Patty's Industrial Hygiene and Toxicology

THIRD REVISED EDITION
In Three Volumes

Volume 1
GENERAL PRINCIPLES

Volumes 2A, 2B, and 2C
TOXICOLOGY

Volume 3
THEORY AND RATIONALE
OF INDUSTRIAL HYGIENE
PRACTICE

Patty's Industrial Hygiene and Toxicology

THIRD REVISED EDITION

Volume 2C
TOXICOLOGY
With Cumulative Index
For Volume 2

GEORGE D. CLAYTON
FLORENCE E. CLAYTON
Editors

Contributors

B. D. Astill
R. R. Beard
G. D. DiVincenzo
Derek Guest
Rolf Hartung

L. W. Hazleton
Gary V. Katz
W. J. Krasavage
S. B. McCollister
R. Montgomery

J. L. O'Donoghue
V. K. Rowe
H. E. Stokinger
Robert J. Weir
Mark A. Wolf (deceased)

1807 1982
175 YEARS OF PUBLISHING

A WILEY-INTERSCIENCE PUBLICATION
JOHN WILEY & SONS
New York • Chichester • Brisbane • Toronto • Singapore

Library of Congress Cataloging in Publication Data:

Patty, Frank Arthur, 1897–
  Patty's Industrial hygiene and toxicology.
  Vol. 3 edited by L. V. Cralley, L. J. Cralley.
  "A Wiley-Interscience publication."
  Includes bibliographical references and indexes.
  CONTENTS: v. 1. General principles.—v. 2.
Toxicology—v. 3. Theory and rationale of
industrial hygiene practice.
  1. Industrial hygiene. 2. Industrial toxicology.
I. Clayton, George D. II. Clayton, Florence E.
III. Battigelli, M. C. IV. Title. V. Title:
Industrial hygiene and toxicology. [DNLM: WA400.3
P322 1978]

RC967.P37 1978   363.1'1   77-17515
ISBN 0-471-16046-6 (v. 1)   AACR1
ISBN 0-471-09258-4 (v. 2c)

Printed in the United States of America

10 9 8 7 6 5 4 3 2 1

Contributors

BERNARD D. ASTILL, Ph.D., Health, Safety, and Human Factors Laboratory, Eastman Kodak Company, Kodak Park, Rochester, New York

RODNEY R. BEARD, M.D., M.P.H., Professor of Preventive Medicine, Emeritus, Stanford University School of Medicine, Stanford, California

GEORGE D. DiVINCENZO, Ph.D., Technical Associate, Health, Safety, and Human Factors Laboratory, Eastman Kodak Company, Kodak Park, Rochester, New York

DEREK GUEST, Ph.D., Health, Safety, and Human Factors Laboratory, Eastman Kodak Company, Kodak Park, Rochester, New York

ROLF HARTUNG, Ph.D., Professor of Environmental Toxicology, School of Public Health, Department of Environmental and Industrial Health, University of Michigan, Ann Arbor, Michigan

LLOYD W. HAZLETON, Ph.D., Retired, formerly President, Hazleton Laboratories, McLean, Virginia

GARY V. KATZ, Ph.D., Health, Safety, and Human Factors Laboratory, Eastman Kodak Company, Kodak Park, Rochester, New York

WALTER J. KRASAVAGE, Senior Toxicologist, Health, Safety, and Human Factors Laboratory, Eastman Kodak Company, Kodak Park, Rochester, New York

SUSAN B. McCOLLISTER, Toxicologist, The Dow Chemical Company, Midland, Michigan

RUTH MONTGOMERY, Haskell Laboratory for Toxicology and Industrial Medicine, E. I. duPont deNemours & Company, Wilmington, Delaware

JOHN L. O'DONOGHUE, V.M.D., Ph.D., Pathology Group Leader, Health, Safety, and Human Factors Laboratory, Eastman Kodak Company, Kodak Park, Rochester, New York

V. K. ROWE, Sc.D. (hon.), Consultant, formerly Director, Toxicological Affairs, Dow Chemical Company, Midland, Michigan

HERBERT E. STOKINGER, Ph.D., Retired, formerly Chief Toxicologist, U.S. Public Health Service, Cincinnati, Ohio

ROBERT J. WEIR, Ph.D., Vice President, Bionetics, Kensington, Maryland

MARK A. WOLF, The Dow Chemical Company, Midland, Michigan (deceased)

Preface

The man who was the catalyst for the *Industrial Hygiene and Toxicology* series died last August, shortly after the manuscripts for the last of the three books of Volume 2 on toxicology were delivered to the publisher. Four years ago, his wife and coeditor Ruth expired. The progenitors of this series, which first came into print in 1948, are no longer here to see what their consanguinity has inspired. We are fortunate to have Frank's last contribution in Volume 1, Chapter 1, third revision. The series will serve as a monument to the memory of Frank Patty's acumen and inquisitive motivation.

Preparation of the third revision of this series was a mammoth undertaking which was begun in 1976: this book, Volume 2C, concludes the volume on toxicology. Volume 3, edited by Cralley and Cralley, was completed in 1979, before the final book of Volume 2.

As editors of Volumes 1 and 2, we found the task frustrating at times but always stimulating and challenging, paralleling the appeal of this profession. The men and women who have contributed their time, knowledge, and expertise for the benefit of others in this and related fields deserve our gratitude for their devotion and perseverance in bringing these fascicles to fruition.

The Preface to Volume 1, third revision, which we prepared in 1977, scrutinized the change of approach to industrial hygiene that had transpired during the period of 1950–1975. We stated that much had been accomplished in the area of recognition and evaluation, but looked to greater strides in controls. Almost five years have elapsed since that time; the achievements in controls which we envisioned have not taken shape to the desired degree. There are still too many people becoming ill or having reduced productive capacity. Although evaluation and recognition techniques have been refined, and despite fantastic expenditures of money, we lag in the improvement and escalation of controls. Remarkable strides have been made scientifically, permitting a better understanding of the chemistry of industrial contaminants and, probably more important, the physiologic effects of these contaminants on humans. With this knowledge we have been able to refine TLVs and carcinogenic and other effects of industrial chemicals. Contrasted with the knowledge available in 1962, the data bank is tremendous.

During the 12 years that have elapsed since the passage of the Occupational Safety and Health Act of 1970 the accomplishments in protecting workers' health have been too few and the pace too slow. Apparently our modus operandi was oblique. The amount of money spent became more important than how it was spent. We were too eager to reach our desired goal quickly. Since our present and past approach has not been as effective as desired, a new approach is essential.

One of the problems facing those entrusted with workers' health protection and community safety lies with our antiquated plants. "Band-Aid" methods, such as local exhaust ventilation on existing machinery and the use of respirators (other than in emergency) for protection from contaminants, are being used.

With the present administration dedicated to improving plants and increasing the productivity of the work force, we should seize the opportunity to chart the right course. What is the right course?

Current practice is for the chief engineer or engineering manager to approve plans for new plant construction that are intended to ensure the integrity of the plant's performance. We believe this is inadequate. To assure that hazards, both in and out of the work place, are "designed out" of the process, the final plans of each project should have the signature not only of the engineer, but also of a qualified industrial hygienist, industrial physician, and industrial toxicologist.

Representatives of these three professions should periodically review the project in its various construction stages. During the start-up phase, detailed examinations of the work place, the air and liquid effluents should be studied and certified as acceptable. In addition, representatives of these groups should prepare a list outlining the medical, toxicologic, industrial hygiene, and environmental hazards associated with the plant. This list should include a study of all raw materials and by-products and should be an integral part of the documents turned over to the operating personnel upon completion of the plant, certifying it for safe occupancy and use.

This procedure would virtually eliminate creation of hazards and would assure the operating personnel a safe place to work and the community a facility that would not adversely affect the environment. To assist in promoting this concept, federal, state, and municipal laws should be changed to accommodate these recommendations. No new major construction should be permitted without the approval of plans by these three professionals, in addition to the presently required engineering approval.

To promote this procedure and achieve this goal, tax incentives might well be considered. In the long run, prevention is much less costly than the ineffectual curing methods to which we have resorted to date.

We *can* achieve the optimum goals of the Occupational Safety and Health Act of 1970 if we have the fortitude to recognize where we have failed to date and the wisdom to select a new course for synergetic action, and if we amplify communication between the scientific and political communities. *Now* is the time.

George D. Clayton
Florence E. Clayton

San Luis Rey, California
March 1982

Contents

Contents
Volume 2A

Contents
Volume 2B

USEFUL EQUIVALENTS AND CONVERSION FACTORS

1 kilometer = 0.6214 mile

1 meter = 3.281 feet

1 centimeter = 0.3937 inch

1 micrometer = 1/25,4000 inch = 40 microinches = 10,000 Angstrom units

1 foot = 30.48 centimeters

1 inch = 25.40 millimeters

1 square kilometer = 0.3861 square mile (U.S.)

1 square foot = 0.0929 square meter

1 square inch = 6.452 square centimeters

1 square mile (U.S.) = 2,589,998 square meters = 640 acres

1 acre = 43,560 square feet = 4047 square meters

1 cubic meter = 35.315 cubic feet

1 cubic centimeter = 0.0610 cubic inch

1 cubic foot = 28.32 liters = 0.0283 cubic meter = 7.481 gallons (U.S.)

1 cubic inch = 16.39 cubic centimeters

1 U.S. gallon = 3.7853 liters = 231 cubic inches = 0.13368 cubic foot

1 liter = 0.9081 quart (dry), 1.057 quarts (U.S., liquid)

1 cubic foot of water = 62.43 pounds (4°C)

1 U.S. gallon of water = 8.345 pounds (4°C)

1 kilogram = 2.205 pounds

1 gram = 15.43 grains

1 pound = 453.59 grams

1 ounce (avoir.) = 28.35 grams

1 gram mole of a perfect gas ≏ 24.45 liters (at 25°C and 760 mm Hg barometric pressure)

1 atmosphere = 14.7 pounds per square inch

1 foot of water pressure = 0.4335 pound per square inch

1 inch of mercury pressure = 0.4912 pound per square inch

1 dyne per square centimeter = 0.0021 pound per square foot

1 gram-calorie = 0.00397 Btu

1 Btu = 778 foot pounds

1 Btu per minute = 12.96 foot-pounds per second

1 hp = 0.707 Btu per second = 550 foot-pounds per second

1 centimeter per second = 1.97 feet per minute = 0.0224 mile per hour

1 footcandle = 1 lumen incident per square foot = 10.764 lumens incident per square meter

1 grain per cubic foot = 2.29 grams per cubic meter

1 milligrm per cubic meter = 0.000437 grain per cubic foot

To convert degrees Celsius to degrees Fahrenheit: °C (9/5) + 32 = °F

To coonvert degrees Fahrenheit to degrees Celsius: (5/9) (°F − 32) = °C

For solutes in water: 1 mg/liter ≏ 1 ppm (by weight)

Atomspheric contamination: 1 mg/liter ≏ 1 oz/1000 cu ft (approx)

For gases or vapors in air a t 25°C and 760 mm Hg pressure:

    To convert mg/liter to ppm (by volume): mg/liter (24,450/mol. wt.) = ppm

    To convert ppm to mg/liter: ppm (mol. wt./24,450) = mg/liter

CONVERSION TABLE FOR GASES AND VAPORS[a]

(Milligrams per liter to parts per million, and vice versa; 25°C and 760 mm Hg barometric pressure)

Molecular Weight	1 mg/liter ppm	1 ppm mg/liter	Molecular Weight	1 mg/liter ppm	1 ppm mg/liter	Molecular Weight	1 mg/liter ppm	1 ppm mg/liter
1	24,450	0.0000409	39	627	0.001595	77	318	0.00315
2	12,230	0.0000818	40	611	0.001636	78	313	0.00319
3	8,150	0.0001227	41	596	0.001677	79	309	0.00323
4	6,113	0.0001636	42	582	0.001718	80	306	0.00327
5	4,890	0.0002045	43	569	0.001759	81	302	0.00331
6	4,075	0.0002454	44	556	0.001800	82	298	0.00335
7	3,493	0.0002863	45	543	0.001840	83	295	0.00339
8	3,056	0.000327	46	532	0.001881	84	291	0.00344
9	2,717	0.000368	47	520	0.001922	85	288	0.00348
10	2,445	0.000409	48	509	0.001963	86	284	0.00352
11	2,223	0.000450	49	499	0.002004	87	281	0.00356
12	2,038	0.000491	50	489	0.002045	88	278	0.00360
13	1,881	0.000532	51	479	0.002086	89	275	0.00364
14	1,746	0.000573	52	470	0.002127	90	272	0.00368
15	1,630	0.000614	53	461	0.002168	91	269	0.00372
16	1,528	0.000654	54	453	0.002209	92	266	0.00376
17	1,438	0.000695	55	445	0.002250	93	263	0.00380
18	1,358	0.000736	56	437	0.002290	94	260	0.00384
19	1,287	0.000777	57	429	0.002331	95	257	0.00389
20	1,223	0.000818	58	422	0.002372	96	255	0.00393
21	1,164	0.000859	59	414	0.002413	97	252	0.00397
22	1,111	0.000900	60	408	0.002554	98	249.5	0.00401
23	1,063	0.000941	61	401	0.002495	99	247.0	0.00405
24	1,019	0.000982	62	394	0.00254	100	244.5	0.00409
25	978	0.001022	63	388	0.00258	101	242.1	0.00413
26	940	0.001063	64	382	0.00262	102	239.7	0.00417
27	906	0.001104	65	376	0.00266	103	237.4	0.00421
28	873	0.001145	66	370	0.00270	104	235.1	0.00425
29	843	0.001186	67	365	0.00274	105	232.9	0.00429
30	815	0.001227	68	360	0.00278	106	230.7	0.00434
31	789	0.001268	69	354	0.00282	107	228.5	0.00438
32	764	0.001309	70	349	0.00286	108	226.4	0.00442
33	741	0.001350	71	344	0.00290	109	224.3	0.00446
34	719	0.001391	72	340	0.00294	110	222.3	0.00450
35	699	0.001432	73	335	0.00299	111	220.3	0.00454
36	679	0.001472	74	330	0.00303	112	218.3	0.00458
37	661	0.001513	75	326	0.00307	113	216.4	0.00462
38	643	0.001554	76	322	0.00311	114	214.5	0.00466

CONVERSION TABLE FOR GASES AND VAPORS (Continued)

(Milligrams per liter to parts per million, and vice versa; 25°C and 760 mm Hg barometric pressure)

Molec-ular Weight	1 mg/liter ppm	1 ppm mg/liter	Molec-ular Weight	1 mg/liter ppm	1 ppm mg/liter	Molec-ular Weight	1 mg/liter ppm	1 ppm mg/liter
115	212.6	0.00470	153	159.8	0.00626	191	128.0	0.00781
116	210.8	0.00474	154	158.8	0.00630	192	127.3	0.00785
117	209.0	0.00479	155	157.7	0.00634	193	126.7	0.00789
118	207.2	0.00483	156	156.7	0.00638	194	126.0	0.00793
119	205.5	0.00487	157	155.7	0.00642	195	125.4	0.00798
120	203.8	0.00491	158	154.7	0.00646	196	124.7	0.00802
121	202.1	0.00495	159	153.7	0.00650	197	124.1	0.00806
122	200.4	0.00499	160	152.8	0.00654	198	123.5	0.00810
123	198.8	0.00503	161	151.9	0.00658	199	122.9	0.00814
124	197.2	0.00507	162	150.9	0.00663	200	122.3	0.00818
125	195.6	0.00511	163	150.0	0.00667	201	121.6	0.00822
126	194.0	0.00515	164	149.1	0.00671	202	121.0	0.00826
127	192.5	0.00519	165	148.2	0.00675	203	120.4	0.00830
128	191.0	0.00524	166	147.3	0.00679	204	119.9	0.00834
129	189.5	0.00528	167	146.4	0.00683	205	119.3	0.00838
130	188.1	0.00532	168	145.5	0.00687	206	118.7	0.00843
131	186.6	0.00536	169	144.7	0.00691	207	118.1	0.00847
132	185.2	0.00540	170	143.8	0.00695	208	117.5	0.00851
133	183.8	0.00544	171	143.0	0.00699	209	117.0	0.00855
134	182.5	0.00548	172	142.2	0.00703	210	116.4	0.00859
135	181.1	0.00552	173	141.3	0.00708	211	115.9	0.00863
136	179.8	0.00556	174	140.5	0.00712	212	115.3	0.00867
137	178.5	0.00560	175	139.7	0.00716	213	114.8	0.00871
138	177.2	0.00564	176	138.9	0.00720	214	114.3	0.00875
139	175.9	0.00569	177	138.1	0.00724	215	113.7	0.00879
140	174.6	0.00573	178	137.4	0.00728	216	113.2	0.00883
141	173.4	0.00577	179	136.6	0.00732	217	112.7	0.00888
142	172.2	0.00581	180	135.8	0.00736	218	112.2	0.00892
143	171.0	0.00585	181	135.1	0.00740	219	111.6	0.00896
144	169.8	0.00589	182	134.3	0.00744	220	111.1	0.00900
145	168.6	0.00593	183	133.6	0.00748	221	110.6	0.00904
146	167.5	0.00597	184	132.9	0.00753	222	110.1	0.00908
147	166.3	0.00601	185	132.2	0.00757	223	109.6	0.00912
148	165.2	0.00605	186	131.5	0.00761	224	109.2	0.00916
149	164.1	0.00609	187	130.7	0.00765	225	108.7	0.00920
150	163.0	0.00613	188	130.1	0.00769	226	108.2	0.00924
151	161.9	0.00618	189	129.4	0.00773	227	107.7	0.00928
152	160.9	0.00622	190	128.7	0.00777	228	107.2	0.00933

CONVERSION TABLE FOR GASES AND VAPORS (Continued)

(Milligrams per liter to parts per million, and vice versa; 25°C and 760 mm Hg barometric pressure)

Molec-ular Weight	1 mg/liter ppm	1 ppm mg/liter	Molec-ular Weight	1 mg/liter ppm	1 ppm mg/liter	Molec-ular Weight	1 mg/liter ppm	1 ppm mg/liter
229	106.8	0.00937	253	96.6	0.01035	277	88.3	0.01133
230	106.3	0.00941	254	96.3	0.01039	278	87.9	0.01137
231	105.8	0.00945	255	95.9	0.01043	279	87.6	0.01141
232	105.4	0.00949	256	95.5	0.01047	280	87.3	0.01145
233	104.9	0.00953	257	95.1	0.01051	281	87.0	0.01149
234	104.5	0.00957	258	94.8	0.01055	282	86.7	0.01153
235	104.0	0.00961	259	94.4	0.01059	283	86.4	0.01157
236	103.6	0.00965	260	94.0	0.01063	284	86.1	0.01162
237	103.2	0.00969	261	93.7	0.01067	285	85.8	0.01166
238	102.7	0.00973	262	93.3	0.01072	286	85.5	0.01170
239	102.3	0.00978	263	93.0	0.01076	287	85.2	0.01174
240	101.9	0.00982	264	92.6	0.01080	288	84.9	0.01178
241	101.5	0.00986	265	92.3	0.01084	289	84.6	0.01182
242	101.0	0.00990	266	91.9	0.01088	290	84.3	0.01186
243	100.6	0.00994	267	91.6	0.01092	291	84.0	0.01190
244	100.2	0.00998	268	91.2	0.01096	292	83.7	0.01194
245	99.8	0.01002	269	90.9	0.01100	293	83.4	0.01198
246	99.4	0.01006	270	90.6	0.01104	294	83.2	0.01202
247	99.0	0.01010	271	90.2	0.01108	295	82.9	0.01207
248	98.6	0.01014	272	89.9	0.01112	296	82.6	0.01211
249	98.2	0.01018	273	89.6	0.01117	297	82.3	0.01215
250	97.8	0.01022	274	89.2	0.01121	298	82.0	0.01219
251	97.4	0.01027	275	88.9	0.01125	299	81.8	0.01223
252	97.0	0.01031	276	88.6	0.01129	300	81.5	0.01227

[a] A. C. Fieldner, S. H. Katz, and S. P. Kinney, "Gas Masks for Gases Met in Fighting Fires," U.S. Bureau of Mines, Technical Paper No. 248, 1921.

Patty's Industrial Hygiene and Toxicology

THIRD REVISED EDITION
In Three Volumes

Volume 1
GENERAL PRINCIPLES

Volumes 2A, 2B, and 2C
TOXICOLOGY

Volume 3
THEORY AND RATIONALE
OF INDUSTRIAL HYGIENE
PRACTICE

CHAPTER FIFTY

Glycols

V. K. ROWE, Sc.D. (Hon.), and
M. A. WOLF*

1 INTRODUCTION

The glycols discussed in this chapter are prepared by a variety of routes and have a wide spectrum of uses. They vary from slightly viscous liquids to waxy solids. Generally they are low in volatility, soluble in water, alcohols, and ketones, and insoluble in hydrocarbons and similar compounds. The properties of several of the more common ones are given in Table 50.1. Industrial exposure is most likely to involve contact with the skin but inhalation of vapors and/or mists may be significant in some instances. The glycols, in general, are not highly toxic substances and some are essentially innocuous. All these factors and others are discussed in the following sections dealing with the individual compounds.

2 ETHYLENE GLYCOL; 1,2-Ethanediol; CAS No. 107-21-1

$$HOCH_2CH_2OH$$

2.1 Source, Uses, and Industrial Exposure

Ethylene glycol historically has been made commercially by the hydrolysis of ethylene oxide. Presently it is also being produced commercially by the oxidation of ethylene in the presence of acetic acid to form ethylene diacetate, which is

* Deceased.

Table 50.1.   Physical and Chemical Properties of Common Glycols (Diols)

Property	Ethylene Glycol	Diethylene Glycol	Triethylene Glycol	Propylene Glycol
CAS No.	107-21-1	111-46-6	112-27-6	57-55-6
Molecular formula	$C_2H_6O_2$	$C_4H_{10}O_3$	$C_6H_{14}O_4$	$C_3H_8O_2$
Molecular weight	62.07	106.12	150.1	76.1
Specific gravity (25/4°C)	1.11	1.12	1.125 (20/20°C)	1.033
Boiling point, °C (760 mm Hg)	197.4	245	287.4	187.9
Freezing point, °C	−13.4	−8.0	−4.3	−31.0
Vapor pressure, mm Hg (25°C)	0.06 (20°C)	<0.01	0.001	0.13
Refractive index (25°C)	1.432		1.456	1.431
Flash point, °F (O.C.)	240	290	330	215–225
Percent in saturated air (25°C)	0.017	0.0013 (20°C)	0.00013 (20°C)	0.038
1 ppm ⇆ mg/m³ at 25°C and 760 mm Hg	2.54	4.35	6.14	3.11
1 mg/l ⇆ ppm at 25°C and 760 mm Hg	365.0	230.7	162.8	321.6

then hydrolyzed to the glycol, with the acetic acid being recycled in the process. Production by 1980 should have approached, if not exceeded, 5 billion lb/year.

More than 25 percent of the ethylene glycol produced is used in antifreeze and coolant mixtures for motor vehicles. It is used in hydraulic fluids and heat exchangers and as a solvent. Large amounts are used as a chemical intermediate in the production of ethylene glycol esters, ethers, and resinous products, particularly polyester fibers and resins.

Contact with the skin and eyes is most likely to occur in industrial handling. Inhalation may be a problem if the material is handled hot or if a mist is generated by heat or by violent agitation. Swallowing is not likely to be an industrial problem unless the material is stored in unmarked or mislabeled containers.

2.2   Physical and Chemical Properties

Ethylene glycol is a colorless, odorless, viscous, hygroscopic liquid with a bittersweet taste. It is miscible with water, lower aliphatic alcohols, aldehydes, and ketones and is practically insoluble in hydrocarbons and similar compounds. Additional properties are given in Table 50.1.

1,3-Propanediol	Dipropylene Glycol	Tripropylene Glycol	1,5-Pentanediol	2-Methyl-2,4-pentanediol	2-Ethyl-1-3-hexanediol	Styrene Glycol
504-63-2	25265-71-8	24800-44-0	111-29-5	107-41-5	94-96-2	93-56-1
$C_3H_8O_2$	$C_6H_{14}O_3$	$C_9H_{20}O_4$	$C_5H_{12}O_2$	$C_6H_{14}O_2$	$C_8H_{18}O_2$	$C_8H_{10}O_2$
76.1	134.2	192.3	104.2	118.2	146.2	138.2
1.055	1.020	1.021	0.9925	0.9216	0.9422	
(20/20°C)			(74°F)	(20/4°C)	(20/4°C)	
210–211	231.8	267.4	238–240	198	244.2	221
		—	−18.0	−50.0	−40.0	64.0
				(sets to glass)	(sets to glass)	
	<0.01	<0.01		0.05	<0.01	
	(20°C)			(20°C)		
	1.439	1.442		1.426	1.4511	
				(20°C)	(20°C)	
	250–280	285	154	210–215	265	
	<0.0013	<0.0013		0.0066		
				(20°C)		
3.11	5.49	7.86	4.26	4.84	5.98	5.65
321.6	181.8	127.1	234.7	206.0	167.2	177.0

2.3   Determination of Presence of Ethylene Glycol

2.3.1   In the Atmosphere

Although numerous methods are available for the determination of ethylene glycol when it is present in substantial amounts, few of the older methods are applicable to the determination of small amounts such as may be of industrial hygiene significance (1–6).

Davis et al. (84) and Spitz and Weinberger (7) have developed gas chromatographic methods for determining ethylene glycol in an aqueous solution. These methods permit the detection of ethylene glycol in the nanogram range, and should be useful in determining atmospheric concentrations because sampling by scrubbing air with water is effective.

Chairova and Dimov (8) and Esposito and Jamison (9) also have developed gas or gas–liquid chromatographic methods for detecting glycols and their derivatives in mixtures.

2.3.2   Determination of Ethylene Glycol in Biologic Systems

Peterson and Rodgerson (10) recommend gas chromatographic methods for serum analysis (sensitivity 0.02 mg/ml). Bolanowska (11) suggests colorimetric methods when evaluating urine. Russell et al. (12) and Rajagopal and Ramak-rishnan (13) also have devised colorimetric methods, the latter mainly for the detection of ethylene glycol in blood. Because mannitol has been used in the treatment of ethylene glycol poisoning, it is important to know that, when present, mannitol gives a false positive response for ethylene glycol (14).

2.4 Physiologic Response

2.4.1 Summary

Ethylene glycol presents negligible hazards to health in industrial handling, except possibly where it is being used at elevated temperatures. It is low in acute oral toxicity, is not significantly irritating to the eyes or skin, is not readily absorbed through the skin in acutely toxic amounts, and its vapor pressure is sufficiently low so that toxic concentrations cannot occur in the air at room temperatures. Mists or aerosols generally are considered to be low in toxicity also, but if exposures to high concentrations occur, more or less serious effects may occur. Ethylene glycol is not considered to be carcinogenic. Available studies suggest that it is not mutagenic. The principal hazard to health of ethylene glycol is associated with the ingestion of large quantities in single doses. Lesser quantities ingested, inhaled, or absorbed through the skin repeatedly over a prolonged period of time can also present a significant hazard to health. Small quantities likely to be encountered in ordinary industrial situations do not pose a serious problem.

Most of the older published articles dealing with the toxicity of ethylene glycol and the other common glycols have been ably reviewed (15–20).

2.4.2 Single-Dose Oral

The toxicity of ethylene glycol for animals has been determined by numerous investigators. Some of the more representative data are summarized in Table 50.2. These data show that ethylene glycol is low in single-dose oral toxicity to laboratory animals.

The single oral dose lethal for humans has been estimated at 1.4 ml/kg (1.56 g/kg) or about 100 ml (111 g) per person (17). It is apparent that ethylene glycol is much more acutely toxic for humans than for the laboratory animals studied.

2.4.3 Repeated-Dose Oral

Morris et al. (19) maintained rats for 2 years on diets containing 1 and 2 percent ethylene glycol. The findings were those of shortened life-span, calcium oxalate bladder stones, severe renal injury, particularly of the tubules, and centrolobular degeneration of the liver at both levels. Blood (30) in a follow-up study found

Table 50.2. Single-Dose Oral Toxicity (of Ethylene Glycol) to Laboratory Animals

	LD_{50} Values (g/kg of Body Weight)				
Reference	Mouse	Rat	Guinea Pig	Rabbit	Dog
Laug et al. (17)	14.6	6.14	8.20		
Smyth et al. (21)		8.54	6.61		
Bornmann (22)	15.28				
Page (23)					>8.81
Pochebyt (24)		5.89			
Plugin (25)	8.0	13.0	11.0	5.0	
Bove (26)		10.0 to 13.4			
Antonyuk (27)		10.88			
Kersting and Nielsen (28)					Minimum lethal dose 7.36; some dogs survived 14.7
Dow (29)		11.3			

that the no-effect level was no more than 0.2 percent in the diet (100 mg/kg/day in a 2-year test). He found calcification of the kidneys and oxalate crystal stones at the 0.5 percent level in male rats; however, females showed only calcification of the kidneys at the dietary level of 1 percent and higher, but oxalate stones at the 4 percent level only. There was increased water consumption and appearance of protein in the urine of males receiving the 1 and 4 percent diets and in the females on the 4 percent diet.

Antonyuk (27) found that the no-effect level in rats fed for 3 months was 1.08 g/kg/day (0.97 ml/kg/day). This level was the highest level fed and represents a dose of one-tenth the single-dose oral LD_{50}. Gaunt et al. (31) also conducted subacute dietary feeding studies on rats. The no-effect level when fed for 16 weeks was found to be 0.1 percent (71 mg/kg/day for male rats and 85 mg/kg/day for female rats). Based upon this study, the acceptable daily intake for a 70-kg person was suggested to be 50 mg ethylene glycol per day. The males given a diet containing 0.25 percent (~178 mg/kg/day) and higher levels developed oxaluria, oxalate crystals in their urine, renal damage, and at the 1 percent level, increased kidney weight and altered renal function. In the females these effects were seen, but to a lesser degree, only in those given the 1 percent level (~850 mg/kg/day).

Nagano et al. (89) fed ethylene glycol to mice in doses ranging from 62.5 to 4000 mg/kg, 5 days/week for 5 weeks and observed no effects upon the testes or blood.

Roze (88) administered ethylene glycol as a 1.0 percent solution in the drinking water for 3 weeks to groups of old and young rats. He observed increased kidney weights, increased blood urea nitrogen, particularly in the older animals, and depressed growth. The simultaneous treatment of a similar group of rats with intramuscular injections, three times a week for 3 weeks, of 1.6 g/kg of polyethylene glycol 6000 as a 20 percent aqueous solution seemed to enhance the severity of the effects of ingesting ethylene glycol. Speculation, but no proof, of the mode of action is offered in this study.

Blood et al. (32) using three rhesus monkeys conducted 3-year feeding studies. The two males were fed a diet containing 0.2 percent ethylene glycol, which equates to an average daily dose between 0.02 and 0.07 g/kg. The one female was fed a diet containing 0.5 percent, which equates to an average daily dose between 0.14 and 0.17 g/kg. No adverse effects were observed and no oxalate crystals were found in the urine. Roberts and Seibold (33) found that a macaque monkey, when given ethylene glycol at a concentration of 0.25 percent (about 0.24 g/kg/day) in its drinking water for 157 days, developed oxalate deposition in the kidneys. Two other monkeys did not show renal oxalate crystals when they were given 0.25 percent (about 0.4 g/kg/day) for up to 60 days. Higher doses, when given even for short periods, caused morphologic tubular changes of the kidney even though oxalate crystals were not detected. A limited number of the monkeys given 10 percent of ethylene glycol in their drinking water for a few days followed by lower concentrations developed calcium oxalate crystals in the brain. Those monkeys given a total dose of less than 15 ml/kg (less than about 0.4 g/kg/day) developed mild glomerular damage and azotemia, even though no calcium oxalate crystals were detected. This suggests that the toxicity of ethylene glycol is not due altogether to the oxalic acid metabolite, but in part to other causes that may be related to the saturation of the normal metabolic pathway by large doses.

Yoshida et al. (34) fed chicks diets containing 5 percent ethylene glycol, the lowest dose given, for 27 days and found that this level was toxic. Riddell et al. (35) found that the median lethal concentration for ethylene glycol was 7.5 percent when given to chickens in their drinking water for 2 weeks. Those fed 27.9 g/l or more developed renal oxalosis. Those fed less than 27.9 g/l developed no adverse effects.

There seems to be little doubt that the primary effect of repeated oral doses of ethylene glycol is that of kidney injury. Such injury may apparently occur even though oxalate crystals are not deposited in the kidney. High doses of ethylene glycol may lead to deposition of oxalate crystals in the brain. Male animals seem to be more susceptible to ethylene glycol than are females. The work of Gaunt et al. (31), and others, suggests that the daily intake of ethylene glycol in the diet of a 70-kg person may be 50 mg/day.

2.4.4 Single-Dose Toxicity by Injection

Lipkan and Petrenko (36) report that the LD_{50} for mice for ethylene glycol when given intraperitoneally was 5.80 g/kg; the nontoxic effect dose was 0.58

g/kg. When given subcutaneously the LD_{50} was 10.0 g/kg with a nontoxic effect level of 1.73 g/kg.

Mason et al. (37) found that the LD_{50} for rats when given subcutaneously was 5.3 g/kg; the $LD_{0.1}$ was 2.66 g/kg.

2.4.5   Repeated Injection

Paterni et al. (38) have reported on studies in which they administered subcutaneously to rats, daily doses of 4 ml ethylene glycol diluted with water. Progressive hemolytic anemia and a variety of changes in the leukocytes were seen. Other findings included renal lesions, changes in the liver and spleen, and iron deposits in all organs. Bornmann (22) also gave the material parenterally to mice and observed a hemolytic effect but concluded that death was due to the narcotic effect and to renal insufficiency, not to the hemolytic effect. Mason et al. (37) found that rats injected with ethylene glycol subcutaneously twice a week for 1 year at a dose level of 1000 mg/kg were unaffected in that they developed no detectable clinical signs.

2.4.6   Eye Irritation

Carpenter and Smyth (39) report that ethylene glycol failed to cause appreciable irritation when introduced once into the eyes of rabbits.

McDonald et al. (40), using the rabbit, placed onto the eye one drop of ethylene glycol (0.05 ml) every 10 min for 6 hr (36 applications). They found that a concentration of ethylene glycol of 4 percent in a balanced salt solution caused mild conjunctival redness, mild chemosis, minor flare, and iritis. A concentration of 0.4 percent caused no detectable effects beyond those seen with the control balanced salt solution. Later work by McDonald et al. (41) showed that concentrations above 4 percent are capable of causing eye injury.

Grant (42) reports that exposure of humans to vapor or spray of ethylene glycol for 4 weeks at a concentration of 17 mg/m^3 produced no ill effects. Levels of 265 mg/m^3 caused no ocular damage to the eyes of chimpanzees; however, rabbits and rats developed severe eye irritation, edema of the lids, and some corneal opacity when exposed several days to 12 mg/m^3.

The data suggest that in humans eye exposure to vapors or to liquid should cause no serious injury although minor and transient discomfort may occur.

2.4.7   Skin Irritation.

Ethylene glycol produces no significant irritant action upon the skin. A slight macerating action on the skin may result from very severe, prolonged exposures, which is comparable to that caused by glycerin under similar conditions.

2.4.8 Skin Absorption

Hanzlik et al. (43) have shown in animal studies that toxic amounts of ethylene glycol can be absorbed through the skin. The data, however, are erratic and difficult to quantitate. Volkmann (44) reports a case of what was believed to be ethylene glycol poisoning resulting from the massive application of an eczema remedy containing ethylene glycol. A comatose condition accompanied by miosis and slowed pulse occurred 4 hr after application but no oxalate was found in the urine.

For industrial hygiene purposes, it would seem prudent to avoid prolonged and repeated contact with skin, particularly contacts that involve extensive areas of skin.

2.4.9 Absorption—Excretion

Marshall and Cheng (246) exposed rats (nose only) to ^{14}C-ethylene glycol vapor at a concentration of 32 µg/l for 30 min. Approximately 61 percent of the amount inhaled was deposited, largely in the nasopharyngeal region. The total amount retained was calculated to be equivalent to 0.9 ± 0.3 g/kg for the males and 0.6 ± 0.05 g/kg for the females. Blood levels of radioactivity were relatively constant for the first 6 hr post exposure, then declined by apparent first-order kinetics with a half-life of 53 hr. The predominant routes of elimination of the radioactivity were via the expired air and the urine, 55 to 70 percent as ^{14}CO$_2$ and 14 to 26 percent, respectively, of the initial body burden.

2.4.10 Inhalation

According to Browning (20), Flury and Wirth exposed rats for 28 hr during 5 days to an atmosphere essentially saturated (0.5 mg/l) with ethylene glycol. No deaths occurred, but the animals reportedly exhibited slight narcosis. Wiley et al. (45) exposed rats and mice to concentrations of 0.35 to 0.40 mg/l (140 to 160 ppm) 8 hr/day during 16 weeks without producing injury. Antonyuk (27) conducted studies in which animals (probably rats) were exposed to the vapor of ethylene glycol for 24 hr/day for 3 months. The "no-effect" level was found to be 0.3 mg/m^3. Growth depression and blood effects were seen at a level of 8.4 mg/m^3. Whether the reported observations were actually due to exposure to ethylene glycol cannot be stated because there is no indication that the controls were subjected to the same extraneous stresses experienced by the experimental animals. Coon et al. (46) report observing no adverse effects when rats, rabbits, guinea pigs, squirrel monkeys, and dogs were exposed to 57 mg ethylene glycol as a vapor/m^3 8 hr/day, 5 days/week for 6 weeks. When a separate set of these animals was exposed continuously 24 hr/day for 90 days, the rats and rabbits experienced moderate to severe eye irritation. Corneal

opacity was seen in the rabbits after 3 days of exposure and in the rats in 8 days. Otherwise, there were no significant effects seen in any of the animals.

Wills et al. (47) in preliminary studies in preparation for human testing found that monkeys tolerated exposures of 500 to 600 mg of ethylene glycol as an aerosol/m^3 for 2 to 3 weeks. Some of the monkeys survived such exposures for as long as 5 to 7 months. In the human studies, 22 volunteers were exposed to aerosols of ethylene glycol (particle size 1 to 5 μm) at concentrations varying from 3 to 67 mg/m^3. The exposures were essentially continuous (20 to 22 hr/day) for 1 month. The volunteers reported some irritation of the nose and throat, and occasionally slight headache and low backache, but there were no other significant adverse effects. Further tests found that aerosol levels of 200 mg/m^3 were intolerable and were very noticeable at levels of 140 mg/m^3. The authors suggest that if irritating levels of aerosols are avoided, there should be no hazard from such exposures.

As a result of anticipated inhalation problems in space flight by humans, the toxicity of ethylene glycol was studied under simulated flight conditions. Harris (48) summarized the work on rats, rabbits, mice, guinea pigs, dogs, and monkeys, conducted over a 2-year period. The animals were exposed in chambers for 3 weeks at a working pressure of 5 psi in an atmosphere consisting of 100 percent oxygen to which was added ethylene glycol aerosol to saturation. The level was 275 mg ethylene glycol/m^3 or about 100 ppm. Careful examination of the animals revealed no adverse effects other than some pulmonary irritation in the rats and mice; no behavioral changes were detected in the monkeys. Felts (49) reported on studies in which four chimpanzees were exposed for 28 days to saturated vapors of ethylene glycol (256 mg/m^3) at 5 psi in an atmosphere consisting of 68 percent oxygen and 32 percent nitrogen, and observed only minor depression of the white blood cell count 2 weeks after termination of exposure. Two other chimpanzees in a similar but separate experiment experienced some signs of minor renal effects but no eye, behavioral, or clinical effects.

Humans were then exposed under ambient conditions 22 hr/day, for 28 days, to 68.5 mg/m^3 of ethylene glycol as a mixture of vapor and aerosol dispersed in the air. The subjects were unable to detect the ethylene glycol and experienced no adverse effects. When the concentration was increased to 137 mg/m^3 the volunteers reported throat and eye irritation and were able to detect the ethylene glycol by taste. A concentration of 205.5 mg/m^3 was considered intolerable because of the irritation of the eyes and throat (48).

From these data it may be concluded that the hazard from even repeated exposure to vapors or aerosols under ordinary room conditions should pose no practical problems. It should be noted, however, that a hazard due to inhalation may exist in circumstances where the ethylene glycol is being handled hot or where agitation or other mechanical operations may create an excessive fog or heavy mist in the air. Even then, if the warning discomfort from excessive

concentrations of vapor and/or aerosol is heeded, no practical problem would be expected.

2.4.11 Teratogenicity

There is no reported information.

2.4.12 Mutagenicity

McCann et al. (50) report that ethylene glycol was found to be nonmutagenic by the salmonella/microsome mutagenicity test. Pfeiffer and Dunkelberg (246) found ethylene glycol to be inactive when tested against S. typhimurium strains TA-98, TA-100, TA-1535, and TA-1537.

2.4.13 Carcinogenicity

McCann et al. (50), as the result of the nonmutagenic effects of ethylene glycol to salmonella, suggest that it is noncarcinogenic. This concept is supported by various workers. Mason et al. (37) found no chemically related tumors when ethylene glycol was injected into Fischer rats subcutaneously twice a week for 52 weeks, then held for 18 months for examination. Blood (30) also found no tumors associated with ethylene glycol when it was fed at 1 percent in the diet to rats for 2 years.

Derse (51) in work directed at the evaluation of cancer production in mice, found no chemically related tumors when they were given ethylene glycol subcutaneously and observed for 15 months. Homburger (52) also saw no tumors related to ethylene glycol when it was injected subcutaneously into mice for 8 to 12 weeks and then the injection sites were excised, homogenized, and injected into 25 mice. He also found that ethylene glycol did not increase lung tumors when injected once intravenously into mice.

These data, along with no suggestion from extensive human experience that tumors have occurred, indicate that the hazard of cancer from exposure to ethylene glycol should be considered nil.

2.4.14 Degradation in Water

Evans and David (53) exposed ethylene glycol to typical river water at 20°C and found that it was completely degraded in 3 days. At temperatures of 8°C, complete or nearly complete degradation occurred in 7 days, depending on the river water used. Pitter (54) in similar tests also showed that ethylene glycol is rapidly degraded.

2.4.15 Metabolism

Many workers have explored the metabolism of ethylene glycol in order to better understand its toxicologic effects and to discover effective methods of treatment. As a result, a number of schemes have been developed (15, 48,

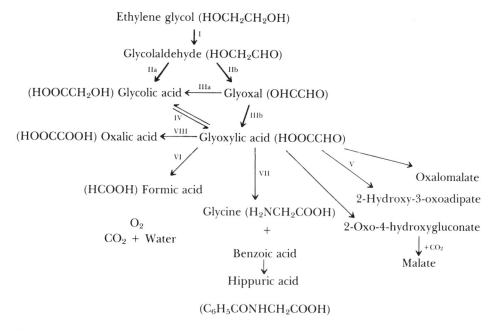

Figure 50.1 Combined scheme of ethylene glycol metabolism. Based on References 15, 56, 60–62, and 64. See text for explanation of the mechanism of reaction as indicated by the Roman numerals.

55–64). Basically, all agree that the main steps in degradation of ethylene glycol are as follows:

Ethylene glycol → glycolaldehyde → glycolic acid → glyoxylic acid

Glyoxylic acid is then metabolized into a number of chemicals that have been identified in expired air, urine, or the blood (see Figure 50.1).

McChesney et al. (61) and Parry and Wallach (63) have summarized the many papers dealing with the postulated mechanisms of metabolism. In their opinion, the data available indicate the following steps, as indicated in Figure 50.1 by Roman numerals:

I. The metabolism of ethylene glycol to glycolaldehyde is mediated by alcohol dehydrogenase.

II. Glycolaldehyde is metabolized to glycolic acid by aldehyde oxidase (IIa), or to a lesser extent to glyoxal (IIb).

III. Glyoxal has been shown to be changed both to glycolic acid in the presence of lactic dehydrogenase, aldehyde oxidase, or possibly both enzymes (IIIb), and to glyoxylic acid via some oxidative mechanism (IIIb).

IV. The main path of the degradation of glycolic acid is to glyoxylic acid. This reaction has been suggested to be caused by lactic dehydrogenase or glycolic acid oxidase.

V–VII.　Once glyoxylic acid is formed, it is apparently degraded very rapidly to a variety of products, a few of which have been observed. Its breakdown to 2-hydroxy-3-oxoadipate is thought to be mediated by thiamine pyrophosphate in the presence of magnesium ions (21). The formation of glycine involves pyridoxal phosphate and glyoxylate transaminase (VII), whereas the formation of carbon dioxide and water via formic acid (VI) apparently involves coenzyme A (CoA) and flavin mononucleotides. Oxalic acid (VIII) formation from glyoxylic acid has been considered to result from the action of lactic dehydrogenase or glycolic acid oxidase.

When ethylene glycol is introduced into the body, it is rapidly distributed into the water of the body. Initially, ethylene glycol is excreted unchanged in the urine for a few hours, accounting for as much as 22 percent of the amount given. Carbon dioxide is also promptly excreted via the lungs and may account for a large amount of the ethylene glycol administered (65). Glycolic acid in the form of salts is also excreted via the urine, being equivalent to as much as 34 to 44 percent of the dose given. Neither glycolaldehyde nor glyoxylic acid have been found in the urine. Oxalic acid, if formed, is also excreted via the urine. The amounts formed depend on the species and on the dose and account for 0.27 to about 3 percent of the dose when given to laboratory animals (62), and possibly up to 2.3 percent of the dose in man (62). Rhesus monkeys and chimpanzees have been shown to excrete larger amounts of oxalic acid salts, as much as 18 to 20 percent of the dose administered. This may have been due to the large doses given (48). The liver and the kidney appear to be the main sites of metabolism (61).

Based on these findings, it has been postulated that ethylene glycol is metabolized at a moderate rate to glycolaldehyde. Glycolaldehyde, however, is apparently metabolized nearly instantaneously to glycolic acid and glyoxal. Glyoxal, too, is rapidly metabolized while glycolic acid is slowly degraded. Glyoxylic acid, because it has not been found in the urine, is also broken down rapidly, apparently to a large extent to carbon dioxide and water via the formic acid route. Its other metabolites are considered to be minor.

2.4.16　Mode of Action

In order to evaluate the mode of action of ethylene glycol, an understanding of the clinical effects it causes is needed. Parry and Wallach (63), using a vast array of data, have summarized the information well.

The effects of ethylene glycol poisoning appear in three fairly distinct stages. The severity of these stages and the advance from one stage to the next depend greatly on the amount of ethylene glycol entering the body.

The initial stage is that considered as "central nervous system effects." This stage usually occurs shortly after exposure, within 30 min or so, and lasts for up to 12 hr.

The second stage is usually labeled the "cardiopulmonary effects" stage and ordinarily lasts from 12 to 36 hr after exposure.

The final stage is known as the "renal failure" stage and occurs if the patient survives the first two stages.

The "central nervous system effects" stage (I) is characterized by signs of drunkenness, although there is no alcoholic odor on the breath; nausea; vomiting; and if large doses are experienced, coma followed by convulsions and death in some cases. Mild hypotension, tachycardia, low grade fever, depressed reflexes, generalized or focal seizures, myoclonic jerks, and tetanic contractions can occur. Ocular signs such as nystagmus, ophthalmoplegia, papilledema, and subsequent optic atrophy have all been reported.

Studies of the blood usually reveal a moderate leukocytosis (10 to 40,000/ mm^3) with predominance of polymorphonuclear cells and a normal hematocrit. Serum sodium bicarbonate levels are often depressed, an anion gap greater than 10 is usually present, and hypercalcemia along with acidosis is often found. Hypocalcemia is often observed also.

Urinanalysis usually shows a low specific gravity with proteinuria, microscopic hematuria, pyuria, and cylindruria. In some cases, calcium oxalate and hippuric acid crystals are seen.

This first stage is usually attributed to the presence of unchanged ethylene glycol which is thought to contribute to the central nervous system effects. The other signs seen are thought to be due to the metabolic products.

The "cardiopulmonary failure" stage (II) is initiated by the onset of coma and is characterized by symptoms such as tachypnea, tachycardia, mild hypotension, and cyanosis. In severe cases, pulmonary edema, bronchopneumonia, cardiac enlargement, and congestive failure are present. Death in this stage, if it occurs, usually comes between 24 and 72 hr after exposure. This stage is usually attributed to acidosis and hypocalcemia caused by the metabolic products.

If the patient survives the first two stages, renal failure in varying degrees becomes the major problem in the final stage. Oxaluria is found in most cases. Costovertebral tenderness may be observed if the patient is alert. The renal damage may vary from mild increase in blood urea nitrogen and creatinine followed by recovery, to complete anuria with acute tubular necrosis that can lead to death.

Originally it was thought that renal failure could be attributed solely to the presence of calcium oxalate. However, much work has been done to show that renal damage can occur at levels of exposure where no or few crystals are detected. Therefore, it is now felt that the various metabolic products, other than oxalate, contribute largely to the renal damage that occurs.

The renal necrosis is characterized by dilated proximal tubules, degeneration of tubular epithelium, and intratubular crystals. Distal tubular degeneration is present usually but is less prominent.

Knowledge of the metabolic process has permitted substantial improvement in therapeutic procedures useful in dealing with cases of poisoning from ingestion of ethylene glycol. Peterson et al. (70) reported that an extract of human liver containing alcohol dehydrogenase actively catalyzed the oxidation of ethylene glycol, that the rate of oxidation of the glycol was slower than that of ethanol, that the purified enzyme oxidized ethylene glycol, and that this could be inhibited by ethanol. They then demonstrated in in vivo experiments in rats and squirrel monkeys that ethanol was effective in reducing the toxicity of ingested ethylene glycol and markedly increased its excretion. The mechanism for this protection is twofold; the ethanol competitively inhibits the oxidation of the glycol to more toxic metabolites and enhances the excretion of the glycol per se.

Since then numerous investigators have extended this information and described effective antidotal treatment for poultry (71), dogs (72–74), cats (75), and swine (83).

Mundy et al. (76) and Van Stee et al. (77) demonstrated that rats and dogs can be protected from ordinarily lethal doses of ethylene glycol by treatment with pyrazole, an alcohol dehydrogenase inhibitor, thus supporting the evidence that alcoholic dehydrogenase is a key material in the mechanism of action. The latter authors (77) suggest that pyrazole and other supportive treatment may be even more effective than ethanol in the treatment of ethylene glycol poisoning, particularly if treatment is delayed.

Parry and Wallach (63) urge that the rapid recognition and immediate treatment of ethylene glycol poisoning is necessary to reduce its toxicity. Ethylene glycol poisoning should be suspected if there is nondiabetic ketoacidosis, an alcohol-like intoxication without the presence of alcohol odor on the breath, coma, and metabolic acidosis coupled with an anionic gap and/or oxalate found in the urine.

2.4.17 Human Experience

When one considers the huge volumes of ethylene glycol that are handled and used industrially, there have, indeed, been few instances of adverse effects. Only one episode that can be considered to be of industrial origin is reported (66), and this involves inhalation of vapors from heated material. One case of alleged absorption through the skin involved a medicinal product (44).

Pons and Custer (85) in their review of 18 human cases stressed effects upon the brain. Grant (86) also noted cerebral injury from acute poisoning and believes it may have caused permanent injury. Ross (87) describes a fatal case in which ¼ to ½ pint of an antifreeze solution was ingested; acute meningoencephalitis followed by anuria and death from renal failure resulted after 12 days.

Troisi (66) reports the results of an investigation of complaints among women workers in an electrolytic condenser factory and attributes the trouble to the inhalation of ethylene glycol vapor. The work involved coating aluminum and paper with a mixture containing 40 percent ethylene glycol, 55 percent boric acid, and 5 percent ammonia at 105°C. Nine of 38 women exposed to the vapors suffered frequent attacks of loss of consciousness and nystagmus, and five of these showed an absolute lymphocytosis. Further examination revealed five additional cases of nystagmus among the other 29 workers. Proper enclosure of the system to prevent inhalation of vapors resulted in complete recovery of all affected individuals, although the two most severely affected were removed to other work.

The lack of reports of adverse human experience from industrial exposures supports the concept that ethylene glycol is low in degree of hazard from industrial handling.

However, numerous poisonings have occurred among humans who have ingested substantial amounts of ethylene glycol, usually as antifreeze, either accidentally or willfully. Judging from the cases where the amount ingested could be well estimated, it appears that the lethal dose for humans is about 1.4 ml/kg (1.56 g/kg). It is apparent, therefore, that the human is far more susceptible to acute ethylene glycol poisoning than most animals.

Haggerty (67), Dreisback (78), and Michelis et al. (79) have estimated that each year 40 to 60 deaths occur that are due to the ingestion of ethylene glycol. Since these reports, nearly each year has produced published reports covering one or more cases of poisoning due to ingestion of antifreeze containing ethylene glycol.

2.4.18   Antidotal Treatment

The treatment of ethylene glycol poisoning has been greatly enhanced by knowledge of the metabolic and pharmacokinetic parameters associated therewith. These were discussed earlier. Several authors (79–82, 87) have described their experience in treating cases of human poisoning. Basically, the suggested treatment involves the infusion of ethanol to competitively inhibit the metabolism of ethylene glycol to oxalic acid, the infusion of alkali to control the profound acidosis typical of such poisoning, the infusion of a diuretic such as mannitol to help prevent or control brain edema, and hemodialysis to remove the ethylene glycol from the circulation and to help manage renal insufficiency. The importance of early diagnosis and treatment cannot be overemphasized.

2.5   Hygienic Standards of Permissible Exposure

The studies of Wiley et al. (45), in which the animals were exposed to an essentially saturated atmosphere of ethylene glycol (140 to 160 ppm) 8 hr/day for 16 weeks, suggest that 100 ppm (274 mg/m^3) in air should be safe for prolonged and repeated 8-hr exposures to humans.

Coon et al. (46) found 57 mg/m^3 to be without adverse effect upon a variety of animal species at exposures of 8 hr/day but quite unacceptable when exposures were essentially continuous over a 6-week period.

Antonyuk (27) recommended a guide for control of 0.3 mg/m³ (0.111 ppm) based on data involving 3 months of continuous exposure of animals.

Wills et al. (47) concluded from controlled studies with human subjects that if irritating levels of the material were avoided, no adverse effects would be expected.

Dubeikovskaya et al. (68) have suggested as a guide for control a level of 5 mg/m³ (1.8 ppm) based upon examination of workers who experienced no ill effects when exposed to air contaminated with ethylene glycol in the range of 17 to 96.2 mg/m³ (median level 44.8 mg/m³).

A threshold limit value of 100 ppm (254 mg/m³) for vapors of ethylene glycol and 10 mg/m³ as particulates has been published (69). These values were established on the basis of long industrial experience and experimental studies. The 100 ppm level may not be low enough to avoid mild irritation in susceptible individuals, however. Therefore, since extraneous conditions may have a bearing on comfort, it is suggested that atmospheric concentrations which do not cause discomfort are not likely to cause adverse effects. Consistent with this position, a threshold limit value for ethylene glycol vapor of 50 ppm with a C designation was established in 1981 (245).

2.6 Odor and Warning Properties

Human volunteers, when exposed to about 28 mg/m³ of ethylene glycol as an aerosol (particulate) for about 4 weeks, reported some complaints of irritation of the throat, mild headache, and low backache. At concentrations of about 140 mg/m³ for part of a day, the complaints became marked; levels of above 140 up to 200 mg/m³ caused reluctance to be exposed for more than a few minutes, and levels of 250 to 300 mg/m³ were intolerable even for one breathing cycle (47). Thus it appears that the TLV of 10 mg/m³ for particulates probably would be without significant odor or other warning properties.

3   DIETHYLENE GLYCOL; Ethanol, 2,2'-Oxybis-; 2,2'-Oxydiethanol; Bis(2-hydroxyethyl) Ether; CAS No. 111-46-6

$$(HOCH_2CH_2)_2O$$

3.1   Source, Uses, and Industrial Exposure

Diethylene glycol is produced commercially as a by-product of ethylene glycol production. It can also be produced directly by reaction between ethylene glycol and ethylene oxide (108). Diethylene glycol is used in gas conditioning and in permanent antifreeze formulations; as a constituent of brake fluids, lubricants, mold release agents, and inks; as a softening agent for textiles; as a plasticizer for cork, adhesives, paper, packaging materials, and coatings; as an intermediate in the production of the explosive diethylene glycol dinitrate; and as an intermediate in the production of certain resins, morpholine, and diethylene glycol esters and ethers.

Diethylene glycol presents practically no hazard from the standpoint of industrial handling. It is quite stable chemically and does not present a hazard due to flammability, except at high temperatures or where mist may be involved. It is not appreciably irritating to the eyes and skin and is not absorbed through the skin in appreciable amounts except possibly under adverse conditions where extensive and prolonged skin contact occurs. Its vapor pressure at room temperatures is so low that toxic concentrations of vapor are impossible. It should be noted, however, that a hazard from repeated prolonged inhalation may exist in operations involving heated material or where mists or fogs are generated. However, any reasonable industrial hygiene control would eliminate this possibility. Although the principal hazard to health presented by diethylene glycol is that of ingestion of a substantial amount, this should not occur in industrial handling unless the material is put in unlabeled or mislabeled containers.

3.2 Physical and Chemical Properties

Diethylene glycol is a colorless, essentially odorless, viscous, hygroscopic liquid. Initially, it has a sweetish taste but its aftertaste is bitter. It is miscible with water, lower aliphatic alcohols, and ketones and is practically insoluble in aliphatic hydrocarbons and fatty oils. Additional properties are given in Table 50.1.

3.3 Determination in the Atmosphere

Methods applicable to the determination of small amounts of diethylene glycol, such as may be of industrial hygiene significance, are few. The chromatographic methods described by Davis et al. (109) and Spitz and Weinberger (7) should be useful for industrial hygiene purposes.

3.4 Physiologic Response

3.4.1 Summary

Diethylene glycol presents negligible hazards to health in industrial handling except possibly where it is being used at elevated temperatures. It is low in acute oral toxicity, it is not irritating to the eyes or skin, it is not readily absorbed through the skin, and its vapor pressure is sufficiently low so that toxic concentrations of vapor can not occur in air at room temperatures. Aerosols generated by violent agitation and/or heat may, however, result in excessive amounts in the air. The principal hazard to health of diethylene glycol is associated with the ingestion of large quantities in single doses.

Prior to 1937, the human toxicologic information on diethylene glycol and other glycols was rather incomplete. However, in 1937 more than 100 deaths were caused by the ingestion of an elixir consisting of sulfanilamide and diethylene glycol as one of the major solvents. As a result of this tragedy, a large number of investigations were conducted in an effort to clarify the toxicologic picture in regard to the various glycols involved.

3.4.2 Single-Dose Oral

Most of the older published articles dealing with the toxicity of diethylene glycol and the other common glycols have been ably reviewed (32). The toxicity of diethylene glycol for animals has been determined by numerous investigators. The findings of Laug et al. (17) and Smyth et al. (21) on small animals are representative. Laug (17) reports the single-dose oral LD_{50} values for rats, guinea pigs, and mice to be 14.8, 7.76, and 23.7 ml/kg (16.6, 8.7, and 26.5 g/kg), respectively, and Smyth et al. (21) report similar values of 20.8 g/kg for rats and 13.2 g/kg for guinea pigs. Plugin (25) has also studied diethylene glycol and reports finding the LD_{50} for mice, rats, rabbits, and guinea pigs to be 13.3, 15.6, 26.9, and 14.0 g/kg, respectively. Laug (17) reports the symptomatology for rabbits, dogs, mice, and guinea pigs to be quite similar; first noted were thirst, diuresis, roughened coat, and refusal of food, followed days later by suppression of urine, proteinuria, prostration, dyspnea, a bloated appearance, coma, lowering of body temperature, and death. As a result of the elixir of sulfanilamide episode the single oral dose lethal for humans has been estimated by Calvery and Klumpp (91) to be about 1 ml/kg.

3.4.3 Repeated-Dose Oral

A comprehensive study of the repeated-dose oral toxicity of diethylene glycol is that reported by Fitzhugh and Nelson (92), who maintained rats for 2 years on diets containing 1, 2, and 4 percent diethylene glycol. Unfortunately the purity of the diethylene glycol was not described and hence some of the effects noted may not, in fact, be attributable to the chemical diethylene glycol. At the dietary level of 1 percent of diethylene glycol, they observed slight growth depression, a few bladder stones identified as calcium oxalate, slight kidney damage, and infrequent liver damage. At the 2 percent dietary level, slight growth depression, a number of bladder stones and bladder tumors, moderate kidney damage, and slight liver damage were noted. At the 4 percent dietary level, there was marked growth depression, slight mortality, frequent bladder stones and bladder tumors, marked kidney damage, and moderate liver damage.

There seems to be little doubt that the bladder stones are directly attributable to the material fed. However, the tumors may well have been a secondary result of mechanical irritation of the bladder by the stones and not the direct effect of a chemical on the cells of the bladder. As a result of this study and an observation previously reported from the U.S. Food and Drug Administration by Morris et al. (19), it was concluded that, in the polyethylene glycol series, the chronic oral toxicity for rats decreased with an increase in molecular weight of the polyglycols.

German workers (93–95) have studied the effects of diethylene and other glycols. Wegener (93) gave rats 1 ml of a 20 percent aqueous solution of diethylene glycol per 100 g of body weight daily over a period of 12 weeks and concluded that it had no influence on the reproductive ability of the animals or on their offspring. Bornmann (94) administered the material to rats in concentrations of 1, 2, 5, 10, and 20 percent in their drinking water and found the material to have a narcotic effect and to cause central paralysis of the respiratory and cardiac centers. Loesser et al. (95) found that concentrations of 5 to 20 percent in the drinking water of rats caused weight loss and death, but that 1 and 2 percent had no such effect. They also report finding 2-naphthol in the urine and bile of animals treated with diethylene glycol; this is difficult to rationalize.

Plugin (25) reports that rats given 3.1 g/kg day for 20 days orally were not affected. He concluded that under these conditions there were no cumulative effects, suggesting that diethylene glycol may be readily metabolized under these conditions.

Weil et al. (96, 97) fed diethylene glycol containing only 0.031 percent of ethylene glycol to weanlings, 2-month-old, and year-old rats for up to 2 years at levels of 4.0 and 2.0 percent in a laboratory chow. Because of the difference in chemical intake, the year-old rats were given a diet containing 6 percent diethylene glycol after 1 month on the 4 percent diet. Although the weanling rats developed more bladder stones than the other groups, the difference was insignificant. The yearling rats developed their bladder stones somewhat earlier. The highest stone formation was 8 in 20 rats at the 4 percent dosage level. None was found in the rats fed the 2 percent level, contrary to the findings of Fitzhugh and Nelson (92). Weil et al. (96, 97) conclude that diethylene glycol substantially free of ethylene glycol does not cause bladder stones, suggesting that it is not metabolized to any great degree to ethylene glycol.

3.4.4 Injection Toxicity

Karel et al. (98) report that when diethylene glycol was given to mice intraperitoneally the LD_{50} was 9.6 g/kg. The toxic effects seen were damage to the spleen, thymus, renal glomeruli, and tubules; high white cell count; and pulmonary congestion. Other LD_{50} values for rats reported are as follows:

intraperitoneal, 6.86 ml/kg (7.7 g/kg); intravenous, 8.0 ml/kg (8.9 g/kg); and subcutaneous, 16.8 ml/kg (18.8 g/kg) (99).

Marchenko (100) injected diethylene glycol intraperitoneally for 3 months to rats at a daily dose of 300 mg/kg and found that it caused brain edema, plethora of brain tissue, petechial hemorrhages, and irregular distribution of cytoplasmic RNA. Histologically, some degenerative changes and vacuolization were seen.

3.4.5   Eye Irritation

Carpenter and Smyth (39) reported that diethylene glycol failed to cause appreciable irritation when introduced into the eyes of rabbits. No cases of injury to human eyes have been reported nor would any be expected.

3.4.6   Skin Irritation

Diethylene glycol produces no significant skin irritation; however, prolonged contact over an extended period of time may produce a macerating action comparable to that caused by glycerol.

3.4.7   Skin Absorption

Hanzlik et al. (43) have shown that commercial diethylene glycol of unknown purity can be absorbed in toxic amounts through the skin of rabbits. The data, however, are erratic and therefore difficult to quantitate. A more definitive LD_{50} value for rabbits is 11.9 ml/kg (13.3 g/kg) (99).

Marchenko (100, 107) applied 2.8 g/kg/day for 2 months to the skin of rats and found that they developed edema of the brain, plethora, and minute brain hemorrhages.

Thus it appears that diethylene glycol may be absorbed through the skin especially upon essentially continuous contact, but that it is not likely to be a problem if reasonable care and caution are practiced. It would seem that the health hazard from skin absorption in anticipated industrial operations would be quite small.

3.4.8   Inhalation

An essentially saturated atmosphere generated at approximately 170°C and a fog generated at about 70°C caused no deaths of rats exposed for 8 hr (99).

Sanina (101) reported that 10 of 16 mice exposed to levels of 4 to 5 mg/m^3 (0.92 ppm) 2 hr/day for 6 to 7 months developed malignant mammary tumors, whereas no tumors developed in the 20 controls. The concentrations stated were generated by heating diethylene glycol at 30 to 35°C in a dish inside the 100-liter chamber housing the mice. The significance of these findings is minimal since the paper gives no indication that the control animals were subjected to the same extraneous stresses, particularly heat, as the experimental animals. Further, there is no indication of the purity of the diethylene glycol tested or of the composition of the atmosphere to which the mice were exposed.

Marchenko (100, 107) exposed mice and rats to about the same concentration as Sanina (101) did for 3 to 7 months and found structural changes in the central nervous system and endocrine and parenchymatous organs. Again the purity of the diethylene glycol was not given and the results appear to be in conflict with the other available data.

Although there appears to be little hazard from an occasional short-term inhalation, the results reported from chronic inhalation, though difficult to interpret, suggest that it would be wise to avoid repeated exposure to the vapor, fog, or mist. Certainly the hazard can be expected to be greater if the substance is handled hot or where agitation or other mechanical operations create a fog or mist in the air.

3.4.9 Mutagenicity

Pfeiffer and Dunkelberg (247) found diethylene glycol to be inactive when tested against S. typhimurium strains TA-98, TA-100, TA-1535, and TA-1537.

3.4.10 Carcinogenicity

Telegina et al. (106) studied the effect of exposure of diethylene glycol on 90 workers who produced aromatic hydrocarbons from crude oil and found that none of the workers exposed for 1 to 9 years experienced any changes in the neoplasm of the skin, or tumors of the nervous system or internal organs. There is, however, no mention of how long after exposure the workers were studied.

Weil et al. (96, 97) in their long-term studies on rats of three different age levels found only one bladder tumor in those fed diets containing 4 percent diethylene glycol. This tumor was found in a rat that also had bladder stones, as was reported by Fitzhugh and Nelson (92). To clarify the question of the cause of the tumor, Weil et al. (96, 97) implanted into the bladders of rats calcium oxalate stones or glass beads. They found that bladder tumors never developed without the presence of a foreign body in the bladder. This led to the conclusion that diethylene glycol essentially free of ethylene glycol is not a primary carcinogen.

Sanina (101) has reported that female mice that inhaled vapors of diethylene glycol developed mammary tumors but the validity of these conclusions is in question because of either serious flaws in the experimental design or inadequate reporting.

There are only limited data on the carcinogenicity of diethylene glycol, but the data available do not suggest that diethylene glycol presents a carcinogenic potential of significance.

3.4.11 Metabolism

Repeated administration to dogs for a week did not lead to consistent increases in urinary oxalate. However, the urinary oxalate was increased in rats maintained on water containing diethylene glycol. Wiley et al. (102) were unable to demonstrate the presence of oxalic acid in the urine following large doses of diethylene glycol to rabbits and dogs. These apparent discrepancies may well be attributable to the purity of the material studied.

3.4.12 Mode of Action

Diethylene glycol in large doses appears to be a depressant to the central nervous system. Deaths from large single doses which occur within 24 hr are believed to result from this action. Acutely toxic doses, not immediately fatal, may exert their effect primarily on the kidney and, to a lesser extent, on the liver. Deaths or serious injuries are associated primarily with renal insufficiency caused by the swelling of the convoluted tubules and a plugging of the tubules with debris. Chronic effects resulting from prolonged and repeated exposure to the commercial product, at least, are most likely to be centered in the kidney and to a lesser degree in the liver. In metabolism studies with the dog, Haag and Ambrose (103) found that a large portion of the diethylene glycol administered was excreted in the urine unchanged.

3.4.13 Human Experience

The human experience in the industrial handling and use of diethylene glycol has been excellent except for the "elixir of sulfanilamide tragedy," in which more than 100 deaths were attributed to its ingestion. Since then, a great number of articles have been published dealing with the clinical and experimental aspects of diethylene glycol poisoning. These have been well summarized by Geiling and Cannon (18), and none seem to reflect the neurologic observations in rats and mice reported by Marchenko (100, 107). A few cases have also been reported from other uses of diethylene glycol in medicinals (104) and from accidents (105). In general, pathology observed in human victims resembles closely that which has been described previously for laboratory animals and consists primarily of degeneration of the kidney with lesser lesions in the liver. Death in practically all these cases was due to renal insufficiency.

3.5 Hygienic Standards of Permissible Exposure

Because of the low vapor pressure of diethylene glycol, its low toxicity when studied by most investigators, and the lack of adverse human experience in the industrial setting, there has been no industrial hygiene standard established. Since significant atmospheric contamination with vapor can occur only under extreme conditions, an industrial hygiene guide of 10 mg/m^3 of particulate material is suggested. Amounts greater than this generally would be considered a nuisance to be avoided.

4 TRIETHYLENE GLYCOL; Ethanol, 2,2'-[1,2-Ethanediyl bis(oxy)]bis-; 2,2'-(Ethylenedioxy)diethanol; Triglycol; Bis(2-Hydroxyethoxy)ethane; CAS No. 112-27-6

$$(CH_2OCH_2CH_2OH)_2$$

4.1 Source, Uses, and Industrial Exposure

Triethylene glycol, like diethylene glycol, is produced commercially as a by-product of ethylene glycol production, its formation being favored by a high ethylene oxide to water ratio (108).

Triethylene glycol is used for many of the same applications as diethylene glycol but it has two distinct properties of importance; it is less volatile and less toxic. It is used as a humectant in tobacco, as a plasticizer, as a dehydrating agent for natural gas, and as a selective solvent. It is a valuable intermediate for the manufacture of plasticizers, resins, emulsifiers, demulsifiers, lubricants, explosives, and many others.

The industrial handling and use of triethylene glycol presents no significant problem from ingestion, skin contact, or vapor inhalation. Its low oral toxicity suggests that it may be considered safe for many applications where intake is limited. Similarly, its negligible skin irritation and absorption properties make it suitable for use to some extent in preparations intended to be applied over appreciable areas of the body. Furthermore, it is stable chemically and does not present a hazard due to flammability, except possibly at high temperatures or where fogs or mists are involved.

4.2 Physical and Chemical Properties

Triethylene glycol is a colorless to pale straw-colored, essentially odorless, viscous, hygroscopic liquid. It is miscible with water and many common solvents. It is practically insoluble in aliphatic hydrocarbons and fats. Additional properties are given in Table 50.1.

4.3 Determination in the Atmosphere

There would seem to be no need for determining atmospheric concentration of triethylene glycol for industrial hygiene purposes. If analysis of the atmosphere were to be made, however, the methods noted in Section 3.3 probably would be useful.

4.4 Physiologic Response

4.4.1 Summary

Triethylene glycol is very low both in acute and chronic oral toxicity, it is not irritating to the eyes or skin, and the inhalation of amounts that conceivably could cause injury does not seem likely.

4.4.2 Single-Dose Oral

Latven and Molitor (110), Smyth et al. (21), Laug et al. (17), and Stenger et al. (111) have studied the single-dose oral toxicity of triethylene glycol and found it to be less toxic than diethylene glycol. Smyth et al. (21) report the oral LD_{50} for rats and guinea pigs to be 22.06 and 14.66 g/kg, respectively. Laug et al. (17) found the LD_{50} values for rats, guinea pigs, mice, and rabbits to be 16.8, 7.9, 18.7, and 8.4 ml/kg, respectively. Stenger et al. (111) state that Woodard found the LD_{50} values for mice, rats, guinea pigs, and rabbits to be 21.0, 18.9, 8.9, and 9.5 g/kg.

4.4.3 Repeated-Dose Oral

The most comprehensive study of the repeated-dose oral toxicity of triethylene glycol is that reported by Fitzhugh and Nelson (92). These investigators fed the material at concentrations of 1.0, 2.0, and 4.0 percent in the diet of rats for 2 years without producing adverse effects. These dosage levels are equivalent to as much as 3 to 4 g/kg/day without effect. Earlier, Lauter and Vrla (112) reported that rats could tolerate 3 percent in their drinking water for 30 days without effect, but 5 percent caused ill effects. These dosages are equivalent to about 5 and 8 g/kg/day.

Lauter and Vrla (112) described the material they used as "commercial grade" but unfortunately Fitzhugh and Nelson (92) give no indication of the quality of the material they studied. Since commercial grade triethylene glycol may contain several percent of diethylene glycol, the possibility that the toxic effect seen by Lauter and Vrla (112) may have been caused by diethylene glycol rather than by triethylene glycol cannot be overlooked.

From these findings, it is apparent that triethylene glycol is very low in repeated-dose oral toxicity, far less than ethylene or diethylene glycols.

4.4.4 Injection Toxicity

Intramuscular.   Lauter and Vrla (112) administered single doses of triethylene glycol to rats intramuscularly by injection and found the LD_{50} to be approximately 8.4 g/kg.

Intraperitoneal.   Karel et al. (113) administered single doses of triethylene glycol to rats intraperitoneally by injection and found the acute LD_{50} dose to be 8.15 g/kg.

Intravenous.   Latven and Molitor (110) report the LD_{50} by intravenous injection to mice to be 7.3 g/kg. Stenger et al. (111) in their studies of triethylene glycol found the LD_{50} values for mice, rats, and rabbits to be 9.5, 11.7, and

10.6, respectively, for male guinea pigs, 1.9, and, for male dogs, greater than 4.5 g/kg.

Repeated daily injections intravenously to dogs of 0.11 and 0.56 g/kg/day for 1 month caused no adverse effects in those given the 0.11 g/kg/day dose and only thrombophlebitis in those receiving 0.56 g/kg/day level (111).

Subcutaneous. Latven and Molitor (110) also report an LD_{50} value by subcutaneous injection of 9.9 g/kg for mice.

Stenger et al. (111) injected triethylene glycol subcutaneously to rats at doses of 1.1, 2.3, and 4.5 g/kg/day for 1 month. Only those rats receiving 4.5 g/kg/day developed a slight decrease in hemoglobin and hematocrit values; otherwise there were no signs of toxicity.

4.4.5 Eye Irritation

Latven and Molitor (110) tested triethylene glycol for its effect upon the rabbit eye and found it to be similar to glycerin and diethylene glycol and less irritating than propylene glycol.

Carpenter and Smyth (39) report that triethylene glycol failed to cause appreciable irritation when introduced into the eyes of rabbits. No cases of injury to human eyes have been reported nor would any be expected.

4.4.6 Skin Irritation

Triethylene glycol produces no significant irritation of the skin. However, prolonged contact over an extended period of time may result in a macerating action similar to that caused by glycerin.

4.4.7 Skin Absorption

No studies have been reported dealing with the skin absorption of triethylene glycol. Although it is possible that, under conditions of very severe prolonged exposures, some of the material may be absorbed through the skin, it is extremely doubtful that a quantity sufficient to produce an appreciable systemic injury would be absorbed.

4.4.8 Inhalation

Interest in the toxicity of triethylene glycol when inhaled was initiated by the observation by Robertson (114) in 1943 and later in 1947 (115) that triethylene glycol was an effective air sterilizer.

During the studies on effectiveness, numerous persons were exposed, and, according to Jennings et al. (116) and Harris and Stokes (117), none were adversely affected. The developments in the field of air sterilization have been well reviewed by Polderman (118).

Also in 1947, Robertson et al. (119) reported extensive experiments with monkeys and rats showing that prolonged inhalation of saturated vapors, as in air disinfection (about 1 ppm), was without any physiologic effect.

Antonyuk (27) exposed test animals (species not given) continuously for 3 months to the vapors of triethylene glycol at concentrations of 5 and 1 mg/m^3 (approximately 0.814 and 0.163 ppm) and found that the 5 mg/m^3 level caused minor effects at most and that the 1 mg/m^3 level caused no effects. An allowable level of 0.3 mg/m^3 was recommended. The significance of this report is nil because there is no indication as to whether the controls were subjected to all the stresses, other than the triethylene glycol, that the experimental animals experienced.

4.4.9   Teratogenicity

Stenger et al. (111) treated female mice, rats, and rabbits during selected periods of pregnancy with triethylene glycol at levels of 2 ml/kg/day (2.25 g/kg/day) by subcutaneous injection, and rats at a level of 4 ml/kg/day (4.50 g/kg/day) by oral administration, and found that there was no evidence of teratogenic effects.

4.4.10   Metabolism

McKennis and co-workers (120) recently studied the fate of ^{14}C-labeled tri-ethylene glycol in rats and of unlabeled material in rabbits. They found in both species that a high percent of small doses was eliminated in the urine unchanged and possibly as the mono- and dicarboxylic acid derivatives of triethylene glycol. In the studies with rats, little if any ^{14}C-oxalate or ^{14}C-triethylene glycol in conjugated form was found in the urine. Small portions of the administered ^{14}C activity were found in the feces (2 to 5 percent) and in expired air (1 percent). Recoveries of the administered ^{14}C activity ranged from 91 to 98 percent.

4.4.11   Human Experience

The human experience in the handling and use of triethylene glycol has been uneventful and without reported cases of any adverse effects.

4.5   Hygienic Standards of Permissible Exposure

It does not seem that a hygienic standard for triethylene glycol is necessary.

5 TETRAETHYLENE GLYCOL; Ethanol, 2,2'-[Oxybis(2,1-ethanediyloxy)]bis-; 2,2'-Oxybis(ethylenedioxy)diethanol; CAS No. 112-60-7

$$(HOCH_2CH_2OCH_2CH_2)_2O$$

5.1 Source, Uses, and Industrial Exposure

Tetraethylene glycol is prepared commercially by adding ethylene oxide to ethylene glycol, diethylene glycol, or water in the presence of a suitable catalyst.

Tetraethylene glycol is used as a plasticizer and solvent where a high boiling point and low volatility are important. It can be an effective coupling agent in formulating water-soluble and water-insoluble materials. It is used as a chemical intermediate in the manufacture of glycol esters and ethers and in certain resins.

Industrial exposure to tetraethylene glycol is mainly topical.

5.2 Physical and Chemical Properties

Tetraethylene glycol is a colorless liquid, miscible with water and many common solvents. It is practically insoluble in aliphatic hydrocarbons and oils. Additional properties are given below.

Molecular formula	$C_8H_{18}O_5$
Molecular weight	194.1
1 ppm $\approx$	7.93 mg/m^3 at 760 mm Hg and 25°C
1 mg/l $\approx$	126.0 ppm at 760 mm Hg and 25°C

5.3 Determination in the Atmosphere

Methods described for ethylene glycol probably can be adapted. The method described by Ramstad et al. (122) for polypropylene glycols may also be applicable.

5.4 Physiologic Response

Summary (99). Tetraethylene glycol is very low in single-dose oral toxicity; the LD$_{50}$ value for rats is 30.8 ml/kg. It is nonirritating to the eyes and skin and is not likely to be absorbed through the skin in toxic amounts, the LD$_{50}$ for rabbits being 20.0 ml/kg. Essentially saturated vapors of this material caused no significant adverse effects when rats were exposed to them for 8 hr.

Yoshida et al. (34) report that chicks fed a diet containing 5 percent of tetraethylene glycol for 27 days as an energy source were unaffected. They postulate that chicks apparently were unable to metabolize it. Weifenbach (121), in his study of the suitability of the use of tetraethylene glycol as a solvent for drugs, found that it caused a slow drop in blood pressure, late production of arrhythmias just before death, and kidney and blood effects when it was infused intravenously into anesthetized rats at a rate of 22 ml/hr. He suggested that it should not be considered an inert solubilizer for drugs.

It appears from these data that tetraethylene glycol should pose no significant health hazard in its handling in industrial operations.

6 POLYETHYLENE GLYCOLS; Poly(oxy-1,2-ethanediyl)α-hydro-ω-hydroxy-, CAS No. 25322-68-3

$$HO(CH_2CH_2O)_n H$$

6.1 Source, Uses, and Industrial Exposure

The polyethylene glycols are prepared commercially by adding ethylene oxide to ethylene glycol, diethylene glycol, or water in the presence of caustic or other catalysts. The molecular weights of the product can be controlled by the proportions of the reactants used (108).

The polyethylene glycols of average molecular weight 600 or less exist as liquids at room temperature. They find primary application as reactive intermediates for the manufacture of fatty acid ester surfactants and as solvents for gas processing.

The polyethylene glycols of average molecular weight 1000 to 2000 exist at room temperature as soft to firm solids with low melting points. They serve primarily as bases for cosmetic creams and lotions, as well as pharmaceutical ointments and toothpaste formulations.

The polyethylene glycols of average molecular weight of 3500 to 20,000 exist at room temperature as firm to hard, brittle, waxlike solids. They are used as binders, plasticizers, molding compounds, stiffening agents, and paper adhesives. Some use of the esters of higher molecular weight polyethylene glycols as cosmetic formulation thickeners has also been noted.

Industrial exposure to the polyethylene glycols is almost entirely limited to topical contact.

6.2 Physical and Chemical Properties

The polyethylene glycols may be represented by the formula $H(OCH_2CH_2)_n OH$. Those having average molecular weights of 200, 300, 400, and 600 are viscous, nearly colorless, odorless, water-soluble liquids having very low vapor pressures.

Table 50.3.   Physical and Chemical Properties of Polyethylene Glycols

Material Designation[a]	Physical State	Mol. Wt. Range	Specific Gravity (25/25°C)	Freezing Range (°C)	Refractive Index (25°C)	Flash Point[b] (°F)	Refs.
200	Liquid	190–210	1.125	Supercools	1.459	340–360	128, 141
300	Liquid	285–315	1.125	−15 to −6	1.463	385–415	128, 141
400	Liquid	380–420	1.125	4–8	1.465	435–460	128, 141
600	Liquid	570–630	1.125	20–25	1.466	475–480	128, 141
1000	Solid	956–1050	1.117	36–40		490–510	128, 141
1450	Solid	1300–1600	1.210	43–46		490	128
1500	Solid	1300–1600	1.21	43–46		490	141
1540	Solid	1300–1600	1.21	43–46		510	141
2000	Solid	1900–2300	1.211	47–50		510	128
4000	Solid	3000–3700	1.204	53–56		520	141
4000	Solid	4200–4800	1.212	54–57		515	128
6000	Solid	6000–7500		60–63		520	141
6000	Solid	7000–8000	1.212	56–59		515	128
9000	Solid	9000–10,000	1.212	60–64		520	128
10,000	Solid						
14,000	Solid						
20,000	Solid						
4,000,000	Solid						

[a] Generally designates average molecular weights.
[b] Cleveland open cup.

The polyethylene glycols having average molecular weights of 1000 and more are nearly colorless, water-soluble, waxy solids at room temperature. All the unstabilized polyglycols are inherently susceptible to oxidative degradation, which occurs with increased rapidity as temperature increases and as the availability of oxygen increases. Their physical and chemical properties are given in Table 50.3.

6.3   Determination in the Atmosphere

The determination of the polyglycols in aqueous solution can be accomplished by the method of Duke and Smith (6) based upon the reaction of alcoholic hydroxyl groups with ammonium hexanitratocerate to form a red product. Infrared spectrophotometry may also be useful. Shaffer and Critchfield (123) have described a method for quantitatively determining the solid polyethylene glycols in biologic materials. Ramstad et al. (122) have described a method for polypropylene glycol which should be adaptable.

6.4 Physiologic Response

6.4.1 Summary

The polyethylene glycols present practically no hazards to health in industrial handling and use. They are not significantly irritating to the eyes, skin, or mucous membranes; they are exceptionally low in oral toxicity; and their vapor pressures are so low that there is no hazard from inhalation.

6.4.2 Single-Dose Oral

All of the polyethylene glycols are very low in single-dose oral toxicity. It is noteworthy that toxicity decreases as molecular weight increases. Representative figures are given in Table 50.4.

6.4.3 Repeated-Dose Oral

Smyth and co-workers (124, 125) summarized the extensive feeding studies they conducted with the polyethylene glycols. The polyethylene glycols having average molecular weights of 400, 1540, and 4000 caused no adverse effect upon dogs when fed in their diet for 1 year at the level of 2 percent (125). When fed to rats for 2 years as a part of their diet, polyethylene glycols 1540 and 4000 had no effect at a level of 4 percent, and polyethylene glycol 400 had no effect at a level of 2 percent (124). The other polyethylene glycols have been studied by dietary feeding techniques using rats, but for shorter periods of 3 to 4 months (29, 125). It appears that several percent of these materials can be tolerated in the diet of rats without appreciable adverse effects, indicating that they are exceptionally low in repeated-dose oral toxicity (see Table 50.5). The hazard from their ingestion would seem to be slight.

6.4.4 Injection Toxicity

The toxicity information from single-dose injection for the polyethylene glycols is summarized in Table 50.6. It is apparent that they are all low in toxicity when injected. As one would expect, the degree of toxicity when given intravenously depends upon the concentration of the compound given and the rate at which it is given (see rabbit data, Table 50.6). This is further substantiated by Weifenback (121), who found that the time required to cause the first signs of toxicity for E-200 depends on the rate of infusion. Both the 6 ml/hr and the 12 ml/hr rates caused light cramps after about 20 min, whereas the rate of 22.5 ml/hr caused strong cramps in about 4.5 min. Further infusion led to a drop in blood pressure, arrhythmia, bloody urine, and finally death.

Roze (135) injected rats intramuscularly with polypropylene glycol 6000 three

Table 50.4. Single-Dose Oral Toxicity of Polyethylene Glycols

	Rats			Guinea Pigs			Rabbits				
Material	M	B	F	M	B	F	M	B	F	Mice	Ref
200		34						20			124
200	34		28			17	14			34	29
300		39			20			21			124
300	30		29	21					21	31	29
400		44			16			27			124
400		44				21	22			36	29
400		33								29	126
600	33		30			28	19			36	29
1,000		42			22						124
1,000	45		32			41			>50	>50	29
1,500		44			29			30			124
1,540		51			37					50	124
2,000	45		>50	>50			>50			>50	29
4,000		59			51			76			124
4,000	>50		>50		50		>50			>50	29
6,000		>50		>50							124
6,000	>50		>50			>50	>50			>50	29
9,000	>50		>50	>50				>50		>50	29
10,000		>50									124
14,000		>32									29
20,000		>16									29
4,000,000		>4.0									126

^{a} M = males; B = both sexes; F = females.

times a week for 3 weeks and observed no adverse effects. However, when this routine was carried out on rats maintained on drinking water containing 1 percent of ethylene glycol, the effects of the ethylene glycol seemed to be enhanced.

A number of these polyethylene glycols have been tested by repeated injection. Table 50.7 summarizes these results.

6.4.5 Eye Irritation

Carpenter and Smyth (39) have reported that the polyethylene glycols do not cause appreciable irritation to the eyes of rabbits. This has been confirmed (128). No cases of injury to human eyes have been reported nor would any be expected.

Table 50.5. Summary of Repeated Oral Dose Toxicity of the Polyethylene Glycols

Mean Molecular Weight	Species	Sex[a]	Duration of Study (months)	Dosage Level (% Concn. in Diet) Without Effect	With Effect	First Sign of Adverse Effect Noted[b]	Ref.
200	Rat	B	3	8	16	LW	125
300	Rat	B	3	4	8	W	125
400	Rat	B	3	8	16	W	125
	Rat	M	24	2	4	W, LW, LP	125
	Rat	F	24	4			125
	Dog	B	12	2			125
600	Rat	B	3	8	16	KW, W	125
1000	Rat	B	3	8	16	W	125
	Rat	B	3	10	15	W	29
1500	Rat	B	3	4	8	W	125
1540	Rat	B	3	4	8	W	125
	Rat	B	24	4	8	LP	125
	Dog	B	12 (Possibly 8)	2			125
2000	Rat	B	4	15			29
4000	Rat	B	3	4	8	W	125
	Rat	B	4	5	10	W, LW	29
	Rat	B	24	4	8	W	125
	Dog	B	12 (Possibly 8)	2			125
6000	Rat	B	3	16	24	KW, W	125
	Rat	M	3	10	15	W	29
	Rat	F	3	15			29
9000	Rat	B	3	15			29
10,000	Rat	B	3	>2.5			99
4,000,000	Rat	B	24	5 (or 2.76)[c]		None	126
4,000,000	Rat	B	3		10	LP, KP	126
4,000,000	Dog	B	24	2 (or 0.56)[c]		None	126

[a] B = both sexes; M = male; F = female.
[b] W = decrease in body weight; LW = increase in liver weight/100 g of body weight; KW = increase in kidney weight/100 g of body weight; LP = slight histologic changes in the liver. KP = slight histologic effects in the kidney.
[c] g/kg/day.

6.4.6 Skin Irritation and Sensitization

Although early reports by Smyth et al. (129) reported that skin sensitization was observed among a few human subjects and in guinea pigs tested with certain polyethylene glycols, later studies (124, 128, 134) show that currently produced materials are without irritating or sensitizing properties. This has been borne out by their very wide application without difficulty in cosmetics.

6.4.7   Skin Absorption—Single Doses

As concluded by Smyth et al. (124), the size of the single dose of the polyethylene glycols required to kill by skin penetration is so large as to defy the establishment of LD_{50} values. Unpublished studies (29), employing essentially the technique of Draize et al. (130), have shown that single doses of 20 g/kg of the various polyethylene glycols ranging from 200 through 9000 were without toxic effects.

6.4.8   Skin Absorption—Repeated Doses

The studies reported by Luduena et al. (131) in 1947 indicated that toxic amounts of polyethylene glycols 200, 400, 1500, and 4000 were quite readily absorbed through the skin of rabbits when applied by inunction 6 days a week for 5 weeks. However, since part of this study was on animals on a deficient

Table 50.6.   Summary of Single-Dose Toxicity by Injection of Polyethylene Glycols

Mean Molecular Weight	Route[a]	Species	Dosage Level	Effects	Ref.
200	IP	Mouse	>4.6 mg/kg	LD_0	127
	IP	Mouse	>1.0 mg/kg	No effect	127
200	SC	Rat	>10 ml/kg (12.5 g/kg)	No effect	99
300	IP	Rat	17.0 g/kg	LD_{50}	99
300	IV	Rat	7.13 g/kg	LD_{50}	99
400	IP	Rat	12.3 g/kg	LD_{50}	99
400	IV	Rat	4.7 g/kg	LD_{50}	99
400	IP	Mouse	14.5 g/kg	LD_{50}	139
	IV	Mouse	8.55 g/kg	LD_{50}	139
	IP	Rat	14.7 g/kg	LD_{50}	139
	IV	Rat	7.3 g/kg	LD_{50}	139
600	IP	Rat	12.9 ml/kg (14.1 g/kg)	LD_{50}	99
1000	IP	Rat	15.6 g/kg	LD_{50}	99
1000	IV slow	Rabbit	>10.0 g/kg	LD_{50}	99
1500	IP	Rat	17.7 g/kg	LD_{50}	99
1500	IV	Rat	8.5 g/kg	LD_{50}	99
1500	IV slow	Rabbit	>10.0 g/kg	LD_{50}	99
1540	IP	Rat	15.4 g/kg	LD_{50}	99
	IV slow	Rabbit	>10.0 g/kg	LD_{50}	99
4000	IP	Rat	9.7 g/kg	LD_{50}	99
4000	IV	Rat	7.5 g/kg	LD_{50}	99
4000	IV slow	Rabbit	> 10.0 g/kg	LD_{50}	99
10,000	IP	Rat	12.6 g/kg	LD_{50}	99
4,000,000	IV	Rat	10 mg/kg as 0.025% solution	No effect	126
4,000,000	IV	Rat	3 mg/kg as 0.10% solution	100% death	126

[a] IP = intraperitoneally; SC = subcutaneously; IV = intravenously.

Table 50.7. Summary of Repeated-Dose Injection Toxicity of Polyethylene Glycols

Average Molecular Weight	Species	Route[a]	Duration	Dosage Level	Effects[b]	Ref.
200	Rabbit	IV	30 day	1 g/kg/day as 5% in H_2O	LP KP	99
300	Rabbit	IV	30 day	1 g/kg/day as 5% in H_2O	No effect	99
400	Rabbit	IV	30 day	1 g/kg/day	LP, KP	99
1000	Rabbit	IV	30 day	1 g/kg/day	LP, KP	99
1500	Rabbit	IV	30 day	1 g/kg/day	No effect	99
1540	Rabbit	IV	30 day	1 g/kg/day	No effect	99
4000	Rabbit	IV	30 day	0.8 g/kg/day as 40% in H_2O	No effect	99
4000	Dog	IV	354 days	90 mg/kg/day	No effect	134
6000	Rabbit	IV	30 day	1 g/kg/day	No effect	99
6000	Rat	IM	3 weeks	1.6 g/kg/day 3× weekly	No effect	135

[a] IV = intravenous; IM = intramuscular.
[b] LP = slight liver histopatholgy; KP = slight kidney histopathology.

diet, and since there were no controls, its significance cannot be evaluated. This is emphasized by the fact that other well-controlled studies by Smyth et al. (125), using similar materials, and by Tusing et al. (132), using a wider spectrum of polyethylene glycols, found no adverse effects from larger doses over a longer period of time.

Schmidt (133) also tested E-300 and E-600 on mice 24 times over an 8-week period. He too found that doses of 0.05 ml to an area of 4 to 6 cm² had no effect.

It is concluded from the data available that there is no hazard from the repeated skin application of currently produced polyethylene glycols.

6.4.9 Inhalation

There would not seem to be any hazard from inhalation of the polyethylene glycols.

6.4.10 Metabolism

The polyethylene glycols in general appear to be slow-acting parasympatho-mimetic-like compounds, according to Smyth et al. (124). When they are given intravenously, they tend to increase the tendency of the blood to clot and if given rapidly cause clumping of the cells, and death occurs from embolism.

6.4.11 Absorption and Excretion

Shaffer and Critchfield (136) in 1947 reported that polyethylene glycols having an average molecular weight of 4000 and 6000 were not absorbed from the rat intestine within 5 hr, whereas lower molecular weight materials (1000 and 1540) were absorbed to a very slight extent. When 1-g doses of materials having average molecular weights of 6000 and 1000 were given intravenously to human subjects, 96 percent of the 6000 molecular weight material and 85 percent of the other were excreted in the urine in 12 hr. When these same two materials in 10-g doses were given orally to five human subjects, none of the 6000 molecular weight material was found in the urine in the following 24 hr, whereas about 8 percent of the 1000 molecular weight material was found.

Shaffer et al. (137) in 1950 reported on studies with human subjects using polyethylene glycol having an average molecular weight of 400. They were able to recover 77 percent in the urine in 12 hr following the administration of 1 g intravenously, and to recover 40 to 50 percent in the urine when a 5 to 10 g dose of the material was given orally. These authors are of the opinion that ethylene glycol is not a metabolite of polyethylene glycol 400.

Principe in 1968 (138) applied polyethylene glycol 400 and 4000 to the skin of horses and assumed that they were not absorbed through the skin because they were not found in the urine. A dose of 4.26 g of polyethylene glycol given orally gave rise to only trace amounts in the urine during the 72-hr test period. If injected intramuscularly the polyethylene glycol 4000 was eliminated in the urine essentially completely in about 30 hr. Chemical studies of the urine indicate that it was excreted essentially unchanged.

Carbon-14-tagged polyethylene glycol 4,000,000 was shown not to be absorbed from the digestive tract of the rat and dog when it was administered orally (126).

Carpenter et al. (134) also tested for the absorption and excretion of polyethylene glycol 4000, and found that when given orally it was excreted via the feces but when it was given by intravenous injection it was excreted via the urine.

It would appear that the high molecular weight polyethylene glycols (1000 and higher) seem to be absorbed from the gut very little or not at all. However, if they are given by injection they seem to be excreted via the urine essentially unchanged. Those with molecular weights of less than 1000 apparently are absorbed from the gut in varying amounts and excreted both in the feces and in the urine essentially unchanged.

6.4.12 Effect on Absorption of Other Materials

Schutz (140) has observed that polyethylene glycol 400 markedly reduces the percutaneous absorption of certain materials which readily penetrate the intact skin in toxic amounts. He found that phenol, dimethylaniline, phenol red, barbital, salicylic acid, and γ-hexachlorocyclohexane were very poorly absorbed from solutions of the glycol.

Polyethylene glycol E-300 (PEG-300) and PEG-300/methylated spirits (PEG-IMS) (6:1 v/v ratio of PEG-300 to methylated spirits (95 percent ethanol, 5 percent methanol)) have been reported to be superior to water when used as decontaminants of the skin of rats exposed to phenol (248, 249). Other similar studies using the skin of swine, believed to be more similar to the skin of humans than is the skin of rats, indicates that a water shower for 15 min is practically equivalent to decontamination with PEG-300 or PEG-IMS. Both treatments were effective in reducing mortality, skin injury, and plasma concentration and retention time of absorbed phenol as compared to animals exposed to phenol but not decontaminated. These data plus the universal availability of water indicates that water is the decontaminant of choice (29).

6.4.13   Human Experience

There are no reported human injuries or adverse effects from the handling of these polyethylene glycols.

6.5   Hygienic Standards of Permissible Exposure

No hygienic standard is believed necessary.

7   PROPYLENE GLYCOL; 1,2-Propanediol; Methyl Ethylene Glycol; 1,2-Dihydroxypropane; CAS No. 57-55-6

$$CH_3CHOHCH_2OH$$

7.1   Source, Uses, and Industrial Exposure

Propylene glycol generally is synthesized commercially by starting with propylene, converting to the chlorohydrin, and hydrolyzing to propylene oxide, which is then hydrolyzed to propylene glycol. It can also be prepared by other methods (108).

Propylene glycol is used in antifreeze formulations, heat exchangers, and brake and hydraulic fluids; in the manufacture of resins, which accounts for a large portion of its use, polypropylene glycols, and propylene glycol ethers and esters; as a solvent in pharmaceuticals, foods, cosmetics, and inks; as a plasticizer for resins and paper; and as a humectant in textiles, tobacco, and pet foods. It is also used in the vapor form as an air sterilizer for hospitals and public buildings.

Industrial exposures are from direct contact, or from inhalation of vapors and of mists where the material is heated or violently agitated. Other exposure is by ingestion resulting from its use in foods and drugs.

7.2 Physical and Chemical Properties

Propylene glycol is a colorless, almost odorless, slightly viscous liquid with a slightly acrid taste. It imparts no odor or taste of its own when used in food colors and flavors. Additional physical and chemical properties are given in Table 50.1.

7.3 Determination in the Atmosphere

The determination of propylene glycol in air can be accomplished by several of the methods noted under ethylene glycol, although it would seem unnecessary for industrial hygiene purposes.

A method for detection of propylene glycol in body fluids is described by Lehman and Newman (142).

7.4 Physiologic Response

7.4.1 Summary

The hazards to health in the industrial handling and use of propylene glycol would seem to be negligible. Its systemic toxicity is especially low and, since 1942, it has been considered a proper ingredient for pharmaceutical products (143). The Food and Drug Administration does not object to its use in food products or in cosmetics (144). The inhalation of atmospheres containing propylene glycol vapor presents no hazard to health. Exposures created by operations producing hot vapors, or by high-speed mechanical action in which a fog of propylene glycol is produced, have not been studied. However, it is difficult to visualize how this condition could create a hazard, since the material is so extremely low in systemic toxicity. The toxicology of propylene glycol has been reviewed extensively (20, 145, 146).

7.4.2 Single-Dose Oral

The single-dose oral toxicity of propylene glycol has been studied by a number of investigators (142, 145, 147, 148, 139, 99, 17). The single-dose oral LD_{50} values range, for rats, from 21.0 to 33.7 g/kg; for mice, from 23.9 to 31.8 g/kg; for guinea pigs, from 18.4 to 19.6 g/kg; for rabbits, from 15.7 to 19.2 g/kg; and for dogs, from 10 to 20 g/kg.

Laug et al. (17) report observing minimal kidney changes from large doses. From one-fourth to one-half of an oral dose given to rats, dogs, or human beings appears unchanged in the urine within 24 hr (142, 148, 149, 150).

7.4.3 Repeated-Oral Dose

Seidenfeld and Hanzlik (151) gave groups of rats drinking water containing 1.0, 2.0, 5.0, 10.0, 25.0, and 50.0 percent propylene glycol over a period of 140 days. Animals receiving water containing either 25.0 or 50 percent propylene glycol died in 69 days, whereas those receiving either 1.0, 2.0, 5.0, or 10.0 percent appeared normal throughout the observation period. The average daily intakes for the latter four groups were calculated to be about 1.6, 3.7, 7.7, and 13.2 g/kg/day of propylene glycol, respectively. Histopathologic examination of the tissues from these animals revealed no renal or other pathologic disturbances. Weatherby and Haag (147) confirmed the fact that rats will tolerate 10.0 percent propylene glycol in the drinking water without physiologic impairment.

Hanzlik and associates (148) found that rats could tolerate up to 30 ml/kg daily of propylene glycol when fed in the diet over a 6-month period. This is equivalent to 1.8 lb daily for a 70-kg person. Morris et al. (19) fed rats 2.45 and 4.9 percent propylene glycol in the diet, allowing, respectively, average daily intakes of 0.9 to 1.77 ml/kg over a 24-month period without significant effect on growth rate. Microscopic examination of the tissues revealed very slight liver damage, but no renal pathology.

Whitlock et al. (152) found that a diet containing 30 percent of propylene glycol was not well tolerated by young rats, and that producing females were unable to bring their young to weaning. Glycerin at 30 percent in the diet was well tolerated. Diets containing 40, 50, or 60 percent of propylene glycol were lethal after a few days.

Gaunt et al. (153) fed both male and female rats propylene glycol in their diet at levels of 6,250, 12,500 and 50,000 ppm for a period of 2 years. They found no significant ill effects based upon mortality, body weight, food consumption, hematology, urinary cell excretion, urine-concentrating ability of the kidneys, organ weights, or histopathology, including tumor incidence. They suggest that the acceptable daily intake for man would be 25 mg/kg/day. (The 50,000 ppm level is equivalent to 2.5 g/kg/day for the rat.)

Van Winkle and Newman (155) showed that propylene glycol, when given in concentrations of 5 or 10 percent in the drinking water of dogs for 5 to 9 months, caused no adverse effects. Criteria employed were liver function, kidney function, and histopathologic examination of the visceral organs. Further, they found no alterations in the serum calcium levels of cats and dogs fed large doses of propylene glycol.

Weil et al. (154) found that male and female dogs fed diets providing propylene glycol at a dose level of 2.0 g/kg for 2 years were unaffected as judged by mortality, body weight change, diet utilization and water consumption, histopathology, organ weights of liver, kidney, and spleen, and measurement of blood, urine, and biochemical parameters. At a daily dose of 5 g/kg, the dogs gained more weight than the controls, especially during the early part of the experiment, owing to the higher caloric intake. An increase in the rate of erythrocyte hemolysis and a slight increase in total bilirubin was also noted. Hemoglobin, packed cell volume, and total erythrocyte count were lowered slightly, whereas the incidence of anisocytes, poikilocytes, and reticulocytes was increased, suggesting that erythrocytes were being destroyed accompanied by accelerated replacement from the bone marrow. This effect was not sufficient, however, even at the 20 percent dietary level to result in any irreversible changes and there was no evidence of damage to the bone marrow or spleen. They state that dogs fed propylene glycol at approximately 8 percent in their diet (equivalent to 2 g/kg/day) can utilize it as a carbohydrate energy source with no adverse effects.

The utilization of propylene glycol as a source of carbohydrate energy in animal feed has led to other dietary feeding studies involving cattle, sheep, chickens, and cats. It is used in pet foods as a humectant as well as a source of energy. In lactating cattle, there is a tendency for the cow to develop hyperketosis which may be due to low carbohydrate stores in the body which are used to produce lactose excreted in the milk. Fisher and co-workers (156, 157) have summarized much of the work done to determine the usefulness of propylene glycol as a dietary supplement to alleviate this condition.

Fisher et al. (156) reported on studies in lactating cows in which propylene glycol was added to food concentrates at levels of 3, 6, and 9 percent. The concentrate was fed to cows for 8 weeks during the early stages of lactation and it was found that there was no consistent effect on feed intake, body-weight change, or efficiency of ration utilization. Propylene glycol, at the 3 and 6 percent levels, appeared to increase the yield of milk but caused a slight decrease in milk fat and an increase in milk lactose content. There was also a significant decrease in hyperketosis in the propylene glycol treated animals as compared to the level in the control animals. Sauer et al. (157), in the study of 120 cows over a 2-year period in which propylene glycol was added to the diet at 3 and 6 percent levels for 8 weeks postpartum, substantiated the suitability of propylene glycol as a food additive for such cattle. It depressed the blood ketone and free fatty acids slightly below the control cattle level when the cows were not stressed by either high lactation yield or low concentrate food intake. However, if the cows were stressed by adverse environmental factors and slightly reduced food concentrate intake, the addition of propylene glycol at 3 and 6 percent to the concentrate ration caused a significant reduction of blood ketones and plasma fatty acids and increased the blood glucose concentration. They suggest that propylene glycol used as a feed additive at 3 and 6 percent of the concentrate should be desirable because of its ability to significantly decrease the incidence of clinical and subclinical ketosis in cows during early lactation when they are most susceptible to such metabolic disorder.

Shiga et al. (158) fed male lambs propylene glycol at levels of 0.5 and 1.0 g/kg in the feed every other day for 104 days. Neither level produced any adverse effects. They did increase the weight gain, especially in the early part of the experiment (about 30 days). Evaluation of the rumen liquor revealed an increase in propionate and total fatty acid concentrations. The weight percentage of wool, dressed carcass, and red meat–fat ratios were also greater than those of the controls.

Propylene glycol has been shown by short-term feeding tests with chicks and chickens to be a suitable source of energy, the suggested level acceptable being in the range of 2.5 to as much as 8 percent in their diet. However, most authors feel that the level should be no more than 2 to 3 percent. Bailey et al. (159) fed chicks from 1 to 26 days of age with diets containing 8 and 16 percent of propylene glycol. The 8 percent level was considered to cause no significant ill effects as judged by live weight gain, food consumption, and carcass analysis. The 16 percent level caused growth depression and lower fat and higher protein content in the carcass. Persons et al. (160) also fed chicks from day 1 for 3 weeks with levels of 5 and 10 percent of propylene glycol in the diet. Both levels showed adverse effects such as depressed body-weight gains at the 5 percent level, and depressed body-weight gains and decreased food efficiency, but no mortality at the 10 percent level. They also fed hens for two periods of 28 days each with diets containing propylene glycol at levels of 2.5 and 5.0 percent. The 2.5 percent level decreased food consumption owing to the hens compensating for the change in energy intake, but there was no effect on egg production or general health. The 5 percent level caused decreased food and energy intake and reduced egg production. The authors suggest that a level of 2.5 percent of propylene glycol is tolerated well by both chicks and hens.

Based upon 21 to 28 day feeding studies on chicks of diets containing 2.5, 5.0, and 10.0 percent of propylene glycol, Waldroup and Bowen (161) found that chicks can utilize 5.0 percent of propylene glycol in their diet without ill effects. The 10 percent level caused depressed body-weight gains, reduced efficiency of feed utilization, diarrhea, and the development of deformed toes.

Yoshida et al. (34) also found that chicks fed 5 percent of propylene glycol in their diet for 27 days were without ill effects. Higher levels caused inferior well being and diarrhea.

Harnisch (162) feels that the no-effect level is in the range of 2 to 3 percent of propylene glycol. This resulted from his 8-week feeding studies on 4- to 5-week-old broilers in which he found that a diet containing 5 percent propylene glycol depressed food intake and reduced feed conversion efficiency.

Because of the use of propylene glycol as a humectant as well as a source of energy in prepared cat food, a study (29) was conducted in which groups of two male cats each were maintained for 94 days on diets containing various amounts of propylene glycol. Two groups of two male cats each served as controls. The average calculated doses of propylene glycol consumed, based on food intake and body weights, were 0, 80, 443, 675, 1763, or 4239 mg/kg/day. The primary treatment-related effect was noted in the red blood cells (RBC), which exhibited Heinz body formation. This effect in the RBC was accompanied by increased amounts of hemosiderin pigment in the Kupffer cells of the liver and reticuloendothelial cells of the spleen. The formation of Heinz bodies and increased hemosiderin occurred in a dose-related manner at doses of 675 mg/kg/day and higher. A daily dose level of 443 mg/kg/day appeared to cause a very slight increase in Heinz body formation without detectable increased hemosiderin present in the liver or spleen when compared to the incidence in the controls.

No treatment-related effects of any type were observed in the group ingesting 80 mg/kg/day. Other hematologic parameters which were evaluated and found to be unaffected by any level of treatment included packed cell volume, RBC count, hemoglobin, RBC morphology to evaluate polychromasia, RBC reticulocyte count, RBC osmotic fragility, methemoglobin, total and differential white blood cell counts, and light microscopic examination of tissue bone marrow. In addition, serum clinical chemistry values of blood urea nitrogen, glutamic pyruvic transaminase activity, alkaline phosphatase activity, glutamic oxaloacetic transaminase activity, glucose concentration, and total bilirubin as well as routine urinalysis were unaffected by treatment. The clinical appearance and demeanor, body weights, organ weights, gross pathology, and histopathology of tissues other than liver and spleen were unaffected by treatment at any of these levels of propylene glycol.

It appears that the cat is much more sensitive than other species to the formation and/or retention of Heinz bodies. Even so, their presence in significant numbers does not seem to adversely affect the cats.

7.4.4 Injection Toxicity

Table 50.8 gives the LD_{50} values by injection of various routes to a number of animal species.

Brittain and D'arcy (165) found that when propylene glycol was injected intravenously in rabbits, there were no effects on blood other than decreased clotting time, increased platelet count, and an increase in polymorphs accompanied by a decrease in lymphocytes. High doses in rats by intramuscular and subcutaneous injections resulted in profound depression, analgesia, and coma (146).

There was no anticholinergic effect in dogs but temporary increased urinary flow, peripheral vasodilatation, and a temporary vasoconstriction of the spleen when propylene glycol was injected intravenously (167).

Repeated subcutaneous injection of propylene glycol to rats at a dose of 2.5 ml/kg (2.6 g/kg) every other day for 1 month caused an increase in oxygen consumption even though there was no kidney or liver histopathology found (167).

7.4.5 Eye Contact

Propylene glycol is not injurious to the eyes of rabbits (39, 168) and has not caused any eye injury in human beings, nor would such be expected, but it may cause transitory stinging, blepharospasm, and lacrimation (108, 139).

7.4.6 Skin Irritation and Absorption

Propylene glycol generally produces no significant irritant action upon the skin. From the results of extensive studies by Warshaw and Herrmann (169) on some

Table 50.8. Acute Toxicity of Propylene Glycol by Injection to Laboratory Animals

Species	Route of Administration	LD_{50} (g/kg)	Ref.
Mouse	Subcutaneous	19.2	146
		15.5	36
	Intraperitoneal	9.73	146
		12.9	139
		13.6	36
		6.8	163
		10.9	145
		11.4	166
	Intravenous	8.3	146
		7.6	139
Rat	Subcutaneous	22.0, 25.0	145
		22.5, 21.7, 29.0	146
	Intraperitoneal	14.7, 13.5	146
	Radiated PG	14.2 (13.7 ml/kg)	164
	Unradiated PG	14.7 (14.2 ml/kg)	164
	Intraperitoneal	13.0	139
		16.8 (16.25 ml/kg)	99
	Intravenous	12.7 (12.3 ml/kg)	99
		6.2	139
		6.8	146
	Intramuscular	14.0, 20.7	146
		15.0, 13.0, 20.0	145
Guinea pig	Subcutaneous	13.0–15.5	146
Rabbit	Intravenous	5.0	145
		6.5	146
	Intramuscular	6.0	145
Dog	Intravenous	25.9	146

866 human subjects with various dermatologic backgrounds, it appears that propylene glycol may cause primary skin irritation in some people, possibly due to dehydration, but the material does not appear to be a sensitizer. Because of the very low systemic toxicity of propylene glycol, no problem from percutaneous absorption can be anticipated. Propylene glycol has been used widely in preparations for topical application and no evidence of systemic injury to humans has been reported. These findings are substantiated by other workers (99, 168, 170–175, 177).

One of the many medicinal uses of propylene glycol is as a solvent in eardrops. Morizono and Johnstone (176) found that a concentration of 10 percent or more or propylene glycol in Ringer's solution when instilled into the inner ear cavity of guinea pigs caused apparent irreversible deafness. They recommend that if propylene glycol is used in eardrops the concentration should be less than 10 percent.

7.4.7 Vapor Inhalation

Robertson et al. (119) exposed sizable groups of rats and monkeys for periods of 12 to 18 months to atmospheres saturated with propylene glycol vapor and produced no ill effects. Human beings also have been exposed to saturated and supersaturated atmospheres for prolonged periods in the air-sterilization program without adverse effect.

The uptake of propylene glycol mist by humans was studied using a 10 percent solution in labeled deionized water nebulized into a mist tent (178). Less than 5 percent of the mist entered the body, and of this 90 percent lodged in the nasopharynx and rapidly disappeared into the stomach. Very little was found in the lungs.

7.4.8 Reproduction

Guerrant et al. (179) checked the reproduction capacity of rats fed up to 30 percent of propylene glycol in their diets through six generations. No adverse effects on reproduction were found when concentrations of propylene glycol in the diet were less than 7.5 percent. At higher levels the rats receiving the propylene glycol consumed less food, grew slower, had young at an older age, produced smaller litters on the average, and weaned fewer young than did the control animals. At the 30 percent level, the females did not breed normally and when they had young they did not feed them properly. They failed to wean third generation young. Emmens (180) fed mice 0.1 ml of a 50 percent water solution of propylene glycol for several days prior to mating and found that it reduced the mating to as little as 30 percent of normal, and litters to 15 percent. The mice swelled visibly with intestinal gases and then recovered.

7.4.9 Teratogenicity

Gebhardt (181) found that 0.05 ml of propylene glycol was not teratogenic when injected into the yolk sac of chick embryos. However, it caused a high mortality of the embryos when injected into the air sac on the fourth day of development and unilateral micromelia in about 20 percent of the survivors. Waldroup and Bowen (161) have reported that chicks fed high levels of propylene glycol in their diets developed a high incidence of toe deformities, 57 out of 168 chicks as compared to the control group, 9 out of 168.

7.4.10 Mutagenicity

Kennedy et al. (182) studied the dominant lethal effects in mice treated with propylene glycol intraperitoneally at a level of 10 mg/kg. The results suggest that it is nonmutagenic at this level. Pfeiffer and Dunkelberg (247) found propylene glycol to be inactive when tested against S. typhimurium strains TA-98, TA-100, TA-1535, and TA-1537.

7.4.11   Carcinogenicity

Dewhurst et al. (183) and Baldwin et al. (184) in studies on the carcinogenicity of other chemicals used propylene glycol as the solvent. As a result they tested propylene glycol alone for carcinogenic activity. Dewhurst et al. (183) used a single injection of 0.2 ml whereas Baldwin et al. (184) gave three to five subcutaneous injections, amount not specified. In neither case were tumors observed over a period of about a year (184) or 2 years (183).

Wallenious and Lecholm (185) applied propylene glycol to the skin of rats three times a week for 14 months but found no tumor formation. Stenback and Shubik (186) confirmed these findings when they applied propylene glycol at undiluted strength and as a 50 and 10 percent solution in acetone to the skin of mice over their lifetime.

No development of tumors has been reported in the lifetime dietary feeding studies (19, 153, 154). In fact, Gaunt et al. (153) specifically states that no tumors were found in the rats. Thus it appears that propylene glycol is without carcinogenic properties.

7.4.12   Metabolism

Ruddick (145), in his review of the toxicity of propylene glycol, summarized the work done to establish its metabolism in the body. It is oxidized to lactic acid or pyruvic acid by two pathways. These two metabolites are then used by the body as a source of energy either by oxidation through the tricarboxylic acid cycle or by generation of glycogen through the glycolytic pathway. The metabolic pathways are thought to be as shown in Figure 50.2.

In ruminants, such as sheep and cattle, research has shown that the metabolism of propylene glycol is carried on to a large extent by the microbial flora in the rumen (187, 158, 156, 188). The metabolite is primarily propionate. In the chick, excessive intake of propylene glycol results in its passage to the cecum, where it is metabolized by bacteria to propionaldehyde (190a).

7.4.13   Mode of Action

Browning (20) states that the studies on propylene glycol indicate that about one-third is excreted via the kidneys as a conjugate with glucuronic acid and the rest is metabolized or excreted in the urine unchanged. This suggests that the organic injury and the central nervous system depressing action is probably due to the excessive presence of the propylene glycol and not to its metabolites or its glucuronide.

7.4.14   Human Experience

There has been no reported injury to humans that resulted from industrial use. However, when used in large doses as a vehicle for repeated medication of a

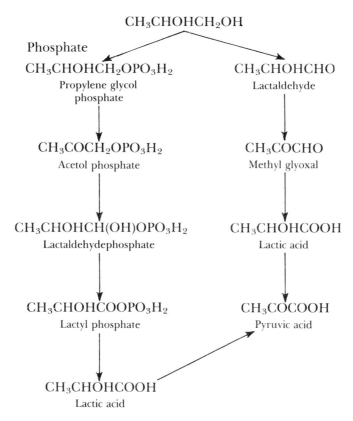

Figure 50.2 Metabolic pathways of propylene glycol.

15-mo-old youngster, it caused adverse signs characterized by hypoglycemia and central nervous system depression. Recovery was prompt upon cessation of treatment (190b). This observation would seem to have no significance as far as the hazard from industrial exposure is concerned.

7.4.15 Pharmacology

Propylene glycol is so low in pharmacological activity that very large doses can be given intravenously, provided it is given slowly. Rapid administration causes death. The material appears to have a sedative-type of effect and is glycogenic (148).

7.5 Hygienic Standards of Permissible Exposure

None appear to be necessary.

8 1,3-PROPANEDIOL; Trimethylene Glycol; CAS No. 504-63-2

$$HOCH_2CH_2CH_2OH$$

8.1 Source, Uses, and Industrial Exposure

1,3-Propanediol is prepared as a by-product in the manufacture of glycerin through the saponification of fat (108).

It is used to lower the freezing point of water and as a chemical intermediate. Industrial exposure is limited and of little concern.

8.2 Physical and Chemical Properties

1,3-Propanediol, an isomer of propylene glycol, is a viscous, colorless, odorless, hygroscopic liquid with a brackish irritating taste (108). Additional properties are given in Table 50.1.

8.3 Determination in the Atmosphere

Methods applicable to the other glycols should be useful.

8.4 Physiologic Response

8.4.1 Summary

Van Winkle (192) found 1,3-propanediol to be about twice as toxic as the 1,2-isomer, to be nonglycogenic, and to cause marked depression in near-fatal doses.

8.4.2 Single-Dose Oral (192)

When 1,3-propanediol was given in single oral doses, the LD_{50} for rats was estimated to be about 15 ml/kg (14.7 to 15.8 g/kg) with a lethal range of 9.5 to 19.0 ml/kg. For cats the LD_{50} was not determined but 3 ml/kg was fatal to cats. Cats, for reasons unknown, appear to be particularly sensitive to single oral doses of the material.

8.4.3 Repeated-Dose Oral

When the material was fed to rats as a part of their diet for 15 weeks, a concentration of 5 percent was without grossly apparent toxic effects but no autopsies were performed. Twelve percent in the diet was not well tolerated, as evidenced by poor growth. A daily dose of 5 ml/kg given by intubation also caused poor growth; a daily dose of 10 ml/kg was fatal to all rats within 5 weeks (192).

1,3-Propanediol was tested as an energy source for chicks as a replacement for corn. It was added to the diet at a level of 10 percent and was fed for 16 days. At this level the chicks experienced extremely poor growth, one-half of that seen in the controls, but no other specific signs (193).

8.4.4 Injection Toxicity (192)

When 1,3-propanediol was given intravenously as a 50 percent aqueous solution, the LD_{50} was found to be 4 to 5 ml/kg (4.2 to 5.3 g/kg) for rabbits and greater than 3 ml/kg (3.17 g/kg) for cats. When it was given intramuscularly the LD_{50} for rats was 6 to 7 ml/kg (6.33 to 7.39 g/kg) and for cats more than 3 ml/kg (3.17 g/kg).

8.4.5 Teratogenicity

Gebhardt (181) found that 1,3-propanediol caused micromelia when injected at the rate of 0.05 ml (0.053 g) into the air sac or yolk sac of fertile eggs on the fourth day or at the beginning of incubation. He suggested that 1,3-propanediol, unlike propylene glycol, may be teratogenic.

8.4.6 Metabolism

Gessner et al. (194) gave 1,3-propanediol to rabbits and analyzed the urine for it and its metabolites; none was found. They suggested that it might have been oxidized completely to carbon dioxide in the body.

8.4.7 Mode of Action

Ulrich and Mestitzova (195) state that 1,3-propanediol is different in its action from 1,2-propanediol (propylene glycol), for 1,3-propanediol harms the kidneys and the liver at much lower doses.

8.5 Hygienic Standards of Permissible Exposure

None seems necessary because of the low vapor pressure at room temperatures and the relatively low toxicity.

9 DIPROPYLENE GLYCOL; Propanol, Oxybis-; CAS No. 25265-71-8

$$CH_3CHOHCH_2OCH_2CHOHCH_3$$

9.1 Source, Uses, and Industrial Exposure

Dipropylene glycol is prepared commercially as a by-product of propylene glycol production. There are three linear isomers possible but these have not been separated and studied, and the exact composition of the commercial product is not known. It is also possible to prepare cyclic isomers such as 2,6-dimethyl-1,4-dioxane and 2,5-dimethyl-1,4-dioxane, but these are not likely to form under conditions employed commercially.

Dipropylene glycol is used for many of the same purposes as the other glycols but mostly in particular applications where its solubility characteristics (greater hydrocarbon solubility) and lower volatility are useful. It is not used in drugs, pharmaceuticals, or food applications because its toxicologic characteristics have not been clearly defined.

Industrial exposure is most likely to be from direct contact and possible inhalation of mist from heated or violently agitated material.

9.2 Physical and Chemical Properties

Dipropylene glycol is a colorless, odorless, slightly viscous liquid. See Table 50.1 for additional properties.

9.3 Determination in the Atmosphere

Dipropylene glycol undergoes the same general reactions as other glycols. The chemical methods for determination are the same nonspecific methods described for other glycols. Fair specificity can be obtained through the use of spectrographic methods. Chromatographic procedures such as described by Ramstad et al. (122) are adaptable.

9.4 Physiologic Response

9.4.1 Summary

Although dipropylene glycol is more active physiologically than propylene glycol, it is still very low in toxicity. The industrial handling and use of dipropylene glycol should present no significant problems from ingestion, skin contact, or vapor inhalation. The information available, however, is not considered adequate to allow an evaluation relative to the suitability of this material for use in foods, drugs, or cosmetics.

9.4.2 Single-Dose Oral

Dipropylene glycol is low in acute oral toxicity. The single dose LD_{50} for rats has been reported to be 14.8 g/kg (148) and 15.0 g/kg (196).

9.4.3 Repeated-Dose Oral

Rats were not affected by 5 percent dipropylene glycol in their drinking water for 77 days (197). In those animals administered a level of 10 percent, some died with hydropic degeneration of kidney tubular epithelium and liver parenchyma. These effects were similar to those of diethylene glycol but less severe and less uniformly produced. Yoshida et al. (34) were able to feed chicks a diet containing 5 percent of dipropylene glycol for 27 days without adverse effects. The chicks were unable to utilize it as an energy source.

9.4.4 Injection Toxicity

Browning (20) reports that the LD_{50} value for rats by intraperitoneal injection is 10.3 ml/kg (10.56 g/kg) and for dogs by intravenous injection 11.5 ml/kg (11.79 g/kg).

9.4.5 Eye Contact

Dipropylene glycol has been shown to cause no significant eye irritation or injury when tested in the eyes of rabbits (20).

9.4.6 Skin Contact

When dipropylene glycol was applied repeatedly to the skin of rabbits for prolonged periods (10 applications in 12 days), it had a negligible irritating action, and there was no indication that toxic quantities were absorbed through the skin (42).

9.4.7 Inhalation

Experimental data are not available on the vapor toxicity of dipropylene glycol. However, it is not likely to produce injury because of its low vapor pressure and low systemic toxicity.

9.4.8 Human Experience

No untoward effects have been reported from the use of dipropylene glycol nor would any be expected.

9.5 Hygienic Standards of Permissible Exposure

Because of the low vapor pressure and the low toxicity of dipropylene glycol, a hygienic standard seems unnecessary.

10 TRIPROPYLENE GLYCOL; Propanol, (1-Methyl-1,2-ethanediyl)bis(oxy)bis-; CAS No. 24800-44-0 and CAS No. 1638-16-0

$$CH_3CHOHCH_2OCH(CH_3)CH_2OCH(CH_3)CH_2OH$$

10.1 Source, Uses, and Industrial Exposure

Tripropylene glycol is made commercially as a by-product of propylene glycol production. It is used much as the other glycols: as an intermediate in the production of ethers, esters, and resins; as a nonvolatile solvent; and as a polyester humectant and a plasticizer. It is not used in drugs, pharmaceuticals, or food applications because its toxicologic properties have not been clearly defined.

Industrial exposure is limited almost entirely to direct contact with the liquid. Vapors or mists may be enountered under unusual conditions.

10.2 Physical and Chemical Properties

Tripropylene glycol is a colorless, slightly viscous liquid. Additional properties are given in Table 50.1.

10.3 Determination in the Atmosphere

Tripropylene glycol undergoes the typical reactions of polyols not having vicinal hydroxyl groups. Chromatographic procedures such as described by Ramstad et al. (122) are adaptable.

10.4 Physiologic Response

Summary. Tripropylene glycol is low in single-dose oral toxicity, the LD_{50} for rats being 11.8 ml/kg (12.0 g/kg) (99) and between 3 and 10 g/kg (29). Yoshida et al. (34) found that chicks were unaffected when they were fed a diet containing 5 percent tripropylene glycol for 27 days. It is nonirritating to the eyes or skin or rabbits (29, 99) and is low in toxicity by absorption through the skin, the LD_{50} for rabbits being greater than 16.0 ml/kg (16.3 g/kg) (99). Essentially continuous contact for 2 weeks resulted in no indication of absorption of toxic amounts through the skin of rabbits (29). Rats exposed to essentially saturated vapors generated at room temperature for 8 hr were unaffected (99); hence it is doubtful that hazardous conditions would occur when it is handled under anticipated industrial conditions.

There have been no reports of human experience.

10.5 Hygienic Standards of Permissible Exposures

None is believed necessary.

11 POLYPROPYLENE GLYCOLS; Poly(propanediols); CAS No. 25322-69-4

$$HO(C_3H_6O)_nH$$

11.1 Source, Uses, and Industrial Exposure

The polypropylene glycols are prepared commercially by reacting propylene glycol or water with propylene oxide.

They are used as lubricants, solvents, plasticizers, softening agents, antifoaming agents, mold release agents, and intermediates in the production of resins, surface active agents, and a large series of ethers and esters. They are widely used in hydraulic fluid compositions.

Industrial exposure is most likely to be direct contact with the skin and eyes. Ingestion should not be a problem except from accident. The very low volatility of these materials makes inhalation improbable except perhaps where mists are formed from violent agitation or high temperatures.

11.2 Physical and Chemical Properties

The polypropylene glycols are clear, lightly colored, slightly oily, viscous liquids having very low vapor pressures. All these materials are quite stable chemically and do not present hazards of flammability except at elevated temperatures. Additional physical and chemical properties are given in Table 50.9. Hereafter in this section the polypropylene glycols are referred to as "P", followed by the average molecular weight.

11.3 Determination in the Atmosphere

The method described by Ramstad et al. (122) should be useful.

11.4 Physiologic Response

11.4.1 Summary

The low molecular weight polypropylene glycols (200 to 1200) have an appreciable single-dose oral toxicity, are mildly irritating to the eyes, are not irritating

Table 50.9. Physical and Chemical Properties of Polypropylene Glycols

Material Designation[a]	Specific Gravity[b]	Pour Point (°C)	Refractive Index (25°C)	Flash Point[c] (°F)	Fire Point[c] (°F)	Ref.
400	1.007	−49	1.445	330		122
				390	405	198
750	1.004	−44	1.447	495	525	203
1200	1.007	−40	1.448	224[d]		202
				460	505	198
2000	1.002	−30	1.449	390[d]		202
				445	510	198
3000	1.001	−29	1.449	440	505	203
4000	1.005	−26	1.450	365[d]		202
				445	515	198

[a] average molecular weight.
[b] 25/25°C.
[c] Cleveland open cup.
[d] Pensky Mertens closed cup.

to the skin, and although they are absorbed through the skin to some extent, skin penetration would not seem to present a serious industrial hazard. The inhalation of mists or vapors from heated material, particularly low molecular weight material, could be hazardous. These materials are not like the low molecular weight polyethylene glycols in their physiologic activity; they are rapidly absorbed from the gastrointestinal tract, are potent central nervous system stimulants, and readily cause cardiac arrhythmias. The higher molecular weight materials with average molecular weights of 2000 or more are very low in toxicity by all routes and do not have the stimulant effect upon the central nervous system typical of the lower molecular weight materials.

11.4.2 Single-Dose Oral

The single-dose oral toxicities of the polypropylene glycols are given in Table 50.10.

The low molecular weight materials (400 to 1200) are rapidly absorbed; excitement and convulsions appear within minutes after administration. With the higher molecular weight material, no excitement or convulsions were observed. Necropsy of animals treated with the largest doses 1 to 8 days after exposure revealed nothing remarkable.

Table 50.10. Single-Dose Oral Toxicities in Rats of Various Polypropylene Glycols

Material (Average Molecular Weight)	Approximate LD_{50} Values (g/kg)		
	Male	Female	Ref.
400	1.6	2.1	202
425	2.91		200
750	0.5	0.3	203
1025	2.15		200
1200	1.4	1.1	202
2000	>15.0	10.3	202
2025	9.76		200
4000	>15.0	>15.0	202

11.4.3 Injection Toxicity

All these materials have been given to animals parenterally (200, 201) and have been found to have essentially the same relative toxicity, one to another, as by the oral route. Table 50.11 shows the toxicity by injection.

11.4.4 Repeated-Dose Oral

Small groups of male rats were maintained for 100 days on diets containing 0.1 and 1.0 percent of P750 and 0.1, 0.3, 1.0, and 3.0 percent of P2000. Those animals that received the diet containing 0.1 percent of P750 were unaffected, as judged by studies of mortality, growth, organ weights, and gross and microscopic examination of the principal internal organs. The animals that

Table 50.11. Single-Dose Toxicity by Injection of Various Polypropylene Glycols

Material (Average Molecular Weight)	Species	Intravenous LD_{50}	Intraperitoneal LD_{50}	Ref.
400	Mice		0.70 g/kg	201
425	Rats	0.41 ml/kg	0.46 ml/kg	200
750	Mice		0.20 g/kg	201
1025	Rats	0.1 ml/kg	0.23 ml/kg	200
1200	Mice		0.11 g/kg	201
2000	Mice		3.60 g/kg	201
2025	Rats	0.71 ml/kg	4.47 ml/kg	200

received the diet containing 1.0 percent of P750, when judged by the same criteria, exhibited only a slight increase in the weight of the livers and kidneys without histologic changes. Hematologic studies on this latter group of animals failed to reveal any abnormalities. One percent of P750 in the diet was well accepted by the rats, and it is especially worthy of note that there was no evidence of any of the pharmacologic signs (excitement, tremors, convulsions) seen in the acutely poisoned animals. It is postulated that the material is readily metabolized or eliminated when absorbed in small doses; this probably accounts for its lack of apparent physiologic effect (203).

The rats that received 0.1, 0.3, and 1.0 percent of P2000 suffered no ill effects as judged by the criteria listed above. Hematologic studies were conducted only at the 1.0 percent level, with all values falling in the normal range. Although the growth of those animals maintained on the diet containing 3.0 percent of P2000 was slightly below normal during most of the test period, there were no other changes attributable to the experimental diets (202).

Polypropylene glycol P1200 was fed to dogs and rats in their diets for 90 days at concentrations of 0.0, 0.1, 0.3, and 1.0 percent. Groups of two dogs of each sex received the medicated diets and a group of three of each sex received unmedicated diet. Groups of 25 rats of each sex constituted all four of the groups. Male dogs on the 1 percent diet received from 317 to 380 mg/kg/day of the P1200 whereas the females received from 275 to 501 mg/kg/day. The average daily dose for the male rats on the 1 percent diet was 526 mg/kg and for the female rats, 810 mg/kg. On the 0.3 percent diet, the male dogs received daily doses averaging from 77 to 99 mg/kg/day; the female dogs, 90 to 123 mg/kg; the male rats, 157 mg/kg; and the female rats, 189 mg/kg.

This study showed no evidence of adverse histopathologic, hematologic, or clinical chemical or other effects from consumption of the chemical with the exception of body weight gains in dogs and rats at the high level (1.0 percent). On completion of the experiment, the high level group of dogs showed net losses in body weight. The high level rats, compared with the nonmedicated controls, showed less gain in body weight (29, 202).

P2025 was fed to rats for 7 days in a diet at levels of 1.58 and 5.0 g/kg/day. There were no effects seen at the 1.58 g/kg/day level; however, there was definite body weight depression at the 5.0 g/kg/day level (99).

11.4.5 Skin and Eye Irritation

Tests conducted on rabbits have indicated that these materials are not significantly irritating to the skin even when exposures are prolonged and repeated (200, 202).

Direct contact with the eyes may cause slight transient pain and conjunctival irritation but no corneal damage. The response is similar to that caused by a mild soap (99, 202).

11.4.6   Skin Sensitization

Not all the polypropylene glycols have been reported as being tested for human skin sensitization properties. However, P2000 was found to cause neither skin irritation nor signs of skin sensitization when applied both continuously and by repeated application to 300 human volunteers (202). P425, P1025, and P2025 caused no responses when applied to 50 human volunteers (99). Thus it would appear that all polypropylene glycols may be considered to be without skin sensitization properties.

11.4.7   Skin Absorption

Acute skin absorption tests conducted by means of a "sleeve" technique similar to that developed by Draize et al. (130) have indicated that the materials are all poorly absorbed through the skin. When single doses of 30.0 ml/kg (30 g/kg) were applied for 24 hr, four of five animals treated with either P400, P750, or P1200 survived, while all six animals so treated with polypropylene glycol 2000 survived (198, 202).

   The LD_{50} values for rabbits by absorption are reported to be 20 g/kg for P425 and more than 20 g/kg for P1025, P1950, and P2025 (99).

   Chronic skin absorption studies have been carried out on P2000. In these studies, the material was bandaged onto the shaved abdomens of groups of five rabbits each, 24 hr/day, five times a week for 3 months. At a dosage level of 1.0 ml/kg (1.0 g/kg) there were no adverse effects as judged by studies of growth, hematology, weights of organs, and gross and microscopic examination of the lungs, heart, liver, kidney, adrenal, testes, stomach, intestine, and skin taken from the site of the prolonged and repeated exposure. Judged by the same criteria, dosage levels of 5.0 and 10.0 ml/kg (5 and 10 g/kg) caused slight depression of growth. At the 10.0 ml/kg (10 g/kg) level, mortality was increased but since the cause was respiratory infection, the significance of the observation is questionable (202).

11.4.8   Pharmacology

Extensive investigation of the pharmacologic activity of these polypropylene glycols has indicated that they are all central nervous system stimulants. P400, P750, and P1200 are quite potent in this respect, whereas P2000 has but little such activity (200, 201).

11.4.9   Metabolism

In the studies by Yoshida et al. (34) in which various glycols including P2000 and P3000 and propylene glycol itself were evaluated as energy sources for chicks, it was found that propylene glycol served as a good energy source. However, neither P2000 nor P3000 were utilized. This led the authors to suggest that chicks were unable to metabolize the polymers.

11.4.10   Human Experience

No cases of toxicity have resulted from the manufacturing, handling, and use of the polypropylene glycols.

11.5   Hygienic Standards of Permissible Exposure

None would seem necessary.

12   BUTANEDIOLS; Butylene Glycols, CAS No. 25265-75-2

$$C_4H_8(OH)_2$$

12.1   Source, Uses, and Industrial Exposure

1,2-Butanediol is a relatively new product and is produced commercially by the hydration of the corresponding 1,2-butylene oxide in a manner similar to that described for other simple glycols (29).

According to Curme and Johnston (108), 1,3-butanediol is prepared commercially by the catalytic reduction of acetaldol, but may be produced by other routes as well. 1,4-Butanediol is produced in Germany by hydrogenation of 2-butyne-1,4-diol but other methods can also be used. 2,3-Butanediol is produced by fermentation, the distribution of optical isomers depending upon the species of bacteria used.

The butanediols are not used extensively commercially but there is considerable interest in them as intermediates in the polyester resins. The 1,3-isomer, because of its low toxicity, has been proposed for cosmetic and pharmaceutical applications.

Exposure is from direct contact in handling and use.

12.2   Physical and Chemical Properties

The commercial preparations of 1,2-, 1,3-, 1,4-, and 2,3-butanediols are clear, viscous liquids miscible with water and alcohol. They all have a molecular weight of 90.12; thus 1 ppm $\approx$ 3.68 mg/m^3 and 1 mg/l $\approx$ 272 ppm at 25°C, 760 mm Hg. The most important physical properties are given in Table 50.12.

Table 50.12. Physical and Chemical Properties of Butylene Glycols (Butanediols)

Butanediol Isomer	CAS No.[a]	Boiling Point (°C) (760 mm Hg)	Flash Point (°F) (O.C.)	Freezing Point (°C)	Refractive Index, 20°C	Vapor Pressure (mm Hg) (20°C)	Specific Gravity (20/20°C)	Ref.
1,2	584-03-2	193.5–195			1.4369 (25°C)		1.0017	29
1,3	107-88-0	207.5	250	<−50	1.4401	0.06	1.0059	108
1,4	110-63-4	230	247	16			1.020	108, 204
2,3	None found	182	185	19 (m.p.)	1.4377	0.17	1.0093	108

[a] General CAS No. is 25265-75-2.

12.3 Physiologic Response

12.3.1 Summary

1,2- and 1,3-Butanediol appear to be very low in oral toxicity when administered in both single and repeated doses. In single oral doses, 1,4-butanediol is much more toxic than either the 1,2- or the 1,3-isomer (about 10 times), and the 2,3-isomer is intermediate. Data on the 1,3- and 1,4-isomers indicate that they are not significantly irritating to the eyes, skin, or mucous membranes, nor are they likely to be absorbed through the skin in hazardous amounts. The undiluted 1,2-isomer is not significantly irritating to the skin but appears to be irritating to the eyes. Dilution to 10 percent with water eliminates the effect upon the eyes.

It would not seem that the butanediols would present any appreciable handling hazards other than possible eye irritation from contact with the 1,2-isomer.

12.3.2 1,2-Butanediol; CAS No. 584-03-2

$$CH_3CH_2CHOHCH_2OH$$

Single-Dose Oral. Single-dose oral toxicity studies have shown this material to be very low in acute oral toxicity for rats, the LD_{50} being about 16 g/kg. In large doses, the material causes narcosis, irritation of the gastrointestinal tract, profound vasodilatation of the visceral vessels, as well as a marked congestion of the kidneys. No hemorrhage was apparent. Deaths that occurred within a few hours are believed to be due to narcosis, and those that were delayed are believed to be due to kidney injury (29).

Repeated-Dose Oral. Schlüssel (205) along with his work on the 1,3-isomer, found that young female rats could tolerate a basic diet in which up to 30 percent of the calories were replaced by 1,2-butanediol, but that 40 percent replacement caused death in 11 to 29 days.

Intravenous Injection. Intravenous injection of up to 1 g/kg in an anesthetized dog failed to cause any noticeable response in blood pressure, heart rate, or respiration (29).

Eye Contact. When applied to the eyes of rabbits, the undiluted liquid was painful, irritating, and injurious, whereas a 10 percent aqueous solution caused no response (29).

Skin Contact. The material was not irritating to the skin of rabbits even when exposures were prolonged and repeated and it was not absorbed through the skin in toxic amounts (29).

Inhalation. Rats were unaffected by a single 7-hr exposure to an atmosphere essentially saturated at 100°C and then cooled to room temperature (29).

Metabolism. Gessner et al. (194) were unable to find any glucuronides or other metabolites in the urine of rabbits given 1,2-butanediol orally. Strack et al. (206) gave 1.0 g/kg of 1,2-butanediol intravenously during a 2-min period to rabbits, and found that it was metabolized slowly, as shown by the fact that the blood level fell slowly and that there was only slow excretion via the urine. That which was excreted in the urine was found either as the glucuronide or unchanged. Tissue analysis showed no accumulation of 1,2-butanediol.

12.3.3   1,3-Butanediol; CAS No. 107-88-0

$$CH_3CHOHCH_2CH_2OH$$

Single-Dose Oral. This material is very low in oral toxicity. When given in single oral doses, Loeser (207), Fischer et al. (208), and Bornmann (94) all report the LD_{50} to be 23.31 ml/kg (23.44 g/kg) for mice and 29.42 ml/kg (29.59 g/kg) for rats. Smyth et al. (209) report the oral LD_{50} for rats to be 22.8 g/kg. Smyth et al. (21) found the LD_{50} for guinea pigs to be 11.0 g/kg.

Repeated-Dose Oral. 1,3-Butanediol also is very low in toxicity when given in repeated oral doses. Loeser (207) reports the feeding of two to three median lethal doses to rats during a 6-week period without organic damage or growth depression. Fischer et al. (208) report that 20 percent in the drinking water of rats for 44 days was without any effect when judged by studies of growth, hematology, liver, kidney, and bladder. However, Bornmann (94) states that 20 percent in the drinking water caused slight depression of growth, but no effect at 10 percent or less. Kopf and co-workers (210) fed rats orally 0.5 or 1.0 ml of 1,3-butanediol twice a week for 45 to 185 days without any effect, and dogs 2.0 ml of a 50 percent aqueous solution twice a week for 5 to 6 months without effect. Schlüssel (205) reports that young rats tolerated a basic diet in which 1,3-butanediol accounted for up to 40 percent of the total calories. Smyth et al. (209) fed groups of 10 rats for 90 days on diets containing sufficient 1,3-butanediol to cause a daily ingestion of up to 5.6 g/kg without any adverse effect as judged by growth, mortality, food consumption, liver and kidney weight changes, and histopathologic examination of the liver, kidney, spleen, and testes. Unfortunately, larger doses were not fed. This was substantiated by

Chagovets et al. (211), who fed 1,3-butanediol at a 10 percent level for 4 weeks without adverse effects. Harris et al. (212) found also that rats fed 47 percent of the calories in their diet for 62 days as 1,3-butanediol experienced no visible adverse effects. However, they found that the levels of glutamate, lactate, pyruvate, and glucose in the brain were below those found in the controls.

Scala and Paynter (213) fed rats 1,3-butanediol at levels in their diets of 1, 3, and 10 percent, and dogs at 0.5, 1, and 3 percent for 2 years. They found the no-effect levels to be 10 percent for the rats and 3 percent for the dogs, the highest levels fed in each case. The criteria evaluated were body weight, food consumption, pharmacological effects, blood studies, urinanalysis, and gross and histopathologic evaluation of 16 tissues in the rat and 18 in the dog.

Miller and Dymsza (214) fed rats a diet for 30 weeks containing 30 percent fats, to which 20 and 30 percent of 1,3-butanediol was added. The 20 percent level was without significant adverse effects, but the 30 percent caused an impairment in the utilization of the diet. The addition of 1,3-butanediol caused an increase in liver phosphohexase and no significant change in serum glucose but an increase in liver glycogen. These findings led the authors to suggest that 1,3-butanediol may not be oxidized via the β-hydroxybutyric acid route.

1,3-Butanediol has also been studied as an energy source for cows (215), chicks (34, 62), and humans in short-term studies (216). Yoshida et al. (215) fed 8-week-old calves a diet containing 5.88 percent of 1,3-butanediol for 6 weeks as a substitute for the usual energy source, and found that the calves were unaffected as judged by growth and digestive ability of the diet. Harnisch (162) and Yoshida et al. (34) both found that young chicks grew well and were unaffected when fed a diet containing 5 percent of 1,3-butanediol for 10 (162) and 27 days (34). However, Harnisch (162) found that the 5 percent level caused depressed food intake when fed over an 8-week period. He suggests that the suitable level of 1,3-butanediol for chicks is 2 to 3 percent. At higher levels, above 5 percent, diarrhea occurred. Tobin et al. (216) found that human volunteers tolerated a diet containing up to 10 percent of 1,3-butanediol for 5 to 7 days with no adverse effects as judged by blood tests; however, there was a slight decrease in blood glucose levels.

Injection Toxicity. When given subcutaneously, the LD_{50} is reported to be 16.51 ml/kg for mice and 20.06 ml/kg for rats (94, 207). Intramuscular injection of 1,3-butanediol was found to cause irritation (217) and inflammatory infiltration (218) that healed with time. Bornmann (94) states that 1,3-butanediol does not cause hemolysis when given parenterally and that acute intoxication results in deep narcosis.

Eye Contact. Grant (42) reports that 1,3-butanediol is capable of causing severe stinging of the eye in humans which is cleared rapidly by irrigation.

There was no irritation or injury. This was consistent with the results of eye tests in rabbits by Carpenter and Smyth (39).

Skin Contact. Smyth et al. (209) report that the material is not irritating to the rabbit skin. This is consistent with the observations of Loeser (207), Husing et al. (219), and Fischer et al. (208), who all conclude that the 1,3-isomer is not irritating to human skin or mucous membranes. This is substantiated by Shelanski (220), who used 200 humans ages 6 to 65. These latter studies also indicate that 1,3-butanediol is not a skin sensitizer.

Inhalation. Smyth et al. (209) exposed rats to saturated vapors of 1,3-butanediol for 8 hr without any adverse effects.

Teratogenicity. Dymsza (221) reports that teratogenic studies in rats and rabbits were found to show no teratologic effects. Gebhardt (181) also found no teratogenic effects in chick embryo tests.

Mutagenicity. Mutagenic studies revealed no chromosome aberrations (221).

Reproduction. Dymsza (221) reported that rats fed a semipurified diet containing 24 percent of 1,3-butanediol for three generations were not adversely affected in their ability to reproduce. Later studies showed that there was no effect on the reproduction of rats fed a diet containing 24 percent of 1,3-butanediol for four generations, nor in dogs fed 20 percent for one generation.

Carcinogenicity. In the 2-year feeding studies on dogs and rats there were no tumors found (213). Thus it appears that 1,3-butanediol is not carcinogenic.

Metabolism. Mehlman and co-workers (223) have studied the metabolism extensively. They conclude that 1,3-butanediol is metabolized by the liver in the following steps (222):

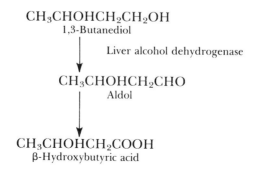

$$CH_3CHOHCH_2CH_2OH$$
1,3-Butanediol

↓ Liver alcohol dehydrogenase

$$CH_3CHOHCH_2CHO$$
Aldol

↓

$$CH_3CHOHCH_2COOH$$
β-Hydroxybutyric acid

β-Hydroxybutyric acid is further metabolized in the tricarboxylic acid cycle to carbon dioxide, which accounts for about 90 percent of the dose administered. In other studies (223) in which rats were fed 1,3-butanediol for 3 to 7 weeks, it was found that the blood level of β-hydroxybutyrate increased significantly. The blood level of acetoacetate, a possible metabolite of beta-hydroxybutyrate, was also higher than normal.

Mode of Action.   Many studies have shown that animals fed 1,3-butanediol develop a significant decrease in adipose tissue (221). This suggests an interference of lipogenesis by either 1,3-butanediol or its metabolites. This lowering of adipose tissue decreases the resistance of the animals to the stress of cold temperatures.

1,3-Butanediol, like other glycols, has narcotic effects at high doses (221). Ayres and Isgrig (224) concluded from their work that 1,3-butanediol at high doses depresses motor activity. They believe that this effect may be due to central nervous system depression (narcotic) and/or muscle relaxation.

12.3.4   1,4-Butanediol; CAS No. 110-63-4

$$CH_2OHCH_2CH_2CH_2OH$$

Single-Dose Oral.   This isomer is about 10 times as toxic when administered to animals as is the 1,2- or the 1,3-isomers. The oral LD_{50} has been found to be 2.14 ml/kg (2.18 g/kg) for mice (208) and 1.78 g/kg for rats (29). Kryshova (230) gives the following LD_{50} values: mice 2.06; rat 1.52; guinea pig 1.20; and rabbit 2.53 g/kg.

Repeated-Dose Oral.   Kryshova (230) fed guinea pigs and rats for 6 months on diets containing 1,4-butanediol at levels of 25 and 30 mg/kg/day. The no-effect level was found to be 25 mg/kg/day. At the 30 mg/kg/day level there were beginning effects, such as blood cholinesterase depression, changes in the protein fraction of the serum, and decrease in the SH-groups of the blood. There were no changes in weight, behavior, or general condition of the rats.

Injection Toxicity.   Zabic et al. (225) report that the LD_{50} value for rats by single intraperitoneal injection is 1.37 g/kg. They noted that rats given the LD_{50} dose developed significant depression of the righting reflex and at the LD_{100} dose, 1.6 g/kg, the righting reflex was lost in less than 20 min, followed by death in 2 to 3 hr.

Rats were given ten daily intraperitoneal injections of either 0.5 g/kg or 1.0 g/kg. Those at the lower dosage were apparently unaffected, whereas those at the higher dose exhibited slight depression of weight gain, some loss of righting reflex, minor changes in plasma and liver free fatty acids, but no increase in liver triglycerides (225).

Eye Contact. Application to the eyes of rabbits showed the material to be but slightly irritating; it caused a very slight conjunctival irritation but no corneal injury (42).

Skin Contact. Repeated application to the rabbits' skin, both intact and abraded, resulted in no appreciable irritation and no evidence of absorption of acutely toxic amounts (29). Judging from these observations, there would seem to be no appreciable hazard from skin contact associated with ordinary industrial handling. Schneider (226), however, reports finding the material highly toxic on the skin. Perhaps this apparent discrepancy can be attributed to the quality of the test material, a factor not to be ignored and one that has been called to attention by others (207).

Teratogenicity. Gebhardt (181) reports that 1,4-butanediol caused no signs of teratogenic action when injected into the yolk sac of fertile chicken eggs.

Metabolism. Gessner et al. (194) report that when butane-1,4-diol was fed to rabbits, most of it appeared to be destroyed; small amounts of succinic acid were found in the urine. Sprince et al. (227), in trying to determine the pharmacologic action of 1,4-butanediol, found its action to be similar to that of γ-hydroxybutyrate and γ-butyrolactone. Maxwell and Roth (228) found that slices of rat brain, liver, kidney, and heart were able to metabolize 1,4-butanediol to γ-hydroxybutyrate, with the liver being the most active. They postulated that the CNS depressing action of 1,4-butanediol is due primarily to the presence of γ-hydroxybutyrate.

Mode of Action. Hinrichs et al. (229) found the material to cause deep narcosis, constriction of the pupils, total loss of reflexes, and kidney injury; they attributed death to paralysis of the vital centers. This has been confirmed (29).

Ulrich and Mestitzova (195) also report that animals experienced accelerated breathing, loss of muscle tonus, drowsiness, and areflexia. In humans given 15 g rectally, miosis, unconsciousness, and coma were seen.

12.3.5 2,3-Butanediol; CAS No. 513-85-9

$$CH_3CHOHCHOHCH_3$$

Summary. Toxicologic data on this isomer seem to be scanty. Fischer et al. (208) report the oral LD_{50} for mice to be 9.0 ml/kg. Gessner et al. (194), in their study of the metabolism of glycols, state that they found glucuronides of 2,3-butanediol in the urine of rabbits equivalent to about 20 percent of the dose given. Gebhardt (181) found no evidence of teratogenic action in chick embryo tests.

12.4 Hygienic Standards of Permissible Exposure

None seems necessary for the butanediols because of their low volatility and low toxicity.

13 POLYBUTYLENE GLYCOLS (29); Poly(oxy-1,4-butanediyl)-α-hydro-ω-hydroxy; CAS No. 25190-06-1

$$HO(C_4H_8O)_nH$$

13.1 Source, Uses, and Industrial Exposure

The polybutylene glycols considered herein are prepared commercially by reacting 1,2-butylene glycol or water with 1,2-butylene oxide. Their number designations indicate their average molecular weights. Other polybutylene glycols can, of course, be made using the other butylene oxides.

Uses for these materials are being developed. Industrial exposure is most likely to be from direct contact with the skin and eyes.

13.2 Physical and Chemical Properties

The polybutylene glycols are clear, viscous, oily, slightly yellow liquids with sweetish tastes. They are less than 1.0 percent soluble in water and greater than 25 percent soluble in methanol and ether. They are quite stable chemically and do not present hazards of flammability except at elevated temperatures.

13.3 Physiologic Response

Limited toxicologic information is available on two polybutylene glycols known as polyglycol B-1000 and polyglycol B-2000, having average molecular weights of 1000 and 2000, respectively. Both appear to be low in single-dose oral toxicity, their oral LD_{50} values for rats being greater than 4.0 g/kg. In such oral doses, however, they do produce marked injury to the kidneys. They are slightly irritating but not damaging to the eyes of rabbits. Prolonged and repeated contact with the skin of rabbits failed to cause any significant topical effect and there was no evidence of absorption of toxic amounts through the skin. It would appear that these materials do not present any appreciable hazards in industrial handling and use. However, until more data become available, care should be taken to avoid ingestion, particularly repeated ingestion.

14 1,5-PENTANEDIOL; 1,5-Pentylene Glycol; Pentamethylene Glycol; CAS No. 111-29-5

$$HOCH_2CH_2CH_2CH_2CH_2OH$$

14.1 Sources, Uses, and Industrial Exposure

1,5-Pentanediol is used largely as a chemical intermediate. Industrial exposure is likely to be from direct contact.

14.2 Physical and Chemical Properties

1,5-Pentanediol is a clear liquid. Additional chemical and physical properties are given in Table 50.1.

14.3 Physiologic Response

Summary (231).   1,5-Pentanediol is low in single-dose oral toxicity; the LD_{50} value for rats is 5.89 g/kg. It is essentially nonirritating to the skin and only very mildly irritating to the eyes. It is not likely to be absorbed through the skin in toxic amounts because the LD_{50} value for rabbits is greater than 20 ml/kg. Rats exposed to essentially saturated vapors generated at room temperature for 8 hr survived. It would appear that the exercise of reasonable care and caution in the handling of 1,5-pentanediol should be adequate to avoid serious toxic effects.

15 2-METHYL-2,4-PENTANEDIOL; Hexylene Glycol; CAS No. 107-41-5

$$(CH_3)_2COHCH_2CHOHCH_3$$

15.1 Source, Uses, and Industrial Exposure

2-Methyl-2,4-pentanediol is prepared commercially by the catalytic hydrogenation of diacetone alcohol (4-hydroxy-4-methyl-2-pentanone) (108).

It is used as a chemical intermediate, a selective solvent in petroleum refining, a component of hydraulic fluids, a solvent for inks, and as an additive for cement (108).

Industrial exposure is likely to be from direct contact or from inhalation, particularly if the material is heated.

15.2 Physical and Chemical Properties (108, 232)

2-Methyl-2,4-pentanediol is a mild-odored liquid. Additional properties are given in Talbe 50.1.

15.3 Determination in the Atmosphere

Determination can be accomplished by the usual methods applicable to polyols. Infrared spectrophotometry and vapor chromatography may be applicable if specificity is necessary.

15.4 Physiological Response

15.4.1 Summary

2-Methyl-2,4-pentanediol is low in single-dose oral toxicity, appreciably injurious to the eyes, slightly irritating to the skin, not readily absorbed through the skin, and sufficiently low in vapor pressure at ordinary temperatures so as not to present appreciable hazard from inhalation. Atmospheres essentially saturated at room temperature (about 66 ppm) are detectable by odor and may be slightly irritating to the eyes. Atmospheric contamination resulting from handling at elevated temperature causes marked irritation of the eyes and hence warning of excessive concentrations. Pharmacologically, the material is a hypnotic. Studies on human subjects show that the material is slowly excreted, largely as the glucuronic acid conjugate.

15.4.2 Single-Dose Oral (232, 233)

When the material is given orally, the LD_{50} for mice is 3.8 ml/kg (3.5 g/kg) and for rats approximately 4.79 g/kg. Spector, according to Larsen (234), reported the LD_{50} value for guinea pigs and rabbits to be from 2.6 to 3.0 g/kg, and for mice and rats, 3.7 to 4.2 g/kg. Hypnosis occurred in mice following single doses of 2.0 ml/kg (1.85 g/kg); with higher doses it was profound. The material caused irritation of the lungs and large intestine, but no gross effects were apparent in the kidney, brain, or heart (232, 233).

15.4.3 Repeated-Dose Oral

According to Larsen (234), Brown et al. administered hexylene glycol to rats for 8 months at a level of 590 mg/kg/day with no detectable adverse effects.

Larsen (234) fed mice orally 20 mg/day in 2 ml of whole milk for up to 81 days and found only minor effects in the kidneys of a few animals. Rats were also fed the material in milk for 4 months at the average rate of 98 and 150 mg/day. The acceptance of the milk containing the hexylene glycol decreased in proportion to the hexylene glycol content. None of the rats showed adverse effects in growth nor were there histopathologic changes in the liver and testes, but there were minor changes in the kidneys. It was also shown that approximately 40 percent of the hexylene glycol was accounted for in the urine, but only 4 per cent of the amount excreted was free glycol; the other 36 percent was conjugated with glycuronic acid. The fertility of male rats given orally an average dose of 148 to 190 mg/day of hexylene glycol for 130 days was unchanged when compared to that of the control group.

15.4.4   Injection Toxicity

Spector, according to Larsen (234), reported the LD_{50} value for mice injected intraperitoneally to be 1.4 g/kg.

15.4.5   Eye Irritation (99, 232, 233)

When introduced into the eyes of rabbits, the undiluted material caused appreciable irritation and corneal injury that was slow to heal.

15.4.6   Skin Irritation

The undiluted material when applied to the uncovered rabbit skin gave rise to minor skin irritation (99).

15.4.7   Skin Absorption

A single 24-hr application of 1.84 g/kg to rabbits caused transitory mild edema and erythema but no deaths (232). The range-finding LD_{50} by cutaneous application to rabbits was found to be 13.3 ml/kg (12.3 g/kg) (232, 233) and 8.56 ml/kg (7.90 g/kg) (99). The inunction of 1 ml/kg/day (0.92 g/kg/day) for 90 days to the skin of rabbits caused no effects; the application of 2 ml/kg/day (1.85 g/kg/day) caused reversible cloudy swelling of the liver.

15.4.8   Inhalation

Rats exposed for 8 hr to air saturated at room temperature all survived (99, 232, 233). The concentration of the vapors in the saturated air was calculated to be approximately 160 ppm (99). Exposure of rats to an atmosphere generated by heating the material to 170°C (estimated to be equivalent to a concentration of 18,000 ppm) for 8 hr also caused no deaths (99).

Rats and rabbits that were exposed to an aerosol at a level of 0.7 mg/l for 7 hr/day for 9 days survived and showed no adverse effects pathologically (99).

15.4.9 Metabolism

Deichmann and Dierker (235) found that the oral administration of hexylene glycol to rats and rabbits resulted in a substantial increase in the amount of hexuronates in the plasma and in the urine. This is substantiated by Larsen (234). Jacobsen (236), in studies on five human subjects, found both free and conjugated hexylene glycol in the urine after single or repeated oral doses. When the daily dose was 600 mg or less, none was detected in the urine. With daily doses up to 5 g/day, substantial amounts of the free hexylene glycol and the conjugated form were found in the urine. Excretion was slow, persisting for up to 10 days after cessation of dosing.

15.5 Hygienic Standards of Permissible Exposure (232)

In the absence of adequate data upon which to establish a hygienic standard, it is suggested that atmospheric concentrations be maintained below those that cause discomfort in the unacclimated individual.

15.6 Odor and Warning Properties (232)

Most human beings exposed 15 min to 50 ppm in the air were able to detect the odor and a few noted eye irritation. At a concentration of 100 ppm, the odor was plain and some noted nasal irritation and respiratory discomfort; at 1000 ppm (4840 mg/m^3), there was irritation of the eyes, nose, and throat, and respiratory discomfort.

16 2,2,4-TRIMETHYL-1,3-PENTANEDIOL; CAS NO. 144-19-4

$$(CH_3)_2CHCHOHC(CH_3)_2CH_2OH$$

16.1 Sources, Uses, and Industrial Exposure

2,2,4-Trimethyl-1,3-propanediol is made by reacting isobutyryl chloride with ethyl isobutyrate as the starting materials. Its main uses are as a component of polyester resins used in food packaging and as an insect repellent. Industrial exposure is likely to be by direct contact with the liquid. Vapors or mists may be encountered under some conditions.

16.2 Physical and Chemical Properties

2,2,4-Trimethyl-1,3-pentanediol is a somewhat volatile liquid with the following properties:

Molecular formula	$C_8H_{18}O_2$
Molecular weight	145.32
1 ppm $\eqsim$	5.94 mg/m^3 at 760 mm Hg and 25°C
1 mg/l $\eqsim$	168.2 ppm at 760 mm Hg and 25°C

16.3 Determination in the Atmosphere

Determination can be accomplished by the usual nonspecific methods used for polyols. Infrared spectrophotometry and gas chromatography may be applicable if specificity is desired.

16.4 Physiologic Response

16.4.1 Summary

2,2,4-Trimethyl-1,3-pentanediol is low in single-dose and repeated-dose oral toxicity, is low in toxicity by injection, is essentially nonirritating to the eye and only slightly to moderately irritating to the skin, is low in toxicity by skin absorption, and is readily metabolized. It would appear that it should pose no significant health hazard in its industrial handling. Because of lack of information on inhalation, it would seem wise to practice reasonable care and caution to avoid prolonged or repeated exposure to its vapors.

16.4.2 Single-Dose Oral

Terhaar et al. (237) report that the LD_{50} for rats, mice, and guinea pigs is near 20 g/kg.

16.4.3 Repeated-Dose Oral (237)

Rats of both sexes were fed for 60 days on diets containing 2.0, 1.0, and 0.5 percent of 2,2,4-trimethyl-1,3-propanediol. The females at the 2 percent level ate less and therefore experienced growth depression. This was accompanied by a slight increase in average weights of the liver, adrenal, kidney, heart, and brain. The males at this level also developed increased liver, adrenal, and kidney weights. Those given the 0.5 percent diet showed no adverse effects by the usual criteria. The group fed the diet containing 1.0 percent of the test material were used for the teratology study discussed later.

16.4.4   Injection Toxicity (237)

Both rats and mice were used to establish the single-dose toxicity by intravenous and intraperitoneal injection. The values reported are near 0.8 and 0.145 g/kg for rats and mice, respectively, by both avenues.

16.4.5   Eye Contact (237)

A 25 percent solution of 2,2,4-trimethyl-1,3-propanediol in 8 percent glycerin and 67 percent ethanol caused moderate but transient irritation to the eyes of rabbits; the response was similar to that caused by the solvent mixture alone.

16.4.6   Skin Irritation (237)

Repeated applications to the skin of rabbits and guinea pigs caused only slight to moderate reddening of the skin. No evidence of the absorption of this material through the skin in acutely toxic amounts was mentioned.

16.4.7   Reproduction (237)

Rats were fed 2,2,4-trimethyl-1,3-propanediol at a dose of 1 percent in their diet through three generations. During each generation the rats were bred two times. The only consistent finding was a decreased pup weight. In the six breedings, pup mortality was greater than in the controls three times, similar twice, and less once. Thus this material at 1 percent in the diet may be considered to have some adverse effect on reproductive capability of rats. More studies are needed to resolve the qualitative and quantitative aspects. The data available suggest that a real hazard from likely industrial exposure would be very small.

16.4.8   Metabolism

Astill and Fassett (238) found that when given orally to rats, 94 to 99 percent of the 2,2,4-trimethyl 1,3-pentanediol was eliminated in the urine in 3 to 4 days, 2 percent in the feces, and less than 0.1 percent as carbon dioxide in expired air. The urinary metabolites were 0.8 to 1.7 percent as unchanged 2,2,4-trimethyl-1,3-pentanediol, 72 to 73 percent as its O-glucuronide, 2.9 to 3 percent as 2,2,4-trimethyl-3-hydroxyvaleric acid, and 4.3 to 4.4 percent as the valeric acid glucuronide.

Absorption studies using rabbits, guinea pigs, and humans were undertaken to establish the fate of this material (238). In rabbits given a single dose of 2.0 g/kg on the skin there was increased urinary glucuronide output accounting for about 30 percent of the dose given. At a dose of 0.18 g/kg, most of the material was absorbed and eliminated via the urine (75 to 82 percent of dose), less than 0.5 percent as carbon dioxide, and 2 percent in the feces. Volatilization at the application site accounted for about 14 percent of the dose given. The guinea pig tests (dose 0.09 mg/kg) showed that the material was absorbed and excreted in a similar manner but to a lesser extent.

Human tests using 3.7 to 4.9 g/person indicated that the material was not absorbed through the skin in appreciable amounts. There was no detectable increase in urinary glucuronides, and this likely was due to the fact that 82 to 93 percent of the dose applied to the skin was recovered from the application site.

16.5 Hygienic Standards of Permissible Exposure

No inhalation data are reported. The toxicologic and metabolic information suggests that a standard may not be necessary.

17 2-ETHYL-1,3-HEXANEDIOL; 1,3-Hexanediol, 2-ethyl-, CAS No. 94-96-2

$$CH_3CH_2CH_2CHOHCH(C_2H_5)CH_2OH$$

17.1 Source, Uses, and Industrial Exposure (108)

2-Ethyl-1,3-hexanediol is produced commercially by the hydrogenation of butyraldol (2-ethyl-3-hydroxycaproaldehyde).

It is used largely as an insect repellent, but it is also used as a solvent for resins and inks, a plasticizer, and a chemical intermediate in the production of polyurethane resins.

Industrial exposure is largely by direct contact. Extensive experience with human beings has been acquired through its extensive use as an insect repellent.

17.2 Physical and Chemical Properties (108)

2-Ethyl-1,3-hexanediol is an oily, colorless, slightly viscous liquid, soluble in alcohol and ether but poorly soluble in water (4.2 percent at 20°C). Additional properties are given in Table 50.1.

17.3 Determination in the Atmosphere

Methods applicable to other polyols having terminal hydroxyl groups should be useful.

17.4   Physiologic Response

17.4.1   Summary

2-Ethyl-1,3-hexanediol is low in single- and repeated-dose oral toxicity, not appreciably irritating to the human skin, somewhat irritating to mucous membranes, and slowly absorbed through the skin. Once absorbed it causes narcosis but little organic injury. It is considered safe for use as an insect repellent.

17.4.2   Single-Dose Oral

Lehman (239) reports that when 2-ethyl-1,3-hexanediol was fed in single oral doses to various species, the LD_{50} values obtained were as follows: rats, 2.5; mice, 4.2; guinea pigs, 1.9; and chicks, 1.4 g/kg. Smyth et al. (209) report the oral LD_{50} for rats to be 2.71 g/kg. In large doses, the material appears to cause deep narcosis, and this is believed to be the cause of death.

17.4.3   Repeated-Dose Oral

Smyth and co-workers (209) report that rats fed for 90 days a diet that provided a daily intake of 0.70 g/kg of the glycol did not grow as well as the controls, but apparently suffered no organic injury. When rats were maintained on a diet that supplied a daily intake of 0.48 g/kg, growth was normal and no detectable adverse effects were noted.

Lehman (239) reports that rats were fed for up to 2 years on diets containing 2.0, 4.0, and 8.0 percent of 2-ethyl-1,3-hexanediol. Growth was depressed at all levels. At the 8.0 percent level, all animals were dead within 18 weeks, death being due to inanition. Those at the 4.0 and 2.0 percent levels survived, and autopsy revealed no organ injury attributable to the glycol.

17.4.4   Eye Contact

Contact with undiluted 2-ethyl-1,3-hexanediol was found to cause moderate irritation and possibly some corneal injury to rabbits (99).

17.4.5   Skin Irritation

2-Ethyl-1,3-hexanediol is somewhat irritating to the skin of rabbits, but human skin appears to be quite resistant (239). Mucous membranes, however, are quite sensitive to the material (239). This is confirmed by Carpenter and Smyth (39) in their studies with the rabbit eye.

17.4.6   Skin Absorption (239)

The single-dose LD_{50} by skin absorption for rabbits is reported to be greater than 10 ml/kg (9.42 g/kg). However, when rabbits were inuncted daily for 90 days, 2 ml/kg (1.88 g/kg) caused about 50 percent mortality, and somewhat greater mortality resulted at the dosage level of 4 ml/kg (3.76 g/kg). Continued contact caused appreciable irritation to the skin of rabbits, and animals that died exhibited moderate liver and kidney injury. Leukocytosis was observed in only one animal treated at the 4 ml/kg (3.76 g/kg) level. Lehman (239) concludes, "The fact that the compound is poorly absorbed through the skin of humans and is nonirritating warrants the conclusion that it may safely be used as a component of an insect repellent product."

17.4.7   Skin Sensitization

Guinea pigs were not sensitized by this material. Two hundred twenty-three human volunteers were patch tested for sensitization with the undiluted product. Three of the volunteers developed mild skin reddening; otherwise no significant signs of sensitization occurred (99). When tested as a repellent in stick formulations containing sodium stearate, none showed signs of irritation, sensitization, or fatiguing of the skin (99).

17.4.8   Inhalation

Rats exposed to essentially saturated vapors generated at room temperature for 2 hr survived; those exposed for 8 hr died. The rats exposed to a mist generated at 170°C for 2 hr also survived whereas those exposed for 8 hr died. Those exposed to a fog generated at 70°C with a nebulizer at a concentration estimated to be 4800 ppm for 8 hr all survived (99).

17.4.9   Metabolism

Gessner et al. (194) report that they were unable to detect any metabolites from 2-ethyl-1,3-hexanediol when it was fed to rabbits. In other studies there was no increase in excretion of ascorbic acid by rats. Rabbit urine was thought to have a mercaptan odor and there was a definite increase in blood urea and blood and urine glucuronates (99).

17.5   Hygienic Standards of Permissible Exposure

None would seem necessary. However, in view of the limited data available, it would seem wise to observe precautions to avoid prolonged or repeated exposures to high concentrations of vapor, mists, or fogs.

18 STYRENE GLYCOL; 1,2-Ethanediol, 1-Phenyl-; Phenyl Glycol; CAS No. 93-56-1

$$C_6H_5CHOHCH_2OH$$

18.1 Sources, Uses, and Industrial Exposure

Styrene glycol is made commercially from styrene oxide by hydrolysis. Styrene glycol is used largely as a chemical intermediate. Industrial exposure is likely to be by direct contact with the solid or solutions of the material. Vapors and mists may be encountered under particular conditions.

18.2 Physical and Chemical Properties

Styrene glycol is a white, practically odorless solid. Some properties are given in Table 50.1.

18.3 Determination in the Atmosphere

Methods for the determination of small amounts of styrene glycol have not been developed. It is believed, however, that chemical methods applicable to other glycols could easily be modified.

18.4 Physiologic Response

18.4.1 Summary

Unpublished range-finding toxicologic studies (29) indicate that styrene glycol is low in oral toxicity and not significantly irritating to the skin. It would not be expected to present an appreciable hazard in ordinary industrial handling and use.

18.4.2 Single-Dose Oral

The LD_{50} for guinea pigs appears to be between 2.0 and 2.6 g/kg.

18.4.3 Repeated-Dose Oral

When given by intubation as a solution in olive oil five times a week for 1 month, dosage levels of 0.5 and 1.0 g/kg were tolerated by rabbits and rats. A dosage level of 1.0 g/kg caused minor liver injury in the rabbit.

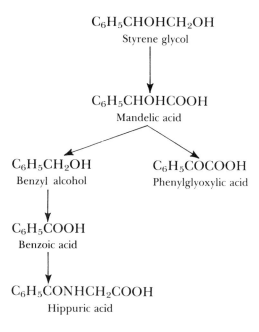

Figure 50.3 Metabolic pathway for styrene glycol.

18.4.4   Skin Contact

Prolonged and repeated contact with a 20 percent solution of styrene glycol in propylene glycol caused no injury to the skin of rabbits nor did it penetrate the skin in toxic amounts.

18.4.5   Mutagenic Activity

Milvy and Garro (240), using a spot test on salmonella strains developed by Ames and his co-workers for screening potential carcinogens, found that styrene glycol was not mutagenic. In addition, it did not inhibit the growth of bacteria. This was confirmed by Vainio et al. (241).

18.4.6   Metabolism

According to Ohtsuji and Ikeda, and Ikeda and Imamura, as quoted by Milvy and Garro (240) styrene glycol could be metabolized as shown in Figure 50.3.

18.5   Hygienic Standards of Permissible Exposure

No inhalation data. Data available, however, would not indicate the necessity for a standard.

19 MIXED POLYGLYCOLS

19.1 Source, Uses, and Industrial Exposure

Mixed polyglycols have been designed to supply products that have properties which are desirable for special needs. Basically they consist of reaction mixtures of ethylene oxide and propylene oxide and sometimes butylene oxide with various polyols.

They have a wide variety of uses, such as antifoam agents, synthetic oils, industrial lubricants, surfactants, mold release agents, cosmetic oil substitutes, pharmaceutical uses, chemical intermediates, and others.

Industrial exposure is expected to be from direct contact. Because of the great variation in these compounds they are discussed by product or groups of related products.

19.2 Polyglycol 11 and 15 Series (29)

The polyglycol 11 series is prepared by reacting glycerol with different amounts of propylene oxide. The polyglycol 15 series is prepared by reacting glycerol with various mixtures of propylene oxide and ethylene oxide.

19.2.1 Physical and Chemical Properties

All these polyglycols are clear viscous liquids with the properties shown in Table 50.13.

Table 50.13. Physical and Chemical Properties of Polyglycols of the 11 and 15 Series

Commercial Designation, Polyglycol	Average Molecular Weight	Specific Gravity (25/25°C)	Refractive Index, 25°C	Flash Point (°F)	Solubility (g/100 g) (25°C)	
					Water	Methanol or Ether
11-100	1030	1.026	1.452	435	<0.1	100
11-200	2700	1.018	1.452	435	<0.1	100
11-300	4000	1.017	1.450	440	<0.1	100
11-400	4900	1.017	1.450	445	<0.1	100
15-100	1100	1.070	1.460	510	Misc.	Misc.
15-200	2600	1.063	1.460	470	Misc.	Misc.
15-500	5000	1.051	1.459	475	Misc.	Misc.
15-1000	9000	1.053	1.458	480	Misc.	Misc.

Table 50.14. Summary of Toxicologic Information on Certain Mixed Polyglycols

Polyglycol	Rat LD_{50} (g/kg)	Eye Irritation (Rabbits)[a]	Effect on Skin Irritation (Rabbits)[b]	Effect on Skin Sensitization (Humans)[c]	Percutaneous Absorption (Rabbits)[d]
11-100	2.0	None	Slight		None
11-200	>4.0[e]	Trace	None		None
11-300	>4.0[e]	Trace	Slight	None	None
11-400	>4.0[e]	None	None	None	None
15-100	31.6	Trace	Very slight		None
15-200	15–20	Trace	Very slight	None[f]	None[g]
15-500	16	Trace	None		None
15-1000	>4.0[e]	None	None	None	None

[a] One to two drops directly in eye; "trace" = conjunctival irritation but no corneal injury.
[b] Contact 24 hr/day for 12 days.
[c] Repeated insult test on 50 human beings.
[d] None apparent from skin irritation test.
[e] No deaths at this, the largest dose fed.
[f] Repeated insult test on 50 human beings plus "Swartz" test on 200 human beings.
[g] LD_{50} by "Draize" test = >30 g/kg.

19.2.2 Physiologic Response

None of the mixed polyglycols in the 11 or 15 series described above presents any handling hazards of significance. The toxicologic information is summarized in Table 50.14.

19.2.3 Hygienic Standards of Permissible Exposure

The low volatility and low toxicity of these materials would seem to make a hygienic standard unnecessary.

19.3 Ucon®* Fluids (242)

50-HB-260
50-HB-5100
25-H-2005
75-H-1400

* Union Carbide Corporation's trade name for polyalkylene glycols and diesters.

19.3.1   Source and Physical Properties

Fluids 50-HB-260 and 50-HB-5100 are water-soluble monobutyl ethers of polymers of ethylene oxide and propylene oxide with approximate mean molecular weights of 940 and 4000, respectively. Fluid 25-H-2005 and 75-H-1400 are polymers with two terminal hydroxyl groups with approximate molecular weights of 4100 and 2200, respectively. The number 25 or 75 in the fluids' names indicate the percentage of units of ethylene oxide in the polymers.

19.3.2   Physiologic Response

Single-Dose Oral.   These products are low in single-dose oral toxicity. The results are summarized in Table 50.15. The only sign of nonfatal effect was that of depression in varying degrees depending on the product.

Repeated-Dose Oral.   Male and female rats and dogs were fed diets containing these materials for a 2-year-period. In the rat studies, the materials were added to the diet so that each product was consumed at a constant level throughout the experiment. The highest levels in all experiments was 0.5 g product/kg/day. The only adverse effect seen in the studies on these four products was that of slight growth depression in the female rats consuming 0.5 g/kg/day of 25-H-2005. The other products caused no adverse effects at this level. Criteria observed were behavior, diet consumption, mortality, life-span, incidence of infections, terminal liver and kidney to body weight ratios, body weight gain, hematocrit, total red cell count, incidence of neoplasms, and gross and histopathology in 20 tissues. Dogs were fed diets containing 0.15 and 1.67 percent of these products. This percentage was kept constant throughout the experiments. These levels were calculated to be an average dose of 0.62, 0.61, 0.62, and 0.50 g/kg/day for 50-HB-260, 50-HB-5100, 25-H-2005, and 75-H-1400, respectively.

The observations made on the dogs were appetite, body weight change, mortality, terminal liver and kidney to body weight ratios, hematocrit, hemo-

Table 50.15.   The Single-Dose Oral Toxicity of Four Ucon Fluids to Various Laboratory Animals

Product	Molecular Weight	LD$_{50}$ Values (ml/kg Body Wt)			
		Rat Male	Rat Female	Mouse Female	Rabbit Male
50-HB-260	940	5.95	4.49	7.46	1.77
50-HB-5100	4000	>64	45.2	49.4	15.8
25-H-2005	4100	14.1	35.9	22.6	35.6
75-H-1400	2200	>64	>16	45.2	35.4

globin, red and white cell total counts, differential white cell counts, serum urea nitrogen, serum alkaline phosphatase, 15 min BSP retention, and gross and histopathology on 18 tissues. The only dose-related response seen was that of granular degeneration of the cytoplasm of the smooth muscle in the intestinal wall in the dogs fed the 1.67 percent level of 25-H-2005. The significance of this finding is unknown. The other products were without adverse effects at the 1.67 percent level.

Carcinogenicity. In addition to the negative carcinogenic results seen in the dietary feeding in rats, none of these products produced papillomas or carcinomas in mice painted on the clipped skin of the back three times a week until death.

Metabolism–Excretion. Rats were fed these products, labeled primarily on the ethylene oxide moiety with ^{14}C, at a dosage of 67 mg/kg of body weight. The urine, feces, carbon dioxide, and carcasses were analyzed for ^{14}C over a 7-day period. The total ^{14}C recovery from all areas ranged from 80 to 95 percent of the dose given. The findings are given in Table 50.16.

These data show that only 50-HB-260 was metabolized to $^{14}CO_2$ in significant amounts. Fluid 50-HB-260 was absorbed from the gut in large amounts, fluid 25-H-2005 in a lesser amount, and fluids 50-HB-5100 and 75-H-1400 in insignificant amounts. The identity of the ^{14}C compounds found in the urine and feces was not determined.

19.3.3 Hygienic Standards of Permissible Exposure

None appears to be needed because of the very low volatility and the low degree of toxicity of these products.

19.4 Pluronic®* Polyols (243)

19.4.1 Physical and Chemical Properties

These products are block polymers made by the condensation of propylene oxide onto a propylene glycol nucleus followed by the addition of ethylene oxide to the ends of the base at levels ranging from 10 to 80 percent of the final molecule. The products have molecular weights ranging from 1000 to over 15,000. The empirical formula is

$$HO(CH_2CH_2O)_x(CH(CH_3)CH_2O)_y(CH_2CH_2O)_xH$$

---

* BASF Wyndotte's trade name for polyoxyalkylene derivatives of propylene glycol.

Table 50.16. Average Percentage of Total Reactivity Found

Product	Molecular Weight	In Urine	In Feces	In Cage Washings	In CO_2	In Carcass	Total Recovered
50-HB-260	940	46.4	26.7	5.0	12.5	0	90.6
50-HB-5100	4000	5.3	89.0	0.6	0	0	94.9
25-H-2005	4100	25.6	49.5	4.5	0.1	0	79.7
75-H-1400	2200	1.6	76.0	2.8	0	0	80.4

The subscripts x indicate that the end moieties are statistically equal. Some of the properties of these products are given in Table 50.17.

These materials are relatively nonhydroscopic, vary from water-insoluble to water-soluble products, and are soluble in common aromatic hydrocarbons, chlorinated hydrocarbons, acetone, and low molecular weight alcohols. They are good solvents for a number of diversified materials such as iodine and perfume oils. Usually those with molecular weights up to 3400 to 3800 are liquids, those in the range of 4000 to 6500 are pastes, and those with molecular weights of greater than 6600 are solids.

19.4.2 Physiologic Response

Summary. The Pluronic polyols discussed can be considered to be low in health hazard in industrial operations. Generally the low molecular weight polyols are possibly very slightly more toxic than those with a higher molecular weight. In the feeding studies, the lower molecular weight products appear to be less tolerated than the higher molecular weight products. They are at most only mildly irritating to the eyes and are not likely to cause corneal injury.

Table 50.17. Some Properties of the Pluronic Polyols

Pluronic Designation	Approximate Molecular Weight	Taste and Odor	Chemical Stability to Acids	Physical Form
L44	2200	Mild bitter taste	Good	Liquid
L61	2000	Mild bitter taste	Good	Liquid
L62	2500	Mild bitter taste	Good	Liquid
L64	2900	Mild bitter taste	Good	Liquid
L101	3800	Mild bitter taste	Good	Liquid
F68	8350	Practically odorless and tasteless	Good	Solid
F108	14,000	Mild bitter taste	Good	Solid
P85	4600	Mild bitter taste	Good	Paste
P127	11,500	Mild bitter taste	Good	Solid

Those tested for skin contact properties were neither irritants nor sensitizers. They appear to be low in toxicity by absorption through the skin.

Single-Dose Oral.   All these products are low in single-dose oral toxicity. The LD_{50} values are given in Table 50.18.

Repeated-Dose Oral.   A number of the Pluronic polyol products have been fed to rats and/or dogs and found to be moderate to low in repeated-dose oral toxicity. The results of the tests are summarized in Table 50.19. In these tests the criteria evaluated were growth, mortality, behavior, hematologic studies and urinalysis, gross pathology and histopathology, and in the dogs, EKG evaluations.

Injection.   Pluronic polyol F68 caused no adverse effects when injected intravenously into dogs at a level of 0.1 g/kg or into rabbits at a level of 1.0 g/kg. The LD_{50} intravenously for mice is 5.5 g/kg and for rats 3.95 g/kg. The LD_{50} when injected intraperitoneally into mice is between 5 and 10 g/kg.

Daily injection of Pluronic polyol F68 intravenously into rabbits and dogs at a dose of 0.5 g/kg/day caused no adverse effects as judged by body weight, hemogram, blood chemistry, and urinalysis.

Eye Contact.   Pluronic polyol F68 when tested as 5 and 10 percent aqueous solution was not irritating to the eyes of rabbits. Pluronic polyols L44, L62, and

Table 50.18.   Single-Dose Oral Toxicity of Pluronic Polyols

Product	Average Molecular Weight	Species	LD_{50} (g/kg)	Signs
L44	2200	Rat	5	Depression and prostration; death and respiratory paralysis probable
L61	2000	Rat	2.1	Salivation, hypoactivity; ruffed fur
L62	2500	Rat	5	Depression and prostration; death and respiratory paralysis probable
L64	2900	Rat	5	Depression and prostration; death and respiratory paralysis probable
F68	8350	Mouse	>15	At 15 g/kg about 20 percent died
		Rat	>15	No effects seen
		Dog	>15	No effects seen other than diarrhea
		Rabbit	>15	No effects seen
		Guinea pig	>15	Mild sedation
P85	4600	Rat	>34	
F127	11,500	Rat	15.4	

Table 50.19. Results of Dietary Feeding of Pluronic Polyol Products to Rats and/or Dogs

Product	Species	Dose (g/kg/day)	Duration	Effect
L33	Dog	0.04	90 days	No effect
	Dog	0.2	90 days	Some emesis; no deaths
	Dog	1.0	90 days	Emesis; growth depression; no deaths
	Rat	0.2	90 days	No effects
	Rat	1.0	90 days	Not palatable, reduced food intake resulting in severe growth depression
L62	Dog	0.04	2 years	No significant effects
	Dog	0.2	2 years	Emesis, some periods of salivation; no significant pathologic changes
	Dog	0.5	2 years	Deaths due to poor nutritional state; 4 of 6 dogs died; no pathology seen
	Rat	0.04	2 years	No effects
	Rat	0.2	2 years	Only growth depression
	Rat	0.5	2 years	Severe growth depression, some deaths due to inanition, no pathology seen
F68	Dog	0.1	6 months	No effect
	Rat	5% in diet	2 years	No effect
	Rat	7.5% in diet	2 years	Decrease in growth rate; no pathology seen
P85	Dog	1.0	90 days	No effect
	Rat	1.0	90 days	No effect
L101	Dog	0.5	90 days	No effect
	Rat	0.5	90 days	No effect
F108	Dog	5.0	90 days	No effect
	Rat	5.0	90 days	No effect

L64 were tested as 25, 50, 75, and 100 percent in the rabbit eye and were found to cause only transient mild to moderate irritation but no corneal injury. The degree of irritation increased with concentration. Pluronic polyols L33, F68, L101, and F127 were found to cause no local anesthetic effects when tested in the eyes of rabbits.

Skin Contact. Pluronic polyols F68, L44, L62, and L64 were tested for skin irritation and skin sensitization using human volunteers. None was a skin irritant or a skin sensitizer. Pluronic polyol F127 at a level of 25 percent in a formulation was found to cause a trace of irritation on rabbit skin but was not considered a primary skin irritant as defined by the Federal Hazardous Substances Act.

Skin Absorption.   The LD_{50} by skin absorption for Pluronic polyol F127 was found to be greater than 2.0 g/kg for rabbits, the highest dose tested. Because of the low single-dose oral toxicity of the these products, a hazard from absorption would not seem likely.

Inhalation.   No inhalation studies have been reported. It would seem that there should be no significant health hazard from inhalation under ordinary industrial operations primarily because of their very low degree of volatility.

Reproduction.   Pluronic polyols L101 and F108 were fed to rats through three generations at levels of 100, 250, and 500 mg/kg for L101 and 300, 1000, and 2500 mg/kg for F108. No adverse effects were seen as judged by growth and the ability to mate and produce viable normal offspring or gross and histopathology of the parents at the end of the test period.

Carcinogenicity.   Only Pluronic polyol L62 was fed to rats for 2 years. In these studies, there was no mention of the finding of tumors during gross or histopathologic examination. Thus it may be assumed that this product is not carcinogenic.

Metabolism.   Pluronic polyol F68 was the only material studied. When given to rats intravenously, essentially all the material given was excreted unchanged via the urine in 24 hr. When given orally, essentially all of it was excreted unchanged in the feces. Therefore it can be assumed that Pluronic F68 is not metabolized in the body.

Human Experience.   No adverse human experience has been reported.

19.4.3   Hygienic Standards of Permissible Exposure

None seems necessary for Pluronic polyols because of their very low degree of volatility and their low degree of toxicity.

19.5   Pluracol®*   V-10, TP-440, and TP-740 Polyols (244)

19.5.1   Source, Uses, and Industrial Exposure

Pluracol V-10 polyol is a polyoxyalkalene derivative of trimethylol propane with an average molecular weight of more than 20,000. It is used as a thickening agent in hydraulic fluids.

* BASF Wyandotte's trade name for polyoxyalkalene derivatives of trimethylol propane.

Pluracol TP-440 and TP-740 polyols are polyoxypropylene derivatives of trimethylol propane. They are used in the urethane industry. The general structure is

$$CH_2(OCH_2CH[CH_3])_x—OH$$

$$CH_3CH_2C—CH_2(OCH_2CH[CH_3])_y—OH$$

$$CH_2(OCH_2CH[CH_3])_z—OH$$

where x, y, and z vary from product to product but are approximately equal for each product.

Industrial exposure to these products is likely to be from direct contact.

19.5.2 Physical and Chemical Properties

The physical and chemical properties are summarized in Table 50.20.

19.5.3 Physiologic Response

Pluracol V-10, TP-440, and TP-740 polyol should present no significant health hazards in anticipated industrial handling. The reported toxicologic information on them is summarized in Table 50.21.

19.5.4 Hygienic Standards for Permissible Exposure

No toxicologic testing has been reported that would permit a suggested hygienic standard. It is believed that none is likely to be needed because of the low degree of toxicity and the low degree of volatility of these Pluracol products.

Table 50.20. Physical and Chemical Properties of Pluracol Polyols V-10, TP-440, and TP-740

Property	V-10	TP-440	TP-740
Form	Pale yellow viscous liquid	Liquid	Liquid
Average molecular weight	>20,000	418	732
Specific gravity	1.089 (60/60°F)		
Viscosity	45,000 cgs at 100°C	625 cgs at 25°C	325 cgs at 25°C
Flash point	510°F		
Solubility in water	up to 75%	0.2%	0.1%

Table 50.21. Summary of Toxicologic Information on Three Pluracol Products

Product	Rat LD_{50} (g/kg)	Eye Irritation, Rabbits	Skin Effects		Skin Absorption, Rabbits
			Irritation	Sensitization	
Pluracol V-10	>10.0[a]		Humans, none	None	
Pluracol TP-440	3.72[b]	Moderate, no corneal injury	Rabbit, trace		>3 g/kg
Pluracol TP-740	2.5[b]				

[a] No adverse effects seen at this level, the highest dose fed.
[b] High doses that caused death resulted in sedation, then convulsions, salivation, lacrimation, tremors, and muscular weakness.

REFERENCES

1. K. G. Bergner and H. Sperlich, *Z. Lebensm. Unters. Forsch.*, **97**, 253 (1953).
2. B. Worshowsky and P. J. Elving, *Ind. Eng. Chem.*, **18**, 253 (1946).
3. R. C. Reinke and E. N. Luce, *Ind. Eng. Chem.*, **18**, 244 (1946).
4. W. H. Evans and A. Dennis, *Analyst*, **98**, 782 (1973).
5. P. Lepsi, *Prac. Lek.*, **25**, 330 (1973).
6. F. R. Duke and G. F. Smith, *J. Ind. Eng. Chem. Anal. Ed.*, **12**, 201 (1940).
7. H. D. Spitz and J. Weinberger, *J. Pharm. Sci.*, **60**, 271 (1971).
8. N. Chairova and N. Dimov, *Heft. Khim.*, **361**, 361 (1973).
9. G. G. Esposito and R. G. Jamison, *Soc. Automot. Eng. J.*, **79**, 40 (1971).
10. R. L. Peterson and D. O. Rodgerson, *Clin. Chem.*, **20**, 820 (1974).
11. W. Bolanowska, *Chim. Anal. (Warsaw)*, **20**, 413 (1975).
12. J. Russell, F. McChesney, and L. Golberg, *Food Cosmet. Toxicol.*, **7**, 107 (1969).
13. G. Rajagopal and S. Ramakrishnan, *Anal. Biochem.*, **65**, 132 (1975).
14. I. J. Gilmour, R. J. W. Blanchard, and W. F. Perry, *N. Engl. Med.*, **291**, 51 (1974).
15. Anonymous, *BIBRA*, **11**, 373 (1972).
16. W. F. von Oettingen, *U.S. Public Health Bull. No. 281* (1943).
17. E. P. Laug, H. O. Calvery, H. J. Morris, and G. Woodard, *J. Ind. Hyg. Toxicol.*, **21**, 173 (1939).
18. E. M. K. Geiling and P. R. Cannon, *J. Am. Med. Assoc.*, **111**, 919 (1938).
19. H. J. Morris, A. A. Nelson, and H. O. Calvery, *J. Pharmacol. Exp. Therap.*, **74**, 266 (1942).
20. E. Browning, *Toxicity and Metabolism of Industrial Solvents*, Elsevier, Amsterdam, 1965.
21. H. F. Smyth, Jr., J. Seaton, and L. F. Fischer, *J. Ind. Hyg. Toxicol.*, **23**, 259 (1941).
22. G. Bornmann, *Arzneim. Forsch.*, **4**, 643 (1954); *Chem. Abstr.*, **49**, 7131 (1955).
23. I. H. Page, *J. Pharmacol.*, **30**, 313 (1955).
24. B. F. Pochebyt, *Zdravookhr. Beloruss*, **4**, 19 (1975); *Chem. Abstr.*, **83**, 73168J (1976).
25. V. P. Plugin, *Gig. Sanit.*, **33**, 16 (1968).
26. K. Bove, *Am. J. Clin. Path.*, **45**, 46 (1966).

27. O. K. Antonyuk, *Gig. Sanit.*, **39**, 106 (1974).

28. E. J. Kersting and S. W. Nielsen, *Am. J. Vet. Res.*, **27**, 574 (1966).

29. The Dow Chemical Company, unpublished data.

30. F. R. Blood, *Food Cosmet. Toxicol.*, **3**, 229 (1965).

31. J. F. Gaunt, J. Hardy, S. D. Gangolli, K. R. Butterworth, and A. G. Lloyd., *BIBRA*, **14**, 109 (1975).

32. F. R. Blood, G. A. Elliott, and M. S. Wright, *Toxicol. Appl. Pharmacol.*, **4**, 489 (1962).

33. J. A. Roberts and H. R. Seibold, *Toxicol. Appl. Pharmacol.*, **15**, 624 (1969).

34. M. Yoshida, H. Hoshii, and H. Morimoto, *Nippon Kakim Gakkaishi*, **6**, 73 (1969).

35. C. R. Riddell. S. W. Nielsen, and E. J. Kersting, *J. Am. Vet. Med. Assoc.*, **150**, 1531 (1967).

36. G. N. Lipkan and V. S. Petrenko, *Fiziol. Akt. Veshchestva*, **4**, 125 (1972).

37. M. M. Mason, C. C. Cate, and J. Baker, *Clin. Toxicol.*, **4**, 185 (1971).

38. L. Paterni, F. Dotta, and M. Sappa, *Folia Med. Naples*, **39**, 242 (1956); *Chem. Abstr.*, **50**, 11524 (1956).

39. C. P. Carpenter and H. F. Smyth, Jr., *Am. J. Ophthalmol.*, **29**, 1363 (1946).

40. T. O. McDonald, M. D. Roberts, and A. R. Borgmann, *Toxicol. Appl. Pharmacol.*, **21**, 143 (1972).

41. T. O. McDonald, K. Kasten, R. Hervey, S. Gregg, A. R. Borgmann, and T. Murchison, *Bull. Parenter. Drug Assoc.*, **27**, 153 (1973).

42. W. M. Grant, *Toxicology of the Eye*, 2nd ed., Charles C Thomas, Springfield, Ill., 1974.

43. P. J. Hanzlik, W. S. Lawrence, J. K. Fellows, F. P. Luduena, and G. L. Lacqueur, *J. Ind. Hyg. Toxicol.*, **29**, 325 (1947).

44. E. Volkmann, *Hippokrates*, **21**, 549 (1950); *Chem. Abstr.*, **47**, 9567 (1953).

45. F. J. Wiley, W. C. Hueper, and W. F. von Oettingen, *J. Ind. Hyg. Toxicol.*, **18**, 123 (1936).

46. R. A. Coon, R. A. Jones, L. J. Jenkins, Jr., and J. Siegel, *Toxicol. Appl. Pharmacol.*, **16**, 646 (1970).

47. J. H. Wills, F. Coulston, E. S. Harris, E. W. McChesney, J. C. Russell, and D. M. Serrone, *Clin. Toxicol.*, **7**, 463 (1974).

48. E. Harris, *Proceedings of the 5th Annual Conference on Atmospheric Contamination in Confined Spaces*, AMRL TR-69-130, paper No. 8, Aerospace Med. Res. Lab. Wright-Patterson Air Force Base, Ohio, 1969, p. 9.

49. M. Felts, *Proceedings of the 5th Annual Conference on Atmospheric Contamination in Confined Spaces*, AMRL TR-69-130, paper No. 9, Aerospace Med. Res. Lab., Wright-Patterson Air Force Base, Ohio, 1969, p. 105.

50. J. McCann, E. Choi, E. Yamasaka, and B. N. Ames, *Proc. Natl. Acad. Sci. US*, **72**, 5135 (1975).

51. P. H. Derse, *U.S. Clearinghouse Fed. Sci. Tech. Inform.*, *P.B. Rep. No. 195153* (1969).

52. F. Homburger, *U.S. Clearinghouse Fed. Sci. Tech. Inform.*, *P.B. Rep. No. 183027* (1968).

53. W. H. Evans and E. J. David, *Water Res.*, **8**, 97 (1974).

54. P. Pitter, *Collect Czech. Chem. Commun.*, **38**, 2665 (1973).

55. P. K. Gessner, D. V. Parke, and R. T. Williams, *Biochem. J.*, **79**, 482 (1961).

56. L. H. Smyth, Jr., T. D. R. Hochaday, M. L. Efron, and J. E. Clayton, *Trans. Assoc. Am. Physicians*, **77**, 317 (1964).

57. L. Hansson, R. Lindfors, and K. Laiko, *Deut. Z. Gerichtl. Med.*, **59**, 11 (1967).

58. J. S. King and A. Wainer, *Proc. Soc. Exp. Biol. Med.*, **128**, 1162 (1968).

59. J. M. Melon and J. Thomas, *Therapie*, **26**, 985 (1971).

60. C. C. Liang and L. C. Ou, *Biochem. J.*, **121**, 447 (1971).

61. E. W. McChesney, L. Golberg, and E. S. Harris, *Food Cosmet. Toxicol.*, **10**, 655 (1972).

62. F. Underwood and W. M. Bennett, *J. Am. Med. Assoc.*, **226**, 1453 (1973).

63. M. F. Parry and R. Wallach, *Am. J. Med.*, **57**, 143 (1974).

64. K. L. Clay and R. C. Murphy, *Toxicol. Appl. Pharmacol.*, **39**, 39 (1977).

65. K. Hagstam, D. H. Ingvar, M. Paatela, and H. Tallquist, *Acta. Med. Scand.*, **178**, 599 (1965).

66. F. M. Troisi, *Br. J. Ind. Med.*, **7**, 65 (1950).

67. R. J. Haggerty, *N. Engl. J. Med.*, **261**, 1296 (1959).

68. L. S. Dubeikovskaya, T. P. Asanova, G. Y. Rozina, L. F. Budanova, E. S. Zenkevich, N. V. Revnova, and L. E. Gorn, *Gig. Tr. Prof. Zabol.*, **17**, 1 (1973).

69. *Threshold Limit Values for Chemical Substances and Physical Agents in the Environment with Intended Changes for 1977*, American Conference of Governmental Industrial Hygienists, Cincinnati, Ohio.

70. D. I. Peterson, J. E. Peterson, M. G. Hardinge, L. Linda, and W. E. C. Wacker, *J. Am. Med. Assoc.*, **186**, 955 (1963).

71. C. Riddell, S. W. Nielson, and E. J. Kersting, *J. Am. Vet. Med. Assoc.*, **150**, 1531 (1967).

72. D. M. Nunamaker, W. Medway, and P. Berg, *J. Am. Vet. Med. Assoc.*, **159**, 310 (1971).

73. S. D. Beckett and R. P. Shields, *J. Am. Vet. Med. Assoc.*, **158**, 472 (1971).

74. J. L. Sanyer, F. W. Oehme, and M. D. McGavin, *Am. J. Vet. Res.*, **34**, 527 (1973).

75. L. Penumarthy and F. W. Oehme, *Am. J. Vet. Res.*, **36**, 209 (1975).

76. R. L. Mundy, L. M. Hall, and R. S. League, *Toxicol. Appl. Pharmacol.*, **28**, 320 (1974).

77. E. W. Van Stee, A. M. Harris, M. L. Horton, and K. C. Back, *J. Pharmacol. Exptl. Therap.*, **192**, 251 (1975).

78. R. H. Dreisbach, in *Handbook of Poisoning*, Lange Medical Publ., Los Altos, Calif., 1974, p. 151.

79. M. F. Michelis, B. Mitchell, and B. B. Davis, *Clin. Toxicol.*, **9**, 53 (1976).

80. W. E. C. Wacker, H. Haynes, R. Druyan, W. Fisher, and J. E. Coleman, *J. Am. Med. Assoc.*, **194**, 173 (1965).

81. J. Pintér, J. Császár and E. Wölfer, *Zeit. Urologie* **59**, 885 (1966) (translation by Sunlone Inc., Arlington Heights, Ill.).

82. H. C. Aquino and C. D. Leonard, *J. Kentucky Med. Assoc.* **70**, 463 (1972).

83. G. D. Osweiler and P. G. Eness, *J. Am. Vet. Med. Assoc.*, **160**, 746 (1972).

84. A. Davis, A. Roaldi, and L. E. Tufts, *J. Gas Chromatogr.* **1964** (Sept.), 306.

85. C. A. Pons and R. P. Custer, *Am. J. Med. Sci.*, **211**, 544 (1946).

86. A. P. Grant, *Lancet*, **263**, 1252 (1952).

87. I. P. Ross, *Br. Med. J.*, **1**, 1340 (1956).

88. C. Roze, *C. R. Acad. Sci., Paris, Ser. D*, **266**(7), 694 (1968).

89. K. Nagano, E. Nakayama, M. Koyano, H. Oobayaski, H. Adachi, and T. Yamada, *Jap. J. Ind. Health*, **21**, 29 (1979) (in Japanese, English summary).

90. W. F. von Oettingen, *U.S. Public Health Bull. No. 281*, (1943).

91. H. O. Calvery and T. G. Klumpp, *Southern Med. J.*, **32**, 1105 (1939).

92. O. G. Fitzhugh and A. A. Nelson, *J. Ind. Hyg. Toxicol.*, **28**, 40 (1946).

93. H. Wegener, *Arch. Exp. Pathol. Pharmakol.*, **220**, 414 (1953); *Chem. Abstr.*, **48**, 2919 (1954).

94. G. Bornmann, *Arzneim. Forsch.*, **4**, 643, 710 (1954), **5**, 38 (1955); *Chem. Abstr.*, **49**, 7131 (1955).

95. A. Loesser, G. Bornmann, L. Grosskinsky, G. Hess, R. Kopf, K. Ritter, A. Schmitz, E. Stürner, and H. Wegener, *Arch. Exp. Pathol. Pharmakol.*, **221**, 14 (1954); *Chem. Abstr.*, **48**, 4698 (1954).

96. C. S. Weil, C. P. Carpenter, and H. F. Smyth, Jr., *Arch. Environ. Health*, **11,** 569 (1965).

97. C. S. Weil, C. P. Carpenter, and H. F. Smyth, Jr., *Ind. Med. Surg.*, **36,** 55 (1967).

98. L. Karel, B. H. Landing, and T. S. Harvey, *Fed. Proc.*, **6,** 342 (1947).

99. Union Carbide Corporation, unpublished data.

100. S. A. Marchenko, *Vrach. Delo*, **2,** 138 (1973).

101. Y. P. Sanina, *Gig. Sanit.*, **33**(2), 36 (1968).

102. F. H. Wiley, W. C. Hueper, D. S. Bergen, and F. R. Blood, *J. Ind. Hyg. Toxicol.*, **20,** 269 (1938).

103. H. B. Haag and A. M. Ambrose, *J. Pharmacol. Exp. Therap.*, **59,** 93 (1937).

104. M. D. Bowie and D. McKenzie, *S. Afr. Med. J.*, **46,** 931 (1972).

105. P. Auzepy, H. Taktak, P. L. Toubas, and M. Deparis, *Sem. Hop. Paris*, **49,** 1371 (1973).

106. K. A. Telegina, N. A. Mustawa, S. Z. Sakaeva, and V. I. Boiko, *Gig. Tr. Prof. Zabol*, **15,** 40 (1971).

107. S. A. Marchenko, *Farmakol. Toksikol.* (*Kiev*), **1973**(8), 1975.

108. G. O. Curme, Jr. and F. Johnston, Eds., *Glycols*, American Chemical Society Monograph Series 114, Reinhold, New York, (1952).

109. A. Davis, A. Roaldi, and L. E. Tufts, *J. Gas Chromatogr.*, **1964**(Sept.), 306.

110. A. R. Latven and H. Molitor, *J. Pharmacol. Exp. Therap.*, **65,** 89 (1939).

111. E. G. Stenger, L. Aeppli, E. Peheim, and F. Roulet, *Arzneim Forsch.*, **18,** 1536 (1968).

112. W. M. Lauter and V. L. Vrla, *J. Am. Pharm. Assoc.*, **29,** 5 (1940).

113. L. Karel, B. H. Landing, and T. S. Harvey, *J. Pharmacol. Exp. Therap.*, **90,** 338 (1947).

114. O. H. Robertson, *Harvey Lect. Ser.*, **38,** 227 (1943).

115. O. H. Robertson, *Wisconsin Med. J.*, **46,** 311 (1947).

116. B. H. Jennings, E. Biggs, and F. C. W. Olson, *Heating, Piping, Air Conditioning*, **16,** 538 (1944).

117. T. N. Harris and J. Stokes, Jr., *Am. J. Med. Sci.*, **209,** 152 (1945).

118. L. D. Polderman, *Soap Sanit. Chem.*, 133 (July 1947).

119. O. H. Robertson, C. G. Loosli, T. T. Puck, H. Wise, H. M. Lemon, and W. Lester, Jr., *J. Pharmacol. Exp. Therap.*, **91,** 52 (1947).

120. H. McKennis, Jr., R. A. Turner, L. B. Turnbull, and E. R. Bowman, *Toxicol. Appl. Pharmacol.*, **4,** 411 (1962).

121. H. Weifenbach, *Arzneim. Forsch.*, **23,** 1087 (1973).

122. T. Ramstad, T. J. Nestrick, and R. H. Stehl, *Anal. Chem.*, **50,** 1325 (1978).

123. C. B. Shaffer and F. H. Critchfield, *Ind. Eng. Chem. Anal. Ed.*, **19,** 32 (1947).

124. H. F. Smyth, Jr., C. P. Carpenter, and C. S. Weil, *J. Am. Pharm. Assoc. Sci. Ed.*, **39,** 349 (1950).

125. H. F. Smyth, Jr., C. P. Carpenter, and C. S. Weil, *J. Am. Pharm. Assoc. Sci. Ed.*, **44,** 27 (1955).

126. H. F. Smyth, Jr., C. S. Weil, M. D. Woodside, J. B. Knaak, L. J. Sullivan, and C. P. Carpenter, *Toxicol. Appl. Pharmacol.*, **16,** 442 (1970).

127. E. G. Worthley and C. D. Scott, *Lloydia* **29,** 123 (1966).

128. The Dow Chemical Company, *Dow Polyethylene Glycols*, 1959.

129. H. F. Smyth, Jr., C. P. Carpenter, C. B. Shaffer, J. Seaton, and L. Fischer, *J. Ind. Hyg. Toxicol.*, **24,** 281 (1942).

130. J. H. Draize, G. Woodard, and H. O. Calvery, *J. Pharmacol. Exp. Therap.*, **82,** 377 (1944).

131. F. P. Luduena, J. K. Fellows, G. L. Laqueur, and R. L. Driver, *J. Ind. Hyg. Toxicol.*, **29,** 390 (1947).

132. T. W. Tusing, J. R. Elsea, and A. B. Sauveur, *J. Am. Pharm. Assoc. Sci. Ed.*, **43,** 489 (1954).

133. O. Schmidt, *J. Soc. Cosmet. Chem.*, **21,** 835 (1970).

134. C. P. Carpenter, M. D. Woodside, E. R. Kinkead, J. M. King, and L. J. Sullivan, *Toxicol. Appl. Pharmacol.*, **18,** 35 (1971).

135. C. Roze, *C. R. Acad. Sci. Paris, Ser. D,* **266,** 694 (1968). *Chem. Abstr.* 85782h (1968).

136. C. B. Shaffer and F. H. Critchfield, *J. Am. Pharm. Assoc. Sci. Ed.*, **36,** 152 (1947).

137. C. B. Shaffer, F. H. Critchfield, and J. H. Nair, *J. Am. Pharm. Assoc. Sci. Ed.*, **39,** 340 (1950).

138. A. H. Principe, *J. Forensic Sci.*, **13,** 90 (1968).

139. W. Bartsch, G. Sponer, K. Dietmann, and G. Fuchs, *Arzneim. Forsch.*, **26,** 1581 (1976).

140. E. Schutz, *Arch. Exp. Pathol. Pharmakol.*, **232,** 237 (1957); *Ind. Hyg. Dig. Abstr.*, **22,** 720 (1958).

141. Union Carbide Chemicals Company, *Physical Properties of Synthetic Organic Chemicals*, 1958.

142. A. J. Lehman and H. W. Newman, *J. Pharmacol. Exp. Therap.*, **60,** 312 (1937).

143. Council on Pharmacy and Chemistry of the American Medical Association, *New and Nonofficial Remedies*, Lippincott, Philadelphia, 1949.

144. Food and Drug Administration, *Food, Drug, Cosmet. Law J.*, **13,** 856 (1958).

145. J. A. Ruddick, *Toxicol. Appl. Pharmacol.*, **21,** 102 (1972).

146. Anonymous, *GRAS (Generally Recognized as Safe) Food Ingredients—Propylene Glycol and Derivatives*, National Technical Information Service, U.S. Dept. of Commerce, Rep. No. FDABF-GRAS-033 (1973).

147. J. H. Weatherby and H. B. Haag, *J. Am. Pharm. Assoc.*, **27,** 466 (1938).

148. P. J. Hanzlik, H. W. Newman, W. Van Winkle, A. J. Lehman, and N. K. Kennedy, *J. Pharmacol. Exp. Therap.*, **67,** 101 (1939).

149. W. Van Winkle, Jr., *J. Pharmacol. Exp. Therap.*, **72,** 344 (1941).

150. H. W. Newman and A. J. Lehman, *Proc. Soc. Exp. Biol. Med.*, **35,** 601 (1936–37).

151. M. A. Seidenfeld and P. J. Hanzlik, *J. Pharmacol. Exp. Therap.*, **44,** 109 (1932).

152. G. P. Whitlock, N. B. Guerrant, and R. A. Dutcher, *Proc. Soc. Exp. Biol. Med.*, **57,** 124 (1944).

153. I. F. Gaunt, F. M. B. Carpanini, P. Grasso, and A. B. G. Lansdown, *Food Cosmet. Toxicol.*, **10,** 151 (1972).

154. C. S. Weil, M. D. Woodside, H. F. Smyth, Jr., and C. P. Carpenter, *Food Cosmet. Toxicol.*, **9,** 479 (1971).

155. W. Van Winkle, Jr. and H. W. Newman, *Food Res.*, **6,** 509 (1941).

156. L. J. Fisher, J. D. Erfle, G. A. Dodge, and F. D. Sauer, *Can. J. Anim. Sci.*, **53,** 289 (1973).

157. F. D. Sauer, J. D. Erfle, and L. J. Fisher, *Can. J. Anim. Sci.*, **53,** 265 (1973).

158. A. Shiga, J. Tsuji, K. Shinozaki, and Y. Kobayashi, *Nippon Chikusan Gakkai-Ho*, **46,** 341 (1975).

159. H. S. Bailey, S. J. Slinger, and J. D. Summers, *Poultry Sci.*, **46,** 19 (1967).

160. J. N. Persons, B. L. Damron, P. W. Waldroup, and R. H. Harms, *Poultry Sci.*, **47,** 351 (1968).

161. P. W. Waldroup and T. E. Bowen, *Poultry Sci.*, **47,** 1911 (1968).

162. S. Harnisch, *Arch. Gefluegelkd.*, **37,** 187 (1973).

163. E. G. Worthley and C. D. Scott, *Lloydia*, **29,** 123 (1966).

164. J. R. Hickman, *J. Pharm. Pharmacol.*, **17,** 255 (1965).

165. R. T. Brittain and P. F. D'arcy, *Toxicol. Appl. Pharmacol.*, **4,** 738 (1962).

166. K. J. Davis and P. M. Jenner, *Toxicol. Appl. Pharmacol.*, **1,** 576 (1959).

167. J. Bost and Y. Ruckebusch, *Therapie*, **17,** 83 (1962).

168. C. S. Weil and R. A. Scala, *Toxicol. Appl. Pharmacol.*, **19,** 276 (1971).

169. T. G. Warshaw and F. Herrmann, *J. Invest. Dermatol.*, **19,** 423 (1952).

170. R. E. Davies, K. H. Harper, and S. R. Kynoch, *J. Soc. Cosmet. Chem.*, **23**, 371 (1972).
171. L. A. Goldsmith and H. P. Baden, *J. Am. Med. Assoc.*, **220**, 579 (1972).
172. F. Morikawa, *Nippon Hifuka Gakkai Zasshi*, **82**, 794 (1972).
173. L. Phillips II, M. Steinberg, H. I. Maibach, and W. A. Akers, *Toxicol. Appl. Pharmacol.*, **21**, 369 (1972).
174. R. N. Shore and W. B. Shelley, *Arch. Dermatol.*, **109**, 397 (1974).
175. F. N. Marzulli and H. I. Maibach, *Food Cosmet. Toxicol.*, **12**, 219 (1974).
176. T. Morizono and B. M. Johnstone, *Med. J. Aust.*, **18**, 634 (1975).
177. J. P. Nater, A. J. M. Baar, and P. J. Hoedemaeker, *Contact Derm.*, **3**, 181 (1977).
178. S. K. Bau, N. Aspin, D. E. Wood, and H. Levison, *Pediatrics*, **48**, 605 (1971).
179. N. B. Guerrant, C. P. Shitlock, M. L. Wolff, and R. A. Dutcher, *Bull. Natl. Formulary Comm.*, **15**, 205 (1947).
180. C. W. Emmens, *J. Reprod. Fert.*, **26**, 175 (1971).
181. D. O. E. Gebhardt, *Teratology*, **1**, 153 (1968).
182. G. L. Kennedy, Jr., D. W. Arnold, M. L. Keplinger, and J. C. Calandra, *Toxicology*, **5**, 159 (1975).
183. F. Dewhurst, D. A. Kitchen, and G. Calcutt, *Br. J. Cancer*, **26**, 506 (1972).
184. R. W. Baldwin, G. J. Cunningham, W. R. D. Smith, and S. J. Surtees, *Br. J. Cancer*, **22**, 133 (1968).
185. K. Wallenious and U. Lecholm, *Odontol. Revy*, **24**, 39, 115 (1973).
186. F. Stenback and P. Shubik, *Toxicol. Appl. Pharmacol.*, **30**, 7 (1974).
187. J. L. Clapperton and J. W. Czerkawski, *Br. J. Nutrit.*, **27**, 553 (1972).
188. D. Giesecke, *Arch. Int. Physiol. Biochim.*, **82**, 645 (1974).
189. D. Giesecke, G. Dirksen, R. Guenzel, and G. Baumer, *Dtsch. Tieraerztl. Wochenschr.*, **82**, 105 (1975).
190. (a) M. Yoshida and H. Ikumo, *Agric. Biol. Chem.*, **35**, 1628 (1971); (b) G. Martin and L. Finberg, *J. Pediat.*, **77**, 877 (1970).
191. G. Martin and L. Finberg, *J. Pediatr.*, **77**, 877 (1970).
192. W. Van Winkle, Jr., *J. Pharmacol. Exp. Therap.*, **72**, 227 (1941).
193. R. D. Creek, *Poultry Sci.*, **49**, 1686 (1970).
194. P. K. Gessner, D. V. Parke, and R. T. Williams, *Biochem. J.*, **74**, 1 (1960).
195. L. Ulrich and M. Mestitzova, *Prac. Lek.*, **7**, 299 (1963).
196. H. E. Christensen, Ed., *Registry of Toxic Effects of Chemical Substances* 1976 ed., U.S. Dept. of Health, Education, and Welfare, National Institute for Occupational Safety and Health, Rockville, Md., 1976.
197. H. D. Keston, N. G. Mulinos, and L. Pomerantz, *Arch. Pathol.*, **27**, 447 (1939).
198. The Dow Chemical Company, *Manufacturer's Check List—Potential Uses for Polypropylene Glycols*, Form No. 111-430-69, 1969.
199. Union Carbide Chemicals Co., *Physical Properties of Synthetic Organic Chemicals*, No. F-6136K, 1957.
200. C. B. Shaffer, C. P. Carpenter, F. H. Critchfield, J. H. Nair, III, and F. R. Frank, *Arch. Ind. Hyg. Occup. Med.*, **3**, 448 (1951).
201. F. E. Shideman and L. Procita, *J. Pharmacol. Exp. Therap.*, **103**, 293 (1951).
202. The Dow Chemical Company, *Polypropylene Glycols from Dow*, Form No. 118-1014-80, 1980.

203. The Dow Chemical Company, *Polypropylene Glycols,* Form No. 125-129-57, 1959.

204. Eastman Kodak Company, unpublished data.

205. H. Schlüssel, *Arch. Exptl. Pathol. Pharmakol.,* **221,** 67 (1954); *Chem. Abstr.,* **48,** 5315 (1954).

206. E. Strack, D. Biesold, and H. Thiele, *Z. Ges. Exp. Med.,* **132,** 522 (1960).

207. A. Loeser, *Pharmazie,* **4,** 263 (1949); *Chem. Abstr.,* **43,** 8558 (1949).

208. L. Fischer, R. Kopf, A. Loeser, and G. Meyer, *Z. Ges. Exp. Med.,* **115,** 22 (1949); *Chem. Abstr.,* **44,** 9070 (1950).

209. H. F. Smyth, Jr., C. P. Carpenter, and C. S. Weil, *Arch. Ind. Hyg. Occup. Med.,* **4,** 119 (1951).

210. R. Kopf, A. Loeser, G. Meyer, and W. Franke, *Arch. Exp. Pathol. Pharmakol.,* **210,** 346 (1950); *Chem. Abstr.,* **45,** 5308 (1951).

211. R. V. Chagovets, N. N. Velikii, P. K. Parklomets, N. Y. Simonova, G. V. Chichkovskaya, T. M. Turganbaeva, and S. P. Chaika, *Ukr. Biokhim. Zh.,* **46,** 275 (1974).

212. R. L. Harris, M. A. Mehlman, and R. L. Veech, *Fed. Proc.,* **31,** 670 (1972).

213. R. A. Scala and O. E. Paynter, *Toxicol. Appl. Pharmacol.,* **10,** 160 (1967).

214. S. A. Miller and H. Dymsza, *J. Nutr.,* **91,** 79 (1967).

215. M. Yoshida, K. Osada, S. Fujishiro, and R. Oda, *Agric. Biol. Chem.,* **35,** 393 (1971).

216. R. B. Tobin, M. A. Mehlman, C. Kies, H. M. Fox, and J. S. Soeldner, *Fed. Proc.,* **34,** 2171 (1975).

217. S. L. Hem, D. R. Bright, C. S. Banker, and J. P. Pogue, *Drug Dev. Commun.,* **1,** 471 (1974–1975).

218. H. Dominguez-Gil and R. Cadorniga, *Farm. Ed. Prat.,* **26,** 535 (1971).

219. E. Husing, R. Kopf, and A. Loeser, *Fette Seifen,* **52,** 45 (1950); *Chem. Abstr.,* **44,** 7999 (1950).

220. M. V. Shelanski, *Cosmet. Perfum.,* **89,** 96 (1974).

221. H. A. Dymsza, *Fed. Proc.,* **34,** 2167 (1975).

222. R. L. Tate, M. A. Mehlman, and R. B. Tobin, *J. Nutr.,* **101,** 1719 (1971).

223. M. A. Mehlman, R. B. Tobin, H. K. J. Hahn, L. Kleager, and R. L. Tate, *J. Nutr.,* **101,** 1711 (1971).

224. J. J. B. Ayres and F. A. Isgrig, *Psychopharmacologia,* **16,** 290 (1970).

225. J. E. Zabik, D. P. VanDam, and R. Maichel, *Toxicol. Appl. Pharmacol.,* **25,** 461 (1973).

226. W. Schneider, *Pharm. Ind.,* **12,** 226 (1950); *Chem. Abstr.,* **45,** 3998 (1951).

227. H. Sprince, J. A. Josephs, Jr., and C. R. Wilpizeski, *Life Sci.,* **5,** 2041 (1966).

228. R. Maxwell and R. H. Roth, *Biochem. Pharmacol.,* **21,** 1621 (1972).

229. A. Hinrichs, R. Kopf, and A. Loeser, *Pharmazie,* **3,** 110 (1948); *Chem. Abstr.,* **42,** 5567 (1948).

230. S. P. Kryshova, *Gig. Sanit.,* **33,** 37 (1968).

231. H. F. Smyth, Jr., C. P. Carpenter, C. S. Weil, U. C. Pozzani, and J. A. Striegel, *Am. Ind. Hyg. Assoc. J.,* **23,** 95 (1962).

232. Shell Chemical Corporation, Ind. Hyg. Bull., *Hexylene Glycol,* SC:57–101 and SC:57–102, 1958.

233. H. F. Smyth, Jr. and C. P. Carpenter, *J. Ind. Hyg. Toxicol.,* **30,** 63 (1948).

234. V. Larsen, *Acta Pharmacol. Toxicol.,* **14,** 341 (1958).

235. W. B. Deichmann and M. Dierker, *J. Biol. Chem.,* **163,** 753 (1946).

236. E. Jacobsen, *Acta Pharmacol. Toxicol.,* **14,** 207 (1958).

237. C. J. Terhaar, W. J. Krasavage, and R. L. Roudabush, *Toxicol. Appl. Pharmacol.,* **29,** 87 (1974).

238. B. D. Astill and D. W. Fassett, *Toxicol. Appl. Pharmacol.,* **29,** 151 (1974).

239. A. J. Lehman, *Assoc. Food Drug Offic. U.S. Quart. Bull.,* **19,** 87 (1955).

240. P. Milvy and A. J. Garro, *Mutat. Res.,* **40,** 15 (1976).

241. C. Vainio, R. Pakkonen, K. Ronnholm, V. Raunio, and O. Pelkonen, *Scand. J. Work Environ. Health,* **2,** 147 (1976).

242. H. F. Smyth, Jr., C. S. Weil, J. M. King, J. B. Knaak, L. J. Sullivan, and C. P. Carpenter, *Toxicol. Appl. Pharmacol.,* **16,** 675 (1970).

243. BASF Wyandotte Corporation, Bulletin No. 05-3012 (765).

244. C. W. Leaf, *Soap and Chemical Specialties,* **1967**(August), 48.

245. *Threshold Limit Values for Chemical Substances and Physical Agents in the Environment with Intended Changes for 1981,* American Conference of Governmental Industrial Hygienists, Cincinnati, Ohio.

246. T. C. Marshall and Y. S. Cheng, *Inhalation Toxicology Research Institute Annual Report,* LMF-84, 493, Dec. 1980.

247. E. H. Pfeiffer and H. Dunkelberg, *Food Cosmet. Toxicol.,* **18,** 115 (1980).

248. D. M. Conning and M. J. Hayes, *Br. J. Ind. Med.,* **27,** 155 (1970).

249. V. K. H. Brown, L. B. Valeris, and D. J. Simpson, *Arch. Environ. Health,* **30,** 1 (1975).

CHAPTER FIFTY-ONE

Derivatives of Glycols

**V. K. ROWE, Sc. D. (Hon.), and
M. A. WOLF***

INTRODUCTION

Many derivatives, ethers, esters, and ether esters, of various glycols are discussed in this chapter. Groupings are based largely on chemical structures where general discussions permit the avoidance of repetition. However, for the most part, individual compounds are discussed separately.

Trade names are given where applicable to aid the reader in associating them with chemical structure. The following appear:

Carbitol®	Union Carbide Corporation's trademark for ethers of diethylene glycol
Cellosolve®	Union Carbide Corporation's trademark for ethers of ethylene glycol
Dowanol®	The Dow Chemical Company's trademark for glycol ethers
Ektasolve®	Eastman Kodak's trademark for solvents
Kodaflex®	Eastman Chemicals Products, Inc. trademark
Oxitol®	Shell Chemical Company's trademark for ethers of ethylene glycol
Pluracol®	BASF Wyandotte's trademark for a series of organic compounds used as chemical intermediates and as constituents of other functional fluids, for example, brake fluids

* Deceased.

Poly-solv®   Olin Corporation's trademark for glycol ether solvents
Propasol®   Union Carbide Corporation's trademark for a series
                    of solvents for water-based enamels
Santicizer®   Monsanto Corporation's trademark for plasticizers
Ucar®   Union Carbide Corporation's trademark for various
                    synthetic organic chemicals

1.1 Determination in the Atmosphere

The choice of the best method for the determination of the glycol ethers varies
with existing conditions.

In many instances where materials are very soluble in water, samples of air
can be taken effectively by scrubbing through water. In instances where water
solubility is limited, alcohol, cold carbon disulfide, or other suitable solvents can
be used as scrubbers. In certain situations the material can be absorbed on
sillica gel or activated alumina, then desorbed with water, methanol, or acetone.
Gas chromatography is likely to be the analytical method of choice for final
analysis but infrared absorption may sometimes be useful. The methods
described by Singliar and Dykyj (1) and by Esposito and Jamison (2) for
mixtures may be useful. Specific conditions should be developed for individual
materials in each laboratory and validated before use.

2 ETHERS OF MONO-, DI-, TRI-, TETRA-, AND POLYETHYLENE GLYCOL

2.1 Source, Uses, and Industrial Exposure

The glycol ethers most commonly encountered industrially are colorless liquids
with mild ethereal odors. These monoalkyl ethers are usually produced by
reacting ethylene oxide with the alcohol of choice, but also may be made by the
direct alkylation of a selected glycol with an alkylating agent such as dialkyl
sulfate. The dialkyl ethers may be prepared by reacting the sodium alcoholate
of the glycol monoether with an alkyl halide. The yields of these ethers can be
varied by changing the mole ratios of reactants and by changing catalysts. For
a more comprehensive discussion of the preparation of these materials and
references, refer to Boese et al. (3).

The miscibility of most of these ethers with water and a large number of
organic solvents makes them especially useful as mutual solvents in many
oil–water compositions. They are used as solvents for various resins, lacquers,
paints, varnishes, dyes, inks, printing pastes, cleaning compositions, liquid soaps,
and even cosmetics. They also are used widely as components of hydraulic
fluids and as chemical intermediates.

Industrial exposure may occur by any of the common routes. Excessive exposure to certain of the glycol ethers may occur from inhalation or by absorption through the skin. Ingestion is not likely to be a factor in industrial handling. The toxicologic information, experience, and hazards of each individual compound are discussed separately.

2.2 Physical and Chemical Properties

The physical and chemical properties of the glycol ethers most commonly encountered are given in Tables 51.1 and 51.2.

2.3 Determination in the Atmosphere

Refer to the introduction of this chapter.

3 ETHYLENE GLYCOL MONOMETHYL ETHER; Ethanol, 2-methoxy-; Methyl Cellosolve® Solvent; Dowanol® EM Glycol Ether; Methyl Oxitol® Glycol; CAS No. 109-86-4

$$CH_3OCH_2CH_2OH$$

3.1 Physical and Chemical Properties

See Table 51.1.

3.2 Physiologic Response

3.2.1 Summary

Ethylene glycol monomethyl ether is low in single-dose oral toxicity, moderate in repeated-dose oral toxicity, not appreciably irritating to the skin, and mildly irritating to the eyes, may be absorbed through the skin but is low in toxicity by this route, and is appreciably toxic when inhaled. Its vapors are irritating in acutely toxic concentrations, but concentrations that may cause serious systemic toxicity upon prolonged and repeated inhalation have negligible warning properties. The material exerts its principal physiologic action upon the brain, blood, and kidneys.

3.2.2 Single-Dose Oral

According to Carpenter and co-workers (4), the single-dose oral toxicity of ethylene glycol monomethyl ether is 3.4 g/kg for rats, 0.89 g/kg for rabbits, and 0.95 g/kg for guinea pigs. Union Carbide (5) lists the LD_{50} for rats as 2.46 g/kg,

Table 51.1. Physical and Chemical Properties of Ethers of Ethylene Glycol

Property	Methyl	Ethyl	n-Propyl	Isopropyl
CAS No.	109-86-4	110-80-5	2807-30-9	109-59-1
Molecular formula	$C_3H_8O_2$	$C_4H_{10}O_2$	$C_5H_{12}O_2$	$C_5H_{12}O_2$
Molecular weight	76.1	90.1	104.1	104.1
Specific gravity 25/4°C	0.962	0.926	0.909	0.900
Boiling point (°C) (760 mm Hg)	124.2	135.0	150-152	140-143
Freezing point (°C)	−85	−100		
Vapor pressure (mm Hg) (25°C)	9.7	5.75	2.9	5.2
Refractive index (25°C)	1.400	1.406	1.412	1.407
Flash point (°F) (open cup)	115	120	125	145
Autoignition temperature (°C)	285	235		
Flammability limits (Vol. % in air)	2.5–19.8	1.82–14.0		
	(125–140°C)	(140–150°C)		
Vapor density (air = 1)	2.6	3.0		3.6
Percent in saturated air (25°C)	1.28	0.76	0.38	0.68
1 ppm ≎ mg/m³ at 25°C, 760 mm Hg	3.11	3.68	4.25	4.25
1 mg/l ≎ ppm at 25°C, 760 mm Hg	322	272	235	235

whereas Saparmamedov (6) found the LD_{50} for mice to be 2.8 g/kg when given in oil solution. If given in water the LD_{50} is slightly higher. In massive doses, the material has a narcotic action but at lower dosage levels deaths are delayed and are accompanied by lung edema, slight liver injury, and marked kidney injury. Hematuria may occur from single doses.

3.2.3 Repeated-Oral Administration

Gross (7) fed rabbits repeated daily doses of ethylene glycol monomethyl ether and found that seven doses of 0.1 ml/kg caused temporary hematuria. Larger doses caused exhaustion, tremors, albuminuria, hematuria, and death. Autopsy revealed severe kidney injury. In studies in which rats were fed diets containing 0.01 to 1.25 percent in their diets for 90 days, the lowest level, 0.01 percent, was found to cause depression of appetite. Beginning pathologic effects were seen at the 0.05 percent level. The 1.25 percent level caused death (8).

Nagano et al. (13) fed mice doses ranging from 62.5 to 4000 mg/kg 5 days a

	Ether						
n-Butyl	Isobutyl	n-Hexyl	2-Methyl-pentyl	2,6,8-Trimethyl-4-nonyl	Diethyl	Phenyl
111-76-2	4439-24-1	112-25-4	29290-45-7	10137-98-1	629-14-1	122-99-6
$C_6H_{14}O_2$	$C_6H_{14}O_2$	$C_8H_{18}O_2$	$C_8H_{18}O_2$	$C_{14}H_{30}O_2$	$C_6H_{14}O_2$	$C_8H_{10}O_2$
118.2	118.2	146.2	146.2	230.4	118.2	138.2
0.898	0.887	0.889	0.888	0.866	0.842	1.104
		(20/20°C)	(20/20°C)	(20/20°C)		
170.8	160.3	208.1	197.1	227	121.4	245.6
				(300 mm Hg)		
−77		−45	<−80	<−50		8.5
			(sets to glass)	(sets to glass)		
0.88	1.6	0.05	0.09	<0.01	12.5	0.0073
		(20°C)	(20°C)	(20°C)		
1.417	1.413		1.428	1.439		1.536
			(20°C)	(20°C)		
165	142	195	88		95	265
244						
1.13–10.6						
(160–180°C)						
4.1	4.1		5.06	7.97	4.1	4.8
0.093	0.21		0.01		1.64	0.00096
4.83	4.83	5.98	5.98	9.42	4.83	5.65
207	207	168	168	106	207	177

week for 5 weeks and observed testicular atrophy and leukopenia, the intensity of which was dose related. Testicular injury from higher doses was reported earlier by Wiley et al. (21).

3.2.4 Injection

The LD_{50} value for rats by intravenous injection is 2.14 g/kg (4).

3.2.5 Eye Irritation

When ethylene glycol monomethyl ether was introduced into the eyes of rabbits, it produced immediate pain, conjunctival irritation, and slight transitory cloudiness of the cornea, which cleared within 24 hr (9). Carpenter and Smyth (11) classify the material along with ethanol in regard to its effect on the rabbit eye.

Grant (12) reports one recorded human eye exposure incident in which complete recovery occurred within 48 hr after exposure.

Table 51.2. Physical and Chemical Properties of Some Common Ethers of Di- and Triethylene Glycol

Property	Ethers of Diethylene Glycol						Ethers of Triethylene Glycol		
	Methyl	Ethyl	Ethyl Vinyl	n-Butyl	n-Hexyl	1,4-Dioxane	Methyl	Ethyl	Butyl
CAS No.	111-77-3	111-90-0	10143-53-0	112-34-5	112-59-4	123-91-1	112-35-6	112-50-5	143-22-6
Molecular formula	$C_5H_{12}O_3$	$C_6H_{14}O_3$	$C_8H_{16}O_3$	$C_8H_{18}O_3$	$C_{10}H_{22}O_3$	$C_4H_8O_2$	$C_7H_{16}O_4$	$C_8H_{18}O_4$	$C_{10}H_{22}O_4$
Molecular weight	120.1	134.2	160.2	162.2	190.3	88.1	164.2	178.2	206.3
Specific gravity 25/4°C	101.9	0.986	0.941 (20/20°C)	0.948	0.935	1.035	1.052	1.018	0.983
Boiling point (°C) (760 mm Hg)	194.1	202.0	191.2	230.4	259.1	101.3	249.2	256.3	279.4
Freezing point (°C)	−85.0	−76.0	−80 (sets to glass)	−68.0	−33.3		−44.0	−18.8	−35.2
Vapor pressure (mm Hg) (25°C)	0.18	0.14	0.2 (20°C)	0.043	<0.01	37	<0.01	<0.01	0.0025
Refractive index (25°C)			1.429 (20°C)			1.422			
Flash point (°F) (O.C.)	200	205	91	200	285	65	245	275	
Autoignition temperature (°C)	~250	~250	~201	~228					
Flammability limits, Vol. % in air	1.5–9.5	1.2–8.5	0.4 (lower limit)	5.58		1.97–22.25			
Vapor density (air = 1)	4.14	4.62	5.54			~3			
Percent in saturated air (25°C)	0.048	0.018	0.03	0.0057		4.75	0.021	0.0026	0.00032
1 ppm ⇌ mg/m³ at 25°C, 760 mm Hg	4.91	5.49	6.55	6.64	7.78	3.60	6.72	7.29	8.44
1 mg/l ⇌ ppm at 25°C, 760 mm Hg	204	188.2	152.6	150.8	128.5	278	148.9	137.2	118.5

3.2.6 Skin Contact—Irritation

Ethylene glycol monomethyl ether in repeated and prolonged contact with the skin of rabbits failed to cause any appreciable irritation (9).

3.2.7 Skin Absorption

Ethylene glycol monomethyl ether is absorbed readily through the skin in toxic amounts. Quantitation by the "sleeve" technique essentially as described by Draize et al. (10) indicated that the LD_{50} was approximately 2 g/kg for rabbits (9). In similar tests LD_{50} values for rabbits were found to be 1.29 g/kg (8). Saparmamedov (6), however, found that when the compound was applied to mouse skin there were no signs of absorption. The signs of intoxication resulting from absorption through the skin are essentially the same as those resulting from other routes of administration.

3.2.8 Inhalation

The toxicity of ethylene glycol monomethyl ether when inhaled has been determined for several animal species. Perhaps the most pertinent studies in this area have been conducted by Gross (7) and by Werner and co-workers (14–16). This work has been well summarized by Smyth (18) and Browning (17). Gross (7) found that a few repeated exposures to 800 or 1600 ppm produced serious systemic intoxication, characterized for the most part by irritation of the respiratory tract and lungs, hematuria, albuminuria, cylindrical casts in the urine, and severe glomerulitis. Werner et al. (14) exposed mice for 7 hr to various concentrations and found the LC_{50} to be 1480 ppm. They concluded that lung and kidney injury was generally the cause of death. Werner et al. (15) exposed groups of rats 7 hr/day, 5 days a week for 5 weeks to a concentration averaging 310 ppm. They noted after 1 week of exposure that there was an increase in the percentage of immature granulocytes in the circulating blood. They observed no changes in the kidneys or lungs. Werner et al. (16) exposed two dogs to a vapor concentration of 750 ppm of ethylene glycol monomethyl ether 7 hr/day, 5 days a week for 12 weeks. Again, the most significant changes were in the blood. The hemoglobin concentration, cell volume, and the number of erythrocytes were decreased. The red cells showed an increased hypochromia, polychromatophilia, and microcytosis. The blood picture, in regard to white cells, was characterized by a greater than normal number of immature forms. These authors point out that the methyl ether, which is significantly less hemolytic in vitro than the other common alkyl ethers of ethylene glycol, produced the greatest alteration in the red cells in the dogs. This is quite a different finding from that which occurs in mice (14) and rats (15), where there appears to be a definite correlation between hematologic effects and hemolytic potency. The site at which these blood changes occur is obscure. The lack of significant amounts of hemosiderin in the spleen suggests that it is not hemolytic. Although no serious damage to the bone marrow was observed in these studies, it is doubtful the studies were sufficiently prolonged to demonstrate that the effect was not centered in the marrow.

Goldberg et al. (19) exposed rats for 4 hr/day for 7 days to 125 ppm and found that they appeared normal in health but they exhibited a decrease in pole climbing response. No tolerance to the exposure was seen; repeated exposure increased the severity of the response, suggesting to the authors an accumulation of ethylene glycol methyl ether or its metabolites. They also found that a single exposure of mice to 125 ppm for 4 hr potentiated the hypnotic effects of barbiturates; at 500 ppm for 4 hr, a decrease in motor activity occurred. Dajani (24) also studied the effect of this material on the behavior of mice and rats. Inhibition of conditioned avoidance response in rats was seen only when lethal or near-lethal exposures occurred, that is, 3317 ppm by inhalation or 1.45 mg/kg orally. At 1656 ppm the inhibition was 10 percent.

3.2.9 Repeated Injection

Gross (7) gave repeated subcutaneous injections of ethylene glycol monomethyl ether to guinea pigs and rabbits. His results show that with the guinea pig, whereas seven daily injections of 0.25 ml/kg caused no symptoms, five injections of either 0.5 or 1.0 ml/kg caused prostration, labored breathing, and death. The response of rabbits was quite similar, except that the rabbit appears to be slightly more resistant than the guinea pig to the effects of the material; deaths occurred only after seven injections of 1.0 ml/kg.

Carpenter et al. (4) found that ethylene glycol monomethyl ether in concentrations of more than 25 percent in 0.75 percent sodium chloride was hemolytic to rat erythrocytes. When such a 25 percent solution was given intravenously to rats, the LD_{50} was found to be 2.7 g/kg; for the undiluted material the LD_{50} was 2.2 g/kg. When the undiluted material was given intraperitoneally to the rat, the LD_{50} was 2.5 g/kg.

3.2.10 Metabolism

Wiley et al. (21) injected (site unspecified but believed to be intramuscular) two dogs with 6 ml and two rabbits with 2 ml daily. One rabbit died after two injections and one dog was anuric after the third (last) injection. The authors were unable to find any increase in oxalic acid in the urine of any of these animals and they found no increase in methanol or formic acid in the urine of the rabbits. Since under similar conditions the injection of ethylene glycol tripled the urinary excretion of oxalic acid, the injection of ethylene glycol monoacetate doubled it, and since they could find no evidence of the methyl group being liberated, they concluded that this glycol ether was not hydrolyzed to the glycol. Quantitatively, they showed ethylene glycol monomethyl ether to be more toxic than ethylene glycol monoacetate, ethylene glycol diethyl ether, ethylene glycol, or diethylene glycol. Clinical examination and autopsy of the animals treated with ethylene glycol monomethyl ether revealed anuria, calcified casts in the urine, irritation of the bladder mucosa, hemorrhage in the gastrointestinal tract, lung edema, and liver and testicular injury.

Nitter-Hauge (23), in evaluating the poisoning in two human cases, noted that the clinical signs were similar to those caused by ethylene glycol and methanol. He suggested that ethylene glycol methyl ether therefore may be hydrolyzed in man with the formation of ethylene glycol and methanol.

Zavon (22) reported that methoxyacetic acid is considered the principal metabolite. Dajani (24) reports that ethylene glycol methyl ether was found equally distributed in brain, plasma, lung, and liver 1 hr after administration and that the half-life in the body was approximately 1 to 2 hr, unless doses were near the lethal level. In inhalation studies the plasma levels of ethylene glycol methyl ether increased nearly linearly from exposures of 1, 2, 4, and 6 hr to 3317 ppm. When the exposure was extended to 8 hr the concentration in the plasma more than doubled that found after the 6-hr exposure, suggesting that metabolic and/or excretory mechanisms were saturated. The material was detected in the urine of rats 30 min after the intraperitoneal administration of 1.0 ml/kg and continued to be present for a total of 7 hr after the injection, 3 hr after it was last seen in the blood.

In work on rats that were given a priming intravenous dose of 310 mg/kg, then an infusion dose of ethylene glycol methyl ether at a rate of 8 mg/kg/min followed with a rate of 16 mg/kg/min, the methyl ether was eliminated via the breath unchanged at a rate of 8 mg/kg/min (25), the same as for the ethyl ether.

Based on these findings it appears that ethylene glycol methyl ether is metabolized very slowly in the body and is eliminated mainly via expired air and in the urine. The metabolites that may be formed still must be identified. The cause of the clinical symptoms has not been determined, but it would seem that they may be caused in large measure by unmetabolized ethylene glycol methyl ether (21).

3.2.11   Human Experience

The only fatal case of poisoning in a human due to the ingestion of ethylene glycol monomethyl ether is that recorded by Young and Woolner (26). The amount of material consumed is speculative, but it is believed that the man consumed about 200 ml of the material mixed with rum. He was admitted to the hospital in a comatose condition and died 5 hr later without regaining consciousness. The urine contained ethanol but no methanol, thus supporting the conclusion of Wiley et al. (21) that the ether is not hydrolyzed. Autopsy revealed hemorrhagic gastritis, marked degeneration of the kidney tubules, and fatty degeneration of the liver. In 1936, Donley (27) described a case of "toxic encephalopathy" suffered by a female who was employed in "fusing" shirt collars by dipping them in a solution composed of ethylene glycol monomethyl ether, isopropanol, and cellulose acetate. She suffered from headache, drowsiness, lethargy, generalized weakness, irregular and unequal pupils, disorientation, and psychopathic symptoms. Two years later Parsons and Parsons (28) described two cases of poisoning resulting from the inhalation of vapors of ethylene glycol monomethyl ether again encountered in the manufacture of "permanently starched" collars. The symptoms experienced by these two men were weakness, sleepiness, headache, gastrointestinal upset, nocturia, loss of weight, burning of the eys, and a complete change of personality from one of sharp intelligence to one of stupidity and lethargy. Clinical examination revealed a macrocytic anemia. Both patients apparently recovered completely. As a result of the two cases reported (28), Greenburg and co-workers (29) examined these 2 and 17 other workers employed in the same factory who were using ethylene glycol monomethyl ether. Actually, the solvent being used was 33 percent ethylene glycol monomethyl ether and 67 percent denatured ethanol; no benzene was present and the denaturants are not suspect. All revealed abnormalities in the blood pictures, suggesting macrocytic anemia, and all suffered some degree of excessive fatigue, abnormal reflexes, and tremors. Examination revealed general immaturity of the leukocytes in every case.

Unfortunately, from the industrial hygiene researcher's viewpoint, ventilation of the operation causing these effects was improved before Greenburg et al. (29) had a chance to measure the concentrations to which the men were actually exposed. After changes in the ventilation system had been accomplished, concentrations ranged from 25 to 76 ppm. Apparently these findings led the investigators to suggest a threshold limit of 25 ppm, even though the concentrations to which the affected persons were exposed were, in all probability, much higher.

Zavon (22) reported five cases of illness in workers. A simulated exposure suggests that these workers were exposed to levels ranging from 61 to 3960 ppm. These men practiced poor hygienic controls and worked 9 to 10 hr/day 6 days a week. Four of the men, those best studied, showed symptoms of central nervous system depression, and in one case cerebral atrophy as shown by ataxia, a positive Romberg test, slurred speech, and tremors. The changes in personality seen were similar to those described by Donley (27) and by Parsons and Parsons (28). All showed anemia, and in one case a hypocellular bone marrow with decrease of the erythroid elements.

Nitter-Hauge (23) reports on two cases of men who drank an estimated dose of 100 ml ethylene glycol methyl ether. The first clinical symptoms occurred after at least 8 to 18 hr and were similar to those reported by Parsons and Parsons (28). Marked acidosis was seen and one patient developed marked oxaluria.

Ohi and Wegman (20) reported on two cases of poisoning that followed the substitution of ethylene glycol methyl ether for acetone in a mandrel cleaning operation. Since vapor concentrations were well below the accepted safe level and there was appreciable skin contact, the authors believe that cutaneous absorption may have been a significant factor. The signs and symptoms in these cases were typical, and it is noteworthy that recovery occurred upon cessation of exposure. The high frequency of mental retardation, neurologic symptoms, drowsiness, fatigue, macrocytic anemia, and the abnormal leukocyte picture presented by persons excessively exposed to ethylene glycol monomethyl ether leaves little doubt that the effects are centered primarily in the brain, blood, and kidneys.

3.3 Hygienic Standards of Permissible Exposure

The American Conference of Governmental Industrial Hygienists (30) in 1980 recommended a threshold limit of 25 ppm for ethylene glycol monomethyl ether. The Occupational Safety and Health Administration also has adopted 25 ppm (80 mg/m^3) as a guide for control (31). The experimental data upon which this value is based seem inadequate for the establishment of a firm industrial hygiene standard, particularly in view of the seriousness of the consequences of excessive exposure.

3.4 Odor and Warning Properties

May (32), under controlled conditions and using human volunteers, found that the odor threshold for ethylene glycol methyl ether was approximately 60 ppm and the level of strong odor 90 ppm. In another study, human volunteers were exposed to levels of 25 and 115 ppm (9) for a full working day. The 25-ppm level was considered to be the maximum tolerated odor level by three of five people, whereas the intolerable level appeared to be more than 115 ppm.

4 ETHYLENE GLYCOL MONOVINYL ETHER; Ethanol, 2-(Ethenyloxy)-; CAS No. 764–48–7

$$HOCH_2CH_2OCH{=}CH_2$$

4.1 Physical and Chemical Properties

Molecular formula	$C_4H_8O_2$
Molecular weight	88.1
1 ppm $\approx$	3.60 mg/m^3 at 25°C, 760 mm Hg
1 mg/l $\approx$	278 ppm at 25°C, 760 mm Hg

4.2 Physiologic Response

Ethylene glycol monovinyl ether is low in single-dose oral toxicity, the LD_{50} value for mice being 2.9 g/kg. Rabbit eye tests resulted in irritation but no apparent corneal injury. The ether was irritating to the skin of rats but it did not cause sensitization of the skin of guinea pigs. The LC_{50} value for inhalation (length of exposure not given) was found to be 29 mg/1 or 8150 ppm. Adverse changes were seen in the liver, kidneys, and lungs (33).

   Gadaskina and Rudi (34) administered 400 mg/kg of ethylene glycol monovinyl ether orally and found that ethylene glycol was excreted in the urine during the next 24 hr. The amount of ethylene glycol represented approximately 24 percent of that which could be formed from complete hydrolysis of the dose given. When both ethylene glycol monovinyl ether and ethyl alcohol were administered simultaneously, the amount of ethylene glycol excreted in the urine increased. The results suggest that the toxic effects of ethylene glycol monovinyl ether may be due mainly to its metabolites such as ethylene glycol and its breakdown products (see Chapter 50, sections 2.4.16, 17, and 18).

5 ETHYLENE GLYCOL MONOETHYL ETHER; Ethanol, 2-Ethoxy-; Cellosolve® Solvent; Dowanol® EE Glycol Ether; CAS No. 110-80-5

$$C_2H_5OCH_2CH_2OH$$

5.1 Physical and Chemical Properties

See Table 51.1.

5.2 Physiologic Response

5.2.1 Summary

Ethylene glycol monoethyl ether is low in oral toxicity, not significantly irritating to the skin, slightly irritating to the eyes and mucous membranes, readily absorbed through the skin but low in toxicity by this route, and somewhat toxic when inhaled. Its vapors are irritating and disagreeable in acutely toxic concentrations but they are not objectionable at levels considered safe for prolonged and repeated daily exposure. The material exerts its action primarily upon the blood and it is believed that changes in the blood picture reflect the first evidence of excessive exposure.

5.2.2 Single-Dose Oral

Carpenter and co-workers (4) found the single-dose oral toxicity of ethylene glycol monoethyl ether to be 5.5 g/kg for rats, 3.1 g/kg for rabbits, and 1.4 g/kg for guinea pigs. The values given for rats, 5.5 g/kg, is somewhat higher than the 3.0 g/kg previously reported by Smyth, et al. (35). Laug and co-workers (36) report the following oral LD_{50} values: 3.46 g/kg for rats, 4.31 g/kg for mice, and 2.79 g/kg for guinea pigs. Others have reported the following LD_{50} values: 2.8 g/kg for rats (9); 4.8 g/kg for mice, 4.45 g/kg for rats, 2.13 g/kg for guinea pigs, and 1.48 g/kg for rabbits (37); and 3.5 g/kg for mice when administered as an oil solution, slightly higher as a water solution (37). Laug et al. (36) observed that the animals displayed no immediate signs of distress. However, they did observe hemorrhage of the stomach and intestine, mild liver injury, severe kidney injury, and hematuria in animals that were seriously affected or died. They concluded as a result of their study that ethylene glycol monoethyl ether should not be used in applications where consumption by human beings could be expected. Others found that reversible liver and kidney injury were seen at levels as low as 0.25 g/kg (9). Stenger et al. (37) report that the animals tested showed signs of dyspnea, somnolence, and ataxia. In addition, both the rats and rabbits exhibited signs of cramps.

5.2.3 Repeated-Dose Oral

Gross (39) fed rabbits repeated daily doses of ethylene glycol monoethyl ether and found that seven doses of 0.1 ml/kg (0.093 g/kg) caused temporary albuminuria, whereas seven doses of 0.25 ml/kg (0.23 g/kg) caused both albuminuria and hematuria after the seventh feeding. When the dosage was increased to 1 ml/kg (0.93 g/kg), albuminuria and hematuria were observed after the seventh day, followed by death on the eight day due to kidney injury. Two doses of 2 ml/kg (1.86 g/kg) caused exhaustion, refusal to eat, albuminuria, cylinders in the urine, and death believed due to kidney injury.

Smyth et al. (35) report maintaining rats for 90 days on drinking water containing ethylene glycol monoethyl ether. They found that the maximum dose having no effect was 0.21 g/kg/day, that a dose of 0.74 g/kg reduced growth and appetite, altered liver and kidney weights, and produced microscopic lesions in these organs, and that mortality was increased when the dosage was 1.89 g/kg/day.

In 90-day dietary feeding studies in rats, a level of 0.25 percent was found to cause no significant ill-effects, whereas a level of 1.25 percent in the diet was found to cause body weight depression (8).

Stenger et al. (37) fed rats ethylene glycol monoethyl ether for 13 weeks at doses of 0.093 to 0.73 g/kg/day. The no-effect level established was 0.093 g/kg/day, whereas the level of 0.185 g/kg/day caused beginning adverse effects, consisting of growth depression, reduced food intake, reduced hemoglobin content and hematocrit values, and histologic changes in the liver, kidney, and testes at the 0.73 g/kg/day level. Stenger et al. also fed dogs for 13 weeks at levels of 0.046 to 0.185 g/kg/day and found reduced hemoglobin levels and hematocrit values after 5 weeks. These dogs also developed pathologic changes in the kidneys and testes similar to those seen in the rats.

Nagano et al. (13) fed mice doses ranging from 62.5 to 4000 mg/kg, 5 days/ week for 5 weeks and observed testicular atrophy and leukopenia, the intensity of which was dose related. They concluded that the ethyl ether of ethylene glycol was less potent than the methyl ether.

Morris et al. (38) report upon studies in which ethylene glycol monoethyl ether was fed in the diet of rats for a 2-year period. At the dosage level of 1.45 percent, equivalent to about 0.9 g/kg/day, they observed only slight kidney damage, but did see appreciable tubular atrophy in the testes, accompanied by marked interstitial edema in about two-thirds of the animals. They did not find any oxalate concretions in the kidneys or bladders, as reported for ethylene glycol.

5.2.4 Parenteral Administration

Gross (39) gave repeated subcutaneous injections of ethylene glycol monoethyl ether to rabbits. The results show that seven doses of 0.25 ml/kg (0.23 g/kg) produced no apparent effect. However, higher doses produced essentially the same response as observed from oral administration, although the intensity of response seemed to be somewhat greater.

Carpenter et al. (4) found ethylene glycol monoethyl ether in concentrations of more than 18 percent in 0.75 percent sodium chloride to be hemolytic to rat erythrocytes. When such an 18 percent solution was given intravenously to rats, the LD_{50} was found to be 3.3 g/kg, and when the undiluted material was given, the LD_{50} was found to be 2.4 g/kg. When given intraperitoneally to rats, the LD_{50} of the undiluted material was found to be 2.14 g/kg.

Von Oettingen and Jirouch (40) concluded that ethylene glycol monoethyl ether given subcutaneously to mice was less toxic than ethylene glycol, which in turn was much less toxic than ethylene glycol monobutyl ether. They found the minimum lethal dose to be about 5.0 ml/kg (4.66 g/kg) for the monoethyl ether, 2.5 ml/kg (2.78 g/kg) for ethylene glycol, and 0.5 ml/kg (0.45 g/kg) for ethylene glycol monobutyl ether. They observed also that the ethyl ether had much less effect upon the central nervous system than did the other ethylene glycol ethers. They observed, however, that large doses were capable of causing severe kidney injury.

Stenger et al. (37) administered ethylene glycol monoethyl ether subcutaneously for 4 weeks and found that doses of up to 0.38 g/kg/day caused no deaths in rats. However, doses of 0.185 and 0.38 g/kg caused dyspnea, somnolence, mild ataxia, some growth depression in the females, and some reduction of hemoglobin levels and hematocrit values. At the 0.38 g/kg level interstitial testicular edema, dissociation of liver parenchyma and tubular lesions of the kidney were observed. Dogs given this product intravenously at levels of 0.093

and 0.46 g/kg/day for 22 days developed local irritation at the injection site and ataxia, but no change in hemoglobin levels or hematocrit values. They also established the folowing LD_{50} values; for subcutaneous injection, rat, 3.16 and rabbit, 1.85 g/kg; for intravenous injection, mouse, 4.8, rat, 4.45, guinea pig, 2.13, and rabbit, 1.48 g/kg.

5.2.5 Eye Irritation

Carpenter and Smyth (11) classified ethylene glycol monoethyl ether along with ethanol. Other studies (43) confirm this and indicate that when the material is introduced directly into the eye, it produces immediate pain, some conjunctival irritation, and a slight transitory irritation of the cornea, which clears within 24 hr. Weil and Scala (41), in summarizing the work of a number of laboratories, state that the eye tests show that only slight transient eye irritation was found. These observations would indicate that the material does not present a serious hazard to the eyes, although it may be painful and uncomfortable.

5.2.6 Skin Contact

Ethylene glycol monoethyl ether, even when in prolonged and repeated contact with the skin of rabbits, failed to cause more than a very mild irritation (9). These findings have been confirmed by others (6, 41, 42).

5.2.7 Skin Absorption

The material is absorbed through the skin of rabbits in acutely toxic amounts. Quantitation by the "sleeve" technique essentially as described by Draize et al. (10) indicated that the LD_{50} was 3.6 ml/kg (3.35 g/kg) (4). When the material was applied by inunction, the LD_{50} was 16.3 ml/kg (15.1 g/kg).

5.2.8 Inhalation Studies

The acute response of guinea pigs to ethylene glycol monoethyl ether in air was studied by Waite et al. (43). They found that guinea pigs could survive exposure intensities of 6000 ppm for 1 hr, 3000 ppm for 4 hr, and 500 ppm for 24 hr without apparent harm. More intense exposures caused injury of the lungs, hemorrhage in the stomach and intestines, and congestion of the kidneys. They concluded that air essentially saturated with the vapor of ethylene glycol monoethyl ether at room temperature was sufficiently disagreeable and irritating to the eyes to provide adequate warning to prevent acute poisoning. Gross (39) reports that the majority of animals repeatedly exposed to 1400 ppm of ethylene glycol monoethyl ether 8 hr/day died after 4 to 12 exposures. Cats were found to be most susceptible, dying 2 days after 4 or 5 days of exposure. One of two mice died after nine exposures but the other survived 12 exposures without evident effects. Two rabbits survived 12 exposures, one dying 7 days later, while two guinea pigs survived 12 exposures without evidence of injury.

Werner and co-workers (14) exposed mice for 7 hr to various concentrations and found the LC_{50} to be 1820 ppm. They attributed death to lung and kidney injury.

Werner et al. (15) exposed groups of rats 7 hr/day, 5 days a week for 5 weeks to a concentration averaging 370 ppm of ethylene glycol monoethyl ether vapor, and noted only a slight effect upon the cellular elements of the blood. Werner et al. (44) exposed two dogs to a vapor concentration of 840 ppm of ethylene glycol monoethyl ether 7 hr/day, 5 days a week for 12 weeks and observed a slight decrease in hemoglobin and red cells. The blood picture was characterized by a greater than normal number of the immature white cells. There was no evidence of kidney injury or of bone marrow injury. There was, however, an increase in the number of calcium oxalate crystals in the urine. The material appeared to be distinctly less toxic than ethylene glycol monomethyl ether or ethylene glycol monobutyl ether.

Goldberg et al. (45) found that rats exposed to the vapors of this material for 4 hr/day, 5 days/week for 10 exposures experienced an early transient weight gain at an exposure level of 2000 ppm. However, after 3 days of exposure, the rats returned to a normal growth pattern. Those exposed to 4000 ppm showed no inhibition of conditioned avoidance–escape behavior but did exhibit consistent reduced weight throughout the exposure period.

Kasparov et al. (46) report an LC_{50} for mice of 77 mg/l (20,900 ppm), the length of exposure not given. In chronic tests using mice, repeated exposures to 3000 mg/m³ (8100 ppm) caused a depression of growth, a lowering of cholinesterase levels, and an increase in protein excretion in the urine.

5.2.9   Teratogenicity

Stenger et al. (37) in a study of possible teratogenic effects of ethylene glycol monoethyl ether, injected it subcutaneously to mice from day 1 to 18 of gestation, at a dose of 0.093 g/kg/day, to rats at the same level from day 1 through day 21, and to rabbits at a level of 0.023 g/kg on days 7 through 16. Under these conditions none of the animals developed any serious malformations indicative of teratogenic effects. However, in the rats only, there was the beginning of skeletal aberrations at this level, which is commonly considered to be an indication of fetal toxicity rather than a teratogenic effect. In the rat tests where the material was fed in graduated doses of 0.026 to 0.38 g/kg/day from day 1 to day 21, there was a dose-dependent increase in skeletal aberrations starting at the 0.093 g/kg/day level. Based on these studies, it would seem that ethylene glycol monoethyl ether is not likely to cause significant teratogenic effects.

5.2.10   Carcinogenicity

In the 2-year dietary feeding studies on rats at a level equivalent to about 0.9 g/kg/day as reported by Morris et al. (38), there were no indications of carcinogenicity.

5.2.11   Metabolism

Very little work has been conducted to study the metabolism of ethylene glycol monoethyl ether. Studies (47) to evaluate the rate of its elimination were conducted in rats using a priming dose of 140 mg/kg intravenously followed by the infusion initially of 8, then of 16 mg/kg/min. Under these conditions it was found that this product was excreted via the lungs at a rate of slightly more than 8 mg/kg/min unchanged. This implies that accumulation is not to be expected.

5.2.12   Human Experience

Human experience in the use of ethylene glycol monoethyl ether has been quite uneventful. According to Browning (17), examinations of workers employed in the manufacture of lacquers and paint by the Factory Department revealed very little evidence of any injury to health from the use of this material. Browning also reports that operators applying ethylene glycol monoethyl ether with a spray gun can work all day without discomfort or ill effects.

Fucik (48) cited a case of a 44-year-old female who drank about 40 ml of this product by mistake. She became vertiginous and became unconscious shortly after exposure. Upon examination she was found to be cyanotic and had tachypneumonary edema, repeated tonic–clonic spasms, and acetone on her breath. She was given oxygen and other supportive treatment which resulted in essentially complete recovery after some 44 days.

5.3   Hygienic Standards of Permissible Exposure

The American Conference of Governmental Industrial Hygienists (30) in 1980 recommended a threshold limit of 100 ppm for ethylene glycol monoethyl ether. However, they are proposing lowering the TLV to 50 ppm with a C designation. This appears to be a reasonable figure but the margin of safety it provides is believed to be small. The Occupational Safety and Health Administration, however, recommends 200 ppm as a guide for control (31).

Goldstein et al (49) have suggested a maximum allowable concentration of 500 mg/m^3 (135 ppm).

5.4 Odor and Warning Properties

Ethylene glycol monoethyl ether, according to May (32), has an odor threshold of about 25 ppm with a strong odor at about 50 ppm. Human volunteers with some work experience in industrial environments reported that levels of 125 ppm were noticeable and that the odor level that would be intolerable was greater than 255 ppm (9). Thus the warning properties should not be relied upon to prevent prolonged daily exposures to concentrations of vapors that could cause adverse effects.

6 ETHYLENE GLYCOL MONO-*n*-PROPYL ETHER; Ethanol, 2-Propoxy-; *n*-Propyl Oxitol® Glycol; CAS No. 2807-30-9

$$C_3H_7OCH_2CH_2OH$$

6.1 Physical and Chemical Properties

See Table 51.1.

6.2 Physiologic Response

6.2.1 Summary

Ethylene glycol mono-*n*-propyl ether is a volatile liquid with a mild ethereal odor and a bitter taste. It is moderate to low in single-dose oral toxicity, the LD_{50} being between 0.5 and 4.4 g/kg. The material is appreciably irritating to the eyes of rabbits, causing injury to the conjunctival membranes and the cornea, and also some iritis; healing appeared to be complete within a week. The material does not appear to be appreciably irritating to the skin but it can be absorbed through the skin in lethal quantities. It is quite toxic when inhaled in high concentrations. The most striking observation in rats treated with toxic doses by any route was the presence of large amounts of blood in the urine.

6.2.2 Single-Dose Oral

According to Smyth (18), the LD_{50} for rats of ethylene glycol mono-*n*-propyl ether is 4.45 g/kg. However, other work (9) indicates that the oral LD_{50} for rats is between 0.5 and 1.0 g/kg. The reason for this difference is unknown. Gross (7) states that a single dose of 1 ml/kg caused the death of a rabbit. In both rabbits and rats, rather large amounts of blood appear in the urine in a matter of hours after feeding 1 g/kg, and death occurs within a matter of a few days after feeding. Saparmamedov (6) reports that the LD_{50} for mice is 2.4 g/kg.

6.2.3 Parenteral Administration

Gross (7) reports that a single subcutaneous injection of 1 ml/kg in the rabbit caused death with serious kidney injury and blood pigments in the urine. A guinea pig given an intraperitoneal injection of 1 ml/kg died and showed considerable kidney injury, but one given 0.5 ml/kg survived. Repeated subcutaneous injections of 0.5 ml/kg were survived by the guinea pig and the rabbit, but when the dosage was 1 ml/kg, the guinea pig died several days after the seventh injection.

6.2.4 Eye Contact

When ethylene glycol mono-*n*-propyl ether was introduced into the eyes of rabbits, Gross (7) found it to cause reddening and swelling of the conjunctiva and the lids and corneal damage. In other studies, these same effects were noted but, in addition, some iritis also was observed (9, 51).

6.2.5 Skin Contact

Gross (7) and Smyth et al. (51) report that the material is not irritating to the skin. Other work (9) confirms this for ordinary exposure but indicates also that if the material is confined to the skin for prolonged periods of time, it may produce appreciable irritation and possibly even a burn. However, the material may be absorbed through the skin in lethal amounts (9). Smyth et al. (51) states that the LD_{50} by absorption in rabbits is 0.87 g/kg. Saparmamedov (6), using mice, reports that no skin irritation nor signs of skin absorption of toxic amounts occurred.

6.2.6 Inhalation Studies

Gross (7) reports that rabbits tolerated exposures lasting 1 or 3 hr to a concentration of 2400 ppm with only irritation of the mucous membranes and no aftereffects. Others (9) have observed that when rats were given single 7-hr exposures to an atmosphere essentially saturated with ethylene glycol mono-*n*-propyl ether that, although all survived, all showed bloody urines within 2 hr after the exposures terminated, together with lung, liver, and kidney injuries. Hematologic studies revealed a reduction in packed cell volume but no evidence of hemolysis. Animals exposed for only 4 hr were normal 7 days later, whereas those exposed for 7 hr exhibited severe kidney injury 2 weeks later. Werner et al. (14) report that the LC_{50} value for rats for a 7-hr exposure is 1530 ppm. They report that the toxic action was that of dyspnea and hemoglobinuria. Carpenter et al. (52) found that some of the rats exposed to 2000 ppm for 4 hr died.

In repeated inhalation experiments, Gross (7) found that mice and guinea pigs were unaffected by twelve 8-hr exposures to 600 ppm, whereas cats and rabbits died.

6.2.7   Human Experience

There is no report of adverse effects in humans, but this may be accounted for by the fact that the material has not been as widely used as some of the other glycol ethers. Certainly, the material produces effects that should cause one to be very cautious and to avoid exposure wherever possible.

6.3   Hygienic Standards of Permissible Exposure

There are no data available that would allow the establishment of a hygienic standard for repeated vapor exposure. In view of the rather serious effects produced by this material it would seem only prudent to avoid inhaling the vapors wherever possible. In addition, it would be wise to take precautions to prevent contact with the eyes and prolonged or repeated exposure to the skin.

7   ETHYLENE GLYCOL MONOISOPROPYL ETHER; Ethanol, 2-(1-Methylethoxy)-; Isopropyl Oxitol® Glycol; Cas No. 109–59-1

$$(CH_3)_2CHOCH_2CH_2OH$$

7.1   Physical and Chemical Properties

See Table 51-1.

7.2   Physiologic Response

7.2.1   Summary

Ethylene glycol monoisopropyl ether is moderate to low in single-dose oral toxicity, the LD_{50} being between 0.5 and 5.6 g/kg. The material is appreciably irritating to the eyes of rabbits, causing injury to the conjunctival membranes, the cornea, and also some iritis. Healing appears to be complete within a week. The material does not appear to be appreciably irritating to the skin but it is readily absorbed through the skin in lethal quantities. It causes severe kidney injury when inhaled in appreciable concentrations. The most striking observation in rats treated with the material by any route was the passing of large amounts of blood in the urine. It appears to be some-what more toxic than the *n*-propyl isomer and other ethers of ethylene glycol.

7.2.2 Single-Dose Oral

The single-dose oral LD_{50} of ethylene glycol monoisopropyl ether for rats is between 0.5 and 1.0 g/kg. It causes severe kidney and liver injury and the passage of large amounts of blood in the urine (9). Smyth et al. (51) report an LD_{50} for rats of 5.66 g/kg. The LD_{50} for mice fed ethylene glycol isopropyl ether in oil is 2.3 g/kg, and in water, 2.18 g/kg (6).

7.2.3 Eye Contact

When the material was instilled into the eyes of rabbits it caused marked conjunctival irritation, marked corneal injury, and some iritis. Healing was essentially complete in about 7 days (9, 51).

7.2.4 Skin Contact

The material is not appreciably irritating to the skin under ordinary conditions of exposure, but if it is confined for prolonged periods, it may cause appreciable irritation, even a burn. In addition, the material may be absorbed through the intact skin in lethal amounts, the LD_{50} being in the range of 1.0 (9) to 1.6 g/kg (51). Saparmamedov (6) reports that mice, when treated on the skin, experienced no significant irritation nor signs of absorption of toxic amounts through the skin.

7.2.5 Inhalation

Smyth *et al.* (51) report that rats exposed for 2 hr to 4000 ppm experienced no deaths whereas a 4-hr exposure caused four of six rats to die. Table 51.3 summarizes further single-dose inhalation data on rats (9).

Table 51.3. Summary of Results of Single Exposures of Rats to Vapors of Ethylene Glycol Monoisopropyl Ether[9]

Concentration	Duration of Exposures (hr)	No. Dying/ No. Exposed	Comments and Observations
Essentially saturated at 100°C and cooled to room temperature	7	4/4	After 3 hr, blood in the urine; 1 animal autopsied after exposure had black and severely enlarged kidneys
As above	4	2/3	Passed bloody urine immediately after exposure; excess urine; 3 days after

Table 51.3. *(Continued)*

Concentration	Duration of Exposures (hr)	No. Dying/ No. Exposed	Comments and Observations
			exposure the kidneys were severely necrotic and dark in color
As above	1	0/3	Bloody urine, slight weight loss, severely injured kidneys
As above	0.5	0/3	Bloody urine, slight weight loss, questionable kidney injury
160 ppm	6.7, 4	0/5	Bloody urine during exposure; slight weight loss; one rat was autopsied 2 days after exposure and the liver and kidneys were pale in color, urine in bladder was clear; the 4 survivors autopsied 15 days after exposure exhibited evidence of slight to moderate kidney injury
80 ppm	7.0	0/5	Slight weight loss; much blood passed in urine; one rat was sacrificed immediately after exposure and no gross pathologic changes noted; another sacrificed the day after exposure had severely affected kidneys, a pale-colored liver, and bloody urine in the bladder; a rat sacrificed 9 days after exposure showed no evidence of gross changes of the kidney
80 ppm	4.0	0/5	Questionable evidence of blood in urine, slight weight loss; 1 rat sacrificed immediately after exposure had severely affected kidneys; 1 sacrificed the day after exposure had slight pathology of the kidneys, but no blood in urine; another sacrificed 9 days after exposure appeared grossly to have slightly injured kidneys

Werner et al. (14), using mice, report that the 7-hr LC_{50} for mice is 1930 ppm.

Only limited repeated exposure inhalation studies have been reported (53). Rats exposed 6 hr/day for 15 days to 1000 ppm experienced initial nasal irritation, lethargy, hemoglobinuria, porphyrinuria, low hemoglobin (fourth day), low reticulosis without histopathologic injury, and congestion of the lungs. Both the blood and urine picture returned to normal. Those exposed to 300 ppm were less affected, whereas those exposed to 100 ppm under the same conditions were unaffected.

It appears that ethylene glycol isopropyl ether is a highly toxic substance when inhaled.

7.2.6 Metabolism

Hutson and Pickering (54) studied the metabolism of ethylene glycol isopropyl ether in both the dog and the rat using ^{14}C tagged material. The metabolism was similar in both animals. It was found to be rapidly metabolized in the rats as 88 percent was excreted from the body in 24 hr; 73 percent via the urine, and 14 percent via the lungs as carbon dioxide. The major urinary metabolites were isopropoxyacetic acid, 30 percent; N-isopropoxyacetylglycine, 46 percent, and ethylene glycol, 13 percent. Based on these results, the authors suggest the metabolic scheme shown in Figure 51.1 for the rat.

7.2.7 Human Experience

No reports of adverse effects from the handling or use of ethylene glycol monoisopropyl ether have been made. However, this may be because of its limited use to date.

7.3 Hygienic Standards of Permissible Exposure

There are no data available that would allow the establishment of a hygienic standard for repeated exposure to vapors. In view of the high toxicity of the material for animals, it would seem only prudent to take particular care to prevent all possible exposure of human beings to this material until such time as its toxicity becomes better evaluated.

8 ETHYLENE GLYCOL MONOBUTYL ETHER; Ethanol, 2-Butoxy-; Butyl Cellosolve® Solvent; Dowanol® EB Glycol Ether; Butyl Oxitol® Glycol; CAS No. 111-76-2

$$C_4H_9OCH_2CH_2OH$$

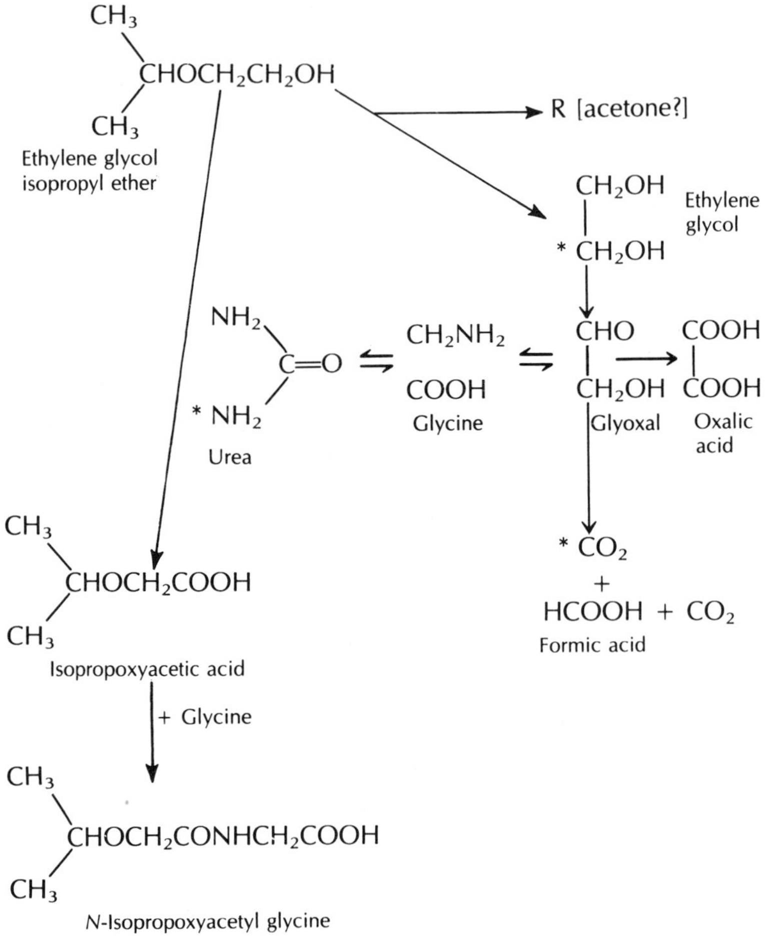

Figure 51.1 Metabolic scheme of ethylene glycol monoisopropyl ether. (The asterisk indicates those compounds excreted by rats.)

8.1 Physical and Chemical Properties

See Table 51.1.

8.2 Physiologic Response

8.2.1 Summary

Ethylene glycol monobutyl ether, according to von Oettingen and Jirouch (40), at first tastes sour but later causes a burning sensation followed by numbness of the tongue, indicating paralysis of the sensory nerve endings. It is moderate to low in single-dose oral toxicity, appreciably irritating and injurious to the eyes, not significantly irritating to the skin, absorbed through the skin in toxic amounts, and moderately toxic when inhaled. Ethylene glycol monobutyl ether is metabolized, at least in part, to butoxyacetic acid and this substance is excreted in the urine of most animal species and of human beings. Animal tests also indicate that ethylene glycol butyl ether is excreted via the lungs. The material itself and its metabolite are hemolytic agents. The presence of butoxyacetic acid in the urine would suggest that exposure to ethylene glycol monobutyl ether had occurred.

Exposure of human beings to high concentrations of ethylene glycol mono-butyl ether vapors, probably in the range of 300 to 600 ppm, for several hours would be expected to cause respiratory and eye irritation, narcosis, and damage to the kidney and liver. Deaths that occur promptly from inhalation in animals are generally attributed to narcosis, but if death is delayed several days it is likely to be caused by pneumonitis and/or kidney injury. The first sign of organic abnormality in man resulting from excessive exposure by any route likely would be an abnormal blood picture characterized by erythropenia, reticulocytosis, granulocytosis, and leukocytosis. Somewhat more intense ex-posure would be likely to cause fragility of erythrocytes and hematuria.

8.2.2 Single-Dose Oral

Gross (7) reports that rabbits tolerated without apparent ill effect single oral doses of 0.1 and 0.5 ml/kg, and that 1.0 and 2.0 ml/kg caused death within 30 or 22 hr, respectively. Carpenter and co-workers (4) fed single doses of ethylene glycol monobutyl ether to young adults of various species and have calculated the following LD_{50} values: rat, 2.5; mouse, 1.2; rabbit, 0.32; and guinea pig, 1.2 g/kg. They noted that, at least in the rat, large old animals are much more susceptible than weanling animals, the LD_{50} values being about 0.55 and about 3.0 g/kg, respectively. Another group (9) found the LD_{50} for young adult female rats weighing 150 to 200 g to be 0.47 g/kg. Recent studies resulted in a calculated LD_{50} value of 0.62 g/kg. Other LD_{50} values reported are 1.48 g/kg for rats (55) and 1.7 g/kg for mice when fed as an oil solution, or 1.17 g/kg when fed as a water solution (9).

Acute or prompt deaths are likely to be due to the narcotic effects of the substance, whereas deaths that were delayed several days usually can be attributed to congested lungs and severely damaged kidneys. Necropsy of animals that died revealed congested lungs, mottled livers, severely congested kidneys, and hemoglobinuria.

8.2.3 Repeated-Dose Oral

Carpenter and co-workers (4) report maintaining rats on diets containing 2.0, 0.5, 0.125, and 0.03 percent of ethylene glycol monobutyl ether. At the top level, growth depression and increased kidney and liver weights were observed but no hematuria and no histopathologic lesions of pertinence were noted. At the 0.5 percent level, growth depression and increased liver weight were observed. No effects of significance were observed at the two lowest dosage levels.

Nagano et al. (13) fed mice doses ranging from 62.5 to 4000 mg/kg, 5 days/week for 5 weeks and observed no significant effects upon the testes or blood.

8.2.4   Toxicity by Injection

Gross (7) gave single subcutaneous injections to rabbits and cats. He observed that rabbits tolerated up to 0.1 ml/kg without reaction, that 0.2 ml/kg caused temporary slight kidney inflammation, and that doses of 0.4 ml/kg were fatal. He found that cats tolerated 1 ml/kg without particular symptoms but that 2 ml/kg caused death 3 days after injection with signs of kidney injury.

Carpenter et al. (4) found that undiluted ethylene glycol monobutyl ether caused hemolysis of rat erythrocytes and that all concentrations above 3 percent in 0.75 percent sodium chloride did likewise. When such a solution was injected intravenously in rats, mice, and rabbits, the LD_{50} values were found to be 0.38, 1.1, and 0.50 g/kg. When the undiluted material was injected intravenously, the LD_{50} values found were 0.34 ml/kg for rats and 0.28 ml/kg for rabbits. When the undiluted material was given intraperitoneally, the LD_{50} value for rats was found to be 0.55 ml/kg.

8.2.5   Eye Contact

Carpenter and Smyth (11) have classified ethylene glycol monobutyl ether along with such materials as acetonyl acetone, dioxane, and isopropanol. Other studies (9, 56) indicate that when the material is introduced directly into the eye it produces marked pain, appreciable conjunctival irritation, and slight transitory injury of the cornea, which heals within a few days.

8.2.6   Skin Contact

Prolonged and repeated contact of ethylene glycol monobutyl ether with the skin of rabbits failed to cause more than a very mild simple irritation. This was confirmed by Saparmamedov (6) in his studies using mice.

8.2.7   Skin Absorption

The material may be absorbed through the skin of rabbits in toxic amounts. According to Carpenter et al. (4) the LD_{50} for rabbits was found to be 0.45 ml/kg (0.40 g/kg) when the material was confined to the skin for 24 hr and 2 ml/

kg (1.8 g/kg) when applied to the skin by gentle massage. Evidence of rapid absorption through the skin also was obtained when these investigators showed that erythrocyte fragility was increased 1 hr after a single 3-min contact with 0.56 ml/kg on 4.5 percent of the total skin surface area. Saparmamedov (6), using mice, reports that this material caused no significant toxic effects by absorption.

8.2.8 Inhalation Studies

Werner and co-workers (14) exposed mice for 7 hr to various concentrations and found the LC_{50} to be 700 ppm. Death was attributed to lung and kidney injury. Carpenter et al. (4) exposed rats of different ages to various concentrations of vapor. They found that 1-year-old rats were more susceptible than young, actively growing rats. At a concentration of 375 ppm, the old adults died after 7 hr whereas the 6-week-old rats survived 8 hr at 500 ppm. Gross (7) reports that the inhalation of 520 ppm 8 hr/day for 8 to 12 days caused the death of cats, rabbits, and guinea pigs but had no adverse effects upon mice. Kidney inflammation was present in all animals that died.

Werner et al. (15) exposed one group of rats 7 hr/day, 5 days a week for 5 weeks to concentrations averaging 320 ppm, and another group to 135 ppm of ethylene glycol monobutyl ether. They noted, after a week, an increase in the percentage of reticulocytes and young granulocytes in the circulating blood, and a decrease in the hemoglobin concentration and erythrocyte count. The blood picture of animals so affected returned to normal shortly after exposure was terminated.

Werner et al. (16) exposed two dogs to a vapor concentration of 415 ppm 7 hr/day, 5 days a week for 12 weeks. Although the animals failed to grow well and displayed the typical reversible blood changes, they did not show evidence of organic injury by histologic procedures. However, it was noted that there was an increase in the number of calcium oxalate crystals in the urine and that there was a moderate retention of urea in the blood throughout the exposure period.

Extensive repeated inhalation studies have been conducted by Carpenter et al. (4). The quantitative results are summarized in Table 51.4. The qualitative observations are as follows. At high concentrations, the rats exhibited hemorrhage of the lungs, congestion of the viscera, liver injury, hemoglobinuria, and a marked erythrocyte fragility. Females were more sensitive than males. At the lower concentrations, increased fragility of the erythrocytes appears to be the most sensitive criterion of effect. This response was transitory, however, since it was apparent during and shortly after exposure, but not the following day.

The guinea pigs were quite resistant to the effects of ethylene glycol monobutyl ether. At high concentrations, congestion and cloudy swelling of the tubules of the kidneys were observed, but no increase in the fragility of the

Table 51.4. Summary of Results of Repeated Inhalation Studies on Ethylene Glycol Monobutyl Ether[4]

Species	Exposure Intensity (No. of 7-Hr Exposures)	Solvent Concn. Causing Some Injury[a] (ppm)	Highest Vapor Concentration (ppm) Not Causing			
			Injury[a]	Increased Erythrocyte Fragility	Hemo-globinuria	Other Hematologic Changes
Rat	30	107	54	<54	107	
Guinea pig	30	203	107	494	494	
Mouse	30	396	200	<112	112	
	60	200	111	<111	111	
	90	401	201	<112	112	
	90 plus 42 days' rest	401	401	401	401	
Dog	90	385	200	100	?	<100
Monkey	90	?	210	<100	?	<100
Human	Two 4-hr periods separated by a 30-min lunch period			>195		
Rat	Two 4-hr periods separated by a 30-min lunch period			<195		
Rat	4 hr			<113		
Human	4 hr			>113		

[a] As judged by mortality, growth, kidney and/or liver weight changes, and gross and/or micropathology of organs.

erythrocytes occurred at any concentration studied. Mice were essentially as resistant as the guinea pigs, with the exception that their erythrocytes were as fragile as those of the rat.

Dogs exposed to high concentrations suffered congestion of the kidneys and lungs, weight loss, increased fragility of the erythrocytes, nasal and eye infections, apathy, anorexia, nausea, and other changes in the circulating blood. The leukocytes were markedly increased whereas the erythrocyte and hemoglobin content was markedly decreased. There was also a marked increase in plasma fibrinogen. At the lowest level of exposure, 100 ppm, the dogs exhibited a transitory increase in the leukocyte count, a marked decrease in erythrocyte count, and a low hematocrit. Twenty-four-hour urine samples contained 94 to 100 mg of butoxyacetic acid.

Monkeys exposed to 200 ppm suffered marked reduction in the number of circulating red blood cells and in hemoglobin concentration. The fragility of the erythrocytes was markedly increased and plasma fibrinogen levels were elevated to about four times normal values. There was, however, no appreciable organic injury observed at necropsy. At the lowest level, 100 ppm, monkeys exhibited typical hematologic effects but all had returned to normal by the end of the exposure period. The female monkey excreted 30 mg of butoxyacetic acid over a 48-hr period after receiving the forty-second exposure.

When human subjects inhaled 200 ppm of ethylene glycol monobutyl ether for 8 hr, there were no objective effects although the presence of butoxyacetic acid in their urines proved that absorption had occurred. The consensus was, however, that this concentration was too high for comfort, with eye, nose, and throat irritation evident. When the concentration was lowered to 100 ppm, similar subjective complaints arose from sensitive persons. It appears that this chemical is one of the few materials to which the human is more resistant than the usual experimental animals. This appears to be due, in part at least, to the fact that humans are more resistant than are most laboratory animals to the hemolytic effects caused by the material itself or its metabolite.

Goldberg et al. (45) exposed avoidance–escape behavior-trained rats 4 hr/day, 5 days a week for 2 weeks to concentrations of ethylene glycol butyl ether vapor of 50, 100, 200, and 400 ppm with no effect on growth or behavior. However, transient hematuria was observed at the 200 and 400 ppm levels.

8.2.9   Metabolism and Mode of Action

Carpenter and co-workers (4) have demonstrated that butoxyacetic acid is a metabolite of ethylene glycol monobutyl ether in the rat, guinea pig, rabbit, dog, rhesus monkey, and man. They have developed a method for estimating its concentration in the urine that can be used to detect the first evidence of absorption of the material by any route. However, it does not appear that the butoxyacetic acid content of the urine can be used as a measure of intensity of exposure. They found 55 mg of butoxyacetic acid in 16 hr urine samples of dogs exposed to 385 ppm. They found 100 and 42 mg in 24-hr urine samples from two dogs exposed to 200 ppm of vapor and 100 and 94 mg in similar urine samples from two dogs exposed to 100 ppm. One of two monkeys exposed to 100 ppm excreted 30 mg of butoxyacetic acid in a 48-hr period and very little was found in a similar urine sample from the other animal. Human beings exposed 8 hr to 195 ppm excreted anywhere from 6 to 300 mg of butoxyacetic acid in a 24-hr period, whereas persons exposed to 98 ppm for 8 hr excreted from 75 to 250 mg in a 24 hr period. At best, it would seem that the presence of butoxyacetic acid in the urine can be considered evidence of exposure to ethylene glycol monobutyl ether.

Carpenter et al. (4) also showed that ethylene glycol monobutyl ether, and to a greater extent its metabolite, butoxyacetic acid, both increase the osmotic fragility of the erythrocyte. This action appears to be greatest in the rat, mouse, and rabbit and distinctly less in the guinea pig, dog, rhesus monkey, and human.

Other studies (7) have shown that ethylene glycol butyl ether in addition to being metabolized to butoxyacetic acid is excreted via the lungs. Rats infused intravenously were shown by gas chromatography to excrete ethylene glycol butyl ether at a rate of 3 to 4 mg/kg/min.

8.2.10   Human Experience

The human experience in the use of ethylene glycol monobutyl ether has been remarkably free of serious complications. Browning (17) states that there is only one possible injury in industrial use of the material, and this was reported by the English Factory Inspection in 1934. A man had two isolated attacks of hematuria at 5-month intervals. The cause, however, was somewhat complicated by the fact that diethylene glycol monobutyl ether was also associated with the process in which he was working. The man also suffered some nasal and eye irritation. Browning (17) also mentions two females who reported irritation of the mucous membranes and headache as a result of working on a process in which the material was used.

8.3.   Hygienic Standards of Permissible Exposure

The American Conference of Governmental Hygienists (30) in 1980 recommended a threshold limit value of 50 ppm for ethylene glycol monobutyl ether. However, they are proposing to lower the TLV to 25 ppm with a skin notation. This figure was also adopted by the Occupational Safety and Health Administration in 1976 (31). On the basis of the data on animals and human beings, it would appear that this is a reasonable figure. There is, however, a hazard other than vapor that must not be overlooked when handling this material— that of possible absorption of toxic quantities through the skin. Because of the low vapor pressure of this substance at room temperature, the hazard from skin absorption could well be greater, or contribute substantially to the overall hazard.

8.4 Odor and Warning Properties

In limited studies using humans with industrial hygiene experience, the lowest concentration of ethylene glycol butyl ether vapor considered to be unpleasant and therefore disagreeable was 40 ppm; the maximum tolerated level was shown to be 60 ppm or slightly higher (9).

9 ETHYLENE GLYCOL MONOISOBUTYL ETHER; Ethanol, 2-(2-Methylpropoxy)-; Ektasolve® EIB Solvent; Isobutyl Cellosolve® Solvent; CAS No. 4439-24-1

$$CH_3CH(CH_3)CH_2OCH_2CH_2OH$$

9.1 Physical and Chemical Properties

See Table 51.1.

9.2 Physiologic Response

9.2.1 Summary

Ethylene glycol isobutyl ether is moderate in single-dose oral toxicity, appreciably irritating and injurious to the eyes, and essentially nonirritating to mildly irritating to the skin, but can be absorbed through the skin in toxic amounts, being moderately to highly toxic by this route. It is moderately toxic by inhalation. Its acute effects seem to be centered in the liver and kidney. Hematuria is a consistent observation.

9.2.2 Single-Dose Oral

Ethylene glycol isobutyl ether has been reported to have an LD_{50} for rats of 0.4 g/kg (57); other data indicate about 1.0 g/kg (9). It has been found that when ingested relatively small amounts may cause liver and kidney injury. High levels caused the development of bloody urine (9). However, mice seem to be more resistant; the LD_{50} for them is greater than 1.6 g/kg (57).

9.2.3 Injection

When this ether was given to rats intraperitoneally, an LD_{50} value of 0.2 to 0.4 g/kg was found (57). For mice the LD_{50} was calculated to be 0.4 g/kg (57).

9.2.4 Eye Contact

Ethylene glycol isobutyl ether is capable of causing considerable eye injury. Instillation into the eyes of rabbits resulted in slight iritis and moderate irritation and corneal injury (9). The eye injury may be slow in healing and even may result in some permanent damage (57).

9.2.5 Skin Irritation

Ethylene glycol isobutyl ether was found to cause moderate skin irritation when held in contact with guinea pig skin for 24 hr (57). However, other data indicate that this material is not significantly irritating to the skin of rabbits even when in essentially continuous contact for 72 hr (9).

9.2.6 Skin Absorption

In the skin tests (9), although there was no significant irritation, the rabbits became prostrate, had difficulty in breathing, and passed blood in their urine. Further tests indicate that the dermal LD_{50} for rabbits is in the range of 0.2 to 0.4 g/kg. The animals that died did so in 2 to 4 days after the 24-hr exposure (9). Guinea pigs, on the other hand, were more resistant to toxicity by skin absorption; the LD_{50} value is reported as 10 ml/kg (8.87 g/kg).

9.2.7 Inhalation

Rats exposed for 0.5 hr to air essentially saturated with the vapor of ethylene glycol isobutyl ether, estimated to be in the range of 2000 ppm, developed liver and kidney injury accompanied by passage of blood in the urine. One-third of the rats died. All the rats exposed for 4.9 hr died, and two-thirds of those exposed for 1 hr died (9). A 6-hr exposure to 200 ppm caused signs of irritation, whereas a similar exposure to 1600 ppm caused death of one-third of the rats tested (57).

9.2.8 Human Experience

There have been no reported cases of adverse effects from exposure to ethylene glycol isobutyl ether. However, this material produces effects in animals; hence it would seem wise to take precautions to avoid eye and skin contact and inhalation.

9.3 Hygienic Standards of Permissible Exposure

There are insufficient data to permit the establishment of a hygienic standard for repeated vapor exposure. However, because of its similarity both in chemical nature and its toxic effects to ethylene glycol butyl ether, it would seem wise to exercise adequate precautions to avoid inhalation of its vapor. Until additional data become available, it certainly would seem prudent to consider using the hygienic standard for ethylene glycol butyl ether as a guide for control of this material.

10   ETHYLENE GLYCOL TERTIARY BUTYL ETHER; Ethanol, 2-*tert*-Butoxy-; CAS No. 7580-85-0

$$(CH_3)_3COCH_2CH_2OH$$

10.1   Physical and Chemical Properties

Molecular formula	$C_6H_{18}O_2$
Molecular weight	118.2
1 ppm $\approx$	4.83 mg/m^3 at 25°C, 760 mm Hg
1 mg/l $\approx$	207 ppm at 25°C, 760 mm Hg

10.2   Physiologic Response

Gage (53) reports that rats exposed to essentially saturated atmosphere (2400 ppm) for 5 hr became comatose, developed hemoglobinuria, and had reduced hemoglobin, followed by death. Those exposed four times to 250 ppm for 6 hr survived but were similarly affected; those exposed 15 times for 6 hr to 100 or 50 ppm developed increased osmotic fragility of the red cells but were otherwise normal; at 20 ppm no adverse effects were seen.

11   ETHYLENE GLYCOL MONO-2-METHYLPENTYL ETHER; Ethanol, 2-(1-Methylpentyl)oxy-; CAS No. 29290-45-7

$$CH_3CH_2CH_2CH(CH_3)CH_2OCH_2CH_2OH$$

11.1   Physical and Chemical Properties

Ethylene glycol mono-2-methylpentyl ether is a clear liquid infinitely soluble in alcohol and many other solvents. Its solubility in water is low, 0.63 percent at 20°C.

11.2   Physiologic Response

Ethylene glycol mono-2-methylpentyl ether is low in single-dose oral toxicity, the LD_{50} for rats being 3.73 ml/kg (58). It is markedly irritating and injurious to the eyes, being rated 6 on a scale of 10 in rabbit eye tests. However, it is only very mildly irritating to the skin; it is rated 2 on a scale of 10 in rabbit skin tests but moderate in toxicity by skin absorption, the LD_{50} for rabbits being 0.44 ml/ kg. No deaths were noted in rats exposed 4 hr to essentially saturated atmospheres. Thus this material appears to present a moderate potential health hazard in anticipated industrial handling. Precautions should be taken to prevent eye exposure and to avoid skin contact. It would seem prudent to avoid prolonged or repeated inhalation of vapors until more information becomes available.

12 ETHYLENE GLYCOL-*n*-MONOHEXYL ETHER; Ethanol, 2-(Hexyloxy)-; *n*-Hexyl Cellosolve® Solvent; CAS No. 112-25-4

$$CH_3(CH_2)_5OCH_2CH_2OH$$

12.1 Physical and Chemical Properties

See Table 51.1.

12.2 Physiologic Response

Toxicologic studies on ethylene glycol-*n*-monohexyl ether (59) indicate that it is low in single-dose oral toxicity; the LD_{50} for rats is 1.48 g/kg. It is severely injurious to the eye, being capable of causing severe rabbit eye irritation and marked to severe corneal damage, but it is essentially nonirritating to the skin of rabbits. However, it is moderate in toxicity by skin absorption; the LD_{50} for rabbits is 0.89 ml/kg (0.79 g/kg). Because of its very low volatility, rats exposed to essentially saturated vapors were not significantly affected. Therefore, ethylene glycol-*n*-monohexyl ether should be handled with great care to prevent eye contact and to avoid skin contact because of its toxicity by absorption. Until more is known it would seem wise also to avoid repeated exposures to its vapor.

13 ETHYLENE GLYCOL MONO-2,4-HEXADIENE ETHER; Ethanol, 2-(2,4-Hexadienyloxy)-; CAS No. 27310-21-0

$$CH_3CH=CHCH=CHCH_2OCH_2CH_2OH$$

13.1 Physical and Chemical Properties

Molecular formula	$C_8H_{14}O_2$
Molecular weight	142.2
1 ppm ≏	5.82 mg/m³ at 25°C, 760 mm Hg
1 mg/l ≏	172 ppm at 25°C, 760 mm Hg

13.2 Physiologic Response

Summary (60). 2-(2,4-Hexadienyloxy)ethanol is low in single-dose oral toxicity, the LD_{50} value for rats being 3.36 ml/kg. It is moderately irritating to the eyes of rabbits, being rated 5 on a scale of 10. However, it is only slightly irritating to the rabbit skin and may be absorbed through the skin in toxic amounts if exposure is prolonged and excessive; the LD_{50} for rabbits is 1.01 ml/kg. Because of its apparent low degree of volatility, none of the rats exposed for 8 hr to essentially saturated vapors generated at room temperature died. Thus the major hazards of 2-(2,4-hexadienyloxy)ethanol probably are from eye and dermal contact; suitable precautions should be taken to avoid such contact. It would seem wise also, until more is known, to avoid repeated prolonged exposure to its vapors, especially if handled hot, or if fogs or mists are generated.

14   ETHYLENE GLYCOL MONO-2,6,8-TRIMETHYL-4-NONYL ETHER; Ethanol, 2-[3,5-Dimethyl-1-(2-methylpropyl)hexyloxy]-; CAS No. 10137-98-1

$$HOCH_2CH_2OCH(CH_2CHCH_3CH_2CHCH_3CH_3)(CH_2CHCH_3CH_3)$$

14.1   Physical and Chemical Properties

Ethylene glycol mono-2,6,8-trimethyl-4-nonyl ether is a clear liquid infinitely soluble in alcohol and many other solvents. Its solubility in water is low, <0.01 percent at 20°C. Other properties are given in Table 51.1.

14.2   Physiologic Response

Summary (58).   Ethylene glycol mono-2,6,8-trimethyl-4-nonyl ether is low in single-dose oral toxicity; the LD_{50} for rats is 5.36 ml/kg. It is moderate to markedly irritating to the eye and may cause corneal injury, being rated 5 on a scale of 10 in rabbit eye tests. It is only very slightly irritating to the skin (rate of 2 on scale of 10) and low in toxicity by skin absorption, the LD_{50} value for rabbits being 3.15 ml/kg. None of the rats died when they were exposed for 8 hr to essentially saturated atmospheres. Thus the main hazard from the handling of this material is that of eye exposure. Precautions should be observed to avoid eye contact; reasonable caution and cleanliness should be adequate to avoid skin and inhalation hazards. No hygienic standard for exposure has been established and it appears that none is necessary.

15   ETHYLENE GLYCOL MONOPHENYL ETHER; Ethanol, 2-Phenoxy-; Dowanol® EPH Glycol Ether; Phenyl Cellosolve® Solvent; CAS No. 122-99-6

$$C_6H_5OCH_2CH_2OH$$

15.1   Physical and Chemical Properties

See Table 51.1.

15.2   Physiologic Response

Summary (9).   Extensive toxicologic data on ethylene glycol monophenyl ether are not available. Range-finding studies on the technical or commercially available product show it to be low in single-dose oral toxicity, the LD_{50} being between 1.0 and 2.0 g/kg. More recent studies indicate that the LD_{50} for rats is about 4.0 g/kg. It is not appreciably irritating to the intact skin and is not readily absorbed through the skin in acutely toxic amounts. In undiluted form, it is severely damaging to the eyes of rabbits. When diluted to 5 percent, it caused only mild irritation of conjunctival membranes. Rats tolerated, without apparent adverse effects, one 7-hr exposure to vapors saturated at 100°C and cooled to room temperature. Reasonable handling precautions, plus particular care to prevent contact with the eyes, should prevent any serious toxic effects.

16 ETHYLENE GLYCOL MONOMETHYLPHENYL ETHER; 2-Methylphenyl Cellosolve®
Solvent; 1-(2-Hydroxyethoxy)methylphenol; (CAS no.—none found)

$$CH_3C_6H_4OCH_2CH_2OH$$

16.1 Physiologic Response

(61) Ethylene glycol monomethylphenyl ether is low in single-dose oral toxicity, the LD_{50} for rats being 3.73 g/kg. It appears to be severely irritating and injurious to the eyes but should cause only minor skin irritation. It is moderate in toxicity by skin absorption; the LD_{50} for rabbits is 0.44 ml/kg by this route. One of six rats exposed for 8 hr to essentially saturated vapors generated at room temperature died. Thus it appears that ethylene glycol monomethylphenyl ether poses a potential health hazard from eye contact and possibly from inhalation. Precautions therefore should be taken to prevent eye contact and, until more information is available, to avoid inhalation, especially from prolonged or repeated exposure.

17 ETHYLENE GLYCOL DIMETHYL ETHER; Ethane, 1,2-Dimethoxy-; Dimethyl
Cellosolve® Solvent; Dimethoxyethane; CAS No. 110-71-4

$$CH_3OCH_2CH_2OCH_3$$

17.1 Physical and Chemical Properties

Molecular formula	$C_4H_{10}O_2$
Molecular weight	90.1
1 ppm ≈	3.68 mg/m³ at 25°C, 760 mm HG
1 mg/l ≈	272 ppm at 25°C, 760 mm Hg

17.2 Physiologic Response

Summary. Goldberg et al. (45) exposed trained female rats for 4 hr/day, 5 days/week for 2 weeks to concentrations of 1000, 2000, 4000, and 8000 ppm of ethylene glycol dimethyl ether vapor. A single exposure to 8000 ppm caused a significant decrease in the avoidance response; however, it had no effect on the escape response. Further exposures decreased both responses, and after five daily 4 hr exposures half the rats died. Autopsy revealed massive hemorrhage of the lung and the gastrointestinal track. Several of those exposed to 4000 ppm for 10 exposures died. They also showed similar gross pathology. Those exposed to 1000 and 2000 ppm all survived but both groups showed significant though slight avoidance response but no escape response. Complete recovery from the behavioral effects occurred within a few days after cessation of exposure. In addition, there was a definite reduction in growth depending on the concentration. Thus it would appear that the toxicity of ethylene glycol dimethyl ether may be considered to be low from limited repeated short exposures. Care should be exercised, however, to avoid prolonged or repeated inhalation of high concentrations of the vapor. Data to establish an acceptable level of prolonged and repeated exposure are not available.

18 ETHYLENE GLYCOL DIETHYL ETHER; Ethane, 1,2-Diethoxy-; Diethyl Cellosolve® Solvent; CAS No. 629-14-1

$$C_2H_5OCH_2CH_2OC_2H_5$$

18.1 Physical and Chemical Properties

See Table 15.1.

18.2 Physiologic Response

18.2.1 Summary

Ethylene glycol diethyl ether vapor is irritating to the eyes and mucous membranes. It is low in oral toxicity, appreciably irritating to the eyes, not appreciably irritating to the skin, but it is moderately toxic when inhaled. The material is a weak narcotic agent. In the dog, at least, its administration does not result in an increase in urinary oxalic acid, but it does exert its toxic action principally on the kidney. The irritating nature of the vapors is probably sufficient to warn of concentrations hazardous to life upon single exposure but probably not adequate to prevent exposure to vapor concentrations hazardous upon prolonged and repeated exposure.

18.2.2 Single-Dose Oral

Smyth et al. (62) fed single doses of ethylene glycol diethyl ether as a 10 percent aqueous solution to rats and guinea pigs and found the LD$_{50}$ to be 4.39 g/kg for the rat and 2.44 g/kg for the guinea pig.

18.2.3 Repeated-Dose Oral

Gross (7) reports feeding a dog and a rabbit 1 ml/kg six times within a week without symptoms, but the urine was not investigated. A cat that received 1 ml/kg four times was seriously intoxicated each time and died after the fourth dose. Necropsy revealed no abnormal findings.

18.2.4   Parenteral Administration

Gross (7) also reported that guinea pigs survived seven subcutaneous injections of 0.5 ml/kg even though they suffered serious weight loss. When the dose was increased to 1 ml/kg, however, death resulted after seven injections. After four injections, the animals showed temporary narcotic symptoms and prostration was apparent just prior to death. Necropsy revealed kidney injury characterized by parenchymatous and interstitial nephritis.

Wiley and co-workers (21) gave two dogs 9.5 ml/day subcutaneously for 7 days and observed no increase in the oxalic acid content of the urine. This treatment reportedly caused no noticeable effects in the intact animal. However, necropsy revealed injury to the vasculature, liver, brain, testes, and particularly to the kidney.

18.2.5   Eye Contact

Carpenter and Smyth (11) have classified ethylene glycol diethyl ether along with ethylene glycol monobutyl ether, acetonyl acetone, isopropanol, and dioxane. Hence the material can be expected to be painful and to produce conjunctival irritation and slight transitory injury of the cornea, which should heal within a few days.

18.2.6   Skin Contact

Gross (7) reports that studies on guinea pigs, rabbits, and dogs revealed that the subject material does not injure the skin.

18.2.7   Inhalation

Gross (7) reports that the inhalation of 10,000 ppm for 1 hr caused irritation of the mucous membranes and a suggestion of narcosis. Cats were more sensitive than were rabbits, guinea pigs, or dogs but all survived the exposure. The same authors report that 12 daily 8-hr exposures of mice, guinea pigs, rabbits, and cats to 500 ppm resulted in the death of one of two rabbits and the two cats but no evident injury to the mice and guinea pigs. Microscopic examination of the tissues of both cats showed definite symptoms of kidney injury and in one of them, a serious purulent inflammation of the trachea, which may well have been of infectious etiology.

18.2.8   Human Experience

There were no reported instances in which ethylene glycol diethyl ether caused any adverse effect in human beings.

18.3 Hygienic Standards of Permissible Exposure

There are not sufficient data upon which to base a standard for permissible vapor exposure. The indications are that such a limit probably should be somewhat less than 100 ppm. The material does not have sufficient warning properties to prevent excessive exposure when contact is prolonged and repeated.

19 DIOXANE; 1,4-Dioxane; p-Dioxane; Diethylene-1,4-dioxide; CAS No. 123-91-1

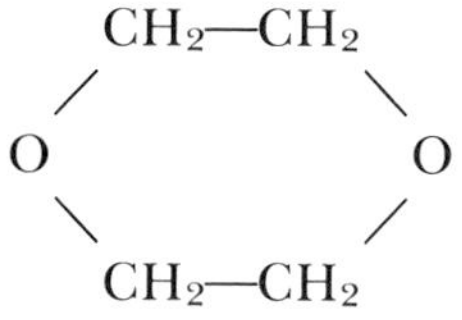

$$
\begin{array}{ccc}
 & CH_2{-}CH_2 & \\
\diagup & & \diagdown \\
O & & O \\
\diagdown & & \diagup \\
 & CH_2{-}CH_2 &
\end{array}
$$

19.1 Source, Uses, and Industrial Exposure

Dioxane can be made by several routes. Probably the most common are by dimerizing ethylene oxide or by dehydration of ethylene glycol. These and other methods are discussed by Curme and Johnston (3). The material is available in large amounts and is used largely in industry as a solvent for lacquers, plastics, varnishes, paints, dyes, fats, greases, waxes, and resins. When perfectly dry it is stable indefinitely. However, it is hygroscopic and because of its ether linkages, it produces peroxides and other degradation products upon standing in the presence of moisture.

19.2 Physical and Chemical Properties

Dioxane is a colorless liquid miscible with water and most organic solvents; it forms a constant boiling mixture with water that contains 81.6 percent dioxane and boils at 87.8°C, 760 mm Hg. Additional properties are given in Table 51.2.

19.3 Determination in the Atmosphere

Satisfactory chemical methods for determining low concentration of dioxane vapor in the air have been developed. It can be determined by means of the interferometer, adsorption, or a combustible gas indicator. It is also reported (63) that dioxane reacts with tetranitromethane with formation of a bright yellow color. It is possible that this reaction can be adapted to determinations in air. Spectrographic techniques and gas chromatography (64) also offer promising possibilities.

19.4 Physiologic Response

19.4.1 Summary

Dioxane is low in single-dose oral toxicity. The liquid is painful and irritating to the eyes, irritating to the skin upon prolonged or repeated contact, and can be absorbed through the skin in toxic amounts. Dioxane vapor has poor warning properties and can be inhaled in amounts that may cause serious systemic injury, principally in the liver and kidney. It is this latter effect that is largely responsible for the hazardous nature of this solvent. Serious and fatal exposures can be experienced without forewarning; illness sometimes becomes apparent hours after exposure. It may also cause cancer in animals, especially when ingested repeatedly in large doses. The toxicity of dioxane, especially from the standpoint of carcinogenicity, has been reviewed extensively (64).

19.4.2 Single-Dose Oral

Laug and co-workers (36) studied the single-dose oral toxicity of dioxane rather thoroughly. They determined the LD_{50} values for mice, rats, and guinea pigs to be 5.66, 5.17, and 3.90 g/kg, respectively. Other LD_{50} values have since been reported:

Species	LD_{50} (g/kg)	Ref.
Rat	6.2	64
Rat	7.35	64
Rat	7.12	65
Rat	5.6	66
Guinea pig	1.27	64
Guinea pig	3.15	65
Rabbit	2.1	64
Mouse	5.7	65
Cat	2.0	65

Symptoms progress from weakness, depression, incoordination, and coma to death. Autopsy reveals hemorrhagic areas in the pyloric region of the stomach, bladders distended with urine, enlarged kidneys, and slight proteinuria, but no hematuria.

Microscopic changes in the liver and kidneys varied in intersity from animal to animal, and generally were of a type seen in diethylene glycol poisoning.

19.4.3   Repeated-Dose Oral

Gross (7) reports that rabbits and guinea pigs fed by gavage 0.1 ml/kg 10 times exhibited dropsical changes in the liver and also that some animals repeatedly fed 0.5 ml/kg (0.52 g/kg/) died after 5, 16, and 20 feedings.

Argus et al. (67) gave rats dioxane at 1 percent in their drinking water for 63 weeks (average total dose 132 g dioxane during the experiment). Six of the 26 experimental animals developed hepatomas; histologic studies also revealed kidney changes resembling glomerulonephritis.

It is reported (64) that Hoch-Ligeti and Argus (94) gave guinea pigs drinking water containing dioxane in the range of 0.5 to 2 percent for 23 months. (The total amount per guinea pig was calculated to be 588 to 635 g dioxane.) The animals were examined histologically within 28 months. Nine of 22 males developed tumors and hyperplasias described as tumorlike, three guinea pigs early hepatomas, two carcinoma of the gall bladder, and one an adenoma of the kidney. Nine guinea pigs developed peri- and intrabronchial epithelial hyperplasia and nodular mononuclear infiltration in the lungs. Two guinea pigs had gall bladder carcinomas, three had early hepatomas, and one had an adenoma of the kidney. Four of ten control animals had some abnormalities in their upper respiratory system. The small number of animals involved makes conclusions uncertain.

Groups of 30 rats were given drinking water containing 0.75, 1.0, 1.4, or 1.8 percent of dioxane for a period of 13 months, then examined histologically at 16 months (Hoch-Ligeti et al. (68). One rat each at the 0.75 and 1.0 percent level developed nasal tumors. Two rats each at the 1.4 and 1.8 percent level developed nasal tumors as well as hepatocellular carcinomas in the liver. In another report by Argus et al. (69), apparently from the same experiment described above, the authors stated that the hepatocarcinogenicity of dioxane in male rats was a function of the total dose administered. The authors also report that in addition to the tumors in six of the rats reported they found incipient or precancerous tumors as follows: four incipient tumors in the livers of the rats given 0.75 percent dioxane, 13 incipient liver tumors and three hepatomas in those receiving 1.4 percent, and 17 incipient liver tumors and 12 hepatomas in those given 1.8 percent. At all doses the rats developed marked kidney alterations characterized by epithelial proliferation of Bowman's capsule, periglomerular fibrosis, and distension of tubules. However, no nasal tumors were seen in these animals.

Argus et al. (69) also performed electron microscopic studies on 10 male rats administered 1.0 percent dioxane in their drinking water for up to 13 months. Half the animals were killed and examined after 8 months, the rest at 13 months. No liver tumors were seen in those killed after 8 months, but two of the five rats sacrificed at 13 months were considered to have precancerous changes in the liver (incipient liver tumors). Based on all this work Argus et al. (69) suggest that dioxane is a weak to moderate hepatocarcinogen.

In the studies of Argus et al. and Hoch-Ligeti et al. (64, 67–69) the level of dioxane administered was considered massive, as much as 1 g/kg/day. Hence there was no information on low levels of intake. To evaluate the effect of the dose response at such lower levels Kociba et al. (70) gave dioxane in drinking water at levels of 0.01, 0.1, and 1.0 percent for 2 years to male and female rats. These levels were calculated to be an average of 9.6 mg/kg/day for male rats, 19.0 mg/kg/day for female rats at the 0.01 percent level, 94 and 148 mg/kg/day at the 0.1 percent level, and 1000 and 1600 mg/kg/day at the 1.0 percent level. The observations made were those of body weight and growth, blood analysis for hemoglobin content, packed cell volume, and total erythrocyte and differential white cell counts, gross and histopathological effects on 27 tissues, and organ weights on five organs.

At the 1 percent level, the rats experienced decreased body weight gains, decreased survival (only one of 60 male rats survived), and decreased water consumption. In addition, hepatocellular and renal tubular epithelial degeneration and necrosis as well as hepatic regeneration were also seen. Nasal carcinomas were observed in 3 of the 66 rats that survived. Ten of the 66 rats developed hepatic tumors as compared to 1 of 106 of the control group.

At the 0.1 percent level, the rats developed variable degrees of renal and hepatic changes similar to those seen at the 1.0 percent level, but there were no treatment-related tumors. The 0.01 percent level was found to be the "no-effect" level as there were no adverse effects detected. Thus the authors concluded that the toxicity of dioxane is related to the dose and that liver injury precedes the development of tumors in the liver.

19.4.4 Injection Toxicity

Single Injections. de Navasquez (71) reports that intravenous injection of dioxane in guinea pigs, rabbits, and cats causes a selective action on the convoluted tubules of the kidney, characterized by acute hydropic degeneration, and liver cell degeneration. Deaths from dioxane were due to uremia caused by intrarenal obstruction and anuria. The LD_{50} values reported are as folows:

Species	Route	LD_{50} (g/kg)	Ref.
Mouse	Intraperitoneal	0.75	64
Rat	Intraperitoneal	5.49 to 6.00	64
Rabbit	Intravenous	1.55	71

19.4.5 Eye Contact

Dioxane, when tested in the eyes of rabbits, was found to cause irritation and transient corneal injury (11). It has caused no serious effects from eye contact in humans (12).

19.4.6 Skin Irritation

Dioxane is not considered to be a skin irritant. However, prolonged and repeated contact can cause eczema, as can any effective fat solvent. King et al. (72) painted the clipped backs of mice three times a week for 60 weeks with 0.2 ml of an unspecified concentration of dioxane in acetone. They report that one suspected skin sarcoma and one malignant lymphoma, but no papillomas, were observed in 47 of the 60 mice treated. Pretreatment of a similar group of mice with one application of 50 μg of 7,12-dimethylbenzanthracene followed by 59 weeks of treatment with dioxane resulted in a significant increase in skin tumors when compared to dioxane or 7,12-dimethylbenzanthracene alone. There were 9 of 59 of the mice that survived the treatment. These results suggest that dioxane is not likely to be a skin cancer producer but that it may be a skin cocarcinogen. Later work (64) by these workers showed that the high degree of cocarcinogenic activity of dioxane was not found as in the earlier studies; hence this aspect needs further study before it can be assessed.

It is reported (64) that Perone et al. (95) applied four grades of dioxane to the skin of mice for 78 weeks using ethyl alcohol as the solvent. There was no evidence of skin tumor formation in any of the mice; hence the authors concluded that dioxane was not likely to cause skin tumors from skin contact.

There is a report (64) of a woman who had experienced a severe skin burn from exposure to isoprene who later developed contact eczema after possible exposure to dioxane. The cause of this response was not clear.

19.4.7 Skin Absorption

Fairley et al. (73) report observing kidney and liver injury in rabbits and guinea pigs as a result of repeated topical application of dixoane. Unsteadiness and incoordination have been shown to occur if sufficient dioxane is administered. It is reported (65) that the dermal LD_{50} for rabbits is 7.6 g/kg (97).

19.4.8 Inhalation

Acute Effects. Yant et al. (74) exposed guinea pigs for 3 hr to concentrations of 1000 to 30,000 ppm. Gross (7) exposed rats, mice, guinea pigs, and rabbits for 8 hr to concentrations ranging from 4000 to 11,000 ppm. At the higher concentrations, marked irritation of the mucous membranes was apparent. Deaths occurring during exposure or shortly afterward were usually due to respiratory failure because of lung edema, but the animals also exhibited congestion of the brain. Delayed deaths were usually due to pneumonia. Liver and kidney injuries were almost always apparent upon microscopic examination in animals dying days after exposure as well as in those apparently recovering if they were killed several days after exposure.

Wirth and Klimmer (93) as cited (64) exposed cats to dioxane at concentrations of 12,000 ppm for 7 hr, 18,000 ppm for 4.3 hr, 24,000 ppm for 4 hr, and 31,000 ppm for 3 hr and found that loss of equilibrium, increased salivation, and lacrimation occurred. Narcotic effects were also seen and the rapidity of development of such effects depended upon the dioxane concentration. Activity of all the cats decreased gradually after exposure, followed by their death. Necropsy findings were those of fatty livers and inflamed respiratory organs. Similar narcotic effects also have been reported for the rabbit (75). The LC_{50} for a 4-hr exposure of rats to dioxane is 14,260 ppm (76).

Chronic Effects.   Fairley et al. (73) exposed rats, mice, guinea pigs, and rabbits 1½ hr a day to concentrations of 10,000, 5000, 2000, and 1000 ppm of dioxane vapor. At the higher levels, mortality was high and deaths were usually due to lung injury. Animals that survived repeated exposures at all levels suffered marked liver and kidney injury.

Gross (7) reports giving cats, rabbits, and guinea pigs (two of each species) 45 daily 8-hr exposures to a concentration of 1350 ppm. Only one cat became ill and had to be sacrificed; it exhibited typical liver and kidney injury. The other animals killed after exposures ceased showed either no or but slight liver and kidney injury. This investigator also exposed a similar group of animals plus two mice to 2700 ppm 8 hr/day. Seven of the 10 animals died after 4 to 26 exposures while the rest survived 34 exposures. The signs were irritation of mucous membranes, emaciation, sometimes cramps, narcosis, and albuminuria, and always severe liver and kidney injury. In some cases, the blood urea nitrogen doubled in concentration.

Goldberg et al. (45) exposed trained female rats to dioxane at levels of 1500, 3000, or 6000 ppm for 4 hr/day, 5 days a week for 2 weeks. The rats were observed for avoidance and escape responses before, during, and 2 hr after exposure each day. No weight changes in the rats were noted until the last day of exposure to 6000 ppm. At the 1500 ppm level only one rat showed variable avoidance responses. At the 3000-ppm level two, and on occasion three, rats showed inhibition of avoidance response. At the 6000-ppm level maximum inhibition was seen on the second day, with all being inhibited in avoidance response and three of eight rats showing inhibition of escape response. Subsequent exposures no longer produced escape responses and there was a reduced but variable effect on avoidance response. This suggests that all inhibition effects on behavior were temporary and reversible.

Torkelson et al. (77) exposed rats to 111 ppm of dioxane 7 hr/day, 5 days/week for 2 years and found that they were unaffected as judged by appearance, demeanor, growth, mortality, hematologic and clinical chemistry studies, organ weights, and gross or histopathologic examination of many tissues. No hepatic or nasal carcinomas were found. These findings support the hypothesis of Kociba et al. (70) that liver injury precedes liver tumor development.

19.4.9 Teratogenicity and Mutagenicity

Schwetz et al. (96) conducted teratogenic studies on both rats and mice given 1,1,1-trichloroethane containing 3.5 percent of dioxane by inhalation. The concentration of dioxane was calculated to be 32 ppm. At this level there was no evidence of maternal, embryonal, or fetal toxicity nor were there any signs of teratogenic response. Although these studies are inconclusive because of the nature of the product tested, the results suggest that dioxane could be without teratogenic activity. However, further work is needed to determine the correctness of this assumption.

The genetic activity of dioxane was evaluated in *Salmonella typhimurium* tester strains TA1535, TA1537, and TA1538 and in *Saccharomyces cerevisiae*, all with and without metabolic activation. The results were negative in all cases (9). Thiess et al. (81) found no chromosomal changes in six workers exposed to dioxane in a manufacturing plant.

19.4.10 Carcinogenicity

Animal studies show that dioxane may be considered a weak to moderate carcinogen to animals at high dosage levels. (69) The types of tumors seen were largely hepatic and nasal tumors that resulted from ingestion of drinking water containing dioxane. Kociba et al. (70) found that the development of tumors was dose related and that hepatic tumor development is preceded by liver organic injury. The data so far indicate that inhalation of dioxane has a low potential to cause tumors in animals at concentrations at or below 100 ppm (77). The data on the production of skin tumors as a result of skin contact suggest that dioxane is not likely to cause such response. Human epidemiologic studies to date have produced no indication that exposures of humans to dioxane in industrial operations have caused an increased incidence of tumors (see Section 19.4.12).

19.4.11 Metabolism

Fairley et al. (85) expressed the opinion that the renal and hepatic injury produced by dioxane is a result of its metabolism to diglycolic and oxalic acids.

However, Wiley et al. (21) questioned this hypothesis when they found that dogs and rabbits fed dioxane repeatedly did not excrete in the urine more oxalic acid than control animals, although animals fed ethylene glycol or ethylene glycol monoacetate did excrete considerably more oxalic acid than did control animals. No attempts to define the metabolites of dioxane by chemical means were reported until Braun and Young (86) determined by various chromatographic techniques along with mass spectrometry and nuclear magnetic resonance spectrometry that the major urinary metabolite of dixoane in rats was β-hydroxyethoxyacetic acid. Shortly thereafter, Wov et al. (87) reported the identification of p -dioxan-2-one as the major urinary metabolite using similar techniques. Although β-hydroxyethoxyacetic acid and p-dioxan-2-one may be interconverted by adjusting the pH and aqueous environment (acidic and dehydrating conditions favoring p-dioxan-2-one), Braun and Young (86) considered p-dioxan-2-one an artifact due to on-column heating during gas chromatography, and did not find p-dioxan-2-one when untreated urine was thin layer chromatographed. Braun and Young (86) also found small amounts of dioxane per se in rat urine and suggested that a second minor urinary metabolite may be diethylene glycol. In human volunteers (88) as well as dioxane plant personnel (88), β-hydroxyethoxyacetic acid was found in urine, indicating that humans metabolize dioxane to the same product as rats. Powar and Mungikar (90) showed that two daily doses of 2 g dioxane/kg to mice caused increases in liver weight, liver microsomal protein, the activities of liver aminopyrine-N-demethylase and acetanilide hydroxylase, components of hepatic microsomal electron transport, and decreases in both NADPH-linked and ascorbate-induced lipid peroxidation. Wov et al. (91) studied the effects of inducers and inhibitors of hepatic mixed-function oxidases on the excretion of p-dioxan-2-one in rats and concluded that these oxidases were involved in the metabolism of dioxane. Further evidence that high doses of dioxane can induce its own metabolism was provided by Young et al. (92) who showed that 1000 mg/kg/day, but not 10 mg/kg/day, caused more rapid excretion of metabolites, and that 1000 mg/kg doses exceeded the capacity of the rat to metabolize dioxane by 62-fold.

19.4.12　Human Experience

In 1933, five cases of fatal industrial poisoning from dioxane were reported in England (78, 79). These reports are summarized and discussed excellently by Browning (17). In general, men working in a synthetic textile factory apparently inhaled excessive amounts of dioxane. The symptoms were irritation of the upper respiratory passages, coughing, irritation of the eyes, drowsiness, vertigo, headache, anorexia, stomach pains, nausea, vomiting, uremia, coma, and death. Autopsy revealed congestion and edema of the lungs and brain, and marked injury of the liver and kidney. Death was attributed to kidney injury. Blood counts on three of the men who died showed no abnormalities other than considerable leukocytosis. The exposure that these men received is still unknown and it is debatable whether the deaths were caused by chronic exposures or whether they were a result of relatively few intense exposures. de Navasquez (71) believes, on the basis of previous experience in this same factory over several months during which time no trouble was encountered, and on the basis that others in the factory at the same time did not show any serious symptoms, that the persons who died were exposed to high concentrations over a relatively short period of time. It is obvious from this incident that dioxane does not have warning properties adequate to prevent short-term exposures dangerous to life.

Johnstone (80) reported the death of a worker who was exposed for 1 week to about 500 ppm of dioxane in the air. In addition, the worker used dioxane to wash glue from his hands, which probably resulted in absorption of some dioxane through the skin. Autopsy revealed damage to kidneys, liver, and brain.

Thiess et al. (81) conducted an epidemiologic study on 74 workers who had been working in a dioxane manufacturing factory from 3 to 41 years. Analysis of the workroom air in the factory in 1974 indicated at that time that the level of exposure was up to 14.24 ppm of dioxane. Of the 74 workers studied, 24 were still working in the plant, 26 were working elsewhere, 15 had retired, and 12 were dead. The study of these workers showed no evidence of adverse effects because of their exposure to dioxane. The studies included extensive medical and physical examinations, chromosome analysis in six of the actively working employees, and a careful analysis of the cause of death of the 12 dead workers.

Buffler et al. [see NIOSH criteria document (64)] conducted a study in 1975 of 165 employees who had worked in a dioxane manufacturing plant in Texas between April 1954 and June 1959. The levels of exposure ranged up to 32 ppm of dioxane. Again, there were no indications of adverse effects due to exposure to dioxane (and in some workers exposure to other solvents) as revealed by careful medical and physical examinations and by evaluation of the causes of death of those who had died. As a part of this study, the urine of five workers was analyzed for dioxane and hydroxyethoxyacetic acid (HEAA). Essentially all of the dioxane was metabolized to HEAA; the ratio of dioxane to HEAA in the urine was 1 to 118. This suggests that at the exposure concentrations the workers experienced, the metabolic pathway in humans was not saturated. This led the authors to believe that exposure to dioxane vapor at low levels (up to 32 ppm) poses no significant health hazard.

Dernehl and Peele [see NIOSH criteria document (64)], in a similar epidemiologic study on workers in a dioxane manufacturing plant involving 80 workers, again found no evidence of adverse effects due to their possible exposure to dioxane vapor as shown by complete physical examinations, chest X-rays, electrocardiograms, a series of liver profile tests, and, in the case of those who had died, a careful review of the cause of their death. Air analysis studies revealed that the exposure levels ranged from 0.05 to 51 ppm.

19.5 Hygienic Standards of Permissible Exposure

For a number of years, official agencies and individuals have recommended an industrial hygiene standard of 100 ppm. In 1972 a level of 50 ppm was tentatively set; in 1974 this level was established as the threshold limit value for dioxane by the American Conference of Governmental Industrial Hygienists and was the level still recommended in 1979 (82). However, in 1980 a notice of intended change to 25 ppm was published (30). The OSHA standard published in 1976 is 100 ppm (31). In the NIOSH criteria document published in 1977 the recommendation was made that the limit of exposure to dioxane should be the lowest concentration reliably measurable over a short period of time, that is, 1 ppm as measured over a 30-min sampling period using a sampling rate of 1 l/min. However, as of early 1980 this suggested guide had not been formally adopted.

19.6 Odor and Warning Properties

The odor of dioxane in low concentrations is faint and generally inoffensive and has been described as being somewhat alcoholic. According to Silverman et al. (83), it was concluded from studies on 12 subjects exposed 15 min to various concentrations of dioxane that 200 ppm was the highest that they considered acceptable; at 300 ppm, it caused irritation of the eyes, nose, and throat; and at 500 ppm, it was objectionable. Even at higher concentrations, the initial irritation to eyes and respiratory passages is transitory and it is certain that the warning properties of dioxane are completely inadequate to prevent short-term exposure to toxic amounts. Yant et al. (74) reported immediate slight burning of the eyes accompanied by lacrimation and slight irritation of the nose and throat from an exposure of 1600 ppm for 10 min; at 5500 ppm eye irritation and a burning sensation in the nose and throat were noted; and at 10,000 ppm or more, pulmonary irritation occurred.

May (32) reported that the odor threshold for dioxane was 170 ppm and 270 ppm was considered a strong–objectionable odor.

Laing (84), however, reports that the odor threshold is about 7 ppm for rats and that the odor detection is about the same in rats and man. This level was substantiated by Thiess et al (81), who reported that the odor recognition level to humans was about 5.6 ppm.

20 DIETHYLENE GLYCOL MONOMETHYL ETHER; Ethanol, 2-(2-Methoxyethoxy)-; Methyl Carbitol® Solvent; Dowanol® DM Glycol Ether; CAS No. 111-77-3

$$CH_3OCH_2CH_2OCH_2CH_2OH$$

20.1 Physical and Chemical Properties

See Table 51.2.

20.2 Physiologic Response

20.2.1 Summary

Diethylene glycol monomethyl ether is low in oral toxicity, painful but not seriously injurious to the eyes, and not irritating to the skin. Although it can be absorbed in toxic amounts through the skin, severe exposure would be required before serious effects would be expected. Hazardous amounts are not likely to be inhaled under ordinary conditions, but where heated material is encountered, care is warranted. No adverse human experience has been reported.

20.2.2 Single-Dose Oral

Smyth et al. (62) fed diethylene glycol monomethyl ether as a 50 percent aqueous solution to rats and guinea pigs and found the LD_{50} values to be 9.21 ml/kg for rats and 4.16 for guinea pigs (9.21 and 4.16 g/kg, respectively). The LD_{50} value for rabbits is 7.19 g/kg (8). Others (9) have fed the material undiluted to groups of 10 male and 10 female rats and have found the LD_{50} value to be between 6.5 and 7.0 ml/kg for males and between 5.5 and 6.0 ml/kg for females (6.5 to 7.0 g/kg for males and 5.5 to 6.0 g/kg for females). Deaths usually occurred within 48 hr or not at all, and were believed due either to profound narcosis or kidney injury.

20.2.3 Repeated-Dose Oral

Smyth and Carpenter (98) report administering commercial diethylene glycol monomethyl ether to rats for 30 days as a part of their drinking water. They found that the maximum dose having no effect was less than 0.19 g/kg based upon microscopic study of the tissues. Animals survived the highest dosage level, 1.83 g/kg.

20.2.4 Eye Contact

Diethylene glycol monomethyl ether is somewhat painful to the eyes and is capable of causing only transitory injury (9, 11). Thus the material presents no serious hazard from eye contact under ordinary industrial handling conditions.

20.2.5   Skin Contact

Diethylene glycol monomethyl ether is not appreciably irritating to the skin (9). Although the material can be absorbed through the skin of rabbits in toxic amounts, extensive and prolonged contact is required to cause serious effects. The LD_{50} is reported to be about 20 ml/kg (20 g/kg) (9). Kligman, as reported by Opdyke (99), found that diethylene glycol methyl ether at a concentration of 20 percent in petrolatum when applied as a closed-patch test for 48 hr caused no irritation or sensitization in 25 human subjects. In view of these observations, there would seem to be no significant hazard from skin contact in ordinary industrial operations.

20.2.6   Inhalation

The only information available (8) states that rats exposed to an essentially saturated atmosphere of diethylene glycol methyl ether generated at room temperature caused no deaths or significant ill effects. Because of the low volatility of this material at normal room temperatures and because of its low oral toxicity, and its apparent low toxicity from single exposures to vapors, the material is believed to present no unusual hazards from inhalation when handled at room temperature. However, vapors of this material generated at elevated temperatures, or when breathed repeatedly over a prolonged period, may well present a hazard from inhalation.

20.2.7   Human Experience

No adverse experience of humans is reported.

20.3   Hygienic Standards of Permissible Exposure

Data adequate to allow the establishment of an hygienic standard are lacking.

21   DIETHYLENE GLYCOL MONOETHYL ETHER; Ethanol, 2-(2-ethoxyethoxy)-; Carbitol® Solvent; Dowanol® DE Glycol Ether; CAS No. 111-90-0

$$C_2H_5OCH_2CH_2OCH_2CH_2OH$$

21.1   Physical and Chemical Properties

See Table 51.2.

21.2 Physiologic Response

21.2.1 Summary

Commercial products may contain an appreciable amount of ethylene glycol and such products *may* be more toxic than the relatively pure ether. This fact must not be overlooked when evaluating the available toxicologic data or when choosing a product for specific use.

It is low in oral toxicity and not appreciably irritating to the eyes or skin, but is readily absorbed in toxic amounts through the skin. Its volatility is sufficiently low that acutely hazardous vapor concentrations do not occur at ordinary temperatures. The material appears to be readily oxidized in the body. Experience with human subjects has been uneventful. It is generally agreed that the relatively pure ether does not present any serious industrial hazards. Reasonable precautions are adequate to ensure safe handling.

21.2.2 Single-Dose Oral

Smyth et al. (62) fed specially purified diethylene glycol monoethyl ether as a 50 percent aqueous solution and found the LD_{50} values to be 8.69 g/kg for the rat and 3.67 g/kg for the guinea pig. The LD_{50} value for rabbits is 3.62 g/kg (8). When a 40 percent aqueous solution of a commercial product was fed, the LD_{50} values found were 9.74 g/kg for the rat and 4.97 g/kg for the guinea pig. Gross (7), working with an industrial product, found that 1 ml/kg was lethal to cats. It caused disturbance of equilibrium, gastrointestinal inflammation, pneumonia, and kidney injury, but no albuminuria. A dog that received a single dose of 2 ml/kg as a 20 percent aqueous solution was unaffected except for slight vomiting 3 hr after feeding. Laug and co-workers (36) found the oral LD_{50} value to be 5.54 g/kg for rats, 6.58 g/kg for mice, and 3.87 g/kg for guinea pigs. They likened the response of animals fed this ether to that caused by ethanol, but also noted pneumonia and kidney injury in treated animals. They concluded that this material is not suitable for use in foods or drugs. Others (9) have found the LD_{50} for rats to be 5.4 g/kg and observed the effects of acutely toxic doses to be characterized by ataxia and depression with little apparent injury to visceral organs.

21.2.3 Repeated-Dose Oral

Smyth and Carpenter (98) report maintaining groups of five rats for 30 days on drinking water containing various amounts of purified diethylene glycol monoethyl ether. Dosage levels ranged from 0.21 to 3.88 g/kg/day. They found that 0.49 g/kg was tolerated without any adverse effect, that 0.87 g/kg caused reduction in appetite, and that 1.77 g/kg caused some organic injury.

Morris et al. (38) maintained rats for 2 years on a diet containing 2.16 percent of a purified diethylene glycol monoethyl ether. This probably is equivalent to slightly more than 1.0 g/kg/day. The only adverse effects they noted were a few oxalate concretions in a kidney of one animal, slight liver damage, and some interstitial edema in the testes. Since the quality of the material tested was not established, the possibility of the concretions being caused by the presence of small amounts of ethylene glycol in the test sample cannot be overlooked.

Hanzlik et al. (100) administered the pure ether to rats at the level of 1.0 percent in their drinking water and to mice at the level of 5.0 percent in their food for a 2-year period without causing significant adverse effects. They also found that when this same ether containing ethylene glycol was fed, kidney injury typical of the glycol occurred. They concluded that the ether was relatively noninjurious.

Hall et al. (101) fed rats diethylene glycol ethyl ether containing 0.4 percent ethylene glycol as an impurity in diets containing 0.25, 1.0, and 5.0 percent for 90 days. They state that the no-effect level was the 1 percent level. At the 5 percent level, they found retarded growth in both sexes, reduced food intake, no hematologic changes, slightly impaired renal function, especially in the males, increase in kidney weights in both sexes and of the testes in the males, degeneration of kidney tubules, fatty changes in the liver, and testicular edema.

Gaunt et al. (102), using diethylene glycol ethyl ether containing less than 0.4 percent ethylene glycol, fed the material for 90 days to rats at dietary levels of 0.5 and 5.0 percent, to mice at dietary levels of 0.2, 0.6, 1.8, and 5.4 percent, and by intubation to guinea pigs at doses of 167, 500, and 1500 mg/kg. Three guinea pigs given 1500 mg/kg/day for 14 to 21 days died showing signs of uremia, while six of the 20 male mice fed the 5.4 percent level died with signs of severe renal damage. Oxaluria developed both in the rats and mice at the highest level. A reduction of hemoglobin developed in all three species at the highest levels. They state that the no-effect levels seen in their studies were about 250 mg/kg/day for rats, 850 to 1000 mg/kg/day for mice, and 167 mg/kg/day for guinea pigs.

Butterworth et al. (103) wanted to study the effect of diethylene glycol ethyl ether on a nonrodent species so they used ferrets in a 9-month study. The material was fed in a diet at levels of 0.5, 1.0, 2.0, and 3.0 ml/kg/day (approximately 0.5, 1.0, 2.0, and 3.0 g/kg/day). No adverse effects such as body and organ weight changes, hematologic effects, or histopathologic injury were seen. Based on this and other data, and using the 100-fold margin of safety, they feel that the acceptable daily intake for man is 2.1 ml/day (approximately 2.1 g/day) for a 70-kg person.

Smyth et al. (104) fed rats through three generations, over a 2-year period, two grades of diethylene glycol ethyl ether; one contained less than 0.2 percent ethylene glycol, the other 29.5 percent. The drinking water levels were 0.01, 0.04, 0.2, and 1 percent (calculated to be approximately 0.01, 0.04, 0.20, and 0.95 g/kg). The adverse effects were in good agreement with those seen in earlier studies. The sample containing the high level of ethylene glycol (29.2 percent) was considerably more toxic than the more pure grade. The maximum safe dosage for the impure material was 0.01 g/kg/day, whereas it was about 0.20 g/kg/day for the more pure sample (less than 0.2 percent ethylene glycol). Thus they consider the cumulative toxicity of diethylene glycol ethyl ether is largely due to the ethylene glycol content. On the basis of their work, the authors suggest that the permissible intake for man may be 1.4 g/day (using a 10 fold margin of safety).

21.2.4   Injection Toxicity

Various workers have reported LD_{50} values (65):

Rat	IV	2.9 g/kg
Mouse	IV	3.9 g/kg
Dog	IV	3.0 g/kg
Rabbit	IV	0.9 g/kg
Mouse	SC	5.5 g/kg
Rabbit	SC	2.0 g/kg

21.2.5   Eye Irritation

Diethylene glycol monoethyl ether is slightly painful but causes no more than a trace of irritation of the conjunctival membranes (9, 11, 40). Thus this material probably does not present a serious hazard from eye contact.

21.2.6   Skin Irritation

Diethylene glycol monoethyl ether is not irritating to the skin of rabbits even upon prolonged and repeated contact (9). Cranch et al. (105) found the material to be neither a primary irritant nor a sensitizer, and no different from wool, fat, or glycerine when applied to the skin of 98 human subjects. Furthermore, they report that 70 percent aqueous material did not retard wound healing. Meininger (106) confirmed the lack of skin irritating effect in humans and was unable to find evidence that it was absorbed through the skin of human subjects. Kligman, according to Opdyke (99), also found that human volunteers showed neither irritation nor signs of sensitization when the material was tested at a 20 percent level in petroleum for a 48-hr closed-patch test.

21.2.7   Skin Absorption

Hanzlik et al. (107) have reported the results of extensive skin absorption studies with diethylene glycol monoethyl ether. They report the LD_{50} to be 8.5 ml/kg when applied by inunction for 2 hr to 100 cm^2 or slightly more. When such applications were repeated daily for 30 days, the LD_{50} was estimated to be 0.32 ml/kg with the no-effect level between 0.04 and 0.08 ml/kg. They report transient dermatitis and both microscopic injury and impairment of kidney function.

Purified diethylene glycol monoethyl ether has been subjected to repeated skin absorption studies by others (9). The material used in these studies had a boiling range (°C) at 760 mm Hg as follows:

First drop	198.7
5%	201.2
50%	202.5
95%	203.7
Dry point	208.0

The material was applied on a 3 × 3 in. cotton pad just heavy enough to hold the dose; this was covered with a 5 × 5 in. impervious film and the whole bandaged onto the clipped abdomen of each rabbit five times a week for 3 months. The dosages used were 0.1, 0.3, 1.0, and 3.0 ml/kg/day (approximately 0.1, 0.3, 1.0, and 3.0 g/kg/day), and five animals were used at each dosage. The bandage was held in intimate contact with the skin for 24 hr, thus allowing essentially all to be absorbed each day; there was no leakage of consequence. The criteria used in judging effect were growth, mortality, hematologic studies, organ weight studies, blood urea nitrogen determinations, and gross and microscopic examination of the treated areas of skin and of the principal organs. The only effects at the top dosage level of 3.0 ml/kg were an increase in blood urea nitrogen and severe kidney injury in one of the four surviving animals. Minor kidney changes were seen in two of the animals and one was unaffected. At the dosage level of 1 ml/kg, moderate kidney changes were seen in three of the four surviving animals and at the lower dosage levels no adverse effects were seen. Thus the no-effect dosage level for repeated exposures over a 90-day period approximates 0.3 ml/kg/day (0.3 g/kg/day). This figure is about 10 times larger than that found by Hanzlik (107) and might well reflect a difference in purity of the material tested.

21.2.8   Inhalation

Gross (7) reports that rabbits, cats, guinea pigs, and mice were not injured by 12 daily exposures to an atmosphere essentially saturated with diethylene glycol monoethyl ether.

21.2.9 Carcinogenicity

The 2-year dietary feeding studies of Morris et al. (38), in which purified diethylene glycol ethyl ether was administered at a level of slightly more than 1.0 g/kg/day to rats, indicate that there were some adverse effects. However, the authors make no mention of the development of tumors. Although this study was not designed specifically for the evaluation of carcinogenic properties, it would seem that the studies suggest that diethylene glycol ethyl ether has little or no carcinogenic potential.

21.2.10 Metabolism

Fellows et al. (108) found that diethylene glycol monoethyl ether is largely destroyed by the body or excreted as the glucuronate. They noted that when given to rabbits orally, or by injection, the urinary content of glucuronic acid increased as it did when propylene glycol was given. Why this should occur with these two materials and not with ethylene and diethylene glycol and glycerol is unknown.

21.2.11 Human Experience

Insofar as is known, no toxic effects have resulted from the industrial use of diethylene glycol monoethyl ether. However, if the material is intended for uses involving prolonged or repeated contact with the skin, it would seem only prudent to choose a product of low ethylene glycol content. The material is not considered suitable for uses where ingestion may be expected.

Browning (17) does report that an alcoholic man who drank a liquid containing 47 percent diethylene glycol ethyl ether (about 300 ml) and less than 0.2 percent methanol developed severe symptoms of central nervous and respiratory injury, thirst, acidosis, and albumin in the urine but no oliguria. He recovered upon symptomatic treatment.

21.3 Hygienic Standards of Permissible Exposure

Data adequate for the establishment of a hygienic standard for diethylene glycol monoethyl ether are not available. However, in view of the low vapor pressure and the low toxicity of the material, such a standard hardly seems necessary. Reasonable and ordinary precautions to avoid inhalation of vapors or mists would seem adequate to prevent excessive vapor exposure.

22 DIETHYLENE GLYCOL MONOBUTYL ETHER; Ethanol, 2(2-Butoxyethoxy)-; Butyl Carbitol® Solvent; Dowanol® DB Glycol Ether; CAS No. 112-34-5

$$C_4H_9OCH_2CH_2OCH_2CH_2OH$$

22.1   Physical and Chemical Properties

See Table 51.2.

22.2   Physiologic Response

22.2.1   Summary

Diethylene glycol monobutyl ether is generally available commercially and is a fairly pure product. It is low in single-dose oral and vapor toxicity, moderately toxic in repeated-dose oral toxicity, moderately irritating and injurious to the eyes, not appreciably irritating to the skin, and not absorbed through the skin in acutely toxic amounts except at large dosage levels. The results of the limited repeated-dose oral work reported suggest that the material may be rather toxic when inhaled or absorbed through the skin in repeated small doses.

22.2.2   Single-Dose Oral

Smyth et al. (62) fed diethylene glycol monobutyl ether as a 50 percent aqueous solution and found the LD_{50} to be 6.56 g/kg for rats and 2.00 g/kg for guinea pigs. Others have found the LD_{50} for rats to be 5.66 g/kg when it was fed undiluted (9). The LD_{50} value for rabbits is 2.2 g/kg (8).

22.2.3   Repeated-Dose Oral

Smyth and Carpenter (98) maintained groups of five rats for 30 days on drinking water containing various amounts of the material. Based on the water consumption, the dosage levels ranged from 0.051 to 1.83 g/kg/day. They found that 0.051 g/kg was the maximum dosage having no effect, 0.094 g/kg caused a reduction in appetite, and 0.65 g/kg/day caused histopathologic injury in either the liver, kidney, spleen, or testes.

22.2.4   Injection

Rats were given 0.126, 0.252, or 0.5 g/kg of diethylene glycol butyl ether by intraperitoneal injection. None of the rats were adversely affected (9). The LD_{50} value for mice when injected intraperitoneally is 0.85 g/kg (65).

22.2.5   Eye Contact

Diethylene glycol monobutyl ether is capable of causing moderate irritation and moderate transient corneal injury (9, 11). This indicates that although the material is not likely to cause impairment of vision, it may cause considerable discomfort.

22.2.6 Skin Contact (9)

Diethylene glycol monobutyl ether is only very slightly irritating to the skin of rabbits even upon prolonged and repeated skin contact. However, when humans were patch-tested with undiluted material, a limited number of the volunteers developed reddening of the skin (8). The material is absorbed through the skin but the acute toxicity by this route is slight, the LD_{50} being about 4 g/kg when exposure was continuous for 24 hr.

22.2.7 Inhalation (9)

Three rats survived a single 7-hr exposure to an atmosphere saturated with the material at 100°C and then cooled to room temperature. The animals appeared normal throughout the exposure and exhibited only a slight transient weight loss. No autopsies were made. These results indicate that there is little hazard from a single vapor exposure.

22.2.8 Human Experience

There has been no adverse human experience that can be attributed definitely to this material (17).

22.3 Hygienic Standards of Permissible Exposure

Data for the establishment of a hygienic standard for diethylene glycol mono-butyl ether are not available. However, the limited repeated-dose data available suggest caution in handling where repeated exposures may be expected.

23 DIETHYLENE GLYCOL ISOBUTYL ETHER; Ethanol, 2-[2(2-Methylpropoxy)ethoxy]-; Dowanol® DiB Glycol Ether; Ektasolve® DIB Solvent; CAS No. 18912-80-6

$$CH_3CH(CH_3)CH_2OCH_2CH_2OCH_2CH_2OH$$

23.1 Physical and Chemical Properties

Molecular formula	$C_8H_{18}O_2$
Molecular weight	162.2
1 ppm ≈	6.64 mg/m^3 at 760 mm Hg and 25°C
1 mg/l ≈	150.8 ppm at 760 mm Hg and 25°C

23.2   Physiologic Response

Summary.   Diethylene glycol isobutyl ether is low in single-dose oral toxicity, the LD_{50} for rats being in the range of 4.0 (9) to 1.6 g/kg (65). It is severely hazardous from eye contact because it causes severe irritation and corneal damage which could result in possible impairment of vision (9). However, it is only very mildly irritating to the skin (9). Skin absorption should present no hazard, since the LD_{50} for rabbits is greater than 4.0 g/kg (9). Because of the low volatility of this material at room temperature, even saturated vapors caused no significant effects to rats exposed for 7 hr. Those exposed to vapors generated at 100°C also survived but the rats passed blood-colored urine during the 7-hr exposure and for a day after exposure (9). All the rats exposed for 6 hr to approximately 6700 ppm also survived (65). Apparently the vapor must have been generated by heating the material.

Based upon these data, special care in the handling of diethylene glycol isobutyl ether should include precautions to prevent eye contact and avoidance of prolonged or repeated inhalation of vapors, mists, or fogs, especially if the material is handled hot, or if the material is agitated or nebulized so as to produce mists or fogs.

24   DIETHYLENE GLYCOL-*n*-HEXYL ETHER; Ethanol, 2-[2-(2-Hexyloxy)ethoxy]-; *n*-Hexyl Carbitol® Solvent; CAS No. 112-59-4

$$CH_3(CH_2)_4CH_2OCH_2CH_2OCH_2CH_2OH$$

24.1   Physical and Chemical Properties

See Table 51.2.

24.2   Physiologic Response

Summary (8).   It is low in single-dose toxicity; the LD_{50} for rats is 4.92 g/kg. It is severely irritating to the eye and may cause marked corneal injury which may be slow in healing. Even a 5 percent solution is capable of causing some slight eye injury. It is mildly to moderately irritating to the skin but should cause no serious injury unless the exposure is prolonged and severe. It is low in toxicity by skin absorption, for the LD_{50} for rabbits is 1.5 g/kg. Rats exposed to essentially saturated vapors generated at room temperature for 8 hr experienced no significant ill effects and all survived.

Diethylene glycol-*n*-hexyl ether, because of its toxicity, should be handled with care and caution to prevent eye contact. It would seem wise also to exercise care to avoid prolonged or repeated inhalation exposure, especially if it is handled hot or if the operations call for agitated or nebulization so that a fog or a mist is generated.

25 DIETHYLENE GLYCOL DIVINYL ETHER; Ethene, 1,1-[oxybis(2,1-ethanediyloxy)bis]-; CAS No. 764-99-8

$$CH_2\!\!=\!\!CHOCH_2CH_2OCH_2CH_2OCH\!\!=\!\!CH_2$$

25.1 Physical and Chemical Properties

Molecular formula	$C_8H_{14}O_3$
Molecular weight	158.1
1 ppm $\approx$	6.47 mg/m^3 at 760 mm Hg, 25°C
1 mg/l $\approx$	154.5 ppm at 760 mm Hg, 25°C

25.2 Physiologic Response

Summary (61). Diethylene glycol divinyl ether is low in single-dose oral toxicity; the LD_{50} for rats is 3.73 ml/kg. It is essentially nonirritating to the eyes and only very slightly irritating to the skin. It is not likely to be absorbed through the skin in toxic amounts, for the LD_{50} value for rabbits by this route is 14.1 ml/kg. Exposure of rats for 8 hr to an essentially saturated atmosphere generated at room temperature resulted in no deaths.

It would appear that the practice of reasonable care and caution in the handling of this compound should prevent the development of serious toxic effects. It would seem wise, until more is known, to practice precautions to avoid repeated prolonged exposure to vapors, mist, or fog. This is especially so if the compound is handled hot or is agitated.

26 DIETHYLENE GLYCOL ETHYL VINYL ETHER; Ethene, 2-[2-(2-Ethoxyethoxy)ethoxy]-; Ethane, 1-(2-Ethoxyethoxy)-2-vinyloxy-; Vinyl ethyl Carbitol® Solvent; CAS No. 10143-53-0

$$CH_2\!\!=\!\!CHOCH_2CH_2OCH_2CH_2OCH_2CH_3$$

26.1 Physical and Chemical Properties

Diethylene glycol ethyl vinyl ether is infinitely soluble in alcohol and many solvents. Its solubility in water is about 9.4 percent at 20°C. Additional properties are given in Table 51.2.

26.2 Physiologic Response

Summary (61). Diethylene glycol ethyl vinyl ether may be considered to be low in health hazard. It is low in single-dose oral toxicity; the LD_{50} value for rats is 11.3 ml/kg. It is essentially nonirritating to the eyes, being rated 1 on a scale of 10 in rabbit eye tests. It is only very mildly irritating to the skin, being rated 2 on a scale of 10 in rabbit irritation tests, and is low in toxicity by skin absorption, the LD_{50} value for rabbits being 8.41 ml/kg. None of the rats exposed for 8 hr to essentially saturated vapors died. The practice of reasonable caution and good cleanliness should be adequate to avoid health problems when handled under anticipated industrial conditions. No hygienic standard has been established and none seems to be necessary.

27 DIETHYLENE GLYCOL MONOMETHYLPENTYL ETHER: Ethanol, 2-[2-(2-Methylpentyl)oxy]-; 2-Methylpentyl Carbitol® Solvent; CAS No. 10143-56-3

$$CH_3CH_2CH_2CH(CH_3)CH_2OCH_2CH_2OCH_2CH_2OH$$

27.1 Physical and Chemical Properties

Molecular formula	$C_{10}H_{22}O_3$
Molecular weight	190.3
1 ppm ≈	7.78 mg/m^3 at 760 mm Hg, 25°C
1 mg/l ≈	128.5 ppm at 760 mm Hg, 25°C

27.2 Physiologic Response

Summary (61). Diethylene glycol monomethylpentyl ether is low in single-dose oral toxicity, the LD_{50} value for rats being 5.66 g/kg. It is severely irritating and injurious to the eyes of rabbits, being rated 6 on a scale of 10, but is only very slightly irritating to the skin. By absorption through the skin, it is low in toxicity, the LD_{50} value for rabbits being 1.58 ml/kg. Thus it may be absorbed through the skin but should not pose a serious health hazard from absorption unless skin exposure is excessive and prolonged. It should not pose a significant health hazard from an occasional exposure to its vapors generated at room temperature, for none of the rats exposed for 8 hr to essentially saturated vapors died. Thus the main health hazard is that of eye contact. Precautions therefore should be taken to prevent eye contact. It would also seem wise, until more is known, to avoid especially prolonged or repeated exposure to hot vapors, mists, or fogs.

28   TRIETHYLENE GLYCOL MONOMETHYL ETHER; Ethanol, 2-[2-(2-Methoxyethoxy)ethoxy]-; Methoxytriglycol; Dowanol® TM Glycol Ether; CAS No. 112-35-6

$$CH_3OCH_2CH_2OCH_2CH_2OCH_2CH_2OH$$

28.1   Physical and Chemical Properties

See Table 51.2.

28.2   Physiologic Response

Summary.   Smyth et al. (61) found that triethylene glycol monomethyl ether is low in single-dose oral toxicity, the LD_{50} for rats being 11.3 ml/kg (11.8 g/kg). The LD_{50} value for rats when injected intravenously is 7.7 ml/kg (8.1 g/kg) and 7.1 ml/kg (7.4 g/kg) when injected intraperitoneally. It is not injurious to the eyes or skin, and is low in toxicity by absorption through the skin, the LD_{50} value for rabbits being 7.1 ml/kg (7.4 g/kg). Exposure of rats to essentially saturated vapor for 8 hr caused no significant effects. Repeated inhalation studies have not been reported. It appears from these data that triethylene glycol monomethyl ether should present no significant health hazard in ordinary industrial handling. However, until more is known, it would seem wise to avoid repeated prolonged exposure to its vapors or mists, especially if it is handled hot.

29   TRIETHYLENE GLYCOL MONOETHYL ETHER; Ethanol, 2-[2-(2-Ethoxyethoxy)ethoxy]-; Ethoxytriglycol, CAS No. 112-50-5

$$C_2H_5OCH_2CH_2OCH_2CH_2OCH_2CH_2OH$$

29.1   Physical and Chemical Properties

See Table 51.2.

29.2   Physiologic Response

Summary.   Smyth and co-workers (98, 109) report that range-finding studies on triethylene glycol monoethyl ether showed the oral LD_{50} to be 10.6 g/kg for rats and the skin absorption LD_{50} to be 8 ml/kg (8.2 g/kg) for rabbits, and that the material is not injurious to the skin or eyes. They also report maintaining groups of 10 rats for 30 days on drinking water containing sufficient material to result in daily intakes ranging from 0.18 to 3.30 g/kg. They found the maximum intake having no effect to be 0.75 g/kg/day, but did not disclose the nature of the effect at higher dosage levels. The effect noted apparently was not increased mortality, decreased food intake, decreased growth, or microscopic changes in the liver, kidney, spleen, or testes, for none of these criteria were affected at the 3.30 g/kg/day level. It appears from these data that the material would present no appreciable hazard in ordinary industrial handling.

30 TRIETHYLENE GLYCOL MONOBUTYL ETHER; Ethanol, 2-[2-(2-Butoxyethoxy)ethoxy]-; Butoxytriglycol; CAS No. 143-22-6

$$\text{C}_4\text{H}_9\text{OCH}_2\text{CH}_2\text{OCH}_2\text{CH}_2\text{OCH}_2\text{CH}_2\text{OH}$$

30.1 Physical and Chemical Properties

See Table 51.2.

30.2 Physiologic Response

Summary (61). Triethylene glycol monobutyl ether has a low single-dose oral toxicity, the LD_{50} for rats 6.73 ml/kg. It should cause no significant irritation to the skin but may cause marked eye irritation and corneal injury, which may be slow in healing. It is low in single-dose skin absorption; the LD_{50} for rabbits is cited as being 3.54 ml/kg. A single 8-hr exposure to rats at essentially saturated vapors resulted in no significant adverse effects. It appears that this material should present no appreciable health hazard from skin contact or from a single exposure to vapors generated at room temperature. However, the eye injury in rabbits indicates that suitable precautions should be taken during industrial handling to avoid all eye contact. Until more is known, it would seem wise to observe suitable precautions to avoid inhalation of vapors or mists, especially when handled hot or when there is a possibility of repeated prolonged exposure.

31 TETRAETHYLENE GLYCOL MONOVINYLETHYL ETHER; 3,6,9,12,15-Pentaoxaheptadec-1-ene; 2-Vinylethoxytetraethylene Glycol; CAS No.—none found

$$\text{CH}_2\text{=}\text{CHCH}_2\text{CH}_2\text{OCH}_2\text{CH}_2\text{OCH}_2\text{CH}_2\text{OCH}_2\text{CH}_2\text{OCH}_2\text{CH}_2\text{OH}$$

31.1 Physical and Chemical Properties

Molecular formula	$C_{12}H_{24}O_5$
Molecular weight	248.3
Boiling point	295.6°C

Freezing point	$-21°C$
1 ppm ≏	10.15 mg/m^3 at 760 mm Hg, 25°C
1 mg/l ≏	98.5 ppm at 760 mm Hg, 25°C

31.2 Physiologic Response

Summary (61). Tetraethylene glycol monovinylethyl ether is low in single-dose oral toxicity; the LD_{50} value for rats is 6.17 ml/kg. It may cause mild eye irritation and only minor skin irritation, being listed 3 and 2, respectively, on a scale of 10 in rabbit tests. It is not likely to pose a health hazard from skin absorption, since the LD_{50} value for rabbits is 6.35 ml/kg. None of the rats exposed for 8 hr to the vapors of this material generated at room temperature died. This may be due in part to its low volatility and to its low degrees of toxicity. There are no data reported on the effects of repeated vapor exposure. The practice of reasonable safety precautions should be adequate in most industrial operations. However, because of the lack of data, it would seem wise to avoid repeated prolonged vapor exposures and to avoid exposures if the material is handled hot or if mists or fogs are generated.

32 TETRAETHYLENE GLYCOL MONOPHENYL ETHER; Ethanol, 2-{2-[2-(2'-phenoxyethoxy)ethoxy]ethoxy}-; CAS No. 36366-93-5

$$C_6H_5OCH_2CH_2OCH_2CH_2OCH_2CH_2OCH_2CH_2OH$$

32.1 Physical and Chemical Properties

Appearance	Light yellow liquid
Molecular formula	$C_{14}H_{22}O_5$
Molecular weight	270.2
1 ppm ≏	11.04 mg/m^3 at 760 mm Hg, 25°C
1 mg/l ≏	90.6 ppm at 760 mm Hg, 25°C

32.2 Physiologic Response

Summary (9). Tetraethylene glycol monophenyl ether is low in single-dose oral toxicity; the LD_{50} for rats is in the range of 1.26 to 5.0 g/kg. It is not significantly irritating to the skin nor is it likely to be absorbed through the skin in toxic amounts. However, eye contact may be hazardous for it is capable of causing irritation, moderate corneal damage, and some impairment of vision in rabbits. Vapor inhalation studies were not conducted because of its very low volatility. Based on these limited data, it would appear that tetraethylene glycol monophenyl ether should be handled with care to prevent eye contact. However, skin contact and ingestion should not pose a problem in industrial handling. Until more is known it would be wise to avoid prolonged or repeated exposure to its vapors or mists, especially if it is handled hot.

33   TETRAETHYLENE GLYCOL DIETHYL ETHER; 3,6,9,12,15-Pentaoxyheptadecane; Diethoxytetraethylene Glycol; CAS No. 4353-28-0

$$CH_3CH_2O(CH_2CH_2O)_4CH_2CH_3$$

33.1   Physical and Chemical Properties

Appearance	Colorless liquid
Molecular formula	$C_{12}H_{26}O_5$
Molecular weight	250.38
1 ppm ≈	10.24 mg/m^3 at 760 mm Hg, 25°C
1 mg/l ≈	97.7 ppm at 760 mm Hg, 25°C

33.2   Physiologic Response

Summary (51).   Tetraethylene glycol diethyl ether is low in single-dose oral toxicity, the LD$_{50}$ for rats being 4.29 ml/kg. It is only slightly irritating to the skin and eyes and is not likely to be absorbed through the skin in acutely toxic amounts, since the LD$_{50}$ value for rabbits by absorption is 6.35 ml/kg. A single 8-hr exposure of rats to the essentially saturated vapors resulted in no significant adverse effects from inhalation. It would appear that tetraethylene glycol diethyl ethers should pose no significant health hazard from ordinary industrial handling. However, it would seem wise to avoid repeated prolonged exposure to vapors of this material, especially when it is handled hot, until more is known concerning the hazard from such repeated exposure.

34.   POLYETHYLENE GLYCOL METHYL ETHERS; CAS No. 9004-74-4

34.1   Physiologic Response

Summary (8).   Studies on polyethylene glycol methyl ethers with average molecular weights of 350, 550, and 750 indicate that they are low in single dose oral toxicity, the LD$_{50}$ value for rats being 22 ml/kg, 39.8 ml/kg, and 39.8 ml/kg respectively. All caused at most only minor irritation, possibly minor transient corneal injury when tested in the eyes of rabbits. Skin irritation tests using rabbits resulted in no more than a trace of irritation. None of these products was absorbed through the skin to any appreciable extent, the LD_{50} by absorption being greater than 20 ml/kg for all three. Because of their very low degree of toxicity and their low volatility, it would seem reasonable to assume that these products are not likely to present a health hazard from inhalation under anticipated industrial handling.

35 ETHERS OF MONO-, DI-, TRI-, AND POLYPROPYLENE GLYCOL

35.1 Source, Uses, and Industrial Exposure

The ethers of mono-, di-, tri-, and polypropylene glycol generally are prepared commercially by reacting propylene oxide with the alcohol of choice in the presence of a catalyst. They also may be prepared by direct alkylation of the selected glycol with an appropriate alkylating agent such as a dialkyl sulfate in the presence of alkali. Preparation under commercial conditions yields products that are mixtures of the alpha and beta isomers, largely alpha.

The methyl and ethyl ethers of these propylene glycols are miscible with water and a great variety of organic solvents. The butyl ethers have limited water solubility but are miscible with most organic solvents. This mutual solvency in water and oils makes some of these materials exceptionally useful as coupling and dispersing agents, as solvents for lacquers, paints, resins, dyes, oils, and greases. The methyl ether of dipropylene glycol is used in various cosmetics as a solvent and dispersing agent. The butyl ethers of polypropylene glycols 400 and 800 are sometimes used as fly repellents.

Industrial exposure may occur by any of the common routes, but the hazards would not seem to be great except under the most adverse conditions. The toxicologic information, experience, and hazards of each individual compound are discussed separately.

35.2 Physical and Chemical Properties

The known physical properties of the most common ethers are given in Tables 51.5 and 51.6. The chemical composition, as far as is known, is described as follows.

The monoalkyl ethers of propylene glycol appear in two isomeric forms, the alpha, or 1-alkyloxy-2-propanol ($ROCH_2CHOHCH_3$), and the beta, or 2-alkyloxy-1-propanol ($HOCH_2CHORCH_3$). A commercial product Dowanol PM glycol ether is a mixture of the two isomers consisting of at least 95 percent alpha, with the balance the beta isomer. The monoalkyl ethers of dipropylene glycol presumably can appear in four isomeric forms. The commercial product Dowanol DPM glycol ether is believed to be a mixture of these but to consist to a very large extent of the isomer in which the alkyl group has displaced the

**Table 51.5.   Physical and Chemical Properties of Some Common Ethers of
Propylene Glycol**

	Ether				
Property	Methyl	Ethyl	Isopropyl	n-Butyl	Phenyl
CAS No. (1-alkoxy isomer)	107-98-2	1569-02-4	3944-36-3	5131-66-8	4169-04-4
CAS No. (alkoxy position unspecified)	1320-67-8	52125-53-8	29387-84-6	29387-86-8	
Molecular formula	$C_4H_{10}O_2$	$C_5H_{12}O_2$	$C_6H_{14}O_2$	$C_7H_{16}O_2$	$C_9H_{12}O_2$
Molecular weight	90.1	104.1	118.2	132.2	152.1
Specific gravity (25/4°C)	0.917	0.896	0.875	0.879	1.059
Boiling point (°C) (760 mm Hg)	119.6	131–134	139–141	169–172	242.7
Freezing point (°C)	− 100				11.4
Vapor pressure (mm Hg) (25°C)	11.8	8.2	5.3	1.4	<1.0
Refractive index (25°C)	1.402	1.404	1.405	1.415	1.522
Flash point (°F) (O.C)	100	130	140	150	264
Vapor density (air = 1)	3.11				5.25
Percent in saturated air (25°C)	1.55	1.08	0.70	0.18	
1 ppm ≏ mg/m³ at 25°C, 760 mm Hg	3.68	4.25	4.83	5.41	6.22
1 mg/1 ≏ ppm at 25°C, 760 mm Hg	272	235	207	185	160.8

hydrogen of the primary hydroxy group of the dipropylene glycol; the internal
ether linkage is between the secondary carbon of the alkyl etherized propylene
unit and the primary carbon of the other propylene unit, thus leaving the
remaining secondary hydroxy group unsubstituted. They are usually designated
by the formula $ROC_3H_6OC_3H_6OH$. The monoalkyl ethers of tripropylene
glycol can appear in eight isomeric forms. The commercial product Dowanol
TPM glycol ether, however, is believed to be a mixture of isomers consisting
largely of the one in which the alkyl group displaces the hydrogen of the
primary hydroxyl group of the tripropylene glycol and the internal ether
linkages are between secondary and primary carbons. They may be represented
by the formula $ROC_3H_6OC_3H_6OC_3H_6OH$.

35.3   Determination in the Atmosphere

Refer to the introduction of this chapter.

Table 51.6. Physical and Chemical Properties of Ethers of Di- and Tripropylene Glycols and Butylene Glycol

Property	Dipropylene Glycol Ethers			Tripropylene Glycol Ethers			Butylene Glycol		
	Methyl	Ethyl	Butyl	Methyl	Ethyl	Butyl	Methyl	Ethyl	n-Butyl
CAS No.	34590-94-8	15764-24-6	29911-28-2	20324-33-8 25498-49-1	20178-34-1	57499-93-1	53778-73-7 111-32-0	111-73-9	None found
Molecular formula	$C_7H_{16}O_3$	$C_8H_{18}O_3$	$C_{10}H_{22}O_3$	$C_{10}H_{22}O_4$	$C_{11}H_{24}O_4$	$C_{13}H_{28}O_4$	$C_5H_{12}O_2$	$C_6H_{14}O_2$	$C_8H_{18}O_2$
Molecular weight	148.2	162.2	190.2	206.3	220.3	248.3	104.1	118.2	146.2
Specific gravity (25/4°C)	0.948	0.927		0.961			0.983 (25/25°C)	0.888 (25/25°C)	0.877 (25/25°C)
Boiling point (°C) (760 mm Hg)	189.6	193–195	214–217	242.8	250±	255±	136	147	180–187
Freezing point (°C)	−80			−60					
Vapor pressure (mm Hg) (25°C)	0.38	0.30	0.06	0.017	0.011	0.008	5.5	3.0	0.62
Refractive index (25°C)	1.419	1.421	1.418	1.428	1.426	1.424	1.408	1.410	1.420
Flash point (°F) (open cup)	185	205		2.60			110	145	160
Percent in saturated air (25°C)	0.051	0.04	0.008	0.003	0.0014	0.001	0.72	0.40	0.081
1 ppm ⇌ mg/m³ at 25°C, 760 mm Hg	6.06	6.64	7.77	8.44	9.00	10.1	4.25	4.83	5.98
1 mg/l ⇌ ppm at 25°C, 760 mm Hg	165	150.7	128.6	118.5	111	98.6	235	207	167

36 PROPYLENE GLYCOL MONOMETHYL ETHER; 2-Propanol, 1-Methoxy-; Dowanol®
PM Glycol Ether; Propasol® Solvent M; Poly-solv® MPM Solvent; CAS Numbers 107-98-2
and 1320-67-8

$$CH_3OCH_2CHOHCH_3$$

(3-Methoxy-1-propanol is an isomer but is of little or no commercial significance.)

36.1 Physical and Chemical Properties

See Table 51.5.

36.2 Physiologic Response

36.2.1 Summary

Commercial propylene glycol monomethyl ether is low in both single- and
repeated-dose oral toxicity, transiently painful to the eyes, not appreciably
irritating to the skin, but can be absorbed through the skin in toxic amounts if
exposure is extensive and prolonged. The vapors are low in toxicity and the
hazard from inhalation also is low because acutely toxic concentrations are
essentially intolerable to humans, and concentrations that might cause effects
from repeated exposures are very disagreeable (irritating to the eyes and
mucous membranes and nauseating to some persons). The primary effect of
the material is that of an anesthetic agent.

The 3-methoxy-1-propanol isomer is similar in toxicity to the other propanol
glycol methyl ethers.

36.2.2 Single-Dose Oral

The single-dose oral toxicity for propylene glycol monomethyl ether has been
reported by several workers (see Table 51.7).

Rowe et al. (110) found that deaths in rats from massive doses were associated
with profound central nervous system depression. Shideman and Procita (111)
attributed the deaths in dogs to respiratory arrest and point out that if the
acute effects, which may last for as long as 48 hr, depending on dosage, are
survived there are no residual effects. Stenger et al. (112) report that the effects
from a single dose by ingestion include dyspnea, somnolence, and ataxia.

It is interesting to note that the alpha and beta isomers, 1-methoxy-2-propanol
and 2-methoxy-1-propanol, respectively, are both low in toxicity and in a similar
range.

Table 51.7. Single-Dose Oral Toxicity for Propylene Glycol Monomethyl Ether

Author	Ref.	LD_{50} (g/kg of Body Weight)				Sample Tested
		Rat	Mouse	Dog	Rabbit	
Rowe et al.	110	6.6				Commercial grade
Smyth et al.	62	7.51				Alpha isomer
		5.71				Beta isomer
Shideman and Procita	111			9.2 ± estimate		
Stenger et al.	112		10.8	4.6 to 5.5	5.3	
Smyth et al.	61	5.20				
Smyth et al.	51	11.9 estimate				3-Methoxy-1-propanol

36.2.3   Repeated-Dose Oral

Rowe et al. (110) report that groups of rats that received 26 doses of 1.0 g/kg or less over a 35-day period showed no ill effects as judged by appearance, growth, organ weights, and histopathologic examination of the organs. Under the same test conditions, 3.0 g/kg of this material produced only minor effects in the liver and kidney.

Stenger et al. (112) administered propylene glycol monomethyl ether five times per week for 13 weeks to rats and for 14 weeks to dogs. The rats received doses of 0.5, 1.0, 2.0, and 4.0 ml/kg/day and the dogs 0.5, 1.0, and 2, and 3 ml/kg/day. Both the dogs and rats experienced mild to severe central nervous system depression in a dose-related manner. In the rat, this caused a growth depression because of reduced food intake. In addition the livers of the rats became enlarged, especially at doses greater than 1.0 ml/kg/day. This was accompanied by cell necrosis in the liver mainly in the peripheral parts of the lobules. Appreciable mortality occurred at the 4.0 ml/kg dosage level. Male dogs developed numerous spermiophages in the epididymis. The meaning of this is not clearly understood. In both the rats and dogs, there was minor kidney injury at the higher doses.

36.2.4   Single-Dose Injection

Unpublished data from Union Carbide (8) give the following toxicity information on propylene glycol monomethyl ether:

Species	Route	LD_{50} (g/kg)
Rat	Intraperitoneal	3.72
Rat	Intravenous	5.66
Rabbit	Intravenous	4–8

Stenger et al. (122) report the following data:

Species	Route	LD_{50} (g/kg)
Mouse	Intravenous	4.9
Rat	Intravenous	3.9
Rabbit	Intravenous	1.1
Dog	Intravenous	1.8–2.3
Rat	Intraperitoneal	3.9
Rat	Subcutaneous	7.2
Rabbit	Subcutaneous	4.6

The effects from intravenous injection in dogs were those of pain at the site of injection, shallow breathing, decreased blood pressure, auricular arrhythmia, and death due to convulsions. Based upon these data, it is apparent that propylene glycol monomethyl ether is low in toxicity even when injected.

36.2.5 Eye Contact

Repeated application of one drop of the undiluted material onto the eyeballs of rabbits for 5 days caused only a mild transitory irritation of the eyelids after each dose (110).

Smyth et al. (61) indicate that eye irritation is mild, having a rating of 3 on a scale of 10. On the other hand, 3-methoxy-1-propanol is reported to be capable of causing moderate eye injury, being rated 5 on a scale of 10 (51). Thus propylene glycol methyl ether should not cause a significant problem from eye contact under the usual conditions of manufacture and use.

36.2.6 Skin Contact (110)

Propylene glycol monomethyl ether, when tested for skin irritation on rabbits, failed to cause more than a very mild simple irritation, and that only after constant contact for several weeks. Smyth et al. (61) also report only minor irritation at most.

36.2.7 Skin Absorption

When the ether was applied to rabbits under a "cuff" as advocated by Draize et al. (10), the LD_{50} value was found to be in the range of 13 to 14 g/kg by both Rowe et al. (110) and Smyth et al. (61). Depression and incomplete anesthesia are signs commonly associated with the absorption of acutely toxic quantities of this material, especially at dosage levels above 10 ml/kg. When measured doses of propylene glycol monomethyl ether were bandaged repeatedly onto the clipped abdomens of rabbits over a 90-day period, a significant amount of absorption through the skin occurred. At the high dosage levels of 7 and 10 ml/kg, the material caused narcosis and an increased mortality. At the lower levels of 2 and 4 ml/kg, only mild narcosis was apparent. The only organic injury noted in any case was a slight increase in kidney weights of the animals at the top dosage level (10 ml/kg) (110).

The above skin contact data show that neither of the isomers of propylene glycol monomethyl ether is likely to cause primary irritation. Although absorption through the skin in toxic amounts can occur, the hazard from this route is very low under the usual conditions of use.

36.2.8   Inhalation (110)

Acute vapor studies conducted on rats and guinea pigs have shown that (a) at a concentration of approximately 5000 ppm, rats and guinea pigs survived single 7-hr exposures; (b) at a concentration of approximately 10,000 ppm, the LD_{50} for rats was 5 to 6 hr and for guinea pigs it was greater than 7 hr; (c) at a concentration of approximately 15,000 ppm (saturated, with some mist present), the LD_{50} for rats was 4 hr and for guinea pigs 10 hr; and (d) deaths resulting from single inhalation exposures appeared to be due to anesthetic action.

Rabbits and monkeys subjected to 132 exposures of propylene glycol monomethyl ether at 800 ppm (2.91 mg/l) over a period of 186 days showed no evidence of adverse effects as judged by gross appearance and behavior, growth, final body and organ weights, hematology, and microscopic examination of tissues. Rats and guinea pigs showed no ill effects by the same criteria when they received 130 exposures in a period of 184 days to a concentration of 1500 ppm (5.46 mg/l). The effects observed at higher concentrations were those of slight growth depression and very slight liver and lung effects. A mild central nervous system depression was noted to result from exposure at the start of the experiments at the 3000-ppm level. However, recovery was rapid after cessation of each day's exposure. The animals developed a tolerance after several weeks so that this response was not observed later.

Goldberg et al. (45) exposed rats to concentrations of 2500, 5000, and 10,000 ppm for 4 hr/day, 5 days/week for 2 weeks in order to study the effect of inhalation on behavior. There was a transient nonspecific depression of behavior for the first several exposures to 5000 and 10,000 ppm. However, there was a rapid development of tolerance. Decreased growth rate was seen at the 10,000-ppm level.

The above data on inhalation, both from single and repeated exposures, show that the vapor toxicity of the material is surprisingly low, and the hazard from vapor exposure is considerably less than that presented by most common solvents.

The hazard of acute poisoning from single exposures is believed to be relatively minor; toxic vapor concentrations are close to saturation values, they are extremely disagreeable, if not intolerable, and any harmful effects consist of functional depression (anesthesia) of the nervous system. Certainly there is no serious organic injury. The hazard of chronic poisoning, likewise, is quite low. Levels that may be toxic upon repeated exposure probably will not be tolerated voluntarily.

36.2.9 Teratogenicity

Stenger et al. (112) administered propylene glycol monomethyl ether both by gavage and by subcutaneous injection repeatedly to mice, rats, and rabbits during the first 18 to 21 days of gestation. Doses varied from 0.04 to 2.00 ml/ kg of body weight. Only the rat fetus showed any effect, a delayed ossification of the skull at the highest dose given (0.8 ml/kg). There were no effects on the number of young born. Thus it would appear that, although propylene glycol monomethyl ether may be capable of causing minor fetotoxic effects, these effects are likely to be seen only when exposure to levels causing other adverse effects occur. In addition, the delayed ossification of the skull can be considered at most a minor effect because it is only a delay in a normal developmental process and is not a teratogenic effect.

36.2.10 Mutagenicity and Carcinogenicity

No reports of studies were found.

36.2.11 Mode of Action

Shideman and Procita (111), in studying the pharmacology of the propylene glycol ethers in anesthetized and artificially respiring dogs, discovered that in proper dosage propylene glycol monomethyl ether caused auricular fibrillation. The significance of this insofar as practical use is concerned is doubtful, for they never observed it in intact animals.

36.2.12 Human Experience

There is no unfavorable human experience known other than from the disagreeableness of the odor.

36.3 Hygienic Standards of Permissible Exposure

The American Conference of Governmental Industrial Hygienists in their 1980 listing of Threshold Limit Values recommend a time-weighted average TLV of 100 ppm (360 mg/m^3) for repeated 8-hr human exposure (30).

36.4 Odor and Warning Properties

Propylene glycol monomethyl ether in concentrations of the order of 3000 ppm or more had a very strong odor and caused marked nasal irritation that was exceedingly difficult for humans to tolerate. Concentrations of 1000 ppm have quite an objectionable odor and are not likely to be tolerated voluntarily.

Stewart et al. (113) in controlled human exposures found that levels of propylene glycol monomethyl ether above 100 ppm were objectionable because of odor. Eye, nasal, and throat irritation became objectionable prior to the first indications of central nervous system impairment, which occurred at a level of 1000 ppm. Breath analysis data showed that propylene glycol monomethyl ether was rapidly excreted via the lungs.

The human volunteers all experienced a rapid development of odor tolerance. Hence, unless prompt action is taken when the objectionable odor is experienced, it cannot be relied upon to prevent exposures that might be hazardous. However, because the odor is readily detected and is objectionable, propylene glycol monomethyl ether vapors are considered to have adequate warning properties, if heeded.

37 PROPYLENE GLYCOL MONOETHYL ETHER; 2-Propanol, 1-Ethoxy-; Dowanol® PE Glycol Ether; CAS No. 1569-02-4 and CAS No. 52125-53-8

$$C_2H_5OCH_2CHOHCH_3$$

(3-Ethoxy-1-propanol is an isomer but is of little or no commercial significance.)

37.1 Physical and Chemical Properties

See Table 51.5.

37.2 Physiologic Response

37.2.1 Summary

Commercial propylene glycol monoethyl ether is low in single-dose oral toxicity, somewhat irritating to the eyes, and not appreciably irritating to the skin. Although it can be absorbed through the skin, this is not likely to be a hazard in ordinary handling and use. Its vapors do not appear to be particularly toxic and concentrations dangerous in short periods (hours) are strongly irritating and narcotic. Data adequate to allow the establishment of safe limits for repeated exposure are not available. The overall hazard associated with handling and use would seem to be slight.

The 3-ethoxy-1-propanol isomer is similar in toxicity to the commercial propylene glycol monoethyl ether except that it appears to be somewhat more injurious to the eyes and more toxic by percutaneous absorption (51).

37.2.2   Single-Dose Oral

Smyth et al. (62) fed the two propylene glycol monoethyl ethers, alpha and beta, as 50 percent aqueous solutions to rats and found the LD_{50}s to be 7.11 and 7.00 g/kg, respectively. Unpublished studies (9) on a commercial product, Dowanol PE glycol ether, yielded an oral LD_{50} value for rats of more than 5.0 g/kg. Marked narcosis and some kidney injury were observed from large doses in both studies.

37.2.3   Repeated-Dose Oral

According to Smyth and Carpenter (98), who administered the beta isomer to rats in their drinking water for 30 days, a daily dose of 0.68 g/kg was without adverse effect, whereas a daily dose of 2.14 g/kg caused reduced growth. Reference to this same article later by Smyth (51) indicates that kidney injury, but no death, also resulted from the 2.14 g/kg daily dosage level.

37.2.4   Eye Contact (9)

When the commercial product was applied to the eyes of rabbits on five consecutive days, it caused moderate conjunctival irritation and some transient cloudiness of the cornea. Healing was essentially complete in 3 to 7 days.

37.2.5   Skin Contact (9)

When the material was confined to the skin of rabbits in a manner essentially the same as described by Draize et al. (10), all of six animals treated survived 5 ml/kg, three of five survived 7 ml/kg, and one of five survived doses of either 10 or 15 ml/kg. The LD_{50} was estimated to be about 9 ml/kg. The material in large doses caused marked central nervous system depression and deaths usually occurred within 48 hr after treatment. No appreciable irritation of the skin resulted.

37.2.6   Inhalation

Gross (7) reports that a mouse, guinea pig, and rabbit tolerated up to 7000 ppm for 1 hr without effect other than irritation of the eyes and respiratory organs. A 2-hr exposure caused more irritation and a rabbit showed signs of kidney injury, transient albumin, and red cells in the urine. All of five rats exposed for

4 hr to a concentration calculated to be 10,000 ppm survived, but they showed marked irritation of the eyes and nares, and they were deeply anesthetized at the end of the exposure (9). Gross (7) also reported exposing cats, guinea pigs, and rabbits 8 hr/day to a concentration of about 1200 ppm. One of two cats and one of two guinea pigs tolerated 12 such exposures without apparent effect but the other cat and guinea pig died after the treatment. The two rabbits succumbed after 3 or 9 days; necropsy revealed pneumonia and kidney injury.

37.2.7   Human Experience

No adverse human experience has been reported.

37.3   Hygienic Standard of Permissible Exposure

Data adequate to establish a hygienic standard for repeated exposure are not available.

38   PROPYLENE GLYCOL-*n*-MONOPROPYL ETHER; 2-Propanol, 1-Propoxy-; Propasol® Solvent P; CAS No. 1569-01-3 (1-Alkoxy Isomer) and CAS No. 30136-13-1 (Alkoxy Position Unspecified)

$$CH_3CH_2CH_2OCH_2CHOHCH_3$$

38.1   Physical and Chemical Properties

Molecular formula	$C_6H_{14}O_2$
Molecular weight	118.2
1 ppm ≈	4.83 mg/m^3 at 760 mm Hg, 25°C
1 mg/l ≈	207 ppm at 760 mm Hg, 25°C

38.2   Physiologic Response

Summary.   Propylene glycol-*n*-monopropyl ether has a low single-dose oral toxicity, for the LD_{50} value for rats is in the range of 2.8 to 3.0 g/kg. At high doses the animals developed central nervous system depression and some evidence of kidney injury. The animals that died did so within 24 hr (9). In the eye tests on rabbits, the product was found to cause moderate to marked irritation and moderate to marked corneal injury and slight iritis. Healing was essentially complete in a week (9). It should cause, at most, only minor irritation when in prolonged or repeated skin contact (8, 9). However, it may cause some local injury to abraded skin if contact is prolonged (9). It is low in toxicity by skin absorption, the LD_{50} for rabbits being 3.55 g/kg (8). Essentially saturated vapors generated at room temperature when inhaled by rats for 8 hr (8) or for 7 hr (9) caused no deaths and no significant adverse effects (estimated concentration 2300 to 2500 ppm). The rats exposed to vapors generated at 100°C likewise showed no significant ill effects and all survived (estimated concentration 6700 ppm) (9).

Based upon these data, propylene glycol-mono-*n*-propyl ether may be considered to present no unusual health hazards in anticipated industrial operations. Precautions should be exercised to avoid eye contact. It would seem wise, until more is known, to avoid repeated inhalation of vapor.

39 PROPYLENE GLYCOL MONOISOPROPYL ETHER; 2-Propanol, 1-Isopropoxy-; CAS No. 3944-36-3 and CAS No. 29387-84-6

$$(CH_3)_2CHOCH_2CHOHCH_3$$

39.1 Physical and Chemical Properties

See Table 51.5.

39.2 Physiologic Response

Summary (9).   Propylene glycol monoisopropyl ether is low in single-dose oral toxicity for rats, the LD_{50} being greater than 2 g/kg and in the neighborhood of 4 g/kg. In the eyes of rabbits, it causes conjunctival irritation, some corneal injury, and iritis which heals within a week. It is only very slightly irritating even upon prolonged and repeated contact with the intact skin, and does not seem to be readily absorbed in acutely toxic amounts. Rats that received a single 7-hr exposure to an essentially saturated atmosphere survived but exhibited drowsiness, labored breathing, temporary weight loss, and mild kidney injury at autopsy.

Based upon these data, propylene glycol monoisopropyl ether may be considered to present no unusual health hazards in anticipated industrial operations. Precautions should be exercised to avoid eye contact. It would seem wise, until more is known, to avoid repeated inhalation of vapor.

40 PROPYLENE GLYCOL MONO-*n*-BUTYL ETHER; 1-Propanol, 2-Butoxy-; Propasol® Solvent P; CAS No. 5131-66-8 and CAS No. 29387-86-8

$$C_4H_9OCH_2CHOHCH_3$$

40.1 Physical and Chemical Properties

See Table 51.5.

40.2 Physiologic Response

40.2.1 Summary

Propylene glycol mono-*n*-butyl ether is low in single-dose oral toxicity, is markedly irritating and somewhat injurious to the eyes, is mildly irritating to the skin, may be adsorbed through the skin but is low in toxicity by this route, and is not particularly hazardous from an occasional exposure from inhalation. However, repeated exposure may lead to liver injury, especially at high levels. It is considered to have no significant teratogenic properties. Thus it appears to present no unusual health hazards from industrial handling.

40.2.2 Single-Dose Oral

The toxicity of propylene glycol mono-*n*-butyl ether has been shown to be low by various reports. The LD_{50} value for rats has been reported to be 1.9 (9), 2.5 (60), and 5.2 g/kg (51).

40.2.3 Eye Contact

When studied on rabbits, it was found to be appreciably irritating to the eyes; one drop in an eye on five consecutive days caused marked conjunctival irritation and corneal cloudiness, which healed within a week (9). Similar findings are reported by Carpenter et al. (60) and Smyth et al. (51).

40.2.4 Skin Contact

Repeated applications (10 in 14 days) to the skin of rabbits resulted in slight irritation and some evidence that toxic amounts were absorbed (9). Both Carpenter et al. (60) and Smyth et al. (51) also report that propylene glycol mono-*n*-butyl ether caused only mild irritation.

40.2.5 Skin Absorption

Single-dose absorption studies conducted essentially as described by Draize et al. (10) showed that all of five animals receiving 1.8 g/kg survived, two of five receiving 2.6 g/kg survived, and none of five receiving 4.4 g/kg survived. When the material was confined under a cuff for 24 hr, as in these studies, rather severe injury to the skin occurred and the animals became deeply narcotized.

Deaths from the larger doses occurred within a few hours after application of the material. All deaths occurred within 24 hr after treatment or not at all (9). Carpenter et al. (60) in their tests found an LD_{50} by absorption for rabbits of 3.1 g/kg; Smyth et al. (51) found 1.4 g/kg.

40.2.6   Inhalation

Carpenter et al. (60) and Smyth et al. (51) have shown that rats exposed for 8 hr to essentially saturated vapor of propylene glycol mono-*n*-butyl ether generated at room temperature were essentially unaffected.

In repeated inhalation studies, rats were exposed to 600 ppm as a metered concentration of vapor for 7 hr/day for 31 days. The only effects seen were increased liver weights in the female rats only (8).

40.2.7   Teratogenicity (9)

Pregnant Swiss Webster mice were treated from day 6 to 15 of gestation either by ingestion or by subcutaneous injection; the highest levels were 226.4 and 151.6 mg/kg/day by ingestion and by injection, respectively. On day 17 the mice were sacrificed and examined. There were no effects in any of the mice as shown by maternal body weight, mortality or behavior, reproduction, fetal body weights, and fetal development as shown by skeletal, internal, and external examinations.

40.2.8   Human Experience

There has been no adverse human experience reported. It would seem, however, that propylene glycol mono-*n*-butyl ether should be handled in such a way that contact with the eyes and prolonged or repeated contact with the skin are prevented.

40.3   Hygienic Standard of Permissible Exposure

Until more extensive data regarding the vapor inhalation toxicity become available, no hygienic standard can be suggested. In the meantime, it would seem prudent to avoid prolonged or repeated repeated inhalation of vapors in concentrations which cause discomfort.

41   PROPYLENE GLYCOL MONOETHYLBUTYL ETHER MIXED ISOMERS; 2-
Ethylbutoxypropanol Mixed Isomers; Propylene Glycol, Monohexyl Ether Mixed Isomers; CAS No. 26447-42-7

$$C_6H_{13}OCH_2CHOHCH_3 \text{ mixed isomers}$$

41.1 Physical and Chemical Properties

Molecular formula	$C_9H_{20}O_2$
Molecular weight	160.2
1 ppm ≈	5.40 mg/m^3 at 760 mm Hg, 25°C
1 mg/l ≈	185 ppm at 760 mm Hg, 25°C

41.2 Physiologic Response

Summary (51).   Propylene glycol monoethylbutyl ether mixed isomers presents a serious hazard from eye contact, but no significant hazard from skin contact, skin absorption, single-dose ingestion, or an occasional exposure to its vapors generated at room temperature. The LD_{50} for rats from a single oral dose is 3.56 ml/kg. Eye tests on rabbits indicate that it can cause severe irritation and damage; it is rated 9 on a scale of 10. Skin irritation tests on rabbits indicate that it is only very slightly irritating, being rated 2 on a scale of 10. The LD_{50} for rabbits by skin absorption is 6.0 ml/kg. None of the rats exposed for 8 hr to essentially saturated vapors died. Precautions therefore should be taken to prevent eye contact, but reasonable care and cleanliness should be adequate to avoid skin difficulties. It would also seem wise, until more is known, to avoid exposure to vapors if the material is handled hot or if mists or fogs are generated.

42 PROPYLENE GLYCOL BUTOXYETHYL ETHER; 2-Propanol, 1-Butoxyethoxy-;
Propasol® Solvent BEP; CAS No. 124-16-3

$$C_4H_9OC_2H_4OCH_2CHOHCH_3$$

42.1 Physical and Chemical Properties

Molecular formula	$C_9H_{20}O_3$
Molecular weight	176.1
1 ppm ≈	7.20 mg/m^3 at 760 mm Hg; 25°C
1 mg/l ≈	138.8 ppm at 760 mm Hg, 25°C

42.2 Physiologic Response

Summary (8).   Propylene glycol butoxyethyl ether is a clear, colorless liquid of low volatility. The toxicologic information available is on a commercial-grade material and is not extensive. The single-dose oral toxicity should be considered low, since the LD_{50} for rats has been reported as 4.0 and 5.66 ml/kg. A single drop when applied to the eyes of a rabbit caused moderate to severe irritation and corneal injury. Skin contact resulted in minor irritation on rabbits. It can be absorbed through the skin in toxic amounts; however, its toxicity is low by this route, since the LD_{50} for rabbits is in the range of 2.83 to 3.00 ml/kg. Concentrated vapors generated at about 19°C killed none of the rats exposed for 8 hr. Studies on the red blood cells of these exposed rats showed no alteration of the fragility of the cells. The estimated level of exposure is about 0.25 mg/l or about 35 ppm.

Based upon these data, propylene glycol-mono-*n*-propyl ether may be considered to present no unusual health hazards in anticipated industrial operations. Precautions should be exercised to avoid eye contact. It would seem wise, until more is known, to avoid repeated inhalation of vapor.

43   PROPYLENE GLYCOL PHENYL ETHER; 2-Propanol, 1-Phenoxy-; Dowanol® PPh Glycol Ether; CAS No.4169-04-4

$$C_6H_5OCH_2CHOHCH_3$$

43.1   Physical and Chemical Properties

See Table 51.5.

43.2   Physiologic Response

Summary (9).   Propylene glycol phenyl ether as tested is a commercial-grade liquid product that is clear and colorless and has a slight odor described as similar to "rotten oranges." The toxicologic information is not extensive. It is considered low in single-dose oral toxicity; the LD_{50} for male rats has been calculated to be 2.83 g/kg, and for female rats 3.73 g/kg. The main effect seen in these rats was central nervous system depression. Death usually occurred in 48 hr or less. When two drops of the liquid were applied to the eyes of rabbits, some initial pain, slight irritation, and slight corneal injury occurred. The eyes healed in several days to one week. Essentially continuous skin contact for 2 weeks under a cloth covering caused only mild irritation to rabbits. Apparently it is low in toxicity by absorption through the skin from a single 24-hr contact, for the LD_{50} for rabbits was found to be greater than 2.0 g/kg, the highest level tested. Exposure of rats for 7 hr to essentially saturated vapor generated at room temperature resulted in no significant adverse effects. When exposed for 7 hr to vapor generated at 100°C, the rats experienced, during exposure, significant nasal and respiratory irritation as shown by difficult and jerky breathing as well as eye irritation. However, none of the rats died as a result of the exposure.

Based upon these data, propylene glycol-monophenyl ether may be considered to present no unusual health hazards in anticipated industrial operations. Precautions should be exercised to avoid eye contact. It would seem wise, until more is known, to avoid repeated inhalation of vapor.

44 DIPROPYLENE GLYCOL MONOMETHYL ETHER; 2-Propanol, 1-(2-Methoxy-2-methylethoxy)-; Dowanol® DPM Glycol Ether; Ucar® Solvent 2LM; Propasol® Solvent DM; Poly-solv® DPM Solvent; CAS No. 34590-94-8

$$CH_3CHOHCH_2OCH_2CH(OCH_3)CH_3$$

44.1 Physical and Chemical Properties

See Table 51.6.

44.2 Physiologic Response

44.2.1 Summary

Dipropylene glycol monomethyl ether is low in single-dose oral toxicity, transiently painful but not damaging to the eyes, and is neither appreciably irritating to the skin nor readily absorbed through the skin of rabbits in toxic amounts when exposures are prolonged and repeated.

It caused neither irritation nor sensitization when tested on human subjects. It is low in toxicity by inhalation. The hazards to health associated with the handling and ordinary use of this material would seem to be minimal.

44.2.2 Single-Dose Oral

The single-dose oral LD_{50} for male and female rats was found to be 5.50 and 5.45 ml/kg (5.22 and 5.18 g/kg), respectively (110). The material produces marked central nervous system depression. Other values reported are 5.66 ml/kg (5.43 g/kg) for rats and 7.5 g/kg for dogs (8). Shideman and Procita (111) estimated the single-dose oral LD_{50} for dogs to be 7.5 ml/kg. They noted that death was due to respiratory failure and usually occurred within 48 hr or not at all.

44.2.3 Repeated-Dose Oral

Browning (17) reports that rats given dipropylene glycol methyl ether at a dosage of 1.0 g/kg for 35 days showed no signs of adverse effects.

44.2.4 Injection Toxicity

The LD_{50} by intravenous injection to anesthesized dogs is 0.35 to 0.5 ml/kg (0.33 to 0.47 g/kg), and for artificially respired dogs 1.3 ml/kg (1.23 g/kg) (17).

44.2.5 Eye Contact (110)

Dipropylene glycol monomethyl ether is not appreciably irritating to the eyes. When one drop of undiluted material was placed in a rabbit's eye on each of five consecutive days, a mild transitory irritation of the conjunctival membranes occurred. Fluorescein staining revealed no corneal damage.

44.2.6 Skin Irritation (110)

Continuous contact of dipropylene glycol monomethyl ether with the skin of numerous rabbits for 90 days caused only a very slight scaliness, far less in intensity than might have been expected on the basis of the solvent properties of the material. In fact, the response was similar to that produced by water alone under the same conditions. When patch-tested on 250 human beings by accepted techniques, it produced no evidence of primary irritation or sensitization of the skin.

44.2.7 Skin Absorption (110)

Single application studies conducted essentially as described by Draize et al. (10) revealed that dipropylene glycol monomethyl ether is not absorbed through the skin in acutely dangerous amounts even when massive doses (20 ml/kg; 19.0 g/kg) are held in continuous contact with a large area of the rabbit's skin for a period of 24 hr. Sufficient absorption did occur, however, to result in transient narcosis. Others have reported LD_{50} values for rabbits of 13 to 14 g/kg (17) and 10.6 ml/kg (10.1 g/kg) (8).

When dipropylene glycol monomethyl ether was applied five times a week for 90 days at dosage levels of 1.0, 3.0, 5.0, and 10.0 ml/kg, the following observations were made (110):

1.   Mortality was high at the 10.0 ml/kg dosage level, slight at 5.0 ml/kg level, and absent at the 1.0 and 3.0 ml/kg levels.
2.   No adverse body weight occurred at any level except just prior to death in those animals that succumbed, presumably to the narcotic effects of the top dosage levels.
3.   No hematologic changes occurred at any dosage level.
4.   The effect of severe (repeated and prolonged) exposure to the skin was slight, being similar to that caused by distilled water under similar conditions.

5. Observations for gross pathology revealed only gastric distension and occasional gastric irritation in those animals dying at the 10 ml/kg dosage level.

6. No significant organ weight changes occurred at any dosage level.

7. The blood urea nitrogen concentration was unaffected in the animals surviving the 3.0 and 5.0 ml/kg dosage level.

8. Histopathologic studies conducted on the liver, lung, spleen, adrenal, heart, testes, and stomach of those animals receiving the 5.0 and 10.0 ml/kg dosage levels revealed no changes. The kidneys of those animals on the 10.0 ml/kg level showed some granular and some hydropic changes; at the 5.0 ml/kg level these same kidney abnormalities were observed but they were of no greater intensity than those observed in some of the controls.

44.2.8 Inhalation (110)

Three groups of three adult male white rats were given single 7-hr exposures to an atmospheric concentration of dipropylene glycol monomethyl ether calculated to be about 500 ppm. This atmosphere was laden with fog and the animals were wet with the material at the end of the exposure. They exhibited only mild narcosis, from which they rapidly recovered.

Groups of rats, rabbits, guinea pigs, and monkeys were exposed 7 hr/day, 5 days a week for periods of from 6 to 8 months to an atmosphere containing about 300 ppm (essentially saturated) of the material. When judged by growth, general appearance, hematologic studies, gross observations at autopsy, and organ weight studies, only the rats exhibited a mild transitory narcosis during the first few weeks of exposure and a statistically significant but very slight increase in the weights of their livers. The other species failed to exhibit any abnormalities. The results of histopathologic examination revealed no evidence of adverse effect except for minor changes in the livers of the female guinea pigs, rabbits, and monkeys. It should be noted that this concentration of vapor is quite disagreeable to human subjects and it is doubtful if it would be tolerated voluntarily.

44.2.9 Pharmacology

Shideman and Procita (111) in 1951 reported rather extensive studies on dipropylene glycol monomethyl ether. In the intact dog, they found the material to be primarily a depressant of the central nervous system with death being due to respiratory failure. In the anesthetized dog, they observed that intravenous injection caused a precipitous drop in blood pressure. This response was unaffected by the prior administration of atropine and by bilateral vagotomy. They also noted that in the anesthetized and artificially respired dog, proper doses of the material intravenously induced auricular, but not ventricular, fibrillation. Rucknagel and Surtskin (114) in 1952 confirmed the observations of Shideman and Procita (111). The practical significance of these findings is questionable since it has never been observed in the intact animal, regardless of the dose or mode of administration.

44.2.10   Human Experience

No injury or adverse effects have been reported from the handling and use of dipropylene glycol monomethyl ether.

44.3   Hygienic Standards of Permissible Exposure

In view of the data and experience available, it would appear that if vapor concentrations were controlled to levels not disagreeable to unacclimated human subjects, there would be no hazard from inhalation. In 1961 The Conference of Governmental Industrial Hygienists recommended a threshold limit value of 100 ppm. This was still the accepted limit in 1980 (30) with a Skin notation. OSHA also recommends 100 ppm as a standard for control (31).

44.4   Odor and Warning Properties

In the inhalation studies (110) the level of 300 ppm was reported to be quite disagreeable to humans. Further studies (9) done with only five unacclimated human volunteers under controlled laboratory conditions suggest the following levels as possible guides for warning properties:

   Probable minimum concentration that may cause minor nasal irritation about 35 ppm.
   Probable minimum concentration that may cause some tolerable eye, throat, and respiratory irritation about 75 ppm.
   Lowest concentration at which odor was rated intolerable about 80 ppm.

Based on these findings, it is possible that an occasional person may find the vapor of dipropylene glycol monomethyl ether intolerable at the recommended TLV.

45   DIPROPYLENE GLYCOL MONOETHYL ETHER; 2-Propanol, 1-(2-Ethoxy-2-methylethoxy)-; CAS No. 15764-24-6

$$CH_3CHOHCH_2OCH_2CH(OCH_2CH_3)CH_3$$

45.1   Physical and Chemical Properties

See Table 51.6.

45.2 Physiologic Response

45.2.1 Summary (9)

Dipropylene glycol monoethyl ether is low in single-dose oral toxicity, the LD_{50} for rats being 4 ml/kg (3.71 g/kg). When one drop of the liquid was introduced into the eyes of rabbits for five consecutive days, it caused a slight transitory conjunctival irritation but no corneal injury. Repeated applications (10 in 14 days) to the skin of rabbits resulted in only a very slight exfoliation and there was no evidence that toxic amounts were absorbed. However, when single doses of the material were confined to the skin for 24 hr under a cuff, essentially as described by Draize et al. (10) all of five animals receiving 5 ml/kg (4.63 g/kg) survived, three of ten receiving 10 ml/kg (9.27 g/kg) survived, and none of five receiving 15 ml/kg (13.9 g/kg) survived. All animals exhibited transient weight loss. Narcosis was apparent in all animals but was profound at the higher dosage levels. The animals were cold to the touch and the skin beneath the cuff was burned. Death occurred in almost all cases within 1 or 2 days and recovery was apparently complete in the survivors within 3 days. Rats were exposed to atmospheres essentially saturated (calculated to be about 400 ppm) for 7 hr. One of 12 animals so exposed died. The others evidenced some degree of irritation of the eyes and nares, transient weight loss, but recovery appeared complete 24 hours later.

45.2.2 Human Experience

There have been no reports of adverse human experience.

45.3 Hygienic Standards of Permissible Exposure

Until more extensive data regarding the vapor inhalation toxicity become available, no hygienic standard can be suggested. It would seem, however, that under reasonable conditions of handling at room temperature, there would be no appreciable hazards.

46 DIPROPYLENE GLYCOL MONO-*n*-BUTYL ETHER; 2-Propanol, 1-(2-Butoxy, 2-methylethoxy)-; CAS No. 29911-28-2

$$CH_3CHOHCH_3OCH_2CH(OCH_2CH_2CH_2CH_3)CH_3$$

46.1 Physical and Chemical Properties

See Table 51.6.

46.2 Physiologic Response

46.2.1 Summary (9)

Dipropylene glycol mono-*n*-butyl ether is low in single oral doses to rats, the LD_{50} being 2 ml/kg. When one drop of the material was introduced into the eye of a rabbit, on five consecutive days, it produced only a transient, slight conjunctival irritation but no corneal injury. Repeated applications (10 in 14 days) to the skin of rabbits resulted in very slight simple irritation that readily cleared; based upon appearance, behavior, and weight change, there was no evidence that absorption of toxic quantities had occurred. No quantitative skin absorption studies or inhalation work were done.

46.2.2 Human Experience

There has been no adverse human experience reported and none would be expected under reasonable conditions of handling.

46.3 Hygienic Standards of Permissible Exposure

Until appropriate data regarding the vapor inhalation toxicity of the material become available, no hygienic standard can be suggested. It would seem, however, that there would be little hazard from inhalation under ordinary conditions of handling and use.

47 TRIPROPYLENE GLYCOL MONOMETHYL ETHER; 2-Propanol, 1-[2-(2-Methoxy-1-methylethoxy)-1-methylethoxy]-; Dowanol® TPM Glycol Ether; CAS No. 20324-33-8 and CAS No. 24498-49-1

$$CH_3CHOHCH_2OCH(CH_3)CH_2OCH(CH_3)CH_2OCH_3$$

47.1 Physical and Chemical Properties

See Table 51.6.

47.2 Physiologic Response

47.2.1 Summary

Tripropylene glycol monomethyl ether is a colorless liquid, low in volatility, with slight, pleasant, ethereal odor and a bitter taste. It is low in single-dose oral toxicity, transiently painful but not damaging to the eye, and not appreciably irritating to the skin unless exposure is severe, but can be absorbed through the skin in toxic quantities if exposure is prolonged and repeated. Its vapor pressure and toxicity are sufficiently low so that no hazard exists from a single vapor exposure and probably none exists even when exposures are repeated. The hazards to health from ordinary handling and use would seem to be negligible.

47.2.2 Single-Dose Oral

When fed in single doses to rats the LD_{50} for the material appeared to be about 3.3 g/kg. The primary effect of the material appears to be narcotic (110). Shideman and Procita (111) have estimated the LD_{50} for dogs to be 5 ml/kg (4.8 g/kg). They assert that the primary effect of the material is central nervous system depression with death from large doses due to respiratory failure.

47.2.3 Repeated-Dose Oral

No data available.

47.2.4 Eye Contact (110)

Repeated instillation of the liquid into the eye of a rabbit failed to cause serious injury. Evidence of transient pain was observed.

47.2.5 Skin Contact (110)

When tripropylene glycol monomethyl ether was applied repeatedly to the skin of rabbits, it caused only a very mild simple irritation. This would indicate that the material would not be likely to produce skin irritation unless exposures were very severe. When applied to rabbits for 24 hr under a cuff, as advocated by Draize (10), a single dose of 20 ml/kg (19.22 g/kg) was survived. Such an observation indicates no practical hazard of systemic intoxication from occasional skin contact.

When measured doses were bandaged repeatedly onto the clipped abdomens of rabbits over a 90-day period, a significant amount of absorption through the skin occurred. At the high dosage levels (5 to 10 ml/kg, 4.8 to 9.65 g/kg), the material caused narcosis and kidney injury. At lower dosage levels, below 5 ml/kg, narcosis was not apparent, but there was some kidney injury even at the lowest dosage level administered, 1 ml/kg/day. These results indicate that prolonged and repeated skin contact with appreciable amounts of the material should be avoided.

47.2.6   Acute Vapor Inhalation (110)

When rats were exposed once for 7 hr to an atmosphere essentially saturated at 25°C with the vapors, no ill effects were observed. This indicates that the material does not present a hazard from acute vapor exposure at ordinary temperatures.

47.2.7   Repeated Vapor Inhalation

No repeated vapor exposures have been conducted.

47.2.8   Pharmacology

Shideman and Procita (111) report that in anesthetized and artificially respired dogs proper intravenous dosage of the material caused auricular fibrillation. The significance of this in practical use is doubtful since it has never been observed in the intact animal, regardless of the dose or mode of administration.

47.2.9   Human Experience

No unfavorable experience is known to us.

47.3   Hygienic Standard of Permissible Exposure

No hygienic standard for tripropylene glycol monomethyl ether would seem to be required, because of its relatively low toxicity and low volatility.

48   TRIPROPYLENE GLYCOL MONOETHYL ETHER; 2-Propanol, 1-[2-(2-Ethoxy-1-methylethoxy)-1-methylethoxy]-; CAS No. 20178-34-1

$$CH_3CHOHCH_2OCH(CH_3)CH_2OCH(CH_3)CH_2OCH_2CH_3$$

48.1   Physical and Chemical Properties

See Table 51.6.

48.2   Physiologic Response

48.2.1   Summary (9)

Tripropylene glycol monoethyl ether is a colorless liquid, low in volatility, with a slight pleasant ethereal odor and a bitter taste. When single oral doses of the material were administered to rats, the LD_{50} was found to be 2 ml/kg. When one drop of the liquid was introduced into the eye of a rabbit on five consecutive days, it produced only a very slight conjunctival irritation and no corneal injury. Repeated applications (10 in 14 days) to the skin of rabbits resulted in very slight simple irritation without evidence that toxic amounts were absorbed. No quantitative skin absorption studies or inhalation studies were conducted.

48.2.2   Human Experience

There has been no adverse human experience reported.

48.3   Hygienic Standards of Permissible Exposure

No hygienic standard for tripropylene glycol monoethyl ether would seem to be required because of its relatively low volatility and low toxicity. There would seem to be little or no hazard associated with ordinary handling and use of this substance.

49   TRIPROPYLENE GLYCOL MONO-*n*-BUTYL ETHER; 2-Propanol, 1-[2-(2-Butoxy-1-methylethoxy)1-methylethoxy]-; CAS No. 57499-93-1

$$CH_3CHOHCH_2OCH(CH_3)CH_2OCH(CH_3)CH_2OCH_2CH_2CH_2CH_3$$

49.1   Physical and Chemical Properties

See Table 51.6.

49.2   Physiologic Response

49.2.1   Summary (9)

Tripropylene glycol mono-*n*-butyl ether is a colorless liquid, low in volatility, with a slight ethereal odor and a bitter taste. When administered in single oral doses to rats, the LD_{50} was found to be 1.84 ml/kg. When one drop of the liquid was instilled into the eye of a rabbit on five consecutive days, no apparent irritation was produced. Repeated applications (10 in 14 days) to the skin of rabbits resulted in only a very slight exfoliation. There was, however, a slight weight loss during the course of the experiment, indicating that perhaps toxic amounts were being absorbed. It would appear, therefore, that prolonged and repeated skin contact with this substance should be avoided.

49.2.2   Human Experience

There has been no adverse human experience reported.

49.3   Hygienic Standards of Permissible Exposure

Because of the very low vapor pressure of this substance and its relatively low toxicity, it does not seem that a hygienic standard would be necessary. There would seem to be very little hazard associated with the handling and use of this substance.

50   MONO-, DI-, AND TRIPROPYLENE GLYCOL ALLYL ETHERS; Dowanol® PA-T
Glycol Ether; CAS No. 1331-17-5

$$CH_2{=}CHCH_2[OCH_2CH(CH_3)]_xOH$$

50.1   Physical and Chemical Properties

Appearance   Clear, colorless liquid
Purity   Mixture of allyl alcohol ethers of propylene glycols with less than 0.1% free allyl alcohol

50.2   Physiologic Response

Summary (9).   Dowanol PA-T glycol ether is a clear colorless liquid which is low in volatility at room temperature. The toxicologic information available was done on a commercial-grade product and is limited in nature. The single-dose oral toxicity is considered to be low; the LD_{50} for female rats is in the range of 1 to 2 g/kg. The rats at those doses where death occurred showed signs of central nervous system depression. Death was frequently preceded by convulsions. Gross autopsy revealed significant irritation of the stomach. Eye contact exposure using rabbits resulted in marked irritation and moderate corneal injury which healed in 3 to 7 days. Essentially continuous contact with the skin for 2 weeks resulted in only minor irritation in rabbits. In these skin tests, the rabbits showed no signs indicative of absorption through the skin of toxic amounts of this material. In inhalation tests using rats, the exposure period was 7 hr. Those exposed to essentially saturated vapor generated at room temperatures developed no adverse effects; however, those exposed to vapor generated at 100°C all died, three during the exposure period, the other three within 24 hr. These rats showed congested lungs and nasal passages. The effect of repeated exposures to vapors of this material have not been studied.

51 MONO-, DI-, AND TRIPROPYLENE GLYCOL, ISOBUTYL ETHERS; Dowanol PIB-T; (A Mixture of about 70% Propylene Glycol, Isobutyl Ether; 20 to 25% Dipropylene Glycol, Isobutyl Ether; and 5 to 10% Tripropylene Glycol, Isobutyl Ether); CAS NO.— None Found

$$(CH_3)_2CHCH_2[OCH_2CH(CH_3)]_xOH$$

51.1 Physical and Chemical Properties

Appearance	Clear, essentially colorless liquid
Boiling Point	170 to 173°C
Specific Gravity	0.882 at 25/25°C

51.2 Physiologic Response

51.2.1 Summary (9)

Dowanol PIB-T glycol ether is low in single-dose oral toxicity. Rabbit eye tests indicate that it should be considered low in hazard from eye contact, probably causing only minor transient irritation. Rabbit skin tests suggest that it is not likely to cause significant skin irritation, nor is it likely to be absorbed through the skin in acutely toxic amounts. It is low in toxicity from occasional or from short-term repeated inhalation exposures. The effects of prolonged repeated inhalation are not known. The hazard to health from industrial handling of this material would seem to be minimal.

51.2.2 Single-Dose Oral

The single-dose oral LD_{50} value for male and female rats is 4.29 and 2.46 g/kg, respectively. The only reported adverse effects noted were sedation and possible minor kidney injury. Based upon these findings the material may be considered low in single-dose oral toxicity and therefore should present no significant health hazard from ingestion in industrial handling.

51.2.3 Eye Contact

The application of two drops of undiluted material into the eyes of rabbits caused only slight pain upon contact followed by transient conjunctival redness in the eyes of rabbits. Thus this product may be considered to be low in hazard from eye contact.

51.2.4 Skin Contact

The daily repeated application of the undiluted material under a cloth covering for 2 weeks to the healthy skin of rabbits resulted in no irritation during the first week of application but slight scaliness during the second week. An essentially continuous contact under a cloth covering for 3 days to abraded skin resulted in the same response.

51.2.5 Skin Absorption

Using a modified "Draize" technique (10) of application under an impervious cuff for 24 hr to rabbit skin, and LD_{50} value by absorption was shown to be about 8.0 g/kg of body weight. Slight skin irritation, edema, and slight necrosis occurred.

51.2.6 Inhalation

Groups of four rats were exposed for 7 hr to essentially saturated atmospheres generated at room temperature and at 100°C. All the rats appeared drowsy and unsteady when removed, the effect being greater in those rats exposed to the vapor generated at 100°C. Visual observation at autopsy indicated no adverse effects to the organs.

Groups of 20 male rats, four male rabbits, and two male beagle dogs were subjected to twenty 7-hr exposures in 28 days to 100 or 200 ppm of the vapor. The rats experienced mild eye irritation during the first 5 days of exposure but this subsided to no irritation for the rest of the test period. Neither the rabbits nor the dogs showed any eye irritation. None of the animals experienced any adverse effects as reflected by hematology, clinical chemistry, pathology, or body and organ weights. Rats exposed for 7 hr/day, 5 days a week for a total of 28 exposures to 600 ppm of vapor were found to develop slight liver injury, which consisted of central lobular granular degeneration with occasional necrosis of parenchymal cells.

Based upon these studies it would appear that eye irritation would likely precede organic injury caused by short-term repeated exposure.

51.2.7 Mutagenicity

Mutagenic effects were studied in the bone marrow cells of rats exposed to 200 ppm of vapors for 7 hr/day, 5 days/week for 4 weeks. No significant effects were seen in the metaphase chromosomes. Although these data are not conclusive, they suggest that this material is not a strong mutagen and that it is entirely possible that it may have no mutagenic properties at all.

51.2.8 Mode of Action

Pharmacologic studies, using rats, indicate that the material reduces blood pressure at a dose of 5 mg/kg when introduced into the jugular vein. Tests with guinea pig cardiovascular preparations show that it produced only smooth muscle relaxation and only when high doses were given.

51.2.9 Human Experience

There has been no adverse human experience reported. The toxicologic data available suggest that the material should be handled with reasonable care, and precautions should be practiced to avoid breathing vapor, especially at concentrations that cause eye irritation or discomfort.

51.3 Hygienic Standards of Permissible Exposure

Adequate inhalation data are not available to permit the suggestion of an hygienic standard for permissible exposure. It would seem wise, therefore, to avoid prolonged or repeated inhalation of the vapors of the isobutyl ethers of a mixute of propylene, dipropylene, and tripropylene glycols in concentrations that cause discomfort or eye irritation.

51.4 Odor and Warning Properties

Limited human tests, done under controlled laboratory conditions using un-acclimated people who were exposed to concentrations of vapor in the range of 10 to 20 ppm indicate that intolerable concentrations probably exceed 20 ppm. However, several of the volunteers reported the detection of odor at levels of 10 ppm.

52 POLYPROPYLENE GLYCOL BUTYL ETHERS; Butoxypolypropylene Glycol CAS No. 9003-13-8

52.1 Physical and Chemical Properties

Property	Butoxypolypropylene Glycol	
	BPG 400	BPG 800
Molecular formula	$C_4H_9O(C_3H_6O)_xH$	$C_4H_9O(C_3H_6O)_yH$
Molecular weight	~400	~800
Specific gravity (25/25°C)	~0.973	~0.990
Vapor pressure mm Hg (25°C)	<0.1	<0.1
Solubility in H_2O (g/100 g at 20°C)	0.2	0.1

52.2 Physiologic Response

Summary. Carpenter et al. (115) have reported on toxicologic studies conducted on two polypropylene glycol butyl ethers having molecular weights approximating 400 (BPG 400) and 800 (BPG 800). Both BPG 400 and BPG 800 are low in toxicity when given either orally or intraperitoneally. The oral LD_{50} values for BPG 400 for male rats, male guinea pigs, and male rabbits were found to be 5.84, 2.46, and 3.30 g/kg, respectively. Similar values for these same species for BPG 800 were 9.16, 6.8, and 23.7 g/kg, respectively. The intraperitoneal LD_{50} values for rats were found to be 0.32 g/kg for BPG 400 and 0.91 g/kg for BPG 800. When large single oral doses were given, the principal effects of both materials were gastrointestinal irritation, congestion of internal organs, and death, usually within 24 hr or not at all. The BPG 800 was less distressing to the animals than the BPG 400 and more likely to cause convulsions and lung hemorrhage.

Dietary feeding studies over a 90-day period showed the no-effect dosage level for BPG 400 to be between 0.16 and 0.67 g/kg/day and for BPG 800 to be less than 0.52 g/kg/day. The liver and/or kidney were the first organs affected when subacute dosage levels were given repeatedly.

Neither material is more than very slightly irritating to the rabbit's skin or eyes. BPG 800 is neither an irritant nor a sensitizer of human skin. Neither is readily absorbed through the skin in acutely toxic amounts but repeated inunction of the rabbit's skin for 30 days showed BPG 400 to be moderate in toxicity and BPG 800 to be low in toxicity by this route, the no-effect levels being less than 0.1 ml/kg/day and 1.0 or more ml/kg/day, respectively. BPG 400 is readily absorbed from the gastrointestinal tract whereas BPG 800 is poorly absorbed. BPG 400 was not stored in the bodies of rats fed large doses for 30 days. Neither material presents a hazard from inhalation under reasonable conditions. Rats exposed to atmospheres essentially saturated at room temperatures for 8 hr were unaffected and they suffered only mild effects when exposed for 8 hr to fogs of the material.

It would seem that under ordinary conditions, neither BPG 400 nor BPG 800 would present any appreciable hazards to health in industrial handling. It is clear, however, that the BPG 800 is distinctly less toxic than BPG 400.

53 ETHERS OF BUTYLENE GLYCOL

53.1 Source, Uses, and Industrial Exposure

The methyl, ethyl, and *n*-butyl ethers of butylene glycol considered herein are prepared by reacting the appropriate alcohol with so-called "straight-chain" butylene oxide consisting of about 80 percent 1,2-isomer and about 20 percent 2,3-isomer in the presence of a catalyst. They are colorless liquids with slight, pleasant odors. The methyl and ethyl ethers are miscible with water but the butyl ether has limited solubility. All are miscible with many organic solvents and oils; thus they are useful as mutual solvents, dispersing agents, solvents for inks, resins, lacquers, oils, and greases. Industrial exposure may occur by any of the common routes.

53.2 Physical and Chemical Properties

See Table 51.6.

53.3 Determination in the Atmosphere

The same methods as described for the ethers of propylene glycol are applicable.

54 BUTYLENE GLYCOL MONOMETHYL ETHER; CAS No. 111-32-0

$$CH_3OC_4H_8OH$$

54.1 Physical and Chemical Properties

See Table 51.6.

54.2 Physiologic Response

54.2.1 Summary (9)

Butylene glycol monomethyl ether is low in single-dose oral toxicity for rats, moderately irritating and injurious to the eyes of rabbits, not appreciably irritating to the skin, and not readily absorbed through the skin of rabbits in toxic amounts. Prolonged inhalation by rats of air essentially saturated with vapors causes drowsiness, unsteadiness, and slight injury of the liver and kidneys; when some fog was present, deaths occurred.

54.2.2 Single-Dose Oral

When single oral doses of butylene glycol monomethyl ether were fed by gavage to rats, doses of 2 g/kg were survived and doses of 4 g/kg were fatal. No pathology was noted upon gross observation of the organs of the survivors.

54.2.3 Eye Contact

The undiluted material when instilled into the eyes of rabbits was painful, and caused moderate conjunctival irritation, moderate corneal injury, and some iritis, all of which cleared within a week.

54.2.4 Skin Contact

The material caused only very slight irritation of the skin of rabbits even though exposures were prolonged and repeated. Evidence of absorption through the skin was not apparent.

54.2.5 Inhalation

An atmosphere essentially saturated (probably 6000 to 7000 ppm) with vapor was not lethal to any of three rats exposed for 7 hr but it did cause drowsiness, unsteadiness, temporary weight loss, and some mild kidney and liver changes. An exposure lasting 4 hr caused drowsiness and unsteadiness but no histopathologic changes. When six rats were exposed for 7 hr to a supersaturated atmosphere (containing some fog), all animals became very drowsy, were unable to stand, breathed with difficulty, and four died. No significant pathologic changes were seen in the organs of the survivors.

54.2.6 Human Experience

No adverse human experience has been noted.

54.3 Hygienic Standards of Permissible Exposure

Data adequate to establish a hygienic standard for repeated inhalation are lacking. Although the butylene glycol monomethyl ether considered herein would not appear to present serious hazards in ordinary handling and use, care should be taken to avoid contact with the eyes. Otherwise, ordinary precautions such as the avoidance of prolonged or repeated contact with the skin and the inhalation of high concentrations of vapors or of mists for prolonged periods would seem to be adequate to ensure safety in industrial handling and use.

55 BUTYLENE GLYCOL MONOETHYL ETHER; CAS No. 111-73-9

$$C_2H_5OC_4H_8OH$$

55.1 Physical and Chemical Properties

See Table 51.6.

55.2 Physiologic Response

55.2.1 Summary (9)

Butylene glycol monoethyl ether is low in single-dose oral toxicity for rats, moderately irritating and injurious to the eyes of rabbits, not appreciably irritating to the skin, and not readily absorbed through the skin in toxic amounts. Air essentially saturated with vapor was only slightly toxic to rats.

55.2.2 Single-Dose Oral

When single oral doses of butylene glycol monoethyl ether were fed to rats, doses of 1 g/kg were survived, doses of 2 g/kg caused the death of one of three, and doses of 4 g/kg caused the death of one of two. Doses of 2 and 4 g/kg caused injury to the lungs, liver, and kidney, but lower dosage levels did not.

55.2.3 Eye Contact

When the material was instilled into the eyes of rabbits it caused pain, marked conjunctival and corneal injury, and slight iritis.

55.2.4 Skin Contact

The undiluted material was not appreciably irritating to the skin of rabbits even when exposures were prolonged and repeated. Evidence of absorption through the skin was not apparent.

55.2.5 Inhalation

Rats exposed for 7 hr to an atmosphere essentially saturated (probably 3000 to 4000 ppm) with vapor survived without serious effects, although some kidney injury was apparent upon gross examination of the organs. However, when a fog was present, six rats exposed for 7 hr exhibited irritation of the eyes and nares, drowsiness, inability to stand, and difficulty in breathing. Although none died, all were ill and gross examination of the organs revealed injury to the lungs, kidneys, and livers.

55.2.6 Human Experience

No adverse human experience has been noted.

55.3 Hygienic Standards of Permissible Exposure

Data adequate to establish a hygienic standard for repeated inhalation are lacking. Although butylene glycol monoethyl ether would not appear to present serious hazards in ordinary handling and use, care should be taken to prevent contact with the eyes. Otherwise, ordinary precautions such as the avoidance of prolonged and repeated skin contact and the inhalation for prolonged periods of time of high concentrations of vapor or of mist would seem adequate to ensure safety in industrial handling and use.

56   BUTYLENE GLYCOL MONO-*n*-BUTYL ETHER; CAS No.—None Found

$$C_4H_9OC_4H_8OH$$

56.1   Physical and Chemical Properties

See Table 51.6.

56.2   Physiologic Response

56.2.1   Summary (9)

Butylene glycol mono-*n*butyl ether is low in single-dose oral toxicity for rats. In studies on rabbits, it was found to be moderately irritating and injurious to the eyes, not appreciably irritating to skin when unconfined but moderately injurious when confined, and not readily absorbed through the skin in acutely toxic amounts. A saturated atmosphere was low in toxicity to rats.

56.2.2   Single-Dose Oral

When single oral doses of butylene glycol mono-*n*-butyl ether as a 20 percent solution in corn oil were fed to rats, doses of 2.0 g/kg were survived and doses of 4.0 g/kg were lethal. High doses caused prostration and labored breathing. Necropsy of surviving animals revealed some kidney injury.

56.2.3   Eye Contact

The undiluted material, when instilled into the eyes of rabbits, was painful, irritating to the conjunctival membranes, injurious to the cornea, and caused some iritis. All these effects disappeared within 7 days after exposure.

56.2.4   Skin Contact

The material was not irritating to the skin of rabbits when applied to the uncovered skin. However, when applied under a bandage for 48 hr, it caused a burn. Acutely toxic amounts were not readily absorbed through the intact skin.

56.2.5   Inhalation

Rats were not noticeably affected by an exposure of 7 hr duration to an atmosphere saturated at 100°C and then cooled to room temperature. Necropsy, however, showed that some kidney injury had resulted.

56.2.6   Human Experience

No adverse human experience has been reported.

56.3   Hygienic Standards of Permissible Exposure

Data adequate to establish a hygienic standard for repeated inhalation are lacking. The limited toxicologic information available does not indicate that butylene glycol mono-*n*-butyl ether presents any unusual hazards. Ordinary precautions to prevent contact with the eyes, prolonged and repeated contact with the skin, and inhalation of vapors would seem adequate to prevent the occurrence of systemic toxicity.

57   ESTERS, DIESTERS, AND ETHER—ESTERS OF GLYCOLS

57.1   Source, Uses, and Industrial Exposure

The common esters and diesters of the common polyols are prepared commercially by esterifying the particular polyol with the acid, acid anhydride, or acid chloride of choice in the presence of a catalyst. Mono- or diesters result, depending upon the proportions of each reactant employed. The ether—esters are prepared by esterifying the glycol ether in a similar manner. Other methods also can be used (3).

The acetic acid esters have remarkable solvent properties for oils, greases, inks, adhesives, and resins. They are widely used in lacquers, enamels, dopes, and adhesives to dissolve the plastics or resins. They are also used in lacquer, paint, and varnish removers.

The nitric acid esters are not typical of the esters or ether esters of organic acids and are considered separately later in this chapter. They are used as explosives, usually in combination with nitroglycerin to reduce the freezing point.

Industrial exposures of consequence are most likely to occur through the inhalation of vapors, although excessive contact with the eyes and skin may also occur. With the dinitrate, a serious hazard exists from absorption through the skin.

57.2   Physical and Chemical Properties

All the esters and ether—esters of organic acids are colorless volatile liquids. Generally the odors are mild, sometimes fruity, and they all have a bitter taste. See Tables 51.8 and 51.9 for additional properties.

Table 51.8. Physical and Chemical Properties of Esters of Various Glycol Esters

| | Ethylene Glycol | | Triethylene Glycol Diacetate | 2-Methyl-2-propene-1,1-diol Diacetate | 2,2-Dimethyl-1,3-propanediol Diacrylate | 2-Methyl-2,4-pentanediol Diacetate | 2,2,4-Trimethyl-1,3-pentanediol Diisobutyrate |
	Acetate	Diacetate					
CAS No.	542-59-6	111-55-7	111-21-7	869-29-4 10476-95-6	2223-82-7	1637-24-7	6846-50-0
Molecular formula	$C_4H_8O_3$	$C_6H_{10}O_4$	$C_{10}H_{18}O_6$	$C_8H_{12}O_4$	$C_{11}H_{16}O_4$	$C_{10}H_{18}O_4$	$C_{16}H_{30}O_4$
Molecular weight	104.1	146.2	234.3	172.2	212.2	202.2	286.4
Specific gravity, 25/4°C	1.106	1.106	1.117 (20/20°C)	1.051 (20/20°C)	1.030 (20/20°C)	1.000 (20/20°C)	0.944
Boiling point (°C) (760 mm Hg)	182	190.8	285.7				280
Freezing point (°C)			sets to glass < −50				−70
Vapor pressure (mm Hg) (25°C)		0.25	<0.01			0.07 (20°C)	
Refractive index (25°C)	1.422	1.416	1.439 (20°C)			1.423 (20°C)	
Flash point (°F) (O.C.)	215	215–255	174	215	253	235	
Vapor density (air = 1)	3.6	5.0	8.1				
Percent in saturated air (25°C)	0.044	0.044	10^{-6}				
1 ppm ⇌ mg/m³ at 25°C, 760 mm Hg	4.25	5.98	9.58	7.03	8.68	8.25	11.68
1 mg/l ⇌ ppm at 25°C, 760 mm Hg	235	168	104.3	142.0	115.2	121.0	85.7

Table 51.9. Physical and Chemical Properties of Some Ether Acetates of Certain Glycols

| | Acetate | | | | | | | |
| | Ethylene Glycol | | Diethylene Glycol | | | Propylene Glycol, Methyl | Dipropylene Glycol, Methyl | Tripropylene Glycol, Methyl |
Property	Methyl	Ethyl	Methyl	Ethyl	Butyl			
CAS No.	110-49-6 32718-56-2	111-15-9	629-38-9	112-15-2	124-17-4	108-65-6	None found	None found
Molecular formula	$C_5H_{10}O_3$	$C_6H_{12}O_3$	$C_7H_{14}O_4$	$C_8H_{16}O_4$	$C_{10}H_{20}O_4$	$C_6H_{12}O_3$	$C_9H_{18}O_4$	$C_{12}H_{24}O_5$
Molecular weight	118.13	132.16	162.2	176.2	204.3	132.1	190.1	248.2
Specific gravity 25/4°C	1.007	0.975	1.04	1.01	0.981	0.957		
Boiling point (°C) (760 mm Hg)	144.5	156.3	194.2	217.4	246.8	146	209	258
Vapor pressure (mm Hg) (20°C)	2.0–3.7	2.8	0.12	0.05	<0.01			
Refractive index (25°C)	1.402	1.406				1.400		
Flash point (°F) (O.C.)	140	138	180	230	240			
Vapor density (air = 1)	4.1	4.7		6.07				
Percent in saturated air (25°)	0.31–0.60	0.21–0.27		0.01	<0.002			
1 ppm ⇋ mg/m³ at 25°C, 760 mm Hg	4.83	5.40	6.63	7.20	8.34	5.40	7.77	10.15
1 mg/l ⇋ ppm at 25°C, 760 mm Hg	207	185	151	139	120	185	128.6	98.5

57.3   Determination in the Atmosphere

The choice of methods for the determination of the esters, diesters, and ether—esters of various glycols vary with existing conditions. Gas chromatography would seem to offer the best means not only of resolving mixtures of vapors but also of identifying the components. Mass spectrophotometry may also be used. Chemical methods such as proposed by Morgan (116) for esters or ether—esters may be useful where spectroscopic equipment is not available.

57.4   Physiologic Response

Summary.   Generally speaking, the fatty acid esters of the glycols and glycol ethers, either in the liquid or vapor state, are more irritating to the mucous membranes than those of the parent glycol or glycol ethers. However, once absorbed into the body, the esters are saponified and the systemic effect is quite typical of the parent glycol or glycol ether. Lepkovski et al. (117), in studies with higher fatty acids of glycols, concluded that the fatty acids were liberated and used nutritionally. Furthermore, they observed that severe injury to the tubular epithelium of the kidneys occurred when the esters of ethylene glycol and diethylene glycol were fed, but not when equivalent amounts of fatty esters of propylene glycol, glycerol, ethyl alcohol, methyl alcohol, or the free fatty acids themselves were fed. Shaffer and Critchfield (118), in studies with polyethylene glycol 400 monostearate, concluded that it was low in toxicity and also utilized nutritionally.

Lest generalizations be too broadly interpreted, it should be noted that the nitric acid esters are highly toxic and exert a physiologic action quite different from that of the parent polyol.

58   ETHYLENE GLYCOL MONOACETATE; 1,2-Ethanediol, Monoacetate; Glycol Monoacetate; Solvent GC; Cellosolve® Acetate; CAS No. 542-59-6

$$HOCH_2CH_2OOCCH_3$$

58.1   Physical and Chemical Properties

See Table 51.8.

58.2   Physiologic Response

58.2.1   Summary

Ethylene glycol monoacetate is low in single-dose oral toxicity, the LD_{50} values reported by Smyth et al. (62) being 8.25 g/kg for rats and 3.80 g/kg for guinea pigs when fed as a 50 percent aqueous solution. It is moderately irritating to the eyes (11), but not appreciably irritating to the skin of animals or humans (7). Dogs were apparently unaffected by 12 feedings of 0.1 or 0.5 ml/kg (7). Seven subcutaneous injections of 0.5 or 1.0 ml/kg did not injure guinea pigs (7). Twelve 8-hr exposures to an atmosphere essentially saturated with vapor at room temperature were survived by cats, guinea pigs, and mice, but caused lung irritation and slight kidney injury. One rabbit treated similarly died (7). Studies by Rosser, (119), quoted by Gross (7), indicate that cats tolerated a single 6-hr exposure to an atmosphere containing mist (28 mg/l) but succumbed from two such exposures. Wiley and co-workers (21) gave repeated daily injections (route unspecified) of ethylene glycol monoacetate to dogs (8.5 ml/day) and rabbits (3.5 ml/day) and observed an increase in urinary oxalic acid similar to that caused by the administration of an equivalent amount of ethylene glycol. Also noted were degenerative changes in the kidneys, testes, and brain.

58.2.2 Human Experience

There have been no reports of untoward effects in human beings.

58.3 Hygienic Standards of Permissible Exposure

The toxicologic data available are inadequate to allow the establishment of a hygienic standard for repeated inhalation.

59 ETHYLENE GLYCOL DIACETATE; 1,2-Ethanediol, Diacetate; CAS No. 111-55-7

$$CH_3COOCH_2CH_2OOCCH_3$$

59.1 Physical and Chemical Properties

See Table 51.8.

59.2 Physiologic Response

59.2.1 Summary

Ethylene glycol diacetate is a colorless, slightly volatile liquid with an acetic, esterlike odor and a bitter taste. The single-dose oral toxicity of this material is low, the LD_{50} values being 6.86 g/kg for rats and 4.94 g/kg for guinea pigs when fed as a 50 percent aqueous solution (62). Yoshida et al. (120) evaluated this material for its nutritive value in the diets of chicks and rats. The energy was well utilized but the diet intake at high dose levels tended to decrease, probably because of the bitter taste. The authors state that this material is somewhat volatile and some free acetic acid was released during storage of the diet. Kesten and co-workers (121) gave the material intravenously and orally to animals and concluded that it did not cause hydropic degeneration of the renal convoluted tubules as do the glycols having ether linkages between the glycol units. Mulinos and co-workers (122) fed the material in water over a prolonged period to rats and rabbits and observed that 1 to 3 percent solutions caused occasional calcium oxalate crystals in the kidneys, and that 5 percent solutions caused large crystalline deposits and death. They did not observe any hydropic changes in the kidneys of either rats or rabbits, confirming the earlier results of Kesten et al. (121).

59.2.2 Human Experience

None reported.

59.3 Hygienic Standard of Permissible Exposure

None can be established on the very limited toxicologic information available. However, it appears that the material should be considered to be like ethylene glycol.

60 ETHYLENE GLYCOL DIACRYLATE; 2-Propenoic acid, 1,2-Diethanediyl Ester; CAS No. 2274-11-5

$$CH_2{=}CHCOOCH_2CH_2OOCCH{=}CH_2$$

60.1 Physical and Chemical Properties

Molecular formula	$C_8H_{10}O_4$
Molecular weight	170.2
1 ppm ≈	6.95 mg/m^3 at 760 mm Hg, 25°C
1 mg/l ≈	143.7 ppm at 760 mm Hg, 25°C

60.2 Physiologic Response

Summary. Carpenter et al. (60) report that ethylene glycol diacrylate is moderate in single-dose oral toxicity, the LD_{50} for rats being 1.07 ml/kg. Eye contact is a serious hazard for the material is capable of causing severe irritation and serious corneal damage, as indicated by rabbit tests (grade 8 on a scale of 10). The material is slightly irritating to the skin of rabbits (grade 3 on a scale of 10). It may be absorbed through the skin in toxic amounts; the LD_{50} for rabbits is 0.57 ml/kg. Rats were not seriously affected by a single 8-hr exposure to an essentially saturated atmosphere, probably because of its low volatility.

It would seem wise to handle ethylene glycol diacrylate with great care to prevent eye contact, to avoid skin contact and to avoid repeated exposure to vapor. Situations where the material is handled hot, or where fogs or mists could be generated, could be hazardous.

61 DIETHYLENE GLYCOL DIACRYLATE; 2-Propenoic acid, Oxydi-2,1-ethanediyl; CAS No. 4074-88-8

$$CH_2{=}CHCOOCH_2CH_2OCH_2CH_2OOCCH{=}CH_2$$

61.1 Physical and Chemical Properties

Molecular formula	$C_{10}H_{14}O_5$
Molecular weight	214.2
1 ppm $\backsim$	8.76 mg/m^3 at 760 mm Hg, 25°C
1 mg/l $\backsim$	114.2 ppm at 760 mm Hg, 25°C

61.2 Physiologic Response

Summary (60). Diethylene glycol diacrylate is a liquid that is moderate in single-dose oral toxicity, since the LD_{50} for rats is 0.77 ml/kg. It is severely injurious to the eyes of rabbits, being capable of causing severe corneal damage (rated 9 on a scale of 10). It is markedly irritating to rabbit skin, being rated 5 on a scale of 10. It is very toxic from skin absorption; the LD_{50} for rabbits by this route is 0.18 ml/kg. Probably because of its very low volatility, it appears to be low in toxicity from inhalation, for rats exposed for 8 hr to an essentially saturated atmosphere, presumably generated at room temperatures, were not seriously affected. Thus diethylene glycol diacrylate must be considered hazardous to health from skin and eye contact. It would seem wise to handle diethylene glycol diacrylate with great care to prevent eye and skin contact and, until more data are available, to avoid prolonged or repeated exposure to vapors, particularly if the material is handled hot or in a manner that results in the generation of a fog or mist.

62 TRIETHYLENE GLYCOL DIACETATE; Ethanol, 2,2′-[1,2-Ethanediylbis(oxy)]bis-, Diacetate; Triglycol Diacetate; CAS No. 111-21-7

$$(CH_2OCH_2CH_2OOCCH_3)_2$$

62.1 Physical and Chemical Properties

See Table 51.8.

62.2 Physiologic Response

Summary (51). Triethylene glycol diacetate is a colorless liquid with a low degree of volatility. It is low in single-dose oral toxicity; the LD_{50} value for rats is reported to be 22.6 ml/kg. It is also low in single-dose toxicity when injected intraperitoneally, since the LD_{50} for rats is 1.41 ml/kg. Skin and eye contact should give rise to no significant irritation, nor is it likely to be absorbed through the skin in toxic amounts, for the LD_{50} for rabbits by absorption is reported to be 8.0 ml/kg. A single 8-hr exposure of rats to essentially saturated vapors generated at approximately 20°C resulted in no significant adverse effects. This suggests that triethylene glycol diacetate should pose no significant health hazard from handling in industrial operations. However, it does seem wise, until more is known, to take precautions to avoid repeated prolonged exposure to its vapors, especially if it is handled hot.

63 TRIETHYLENE GLYCOL DIVALERATE; Pentanoic acid, 1,2-Diethanylbis(oxy-2,1-ethane) Diyl Ester; Valeric Acid, Triethylene Glycol Diester; CAS No.—None Found

$$(CH_2OCH_2CH_2OOC(CH_2)_3CH_3)_2$$

63.1 Physical and Chemical Properties

Molecular formula	$C_{16}H_{30}O_6$
Molecular weight	318.4
1 ppm ≈	13.05 mg/m$_3$ at 760 mm Hg, 25°C
1 mg/l ≈	0.999 ppm at 760 mm Hg, 25°C

63.2 Physiologic Response

Summary (60). Triethylene glycol divalerate is a liquid that is low in single-dose oral toxicity, the LD_{50} for rats being 14.3 ml/kg. It is essentially nonirritating to the eyes of rabbits, being grade 1 on a scale of 10 and only very mildly irritating to rabbit skin with a grade of 2 on the scale of 10. Rabbit skin absorption tests indicate that it is very low in toxicity by absorption, the LD_{50} being >16 ml/kg. A single 8-hr exposure of rats to an essentially saturated atmosphere of the vapors of this material resulted in no deaths of the rats; hence it seems to be low in toxicity from an occasional vapor exposure. Thus the industrial handling of triethylene glycol divalerate should pose no significant health hazard owing to its low degree of toxicity. However, until more is known, it would seem wise to avoid exposure from inhalation, especially if it is handled hot, if mists or fogs are generated, or if repeated vapor exposures are likely.

64 PROPYLENE GLYCOL MONOACRYLATE; 1,2-Propanediol, 1(or 2)-Acrylate; Acrylic Acid, Hydroxypropyl Ester; CAS No. 25584-83-2, CAS No. 999-61-1 (1-Acrylate), and CAS No. 2918-23-2 (2-Acrylate)

$$CH_2\!\!=\!\!CHCOOCH_2CHOHCH_3 \quad \text{and} \quad CH_2\!\!=\!\!CHCOOCH(Ch_3)CH_2OH$$

64.1 Physical and Chemical Properties

Molecular formula	$C_6H_{10}O_3$
Molecular weight	130.1
1 ppm ≎	5.32 mg/m^3 at 760 mm Hg, 25°C
1 mg/l ≎	188.1 ppm at 760 mm Hg, 25°C

64.2 Physiologic Response

64.2.1 Summary

Propylene glycol monoacrylate is a clear to light yellow, very slightly volatile liquid with a sweetish solvent odor. It is moderate to low in single-dose oral toxicity. The liquid is markedly painful and injurious to the eyes, resulting in possible impairment of vision. Prolonged or repeated skin contact can cause marked irritation, even a burn; however, it is not considered to be corrosive by the DOT test. It can be absorbed readily through the skin in acutely toxic amounts and should be considered a possible weak skin sensitizer. The vapor of propylene glycol monoacrylate causes eye, nasal, and respiratory irritation. However, these warning properties should not be relied upon as adequate to prevent adverse effects to the lungs and to the upper respiratory system from inhalation exposures.

64.2.2 Single-Dose Oral

Propylene glycol monoacrylate has a moderate to low single-dose oral toxicity. The LD$_{50}$ for rats has been reported to range from 0.25 to 0.50 g/kg, (9) 0.59 g/kg (123), and 1.3 g/kg (51). The rats died overnight and gross autopsy revealed that there were no significant pathologic effects (9).

64.2.3 Eye Contact

Propylene glycol monoacrylate is markedly injurious to the eye. The application of approximately 0.1 ml to the eye of rabbits caused marked to severe irritation and corneal injury sufficient to cause impairment of vision (9, 51). In one report the impairment was of such a nature that it could possibly be permanent in nature (9).

64.2.4   Skin Contact

A single 24-hr contact has been reported to cause marked irritation and some damage to the skin of rabbits (51). Even a 24 hr exposure under a cloth covering to a 10 percent water solution has been observed to cause a moderate burn resulting in a scar (9). However, the same dose, when applied nine times to the uncovered skin in a 2-week period, was observed to cause no significant irritation. The application of the undiluted material to the covered skin of rabbits resulted in the death of the rabbits in 24 hr.

Propylene glycol monoacrylate was tested for corrosivity as directed by the Department of Transportation and was found to be noncorrosive (9).

64.2.5   Skin Absorption

Propylene glycol monoacrylate has been shown by rabbit skin absorption tests to be highly toxic, the LD_{50} for rabbits being 0.17 g/kg (51) and in the range of 0.25 g/kg (9). Death occurred overnight. The animals that survived developed severe irritation, moderate edema, and moderate to severe necrosis (9).

64.2.6   Skin Sensitization

Male and female guinea pigs were given six applications in 3 weeks of a 0.5 percent solution of this material in a solvent made of Tween 80 (1 part) and dipropylene glycol methyl ether (9 parts). Then they were given a challenge dose after a 2-week rest period. Four of the 10 guinea pigs displayed weak sensitization reactions. Thus propylene glycol monoacrylate may be capable of causing sensitization in susceptible persons.

64.2.7   Inhalation—Single Exposure

Rats exposed for 7 or 8 hr to vapors of propylene glycol monoacrylate generated at room temperature to produce an essentially saturated atmosphere survived the exposure with no significant adverse effects (9, 51). All the rats exposed for 4.25 hr to the vapor generated at 100°C died either during the exposure or within 1 hr after exposure terminated. During the exposure, marked signs of nasal, respiratory, and eye irritation were observed (9). Rats exposed for 1 hr to vapor generated at 100°C survived but gross observation revealed some slight kidney and liver injury. The concentration of the vapors generated at 100°C was estimated to be in the range of 650 ppm.

64.2.8 Inhalation—Repeated Exposure (9)

Dogs, rats, rabbits, and mice were exposed to 5 ppm (the lowest concentration tested) of propylene glycol monoacrylate vapor for 6 hr/day for 5 days/week for a total of 20 or 21 exposures in a month. During the exposure period both the dogs and rabbits exhibited signs of nasal and respiratory irritation. The rats, however, showed no visual signs but pathologically the irritation was observed. This was seen in the dogs and rabbits as well. Both the dogs and rabbits showed signs of eye irritation. There were no treatment-related effects in any of the animals as shown by organ to body weight ratios, hematology, clinical chemistry, and urinalysis. However, microscopic testicular changes were seen in three of the four treated dogs. These changes were not observed in the other animals. The pathologic findings indicate that the changes may have been due to the respiratory system effects observed, and/or perhaps to the fact that the dogs were sexually immature.

Based upon these findings the no-effect level for repeated inhalation exposure to animals must be less than 5 ppm.

65 1,3-BUTANEDIOL DIACRYLATE; 2-Propenoic Acid, 1-Methyl-1,3-Propanediyl Ester; 1,3-Butylene Glycol Diacrylate; Acrylic Acid, 1,3-Butylene Glycol Diester; CAS No. 19485-03-1

$$CH_2=CHCOOCH_2CH_2CH(OOCCH=CH_2)CH_3$$

65.1 Physical and Chemical Properties

Molecular formula	$C_{10}H_{14}O_4$
Molecular weight	198.2
1 ppm ≈	8.10 mg/m$_3$ at 760 mm Hg, 25°C
1 mg/l ≈	123.4 ppm at 760 mm Hg, 25°C

65.2 Physiologic Response

Summary (60). 1,3-Butanediol diacrylate is low in single-dose oral toxicity, the LD$_{50}$ for rats being 3.54 ml/kg. It is markedly irritating and damaging to the eye, being rated 9 on a scale of 1 to 10. It is markedly irritating to the skin and may cause a burn upon prolonged or repeated contact. It is also moderate in toxicity by skin absorption; the LD$_{50}$ for rabbits is 0.45 ml/kg. Because of its low volatility, it is not likely to be a problem from inhalation from a single or occasional exposure. Rats exposed to essentially saturated vapors for 8 hr survived. However, it would seem wise to avoid inhalation of vapors, especially when it is handled hot or when mists may be produced.

66 BUTYLENE GLYCOL ADIPIC ACID POLYESTER; Santicizer® 334F; CAS No. 9080-04-0

$$[RCO]_2[OCH(CH_3)CH_2CH_2O]_xOC(CH_2)_4OC-]_{x-1}$$

$R = C_{13}H_{27}, C_{15}H_{31},$ and $C_{17}H_{35}$

$x = $ a number such that the molecular weight
of the molecule is in the range of
1700 to 2200

66.1 Physiologic Response

Summary (124).   Both rats and dogs were fed for 2 years on a diet containing 1000, 5000 or 10,000 ppm of butylene glycol adipic acid polyester. The no-effect level for both dogs and rats was found to be 10,000 ppm as judged by body weight, organ weights, food consumption, mortality, behavior, clinical observations of the blood, and gross and microscopic pathology. Rat reproduction was likewise unaffected by feeding 10,000 ppm of the chemical for three generations.

67 2-METHYL-2-PROPENE-1,1-DIOL DIACETATE; Methyl Allylidene Diacetate; CAS No. 869-29-4 and CAS NO. 10476-95-6

$$CH_2{=}C(CH_3)CH(OOCH_3)_2$$

67.1 Physical and Chemical Properties

See Table 51.8.

67.2 Physiologic Response

Summary (51).   2-Methyl-2-propene-1,1-diol diacetate is a seriously hazardous material. It is moderate in single-dose oral toxicity; the LD_{50} for rats is 0.44 ml/kg. Rabbit eye studies showed the material to cause severe irritation and corneal injury, being rated 9 on a scale of 10. It is markedly irritating to the skin and can cause a burn, but the greater hazard is that of skin absorption, the LD_{50} for rabbits being 0.044 ml/kg. It is hazardous from inhalation also. All the rats exposed for 1 hr to essentially saturated vapors died, and five of six rats exposed to 62.5 ppm for 4 hr died. When handling this material special precautions should be taken to prevent eye and skin contact and inhalation of vapor.

68 2,2-DIMETHYL-1,3-PROPANEDIOL DIACRYLIC ESTER; 2-Propenoic Acid, 2,2-Dimethyl-1,3-propanediyl Ester; Neopentyl Glycol Diacrylate; CAS No. 2223-82-7

$$CH_2{=}CHCOOCH_2C(CH_3)_2CH_2OOCCH{=}CH_2$$

68.1 Physical and Chemical Properties

See Table 51.8.

68.2 Physiologic Response

Summary (60). 2,2-Dimethyl-1,3-propanediol diacrylate is low in single-dose oral toxicity; the LD_{50} value for rats is 6.73 ml/kg. It is moderately irritating to the eyes and may cause some corneal injury; it is rated 4 on a scale of 10 in rabbit eye tests. It is also moderately injurious to the skin, being rated 5 on a scale of 10 in rabbit skin tests. It is also moderately toxic from skin absorption, the LD_{50} for rabbits being 0.40 ml/kg. However, it should be low in hazard from an occasional exposure to vapors, for rats exposed 8 hr to an essentially saturated vapor did not die. No data involving repeated exposure are available. Precautions should be taken to avoid skin and eye contact, and, until more is known, to avoid repeated prolonged exposure to vapors or to mists or fogs if generated. Precautions should also be observed if the material is handled hot.

69 2-METHYL-2,4-PENTANEDIOL DIACETATE; Acetic Acid, 2-Methyl-2,4-pentanediol Diester; Hexylene Glycol Diacetate; CAS No. 1637-24-7

$$(CH_3)_2CO(OCCH_3)CH_2CHO(OCCH_3)CH_3$$

69.1 Physical and Chemical Properties

See Table 51.8.

69.2 Physiologic Response

Summary (51). 2-Methyl-2,4-pentanediol diacetate is low in single-dose oral toxicity; the LD_{50} rats is 3.36 ml/kg. It is only very mildly irritating to the eyes, being rated 2 on a scale of 10 in rabbit eye tests, and mildly irritating to the skin, being rated 3 on a scale of 10 in rabbit skin tests. It is very low in toxicity by skin absorption; the LD_{50} value for rabbits is 16.0 ml/kg by this route. It is low in toxicity from single exposure to its vapors, for none of the rats exposed for 8 hr to essentially saturated vapors died. Thus this material presents no unusual health hazards and may be handled safely under most industrial conditions of reasonable care and cleanliness are practiced. However, until more is known, it would seem wise to avoid repeated prolonged vapor exposures or exposure to vapors generated when handled hot or as mists or fogs.

70 2,2,4-TRIMETHYL-1,3-PENTANEDIOL DIISOBUTYRATE; Propanoic acid, 2-Methyl-, 2,2-Dimethyl-1-(1-MethylEthyl)-1,3-propanediyl Ester; Kodaflex® TXIB Plasticizer; CAS No. 6846-50-0

$$(CH_3)_2CHCOOCH_2C(CH_3)_2CH(OOCCH\ (CH_3)_2)CH(CH_3)_2$$

70.1 Sources, Uses, and Industrial Exposure

2,2,4-Trimethyl-1,3-pentanediol diisobutyrate is made by reacting 2,2,4-trimethyl-1,3-pentanediol with isobutyryl anhydride. Its main potential use is as a plasticizer in food packaging. Industrial exposure is likely to be by direct contact with the liquid and by exposure to the vapor or mist under some conditions.

70.2 Physical and Chemical Properties

2,2,4-Trimethyl-1,3-pentanediol diisobutyrate is a clear, somewhat viscous liquid and miscible with water. For additional properties see Table 51.8.

70.3 Determination in the Atmosphere

Both gas chromatographic and infrared methods may be useful (125).

70.4 Physiologic Response (125)

70.4.1 Summary

Astill et al. (125) report that 2,2,4-trimethyl-1,3-pentanediol diisobutyrate is low in single-dose and repeated-dose oral toxicity, is only slightly irritating to the skin, is not absorbed through the skin in acutely toxic amounts, is not a skin sensitizer to guinea pigs and is low in toxicity by inhalation. It would not be expected to present an appreciable health hazard in anticipated industrial handling or in its use as a plasticizer.

70.4.2 Single-Dose Oral

2,2,4,-Trimethyl-1,3-pentanediol diisobutyrate has an LD_{50} value of greater than 3.2 g/kg for rats and 6.4 g/kg for mice.

70.4.3 Repeated-Dose Oral

Rats were fed diets containing 2,2,4-trimethyl-1,3-pentanediol diisobutyrate at levels of 0.1 and 1.0 percent for 100 days with no adverse effects as shown by behavior, feed intake, weight gain, or histology. Similar findings were seen in dogs fed at levels of 0.1, 0.35, and 1.0 percent of this material for 13 weeks.

70.4.4 Injection Toxicity

When this material was injected intraperitoneally at a dose of 3.2 g/kg, one-half the rats, but none of the mice died. The adverse effects seen were weakness and some vasodilation.

70.4.5 Skin Irritation

The application of up to 200 ml/kg of the material on the skin of guinea pigs under an occluded patch gave rise to slight irritation. When it was applied by gentle inunction, without covering, only very slight irritation developed.

70.4.6 Skin Absorption

In the above tests, no systemic toxic effects were detected.

70.4.7 Skin Sensitization

The results of skin sensitization studies using guinea pigs indicated that it was not a sensitizer.

70.4.8 Inhalation

Exposure of rats to concentrations of as much as 5.3 mg/l of vapor for 6 hr caused only vasodilation but no deaths. To attain this high concentration the compound was heated to 100°C.

70.4.9 Metabolism

Single-dose oral feeding of 2,2,4-trimethyl-1,3-pentanediol diisobutyrate shows that from one-third to one-half of the compound was excreted in the feces as monoisobutyrate of the parent glycol and that the rest was absorbed from the gut and eliminated via the urine with little or none deposited in tissues or organs. The metabolites found in the urine consist of unchanged 2,2,4-trimethyl-1,3-pentanediol and its metabolites, such as its O-glucuronide, O-sulfate, and 2,2,4-trimethyl-3-hydroxyvaleric acid and its glucuronide, all of which are physiologically inert. Thus it appears that this compound is essentially physiologically inert at intake levels likely to be encountered in use.

70.5   Hygienic Standards for Permissible Exposure

The data suggest that there should be no necessity for setting a standard.

71   ETHYLENE GLYCOL MONOMETHYL ETHER ACETATE; Ethanol, 2-Methoxy-, Acetate; 2-Methoxyethyl Acetate; Methyl Cellosolve® Acetate; Methyl Glycol Acetate; CAS No. 110-49-6 and CAS No. 32718-56-2

$$CH_3OCH_2CH_2OOCCH_3$$

71.1   Physical and Chemical properties

See Table 51.9.

71.2   Physiologic Response

71.2.1   Summary

Ethylene glycol monomethyl ether acetate is low in single-dose oral toxicity, not significantly irritating to the eyes or skin, poorly absorbed through the skin, and moderately toxic when inhaled. Its effects are similar to those of ethylene glycol monomethyl ether, being centered in the blood, kidneys, and brain.

71.2.2   Single-Dose Oral

The LD_{50} values of 1.25 g/kg for guinea pigs and 3.93 g/kg for rats reported by Smyth et al. (62) were determined by feeding 50 percent aqueous solutions. Later, Smyth and Carpenter (98) reported the LD_{50} for rats to be 3.39 g/kg, but did not specify in what form it was fed.

71.2.3   Repeated-Dose Oral

Gross (7) reports that rabbits died after receiving three daily doses of either 0.5 or 1.0 ml/kg, and all showed kidney injury with albumin and granular casts in the urine.

Nagano et al. (13) fed mice doses ranging from 62.5 to 4000 mg/kg 5 days/week for 5 weeks and observed testicular atrophy and leukopenia, the intensity of which was dose related. The response was essentially equivalent to that observed with the unesterified ether.

71.2.4   Subcutaneous Injections

Browning (17) reports that the lethal dose for mice is 5.0 ml/kg (5.04 g/kg) and 4.6 ml/kg (4.63 g/kg). Seven injections of 0.5 ml or four injections of 1.0 ml in guinea pigs caused their death 1 to 5 days after treatment. Kidney injury was apparent (7).

71.2.5   Eye Contact

The material is mildly irritating to the eyes (11).

71.2.6   Skin Contact

The material is not significantly irritating to the skin (7).

71.2.7   Skin Absorption

It can be absorbed through the skin if exposure is prolonged. Using the "cuff" procedure on rabbits Smyth and Carpenter (98) estimated the LD_{50} to be 5.25 ml/kg (5.29 g/kg).

71.2.8   Sensitization

Jordan and Dahl (126) report that a 58-year-old woman developed dermatitis on the nose due to her eyeglasses. Patch testing revealed that the causative agent could have been ethylene glycol monomethyl ether acetate and/or ethyl acetate. They state that ethylene glycol monomethyl ether acetate should be considered only as a rare cause of allergic dermatitis, for this was the first incident recorded in the literature. Even this qualification seems conservative.

71.2.9   Inhalation

Gross (7) reports the results of several inhalation studies; they may be summarized as follows: (1) in an essentially saturated atmosphere, 22 mg/l (about 4554 ppm), mice and rabbits tolerated single exposures for 3 hr with only irritation of the mucous membranes; guinea pigs survived exposure of 1 hr, but succumbed days later. (2) Cats died after receiving one 9-hr exposure to 2500 ppm and survived one 7-hr exposure to 1500 ppm but did show an increase in blood clotting time and showed changes in the brain. (3) Cats, guinea pigs, rabbits, and mice were given repeated 8-hr exposures to 500 and 1000 ppm; at 500 ppm the cats showed slight narcosis and died, but the others lived. At 1000 ppm deaths occurred in all species. With both concentrations kidney injury occurred. (4) Cats tolerated repeated 4- to 6-hr exposures to 200 ppm, but decreases in blood pigments and in numbers of red cells were noted. Smyth et al. (98) report that a single 4-hr exposure to 7000 ppm killed two of six rats.

71.2.10   Metabolism

Browning (17) states that this material must be saponified by the body because it gives rise to toxic signs similar to those caused by ethylene glycol methyl ether. Some acetaldehyde may also be formed.

71.2.11   Human Experience

The only adverse human experience reported is that by Jordan and Dahl (126), which is of a questionable nature (see Section 71.2.8).

71.3   Hygienic Standard of Permissible Exposure

The American Conference of Governmental Industrial Hygienists (30) has recommended a threshold limit value of 25 ppm (1980). Since the data on this ester are not adequate to permit such a recommendation, this value must be based upon analogy with ethylene glycol monomethyl ether. Since the ester is converted to the ether in the body, this seems justifiable but subject to the same limitations as expressed in the discussion of the ether (see Section 3.3). The value of 25 ppm has also been adopted by the Occupational Safety and Health Administration (31).

72   ETHYLENE GLYCOL MONOETHYL ETHER ACETATE; Ethanol, 2-Ethoxy-, Acetate; 2-Ethoxyethyl Acetate; Cellosolve® Acetate; Poly-solv® EE Acetate; CAS No. 111-15-9

$$C_2H_5OCH_2CH_2OOCCH_3$$

72.1   Physical and Chemical Properties

See Table 51.9.

72.2   Physiologic Response

72.2.1   Summary

High concentrations of vapor are irritating to the eyes and nose. The material is fairly low in single-dose oral toxicity, somewhat irritating to the eyes, not appreciably irritating to the skin, poorly absorbed through the skin, and not expecially toxic when inhaled in amounts likely to be encountered under ordinary conditions. It is capable of causing central nervous system depression, blood changes, and lung and kidney injury.

72.2.2 Single-Dose Oral

Smyth et al. (62) report LD_{50} values of 5.10 g/kg for rats and 1.91 g/kg for guinea pigs when the material was fed as a 50 percent aqueous suspension in 1 percent Tergitol 7 surfactant. Carpenter (127) reports the oral LD_{50} for rabbits to be 1.95 g/kg. The toxic effects seen were some degree of gastrointestinal tract irritation and kidney injury, and to a lesser degree, injury of the liver. Rarely was bloody urine passed, but the bile was often orange or red. Narcosis occurred mostly in those animals at or above the LD_{50} dose level.

72.2.3 Repeated-Dose Oral

Nagano et al. (13) fed mice doses ranging from 62.5 to 4000 mg/kg 5 days/week for 5 weeks and observed testicular atrophy and leukopenia, the intensity of which was dose related. They concluded that the response was similar to that of the unesterified ethyl ether.

72.2.4 Eye Irritation

Ethylene glycol monoethyl ether acetate is somewhat irritating to the eyes of rabbits (11).

72.2.5 Skin Irritation

The liquid is not significantly irritating to the skin unless exposure is prolonged or frequently repeated (127).

72.2.6 Skin Absorption

Absorption of the material through the intact skin can occur but the lethal dose is large. Carpenter (127) reports the LD_{50} for rabbits to be 10.3 g/kg.

72.2.7 Injection

Von Oettingen and Jirouch (40) report the fatal dose for mice to be 5.0 ml/kg. Gross (7) reports that single doses of 0.5 or 1.0 ml/kg were well tolerated when injected intraperitoneally in guinea pigs. When injected subcutaneously seven times, doses of 0.5 and 1.0 ml/kg caused temporary ill effects but no deaths.

72.2.8   Inhalation

According to Gross (7), a 1-hr exposure to an atmosphere essentially saturated
with vapor (<4000 ppm) was survived by guinea pigs. Cats exposed once for
2 to 6 hr to an atmosphere laden with fog survived, but two such exposures
caused vomiting, paralysis, albumin in the urine, and death. Mice, guinea pigs,
and a rabbit were unaffected by twelve 8-hr exposures to a concentration of
450 ppm, but another rabbit and two cats died before the end of the exposure
period. Albumin occurred in the urine, and the kidneys of the animals that
died were injured.

Carpenter (127) states that rats died as a result of a 2-hr exposure to an
atmosphere (probably fog-laden) generated by bubbling the air through the
boiling liquid, but survived a 4-hr exposure when the air was bubbled through
the liquid at room temperature (1500 ppm). Under the latter conditions deaths
resulted from an 8-hr exposure. Dogs survived 120 daily 7-hr exposures to a
concentration of 600 ppm without apparent injury. Carpenter was unable to
detect methemoglobin in the blood, other hematologic changes, any effect upon
numerous clinical tests or any histopathologic changes in the tissues. This is a
bit unexpected when one considers the effects caused by ethylene glycol
monoethyl ether and the ether—ester on rats, but it may be explained by the
difference in susceptibility of the two species and perhaps the dosage. He
concludes that ethylene glycol monoethyl ether acetate is in the same range of
toxicity as methyl ethyl ketone, propylene dichloride, and tetrachloroethylene,
but its hazards are believed to be less because its vapor pressure is substantially
lower.

72.2.9   Human Experience

There are no records of adverse human experience that can be attributed to
ethylene glycol monoethyl ether acetate. Perhaps the reason for this is that the
vapors are objectionable in concentrations necessary to cause adverse effects.

72.3   Hygienic Standard of Permissible Exposure

The American Conference of Governmental Industrial Hygienists (30) has
recommended (1980) a threshold limit value of 100 ppm, with a notice of
intention to lower the value to 50 ppm with a C designation. Since the systemic
effects of the material are in all probability directly related to the metabolic
product, ethylene glycol monoethyl ether, the margin of safety provided by this
value may be small. The Occupational Safety and Health Administration has
also accepted the 100 ppm level as a guide for control (31).

73   DIETHYLENE GLYCOL MONOMETHYL ETHER ACETATE; Ethanol, 2-(2-Methoxyethoxy)-, Acetate; 2-(2-Methoxyethoxy)ethyl Acetate; Methyl Carbitol® Acetate; CAS No. 629-38-9

$$CH_3OCH_2CH_2OCH_2CH_2OOCCH_3$$

73.1   Physical and Chemical Properties

See Table 51.9.

73.2   Physiologic Response

Summary.   Diethylene glycol monomethyl ether acetate has a low single-dose oral toxicity. The LD_{50} values reported by Smyth et al. (62) are 11.96 g/kg for rats and 3.46 g/kg for guinea pigs when the material is fed as a 50 percent aqueous solution. Carpenter and Smyth (11) found the material to be appreciably irritating to the rabbit eye. There has been no adverse human experience reported nor would any be expected in ordinary industrial handling and use.

73.3   Hygienic Standard of Permissible Exposure

None has been suggested nor would one seem necessary in view of the low volatility and the nature of the material.

74   DIETHYLENE GLYCOL MONOETHYL ETHER ACETATE; Ethanol, 2-(2-Ethoxyethoxy)-, Acetate; Carbitol® Acetate; CAS No. 112-15-2

$$C_2H_5OCH_2CH_2OCH_2CH_2OOCCH_3$$

74.1   Physical and Chemical Properties

See Table 51.9.

74.2   Physiologic Response

Summary.   Diethylene glycol monoethyl ether acetate is low in single-dose oral toxicity. The LD_{50} values reported by Smyth et al. (62) are 11.0 g/kg for rats and 3.93 g/kg for guinea pigs when the material is administered as a 50 percent aqueous solution. The LD_{50} for rabbits is 4.4. g/kg (8). Carpenter and Smyth (11) found the material to be only very slightly irritating to the rabbit eye. The skin absorption LD_{50} for rabbits is 15.0 g/kg (8), but it is only very slightly irritating to the skin. Human patch tests, using undiluted material, resulted in a limited number of people developing mild skin irritation (8). Rats and guinea pigs exposed to essentially saturated atmospheres at room temperature for 8 hr all survived, but gross autopsy revealed injury to the lungs and kidneys (8). No adverse human experience has been reported nor would any be expected under ordinary conditions of industrial handling and use.

74.3 Hygienic Standard of Permissible Exposure

None has been suggested. However, in light of the limited data available, it would seem wise to avoid prolonged or repeated exposures to vapors, mists, or fogs of diethylene glycol ethyl ether acetate.

75 DIETHYLENE GLYCOL MONOBUTYL ETHER ACETATE; Ethanol, 2-(2-Butoxyethoxy)-, Acetate; Butyl Carbitol® Acetate; CAS No. 124-17-4

$$C_4H_9OCH_2CH_2OCH_2CH_2OOCCH_3$$

75.1 Physical and Chemical Properties

See Table 51.9.

75.2 Physiologic Response

75.2.1 Summary

Diethylene glycol monobutyl ether acetate was at one time used widely as an insect repellent. The material is low in single-dose oral toxicity and not appreciably irritating to the eyes or skin, but it can be absorbed through the skin in toxic amounts when contact is extensive and prolonged.

75.2.2 Single-Dose Oral

Smyth and co-workers (62) fed the material as a 50 percent suspension in 1 percent Tergitol 7 surfactant, and found the LD_{50} to be 11.92 g/kg for rats and 2.34 g/kg for guinea pigs. The LD_{50} for rabbits is 2.26 g/kg (8). Others (9) have fed the material emulsified with 5 percent gum arabic and found it to be more toxic, the LD_{50} for rats being estimated at 7 ml/kg. In doses slightly below the lethal level, it causes marked narcosis. Draize et al. (128) report LD_{50} values for several species as follows: rabbits, 2.8; guinea pigs, 2.7; rats, 7.1; mice, 6.6; and chicks, 5.0 ml/kg.

75.2.3 Eye Contact

Carpenter and Smyth (11) have found the material to be only slightly irritating to the eys of rabbits.

75.2.4 Skin Contact

Repeated and prolonged contact with the skin causes mild erythema and exfoliation in rabbits and humans, particularly when sweating (128).

75.2.5 Skin Absorption

The material is absorbed through the skin of rabbits in toxic quantities. Draize et al. (128) estimate the single-dose LD_{50} by skin absorption for the rabbit to be 5.5 ml/kg and the repeated dose (90-day) LD_{50} to be 2 ml/kg. Others (8) report an LD_{50} for rabbits of 14.8 ml/kg (14.5 g/kg). In rabbits inuncted repeatedly, they observed hematuria with degenerative changes in the kidney. Hemoglobinuria was severe when the dose was 4 ml/kg.

75.2.6 Inhalation

A single 8-hr exposure to essentially saturated vapors caused the death of one of six rats.

75.2.7 Human Experience

Diethylene glycol monobutyl ether acetate has been used by many human being as an insect repellent. This has resulted in rather extensive and intimate contact with the skin and undoubtedly some inhalation of vapors. Hoehn (129) reports one case of a 3-year-old child who allegedly suffered nephrosis as a result of such use. As a result, and because this conceivably could have occurred, the material was withdrawn from this market. It would seem that the material would present no appreciable hazard in industrial handling and use.

75.3 Hygienic Standard of Permissible Exposure

None has been suggested. Until more is known about the toxicity by inhalation, it would seem wise to avoid prolonged or repeated exposure.

76 PROPYLENE GLYCOL MONOMETHYL ETHER ACETATE; CAS No. 108-65-6.
DIPROPYLENE GLYCOL MONOMETHYL ETHER ACETATE; CAS No.—None Found.
TRIPROPYLENE GLYCOL MONOMETHYL ETHER ACETATE; CAS No.—None Found

76.1 Physical and Chemical Properties

See Table 51.9.

76.2 Physiologic Response

Summary (9). All these compounds are low in single-dose oral toxicity for rats. When fed undiluted, single doses of 3 ml/kg were survived by all animals in each case; doses of 10 ml/kg caused the death of three of five fed the propylene glycol derivative, and all of five fed the di- and tripropylene glycol derivatives. The propylene glycol derivative is somewhat painful and irritating to the eyes, but the others are not. None of them is appreciably irritating to the skin of rabbits, nor are they absorbed through the skin in significant amounts, even when applied repeatedly for a 2-week period of time. It would not seem that any of them would present any serious hazards in ordinary industrial handling and use.

77 ETHYLENE GLYCOL DINITRATE; 1,2-Ethanediol, Dinitrate; Nitroglycol; Dinitroglycol; EGDN; EGN; Ethylene Nitrate; CAS No. 628–96–6

$$O_2NOCH_2CH_2ONO_2$$

77.1 Physical and Chemical Properties

Ethylene glycol dinitrate is a yellow oily liquid as ordinarily produced, but when pure it is colorless. It is soluble in many solvents, such as carbon tetrachloride, ether, benzene, toluene, and acetone, limited in solubility in common alcohols, and only very slightly soluble in water. It decomposes violently upon heating or impact with a force similar to that of nitroglycerin. Additional physical and chemical properties are given in Table 51.10.

77.2 Determination in the Atmosphere

In accordance with NIOSH Method No. P&CAM 203 (161) and No. S216 (162), a known volume of air is drawn through a tube packed with Tenax GC to trap the vapors present. The analyte is desorbed with ethyl alcohol. Analysis is made using gas chromatography employing an electron capture detector.

77.3 Physiologic Response

77.3.1 Summary

Ethylene glycol dinitrate is moderate in toxicity to rats when given in single oral doses. It is not likely to be significantly irritating upon eye or skin contact, but it is readily absorbed through the skin in toxic amounts. Even minor skin exposures can cause adverse clinical effects. Inhalation of as little as 0.02 ppm

Table 51.10. Physical and Chemical Properties of Some Glycol Dinitrates

Property	Ethylene Glycol Dinitrate	Diethylene Glycol Dinitrate	Triethylene Glycol Dinitrate	Propylene Glycol Dinitrate
CAS No.	628-96-6	693-21-0	111-22-8	6423-43-4
Common designation	EGDN	DEGDN	TEGDN	PGDN
Molecular formula	$C_2H_4N_2O_6$	$C_4H_8N_2O_7$	$C_6H_{12}N_2O_8$	$C_3H_6N_2O_6$
Molecular weight	152.1	196.1	240	166
Specific gravity (25/4°C)	1.491			1.4
Boiling point (°C) (760 mm Hg)	Explodes			92 at 10 mm Hg
Vapor pressure (mm Hg) (25°C)	0.072			0.07 (22.5°C)
Vapor density (air = 1)	5.24			
Percent in saturated air (25°C)	0.0095			
1 ppm ⇌ mg/m³ at 25°C, 760 mm Hg	6.24	8.02	9.82	6.79
1 mg/l ⇌ ppm at 25°C, 760 mm Hg	160.9	124.7	101.9	147.3

has been reported to have caused adverse effects but absorption through the skin may have contributed significantly to the total exposure. Repeated exposures to vapor may lead to a tolerance as long as daily exposures occur, but this is lost in 24 to 60 hr. The major clinical effects are lowered blood pressure due to vasodilatation, increased pulse rate, headache, dizziness, nausea, vomiting, hypotension, tachycardia, peripheral paresthesis, and chest pain. Anginal-type attack and, in some cases, sudden deaths have occurred upon reexposure after a few days of nonexposure; a limited number have been reported in the absence of reexposure.

77.3.2   Oral—Single Dose

The single-dose oral LD_{50} for rats of ethylene glycol dinitrate has been reported to be 616 mg/kg. This value is similar to that for propylene glycol dinitrate (65). The chief effect is blood vessel dilation (130).

77.3.3   Injection—Single Dose

Gross et al. (133) report that the LD_{100} values for ethylene glycol dinitrate, when injected subcutaneously to the rabbit and cat, are 300 mg/kg and 100 mg/kg, respectively. According to Jones et al. (134), Kylin and his co-workers reported that the toxicity of ethylene glycol dinitrate was very similar to that of propylene glycol dinitrate when given intraperitoneally, the LD_{50} values of which range from 0.4 to 1.05 g/kg, depending on the species.

Sugawara (141) gave rabbits 150 mg/kg subcutaneously without significant effects.

77.3.4 Injection—Repeated Dose

Intramuscular. Cavagna et al. (135) report that there was an increase in catecholamine in myocardial tissue reaching a maximum concentration at the end of each week during which rats were given 5 mg/kg every 12 hr (11 times/week) for 15 weeks. However, there were no signs of long-term accumulation. At the end of each week and at the end of the experiment, the catecholamines peak in 24 hr but returned to normal in 48 hr. Vigliani et al. (136) also saw an increase in catecholamines in the heart when rats were injected intramuscularly with 60 mg/kg/day in three divided doses for 5 days.

Subcutaneous. Kalin et al. (137) report that rats given 75 mg/kg/day for 14 to 28 days of ethylene glycol dinitrate developed no effect on the uptake, storage, release or resynthesis of noradrenaline in the adrenergic nerves. These findings are contrary to those found by Minami et al. (138), Clark (139), and Clark and Litchfield (140), who report varying effects, such as increased catecholamines (138) at a dose of 65 mg/kg/day for 5 days/week for 8 weeks, increase in plasma corticosterone (130) at the same dose, and a fall in blood pressure in dogs given 10 mg/kg/day, 5 days/week for 5 weeks, and in rats given 65 mg/kg/day, 5 days/week for 10 weeks. All (138–140) report that the effects disappeared in 2 to 3 days after cessation of exposure.

77.3.5 Eye Contact

There are no reports of eye tests conducted on ethylene glycol dinitrate. Extensive human experience suggests that eye contact with it is not likely to cause significant irritation. This is consistent with the observation that propylene glycol dinitrate causes at most only mild irritation. However, the implied toxicity of this material by absorption indicates that it would be wise to use suitable eye protection where eye exposure might occur.

77.3.6 Skin Contact

No animal tests for skin effects have been reported. Extensive human experience, however, indicates that ethylene glycol dinitrate is not likely to give rise to skin irritation.

77.3.7 Skin Absorption

Ethylene glycol dinitrate is readily absorbed through the skin in toxic amounts, more readily than nitroglycerine (130). The minimum dose, applied as a 1 percent solution in alcohol, which causes headaches in man, was reported as 18 to 35 mg. Einert et al. (131), in reviewing the problems encountered in dynamite plants where nitroglycerin and ethylene glycol dinitrate are handled, quote Ebright, who states that headaches could result from merely "shaking hands with persons who handle dynamite." They also quote Gross and co-workers, who showed that practically all of the ethylene glycol dinitrate was absorbed in 8 days, with up to one-third absorbed in the first 4 to 12 hr. In their studies Einert et al. (131) showed that skin exposure contributed significantly to the clinical symptoms. They considered the risk from absorption from gloves to be low if less than 0.25 mg of the material was absorbed in 8 hr, moderate if 0.25 to 0.75 mg was absorbed, and high if more than 0.75 mg was absorbed. These figures emphazie the high toxicity of ethylene glycol dinitrate by percutaneous absorption.

Morikawa et al. (132), in their study using plethysmographic measurements, found that nonexposed volunteers who wore used protective gloves after they had been washed with detergents experienced characteristic changes in the pulse wave and a drop in systolic blood pressure, thus indicating that it is very difficult to remove ethylene glycol dinitrate from contaminated gloves. It is expected that the same would be the case with other wearing apparel, particularly shoes.

77.3.8 Inhalation

Cats that inhaled an average concentration of 2 ppm of ethylene glycol dinitrate for a total of 1000 8-hr exposures exhibited only temporary blood changes. If exposed to 21 ppm, there were marked blood changes. Higher levels killed the cats in a short time (133). Rats and guinea pigs exposed to 80 ppm for 6 months experienced drowsiness and Heinz body formation in the red blood cells. The pathologic effects from chronic exposure were fatty changes in the liver, heart muscle, and kidney, with pigment deposits in the liver and spleen similar to those seen in anemia.

Human experience has indicated that those consuming alcohol are especially susceptible to ethylene glycol dinitrate poisoning (142).

Carmicheal and Lieben (143), in a survey of sudden deaths to people working in dynamite plants, found that the deaths occurred in persons exposed to levels considerably greater than those considered safe. Forssman et al. (144), in a study of dynamite workers in Sweden, reported that no chronic effects were seen from repeated exposures to a mixture of nitroglycerin and ethylene glycol dinitrate up to 0.5 ppm (5 mg/m^3). Einert et al. (131), in medical and industrial hygiene surveys in California, found levels of 0.013 to 0.44 ppm of a mixture of nitroglycerin and ethylene glycol dinitrate. Fourteen to 48 percent of the employees developed adverse clinical symptoms. The authors further studied the possible effects of skin absorption and showed that this type of exposure was a significant factor in the cause of adverse effects. Unacclimated human volunteers developed mild headaches and a drop in blood pressure in 25 min when exposed to 0.054 ppm (145).

Morikawa et al. (132), using plethysmography, report that up to 100 employees exposed frequently to levels of 0.5 ppm or more suffered varying degrees of clinical effects, primarily in the summer. In 1963 the control of vapor in the explosive factory was improved so that exposure levels ranged from 0.02 to 0.066 ppm, mostly ethylene glycol dinitrate. Even at these levels, a few of the employees experienced headaches and abnormal pulse waves similar to those caused by ethylene glycol dinitrate. These authors recognize that skin absorption is an important factor in industrial handling and it is possible that this route contributed to the observed effects.

Workers in a Japanese dynamite plant developed no methemoglobin in their blood when exposed repeatedly to 0.2 to 2 ppm of ethylene glycol dinitrate for about 6 hr/day (146).

77.3.9   Mutagenicity

Kanonova et al. (147) evaluated a number of alkyl nitrates for mutagenic effects using the extracellular bacteriophage T4B of *Escherichia coli* and the alkylation of the DNA of the phage. Ethylene glycol dinitrate was found to cause mutagenic changes whereas nitroglycerin did not.

77.3.10   Metabolism

Numerous workers have conducted studies to elucidate both the metabolism and the pharmacologic action of ethylene glycol dinitrate because of its use in pharmacology and its adverse human exposure effects. These studies have been summarized by Litchfield (148, 149). Basically, the studies show that when animals and humans are exposed to ethylene glycol dinitrate the material appears in the blood immediately, peaking in concentration in about 30 min. Apparently it is rapidly metabolized, for it disappears in about 4 hr with very little found excreted in the urine. When it was given orally a trace amount was found in the feces. Blood analysis shows that inorganic nitrite, inorganic nitrate, and ethylene glycol mononitrate are the main metabolites.

The inorganic nitrite concentration in the blood peaked in about 2 hr and returned to normal in 12 hr. During this period, very little or no inorganic nitrites were found in the urine. This implies that the inorganic nitrite was either metabolized to inorganic nitrate or incorporated into the body.

Inorganic nitrate levels in the blood peaked in about 3.5 hr, then returned to normal levels in 12 hr. During this period, sufficient inorganic nitrate was excreted in the urine to account for approximately 60 percent of the amount of ethylene glycol dinitrate administered. The fate of the other 40 percent of inorganic nitrate is not known.

The ethylene glycol mononitrate found in the blood has been shown to be rapidly metabolized to ethylene glycol and inorganic nitrate and inorganic nitrite. Less than 0.5 percent of the ethylene glycol mononitrate was found to be excreted in urine.

Very little ethylene glycol has been seen in the urine after administration of either ethylene glycol dinitrate or mononitrate; however, in the blood, ethylene glycol reaches a peak level in 2 to 3 hr; hence it appears that it is metabolized rapidly (152).

Based upon these facts, the scheme of metabolism of ethylene glycol dinitrate shown in Figure 51.2 has been proposed (149). The mechanism of step I is considered to proceed mainly to ethylene glycol mononitrate and inorganic nitrites via organic nitrate reductase accompanied by reduced glutathione. However, Tsuruta and Hasegawa (151) found that there are two enzymes involved, a nitrate- as well as a nitrite-forming enzyme.

Step II is accomplished by the addition of $(OH)^-$, but the mechanism has not been suggested. Step III occurs as a result of reductive hydrolysis whereas step IV is the result of direct hydrolysis. Inorganic nitrites are known to be metabolized to inorganic nitrates by oxidation. Because of the ability of inorganic nitrites to cause the production of methemoglobinemia, it was suggested that some of the adverse effects of ethylene glycol dinitrate might be due to the presence of methemoglobin. Tests in rabbits by Hasegawa and Sato (153, 154) showed that methemoglobin is in fact produced by ethylene glycol dinitrate and that the amounts produced parallel the amount of free ethylene glycol

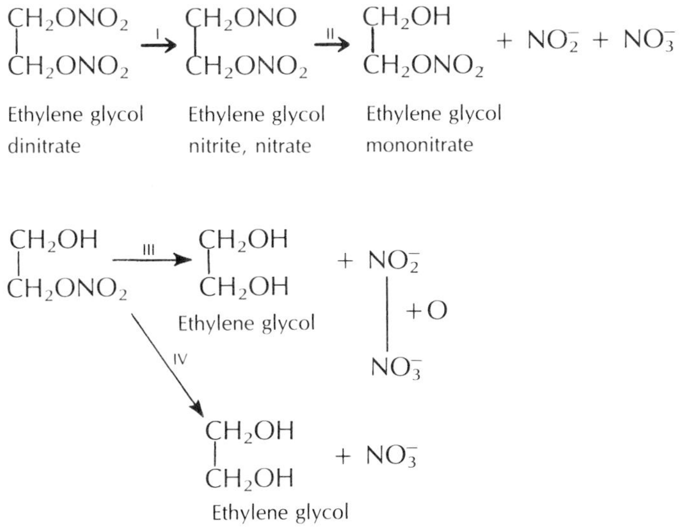

$$
\begin{array}{ccc}
\underset{|}{CH_2ONO_2} & \underset{|}{CH_2ONO} & \underset{|}{CH_2OH} \\
CH_2ONO_2 & \xrightarrow{\ I\ } CH_2ONO_2 & \xrightarrow{\ II\ } CH_2ONO_2 \quad + NO_2^- + NO_3^-
\end{array}
$$

Ethylene glycol　　　　Ethylene glycol　　　Ethylene glycol
dinitrate　　　　　　　nitrite, nitrate　　　mononitrate

$$
\underset{CH_2ONO_2}{\overset{CH_2OH}{|}} \xrightarrow{\ III\ } \underset{CH_2OH}{\overset{CH_2OH}{|}} \quad + NO_2^-
$$

Ethylene glycol

$\xrightarrow{\ IV\ }$

$$\underset{CH_2OH}{\overset{CH_2OH}{|}} \quad + NO_3^-$$

Ethylene glycol

$$+O \to NO_3^-$$

Figure 51.2 Scheme of ethylene glycol dinitrate metabolism.

dinitrate in the blood (153). In their study of the mechanism of production of methemoglobin, they found it was produced only in the presence of oxygen. On this basis, they suggest that methemoglobin is formed in the following manner:

$$HB(O_2)_4 + \begin{matrix} CH_2ONO_2 \\ | \\ CH_2ONO_2 \end{matrix} + 3H_2O \rightarrow Hb(OH)_4 + \begin{matrix} CH_2OH \\ | \\ CH_2ONO_2 \end{matrix} + NO_3^- + 3O_2$$

Hemoglobin   Methemo-   Ethylene
oxygenated   globin   glycol mono-
                                                    nitrate

This scheme, however, does not follow the postulated scheme for the metabolism of ethylene glycol dinitrate in which inorganic nitrites are formed. Hence it is possible that the production of inorganic nitrites may also be involved in the production of methemoglobin as well.

Hasegawa and Sato (154) and Hasegawa et al. (146) as well as others report that humans exposed to ethylene glycol dinitrate do not develop methemoglobin; hence the adverse effects seen apparently are not related to methemoglobin formation.

77.3.11   Mode of Action

The clinical effects due to exposure to ethylene glycol dinitrate have been reported by many medical observers. They have been summarized by Litchfield (149) and Plunkett (155) to be lowered blood pressure because of vasodilatation, increased pulse rate, headache, dizziness, nausea and vomiting, hypotension, tachycardia, peripheral paresthesia, and chest pain. Anginal-type attacks, and in some cases deaths, that occur in workers usually result 48 to 60 hr after cessation of exposure. Exposed workers become tolerant during the work week but lose this tolerance over the weekend or when away from exposure. Thus they experience the above mentioned effects when exposure occurs again. This is known as "Monday headache."

Gobbato and DeRosa (156) exposed humans to ethylene glycol dinitrate by inhalation and report that it causes severe cranioencephalic dilation as shown by an increase in systolic blood pressure, in cerebral flow, and in vasoelasticity with a reduction in local capillary resistance. The changes began almost immediately upon exposure and reached a maximum in 4 to 8 min and disappeared in 15 to 20 min. This corresponds with the presence of ethylene glycol dinitrate in the blood. This suggests that the vasodilatation seen is due to the ethylene glycol dinitrate and not to its metabolites. Clark and Litchfield

(152) report that the initial drop in blood pressure due to ethylene glycol dinitrate exposure returns to normal. This is then followed by a longer period of decrease in blood pressure which paralleled the presence of the metabolite, ethylene glycol mononitrate.

The causes of "Monday headaches" and/or "Monday deaths" have also been explored. Vigliani et al. (136) feel that the causes of the sudden deaths associated with exposures to ethylene glycol dinitrate are very similar to those resulting from nitroglycerin exposure. Thus the causes of death from nitroglycerin may also be considered possible causes of death from exposure to ethylene glycol dinitrate. They are as folows:

1.   Peripheral vascular collapse due to prolonged vasodilatation.
2.   Coronary insufficiency due to vascular coronary spasm resulting from prolonged vasodilatation.
3.   Coronary insufficiency due to lowered pulse pressure, increased diastolic pressure, and high pulse rate.
4.   Sensitization to cardioinhibitory trigeminovagal or senocarotid reflexes.
5.   Coronary insufficiency due to increased consumption of oxygen by the myocardium, caused by prolonged inhibition.
6.   Peripheral vascular collapse due to vasomotor paralysis resulting from hypothalmic alterations.
7.   Coronary insufficiency due to vascular sclerosis.

Examination of these hypotheses show that there are two major concepts: coronary insufficiency and serious changes in the dynamics of the cardiovascular system. However, the exact cause or causes have not been explained satisfactorily.

Vigliani et al. (136) in their work found that there was an increase in the catecholamine content of the heart because of the biologic action of ethylene glycol dinitrate. This suggests that some of the adverse effects due to exposure to ethylene glycol dinitrate may be explained by the storage, then sudden release, of large amounts of catecholamines.

Minami et al. (138) postulate that ethylene glycol dinitrate affects not only the catecholamine metabolism of the heart but also that of the brain and adrenals. Kalin et al. (137) report that they found that ethylene glycol dinitrate had no clear effect upon the uptake, storage, release, or resynthesis of noradrenaline in the adrenergic nerves of the rat.

Suwa et al. (157) also studied the effect of ethylene glycol dinitrate on the heart. They found evidence that the heart continued to be affected even after ethylene glycol dinitrate was no longer detectable in the blood. They feel that the supersensitivity of the heart might be due to adverse effects on the nervous system, not to electrolyte changes in the myocardium.

Gobbato and DeRosa (156), from studies in humans, suggested that the adverse effects from exposure were due to circulatory collapse resulting from systemic vasodilatory effects.

Clark (139), in his study of the effects of ethylene glycol dinitrate on pituitary-adrenocortical function, concluded that this material apparently increased the level of corticosterone in the plasma. This may be due to hypotension which could stimulate the pituitary gland.

77.3.12   Human Experience

There are many reports in the literature of human exposure to ethylene glycol dinitrate either alone or in association with nitroglycerin, mainly in dynamite plants. Carmicheal and Lieben (143) and von Oettingen (142) have summarized them.

Most authors agree that exposures in industrial plants are the result of inhalation of vapors and/or by absorption through the skin. Because of the greater volatility of ethylene glycol dinitrate as compared to nitroglycerin, it is felt that the inhalation effects in operations where both are handled are due mainly to ethylene glycol dinitrate.

Single occasional exposures are more likely to cause the clinical signs described. Most authors agree that repeated daily exposure gives rise to a tolerance which reduces the clinical effects, and this continues during the work week. However, over the weekend, this tolerance is lost so that on return to work and exposure the clinical effects again occur. In a limited number of workers these effects can be severe, resulting in sudden death due to angina-type effects, usually in the summer months. These phenomena have come to be known as "Monday headache," or in the case of sudden death, "Monday morning death." Symanski (158) in 1952 summarized the reports of sudden deaths occurring worldwide and found 44 deaths. Carmicheal and Lieben (143), in a survey of such deaths in Pennsylvania in 1961, listed 12 such deaths and Forssman et al. (159) also listed a number of such deaths.

Forssman et al. (159) report that there were no chronic effects in workers exposed repeatedly to 5 mg/m^3 (0.5 ppm) to nitroglycerin and ethylene glycol dinitrate. Their criteria were clinical, X-ray of heart and lung, electrocardiographic studies, and blood examinations. However, Trainor and Jones (145) found that human volunteers exposed to 2 mg/m^3 (0.2 ppm) of a mixture of ethylene glycol dinitrate and nitroglycerin experienced a fall in blood pressure and marked headaches in 3 min. Even at a level of 0.7 and 0.5 mg/m^3 (0.08 and 0.05 ppm) the same effects were seen after 25 min.

Morikawa et al. (132) found that a significant number of dynamite plant workers developed slight to moderate abnormal pulse waves typical of those caused by ethylene glycol dinitrate when tested by plethysmography. These workers experienced vapor exposures exceeding 0.1 ppm of ethylene glycol dinitrate in 6 of 104 measurements. Cutaneous absorption from contaminated rubber gloves also occurred.

Hematologic and clinical chemical studies on 35 subjects exposed to ethylene glycol dinitrate showed no abnormalities in blood electrolytes, liquids, cholesterol levels, and various enzyme activities (160).

Hasegawa et al. (146) report that dynamite workers exposed to 0.2 to 2 ppm of vapors 6 hr/day repeatedly developed no detectable methemoglobinemia. This was thought to be due to the fact that man is the lowest producer of methemoglobin when exposed to ethylene glycol dinitrate of the animals they studied.

Humans consuming alcohol are believed to be especially susceptible to this material (142).

77.4 Hygienic Standard of Permissible Exposure

Both the Conference of Governmental Industrial Hygienists (30) and Occupational Safety and Health Administration (31) have suggested 0.2 ppm both as a ceiling level and as a guide for repeated daily exposures of 8 hr/day. The 1980 TLV list gives notice of lowering the TLV to 0.02 ppm. The Occupational Safety and Health Administration includes a guide of 1 mg/m^3 (1.61 ppm) as well. It should be noted that ethylene glycol dinitrate frequently is handled along with nitroglycerin, and since they both have essentially the same physiologic activity, the control value of 0.2 ppm (or 0.02) should be used for either material alone or as the sum of the concentrations of each.

78 DIETHYLENE GLYCOL DINITRATE; Ethanol, 2,2'-Oxybis-, Dinitrate; DEGDN; CAS No. 693-21-0

$$O_2NOCH_2CH_2OCH_2CH_2ONO_2$$

78.1 Physical and Chemical Properties

See Table 51.10.

78.2 Determination in the Atmosphere

No official methods have been published but it seems likely that NIOSH Method No. P&CAM 203 (161) or No. S216 (162) recommended for ethylene glycol dinitrate and nitroglycerin would we adaptable.

78.3 Physiologic Response

78.3.1 Summary

Diethylene glycol dinitrate is a colorless or yellowish liquid of low volatility. Depending on purity of the product, it gives water a bitter-astringent taste at concentrations of 130 mg/l. The odor threshold in water is about 650 mg/l. It is low in single-dose oral toxicity and high in repeated-dose oral toxicity. The main adverse effects are those of central nervous system injury and hypotension. Although it apparently does cause the production of methemoglobinemia, it does so only as fairly high doses. Chronic exposure has been shown to cause a drop in blood pressure. No information was found concerning the effects of eye or skin contact or of inhalation.

78.3.2 Single-Dose Oral (163)

Diethylene glycol dinitrate has a low single-dose oral toxicity, the LD_{50} for mice, rats, and guinea pigs being 1250, 1180, and 1250 mg/kg, respectively. The signs seen were acute cyanosis and central nervous system injury. Intravenous injection of 0.4 mg/kg to rabbits showed that this material also reduces blood pressure and induces prolonged hypotension.

78.3.3 Repeated-Oral (163)

Rats were given by gavage diethylene glycol dinitrate in vegetable oil six times a week for 6 months at doses of 0.05, 0.5, and 5 mg/kg. These doses did not cause the production of methemoglobin. However, those given doses of 5 and 0.5 mg/kg showed signs of changes in conditioned reflex activity, CNS effects, and the immunologic condition. The 5 mg/kg dose decreased the blood pressure by the fifth to the sixth month as well as a change in the mitotic activity of the bone marrow and cardiovascular effects. Hematologic and histopathologic studies revealed no changes. The no-effect level was determined to be 0.05 mg/kg.

78.3.4 Mode of Action

The main effects seem to be those of central nervous system injury and hypotension, but unlike the other dinitrates, it seems to be low in its ability to cause methemoglobinemia especially at small doses. The hypotensive effect was not accompanied by changes in ECG readings, suggesting that this material influences vascular tonus without impairing the activity of the myocardium.

79   TRIETHYLENE GLYCOL DINITRATE; Ethanol, 2,2'-[1,2-Ethanediylbis(oxy)]bis-, dinitrate; TEGDN; CAS No. 111-22-8

$$O_2NOCH_2CH_2OCH_2CH_2OCH_2CH_2ONO_2$$

79.1   Physical and Chemical Properties

See Table 51.10.

79.2   Determination in the Atmosphere

No official methods have been published but it seems likely that NIOSH Method No. P&CAM 203 (161) or No. S216 (162) recommended for ethylene glycol dinitrate and nitroglycerin would be adaptable.

79.3   Physiologic Response

Summary (164).   Triethylene glycol dinitrate is low in single-dose toxicity by various routes in several species, as shown by the following data:

Species	Sex	Route of Administration	LD_{50} (mg/kg)
Mouse	M	IP	945
Guinea pig	M	IP	700
Rat	M	IP	796
Rat	M	SC	2520
Rat	M	Oral	1000

The effects of exposure were those of methemoglobinemia and hypotension in rats. Unlike both ethylene and propylene glycol dinitrate-treated rats, the rats treated with triethylene glycol dinitrate developed tremors, convulsions, then death apparently due to respiratory arrest. Using a phrenic nerve diaphram preparation, triethylene glycol dinitrate blocked nerve-stimulated contraction, suggesting that this product interferes with nerve-muscle communication which may be the cause of the tremors, convulsions, and respiratory arrest. Daily dermal applications of 21 mmole/kg (5.0 g/kg approximately) caused death of 9 of 11 rabbits treated in 2 to 3 weeks. All the rabbits lost weight and those that died developed the same signs as seen when this material was given by ingestion or by injection. As with the other glycol dinitrates, triethylene glycol dinitrate is apparently metabolized in the red cells in the presence of hemoglobin to mononitrate, inorganic nitrate, and inorganic nitrite.

80   PROPYLENE GLYCOL DINITRATE; 1,2-Propanediol, Dinitrate; PGDN; CAS No. 6423-43-4

$$CH_3CHONO_2CH_2ONO_2$$

80.1   Physical and Chemical Properties

See Table 51-10.

80.2   Determination in the Atmosphere

No official methods have been published but it seems likely that NIOSH method No. P&CAM 203 (161) or No. S216 (162) recommended for ethylene glycol dinitrate and nitroglycerin would be adaptable.

80.3   Physiologic Response

80.3.1   Summary

Propylene glycol dinitrate is a red-orange or colorless liquid depending upon the purity of the product. It is moderate to low in single-dose oral toxicity and moderately toxic by injection. Eye contact should cause at most only mild transient irritation, whereas skin contact may cause only mild irritation upon prolonged or repeated contact. It is readily absorbed through the skin. Repeated skin exposures readily can lead to adverse effects. A single short vapor exposure to concentrations of 1 to 1.5 ppm or more can lead to headaches.

The main effects are circulatory collapse and the development of methemoglobinemia. The symptoms of overexposure are headaches, nasal congestion, dizziness, eye irritation, vasomotor collapse, unconsciousness, and death, depending upon degree of exposure.

80.3.2   Single-Dose Oral

The single-dose oral toxicity is moderate to low for propylene glycol dinitrate, the LD_{50} values reported for female rats are 1.19 g/kg (165) and 0.86 g/kg (134), and for male rats, 0.25 g/kg (164). The adverse signs seen in the rat were prostration, lethargy, anoxia, and signs of methemoglobinemia (165). Jones et al. (134) report in addition reduced response to external stimuli and mild convulsions prior to death.

80.3.3 Single-Dose Injection

The reported LD_{50} values and route of administration are as follows:

Species	Sex	Route of Injection	LD_{50} (g/kg)	Ref.
Rat	F	SC	0.463	165
	M	SC	0.524	165
	M	SC	0.53	164
Mouse	F	SC	1.21	165
Cat	F	SC	0.2 to 0.3	165
Mouse		IP	0.93	165
Mouse	M	IP	1.05	164
Guinea pig	M	IP	0.40	164
Rat	M	IP	0.48	164

The signs seen after injection were similar to those seen after ingestion. Methemoglobinemia was shown to be directly related to the dose administered. Rats receiving doses that proved fatal to a significant number experience an increase in methemoglobin levels of up to 70 to 85 percent (164).

Clark and Litchfield (165) reported that the hemoglobin in rats given an LD_{50} dose subcutaneously was nearly all converted to methemoglobin. They postulate that death is the result of this methemoglobin formation.

Clark and Litchfield (165) also followed the blood pressure of rats given propylene glycol dinitrate subcutaneously and found again that the drop in blood pressure reflected the dose administered. A dose of 0.16 g given subcutaneously caused an average drop of about 60 percent.

80.3.4 Eye Contact

Jones et al. (134) found that propylene glycol dinitrate is only very slightly irritating to the eyes of rabbits. Application of 0.1 ml to the eye caused only mild irritation but no corneal response. The irritation disappeared entirely within 24 hr.

80.3.5 Skin Contact

Skin tests on rabbits showed that this material caused no observable skin irritation when the method similar to that of Draize was used (165). Jones et al. (134) report that the application of 1 g/kg to the skin of rabbits repeatedly caused slight irritation.

80.3.6   Skin Absorption

Clark and Litchfield (165) applied a 10 percent solution of propylene glycol dinitrate to the skin of rats at a dose of 50 mg/kg and found that the rats experienced a detectable fall in blood pressure. This suggests that this material is absorbed through the skin.

Jones et al. (134) administered this material to the skin of rabbits for 2 hr/day for 20 days in doses of 1.0, 2.0, and 4.0 g/kg. At the 2 g/kg level the rabbits all developed initially signs of weakness, cyanosis, and shallow, rapid breathing. One of the five animals died. Those that survived gradually improved so that at the end of the experiment they appeared normal. At the 4 g/kg level 13 of 14 rabbits died after the fifth application. Autopsy revealed the internal organs were dark blue-gray in color and the urinary bladder was markedly distended. The hemoglobin and hematocrit values were depressed and urinary nitrates accounted for approximately 7 percent of the propylene glycol dinitrate given at the 4 g/kg level.

80.3.7   Inhalation—Single Exposure

Jones et al. (134) exposed six rats for 4 hr to a mist of propylene glycol dinitrate (estimated concentration 1350 mg/m^3) and found that the rats showed no toxic signs and there was no mortality. However, the mean methemoglobin level immediately after exposure had risen to 23.5 percent. Both a rabbit and a squirrel monkey were exposed 23 hr/day to concentrations of 240 and 415 mg/m^3, respectively. The rabbit died on the fourth day, the monkey on the third day. Both experienced increased methemoglobin levels, the rabbit 18.2 and the monkey 40.2 percent upon death. Both animals became cyanotic during the experiment.

80.3.8   Inhalation—Repeated Exposure

Jones et al. (134) also exposed rats for 7 hr/day, 5 days/week for a total of 30 exposures at a level of 65 mg/m^3 (about 10 ppm). These rats experienced no mortality nor toxic effects as shown by hematology and histopathologic examination.

In further studies, rats of both sexes, guinea pigs of both sexes, male squirrel monkeys, and male beagle dogs were exposed to 67 ± 8, 108 ± 11, and 236 ± 24 mg/m^3 of propylene glycol dinitrate continuously for 90 days. These levels are estimated to be 10, 16, and 35 ppm, respectively. None of the animals died due to the exposures. The adverse effects seen were anemia, pigment deposition in various organs, fatty changes in the liver, methemoglobin formation, and increased serum and urinary nitrates at the highest exposure level. Dogs and monkeys exposed to about 10 ppm exhibited fatty changes and pigment deposits, and the guinea pigs developed some minor lung hemorrhages at the 16 ppm level. In these tests three of the monkeys at the 35 ppm level were removed from exposure for 2 hr once a week to evaluate avoidance behavior. Their behavior was unaffected.

Based on these data the authors (134) recommend an industrial hygiene standard of 1.2 mg/m^3 (0.2 ppm) for a 40 hr/week exposure and a zero level for those in confined spaces such as a space capsule.

Mattsson et al. (167) exposed rhesus monkeys to the vapor of propylene glycol dinitrate for 4 hr at concentrations of 2 to 33 ppm. At the 2 ppm level the visual-evoked response was altered (increased 20 percent) but behavioral response was unaltered even at the 33 ppm level.

Young et al. (168) found that primates (*Macaca mulatta*) when exposed to 28.2 mg/m^3 (4 ppm) of propylene glycol dinitrate for 23 hr/day for 14 days showed no significant disruption of avoidance behavior, motor coordination, or sensory function.

Stewart et al. (166) exposed human volunteers to levels of propylene glycol dinitrate in air of 0.03 up to 1.5 ppm for up to 8 hr. In addition, some were exposed 8 hr/day for 5 days to 0.2 ppm. The detectable level of propylene glycol dinitrate vapor in air was found to be in the range of 0.2 to 0.26 ppm by most of the subjects.

A single 8-hr exposure to 0.2 ppm or more caused disruption of the organization of the visual-evoked response and headache in a majority of the subjects. Marked impairment of balance occurred in those exposed for 6.5 hr to 0.5 ppm. Those exposed for 40 min to 1.5 ppm developed eye irritation in addition.

Repeated 8-hr exposures for 5 days to 0.2 ppm resulted in tolerance to the induction of headaches; however, the alteration in the visual-evoked response appeared to be cumulative. These findings led the authors (166) to suggest that the guide for the control to 0.2 ppm may not be adequate to prevent adverse responses, especially where repeated exposures may occur.

80.3.9 Metabolism

Clark and Litchfield (165) studied the metabolism of propylene glycol dinitrate in rats. After a subcutaneous dose of 65 mg/kg it was found that propylene glycol dinitrate rapidly enters the bloodstream where it is readily metabolized to propylene glycol mononitrate, inorganic nitrate, and a small amount of inorganic nitrite. The propylene glycol mononitrate apparently is readily metabolized, for the blood level decreases with time and none appears in the urine. Urinanalysis showed that inorganic nitrates are the major product; they represented about 56 percent of the dose administered. Essentially no propylene glycol dinitrate or propylene glycol mononitrate and very little inorganic nitrites were excreted. Studies using whole blood and serum indicate that the metabolism of propylene glycol dinitrate occurs in the red blood cells (169). Based on these findings the metabolism of propylene glycol dinitrate may be as follows:

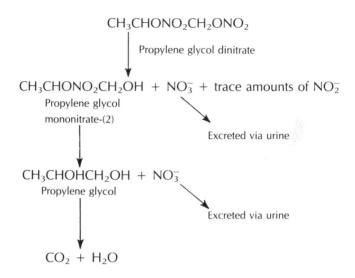

$$CH_3CHONO_2CH_2ONO_2$$
Propylene glycol dinitrate

$$CH_3CHONO_2CH_2OH + NO_3^- + \text{trace amounts of } NO_2^-$$
Propylene glycol
mononitrate-(2)

Excreted via urine

$$CH_3CHOHCH_2OH + NO_3^-$$
Propylene glycol

Excreted via urine

$$CO_2 + H_2O$$

The mechanisms likely to be involved in the various steps of the metabolism scheme are likely to be similar to those postulated in the metabolism of ethylene glycol dinitrate.

80.3.10   Mode of Action

The presence of propylene glycol dinitrate in the blood has been shown to lead to a correlatable drop in blood pressure. As the material is metabolized there is a correlatable increase in the formation of methemoglobin. In rats it has been shown that the hemoglobin is essentially completely converted to methemoglobin at doses that result in the death of the animals (165). On this basis it may be postulated that death is the result of circulatory collapse and anoxia due to the high level of methemoglobin in the blood.

80.3.11   Human Experience

Stewart et al. (166) report that human exposures to Otto Fuel II, which is largely propylene glycol dinitrate, caused headaches, nasal congestion, dizziness, eye irritation, vasomotor collapse, and unconsciousness. However, the number of people involved was not given.

80.4 Hygienic Standards of Permissible Exposure

Jones et al. (134), on the basis of their work, have recommended that the guide for the control of propylene glycol dinitrate in air should be 0.2 ppm for daily 8-hr exposures. This is being followed by the U.S. Navy, and is substantiated by the fact that human experience at such levels of exposure resulted in no untoward symptoms. However, the work of Stewart et al. (166) indicates that prolonged or repeated exposures to 0.2 ppm may cause headaches in a significant number of human volunteers; hence the 0.2 ppm level may not provide an appropriate margin of safety if headaches are to be avoided.

80.5 Odor and Warning Properties

The work of Stewart et al. (166) indicates that the detection of propylene glycol dinitrate in air by humans occurs in the range of 0.2 to 0.26 ppm. Repeated exposures at this level gave rise to headaches and visual-evoked responses. Thus it may be that the warning properties of propylene glycol dinitrate may not be adequate to prevent some adverse responses, especially if repeated exposures occur.

REFERENCES

1. M. Singliar and J. Dykyj, *Collect. Czech. Chem. Commun.*, **34**(3), 767 (1969).
2. G. G. Esposito and R. G. Jamison, *Soc. Automot. Eng. J.*, **79**(1), 40 (1971).
3. A. B. Boese, Jr., C. K. Kink, and H. G. Goodman, Jr. in G. O. Curme, Jr., and F. Johnston, Eds., *Glycols*, American Chemical Society Monograph Series 114, Reinhold, New York, 1952.
4. C. P. Carpenter, U. C. Pozzani, C. S. Weil, J. H. Nair, III, G. A. Keck, and H. F. Smyth, Jr., *AMA Arch. Ind. Health*, **14,** 114 (1956).
5. Union Carbide Corporation, Industrial Brochure F-4765E.
6. E. Saparmamedov, *Zdravookhr. Turkm.*, **18,** 26 (1974).
7. E. Gross, in K. B. Lehmann and F. Flury, *Toxicology and Hygiene of Industrial Solvents* (Transl. by E. King and H. F. Smyth, Jr.), Springer, Berlin, 1938.
8. Union Carbide Corporation, unpublished data.
9. The Dow Chemical Company, unpublished data.
10. J. H. Draize, G. Woodard, and H. O. Calvery, *J. Pharmacol. Exp. Therap.*, **82,** 377 (1944).
11. C. P. Carpenter and H. F. Smyth, Jr., *Am. J. Ophthalmol.*, **29,** 1363 (1946).
12. W. M. Grant, *Toxicology of the Eye*, 2nd ed., Charles C Thomas, Springfield, Ill., 1974.
13. K. Nagano, E. Nakayama, M. Koyano, H. Oobayashi, H. Adachi, and T. Yamada, *Jap. J. Ind. Health*, **21**, 29 (1979) (in Japanese, English summary).
14. H. W. Werner, J. L. Mitchell, J. W. Miller, and W. F. von Oettingen, *J. Ind. Hyg. Toxicol.*, **25,** 157 (1943).

15. H. W. Werner, C. Z. Nawrocki, J. L. Mitchell, J. W. Miller, and W. F. von Oettingen, *J. Ind. Hyg. Toxicol.,* **25,** 374 (1943).

16. H. W. Werner, J. L. Mitchell, J. W. Miller, and W. F. von Oettingen, *J. Ind. Hyg. Toxicol.,* **25,** 409 (1943).

17. E. Browning, *Toxicity and Metabolism of Industrial Solvents,* Elsevier, Amsterdam, 1965.

18. H. F. Smyth, Jr. in G. O. Curme, Jr. and F. Johnson, eds., *Glycols,* American Chemical Society Monograph Series 114, Reinhold, New York, 1952.

19. M. E. Goldberg, C. Haun, and H. F. Smyth, Jr., *Toxicol. Appl. Pharmacol.,* **4,** 148 (1962).

20. G. Ohi and D. H. Wegman, *J. Occup. Med.,* **20,** 675 (1978).

21. F. H. Wiley, W. C. Hueper, D. S. Bergen, and F. R. Blood, *J. Ind. Hyg. Toxicol.,* **20,** 269 (1938).

22. M. R. Zavon, *Am. Ind. Hyg. Assoc. J.,* **24,** 36 (1963).

23. S. Nitter-Hauge, *Acta Med. Scand.,* **188,** 277 (1970).

24. E. Z. Dajani, *Diss. Abst. Int. B.,* **30,** 1819 (1969).

25. Shell Research Limited , London unpublished data, 1970.

26. E. G. Young and L. b. Woolner, *J. Ind. Hyg. Toxicol.,* **28,** 267 (1946).

27. D. E. Donley, *J. Ind. Hyg. Toxicol.,* **18,** 134 (1936).

28. C. E. Parsons and M. E. M. Parsons, *J. Ind. Hyg. Toxicol.,* **20,** 124 (1938).

29. L. Greenburg, M. R. Mayers, L. J. Goldwater, W. J. Burke, and S. Moskowitz, *J. Ind. Hyg. Toxicol.,* **20,** 134 (1938).

30. American Conference of Governmental Industrial Hygienists, *Threshold Limit Values for Chemical Substances and Physical Agents in the Environment with Intended Changes for 1980.*

31. U.S. Government, 29 CFR, 1910.1000, July 19, 1978.

32. J. May, *Staub,* **26,** 385 (1966).

33. F. A. Rudi, *Gig. Sanit.,* **11,** 94 (1974); *Chem. Abstr.,* **82,** 133649H (1974).

34. L. D. Gadaskina and F. A. Rudi, *Gig. Tr. Prof. Zabol.,* **2,** 31 (1976).

35. H. F. Smyth, Jr., C. P. Carpenter, and C. S. Weil, *Arch. Ind. Hyg. Occup. Med.,* **4,** 119 (1951).

36. E. P. Laug, H. O. Calvery, H. J. Morris, and G. Woodard, *J. Ind. Hyg. Toxicol.,* **21,** 173 (1939).

37. E. G. Stenger, L. Aeppli, J. Muller, E. Pehein, and P. Thomann, *Arzneim. Forsch.,* **21,** 880 (1971).

38. H. J. Morris, A. A. Nelson, and H. O. Calvery, *J. Pharmacol. Exp. Therap.,* **74,** 266 (1942).

39. E. Gross in K. B. Lehmann and F. Flury, *Toxicology and Hygiene of Industrial Solvents* (transl. by E. King and H. F. Smyth, Jr.), Springer, Berlin, 1938.

40. W. F. von Oettingen and E. A. Jirouch, *J. Pharmacol. Exp. Therap.,* **42,** 355 (1931).

41. C. S. Weil and R. Scala, *Toxicol. Appl. Pharmacol.,* **19,** 276 (1971).

42. P. T. Kan, M. A. Simetskii, and V. I. Tlyashohanko, *Vses. Nauchn. Issled. Inst. Vet. Sanit. Tr. Moscow,* **39,** 369 (1971).

43. C. P. Waite, F. A. Patty, and W. P. Yant, *U.S. Public Health Rep.,* **45,** 1459 (1930).

44. See Reference 16.

45. M. E. Goldberg, H. E. Johnson, U. C. Pozzani, and H. F. Smyth, Jr., *Am. Ind. Hyg. Assoc. J.,* **25,** 369 (1964).

46. A. A. Kasparov, N. V. Marinenko, and E. I. Marinenko, *Nauch. Tr. Irkutsk. Med. Inst.,* **115,** 26 (1972).

47. Shell Research Limited, Tunstall Laboratory, Sittingbourne, England, unpublished data.

48. J. Fucik, *Prac. Lek.,* **21,** 116 (1969).

49. I. Goldstein, E. Dumitru, and V. David, *Igiena,* **19,** 209 (1970).

50. G. O. Curme, Jr., and J. Johnston, *Glycols, American Chemical Society Monograph Series* 114, Reinhold, New York, 1952.

51. H. F. Smyth, Jr., C. P. Carpenter, C. S. Weil, U. C. Pozzani, J. A. Striegel, and J. S. Nycum, *Am. Ind. Hyg. Assoc. J.,* **30,** 470 (1969).

52. C. P. Carpenter, H. F. Smyth, Jr., and U. C. Pozzani, *J. Ind. Hyg. Toxicol.,* **31,** 343 (1949).

53. J. C. Gage, *Br. J. Ind. Med.,* **27,** 1 (1970).

54. D. H. Hutson and B. A. Pickering, *Xenobiotica,* **1,** 105 (1971).

55. H. E. Christensen, Ed., *Registry of Toxic Effects of Chemical Substances,* 1976 ed., U.S. Dept. Health, Education, and Welfare, NIOSH, Rockville, Md.

56. A. P. Grant, *Lancet,* **263,** 1252 (1952).

57. Eastman Kodak Company, Publication M-165, 1969.

58. H. F. Smyth, Jr., C. P. Carpenter, C. S. Weil, U. C. Pozzani, J. A. Striegel, and J. S. Nycum, *Am. Ind. Hyg. Assoc. J.,* **30,** 410 (1969).

59. H. F. Smyth, Jr., C. P. Carpenter, C. S. Weil, and U. C. Pozzani, *Arch. Ind. Hyg. Occup. Med.,* **10,** 61 (1954).

60. C. P. Carpenter, C. S. Weil, and H. F. Smyth, Jr., *Toxicol. Appl. Pharmacol.,* **28,** 313 (1974).

61. H. F. Smyth, Jr., C. P. Carpenter, C. S. Weil, U. C. Pozzani, and J. A. Striegel, *Am. Ind. Hyg. Assoc. J.,* **23,** 95 (1962).

62. H. F. Smyth, Jr., J. Seaton, and L. Fischer, *J. Ind. Hyg. Toxicol.,* **23,** 259 (1941).

63. E. W. Reid and H. E. Hoffman, *Ind. Eng. Chem.,* **21,** 695 (1929).

64. U.S. Department of Health, Education and Welfare, Public Health Service, Center for Disease Control, National Institute for Occupational Safety and Health, *NIOSH Criteria for a Recommended Standard Occupational Exposure to Dioxane,* September, 1977.

65. See Reference 55.

66. M. P. Argus, R. S. Sohal, G. M. Bryant, C. Hoch-Ligeti, and J. C. Arcos, *Eur. J. Cancer,* **9,** 237 (1973).

67. M. F. Argus, J. C. Arcos, and C. Hoch-Ligeti, *J. Natl. Cancer Inst.,* **35,** 949 (1965).

68. C. Hoch-Ligeti, M. F. Argus, and J. C. Arcos, *Br. J. Cancer,* **24,** 164 (1970).

69. M. F. Argus, R. S. Sohal, G. M. Bryant, C. Hoch-Ligeti, and J. C. Arcos, *Eur. J. Cancer,* **9,** 231 (1973).

70. R. J. Kociba, S. B. McCollister, C. Park, T. R. Torkelson, and P. J. Gehring, *Toxicol. Appl. Pharmacol.,* **30,** 275 (1974).

71. S. de Navasquez, *J. Hyg.,* **35,** 540 (1935).

72. M. E. King, A. M. Shefner, and R. R. Bates, *Environ. Health Perspect.,* **5,** 163 (1973); DHEW Publication No. (NIH) 74-218.

73. A. Fairley, E. C. Linton, and A. H. Ford-Moore, *J. Hyg.,* **34,** 486 (1934).

74. W. P. Yant, H. H. Schrenk, F. A. Patty, and C. P. Waite, *U.S. Public Health Rep.,* **45,** 2023 (1930).

75. N. Nelson, *Med. Bull.,* **11,** 226 (1951).

76. U. C. Pozzani, C. S. Weil, and C. P. Carpenter, *Am. Ind. Hyg. Assoc. J.,* **20,** 364 (1959).

77. T. R. Torkelson, B. K. J. Leong, R. J. Kociba, W. A. Richter, and P. J. Gehring, *Toxicol. Appl. Pharmacol.,* **30,** 287 (1975).

78. H. Barber, *Guy's Hosp. Rep.,* **84,** 267 (1934).

79. S. A. Henry, Annual Report of Chief Inspector of Factories, H. M. Stationery Office, London, 1934.

80. R. T. Johnstone, *Am. Arch. Ind. Health*, **20**, 445 (1959).

81. A. M. Thiess, E. Tress, and I. Fleig, *Arbeitmed. Sozialmed. Praventivmed.*, **11**, 35 (1976).

82. American Conference of Governmental Industrial Hygienists, *Threshold Limit Values for Chemical Substances and Physical Agents in the Workroom Environment with Intended Changes for 1979.*

83. L. Silverman, H. F. Schulte, and W. W. First, *J. Ind. Hyg. Toxicol.*, **28**, 262 (1946).

84. D. G. Laing, *Chem. Senses Flavor*, **1**, 257 (1975).

85. A. Fairley, E. C. Linton, and A. H. Ford-Moore, *J. Hyg.*, **36**, 341 (1936).

86. W. H. Braun and J. D. Young, *Toxicol. Appl. Pharmacol.*, **39**, 33 (1977).

87. Y. Wov, J. C. Arcos, and M. F. Argus, *Biochem. Pharmacol.*, **26**, 1535 (1977).

88. J. D. Young, M. B. Chenoweth, W. H. Braun, G. E. Blau, and L. W. Rampy, *J. Toxicol. Environ. Health.*, **3**, 507 (1977).

89. J. D. Young, W. H. Braun, P. J. Gehring, B. S. Horvath, and R. L. Daniel, *Toxicol. Appl. Pharmacol.*, **38**, 643 (1976).

90. S. S. Powar and A. M. Mungikar, *Bull. Environ. Contam. Toxicol.*, **15**, 762 (1976).

91. Y. Wov, M. F. Argus, and J. C. Arcos, *Biochem. Pharmacol.*, **26**, 1539 (1977).

92. J. D. Young, W. H. Braun, and P. J. Gehring, *J. Toxicol. Environ. Health*, **4**, 709 1978.

93. W. Wirth and O. Klimmer, *Arch. Gewerbepathol. Gewerbehyg.*, **17**, 192 (1936).

94. C. Hoch-Ligeti and M. F. Argus, in P. Nettesheim, M. G. Hanna, Jr., and J. W. Deatherage, Jr., Eds., Conference on the Morphology of Experimental Respiratory Carcinogenesis, Gatlinburg, Tenn. 1970; *AEC Symposium Series No. 21*, Springfield, Va. National Technical Information Service, 267–279; *IARC*, **11**, 247 (1976).

95. V. B. Perone, L. D. Scheel, and W. P. Tolov, National Institute for Occupational Safety and Health, unpublished report, 1976.

96. B. A. Schwetz, B. J. K. Leong, and P. J. Gehring, *Toxicol. Appl. Pharmacol.*, **32**, 84 (1975).

97. Union Carbide Corporation, Data Sheet, 1971.

98. H. F. Smyth, Jr., and C. P. Carpenter, *J. Ind. Hyg. Toxicol.*, **30**, 63 (1948).

99. D. L. J. Opdyke, *Food Cosmet. Toxicol.*, **12**, 517 (1974).

100. P. J. Hanzlik, W. S. Lawrence, and G. L. Laqueur, *J. Ind. Hyg. Toxicol.*, **29**, 233 (1947).

101. D. E. Hall, F. S. Lee, P. Austin, and F. A. Fairweather, *Food Cosmet. Toxicol.*, **4**, 263 (1966).

102. I. F. Gaunt, J. Colley, P. Grasso, A. B. G. Lansdown, and S. D. Gangolli, *Food Cosmet. Toxicol.*, **6**, 689 (1968).

103. K. B. Butterworth, I. F. Gaunt, and P. Grasso, *BIBRA*, **15**, 115 (1976).

104. H. F. Smyth, Jr., C. P. Carpenter, and C. B. Schaffer, *Food Cosmet. Toxicol.*, **2**, 641 (1964).

105. A. G. Cranch, H. F. Smyth, Jr., and C. P. Carpenter, *Arch. Dermatol. Syphilol.*, **45**, 553 (1942).

106. W. M. Meininger, *Arch. Dermatol. Syphilol.*, **58**, 19 (1948).

107. P. J. Hanzlik, W. S. Lawrence, J. K. Fellows, F. P. Luduena, and G. L. Laqueur, *J. Ind. Hyg. Toxicol.*, **29**, 325 (1947).

108. J. K. Fellows, F. P. Luduena, and P. J. Hanzlik, *J. Pharmacol. Exp. Therap.*, **89**, 210 (1947).

109. H. F. Smyth, Jr., C. P. Carpenter, and C. S. Weil, *Arch. Ind. Hyg. Occup. Med.*, **4**, 119 (1951).

110. V. K. Rowe, D. D. McCollister, H. C. Spencer, F. Oyen, R. L. Hollingsworth, and V. A. Drill, *Arch. Ind. Hyg. Occup. Med.*, **9**, 509 (1954).

111. F. E. Shideman and L. Procita, *J. Pharmacol. Exp. Therap.*, **102**, 70 (1951).

112. E. G. Stenger, L. Aeppli, L. Machemer, D. Mullen, and J. Trokan, *Arzneim. Forsch.*, **22**, 569 (1972).

113. R. D. Stewart, E. D. Baretta, H. C. Dodd, and T. R. Torkelson, *Arch. Environ. Health*, **20**, 218 (1970).

114. D. L. Rucknagel and A. Surtskin, *Proc. Soc. Exp. Biol. Med.*, **80**, 584 (1952).

115. C. P. Carpenter, F. N. Critchfield, J. H. Nair, III, and C. B. Shaffer, *Arch. Ind. Hyg. Occup. Med.*, **4**, 261 (1951).

116. P. W. Morgan, *Ind. Eng. Chem. Anal. Ed.*, **18**, 500 (1946).

117. S. Lepkovski, R. A. Over, and H. M. Evans, *J. Biol. Chem.*, **108**, 431 (1935).

118. C. B. Shaffer and F. H. Critchfield, *Fed. Proc.*, **7**, 254 (March, 1948).

119. E. Rosser, in K. B. Lehmann and F. Flury, *Toxicology and Hygiene of Industrial Solvents* (transl. by E. King and H. F. Smyth, Jr.), Springer, Berlin, 1938.

120. M. Yoshida, H. Hoshii, and M. Matsui, *Agric. Biol. Chem.*, **36**, 2473 (1972).

121. H. D. Kesten, M. G. Mulinos, and L. Pomerantz, *Arch. Pathol.*, **27**, 447 (1939); *Chem. Abstr.*, **33**, 4659 (1939).

122. M. G. Mulinos, L. Pomerantz, and M. E. Lojkin, *Am. J. Pharm.*, **115**, 51 (1943); *Chem. Abstr.*, **37**, 4136 (1943).

123. W. E. Rinehart, M. Kaschak, and E. A. Pfitzer, *Chem-Toxicol. Ser. Bull.*, **6**, Industrial Hygiene Foundation of America Inc. (1967).

124. O. E. Fancher, G. L. Kennedy, Jr., J. B. Plank, D. C. Lindberg, W. H. Hunt, and J. C. Calandra, *Toxicol. Appl. Pharmacol.*, **26**, 58 (1973).

125. B. S. Astill, C. J. Terhaar, and D. W. Fassett, *Toxicol. Appl. Pharmacol.*, **22**, 387 (1972).

126. W. Jordan and M. Dahl, *Arch. Dermatol.*, **104**, 524 (1971).

127. C. P. Carpenter, *J. Am. Med. Assoc.*, **135**, 880 (1947).

128. J. H. Draize, E. Alvarez, M. F. Whitesell, G. Woodard, E. C. Hagen, and A. A. Nelson, *J. Pharmacol. Exp. Therap.*, **43**, 26 (1948).

129. D. Hoehn, *J. Am. Med. Assoc.*, **128**, 513 (1945).

130. Anonymous, *Am. Ind. Hyg. Assoc. J.*, **27**, 574 (1966).

131. C. Einert, W. Adams, R. Crothers, H. Moore, and F. Ottoboni, *Am. Ind. Hyg. Assoc. J.*, **24**, 435 (1963).

132. Y. Morikawa, K. Muraki, Y. Ikoma, T. Honda, and H. Takamutsu, *Arch. Environ. Health*, **14**, 614 (1967).

133. E. Gross, M. Bock, and F. Hellrung, *Arch. Exp. Path. Pharmakol.*, **200**, 271 (1942).

134. R. A. Jones, J. A. Strickland, and J. Siegel, *Toxicol. Appl. Pharmacol.*, **22**, 128 (1972).

135. G. Cavagna, G. Locati, and M. Capizzi, *Med. Lav.*, **59**, 772 (1968).

136. E. C. Vigliani, G. Cavagna, G. Locati, and V. Foa, *Arch. Environ. Health*, **16**, 477 (1968).

137. M. Kalin, B. Kylin, and T. Malmfors, *Arch. Environ. Health*, **19**, 32 (1969).

138. M. Minami, A. Okada, A. Takizawa, and J. Kubota, *Br. J. Ind. Med.*, **29**, 321 (1972).

139. D. G. Clark, *Toxicol. Appl. Pharmacol.*, **21**, 355 (1972).

140. D. G. Clark and M. H. Litchfield, *Br. J. Ind. Med.*, **26**, 150 (1969).

141. N. Sugawara, *Bull. Yamaguchi Med. Sch.*, **20**, 99 (1973).

142. W. F. von Oettingen, Natl. Insts. Health Bull. No. 186 (1946).

143. P. Carmichael and J. Lieben, *Arch. Environ. Health*, **7**, 424 (1963).

144. S. Forssman, N. Nasretiez, G. Johansson, G. Sundell, O. Wilander, and G. Bostrom, Swedish Employers Confederation, Stockholm, Sweden.

145. D. C. Trainor and R. C. Jones, *Arch. Environ. Health,* **12,** 231 (1966).

146. H. Hasegawa, M. Sato, and H. Tsuruta, *Ind. Health,* **8,** 153 (1970).

147. S. D. Kanonova, A. M. Korolev, L. T. Cremeko, and L. L. Gumanov, *Genetika,* **8,** 101 (1972).

148. M. H. Litchfield, *J. Pharma. Sci.,* **60,** 1599 (1971).

149. M. H. Litchfield, *Drug Metab. Rev.,* **2,** 239 (1973).

150. H. Tsuruta and H. Hasegawa, *Ind. Health,* **8,** 99 (1970).

151. H. Tsuruta and H. Hasegawa, *Ind. Health,* **8,** 119 (1970).

152. D. G. Clark and M. H. Litchfield, *Br. J. Ind. Med.,* **24,** 320 (1967).

153. H. Hasegawa and M. Sato, *Ind. Health,* **1,** 20 (1963).

154. H. Hasegawa and M. Sato, *Ind. Health,* **8,** 88 (1970).

155. E. R. Plunkett, *Handbook of Industrial Toxicology,* Chemical Publishing Company, Inc., New York, 1976.

156. F. Gobbato and E. DeRosa, *Folia Med.,* **52,** 537 (1969).

157. K. Suwa, M. Sato, and H. Hasegawa, *Ind. Health,* **2,** 80 (1964).

158. H. Symanski, *Arch. Hyg. Bakt.,* **136,** 139 (1952).

159. S. Forssman, N. Nasretiez, G. Johannsson, G. Sundell, O. Wilander, and G. Bostrom, *Arch. Gewerbepathol. Gewerbehyg.,* **16,** 157 (1958).

160. P. Martini and L. Massari, *Med. Lav.,* **56,** 62 (1965).

161. U.S. Dept. of Health, Education and Welfare, *NIOSH Manual of Analytical Methods,* Vol. 1, 2nd ed., P&CAM 203, 1977.

162. U.S. Dept. of Health, Education and Welfare, *NIOSH Manual of Analytical Methods,* Vol. 3, 2nd ed., S216, 1977.

163. G. N. Krasovsky, A. A. Korolav, and S. A. Shigan, *J. Hyg. Epidemiol. Microbiol. Immunol.,* **17,** 114 (1973).

164. M. E. Andersen and R. G. Mahl, Am. Ind. Hyg. Assoc. J., **34,** 526 (1973).

165. D. G. Clark and M. H. Litchfield, *Toxicol. Appl. Pharmacol.,* **15,** 175 (1969).

166. R. D. Stewart, J. E. Peterson, P. E. Newton, C. L. Hake, M. J. Hosko, A. J. Lebrun, and G. M. Lawton, *Toxicol. Appl. Pharmacol.,* **30,** 377 (1974).

167. J. L. Mattsson, J. W. Crock, Jr., and L. J. Jenkins, Jr., U.S. NTIS Rep. AD-A023770, 1975.

168. R. W. Young, C. R. Curran, C. G. Franz, and L. J. Jenkins, Jr., US NTIS Rep. AD-A026841, 1976.

169. M. E. Andersen and R. A. Smith, *Biochem. Pharmacol.,* **22,** 3247 (1973).

CHAPTER FIFTY-TWO

Inorganic Compounds of Oxygen, Nitrogen, and Carbon

RODNEY R. BEARD, M.D., M.P.H.

1 OXYGEN, O_2

1.1 Occurrence and Uses

Oxygen is a normal constituent of the air we breathe (20.95 percent by volume). Men and animals depend on oxygen and can live only a few minutes in its absence. Oxygen is required by the body for combustion in the tissues in amounts proportional to energy expenditures. Henderson and Haggard (1) give the approximate energy expenditures, oxygen consumption, and volume of breathing of an average (70 kg) man (see Table 52.1). A table giving greater detail is available (2).

The capacity to supply oxygen to the tissues declines with age, as indicated in Table 52.2.

1.2 Physical and Chemical Properties

Physical state	Colorless, odorless gas
Boiling point	$-183°C$

4053

Density	1.429 g/l at 0°C, or 1.105 (air = 1)
Solubility in water	0.0694 g/kg at 0°C; 0.0325 at 37°C
1 mg/m^3 ≎	0.75 ppm
1 ppm ≎	1.33 mg/m^3

1.3 Oxygen Deficiency

Oxygen deficiency can be encountered in numerous situations. Tanks, vats, holds of ships, silos, mines, or any poorly ventilated area where the air may be diluted or displaced by gases or vapors of volatile materials, or where oxygen may be consumed by chemical or biologic reactions, are potentially dangerous. A worker entering any such place should wear a lifeline that is continually attended by another person who has no other duty.

The normal 21 percent of oxygen in the air produces a partial pressure of about 160 torr at sea level. If the oxygen partial pressure is reduced to 120 torr, impairment of mental performance is soon detectable.

A general review of effects of hypoxia with discussion of the problems of prolonged exposure and work appears in *Physiological Responses and Adaptations to High Altitude* by Lahiri (3a). Another recent review appears in *Aviation Medicine* by Dhenin, et al (3b). Hypoxia is almost invariably associated with hypocapnia, a subnormal amount of carbon dioxide in the blood, and this complicates matters.

Effects of hypoxia are not usually detectable when oxygen pressure is greater than 120 torr, but there may be some interference with adaptation of the eye to darkness. From 160 to 120 torr, the uptake of oxygen decreases linearly with the decrease in oxygen pressure. At less than 120 torr, chemical receptors in the aorta and the carotid arteries detect the declining blood saturation and stimulate increased breathing. This partially compensates for the lessened

Table 52.1. Energy Expenditure, Oxygen Consumption, and Volume of Breathing of Man[a]

Activity	Energy Used (cal/min)	O$_2$ consumption (l/min) (0°C, 760 mm)	Vol. of Air Breathed, (l/min) (20°C)
Rest in bed, fasting	1.15	0.240	6
Sitting	1.44	0.300	7
Standing	1.72	0.360	8
Walking, 2 mph	3.12	0.650	14
Walking, 4 mph	5.76	1.200	26
Slow run	9.60	2.000	43
Maximum exertion	14–20	3.000–4.000	65–100

[a] From F. Patty (1), after Y. Henderson and H. W. Haggard.

Table 52.2.   Age and Maximal Oxygen Intake[a]

Mean Age (years)	Number of Subjects	Mean Maximal O_2 Intake (l/min)	O_2 (cm^3/kg of Body wt/min)
17.4	11	3.61	52.8
24.5	11	3.53	48.7
35.1	10	3.42	43.1
44.3	9	2.92	39.5
51.0	7	2.63	38.4
63.1	8	2.35	34.5
75.0	3	1.71	25.5

[a] From F. Patty (1) after T. Sollman.

supply of oxygen, but at the same time causes excessive loss of carbon dioxide. There is considerable variation among individuals in their sensitivity to hypoxia. During exercise, increased respiration may occur when oxygen pressure is about 140 torr. With further lowering of oxygen pressure, respiration increases more, and the output of blood from the heart also increases. This results in increased oxygen use, as a result of the increased work. However, the increase in ventilation is not sufficient to cause awareness of hard breathing, and the increase in oxygen pressure in the blood is not great. At the same time, there is a constriction of the peripheral arteries, increasing the supply of blood to the brain and other organs. However, the decrease in carbon dioxide pressure stimulates constriction of the brain arteries, and the net result of an exposure to moderate hypoxia is probably an initial reduction of cerebral flow. As the hypoxia increases, oxygen lack becomes the stronger stimulus, the hypocapnic vasoconstriction is overcome, and the net cerebral blood flow is increased (4a). A gradual increase of heart rate is first seen at an oxygen pressure of 137 torr, the increase becoming greater as oxygen pressure goes down to 67 torr, which is the minimum that can be sustained by an unacclimatized person. Were it not for the increased blood flow, consciousness would be lost at a higher ambient O_2 pressure. The brain is the organ most susceptible to the effects of hypoxia because it has a high metabolic rate, has no reserve store of oxygen, and is incapable of anaerobic metabolism. An important criterion of the effects is the level of hypoxia that can be tolerated without loss of function. Table 52.3 shows the results of studies on men in a low pressure chamber. There would probably not be much difference if the hypoxia had been induced by breathing air diluted with nitrogen, but this is not certain.

Boismare, et al., (4b) reviewed the biochemical changes in catecholamines and discussed the action of apomorphine in preventing interference with a conditional avoidance reaction by hypoxia.

Symptoms and signs of oxygen deficiency are summarized in Table 52.4.

Table 52.3. Maintenance of Useful Consciousness in Hypoxic Hypoxia[a]

Altitude		Oxygen Partial	Subjects at Critical
m	(ft)	Pressure (torr)	Threshold n
5000	(16,400)	85	3
6000	(19,685)	74	18
7000	(22,966)	65	55
8000	(26,247)	48	20
9000	(29,528)	42	4

[a] Distribution of 100 subjects according to their critical threshold during low pressure chamber ascent at 309 m/min with 5 min stay at each 1000 m. Critical threshold was failure to pass a comprehension and writing test. Modified from Luft, cited by Randel (4).

1.4 Oxygen Toxicity

Oxygen poisoning was described by Paul Bert in 1878 (5). He observed convulsions and other effects in animals exposed to pure oxygen at 3 to 4 atm pressure, and also saw decreased oxidation in tissues at a pressure as low as 1 atm. He concluded that oxygen at pressures above 1 atm is a poison.

Although it appears that all living cells can be damaged by oxygen, the respiratory system and the central nervous system are most conspicuously affected by its toxic actions. The respiratory system is more susceptible to injury, but the effects on the central nervous system are more dramatic. Animals or

Table 52.4. Response of Man to the Inhalation of Atmospheres Deficient in Oxygen[a]

Stage	Oxygen Vol. (%)	Symptoms or Phenomena
1	12–16	Breathing and pulse rate increased, muscular coordination slightly disturbed
2	10–14	Consciousness continues, emotional upsets, abnormal fatigue upon exertion, disturbed respiration
3	6–10	Nausea and vomiting, inability to move freely, loss of consciousness may occur; may collapse and although aware of circumstances be unable to move or cry out
4	Below 6	Convulsive movements, gasping respiration; respiration stops and a few minutes later heart action ceases

[a] From F. Patty (1), after F. A. D. Alexander and H. E. Himwick.

men exposed to a 4-atm pressure of oxygen or more are likely to show muscular twitching or general convulsions (like those of epilepsy) within 1 hr. Pure oxygen at atmospheric pressure, or even less, can cause pulmonary irritation and edema after 24-hr exposures; higher pressures cause trouble in a shorter time. Half an atmosphere of pure oxygen can probably be tolerated indefinitely without lung damage; 3 atm is probably safe for healthy adults for 1 hr. Intermittent breathing of air or another gas mixture containing oxygen at the accustomed $\frac{1}{5}$ atm pressure lessens the risk of lung injury.

Industrial exposures to high oxygen pressure are uncommon. Sea diving is probably the most frequent, and the techniques of using special gas mixtures of helium, oxygen, and nitrogen are highly developed. Caison workers and tunnel makers may also be exposed to pressures that are high enough to cause lung damage. This is discussed in Chapter 9, "Physiological Effects of Abnormal Atmospheric Pressures," in Volume 1 of this series.

There is a temptation to use undiluted oxygen for diving; it is available, and it may seem wasteful to carry a lot of inert gas when it is only the oxygen that is used. Self-contained respirators commonly provide undiluted oxygen, but they often require rebreathing of the gas, so it is diluted with nitrogen that comes out of the body. Also, they are not often used for as much as 1 hr without an intervening spell of breathing normal air. Aviators usually use systems that provide diluted oxygen, but they can get pure oxygen when they desire it. At altitudes below 5500 m (18,000 ft), they can be exposed to more than $\frac{1}{2}$ atm oxygen pressure, but this is not likely to be continued over a long period of time. Caisson workers may be exposed to a partial pressure of oxygen exceeding $\frac{1}{2}$ atm, if the working pressure is more than $2\frac{1}{2}$ atm (0.25 kPa or 37.5 psia), but we have seen no reports of oxygen poisoning from this occupation.

1.5 Absorption, Excretion, and Metabolism

Oxygen is absorbed almost exclusively through the lungs, although it may be taken up through mucous membranes of the gastrointestinal tract, the middle ear, and the accessory sinuses. It diffuses through the lining of the lung alveoli into the blood capillaries, is dissolved in the blood plasma, diffuses into the red blood cells, and is bound to the hemoglobin that they contain. The red cells are transported to the body tissues by the flowing blood. Where the partial pressure of oxygen is low, the oxygen dissolved in the plasma diffuses out of the capillaries. As the plasma concentration of oxygen diminishes, it is replaced by that contained in the red cells. The blood is returned to the lungs for a new supply of oxygen.

Within the tissue cells, the oxygen is used mainly by structures called mitochondria, in which the chemical reactions that supply energy for the functions of the cell take place. Most of the oxygen is combined with carbon to form carbon dioxide, or with hydrogen to form water. A small part is combined into various compounds that form cellular structures or cellular products. The carbon dioxide that has been formed is transported by the blood to the lungs, where it diffuses out into the alveoli and is breathed out. The water is carried to the kidneys where it is separated from the plasma and excreted in the urine.

1.6 Physiologic and Pathologic Effects in Animals

Mustafa and Tierney (6a) published a thorough review of biochemical and metabolic aspects of oxygen poisoning of animal lungs in 1978, with 233 references. They show the similarity of the effects of oxygen, ozone, and nitrogen dioxide injury. They believe the basic mechanisms to be similar for all three, but not identical.

Although oxygen is essential for many forms of life, it would probably be lethal to respiring organisms if it were not for biochemical defense mechanisms that destroy toxic intermediates. These defense mechanisms appear to be adequate when O_2 is present in air at sea level, but are inadequate when the PO_2 is 3 to 5 times greater. It is tempting to speculate that in some humans the defense mechanisms may not be totally adequate throughout many years; their lungs may be particularly sensitive and have chronic injury secondary to breathing O_2 in air.

Oxygen itself is probably not toxic to most biologic systems, and even the intermediates produced during oxidation in mitochondria by the cytochrome c oxidase reaction are highly compartmentalized and apparently have no toxic effects. The reaction of an O_2 molecule with cytochrome c oxidase involves transfer of 4 electrons (a tetravalent attack on the O_2 molecule) and is responsible for most O_2 utilized by aerobic cells. However, other biologic oxidations may involve a single electron transfer to an O_2 molecule. This univalent reduction produces the superoxide anion, O_2^-, a free radical. This anion may be toxic by itself, and if it reacts with hydrogen peroxide, it may generate the hydroxyl radical $OH\cdot$, which is considered extremely toxic. The superoxide anion can probably attack relatively few molecules, but the hydroxyl radical is powerful enough to attack virtually every type of organic molecule. The major mechanism of defense against this catastrophic cascade of free radical reactions is the activity of the enzyme superoxide dismutase, which catalyzes the dismutation of superoxide anion to hydrogen peroxide. In the absence of superoxide anion, the hydroxyl radical is not formed, and the toxicity of the hydrogen peroxide is relatively low.

Other reviews of the toxicity of oxygen have been published recently by Jacquez (6b) and by Deneke and Fanburg (6c).

Stevens and Autor (7) demonstrated that neonatal rat lung tissue responds to high oxygen pressure by increased production of superoxide dismutase, both in vivo and in vitro. The increase was accounted for solely by a change in mitochondrial manganosuperoxide dismutase, and it could be abolished by any of several materials that inhibit protein synthesis.

Yam et al. (8) compared lung enzyme activities of neonatal rats with those of adult rats after both had been exposed to 96 to 98 percent oxygen. Up to about 36 days of age, most of the rats survived a 72-hr exposure, and rats older than 50 days mostly died. Newborn rats sampled day by day showed significant increases of superoxide dismutase activity (144 percent) reduced glutathione (176 percent), and glucose-6-phosphate dehydrogenase (151 percent). Adult rats showed an increase only after 3 days and only in glucose-6-phosphate dehydrogenase activity, the other enzymes remaining the same as in air-exposed animals.

Liu et al. (9) demonstrated that limitations of oxygen supply, whether by reason of low ambient oxygen pressure or limitation of oxygen supply to specific tissues, are associated with reductions in superoxide dismutase activity, consistent with the thesis that superoxide dismutase plays an important role in protection against oxygen toxicity.

The metabolic responses to hyperbaric oxygen exposure are complex and not well understood. It appears that mitochondria are damaged by oxygen and the rate of oxygen consumption of lung tissues is diminished after acute exposures of animals to undiluted oxygen. On the third day of exposure, a rebound of oxygen use has been observed, and it is suggested that this reflects the replacement of epithelial, endothelial, and interstitial lung cells that were injured in the earlier period. This may be an adaptive response that enables continued lung function despite the injury. Glucose metabolism by lung tissues is increased two- or threefold by exposure to undiluted oxygen for 2 or 3 days, with increased lactate production, an effect like that of hypoxia. This paradox is explained by the reduced mitochondrial activity that may be caused either by an insufficient supply of oxygen or by the injury due to an excess of oxygen.

Effects of oxidants on lipid metabolism are important because they influence the secretion or inhibition of pulmonary surfactant, the material that reduces surface tension of the fluid that coats the interior of the lungs and thereby enables them to be inflated. Dipalmitoyl phosphatidylcholine, an essential ingredient of surfactant, was not diminished by oxygen exposure for 3 days, but it increased after the animals were returned to normal air.

Sheffield (10) observed that pulmonary surfactant activity was markedly less in mice dying in respiratory distress during exposure to increased oxygen pressure than it was in control mice in air or in mice that survived the exposure.

Nucleic acid content and indices of protein synthesis in lung tissues also change as if there were first a depression of activity, then, after several days, an increase. The increase appears to be concurrent with proliferation of type 2 epithelial cells. These cells can subsequently transform to the commonly more numerous flat type 1 cells that normally line the alveoli, and also to other types of epithelial cells that are destroyed by oxygen. There is evidence of increased collagen production, consistent with the occurrence of pulmonary fibrosis after prolonged exposure to hyperbaric oxygen.

One of the earliest changes detectable after oxygen exposure is a reduction of the capacity of the lung tissue to perform its normal function of removal of 5-hydroxytryptamine. This may signal a more general depression of an important physiologic activity, the control of the concentration of metabolically active compounds that are continuously produced by other parts of the body.

Sackner et al. (11) studied the effects of graded concentrations of oxygen on tracheal mucus velocity and cellular changes in the tracheas of anesthetized dogs. One hundred percent oxygen caused 45 percent reduction of mucus velocity after 2 hr of exposure; 75 percent oxygen caused a 42 percent reduction after 9 hr; 50 percent oxygen caused an initial increase in velocity that was reversed after 12 hr, and ultimately a 50 percent reduction after 30 hr. Acute tracheobronchitis was seen in the animals exposed to 100 and 75 percent oxygen.

1.6.1   Factors that Modify Oxidant Injury of the Lung

Marked differences in susceptibility to oxygen poisoning have been reported between mice and rats; little is known about comparisons among other species. Larger animals such as dogs and monkeys are less susceptible than smaller ones, especially rodents. Older rats are more sensitive than younger ones, and consequently, heavier rats are more sensitive than lighter ones; but rats reduced in weight by dietary restrictions are also less sensitive than heavier ones of the same age. Northway et al. (12) and Bonikos et al. (13) observed that newborn mice are notably less susceptible to the lethal effect of undiluted oxygen than adult animals. Frank et al. (14) exposed newborn and adult animals of five species to 95 + percent oxygen for 7 days. Measurements of antioxidant enzyme activity in the lungs (superoxide dismutase, catalase, glutathione peroxidase) showed no significant responses in adult animals to a 24-hr exposure to 95 + percent oxygen, nor were these enzymes increased in newborn guinea pigs or hamsters. Newborn mice, rabbits, and rats showed significant increases in all the enzymes ranging from 12 to 36 percent (except that mice showed only a 5 percent increase in catalase). The exposure time to cause 50 percent mortality was much greater for the newborn mice, rats, and rabbits than for the adults, but the newborn guinea pigs and hamsters only lived as long as the adults of the same species.

Nutritional factors influence susceptibility. Restriction of vitamin E intake increases susceptibility to oxygen injury in several species; a moderate intake is as effective in preventing injury as a large dose. "It appears that optimal protection can be achieved with 2 to 4 times the recommended daily dietary allowance." (The current National Research Council recommendation is 4 to 15 international units per day for infants, children, and adult humans.)

Hall et al. (15) reported that acclimatization to hypobaric anoxia, induced by keeping female mice in simulated altitudes of 0, 5,000, or 15,000 ft for 1, 2, 4, or 8 weeks, prior to exposure to oxygen at high pressure (4, 6, or 9 atm) enhanced their susceptibility to convulsions and to death. Duration of acclimatization at a given altitude had no effect on the time to convulsion, but did decrease the time to death. The times to convulsion and death showed a log-linear relationship to both altitude and oxygen pressure. They warn that previous exposure to low oxygen tension should be taken into account in hyperoxic therapy or other conditions in which a person may be exposed to increased oxygen pressure.

The work of Winter et al. (16) tends to confirm their conjecture that any kind of diffuse injury to the lung enhances resistance to injury by oxygen for a few days, as a result of modifications in pulmonary morphology, particularly the proliferation of granular (type 2) pneumocytes. They caused pulmonary irritation by the intravenous injection of oleic acid in rabbits. Animals prepared in this way survived exposure to 1 atm pressure of oxygen twice as long as those not previously injured.

Frank and Roberts (17) discuss this effect more broadly and use a less specific experimental approach. They comment that more than 50 agents have been reported to have some efficacy in the prevention of oxygen toxicity, but most of them had been tested for their protection from high pressures where central nervous effects are prominent. They had found that bacterial endotoxin had a protective effect against pulmonary reactions to oxygen, and in the paper cited, describe extensive experiments to confirm this point and to refine the data. The mechanism by which endotoxin acts in this context is unknown; it has many biologic actions that affect every organ system of the body; however, it has been demonstrated to cause increased activities of antioxidants in the lung. They gave rats small doses of purified bacterial endotoxin by intraperitoneal injection, then exposed them to 95 percent oxygen for 72 hr. Ninety-seven percent of the endotoxin-treated rats survived, compared to 32 percent of control animals. The lungs of survivors showed less collagen and reticular fiber deposition in the rats that had been treated with endotoxin. Endotoxin treatment was associated with increased antioxidant enzyme activities of the lung tissues in a dose-dependent manner.

Northway et al. (12) report three phases of reaction in the newborn: first, inhibition of lung DNA synthesis, lasting 96 hr; an intermediate phase from 96 to 144 hr, characterized by a sharp increase in mortality and a minimum lung DNA-to-body weight ratio; and a third phase of increased DNA synthesis, lung and body growth, and a return to control levels of lung DNA-to-body weight ratio. In the same experimental exposures, Ludwin et al. (18) established that mitochondria and cell membranes of ciliated bronchial cells were the earliest affected lung tissues, with changes after 3 days, and that severe bronchiolitis was seen after 7 days of exposure. These changes were in addition to the effects on the capillaries, described by others.

Bonikos et al. (19) exposed newborn mice to oxygen at 1 atm pressure for up to 6 weeks, and saw progressive evolution of dense fibrous tissue deposition, chronic bronchitis, and bronchiolitis. Eighteen percent of 40 (or more) animals survived the 6-week exposure, whereas almost total mortality of adult mice would be expected within 8 days of such exposure. Early changes included loss of cilia in the lining cells of large bronchi and patchy necrosis of the bronchial and bronchiolar lining, and proliferation of granular type 2 pneumocytes in the alveoli. These changes were more severe after 3 weeks. In the later weeks, development of fibrosis and breakdown of alveoli supervened, leading to an emphysema-like appearance.

Hughson et al. (20) observed heart lesions in rats exposed to 5 atm pressure of oxygen for 75 min. Early mitochondrial swelling was followed by progressive degeneration of myofibrillar structure, becoming more severe over the 4 hr during which the animals were observed. All animals showed labored respirations or convulsions, or both, during the high pressure oxygen exposure, and 6 out of 30 died. Compounds such as ascorbic acid (vitamin C), p-aminobenzoic acid, glutathione, and butylated hydroxytoluene also have been used as antioxidants to protect biologic systems.

Pratt reviewed the pathologic changes of oxygen poisoning in the lung in 1974 (21). The type 2 pneumocyte is the lung cell most resistant to injury. The type 1, membranous pneumocyte is killed by oxygen at 1 atm pressure within 1 to 4 days, and capillary endothelial cells are also highly susceptible. In animals that survive, the type 2 pneumocytes proliferate and replace the destroyed type 1 cells. Subsequently, the type 2 cells may transform to type 1, and the lung structure returns nearly to normal.

Robinson et al. (22) looked for biochemical responses in a primate, the baboon, upon exposure to 95 percent oxygen at atmospheric pressure for 120 hr (except 2 out of 10 that became dyspneic and cyanotic, and were removed after only 96 hr). Superoxide dismutase in lung tissue decreased to 68 percent of control activity immediately after exposure; catalase showed no significant change; glucose-6-phosphate dehydrogenase rose to 172 percent of control activity; cytochrome oxidase decreased to 64 percent of control activity; oxygen use by lung tissue decreased by 37.5 percent; lactic acid production by the lung increased by 612 percent. All the changed values reverted to normal after 2 weeks of recovery. Morphologic changes in the lung were similar to those described above. A striking observation was that the arterial oxygen pressure fell from 90 to 51 torr after 5 days of oxygen exposure, an average decrease of 44 percent, indicating the severity of impairment of lung function.

Wolfe et al. (23) exposed baboons to 95 percent oxygen for 5 days. Lung biopsies were performed before and after oxygen exposure. Pulmonary function was evaluated before exposure, at the end of exposure, and after 7 and 14 days. Six baboons removed from oxygen environment after 96 to 110 hr and exposed to room air died within 3 to 20 hr because of profound hypoxia (arterial oxygen pressure 40 torr). The remaining 12 baboons were "weaned" to room air over a 3-day period. Electron microscopy showed endothelial swelling, edema of interstitial alveolar membranes, and increased numbers of type 2 pneumocytes. Lung volume measurements showed significant decreases in total lung capacity, vital capacity, the ratio of closing capacity to total lung capacity, and dynamic compliance. Biochemical studies indicated a shift toward anaerobic metabolism. These changes were all reversible in the surviving animals.

Raffin et al. (24) exposed tissue cultures of rat lung macrophage cells to 40 or 60 percent oxygen at atmospheric pressure for 48 hr and observed impairment of ability to take up foreign particles (phagocytosis). After reexposure to air for 48 hr, the cells regained their phagocytic abilities. Superoxide dismutase activity was markedly increased in the cells exposed to oxygen. Preexposure to 40 or 60 percent oxygen increased the susceptibility of these cell cultures to injury by 95 percent oxygen in a dose-related manner.

Simon et al. (25a) demonstrated that cell cultures, derived from type 2 pneumocytes, when exposed to 95 percent oxygen for 96 hr showed a marked decrease in oxygen consumption, suggesting impairment of mitochondrial energy conversion. In addition, an increase in lactate production and a shift to glycolytic energy production were seen.

Heino, et al. (25b) observed changes in mitochondria and dilatation of the Golgi apparatus within type 2 pneumocytes of rats after 24 hr exposure to 100 percent oxygen, but not after 12 hr exposure.

Torbati et al. (26) state that the most clearly defined electrical manifestation of oxygen poisoning in the brain is the occurrence of electrical discharges consisting of generalized, high amplitude spike activity developing simultaneously in many brain structures. They suggest that this may be the result of toxic action on the neurons, or a result of decreased cerebral blood flow owing to constriction of the blood vessels, a phenomenon known to result from exposure to high oxygen pressure.

1.7   Effects in Man

Human poisoning by oxygen has been the subject of several recent reviews. Among them are those by Miller and Winter (27), Raffin (28), Wood (29), and Edwards et al. (30).

Raffin et al. (31) present an illuminating discussion of the etiology of oxidant injury to lung tissues, the role of free radicals, and the protective role of superoxide dismutase.

Pulmonary symptoms of oxygen poisoning generally precede measurable decrements in pulmonary function. They usually begin with a sensation of tightness in the chest on deep inspiration and discomfort under the sternum (breastbone). These increase in severity and become continuous, and coughing appears. Intense burning pain and uncontrollable coughing spasms develop with prolonged exposure.

Reduction of vital capacity is the earliest measurable sign of oxygen poisoning.

Symptoms may appear after about 36 hr in persons exposed to ½ atm pressure of oxygen, in about 10 hr at 1 atm, and in 4 to 6 hr at 2 atm, as shown in Figure 52.1. Reviewers of oxygen toxicity in humans have generally adopted the theory that free radical production is the underlying biochemical mechanism of injury. Pathologically, "One of the earliest responses of the lung is accumulation of water in the interstitial space and within the pulmonary pneumocytes. This widens the alveolar-capillary membrane, increasing the diffusing distances, but not enough to interfere significantly with gas exchange. . . . The thickened alveolar-capillary membrane is necessarily stiffer and hence somewhat less compliant. Vital capacity, therefore, decreases with the decrease in lung compliance. With intracellular swelling of the surfactant-producing type 2 pneumocytes, spatial derangements of phospholipid building blocks in the ribosomes of the endoplasmic reticulum may occur. Thus, the necessary order for the assembly of phospholipids in the production of surfactant granules may become random, resulting in the production of abnormal surfactant. With swelling and blunting of the microvilli of type 2 pneumocytes, the extrusion and layering of surfactant may be grossly uneven . . ." (29).

The development of pathologic changes in man has not been followed in great detail, but it appears to be similar to the evolution of oxygen poisoning as seen in other primates and not much different from the sequence seen in lower animals. Two phases can be recognized, an exudative phase and a proliferative phase. At the onset, injury to the lining of the trachea and bronchi has been demonstrated.

Sackner et al. (32) made direct visual observations of the interior of the trachea and measured the rate of mucus transport in 10 human volunteers who

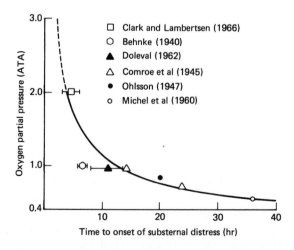

Figure 52.1 Time of onset of pulmonary distress with increasing oxygen pressure. From Wood (29).

breathed oxygen through a face mask for 6 hr. The oxygen concentration was 90 to 95 percent. Observations were made after 3 and 6 hr. Tracheal mucus velocity was decreased after 3 hr and all subjects showed visual evidence of tracheal irritation after 6 hr. Tracheal mucus flow was restored to control levels 15 min after administration of a beta adrenergic stimulant, terbutaline, by subcutaneous injection. Three subjects had fever and two had symptoms of acute bronchitis several hours after the study was completed; one had sinusitis and conjunctivitis. These phenomena might be related to the procedure of passing a fiberoptic bronchoscope into the trachea. Two others became nauseated and vomited during the evening of the study. All were extremely fatigued. None complained of substernal pain or other symptoms of tracheal or bronchial irritation during the oxygen exposure.

In the deeper levels of the lung, the exudative phase is characterized by injury to the endothelial lining cells of the blood capillaries and to the pavement-like type 1 pneumocytes that line the alveoli. Swelling of the cells and of interstitial structures, hemorrhage into the interstitial area and the alveoli, fibrinous exudate in the alveoli, and formation of a hyaline membrane may ensue. The proliferative phase, beginning a week or so after the onset, is characterized by infiltration of macrophage cells that scavenge dead tissue and foreign materials and by proliferation of type 2 pneumocytes, rounded cells that have the potential to transform into type 1 pneumocytes and other cell types, and proliferation of fibroblasts that make scar tissue. The extent of scarring depends on the severity of the injury. There is some speculation that chronic, prolonged, low level injury can produce severe fibrotic changes in the lung.

Hendricks et al. (33) exposed five healthy men to 2 atm oxygen pressure alternating with nitrogen–oxygen mixture. The cycle was 20 min on oxygen, 5 min on normoxic nitrogen–oxygen mixture, repeated until symptoms or signs of toxicity appeared. Compared with the observations of Clark and Lambertson (34), the intermittent return to normal oxygen pressure more than doubled the tolerance to high pressure oxygen, as measured either by symptoms or by the effect on the vital capacity.

The central nervous effects of oxygen are reviewed by Wood (29) and Edwards (30). As shown in Figure 52.2, these effects come on more quickly than the pulmonary symptoms, but only under a higher oxygen pressure. Current knowledge of the progressive neurologic symptoms of oxygen poisoning has been aptly expressed by Lambertson (35a): "The convulsions are usually but not always preceded by the occurrence of localized muscular twitching, especially about the eye, mouth, and forehead. Small muscles of the hands may be involved and incoordination of diaphragm activity in respiration may occur. These phenomena increase in severity over a period which may vary from a few minutes to nearly an hour with essentially clear consciousness being retained. Eventually an abrupt spread of excitation occurs and the rigid tonic

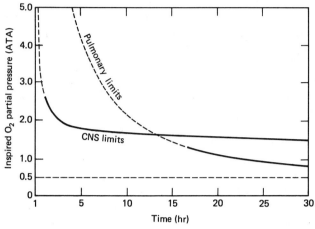

Predicted human pulmonary and CNS tolerance to high pressure

Figure 52.2   Time to onset of pulmonary or central nervous symptoms with increasing oxygen pressure. From Edwards et al. (30).

phase of the convulsion begins. The tonic phase lasts for about 30 sec and is accompanied by an abrupt loss of consciousness. Vigorous clonic contractions of the muscular groups of head and neck, trunk and limbs then occur becoming progressively less violent over about one minute." The seizures have been likened to those seen in grand mal epilepsy. Several reviewers state that central nervous effects are seen in men exposed to 2 atm oxygen pressure, or more. This is the pressure encountered by divers at a depth of about 33 ft or 10 m. The U.S. Navy has published the following guide for safe diving with 100 percent O_2:

Depth (m)	Time (min)
3	240
6	105
9	45
12	10

These exposures are for "nonworking" dives. Exercise increases the likelihood of convulsions. Divers are more likely to have trouble than persons exposed to the same oxygen pressure in a hyperbaric chamber. Other estimates of safe exposure limits indicate that "Men do not run any great risk of convulsions when exposed in a dry chamber and at rest to 2.3 or 3 ATA (atmospheres pressure absolute) oxygen for periods of 2 hours or 1 hour respectively, but that increasing the pressure above this level greatly increases the chance of oxygen toxicity. An exposure of 2 hours at 3 ATA oxygen may be the practical safe limit under these conditions although additional studies at this exposure are necessary to clarify the situation" (29).

Medical workers in hyperbaric chambers may be exposed to this risk and should be carefully supervised to avoid the hazard.

Banks, et al. (35b) studied speed of recognition of visual symbols by eight healthy young men. After short exposures to air pressures up to 7 times atmospheric, they found significant deterioration at 3 atm (ATA); this increased with higher pressures. Performance was more consistent with a digital display than with a clock face. This (35b) is an interesting review of the subject.

The mechanisms of central nervous injury are not clear. Carbon dioxide accumulation has been considered to be only a minor factor. The rapid reversibility of the convulsive state induced by oxygen under high pressure argues against peroxidation of lipids in cellular membranes as a major factor. There is considerable support for the idea that hyperoxia causes destruction of neurotransmitter compounds such as γ-aminobutyric acid (GABA). This has been demonstrated in animals.

There has been interest in protective measures that might prevent convulsions and other central nervous effects (4b). Anesthetic and sedative agents such as barbiturates are effective in animals, but as Wood (29) points out, they must be used with great caution because they may merely suppress the convulsions, and "hidden" damage to the central nervous system or other tissues may still occur.

1.8 Determination in Air

Several direct reading instruments are available at present from equipment manufacturers. For the most current information on such instruments the reader should contact the National Institute of Occupational Safety and Health, R. A. Taft Laboratories, 4676 Columbia Parkway, Cincinnati, Ohio; or the U.S. Environmental Protective Agency, Research Triangle Park, North Carolina.

2 OZONE, O_3

2.1 Occurrence and Uses

Ozone is triatomic oxygen, O_3. It is a bluish gas with a slightly pungent odor, about 1.6 times as heavy as air, and it is chemically highly reactive. It occurs naturally in the atmosphere, arising from the action of ultraviolet light upon oxygen at high altitudes. The maximal concentration is observed at an altitude of about 23 km (75,000 ft), but it can be detected at all altitudes up to about 90 km (300,000 ft). A second major source is the photochemical reaction of oxides of nitrogen and hydrocarbons from natural sources or from human activities.

A third, minor source is from natural electrical discharges. The latter are most conspicuous as mountain lightning storms, and this association probably led to the idea that an ozone-rich atmosphere is pleasant and healthful, and the adoption of place names such as "Ozone Park."

There was little commercial use of ozone until recently, when there has been a surge of interest in using it for water and sewage treatment and for the stabilization of industrial wastes. Should the current concern over health effects of long-term exposure to chlorine and chlorinated hydrocarbons increase, the use of ozone might expand very rapidly.

The most likely sources of exposures to ozone in industry are leakage from ozone-using processes and high voltage electrical equipment and from electric arc welding. The latter is more potently a source of nitrogen oxides. Aircraft flying at altitudes greater than 10,000 m (33,000 ft) may take significant quantities of atmospheric ozone into their cabin ventilators (36–38a).

Industrial exposures take place against a background of community exposure, although the ozone level in the ambient atmosphere is somewhat attenuated as it infiltrates a building. Indeed, significant protection from high atmospheric levels of ozone and other photochemical oxidants can be had simply by remaining indoors. Recent reviews of community ozone exposures have been prepared by the National Academy of Science (36), the Environmental Protection Agency (37) and by Whittemore (38b).

Ozone and associated photochemical oxidants are products of extremely complex chemical reactions for which the principal ingredients are a wide variety of hydrocarbons, oxides of nitrogen, and sunlight. The raw materials of photochemical smog arise from many sources. One of the principal classes is organic chemicals, consisting mainly of hydrocarbons (see Figure 52.3). Some aliphatic compounds have little reactivity in the photochemical reaction, and in air pollution control work, it is the practice to disregard methane.

In oversimplified form, the major elements of the photochemical oxidant-forming reactions appear to be as follows (Figure 52.3): (1) nitrogen dioxide (NO_2) absorbs light and breaks down to nitric oxide (NO) and atomic oxygen (O); (2) atomic oxygen combines with molecular oxygen (O_2) to form ozone (O_3); (3) nitric oxide combines again with ozone to form nitrogen dioxide and molecular oxygen with release of thermal energy. This cycle results in little accumulation of ozone. However, the introduction of any of a number of other substances can interfere with this cycle. Most notably, nitric oxide (NO) reacts preferentially with peroxy radicals (RO_2) which can be formed from various hydrocarbons. This reduces the availability of nitric oxide for the reaction with ozone, and the ozone concentration of the mixture builds up. Figure 52.4 shows the chemical changes that take place during photoirradiation of the system.

As stated in the National Research Council report on ozone (1977) (36), "In summary: the concentration of ozone in the polluted atmosphere is controlled by the intensity of sunlight and the ratio of nitrogen dioxide to nitric oxide.

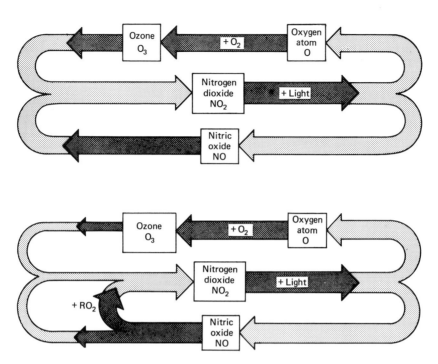

Figure 52.3   The supposed effect of added hydrocarbon on the concentration of ozone in the lower atmosphere. In the upper diagram, the dissociation of NO_2 by sunlight yields equal numbers of molecules of NO and O; the latter react with O_2 to give O_3. NO reacts with O_3 to form NO_2 again; this process gives rise to a small, stable O_3 concentration. In the lower diagram, the effect of hydrocarbon is to form peroxy radicals, RO_2, that react with NO to form NO_2; this diminishes the supply of NO, so the O_3 builds up. (Modified from Fig. 2.4 of Reference 36.)

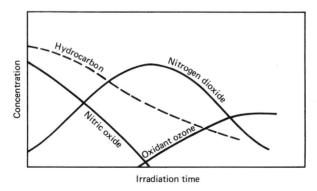

Irradiation time

Figure 52.4.   Chemical changes occurring during photoirradiation of hydrocarbon–nitrogen oxide–air systems. (From Fig. 4.1, Reference 37.)

Hydrocarbons and other pollutants—such as aldehydes, ketones, chlorinated hydrocarbons, and carbon monoxide—react to form peroxy radicals. These in turn, react with nitric oxide, causing the ratio [NO_2:NO] to increase. As a consequence . . ., the ozone concentration also increases."

The study of ozone toxicity has been complicated by problems of measurement of low concentrations of the gas in the atmosphere. These have been resolved since the development of instruments using specific ultraviolet absorption or chemiluminescence reactions. However, comparisons among different reports on ozone effects are impaired by uncertainties about the modes of calibration that have been employed for analytical instruments, which introduce differences of a quarter or a third (see Section 2.6).

2.2 Physical and Chemical Properties

Physical state	Bluish gas
Molecular weight	48
Boiling point	$-112°C$
Melting point	$-193°C$
Density	1.6 (air = 1)
Solubility in water	0.49 ml/100 ml at 0°C
1 mg/m^3 ⇌	$\frac{1}{2}$ ppm at 20°C, 760 torr
1 ppm ⇌	2 mg/m^3 at 20°C, 760 torr

2.3 Absorption, Excretion, and Metabolism

Major factors in the respiratory uptake of ozone, aside from the concentration of the gas in the air, are the structure of the respiratory tract, the influence of the mucus that covers the air passages, the flow of air, and the physical and chemical properties of ozone (36).

Ozone is rather sparingly soluble in water, so a considerable part of the inhaled gas reaches the smallest bronchioles and the alveoli. As might be expected, its effects are observed throughout the respiratory tract.

Experiments in beagle dogs, rabbits, and guinea pigs indicate that more than half the ozone inhaled is taken up by the nasal and pharyngeal mucosa, with exposures to less than 2 ppm. Tracheobronchial absorption has not been measured, but mathematical modeling predicts the maximum dose to be taken up by the respiratory bronchioles, a result that agrees well with experimental observations in various animal species (37).

Ozone is among the most reactive chemicals, and it should be expected that it would be rapidly used up near the point where it is absorbed. This appears to be the case; there has been no demonstration of the presence of inhaled ozone outside of the respiratory organs. However, its effects are not limited to the lungs and airways. Changes in blood morphology, in central nervous system functions, and in liver enzyme concentrations have been demonstrated (39–41). The metabolism of ozone is mainly through the oxidation of lung lipids (37). Reaction products that may be excreted include malonaldehyde, ethane, and pentane (see Section 2.4).

2.4 Physiologic and Pathologic Responses in Animals

2.4.1 Morphologic Changes

Extensive work on the effects of ozone has been done using rats and mice, and some with rabbits, dogs, hamsters, and monkeys. Reports up to 1977 were reviewed in the Environmental Protection Agency Criteria Document of 1978 (37). Goldstein reviewed details of toxic mechanisms up to 1977 (42). Stokinger and Coffin (43) summarized acute toxicity studies in 1968, stating that the lethal concentration for half the exposed animals (LC_{50}) for a 4-hr exposure was about 6 ppm for rats and mice. Cats, rabbits, and guinea pigs were less susceptible, and dogs were relatively resistant. The animals died of shock, with pulmonary edema and hemorrhage.

Stokinger reported that acute exposures of 1 to about 3 hr to concentrations in the magnitude of 5 ppm cause death and peeling off of the lining cells of the lower bronchi and bronchioles, and severe swelling of the lung tissues. Large amounts of fluid (mainly water) exude and displace the air from the alveoli. The animals die from want of oxygen and from the impediment to the flow of blood through the lungs, which results in heart failure (39).

Scheel et al. (44) demonstrated lethal effects at much lower levels. Sublethal exposures cause impaired breathing and reduced oxygen consumption. It appears certain that severe damage to pulmonary structures occurs in small animals with exposures to about 1 ppm for 3 hr or less.

Prolonged exposures at concentrations around 0.3 ppm, near the range observed in polluted community air, have led to lesions of the terminal bronchioles and adjacent alveoli. There is destruction of the lining cells of these structures, followed in a few hours or days by proliferation of so-called type 2 cells. The increase in type 2 cells causes thickening of the bronchiolar walls and alveolar septa, and this may be presumed to interfere with the flow of air (45–47). Cilia become swollen and clumped throughout the airways, small round bodies appear on the surface, and in some areas a pseudomembrane develops (48).

Mellick et al. (46) reported that in monkeys, ozone appears to be absorbed along the entire respiratory tract, penetrating deeply into the peripheral, nonciliated airways, and causing its most conspicuous lesions in the respiratory bronchioles and alveolar ducts. In their studies the extent of lesions deep in the lung increased as the concentration of ozone went from 0.2 to 0.8 ppm. These were prolonged exposures, 8 hr/day for seven successive days.

The lower respiratory structures of monkeys are anatomically similar to those of man; these differ significantly from those of rodents (46).

Plopper et al. (49) exposed 2-month-old rats to 0.1 or 0.2 ppm continuously for 7 days, and saw lung changes in all at 0.2 ppm and in 4 of 11 at 0.1 ppm. Significant biochemical changes also were observed at 0.1 ppm.

2.4.2 Biochemical Changes in the Lung

The mechanism of ozone toxicity at the cellular level is not clear. Several theories have been advanced. The oxidation of sulfhydryl groups or their precursors, the oxidation of polyunsaturated lipids, the formation of toxic compounds by oxidation of lipids, and the formation of free radicals, probably in combination, are among the more persuasive (37, 50).

The superoxide radical is a highly reactive oxidant. It can be changed to a less reactive form by the action of the enzyme superoxide dismutase. Seven-day exposures of rats to 0.2 ppm or higher concentrations of ozone were followed by dose-related increases of superoxide dismutase in lung tissue (50). Malonaldehyde, a product of lipid peroxidation, has been observed to increase with exposure to low levels of ozone.

Several investigators have reported increased susceptibility to ozone effects in vitamin E-depleted animals, or protective effects by preliminary treatment with this antioxidant vitamin (37). The peroxidation of lipids and free radical formation, which leads to further peroxidation, are impeded by vitamin E and other antioxidants (37). Menzel (50) discussed this effect in detail, and further stated that the level of vitamin E needed to protect against lipid peroxidation toxicity may be much greater than the present U.S. dietary intake. Dumelin et al. (51) reported that lipid peroxidation in vivo increases the amount of pentane and ethane in the expired air of rats. They used this to measure the protection afforded by vitamin E.

Other naturally occurring antioxidants in the cells include reduced nicotinamide adenine dinucleotide (NADH), reduced nicotinamide dinucleotide phosphate (NADPH), and sulfhydryl compounds, particularly reduced glutathione (GSH) and related enzymes. In vitro studies showed these compounds to be susceptible to oxidation by ozone. GSH in the lung is diminished by rather high (4 to 8 ppm) exposures of ozone, but not by 1.5 ppm or less. However, GSH content was observed to *increase* following low level ozone exposures. Other enzymes, glucose-6-phosphate dehydrogenase (G-6-PD), 6-phosphogluconate dehydrogenase (6-P-GD), and lactate dehydrogenase (LDH), showed similar increases. These increases are likely to be reflections of increased cellular replacement as part of the process of repair of ozone-damaged tissues. Mitochondrial enzymes showed similar reductions in activity after severe, acute exposure, with increased levels during recovery, about 7 days.

The oxygen uptake of ozone-exposed lung tissue was also observed to decrease immediately after acute exposures and to increase during recovery.

Although most of these studies were done in rats and mice, one excellent experiment using two species of monkeys showed similar results (37).

Prolonged exposures to low concentrations of ozone led to increased enzyme activity, and increased protein production and oxygen consumption by lung tissue. Increased lung protein synthesis was seen on the second through seventh days of ozone exposure at 0.2 ppm. Collagen (the protein of supporting structural tissue) increased more than other protein, suggesting the possibility of inducing lung fibrosis. The secretion of glycoprotein was reduced by exposures to 0.2 ppm.

Clark et al. (52) exposed squirrel monkeys to 0.75 ppm for 4 hr/day on four successive days, with 30°C, 50 percent relative humidity, and intermittent exercise. Lung tissues showed markedly increased lipid peroxidation and decreased vitamin E. Tissue LDH and G6PDH were increased. No changes in superoxide dismutase or malic acid dehydrogenase were observed.

Lee edited a review describing other biochemical responses to ozone in 1977 (53). Mustafa and Tierney published a review of biochemical and metabolic responses in the lung in 1978 (6).

2.4.3 Blood Changes

After exposure to ozone in vivo, red blood cells tend to lose their disklike shape and to become spherical. They can be hemolyzed more easily. Hemoglobin characteristics and oxygen transport appear to be unaffected. The acetylcholinesterase (AChE) bound to red cell membranes is decreased. Glutathione and related enzymes appear to be unaffected.

Leukocytes appear to be unaffected by ozone exposures up to 2 ppm in vivo.

The serum albumin diminishes and the alpha and gamma globulins increase, whereas total protein remains unchanged after prolonged exposures of rabbits to 0.4 or 1.0 ppm. Several other blood constituents may be affected, but their significance is not obvious (37). Chow et al. (54) reported that several enzymes in the blood were augmented in rats after 7-day exposures to 0.7 ppm and that animals on a vitamin E-deficient diet were more greatly affected. Also, although both sets of animals showed lung lesions, the lesions were more severe in the rats with deficient vitamin E intake. Clark et al. (52) reported increased fragility, diminished acetylcholinesterase, and diminished glutathione in red blood cells, but no significant changes of lactic acid dehydrogenase (LAD) in the blood of squirrel monkeys exposed to 0.75 ppm of ozone 4 hr/day for four successive days, with intermittent exercise and elevated temperature.

2.4.4 Infection

Stokinger (55) noted, "Exposures to ozone bring out or activate latent, subclinical respiratory infections in animals, resulting in full-blown respiratory disease that often ends fatally. The concentrations of ozone responsible for this exacerbation were around 1 ppm. . . ." Recent observations in several laboratories have confirmed that mice become less resistant to several kinds of bacterial infections after ozone inhalations.

The Environmental Protection Agency Criteria Document (37) in its summary indicates that the most sensitive indicator of ozone injury is increased susceptibility to infection, observed in mice after an exposure to 0.08 ppm for 3 hr. The effect persisted for only a few hours. Ozone inhalation after the exposure to infection also decreased resistance. Other pollutants (nitrogen dioxide or sulfuric acid) have additive effects. Exposures of 0.1 to 0.5 ppm for 2½ to 4 hr caused alterations of the functions of alveolar macrophages, cells that work to remove bacteria and other particles from the lungs. Decreased bactericidal activity in the lung was shown with 0.6 ppm in the same exposure time. Goldstein et al. (56) showed that impairment of bactericidal capacity in macrophages was associated with reduced activity of lysosomal enzymes.

2.4.5 Tolerance and Adaptation

Stokinger (39) reported striking development of tolerance for ozone upon repeated exposures of animals, such that they could withstand multilethal exposures for 4 to 6 weeks afterward. A single 1-hr preexposure at a nominal level of 1 ppm was sufficient to provide some protection. Repeated exposures were more effective. Tolerance was sustained for 1 month or longer in rats, and up to 14 weeks in mice. This phenomenon has been observed repeatedly by others (40, 57). Stokinger also showed cross protection from ozone by previous exposures to nitrogen dioxide, and vice versa. Douglas found no cross protection between ozone and oxygen (57). Chow observed that several lung enzymes were maintained at higher levels in preexposed rats after an ozone challenge and suggested that the glutathione peroxidase system might have a role in the development of tolerance (58a).

Veninga et al. (58b) reported adaptive reactions of hepatic reduced ascorbic acid content and plasma creatine phosphokinase activity in mice in response to low doses of ozone given in periods up to 4 hr, the concentrations ranging up to 0.8 ppm.

2.4.6 Exercise

Stokinger (59) noted that 1 ppm of ozone was without evident acute effects in resting rats, but was lethal for rats made to exercise. Fukase et al. (60) recently confirmed this, showing that mice made to exercise during three daily 4-hr exposures to 0.2 to 1.0 ppm showed a dose-related increase of lung weight and of reduced glutathione in the lungs, as compared to nonexercised controls.

2.4.7   Age

The effects of age upon the ozone response are uncertain. It is widely believed that infants and immature animals are more susceptible. Stokinger (59) observed that half of his young mice died after a 4-hr exposure to 4 ppm which was tolerated by mature animals; Bils (61) also saw more pronounced lung changes in 4-day-old mice, as compared to 4- and 8-week-olds after 1- or 2-day exposures to 1 ppm. Stephens et al. (62), however, found that rats, from birth to weaning, were resistant to injury by ozone exposures of up to 72 hr at 0.85 ppm, whereas older animals showed severe changes of the epithelium of the terminal airways. Infant mice were similarly resistant. Richkind and Hacker (63) reported that wild deer mice less than 1 year old were remarkably more susceptible than older animals. This was also true of the laboratory-reared progeny of wild ancestors. Guerrero et al. (69) observed that older cells in tissue culture were more susceptible to ozone than younger cells.

2.4.8   Genetic Effects

Ozone can alter the structure of nucleic acids (65).

The similarity of ozone and ionizing radiation in producing free radicals and peroxidation of fatty acids has been remarked by Stokinger (39) and more recently by Menzel (66). Ability to cause chromosome breaks is associated in this similarity. Since Fetner (67) observed chromosome breakage in growing roots of beans after ozone exposure in 1958, numerous studies of this radiationlike effect have been reported (36, 37). Zelac et al. (68) demonstrated increased chromosome breakage in lymphocytes of Chinese hamsters after 5-hr exposures to 0.2 ppm of ozone and compared this with the effect of 230 rads of X-rays, alone and in combination, and found the ozone more potent. There was no suggestion of synergism; the combined effect was less than the sum of those from the independent exposures. They concluded that the risks of permissible industrial ozone exposures (at 0.1 ppm) might be greater than those from permissible X-ray exposures (3 rems per quarter). Gooch et al. (69), however, found only minor effects from heavy in vitro exposures of human peripheral leukocytes, and no changes in Chinese hamsters exposed to 5.2 ppm for 6 hr, or in mice exposed to 0.99 ppm for 1 hr.

Genetic resistance and susceptibility to ozone injury has been shown in plants, and selective survival of resistant strains is observed. A number of investigators have induced genetic changes in bacteria. Cell cultures, both animal and human, showed chromosome aberrations. Mice showed increased neonatal deaths and various body abnormalities in offspring of mothers exposed to ozone during gestation (37).

Kavlock et al. (70) exposed pregnant rats to several concentrations of ozone for several days at different stages of gestation. Exposures of 0.44 ppm 8 hr/

day throughout the period of organogenesis produced no significant effects. Higher concentrations, up to 1.97 ppm, appeared to cause minimal teratogenesis. In midgestation, administration of 1.26 ppm was followed by increased rates of fetal absorption.

Von Nieding (71) reviewed the field in 1978 and found there was some evidence for mutagenicity but no experimental showing of carcinogenesis. Nitrosamine production by ozone exposures suggests a potential hazard.

Many of the reports cited in this field are lacking in details, and the ozone measurement methods are not specified, or are of doubtful accuracy at the low concentrations described. Moreover, the techniques of detection of chromosome damage are difficult, and the interpretation of the observations requires unusually skilled judgments.

2.4.9   Central Nervous System Effects

Mice avoid exposure to ozone at 0.6 ppm or more. They show depression of gross motor activity at levels of 0.1 ppm and higher. Subhuman primates show increased reaction time after 30 min exposure to 0.5 ppm. Rats showed an 84 percent reduction in voluntary activity when exposed for 7 days to 1.0 ppm. Rats exposed to 0.056 ppm for 93 days showed no behavior changes. Encephalographic studies in rats showed changes associated with exposures to 0.5 to 1.0 ppm for 1 hr (37).

Ozone has a depressant effect in several species. For example, reduced voluntary running activity in mice is a very sensitive biologic indicator of its action, manifest at levels of 0.2 and even 0.1 ppm. The mechanism of this action is not clear, although diminished concentration of catecholamines in the cerebral cortex has been observed (36).

2.5   Effects in Man

The systemic effects of ozone appear to be almost entirely the consequences of irritation of the respiratory tract. Inasmuch as ozone is highly reactive and only moderately soluble, one might expect to see signs of action at all levels. In fact, it is only with relatively high concentrations that affects are noted in the upper respiratory tract or in the trachea or bronchi.

As in other sections of this discussion of ozone, statements of ozone concentration must be viewed skeptically, unless a reliable method of measurement (see Section 2.7) has been specified.

The odor of ozone can be detected at 0.05 ppm (72) or possibly at even 0.02 ppm (73), but some apparently normal people do not detect 0.3 ppm.

2.5.1   Industrial Exposures

Early reports indicated that 1.5 ppm was intolerable and as little as 0.4 ppm was uncomfortable. In 1931, Flury and Zernik described mucous membrane irritation following 0.47 ppm, sleepiness and headache after 1 hr at 0.94 ppm, and increased pulse rate, sleepiness, and prolonged headache after higher concentrations (74). Several studies in welders (usually in mixed exposures, not limited to ozone) (75–79) showed no effects from short exposures at 0.2 or 0.25 ppm, sensations of chest constriction and throat irritation at 0.3 to 0.8 ppm, dry mouth and throat and irritation of nose and eyes at 0.8 to 1.7 ppm, and severe headaches, throat irritation, cough, choking, painful breathing, and signs of pneumonia or pulmonary edema at 9.2 ppm. An exposure at concentrations as high as 11.2 ppm for 2 hr caused perspiration, coughing, and collapse; oxygen inhalation relieved the symptoms and in 2 days all symptoms disappeared. Challen et al. (77) observed that when ozone concentrations were reduced to the region of 0.2 ppm there were no further reports of symptoms. Long-term exposures of smokers to 0.2 to 0.3 ppm in a welding shop were associated with small decreases in some tests of pulmonary functions in a few subjects.

Akbarkhanzadeh (80) observed respiratory symptoms in 51 percent of long-term welders, compared to 26 percent of suitable control subjects. The welders also showed more impairment of pulmonary function than the controls.

Current studies of exposures in high-flying aircraft which sometimes exceed 0.1 ppm and are rarely as great as 0.6 ppm suggest that passengers and sedentary workers rarely complain of ozone symptoms, but that physically active flight attendants experience chest pain, coughing, and excessive fatigue, and that impairment of some pulmonary functions may result from repeated exposures; causes other than ozone have not been excluded. The pressure altitude of these exposures was over 1500 m (5000 ft). The ozone analyses may have underestimated the true values slightly (81).

Lategola et al. (82) studied symptoms and cardiopulmonary functions in two experiments conducted in an altitude-simulating chamber at a pressure altitude of 1029 m (6000 ft). In the first experiment, 15 men and 12 women were exposed to 0.2 ppm for 4 hr, with treadmill exercise in the last 10 min of each hour. In the second experiment, 14 men and 14 women were similarly exposed to 0.3 ppm. The subjects were between 20 and 30 years old, and were "considered to be representative of the flight attendant population." No symptoms or group decrements of pulmonary function were observed at a significance level less than 90 percent in the experiment with 0.2 ppm ozone. With 0.3 ppm, there was an increase of symptoms, significant at the 95 percent level, and a decrease of all pulmonary functions tested, at a significance level of 98.4 percent or greater.

Allen et al. (83) observed measurable increases in ozone from photocopying machines in poorly ventilated rooms. Selway et al. (84) measured ozone production by 10 photocopying machines of eight models from four makers, and found levels from 0.002 to 0.153 ppm at the operators' breathing zones. Cleaning the machines reduced ozone emissions markedly, but they returned to precleaning levels within a few weeks. Excessive concentrations were observed only under abnormal operating conditions, but the use and neglect of such equipment is so widespread that it is an ozone source worthy of concern.

Protection from Ozone.   Protection from intoxication can be accomplished by application of the principles of industrial hygiene. Increases of ozone over naturally occurring levels in the open air of the community should be avoided by limitations on the emission of air pollutants which are converted to ozone, namely, reactive hydrocarbons and oxides of nitrogen. In industry, ozone-producing processes should be replaced where this is possible. Removal of ozone by ventilation is valuable where there are concentrated sources, and dilution can be applied where the sources are diffuse.

The observation that antioxidants can neutralize ozone in vitro, and, more importantly, that Vitamin E gives substantial protection to experimental animals suggests that workers who may be exposed should give attention to their dietary intake of this vitamin. The most abundant sources are vegetable oils, nuts, wheat germ, milk, eggs, muscle meats, fish, cereals, and leafy vegetables (85a). Hence the basic nutritional advice to take a highly varied food intake is sound. Supplementation of the diet with vitamin E concentrate may be attractive as a way to combat ozone poisoning, and competent chest disease specialists have recommended that people in Los Angeles should take 100 international units of vitamin E daily to protect them from the excessive oxidant levels sometimes encountered there. Menzel suggests that the level of vitamin E afforded by the usual United States dietary intake is much below the level needed to protect against peroxidation toxicity (50). Roberts (85b) warns against overdosage with this vitamin, but apparently would not be alarmed by the use of 100 units of all-rac-α tocopherol per day by an adult.

Posin et al. (86) tested the protection afforded against ozone by heavy dosage (18,000 IU) of vitamin E in 11 young men. They found no significant differences in comparison with 11 control subjects. The exposures were at 0.5 ppm with intermittent exercise, sufficient to induce significant blood alterations.

Dependence on pills is generally to be discouraged, and the use of prophylactic medication should never take the place of sound industrial hygiene practices that can eliminate hazards.

2.5.2   Experimental Observations

Hackney summarized the requirements to be considered in the design of experiments on human health effects of air pollutants in 1975 (87). These requirements are demanding, but they must be observed if experimental efforts are not to be wasted and if confusion is to be avoided. They should be reviewed by anyone planning to experiment in this field, or undertaking a critical review of reported studies.

"Air Quality Criteria for Ozone and Other Photochemical Oxidants" (37) summarizes more than 20 reports of experiments with humans. The most significant of these are discussed here.

Von Nieding et al. (88) observed reduction of arterial oxygen pressure and increased resistance to airflow in the lungs of humans after a 2-hr exposure to 0.2 ppm of ozone.

Horvath et al. examined changes in pulmonary functions following resting exposures to 0.25, 0.50, and 0.75 ppm of ozone for 2 hr in eight 21-year-old male and seven 22-year-old female subjects. Immediately following exposures to 0.5 and 0.75 ppm, several indices of pulmonary function decreased significantly, by about 10 percent for the higher exposure and 5 percent for the lesser. Exposures of the same subjects to 0.25 ppm were not followed by significant changes in pulmonary function tests (89).

That increasing physical activity should increase the effects of ozone exposure appears to be obvious; increased activity causes increased volume, rate, and depth of respiration; for any given concentration of ozone, the inspired dose is increased. This has been demonstrated experimentally in humans by several investigators. Bates et al. (90) recorded significant reductions in some pulmonary functions after 2 hr of exposure to 0.75 ppm of ozone. In the same laboratory, Hazucha et al. (91) followed this with tests for 2 hr at 0.75 and 0.37 ppm, combined with intermittent exercise sufficient to double the rate of ventilation. Some of the 12 subjects noticed throat dryness, slight cough, and throat irritation after 30 min; after 2 hr, all complained of substernal soreness, chest tightness, and cough, and a few had difficult breathing and wheezing. There were significant reductions in lung function tests related to the maximal expiratory flow rate from the lungs after the exposures to 0.37 ppm. The higher concentration, 0.75 ppm, led to greater effects. The total lung capacity was not affected, but the residual volume increased, probably as a result of impaired flow in the small airways (91). Hackney et al. (92) in carefully designed experiments, exposed four healthy men 36 to 49 years of age to ozone alone and in combination with nitrogen dioxide and carbon monoxide, for periods over 4 hr, with the exposure chamber at a "warm day" temperature of 31°C and relative humidity of 35 percent. Ozone concentrations were 0.25, 0.37, or 0.5 ppm. An extensive range of sensitive pulmonary function tests were done. A few statistically significant effects associated with ozone were observed, but they were small and inconsistent. Addition of nitrogen dioxide at the relatively low level of 0.3 ppm, and of carbon monoxide at 30 ppm, had no apparent effects. A second group of four subjects was tested similarly. These subjects were also in good health and had normal pulmonary function tests, but had prior histories of cough, chest discomfort, or wheezing associated with allergy or exposure to polluted air. Marked effects developed among some of these on their first exposure to ozone, so the duration of the exposures was cut to 2 hr on subsequent tests. Pulmonary functions related to expiratory flow were all consistently reduced in this sensitive group, and they all had symptoms of chest pain or cough. The protocol was further modified by reducing the ozone concentration to 0.25 ppm for two exposures, followed by 0.37 ppm for two exposures a week later. The "hyperreactive" subjects had no symptoms after the 0.25 exposure. They had marked symptoms following the 0.37 ppm exposure. They showed little change in pulmonary function tests at the lower exposure levels. In a subsquent study extending this line of work, Hackney's group repeated exposures to 0.25, 0.37, and 0.5 ppm of ozone, with and without addition of 0.3 ppm nitrogen dioxide and 30 ppm carbon monoxide. These studies were done with intermittent exercise sufficient to double the ventilatory rate (93). Ozone exposure associated with exercise led to symptoms of cough, throat irritation, and substernal pain even in subjects with no prior history of sensitivity to the effects of "smog"; similar exposures without exercise had produced minimal symptoms in these "normal" subjects, although "hyper-reactive" subjects had felt these symptoms severely. Biochemical effects on blood cells and plasma were also observed (see Section 2.4.3). On several grounds, Hackney concluded that under Los Angeles summer conditions, impairment by ozone concentrations above 0.25 to 0.3 ppm would be implied, and that some individuals would be at risk at considerably lower levels.

Bates et al. (94) reported in 1970 that preliminary studies showed reductions of dynamic pulmonary compliance and of gas diffusion rates in the lungs of eight subjects exposed to 0.75 ppm of ozone for 2 hr. Hazucha et al., in the same laboratory, reported in 1973 (91) that in four nonsmoking subjects, 2-hr exposure to 0.37 ppm ozone with intermittent light exercise led to a 7.4 percent reduction in forced expiratory volume, and for smokers, to a 4.5 percent reduction. With 0.75 ppm of ozone, the reductions were 24.6 and 15.6 percent, respectively. These differences were statistically significant at better than the 95 percent level. These differences were attributed to an increase in the residual volume, which in turn was thought to be related to either a decrease in the static recoil of the lung or to small airway obstruction or both. The "closing capacity" of the lung, reputed to be a sensitive indicator of small airway obstruction, was similarly affected. They concluded that a concentration of 0.37 ppm for 2 hr is unacceptably high if impairment of pulmonary function is to be avoided in normal, active people.

Silverman et al. (95) studied exposures to 0.37, 0.5, or 0.75 ppm of ozone, under conditions of rest or intermittent light exercise in 28 healthy young persons (95). Ten smokers in the group were not considered separately in the analysis. Expiratory pulmonary function tests showed significant decreases in exercised subjects with all concentrations, but only the 0.75-ppm exposure was followed by significant changes in those at rest. The pulmonary function decrements were related to the ozone dosage, though there was said to be some indication that a high-concentration, short-time exposure was more damaging than an equal dose in lower concentration for a longer time.

Kerr et al. (96) exposed 20 healthy adults, 10 of whom were smokers, to 0.5 ppm ozone for 6 hr, with intermittent light exercise. Those who experienced chest discomfort and cough were most likely to show decrements on pulmonary function tests. Smokers showed much less severe effects than nonsmokers. Significant expiratory obstruction was observed, as indicated by pulmonary function tests. Diffusing capacity, pulmonary compliance, and nitrogen elimination were unaffected.

2.5.3   Effects on Exercise Capacity

Folinsbee et al. (97) exposed 13 healthy young men to 0.75 ppm of ozone with light exercise for 2 hr, then tested their maximum capabilities, with suitable controls. The maximum oxygen consumption and the attained work load were both 10 percent less after ozone exposure than after exposure to air, the maximum ventilation rate was 16 percent less, and the maximum heart rate was 6 percent less. Decreases in vital capacity and expiratory airflow were noted, as well as cough and chest discomfort. DeLucia and Adams (98) sought to demonstrate changes in pulmonary function related to a low ozone concentration coupled with exercise graded from light to heavy. One-hour exposures of six subjects (one with a prior history of asthma) were used, with ozone concentrations of 0.15 and 0.3 ppm. All subjects showed some signs of toxicity when working at 45 or 65 percent of maximal capacity with ozone. Respiratory exhalation indices showed decrements of 9 to 21 percent with the heavier work schedule, and higher ozone concentration, but consistent effects were not seen with less severe exposures. Distinct differences in individual sensitivity to ozone were manifest in exercise performance. The two most sensitive subjects were unable to complete 1 hr at 65 percent of maximal work rate while breathing 0.3 percent ozone, and they were said to show marked effects at lower concentrations. Base-line pulmonary function tests with exercise in the absence of ozone were done, but the results were not published.

Savin and Adams (99) subsequently tested nine trained male athletes in exposures to 0.15 and 0.3 ppm of ozone (reliable values), using incremental work loading on a bicycle ergometer to the maximal level of voluntary exertion. Exposure durations averaged just over 30 min. No significant effects were noted on exercise capacity nor on pulmonary function tests. Dillard et al. (100) exposed 10 subjects to 0.24 ppm (reliably measured) of ozone while exercising at 50 percent of maximal oxygen consumption for 1 hr. They observed throat tickling in all, and several complained of cough, chest tightness, pain on deep inspiration, congestion, wheezing, or headache. All showed increased residual lung volume and lessened vital capacity, as compared with control tests, seven showed diminished airflow rates, and eight showed decreased 1-sec forced expiratory volumes. Horvath et al. (89) examined the exercise performance of eight young men and five young women after exposures to 0.25, 0.50, and 0.75 ppm of ozone for 2 hr. A significant increase in the expiratory exchange ratio (respiratory quotient, RQ) was observed after 30 min at 0.75 ppm, and this elevation persisted. The ventilatory equivalent (volume of air breathed per unit of oxygen extracted) was also elevated, and there was a significant decrease in oxygen uptake in all concentrations of ozone as compared with filtered air. However, on treadmill tests of maximum work capacity, none of the pollutant conditions reduced subsequent maximum exercise performance (oxygen consumption, heart rate, or total performance time).

2.5.4 Aggravation of Chronic Lung Disease

It has been assumed, reasonably enough, that persons with chronic pulmonary disease should be unusually susceptible to the effects of pollutants. However, quantitative evidence of this is very sparse. In 1961, Schoettlin and Landau (101) reported that asthmatic patients in the Los Angeles area showed a low correlation of asthma attacks with oxidant air pollution. However, they observed also that there were significantly more asthma attacks on days when the oxidant level was over 0.25 ppm (true value probably lower).

Motley et al. (cited in Reference 37) reported that persons with pulmonary emphysema showed improved respiratory functions after going from "smoggy" air into a room with clean air. Remmers and Balchum (cited in Reference 37) similarly studied patients with chronic pulmonary disease by putting them in a filtered air room, and found a small degree of correlation between their rate of oxygen consumption and their exposure to oxidant pollution, with smokers affected more strongly than nonsmokers. Several other inconclusive studies have been reported.

Kagawa and Toyama reported effects of photochemical air pollution on lung function in 20 schoolchildren who were observed weekly from June through December. Ozone was significantly associated with increased airway resistance (102).

Linn et al. (41) performed careful laboratory studies in 22 Los Angeles asthmatics, exposing them to 0.2 ppm ozone for 2 hr with intermittent light exercise. Comparisons were made with filtered air and sham ozone exposures. There was a slight increase in symptoms reported with the ozone exposure, but there were no perceptible consistent effects on sensitive tests of pulmonary function. Silverman (103) observed that 11 out of 17 adult (?) asthmatics showed impairment of airflow after exposure to 0.25 ppm for 2 hr.

2.5.5 Adaptation in Humans

Adaptation in humans was tested by Hackney et al. (104). They compared the reactions of nine newcomers to those of six long-standing Los Angeles residents when exposed to 0.4 ppm of ozone with temperatures and humidity simulating ambient pollutant exposures. The newcomers showed more symptoms and greater loss of pulmonary functions. Hackney et al. also tested the effect of repeated exposures of humans with known respiratory hyperreactivity (105). Six subjects were exposed to 0.5 ppm for 2 hr/day on four successive days, with intermittent light exercise, temperature 31°C, and relative humidity 35 percent. Five of the six showed decrements of pulmonary function on the first three days, with return nearly to base line on the fourth day.

In another experiment in Los Angeles, Hackney et al. (93) compared responses to ozone exposure in four Los Angeles residents with those in four eastern Canadians, who had been studied previously in Montreal by Bates. All eight were exposed to 0.37 ppm of ozone for 2 hr, with intermittent light exercise. Although the differences were not deemed statistically significant, the Canadians showed decrements in several lung function tests, whereas the Californians did not. Buckley et al. (106) reported biochemical changes in the same subjects; the Canadians showed somewhat greater effects than the Californians. This may be evidence of adaptation to oxidant exposures, or to a selective process in which susceptible persons have moved away from Los Angeles. Hazucha et al. (107) studied several sequences of repeated ozone exposures with intervals of 1 to 2 months between them and observed that only the four subjects exposed to 0.6 ppm initially, with exposure to a 0.4 ppm reinforcing exposure after a 1- or 2-month interval, were resistant to a subsequent exposure to 0.6 ppm. Initial exposures of 0.4 or 0.2 ppm, and reinforcing exposures of less than 0.4 ppm, appeared to be ineffectual.

2.5.6 Central Nervous System Effects

Headache is sometimes reported as a symptom of ozone poisoning, and there is evidence of central nervous system depression in animals, but there has been very limited investigation of central nervous system effects in humans. Lagerwerff (108) reported decreased dark-adapted visual acuity, increase in peripheral vision, and changes in eye muscle balance associated with exposures to 0.2, 0.35, or 0.5 ppm for 3 or 6 hr, with some evidence of a dose-related response. Gliner et al. (109) studied effects of ozone on sustained visual and auditory attention tasks (vigilance performance) under ozone exposures of 0.25, 0.5, and 0.75 ppm (ozone concentrations probably reliable). Eight men and seven women aged 19 to 27 years were the subjects. The 0.75-ppm level of ozone was associated with impaired performance on tests with a high ratio of signals to nonsignals, (1:6), but not with a lower ratio (1:30). The authors suggest that ozone, which caused symptoms of upper respiratory irritation at this concentration, gave rise to overarousal. We observed a somewhat similar result when we exposed human subjects to 0.3 ppm; the average response latency (reaction time) was shortened, as compared with control sessions.

2.5.7 Epidemiology

There have been no epidemiologic studies relating to ozone alone. A few reports on the effects of "oxidant measured as ozone" are available. There are several complicating factors that are hard to evaluate in considering these reports. Among these are the presence of other oxidants that may influence both the air analysis and the toxic potential of the mixture, such as nitrogen dioxide and peroxyacetyl nitrate (PAN). Other irritants such as aldehydes, and airborne particles, including sulfates, may influence the reactions of exposed humans. The effects of high atmospheric temperatures and humidity should be taken into account. The characteristics of the population sample should be identified, including the range of physical activities. It has been impossible to satisfy all these requirements for the few retrospective studies in which accumulated data on environmental exposures are correlated with health effects, but the dearth of information makes it necessary to look at a variety of unsatisfactory data and to glean as much information as can be extracted.

Goldsmith and Friberg (102) reviewed the epidemiologic evidence up to 1975 and concluded that although photochemical oxidant could have been a factor in mortality in some areas, no role had been proved.

Several other efforts to detect an association of exposures to photochemical oxidant and increased death rates have been reported, mainly from the Los Angeles area. The periods when high mortality was observed along with high oxidant were also periods of excessive heat, and it appears that the high temperatures were more important, by far, than the oxidant as a cause of death (37, 110). Aggravation of preexistent asthma by air pollution was studied by Schoettlin and Landau (101) with somewhat equivocal results, suggesting that a minority of asthmatics experienced aggravation by oxidant at about 0.2 ppm, hourly average level. This report has been very widely cited as a basis for setting air quality standards.

Hammer et al., cited in Reference 37, studied the association of headache, eye discomfort, cough, and chest discomfort with air pollutant concentrations in a population of student nurses. They found that symptoms were more related to oxidant than to other air pollutants, and that eye discomfort was reported most frequently on days when the oxidant level exceeded 0.15 to 0.19 ppm (maximum hourly average), headaches and chest discomfort were more frequent above 0.25 to 0.29 ppm, and coughs were more frequent above 0.3 to 0.39 ppm.

Athletic performance appears to be affected by oxidant. Wayne et al. (111) compared the performance of high school track competitors under varied oxidant exposures and found that failure to improve performance, as the season progressed, correlated with increased oxidant levels in the hour preceding the run. Levels less than 0.2 ppm did not show a significant effect.

2.5.8 Summary of Effects in Humans

In summary, it appears that healthy persons engaged in light exercise are likely to experience chest discomfort and cough, and to show measurable impairment of expiratory pulmonary functions when exposed for 2 hr or less to ozone concentrations lower than 0.37 ppm. Some persons who appear to be healthy show significant responses to 1-hr exposures of 0.15 ppm or less, with light exercise. Heavier exercise increases the responses materially. Repeated low level exposures, insufficient to evoke symptoms, lead to a small degree of tolerance for ozone; thus newly exposed persons are likely to feel the effects of an increased ozone level more than "old hands." The concentrations cited in this section were generally measured by methods that used the neutral buffered potassium iodide calibration method, and the true values of ozone were probably lower than the values cited. Exposures to much higher concentrations sometimes encountered in industry are likely to cause severe chest pain, pulmonary edema, and collapse. Although experiments on animals suggest that ozone may initiate destruction of lung tissue leading to emphysema, the few reports on follow-up observations of men days or weeks after severe exposures indicate complete recovery without disability, unless complications such as infectious pneumonia or cardiac injury occur.

The effects of ozone and photochemical oxidant in humans are summarized in Tables 52.5 and 52.6.

A valuable review of the health effects of low concentrations of ozone and other air pollutants by Ferris was published in 1978 (110).

The reader who would familiarize himself with the fascinating, complex interactions involved in the transport of oxygen and other gases in the blood will benefit from Dejours' book, *Comparative Physiology* (112).

2.5.9 Mutagenic Effects

Studies of mutagenic effects in man have been equivocal. McKenzie et al (113) exposed 30 men to 0.4 ppm of ozone for 4 hr and studied lymphocyte cultures taken before and at intervals of zero, 3 days, 2 weeks, and 4 weeks after exposure. They found no trend toward significant changes in chromosomes.

Table 52.5. Effects of Ozone in Humans—Occupational Exposures

Ozone Concentration	Subjective Complaints	Clinical Manifestations	Comments	Ref.
0.25	None	—	Inert gas shielded arc welding	Kleinfeld et al. (75)
0.3–0.8	Throat irritation, chest	None	Inert gas shielded arc welding. Two affected among four	(75)
0.9 peak	Throat irritation, lassitude, headache	Pulmonary edema in one of three	Inert gas shielded arc welding. Trichloroethylene also present	(75)
0.8–1.7	Dry mouth and throat irritation of nose and eyes, chest tightness	—	Welding shop	Challen (77)
0.47	Mucosal irritation	—	—	Flury and Zernik (74)
0.94	Severe mucosal irritation, sleepiness			(74)
>0.94	Pulse increase, sleepiness, headache			(74)
0.1–0.6	Substernal pain, cough, mucous membrane irritation	None	Aircraft flight attendants, moderate physical activity.	Reed (81)
	None		Aircraft crew, sedentary	

Table 52.6. Effects of Ozone in Man—Selected Experimental Observations[a]

Ozone Conc. (ppm)	Exposure Time	No. and Sex of Subjects	Ages	Smokers	Subjective Complaints	Objective Measurements	Other Comments	Refs.
1.0	1 hr	4 M			Throat irritation, cough	Airway resistance increased >20% in 3 of 4 Ss		Goldsmith and Nadel (36, 37)
0.9	5 min	4 M				Decreased airway conductance	Exercise during exposure	Kagawa and Toyama (36, 37)
0.6–0.8	2 hr	10 M 1 F			Substernal soreness, tracheal irritation 6–12 hr after exposure	Decreased vital capacity, airflow and diffusion rate		Young et al. (79)
0.75	2 hr	10 M	25–33	2	Cough, chest tightness, substernal soreness in most Ss; some had pharyngitis, dyspnea, and wheezing	Decreased pulmonary airflow, increased airway resistance	2 Ss exercised intermittently and showed worse effects	Bates et al. (94)
0.37–0.75	2 hr	12 M	Young	6	As above	Decreases in FVC and airflow	Alternating rest and exercise	Hazucha et al. (91)
0.37, 0.5, 0.75	2 hr	20 M 8 F	Young	10	Throat irritation, tracheal soreness, cough correlated with intensity of exposure; mod. severe at 0.75 ppm with exercise	Dose-related increase of breathing rate, decline of tidal volume	Rest, or exercise to 40, 60, or 75% of maximal tolerance	Folinsbee et al. (97)
0.75	2 hr	13 M	19–30 mean 24.6	4	Throat irritation, discomfort on deep breath, cough	Maximum tolerated work load, heart rate, minute volume, tidal volume, O$_2$ uptake all declined	Alternate rest and 50 W work, 15 min periods. 2 Ss unable to complete 2 hr at 0.75 ppm	Folinsbee et al. (97)

4087

Table 52.6. (Continued)

Ozone Conc. (ppm)	Exposure Time	No. and Sex of Subjects	Ages	Smokers	Subjective Complaints	Objective Measurements	Other Comments	Refs.
0.25, 0.50, 0.75	2 hr	8 M 5 F	21 22		Not described	No effect at 0.25, decreases in FVC at 0.5 and 0.75 ppm	No impairment of exercise capacity after exposure	Horvath et al. (89)
0.5	4–5 hr	4 M	36–49		None	None	Ss with normally reactive airways. Alternating rest and exercise; 31 C, 35% RH	Hackney et al. (92)
0.5	4–5 hr	4 M	29–41		Uncomfortable, unable to complete protocol	Marked changes in pulmonary function tests	Ss with hyperreactive airways	Hackney et al. (92)
0.5	2 hr 2 days	7 M	25–36			Minimal effects first day, 5 of 7 showed effects second day	Ss with normally reactive airways	Hackney et al. (93)
0.37, 0.25	4–5 hr	4 M	29–41			Slight effects at 0.37 No effects at 0.25	Ss with hyperreactive airways	Hackney et al. (93)
0.37, 0.35	2 hr	20 M 8 F	19–29	10	Dryness of nose and throat, throat irritation, cough, discomfort on deep breath, nausea, all dose related	High correlation (0.7–0.95) of impairment of VC and FEF with ozone dose	Alternate rest and exercise to increase ventilation 2½ ×	Silverman et al. (95)
0.37	2 hr	5 M	27–41			No important changes	Subjects with normally reactive airways	Hackney et al. (92)
0.3	1 hr	10			Throat tickle; few had chest pain, cough	Diminished VC, increased airway resistance and residual volume	Exercise at 50% of maximal O_2 consumption	Dillard et al. (100)
0.15,	½ hr	9 M	26.4	0	Dose related, but not	Nonsignificant trend	Trained athletes tested	Savin and

4088

Conc.	Duration	Subjects	mean	Smokers	specified	Effects	for work capacity	Reference
0.3						toward reduced maximum O_2 uptake; no change in work capacity. Max. ventilation increased		Adams (99)
0.25	2 hr 2 days	7	22–41			No obvious effects	3 of 7 subjects with hyperreactive airways exposed to O_3 alone or combined with NO_2 or NO_2 and CO	Hackney et al. (93)
0.25	2 hr	5 M 12 F	20–71 41 mean	0 7 ex-	Not reported	Increased airway resistance in 6 Ss, no effect in 11	Chronic asthmatics, longstanding. Tested at rest	Silverman et al. (95)
0.2	2 hr	20 M 2 F	19–59	6 9 ex- 7 non	Nonsignificant increases	Small increases in blood enzymes; no changes in pulmonary function	Chronic asthmatics, longstanding. Intermittent exercise, 31°C	Linn et al. (41)
0.15	1 hr	6 M			Congestion, wheezing, and headache with more severe exposures	Ventilation and O_2 uptake unchanged; decreased airway conductance after 65% maximal workload	Graded exercise to 25, 45, and 65% of maximum tolerance	DeLucia and Adams (98)
0.1	2 hr	11 M	22–38	2	Not described	Reduction of arterial O_2 pressure and increase of airway resistance		von Nieding et al. (88)
0.1	1 hr	4 M				Airway resistance increased 3.3% average, 45% in one S	One S with history of asthma	Goldsmith and Nadel (37)

[a] Abbreviations used in table: Ss = subjects; FVC = forced vital capacity; FEF = forced expiratory flow rate; VC = vital capacity.

Guerrero et al. (114) used a sister chromatid exchange analysis, said to be a more sensitive technique, and found a dose-related effect in human fetal lung diploid cells which were exposed in vitro to 0.025 to 1.0 ppm of ozone for 1 hr. However, they saw no effects in cultures from the bloods of 31 humans exposed to 0.5 ppm of ozone for 2 hr with intermittent exercise, 31°C ambient temperature, and 35 percent relative humidity.

2.6 Effects of Mixtures of Ozone with Other Gases

Information is available about the interactions of ozone with nitrogen dioxide (NO_2) and with sulfur oxides, but it is not definitive.

In studying the combined action of ozone and nitrogen dioxide, it must not be forgotten that these gases react rapidly to form the nitrate radical.

Goldstein (115) observed this precaution in testing the effects of the combination in vitro on human red blood cells. Four kinds of effects were studied, and in general the responses to the two gases were additive. However, a slight degree of synergism was apparent at lower concentrations for lipid peroxidation, and a degree of antagonism at higher concentrations. Conversion of hemoglobin to methemoglobin by nitrogen dioxide was potentiated by ozone. Osmotic fragility, acetylcholine esterase, and reduced glutathione measurements showed only additive effects. Goldstein (116) reported that the immediate effects of ozone and nitrogen dioxide in rat lungs are dissimilar with respect to lipid peroxidation, lung protein, or nonprotein sulfhydryl levels, or the agglutinability of alveolar macrophages by concanavalin A. Ehrlich et al. (117) studied effects on the resistance of mice to streptococcal pneumonia with 1.5 to 5.0 ppm of nitrogen dioxide, and 0.05 to 0.5 ppm of ozone. Daily 3-hr exposures up to 20 days were used. They found a suggestion of a synergistic effect, but it was not clearly demonstrated.

Freeman et al. (118) studied effects of ozone and nitrogen dioxide mixtures on the lungs of month-old rats, in concentrations of 0.9 ppm each, or 2.5 ppm nitrogen dioxide and 0.25 ppm ozone, for 60 days. The higher concentrations caused emphysema-like lesions. Ozone was mainly responsible. No potentiation of one gas by the other was apparent.

Hyde et al. (119) reported on beagle dogs exposed to diluted auto exhaust, diluted irradiated auto exhaust, nitrogen dioxide (NO_2), and nitric oxide (NO) in various concentrations for 68 months. The concentrations were oxidant, as ozone, 0.2 ppm; nitrogen dioxide, 0.046 to 0.92 ppm; nitric oxide, 0.18 to 1.6 ppm; and sulfur dioxide, 0.41 ppm. (Carbon monoxide, hydrocarbon, and sulfuric acid were also measured.) These exposures were followed by 32 to 36 months in clean air. Gross pathologic and microscopic examinations, including scanning and transmission electron microscopy, showed no significant gross lesions. Microscopically, there were enlargement of certain air spaces, with increased numbers and sizes of interalveolar pores in some, and hyperplasia of nonciliated bronchial epithelial cells. No evidence of interaction among the gases was seen.

Von Neiding et al. (88) exposed 11 healthy men to nitrogen dioxide, ozone, and sulfur dioxide alone or in two- or three-gas mixtures for 2 hr. The concentrations were nitrogen dioxide, 5 ppm; ozone, 0.1 ppm; and sulfur dioxide, 5 ppm. Several measures of lung functions were used. No evidence of interaction was found, except that following exposure to the three-gas mixture, there was an increased reaction to inhalation of an acetylcholine aerosol. The acetylcholine challenge was not used with the two-gas mixtures.

Sulfur oxides are prominent among the community air pollutants which arise from several widely used fuels, and there has been concern about their interactions with ozone. As indicated in the preceding paragraph, one experiment showed no long-term effects in dogs.

Amdur et al. (120) showed that inhalation of ozone at 0.2, 0.4, and 0.8 ppm caused increased respiratory frequency in guinea pigs, and the two higher concentrations caused a decrease in compliance. Sulfur dioxide at the same concentrations caused no changes, and mixtures of ozone and sulfur dioxide produced the same results as ozone alone. Juhos et al. (121) exposed rats to sulfuric acid aerosol at 2 mg/m^3 for 8 hr/day, or to ozone at 0.9 ppm for 4 hr/day, or to a mixture of the two for 8 hr/day, for 82 days. The sulfuric acid caused only slight changes in the respiratory tract, whereas ozone caused definite hypertrophy and hyperplasia of respiratory bronchiolar and adjacent alveolar epithelium. The mixture was only questionably more effective than ozone alone. Cavender et al. (122) exposed rats and guinea pigs to ozone at 0.5 ppm, sulfuric acid mist at 10 mg/m^3, or their combination, for 6 hr/day, 5 days/week, for 6 months. Only two of 70 rats showed hyperplasia of type II cells and focal alveolitis after the ozone exposure, and only two of the rats exposed to the mixture showed slight bronchiolar changes. Guinea pigs exposed to ozone or the mixture characteristically had lesions near the terminal bronchioles. There was no evidence of potentiation by the sulfuric acid. Cavender et al. (123) also studied exposures of 2 or 7 days, with ozone at 1.0 or 2.0 ppm and sulfuric acid mist at 5.0 or 10.0 mg/m^3, and saw no synergism. Last and Cross (124) exposed rats to 0.5 ppm ozone and 1.1 mg/m^3 of sulfuric acid mist, and observed a pronounced interaction, as measured by metabolic changes in tracheal explant cultures and in lung homogenates.

The complexities of efforts to evaluate the effects of combined exposures of pollutants in man are shown by a sequence of experiments in the laboratories of Bates in Montreal and Hackney in Los Angeles. Bates and Hazucha in Montreal (125) observed significant disturbances of lung functions with exposures to 0.37 ppm ozone; these were sharply increased by the addition of 0.37 ppm sulfur dioxide, although sulfur dioxide alone produced no evident effects. Hackney et al. in Los Angeles (93) reported no effects in similar exposures of healthy subjects, but observed that subjects judged to have hypersensitive airways were strongly affected by an ozone–sulfur dioxide mixture. Bates's and Hackney's groups collaborated in an unusually well designed, clearly described set of studies at Hackney's Rancho Los Amigos laboratory in Los Angeles. These consisted of experiments on eight Los Angeles subjects—four of whom were known to be sensitive to oxidant irritation—and four of the Montreal subjects. The ozone and sulfur dioxide concentrations were both 0.37 ppm. The Los Angeles subjects showed significant but less severe reactions than had been observed in Montreal. The responses with sulfur dioxide added were only slightly greater than with ozone alone. Two Canadians who showed severe reactions in Montreal showed only moderate responses in Los Angeles, and two who showed moderate responses in Montreal were about the same in Los Angeles. Later, it was shown that the Montreal exposure chamber had significant amounts of sulfates present, and this was suggested as the explanation for the more severe responses seen there. It was also suggested that Los Angeles residents may have a degree of adaptation to oxidant (126a).

Kleiman, et al. (126b) reviewed the effects of combinations of O_3, SO_2 and H_2SO_4 and reported on experiments with these irritants. The effect of ozone alone at 0.37 ppm was almost as great as that of ozone, 0.37 ppm plus SO_2, 0.37 ppm, and H_2SO_4 aerosol, 100 $\mu g/m^3$. There were 19 subjects, including 5 females, ranging from 19 to 54 years of age, 3 current smokers, 7 ex-smokers, and 9 non-smokers, and 4 with allergy histories, but no asthmatics.

Bedi et al. (127a) exposed nine humans to 0.4 ppm O_3 and 0.4 ppm sulfur dioxide for 2 hr with intermittent light exercise at 25°C and 45 percent relative humidity. Ozone alone caused significant decreases in maximum expiratory flow, forced vital capacity, and inspiratory capacity. Addition of sulfur dioxide caused no additional effects.

Folinsbee et al., (127b) review the evidence for interaction, synergism, of O_3 with NO_2 and SO_2 and report their own experimental work on O_3 + NO_2 at 0.5 ppm each for 2 hr with several conditions of elevated temperature, relative humidity, and exercise. They found considerable, statistically significant decreases of vital capacity, inspiratory capacity, and several measures of air flow, but no evidence of synergism. (Their O_3 concentration was probably closer to 0.4 ppm, since they did not take into account measurement factors mentioned in Section 2.7. Also, in their Table 4, the references should be 5 and 6, instead of 14 and 13.) There is no significant degree of interaction of ozone with NO_2 or SO_2.

2.7 Determination in Air

Five basic methods for analysis of ozone in air are available: wet chemistry reduction of potassium iodide (using colorimetric or electrometric evaluation), chemiluminescence, ultraviolet photometry, infrared photometry, and gas phase titration with nitric oxide.

A basic difficulty arises from the instability of ozone, which makes it impossible to distribute standard samples prepared by a central laboratory. Consequently, instruments for field use must be calibrated against an indirect reference standard. The preferred method is the use of an ozone generator that has been calibrated by comparison against gas-phase titration, long-path ultraviolet absorption, or titration with neutral buffered potassium iodide. The ozone generator is a mercury vapor lamp which acts on a current of clean air flowing through a quartz tube. Such generators have shown satisfactory stability.

Gas-phase luminescence has been designated as the principal reference method by the U.S. Environmental Protection Agency. Instruments using gas-solid luminescence and ultraviolet photometry have been designated as equivalent methods. The calibration based on neutral buffered potassium iodide that has been the accepted standard will undoubtedly be replaced by more precise methods using ultraviolet photometry or gas-phase titration (36, 37).

Prior to about 1975, ozone measurements were made by instruments that used the reduction of potassium iodide solutions. These were subject to interference by several substances, most notably by nitrogen dioxide, which added to the effect of ozone, and by sulfur dioxide, which subtracted from the ozone. In some areas where ozone levels were high and the levels of interfering substances were low, such instruments gave useful information, but the values were not properly comparable with determinations made under other circumstances. Worse, it was found that the standard calibration procedures used by several government agencies introduced systematic differences.

In consequence of these uncertainties of measurement, it is difficult to interpret toxicologic reports and epidemiologic studies, since the systematic errors must be appraised and taken into account, if possible. The following relationships among several methods of calibration have been reported (36):

$$O_3 \text{ (CARB)} = 1.29 \, O_3 \text{ (UV)} - 0.005$$

$$O_3 \text{ (LAAPCD)} = 0.960 \, O_3 \text{ (UV)} - 0.032$$

$$O_3 \text{ (EPA)} = 1.11 \, O_3 \text{ (UV)} - 0.035$$

where CARB indicates California Air Resources Board, 2 percent neutral buffered KI method; LAAPCD indicates Los Angeles Air Pollution Control District, 2 percent unbuffered KI method; EPA indicates Environmental Protection Agency, 1 percent neutral buffered KI method, and O_3 (UV) indicates the ozone level found by ultraviolet photometry.

Unfortunately, too few of the published reports concerning the effects of ozone at measured concentrations indicate the method of calibration that was used.

Que Hee (127c) recently reviewed methods for calibrating ozone measuring instruments and described a new method based on the evolution of formaldehyde by the reaction of ozone with ethylene.

2.8 Threshold Limit Values and Maximum Allowable Exposures

The American Conference of Governmental Industrial Hygienists' (ACGIH) (128) threshold limit value (TLV) for ozone for 1981 was 0.1 ppm. The American National Standards Institute-American Society for Testing Materials (ANSI/ATPM) standard for 1977 (129) was also 0.1 ppm. Both of these were for time-weighted averages over 8 hr. A short-term exposure limit (STEL) of 0.3 ppm was set by ACGIH. The ANSI standard said no worker should be exposed to a ceiling concentration in excess of 0.3 ppm, as measured over a sampling time of 10 min.

The Environmental Protection Agency set a primary air quality standard for oxidant, measured as ozone, of 0.08 ppm, 1 hr average, in 1971. This was a community air standard. In 1979, this was changed to an ozone standard of 0.12 ppm (130).

In other parts of the world, the maximum allowable concentration (MAC) or permissible limit in industry in 1970 was 0.1 ppm in most of the countries where a standard had been set. Rumania, the Soviet Union, and Yugoslavia had a MAC of 0.1 mg/m^3, equal to 0.05 ppm. We believe these values were unchanged in 1980 (131).

3 NITROGEN, N_2

3.1 Occurrence and Uses

Nitrogen is the main component of air, comprising 78 to 79 percent by volume. It is used in industry in its gaseous form to displace oxygen or explosive gases from enclosed spaces, including not only fuel tanks and other large volumes, but also communications cables and small packages of materials that would be damaged by oxidation. Compressed nitrogen is mixed with oxygen, helium, or other gases for deep-sea diving. Liquefied nitrogen is widely used for chilling metals to alter their physical characteristics and it is used in medicine and biology for quick freezing of tissues and microorganisms. Skin contact with even a small amount of liquid nitrogen can cause a serious burn, and it should be handled with great care.

3.2 Physical and Chemical Properties

Physical state	Colorless, odorless gas
Molecular weight	28.0134
Boiling point	−195.8°C
Melting point	−200.9°C
Density	1.2506 g/l at 0°C, or 0.9673 (air = 1)

Solubility in water	0.0294 g/kg at 0°C; 0.0146 at 37°C
1 mg/m³ ≈	0.79 ppm
1 ppm ≈	1.27 mg/m³

3.3 Nitrogen Toxicity

In industrial hygiene, nitrogen has its main importance for what it is not—it is not oxygen and does not support the chemical reactions needed for maintenance of life. It is likely to cause trouble because it displaces oxygen, leading to hypoxic asphyxia. However, it has recently been recognized as having direct toxic actions of its own, as seen in sea diving.

The occurrence of nitrogen poisoning in divers was recognized in the 1920s, but recent developments have brought much greater attention to it. Interest in sea diving was centered in militaty organizations for many years, and most of our knowledge of this field was developed by the U.S. Navy and other naval organizations. The *U.S. Navy Diving Manual* (132) is one of the best sources of information. Bennett and Elliott (133), Edmonds et al. (134), and Strauss (135) have reviewed the field recently. Dueker (136) provides an excellent discussion addressed to nontechnical readers. See Chapter 9 of Volume 1 of this series (137).

At a depth of about 30 m (100 ft) a diver breathing air is likely to become disoriented and to behave irrationally. "A diver may remove his mouthpiece, take off his equipment, or dive aimlessly to the point of unconsciousness or exhaustion of his air supply. To the narcotized diver, the sea world is pleasant indeed: he is all-powerful; there are no dangers; everything seems amusing. The fish around him do not have air bottles, so why should he?" These symptoms come on in a matter of minutes (136).

Apparently the mechanism of this narcosis is similar to that of various inhalation anesthetics, for which the theory is that any foreign substance dissolved in brain cells leads to narcosis. Under increased pressure, sufficient nitrogen dissolves in the fat-containing brain cells to produce symptoms. Other gases such as nitrous oxide or xenon are more soluble in fats, and they cause narcosis at normal atmospheric pressure. For most persons, the nitrogen partial pressure must be increased to more than 1200 torr to cause symptoms. There is marked individual variation, however.

Prompt reduction of the nitrogen partial pressure gives relief. The main problem is to prevent the diver from doing something foolish before he can be returned to normal pressure.

For shallow dives, the problem is avoided by increasing the proportion of oxygen in the breathing mixture, but this introduces the problem of oxygen toxicity (see Section 1.4). For deep dives, carefully balanced mixtures of nitrogen, oxygen, and helium are used.

Another problem arises from the solution of oxygen in body tissues under increased pressure of air, that of decompression sickness, or "bends," so-called because of the contorted postures of the agonized sufferers.

Decompression sickness is attributed to the formation of bubbles of gas, mainly nitrogen, in the tissues and in the blood vessels of persons who have been exposed to increased pressure for a time and who are suddenly released from the pressure. The decompressed gases come out of solution like the carbon dioxide in a freshly opened bottle of soda water. This can be avoided by preliminary breathing of oxygen (to wash the dissolved nitrogen out of the body), the use of breathing mixtures that contain the least soluble gases (helium, neon), and by very slow decompression. In very deep sea work, at least the last two are used; decompression may take several days.

3.4 Threshold Limit Value or Maximal Allowable Concentration

Several naval organizations have developed elaborate schedules for the control of diving and work under increased atmospheric pressure. The Diving Manual of the U.S. Navy is a good example (132).

4 NITROUS OXIDE, N_2O

4.1 Occurrence and Uses

Nitrous oxide, N_2O, is the most abundant of the nitrogen compounds in the atmosphere. It arises mainly from bacterial decomposition of organic nitrogen compounds in soil. Its concentration is about 0.25 to 0.29 ppm (138).

Nitrous oxide has little industrial use. It is used as a foaming agent for whipped cream, as an oxidant for organic compounds, to make nitrates of alkali metals, and in some rocket fuel combinations (139). It is used in medical practice as an anesthetic.

4.2 Physical and Chemical Properties (139, 140)

Physical state	Colorless gas, slightly sweet odor and taste
Molecular weight	44.02
Melting point	$-90.81°C$
Boiling point	$-88.46°C$
Density (air = 1)	1.53
Solubility in water at 25°C (ml/l)	600

4.3 Absorption, Excretion, and Metabolism

Nitrous oxide enters the body by inhalation and is readily absorbed through the lungs and transported throughout the body. It appears not to be metabolized, and is excreted through the lungs when the exposed person or animal breathes clean air.

4.4 Physiologic and Pathologic Responses in Animals and Man

Nitrous oxide is weakly narcotic, and when mixed with oxygen in excess of 80:20 can induce deep narcosis. It is an asphyxiant, and can cause death or permanent brain injury by hypoxia. Doses that do not cause hypoxic damage cause no apparent lasting effect. Brain damage from repeated abuse of nitrogen dioxide by persons willfully inducing disorientation has been suggested. Cohen et al. (141) observed that dentists and their assistants, chronically exposed to nitrous oxide, suffer increased rates of neurologic disease, kidney disease, liver disease, cancer, and spontaneous abortion.

5. NITRIC OXIDE, NO, AND NITROGEN DIOXIDE, NO$_2$

These compounds are considered together because they are encountered together in dynamic equilibrium.

Nitric oxide toxicity has been studied on a few occasions in animals, but at concentrations much higher than those likely to be encountered in practice (142). It has been established that nitrogen dioxide is much more toxic than nitric oxide, so analytic discrimination of the two in air is important for health protection.

5.1 Occurrence and Uses

Nitric oxide and nitrogen dioxide both occur naturally as a result of bacterial action on nitrogenous compounds and to a lesser extent from volcanic action, and from fixation by lightning. In the absence of man-made sources, the nitrogen dioxide concentration is about 0.0002 to 0.005 ppm, and that of nitric oxide is considerably less. In urban areas, the annual mean nitrogen dioxide concentration usually ranges from 0.01 to 0.05 ppm, but with marked variations. Shorter term averages to as high as 0.45 ppm for an hour (142).

The main source of urban nitric oxide and nitrogen dioxide is combustion of fossil fuels (138, 142). Escape of these gases from industrial processes where nitric acid is made or used, or from fertilizer or explosives factories, can be important in local areas and in the plants themselves. In general, higher combustion temperatures yield more nitrogen oxides. This topic is reviewed in a National Research Council report (138).

Industrial exposures can take place wherever nitric acid is made or used and has occurred most commonly where metals are dipped in acid baths. Electric arc welding, and to a lesser extent gas welding, can generate hazardous concentrations. The fermentation of silage produces high concentrations, and poisonings of farmers have occurred.

Nitric oxide is prepared commercially by passing air through an electric arc or by oxidation of ammonia. It is shipped as a compressed gas in steel cylinders. It is used extensively in the production of nitric acid, as a bleach for rayon, and as an oxidant in organic chemical reactions. Nitrogen dioxide is prepared commercially by the oxidation of nitric oxide.

Nitrogen dioxide and its polymer, nitrogen tetroxide, are always found together at normal environmental temperatures. Nitrogen dioxide can be used to nitrate benzene, anthracene, and naphthalene at 20 to 60°C but has not been extensively applied. It is a by-product of many operations and results whenever nitric acid acts upon metals, as in bright dipping, pickling, and etching, or upon organic material, as in the nitration of cotton or other cellulose. It is a by-product of the manufacture of many chemicals including explosives, dyes, lacquers, and celluloid, It also results, in significant amounts, from the slow burning of explosives or the detonation of explosives having a high oxygen balance; from electric arcs and electric- and gas-welding or gas-shrinking operations in confined and unventilated areas; from the burning of nitrocellulose; from the accidental spillage of nitric acid; during operations incidental to the manufacture or recovery of nitric acid; from the reduction of nitrates as, for instance, in the accidental pollution of a molten nitrate salt, heat-treat bath with some readily oxidizable matter. Nitrogen dioxide has been used as a component of fuel for jet propulsion.

The brown mixture arising from the bright dipping of copper or brass or from nitration reactions is essentially a pure mixture of NO_2 and N_2O_4. So far as industrial exposures are concerned, it matters little whether the nitrogen oxides enter the air as NO, NO_2, or N_2O_4, since the NO gradually changes to NO_2. This mixture has frequently been erroneously referred to as nitrous fumes, but it is not nitrous oxide, N_2O, and it is a gaseous mixture and not a fume. It may, for convenience, more properly be referred to as NO_x (pronounced "nox").

Fatal concentrations have been reported in farm silos shortly after filling.

5.2 Physical and Chemical Properties (139, 140)

Property	Nitric Oxide	Nitrogen Dioxide
Physical state	Colorless gas	Brown gas, yellow liquid
Molecular weight	30.01	46.01
Density (air = 1)	1.04	1.58
Melting point	−163.6°C	−11.2°C
Boiling point	−151.8°C	+21.2°C
Solubility in water (ml/l)	73.4	Slight
1 ppm ≂	1.23 mg/m³	1.88 mg/m³
1 mg/m³ ≂	0.813 ppm	0.532 ppm

5.3 Absorption, Excretion, and Metabolism

Nitric oxide is somewhat soluble in water; nitrogen dioxide decomposes to nitrous and nitric acids in solution. On inhalation, a considerable proportion of these gases is absorbed in the upper respiratory tract. Yokoyama (143) found that isolated upper airways of dogs and rabbits removed 42 percent of nitrogen dioxide. Dalhamn and Sjöholm (144) found that about 50 percent of nitrogen dioxide was absorbed in passage through the nose and throat of a rabbit. Goldstein et al. (145) observed that 50 to 60 percent of nitrogen dioxide was absorbed by monkeys exposed to 0.3 to 0.9 ppm of nitrogen dioxide. The radioactive tracer they used remained in the lung for prolonged periods, with some spread to other tissues. Nitrates and nitrites have been detected in the urine of animals after the inhalation of nitrogen dioxide. According to Flury and Zernik, nitric oxide combines with hemoglobin to form methemoglobin (74). The nitrite ions resulting from the reaction of nitric oxide would be expected to produce this result.

A World Health Organization review of nitrate and nitrite poisoning from water and food (146) concludes that available information does not permit the establishment of a quantitative dose–response relationship for human exposure to nitrates in water and food. However, it was observed that only infants who received over 10 mg/kg/day of nitrate showed methemoglobin levels above normal. Assuming a weight of 4 kg and a respiratory volume of 4 m^3/day for the infant, such a dose could be inhaled if the nitrate concentration in air were 8.75 mg/m^3 of nitrogen dioxide, equivalent to 3.5 ppm, a level exceeding by almost 400-fold the average values observed in the country's most polluted areas.

5.4 Physiologic and Pathologic Responses in Animals

The National Research Council review (138) cites a threshold for acute mortality of 40 to 50 ppm of NO_2 for 1-hr exposures of several species. This is shown graphically in Figure 52.5. It is noted that primates are more susceptible than rodents.

All animal species tested survive long-term exposures to concentrations of nitrogen dioxide that exceed those encountered in community air pollution, as shown in Table 52.7.

The recent review by a World Health Organization task force (142) indicates that the most prominent morphologic changes in several animal species occur in the epithelia of the terminal bronchioles and the alveolar ducts. Exposure to 0.25 to 1.0 ppm leads to bronchitis, bronchopneumonia, atelectasis, protein leakage into the alveoli, changes in collagen and elastin and in the mast cells of the lung, reduction or loss of cilia, and adenomatous changes

The effects are more pronounced at 2.0 to 25 ppm. Ciliated bronchiolar cells and the flat type 1 alveolar lining cells are injured first. They are replaced by nonciliated bronchiolar cells and type 2 alveolar cells. Prolonged exposure

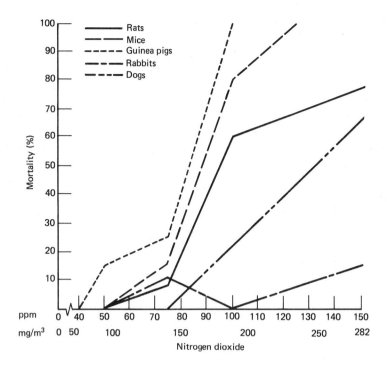

Figure 52.5 Mortality of different animal species after a 1-hr exposure to nitrogen dioxide. From Reference 138.

results in narrowing of the small airways by exudate, hypertrophy of the respiratory epithelium, and swelling of the basement membrane.

A point not emphasized by the WHO review is that several authors have reported lung lesions in animals resembling those of human emphysema, a very troublesome lung disease that has appeared to be increasing in prevalence. Hypertrophy of bronchiolar epithelium was seen in monkeys exposed continuously at 2 ppm for 14 months. Exposure of rats to 10 or 25 ppm for 26 or 13 weeks respectively was followed by development of large, distended lungs that did not collapse under atmospheric pressure when removed from the thorax. The lungs appeared grossly emphysematous and the thoracic cage was enlarged, with dorsal kyphosis, resembling changes seen in emphysematous humans.

The review by the U.S. National Research Council comments on these reports that the lesions induced in rats lack an essential characteristic of human emphysema. "Destructive bullous lesions are the *sine qua non* of emphysema, but these bullae do not develop in rodent models, even in rats exposed for a lifetime to nitrogen dioxide concentrations of 1.5 and 3.8 mg/m³ (0.8 and 2.0 ppm). The lungs ... are grossly normal; ... only minor ciliary loss, epithelial hyperplasia and 'cytoplasmic blebbing' [are seen]. These animals have a normal lifespan and die of diseases unrelated to nitrogen dioxide exposures" (138).

Bils (156) observed thickening and numerous fenestrations of alveolar septa in squirrel monkeys exposed to 3 ppm for 4 hr/day for 4 days.

Port et al. (157) exposed several species to 0.1 ppm continuously for 6 months, with added daily peaks to 1.0 ppm for 2 hr. Occasional foci of distended alveoli occurred, and an increase in the size and number of alveolar pores was seen. Holt et al. (158) observed emphysematous lesions in mice exposed to 10 ppm of NO_2 for 2 hr/day, 5 days a week for various periods up to 30 weeks. However, mice similarly exposed to 10 ppm of nitric oxide showed much more pronounced emphysematous changes. There is no explanation for this paradoxical reversal of the usual levels of toxicity, and the methods of measurement of the gases are not specified.

Drozdz et al. (159) observed bronchiolar damage and emphysematous changes in the lungs of guinea pigs exposed to 2 mg/m^3 of nitrogen dioxide (1 ppm), along with some skin changes and destruction of collagen. Hyde et al. (160) exposed 104 beagle dogs to various combinations of "raw" or photochemically reacted automobile exhaust, oxides of sulfur, or oxides of nitrogen for 16 hr/day for 68 months, then to clean air for an additional 30 to 36 months, then did very sophisticated pathologic studies. Enlargement of air spaces in proximal acinar regions, with and without increases in the number and size of interalveolar pores, and hyperplasia of nonciliated bronchiolar cells were conspicuous changes. These were most strongly associated with exposure to nitrogen dioxide at 0.64 ppm. They were less conspicuous in animals exposed to irradiated automobile exhaust that contained 0.92 ppm of nitrogen dioxide and 0.18 ppm of nitric oxide. Addition of sulfur oxides appeared to make the lesions worse.

Table 52.7. Survival of Animals Exposed Chronically to High Concentrations of Nitrogen Dioxide (138)

Species	Concentration mg/m^3	ppm	Duration of Exposure	Fatalities Attributed to Exposure	Ref.
Mice	0.94	0.5	12 months	None reported pneumonitis	147
Rats	1.52	0.8	Lifetime	None	148
	3.76	2.0	Lifetime	None	149
	23.50	12.5	213 days	11% fatality	150
Guinea pigs	7.52	4.0	4 hr/day, 5 days/week for 6 months	None	151
	28.20	15.0	6 months	None	151
Squirrel monkeys	1.88	1.0	16 months	None	152
Stump-tailed macaque	3.76	2.0	2 years	None	153
Dogs	9.40	5.0	15 months	None	154
Rabbits	2.44	1.3	17 weeks	None	155

Exposure to 1.64 ppm of nitric oxide and 0.14 ppm of nitrogen dioxide was most effective in causing increases in alveolar pores.

Damage to pulmonary function is described in the WHO review (142). Rapid breathing throughout life characterized rats exposed continuously to 0.8 ppm. Beagles exposed daily for 6 months to 0.64 ppm nitrogen dioxide with 0.25 ppm nitric oxide showed reduction of diffusion capacity and of peak expiratory flow rates. But another experimenter, using somewhat heavier exposures for 72 weeks, found no changes. Rats that showed rapid breathing with exposure to 2.0 ppm for 2 years showed no changes in transpulmonary resistance nor compliance. Nonhuman primates also breathed rapidly when exposed to 2 ppm for 7 years, or to 5 ppm for 2 months, and in the latter study showed reduced tidal volumes. Guinea pigs exposed to 5 ppm for 5½ months showed no changes in expiratory flow resistance, but guinea pigs exposed to 5.2 ppm for only 4 hr showed increased respiratory rates and decreased tidal volumes; these returned to normal after cessation of the exposure. Rats exposed to 2.9 ppm for 5 days/week for 9 months showed a 13 percent decrease in lung compliance, but rabbits exposed to 5 ppm 6 hr/day for 18 months showed no changes. Higher concentrations (8 to 12 ppm) evoked changes.

Abraham et al. (161) exposed sheep to 7.5 or 1.5 ppm of nitrogen dioxide, and observed that a 2-hr exposure to 7.5 ppm can produce bronchial hyperactivity to a bronchial constrictor drug and that a 2-hr exposure to 15 ppm induces airway hyperactivity, depression of tracheal mucus movement, and in some animals, increased pulmonary resistance. Also, a 4-hr exposure to 15 ppm causes a transient increase in pulmonary resistance without an effect on the blood circulation.

Biochemical changes described in the WHO and NRC reviews include alterations in several lung enzymes, in the lipid content of the lungs, and in the stability of pulmonary surfactant, and a decrease in lung glutathione. A dose-related response is observed from 6.0 to 40.0 ppm.

The biochemical mechanisms of nitrogen dioxide cellular injury are under very active study. Two theories are emerging: one emphasizes lipid peroxidation, the other the oxidation of low molecular weight reducing substances and proteins. These need not be mutually exclusive.

Nitrogen dioxide appears to act in the same way as ozone as an oxidant of unsaturated fatty acids in vitro (162). Both nitrogen dioxide and ozone initiate oxidation through free radicals, and both show an induction period that can be prolonged by vitamin E or other free radical scavenging agents. Polar nitrogen-containing compounds and peroxides are produced.

Rats exposed to 10 to 33 ppm of nitrogen dioxide showed increased mortality and decreased unsaturated fatty acids in lung washings (163–165), and the effect was enhanced in vitamin E-deficient animals. However, another experiment found no reduction of lung phospholipids in guinea pigs exposed to 0.05 or 0.5 ppm of nitrogen dioxide with or without an equal concentration of ammonia for 122 days, 8 hr/day (166).

Exposure of rabbits to 1.0 ppm for 2 weeks depressed lecithin syntheses. The effect appeared to decline during the second week (167).

Menzel (50) described dietary experiments with mice exposed to 0.97 ppm of ozone and fed saturated or unsaturated fats and deprived of vitamin E or given a supplement, up to 3 IU/day. He observed that there was a pronounced delay in mortality in the animals given the vitamin E. This is given as evidence that the mechanism of ozone toxicity is the initiation of lipid peroxidation and the formation of free radicals. Vitamin E is an effective scavenger of free radicals. Nitrogen dioxide, given in the same experiment, was less toxic than ozone, causing no mortality at 13 ppm, 8 hr/day for 90 days. There was, however, evidence of pulmonary edema, and this was least pronounced in the mice that received a high dose of vitamin E. Menzel discusses the chemical reactions involved at some length, and concludes that there may be a need for a marked increase in the vitamin E intake of the population of the United States.

Goldstein had reported lipid peroxidation by nitrogen dioxide in an in vitro system (115), but in vivo tests with nitrogen dioxide at 2 to 20 ppm (lethal) for 2 hr showed no evidence of this effect, nor of sulfhydryl oxidation (116). However, the assays were made immediately after exposure, and it has been shown that products of lipid peroxidation are maximal after a delay of several hours.

It now appears that lipid peroxidation causes severe damage to cell membranes and this mechanism adequately explains the effects of nitrogen dioxide, ozone, and other oxidants on the lung, with differences arising because of variations in solubility and reaction speed. Much research in this field is in progress and rapid elucidation of details should occur.

Other mechanisms have also been shown for ozone, but are less well known for nitrogen dioxide. Oxidation of sulfhydryl compounds is an example.

The most conspicuous effect of nitrogen dioxide poisoning is pulmonary edema, and this may be responsible for the appearance of reduced concentrations of lung enzymes and proteins. Also, changes in cell populations may influence concentrations of tissue constituents.

Since preparation of the review articles referred to, a report by Menzel et al. (168) was released in which it was observed that after 7 days exposure to 0.5 ppm, serum lactic acid dehydrogenase (LDH), creatine phosphokinase (CPK), glutamic–oxalacetic transaminase (SGOT), and glutamic–pyruvic transaminase (SGPT) were elevated, whereas lung glutathione peroxidase (GSH peroxidase) and acid phosphatase were not affected. Lung and plasma lysosyme levels were elevated, in contrast to the earlier observation of lowered levels reported by Chow et al. (169).

Ayaz and Csallany (169a) found that exposure of female mice for 17 months to 1.0 ppm caused suppression of glutathione peroxidase activity, more pronounced in vitamin E-deficient animals, whereas exposure to 0.5 ppm did not produce this effect.

Menzel et al. (168) found reduction of red blood cell GSH peroxidase after 7 days exposure to 0.5 ppm, but not of lung GSH peroxidase; however, the level of this enzyme in both lungs and red cells was normal after 4 months exposure. This is distinctly different from the observation with ozone.

A conspicuous effect of nitrogen dioxide exposure is decreased resistance to pulmonary infection, as reported in review documents (138, 142). Protection from infection of the lung is accomplished by three main mechanisms: the mucociliary system, which extends from the nose to the terminal bronchioles, removes 50 to 90 percent of particles that are deposited; particle-ingesting cells, macrophages, take up and kill many bacteria and may inactivate virus particles; antibodies in blood, cellular, and extracellular fluids join in the defense system. Toxic agents can interfere with any of these processes. Nitrogen dioxide affects all of them (138, 142).

Gardner et al. (170) recently reaffirmed the effectiveness of nitrogen dioxide in increasing susceptibility of mice to infection with *Streptococcus pyogenes* in concentrations as low as 0.5 ppm. They showed progressively greater effects with exposures of 7 days to 1 year. Similar results were seen with higher concentrations, up to 28 ppm, with durations from 10 to 35 min. They emphasize that any specific dose is more effective if delivered in a shorter time; that is, a high concentration for a short time is worse than a low concentration for a long time. They also demonstrated that continuous exposure is worse than intermittent exposure for a few hours, but not for several days or longer. They suggest that this results from early effects upon macrophages, and later effects on other components of the defense system.

Ehrlich et al. (171) reported increased mortality from *S. pyogenes* in mice after exposure to nitrogen dioxide at 2.0 ppm for 3 hr. Mice made to exercise during exposure to 1.0, 2.0, or 3.0 ppm for 3 hr showed less resistance to infection than controls unexposed to nitrogen dioxide.

Schiff (172) challenged hamster tracheal ring organ cultures with influenza virus after 1 or 2 weeks of exposure to 2.0 ppm of nitrogen dioxide for 2 hr/day, 5 days/week. The infection caused decreased ciliary activity in exposed cultures, as compared to controls held in filtered air. After 2 weeks, there was decreased ciliary activity and other changes in noninfected, nitrogen dioxide-exposed cultures.

Chronic exposures to 0.5 ppm or more for 30 days or longer led to increased susceptibility to bacterial challenge by *Klebsiella pneumoniae* in mice or squirrel monkeys (138).

Mice exposed to 10 ppm for 1 to 5 days showed diminished resistance to mouse adapted influenza virus, but lower concentrations were ineffective. However, in another experiment exposures to 38 ppm for 6 weeks appeared to enhance resistance in mice. The results of studies on antibody production are also confusing, sometimes indicating suppression, and sometimes enhancement, by nitrogen dioxide exposures (138, 142).

Several experiments demonstrated that high concentrations of nitrogen dioxide impaired mucociliary clearance (138). According to Giordano and Morrow (173), concentrations over 5 ppm decrease mucociliary transport.

Several studies demonstrated the susceptibility of macrophages to nitrogen dioxide injury (138), and this probably accounts for much of the increased susceptibility to infection that has been described.

Aranyi et al. (174) observed morphologic changes of lung macrophages after exposure of mice to 2.0 ppm continuously, or 0.5 ppm with daily 1-hr peaks of 2 ppm 5 days/week, for 21 weeks. No effect was seen with 0.5 ppm continuously, or 0.1 ppm with 1 ppm peaks.

Green and Schneider (175) observed that exposure of baboons to 2 ppm nitrogen dioxide for 6 months caused changes in the responsiveness of their pulmonary macrophages to an immunologic factor.

Blood changes have been reported from exposures as low as 0.3 to 2.0 ppm. Polycythemia (increased numbers of red cells) and leukocytosis (increased numbers of white cells) were observed. However, several studies at levels up to 5 ppm showed no effects on blood cells or hemoglobin; 10 ppm caused an increase of methemoglobin after 1 hr.

The principal effects of low level exposures of experimental animals to nitrogen dioxide are summarized in Table 52.8, modified from one published in the WHO report (142).

5.5 Effects in Man

Nitrogen dioxide has been recognized as an industrial poison since the early years of the century at least; the condition is usually called "nitrous fume inhalation." The WHO review (142) has a good summary of the effects of acute exposures (see Table 52.9). Milne (175a) wrote an interesting review in 1969, calling attention to the occurrence of an effect that could be delayed for a month or more, called *bronchiolitis obliterans*. Proctor and Hughes (176), in their brief summary, say that bronchiolitis obliterans is a rare complication. The National Institute of Occupational Safety and Health has published criteria for occupational exposures (177).

Brief exposure to 250 ppm or more causes coughing, production of frothy or mucoid sputum, and increasingly difficult breathing. Within 2 hr, severe pulmonary edema with its consequent impairment of oxygen uptake and of heart function may develop, and death may soon follow. In other cases, there may be a latent period of 8 hr or more in which the person feels well and may wish to return to work, but after this interval, the onset of pulmonary edema is rapid; it is widely accepted that the severity of the pulmonary reaction is lessened if the person avoids exertion during the latent period. In still other cases, there may be mild or moderate signs and symptoms of injury right after the exposure; these subside over a period of 2 to 3 weeks, only to be followed

Table 52.8. Experimental Animal Studies of Nitrogen Dioxide Toxicity[a]

I. Local Effects on the Respiratory System

NO₂ Concn (ppm)	Length of Exposure days	hr/day	Effects	Responses[b]	Species	Number of Animals
1.0	4	24	Glutathione peroxidase activity increased in lung		Rat	5(11)
1.0	1	4	Peroxidation of lung lipids		Rat	6(10)
0.8	990	24	Elevated respiratory rates	9/9(0/12)	Rat	6(10)
0.8	990	24	Minimal bronchiolar epithelial hypertrophy	9/9(0/12)	Rat	9(12)
0.8	5	24	Decrease in lung reduced glutathione		Mouse	10(10)
0.5–0.8	30	24	Degeneration and desquamation of mucous membrane	10/10(0/5)	Mouse	10(5)
0.5–0.8	30	24	Ciliated cell damage; edema of alveolar epithelia; epithelial proliferation, peripheral bronchus (adenomatous)	6/10(0/5)	Mouse	10(5)
0.64 plus 0.25 NO	61 months	16	Reduction in pulmonary diffusion capacity	6/11(3/8)	Dog	11(18)
0.64 plus 0.25 NO	61 months	16	Decreased peak expiratory flow rate	6/11(1/8)	Dog	11(18)
0.64 plus 0.25 NO	61 months	16	Enlargement of proximal alveolar air spaces		Dog	12(20)
0.5 plus 0.2 NO	18 months	16	No changes in CO diffusion capacity, compliance, or total expiratory resistance	0/12(0/20)	Dog	12(20)
0.5	90–360	6, 18, 24	Evidence of focal emphysema	12/12(0/4)	Mouse	12(4)
0.5	1	4	Reduction in mitochondria of alveolar cells and degradation of mast cells	6/6(0/6)	Rat	6(6)

Dose	Duration	No.	Effect	Species		
0.5	4 months	8	Decrease in plasma cholinesterase; rbc or lung GSH peroxidase unchanged. Increase in lung acid phosphatase and in plasma and lung lysozyme. Reduction of rbc glutathione peroxidase, later returning to normal	Guinea pig		
0.4	7	24	Increase in protein content of lung lavage fluid	Guinea pig		
0.32	90	24	Bronchitis, peribronchitis, light pneumosclerosis; no effect at 0.8 ppm	Rat		15(15)
0.1 plus	180	24	Alveolar distention, increased septal pores	Several		
0.25	24–36	4	Lung collagen fiber changes	Rabbit	2/3(0/1)	3(1)
0.05	90	24	No pathological or histological changes	Rat	0/10(0/10)	10(10)

II. Other Effects

Dose	Duration	No.	Effect	Species		
1.0	16 months	24	Increased serum neutralizing antibody titers	Squirrel monkey	5	
1.0	16 months	24	Decreased aminotransferase activity in brain and liver, increased activity in blood serum and heart	Guinea pig	30(50)	
0.9	30	24	Reduced antibody production in spleen	Mouse	9(9)	
0.7–0.8	1 month	24	No change in growth rate	Mouse	20(20)	
0.5 plus 1 hr at 2.0 daily	3 months	24	Changes in circulating immunoglobulins; depression in serum neutralizing antibody titers	Mouse	112–160 (112–160)	
0.5–0.8 plus 50 CO 1–1.5 months		24	No change in blood carboxyhemoglobin	Mouse	94(49)	
0.32	90	24	Changes in conditioned reflexes	Rat	15(15)	
0.3–0.4	1	2	Increase in liver ascorbic acid	Mouse	20(114)	

Table 52.8. (Continued)

| NO$_2$ Concn (ppm) | Length of Exposure | | II. Other Effects | Responses[b] | Species | Number of Animals |
	days	hr/day	Effects			
0.05	90	24	No effects on weight gain, central nervous system activities of cholinesterase, catalase, or of SH groups in blood, hemoglobin, or erythrocytes		Rat	10(10)
			III. Interaction with Infectious Agents			
0.5–1.0	39	24	Adenomatous proliferation of bronchial and bronchiolar epithelium	7/12(1/12)	Mouse	12(12)
0.5	12 months	24	Increased susceptibility to pulmonary infection; reduced lung clearance of inhaled microbes		Mouse	4(4)
0.3–0.5	3 months	24	Adenomatous proliferation of peripheral bronchial cells; not increased by 3 months additional exposure		Mouse	12(8)

[a] Modified from Reference 142.
[b] Number of animals showing effects/total number of animals; numbers in parentheses refer to control groups.

by abrupt onset of fever, chills, and difficult breathing that may end in death. When death occurred shortly after poisoning, autopsy showed an inflammatory reaction of the respiratory tract and lung edema. In cases where death occurred after several weeks of apparent recovery, the small bronchi showed fibrotic narrowing. The chest X-rays in the first type showed widespread replacement of air-filled lung by edema; in the late-onset cases, many fine nodules, corresponding to bronchiolar fibrosis, were seen.

Wherever there is exposure to unusual amounts of nitrogen dioxide, the exposed persons should be under medical supervision for a period of 72 hr, to detect the earliest signs of pulmonary edema. Treatment may include administration of oxygen under pressure. If there is eye irritation at the time of exposure, the eyes should be flushed with water.

A review by the American Conference of Governmental Industrial Hygienists (178) suggests that a 60-min exposure of humans to 100 ppm leads to pulmonary edema and death; 50 ppm to pulmonary edema with possible subacute or chronic lesions in the lungs; and 25 ppm to respiratory irritation and chest pain. Fifty parts per million is moderately irritating to the eyes and nose; 25 ppm is irritating to some people (177).

As with ozone, occupational exposures to nitrogen dioxide must be considered against the background of community exposure that continues throughout the day, without rests or holidays. However, community nitrogen dioxide concentrations do not closely approach the levels that have been recognized as hazardous in industry. Yet it may be found that the levels now acceptable in industry may have to be revised downward as we learn more about nitrogen dioxide and chronic disease.

The World Health Organization review (142) summarizes information on health effects in humans up to 1977, and the National Research Council report (138) does so up to 1974. These reviews are mainly addressed to the need for control of community exposures.

The odor of nitrogen dioxide is perceptible for some persons at 0.11 ppm, and for most at 0.22 ppm. Dark adaptation, the ability to perceive dim lights, is impaired by as little as 0.074 ppm. These responses are immediately reversible.

Exposure to 0.7 to 2.0 ppm for 10 min causes increased resistance to the flow of air in the respiratory tract. A number of authors have reported that 4 to 5 ppm produced an increase in airway resistance and a decrease in the rate of diffusion of oxygen or carbon monoxide through the lung membranes. A majority of asthma patients showed an increased response to a bronchial constrictor agent after exposure to 0.1 ppm for 2 hr. Normal persons also showed increased responses to bronchoconstrictors when exposed for 2 hr to nitrogen dioxide at 0.05 ppm and sulfur dioxide at 0.1 ppm. See Table 52.9.

Nitrogen dioxide alone at 3 or 6 ppm caused 16 and 34 percent increases in airway resistance; addition of sodium chloride aerosol of 0.95 μg average particle size, 1.4 mg/m^3, doubled the resistance. Particles of 0.22 μg failed to cause an increase.

Table 52.9. Controlled Studies on Humans—Effects of Nitrogen Dioxide on Lung Function[a]

NO$_2$ Concn. (ppm)	Length of Exposure		Effects[b]	Subjects
	Days	hr/day		
5	1	2	Increase of R_{aw}; decrease of AaD_{O_2}	11 healthy Ss
5	1	15 min	PA_{O_2} before, during and after exposure unchanged, but Pa_{O_2} decreased; AaD_{O_2} increased	14 chronic bronchitis patients
5	1	15 min	DL_{CO} decreased	16 healthy Ss
4–5	1	10 min	Decrease in lung compliance with corresponding increases in expiratory . and inspiratory flow resistance	5 healthy Ss
1.6–2	1	15 min	R_{aw} increase	15 patients with chronic bronchitis
1.0	2	2	Slight reduction of FVC second day	20 healthy Los Angeles Ss
			Reduction of erythrocyte acetylcholinesterase second day	10 of foregoing Ss
0.7–2	1	10 min	Increase in inspiratory and expiratory flow resistance	10 healthy Ss
0.1	1	1	A slight but significant increase in initial SR_{aw} and enhancement of bronchoconstrictor effect of carbachol in 13 Ss	20 asthmatics
0.05 plus	1	2	No effect on R_{aw} and AaD_{O_2}; sensitivity of the bronchial tree to acetylcholine increased compared with preexposure level	11 healthy Ss

[a] Modified from Reference 142.
[b] Abbreviations: Ss = Subjects; R_{aw} = airway resistance; SR_{aw} = specific airway resistance, i.e., the product of airway resistance and thoracic gas volume; AaD_{O_2} = alveolar to arterial oxygen pressure difference; DL_{CO} = diffusing capacity of the lung for carbon monoxide; PA_{O_2} = alveolar partial pressure of oxygen; Pa_{O_2} = arterial partial pressure of oxygen.

Epidemiologic studies of the effects of air pollutants are very difficult under the best of circumstances. Studies of nitrogen dioxide are usually complicated because it always occurs in combination with other pollutants; usually the measurements are with other pollutants, and the measurements are usually uncertain. It is not possible on the basis of available epidemiologic knowledge to ascribe a health hazard to nitrogen dioxide alone.

Epidemiologic studies of pulmonary function have sometimes shown changes consistent with the hypothesis of damage by nitrogen dioxide, but other causes have also been present. Evidence of effects on susceptibility to respiratory infection has been sought, and the results are similarly equivocal. Several studies have suggested an association of chronic disease with nitrogen dioxide exposures, but the observations are not conclusive. See Table 52.10.

Hackney et al. (178a) exposed 20 healthy subjects, residents of Los Angeles, 23 to 48 years old, including two current and two former smokers, to 1 ppm nitrogen dioxide, with light exercise, at 31°C and 35 percent humidity. An extensive array of pulmonary function, clinical, and biochemical tests were done during and after the exposure. Only a marginal reduction of forced vital capacity was seen after two successive days of exposure. There was a slight increase of symptoms of respiratory irritation.

Folinsbee et al. (179) studied cardiovascular and metabolic responses in 15 healthy males aged 20 to 25 years, exposed to 0.62 ppm of nitrogen dioxide, with varying amounts of exercise at 40 percent of maximal oxygen consumption. They found no differences in cardiac or pulmonary function with increasing nitrogen dioxide exposure.

Posin et al. (180) found decreases of erythrocyte membrane acetylcholinesterase after exposures to 1 or 2 ppm nitrogen dioxide for 2½ or 3 hr, repeated

Table 52.10.   Epidemiologic Studies of Community Exposure to Nitrogen Dioxide

Concentrations and Population	I. Pulmonary Function Averaging time	Effect and/or Response
Pulmonary function tests on 20 normal 11-year-old schoolchildren were made once or twice a week for 17 months. The concentrations of NO_2 in the higher temperature season at the time of measurement ranged from approx. 0.02 to 0.19 ppm	1 hr	Association with decrease in specific airway conductance and V_{max} at 50% FVC during the high temperature season; in one subject, V_{max} at 50% FVC decreased steeply at NO_2 levels of approximately 0.04 ppm; the observed effect is not associated with NO_2 alone, but with combined exposure to NO_2, SO_2, particulates, and O_3

Table 52.10. (*Continued*)

II. Acute Respiratory Diseases

Exposed	Control	Averaging Time	Effect and/or Response
0.08–0.15 ppm NO_2 with 4–7 $\mu g/m^3$ nitrate, 10–13 $\mu g/m^3$ under 0.01 ppm SO_2, 63–96 $\mu g/m^3$ particulates; exposure to H_2SO_4 and HNO_3 fumes also present but not measured	0.03–0.06 ppm NO_2 with 2–3 $\mu g/m^3$ nitrates, 10 $\mu g/m^3$ sulfates under 0.01 ppm SO_2, 62–72 $\mu g/m^3$ particulates	1 year	Increased incidence of acute respiratory disease in school children and parents in Chattanooga. Increased incidence of lower respiratory disease in Chattanooga infants and school children
Over 0.5 ppm NO_2 $\frac{1}{2}$–1 hr peak indoor concentration	Under 0.05 NO_2	1 hr	No evidence of increased acute respiratory disease in housewives cooking with gas stoves compared with those using electric stoves

III. Chronic Respiratory Disease

Exposed	Control	Averaging Time	Effect
0.055 ppm NO_2 with 0.035 ppm SO_2	0.04 ppm NO_2 with 0.01 ppm SO_2	1 year	No significant increase in chronic respiratory symptoms among central city traffic police officers in Boston
0.05 ppm NO_2 with 0.01 ppm SO_2, 120 $\mu g/m^3$ particulates, 0.14 ppm oxidants (mean of daily 1-hr maxima)	0.023 ppm NO_2 with 0.01 ppm SO_2, 78 $\mu g/m^3$ particulates, 0.074 ppm oxidants (mean of daily 1-hr maxima)	1 year	No effect on prevalence of chronic respiratory symptoms or on lung functions of nonsmoking subjects living in Southern California
Instantaneous levels up to 0.74 ppm in homes with gas stoves	Other heat for cooking		Reduced pulmonary function in 6–9 year old children

[a] Modified from Reference 142.

on two successive days; peroxidized red blood cell lipids only after 2 ppm; elevated glucose-6-phosphate dehydrogenase (G-6-PD) only after a second exposure to 2 ppm. Small decreases of hemoglobin and hematocrit occurred after both levels. However, similar but smaller changes in acetylcholinesterase, hemoglobin, and hematocrit followed similar exposures to clean air. The subjects were 10 healthy adult males who were exposed in a manner similar to that described above for Hackney's pulmonary function experiments.

Von Nieding and Wagner (181) exposed 111 subjects with chronic nonspecific lung disease to 0.5 to 8 ppm of nitrogen dioxide for 15 to 60 min. Significant increases of airway resistance were seen with concentrations as low as 1.6 ppm (mean of 15 subjects). They concluded, ". . . NO_2 may act by release of histamine, causing a bronchiolar, alveolar, and interstitial edema, thus differing from irritant air pollutants like SO_2, where reflex bronchoconstriction causes in some bronchitics dramatic increases of airway resistance at similar low concentrations."

On the question of the effects of exposures in the home, there are conflicting reports. Mitchell et al. (182) studied 232 households that used electric stoves for cooking and 209 households that used gas, and found no significant differences in the incidence of respiratory illnesses. The average nitrogen dioxide level in a sample of electric homes was 2 pphm (parts per hundred million), with a range of 0 to 6 pphm; in gas homes, the average was 5 pphm with a range of 0.5 to 11 pphm, whereas the outdoor average was 3 pphm with a range of 1.5 to 5 pphm. Peak levels in gas cooking households were generally eight times higher than the average and sometimes exceeded 100 pphm. Ferris et al. (183) reported that in homes where gas was used for cooking, the nitrogen dioxide concentration was higher than out of doors, with instantaneous kitchen levels up to 0.74 ppm, and that pulmonary function was lower in 6- to 9-year-old children in gas-using homes.

5.6 Determination in Air

Nitrogen dioxide and nitric oxide can be measured separately or together by several manual or automatic methods. "Grab" samples or integrated samples collected over various periods of time can be used.

Among the manual methods, references to the Griess–Saltzman technique are most frequent. Nitrogen dioxide reacts with sulfanilamide to form a diazonium salt that is coupled with a naphthylethylenediamine compound to give a red dye, which is estimated colorimetrically. Color fading is a limitation of the method, and it requires a high level of skill. It has been adapted for use in continuous recorders.

The Jacobs–Hochheiser method is a modification that uses sodium hydroxide as an absorbing medium. It has been the subject of controversy, and the Environmental Protection Agency abandoned it as a reference method. It overestimates low concentrations and underestimates high concentrations. Reports of studies that used this method are hard to evaluate.

Several modifications said to be superior to the Saltzman method include the arsenite or Christie method, the triethanolamine or Levaggi method, and the TG-S-ANSA method.

All the foregoing methods measure nitrogen dioxide, and nitric oxide must be oxidized to nitrogen dioxide to be measured. The efficiency of this oxidation has given some problems.

The chemiluminescent method depends on the reaction between nitric oxide and ozone, which emits light that can be measured. Nitrogen dioxide can be measured after reduction to nitric oxide; this reduction has been the source of some difficulties. Commercial instruments have been developed that are stable, sensitive, and durable. The accuracy of calibration is very important. Gas-phase titration is used to attain satisfactory precision. Jones and Ridgik (184) published a method for field calibration of these meters. Palmes and Tomczyk (185) described a personal sampler for oxides of nitrogen for monitoring workplace air.

5.7 Threshold Limit Values and Allowable Exposures

The American Conference of Governmental Industrial Hygienists (186) established a threshold limit value (TLV) of 3 ppm for nitrogen dioxide in 1981. A short-term exposure limit (STEL) of 6 ppm was adopted.

The TLV for nitric oxide is 25 ppm and the STEL 35 ppm, as of 1981.

In 1971 the U.S. Environmental Protection Agency (187) established a national ambient air quality standard for nitrogen dioxide of 100 μg/m^3 (0.053 ppm), averaged annually. The German Federal Republic maximum allowable concentration for industrial exposures was 5 ppm in 1979 (181).

6 CARBON MONOXIDE, CO

Of all the gases that have poisonous effects upon humans and animals, carbon monoxide is the most widely encountered. It exerts its effects by combining with the hemoglobin of the blood and interrupting the normal oxygen supply to the body tissues. Although the resultant oxygen deficiency is a reversible chemical asphyxia, nevertheless the damage done by severe asphyxia from any cause may not be reversible.

6.1 Sources of Carbon Monoxide

Exposure to carbon monoxide in industry, or even in private life, may occur whenever carbonaceous material, such as coal, wood, paper, oil, gas, gasoline, or any other organic material, is burned. Carbon monoxide is a product of incomplete combustion, and is not likely to result where a flame burns in an abundant air supply without contacting any surface. Whenever a flame touches a surface that is cooler than the ignition temperature of the gaseous part of the flame, carbon monoxide may result. Notorious in this respect are water heaters: the temperature of the water-filled coils cannot rise appreciably above the boiling point of water at the pressure involved; if a flame is allowed to play on the coils, a substantial amount of carbon monoxide is produced, and where the heater is not effectively vented to the exterior, contamination of the room atmosphere results.

Gas or coal heaters in the home and gas space heaters in industry have been frequent sources of carbon monoxide when not provided with effective vents. Gas heaters, although they may be properly adjusted when installed, may become hazardous sources of carbon monoxide if not correctly maintained. Automobile exhaust gas in garages, especially small private garages, is perhaps the most familiar source of carbon monoxide exposures. "Clean air" adjustments to automobile engines, and the use of catalytic afterburners, reduce carbon monoxide emissions, but they should not be relied upon to provide assurance of safety from poisoning by exhaust gases in garages or other enclosed spaces.

Major industrial sources are the cupolas of iron foundries, catalytic cracking units in petroleum refineries, lime kilns and kraft recovery furnaces in kraft paper mills, and the sintering of blast furnace feed in sintering plants (189a).

Additional potential exposures occur in the manufacture and use of illuminating gas, or "manufactured gas" (although this hazard is less frequently encountered than when the earlier editions of this book were written); the manufacture of synthetic methane or other organics from carbon monoxide; carbide manufacture; the distillation of coal or wood; operations near furnaces, ovens, stoves, forges, and kilns, which are especially likely to produce excessive carbon monoxide during the period in which they are being brought to normal operating temperatures after a period of idleness; controlled atmosphere heat-treating of metals; fire fighting; mines, following fires or the use of explosives; testing internal combustion engines; and many other sources. Portable stoves, formerly called "salamanders," when used to heat buildings under construction may be dangerous sources of carbon monoxide. Other sources that may give cause for concern are faulty exhaust equipment on automobiles, buses, airplanes, and cabin cruisers; improperly located air inlets for automobile ventilation; and compressed air for respiratory devices such as supplied-air respirators or "scuba" diving equipment, when supplied from reciprocating compressors, in which carbon monoxide may be produced by overheating of lubricating oil. Production of liquid fuels from coal may be a source in the future.

Rapid increases in the price of crude oil have accelerated a shift in the raw material base for chemical feedstocks to coal and the replacement of ethylene as a feedstock by synthesis gas, a mixture of hydrogen and carbon monoxide which can be produced directly from many carbonaceous sources. Organics likely to be produced from synthesis gas include ethanol, ethylene glycol and vinyl acetate (188b). This shift will increase opportunities for exposure to carbon monoxide, unless great care is taken in the planning, construction and operation of facilities for the use of synthesis gas.

The extent of room atmospheric pollution by carbon monoxide from any source, as for instance an automobile, can be computed from the rate of production of carbon monoxide and the amount of general ventilation by use of the formula (189)

$$C = \frac{10^6 K(1 - e^{-Rt})}{RV}$$

where C = carbon monoxide concentration in parts per million in a room after a given time, t; R = air changes per hour; t = time in hours; V = volume of room (in any convenient units); K = volume of carbon monoxide liberated per hour (in the same units); and e = base of the natural system of logarithms.

Inspection of the formula shows that as R or t increases, the factor $(1 - e^{-Rt})$ approaches 1 and equilibrium is reached—more quickly for larger values of R. Where V is relatively small, the concentration builds up rapidly.

$$\frac{K}{RV} = \text{equilibrium concentration}$$

This formula can be used in computing general ventilation rates, the concentration of any gas or vapor in the air, or the rate of admitting any gas or vapor to the room, providing all the other factors are known.

The quantity of carbon monoxide produced by an automobile gasoline engine varies with the air/fuel ratio; speed; temperature of the combustion chamber, cylinder, and cylinder walls; compression ratio; spark advance; piston displacement; the presence of air injection into the exhaust manifold, or a catalytic afterburner; and other factors. An idling engine with a rich fuel/air ratio may produce an exhaust with upward of 7 percent carbon monoxide, while the same engine with carburetor adjusted for efficient operation, and at a speed of 60 km (35 miles) or more per hour, may produce less than 0.5 percent of carbon monoxide, or an even lower concentration with exhaust manifold air injection and an afterburner. In general, combustion products having a high percentage of carbon monoxide have a low percentage of nitrogen oxides, and vice versa.

Engineering improvements in automobile engines have lowered the carbon monoxide content of vehicle combustion products appreciably over the years, but there are still occasions when the carbon monoxide concentration of city streets exceeds 100 ppm.

The relative proportions of carbon monoxide and nitrogen oxides and their relative toxicities are of prime importance in the control of vehicle combustion products in tunnels and mines, or wherever quantities of ventilating air are regulated according to the concentration of carbon monoxide. Where the traffic is moving at less than 40 km (25 miles) per hour the ratio of carbon monoxide to nitrogen oxides emitted by automobiles may be as much as 1000 : 1, but where the traffic moves at higher speeds, the ratio may approach 1 : 1.

An unexpected source of carbon monoxide exposure follows the inhalation of methylene chloride, used as a paint stripper, among other uses. Methylene chloride is metabolized to carbon monoxide (190), and a 3-hr exposure, even in a well ventilated room, can result in a carbon monoxide hemoglobin saturation of 16 percent (191).

Carbon monoxide is a serious factor in community air pollution. Ayres and Aronow, and Anderson et al. (cited in Reference 192) have shown that coronary artery disease that causes heart pain is aggravated when the proportion of hemoglobin bound by carbon monoxide reaches or exceeds about 5 percent. There are contradictory reports concerning the effects of small doses on central nervous functions. There is general agreement that the carbon monoxide hemoglobin saturation should be kept below 5 percent, and allowing for a margin of safety, the Environmental Protection Agency set 1½ percent as a goal.

A small amount of carbon monoxide is produced normally in the body. This *endogenous* carbon monoxide is sufficient in amount to maintain a carbon monoxide hemoglobin saturation of about 0.4 to 0.7 percent. In some persons with blood disease, such as hemolytic anemia, the carbon monoxide saturation may reach 6 percent.

A major source of carbon monoxide for many people is tobacco smoking. Cigarette smoke contains over 2 percent carbon monoxide, but the average concentration in the smoke that reaches the lungs is about 400 ppm (193). Smokers of one pack per day of cigarettes usually have 5 or 6 percent carbon monoxide hemoglobin saturation during their waking hours; smokers of two or three packs per day may show 7 to 9 percent, and heavy cigar smokers may reach 20 percent.

6.2 Physical and Chemical Properties

Physical state	Colorless, odorless gas
Molecular weight	28
Specific Gravity	Essentially the same as air
Melting point	$-207°C$
Boiling point	$-190°C$
Solubility	3.5 ml/100 ml of water at 0°C, 2.3 ml/100 ml at 20°C, 1.5 ml/100 ml at 60°C

Flammability   Flammable range 12.5 to 74.2 percent; ignition temper-
                             ature, 610°C (1130°F)

6.3 Determination in the Atmosphere

Carbon monoxide can readily be measured by indicator tubes, or more
accurately by an electrochemical reaction that activates a small, rugged, portable
meter. A catalytic oxidation indicator, in use for many years, may still be found
in operation. For greater precision, infrared spectroscopy or gas chromatography can be used, and volumetric analysis also remains possible.

6.4 Determination in Blood

Carbon monoxide hemoglobin has distinctive light absorption peaks at 568 and
539 mμ, contrasting with oxyhemoglobin peaks at 576 and 540 mμ. Also, there
are peaks in the Soret region at approximately 420 mμ for carboxyhemoglobin
and 414 mμ for oxyhemoglobin. These have been the foundations for several
methods of measurement of the carboxyhemoglobin concentration in drop-size
blood samples. The CO-Oximeter® (194) is widely used in medical pulmonary
function laboratories and is sufficiently accurate and sensitive for clinical
purposes. It is not reliable at concentrations below about 2 percent. The method
described by Small et al. (195) using spectrophotometric omparisons in the
Soret band is convenient and precise for low concentrations, using small samples
(100 μl).

Gas extraction, with volumetric measurement, or by infrared spectrophotometry, or by gas chromatography, has been the basis of precise measurements.
The method of Coburn et al. (196) is satisfactory, but requires 2 to 3 ml of
blood. Dahms and Horvath (197) described a similar method that uses a smaller
blood sample (250 μl or less), takes but 3 min, and is as precise as any other.

Estimation of carboxyhemoglobin concentration by analysis of expired breath
is a simple and practical technique. After breathing deeply of carbon monoxide-
free air, the person holds his breath for 20 sec or more, then exhales half of it;
the remainder, consisting of air from the alveoli (assuming he has normal lungs)
is collected in a plastic bag. The carbon monoxide concentration is measured.
A portable instrument such as the electrochemical analyzer can be used, or
even an indicator tube. The relationship between alveolar carbon monoxide
and carboxyhemoglobin is as follows (198, 199), when expired carbon monoxide
is 25 to 75 ppm: % COHb = ppm/5 + 0.5; when expired CO is 75 to 120 ppm,
% COHb = ppm/5.5; when expired CO is 120 to 250 ppm, % COHb =
ppm/7.

Applying this method, nonsmoking humans can be used as integrating
samplers to determine the average carbon monoxide exposure over a period of
time.

6.5 Physiologic Responses

6.5.1 Acute

J. S. Haldane, C. G. Douglas, J. B. S. Haldane, R. R. Sayers, W. P. Yant, Y. Henderson, and H. W. Haggard were notable early investigators who described the symptoms and signs of carbon monoxide poisoning in detail. Stewart (191) recently summarized their observations and some more recent ones. These and other observations are shown in Table 52.11.

There have been conflicting statements about the effects of small doses of carbon monoxide on cerebral functions and human behavior. Laties and Merigan (200) recently reviewed this field, and, although the reports of effects with carboxyhemoglobin levels below 5 percent are not fully convincing, they

Table 52.11. Human Response to Various Concentrations of Carboxyhemoglobin[a]

Blood Saturation COHb (%)	Response of Healthy Adult[b]	Response of Patient Ill with Severe Heart Disease
0.3–0.7	Normal range due to endogenous CO production; no known detrimental effect	
1–5	Selective increase in blood flow to certain vital organs to compensate for reduction in O_2-carrying capacity of the blood	Patient with advanced cardiovascular disease may lack sufficient cardiac reserve to compensate; diminished vigilance in some circumstances
5–9	Visual light threshold increased	Less exertion required to induce chest pain in patients with angina pectoris
16–20	Headache; visual-evoked response abnormal	May be lethal for patients with severely compromised cardiac function
20–30	Throbbing headache; nausea; fine manual dexterity abnormal; judgment and calculation impaired	
30–40	Severe headache; nausea and vomiting; syncope	
50–60	Coma; convulsions	
67–70	Lethal if not treated	

[a] Modified from reference 191.

[b] Exposure to CO in concentrations in excess of 50,000 ppm can result in a fatal cardiac arrhythmia and death before the carboxyhemoglobin saturation is significantly elevated.

cannot be dismissed without further attention. It is hard to conceive that so small a reduction in the oxygen capacity of the blood would produce a hypoxic effect. The possibility of interference with enzyme activities or other direct toxic action on the cells has been considered, but no evidence of a mechanism has been found. Cytochrome oxidase and cytochrome P450 are inhibited by carbon monoxide in high concentrations, but that doesn't explain a low level effect.

These low level effects on brain function can be demonstrated only by very sensitive tests of unusual visual functions or exacting tests of vigilance. They appear to have little practical importance. Significant deterioration of brain functions and on performance of practical tests appears when carbon monoxide hemoglobin is over 20 percent in healthy young individuals.

These symptoms are primarily, if not exclusively, the consequences of hypoxia, an inadequate oxygen supply to tissues of the body, especially to the brain. Oxygen is carried from the lungs to the tissues where it is used by the blood. A small amount is dissolved in the watery part of the blood, the plasma, while the greater part is carried in the red blood cells, or erythrocytes, in chemical combination with the red pigment hemoglobin. Hemoglobin is an iron-containing molecule that has four points of attachment (ligands) at which oxygen molecules can attach in a reversible reaction. The reaction velocities for each of these ligands differ so there are different equilibrium constants for the four ligands, as shown in Table 52.12 (188).

The binding of carbon monoxide to hemoglobin is 220 to 290 times stronger than that of oxygen, as expressed in the equation

$$\frac{[\text{COHb}]}{[\text{O}_2\text{Hb}]} = \frac{M \times P_{\text{CO}}}{P_{\text{O}_2}}$$

where [COHb] = carboxyhemoglobin concentration
$\quad\quad$ [O_2Hb] = oxyhemoglobin concentration
$\quad\quad\quad$ P_{CO} = partial pressure of carbon monoxide
$\quad\quad\quad$ P_{O_2} = partial pressure of oxygen
$\quad\quad\quad\quad$ M = affinity constant of hemoglobin for carbon monoxide as compared to that for oxygen

Since carbon monoxide is absorbed through the lungs, the rate of uptake is speeded by greater pulmonary ventilation, as in vigorous exercise, and is slowed by impaired diffusion through the lung, as in various pulmonary diseases, or by impaired blood circulation, or by deficient hemoglobin, as in anemia. However, the effects of deficient hemoglobin are aggravated by carbon monoxide.

Although the partial pressure of carbon monoxide in the lungs may be only 10^{-7} atm, the partial pressure in the red blood cells is almost nil, so a very large part of the inhaled gas is taken up by the blood. Coburn el al. (201) worked out a satisfactory equation for predicting the buildup of carbon monoxide hemoglobin. Stewart (191) demonstrated the accuracy of the computation by empirical

Table 52.12.   Reactions of Hemoglobin with Oxygen or Carbon Monoxide

$$Hb_4 + X \underset{k_1}{\overset{k'_1}{\rightleftharpoons}} XHb_4 \qquad K_1 = \frac{k'_1}{k_1}$$

$$XHb_4 + X \underset{k_2}{\overset{k'_2}{\rightleftharpoons}} (X)_2Hb_4 \qquad K_2 = \frac{k'_2}{k_2}$$

$$(X)_2Hb_4 + X \underset{k_3}{\overset{k'_3}{\rightleftharpoons}} (X)_3Hb_4 \qquad K_3 = \frac{k'_3}{k_3}$$

$$(X)_3Hb_4 + X \underset{k_4}{\overset{k'_4}{\rightleftharpoons}} (X)_4Hb_4 \qquad K_4 = \frac{k'_4}{k_4}$$

In these equations, X represents either oxygen or carbon monoxide. In either case, the reactions proceed in step by step fashion, with different velocities and equilibrium constants for each step. The value of K_4 is much greater than that of K_1, K_2, or K_3, because k'_4 is much greater than k_4. Consequently, the last ligand that binds oxygen or carbon monoxide dissociates very readily. This makes oxygen very easily available when the partial pressure drops to about 40 torr, the pressure in mixed venous blood. Thus the tissue cells are supplied. The other ligands maintain a reserve that can be drawn upon if the tissue concentration falls even lower.

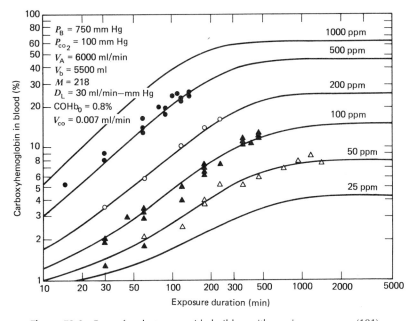

Figure 52.6   Rate of carbon monoxide buildup with varying exposures (191).

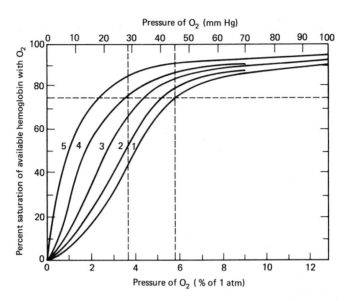

Figure 52.7 Dissociation of oxyhemoglobin in the presence of various quantities of carbon monoxide hemoglobin: 1 = 0 percent, 2 = 10 percent, 3 = 25 percent, 4 = 50 percent, 5 = 75 percent.

tests, shown in Figure 52.6. This figure demonstrates the rate of buildup with several levels of carbon monoxide exposure.

The effect of carbon monoxide on the oxygen-carrying power of the blood is greater than would occur if a person lost the use of an equal amount of hemoglobin by other means, such as hemorrhage. Aside from compensatory mechanisms that would counteract the effects of blood loss, carbon monoxide acts specifically to cause hemoglobin to cling more tightly to its oxygen content than would otherwise occur. This is shown in the curves of oxyhemoglobin dissociation in Figure 52.7.

The dissociation curves for oxyhemoglobin in the presence of various concentrations of carboxyhemoglobin illustrate why a normal person with 50 percent saturation with carbon monoxide is unable to undergo physical exertion comparable to that of an anemic person with only 50 percent of hemoglobin.

If the oxygen saturation of the available hemoglobin of the venous blood of each person is reduced say, to 75 percent at rest, the anemic person has a remaining oxygen tension of 44 torr, whereas the person with 50 percent carbon monoxide saturation has only 28 torr of oxygen tension. Then, if each person exercises, the venous blood becomes less saturated with oxygen—say, 60 percent—and the anemic person retains a partial pressure of about 36 torr of oxygen, which is still ample to support oxidation in the tissues, but the person with 50 percent carbon monoxide hemoglobin has suffered a drop in oxygen tension to less than 20 torr, which is probably insufficient to prevent fainting.

Other factors influence the dissociation of both oxyhemoglobin and carboxy-hemoglobin: carbon dioxide, for instance, enhances the dissociation of each.

Almost all of the carbon monoxide that has been inhaled is eliminated through the lungs when the previously exposed person enters an atmosphere free of carbon monoxide. The time required to eliminate half of the gas is 3 to 5 hr, depending on the amount of respiration, which acts to wash it out of the body. Increased oxygen pressure helps to dislodge it from the hemoglobin. One hundred percent oxygen given at atmospheric pressure reduces the half-elimination time to about 80 min. Hyberbaric oxygen at 3 atm pressure absolute (3 ATA) reduces the half time to about 25 min.

Adaptation. Stewart (191) reported that smokers, and even nonsmokers in areas with dense air pollution, showed a compensatory increase in hemoglobin, although this has been questioned by others. In any case, the increase is insufficient to give any substantial measure of protection.

Emergency Treatment. Prompt administration of oxygen under increased pressure may restore life and health to a person poisoned by carbon monoxide (202). Quick transportation to a hyperbaric chamber for this treatment should be the first order of business for anyone in whom this is suspected. Meanwhile, artificial respiration should be applied if the person is not breathing. Oxygen should be given at the highest possible concentration, except that it is beneficial for it to contain 3 to 5 percent carbon dioxide as a stimulant to respiration.

It will be recalled that a high partial pressure of oxygen speeds the release of carbon monoxide from hemoglobin, and that increased respiration helps to keep oxygen pressure high and carbon monoxide pressure low in the alveoli.

The dramatic effects of hyperbaric treatment with oxygen arise from two sources. Oxygen at 3 atm absolute pressure (3 ATA) dissolves in blood plasma to the extent of 6.4 ml/100 ml of blood. This is sufficient to satisfy the needs of the body even if all the hemoglobin were nonfunctioning. Second, the increased oxygen pressure causes rapid clearance of carbon monoxide from the hemoglobin, reducing the half-release time to about 25 min.

6.5.2 Chronic Poisoning

Grut (203a) described a syndrome that was called chronic carbon monoxide poisoning, but it was not clear that carbon monoxide was the sole cause, or even an essential one, for the development of the symptoms that were described. There is some difficulty in making a distinction between frequently repeated episodes of acute poisoning, such as have been observed in garage workers, and a truly chronic disease, in which different or more extensive changes take place, and different signs and symptoms arise. Convincing evidence of chronic poisoning has not be presented.

Effects in Pregnancy.   Advancing knowledge of the causes of damage to the fetuses in the wombs of their mothers has stimulated questions about the possible effects of CO. Longo reviewed this topic in 1977 (203b). It is established that CO crosses the placenta from the mother's blood to the fetus. The fetal COHb concentration lags behind that of the mother. Whereas it takes about 2 hr for the maternal COHb to reach half the final equilibrium level after an increase in ambient CO, it takes about 7 hr for the fetal COHb to reach half of the steady state value. The excretion of CO by the fetus is similarly delayed.

There is no experimental evidence that the fetal COHb levels that might be reached in an industry that observes current good hygienic practices (about 8 percent) would be harmful, but higher concentrations have profound effects in animals. Obstetricians show interest in the improved oxygenation of the fetus that can be attained by stopping the mother's smoking for 2 days before delivery. Smoker mothers have smaller babies, on the average. These observations suggest that the current TLV of 50 ppm is not adequate to protect fetuses, but more knowledge is needed to decide what stages of pregnancy might be affected.

6.5.3   Sequelae

Sequelae of severe, acute poisoning are well known. In almost all cases, acute, severe poisoning results in either death or complete recovery. In a few cases in which prolonged coma occurred, residual neurologic symptoms and signs appeared (204, 205). These have been extremely varied, ranging from mild memory lapses to various palsies and to nearly complete loss of cerebral function. It is impossible to predict the ultimate disability when a person has been in hypoxic coma for several hours. Some make amazing recoveries.

Neurologic textbooks usually state that carbon monoxide poisoning is a cause of the Parkinson syndrome, also known as paralysis agitans. However, they leave an erroneous impression that minor carbon monoxide poisonings may cause this disorder after a long latent period in which the person appears well. We reviewed the case reports cited by a number of neurology texts, and all were severe poisonings accompanied by coma, and the neurologic signs came on within a few days after the episodes. The opinion that neurologic sequelae only follow poisonings that cause coma has been expressed by Kuroiwa (206), a neurologist with very extensive extensive experience with carbon monoxide poisonings.

6.5.4   Hygienic Standards of Permissible Exposure

The American Conference of Governmental Industrial Hygienists Threshold Limit Value (TLV) is 50 ppm; the Short Term Exposure Limit (STEL) is 400 ppm—both as of 1981 (186). The U.S. national ambient air quality standard is, for 8-hr average, 10 mg/m^3; and for 1 hr average, 35 mg/m^3 (130).

7 CARBON DIOXIDE, CO_2

7.1 Occurrence and Uses

Carbon dioxide is a normal constituent of the atmosphere, about 0.03 percent by volume above the ocean and up to 0.06 percent in urban areas; the exhaled breath contains up to 5.6 percent; the gas is also widely encountered in industry in harmless concentrations. Carbon dioxide may be recovered from lime or cement kilns, from flue gases, from fermentation processes, and from some natural gas wells. It is first purified, then dehydrated, and compressed. If solid carbon dioxide, dry ice, is desired, it is then manufactured from the compressed liquid. When the pressure upon this compressed, liquefied carbon dioxide is suddenly released, a portion of it solidifies to "snow" as the balance expands to a gas again and is drawn off and recompressed. The snow is then pressed into 220-lb blocks of solid dry ice. The compressed and bottled gas is used for carbonating beverages and the dry ice is used for preserving foods, especially during transportation. Other incidental uses are for chilling aluminum rivets and shrinking cylinder liners or bearing inserts. Industrial exposures may occur in mines, caves, tunnels, wells, the holds of ships, as well as tanks, vats, or any place where fermentation processes may have depleted the oxygen with formation of carbon dioxide. The manufacture, storage, and use of dry ice also offer exposures, as do carbon dioxide fire extinguishers when operated in confined areas.

It is important to remember that carbon dioxide is heavier than air, and that high concentrations can persist in open pits or tanks; they should be cleared by forced ventilation or supplied air, or oxygen respirators should be worn when entering them.

7.2 Physical and Chemical Properties

Physical state	Colorless gas
Molecular weight	44.01
Boiling point	$-78.5°C$
Melting point	Sublimes at atmospheric pressure
Density	1.997 at O°C, or 1.53 (air = 1)
Solubility in water	3.35 g/kg at 0°C; 1.07 g/kg ar 37°C; 171 ml/100 ml at 0°C; 90 ml/100 ml at 20°C
$1\ mg/m^3 \approx$	0.5 ppm
1 ppm $\approx$	$2\ mg/m^3$

7.3 Toxicity of Carbon Dioxide

Acute Effects. Except as a contributor to oxygen deficiency, a very real danger, carbon dioxide does not offer serious industrial exposures. The initial effect of inhalation of excessive carbon dioxide is noticed in concentrations of about 2 percent, 20,000 ppm, when the breathing becomes deeper and the tidal volume is increased (207–210). The depth of respiration is markedly increased at 4 percent; at 4.5 to 5 percent breathing becomes labored, and distressing to some individuals. Concentrations of 8 to 10 percent have been inhaled by men for periods up to 1 hr with no evident, harmful effects. The role of carbon dioxide in oxygen deficiency need not be elaborated upon, for it acts similarly to any other diluent gas. It is worth noting that in many instances the carbon dioxide may have been formed by processes, such as combustion or fermentation, that were at the same time depleting the oxygen supply in the air.

7.4 Threshold Limit Value

The threshold limit value set by the American Conference of Government Industrial Hygienists has been maintained at 5000 ppm for many years (128, 178). The Short Term Exposure Limit (STEL) is 15,000 ppm.

8 PHOSGENE, $COCl_2$

8.1 Occurrence and Uses

Phosgene is used in the manufacture of a wide variety of organic chemicals, including dyestuffs, isocyanates and their derivatives, carbonic acid esters (polycarbonates), and acid chlorides. It is also used in metallurgy to separate ores by chlorination of the oxides and volatilization. It is produced commercially by the catalytic chlorination of carbon monoxide and supplied in liquid form in steel cylinders. Its chief importance in industrial hygiene, however, lies in its occurrence as one of the products of combustion whenever a volatile chlorine compound, such as chlorinated solvent or its vapor, comes in contact with a flame or very hot metal. This ordinarily does not produce a serious threat to health except where ventilation is not satisfactory, the area is confined, or considerable quantities of chlorinated vapors are involved. It may be encountered in the use of carbon tetrachloride (211) for extinguishing fires in confined spaces.

8.2 Physical and Chemical Properties

Physical state	Colorless gas
Molecular weight	98.92
Boiling point	8.3°C
Melting point	− 104°C
Density	1.392 g/l, or 1.08 (air = 1)

Solubility	Decomposes in water, alcohol, very soluble in benzene, toluene
$1 \ mg/m^3 \approx$	0.25 ppm
1 ppm $\approx$	$4.1 \ mg/m^3$

8.3 Toxicity of Phosgene

Phosgene is mildly irritant to mucous membranes in concentrations below 10 ppm, and very irritant to the entire respiratory tract in considerably higher concentrations. A single, shallow breath of a moderately high concentration causes a rasping, burning sensation in the nose, pharynx, and larynx that is not readily forgotten.

The irritant properties of phosgene are not sufficient to give warning of hazardous concentrations. One-half part per million can be recognized through the sense of smell by normal persons acquainted with its odor, but olfactory fatigue or adaptation can cause a gradually increasing concentration to go unnoticed. The least concentration that can cause immediate throat irritation is 3 ppm; 4 ppm causes immediate irritation of the eyes; 4.8 ppm causes cough. Brief exposure to 50 ppm may be rapidly fatal (212).

The most serious effect of phosgene is lung irritation. Only a relatively small portion of the inhaled gas hydrolyzes in the respiratory passages, but in the moist atmosphere of the terminal spaces of the lungs complete hydrolysis occurs with irritant effects upon the alveolar walls and blood capillaries (74). The result of this action is a gradually increasing edema, until as much as 30 to 50 percent of the total blood plasma has accumulated in the lungs, causing "dry land drowning." The air spaces grow less and less; the blood is thickened by loss of plasma, which results in slowed circulation; oxygen exchange is slowed; and the overworked heart, with insufficient oxygen, weakens. The end result may be either asphyxiation or heart failure, and this may be delayed. High concentrations of phosgene are immediately corrosive to lung tissue and result in sudden death by suffocation (213, 214).

Diller (215) discusses the clinical problems arising in cases of phosgene exposure. In the absence of special detector badges worn by workers, there is no way of knowing the extent of phosgene exposure. But if one waits for the appearance of symptoms, pulmonary edema may be lethal. Consequently, any exposed person must be treated as if the exposure is life threatening. The person should be kept at rest and given a glucocorticoid anti-inflammatory medication, and should be given oxygen-enriched air. Medical observation for the detection of onset of lung edema should be maintained. After 8 hr, a chest X-ray should be taken; if there are no signs of edema, the patient may be discharged. If the exposure is thought to have been severe, especially if there has been any indication of eye, nose, or throat irritation, the glucocorticoid should be given intravenously and positive pressure breathing should be given for 15 min in each hour, as well as oxygen enrichment. X-Rays should be made at 2, 4, and 8 hr, and additional supportive measures introduced, if necessary.

Recovery from phosgene injury of the lungs is likely to be prolonged for weeks or even months, and convalescence should be observed by the use of pulmonary function tests.

The normal responses to slight gassing are, besides a dryness or burning sensation in the throat, numbness, vomiting, pain in the chest, bronchitis, and possibly dyspnea. There is sometimes a latent effect: the period between inhaling low concentrations of the gas and the appearance of dyspnea may be several hours, almost free of symptoms. The action of phosgene and its sequelae resemble in some respects those of nitrogen dioxide. Flury and Zernick recorded the response to various concentrations of phosgene as follows:

Response	Concentration (ppm)
Maximum amount for prolonged exposure	1
Dangerous to life, for prolonged exposure	1.25–2.5
Cough or other subjective symptoms within 1 min	5
Irritation of eyes and respiratory tract in less than 1 min	10
Dangerous to life in 30 to 60 min	12.5
Severe lung injury within 1 to 2 min	20
Dangerous to life for as little as 30 min	25
Rapidly fatal (30 min or less)	90

Splashes of phosgene in the eye produce severe irritation, and phosgene on the skin can cause severe burns (213).

8.4 Threshold Limit Value

The threshold limit value established by the American Conference of Governmental Industrial Hygienists, designed to prevent the occurrence of pulmonary edema, is 0.1 ppm (1981) (128, 178).

REFERENCES

1. Y. Henderson and H. W. Haggard, *Noxious Gases*, Reinhold, New York, 1943.
2. J. F. Parker and V. R. West, Eds., *Bioastronautics Data Book*, 2nd ed., National Aeronautics and Space Administration, Washington, 1973.
3a. S. Lahiri, "Physiological Responses and Adaptation to High Altitude," *Internat. Rev. Physiol., Experimental Physiol. II*, Vol. 15, D. Robertshaw, ed., University Park Press, Baltimore, 1977.
3b. G. Dhenin, G. R. Sharp, and J. Ernsting, Eds., *Aviation Medicine*, TriMed, London, 1978.
4a. H. A. Randel, Ed., *Aerospace Medicine*, Williams and Wilkins, Baltimore, 1971.
4b. F. Boismare, C. Saligaut, N. Moore, and J. P. Raoult, "Avoidance Learning and Mechanism

of the Protective Effect of Apomorphine Against Hypoxic," *Aviation Space Environ. Med.*, **52**(5), 299–303, 1981.

5. P. Bert, *La pression barometrique*, G. Masson, Paris, 1878. English Translation by M. A. Hitchcock and F. A. Hitchcock, College Book Co., Columbus, Ohio, 1943.

6a. M. G. Mustafa and D. F. Tierney, "Biochemical and Metabolic Changes in the Lung with Oxygen, Ozone, and Nitrogen Dioxide Toxicity," *Amer. Rev. Resp. Dis.*, **118**, 1061–1090 (1978).

6b. J. A. Jacquez, "Toxic Effects of Respiratory Gases," *Respiratory Physiology*, Hemisphere, Washington, D.C., 1979, Ch. 10.

6c. S. M. Deneke and B. L. Fanburg, "Normobaric Oxygen Toxicity of the Lung," *New Eng. J. Med.*, **303**(2), 76–86, 1980.

7. J. B. Stevens and A. P. Autor, "Induction of Superoxide Dismutase by Oxygen in Neonatal Rat Lung," *J. Biol. Chem.*, **252**(10), 3509–3514 (1977).

8. J. Yam, L. Frank, and R. J. Roberts, "Oxygen Toxicity: Comparison of Lung Biochemical Responses in Neonatal and Adult Rats," *Pediatr. Res.*, **12**, 115–119 (1978).

9. J. Liu, L. M. Simon, J. R. Phillips, and E. D. Robin, "Superoxide Dismutase (SOD) Activity in Hypoxic Mammalian Systems," *J. Appl. Physiol.*, **42**(1), 107–110 (1977).

10. P. J. Sheffield, "Effects of Long-Term Exposures to 100% Oxygen at Selected Simulated Altitudes on the Pulmonary Surfactant in Mice," *Aviat. Space Environ. Med.*, **46**(1), 6–10 (1975).

11. M. A. Sackner, J. A. Hirsch, S. Epstein, and A. M. Rywlin, "Effect of Oxygen in Graded Concentrations upon Tracheal Mucous Velocity," *Chest*, **69**(2), 164–167 (1976).

12. W. H. Northway, Jr., L. Rezeau, R. Petricks, and K. G. Bensch, "Oxygen Toxicity in the Newborn Lung: Reversal of Inhibition of DNA Synthesis in the Mouse," *Pediatrics,* **57,** 41–46 (1976).

13. D. S. Bonikos, K. G. Bensch, S. K. Ludwin, and W. H. Northway, Jr., "Oxygen Toxicity in the Newborn; The Effect of 100% O_2 Exposure on the Lungs of Newborn Mice," *Lab. Invest.,* **32**(5), 619–635 (1975).

14. L. Frank, J. R. Bucher, and R. J. Roberts, "Oxygen Toxicity in Neonatal and Adult Animals of Various Species," *J. Appl. Physiol.,* **45**(5), 699–704 (1978).

15. P. Hall, C. L. Schatte, and J. W. Fitch, "Relative Susceptibility of Altitude-Acclimatized Mice to Acute Oxygen Toxicity," *J. Appl. Physiol.,* **38**(2), 279–281 (1975).

16. P. M. Winter, G. Smith, and R. F. Wheelis, "The Effect of Prior Pulmonary Injury on the Rate of Development of Fatal Oxygen Toxicity," *Chest* **66**(1, Suppl., Part 2), 1S–4S (1974).

17. L. Frank and R. J. Roberts, "Endotoxin Protection Against Oxygen-induced Acute and Chronic Lung Injury," *J. Appl. Physiol.* **47**(3), 577–581 (1979).

18. S. K. Ludwin, W. H. Northway, Jr., and K. G. Bensch, "Necrotizing Bronchiolitis in Mice Exposed to 100% Oxygen," *Lab. Invest.,* **31**(5), 425–435 (1974).

19. D. S. Bonikos, K. G. Bensch, and W. H. Northway, Jr., "Oxygen Toxicity in the Newborn," *Am. J. Pathol.,* **85,** C23–650 (1976).

20. M. Hughson, J. D. Balentine, and H. B. Daniell, "The Ultrastructural Pathology of Hyperbaric Oxygen Exposure—Observations on the Heart," *Lab. Invest.* **37**(5), 516–525 (1977).

21. P. C. Pratt, "Pathology of Pulmonary Oxygen Toxicity," *Am. Rev. Resp. Dis.* **110**(6 Part 2), 51–57 (1974).

22. L. A. Robinson, W. G. Wolfe, and M. L. Salin, "Alterations in Cellular Enzymes and Tissue Metabolism in the Oxygen Toxic Primate Lung," *J. Surg. Res.,* **24**(5), 359–365 (1978).

23. W. G. Wolfe, L. A. Robinson, J. F. Moran, and J. E. Lowe, "Reversible Pulmonary Oxygen Toxicity in the Primate," *Ann. Surg.,* **188**(4), 530–543 (1978).

24. T. A. Raffin, L. M. Simon, D. Braun, J. Theodore, and E. D. Robin, "Impairment of

Phagocytosis by Moderate Hyperoxia (40 to 60% Oxygen) in Lung Macrophages," *Lab. Invest.*, **42**(6), 622–626 (1980).

25a. L. M. Simon, T. A. Raffin, W. H. J. Douglas, J. Theodore, and E. D. Robin, "Effects of High Oxygen Exposure on Bioenergetics in Isolated Type II Pneumocytes," *J. Appl. Physiol.*, **47**(1), 98–103 (1979).

25b. M. E. Heino, L. A. Laitenan and T. Tervo, "Early Pulmotoxic Effects of Oxygen on the Rat Alveolar Type II Epithelial Cell," *Aviation Space Environ. Med.*, **52**(5), 294–298 (1981).

26. D. Torbati, D. Parolla, and S. Lavy, "Changes in the Electrical Activity and P_{O_2} of the Rat's Brain under High Oxygen Pressure," *Exp. Neurol.*, **50**, 439–447 (1976).

27. J. M. Miller and P. M. Winter, "Clinical Manifestations of Pulmonary Oxygen Toxicity," in J. B. Brodsky, Ed., "Clinical Aspects of Oxygen," *Int. Anesthesiol. Clinics*, **19**(3), 179–199 (1981).

28. T. A. Raffin, "Oxygen Toxicity—Etiology," in J. B. Brodsky, Ed., "Clinical Aspects of Oxygen," *Int. Anesthesiol. Clinics*, **19**(3), 169–177 (1981).

29. J. D. Wood, "Oxygen Toxicity," in P. B. Bennett and D. H. Elliott, Eds., *The Physiology and Medicine of Diving and Compressed Air Work*, 2nd ed., Balliere Tindall, London, 1975.

30. C. Edwards, C. Lowry, and J. Pennefather, *Diving and Subaquatic Medicine*, Diving Mecical Centre, Sydney, 1976.

31. T. A. Raffin, E. D. Robin, and J. Pickersgill, "Paraquat Ingestion and Pulmonary Injury," *Western J. Med.*, **128**, 26–34 (1978).

32. M. A. Sackner, J. Landa, J. Hirsch, and A. Zapata, "Pulmonary Effects of Oxygen Breathing," *Ann. Intern. Med.*, **82**(1), 40–43 (1975).

33. P. L. Hendricks, D. A. Hall, W. L. Hunter, Jr., and P. J. Haley, "Extension of Pulmonary O_2 Tolerance in Man at 2 ATA by Intermittent O_2 Exposure," *J. Appl. Physiol.*, **42**(4), 593–599 (1977).

34. J. M. Clark and C. J. Lambertson, "Rate of Development of Pulmonary Oxygen Toxicity in Man during O_2 Breathing at 2.0 ATA," *J. Appl. Physiol.*, **30**, 739–752 (1971) [cited by Hendricks et al. (33)].

35a. C. J. Lambertson, "Effects of Oxygen at High Partial Pressure," in W. O. Fenn and H. Rahn, Eds., *Handbook of Physiology, Section 3, Vol. II*, American Physiological Society, Washington, 1965, pp. 1027–1046.

35b. W. W. Banks, T. E. Berkhage, and D. M. Heaney, "Visual Recognition Thresholds in a Compressed Air Environment," *Aviation Space Environ. Med.*, **50**(10), 1003–1006 (1979).

36. Committee on Medical and Biological Effects of Environmental Pollutants, National Academy of Sciences, *Ozone and Other Photochemical Oxidants*, National Academy of Sciences, Washington, D.C., 1977.

37. U.S. Environmental Protection Agency, *Air Quality Criteria for Ozone and Other Photochemical Oxidants*, EPA-600/8-78-004, U.S. EPA, Washington, D.C., 1978.

38a. S. Van Heusden and L. G. J. Mans, "Alternating Measurement of Ambient and Cabin Ozone Concentrations in Commercial Jet Aircraft," *Aviat. Space Environ. Med.*, **49**(9), 1056–1061 (1978).

38b. A. S. Whittemore, "Air Pollution and Respiratory Disease," *Ann. Rev. Public Health*, **2**, 397–429 (1981).

39. H. E. Stokinger, "Ozone Toxicology," *Arch. Environ. Health*, **10**, 719–731 (1965).

40. C. E. Cross, A. J. Delucia, A. K. Reddy, M. Z. Hussain, C. K. Chow, and A. M. G. Mustafa, "Ozone Interactions with Lung Tissue, Biochemical Approaches," *Am. J. Med.* **60**(7), 929–935 (1976).

41. W. S. Linn, R. D. Buckley, C. E. Spier, J. D. Hackney, et al., "Health Effects of Ozone Exposure in Asthmatics," *Am. Rev. Resp. Dis.*, **117**(5), 835–843 (1978).

42. B. D. Goldstein, "The Pulmonary and Extrapulmonary Effects of Ozone," in *Oxygen Free Radicals and Tissue Damage*, Ciba Foundation Symposium 65 (New Series), Excerpta Medica, New York, 1979.

43. H. E. Stokinger and D. L. Coffin, "Biologic Effects of Air Pollutants," in A. C. Stern, Ed, *Air Pollution*, Vol. 1, *Air Pollution and Its Effects*, 2nd ed., Academic Press, New York, 1968.

44. L. D. Scheel, O. J. Dobrogorski, J. T. Mountain, J. L. Svirbely, and H. E. Stokinger, "Physiologic, Biochemical, Immunologic and Pathologic Changes Following Ozone Exposure," *J. Appl. Physiol.*, **14**, 67–80 (1959).

45. G. Freeman, L. Juhos, et al., "Pathology of Pulmonary Disease from Exposure to Interdependent Ambient Gases (Nitrogen Dioxide and Ozone)," *Arch. Environ. Health*, **29**, 203–210 (1974).

46. P. W. Mellick, D. L. Dungworth, L. W. Schwartz, and W. S. Tyler, "Short Term Morphologic Effects of High Ambient Levels of Ozone on Lungs of Rhesus Monkeys," *Lab. Invest.*, **36**(1), 82–90 (1977).

47. M. G. Mustafa and S. D. Lee, "Pulmonary Biochemical Alterations Resulting from Ozone Exposure," *Ann. Occup. Hyg.*, **19**(1), 17–26 (1976).

48. S. Sato, M. Kawakami, S. Maeda, and T. Takashima, "Scanning Electron Microscopy of the Lungs of Vitamin E Deficient Rats Exposed to a Low Concentration of Ozone," *Am. Rev. Resp. Dis.* **113**, 809–821 (1976).

49. C. G. Plopper, D. L. Dungworth, W. S. Tyler, and C. K. Chow, "Pulmonary Alterations in Rats Exposed to 0.2 and 0.1 ppm Ozone: A Correlated Morphological and Biochemical Study," *Arch. Environ. Health*, **34**(6), 390–395 (1979).

50. D. B. Menzel, "Nutritional Needs in Environmental Intoxication: Vitamin E and Air Pollution: An Example," *Environ. Health Perspect.*, **29**, 105–114 (1979).

51. E. E. Dumelin, C. J. Dillard, and A. L. Tappel, "Effects of Vitamin E and Ozone on Pentane and Ethane Expired by Rats," *Arch. Environ. Health*, **33**(3), 129–135 (1978).

52. K. W. Clark, C. I. Posin, and R. D. Buckley, "Biochemical Response of Squirrel Monkeys to Ozone, *J. Toxicol. Environ. Health*, **4**, 741–753 (1978).

53. S. D. Lee, *Biochemical Effects of Environmental Pollutants*, Ann Arbor Science, Ann Arbor, 1977.

54. C. K. Chow, C. G. Plopper, and D. L. Dungworth, "Influence of Dietary Vitamin E on the Lungs of Ozone-Exposed Rats," *Environ. Res.*, **20**, 309–317 (1979).

55. H. E. Stokinger, "Factors Modifying Toxicity of Ozone," in *Ozone Chemistry and Technology*, American Chemical Society, Washington, D.C., 1959.

56. E. Goldstein, H. C. Bartleman, M. van der Ploeg, P. Vanduijn, J. G. van der Stap, and W. Lippert, "Effect of Ozone on Lysosomal Enzymes of Alveolar Macrophages Engaged in Phagocytosis and Killing of Inhaled *Staphylococcus aureus*," *J. Infect. Dis.*, **138**(3), 299–311 (1978).

57. J. S. Douglas, G. Curry, and S. A. Geffkin, "Superoxide Dismutase and Pulmonary Ozone Toxicity," *Life Sci.*, **20**(7), 1187–1192 (1977).

58a. C. K. Chow, "Biochemical Responses in Lungs of Ozone-Tolerant Rats," *Nature*, **260**(5553), 721–722 (1976).

58b. T. S. Veninga, J. Wagenaar and W. Lenistra, "Distinct Enzymatic Responses in Mice Exposed to a Range of Low Doses of Ozone," *Environ. Health Perspect.*, **39**, 153–157 (1981).

59. H. E. Stokinger: "Evaluation of the Hazards of Ozone and Oxides of Nitrogen," *AMA Arch. Ind. Health*, **15**, 181–189 (1957).

60. O. Fukase, H. Watanabe, and K. Isomura, "Effects of Exercise on Mice Exposed to Ozone," *Arch. Environ. Health*, **33**, 198–220 (1978).

61. R. F. Bils: "Ultrastructural Alterations of Alveolar Tissue of Mice. III. Ozone," *Arch. Environ. Health*, **20**, 468–480 (1970).

62. R. J. Stephens, M. F. Sloan, D. G. Groth, B. S. Negi, and R. D. Lunan, "Cytologic Response of Postnatal Rat Lungs to O_3 or NO_2 Exposure," *Am. J. Pathol.,* **93,** 183–200 (1978).

63. K. E. Richkind and A. D. Hacker, "Responses of Natural Wildlife Populations to Air Pollution," *J. Toxicol. Environ. Health,* in press, 11 Sept. 1979.

64. R. R. Guerrero, D. E. Rounds, J. Booher, R. S. Olson, and J. D. Hackney, "Ozone Sensitivity in Aging WI-38 Cells Based on Acid Phosphatase Content," *Arch. Environ. Health,* **34,** 407–412 (1979).

65. E. Christensen and A. C. Giese, "Changes in Absorption Spectra of Nucleic Acids and their Derivatives following Exposure to Ozone and Ultraviolet Radiation," *Arch. Biochem. Physiol.,* **51,** 208–216 (1954).

66. D. B. Menzel, "Toxicity of Ozone, Oxygen, and Radiation," *Ann. Rev. Pharmacol.,* **10,** 379–394 (1970).

67. R. H. Fetner, "Chromosome Breakage in *Vicia faba* by Ozone," *Nature,* **181,** 504–505 (1958).

68. R. E. Zelac, H. L. Cromroy, W. E. Bolch, Jr., B. G. Dunavant, and H. A. Bevis, "Inhaled Ozone as a Mutagen," *Environ. Res.,* **4,** 262–282, 325–342 (1971).

69. P. C. Gooch, D. A. Creasia, and J. G. Brewer, "The Cytogenetic Effects of Ozone: Inhalation and *in vitro* Exposures," *Environ. Res.,* **12**(2), 188–195 (1976).

70. R. Kavlock, G. Daston, and C. T. Grabowski, "Studies on the Developmental Toxicity of Ozone. I. Prenatal Effects," *Toxicol. Appl. Pharmacol.,* **48**(1), 19–28 (1979).

71. G. von Nieding, "Possible Mutagenic Properties and Carcinogenic Action of the Irritant Gas Pollutants NO_2, O_3, and SO_2," *Environ. Health Perspect.,* **22,** 91–92 (1978).

72. H. E. Stokinger, "Ozone," in *Encyclopedia of Occupational Health & Safety,* International Labour Office, Geneva, 1972.

73. A. Henschler, H. Stier, H. Beck, and W. Neumann, *Arch. Gewerbepathol. Gewerbehyg.* **17,** 547, 1960, cited in A. Stern, Ed., *Air Pollution,* 3rd ed., Academic Press, New York, 1977, p. 498.

74. F. Flury and F. Zernik, *Schädliche Gase, Nebel, Rauch und Staubarten,* Springer, Berlin, 1931.

75. M. Kleinfeld, C. Giel and I. R. Tabershaw, "Health Hazards Associated with Inert-Gas-Shielded Metal Arc Welding," *AMA. Arch. Ind. Health,* **15**(1), 27–31 (1957).

76. M. Kleinfeld, "Acute Pulmonary Edema of Chemical Origin," *Arch. Environ. Health,* **10,** 942–946 (1965).

77. P. J. R. Challen, D. E. Hickish, and J. Bedford, "An Investigation of Some Health Hazards in an Inert-Gas Tungsten-Arc Welding Shop," *Br. J. Ind. Med.,* **15**(3), 276–282 (1958).

78. F. J. Kelly and W. E. Gill, "Ozone Poisoning: Serious Human Intoxication," *Arch. Environ. Health,* **10,** 517–519 (1965).

79. W. A. Young, D. B. Shaw, and D. V. Bates, "Pulmonary Function in Welders Exposed to Ozone," *Arch. Environ. Health,* **7,** 337–340 (1963).

80. F. Akbarkhanzadeh, "Long-Term Effects of Welding Fumes upon Respiratory Symptoms and Pulmonary Function," *J. Occup. Med.,* **22**(5), 337–341 (1980).

81. D. Reed, personal communication, and unpublished data from National Aviation and Space Administration, 1979.

82. M. F. Lategola, C. E. Melton, and E. A. Higgins, "Effects of Ozone on Symptoms and Cardiorespiratory Function in a Flight Attendant Surrogate Population," *Aviation, Space Environ. Med.,* **51**(3), 237–246 (1980).

83. R. J. Allen, R. A. Wadden, and E. D. Ross, "Characterization of Potential Indoor Sources of Ozone," *Am. Ind. Hyg. Assoc. J.,* **39**(6), 466–471 (1978).

84. M. D. Selway, R. J. Allen, and R. A. Wadden, "Ozone Production from Photocopying Machines," *Am. Ind. Hyg. Assoc. J.,* **41**(6), 455–459 (1980).

85a. S. R. Williams, *Essentials of Nutrition and Diet Therapy,* Mosby, St. Louis, 1978.

85b. H. J. Roberts, "Perspective on Vitamin E as Therapy," *J. Am. Med. Assoc.,* **246,** 129–130 (1981).

86. C. I. Posin, K. W. Clark, M. P. Jones, R. D. Buckley, and J. D. Hackney, "Human Biochemical Response to Ozone and Vitamin E," *J. Toxicol. Environ. Health,* **5,** 1049–1058 (1979).

87. J. D. Hackney, W. S. Linn, R. D. Buckley, E. E. Pedersen, S. K. Karuza, D. C. Law, and D. A. Fischer, "Experimental Studies on Human Health Effects of Air Pollutants," *Arch. Environ. Health,* **30,** 373–378 (1975).

88. G. von Nieding, H. M. Wagner, H. Krekeler, H. Lollgen, W. Fries, and A. Beuthan, "Controlled Studies of Human Exposure to Single and Combined Action of Nitrogen Dioxide, Ozone and Sulphur Dioxide," *Int. Arch. Occup. Environ. Health,* **43,** 195–210 (1979).

89. S. M. Horvath, J. A. Gliner, and J. A. Matsen-Twisdale, "Pulmonary Function and Maximum Exercise Responses Following Acute Ozone Exposure," *Aviat. Space Environ. Med.,* **50**(9), 901–905 (1979).

90. D. V. Bates, G. M. Bell, C. D. Burnham, M. Hazucha, J. Mantha, L. D. Pengelly, and F. Silverman, "Short-term Effects of Ozone on the Lung," *J. Appl. Physiol.,* **32**(2), 176–181 (1972).

91. M. Hazucha, F. Silverman, C. Parent, S. Field, and D. V. Bates, "Pulmonary Function in Man after Short-Term Exposure to Ozone," *Arch. Environ. Health,* **27,** 183–188 (1973).

92. J. D. Hackney, W. S. Linn, J. G. Mohler, E. E. Pedersen, P. Breisacher, and A. Russo, "Experimental Studies on Human Health Effect of Air Pollutants," *Arch. Environ. Health* **30,** 379–384 (1975).

93. J. D. Hackney, W. S. Linn, S. K. Karuza, H. Greenberg, R. D. Buckley, and E. E. Pedersen, "Experimental Studies on Human Exposure to Ozone Alone and in Combination with Other Pollutant Gases," *Arch. Environ. Health,* **30**(8), 385–390 (1975).

94. D. V. Bates, G. Bell, C. Burnham, M. Hazucha, J. Mantha, L. D. Pengelly, and F. Silverman, "Problems in Studies of Human Exposure to Air Pollutants," *Can. Med. Assoc. J.,* **103**(8), 833–837 (1970).

95. F. Silverman, L. J. Folinsbee, J. Barnard, and R. J. Shephard, "Pulmonary Function Changes in Ozone—Interaction of Concentration and Ventilation," *J. Appl. Physiol.* **41**(6), 859–864 (1976).

96. H. D. Kerr, T. J. Kulle, M. L. Mcilhany, and P. Snidersky, "Effects of Ozone on Pulmonary Function in Normal Subjects, An Environmental-Chamber Study," *Am. Rev. Respir. Dis.,* **111**(6), 763–773 (1975).

97. L. J. Folinsbee, F. Silverman, and R. J. Shepard, "Decrease of Maximum Work Performance Following Ozone Exposure," *J. Appl. Physiol.,* **42**(4), 531–536 (1977).

98. A. J. Delucia and W. C. Adams, "Effects of Ozone Inhalation During Exercise on Pulmonary Function and Blood Chemistry," *J. Appl. Physiol.* **43**(1), 75–81 (1977).

99. W. M. Savin and W. C. Adams, "Effects of Ozone Inhalation on Work Performance and Maximum Oxygen Uptake," *J. Appl. Physiol.,* **46**(2), 309–314 (1979).

100. C. J. Dillard, R. E. Litov, W. M. Savin, E. E. Dumelin, and A. L. Tappel, "Effect of Exercise, Vitamin E, and Ozone on Pulmonary Function and Lipid Peroxidation," *J. Appl. Physiol.,* **45**(6), 927–932 (1978).

101. C. E. Schoettlin and E. Landau, "Air Pollution and Asthmatic Attacks in the Los Angeles Area," *Public Health Rep.,* **76**(6), 545–548 (1961).

102. J. R. Goldsmith and L. T. Friberg, "Effects of Air Pollution on Human Health," in: A. Stern, Ed., *Air Pollution,* 3rd ed., Vol. II, Academic Press, New York, 1977.

103. F. Silverman, "Asthma and Respiratory Irritants (Ozone)," *Environ. Health Perspect.,* **29,** 131–136 (1979).

104. J. D. Hackney, W. S. Linn, R. D. Buckley, and H. J. Hislop, "Studies in Adaptation to Ambient Oxidant Air Pollution: Effects of Ozone Exposure in Los Angeles Residents vs. New Arrivals," *Environ. Health. Perspect.,* **18,** 141–146 (1976).

105. J. D. Hackney, W. S. Linn, J. G. Mohler, and C. R. Collier, "Adaptation to Short-Term Respiratory Effects of Ozone in Men Exposed Repeatedly," *J. Appl. Physiol.* **43**(1), 82–85 (1977).

106. R. D. Buckley, J. D. Hackney, K. Clark, and C. Posin: "Biochemical Responses of Humans to Gaseous Pollutants," in *Biochemical Effects of Environmental Pollutants,* S. D. Lee, Ed., Ann Arbor Science, 1977.

107. M. Hazucha, C. Parent, and D. V. Bates, "Development of Ozone Tolerance in Man," in B. Dimitriades, Ed., *International Conference on Photochemical Oxidant Pollution and Its Control, Proceedings,* Vol. 1, EPA-600/3-77-001a, U.S. Environ. Protection Agency, Research Triangle Park, N.C., Jan. 1977.

108. J. M. Lagerwerff, "Prolonged Ozone Inhalation and Its Effects on Visual Parameters," *Aerosp. Med.,* **34**(6), 479–486 (1963).

109. J. A. Gliner, J. A. Matsen-Twisdale, and S. M. Horvath, "Auditory and Visual Sustained Attention During Ozone Exposure," *Aviat. Space Environ. Med.,* **50**(9), 906–910 (1979).

110. B. F. Ferris, Jr., "Health Effects of Exposure to Low Levels of Regulated Air Pollutants, a Critical Review," *J. Air Pollut. Control Assn.* **28**(5), 482–497 (1978).

111. W. S. Wayne, P. Wehrle, and R. E. Carroll, "Pollution and Athletic Performance," *J. Am. Med. Assoc.* **199,** 901–904 (1967).

112. P. Dejours, *Principles of Comparative Respiratory Physiology,* North Holland/American Elsevier, New York, 1975.

113. W. H. McKenzie, J. H. Knelson, N. J. Rummo, and D. E. House, "Cytogenic Effects of Inhaled Ozone in Man," *Mutation Res.,* **48**(1), 95–102 (1977).

114. R. R. Guerrero, D. E. Rounds, R. S. Olson, and J. D. Hackney, "Mutagenic Effects of Ozone on Human Cells Exposed *in vivo* and *in vitro* Based on Sister Chromatid Exchange Analysis," *Environ. Res.,* **18**(2), 336–346 (1979).

115. B. D. Goldstein, "Combined Exposure to Ozone and Nitrogen Dioxide," *Environ. Health Perspect.,* **13,** 107–110 (1976).

116. B. D. Goldstein, "Combined Exposure to Ozone and Nitrogen Dioxide," *Environ. Health Perspect.,* **30,** 87–89 (1979).

117. R. Ehrlich, J. C. Findlay, J. D. Fenters, and D. E. Gardner, "Health Effects of Short-Term Inhalation of Nitrogen Dioxide and Ozone Mixtures," *Environ. Res.,* **14**(2), 223–231 (1977).

118. G. Freeman, L. T. Juhos, N. J. Furiosi, R. Mussenden, R. J. Stephens, and M. J. Evans, "Pathology of Pulmonary Diseases from Exposure to Interdependent Ambient Gases (Nitrogen Dioxide and Ozone)," *Arch. Environ. Health,* **29**(4), 203–210 (1974).

119. D. Hyde, J. Orthoefer, D. Dungworth, E. Tyler, R. Carter, and H. Lum, "Morphometric and Morphologic Evaluation of Pulmonary Lesions in Beagle Dogs Chronically Exposed to High Ambient Levels of Air Pollutants, *Lab. Invest.,* **38**(4), 455–469 (1978).

120. M. O. Amdur, V. Ugro, and D. W. Underhill, "Respiratory Response of Guinea Pigs to Ozone Alone and with Sulfur Dioxide," *Am. Ind. Hyg. Assoc. J.,* **39**(12), 958–961 (1978).

121. L. T. Juhos, M. J. Evans, R. Mussenden-Harvey, N. J. Furiosi, C. E. Lapple, and G. Freeman, "Limited Exposure of Rats to H_2SO_4 with and without O_3," *J. Environ. Sci. Health,* **C13**(1), 33–47 (1978).

122. C. L. Cavender, B. Singh, and B. Y. Cockrell, "Effects in Rats and Guinea Pigs of Six-Month Exposure to Sulfuric Acid Mist, Ozone, and Their Combination," *J. Toxicol. Environ. Health,* **4**(5–6), 845–852 (1978).

123. F. L. Cavender, W. H. Steinhagen, C. E. Ulrich et al., "Effects in Rats and Guinea Pigs of Short-Term Exposures to Sulfuric Acid Mist, Ozone, and Their Combination," *J. Toxicol. Environ. Health*, **3**(3), 521–533 (1977).

124. J. A. Last and C. E. Cross, "A New Model for Health Effects of Air Pollutants: Evidence for Synergistic Effects of Mixtures of Ozone and Sulfuric Acid Aerosols on Rat Lungs," *J. Lab. Clin. Med.*, **91**(2), 328–339 (1978).

125. D. V. Bates and M. Hazucha, *The Effects of Low Levels of SO and Ozone in the Same Atmosphere on Human Pulmonary Function*, Commission of the European Communities, EPA, WHO International Symposium Proceedings, Recent Advances in the Assessment of the Health Effects of Environmental Pollution, Vol. IV, Paris, 24–28 June 1974, pp. 1977–1988.

126a. K. A. Bell, W. S. Linn, M. Hazucha, J. D. Hackney, and D. V. Bates, "Respiratory Effects of Exposure to Ozone plus Sulfur Dioxide in Southern Californians and Eastern Canadians," *Am. Ind. Hyg. Assoc. J.* **38**(12), 696–706 (1977).

126b. M. T. Kleinman, R. M. Bailey, Y-T. C. Chang, K. W. Clark, M. P. Jones, W. S. Linn, and J. D. Hackney, "Exposures of Human Volunteers to a Controlled Atmospheric Mixture of Ozone, Sulfur Dioxide and Sulfuric Acid," *Amer. Ind. Hyg. J.*, **42**(1), 61–69 (1981).

127a. J. F. Bedi, L. J. Folinsbee, S. M. Horvath, and R. S. Ebenstein, "Human Exposure to Sulfur Dioxide and Ozone: Absence of a Synergistic Effect," *Arch. Environ. Health*, **34**(4), 233–239 (1979).

127b. L. J. Folinsbee, J. F. Bedi and S. M. Horvath, "Combined Effects of Ozone and Nitrogen Dioxide on Respiratory Function in Man," *Amer. Ind. Hyg. Assn. J.*, **42**(7), 534–541 (1981).

127c. S. S. Que Hee, "Formaldehyde Evolution from an Ethylene-Using Ozone Chemiluminescence Meter: A New Method of Calibration," *Amer. Ind. Hyg. Assoc. J.*, **42**(7), 510–514 (1981).

128. American Conference of Governmental Industrial Hygienists, *TLVs® Threshold Limit Values for Chemical Substances and Physical Agents in the Workroom Environment with Intended Changes*, ACGIH, Cincinnati, 1981.

129. American Society for Testing Materials, *Standard Practice for Safety and Health Requirments Relating to Occupational Exposure to Ozone*. Annual Book of ASTM Standards, ASTM, New York, 1977.

130. Environmental Protection Agency, "National Primary and Secondary Air Quality Standards," *Fed. Reg.*, **44**(28), 8202 (1979).

131. ILO/WHO Joint Committee on Occupational Health, *Permissible Levels of Toxic Substances in the Working Environment*, International Labour Office, Geneva, 1970.

132. U.S. Naval Ships Systems Command, *U.S. Navy Diving Manual*, Change 2, U.S. Government Printing Office, Washington, D.C., 1978.

133. P. B. Bennett and D. H. Elliott, Eds., *The Physiology and Medicine of Diving and Compressed Air Work*, 2nd ed., Balliere Tindall, London, 1975.

134. C. Edmonds, C. Lowry, and J. Pennefather, *Diving and Subaquatic Medicine*, Diving Medical Center, Sydney, 1976.

135. R. H. Strauss, Ed., *Diving Medicine*, Grune and Stratton, New York, 1976.

136. C. W. Dueker, *Medical Aspects of Sport Diving*, 2nd ed., A. S. Barnes, San Diego (in prep.).

137. A. R. Behnke, Jr., "Physiological Effects of Abnormal Atmospheric Pressures," in G. D. Clayton and F. E. Clayton, Eds., *Patty's Industrial Hygiene & Toxicology*, 3rd rev. ed., Vol. 1, John Wiley & Sons, New York, 1978, Ch. 9.

138. Committee on Medical and Biological Effects of Environmental Pollutants, National Academy of Sciences, *Nitrogen Oxides*, National Academy of Sciences, Washington, D.C., 1977.

139. M. Windholz, Ed., *The Merck Index*, 9th ed., Merck, Rahway, N.J., 1976.

140. R. C. Weast, Ed., *Handbook of Chemistry & Physics*, 58th ed., CRC Press, Cleveland, 1977.

141. E. N. Cohen, B. W. Brown, et al., "Occupational Disease in Dentistry and Exposure to Anesthetic Gases," unpublished report, Dept. of Anesthesia, Stanford University, Stanford, Calif., 1980.

142. World Health Organization, *Environmental Health Criteria. 4. Oxides of Nitrogen,* WHO/UN Environmental Programme, Geneva, 1977.

143. E. Yokoyama, "Uptake of SO_2 and NO_2 by the Isolated Upper Airways," *Bull. Inst. Public Health,* **17,** 302–306 (1968).

144. T. Dalhamn, and J. Sjöholm, "Studies on SO_2, NO_2 and NH_3: Effect on Ciliary Activity in Rabbit Trachea of Single *in vitro* Exposure and Reabsorption in Rabbit Nasal Cavity," *Acta Physiol. Scand.,* **58,** 287–291 (1963).

145. E. N. Goldstein, N. F. Peele, N. J. Parks, H. H. Hines, E. P. Steffey, and B. Tarkkington, "Fate and Distribution of Inhaled Nitrogen Dioxide in Rhesus Monkeys," *Amer. Rev. Respir. Dis.* **115,** 403–412 (1977).

146. World Health Organization, *Environmental Health Criteria. 5. Nitrates, Nitrites and n-Nitroso Compounds,* WHO/UN Environmental Programme, Geneva, 1978.

147. R. Ehrlich and M. C. Henry, "Chronic Toxicity of Nitrogen Dioxide. I Effect on Resistance to Bacterial Pneumonia," *Arch. Environ. Health,* **17,** 860–865 (1968).

148. G. Freeman, N. J. Furiosi, and G. B. Haydon, "Effects of Continuous Exposure to 0.8 ppm NO_2 on Respiration of Rats," *Arch. Environ. Health,* **13,** 454–456 (1966).

149. G. Freeman, R. J. Stephens, S. C. Crane, and N. J. Furiosi, "Lesion of the Lung in Rats Continuously Exposed to Two Parts per Million of Nitrogen Dioxide," *Arch. Environ. Health,* **17,** 181–192 (1968).

150. G. Freeman and G. B. Haydon, "Emphysema after Low Level Exposure to NO_2," *Arch. Environ. Health,* **8,** 125–128 (1964).

151. O. J. Balchum, R. D. Buckley, R. Sherwin, and M. Gardner, "Nitrogen Dioxide Inhalation and Lung Antibodies," *Arch. Environ. Health,* **10,** 274–277 (1965).

152. J. D. Fenters, J. C. Findlay, C. D. Port, R. Ehrlich, and D. L. Coffin, "Chronic Exposure to Nitrogen Dioxide. Immunologic, Physiologic and Pathologic Effects in Virus-Challenged Squirrel Monkeys," *Arch. Environ. Health,* **27,** 85–89 (1973).

153. G. Freeman, S. C. Crane, R. J. Stephens, and N. J. Furiosi, "The Subacute Nitrogen Dioxide-Induced Lesion of The Rat Lung," *Arch. Environ. Health,* **18,** 609–612 (1969).

154. W. D. Wagner, B. R. Duncan, P. G. Wright, and H. E. Stokinger, "Experimental Study of the Threshold Limit of NO_2," *Arch. Environ. Health,* **10,** 455–466 (1965).

155. L. S. Mitina, "The Combined Effect of Small Concentrations of Nitrogen Dioxide and Sulfur Dioxide Gases," *Gig. Sanit.,* **27**(10), 3–8 (1962).

156. R. F. Bils, "The Connective Tissues and Alveolar Walls in Lungs of Normal and Oxidant-Exposed Squirrel Monkeys," *J. Cell. Biol.* **70,** 318 (1976) (Abstr).

157. C. D. Port, D. L. Coffin, and P. Kane, "A Comparative Study of Experimental and Spontaneous Emphysema," *J. Toxicol. Environ. Health,* **2,** 589–604 (1977).

158. P. G. Holt, L. M. Finlay-Jones, D. Keast, and J. M. Papadimitrou, "Immunological Function in Mice Chronically Exposed to Nitrogen Oxides," *Environ. Res.,* **19**(1), 154–162 (1979).

159. M. Drozdz, E. Kucharz, and J. Szyja, "Effect of Chronic Exposure to Nitrogen Dioxide on Collagen Content in Lung and Skin of Guinea Pigs," *Environ. Res.,* **13,** 369–377 (1977).

160. D. Hyde, J. Orthoefer, D. Dungworth, W. Tyler, R. Carter, and H. Lum, "Morphometric and Morphologic Evaluation of Pulmonary Lesions in Beagle Dogs Chronically Exposed to High Ambient Levels of Air Pollutants," *Lab. Invest.,* **34**(4), 455–469 (1978).

161. W. M. Abraham, M. Welker, W. Oliver Jr., M. Mingle, A. J. Januskiewicz, et al., "Cardiopul-

monary Effects of Short-Term Nitrogen Dioxide Exposure in Conscious Sheep," *Environ. Res.*, **22**, 61–72 (1980).

162. J. N. Roehm, J. G. Hadley, and D. B. Menzel, "Oxidation of Unsaturated Fatty Acids by O_3 and NO_2: a Common Mechanism of Action," *Arch. Environ. Health*, **23**, 142–148 (1971).

163. B. L. Fletcher and A. L. Tappel, "Protective Effects of Dietary Tocopherol in Rats Exposed to Toxic Levels of Ozone and Nitrogen Dioxide," *Environ. Res.*, **6**, 165–175 (1973).

164. D. B. Menzel, J. N. Roehm, and S. B. Lee, "Vitamin E: The Biological and Environmental Antioxidant," *J. Agric. Food Chem.*, **20**, 481–486 (1972).

165. H. V. Thomas, P. K. Mueller, and R. L. Lyman, "Lipoperoxidation of Lung Lipids in Rats Exposed to Nitrogen Dioxide," *Science*, **159**, 532–534 (1968).

166. H. I. Trzeciak, S. Kosmider, K. Kryk, and A. Kryk, "The Effects of Nitrogen Oxides and Their Neutralization Products with Ammonia on the Lung Phospholipids of Guinea Pigs," *Environ. Res.*, **14**, 87–89 (1977).

167. K. Seto, M. Kon, M. Kawakami, S. Yagashita, K. Sugita, and M. Shisido, "Effect of Nitrogen Dioxide Inhalation on the Formation of Protein in the Lung," *Igaku To Seibutsugaku*, **90**, 103–106 (1975) (in Japanese; English abstract).

168. D. B. Menzel, M. D. Abou-Donia, C. R. Roe, R. Ehrlich, B. E. Gardner, and D. L. Coffin, "Biochemical Indices of Nitrogen Dioxide Intoxication of Guinea Pigs Following Low-Level Long-Term Exposure," in B. Dimitriades, Ed., *Proceedings International Conference Photochemical Oxidant Pollution and Its Control*, Vol. II, Sept. 1973, EPA 600/3-77-001b, U.S. Environmental Protection Agency, Research Triangle Park, N.C., (1977), pp. 577–587.

169. C. K. Chow, C. J. Dillard, and A. L. Tappel, "Glutathione Peroxidase System and Lysozyme in Rats Exposed to Ozone or Nitrogen Dioxide," *Environ. Res.* **7**, 311–319 (1974).

169a. K. L. Ayaz and A. F. Csallany, "Long Term NO_2 Exposure of Mice in the Presence and Absence of Vitamin E. II. Effect on Glutathione Peroxidase," *Arch. Environ. Health*, **33**, 292–296 (1978).

170. D. E. Gardner, F. J. Miller, E. J. Blommer, and D. L. Coffin, "The Influence of Exposure Mode on the Toxicity of NO_2," *Environ. Health Perspect.*, **30**, 23–29 (1977).

171. R. Ehrlich, J. C. Findlay, J. D. Fenters, and D. E. Gardner, "Health Effects of Short-Term Exposures to Inhalation of NO_2-O_3 Mixtures," *Environ. Res.*, **14**, 223–231 (1977).

172. L. J. Schiff, "Effect of Nitrogen Dioxide on Influenza Virus Infection in Hamster Trachea Organ Culture," *Proc. Soc. Exp. Biol. Med.*, **156**, 546–549 (1977).

173. A. M. Giordano and P. E. Morrow, "Chronic Low-Level Nitrogen Dioxide Exposure and Mucociliary Clearance," *Arch. Environ. Health*, **25**, 443–449 (1972).

174. C. Aranyi, J. Fenters, and R. Ehrlich, "Scanning Electron Microscopy of Alveolar Macrophages After Exposure to O_2, NO_2, and O_3," *Environ. Health Perspect.*, **16**, 180 (1976).

175. N. D. Green and S. L. Schneider, "Effects of NO_2 on the Response of Baboon Alveolar Macrophages to Migratory Inhibitory Factor," *J. Toxicol. Environ. Health*, **4**, 869–880 (1978).

175a. J. E. H. Milne, "Nitrogen Dioxide Inhalation and Bronchiolitis Obliterans. A Review of the Literature and Report of a Case," *J. Occup. Med.*, **11**, 538–547 (1969).

176. N. H. Proctor and J. P. Hughes, *Chemical Hazards of the Workplace*, Lippincott, Philadelphia, 1978.

177. U.S. Dept. of Health, Education and Welfare, National Institute for Occupational Safety and Health, *Criteria for a Recommended Standard—Occupational Exposure to Oxides of Nitrogen*, HEW Publ. No. (NIOSH) 76-149, Washington, D.C., 1979.

178. American Conference of Governmental Industrial Hygienists, *Documentation of the TLVs for Substances in Workroom Air*, 3rd ed., Cincinatti, Ohio 1976.

178a. J. D. Hackney, F. C. Thiede, W. S. Linn, E. E. Pederson, C. E. Spier, D. C. Law, and D. A. Fischer, "Experimental Studies on Human Health Effects of Air Pollutants. IV. Short-Term Physiological and Clinical Effects of Nitrogen Dioxide Exposure," *Arch. Environ. Health,* **33,** 176–181 (1978).

179. L. J. Folinsbee, S. M. Horvath, J. F. Bedi, and J. C. Delehunt, "Effects of 0.62 ppm NO_2 on Cardiopulmonary Function in Young Male Nonsmokers," *Environ. Res.,* **15**(2), 199–205 (1978).

180. C. Posin, K. Clark, M. P. Jones, J. V. Patterson, R. D. Buckley, and J. D. Hackney, "Nitrogen Dioxide Inhalation and Human Blood Chemistry," *Arch. Environ. Health,* **33**(6), 318–324 (1978).

181. G. von Nieding and H. M. Wagner, "Effect of NO_2 on Chronic Bronchitics," *Environ. Health Perspect.,* **29,** 137–142 (1979).

182. R. I. Mitchell, R. Williams, R. W. Cote, R. R. Lanese, and M. D. Keller, "Household Survey of the Incidence of Respiratory Disease in Relation to Environmental Pollutants," in Commission of the European Communities International Symposium Proceedings, *Recent Advances in the Assessment of Health Effects of Environmental Pollution,* Vol. I, Paris, 24–28 June 1974.

183. B. G. Ferris, Jr., F. E. Speizer, Y. M. M. Bishop, and J. D. Spengler, "Effects of Indoor Environment on Pulmonary Function of Children 6–9 Years Old," *Am. Rev. Respir. Dis. (Suppl.),* **119**(4), 214 (1979).

184. W. Jones and T. A. Ridgik, "Nitric Oxide Oxidation Method for Field Calibration of Nitrogen Dioxide Meters," *Am. Ind. Hyg. Assoc. J.,* **41,** 433–436 (1980).

185. E. Palmes and C. Tomczyk, "Personal Samplers for NO_x," *Am. Ind. Hyg. Assoc. J.,* **40,** 588–591 (1979).

186. American Conference of Governmental Industrial Hygienists, *TLVs® Threshold Limit Values for Chemical Substances and Physical Agents in the Workroom Environment with Intended Changes,* ACGIH, Cincinnati, 1981.

187. U.S. Environmental Protection Agency, *National Primary and Secondary Air Quality Standards, Fed. Reg.,* **36,** 8186–8210 (1971).

188a. U.S. National Institute for Occupational Safety and Health: Criteria for a Recommended Standard, *Occupational Exposure to Carbon Monoxide,* HSM 73-11000, U.S. Government Printing Office, Washington, D.C., 1972.

188b. R. L. Pruett, "Synthesis Gas: A Raw Material for Industrial Chemicals," *Science,* **211,** (4447) 11–16 (1981).

189. G. W. Jones, L. B. Berger, and W. F. Holbrook, *U.S. Bur. Mines Tech. Paper No.,* **337** (1923).

190. M. W. Kubic, *Drug Metab. Disposition,* **2,** 53–57 (1974). Cited in Reference 4.

191. R. D. Stewart, "The Effect of Carbon Monoxide on Humans," *Ann. Rev. Pharmacol.,* **15,** 409–423 (1975).

192. J. R. Goldsmith, "Carbon Monoxide and Coronary Heart Disease: A Review," *Environ. Res.,* **10,** 236–248 (1975).

193. J. R. Goldsmith and S. A. Landaw, "Carbon Monoxide and Human Health," *Science,* **162,** 1352–1355 (1968).

194. A. H. J. Maas, M. L. Hamelink, and R. J. M. de Leeuw, "An Evaluation of the Spectrophotometric Determination of HBO_2, HbCO and Hb in Blood with the CO-Oximeter IL 182," *Clin. Chim. Acta,* **29,** 303–309 (1970).

195. K. A. Small, E. P. Radford, J. M. Frazier, F. L. Rodkey, and H. A. Collision, "A Rapid Method for Simultaneous Measurement of Carboxy- and Methemoglobin in Blood," *J. Appl. Physiol.,* **31**(1), 154–160 (1971).

196. R. F. Coburn, G. K. Danielson, W. S. Blakemore, and R. E. Forster, II, "Carbon Monoxide in Blood: Analytical Method and Sources of Error," *J. Appl. Physiol.,* **19**(3), 510–515 (1964).

197. T. E. Dahms and S. M. Horvath, "Rapid, Accurate Technique for Determination of Carbon Monoxide in Blood," *Clin. Chem.,* **20,** 533–537 (1974).

198. Anonymous, "Carbon Monoxide Exposure Level Evaluation by Expired Air Analysis," *Mich. Occup. Health,* **22**(3), 1 (1978).

199. R. D. Stewart, R. S. Stewart, W. Stamm, and R. P. Seelen, "Rapid Estimation of Carboxyhemoglobin Level in Fire Fighters," *J. Am. Med. Assoc.,* **235**(4), 390–392 (1976).

200. V. G. Laties and W. H. Merigan, "Behavioral Effects of Carbon Monoxide on Animals and Man," *Ann. Rev. Pharmacol. Toxicol.,* **19,** 357–392 (1979).

201. R. F. Coburn, R. E. Forster, and P. B. Kane, "Consideration of the Physiological Variables that Determine the Blood Carboxyhemoglobin Concentration in Man," *J. Clin. Invest.,* **44,** 1899–1910 (1965).

202. B. D. Dinman, "The Management of Acute Carbon Monoxide Intoxication," *J. Occup. Med.,* **16**(10), 662–664 (1974).

203a. A. Grut, *Choric Carbon Monoxide Poisoning, A Study in Occupational Medicine,* Einar Munksgaard, Copenhagen, 1949.

203b. L. D. Longo, "The Biological Effects of Carbon Monoxide on the Pregnant Woman, Fetus, and Newborn Infant," *Amer. J. Obstet. and Gynecol.,* **129**(1), 69–103 (1977).

204. F. H. Shillito, C. K. Drinker, and T. J. Shaughnessy, "The Problem of Nervous and Mental Sequelae in Carbon Monoxide Poisoning" *J. Am. Med. Assoc.,* **106**(9), 669–674 (1936).

205. J. W. Meigs and J. P. W. Hughes, "Acute Carbon Monoxide Poisoning," *Arch. Ind. Hyg. Occ. Med.,* **6,** 344–356 (1952).

206. Y. Kuroiwa, personal communication, 1980.

207. T. Sollman, *A Manual of Pharmacology,* 6th ed., Saunders, Philadelphia, 1944.

208. National Institute for Occupational Safety and Health, *Criteria for a Recommended Standard for Occupational Exposure to Carbon Dioxide,* (NIOSH) 76-194, U.S. Government Printing Office, Washington, 1976.

209. D. J. Cullen and E. L. Eger, "Cardiovascular Effects of Carbon Dioxide in Man," *Anesthesiology,* **41,** 345 (1974).

210a. American Industrial Hygiene Association, "Hygienic Guide Series: Carbon Dioxide," *Am. Ind. Hyg. Assoc. J.,* **25,** 519 (1964).

210b. American Conference of Governmental Industrial Hygienists, *Threshold Limit Values for Chemical Substances and Physical Agents in the Workroom Environment with Intended Changes for 1981,* ACGIH, Cincinatti, Ohio, 1981.

211. W. P. Yant, J. C. Olsen, H. H. Storch, J. B. Littlefield, and L. Scheflan, *Ind. Eng. Chem., Anal. Ed.,* **8,** 20 (1936).

212. S. A. Cucinell, "Review of the Toxicity of Long-Term Phosgene Exposure," *Arch. Environ. Health,* **28,** 272 (1974); cited by N. H. Proctor and J. P. Hughes, Eds., *Chemical Hazards of the Workplace,* Lippincott, Philadelphia, 1978.

213. American Industrial Hygiene Association, "Hygienic Guide Series, Phosgene," *Am. Ind. Hyg. Assoc. J.,* **29,** 308 (1968).

214. National Institute for Occupational Safety and Health, *Criteria for a Recommended Standard. Occupational Exposure to Phosgene,* NIOSH 76-137, U.S. Government Printing Office, Washington, 1976.

215. W. F. Diller, "Medical Phosgene Problems and Their Possible Solution," *J. Occup. Med.,* **20**(3), 189–193 (1978).

CHAPTER FIFTY-THREE

Aliphatic Nitro Compounds, Nitrates, Nitrites

HERBERT E. STOKINGER, Ph.D.

1 ALIPHATIC NITRO COMPOUNDS

1.1 Introduction

Nitroparaffins, or nitroalkanes, are derivatives of alkanes with the general formula C_nH_{2n+1}, in which one or more hydrogen atoms are replaced by the electronegative nitro group ($-NO_2$) and which is attached directly to carbon ($-C-NO_2$), as opposed to the nitrates and nitrites, which are nitric acid esters of mono- and polyvalent aliphatic alcohols bearing the structure $-C-O-NO_x$.

Nitroparaffins are classed as primary, RCH_2NO_2, secondary, R_2CHNO_2, and tertiary, R_3CNO_2, using the same convention as for alcohols. At this time, only primary nitroparaffins are available in commercial quantities with the exception of 2-nitropropane (1). Others available are nitromethane, nitroethane, 1-nitropropane, and tetranitromethane.

1.2 Source and Production

The nitroparaffins, nitromethane, nitroethane, 1-nitropropane, and 2-nitropropane, are produced in large commercial quantities by direct nitration of propane with nitric acid, and the individual compounds are obtained by fractional distillation. The nitroparaffins are produced only by Commercial Solvents Corporation, a division of International Minerals. No production figures appear to be available (2), but estimates run as high as a multimillion pounds annually.

1.3 Uses and Industrial Hazards

1.3.1 Uses

The uses of the nitroparaffins depend on their strong solvent power for a wide variety of substances including many coating materials, waxes, gums, resins, dyes, and numerous organic chemicals. Tables of substances soluble and insoluble in 1- and 2-nitropropane may be found in Reference (3).

Another important use is the production of derivatives, such as nitroalcohols, alkanolamines, and polynitro compounds. In potential range of usefulness for chemical synthesis, the four nitroparaffins are "unequaled" by any other group of organic chemicals. In some cases, they provide better methods of manufacturing well-known chemicals, chloropicrin and hydroxylamine, for example (3).

1.3.2 Industrial Hazards

In 1976 NIOSH estimated that the number of workers exposed to 1-nitropropane was 5400, and to 2-nitropropane, 203,280. The chief industrial hazard from exposure to nitromethane and 1- and 2-nitropropane is inhalation of their vapors (4), the vapor pressures at 25°C being, respectively, 35.5, 13, and 20 mm Hg, which are sufficient to produce high vapor levels in the workplace unless controlled. Although no injuries from inhalation of nitromethane have been reported, mild dermal irritation has occurred as a result of its solvent action. The more highly toxic 2-nitropropane has produced headache, dizziness, nausea, vomiting, and diarrhea, with complaints of respiratory tract irritation. These signs and symptoms resulted from exposures at concentrations ranging from 30 to 300 ppm. 1-Nitropropane is rated somewhat toxic.

Fire and explosion hazards from nitromethane are considered minimal, for it has relatively high flash points of 120 and 103°F, respectively (TOC). The lower limit of flammability is greater than nitromethane, being 2.6 percent. Like nitromethane, these intropropanes react with inorganic bases to form salts that are explosive when dry.

1.4 Physical and Chemical Properties

1.4.1 Physical and Chemical Constants

The constants of six mononitroparaffins and tetranitromethane are given in Table 53.1. Of the seven nitroparaffins listed, only five, nitromethane, nitroe-

Table 53.1. Properties of the Mononitroparaffins and Tetranitromethane (3)

Name	Formula	Mol. Wt.	B.p. (°C)	Sp. Gr.	Solubility in H$_2$O at 20°C (% by Vol.)	Vapor Pressure (mm Hg) (°C)	Vapor Density (Air = 1)	Flash Point (°F)	Conversion Units 1 mg/l (ppm)	1 ppm (mg/m^3)	Oral Lethal Dose, Rabbits (g/kg)
Nitromethane	CH$_3$NO$_2$	61.04	101.2	1.139 (20/20°C)	9.5	27.8 (20)	2.11	110	400.7	2.495	0.75–1.0
Nitroethane	CH$_3$CH$_2$NO$_2$	75.07	114.8	1.052 (20/20°C)	4.5	15.6 (20)	2.58	105	325.7	3.07	0.5–0.75
1-Nitropropane	CH$_3$CH$_2$CH$_2$NO$_2$	89.09	131.6	1.003 (20/20°C)	1.4	7.5 (20)	3.06	120	274.7	3.04	0.25–0.50
2-Nitropropane	CH$_3$CH(NO$_2$)CH$_3$	89.09	120.3	0.992 (20/20°C)	1.7	12.9 (20)	3.06	100	274.7	3.64	0.50–0.75
1-Nitrobutane	CH$_3$CH$_2$CH$_2$CH$_2$NO$_2$	103.12	151	0.9774 (15.6/15.6°C)	0.5	5(25)	3.6		237.1	4.21	0.50–0.75
2-Nitrobutane	CH$_3$CH$_2$CH(NO$_2$)CH$_3$	103.12	139	0.9728 (15.6/15.6°C)	0.9	8(25)	3.6		237.1	4.21	0.50–0.75
Tetranitromethane	C(NO$_2$)$_4$	196.04	125.7	1.650 (13/4°C)	Insoluble	13 (approx., 25)	0.8		124.7	8.02	—

thane, 1- and 2-nitropropane, and tetranitromethane, are of present commercial interest (1); 1- and 2-nitrobutane are included because toxicity data generated in the past anticipated their eventual use.

Nitroparaffins are colorless, oily liquids with relatively high vapor pressures. Their solubility in water decreases with increasing hydrocarbon chain length and number of nitro groups, tetranitromethane being completely insoluble. As expected, their boiling and flash points are higher than their corresponding hydrocarbons. Although flash points are relatively high, under certain conditions of temperature, chemical reaction, and confinement, shock explosion can result, as noted in Section 3.2.

Most organic compounds, including aromatic hydrocarbons, alcohols, esters, ketones, ethers, and carboxylic acids, are miscible with the nitroparaffins, which property is the basis for much of their industrial use (Section 1.3.1).

1.4.2 Chemical Properties (3)

The nitroparaffins are acidic substances, and the polynitro compounds even stronger acids than the corresponding mononitroparaffins; thus trinitromethane is a typical strong acid with an ionization constant of the range 10^{-2}–10^{-3} as for strong inorganic acids. They are thus rapidly neutralized with strong bases and readily titrated.

Tautomerism, the general property of primary and secondary mononitroparaffins, gives rise to a more acidic, "aci" form, or nitronic acid, thus:

$$CH_3CH_2N{\overset{\nearrow O}{\underset{\searrow O}{}}} \xrightarrow{H_2O} CH_3CH{=}N{\overset{\nearrow O}{\underset{\searrow OH}{}}}$$

Nitroethane   Ethanenitronic acid

This tautomerism forms the basis of a number of important chemical reactions through the formation of *nitroparaffin salts*. Mercury fulminate, $Hg(ON{=}C)_2$, is one of the better known compounds which is derived from the mercury salt of nitromethane, $(CH_2{=}NO_2)_2Hg$. The production of *chloronitroparaffins* makes use of this tautomerism, as does the important intermediate methazonic acid,

$$HON{=}CHCH{=}N{\overset{\nearrow O}{\underset{\searrow OH}{}}}$$

which serves as a starting product for a number of well-known compounds, for example, nitroacetic acid and glycine. Chloropicrin, trichloronitromethane

(CCl_3NO_2), however, is an exception; this compound in turn, like other nitro derivatives, is a source of guanidine [$H_2NC(=NH)NH_2$] by reaction with ammonia.

Primary amines are readily formed by hydrogen reduction of the lower nitroparaffins; milder reduction by zinc dust yields hydroxylamines, which are in turn good reducing agents.

Nitrohydroxy compounds result from basic catalytic condensation with formaldehyde (HCHO) to produce aliphatic derivatives, and with aromatic aldehydes to form aromatic nitrohydroxy compounds, for example, $C_6H_5CHOHC(CH_3)NO_2CH_3$. *Aminohydroxy compounds* are formed by direct reduction of both these aliphatic and aromatic nitrohydroxy compounds.

Aromatic amines (with no hydroxy group) can be made by dehydrating aromatic nitrohydroxy compounds followed by reduction of the nitro group, for example, ($C_6H_5CH_2CHNH_2C_2H_5$). Compounds of this general structure have valuable pharmaceutic properties.

A number of other useful products such as β-dioximes, which in turn can yield isoxazoles on hydrolysis, are also among the armamentarium of the nitroparaffin reaction product possibilities.

The unexcelled solvent power of the nitroparaffins has been noted in Section 1.3.1.

1.5   Analytic Determination

Early methods (1940–1959) of determining nitroparaffins used colorimetric procedures. These have now been replaced by instrumental methods (since 1970). Reference to the colorimetric procedures are included here because these procedures were used for monitoring animal exposures which provide the toxicity data in Table 53.2. Instrumental methods of mass spectroscopy and gas chromatography are used routinely by the manufacturer of nitroparaffins, and infrared has been used in animal exposure studies of nitromethane and 2-nitropropane (7).

1.5.1   Determination in Air—Colorimetric Methods

The primary mononitroparaffins, with the exception of nitromethane, on which toxicity data were determined, nitroethane, 1-nitropropane, and 1-nitrobutane, were analyzed colorimetrically by measuring the color developed from an HCl-acidified alkaline solution containing $FeCl_3$. Reproducible results were obtained down to 0.5 mg/25 ml (6a). Analytic data on the secondary nitroparaffins were determined by measuring in a Beckman spectrophotometer, UV radiation at wavelength 2775 Å through an alcoholic solution of the 2-nitropropane (6b).

A more sensitive spectrophotometric determination of primary nitroparaffins utilizes the coupling reaction with *p*-diazobenzenesulfonic acid (6c). Simple,

Table 53.2. Results of Inhalation Experiments (11a)[a]

	Nitromethane						Nitroethane						1-Nitropropane				
Exp. No.	Concn. (%)	Time (hr)	Concn. × Time	Rabbits, No. Killed	Guinea Pigs, No. Killed		Exp. No.	Concn. (%)	Time (hr)	Concn. × Time	Rabbits, No. Killed	Guinea Pigs, No. Killed		Exp. No.	Concn. (%)	Time (hr)	Concn. × Time	Rabbits, No. Killed	Guinea Pigs, No. Killed
6	1.0	6	6	2	2		28	2.5	2	5	2	0							
3	3.0	2	6	2	2		16	3.0	1.25	3.75	2	2							
2	5.0	1	5	2	2		15	3.0	1	3	1	1		29	1.0	3	3	2	2
							17	1.0	3	3	2	1							
4	3.0	1	3	0	2														
5	1.0	3	3	0	2		19	3.0	0.5	1.5	1	0		31	0.5	3	1.5	2	2
7	0.5	6	3	1	1		20	0.5	3	1.5	2	0							
8	0.25	12	3	2	1		25	0.1	12	1.2	1	0							
12	0.10	30	3	0	2		21	0.5	2	1	1	0		30	1.0	1	1	0	1
1	2.25	1	2.25	0	1		18	1.0	1	1	1?	0							
9	3.0	0.5	1.5	0	1		22	0.25	3	0.75	0	0							
32	0.5	3	1.5	0	1		24	0.1	6	0.6	0	0							
11	1.0	1	1	0	0		26	0.05	30	1.5	0	0[b]							
10	3.0	0.25	0.75	0	0		27	0.05	140	7.0	0	0[c]							
13	0.05	140	7.0	0	0[c]														
14	0.1	48	4.8	1[b]	1[c]														

[a] (Two rabbits and two guinea pigs exposed in each experiment).
[b] No animals exposed.
[c] One monkey exposed.

secondary nitroparaffins do not interfere as do some complex secondary nitro alcohols. Nitromethane can be determined at 440 mμ up to 50 μg/ml, and nitroethane and 1-nitropropane at 395 mμ up to 80 and 100 μg/ml, respectively.

Tetranitromethane has been measured by collection in reagent-grade methanol followed by reading at 240 mμ on a Beckman spectrophotometer and comparison to reference calibration curves (7).

Chloronitroparaffins have also been determined by colorimetric procedures. Using alkaline resorcinol for 1,1-dichloro-1-nitroethane, as for *chloropicrin*, (8), color density is read at 480 mμ in a photoelectric spectrophotometer. Color density is linear between 60 and 650 μg/25 ml. For *1-chloro-1-nitropropane*, phenylenediamine in concentrated sulfuric acid is used; the curve is linear between 75 and 250 μg/l at readings at 540 mμ. A colorimetric method for *2-nitro-2-methyl-1-propanol* using chromotropic acid is given in Reference 9 (p. 444). It has an accuracy of 2 percent and a precision of 1 percent.

1.5.2 Determination in Air—Instrumental Methods (9)

Mass spectrography, gas chromatography, and infrared spectroscopy are the current methods of choice for the determination of the nitroparaffins.

Mass spectra have been determined on eight C_1–C_4 mononitroparaffins (9, p. 418). Nitromethane uniquely is the only member to have a major peak at its mass weight, 61. As expected, the two isomers, 1- and 2-nitropropane, have similar spectra, but can be separated with good precision and accuracy in C_1–C_3 nitroparaffin mixtures by the *M/e* 42–43 ratio. Conditions for analytic determination are given on pages 424 and 425 of this reference.

The different conditions recommended for analysis of nitroparaffin mixtures by gas chromatography along with seven systems with their chromatograms for nitroparaffin mixtures, including chloronitroparaffins, are given on pp. 425–429 and 430–433, and for nitroalcohols, on pp. 441–443 (9).

Infrared absorption spectroscopy has been used to monitor animal inhalation chamber concentrations of nitromethane and 2-nitropropane (10). Conditions of use and calibration of a Wilks MIRAN are described. MIRAN was connected to an automatic sampler for hourly analysis of chamber air. Over the 21 weeks exposure concentrations for nitromethane averaged 97.6 ± 4.6 ppm and 745 ± 34 ppm for the two exposure groups; for 2-nitropropane, 27.2 ± 3.1 ppm and 207 ± 15 ppm, respectively.

1.6 Physiologic Response

It should be noted that the inhalation toxicity data on the nitroparaffins gathered in the late 1930s (11a) and summarized below lacked some of the refinements of later work with these compounds (10). Exposure "chambers" consisting of steel drums of 233-liter capacity lacked the space to expose what is now considered an adequate size complement of animal species (11b). The exposure concentrations at levels of 5000 to 50,000 ppm must be considered "nominal," because of measurement by interferometer, the chamber airflow characteristics, and fan circulation of air. Lower exposure levels, although measured colorimetrically (Section 1.5.1), lacked the precision of later day instrumental methods (Section 1.5.2).

1.6.1 Comparative Animal Toxicities (Table 53.2)

Acute Toxicity. The nitroparaffins act chiefly as moderate irritants when inhaled (11a). Animals exposed at levels greater than 10,000 ppm give evidence of restlessness, discomfort, and signs of respiratory-tract irritation, followed by eye irritation, salivation, and later central nervous system symptoms consisting of abnormal movements with occasional convulsions. 2-Nitropropane appears to be more irritating than nitroethane (6b), which is more irritating than nitromethane. Anesthetic symptoms are generally mild, and appear late; most animals that manifest anesthesia die later. This is more marked with nitroethane and nitropropane than with nitromethane inhalation. Incomplete information on the polynitroparaffins indicates that an increased number of nitro groups results in increased irritant properties. The *chlorinated nitroparaffins* are more irritating than the unchlorinated compounds (8). This reaches a severe degree with trichloronitromethane (chloropicrin). Unsaturation of the hydrocarbon chain in the *nitroolefins* also results in an increase in the irritant effects (12).

The primary nitroparaffins fail to show significant pharmacologic effects on blood pressure or respiration (13). Oral doses result in symptoms similar to those produced by inhalation except for the additional evidence of gastrointestinal tract irritation. They are less potent methemoglobin formers than the aromatic nitro compounds, but 2-nitropropane at 80 ppm induced Heinz bodies in 5 to 15 percent of erythrocytes of cats, the most susceptible of species; during prolonged exposure at this concentration, 30 to 35 percent of the erythrocytes contained these abnormal structures (6b).

The nitroparaffins show no significant percutaneous absorption as judged by lack of any systemic effects or weight loss after application of the pure compounds in five daily treatments to the clipped abdominal skin of rabbits (11a).

Animals dying following brief inhalation of the nitroparaffins show general visceral and cerebral congestion. After exposure at high concentrations there is pulmonary irritation and edema, the latter inadequate to be the sole cause of death. Inhalation of 2-nitropropane results in general vascular endothelial damage in all tissues as well as pulmonary edema, hemorrhage, and brain and liver damage (6b). Microscopically the liver shows severe parenchymal degeneration and focal necrosis. Sublethal concentrations of nitromethane produce severe liver changes in dogs consisting of infiltration with chronic inflammatory cells, fatty changes, congestion, and some hemorrhage and necrosis (19). Toxic damage to the kidneys and heart is less prominent. Inhalation of trichloronitromethane produces severe injury of the respiratory tract consisting of inflammation and necrosis of the bronchi, and edema and congestion in the alveoli. The chlorinated nitroparaffins and nitroolefins produce gastrointestinal tract irritation and damage when given by mouth. The *monochloronitroparaffins* are not markedly irritating to the skin or eyes but the *dichloro compounds* (8) and particularly the *nitroolefins* are strong skin and eye irritants (12, 18).

Acute toxicologic and pharmacologic studies of 12 nitroolefins from 2-nitro-2-butane to 3-nitro-3-nonene by Deichmann et al. (12a) by oral, intraperitoneal, and dermal routes showed those of the series from C_4 to C_8 to be highly toxic to rats orally, and those from C_4 through C_9 intraperitoneally, with toxicities tending to decrease with increasing carbon chain length. Approximate oral lethal doses for rats ranged from 280 to 620 mg/kg, and corresponding intraperitoneal doses from 80 to 280 mg/kg. Corresponding percutaneous doses for the rabbit showed no such regularity; doses for the entire series ranged from 420 mg/kg for the C_9 compound, 2-nitro-3-nonene, to 1400 mg/kg for 2-nitro-2-hexene and 3-nitro-3-heptene.

Absorption from the respiratory or gastroenteric tract, peritoneal cavity, or skin is very rapid, and signs of systemic intoxication appear promptly, including hyperexcitability tremors, clonic convulsions, tachycardia, and increased rate and magnitude of respiration, followed by a generalized depression, ataxia, cyanosis, and dyspnea. Death is initiated by respiratory failure and associated with asphyxial convulsions. Pathologic changes were most marked in the lungs, regardless of the mode of administration of a compound.

Altered function in animals inhaling nitroolefins was reported by Murphy et al. (12b) as increased total pulmonary flow resistance and tidal volumes and decreased respiratory rates of guinea pigs, and decreased voluntary activity of mice occurred during inhalation of the vapors of nitroolefins at concentrations near or below the threshold for human, sensory detection (0.1 to 0.5 ppm).

Increasing concentrations increased the magnitude of the effects. Comparison of the effects of 2-nitro-2-butene, 3-nitro-3-hexene, and 4-nitro-4-nonene indicated that the effect on pulmonary function was inversely related to the carbon chain length. However, 4-nitro-4-nonene was slightly more active than the butene and hexene in producing depression of mouse activity. At the low concentrations tested, the effects of nitroolefins were reversible when the animals were returned to clean air. Injection of atropine sulfate overcame the increased pulmonary flow resistance induced by 4-nitro-4-nonene.

1.6.2 Human Experience

Published reports of human experience are scanty. Anorexia, nausea, vomiting, and intermittent diarrhea and headaches in men exposed at 20 to 45 ppm of *nitropropane* during a dipping process ceased to appear when methyl ethyl ketone was substituted (15). Nasal irritation, burning eyes, dyspnea, cough, chest oppression, and dizziness in men handling crude TNT have been attributed to their *tetranitromethane* exposure (16). Headache, methemoglobinemia, and a few deaths have also been attributed to similar exposures. There has been more human experience with trichloronitromethane (*chloropicrin*) because it was used as a war gas often in mixtures with chlorine or phosgene. Chloropicrin is a lacrimator which produces a peculiar frontal headache, coughing, nausea, and vomiting and severe injury of the respiratory tract resulting in pulmonary edema. It has been noted that individuals who have been injured with chloropicrin appear to become more susceptible so that concentrations not producing symptoms in others cause them distress (17).

The 2-nitro-2-olefins, 2-nitro-2-butene, 2-nitro-2-hexene, and 2-nitro-2-nonene, those that have been tested, produced distinct eye irritation at low concentrations (18); the butene and hexene derivatives produced irritation within 3 min at concentrations between 0.1 and 0.5 ppm. For the corresponding nonene, irritation occurred only at concentrations above 1.0 ppm. As in the case of chloropicrin, individuals who had been repeatedly exposed to nitroolefins over a sustained period became increasingly sensitive to the eye-irritating effects of these compounds. Temperature, within the limits tested of 65° to 90°F, appeared to have no influence on the sensitivity of the eye to a nitroolefin, but brief UV irradiation rapidly destroys the lacrimator.

Four fatalities and one near-fatal case have been assembled by Hine et al. (19) of workmen overexposed to solvent compounds. The one agent common to all was 2-nitropropane (2-NP), which occurred in the solvent mixtures in concentrations of 11 to 28 percent. All patients showed typical signs of 2-NP overexposure of headache, nausea, vomiting, diarrhea, and chest and abdominal pains, but the prodromal signs are somewhat nonspecific and similar to those from overexposure to any variety of solvents. Furthermore, 2-NP has no distinctive warning properties in solvent mixtures, and no organoleptic recognition until concentrations well above its TLV are reached.

The characteristic lesion in the fatal cases was destruction of the hepatocytes. In all cases, liver failure was the primary cause of death. This was well-documented by antemortem findings of elevations in the enzymes and post-mortem findings of microscopic evidence of liver changes. The antemortem observation of hemorrhage from the gastrointestinal tract would not have been expected on the basis of animal studies, with the exception of a report of destruction of vascular epithelium caused by the compound. In one of the fatal cases, cerebral necrosis was observed; this finding was predictable on the basis of findings from experiments in animals. No organ damage was observed in the nonfatal case.

Survival time was from 6 to 10 days in the four fatal cases. The failure to detect any solvent in the blood or liver can be explained by the elimination from the body of all but trace amounts during the period between exposure and death. In none of the cases were Heinz bodies observed or methemoglobin detected, but the blood was reported to be darker than normal in the fatal cases.

The near-fatal case, a 23-year-old male printer, had been exposed for about 30 min to a floor spill of 2-NP. Several hours later vomiting and extreme colic-like abdominal pain developed, but no diarrhea or loss of consciousness. The patient was treated with an infusion of 5 percent glucose, Novalgin®, and Torecan®. The vomiting regressed rapidly following this treatment. Subsequent laboratory tests revealed no significant organ damage. The patient was discharged from the hospital 4 days after admission.

The disturbing aspect of these cases is that they occurred almost 40 years after the toxicity of 2-NP had been reported in considerable detail in industrial hygiene journals. The difficulty seems to lie in the lack of recognition of the presence of 2-NP in solvent mixtures, and failure to provide adequate warning labels on such products.

1.6.3 Metabolism

The nitroparaffins are absorbed through the lungs and from the gastrointestinal tract. Applications to the skin give no evidence of sufficient absorption to result in systemic injury. Distribution studies of nitroethane in animals showed rapid disappearance from the body; within 3 hr only 14 percent of the dose was recovered in rats, and by 30 hr, essentially all the dose had been cleared from the tissues, the blood, lungs, liver, and muscle (20a). Partial excretion of nitroethane was via the lungs.

By either inhalation or oral administration nitroethane was shown to be metabolized to aldehyde and nitrite, with the end product the eventual oxidation to nitrate (20b), the overall reactions being

$$CH_3CH_2NO_2 \rightarrow CH_3\overset{O}{CHO} + HNO_2 + H_2O \rightarrow NO_3^-$$

Nitrite, which can be found in blood and urine, follows the administration of nitroethane, nitropropanes, and nitrobutanes, but not after the nitromethane or 2-nitro-2-methylpropane, a tertiary nitroparaffin.

The probable way in which this oxidative denitrification of the nitroparaffins occurs to yield nitrite appears to be through the microsomal cytochrome P-450-dependent monooxygenase system (21). This reaction was shown to occur with both 1- and 2-nitropropanes, using the liver microsomal fraction from male Sprague-Dawley rats, the same species and sex used to demonstrate hepatic carcinoma from 2-NP (see Section 1.6.4). Oddly, nitromethane, which was not carcinogenic in either rats or rabbits, did interact with reduced rabbit liver microsomes (22). Although the reason for the different result is not clear, it is surmised (22) that as the affinity of nitromethane for the heme protein decreases with the increase in number of binding sites, this allows for relative ineffectiveness of the binding, and hence oxidative cleavage of nitromethane.

The chloronitroparaffins appear not to show appreciable percutaneous absorption, as judged by lack of apparent systemic effects (8), but 1,1-dichloro-1-nitroethane induced swelling and irritation after only two applications. The monochlor derivative of 1-nitropropane, however, produced only slight erythema after 10 applications.

Absorption of nitrooleffins from the respiratory or gastrointestinal tract, peritoneal cavity, or skin is very rapid, giving signs of prompt systemic toxicity (12a).

1.6.4   Mode of Action—Animal Carcinogenicity

2-Nitropropane (2-NP) has been found to be a hepatic carcinogen for male, but not female rats of the Sprague-Dawley derived strain following a 6-month daily 7-hr exposure at 207 ppm. A concurrently exposed group of male New Zealand strain of white rabbits showed no such response. Similarly, a parallel exposure of these two species to nitromethane (NM) did not result in any neoplastic changes, although exposure concentrations as high as 745 ± 34 ppm were experienced (10).

Multiple hepatic carcinomas and numerous neoplastic nodules were present in the livers of all 10 rats exposed at 207 ppm of 2-NP for 6 months, but not at approximately one-tenth this level. The hepatocellular carcinomas exhibited a variety of histologic characteristics, and in many cases the normal hepatic parenchyma was destroyed.

These neoplasms were composed of anaplastic hepatocytes, sometimes forming broad sheets and at other times forming trabeculae several cells thick. Blood-filled cysts were occasionally seen in the neoplasm, and mitotic figures were frequently present. The hepatocellular carcinomas appeared to be rapidly growing and severely compressing the surrounding parenchyma. In some instances they had broken through the capsule, and a slight amount of perihepatitis was evident in the capsule. No metastatic hepatocellular carcinomas were seen in any of the other tissues examined.

The lack of finding of NM to be an animal carcinogen in the present study confirms the earlier work of E. Weisburger (23) and, for what it's worth, NM was nonmutagenic for *S. typhimurium* TA98 or TA100 in the Ames test (24).

1.7   Hygienic Standards of Permissible Exposure

Adopted TLVs and tentative STELs have been recommended for five nitroparaffins as listed in Table 53.3. Some official OSHA standards differ because they were adopted from the TLV list of 1968 without subsequent change. Later

Table 53.3. Threshold Limit Values of Nitroparaffins

Compound	TLV[a]		STEL[b] Tentative Values		OSHA Standard	
	ppm	mg/m³	ppm	mg/m³	ppm	mg/m³
Nitromethane	100	250	150	375	100	250
Nitroethane	100	310	150	465	100	310
1-Nitropropane[c]	15	55	25	90	25	90
2-Nitropropane	C25A2[d]	90A2	—	—	25	90
Tetranitromethane	1	8	—	—	1	8

[a] From 1980 list.
[b] Short-term exposure limit as defined in preface, TLV booklet.
[c] In *Notice of Intended Changes.*
[d] Industrial substances suspect of carcinogenic potential for man.

information has prompted changes in the TLVs of 1- and 2-nitropane. The basis for these changes is summarized in *Documentation of TLVs 1980* and in Section 1.6.4.

No TLVs have been established for any nitrooleffin or nitroalcohol because of little or no present industrial interest. Among the chloronitroparaffins, chloropicrin, trichloronitromethane, has a TLV of 0.1 ppm (0.7 mg/m³) and a STEL of 0.3 ppm and 2 mg/m³. For the basis of these limits, see *Documentation of TLVs.*

1.8 Specific Compounds

1.8.1 Nitromethane, CH_3NO_2

The physical and chemical properties of nitromethane (NM) are given in Table 53.1, its analytic determination in Section 1.5, and its toxicity relative to that of other nitroparaffins in Section 1-6.

Acute Toxicity. Animal toxicity has been studied by Machle and co-workers and by Weatherby (14). This information is summarized in Table 53.4. Animals exposed at 30,000 ppm in air for longer than 1 hr developed pronounced nervous system symptoms. At 10,000 ppm nervous system symptoms did not appear until after 5 hr. During exposure at lower concentrations, there was slight irritation of the respiratory tract without evidence of eye irritation. This was followed by mild narcosis, weakness, and salivation. Rabbits, guinea pigs, and monkeys all survived repeated exposures for a total of 140 hr at concentrations of 500 ppm. A single monkey exposed at 1000 ppm for eight 6-hr exposures died. No remarkable changes in either blood pressure or respiration

Table 53.4. Acute Toxicity of Nitromethane (11a, 14)

Route	Animal	Dose		Mortality
Oral	Dog	0.125^a		0/2
		$0.25-1.5^a$		12/12
	Rabbit	$0.75-1.0^a$		Lethal dose
	Mouse	1.2^a		1/5
		1.5^a		6/10
Subcutaneous	Dog	$0.5-1.0^b$		MLD
Intravenous	Rabbit	0.8^a		2/6
		1.0^a		2/6
		$1.25-2.0^a$		9/9
Inhalation	Rabbit	30,000 ppm,	<2 hr.	0/6
			2 hr.	2/2
		10,000 ppm,	6 hr.	3/2
			1-3 hr.	0/4
		5,000 ppm,	6 hr.	1/2
			3 hr.	0/2
		500 ppm,	140 hr.	0/2
	Guinea pig	30,000 ppm,	1-2 hr.	4/4
			30 min.	1/2
			15 min.	0/2
		10,000 ppm,	3-6 hr.	4/4
			1 hr.	0/2
		1,000 ppm,	30 hr.	2/3
		500 ppm,	140 hr.	0/3
	Monkey	1,000 ppm,	48 hr.	1/1
		500 ppm,	140 hr.	0/1

^{a} Dose, g/kg.
^{b} Dose, ml/kg.

followed intravenous injection of NM in anesthetized dogs. The histopathologic changes observed following acute poisoning by all routes were chiefly confined to the liver and kidneys with the liver showing the most prominent injury. Subcapsular damage, focal necrosis, both periportal and midzonal fatty infiltration, congestion, and edema were observed. Three of 10 rats given 0.25 percent NM in their drinking water for 15 weeks and 4 of 10 rats given 0.1 percent died during the course of the experiment. The surviving animals failed to gain weight normally. Histopathologic examination showed mild but definite liver abnormalities. Methemoglobinemia has not been observed in rabbits or rats. Nitromethane is apparently metabolized by a different mechanism from nitroethane and nitropropane in that negligible amounts of nitrites are found in the blood following intravenous injection of 1 mM in rabbits. Skin application of NM does not produce irritation or death in animals.

Subchronic Toxicity.   Six-months inhalation studies of NM in rats or rabbits showed widely differing responses in the two exposed species (10).

During six months of exposure at 98 or 745 ppm of NM, only mild to moderate symptoms of toxicity were observed in rats and rabbits. A reduction in body weight gain was observed in rats exposed at 745 ppm of NM. Hematocrit and hemoglobin levels in rats were slightly depressed from 10 days through 6 months of exposure to 745 NM. Rabbits also provided a suggestion of depression in hemoglobin levels. Other hematologic parameters such as prothrombin time and methoglobin concentration were unaffected in both rats and rabbits. Ornithine carbamyl transferase in rabbits was elevated after 1 and 3 months, but not 6 months of exposure at 745 ppm of NM. No apparent effects on glutamic–pyruvic transaminase in both species or serum T_4 activity in rats were observed. Serum T_4, however, was statistically significantly depressed in rabbits exposed at either 98 or 745 ppm NM at the 6-month testing period as well as at the 1-month sacrifice for rabbits exposed at 745 ppm. Weights of all organs evaluated were comparable to controls in rats and rabbits except for thyroid weights in rabbits after 6 months of exposure at 745 ppm of NM. The increased thyroid weights after 6 months and decreased thyroxin levels at all testing intervals, for both concentrations, indicated an effect on the thyroid in rabbits by NM.

Histopathologic evaluation indicated no exposure-related abnormalities in rats due to exposure to NM at 98 or 745 ppm for up to 6 months. Some evidence of pulmonary edema and other pulmonary abnormalities was observed in rabbits exposed to both levels of NM for 1 month.

The most important observations are that inhalation of NM produces mild irritation and toxicity before narcosis occurs and that liver damage can result from repeated administration at levels in excess of 1000 ppm.

Odor and Warning Properties.   The odors of nitroparaffins are easily detectable, and concentrations below 200 ppm are disagreeable to most observers (11a). The odor and sensory symptoms are not dependable warning properties. The threshold limit value of 100 ppm can be interpreted from the acute and repeated vapor exposures of animals. No injuries in man have been reported, and industrial experience has been good (3).

1.8.2   Tetranitromethane, $C(NO_2)_4$

Tetranitromethane (TNM) is a colorless oily fluid with a distinct, pungent odor. Its physical and chemical properties are given in Table 53.1. It is explosive and is more easily detonated than TNT. Its explosive power is less than TNT except when mixed with hydrocarbons, the mixtures being more powerful explosives and very sensitive to shock. Accidental explosions have occurred in handling and manufacture. It occurs as a contaminant of crude TNT. It can be prepared by the nitration of acetylene with nitric acid to form trinitromethane, the mixture of trinitromethane and nitric acid being converted to tetranitromethane by sulfuric acid at elevated temperatures. It is of interest for use as a propellant and as a fuel additive.

Animal Toxicity–Acute.   A summary of the data on the response of animals to inhalation of various concentrations of tetranitromethane appears in Tables 53.5 and 53.6. In all reported experiments, exposed animals have exhibited similar symptoms, chiefly those of respiratory tract irritation. The first signs are increased preening, change in the respiratory pattern, and evidences of eye irritation followed by rhinorrhea, gasping, and salivation. The symptoms progress to cyanosis, excitement, and death at higher concentrations. Methemoglobinemia occurred in exposed cats. It should be noted that Sievers et al. (16) exposed their animals to tetranitromethane from crude trinitrotoluene, and although the concentrations recorded for TNM were determined by sampling and analysis, other unknown contaminants could have been present. These investigators found that animals exposed at 3 to 9 ppm for 1 to 3 days developed pulmonary edema. Lower concentrations (0.1 to 0.4 ppm) produced only mild irritation.

The results of pathologic examinations on animals dying from acute exposures were all similar. There was marked lung irritation with destruction of epithelial cells, vascular congestion, pulmonary edema, and emphysema with tracheitis

Table 53.5.   Response to Various Concentrations of Tetranitromethane (TNM)

Animal	Concentration (ppm)	Duration of Exposure	Response
1 cat	100	20 min	Death in 1 hr[a]
1 cat	10	20 min	Death in 10 days[a]
5 cats	7–25	2½–5 hr	Death in 1–5½ hr[b]
2 cats	3–9	6 hr × 3	Severe irritation[b]
2 cats	0.1–0.4	6 hr × 2	Mild irritation[b]
20 rats	1230	1 hr	All died in 25–50 min[c]
20 rats	300	1½ hr	All died in 40–90 min[c]
20 rats	33	10 hr	All died in 3–10 hr[c]
19 rats	6.35	6 months	11 deaths[c]
2 dogs	6.35	6 months	Mild symptoms[c]

[a] F. Flury and F. Zernik, *Schädliche Gase*, Springer, Berlin, 1931.
[b] Reference 16.
[c] Reference 7a.

Table 53.6. Acute Toxicity of TNM to Rats and Mice

Test[a]	Rats	Mice
Oral LD_{50} (95% C.L.)	130 (83–205) mg/kg	375 (262–511) mg/kg
Intravenous LD_{50} (95% C.L.)	12.6 (10.0–15.9) mg/kg	63.1 (45.0–88.7) mg/kg
4-Hr inhalation (95% C.L.)	17.5 (16.4–18.7) ppm	54.4 (48.0–61.7) ppm

[a] C.L. = confidence limits.

and bronchopneumonia. Nonspecific changes in the liver and kidney were observed in some animals.

Animal Toxicity—Subchronic. Horn (7a) exposed two dogs and 19 rats to 6.35 ppm for 6 hr/day, 5 days/week for 6 months. Eleven of the 19 rats died in the course of the exposure with evidence of pulmonary irritation, edema, and pneumonia. Some initial anorexia was observed in the dogs. Repeated examinations did not reveal anemia, Heinz bodies, methemoglobinemia, or biochemical disturbances. Rats surviving 6.35 ppm for 6 months developed pneumonitis and bronchitis of a moderate degree, whereas those dying developed more severe pneumonia. Histopathologic examination of two dogs surviving the same concentration for 6 months revealed no evidence of injury (7a).

Human Toxicity. The lowest lethal dose, LD_{LO}, for man by inhalation is given as 500 mg/kg (approx. 35 g) (25). A few deaths and intoxications that have occurred during handling of heated contaminated TNT have been attributed to TNM (11a). Symptoms experienced in the laboratory production of TNM were irritation of eyes, nose, and throat from acute exposures and, after more prolonged inhalation, headache and respiratory distress (26). Skin irritation does not result from repeated contact in man or animals.

Odor and Warning Properties. TNM can be recognized by its characteristic acrid biting odor. It can be determined in air by the methods used by Horn (7a) or Vouk and Weber (27). Sievers et al. (16) felt that the safe working level should be below 5 ppm. The American Conference of Governmental Industrial Hygienists suggests a threshold limit value of 1 ppm. Human experience is lacking, but the animal experiments suggest that this level may not be low enough to prevent all irritation.

1.8.3 Nitroethane, $CH_3CH_2NO_2$

Nitroethane (NE) is an oily, colorless liquid with a somewhat pleasant odor. It represents a lesser explosive hazard than nitromethane and tetranitromethane.

Unconfined quantities are not exploded by heat or shock. Since under appropriate conditions of confinement or contamination with other materials explosions could result, safe handling procedures have been recommended in detail (5a).

Animal Toxicity. The response of rabbits and guinea pigs to inhalation of NE as determined by Machle and co-workers (11a) appears in Table 53.7. It is a moderate respiratory tract irritant. There was more respiratory tract irritation and less narcosis with NE than was observed with NM. Except for this, the symptomatology and pathologic findings were similar to NM. They found no evidence of skin irritation or skin absorption. Scott found increasing nitrite concentrations in the blood of rabbits during inhalation of NE (20b). Nitroethane is excreted via the lungs, is rapidly metabolized, and is completely eliminated in 30 hr (20a).

Hygienic Standard. OSHA adopted the 1968 TLV of 100 ppm (approx. 310 mg/m^3) in 1978. Since that time a STEL of 150 ppm (approx. 465 mg/m^3) has been added by the TLV Committee. The TLV is considered a level that provides a reasonable degree of safety and comparative freedom from irritation.

1.8.4   2-Nitropropane, 2-CH$_3$CHNO$_2$CH$_3$

2-Nitropropane (2-NP) is a colorless, oily liquid. Its physical and chemical properties are given in Table 53.1. The nitropropanes are less of an explosive hazard than the nitromethanes.

Table 53.7.   Response to Inhalation of Nitroethane (11a)[a]

Concentration (ppm)	Time (hr)	Mortality	
		Rabbit	Guinea Pig
30,000	1.25	2	2
	1	1	1
	0.5	1	None
10,000	3	2	1
	1	1	None
5,000	3	2	None
	2	1	None
2,500	3	None	None
1,000	2	1	None
	6	None	None
500	30	None	
	140	None	None[b]

[a] (Two rabbits and two guinea pigs in each experiment).
[b] One monkey exposed, not fatal.

Table 53.8.   Response to Inhalation of 2-Nitropropane (6b)

Animal	Highest Tolerable Concentration (ppm)			Lowest Lethal Concentration (ppm)		
	1 hr	2.25 hr	4.5 hr	1 hr	2.25 hr	4.5 hr
Rat	2353	1372	714[a]	3865	2633	1513
Guinea pig	9523	4313	2381		9607	4622[b]
Rabbit	3865	2633	1401	9523	4313	2381
Cat	787	734	328	2353	1148	714

[a] Time, 7 hr.
[b] Time, 5.5 hr.

Animal Toxicity—Acute.   As seen in Table 53.8, considerable differences in species response were observed. Cats were the most susceptible and guinea pigs the least. High concentrations of 2-NP produced dyspnea, cyanosis, prostration, some convulsions, lethargy, and weakness, proceeding to coma and death. Some animals surviving the acute exposure died 1 to 4 days later. These high concentrations of 2-NP caused pulmonary edema and hemorrhage, selective disintegration of brain cells, and hepatocellular damage, with general vascular endothelial injury in all tissues.

Animal Toxicity—Subchronic.   Cats, rabbits, rats, guinea pigs, and monkeys were exposed repeatedly at 328 and 83 ppm for 7 hr/day. Cats died following several days exposure at 328 ppm but rabbits, rats, and guinea pigs survived 130 exposures and a monkey survived 100. No signs or symptoms were observed in any animals during 130 exposures at 83 ppm (6b).

Cats that died following several exposures at 328 ppm had microscopic evidence of focal necrosis and parenchymal degeneration in the liver and slight to moderate degeneration of the heart and kidneys. The lungs showed pulmonary edema, intraalveolar hemorrhage, and interstitial pneumonitis. The other species exposed at this concentration did not exhibit these findings. Except for one cat, no microscopic tissue changes were observed in the animals exposed at 83 ppm. Inhalation of 2-NP induced methemoglobin formation in cats and to a lesser extent in rabbits. Cats developed 25 to 35 percent methemoglobin when exposed at 750 ppm for 4½ hr and about 15 to 25 percent methemoglobin during repeated, daily, 7-hr exposures at 280 ppm. Heinz bodies appeared in the erythrocytes of cats and rabbits at even lower concentrations.

A later exposure of rats and rabbits to 6 months inhalation of 2-NP at 207 and 27 ppm showed very few classical signs of nitroparaffin toxicity. Ornithine carbamyl transferase was elevated in rabbits after 1 and 3 months of exposure to 207 ppm 2-NP, but not in rats. No effects on body weight gain or hematology were observed. Liver weights from rats exposed at 27 ppm of 2-NP for up to 6 months were comparable to those of controls. However, severe neoplastic changes were observed in the livers of male rats exposed to 207 ppm 2-NP for 6 months (see Section 1.6.4.) (10).

These later studies suggest that hematologic changes do not occur until concentrations above 207 ppm are reached. The focal necrosis of the liver, seen in the earlier study, could have presaged the hepatic carcinoma found in rats after 6 months of exposure, had the 3-week study been of longer duration.

Human Experience.   The first report of worker response from exposure to 2-NP solvent mixtures was made by Skinner (15), then director of the Division of Occupational Hygiene for the Massachusetts Department of Labor.

Workers dipping forms into a solvent mixture of xylene and 2-NP at 110 to 120°F developed anorexia, nausea with vomiting, and occasionally diarrhea. They felt well the following morning, but the symptoms recurred during exposure. Concentrations of 20 to 45 ppm of 2-NP were found in the work area (xylene concentrations, not given). Substitution of methyl ethyl ketone for 2-NP in the solvent mixture relieved the symptoms. A spraying operation utilizing a lacquer with 20 percent 2-NP did not produce symptoms in personnel exposed not more than 4 hr/day for 3 days/week (15). Concentrations found ranged from 10 to 30 ppm. On this basis Skinner suggested 25 ppm as the maximum allowable concentration. As worker exposure without adverse effects did not exceed 4 hr/day more than 3 days/week, a TC_{LO} (lowest toxic concentration) was inferred to be 15 ppm (25).

Hygienic Standards.   The TLV Committee recommended in 1959, and the American Conference of Governmental Industrial Hygienists adopted, 50 ppm (approx. 180 mg/m^3) for 2-NP on the basis of animal studies by Treon and Dutra (6b). However, worker responses in industry and continuing industry hygiene surveys (28a, b) indicated lowering of the TLV to 25 ppm (see *Documentation, TLVs,* 1971 ed.). This is the adopted OSHA standard.

The NIOSH-supported study, which found 2-NP to be an hepatic carcinogen for rats (10), indicated placing 2-NP in the A category for "Industrial Substances Suspect of Carcinogenic Potential for Man," with a *ceiling* value of 25 ppm (1978–1979 TLV lists).

Odor and Warning Properties.   The limits of odor detectability only bracket points of detection and nondetection; odor is reported (6b) to be detectable at 1.05 mg/l (294 ppm), but not at 0.297 mg/l (83 ppm). Because the level of detection lies somewhere between 83 and 294 ppm, odor detection cannot serve as a warning device for overexposure.

1.8.5   1-Nitropropane, $CH_3CH_2CH_2NO_2$

Animal Toxicity.   The acute inhalation toxicity of 1-nitropropane (1-NP) is not greatly different from that of 2-NP (see Tables 53.2 and 53.9); exposures at 5000 ppm of 1-NP propane for 3 hr killed rabbits and guinea pigs, whereas the lowest lethal concentrations of 2-NP for these animals after a 2.25-hr exposure were 4313 and 9607 ppm, respectively. The symptoms and gross pathologic changes observed in the exposed animals were similar to those exposed to nitroethane.

Human Response.   In a limited organoleptic test of 1-NP, human volunteer subjects found concentrations exceeding 100 ppm irritating after brief periods of exposure (29).

Industrial experience with 1-NP has been virtually without incident (5b, 3); a few field reports of warning symptoms such as headache and nausea resulting from prolonged exposure to excessive amounts have been made, but these appear well in advance of systemic injury, which has not occurred.

Hygienic Standards.   The first TLV for 1-NP was established at 25 ppm (approx. 90 mg/m³) in 1963, 5 years after that of 2-NP, the delay being caused by a relative lack of comparative toxicity data and industrial experience. What data existed at that time indicated that 1-NP was somewhat more toxic than other nitroparaffins (30) and 2-NP (11a).

In 1979, 1-NP was placed on the tentative list at 15 ppm as a time-weighted average, in comparison with a ceiling value of 25 ppm for 2-NP. The reduction to 15 ppm was based on the knowledge that its toxicity was greater than that of 2-NP, although it is not considered to be an animal carcinogen as is 2-NP (10).

1.8.6   Nitrobutanes

The formulas and their physical constants are given in Table 53.1.

The toxicology of 1-nitrobutane and 2-nitrobutane has not been studied

Table 53.9.   Response to Inhalation of 1-Nitropropane

Concentration (ppm)	Time (hr)	Mortality[a] Rabbit	Mortality[a] Guinea pig
10,000	3	2	2
10,000	1	None	1
5,000	3	2	2

[a] Two rabbits and two guinea pigs in each experiment.

beyond that reported by Machle et al. (11) (Table 53.2). The effects following oral administration in rabbits were similar to those produced by the other nitroparaffins and the lethal dose range was the same as for 2-NP and NE. As with these materials, no skin irritation or systemic symptoms were observed after five daily open applications to rabbit skin. Less nitrite can be recovered from rabbit blood following intravenous injection of the nitrobutanes than after an injection of equivalent doses of the nitropropanes or nitroethanes. It would be expected that the nitrobutanes would present hazards qualitatively similar to those of 2-NP.

1.8.7 Chlorinated Mononitroparaffins

The formulas of five of these substances, their physical constants, and the oral, lethal doses of four for rabbits appear in Table 53.10.

The chlorinated nitroparaffins are of particular interest in the manufacture of highly accelerated rubber cements and insecticides and in chemical synthesis. The 1,1 derivatives are prepared by the chlorination of the sodium salt of 2-nitropropane; the 1,2 derivatives are prepared by chlorination under anhydrous conditions and strong light (3).

Animal Toxicity. Comparison of the acute, oral, lethal doses for rabbits shows that, with the exception of 2-chloro-2-nitropropane, the chlorinated mononitroparaffins were five times more toxic than the unchlorinated (Table 53.10 versus Table 53.1). The same toxicity difference holds for the respiratory route (Tables 53.11, 53.12). 1,1-Dichloronitroethane is considerably more irritating to skin and mucous membranes than 1-chloro-1-nitropropane. It exhibits greater toxicity by inhalation (see Tables 53.11 and 53.12). Both materials are lung irritants. They cause pulmonary edema and death within 24 hr following exposure at high concentrations. The chief site of injury is the lungs, but damage is also observed in the heart muscle, liver, and kidneys after lethal exposures. Although 1-chloro-1-nitroethane and 2-chloro-2-nitropropane have not been studied in detail, it would be expected that their inhalation toxicity would be qualitatively and quantitatively similar to that of 1-chloro-1-nitropropane.

Very similar findings were made by Soviet scientists for the acute toxicities of 1-chloro-1-nitroethane and propane, and 2-chloro-2-nitropropane and butane (8b).

Hygienic Standards. In 1959 the TLV Committee proposed a limit for 1-chloro-1-nitropropane of 20 ppm (approx. 100 mg/m^3) and 10 ppm (approx. 60 mg/m^3) for 1,1-dichloronitroethane, on the basis of acute toxicity determinations in animals (8), already detailed in Section 1.8.7 and in Tables 53.10 and 53.11.

Table 53.10. Physical and Chemical Properties of the Chlorinated Mononitroparaffins

Name	Formula	Mol. Wt.	B. p. (°C)	Sp. Gr.	Solubility in H₂O (ml/100 ml) (20°C)	Vapor Pressure (mm Hg) (25°C)	Vapor Density (Air = 1)	Flash Point (°F)	Conversion Units		Oral Lethal Dose, Rabbits[a] (g/kg)
									1 mg/l (ppm)	1 ppm (mg/m³)	
Trichloronitromethane (chloropicrin)	CCl_3NO_2	164.38	111.84	1.656 (20/4°C)	Insoluble	16.9 (20°C)	5.7		148.8	6.72	
1-Chloro-1-nitroethane	$CH_3CHClNO_2$	109.51	127.5	1.2860 (20/20°C)	0.4	11.9	3.6	133	237	4.21	0.10–0.15
1,1-Dichloro-1-nitroethane	$CH_3CCl_2NO_2$	143.9	124	1.4271 (20/20°C)	0.25	16	5.0	168	169.9	5.89	0.15–0.20
1-Chloro-1-nitropropane	$CH_3CH_2CHClNO_2$	123.5	139.5–143.3	1.209 (20/20°C)	0.5	5.8	4.3	144	198	5.05	0.05–0.10
2-Chloro-2-nitropropane	$CH_3CHClNO_2CH_3$	123.5	133.6	1.197 (20/20°C)	0.5	8.5	4.3	135	198	5.05	0.5–0.75

[a] From Reference 8.

Table 53.11.   Response to Inhalation of 1,1-Dichloronitroethane (8a)

Average Concentration (ppm)	Duration of Exposure	Mortality[a]	
		Rabbit	Guinea Pig
4910	30 min	2	2
985	$3\frac{1}{2}$ hr	2	1
594	$2\frac{1}{2}$ hr	1	None
254	1 hr	None	None
169	2 hr	1	1
100	6 hr	2	2
60	2 hr	None	None
52	18 hr, 40 min	2	None
34	4 hr	None	None
25	204 hr	None	None

[a] Two rabbits and two guinea pigs in each experiment.

A later review of the data resulted in the reduction of the TLVs for both compounds to the same limit of 2 ppm (1980 TLV list).

1.8.8   Trichloronitromethane, CCl_3NO_2 (Chloropicrin)

The physical constants of chloropicrin are given in Table 53.10. Its uses have included dyestuffs (crystal violet), organic syntheses, fumigants, fungicides, insecticides, rat exterminator, and poison war gas.

Toxicity.   Data on exposures of man and animals to various concentrations of chloropicrin, largely obtained during World War I, have been summarized in

Table 53.12.   Response to Inhalation of 1-Chloro-1-nitropropane (9)

Average Concentration (ppm)	Duration of Exposure	Mortality[a]	
		Rabbit	Guinea Pig
4950	60 min	2	1
2574	2 hr	2	None
2178	1 hr	None	1
1069	1 hr	None	None
693	2 hr	None	None
393	6 hr	1	None

[a] Two rabbits and two guinea pigs in each experiment.

Table 53.13. Response to Various Concentrations of Trichloronitromethane—Animals

Animal	Concentration		Duration of Exposure (min)	Response
	mg/l	ppm		
Dog	1.05	155	12	Became ill
	0.08–0.95	117–140	30	Death of 43% of the animals; survival of remainder
Mouse	0.85	125	15	Death in 3 hr to 1 day
Cat	0.51	76	25	Death usually in 1 day
Mouse	0.34	50	15	Death after 10 days
Dog	0.32	48	15	Tolerated
Cat	0.32	48	20	Death after 8 to 12 days
	0.26	38	21	Survived 7 days
Mouse	0.17	25	15	Tolerated

Tables 53.13 and 53.14 (31–33). Chloropicrin is both a lacrimator and a lung irritant. It is intermediate in toxicity between chlorine and phosgene. Nominal lethal concentrations (mg/l) after 30 min exposure are chloride, 3.0; chloropicrin, 0.8; and phosgene, 0.36. Flury and Zernik state that exposure to 4 ppm for a few seconds renders a man unfit for combat, and 15 ppm for approximately the same period of time results in respiratory tract injury (31). Whereas chlorine in fatal concentrations produces more injury of the upper respiratory tract, trachea, and larger bronchi than in the alveoli, and phosgene acts primarily on the alveoli, chloropicrin produces more injury to medium and small bronchi than to the trachea and large bronchi. The alveolar injury is less than with phosgene but pulmonary edema occurs and is the most frequent cause of early deaths. Exposure to chloropicrin produces more coughing than does phosgene

Table 53.14. Response to Various Concentrations of Trichloronitromethane—Man

Concentration		Duration of Exposure (min)	Response
mg/l	ppm		
2.0	297.6	10	Lethal concentration
0.8	119.0	30	Lethal concentration
0.1	15.0	1	Intolerable
0.050	7.5	10	Intolerable
0.009	1.3		Lowest irritant concentration
0.0073	1.1		Odor detectable
0.002–0.025	0.3–3.7	3–30 sec	Closing of eyelids according to individual sensitivity

and there is less delay in onset of pulmonary edema. Late deaths may occur from secondary infections, bronchopneumonia, or bronchiolitis obliterans. During World War I chloropicrin was noted for its tendency to cause nausea and vomiting. It has been stated that individuals injured by inhalation of chloropicrin become more susceptible, so that concentrations of the gas not producing symptoms in others cause them distress. Chloropicrin is also a potent skin irritant.

Tests for carcinogenicity sponsored by the National Cancer Institute have proved indefinite (1980).

Hygienic Standards.   The original TLV for chloropicrin of 1.0 ppm (approx. 7.0 mg/m³) recommended in 1956 and adopted by the American Conference of Governmental Industrial Hygienists in that year, was based on the information compiled by Vedder (17) and Fries and West (33), and quoted by Flury and Zernik (31), who reported that concentrations of 0.3 to 0.37 ppm resulted in eye irritation in 3 to 30 sec, depending on individual susceptibility. A concentration of 15 ppm could not be tolerated longer than 1 min even by individuals accustomed to chloropicrin.

In 1959 the TLV was reduced to 0.1 ppm in order "to provide greater protection from eye irritation in all workers and to insure against potential pulmonary changes" (*TLV Documentation,* 1st ed., 1962.). It is below the levels detectable by odor or irritation.

OSHA adopted the reduced TLV of 1968, making it an official U.S. standard.

1.8.9   Nitroolefins

Toxicologic interest in the nitro derivatives of straight-chain olefins has stemmed from human health concerns (eye irritation and carcinogenicity) from their presence in urban air pollution derived from automobile exhausts. Deichmann et al. (12a) have shown that nitroolefins are indeed emitted in the exhausts from gasoline engines. For this reason Deichmann et al. studied the toxicologic and pharmacologic actions of a series of 21 straight-chain olefins in experimental animals and human volunteers (12a). Unfortunately, no physical or chemical constants of any member of the series appear to have been published (34). Unfortunately also, for the great amount of effort expended, it was learned too late that, because of their high reactivity, the nitroolefins "decompose readily in the presence of sunlight" (12a), so that only on sunless days would significant amounts be present temporarily in smog to contribute to effects on urban populations. As industrial chemical entities also, no nitroolefin appears at present to have any commercial interest (1). Accordingly, the vast amount of data accumulated by Deichmann and associates over a 10-year period, 1955 to 1965, is summarized only; detailed tabular data can be found in Reference 12a and in previous references. Table 53.15 gives the comparative toxicities of 12 nitroolefins by four routes of exposure.

Table 53.15. Acute Effects of Nitroolefins (12a)

| Name | Formula | Vapor Exposure (5 hr) | | Oral Toxicity, Rats, Single Dose, Undiluted (Approx. Lethal Dose) | | Intraperitoneal Toxicity, Rats, Single Dose, Undiluted (Approx. Lethal Dose) | | Percutaneous Toxicity, Rabbits, open, 5-hr Single Dose, (Approx. Lethal Dose) | |
		Concentration (ppm)	Survival Time, Rats, 47% Humidity	g/kg	mM/kg	g/kg	mM/kg	g/kg	mM/kg
2-Nitro-2-butene	$CH_3C(NO_2):CHCH_3$	1400	100 min	0.28	2.8	0.08	0.8	0.62	6.1
2-Nitro-2-pentene	$CH_3C(NO_2):CHCH_2CH_3$	240	240 min	0.28	2.4	0.08	0.7	0.94	5.4
		55	Survived						
3-Nitro-3-pentene	$CH_3CH_2C(NO_3):CHCH_3$	268	280 min	0.42	3.7	0.05	0.4	0.62	8.2
2-Nitro-2-hexene	$CH_3C(NO_2):CH(CH_2)_2CH_3$	515	50–85 min	0.42	3.3	0.12	0.9	1.40	7.3
		152	Survived						
3-Nitro-3-hexene	$CH_3CH_2C(NO_2):CHCH_2CH_3$	557	30–70 min	0.42	3.3	0.08	0.6	0.94	10.9
		50	Survived						
2-Nitro-2-heptene	$CH_3C(NO_2):CH(CH_2)_3CH_3$	308	3–18 hr	0.94	6.6	0.28	2.0	0.94	6.6
		135	Survived						
3-Nitro-3-hepten	$CH_3CH_2C(NO_2):CH(CH_2)_2CH_3$	54	24 hr	0.62	4.3	0.28	2.0	1.40	9.8
2-Nitro-2-octene	$CH_3C(NO_2):CH(CH_2)_4CH_3$	47	Survived	1.4	9.0	0.28	1.8	0.62	4.0
3-Nitro-3-octene	$CH_3CH_2C(NO_2):CH(CH_2)_3CH_3$	142	18–24 hr	0.62	4.0	0.18	1.2	0.94	6.0
		72	Survived						
3-Nitro-2-octene	$CH_3CH:C(NO_2)(CH_2)_4CH_3$	141	18–72 hr	0.62	4.0	0.18	1.2	0.62	4.0
		44	Survived						
2-Nitro-2-nonene	$CH_3C(NO_3):CH(CH_2)_5CH_3$	64	Survived	2.1	12.3	0.28	1.6	0.62	3.6
3-Nitro-3-nonene	$CH_3CH_2C(NO_2):CH(CH_2)_4CH_3$	59	24 hr	2.1	12.3	0.42	2.5	0.42	2.5
		10	Survived						

Human Eye Irritation. A major quest of the studies of toxicologic and pharmacologic actions of the series of the conjugated, linear nitroolefins was the determination of their irritation to the human eye, in an effort to identify the causative agent(s) in Los Angeles smog. This phase of the work was confined to three nitroolefins, 2-nitro-2-butene, 2-nitro-2-hexene, and 2-nitro-2-nonene, and was conducted jointly by members of the Los Angeles County Air Pollution Control District and the staff at the University of Miami. Both groups found the eye irritation produced by 2-nitro-2-butene and 2-nitro-2-hexene to be of the same order of magnitude. At the University of Miami, the threshold for eye irritation was found to be 0.2 to 0.4 ppm for the nitrohexene. The California group established the following levels: 0.1 ppm nitrohexene and 0.1 ppm nitrobutene. Both groups found the threshold for 2-nitro-2-nonene to be considerably higher.

After repeated exposures, subjects became increasingly sensitive. Temperature, within the limits of 18 to 32°C (65 to 90°F), had no significant influence upon the sensitivity of the eye to a nitroolefin.

Acute Toxicity—Animals. Each of the compounds was investigated by inhalation, oral, intraperitoneal, and cutaneous routes, using rabbits, guinea pigs, rats, mice, chicks, and dogs. The subacute inhalation toxicity of each of four nitroolefins representative of the series was studied, using rabbits, guinea pigs, rats, and mice.

All nitroolefin compounds were found to be highly toxic as well as irritant. Absorption from the respiratory or gastroenteric tract, peritoneal cavity, or skin, was very rapid. Signs of systemic intoxication appeared promptly, including hyperexcitability, tremors, clonic convulsions, tachycardia, and increased rate and amplitude of respiration, followed by a generalized depression, ataxia, cyanosis, and dyspnea. Death was initiated by respiratory failure and associated with asphyxial convulsions. Pathologic changes were most marked in the lungs, regardless of the mode of administration.

Inhalation toxicity showed no definite relationship to chain length, but the acute oral and intraperitoneal toxicities decreased with increasing length of the carbon chain.

Chronic Toxicity and Carcinogenicity. The most significant finding in an 18-month chronic inhalation study using dogs, goats, rats, and mice, was five instances of primary adenocarcinoma of the lung in a group of 27 Swiss mice exposed at 0.2 ppm of vapors of 3-nitro-3-hexene. No such changes were observed in the 21 control mice.

A second chronic inhalation study was performed at 1.0 and 2.0 ppm 3-nitro-3-hexane, in which rats were exposed for 36 months, and dogs for 42 months. The histopathologic examination of tissues from the rats revealed primary malignant lesions (undifferentiated carcinoma) in the lungs of 6 of 100 CFN rats exposed to 1.0 ppm of 3-nitro-3-hexene, and in the lungs of 11 of 100 CFN rats exposed at 2.0 ppm. The male rat appeared more susceptible than the female; of those exposed at 1.0 ppm, there were four males and two females with tumors, and at 2.0 ppm there were seven males and four females with tumors. In this latter group, the lesions in the females were described as "early carcinoma." There were no primary malignant lesions in the lungs of the 100 control rats.

Effects on Skin and Eye.   Open application to the skin of the rabbit (for 5 hr) resulted in intense local irritation, pain, erythema, edema, and later necrosis. One drop in the eye produced marked irritation and corneal damage.

2 ALIPHATIC NITRATES

2.1 General Considerations

The aliphatic nitrates are nitric acid esters of mono- and polyhydric aliphatic alcohols. The nitrate group has the structure $-C-O-NO_2$, where the N is linked to C through O, as contrasted to the nitroparaffins in which N is linked directly to C.

The nitric acid esters of the lower mono-, di-, and trihydric alcohols are liquids (methyl nitrate, ethyleneglycol dinitrate, trinitroglycerin) whereas those of the tetrahydric alcohols (erythritol tetranitrate, pentaerythritol tetranitrate) and the hexahydric alcohol (mannitol hexanitrate) are solids. They are generally insoluble, or only very slightly soluble in water, but are more soluble in alcohol or other organic solvents.

2.2 Source, Production, Use

The sources of the aliphatic nitrates are the corresponding alcohols. They are produced by esterification with nitric acid, with some, ethyl and methyl nitrates, in the presence of urea or urea nitrate, followed by isolation from related nitrate esters; for example, ethyl ether can be nitrated in the vapor phase to yield ethyl nitrate, methyl nitrate, and 2-nitroethyl ethyl ether.

Uses of aliphatic nitrates are chiefly as explosives and blasting powders [nitroglycerin, (NG), pentaerythritol tetranitrate (PETN)], with some limited use as a vasodilating agent (NG). Amyl nitrate has been used to increase the cetane number of diesel fuels. The use of NG and PETN as military explosives waxes and wanes with military exigencies (e.g., in the Vietnam War); military production for ordnance peaked in 1968 as judged by the use of 552,000 short tons of HNO_3, waning gradually to <50,000 tons by 1980 (2). The several aliphatic nitrates tested as promising candidate rocket fuels, methyl, ethyl, and propyl nitrate, appear to be used no longer (1).

2.3 Pharmacology, Symptomatology, and Mode of Action

Most of the information in these areas on the nitrate esters of the monohydric aliphatic alcohols was developed in the late 1930s to the late 1950s; accordingly, little new information is included here beyond that appearing in the first edition of this book (1962).

A prominent exception is a review article on NG and other nitrate esters (35a, 7 pp., 127 refs.), which shows that after more than 100 years of use for relief of angina pectoris, NG is finding wider application in congestive heart failure, in limiting myocardial "infarct size," in long-term angina prophylaxis, and as a diagnostic test for the presence of myocardial ischemia. Not only is its clinical potential now better recognized, but also its mode of action in normal hearts and in myocardial ischemia.

Because one of the shortcomings of NG is its brief action, long-sought nitrate preparations with more sustained effect have now been found in related nitrate esters, erythritol tetranitrate, pentaerythritol tetranitrate, and isosorbide dinitrate. The last, the oldest and best-studied, produces significant reduction in arterial and capillary pressures for the first hour after ingestion of a small dose, and decrease in cardiac output for up to 4 hr, as well as providing effective prophylaxis in angina.

The chief effects of the aliphatic nitrates are dilatation of blood vessels and methemoglobin formation. The vascular dilatation accounts for the characteristic lowering of blood pressure and headache. The individual members in the series differ in intensity and duration of these effects. Animals given effective doses by mouth or parenterally exhibit such signs and symptoms as marked depression in blood pressure, tremors, ataxia, lethargy, alteration in respiration (usually hyperpnea), cyanosis, prostration, and convulsions. Death, when it occurs, is either from respiratory or cardiac arrest. Animals surviving the acute exposure recover promptly.

Nitroglycerin (NG) and erythritol tetranitrate (ETN) are capable of producing approximately the same degree of hypotension in man but the effect of erythritol tetranitrate is more prolonged and requires a larger dose. The maximum blood pressure depression from NG occurs at approximately 4 min, whereas that from ETN occurs at approximately 20 min. Pentaerythritol tetranitrate (PETN) is less effective as a hypotensive agent than ETN. Methyl nitrate causes little depression in blood pressure (35b). The outstanding symptom produced in man is headache. This is usually described as very severe and throbbing, and is often associated with flushing, palpitation, and nausea and, less frequently, with vomiting and abdominal discomfort. Temporary tolerance develops from continued or repeated daily exposures. Doses resulting in headaches in man are NG, 18 mg (skin); ethylene glycol dinitrate (EGDN), 35 mg (skin); and ETN, 45 mg (oral) (35b). PETN in doses of 64 mg orally does not produce headache (36). Thus PETN is the least effective and NG the most effective. Other pharmacologic consequences of vasodilatation are increased pulse rate, an increase in cardiac stroke volume, variable cardiac dilatation and cardiac output, and a shift in blood distribution with increased stasis and pressure in pulmonary arteries (35b).

For some members of the series, the ease of hydrolysis to the alcohol and nitrate and the degree of blood pressure lowering are parallel. Krantz and co-workers, however, feel that there is little evidence that hydrolysis to produce nitrite is necessary for hypotensive action (37–39). They present evidence that the intact nitrate molecule (e.g., isomannide dinitrate) can act directly to lower blood pressure. It appears that this effect of the nitrate esters does not depend exclusively on the liberation of nitrite groups. Dilatation can occur without measurable nitrite in the blood, or when the amount measured is not sufficient to account for the effect observed. The in vivo formation of nitrite is commonly assumed to be the explanation for the methemoglobin-forming properties of the aliphatic nitrates (40). The mechanism of formation of nitrite is not clear (41). It is possible that reduction to nitrite occurs before hydrolysis as follows (42):

$$RONO_2 \xrightarrow{+2H} RONO \xrightarrow{+H_2O} ROH + NO_2^-$$

Ethyl nitrate, EGDN, NG, propyl nitrate, and amyl nitrates are known to cause methemoglobin formation in experimental animals. Ethyl nitrate is a weak methemoglobin former. Nitroglycerin is a moderately active methemoglobin former but EGDN is considerably more effective (approximately four times). Ethylene glycol mononitrate, on the other hand, is not very active in this respect (35b).

Ethyl, propyl, and amyl nitrate, as well as EGDN and NG, induce Heinz body formation in animals. Although ethyl nitrate induces Heinz body formation, ethyl nitrite does not. Ethylene glycol dinitrate is more effective than NG, which is more effective than ethyl nitrate. Toxicity data are not sufficiently complete to permit further comparisons of the effectiveness of the nitrates in producing Heinz body formation. The precise nature of these small, rounded inclusion bodies in the red blood cells, described by Heinz in 1890, is not clear. They have been observed in man and animals after absorption of a variety of chemical compounds, the most prominent of which are the aromatic nitrogen compounds, inorganic nitrites, and the aliphatic nitrates. Their appearance is commonly associated with anemia and the production of methemoglobin. Some evidence indicates that they are proteins in nature, possibly hemoglobin degradation products. Red blood cells containing the inclusion bodies have a shorter life-span and are removed from the circulation by the spleen. Special stains are required to satisfactorily demonstrate their presence. In the case of the aliphatic nitrates, erythrocytes containing Heinz bodies disappear from the circulating blood more slowly than methemoglobin (42–44).

The alkyl nitrates are all absorbed from the gastrointestinal tract (35b). Pentaerythritol tetranitrate is relatively slowly absorbed by this route. Ethylene glycol dinitrate and NG are absorbed through the skin but the absorption of ETN and PETN by this route is slow or absent. The nitric acid esters of the monovalent alcohols are rapidly absorbed from the lung. The absorption of ETN through the lungs is slower than with EGDN. Nitroglycerin is more slowly absorbed from the lungs than EGDN and PETN is more slowly absorbed than ETN.

Trimethylenetrinitramine (cyclonite, RDX) has been included in this section because it is also used as a high explosive. The nitramines contain the grouping —$NHNO_2$. Little is known about the action of the nitramines. However, neither methyl nitramine nor trimethylenetrinitramine exhibits nitrite or nitratelike actions (45).

2.4 Pathology in Animals

Pathologic examinations of animals dying following acute intoxication have either been negative or have revealed only slight nonspecific pathologic changes consisting of congestion of internal organs. Hueper and Landsberg (45d) have described degenerative vascular and parenchymatous lesions in the heart, kidneys, lungs, brain, and testes following several months' administration of large doses of ETN to young rats. They felt that these changes were induced by inadequate nutrition of the tissues following hypoxia and stagnation of the organs' blood supply associated with vasodilatation. On the other hand, von Oettingen et al. were unable to confirm these findings after chronic administration of ETN and PETN at lower doses (45a). Evidence of injury to these organ systems in men chronically exposed to NG, ETN, and PETN is lacking.

2.5 Industrial Experience

No injuries to workers from exposure to any of the monohydric alcohol esters of nitric acid (methyl, ethyl, *n*-propyl, amyl isomers) have appeared in the published literature apparently because industrial interest in these esters was short-lived, these esters having been replaced with polynitrated compounds of greater effectiveness.

Industrial experience has been far different with the polynitrated esters, NG, and its congener EGDN; sudden death has been reported among dynamite workers exposed to mixtures of NG containing up to 80 percent EGDN in U.S. manufacturing plants (46). Reports of similar experiences had already come from Canada, Germany, Italy, Sweden, and the Soviet Union (1942–1960). The occurrence of characteristic and severe headaches in workers handling these nitrate esters was so frequent that such acquired names as "dynamite head" and "powder headache" were common. Hypotension and peripheral vascular collapse were described as associated with these headaches.

McConnell and his associates have summarized the experience with occupational disease and industrial hygiene in government-owned ordnance explosives plants in the United States during World War II (47). These included chemical and explosives manufacturing, munitions and munition loading, and proving and testing areas. Trinitrotoluene and tetryl were common to most areas but NG, PETN, and trimethylenetrinitramine (RDX) were also handled. In 915,000 man-years of experience no fatalities due to the aliphatic nitrates or RDX occurred. Most of the illness and 21 deaths were attributed to trinitrotoluene. Some of the 93,000 cases of mild illness or dermatitis were attributed to PETN, NG, or RDX manufacture (47).

2.6 Analytic Determination

Early methods (1955–1963) used for determining aliphatic nitrate esters consisted of various colorimetric or spectrophotometric procedures. These have now been largely replaced by instrumental methods (since mid 1960). References to the colorimetric procedures are included here (48a, b, 49, 50–55) because they were used for monitoring animal or worker exposures in work summarized later in the sections on specific compounds.

2.6.1 Colorimetric Methods

Colorimetric procedures used for the determination of "traces" of polynitrate esters, NG and EDGN, were reported in the mid 1930s (50), and for PETN and NG a few years later (51). These procedures appear relatively crude by present-day standards, for they were based on the nitration of *m*-xylene by the aliphatic nitrate under determination.

Determination in Air. More precise methods were developed later such as that for amyl nitrate in air by Treon et al. (52), who used it to monitor animal exposures. The procedure utilizes the transmission of ultraviolet radiation at 236 mμ through a 10-ml quartz cell. The curve was linear over a range of 25 to 600 μg/ml.

Colorimetric methods were still being used as late as 1966 for the determination of NG and EGDN in workplace air (53). Following collection in 95 percent alcohol, samples were analyzed by the old phenoldisulfonic acid method for nitrates. Determination of these nitrate esters was also made by hydrolyzing the ethanol or ether samples with alcoholic KOH to nitrite and determining spectrophotometrically the colored product formed by diazotizing sulfanilic acid with 1-naphthylamine at 525 mμ. As little as 0.3 mg NG or EGDN was claimed to be determined from a 10-liter air sample (54).

Determination in Biologic Fluids. A colorimetric method for determining NG and EGDN in blood and urine was reported by Zurlo et al. (55). Nitric esters are extracted with ether from blood or urine (5 ml blood, 10 to 30 ml urine), hydrolyzed with KOH in alcohol, and the nitrous acid liberated, extracted with water, and estimated at 490 mμ after adding α-naphthylamine and p-nitroaniline (EGDN). The degree of accuracy claimed is ± 10 percent; the sensitivity, up to 5 μg. NG is assayed using 0.1 percent KOH; with 2.5 percent KOH, both NG and EDGN can be determined.

2.6.2 Instrumental Methods

Infrared spectrography is generally satisfactory for the identification of aliphatic nitrate esters (48a, b). Spectral correlations have been compiled by Pristera et al. using band assignments at 6.0, 7.8, and 12.0 μm.

A chromatograph procedure has been used for determining EGDN and NG in blood and urine (49).

These methods have the advantage of greater precision and ease of manipulation of samples, and have thus replaced largely the colorimetric methods of the past.

2.7 Mechanism of Tolerance

From Section 2.3 and those that follow, it is seen that considerable progress has been made since the second edition in 1962 in our understanding of the way aliphatic nitrates produce their several effects of vasodilation, tremoring, and sudden death. New insights have been gained into (1) the mode of action of the nitrate esters generally to produce vasodilation through their metabolic conversion to nitrite (Section 2.3); (2) the specific involvement of the thyroid in the acute action of dynamite mix, and plasma corticosterone after EGDN administration, and the inhibition of monoamine oxidase after its chronic administration (Section 2.8.8 EGDN); (3) the mechanism of oxidation of hemoglobin to methemoglobin by PGDN; and (4) the blocking of the phrenic nerve-muscle communication as the cause of the tremoring response and respiratory arrest in triethylene glycol dinitrate intoxication (Section 2.8.8 PGDN).

A mechanism for tolerance development, a common response to organic nitrates, generally has also been elucidated (45b). The study was performed because the treatment of angina is hindered by the development of tolerance to the vasodilator action of these esters. It involved the interaction of NG on aortic tissue sulfhydryl in vitro and in the intact animal made tolerant to NG, on the hypothesis that NG oxidizes a critical sulfhydryl group in the NG "receptor." It was found that in tolerant aorta, sensitivity to NG decreased whereas nitrite formation increased, converting the nitrate receptor to the disulfide form which has a lower affinity for NG. Confirmation of the hypothesis was obtained by reversing the tolerance with dithiothreitol, a disulfide reducing agent, thus demonstrating that tolerance involved chemical alteration of the receptor by NG. Cross tolerance was also demonstrated to other organic nitrates, but lack of it to isoproterenol, inorganic nitrite, and papaverine.

A similar involvement of the sulfhydryl–disulfide system was found in the author's laboratory to be the mechanism for the development of tolerance and cross tolerance to the acute effects of ozone and nitrogen dioxide (45c).

2.8 Specific Compounds

2.8.1 Methyl Nitrate, CH_3ONO_2

Methyl nitrate (MN) presently is of interest to the military as a munition (56). It appears to be of no commercial or industrial interest (1).

Physical and Chemical Properties

Molecular weight   77.04
Boiling point   66°C (explodes)
Specific gravity   1.217 (15°C)
Vapor density   2.66 (air = 1)
Solubility   Slightly soluble in water, soluble in alcohol and ether
    1 mg/l ≈ 317 ppm, 1 ppm ≈ 3.15 mg/m^3 at 25°C, 760 mm Hg

Physiologic Response. The acute toxicity of MN exhibits a peculiar pattern as determined in small laboratory animals; the 4-hr inhalation LC_{50} for the rat was found to be 1275 ppm (C.L. 1200 to 1355), whereas for the mouse, the LC_{50} was 5942 (C.L. 5827 to 6509), a toxicity indicating widespread differences between two closely related species, a difference of almost fivefold in greater susceptibility of the rat (56). Again, orally the rat was more than five times more susceptible than the mouse; rat LD_{50}, 344 mg/kg; mouse, 1820 mg/kg; with the guinea pig showing an intermediate response, LD_{50}, 548 mg/kg. These acute animal data thus provide no good indication of the acute toxicity for man.

Signs and Symptoms (56). Responses of animals to single, lethal doses of MN by inhalation were dose related and followed a general pattern of lethargy, decreased respiratory rate, and cyanosis. All animals were inactive throughout the 4-hr exposure period. Similar responses followed oral administration; both rats and mice were inactive following dosing. Rats also exhibited labored breathing and gasping at the higher dose levels. Death was seldom delayed, with most occurring during the 12-hr period following dosing.

Gross examination of the guinea pigs that died following the single oral dose revealed chocolate-brown discoloration of the blood and lungs indicating severe methemoglobinemia. Except for the livers appearing slightly pale, no other lesions were observed. Gross examinations of the animals surviving the 14-day observation period revealed no treatment-related lesions.

Doses of 12.5 mg/kg have practically no effect on the blood pressure and pulse rate of rabbits. Doses of 52 mg/kg have a slight transient effect. Methyl nitrate is considerably less effective in these respects than NG. The minimal dose causing headache in man is between 117 and 470 mg. As has been observed with the other nitric acid esters, fractional doses produce tolerance that lasts for several days (35b).

No injuries from exposure to MN by handlers appear to have been reported in the published literature, and no TLV has been established.

2.8.2   Ethyl Nitrate, $CH_3CH_2ONO_2$

Uses.   Ethyl nitrate (EN) is a colorless flammable liquid with a pleasant odor and sweet taste. It has been used in organic syntheses of drugs, perfumes, and dyes, and as rocket propellant.

Physical and Chemical Properties

Molecular weight	91.07
Specific gravity	1.116 (20/4°C)
Vapor density	3.14 (air = 1)
Boiling point	87.6°C
Melting point	112°C
Flash point	50°F (CC)
Explodes at	185°F
Solubility	1.3 g in 100 ml of water at 55°C; miscible in all proportions in alcohol and ether

1 mg/l $\backsim$ 269 ppm, 1 ppm $\backsim$ 3.72/m^3 at 25°C, 760 mm Hg

Physiologic Response.   No industrial intoxications have been recorded from EN. It is said that EN has anesthetic properties and on inhalation causes headache, narcosis, and vomiting (35b). In cats, 400 mg/kg in olive oil intraperitoneally produces unconsciousness, increased respiratory rate, dilatation, and fixation of pupils, followed by death in 90 min; 300 mg/kg intraperitoneally is followed by recovery. Moderate methemoglobinemia and Heinz body formation are observed after doses of 125 to 250 mg/kg (35b). Thus the effects of EN resemble those of the other aliphatic nitrates that have been studied.

No TLV has been established for EN.

2.8.3 Propyl Nitrate, $C_3H_7ONO_2$

Uses. *n*-Propyl nitrate (PN) is a pale yellow liquid with a sweet, sickening odor. It has been tested as a fuel ignition promoter, and as a liquid rocket monopropellant, but appears no longer used for these purposes (1).

Propyl nitrate is a strong oxidizing agent, flammable, and a dangerous fire and explosion risk (see below).

Physical and Chemical Properties

Molecular weight	105.09
Boiling point	110.5°C
Freezing point	-100°C
Specific gravity	1.058 (20/4°C)
Vapor pressure	Approx. 16 mm Hg, 25°C
Vapor density	3.62 (air = 1)
Flash point	68°F
Autoignition temperature	350°F
Explosive limits in air	2–100%
Solubility	Very slightly soluble in water, soluble in alcohol and ether

1 mg/l $\approx$ 233 ppm, 1 ppm $\approx$ 4.30 mg/m^3 at 25°C, 760 mm Hg

Physiologic Response. The *acute toxicity* of *n*-propyl nitrate (*n*-PN) vapor is relatively low for small laboratory animals, rat, mouse and guinea pig, according to studies of Rinehart et al. (57) but moderate for the dog. They estimated the 4-hr LC_{50} values to be between 9000 and 10,000 ppm for the rat, 6000 and 7000 ppm for the mouse, and 2000 and 2500 ppm for the dog.

From *subacute*, 8-week, daily, 6-hour exposures, the dog proved still to be the most susceptible of the species tested, 560 ppm constituting an LC_{50}, whereas all guinea pigs exposed for the same period survived 3235 ppm. This level proved to be an LC_{25} for the rat, ranking this species with intermediate, short-term susceptibility between the dog and guinea pig.

Compared to MN, it should be noted that the species susceptibility ranking represents a reversal of that for MN where the rat was almost five times more susceptible than the mouse. This difference possibly indicates a widely different rate of metabolism of *n*-PN from that of MN in the two species.

Table 53.16 shows the acute response of four laboratory species when *n*-PN is administered orally, dermally, and intravenously. It may be seen that percutaneous toxicity is essentially nil, and oral toxicity is very low compared with intravenously administered doses, in which mg/kg doses were lethal compared with g/kg doses orally.

Table 53.16. Acute Response to *n*-Propyl Nitrate—Animals (53, 59)

Animal	Dose (g/kg)	Route	Effect
Rat	7.5	Oral	Approximate lethal dose (sample I)
Rat	5.0	Oral	Approximate lethal dose (sample II)
Rat	1.0	Oral	Weakness, incoordination, cyanosis
Rats	1.5×10	Oral	Weakness, cyanosis, weight loss (first week)
Rabbit	11,17	Skin	Essentially none
Rabbit	0.2–0.25	IV	Approximate LD_{50}
Dog	0.005	IV	Slight fall in blood pressure
	0.050	IV	Hypotension, cyanosis
	0.2–0.25	IV	Death in respiratory arrest
Cat	0.1–0.25	IV	6/7 died in 1 min
	0.025–0.075	IV	Hypotension, methemoglobinemia, survived

Chronic, 26-week inhalation exposures of the dog, guinea pig, and rat showed again that the dog was the most susceptible, 260 ppm being the highest dose tested in which all dogs survived versus 2110 ppm, a level of complete survival of guinea pigs, but which constituted an approximate LC_{50} for the rat (57). Thus the same order of susceptibility holds for these species chronically and acutely.

Signs and symptoms resulting from *n*-PN exposure differed in kind and degree according to species susceptibility (57). The main effect in rodents is anoxia, caused by lowered blood oxygen content resulting from methemoglobin production, whereas dogs developed hemoglobinuria and hemolytic anemia together with methemoglobin production and resultant much lower oxygen-carrying capacity than rodents, which did not develop anemia. The fact that blood levels in dogs returned to normal or near normal on continued exposure, and did not show appreciable development or accumulation of methemoglobin from continuous daily exposures at levels below 900 ppm, points clearly to development of *tolerance* to chronic, low level effects.

Oddly, examination of tissues from repeatedly exposed dogs and rodents showed no pathologic damage except an increase in pigment in spleen and liver, presumably from hemolysis and increased hematopoietic activity.

Toxic signs following acute exposures at high levels are obviously more severe, consisting of cyanosis, methemoglobinemia, and uria, hemolytic anemia, vomiting, convulsions, and death in the dogs, and cyanosis, lethargy, convulsions, and death in the rodents.

What significance these experimental animal responses have for the industrial worker can only be guessed. In such cases, prudence dictates reliance on the responses of the largest and most susceptible species tested, the dog. This has been done in setting hygienic standards of exposure.

Permissible limits of exposure to *n*-PN have been deduced from the experimental work in animals summarized above. A TLV of 25 ppm was recommended by the TLV Committee in 1960 and adopted by ACGIH in 1962. The odor, "presumably detectable at concentration levels of 50 ppm and above" (57), while "resulting in discomfort—in the form of irritation, headache, or nausea," would offer less than the usually desirable warning if the TLV excursion factor for 25 ppm is 1.5. But in the absence of on-the-spot monitoring device, these responses should be persuasive.

2.8.4 Isopropyl Nitrate, $(CH_3)_2CHONO_2$

In contrast to *n*-PN, only preliminary acute toxicity determinations have been made on isopropyl nitrate (isoPN), but these have demonstrated that isoPN is qualitatively like *n*-PN as far as their acute and subacute toxic effects are concerned (58). Both nitrates show low toxicity orally or absorbed through the skin, and are nonproductive of eye injury on single administrations in small laboratory animals. On the other hand, repeated contact with the skin causes irritation, and inhalation of the vapor produces cyanosis, methemoglobinemia, and even death. Combustion products were more toxic than the vapor itself.

Comparative approximate oral lethal doses for the male albino rat were 3.4 g/kg versus 5.0 g/kg for *n*-PN; 17 g/kg on the rabbit skin caused only inflammation from the application of both nitrates, but no overt systemic effects (Table 53.16). The 6-hr exposure to isoPN vapor at 8500 ppm was an approximate LC_{50} for the rat, compared with 9000 to 10,000 ppm as an estimated 4-hr LC_{50} of *n*-PN for this species.

Comparable subacute oral dosing schedules for isoPN and *n*-PN of five times/ week for 2 weeks, at approximately one-fifth the LD (680 mg/kg, isoPN, 1500 mg/kg, *n*-PN) resulted in temporary weakness from *n*-PN, cyanosis and weight loss, methemoglobinemia, and congested spleens; symptoms became less severe on continued treatment, and 10 days after treatment, weight gains occurred. Rats dosed with isoPN showed no overt signs of toxicity. Subacute rabbit skin tests, both at the same dosing schedule (7.5 g/kg, five times/week for 2 weeks), showed no systemic effects from either nitrate, but showed equally inflammation and staining and thickening of the skin.

Thus, from the standpoint of comparative acute and subacute toxicity, isoPN exhibits a closely similar degree of toxicity for small laboratory animals, as far as can be judged from approximate determinations and estimates therefrom.

2.8.5 Amyl Nitrate, $C_5H_{11}ONO_2$

Uses. As used in the military and as tested toxicologically, amyl nitrate (AN) is a mixture of several primary, normal, and branched chain amyl nitrates containing only a trace of amyl alcohol. It has been used to increase the cetane number of diesel fuels, and is a component of Otto fuel II. As studied by Treon et al. (52), it was a clear, slightly yellow liquid with a sickening sweet odor.

Physical and Chemical Properties

Molecular weight	133.15
Vapor density	0.997 (20/4°C)
Boiling range	150–155°C (unstable)
Vapor pressure	(Isoamyl nitrate) 5 mm Hg, 28.8°C
Solubility	0.3 ml in 100 ml of water; miscible with hydrocarbons
Flash point	42°C (closed cup)

1 mg/l ≂ 184 ppm, 1 ppm ≂ 5.44 mg/m^3 at 25°C

Physiologic Response. Treon et al. (52) exposed cats, guinea pigs, rabbits, rats, and mice to measured concentrations of amyl nitrates in air (Section 2.6.1). Selected data from their acute and subacute animal exposures are given in Table 53.17.

As far as can be determined from the noncongruent levels and duration of exposures, toxicity of the AN isomers is greater than that of n-PN by a factor of two- or threefold. Evidently, the diluting-out effect of increasing carbon chain length on toxicity is insufficient to overcome the strong toxic effect of the ONO$_2$ group, or metabolism differs with increasing chain length; higher

Table 53.17. Response to Inhalation of Various Concentrations of Amyl Nitrates (52)

Concentration		Duration	Mortality[a]			
mg/m^3	ppm	(hr)	Guinea Pigs	Rabbits	Rats	Mice
19,900	3730	7	2/2	2/2	3/4	5/5
19,170	3593	3.5	0/2	2/2	1/4	4/5
17,220	3227	1	0/2	0/2	0/4	0/5
16,390	3072	3 × 1	2/2	2/2	4/4	5/5
14,800	2774	3.5	0/2	0/2	0/4	2/4
13,600	2549	0.33	6/2	0/2	0/4	0/5
12,700	2380	2 × 7	2/2	2/2	0/4	5/5
12,300	2305	1	0/2	0/2	0/4	5/5
9,600	1807	7	0/2	1/2	0/4	4/4
9,090	1703	3 × 7	2/2	1/2	2/4	5/5
8,600	1612	7	0/2	0/2	0/4	0/5
3,200	599	9 × 7 + 6.25	0/2	0/2	0/4	0/2
1,400	262	20 × 7	0/2	0/2	0/3	0/5

[a] No cats died following any of these exposures.

members of the aliphatic nitrites, at least, split off the nitrous group, whereas lower members of the series remain intact to exert toxicity (35b) (see Section 3).

Again, there was no similarity to n-PN in species susceptibility; in order of decreasing susceptibility, mouse > rat and rabbit > guinea pig > cat. Except for the cat, which survived all exposures including the highest, 3730 ppm for 7 hr, the differential susceptibility was small; lethal concentrations were for the mouse, 2300 ppm; rat and rabbit, 3600 ppm; and guinea pig, about 3700 ppm.

All species, however, survived exposures of 600 ppm for 10 days at 7 hr/day, or 260 ppm for 20 days at 7 hr/day. Treon et al. pointed out a striking difference in species susceptibility to the toxicity by inhalation of AN; the cat, which was most resistant to AN, was the most susceptible to 2-nitropropane! Again, this is probably attributable to differences in cat metabolic capacities.

At high lethal levels *signs and symptoms* were characterized by tremors, ataxia, alterations in respiration, lethargy, cyanosis, convulsions, coma, and deaths. All exposures except 262 ppm produced signs or symptoms in cats, guinea pigs, and rats, and some alterations in respiration were observed in rabbits and mice at the lowest concentration, 262 ppm. A cat exposed at 599 ppm developed methemoglobin levels up to 59.5 percent after the seventh exposure. Cats exposed at concentrations ranging from 1700 to 3700 ppm of the amyl nitrates showed Heinz body formation, which reached a peak several days after exposure and disappeared slowly after a period of 1 to 3 weeks.

Animals dying during exposure had diffuse degenerative changes in the liver, kidneys, and brain with hyperemia and edema of the lungs. Those sacrificed at varying intervals after exposures had normal findings on pathologic examination.

Persons exposed in the laboratory during these studies developed nausea and headache. No other illness was observed (52).

Thus the effects of mixed amyl nitrates in animals are qualitatively similar to those of the other alkyl mononitrates. The acute and subacute inhalation toxicity for guinea pigs, rats, and mice is greater than that of n-PN nitrate, but the higher boiling point and lower vapor pressure of the amyl nitrates tend to reduce this differential as far as the industrial hazard is concerned.

No observations on men exposed in industry have been recorded and no maximal allowable concentrations have been proposed.

2.8.6   Ethylene Glycol Dinitrate, $C_2H_4(ONO_2)_2$

Uses.   Ethylene glycoldinitrate (EGDN), a colorless liquid produced by the nitration of ethylene glycol in the presence of a dehydrating agent, sulfuric acid, is used in conjunction with NG in the manufacture of low-freezing dynamites, for ethylene glycol is much cheaper than glycerin.

Physical and Chemical Properties

Molecular weight	152.07
Specific gravity	1.483 (8°C)
Boiling point	125°C (50 mm); explodes at 114–116°C
Freezing point	22.8°C
Vapor density	5.24 (air = 1)
Vapor pressure	0.045 mm Hg (20°C)
Ignition temp	195–200°C (760 mm)
Solubility	0.52 g/100 g H_2O (25°C);
	0.85 g/100 g H_2O (60°C);
	Soluble alcohol and dilute alkali

1 mg/l ≈ 161 ppm, 1 ppm ≈ 6.24 mg/m^3 at 25°C, 760 mm Hg

Saturated air contains 10 ppm

In the presence of more than slight traces of free acid, EGDN is unstable with a melting point of −21°C. The melting point of the stable forms is −18.5°C.

Physiologic Response. Early *acute toxicity* tests in animals by Gross et al. (60) were confined to subcutaneous administration, 400 mg/kg being a fatal dose for the rabbit and 100 mg/kg for the cat, a species more susceptible to methemo-globin (MHb) formation. A subcutaneous dose of 60 mg/kg in this species produced 45 percent MHb as well as Heinz bodies. In the rabbit, a dose of 12.5 mg/kg resulted in hypotension, but no MHb (61). Smaller doses of 0.6 mg/kg caused a rapid but transient drop in pulse pressure. These hypotensive effects of EGDN were found repeatedly to be more marked than those from NG, and to be twice as toxic as NG. In Heinz body production, Wilhelmi (62) found EGDN to be four times more effective than NG, and 20 times more than EN in the cat.

In man, acute exposures to EGDN resulted in headache, nausea, vomiting, lowering of blood pressure, increase in pulse rate, and cyanosis. The minimal dose causing headache when applied to the skin is 1.8 to 3.5 ml of a 1 percent alcoholic solution, according to Leake et al. (63). When it was applied in fractional doses totaling 170 mg, tolerance developed in 24 to 36 hr and lasted for 10 to 13 days.

Chronic exposure to the vapors of EGDN, 8 hr daily at 2 ppm for 1000 days caused moderate, temporary blood changes without any clinical aftereffects (cats). Similar exposures of these animals at 10 times the level resulted in marked blood changes, but otherwise showed no organic changes and behaved normally (60).

Ethylene glycol dinitrate readily penetrates the skin and is absorbed through the lungs and gastrointestinal tract, causing chronic poisoning and anemia in contrast to NG (60). Thus exposures from direct skin contact and inhalation of its vapors give rise to symptoms of headache, weakness, dizziness, shortness of breath with chest pain, and general fatigue (64).

Metabolism. In a study of the metabolism of EDGN and its influence on blood pressure of the rat, Clark and Litchfield (65) found that the breakdown of EDGN in blood results in the liberation of inorganic nitrite and nitrate, and ethyleneglycol mononitrate (EGMN). Free EGDN in blood reached a peak in 30 min and fell to zero 8 hr later. Inorganic nitrite was maximal in 1 to 2 hr, falling to zero at 12 hr, whereas nitrate rose more slowly to its maximum in 3 to 5 hr, reaching preinjection levels (approximately 1.0 μg/ml) 12 hr after injection, with no further changes in the blood levels of any of these constituents up to 96 hr.

A marked fall in blood pressure occurred directly after the injection, reaching its lowest value in 30 min. This was followed at 2 to 3 hr by a significant secondary fall, followed by a steady rise to preinjection levels at 12 hr. Ethylene glycol dinitrate was more effective in this respect than EGMN.

Breakdown of EGMN in the blood was demonstrated by the recovery of less than 0.5 percent in the urine. Nitrite is released from the reduced EGDN ester, later oxidized to nitrate, and excreted in the urine, accounting for 57 percent of the injected dose.

Industrial Experience. An essentially worldwide literature on worker exposure to EGDN dates back to the late 1950s, when up to 80 percent EGDN was added to NG to form a lower freezing dynamite. Reports of the effects of NG on those engaged in its manufacture appeared as early as 1914 (66).

Acute effects from exposure to EGDN by either inhalation of its vapors or by skin contact consist of fall in blood pressure and headache; four of five volunteer workers exposed at the then TLV of 2.0 mg/m^3 (approximately 0.2 ppm, measured as NG) experienced a fall of from 30/20 to 10/8 mm Hg, with severe headaches developing in 1 to 3 min; one worker was refractory, exhibiting no change in blood pressure or experiencing headache (53); at 0.7 mg/m^3, blood pressure drop was from 30/20 to 0/0 in 10 volunteers who experienced slight headache or merely slight dullness in the head; at 0.5 mg/m^3, when seven volunteers were tested, only three experienced slight or transitory headache, although blood pressure depressions were commensurate with those from higher exposures. Thus a dose response from short-term inhalation exposures to EGDN has been obtained.

The investigators noted that, being far more volatile than NG, EGDN was responsible for essentially all the observed effects, and quoted Rabinowitz (67) as stressing that alcohol accentuates the severity of the headaches.

In addition to headache, "Pains in the chest, abdomen, and extremities, and symptoms of general fatigue may appear as a result of the so-called acute effect" (46).

A quite different pattern of response develops from repeated, long-term, chronic exposures involving periods of years. Of foremost and arresting attention are the oft-quoted "Monday morning fatalities" and angina. Carmichael and Lieben (46) have assembled a table of at least 38 sudden deaths in dynamite workers that occurred 30 to 48 hr after absence from work over the period from 1926 to 1961. These deaths were reported from Germany, Scotland, and the United States in men from the ages 27 to 58. In the three Pennsylvania plants surveyed over a 5-year period, the 10 deaths occurred in men employed in the mixing and cartridge-filling operations (46). The pattern in the deaths was fatal heart attacks that occurred during the weekend or on a Monday morning in those with similar occupational exposure, and clinical diagnosis of acute infarction with little evidence of definite coronary occlusion. Narrowed coronary lumen and thickened sclerotic arterial walls were found at postmortem examination. These cases develop the symptomatology of cardiac ischemia, which may be foreshadowed by changes in periodic blood pressure and pulse rate determinations.

The importance of measuring not only exposures from the air but those from skin contact as well has been emphasized in a report by Einert et al. (54). Having noted past reports that (1) headache could result from merely "shaking hands with persons who handled dynamite" (66), (2) the minimal effective dose of EGDN when applied to the skin was between 1.8 and 3.5 ml of a 1 percent alcoholic solution (68), and (3) EGDN is more readily absorbed through the skin than NG (69a), Einert et al. estimated skin exposures of operators at the several work sites in an explosives plant by extracting the gloves worn by operators for EGDN. Values obtained varied from less than 0.1 mg for a 2-hr morning exposure in the mix house, to 1 mg for a 4-hr exposure at a semiautomatic cartridge-filling machine. These skin values were found comparable to the measured exposures from air inhaled on an 8-hr shift, and 20 percent represents the amount of EGDN retained from that inhaled (69b), certainly an amount to be reckoned with in any EGDN exposure.

The authors recommended the use of lining gloves as a collection method for estimating skin exposures when combined with the simple clinical methods of pulse and blood pressure measurements before and after work.

Mechanism. Ever since "sudden death" became generally recognized as a disturbing sequel from exposure to EGDN, investigators have made repeated attempts to determine the way this is brought about.

One such attempt has been made by Phipps (70) to understand the reason for the "Monday morning" headaches that at times are preceded by severe anginal attacks, often ending in sudden death. Phipps's approach in the use of thyroid hormone was influenced by an old report that thyroidectomy gives dramatic relief from angina pectoris, the explanation being that the hormone sensitizes the myocardium to epinephrine, and therefore removal of the thyroid gland lowers the level of hormone and the resultant desensitization reduces the risk of anginal attacks.

Using this as background, Phipps found that rats that had been pretreated with thyroid hormone showed increased sensitivity to dynamite mix (85 percent EGDN, 17 percent NG) to the point that an LD_{50} became LD_{95}. Conversely, rendering the animals hypothyroid, with thyroidectomy followed by a depletion period, made the rats resistant. There was 100 percent protection; thus the LD_{50} became LD_0! Therefore, hyperthyroidism potentiates the mix; hypothyroidism antagonizes.

Subsequent attempts to investigate the underlying cause of "sudden death" from dynamite mix using rats as a model were abandoned when a change of seasons (from summer to winter) response to treatment altered, presumably due to altered rat metabolism and hormone release. A change in NIOSH policy from basic to applied research completely terminated the effort.

An investigation with similar purpose was reported by Clark (71). Rats dosed with EGDN (65 mg/kg) showed marked increase in plasma corticosterone rising to a maximum in 15 min and persisting for more than 2 hr. It had been known that systemic hypotension is a potent stimulator of 17-hydroxycorticosteroid secretion, a response dependent on the pituitary. Repeated injections of EGDN lead to a decrease in corticesterone response, as did an injection of EGDN given 24 or 72 hr after the last series of injections. The authors suggest, without investigating the possibilities, that the reduced cortisone response to EGDN is due partly to tolerance to the induced hypotension, and partly to some deficiency in the hypothalamopituitary-adrenal axis.

Investigations have also been made to determine the pharmacologic basis of chronic effects. Chronic intoxication was reasoned by Kalin and Kylin (72) conceivably to have its explanation in the development of a mechanism that attempts to compensate for the effect of EGDN, and when this is interrupted, can exert a pathologic effect. In experimentally testing this hypothesis in rabbit liver homogenates, they showed that EGDN was a competitive inhibitor of monoamine oxidase (MAO), but the EGMN displayed no such effects. Admittedly, the overall effect of MAO in vivo is difficult to assess; on the one hand, there is the increase of monoamines in organs where inhibition occurs, but there is the contrary fact of reduced susceptibility to adrenalin in rats intoxicated with EGDN, although increased myocardial circulation could account for this. And there it stands, hypotheses bolstered by some pharmacologic findings, but seemingly not yet able to account for the ultimate effect, sudden death.

Hygienic Standards of Permissible Exposure. The limits of permissible exposure (TLVs) to EDGN have undergone a succession of revisions since the first recommendation of the Committee in 1962. The first listing appeared as 0.2 ppm (approximately 1.2 mg/m^3) with the "Skin" notation, indicating a potential exposure by this route which should be taken into consideration in evaluating worker exposure. In 1966 the 0.2 ppm was recommended as a "Ceiling" limit as reports of headache reached the Committee, apparently from permitted excursions above the time-weighted average of 0.2 ppm. In 1963, the TLV was listed for the first time with NG as EGDN and/or NG, for the major exposure was now from a dynamite mix with NG. A notation of 0.02 ppm for intermittent exposures was added, again from reports that 0.2 ppm resulted in headaches in those intermittently exposed. These values were retained until 1979 when EGDN was proposed separately from NG as 0.02 ppm as a TWA. The OSHA standard is C ethylene glycol dinitrate and/or nitroglyceıin—Skin, 0.2 ppm, 1 mg/m^3.

2.8.7 Nitroglycerin, $C_3H_2(ONO_2)_3$(Glyceryl Trinitrate)

Uses and Industrial Hazards.   Nitroglycerin (NG) is a colorless, oily liquid that explodes violently from shock or when heated to about 260°C and thus is a severe explosion risk.

Its major use is in explosives and blasting gels, as in the production of low-freezing dynamite in mixture with EGDN, commonly in proportions of 20 and 80 percent. Other explosive uses are in cordite in mixture with nitrocellulose and petroleum, and in blasting gelatin with 7 percent nitrocellulose. In medicine, it has expanding applications in congestive heart failure, in the reduction of infarct size, in myocardial infarction, and in the long-term prophylaxis of angina (35a).

Although NG shares the EGDN most of its toxicologic and pharmacologic properties, it should be noted that when used in its usual dynamite mix of 20 percent NG, 80 percent EGDN, its exposure hazard is essentially nil, as can be calculated from its vapor pressure relative to that of EGDN, yielding a ratio of 1 part in 720 (54). When not used in admixture, its exposure hazard is still far less than that of EGDN, 0.00025 versus 0.045 mm Hg (see below).

Physical and Chemical Properties

Molecular weight	227.1
Specific gravity	1.601
Melting point	13°C
Vapor density	7.8 (air = 1)
Vapor pressure	0.00025 mm Hg, 20°C
Explosion point	260°C
Autoignition point	518°F

Solubility   Slightly soluble in water, partly soluble in alcohol; miscible with ether and chloroform

1 mg/l ≅ 108 ppm, 1 ppm ≅ 9.29 mg/m^3 at 25°C, 760 mm Hg

Physiologic Response. Nitroglycerin is a potent vasodilator of both arterial and venous vascular smooth muscle. Its action is a matter of minutes whether exposure is via the lungs, skin, or mucous membranes. The role of skin in the absorption of the liquid is particularly significant in view of the small air concentrations resulting from its low vapor pressure (54). However, it takes relatively very small amounts to produce an intense throbbing headache, often associated with nausea, and occasionally with vomiting and abdominal pain. Rabinowitz (67) in discussing tolerance and habituation to NG, reports that as little as 0.001 ml NG is capable of producing a severe headache. The tolerance is developed if the exposure to nitroglycerin is maintained in powder workers. In most cases this is transient and the headache may reappear after a weekend or holiday. Considerable variability has been pointed out, however. Rabinowitz reports that one worker was engaged in an exposure situation in the manufacture of explosives for more than 30 years without experiencing a nitroglycerin headache. The powder headache has been described as preceded by a sensation of warmth and fullness in the head which starts at the forehead and moves upward towards the occiput. It may remain for hours or several days and may extend to the back of the neck.

Larger amounts may result in hypotension, depression, confusion, occasionally delirium, methemoglobinemia, and cyanosis. Aggravation of these symptoms and the occurrence of maniacal manifestations after alcohol ingestion have been repeatedly observed. A temporary tolerance to headache develops from repeated exposures but this is usually lost after a few days without exposure. Workers have utilized this phenomenon by placing a small amount of nitroglycerin in their hat bands to ensure continued absorption and to prevent "Monday headache." The occurrence of typical headache, hypotension, palpitation, and flushing have been observed in explosives workers (66, 73, 74), dynamite handlers (77), and those handling cordite (75). The headache is presumably due to cerebral vasodilatation and resembles clinically that produced by histamine. Temporary relief can be obtained from adrenalin or from the administration of ergotamine tartrate. Fatalities from industrial intoxication are uncommon.

Medical studies of explosives workers with combined NG and EGDN exposures have not given evidence of chronic intoxication or injury despite the occurrence of transient symptoms. An extensive study of 276 workers with long exposure to NG and EGDN in three Swedish explosives factories gave no evidence of permanent deterioration in health (76). The average air concentrations of NG-EGDN for most operations were below 5 mg/m^3, usually 2 to 4 mg/m^3. In the group with exposures at concentrations generally below 3 mg/m^3, symptoms such as fatigue and alcohol intolerance were less frequent, but like EGDN, there was little difference in the frequency of headaches. Headache and fatigue were the predominant symptoms. Vomiting, dyspnea, and alcohol intolerance were less frequent and chest pain was least frequent.

Nitroglycerin appears to induce a shift of blood flow from relatively well-perfused myocardium to less-adequately nourished endocardium. It also has hypotensive effects largely due to reductions in diastolic pressure distending the relaxed ventricular wall. Since 1969 it has been generally believed that NG relieves angina by favorably altering the imbalance between myocardial oxygen supply and demand (35a).

Hygienic Standards of Permissible Exposure.   The TLV for nitroglycerin as proposed in the tentative list for 1980 is 0.02 ppm (approximately 0.2 mg/m^3), the STEL, 0.04 ppm (approximately 0.4 mg/m^3). Nitroglycerin was, however, listed among the first TLVs adopted in 1949 with a value of 0.5 ppm, a value taken from the U.S. Public Health Service guide for occupational disease control. This TLV was revised in 1962 to 0.2 ppm (approximately 2 mg/m^3) bearing a "Skin" notation to emphasize the important role of this route of absorption. In 1963 it was first listed with EGDN as C, 0.2 ppm, 2 mg/m^3 Skin, with the notation of 0.02 ppm for the intermittent exposure to avoid headache (see Section 2.8.6) at which value it remained until 1979 when a proposal was made to list NG separately. The OSHA standard for NG is 0.2 ppm 2 mg/m^3 Skin.

2.8.8   Propylene Glycol 1,2-Dinitrate, $CH_3CHONO_2CH_2ONO_2$

Preparation and Use.   Propylene glycol 1,2-dinitrate (PGDN) is prepared by nitrating the corresponding glycol.

A major use of PGDN is as a constituent of Otto fuel II, a torpedo propellant, used by the U.S. Navy.

Physical and Chemical Properties.   PGDN is a volatile, red-orange liquid with a disagreeable odor. Under ordinary conditions it is unstable; accordingly it is stabilized by small additions of 2-nitrodiphenylamine and di-N-butyl sebacate, substances that have been shown to have no overt toxic effects at 50 times the maximal animal test dose. It has the following physical constants:

Freezing point	$-18.4°F$
Boiling point	Decomposer above 250°F (121°C)
Vapor pressure	0.877 mm Hg (77°F)
Density	1.232 g/ml (77°F)
Flash point	260°F
Heat capacity	0.44 B.U./ml

Physiologic Response. Considerable information in this area is available ranging from acute, subacute, and chronic toxicity in animals, to experimental human responses, to highly sophisticated neurophysiologic measurements in workers performing routine maintenance procedures for prolonged periods.

Acute Toxicity—Animals. In a comparative toxicity study with triethyleneglycol dinitrate (TEGDN), LD_{50} values in the mouse, guinea pig, and rat, by oral, subcutaneous, and intraperitoneal injection showed that PGDN was 4.0 times more toxic orally for the rat than TEGDN (LD_{50}, 250 mg/kg); 4.8 times, subcutaneously (LD_{50}, 530 mg/kg); and approximately 1.7 times, intraperitoneally in the guinea pig and rat (LD_{50}, 402 and 479 mg/kg, respectively). Only by the oral route in the mouse was PGDN, for some unexplained reason, slightly less toxic than TEGDN (LD_{50}, 1047 mg/m³), considerably out of line with the LD_{50} values for the rat and guinea pig (78).

Both PGDN and TEGDN produced methemoglobin in the rat, with PGDN producing it at a far faster rate, while exhibiting ataxia, lethargy, and respiratory depression. Rats on TEGDN, on the other hand, were hyperactive to auditory and tactile stimulation. Methemoglobin as a contributing cause of death was shown by pretreating rats with methylene blue, in which case the time to death for PGDN was extended 224 min beyond the 197 for the average time to death without methylene blue (78). Of four enzymes tested for altered activity, after PGDN treatment, only alkaline phosphatase and creatine kinase showed moderate increase in activity.

A 4-hr inhalation exposure of rats to PGDN mist at 1350 mg/m³ (approximately 200 ppm) resulted in no deaths and no overt signs of toxicity after 14 days, but methemoglobin values (mean of 6) reached 23.5 percent (79).

In ocular and primary skin irritation experiments in rabbits, no immediate reaction occurred after instillation of 0.1 ml PGDN, but redness of the conjunctiva was noted after 5 min. The iris and cornea were not involved, and the redness gradually abated and disappeared within 24 hr (79), indicating PGDN to have a low index of irritation to the eye and skin.

Acute Toxicity—Man. Because early signs of toxicity from PGDN involve organoleptic sensations, which small laboratory animals are incapable of registering, investigators turned to human subjects for more definitive physiologic responses (80).

Twenty human volunteers, 17 of whom were aged 22 to 25, plus 2 males, ages 45 and 51, and 1 female, age 24, served as subjects for inhalation exposures of the vapor of PGDN, in a series of 11 experiments involving concentrations ranging from 0.01 ppm (approximately 0.075 mg/m³) to 1.5 ppm for durations of from 1 to 8 hr.

Using headache, mild and severe, and eye irritation as measures of response, the first level definitely to develop a frontal headache was 0.1 ppm after a 6-hr exposure. This occurred in one of three subjects and persisted for several hours. At 0.2 ppm, only 2 of 12 subjects did not experience headache; two, mild within 2 hr, and five, within 8 hr; within 8 hr, however, 6 of the 12 experienced severe headache, with one complaining of eye irritation. Repeated daily exposures at this level resulted in dramatic decrease in intensity of headache. Odor of mild intensity was detected immediately, but after 5 min, the odor was no longer detected. At the 0.35 ppm level, the subjects reported mild headache after 2 hr or more, becoming severe after 8 hr, and one of three of the 2-hr subjects reported eye irritation after 5 min; this subsided completely shortly after exposure.

The 0.5-ppm level was the first to show neurologic and vascular changes as shown by abnormal modified Romberg and heel-to-toe tests, and a mean elevation of diastolic pressure of 12 mm. Headache, initially mild, became progressively worse and throbbing in nature, with dizziness and nausea after 6-hr exposure.

At the highest level, 1.5 ppm (approximately 11.25 mg/m^3), that could be sustained by the subjects for short periods (1.2 and 3.2 hr), headache pain was almost incapacitating, causing termination of the exposure after 3 hr; coffee ameliorated the pain, which persisted for 1 to 7.5 hr postexposure. The Flanagan coordinates test was abnormal.

Exhaled breath concentrations of PGDN after a 1-hr exposure at 1.5 ppm measured 20 to 35 ppb, and remained at this level for the rest of the exposure. At 5 min postexposure only a trace (5 ppb), the limit of sensitivity of the method, was detected, and none 15 min postexposure. Only trace amounts of PGDN were found in the blood of subjects exposed at the higher levels, and no levels of exposure were sufficient to elevate blood nitrate or methemoglobin above control values.

Monitored central nervous system responses showed that levels of 0.2 ppm and above produced disruption of the organization of the visual-evoked response (VER). Subjects repeatedly exposed at 0.2 ppm for 8 hr on a daily basis developed a tolerance of the induction of headaches, but the alteration in VER morphology appeared cumulative. Marked impairment in balance became manifest after exposure at 0.5 ppm for 6.5 hr, and 40 min of exposure at 1.5 ppm added eye irritation to the list of symptoms.

Subacute and Subchronic Response—Animal. As the Navy program of filling torpedos with propellent PGDN progressed without benefit of toxicologic information by inhalation or skin, routes with known potential to cause untoward effects, animal studies were performed by the Navy to fill both these gaps (78, 79).

The inhalation studies performed (79) were designed to determine the minimal dose at which toxic signs could be observed, and accordingly were performed at levels considerably in excess of those permitted in workroom air

(31.4, 14.5, and 9 ppm). Four species of animals, monkeys, dogs, rats and guinea pigs, were exposed continuously for 90 days at these levels, after it had been found that no toxic effects or mortality occurred in rats after 30 exposures at approximately 8.7 ppm on a daily 7-hr, five times weekly basis.

Apart from one monkey that died on day 31 from exposure at the highest level, 31.4 ppm, the death probably complicated by a parasitic infection, there were no other deaths or visible signs of toxicity in any of the other animals in the three exposures, other than those noted below. The rate of body weight gain in all exposures was essentially the same as that of controls, and postexposure hematologic values were all within normal limits for all species except the dogs exposed at 31.4 ppm. These animals showed decreases of 63 and 37 percent in their hemoglobin and hematocrit values. Methemoglobin values increased in all exposed species, being most marked in dogs (23.4 percent maximal) and monkeys, 17 percent. Serum inorganic nitrate determined in monkeys rose as high as 375 µg/ml above controls, and as high as 172 µg/ml in dogs, the only two species examined.

The liver from both dogs in the exposure showed hemosiderin deposits in the sinusoids, bile canaliculi, Kupffer cells, and liver plate cells. Similar pigment accumulations were seen in the cytoplasm of the epithelial cells lining the proximal convoluted tubules in the kidneys of the dogs and some of the rats. Fatty changes were noted in some of the guinea pig and rat livers.

In the 14.5 ppm exposure, guinea pigs consistently showed foci of pulmonary hemorrhage. Fatty changes similar to those seen in the 9 ppm exposure were also seen in the livers of dogs and monkeys. Monkeys had iron-positive granules present in the central areas of the liver and in some kidney sections. Liver sections from dogs revealed heavy iron-positive deposits in the sinusoids and Kupffer cells.

Heavy iron-positive deposits were also present in the liver, spleen, and kidney sections of dogs and monkeys exposed at 236 mg/m^3. Hepatic iron-positive deposits were commonly associated with vacuolar change, mononuclear cell infiltrates, and focal necrosis. Female rats showed focal necrosis of the liver and acute tubular necrosis of the kidney that appeared to be related to the test material; male rats appeared normal. Vacuolar changes noted in the liver of all guinea pigs and in four of nine monkeys were also attributed to the exposures. No changes were noted in any of the other tissue sections examined from the three exposures. Control rats and guinea pigs were normal.

Squirrel monkeys exposed to 14.5 and 31.4 ppm PGDN had elevated serum urea nitrogen and decreased serum alkaline phosphatase levels, indicating the possibility of kidney change in this species. Elevation in the bromsulphalein retention values in one dog exposed at 31.4 ppm was noted. Serum asparate aminotransferase and liver alkaline phosphatase determinations were in close agreement with controls.

Behavioral studies on rhesus monkeys trained to perform in a visual discrimination test (VDT) and in a visual acuity threshold test (VATT), and later exposed continuously for 90 days at 31.4 ppm PGDN, produced no changes in the avoidance behavioral pattern as indicated by the VDT and VATT tests.

Jones et al. (79), noting the similarity in toxicologic effects of EGDN and PGDN, reasoned that circulatory collapse, "sudden death," seen in EGDN exposures could also occur from PGDN and that similarly, urinary nitrate determinations could be useful in determining PGDN absorption.

Dermal applications of PGDN on a subacute schedule in rabbits showed high absorption by this route (79), and hence raise a potential toxic hazard to workers handling PGDN.

In *subacute dermal toxicity* tests in which doses of 1, 2, and 4 g/kg were applied daily for 90 days to the backs of rabbits, 13 of 14 rabbits died after the fifth application at the highest dose. Internal organs took on a dark, blue-gray appearance. At the 2 g/kg dose, weakness and slight cyanosis was seen at the start with one rabbit dead after the sixth application. This was followed by steady physical improvement, except for slight wrinkling and scaling of the skin in the area of application, and the animals appeared normal and showed a weight gain of 15 percent on day 20. In the lowest dosed group, only minor irritation and roughening of the skin was noted. This cleared by the fifth day (79). These findings are in agreement with a previous 3-week dermal application of about 3.5 g/kg by Andersen and Mehl (78); 6 of 11 rabbits died on this regimen, with a mean time to death of 16 days.

Long-Term Effects. The need for further information on the long-term effects at low levels of PGDN on performance of workers under actual conditions of brief, intermittent, and peak airborne concentrations prompted a behavioral toxicologic study involving quantitative eye-tracking and ataxia tests (81).

A study group consisted of 115 active duty (Navy) and civilian personnel, 87 of whom were designated as "chronically exposed." Of these, 29 were tested before and immediately after PGDN exposure during routine torpedo maintenance procedure, designated "turnarounds." Nonexposed controls, numbering 28, were similar in sex distribution, race, smoking habits, and caffeine intake, whereas Otto fuel workers as a group consumed more than twice as much alcohol as did controls, a substance regarded as aggravating the toxicity of aliphatic nitrates.

The mean age of both exposed and control groups, predominantly male, was 33, with similar distribution in all age groupings by decade. A subgroup of 28 with 5 or more years total exposure, in addition to greater alcohol consumption, differed significantly ($P = 0.004$) from controls, 40.6 versus 33.3 years), and in greater intake of caffeine (957.6 versus 484.8 mg/day, $P = 0.015$), and cigarettes (17.3 versus 8.0 packs/year, $P = 0.018$).

The duration of exposure of the entire group averaged 47.4 months, with a range of 1 to 132; subgroups, 91.8 months average, with a range of 60 to 132;

the total exposure indices were 719 and 2017 turnaround months, respectively. A mean of 14 grab samples were taken during each turnaround procedure, where concentrations ranged from 0.00 to 0.22 ppm with a mean concentration for the 29 TAs of 0.03 ppm. Only one sample exceeded the current TLV of 0.2 ppm, and 87.5 percent of all peak concentrations were equal to or less than one-half the TLV.

Neurologic Test Results. Of the two neurologic tests performed on the chronically exposed Otto fuel workers, the oculomotor (quantitative eye-tracking) and quantitative ataxia tests, no statistically significant differences from controls appeared in the mean scores, with the exception of the subgroup which scored *better* on the Sharpened Romberg test. These findings held also for another group of workers with a longer mean exposure duration of almost 8 years with a range of 5 to 11 years.

Eye tests on 29 turnarounds, however, showed a significant decrease ($P = 0.03$) in velocity of eye movement and in latency ($P = 0.04$) when tested before and directly after exposure. Apparent alterations in standing behavior (on one leg for 30 sec) were of questionable significance.

Accordingly, the investigators concluded (1) PGDN intermittent exposures during routine maintenance procedures (turnarounds) can produce subclinical neurologic alterations as measured by oculomotor function tests; (2) these changes can occur at air levels less than half of the TLV of 0.2 ppm; and (3) there is no evidence that repeated, daily exposures for up to 11 years result in permanent impairment of the oculomotor system and its brain control centers (81).

These oculomotor function tests are apprently far more sensitive than the conditioned avoidance behavior test; continuous, 23-hr/day exposures of primates (*Macaca mulatta*) for periods up to 35 days at levels far in excess of those of the Otto fuel workers (0.24 to 3.5 ppm) had no discernible effects on avoidance behavior (82).

Cardiovascular Measurements. Recognizing the seriousness of the cardiovascular effects of the nitrate esters, diastolic hypertension, and angina pain with sudden death (46, 54), measurements were made of systolic and diastolic and blood pressure rate in Otto fuel workers (81). Although statistically significant decrease in diastolic pressure and increase in pulse was found, the changes were of borderline significance ($P = 0.044$), owing to confounding effects of age, alcohol, nicotine and caffeine intake, recent physical activity, and psychologic stress. Thus the investigators concluded it was impossible to draw positive inferences on the relation between PGDN exposure and cardiovascular effects.

Metabolism. The metabolism of PGDN as determined in vitro in blood, and in vivo in specific pathogen-free rats, showed that 50 percent was broken down in 1 hr, and 50 percent of the remainder in the following hour (83). Small concentrations of inorganic nitrite were produced during the incubation in blood, while inorganic nitrate accumulated. At the end of 3 hr, the first time it was measured, there were large amounts of propylene glycol 2-mononitrite, PGMN-2, together with small amounts of PGMN-1. The summed quantities of mononitrates, inorganic nitrate, and nitrite represented 95 percent of the initial amound added to blood. The breakdown occurred in the erythrocytes.

In the intact rat, unlike in the blood, the mononitrates undergo further degradation to nitrogen compounds other than the mononitrates and inorganic nitrate, as only 56 percent of the administered PGDN appeared in the urine as inorganic nitrate. Thus there is qualitatively little to distinguish the in vitro and in vivo metabolism of PGDN from that of EGDN (65). The only point of difference is that PGDN gives rise to two mononitrates with the 2-isomer predominant, whereas EGDN gives rise to only EGMN. Quantitatively, there is less dinitrate and inorganic nitrite in the bloodstream after subcutaneous injection from PGDN than from a comparable injection of EGDN. Excretion was complete in 24 hr following a 65 mg/kg PGDN subcutaneous injection in rats (83).

Mechanism. The mechanism of only one aspect of the physiologic action of PGDN has thus far been elucidated, that of its oxidation of hemoglobin (84). No attempt was made to *explore the mechanism for the most frequently reported* symptom (65 percent), headache.

It had previously been noted by Andersen and Mehl (78) that PGDN produces more methemoglobin in vivo than equivalent doses of TEGDN. Accordingly, effort was made to understand better this oxidative process and perhaps throw some light on the role of hemoglobin in detoxifying the nitrate esters.

The reaction was found to be not enzymatic, but molecular, and was first order in dinitrate and in O_2Hb. The rate of oxidation proceeds linearly with dinitrate concentration, and does not approach a limit as would be the case if enzymatically driven. The rate of oxidation is related complexly to the oxygen concentration; no oxidation occurs at zero oxygen concentration and none at very high concentrations. The stoichiometry was thought to be 1.5 hemes oxidized per ester bond broken in hemolysates, and 1.9 to 2.3 per mole reacted ester in whole cells. From these studies, it was reasoned that hemoglobin would fulfill an important role in detoxifying the effects caused by the intact dinitrates. Hemoglobin in vivo, with the methemoglobin reductase system, acts catalytically to metabolize dinitrates to nitrite and nitrate, and the mononitrates are further degraded by the denitrifying tissue enzymes.

Hygienic Standards of Permissible Exposure. The TLV Committee proposed a ceiling value of 0.2 ppm (approximately 2 mg/m^3) in 1975 after the report of Stewart et al. (80) appeared, showing that human volunteers repeatedly exposed at 0.2 ppm PGDN and above experienced headache and disruption in visual evoked response, and at higher levels, 0.5 ppm, marked impairment of balance. In 1979, the TLV was reduced to 0.02 ppm without the ceiling notation from advanced information of adverse effects in workers with PGDN "at air levels generally less than 0.1 ppm" (81).

2.8.9   Pentaerythritol Tetranitrate, $C(CH_2ONO_2)_4$

Production and Use. Pentaerythritol tetranitrate (PETN) is produced by nitrating erythritol, which is obtained by reacting formaldehyde with acetaldehyde. The PETN, after precipitation in water, dissolved in acetone, and reprecipitated, is used as a water-wet product of 40 percent water content, the only state in which it can be shipped.

PETN is used as a filling in detonating fuse and detonators. In admixture with TNT it is used for loading small caliber projectiles and grenades as well as booster charges.

Physical and Chemical Properties

Molecular weight	316.15
Melting point	141.3°C
Boiling point	180°C (50 mm Hg)
Specific gravity	1.765
Solubility	4.3 mg/100 g H_2O (25°C); 25.4 g/100 g acetone (20°C); slightly soluble in most organic solvents

Pentaerythritol tetranitrate has an explosion temperature of 225°C, near that of NG, but is less sensitive to impact and friction. It is extremely sensitive to initiation of explosion by lead azide and other initiating agents, much more so than TNT or tetryl, and its explosive strength is at least 50 percent greater than TNT. Although PETN safely withstands storage for 18 months at 65°C, continued storage has marked effects of instability; the presence of as little as 0.01 percent free acid or alkali in PETN markedly accelerates its deterioration. It is the least stable of the standard military bursting-charge explosives.

Physiologic Response. Pentaerythritol tetranitrate is absorbed only slowly from the gastrointestinal tract and lung, but not appreciably through the skin, owing to its low water solubility. Its physiologic effects are similar to those of the other aliphatic nitrates, although considerably less potent as a vasodilator than, for example, NG. Doses of 5 mg/kg by mouth in dogs result in a fall in blood pressure but no effect is observed in man after 64 mg orally (36). The daily oral administration of 2 mg/kg for 1 year caused no effects on growth, hematology, or pathology in rats. Patch tests in 20 persons gave no evidence of skin irritation or sensitization. Although some cases of mild illness and dermatitis have been attributed to contact with PETN in ordnance plants (47), it is apparent that PETN is relatively nontoxic and the controls and good housekeeping necessary to prevent explosions from this shock-sensitive material should be adequate to prevent injurious effects in workers.

No TLV has been established for PETN because of insufficient data.

2.8.10 Trimethylenetrinitramine, Cyclotrimethylenetrinitramine, Cyclonite, Hexogen, T_4, RDX

Production and Use. Cyclonite, like PETN, can be produced when only coal, water, air, and electrical energy are available. For this reason, Germany and Italy emphasized its production and use it under the names Hexogen and T_4. The British also used it under the name RDX, which is its military name in the United States.

Cyclonite can be produced by a number of processes. The two most important are the nitration of hexamethylenetetramine (HMX) and a similar nitration in the presence of ammonium nitrate and acetic anhydride. It forms colorless orthorhombic crystals. It has been widely used as a base charge for detonators and as an ingredient of bursting-charge and plastic explosives by the military.

Physical and Chemical Properties

Melting point	(Pure; 204.1°C; impure military grades containing ca. 10 percent HMX, about 190°C
Specific gravity	1.816
Solubility	Insoluble H_2O; slightly soluble most organic solvents; readily soluble hot aniline, phenol, warm concentration HNO_3

The sensitivity to impact of cyclonite approximates that of tetryl (trinitrophenylmethylnitramine) but cyclonite is somewhat more sensitive to friction and rifle-bullet impact. Both are equally sensitive to heat. Cyclonite is more sensitive than tetryl to initiation by lead azide.

Cyclonite is of the same order of brisance and explosive strength as PETN and distinctly superior to tetryl as determined by the sand, rate of detonation, ballistic pendulum, and Trauzl lead-block tests. It may be considered, therefore, to be at least the equal of, if not superior to, any of the solid bursting-charge explosives available in quantity.

The stability of cyclonite is considerably superior to that of PETN and nearly equal to that of TNT, as indicated by the vacuum-stability test. It withstands storage at 85°C for 10 months or at 100°C for 100 hr without measurable deterioration; hence from the viewpoint of stability, cyclonite must be considered highly satisfactory.

Physiologic Response. RDX does not exhibit pharmacologic effects similar to the nitrites or nitrates. Chronic intoxication is characterized by the occurrence of repeated convulsions. It is slowly absorbed from the stomach and apparently from the lungs, but there is no evidence of skin absorption.

The minimum lethal dose as determined in rats by single oral doses of a 4 percent solution was approximately 200 mg/kg (45). There was one death in 35 rats given 15 mg/kg daily for 10 weeks by mouth. Deaths occurred in 17 of 35 rats given 50 mg/kg daily and 15 of 35 on daily doses of 100 mg/kg. The animals lost weight, became increasingly irritable and vicious, and developed frequent convulsions. Gross pathology in those dying during exposure showed lung and gastrointestinal tract congestion. Those surviving had no pathologic changes. Seven dogs given 50 mg/day, 6 days/week for 6 weeks showed no blood changes and no methemoglobinemia. A few hours after the first dose they became excited and irritable. As dosing continued, reflexes became hyperactive and within the first week the animals had generalized convulsions characterized by hyperexcitability and increased activity, followed by clonic movements and salivation, then tonic convulsions and collapse. There was weight loss in all animals and death in one. No microscopic pathology was observed (45).

Epileptiform seizures have occurred in workers manufacturing trimethyle-netrinitramine (T_4) in Italy (85). The convulsions occurred either without warning or after 1 or 2 days of insomnia, restlessness, and irritability. They were generalized tonic–clonic convulsions resembling in all clinical respects the seizures seen in epilepsy but occurring in individuals without a previous history of seizures. They were most frequent in persons doing the drying, sieving, and packing, where the dust could be inhaled. The attacks disappeared when the workers were removed from contact with T_4. The seizures were followed by temporary postconvulsive amnesia, malaise, fatigue, and asthenia, but there was eventually complete recovery.

Similar evidence of systemic intoxication was not observed in a major RDX manufacturing plant during World War II. In this operation there was little or no dusting since the material was handled in a moist state. However, primary irritant and sensitization dermatitis, particularly of the face and eyelids, was encountered during the nitration operation. Studies indicated that an unidentified component in the fumes from the reaction mixture was responsible (86).

Although McConnell (47) attributed some dermatitis to RDX manufacture, this probably was due to intermediates since significant dermatitis was not observed in individuals handling the final purified material (86). This observation is corroborated by von Oettingen's finding that patch testing with the moistened solid did not produce irritation (45a).

Five cases of convulsions and/or unconsciousness occurred among about 26 workers engaged in pelletizing RDX as late as 1962 in the United States (87). Exposure occurred from the release of dust in the workroom air during dumping of the dried RDX powder in an unventilated room and sweeping the RDX that had spilled on the floor; the dust covered the clothing and arms of the workers. The typical symptoms of RDX intoxication occurred either at work or several hours later at home, with few prodromal signs of headache, nausea, and vomiting. Unconsciousness lasted several minutes to 24 hr with varying periods of stupor, nausea, vomiting, and weakness. Recovery was complete with no sequelae, but two men reexposed to RDX had recurrences of illness. When control measures were installed, illnesses disappeared.

Hygienic Standards of Permissible Exposure. The TLV for RDX recommended by The Committee was adopted by ACGIH in 1969 as 1.5 mg/m^3 with the "Skin" notation, and has remained at this value since that time. The value was based on its analogy to TNT and its demonstrated effectiveness in the prevention of injury in an AEC establishment when RDX was maintained below 1.5 mg/m^3 (88). OSHA later adopted the limit. The TLV booklet for 1980 lists RDX as cyclonite without change in TLV but with the addition of an STEL of 3 mg/m^3.

3 ALKYL NITRITES

3.1 General Considerations

The alkyl nitrites are aliphatic esters of nitrous acid. The nitrite group has the structure—CONO. Except for methyl nitrite, which is a gas, the lower members of the series are volatile liquids. In general they are insoluble or only very slightly soluble in water but are soluble or miscible with alcohol and ether in most proportions. They tend to decompose to oxides of nitrogen with exposure to light or heat. Violent decomposition can occur. As a group, they tend to be flammable and potentially explosive. They are oxidizing materials which present the possibility of violent reactions from contact with readily oxidized compounds. The physical and chemical properties of the alkyl nitrites are given in Table 53.18.

The aliphatic nitrites have been of interest mainly because of their pharmacologic properties and therapeutic use, but they are also used to a limited extent

Table 53.18. Physical and Chemical Properties of the Alkyl Nitrites

Name	Formula	Molecular Weight	Physical State	Boiling Point (°C)	Specific Gravity
Methyl nitrite	CH_2ONO	61.04	Gas	−12	0.991 (15°C)
Ethyl nitrite	CH_3CH_2ONO	75.07	Colorless liquid	17	0.900 (15.5°C)
n-Propyl nitrite	$CH_3CH_2CH_2ONO$	89.09	Liquid	57	0.935
Isopropyl nitrite	$(CH_3)_2CHONO$	89.09	Pale yellow oil	45	0.844 (25/4°C)
n-Butyl nitrite	$CH_3(CH_2)_2CH_2ONO$	103.12	Oily liquid	78.2	0.9114 (0/4°C)
Isobutyl nitrite	$(CH_3)_2CHCH_2ONO$	103.12	Colorless liquid	67	0.8702 (20/20°C)
sec-Butyl nitrite	$CH_3CH_2CH(CH_3)ONO$	103.12	Liquid	68	0.8981 (0/4°C)
tert-Butyl nitrite	$(CH_2)_3CONO$	103.12	Yellow liquid	63	0.8941 (0/4°C)
n-Amyl nitrite	$CH_3(CH_2)_4ONO$	117.15	Pale yellow liquid	104	0.8528 (20/4°C)
Isoamyl nitrite	$(CH_3)_2CHCH_2CH_2ONO$	117.15	Transparent liquid	97–99	0.872
n-Hexyl nitrite	$CH_3(CH_2)_5ONO$	131.17	Liquid	129–130	0.8851 (20/4°C)
n-Heptyl nitrite	$CH_3(CH_2)_6ONO$	145.20	Yellow liquid	155	0.8939 (0/4°C)
n-Octyl nitrite	$CH_3(CH_2)_7ONO$	159.23	Greenish liquid	174–175	0.862 (17°C)

in industry as intermediates in chemical syntheses. *n*-Butyl nitrite has been used in the manufacture of rare earth azides. *n*-Propyl nitrite, isopropyl nitrite, and *tert*-butyl nitrite have been used as jet propellants and for the preparation of fuels. They are usually prepared by the action of sodium nitrite on a mixture of the alcohol and sulfuric acid.

The pharmacologic and toxicologic effects of the aliphatic nitrites are chiefly characterized by vasodilatation resulting in a fall in blood pressure and tachycardia. Methemoglobin is produced by larger doses. In these respects the alkyl nitrites resemble closely the inorganic nitrites (sodium nitrite) and the aliphatic nitrates. Inhalation by animals and man results in smooth muscle relaxation, vasodilatation, increased pulse rate, and decreased blood pressure progressing to unconsciousness with shock and cyanosis. Headache is often a prominent symptom and may be due to meningeal congestion and vascular dilatation. The development of tolerance has been observed in the therapeutic use of amyl nitrite for angina pectoris. This disappears after a week or so of "nonexposure." The branched chain compounds are more effective than the straight chains in lowering blood pressure. Isopropyl nitrite is considerably more effective than *n*-propyl nitrite and isobutyl nitrite more than *n*-butyl. The secondary and tertiary butyl compounds also have a more pronounced hypotensive effect than normal butyl nitrite. Methyl nitrite is more effective than are ethyl and propyl nitrites, and amyl nitrite is more effective than ethyl nitrate. As far as the duration of the hypotensive effect is concerned, methyl and ethyl nitrites are most effective, *n*-propyl is the least effective of the lower alkyl nitrites, and the iso derivatives of propyl and butyl nitrite are more effective than the normal compounds (35b). The hypotensive effects are very transient. Amyl nitrite, for example, produces a rapid fall in blood pressure, which lasts only a few minutes after inhalation. Krantz, Carr, and associates have conducted extensive studies on the pharmacology of the alkyl nitrites (37–39). They found that when dogs were exposed by administering the vapor of 0.3 cm^3 through an aspirating bottle into the trachea, the degree of hypotension produced decreased from *n*-hexyl (58 percent fall in blood pressure) through *n*-heptyl (47 percent) and *n*-octyl (30 percent) to *n*-decyl (16 percent). Alkyl nitrites with 11 to 18 carbon atoms in their chain showed slight or no effect on blood pressure under these conditions. However, if injected, they produced hypotension. With chains longer than 2-ethyl-*n*-hexyl-1-nitrite, the duration of action became shorter. Cyclohexyl nitrite produced a fall in blood pressure equivalent to ethyl nitrite or amyl nitrite but the duration was longer. In man it produced severe headache. They felt that the major effects are related to the relaxing action of the nitrites on smooth muscle.

Methemoglobin formation has been repeatedly observed following administration to man and animals. The aliphatic nitrites act as direct oxidants of hemoglobin. One molecule of nitrite and two molecules of hemoglobin can react to form two methemoglobin molecules under appropriate conditions. Side reactions to form nitrosohemoglobin and nitrosomethemoglobin may occur. The amount of methemoglobin formed in cats is directly proportional to the intravenous dose (40). The longer chain compounds induce more methemoglobin formation relative to their hypotensive effect (37). The therapeutic usefulness of methylene blue in acute intoxications accompanied by methemoglobinemia remains controversial even though support for its effectiveness in severe methemoglobinemia continues to appear (40). Although methemoglobinemia is a prominent effect of nitrite absorption, the action of the alkyl nitrites on the vascular system is the major determinant in their toxicity.

The lower aliphatic nitrites are promptly absorbed from the lung. Amyl nitrite is ineffective by mouth since it is destroyed in the gut. It is less effective by injection than by inhalation. Octyl nitrite (2-ethyl-*n*-hexyl-1-nitrite) is not absorbed through the mucous membranes and is ineffective sublingually. It appears that the nitrites are hydrolyzed in vivo to nitrite and the corresponding alcohol, which is then partly oxidized and partly exhaled unchanged.

Although the inorganic nitrites, particularly sodium nitrite, have produced many accidental poisonings by ingestion, reports of industrial intoxications appear to be limited to methyl (89) and ethyl nitrite (90). Animals given increasing doses of amyl nitrite become ataxic, vomit, and have convulsions severe enough to terminate in death (91). In dogs given repeated inhalations at 20- to 90-sec intervals, for up to 7 min, gagging, involuntary urination and defecation, and twitching muscular movements were noted.

No reports of death, however, occurring among "snappers" or "poppers" of pearls of amyl or isobutyl nitrites have appeared, presumably because of tolerance development from frequent use (91).

Methods for the determination of nitrites in air and in biologic fluids have been described (35b).

3.2 Specific Compounds

Except for methyl and ethyl nitrite and amyl nitrite, specific information on the effects of the alkyl nitrites in man is lacking. However, the pharmacologic properties as determined in animals are so uniform within the group that information on these nitrites can be taken as illustrative of the effects and potential hazards of the other members of the series (see Tables 53.18 and 53.19).

3.2.1 Methyl Nitrite, CH$_3$ONO

Methyl nitrite is a gas with a boiling point of $-12°C$, a melting point of $-17°C$, and a specific gravity of 0.991 at 15°C. It is a severe explosion risk when shocked or heated. It has uses in the synthesis of nitrite and nitroso esters. It is formed

Table 53.19. Comparative Toxicity Data on Aliphatic Nitro Compounds, Nitrates, and Nitrites

Chemical Group	Skin Absorption	Irritation	Vascular Dilatation	Methemoglobin Formation	Industrial Experience
Aliphatic nitro compounds —CNO$_2$					
Nitroparaffins	None	Moderate	None	Positive	Irritation, systemic symptoms
Chlorinated nitroparaffins	None	Marked	None	Unknown	Lung injury
Nitroolefins	Positive	Marked	Unknown	Not observed	None
Aliphatic nitrates —CONO$_2$	Positive	None	Marked	Positive	Systemic symptoms, possible deaths
Aliphatic nitrites —CONO	Unknown	None	Marked	Positive	Systemic symptoms, 1 fatality
Trimethylenetrinitramine —CNHNO$_2$	None	None	None	None	Convulsions

as a by-product in the synthesis of a rubber antioxidant when methyl alcohol, $NaNO_2$, and HCl are added to the reaction mixture (92).

Methyl nitrite is a relatively toxic compound by inhalation for both animals and man; seven of eight rats died after 24-hr exposure at 250 ppm, but all eight rats survived thirteen 6-hr exposures at 110 ppm with methemoglobin levels of 30 to 40 percent; organs were grossly normal at autopsy. The 25 ppm level for fifteen 6-hr exposures was apparently a gross "no-effect" level for rats; no toxic signs were observed and organs were normal; however, 35 ppm produced 6 percent methemoglobin in a cat after a 6-hr exposure (93).

Methyl nitrite was found to be a potent cyanosing agent for workers synthesizing a rubber antioxidant (89). Six cases of methyl nitrite intoxication are described, consisting initially of dizziness and later headache and palpitation, the last more pronounced in two workers who consumed alcohol after exposure at work. All men responded satisfactorily to bed rest for 12 hr and the inhalation of oxygen for about 2 hr.

Atmospheric concentrations in the plants where the men had been affected, simulating the conditions at the time of the intoxications, indicated that "50 ppm is the uppermost limit of safety" (89).

3.2.2   Ethyl Nitrite, CH_3CH_2ONO

Ethyl nitrite is a volatile, flammable, colorless liquid. It has a boiling point of 17°C and a specific gravity of 0.900 at 15.5°C. Its flash point is $-31°F$ and the explosive limits in percent by volume in air are 3.01 to 50. The autoignition temperature of the liquid is 195°F. Thus it has a high potential fire and explosion hazard. It decomposes readily to form oxides of nitrogen.

Inhalation of ethyl nitrite by dogs results in as much as 70 mm Hg depression in the blood pressure. This lasts approximately 2 min after a single inhalation. Methemoglobin is formed but Heinz bodies have not been found (35b). Mice and cats exposed for 15 min to 15 ppm did not show recognizable effects. Industrial intoxications characterized by headache, tachycardia, and methemoglobinemia have occurred. A fatality has been described following the inhalation of ethyl nitrite after accidental breakage of a 4-liter bottle containing spirits of ethyl nitrite (24 percent ethyl nitrite in alcohol) (31).

Three cases of ethyl nitrite intoxication have been reported from Czechoslovakia (90) during a synthesis of hydantoin (glycolylurea). Symptoms of both "nitrite" effect and methemoglobinemia were noted. In addition, there was a vasodilator effect upon the blood vessels of the sclerae, producing a peculiar redness of the eyes, a hitherto undescribed manifestation of the nitrite effect. No Heinz bodies were seen, as is to be expected.

3.2.3   Amyl Nitrite, $(CH_3)_2CHCH_2CH_2ONO$ (Isoamyl Nitrite)

Isoamyl nitrite is a light yellow, transparent liquid with a pleasant, fragrant, fruity odor. Its physical properties are molecular weight, 117.15; boiling range,

97 to 99°C; and specific gravity, 0.872. It decomposes upon exposure to air and sunlight. It is flammable and explosive. It is prepared by the addition of sodium nitrite to a mixture of isoamyl alcohol and sulfuric acid followed by distillation.

Amyl nitrite was introduced to medicine in 1859 and has received considerable pharmacologic investigation since that time. Its major use was for the treatment of angina pectoris through its vasodilatory effect on the coronary arteries. However, this effect is transient and it has largely been replaced by nitroglycerin and longer acting nitrates. Amyl nitrite has been most helpful in clarifying the differential diagnosis of murmurs due to left ventricular outflow tract obstruction (which increase following amyl nitrite) from those of mitral regurgitation (which decrease); the apical diastolic rumble of mitral stenosis (which increases) from the Austin-Flint rumble (which decreases); ventricular septal defect (which decreases) from pulmonic stenosis (which increases); and acyanotic tetrology of Fallot (which decreases) from isolated valvular pulmonary stenosis (which increases) (93).

The symptoms following inhalation of large doses by man are flushing of the face, pulsatile headache, disturbing tachycardia, cyanosis (methemoglobinemia), weakness, confusion, restlessness, faintness, and collapse, particularly if the individual is standing. These symptoms are usually of short duration. Industrial intoxications have not been reported (35b).

4 SUMMARY

Although the aliphatic nitro compounds and the nitrates and nitrites have several features in common (nitrogen–oxygen grouping, explosiveness, methemoglobin formation), there are significant differences. Some of their attributes are summarized in Table 53.19. The esters of nitric and nitrous acid, with the nitrogen linked to the carbon through oxygen, are very similar in their pharmacologic effects. Both produce methemoglobinemia and vascular dilatation, with hypotension and headache. These effects are transient. None of the series has appreciable irritant properties. Pathologic changes occur in animals only after high levels of exposure and are generally nonspecific and reversible. The nitric acid esters of the monovalent and lower polyvalent alcohols are absorbed through the skin. Information is not available on the skin absorption of the alkyl nitrites. Members of both groups are well absorbed from the mucous membranes and lungs. Heinz body formation has been observed with the nitrates but not with the nitrites.

The nitroparaffins, like the nitrates and nitrites, cause methemoglobinemia in animals. Heinz body formation parallels this activity within the series. Although some members are metabolized to nitrate and nitrite, there is no significant effect on blood pressure or respiration. As with the lower nitrates and nitrites, anesthetic symptoms are observed in animals during acute exposures, but these occur late. The prominent effect is irritation of the skin, mucous membranes, and respiratory tract. This is most marked with the chlorinated nitroparaffins and the nitroolefins. In addition to the respiratory tract injury, cellular damage may be observed in the liver and kidneys. Except for the nitroolefins, skin absorption is negligible.

The nitramine, trimethylenetrinitramine, has entirely different activity. It is a convulsant for man and animal. Skin absorption, irritation, vasodilatation, methemoglobin formation, and permanent pathologic damage after repeated doses are either insignificant or absent.

Transient illness has been associated with the industrial use or manufacture of these materials but fatalities have been rare and chronic intoxication has been uncommon. Some members of each group present extremely high fire and explosion hazards.

REFERENCES

1. *Chemical Buyer's Directory,* 1980, 1981, Schnell Publishing Co., New York.

2. *Chemical Economic Handbook,* Stanford Research Institute, September 15, 1977.

3. *C.S.C. Tech. Data Sheet No. 1,* "Nitroparaffins", Commercial Solvents Corp. New York, 1961.

4. *Nitromethane, Nitropropanes,* Hygienic Guide Series, American Industrial Hygiene Association, Akron, Ohio, 1960–1961.

5. (a) *C.S.C. Tech. Data Sheet No. 2,* "Storage and Handling of Nitromethane," Commercial Solvents Corp., New York, 1961; (b) *C.S.C. Tech. Data Sheet No. 7,* 2nd ed., "Storage and Handling of 2-Nitropropane," Commercial Solvents Corp., Hillside, Il.

6. (a)E. W. Scott and J. F. Treon, *Ind. Eng. Chem., Anal. Ed.,* **12,** 189 (1940); (b) J. F. Treon and F. R. Dutra, *Arch. Ind. Hyg. Occup. Med.,* **5,** 52 (1952); (c) I. R. Cohen and A. P. Altshuller, *Anal. Chem.,* **31,** 1638 (1959).

7. (a) H. J. Horn, *Arch. Ind. Hyg. Occup. Med.,* **10,** 213 (1954); (b) E. H. Vernot, J. D. MacEwen et al., unpublished results, Wright-Patterson A.F.B., Ohio, October 1976.

8. (a)W. Machle et al., *J. Ind. Hyg. Toxicol.,* **27,** 95 (1945); (b) I. K. Neklesova and M. A. Kudrina, *Hyg. Sanit.,* **34,** 429 (1969).

9. *Encyclopedia of Industrial Chemical Analysis,* F. D. Snell and L. S. Ettre, Eds., Vol. 16, John Wiley and Sons, New York, pp. 417–448.

10. T. R. Lewis, C. E. Ulrich, and W. M. Busey, *J. Environ. Pathol. Toxicol.,* **2,** 233 (1979).

11. (a) W. Machle, E. W. Scott, and J. F. Treon, *J. Ind. Hyg. Toxicol.,* **22,** 315 (1940); (b) W. Machle, E. W. Scott, and J. F. Treon, *J. Ind. Hyg. Toxicol.,* **21,** 72 (1939).

12. (a) W. B. Deichmann, W. E. MacDonald, et al., *Ind. Med. Surg.,* **34,** 800 (1965); (b) S. D. Murphy et al., *Toxicol. Appl. Pharm.,* **5,** 319 (1963).

13. W. Machle and E. W. Scott, *Proc. Soc. Exp. Biol. Med.,* **53,** 42 (1943).

14. J. H. Weatherby, *Arch. Ind. Health,* **11,** 103 (1955).

15. J. B. Skinner, *Ind. Med.,* **16,** 441 (1947).

16. R. F. Sievers et al., *U.S. Pub. Health Rep.,* **62,** 1048 (1947).

17. E. B. Vedder, *The Medical Aspects of Chemical Warfare,* Williams and Wilkins, Baltimore, 1925.

18. K. F. Lampe, T. J. Mende, and W. B. Deichmann, *Ind. Med. Surg.,* **27,** 375 (1958).

19. C. H. Hine et al., *J. Occup. Med.*, **20**, 333 (1978).

20. (a) W. Machle et al., *J. Ind. Hyg. Toxicol.*, **24**, 5 (1942); (b) E. W. Scott, *J. Ind. Hyg. Toxicol.*, **25**, 20 (1943).

21. U. Volker et al., *Biochem. Pharmacol.*, **27**, 2301 (1978).

22. A. E. Wade et al., *Biochem. Pharmacol.*, **26**, 963 (1977).

23. Z. Hadidian, E. K. Weisburger, J. H. Weisburger, et al., *J. Natl. Cancer Inst.*, **41**, 985 (1968).

24. Chung Wai Chiu et al., *Mutat. Res.*, **58**, 11 (1978).

25. *NIOSH Registry of Toxic Effects of Chemical Substances*, U. S. D. H.E.W., Rockville, Md., 1976.

26. K. F. Hager, *Ind. Eng. Chem.*, **41**, 2168 (1949).

27. V. B. Vouk and O. A. Weber, *Br. J. Ind. Med.*, **9**, 32 (1952).

28. (a) Massachusetts Division of Occupational Hygiene, unpublished data; (b) M. Gaultier et al., *Arch. Mal. Prof.*, **25**, 425 (1964).

29. L. Silverman, H. F. Schulte, and M. W. First, *J. Ind. Hyg. Toxicol.*, **28**, 262 (1946).

30. E. Browning, *Toxicity of Industrial Organic Solvents*, Chemical Publishing Co., New York, 1953, p. 374.

31. F. Flury and F. Zernik, *Schädliche Gase*, Springer, Berlin, 1931.

32. A. M. Prentiss, *Chemicals in War*, McGraw-Hill, New York, 1937.

33. A. A. Fries and C. J. West, *Chemical Warfare*, McGraw-Hill, New York, 1921, p. 143.

34. Weast, Ed., *Handbook of Chemistry and Physics*, C.R.C. Press, Cleveland, Ohio.

35. (a) S. E. Warren and G. S. Francis, *Am. J. Med.*, **65**, 53 (1978): (b) W. F. von Oettingen, "The Effects of Aliphatic Nitrous and Nitric Acid Esters on the Physiological Functions with Special Reference to Their Chemical Constitution," *Nat. Inst. Health Bull. No.* **186** (1946).

36. W. F. Von Oettingen, D. D. Donahue, A. H. Lawton, A. R. Monaco, H. Yagoda, and P. J. Valaer, "Toxicity and Potential Dangers of Pentaerythritol-Tetranitrate (PETN)," *U.S. Public Health Bull. No.* **282** (1944).

37. J. C. Krantz, C. J. Carr, and S. E. Forman, *Proc. Soc. Exp. Biol. Med.*, **42**, 472 (1939).

38. J. C. Krantz, C. J. Carr, S. E. Forman, and N. Cone, *J. Pharmacol. Exp. Therap.*, **70**, 323 (1940).

39. M. Rath and J. C. Krantz, *J. Pharmacol. Exp. Therap.*, **76**, 33 (1942).

40. O. Bodansky, *Pharmacol. Revs.*, **3**, 144 (1951).

41. R. T. Williams, *Detoxication Mechanisms*, 2nd ed., Wiley, New York, 1959.

42. P. Rofe, *Brit. J. Ind. Med.*, **16**, 15 (1959).

43. J. B. Hughes and J. F. Treon, *Arch. Ind. Hyg. Occup. Med.*, **10**, 192 (1954).

44. J. F. Treon, F. P. Cleveland, and J. Duffy, *Arch. Ind. Health*, **11**, 290 (1955).

45. (a) W. F. von Oettingen, D. D. Donahue, H. Yagoda, A. R. Monaco, and M. R. Harris, *J. Ind. Hyg. Toxicol.*, **31**, 21 (1949); (b) P. Needleman and E. M. Johnson, Jr., *J. Pharm. Exp. Therap.*, **184**, 709 (1973); (c) E. J. Fairchild II, *Arch. Environ. Health*, **14**, 111 (1967); (d) W. O. Hueper and J. W. Landsberg, *Arch. Pathol.*, **29**, 633 (1940).

46. P. Carmichael and J. Lieben, *Arch. Environ. Health*, **7**, 424 (1963).

47. W. J. McConnell, R. H. Flinn, and A. D. Brandt, *Occup. Med.*, **1**, 551 (1946).

48. (a) F. Pristera et al., *Anal. Chem.*, **32**, 495 (1960); (b) F. Pristera et al., "Compilation of Infrared Spectra of Ingredients of Propellants and Explosives," Tech. Memo. Rept. 1889, AMCMS Code H 30.11.1161.1, Picatiny Arsenal, Dover, N.J. (no date given).

49. A. F. Williams et al., *Nature*, **210**, 816 (1966).

50. J. H. Foulger, *J. Ind. Hyg. Toxicol.*, **18**, 127 (1936).

51. H. Yagoda and F. H. Goldman, *J. Ind. Hyg. Toxicol.*, **25**, 440 (1943).

52. J. F. Treon, F. P. Cleveland, and J. Duffy, *Arch. Ind. Health,* **11,** 290 (1955).

53. D. C. Trainor and R. C. Jones, *Arch. Environ. Health,* **12,** 231 (1966).

54. C. Einert et al., *Am. Ind. Hyg. Assoc. J.,* **24,** 435 (1963).

55. N. Zurlo et al., *Med. Lav.,* **54,** 166 (1963).

56. E. R. Kinkead, J. D. MacEwen, and E. H. Vernot, "Toxic Hazard Evaluation of Five Atmospheric Pollutant Effluents from Ammunition Plants," AMRL Rept. TR-76-XX Oct. 1976.

57. W. E. Rinehart et al., *Am. Ind. Hyg. Assoc. Quart.,* **19,** 80 (1958).

58. D. B. Hood, Haskell Laboratory for Toxicology and Industrial Medicine, E. I. du Pont de Nemours & Co., unpublished data, Rept. No. 21-53, 1953.

59. E. F. Murtha, D. E. Stabile, and J. H. Wills, *J. Pharmacol. Exp. Therap.,* **118,** 77 (1956).

60. E. Gross et al., *Arch. Exp. Pathol. Pharmakol.,* **200,** 271 (1942).

61. J. B. Bradbury, *Br. Med. J.,* **2,** 1213 (1895).

62. H. Wilhelmi, *Arch. Exp. Pathol. Pharmakol.,* **200,** 305 (1942).

63. C. D. Leake et al., *J. Pharm. Exp. Therap.,* **35,** 143 (1931).

64. R. A. Lange et al., *Circulation,* **46,** 666 (1972).

65. D. G. Clark and M. H. Litchfield, *Br. J. Ind. Med.,* **24,** 320 (1967).

66. G. E. Ebright, *J. Am. Med. Assoc.,* **62,** 201 (1914).

67. I. M. Rabinowitz, *Can. Med. Assoc. J.,* **50,** 199 (1944).

68. L. A. Crandell et al., *J. Pharm. Exp. Therap.,* **41,** 103 (1931).

69. (a) E. Gross et al., *Arch. Exptl. Pathol. Pharmakol.,* **200,** 271 (1942); (b) E. Gross et al., *Arch. Toxicol.,* **18,** 200 (1960).

70. F. C. Phipps, *Proc. Soc. Exp. Biol. Med.,* **139,** 323 (1972).

71. D. G. Clark, *Toxicol. Appl. Pharmacol.,* **21,** 355 (1972).

72. M. Kalin and B. Kylin, prepublication report to author, April 1968.

73. A. J. Flemming, C. A. D'Alonzo, and J. A. Zapp, *Modern Occupational Med.,* Lea and Febiger, Philadelphia, 1954.

74. I. Maccherini and E. Camarri, *Med. Lav.,* **50,** 193 (1959).

75. J. S. Weiner and M. L. Thomson, *Br. J. Ind. Med.,* **4,** 205 (1947).

76. S. Forssman, *Arch. Gewerbepathol. Gewerbehyg.,* **16,** 157 (1958).

77. A. M. Schwartz, *New Engl. J. Med.,* **255,** 193 (1959).

78. M. E. Andersen and R. G. Mehl, *Am. Ind. Hyg. Assoc. J.,* **34,** 526 (1973).

79. R. A. Jones et al., *Toxicol. Appl. Pharmacol.,* **22,** 128 (1972).

80. R. D. Stewart et al., *Toxicol. Appl. Pharmacol.,* **30,** 377 (1974).

81. E. P. Horvath, Jr., T. N. Markham, and R. A. Ilka, prepublication report to author by senior investigator, December 1980.

82. R. W. Young, C. R. Curran, G. G. Franz, and L. J. Jenkins. Jr., Effects of Chronic Inhalation of Propylene Glycol 1,2-Dinitrate on Conditioned Avoidance Behavior," Armed Forces. Radiation Biology Research Institute, June, 1976.

83. D. G. Clark and M. H. Litchfield, *Toxicol. Appl. Pharmacol.,* **15,** 175 (1969).

84. M. E. Andersen and R. A. Smith, *Biochem. Pharmacol.,* **22,** 3247 (1973).

85. M. Barsotti and G. Crotti, *Med. Lav.,* **40,** 107 (1949).

86. J. H. Sterner, Eastman Kodak Co., personal communication, 1961.

87. A. S. Kaplan, G. F. Berghout, and A. Peczenik, *Arch. Environ. Health,* **10,** 877 (1965).

88. E. C. Hyatt and M. F. Milligan, *Am. Ind. Hyg. Quart.,* **14,** 289 (1953).

89. W. G. F. Adams, *Trans. Assoc. Ind. Med. Officers,* **14,** 24 (1964).

90. T. Beritic, *Arch. Hig. Rada,* **8,** 333 (1957); *Ind. Hyg. Digest Abstr.* **55** (Jan. 1959).

91. S. Cohen, *J. Am. Med. Assoc.,* **241,** 2077 (1979).

92. J. C. Gage, *Br. J. Ind. Med.,* **27,** 1 (1970).

93. P. T. Cochran, *Am. Heart J.,* **98,** 141 (1979).

CHAPTER FIFTY-FOUR

Polymers

RUTH R. MONTGOMERY

Generally high polymers are substances of great chemical inertness and physiological and toxicological effects are slight or totally absent as may be expected. Although this is generally correct ... and there is a difference between the inert macromolecule and the active monomer, in practice matters are not as simple, and we will see in fact that the macromolecule behaves differently according to the mode of introduction into the body. ... What degradation means needs to be stated precisely. It is important to know not only its causes, but also to know what substances are formed during degradation, and more particularly during thermal degradation of macromolecules. This can, in fact, give rise to a question of industrial toxicology ... (Lefaux, Reference 1).

1 GENERAL CONSIDERATIONS

1.1 Introduction

The word polymer is derived from the Greek words *poly*, or many, and *meros*, or parts. Natural polymers include such diverse materials as proteins, polypeptides, polysaccharides, DNA, wood, wool, and silk. Their chemical structure may be specific or variable and undefined. Frequently they are placed in the more general category of macromolecules, which also includes mineral compounds or derivatives such as silica and glass.

Synthetic polymers are usually very large organic molecules that consist essentially of repeating structural units. The dividing line between polymers of low molecular weight (oligomers) and high polymers is usually placed in the molecular weight range of 10,000 (2) but sometimes as low as 5000 (3). Individual materials discussed in this review are essentially prototype polymers. (Epoxy resins are reviewed in Chapter 32, Volume 2A.) The products as

4209

generally sold frequently contain other ingredients such as curing agents, stabilizers, fillers, reinforcing agents, colorants, and plasticizers. Table 54.1 gives a historical overview and Table 54.2 indicates the relative commercial importance by major class type.

Polymerization describes any process, usually conducted with heat, light, or catalyst, that involves formation of a chain-link macromolecule. Most commercial processes use addition polymerization or condensation polymerization techniques. Addition or *chain reaction* polymerization starts with initiation by free radicals, from decomposition of materials such as benzoyl peroxide or azobisisobutyronitrile, that open the double bond of a monomer; this step usually controls the polymerization rate. The second step, chain growth, where additional monomer units are added to the growing chain, is extremely rapid. The third step, termination, occurs when two free radicals react, when chain transfer starts a new chain as the growing chain is terminated, or by disproportionation of the free radical end of the growing chain. Termination may occur naturally or by deliberate intent from chain transfer agents. Polyethylene and other olefin polymers are typical examples.

Step reaction polymerization is somewhat analogous to condensation in compounds of low molecular weight. Two chemicals of different type interact, frequently with the elimination of fragments or small molecules such as water. Typical examples are the polyesters and polyamides.

Linear or moderately branched polymers* such as polyvinyl chloride and nearly all polyethylene are *thermoplastic*: they soften, flow, or change shape with the application of sufficient heat and/or pressure. *Thermosetting* or network polymers, typically melamine–formaldehyde resins, are invariably cross-linked; during the polymerization process they change irreversibly into infusible, insoluble materials with a physical structure that is thermally stable until degraded. The term "plastic" strictly describes a polymer that is capable of being shaped or molded, but it often identifies a thermoplastic or thermoset polymer that has been formed under heat or pressure and combined with fillers, reinforcing agents, plasticizers, colorants, etc.

In general, organic polymers are useful in a limited temperature range compared to structural metals. The different grades and families of commercial synthetic polymers vary widely in their elevated temperature capabilities; these differences distinguish certain more costly polymers—for example, engineering materials such as polycarbonate or thermoplastic polyester and specialty materials such as polytetrafluoroethylene or polyimide—from lower temperature high volume materials, such as polyethylene, polyvinyl chloride, and polystyrene. Commercial synthetic polymers consist for the most part of chains of carbon,

---

* Branching implies side chains where the individual molecules are still discrete (5). Branched polymers may or may not contain cross-links. To some extent this varies with the polymer; branched polyethylene is typically linear but branched polyesters are typically thermoset.

Table 54.1. Some Commercial Polymers and Approximate Year of Introduction[a]

Date	Material	Typical Application
1868	Cellulose nitrate (polymer)	Knife handles, table tennis balls
1900	Viscose rayon	Linings in clothing, tablecloths, curtains
1909	Phenol–formaldehyde	Coatings, electrical insulators, machine parts
1926	Alkyds	Exterior paints, electrical insulators, distributor caps
1927	Cellulose acetate	Packaging films, lacquers
1927	Polyvinyl chloride	Pipe, siding, flooring, wall coverings
1929	Urea–formaldehyde	Wood adhesives, wood products, electrical switches and parts
1931	Polychloroprene	Industrial hoses, extrusions, wire and cable, footwear
1935	Ethyl cellulose	Moldings, coatings
1936	Polymethyl methacrylate	Display signs
1937	Polyvinyl acetate	Adhesives
1937	Styrene–butadiene copolymers	Tires, footwear, molded items
1938	Cellulose acetate butyrate	Sheets, tubing, lacquers
1938	Polystyrene	Kitchenware, toys
1938	Polyamides (nylons)	Fibers, films
1939	Melamine–formaldehyde	Tableware
1939	Polyvinylidene chloride	Films, paper coatings
1942	Polyesters (cross-linkable)	Boat hulls
1942	Polyethylene (low density)	Squeeze bottles, films
1943	Silicones	Rubber goods
1943	Fluoropolymers	Industrial gaskets, slip coatings
1943	Polyurethane	Foam goods, adhesives, binders, coatings, skateboards
1943	Butyl rubber	Automobile tubes, curing bags, molded items
1948	Acrylonitrile–butadiene–styrene copolymers	Pipe and fittings, appliances, luggage
1950	Polyacrylonitrile (fiber derivatives)	Sweaters and knitwear, blankets
1950	Chlorosulfonated polyethylene	Automotive hoses, wire and cable
1952	Polyethylene terephthalate and related polyesters	Clothing, filling (fiberfill), sailcloth, films
1955	Linear high density polyethylene	Detergent bottles
1956	Acetal resin	Auto parts
1956	Fluoroelastomers	High temperature seals and gaskets
1957	Polypropylene	Carpet fiber, moldings
1957	Polycarbonate	Appliance parts
1958	Fluorinated ethylene propylene	Wire insulation, films, and moldings

Table 54.1. (*Continued*)

Date	Material	Typical Application
1961	Polyisoprene (high cis)	Tires, footwear, medical items
1962	Ethylene–propylene saturated copolymer	Tires, general purpose rubber
1964	Ionomer resins	Moldings
1964	Polyphenylene oxide	High temperature moldings
1965	Polyimides	High temperature films and wire coatings
1965	Polybutene	Films, pressure resistant pipe
1965	Polysulfone	High temperature thermoplastic
1965	Poly(4-methyl-1-pentene)	Clear moldings
1968	Phenylene ether sulfone	High temperature films and moldings
1970	Ethylene–chlorotrifluoroethylene copolymer	Wire insulation
1970	Moldable elastomers	Molded rubber goods
1971	Hydrogels (hydroxy acrylates)	Contact lenses
1972	Moldable polyesters	Engineering thermoplastic
1972	Perfluoroalkoxy polymers	Moldings, films, chemical coatings
1973	Modified ethylene–tetrafluoroethylene copolymer	Wire insulation
1974	Aromatic polyamides (aramid fibers)	Clothing with high temperature resistance, high strength reinforcement of tires and plastic composites

[a] Adapted with permission from a compilation by H. F. Mark and S. Atlas, "Introduction to Polymer Science," in H. S. Kaufman and J. J Falcetta, Eds., *Introduction to Polymer Science and Technology*, Copyright ©1977 by John Wiley and Sons, Inc., New York.

hydrogen, oxygen, and nitrogen, sometimes with halogen substituents (typically chlorine and fluorine). In some cases, the presence of phosphorus, sulfur, boron, or silicon may result in distinctive properties. Silicon and oxygen form a stable backbone (siloxane) that can act as a macromolecular skeleton.

1.2 Basis of Distinctive Polymer Properties

Polymers are notable for their characteristic mechanical properties and low density. Particular polymers may have low thermal conductivity or good chemical resistance. These various properties are a function of the repeating structural units, their arrangement in the chain, and the total size of the chain.

Table 54.2. Production of Certain Synthetic and Natural Polymers

A. U.S. Production of Plastics, Resins, and Elastomers (1980)	
Polymer	1000 lbs Dry Basis
Thermoplastic Resins[a]	
Acrylic resins (includes thermosetting acrylics)	1,028,154
Engineering plastics[b]	580,210
Polyamide resins (nylon and nonnylon types)	315,089
Polyester resins, saturated	607,705
Polyethylene resins	11,719,893
Polypropylene resins	3,698,987
Polytetrafluoroethylene (PTFE) resins	21,591
Acrylonitrile–butadiene–styrene (ABS) resins	981,068
Straight polystyrene	2,264,773
Rubber modified polystyrene	1,183,366
All other types of styrene plastic materials	1,110,858
Polyvinyl acetate	668,343
Polyvinyl alcohol	156,092
Polyvinyl chloride and copolymers	5,485,371
Polyvinylidene chloride latex resins	26,027
All other vinyl and vinylidene resins	380,904
All other thermoplastic resins	893,315
Total	31,121,746
Thermosetting Resins[a]	
Alkyd resins	702,799
Furfuryl type resins	24,630
Melamine–formaldehyde resins	186,030
Phenolic and other tar acid resins	1,744,928
Polyester resins, unsaturated[c]	952,469
Polyether and polyester polyols for urethanes[d]	1,381,824
Polyurethane elastomers[e] and plastic products	253,544
Silicone resins	12,381
Urea–formaldehyde resins	1,310,880
All other thermosetting resins	494,759
Total	7,064,244
Synthetic Elastomers[f] (includes latex)	Metric Tons
Styrene–butadiene	1,074,232
Polybutadiene	311,306
Nitrile	62,525
Ethylene–propylene	144,485
Polychloroprene	150,544
Other synthetic elastomers (includes polyisoprene and butyl rubber but excludes polyurethane rubber)	265,951
Total	2,009,043

Table 54.2. (*Continued*)

B. World Production of Certain Textile Fibers (1980)[g]

	Thousand Metric Tons
Natural Fibers	
Raw cotton	14,137
Raw wool	1,581
Raw silk	56
Total	15,774
Percent of World Total	*53*
Man-Made Fibers Except Olefin and Textile Glass	
Rayon and acetate	3,244
Acrylic and modacrylic	2,083
Nylon and aramid	3,125
Polyester	5,132
Certain other noncellulosic fibers	147
Total	13,731
Percent of World Total	*47*

[a] Adapted from data in *Synthetic Organic Chemicals*, USITC Publ. 1183, U.S. International Trade Commission, Washington, D.C., 1981. This reference should be consulted for data on *additional polymers*. These data "do not necessarily coincide with that reported to the Society of Plastics Industry (SPI) because of differences in both reporting instructions and in the coverage of certain resins." Quantities are given on a dry weight basis unless the system is specified as latex. Dry weight basis is the total weight of the materials including resin and additives but excluding water and other liquid diluents unless they are an integral part of the materials.

[b] Engineering plastics include acetal, polycarbonate, polyimide and amide–imide polymers, polysulfone, polyphenylene oxide, and polyphenylene sulfide. (ABS resins and nylon resins are listed separately.)

[c] Polyester resins are unsaturated resins, later to be copolymerized with a monomer, and polyallyl resins. Data are on an "as sold" basis, including monomer if part of the resin system.

[d] The other principal starting materials used in the production of urethane products are the isocyanic acid derivatives, mainly the 80:20 mixture of toluene-2,4- and -2,6-diisocyanate.

[e] Data on urethane elastomers believed not fully representative of the total urethane market.

[f] The term "elastomers" may be defined as substances in bale, crumb, powder, latex, and other crude forms that can be vulcanized or similarly processed into materials that can be stretched at 68°F to at least twice their original length and, after having been stretched and the stress removed, return with force to approximately their original length (*Synthetic Organic Chemicals*, USITC Publ. 920, U.S. International Trade Commission, Washington,

4214

Synthetic polymers characteristically are a mixture of many chains of different molecular weight. Different manufacturing processes result in trace amounts of different ingredients that can have a major effect on product behavior. Polymers of the same fundamental unit structure but of significantly different molecular weights may have some quite different properties.

The polymer molecular weight or chain conformation is essentially an *average* that varies somewhat with different methods of measurement. The molecular weight of synthetic polymers is generally expressed as the number average, the weight average, or the viscosity average molecular weight (M_n, M_w, or M_v, respectively). When all molecules are within a narrow weight range, then $M_w = M_n$ and the system is monodisperse; otherwise, and more commonly, M_w exceeds M_n. The ratio M_w/M_n, known as the polydispersity index, provides a measure of the molecular weight distribution in a given sample. [For details, see Stevens (4) and Billmeyer (5); Moncrieff (6) provides a stepwise example of M_n and M_w determinations.]

Each type of synthetic polymer requires a characteristic number of monomeric units to acquire desirable mechanical properties such as tensile strength, impact strength, or elasticity. Above a characteristic higher number, these desirable properties become limited, or processing becomes impractical. Consequently, most useful polymers have an approximate weight range of 20,000 to 200,000; this generalization has a great many exceptions such as high molecular weight polyethylene and many elastomers.

Stereoisomeric configurations within the polymer chain can influence polymer properties. These can be the stereoregular syndiotactic and isotactic configurations as well as the random atactic configurations (5). Polymer chains may be amorphous ["cooked spaghetti" (4)] or crystalline with orderly packing of molecular chains; even when crystalline they have some noncrystalline regions. Crystalline regions tend to give stiffness and strength whereas noncrystalline regions are usually responsible for flexibility and toughness. Crystalline polymer tends to be more resistant to solvents and have a higher melting point than amorphous polymer. The crystalline (or crystal) melting point is considered the temperature at which crystallinity disappears (4, 7).

D.C. 1978). Production figures are based on the *RMA Industry Rubber Report*, December 1980, issued March 13, 1981, and are reprinted by permission of the Rubber Manufacturers Association, Washington, D.C.

[g] As listed in the *Textile Organon*, **52** (6), 81, 83 (1981). Data are reprinted by permission of the Textile Economics Bureau, Inc., Roseland, N.J. "The silk and man-made fiber data are on a calender-year basis, while the figures for cotton and wool are on a seasonal basis."

In the classic definition, the glass transition temperature describes a temperature below the melting point where the properties of a polymer change, often dramatically, from those of a brittle solid or glass to those of a rubber. The glass transition temperature of a polymer often characterizes its potential use. The glass transition temperature of fibers should be well above room temperature and that of rubbers well below room temperature. With typical "plastics," all mechanical properties except impact are adversely affected in varying degrees by the onset of the glass transition. In general, the glass transition temperature tends to be in the order of one-half to two-thirds the value of the melting point in degrees Kelvin (4). Heating partially crystalline homopolymer above the glass transition temperature but below the melting point causes softening only in the amorphous regions; the polymer becomes more flexible but remains relatively intact until heating exceeds the melting point.

The glass transition temperature is now viewed as one characteristic phenomenon in a series of transitions and relaxations affecting polymers. It can be measured by many methods (8). Figure 54.1 shows that with atactic polymers the reported spread in glass transition temperature is relatively small and within reasonable limits. A similar situation exists with semicrystalline polymers (nominal 30 to 60 percent crystallinity) that can be quenched to a completely amorphous state. Crystalline polyethylene has a very wide range of reported values (8, 9) and its glass transition temperature has comparatively little physical significance (10).

Potential end use requiring particular properties often determines the selection of polymerization procedure. Some typical techniques used to polymerize thermoplastics are described in the discussions of polyethylene, polyvinyl chloride, and polystyrene. In many cases the plasticizers, fillers, stabilizers, reinforcing agents, and other additives added to polymers significantly alter their physical, chemical, electrical, and mechanical properties. Obviously these additives may affect the suitability of particular commercial products in one or more applications.

The *Polymer Handbook* (11) gives the single most extensive compilation of T_g, T_m, and other physical and chemical properties. Mechanical properties of polymers are often characterized in terms of test behavior (12). Handbooks and monographs (13–17) provide concise summaries. On-line information retrieval systems provide access to current literature and now include the capability for structure searching that can be linked to literature citations (18).

1.3 Analysis of High Polymers

In the general sense, analysis of polymers includes chemical or physical structure as well as mechanical, thermal, electrical, and other physical behavior. In the fields of toxicology and industrial hygiene the focus is primarily on (1) residual

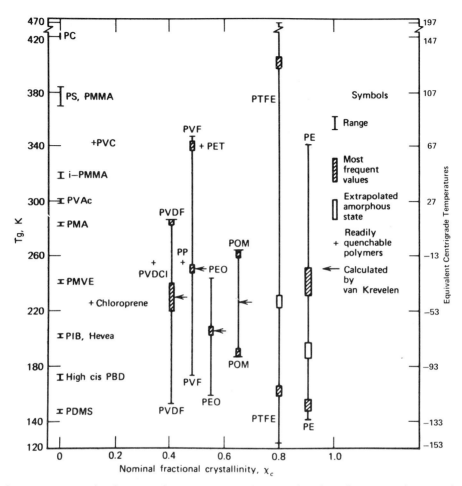

Figure 54.1 Spread in literature glass transition temperature values for polymers as a function of increasing crystallinity. Where there is spread, vertical bars indicate both the total range of values and one or two that are more common. Adapted with permission from Boyer (8), who gives preferred values for polyvinyl fluoride (PVF), polyvinylidene fluoride (PVDF), polyethylene oxide (PEO), polyoxymethylene (POM), and polyethylene (PE).

monomer or monomers; (2) other residues from the manufacture of the raw polymer; (3) dust and volatile emissions during processing of the raw polymer or end use of finished articles; and (4) thermal degradation products. Analysis of high polymers has become increasingly sophisticated over the past decade and this general trend is expected to increase in the years ahead. Interactions between polymers and solvents, diffusion of gases in polymers, oxidation reactions, and degradation reactions can now be studied in detail (19). Techniques used to examine polymer structure and properties include various types of chromatography and spectroscopy, optical methods, light scattering, X-ray diffraction, and neutron scattering.

The key to the utility of plastics is the ease with which they can be converted from the raw resin to the final product (13). Many processing operations with plastics can often be run automatically. The molecular phenomena that permit this relative ease of processing can, if uncontrolled, also facilitate the development of products with a propensity to failure in a needed type of behavior such as tensile strength or impact strength. A complicating factor is that testing in "actual use" conditions is not always comparable to a field test: (1) environmental conditions may cause a change in the product and (2) multiple exposure conditions are not always comparable to one-time test procedures.

The biennial articles sponsored by *Analytical Reviews* (20, 21, and earlier references) outline recent technical literature. Each year the American Society for Testing and Materials reviews many tests designed to characterize the behavior of polymers under stated conditions (see Table 54.3). Federal, state, and local agencies as well as professional associations also develop specifications either generally or for specific purposes.

1.4 Toxicologic Characteristics

The biologic effects of synthetic polymers may be attributable to residual monomer; oligomers or low molecular weight by-products, sometimes cyclic, that become incorporated into the polymer; additives; or molecular changes from the curing process. Trace amounts of residual monomer and other adventitious ingredients are often difficult to remove and can be defined only according to the sensitivity of available analyses. Complex coordination catalysts* that give polymers the desired stereospecificity may, depending on their composition and fate, pose questions of trace metal toxicity. Detailed discussion of these various materials may be found in other chapters or, alternatively, other sources.

Natural polymers may also contain trace ingredients that require scrutiny and toxicologic evaluation. For example, the low molecular weight materials in some tropical woods may cause allergic dermatitis (see Section 7.2).

---

* This term (4) includes the Ziegler–Natta catalysts originally developed in the 1950s, reduced metal oxides, and alfin catalysts. A Ziegler–Natta catalyst is defined as a combination of a transition metal compound (Groups IV to VIII) with an organometallic compound (Groups I to III); the most thoroughly studied have been the combinations of titanium halides with trialkylaluminum compounds but halides or oxyhalides of vanadium, chromium, molybdenum, and zirconium are also common catalyst components. [Ziegler catalysts and Natta catalysts can also be distinguished separately (3).] An alfin catalyst describes a class of heterogeneous catalysts consisting of an alkenyl sodium compound, an alkoxide, and an alkali metal halide (4). The name reflects the *alcohol–olefin* ingredients; typically, amylsodium is reacted with isopropyl alcohol or isopropyl ether and propylene.

Table 54.3. Synopses of Selected Standards Circulated by the American Society for Testing and Materials (ASTM)[a]

A. Detection of Residual Monomer or Low Molecular Weight Component

Method	Part	Title/Subject	Scope/Focus
ANSI/ASTM D 3680-78	35	Residual vinyl chloride in polyvinyl chloride and copolymers by solution injection technique	Under optimum conditions approximately 1 ppm vinyl chloride can be detected in resins
ANSI/ASTM D 3749-78	35	Residual vinyl chloride in polyvinyl chloride homopolymer resins by gas chromatographic headspace technique	Under optimum conditions the level of detection in the headspace vapor is equivalent to about 0.02 ppm in the resin
ANSI/ASTM D 2124-70 (Reapproved 1979)	35	Analysis of components in polyvinyl chloride by infrared spectrophotometry	Identification, in many cases quantitative, of certain resins, plasticizers, stabilizers, and fillers in polyvinyl chloride formulations
ANSI/ASTM D 834-61 (Reapproved 1979)	35	Free ammonia in phenol–formaldehyde molded materials (single stage)	Likelihood that such free ammonia might contaminate foodstuffs in proximity with the molded part or that metal parts might become corroded
ASTM D 1312-56 (Reapproved 1974)	28	Apparent free phenols in synthetic phenolic resins or solutions used for coating purposes	"Applies to all the commonly used resins except those containing p-phenylphenol"
ASTM 1597-60 (Reapproved 1975)	28	Melamine content of nitrogen resins	Method "covers the spectrophotometric determination of melamine in butylated or caprylated melamine–formaldehyde resins and in mixtures of such melamine and urea resins, but does not cover analogous resins made from substituted melamine derivatives"
ASTM D 1615-60 (Reapproved 1975)	28	Glycerol, ethylene glycol, and pentaerythritol in alkyd resins	Applies to determinations in alkyd resins and resin solutions; certain other polyhydric alcohols, urea, melamine, or phenolic resins cause interference

Table 54.3. (Continued)

A. Detection of Residual Monomer or Low Molecular Weight Component

Method	Part	Title/Subject	Scope/Focus
ASTM D 1727-62 (Reapproved 1976)	28	Urea content of nitrogen resins	"Spectrophotometric determination of urea in butylated urea–formaldehyde resin solutions and in mixtures of such urea and melamine resins"
ANSI/ASTM D 2455-69 (Reapproved 1974)	28	Identification of carboxylic acids in alkyd resins	"Method covers the qualitative identification of carboxylic acids in alkyd resins, including resin-modified alkyds. It may be used for analyzing polyesters but additional peaks may appear from monomers such as styrene"
ASTM D 2572-70 (Reapproved 1974)	28	Isocyanate groups in urethane materials or prepolymers	"Method covers the determination of the isocyanate group (NCO) content of a urethane intermediate or prepolymer." Interlaboratory difference should not exceed 0.40 wt. %
ASTM D 3432	29	Free toluene diisocyanate in urethane prepolymers and coating solutions	Gas chromatographic method intended to replace D 2615 (distillation method, Part 28)
ASTM D 2690-80	28	Isophthalic acid in alkyd and polyester resins	Method "covers the gravimetric determination of isophthalic acid content of alkyd resins and polyesters [without] interference from styrene monomer or polymer or from other dicarboxylic acids except terephthalic acid"
ASTM D 2921-70 (Reapproved 1979)	27	Qualitative tests for the presence of water repellents and preservatives in wood products	"Method describes simple qualitative field or laboratory tests to determine water repellency or the presence of [pentachlorophenol, tetrachlorophenol, and other chlorinated phenols] in wood products that are specified to be water repellent preservative treated"

4220

Standard	Page	Title	Description
ANSI/ASTM D 2998-71 (Reapproved 1976)	28	Polyhydric alcohols in alkyd resins	"Method covers the qualitative and quantitative determination of the polyols and alkyd resins, including resin- and polymer-modified alkyds. Quantities as low as 0.5 percent may be detected and measured"
ANSI/ASTM D 3733-78	28	Silicon content of silicone polymers and silicone-modified alkyds by atomic absorption	"Method covers the determination of the silicon content of silicone polymers and silicone-modified alkyds when present in the nonvolatile portion of polymers, resins, or liquid coatings to the extent of 1% or more"

B. Chemical Resistance

Standard	Page	Title	Description
ANSI/ASTM C 581-74 (Reapproved 1978)	35	Chemical resistance of thermosetting resins used in glass fiber-reinforced structures	Relatively rapid test; standard reagents include acids, base, organic solvents, and water
ANSI/ASTM G 20-77	27	Chemical resistance of pipeline coatings	Specimens inspected for visible signs of chemical attack; reference to tests to determine if specimens have undergone loss of mechanical or bonding properties

C. Thermal Effects and Related Tests

Standard	Page	Title	Description
ANSI/ASTM D 2117-64 (Reapproved 1978)	35	Melting point of semicrystalline polymers	Normal operating temperature range 30–200°C; maximum temperature of 350°C with special equipment
ANSI/ASTM D 3045-74 (Reapproved 1979)	35	Heat aging of plastics without load	"Guide to compare thermal aging characteristics of materials as measured by the change in some property of interest"
ANSI/ASTM D 2863-77	35	Minimum oxygen concentration to support candle-like combustion of plastics (oxygen index)	Intended to measure properties under controlled laboratory conditions; not to be used for the description or appraisal of fire hazard
ASTM D 1525-76	35	Vicat softening temperature of plastics	Determination of the temperature at which specified needle penetration occurs under specified test conditions; useful for many thermoplastics but not recommended for materials having a wide softening range

Table 54.3. (Continued)

Method	Part	D. Miscellaneous Title/Subject	Scope/Focus
ANSI/ASTM D 570-77	35	Water absorption of plastics	Serves as a guide to the proportion of water absorbed in relation to effects on properties and also as a control test on the uniformity of a product
ANSI/ASTM D 1434-75	35	Gas transmission rate of plastic film and sheeting	"Determination of the steady-state rate of transmission of a gas through plastics in the form of film sheeting, laminates, and plastic-coated papers or fabrics"
ANSI/ASTM G 21-70 (Reapproved 1980)	35	Determining resistance of synthetic polymeric materials to fungi	Determination of the effect of fungi on the properties of synthetic polymeric materials in the form of molded and fabricated articles, tubes, rods, sheets, and film materials
ANSI/ASTM G 22-76 (Reapproved 1980)	35	Determining resistance of plastics to bacteria	As directly above with reference to bacteria
ANSI/ASTM D 2654-76	32	Moisture content and moisture regain of textiles	Procedures generally applicable to all fibers
AATCC Test Method 30-1974	32	Mildew and rot resistance of textiles; evaluation of fungicide effect	A severe exposure for textile products

[a] Summarized or quoted directly from the tests described in Ref. 12 and printed by permission of the American Society for Testing and Materials, Philadelphia, Pa. Certain standards have also been approved by the American National Standards Institute (ANSI). The American Association of Textile Chemists and Colorists developed AATCC Test Method 30-1974.

The magnitude and type of biologic response vary with the nature of the particular polymer, the amount of the dose, and the route of administration. In the general population the main routes of exposure can be grouped as oral (usually indirectly), cutaneous, and inhalant. Skin reactions are usually traceable to allergy to low molecular weight components or additives.

Acute oral LD_{50} values of polymers are often above the limits of practicable experimental testing; the sheer bulk of the required dose of relatively inert polymer may overshadow the toxic potential (22). Some polymers intended as direct food additives have been assessed in terms of the acceptable daily intake (23). For polymers intended for food contact applications the toxic potential is more often assessed in terms of any possible migration of low molecular weight components into food (24). Sensitive methods such as headspace gas chromatography may be used to determine the relationship between residual monomer (as vinyl chloride) in polymer containers and the content of monomer in foods stored in these containers for extended periods (25). Work with beverage containers manufactured from acrylonitrile has shown that a linear relationship from diffusion is not necessarily present at very low levels of residual monomer (26).

Inhalation toxicity comprises exposure to polymer dust, volatile ingredients generated during processing of polymers, or thermal degradation as discussed in the next subsection. Industrial exposure limits for the airborne dust of specific polymers or their decomposition products per se have been defined or evaluated in only a relatively few cases (27, 28). The biologic effect from particles as a function of physical form is often the outstanding result of gross overexposure to polymer dust. The general guidelines (27) for nuisance dusts—cellulose is considered an example—should always be regarded as the most lenient standard that might be applicable.

A relatively small part of the population receives parenteral exposure as a result of the increasing use of polymers for implants that are essentially tissue replacements. These polymers can produce local sarcomas when implanted into animals as intact films or pellets of sufficient size but cause few if any tumors when implanted as perforated sheets, textile fabrics, or powders. Such sarcomas are attributable to a foreign body process that is "conditioned by the smoothness and size of the implant" (29). The term "solid state carcinogenesis" describes this process and was introduced to replace the term "smooth surface carcinogenesis" because the smooth surfaces of liquids are not locally carcinogenic. Tumor production at smooth round surfaces exceeding a certain critical size appears to result after an initial avascular response with little cellular infiltration. Medical records show that tumors may develop in certain cases after years of exposure to implanted or accidentally embedded inorganic solid surfaces (29).

A similar pattern may develop with implants of organic polymers. Extended irritation from a granulomatous type of response provides an anticarcinogenic influence. The critical point appears to be whether or not tissue homeostasis is maintained. Two-centimeter Millipore® filter disks* with pore sizes ≥0.22 μm induced cellular invasion of the filters with macrophage/phagolysosome activity and were nontumorigenic in mice that had received subcutaneous implants; the same disks with pore sizes ≤0.1 μm appeared to be impenetrable by cells or cytoplasmic processes, induced more fibrosis, and were locally tumorigenic (30). In its review of polyethylene, the International Agency for Research on Cancer noted that "subcutaneous implantation of all forms except the powder resulted in local sarcomas, the incidence of which varied with the size and form of the implants" (31).

Today exposure to synthetic polymers from applications in medicine and allied fields is increasing widely. Polymers are used extensively in implanted prosthetic devices and adhesives, external prosthetic devices, packaging, and disposable equipment (32, 33). Polymeric drugs and composites, where the polymer provides a time-released depot of chemotherapeutic agent, appear particularly promising for anticancer therapy (34, 35). Acrylic polymers are widely used in dentistry (see Section 6.5).

1.5 Flammability and Related Phenomena

Nearly all polymers burn in the presence of sufficient heat and oxygen. The basic lethal factors in uncontrolled fires are toxic gases from the thermal decomposition of the fuel, oxygen deficiency, and heat (36). The type of toxic gases evolved from a given polymeric material during a conflagration depends on the chemical structure of the polymer base, any additive or naturally occurring chemicals contained therein, and environmental conditions.

1.5.1 Types of Structure

Polymer bases have been broadly grouped (37) as those consisting of only carbon plus hydrogen and/or oxygen; those containing nitrogen atoms; and polymers containing halogen or sulfur. Polymers containing only carbon and hydrogen (or carbon, hydrogen, and oxygen) typically form carbon monoxide as the major toxicant with incomplete combustion, which may also produce gaseous and condensed hydrocarbon products. Hydrogen cyanide, ammonia, and nitriles may form with incomplete combustion of a nitrogen-containing polymer; complete combustion may produce nitrogen oxides. Polymers containing halogen or sulfur heteroatoms form acid gases with complete combustion but may form organic halogen and sulfur compounds with incomplete combustion.

Essentially two avenues are open to the preparation of synthetic polymers with improved fire performance. New polymers can be developed with fire

---

* These filter disks are composed of mixed cellulose esters (acetate and nitrate) and may be cast on nylon for microweb reinforcement.

safety characteristics inherent in their structure or fire retardants can be added to existing polymers (38). Depending on the type of polymer, both routes are technologically and commercially important; the second is the only approach available for natural and many synthetic polymers.

1.5.2 Fire Behavior of Polymers

The fire behavior of polymeric materials involves fire dynamics and the adjacent environmental systems in addition to chemical composition. This behavior is not an intrinsic property that can be measured in definitive terms by a simple test (39). The polymer burning process tends to be self-sustaining as long as fuel and oxygen are present. Polymer combustion mechanisms can be characterized as micro, macro, or mass (40). As conflagrations proceed from one scale to another the relative importance of the various factors involved can change dramatically. Realization of these complex phenomena has led to the concept of "practical fire performance" rather than laboratory scale results as a means of evaluating the combustibility of polymers. This concept of practical fire performance is admittedly extremely difficult to define (41).

Assessment of the fire hazard with a given polymer generally involves a complex group of factors from several sets: ease of ignition, rapidity of burning once ignited, and severity of conditions produced by burning (the heat load, the smoke, and the nature and the amount of toxic gases produced). Tests for flammability and fire performance stated in terms of these specific factors provide a measure of response to given combustion conditions but they are often designed under constraints. Emphasis is frequently placed on reproducibility, simplicity, and ease of performance rather than significance of their respective scope (42). Many tests and standard methods are now undergoing critical examination (39).

1.5.3 Limiting Oxygen Index, Ignition Temperature, and Smoke Formation

The limiting oxygen index is the minimum oxygen concentration that a polymer requires to sustain ignition and combustion under specified test conditions. In the standard test (see Table 54.3, Method D 2863) the sample (solid, cellular, or film) is ignited at the top and burns downward with a small candlelike flame. This test is of value in product development and quality control but does not indicate polymer behavior under the turbulent conditions encountered in real fires (39).

The *flash temperature* for a specimen is the lowest temperature of adjacent air at which it ignites when flame or an external heat source is supplied. The lowest air temperature around the specimen at which ignition occurs without a direct heat source is the *self-ignition temperature* (43).

Smoke formation is an aspect of polymer flammability that depends highly on the conditions of combustion. Some materials are recognized as having a greater propensity for smoke formation than others, but in general this formation appears to be less a property of the material than of the fire (44). Smoke from pyrolyzing or actively burning materials is generally evaluated in terms of its opacity or light attenuation and polymers may be compared on a relative basis (45–47). Polymers can also be compared in terms of the type of soot (48, 49) or the amount of char (50, 51).

In general, polymers with unsubstituted aliphatic hydrocarbon backbones are very flammable, but their tendency to generate smoke is minimal. The addition of flame retardants, especially halogen compounds, reduces their tendency to burn and in some, but not all, cases increases the evolution of smoke. Some highly chlorinated polymers can have high limiting oxygen indices and also display high smoke generation. Polymers containing an aromatic group in the side chain, such as polystyrene, may burn readily and produce copious smoke. However, polymers with the aromatic group in the main chain, such as polycarbonate and polyphenylene oxide, tend to be intermediate in both limiting oxygen index and smoke generation (52).

Table 54.4 summarizes some typical oxygen index values and ignition temperatures for polymers.

1.5.4 Physical Form

The physical shape of polymers may facilitate oxygen availability or heat transfer and in some cases may contribute more to flammability than chemical structure. The main physical forms are listed in Table 54.5.

1.5.5 Explosibility

Many polymers are difficult to ignite in solid massive form but nearly all burn in the form of dust (56). Not all combustible dusts are explosible (57). However, dust dispersions can often be ignited explosively by sparks, flame, or metal surfaces above 375°C (56). Explosibility hazards of polymeric dusts are often compared with Pittsburgh coal dust (<74 μm), which is given values of unity for ignition sensitivity, explosion severity, and index of explosibility (see Chapter 27, Volume 1).

Molding powders for injection or extrusion molding are generally cubes or cylindrical pellets approximately 0.1 in. in diameter. They are not dusty and pose no particular explosive hazard although their grinding or processing may generate fine dust that requires consideration of such hazard (56).

Table 54.6 gives data on explosion characteristics of polymeric dusts as compiled by the National Fire Protection Association and values from tests conducted by the Joint Fire Research Organization in England.

Table 54.4. Limiting Oxygen Indices and Ignition Temperatures for Various Polymers[a]

Polymer	Range of Limiting Oxygen Indices[b]	Ignition Temperature (°C) Flash Ignition	Ignition Temperature (°C) Self-Ignition
Polyethylene	17.4–30.2	341	349
Polypropylene	17.4–28.2		570
Natural rubber	17.2		
Polybutadiene	17.1–18.3		
Styrene–butadiene	16.9–19.0		
Neoprene	26.3		
Chlorosulfonated polyethylene	25.1		
Polyvinyl chloride	20.6–80.7	391	454
Chlorinated polyvinyl chloride	55.0		
Polyvinylidene chloride	60.0	532	532
Polyvinyl fluoride	16–22.6		
Polyvinylidene fluoride	43.4–43.7		
Polytetrafluoroethylene	95		
Polyvinyl alcohol	21.6–22.5		530
Polystyrene	17–25.2	296–360	488–496
Acrylic and modacrylic	16.7–29.8		560 (acrylic fiber)
Styrene–acrylonitrile	18–28		
Acrylonitrile–butadiene–styrene	18–39.0		
Wood	22.4–25.4	228–264	260 (white pine shavings)
Cotton	18.6–27.3	230–266	254
Rayon	18.7–19.7		
Cellulose acetate	16.8–27	305	475
Cellulose nitrate	23.8–25.0	141	141
Wool	20–21.5		
Nylon 66 fabric	20–21.5		
Nylon 6 fabric	20–21		
Polyethylene terephthalate fabric	20–21		
Bisphenol A polycarbonate	32.0–33.0		
Polyoxymethylene	15.3–15.7		
Polyphenylene sulfide	48.0–50.0		
Polyether sulfone	30.0–40.0		

[a] Summarized from Refs. 40, 41, and 53–55. Additional ignition temperatures are summarized in Table 54.6.
[b] The limiting oxygen index can also be expressed in terms of the volume of oxygen in the atmosphere where n is a decimal and equals 0.95 for polytetrafluoroethylene (53).

Table 54.5. Most Common Physical Forms of Polymers and Their Relation to Flammability[a]

Category	Products	Definition	Heat Transfer
Film	Films, sheeting, coatings, and adhesives	Strictly, thickness ≤0.01 in.; generally, thickness very small relative to length and width	Absorption depends on adjacent material—air and solid (coating); air (sheeting exposed to air on both sides); or solids (unexposed adhesive)
Formed polymer	Rods, blocks, castings, extrusions, and moldings	Significant thickness and mass in product	Formed product
Fiber	Fabrics, textiles	Cylindrical rods of small diameter arranged in a matrix	Polymeric fibers and matrix (air usually intervening material)
Particles (fines)	Powders; powder coating resins (not a common use of polymers generally)	Large surface area per unit volume	Surrounding atmosphere
Filled polymer	Reinforced polymers, other filled polymers	Solid material other than reference polymer distributed more or less evenly	Polymer and filler or reinforcing material
Foam	Cellular plastics (honeycomb structures)	Gas distributed more or less evenly through polymer	Polymer, contained gas, distribution of polymer and gas

[a] Data summarized from Ref. 54.

Table 54.6. Explosion Parameters of Polymeric Dusts

A. Parameters of Selected Polymeric Dusts as Compiled by the National Fire Protection Association[a]

Type of Dust	Explosibility Index	Ignition Sensitivity	Explosion Severity	Maximum Explosion Pressure (psig)	Maximum Rate of Pressure Rise (psi/sec)	Ignition Temperature		Minimum Cloud Ignition Energy (joules)	Minimum Explosion Concn. (oz/ft³)
						Cloud (°C)	Layer (°C)		
Agricultural dusts									
Cellulose	2.8	1.0	2.8	130	4,500	480	270	0.080	0.055
Cork dust	>10	3.6	3.3	96	7,500	460	210	0.035	0.035
Cotton linter, raw	<0.1	<0.1	<0.1	73	400	520	—	1.92	0.50
Wood, birch bark ground	6.7	3.7	1.8	103	7,500	450	250	0.060	0.020
Wood, flour, white pine	9.9	3.1	3.2	113	5,500	470	260	0.040	0.035
Thermoplastic resins and molding compounds									
Acetal, linear (polyformaldehyde)	>10	6.5	1.9	113	4,100	440	—	0.020	0.035
Methyl methacrylate polymer	6.3	7.0	0.9	84	2,000	480	—	0.020	0.030
Acrylamide polymer	2.5	4.1	0.6	85	2,500	410	240	0.030	0.040
Acrylonitrile polymer	>10	8.1	2.3	89	11,000	500	460	0.020	0.025
Cellulose acetate	>10	8.0	1.6	85	3,600	420	—	0.015	0.040
Cellulose triacetate	7.4	3.9	1.9	107	4,300	430	—	0.030	0.040
Cellulose acetate butyrate	5.6	4.7	1.2	85	2,700	410	—	0.030	0.035
Tetrafluoroethylene polymer (micronized)	≪0.1	≪0.1		b		670	570	b	c
Nylon (polyhexamethylene adipamide) polymer	>10	4.7	1.8	95	4,000	500	430	0.020	0.030
Polycarbonate	8.6	4.5	1.9	96	4,700	710	—	0.025	0.025
Polyethylene, high pressure process	>10	7.5	1.4	81	4,000	450	380	0.030	0.020
Polyethylene, low pressure process	>10	22.4	2.3	80	7,500	450	—	0.010	0.020

Table 54.6. (Continued)

A. Parameters of Selected Polymeric Dusts as Compiled by the National Fire Protection Association[a]

Type of Dust	Explosibility Index	Ignition Sensitivity	Explosion Severity	Maximum Explosion Pressure (psig)	Maximum Rate of Pressure Rise (psi/sec)	Ignition Temperature Cloud (°C)	Ignition Temperature Layer (°C)	Minimum Cloud Ignition Energy (joules)	Minimum Explosion Concn. (oz/ft³)
Polyethylene wax, low molecular weight	5.8	7.2	0.8	74	3,000	400	—	0.035	0.020
Polypropylene (contains no antioxidant)	>10	8.0	2.0	76	5,500	420	—	0.030	0.020
Rayon (viscose) flock, 1.5 denier, 0.020 in. maroon	0.2	0.3	0.8	107	1,700	520	250	0.240	0.055
Polystyrene molding compound	>10	6.0	2.0	77	5,000	560	—	0.040	0.015
Polystyrene latex, spray-dried, contains surfactants	>10	13.4	3.3	100	7,000	500	500	0.015[d]	0.020
Styrene–acrylonitrile (copolymer 70:30)	1.9	3.8	0.5	71	1,400	500	—	0.030	0.035
Styrene–butadiene latex copolymer, >75% styrene, alum coagulated	>10	7.3	1.7	92	3,900	440	—	0.025	0.025
Polyvinyl acetate	0.2	0.6	1.2	69	1,000	550	—	0.160	0.040
Polyvinyl butyral	>10	25.8	0.9	84	2,000	390	—	0.010	0.020
Polyvinyl chloride, fine	≪0.1	≪0.1	<0.1	28	200	660	400	[b]	[d]
Thermosetting resins and molding compounds									
Alkyd molding compound, mineral filler, not self-extinguishing	<0.1	0.2	<0.1	40	300	500	270	0.120	0.155
Allyl alcohol derivative, CR-39 (from dust collector)	>10	5.6	3.6	91	7,500	510	—	0.020	0.035

Table (continued):

	Minimum Ignition Temperature (°C)	Minimum Explosible Concentration (kg/m^3)	Minimum Ignition Energy (mJ)	Maximum Explosion Pressure lbf/in^2	Maximum Explosion Pressure kN/m^2	Maximum Rate of Pressure Rise lbf/in^{2}s	Maximum Rate of Pressure Rise kN/m^{2}s	Maximum Oxygen Concentration to Prevent Ignition (% by vol.)
Allyl alcohol derivative, CR-149 glass fiber mixture (65–35)	<0.1	<0.1	60	1,000	540	—	1.60	0.345
Melamine formaldehyde, unfilled laminating type, no plasticizer	<0.1	0.1	81	800	810	—	0.320	0.085
Urea formaldehyde molding compound, Grade II, fine	1.0	0.6	89	3,600	460	—	0.080	0.085
Urea formaldehyde–phenol formaldehyde molding compound, wood flour filler	0.2	0.4	84	1,700	530	240	0.120	0.085
Polyethylene terephthalate	7.5	2.9	98	5,500	500	—	0.035	0.040
Polyurethane foam (toluene diisocyanate–polyhydroxy with fluorocarbon blowing agent, fire retardant	>10	9.8	96	3,700	550	390	0.015	0.025
Rubber								
Rubber, crude, hard	7.4	4.6	80	3,800	350	—	0.050	0.025
Rubber, synthetic, hard, contains 33% sulfur	>10	7.0	93	3,100	320	—	0.030	0.030
Rubber, chlorinated	≪0.1	0.1	—	No ignition	940	290	b	d

B. Parameters of Selected Polymeric Dusts as Tested at the Joint Fire Research Station[e]

	Minimum Ignition Temperature (°C)	Minimum Explosible Concentration (kg/m^3)	Minimum Ignition Energy (mJ)	Maximum Explosion Pressure lbf/in^2	Maximum Explosion Pressure kN/m^2	Maximum Rate of Pressure Rise lbf/in^{2}s	Maximum Rate of Pressure Rise kN/m^{2}s	Maximum Oxygen Concentration to Prevent Ignition (% by vol.)
Acrylonitrile–butadiene–styrene copolymer	400	—	—	—	—	—	—	—
Acrylonitrile–vinylidene chloride copolymer	—	0.05	70	—	—	—	—	8

Table 54.6. (Continued)

B. Parameters of Selected Polymeric Dusts as Tested at the Joint Fire Research Station[e]

	Minimum Ignition Temperature (°C)	Minimum Explosible Concentration (kg/m³)	Minimum Ignition Energy (mJ)	Maximum Explosion Pressure		Maximum Rate of Pressure Rise		Maximum Oxygen Concentration to Prevent Ignition (% by vol.)
				lbf/in²	kN/m²	lbf/in²s	kN/m²s	
Casein	460	—	—	89	610	1200	8300	—
Cellulose, bleached	410	—	—	—	—	—	—	—
Cellulose acetate	340	—	—	111	770	5400	37300	5
Cellulose acetate, fibres	430	—	—	—	—	—	—	—
Cellulose acetate butyrate	380	—	—	—	—	—	—	—
Cellulose triacetate	390	—	—	—	—	—	—	—
Charcoal, wood	470	0.03	—	—	—	—	—	—
Coal, Pittsburgh <74 µm	530	—	—	—	—	—	—	—
Coal, 30% volatile	530	—	—	—	—	—	—	—
Hydroxyethyl cellulose	420	—	—	—	—	—	—	—
Hydroxyethyl methyl cellulose	410	—	—	—	—	—	—	—
Maize husk	430	—	—	—	—	—	—	—
Melamine–formaldehyde resin	410	0.02	68	87	600	900	6200	—
Methyl cellulose	480	—	—	—	—	—	—	—
Nitrocellulose	—	—	30	>256	>1800	>20900	>144200	—
Nylon, ground flock	450	—	—	—	—	—	—	—
Paper	400	0.03	—	80	550	500	3500	—
Phenol–formaldehyde resin	450	0.015	—	107	740	6500	44900	—
Polyester resin <1400 µm	400	—	—	—	—	—	—	—
Polyethylene	390	0.02	38	87	600	2900	20000	—
Polyethylene glycol	320	—	—	—	—	—	—	—
Polypropylene	380	—	43	45	310	300	2100	—
Polyurethane	460	—	—	—	—	—	—	—
Polyvinyl acetate	450	—	—	—	—	—	—	—
Polyvinyl chloride	510	—	—	Nil	Nil	Nil	Nil	—

Material								
Polyvinylidene chloride[f]	670	—	—	—	—	—	—	—
Rayon, viscose	420	—	—	—	—	—	—	—
Resin, rubber	400	—	—	—	—	—	—	—
Resin, synthetic	400	—	—	—	—	—	—	—
Rubber	380	—	—	—	—	—	—	—
Rubber, latex	450	—	—	—	—	—	—	—
Rubber, synthetic	410	—	—	—	—	—	—	—
Rubber, accelerator	310	—	—	—	—	—	—	—
Rubber crumb	440	—	—	84	580	6800	46900	—
Sodium carboxymethyl cellulose	320	1.1	440	49	340	400	2800	5
Urea–formaldehyde molding powder	450	0.04	—	82	570	700	4800	—
Urea–formaldehyde molding powder, paper filled	430	0.07	49	98	680	600	4100	—
Urea–formaldehyde molding powder, wood filled	430	0.025	40	87	600	1200	8300	—
Urea–formaldehyde resin	430	0.02	34	110	760	1600	11000	5
Wood	360	—	—	90	620	5700	39300	5
Wood, ground fluffed	450	—	—	—	—	—	—	—
Wood flour	380	—	—	—	—	—	—	—
Wood pulp, dehydrated	450	—	—	—	—	—	—	—
Wood pulp, flock	470	—	—	—	—	—	—	—

[a] Based on data developed by the U.S. Bureau of Mines and summarized in Ref. 58. Selected data are reproduced by permission from NFPA 654, *Standard for the Prevention of Dust Explosions in the Plastics Industry*, Copyright ©1975, National Fire Protection Association, Boston, Mass. "To facilitate evaluating the explosibility of dusts and to provide a numerical rating for the relative hazard, an empirical index has been developed by the Bureau of Mines. . . . The potential hazard of a dust is related to its ignition sensitivity and to the severity of the subsequent explosion. . . . The index of explosibility is the product of the ignition sensitivity and the explosion severity. This index is a dimensionless quantity having a numerical value of 1.0 for a dust equivalent in explosibility to the standard Pittsburgh coal. An index greater than 1.0 indicates a hazard greater than that for the coal dust. The notation ≤1.0 designates materials presenting primarily a fire hazard as ignition of the dust cloud is not obtained by spark or flame, but by a surface heated to a relatively high temperature. . . . The relative explosion hazard of a dust may be further classified by ratings of

Table 54.6. (Continued)

weak, moderate, strong, or severe. These ratings are correlated with the empirical index as follows:

Relative Explosion Hazard Rating	Ignition Sensitivity	Explosion Severity	Index of Explosibility
Weak	<0.2	<0.5	<0.1
Moderate	0.2–1.0	0.5–1.0	0.1–1.0
Strong	1.0–5.0	1.0–2.0	1.0–10
Severe	>5.0	>2.0	>10

Explosibility of dusts increases with decrease in particle size. The relative particle diameter is the ratio of the mean particle size of the dust to the mean size of the through No. 200 sieve sample; and the relative explosibility index is the ratio of the index computed for the dust to that computed for a through No. 200 sieve sample. . . . The explosibility index is approximately inversely proportional to the cube of the relative mean particle diameter. . . . For this reason it is advisable to reduce the production of fines as much as practicable. . . ."

[b] No ignition to 8.32 joules, the highest tried.

[c] No ignition to 2 oz/ft^3, the highest tried.

[d] Ignition denoted by flame; all others not so marked, denoted by a glow.

[e] Selected data from Ref. 59. These data are reproduced by permission from Fire Research Technical paper No. 21, *Explosibility Tests for Industrial Dusts*, ©Crown copyright 1975, Borehamwood, Hertfordshire, England.

The classification of explosibility is as follows:

"(a) Dusts which, when tested, ignited and propagated flame in the test apparatus, and
(b) Dusts which, when tested, did not ignite and propagate flame in the test apparatus.

Dusts are classified as Group (a) if they ignite and propagate flame in any of the three tests with a small source of ignition, either 'as received' or after sieving and drying the sample. The explosibility of dusts may be modified by factors such as large particle size, or high moisture content which is not removed by the usual drying methods, or if the sample is a mixture and the explosible proportion is inerted by the other ingredients. With some dusts, although flame propagation has been observed in the tests, and they are therefore classified as explosible, the explosion pressure that the dust can develop under industrial conditions may be very small.

The classification only applied to conditions where a dust is dispersed at or near ordinary atmospheric temperatures. Where a dust cloud is in a heated environment other considerations may apply."

[f] Indicates a Group (b) dust.

1.5.6 Pyrolysis/Combustion Toxicity Tests

The purpose of these tests is to facilitate the evaluation of potential hazards that might arise from the use of a given polymer in a product or system (39). Among the lethal factors identified at the beginning of this section, toxic gases (and smoke) are considered more important in building fires than heat or flames. Carbon monoxide, derived largely from natural cellulosic polymers, is generally recognized as the chief toxicant at present.

Testing to define possible hazards from the increasing use of synthetic polymers has for several decades been accepted as desirable. Zapp in 1960 stated the basic principle that underlies evaluation of their combustion toxicity: "in judging the safety of a synthetic resin for a proposed use, the hazard from combustion or thermal decomposition should be compared under equivalent conditions with that of alternative materials which have, if possible, a history of similar use (60)." However, these hazards are highly complex and the technical implementation of this principle may be even more so. Tests devised in an attempt to comply with a "virtually meaningless" (39) building code statement that the polymer in question should be no more toxic than wood are inherently limited by the rationale. Combustion/pyrolysis tests preferably depend on a given means of degradation and are based on a parameter such as degradation temperature, type or quantity of clinical effect, or a combination of these parameters (61–72). Like tests for flammability and fire performance (see Section 1.5.2), these tests provide a measure of response and suffer from experimental constraints.

The three principal means of thermal degradation—oxidative pyrolysis ("smoldering"), anaerobic pyrolysis, and flaming combustion (73–75)—all can and probably do occur at some instant in an uncontrolled fire, but *they change in importance as the fire proceeds*. The majority of fires neither start nor progress anaerobically, nor is an anaerobic atmosphere compatible with human life. Anaerobic pyrolysis is therefore essentially unsuitable (73) for the direct evaluation of thermal degradation toxicity even though it often yields useful information. Although temperatures as high as 800 to 900°C have been recorded in large-scale test fires, the use of such temperatures in bioassay tests imposes constraints to protect the animals from heat and is in some respects academic. At such extremely high temperatures heat becomes an overpowering toxic stress for an extensive area around the fire center. In general, temperatures extending from ≤400 to 600°C permit coverage of the main features of the decomposition process.

Table 54.7 summarizes results from a series of toxicity tests where rats were exposed to the products from polymers heated to specified temperatures. Table 54.8 gives the relative time to death or incapacitation in mice from polymer degradation products that were generated by a regulated increase in temperature to 800°C.

Table 54.7. Acute Inhalation Toxicity of Thermal Degradation Products of Plastics and Natural Materials When Administered Over 30 Minutes to Rats, According to the German Protocol DIN 53436[a]

Material	300°C CO-Hb %	300°C Exitus	400°C CO-Hb %	400°C Exitus	500°C CO-Hb %	500°C Exitus	600°C CO-Hb %	600°C Exitus
Rigid expanded polystyrene	<15	0/12	<15	0/12	<15	0/18	80	+11/12
Polystyrene	<15	0/12	<15	0/12	<15	0/12	74	+12/12
Nylon 66		0/12	<10	0/12	30	+12/12	22	+12/12
Nylon 6		0/12	<10	0/12	53	+12/12	34	+12/12
Rigid expanded polyurethane	<10	0/12	22	0/12	63	+12/12	76	+12/12
Unsaturated polyester resin and fiber-glass	<15	0/12	23	+3/12	54	+15/18	63	+1/12
Insulating cork	17	0/12	69	+5/18	86	+12/12	87	+12/12
Acrylonitrile–styrene copolymer		0/12	<15	+10/12	17	+16/18	27	0/18
Unsaturated polyester resin and fiber-glass (with flame-retardant finish)	<10	0/18	28	+12/12	76	+18/18	85	+12/12
Polyethylene	<15	0/12	79	+12/12	85	+12/12	82	+12/12
Polypropylene	18	0/12	58	+12/12	84	+12/12	81	+12/12
Sheep's wool	<15	+2/12	19	+12/12	27	+12/12	31	+12/12
Pinewood	32	+3/12	86	+12/12	85	+12/12	85	+12/12
Polyvinyl chloride	25	+10/12	45	+11/12	70	+12/12	76	+12/12
Leather	67	+12/12	69	+12/12	63	+17/18	54	+11/12
Celluloid	59	+12/12	72	+12/12	73	+12/12	75	+12/12

[a] Adapted from H. T. Hofmann and H. Sand, "Further Investigations into the Relative Toxicity of Decomposition Products Given Off from Smoldering Plastics," *J. Combust. Toxicol.*, **1**, 250–258 (1974). Weight of specimens, 5 g; air stream, 100 l/hr; dilution, 1:1. Data are reproduced courtesy of the Journal of Combustion Toxicology, Technomic Publishing Co., Inc., 265 Post Road West, Westport, Conn. 06880.

1.6 Degradation at Ambient Temperature

Like natural polymers, synthetic polymers may degrade owing to physical, chemical, or biologic stress. Synthetic polymers generally have a homogeneous structure; degradation or its lack is relatively uniform throughout the polymer.

Chemical degradation generally occurs as a result of weathering or aging, particularly photodegradation, or solvent action. Aging, impact, or mechanical failure may bring about physical degradation. Biodegradation may describe various types of environmental degradation but is preferably reserved for the degradation brought about by living organisms, usually microorganisms (76, 77), and includes changes brought about by soil burial. Biodegradation of polymers may be intentionally controlled by biocides (78) during use but is considered a desirable attribute in treatment of waste (79).

1.7 Polymer Names and Trade Names

Trade names for polymers may refer to a specific product or a line of products with similar properties. Only a very limited number of product names have been included in this chapter. Dictionary (3), index (80, 81), or review sources (31, 56) provide much more comprehensive information. See also commercial directories and on-line information data bases.

2 OLEFIN RESINS

2.1 Overview

Commercial development of polyethylene began (10) in the 1930s and was followed by full-scale production of the major olefin resins during the 1940s through the 1960s. Polyethylene and other olefin resins are closely related to the paraffins. As a class, these resins have very low water absorption, moderate to high gas permeability, good toughness and flexibility at low temperatures, and relatively low heat resistance. Environmental stress cracking is greater with polyethylene, whereas polypropylene is more susceptible to oxidation. Polyethylene cross-links on oxidation whereas polypropylene degrades to form lower molecular weight products (10). Table 54.9 gives typical data for properties. Amorphous olefins with rubbery characteristics are discussed in Section 3.

2.1.1 Production and Processing

Polyethylene alone presently accounts for more than 30 percent of the total annual U.S. plastics production, and polypropylene is also a major commercial product (Table 54.2). The homopolymer resins are sold by the manufacturer

Table 54.8. Relative Acute Toxicity of Thermal Degradation Products of Polymers as Determined by Time to Incapacitation and to Death in Mice[a]

Films, Forms, Elastomers, and Miscellaneous Polymers	No. of Samples	Time (min) to Death	Time (min) to Incapacitation	No. of Tests
Polyether sulfone	3	12.30 ± 2.08	11.25 ± 1.93	7
Polyphenylene sulfide	4	13.21 ± 3.80	11.53 ± 2.68	9
Polyaryl sulfone	2	13.48 ± 3.17	10.31 ± 0.42	5
Polyurethane, flexible foam	29	14.15 ± 2.84	10.45 ± 1.36	81
Polyamide (nylon)	3	14.36 ± 1.71	12.35 ± 1.44	7
Polyphenyl sulfone	1	15.46	13.32	2
Polyurethane, rigid foam	7	15.49 ± 4.06	11.77 ± 2.95	17
Polymethyl methacrylate (PMMA)	1	15.58	12.61	2
Polyvinylidene fluoride	1	15.86	6.50	2
Polyvinyl chloride (PVC)	2	16.60 ± 0.33	9.32 ± 4.77	4
Acrylonitrile–butadiene–styrene (ABS)	3	17.13 ± 2.45	11.82 ± 1.52	9
Polyethylene, including foam	5	17.31 ± 3.73	10.67 ± 3.65	11
Acrylonitrile rubber (NBR)	3	19.13 ± 2.89	12.46 ± 2.56	9

Fibers and Fabrics	No. of Samples	Time (min) to Death	Time (min) to Incapacitation	No. of Tests
Wool 100%	4	7.65 ± 1.29	5.31 ± 0.48	10
Wool 85–90%/nylon	4	8.87 ± 1.01	5.68 ± 1.06	10
Nylon 100%	9	16.78 ± 3.49	12.01 ± 1.74	18
Silk 100%	2	9.18 ± 0.35	6.70 ± 1.22	4
Silk 70%/rayon 30%	2	12.33 ± 0.58	8.94 ± 0.01	4
Rayon 100%	10	15.40 ± 2.41	9.22 ± 0.85	20
Polyester 100%	3	10.70 ± 2.25	8.15 ± 0.74	6
Polyester 65–87%/cotton	3	10.45 ± 0.26	8.28 ± 0.78	7
Cotton 59–70%/polyester	2	15.66 ± 1.17	9.14 ± 0.13	4
Cotton 100%	10	13.08 ± 2.14	8.39 ± 1.40	37
Cotton 100% FR	6	14.19 ± 3.64	8.85 ± 0.85	13
Aromatic polyamide 100%	3	14.71 ± 1.13	12.58 ± 1.12	7
Cotton 82–86%/rayon	2	12.00 ± 0.16	9.04 ± 0.81	4
Cotton 52–75%/rayon	8	14.53 ± 2.06	9.14 ± 1.30	16
Rayon 54–75%/cotton	18	12.70 ± 2.49	8.55 ± 1.56	37
Rayon 56–73%/polypropylene	2	14.08 ± 2.16	10.68 ± 1.66	4
Polypropylene 100%	4	16.64 ± 2.76	9.31 ± 1.55	8

Material	n			n
Polyphenylene oxide, modified	1	19.96	8.65	2
Bisphenol A polycarbonate	3	20.40 ± 3.77	14.71 ± 1.68	24
Polyvinyl fluoride	1	20.50	16.94	2
Ethylene propylene diene terpolymer (EPDM)	2	20.69 ± 0.04	12.97 ± 3.04	5
Chlorosulfonated polyethylene	2	20.88 ± 2.07	12.64 ± 6.25	7
Polyisocyanurate, rigid foam	2	21.68 ± 1.38	18.90 ± 1.55	4
Polyisoprene, natural rubber	1	22.13	15.35	3
Chlorinated polyvinyl chloride	2	22.25 ± 0.69	7.64 ± 1.92	4
Polychloroprene, including foam	6	22.33 ± 3.80	13.61 ± 1.69	21
Polystyrene	2	23.10 ± 4.33	17.11 ± 2.73	4
Styrene–butadiene rubber (SBR)	1	24.11	15.73	3
Chlorinated polyethylene	2	26.08 ± 1.80	9.31 ± 2.55	4
Wood	12	14.03 ± 1.48	9.92 ± 1.09	33
Cellulosic board	8	16.57 ± 3.54	10.10 ± 1.35	28

Material	n			n
Rayon 52–72%/nylon	3	15.60 ± 3.86	10.99 ± 3.07	6
Nylon 57–62%/rayon	2	15.62 ± 0.15	12.18 ± 0.26	4
Nylon 100%	9	16.78 ± 3.49	12.01 ± 1.74	18

[a] Thermal degradation products were generated in experiments conducted according to a rising temperature program, 200–800°C. Other experiments (data not shown) were conducted according to a fixed 800°C temperature program. The above data are summarized from C. J. Hilado, J. J. Cumming, and C. J. Casey, "Toxicity of Pyrolysis Gases from Natural and Synthetic materials," *Fire Technol.*, **14**, 136–146 (1978) and are reproduced by permission from Fire Technology, National Fire Protection Association, Boston, Mass.

Table 54.9. Selected Properties of Polyethylene, Propylene, and Polybutylene[a]

Property	Low Density Polyethylene	High Density Polyethylene	Polypropylene	Polybutylene	Reference[a]
Melting point (°C)	110–130	120–140	Isotactic: 165–208 Syndiotactic: 138–161	124–135	10, 14, 82–85
Glass transition temperature (°C)	∼ −80 to −125; conflicting interpretations of data with several identified transitions		∼ −3 to −35; conflicting data	−34; wide spread in reported values	10, 14, 83, 84, 86
Softening point (°C)	60–80 (waxes, 20–60)	110–125	85–105 (Vicat, 5-kg load); fibers, 145–150		6, 14, 83, 84
Brittle point	< −70 for high polymers; oligomers, brittle at −30 to ambient temperature		∼0		10,14
Density	0.910–0.925[b]	0.941–0.965[b]	0.89–0.94	0.87–0.95	14, 82, 85
Refractive index	1.51–1.545, increases with density (various amorphous and crystalline forms cited as 1.49–1.58)		1.49		14, 83, 84, 87
Solvents	As high density polyethylene, but temperature 10–30°C lower, depending on degree of branching	Above 80°C: aliphatic and aromatic hydrocarbons, particularly when chlorinated. Generally insoluble below ∼60–80°C. Swelling occurs at temperatures below solubility	Atactic: hydrocarbons and chlorinated hydrocarbons at ambient temperature, isoamyl acetate, diethyl ether. Isotactic: as polyethylene. Swelling at ambient temperatures from aromatic hydrocarbons, chlorinated hydrocarbons, esters, ethers, and oxidizing agents	n-Alkylacetates; aromatic and chlorinated hydrocarbons at elevated temperatures. (Isotactic: as polyethylene)	14, 83, 84, 88

Nonsolvents	As high density polyethylene	Many organic liquids and aqueous solutions at ambient temperatures. Insoluble in more polar organic solvents with small hydrocarbon group even at elevated temperatures	10, 14, 88
	All common organic solvents at ambient temperature, inorganic solvents		
Decomposed by	Strong oxidizing agents, as nitric and sulfuric acids; slowly attacked by halogens and chlorinating agents	Similar to polyethylene, but polypropylene more susceptible to oxidation; rate of attack and range of reagents increasing with rise in temperature, slowly weakened (as fall in fiber strength) by concentrated alkali	10, 14

[a] Data are summarized mainly as a range of values from Roff and Scott (14), *Modern Plastics Encyclopedia* (82), and *Polymer Handbook* compilations (83–88), with emphasis placed on the *Handbook*. Values in °K (86) have been converted to °C. Properties related to physical form are often summarized in specific reviews, such as Refs. 89–91 for fibers. The above listing should be viewed as a general guide. Specific values for properties of a commercial product should be obtained from the manufacturer prior to significant, prolonged, or extensive use. Commercial products, whether or not identified by a particular trade name, may often be prepared by various techniques and/or with various comonomers to give products with somewhat different properties. In the latter event, other CAS registry nos. may be appropriate to identify the products under consideration.

[b] Medium density polyethylene occupies an intermediate range between low and high density; see Section 2.2.1. Cross-linking of polyethylene increases density, decreases crystallinity. With polypropylene, density increases with increasing crystallinity; this direct correlation is more common (14).

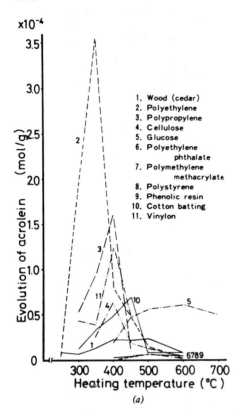

1. Wood (cedar)
2. Polyethylene
3. Polypropylene
4. Cellulose
5. Glucose
6. Polyethylene
       phthalate
7. Polymethylene
       methacrylate
8. Polystyrene
9. Phenolic resin
10. Cotton batting
11. Vinylon

Figure 54.2 Evolution of (a) acrolein and (b) formaldehyde from various materials, based on a sample weight of 500 mg. From Morikawa (103). Reproduced by courtesy of *Journal of Combustion Toxicology,* Technomic Publishing Co., Inc., Westport, Conn.

primarily in dry solid form, as granules, pellets, fibers, or sheets. Most end use applications require additives. Antioxidants can be added during processing of the raw polymer and are essential with polypropylene. Ultraviolet stabilizers or carbon black protect against sunlight, which gradually causes embrittlement. Low density polyethylene intended for film may have slip additives to decrease friction and antiblock additives to prevent layers from blocking or sticking. Additives can also add color or modify electric insulating properties. Infrequently stabilizer additives may be deliberately omitted; for example, unstabilized polyethylene plastic mulch disintegrates more rapidly after the growing season.

2.1.2 Analysis and Specifications

Specifications and related analyses pertaining to potential food contact uses of olefin resins have been delineated (92) by the Food and Drug Administration.* Polyethylene and oxidized polyethylene have been cleared for a number of

* This agency has sponsored studies to assess the migration of butylated hydroxytoluene from high density polyethylene into foods and food simulants (24).

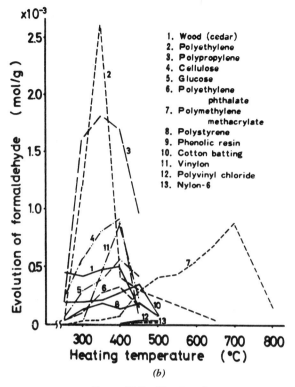

Figure 54.2 *(Continued)*

indirect food additive uses, among which clearance in packaging is probably the most important. Additional specifications pertain to the use of polyethylene in pressure-sensitive adhesives, in coatings, as a food lubricant, and as a replacement for roughage (direct additive) in animal feed. Polyethylene with a molecular weight of 2000 to 21,000 and polyisobutylene with a minimum molecular weight of 37,000 may be used as masticatory substances in chewing gum base. Polypropylene has also been cleared for specified food contact uses including coatings, gaskets, and packaging. Considerations pertaining to the use of polyethylene and polypropylene in containers for pharmaceuticals have been reviewed under the auspices of the World Health Organization (93).

Specifications concerning mechanical and other types of performance depend on end use (see Section 1.3).

2.1.3 Toxicologic Potential

Monomer residue has not been considered a problem. The ethylene, propylene, and 1-butene monomers act toxicologically as asphyxiant gases. Ethylene is a natural constitutent of apples and other fruit and is used commercially to accelerate ripening of fruit. Low molecular weight additives or impurities may be of concern if significant migration of these materials into food or water is possible; in practice the specifications discussed above act as a control.

Table 54.10 indicates that combustion typically yields carbon monoxide as the primary toxic decomposition product. Toxicologically significant quantities of aldehydes may also occur under certain conditions of temperature and air supply; see Figure 54.2.

These olefin resins are typically resistant to microbiologic degradation although they can be treated to increase susceptibility (see Sections 2.1 and 2.2). Depending on formulation of the finished product, they *may* be subject to deterioration of desirable properties from adverse environmental conditions or aging.

2.1.4   Workplace Practice and Standards

Inhalation of dust should be minimized in accordance with general principles of good industrial hygiene practice (Volume 1, Chapter 7). Precaution should be taken against inhalation of thermal decomposition products. Incineration with adequate air supply is generally the preferred method of disposal, but sanitary landfill is acceptable.

2.2   Polyethylene

Polyethylene (PE) has CAS No. 9002-88-4. Trade names include Alathon®, Alkathene®, Lupolen®, and Tyvek® spunbonded olefin sheet; polyethylene is also available in the Bakelite®, Marlex®, Tenite®, and many other product series. Figure 54.3 (110) shows the formula and difference in chain structure relative to density of the more common thermoplastic forms. Polyethylene may also be deliberately cross-linked to provide resistance to creep at high temperatures or to improve resistance to environmental stress cracking (3, 7).

2.2.1   Preparation

At least five distinct processes have been identified (10, 110). High pressure polymerization methods generally use pressure, heat, and a free radical initiator to produce low density polyethylene (d = 0.915 to 0.94 g/cm^3) with molecular weight below 50,000. The Phillips process and Standard Oil (Indiana) process use a hydrocarbon solvent and metal oxide catalyst to produce high density polyethylene (d = 0.96 g/cm^3). The Ziegler polymers are intermediate in density—about 0.945 g/cm^3—between high pressure polyethylenes and those produced by the other two processes (10). More recently a low presssure gas-phase process has been used to prepare both high density and low density polyethylene.

2.2.2 Molecular Weight, Properties, and Applications

Brydson (10) describes the most common commercial grades of polyethylenes as having a number average molecular weight approximately 10,000 to 40,000 (corresponding weight average molecular weight 50,000 to 300,000). Waxlike polyethylenes, about 1000 to 10,000 molecular weight, are classed as oligomers rather than high polymers. The more highly crystalline, high density materials tend to be more brittle. Irradiation increases cross-linking and decreases crystallinity.

Polyethylene free from catalyst and formulated to retard oxidation is a nonpolar material that offers excellent electrical insulating properties. Under load polyethylene gradually deforms or creeps. Structural changes can occur with oxidation at 50°C and with ultraviolet light at room temperature (10). See Table 54.9 and Figure 54.1 for comparative data.

More than half of all low density polyethylene is used in film, particularly in packaging and lining for food, clothing, consumer, and industrial products. Other major applications of low density homopolymers and copolymers are in hot melt coating and adhesive formulations. Powders are used as coatings and to make hollow parts by rotational molding. High density polyethylene is widely used in various molding processes to make containers or pipes where rigidity, impact resistance, and light weight are required. Injection molded structural foam of high density polyethylene offers a high stiffness-to-weight ratio. Cross-linked high density polyethylene shows promise in solar absorption air-conditioning systems (111). Spun-bonded polyethylene can be formulated soft and conformable or stiff and paperlike (112). Applications include disposable apparel, utility and construction sheeting, posters, charts or maps, wall coverings, and packaging.

Ultra high molecular weight polyethylene describes high density polymers with an average weight range of 2 to 6 million (113). The base resin, a fine powder, can be processed into products that offer unusually high abrasion resistance and corrosion resistance.

2.2.3 Oral Toxicity

Rats survived acute doses of 7.95 g/kg of "Marlex" 50* without evidence of adverse effects (114, also cited in Reference 1). Mice receiving a single oral dose of 2500 mg/kg powdered unstabilized and stabilized high density polyethylene also showed no toxic effects (115).

---

* Identified as homopolymer with density of 0.96.

Table 54.10. Overview of Inhalation and Thermal Degradation Data on Polyethylene and Polypropylene[a]

Polymer	TLV	Type of Study				Description of Material	Procedure	Observations/Conclusions	Reference
		Ambient Temperature	Elevated Temperature	Effluent Analysis	Species Exposed				
Polyethylene	No specific TLV; see Section 1.4[b]	X			Rat	High density polyethylene	Exposure 30 min, decomposition products generated at 100°C increments (DIN 53436 protocol)	Products from 400°C degradation lethal, caused 79% carboxyhemoglobin; see Table 45.7	62
			X	X	Mouse	Low density polyethylene generally, one sample high density polyethylene	Exposure 30 min. Polyethylene pyrolyzed without forced air flow, at 800°C (fixed temperature program) or increased at 40°C/min from 200 to 800°C (rising temperature program). Later tests with or without forced air flow at increments of 50 or 100°C	Time to death longer than median when compared to other polymers in series tested (fixed and also rising temperature programs); see Table 45.8. Carbon monoxide concentration/exposure time (800°C fixed temperature program) considered sufficiently high to have been sole toxicant; methane also found. Later tests (incremental temperatures) indicated that 1 liter air/min decreased CO levels; thermal decomposition at 450°C may favor CO production	66, 94–97
			X	X	Mouse	Flexible foam	Exposure 30 min, rising temperature program as above	Carbon monoxide considered "an important if not the principal toxicant"	98
			X	X	Rat	Pellets and semiflexible foam	Exposure 30 min, rats exposed to products generated by flaming or nonflaming conditions 345–385°C	Carbon monoxide and acrolein in toxicologically significant concentrations; CO proportionately greater toxicant from flaming conditions, C_3H_4O proportionately greater from nonflaming conditions just below autoig-	99

					nition. Earlier comparable work indicated no toxicologically significant concentration of formaldehyde	
X	X	Rat (also human experience)	Semiflexible foam	Large-scale fire test	Carbon monoxide and temperature considered limiting toxicants; levels of acrolein considered, in some cases, transiently irritating but not approaching a fatal level	100
X	X	Mouse	"Granular"	Exposure 15 min, decomposition products generated at 350, 500, or 750°C	350 or 500°C degradation lethal; at 750°C, polymer burst into flames and mice survived	101
X	X		Powders, both low density and high density polyethylene; pellets of high density polyethylene	Degradation <500°C, under different conditions of air flow, and heating rate	Varying amounts of carbon monoxide and carbon dioxide, plus smaller amounts of methane, ethylene, ethane, and C_3 to C_6 hydrocarbons	37, 102
	X			"Smoldering combustion" at 300–400°C	Polyethylene yielded largest amounts of acrolein, formaldehyde, and fatty acids when compared with 10–12 other materials; see Figure 54.2	103
X	X			"Flaming combustion" at 700°C, 50–100 l/hr air flow	Production of CO, C_2H_4, CH_4, and acetylene (listed in order of decreasing analysis concentration)	104
X	X			Degradation, 335–450°C	Continuous spectrum of saturated and unsaturated hydrocarbons from C_2 to C_{90}, with lower temperature favoring larger fragments	As cited in 105
X	X			"Thermo-oxidation" at 350°C (also combustion, pyrolysis under helium)	C_3–C_{15} aldehydes represented 48% of the chromatogram area, ketones 3%, olefins 25%, paraffins 12%; CO not analyzed	106

Table 54.10. (Continued)

Polymer	TLV	Ambient Temperature	Elevated Temperature	Effluent Analysis	Species Exposed	Description of Material	Procedure	Observations/Conclusions	Reference
			X	X			Pyrolysis under vacuum at 1000°C	Gaseous products determined, mainly hydrogen, ethane, smaller amounts acetylene and miscellaneous hydrocarbons	107
Polypropylene	No specific TLV; see Section 1.4		X		Rat		See above, this table; Ref. 62	Products of 400°C degradation lethal, caused 58% carboxyhemoglobin; see Table 45.7	62
			X		Mouse	Fabric, resin	Fixed 800°C temperature program and rising temperature programs. See above, this table; Ref. 66	Time to death among the longest of fabrics tested in series without forced air flow; see Table 45.8. Decomposition products not significantly different from or produced less response (longer time to death) than those from polyethylene. Carbon monoxide and CH$_4$ concentrations in effluent increased with temperature at 400–800°C; CO may have been principal toxicant at 500–800°C but not at 400°C	66, 95, 108
		X	X		Mouse		As described above, this table		101
		X	X		Rat	"Granular"	Polymer temperature increased 5°C/min to about 700°C, rats exposed head only to airstream of diluted effluent ≤140 minutes	Polypropylene products among least toxic when series of 10 polymers degraded under slow combustion conditions, but among most toxic when they were degraded	109

while pyrolysis occurring (slow pyrolysis conditions), or temperature increased to 700°C in 1–2 min, maintained ≤10 min, rats exposed 4 hr (rapid combustion conditions)

under rapid combustion conditions

Isotactic polypropylene	X	X	See above, this table	Varying amounts of CO and CO$_2$; varying amounts of propylene, also of CH$_4$ and miscellaneous C$_3$ to C$_6$ hydrocarbons	102
	X		See above, this table	Polypropylene yielded less acrolein, formaldehyde, and fatty acids than polyethylene, but more of these products than the 9–11 other materials in series; see Figure 54.2	103
	X	X	Degradation under nitrogen, 328–1200°C	Saturated and unsaturated hydrocarbons from C$_2$ upwards; as temperature raised, proportionately greater yield of small fragments	As cited in 105
	X		See above, this table	Methyl alkyl ketones represented 57% of the chromatogram area; CO not analyzed	106
	X		As described above, this table		107

[a] These listings are generally selective. Where feasible, data from several articles of one investigative team are collated and summarized as one entry.

[b] The USSR has an exposure limit of 10 mg/m³ for "polyethylene (low presure)" and 5 mg/m³ for "polypropylene (not stabilized)" (28).

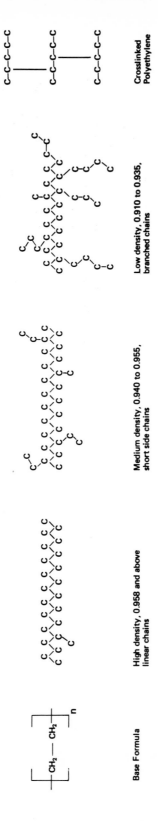

Base Formula

High density, 0.958 and above linear chains

Medium density, 0.940 to 0.955, short side chains

Low density, 0.910 to 0.935, branched chains

Crosslinked Polyethylene

Figure 54.3 Molecular chain structure of polyethylene. From Flesher (110). The numerical definitions of density are not exact; see Roff and Scott (14). Adapted from *Modern Plastics Encyclopedia 1978–79*, copyright McGraw-Hill Inc.

Dietary levels of 1.25, 2.5, or 5 percent "Marlex" 50* for 90 days produced no adverse effects in rats (114, also cited in Reference 1). In other 90-day tests (116, 117), rats and dogs were fed an extract of low molecular weight polyethylene film; the film had been extracted with isooctane to yield 568 mg extract/100 g of film. Rats fed at a level of 13,500 ppm film extract showed liver changes (fat droplets, cloudy swelling, and increased liver weight) that were considered reversible in all cases. Rats fed at levels of 2700 and 540 ppm and dogs fed 2700 ppm showed no adverse effects.

Swine fed 20 percent shredded polyethylene for 65 or 77 days in a program to control weight gain showed no adverse effects on the digestive system (118). Feeding animals aqueous extracts of stabilized polyethylene for 16 to 19 months produced only insignificant, generally transient effects on body weight or conditioned reflexes but no pathologic changes (115).

2.2.4 Inhalation Toxicity and Thermal Degradation

Overall, carbon monoxide and heat are considered the primary or limiting toxicants (100) from thermal decomposition of polyethylene. Under conditions of limited air supply the optimal temperature for the formation of carbon monoxide may be approximately 450°C (96). More efficient conversion to carbon monoxide at 600 compared to 800°C was suggested as the reason for the more rapid time to death of mice exposed to products generated at the lower temperature without direct air supply (119). Other decomposition products that have been identified under varying conditions include carbon dioxide, acrolein, formaldehyde, other aldehydes, ketones, fatty acids, methane, ethane, and acetylene. Probably the most attention has been given to the formation of acrolein, which can be evolved in toxicologically significant amounts.

Incineration of "pure" polyethylene in waste disposal presents a comparatively minor air pollution problem (120). Polyethylene containing filler material with mesh size less than about 3 μm in diameter may be more hazardous. Polyethylene under flaming conditions (44) is considered relatively low in smoke release, whereas under smoldering conditions polyethylene smokes markedly.

Table 54.10 provides an entry to the extensive literature. See also Figure 54.2.

2.2.5 Carcinogenic Potential/Implant Studies

As indicated in Section 1.4, tumors in animals from implanted polyethylene are generally considered consistent with solid state carcinogenesis (29; see also Section 1.4, References 121, 122). Tumors observed in women using intrauterine contraceptive devices of polyethylene are also consistent with this rationale. In an apparent exception, intraperitoneal implants of polyethylene particles <0.4 cm did produce tumors in rats, but it was "suggested that the critical size of particles implanted in the abdominal cavity may be much less for solid-state carcinogenesis to be operative" (122).

* Identified as homopolymer with density of 0.96.

A biochemical correlation with physical factors comes from the tests of Matlaga and his associates (123) with six "medical-grade polymers" including polyethylene. They found that triangular-shaped implants of the respective polymers showed the highest level of cellular lysosomal phosphatase enzyme activity, whereas pentagonal shapes showed less activity and circular shapes the lowest enzyme activity.

2.2.6   Environmental Deterioration

Potts and co-workers (71, 124, 125) reported that a commercial polyethylene-based packaging wrap was biodegradable according to ASTM Method D-1924-63; this biodegradability was presumed to be due to an additive since it did not occur after solvent extraction. The results of tests with 10 homopolymers of varying weights are consistent with their thesis that biodegradability of the polymer correlates with a sufficient number of low molecular weight species (under about 500) rather than the average molecular weight measurement. Polyethylene of high molecular weight can be rendered biodegradable if pyrolyzed to reduce molecular weight; in stepwise tests, the microbial growth rating varied directly with the reduction in molecular weight.

Photodegraded polyethylene appears to be rather readily biodegraded (126). Photodegradation can be accelerated by the use of additives (127–129) such as unsaturated esters (vinyl acetate, alkyl acrylates, or methacrylates) or 2 percent partially degraded polymer.

2.3   Polypropylene

Polypropylene (PP) has CAS No. 9003-07-0*. Plastics are commercially available as Profax® resins and films and in the Amco®, Amerfil®, and other product lines; textiles include Herculon® olefin fiber† and Typar® spun-bonded polypropylene sheet.

$$-\!\left(CH_2CH\right)_n\!-$$
$$\qquad\ \ \ \ \ \ |$$
$$\qquad\ \ \ \ \ \ CH_3$$

2.3.1   Preparation

Polypropylene is commonly prepared by polymerizing propylene with a Ziegler–Natta catalyst system in a diluent such as naphtha. Polymer molecular weight can be controlled by polymerization temperature or pressure, modifi-

---

* Isotactic, syndiotactic, and atactic forms are possible; the isotactic form (CAS No. 25085-53-4) accounts for 90 to 95 percent of commercial polymers (10).

† Olefin fiber is defined (130) as a manufactured fiber in which the fiber-forming substance is any long chain synthetic polymer composed of at least 85 percent by weight of ethylene, propylene, or other olefin units (except amorphous olefins considered as rubber). Commercial olefin fibers per se are primarily polypropylene.

cation of the catalyst system, or the use of hydrogen as a chain transfer agent (10).

2.3.2 Molecular Weight, Properties, and Applications

Number average molecular weights of commercial polypropylenes can range (10) from 38,000 to 60,000, and weight average molecular weights from 220,000 to 700,000. Higher molecular weight resins tend to be softer and less stiff than lower molecular weight resins, which tend to be more brittle. Articles made of polypropylene can withstand boiling water and steam sterilization. "Springy polypropylene" is under study as an elastomeric biomaterial (131). See Table 54.9 and Figure 54.1.

Polypropylene is fabricated by molding (including foams), by extrusion, as film, and as fibers. Injection molding applications and fibers currently are the dominant end uses and account for approximately two-thirds of U.S. consumption (132). Injection moldings are used in the transportation, appliance, housewares, medical, and packaging fields. Polypropylene fibers are light, strong, and resistant to abrasion, and have little affinity for water; they are widely used in carpet face, carpet backing, upholstery, industrial ropes, sacking, and fishing nets.

2.3.3 Oral and Skin Toxicity

Mice given an acute oral dose of 8 g/kg showed no noticeable toxic effects. Feeding rats and mice aqueous extracts of polypropylene (extraction temperatures 20 and 60°C) for 15 months produced no significant changes (133). Rats fed edible-oil solutions of extractable material from a typical but ^{14}C-tagged polypropylene excreted radioactive material in the feces within 2 days. "This accounted for about 93% of the amount fed" (134; also cited in Reference 1). Two-year tests with oligomeric material, molecular weight 800, showed no effects in rats fed a diet as high as 20,000 ppm and dogs fed up to 1000 mg/kg (135).

Polypropylene microfoam sheeting produced no evidence of skin irritation or sensitization when tested on 20 humans (136).

2.3.4 Carcinogenic Potential/Implant Studies

The data from one experiment with subcutaneous implants in rats (137) suggest that the response of local fibrosarcomas observed with disks, and to a relatively small extent with powder, is essentially similar to that observed with polyethylene (see Section 1.4 and Section 2.2, Part 5). As with polyethylene, polypropylene implants of triangular, pentagonal, or circular shape provoked varying levels of cellular lysosomal acid phosphatase enzyme activity (121). In a 6-month test with subcutaneous implants in hamsters (138), sequential examination of polypropylene filaments with and without an antioxidant showed that in vivo degradation could be effectively retarded by the use of an antioxidant.

2.3.5   Inhalation and Thermal Degradation Products

In general the pyrolysis/combustion toxicity appears to be approximately similar to that of polyethylene. Data are summarized in Table 54.10; see also Figure 54.2.

2.3.6   Photodegradation

Addition of partially degraded polymer such as heat-treated polypropylene promotes environmental deterioration (129). Polypropylene can also be rendered degradable by additives (139). A Japanese study (140) showed that polypropylene weathered outdoors for several years became brittle and degraded to low molecular weight products, particularly acetic acid.

2.4   Polybutylene

Polybutene (CAS No. 9003-28-5)* is available as Vestolen® BT and Witron® polybutylene.

$$-\!\!\left[CH_2CH\right]_n\!- \atop \quad\;\; C_2H_5$$

2.4.1   Preparation

Polybutylene is synthesized from butene-1 in a Ziegler–Natta system (10, 141). The initial crystalline form on cooling from the melt is tetragonal; after 5 to 7 days this transforms to a stable rhombohedral form with higher density and melting point.

2.4.2   Molecular Weight, Properties, and Applications

The weight average molecular weight is given as 10^5 to 10^6 (14). The outstanding property of polybutene-1 is its resistance to creep, which is very high for an aliphatic polyolefin (10). Piping and high performance films are the major markets. Other applications are as electrical insulation elastic and high tenacity fibers, as well as blends with other polymers (141). See also Table 54.9.

2.4.3   Oral Toxicity

Rats fed a diet of 10 or 1 percent polybutene-1 for 6 months showed no adverse effects (142).

---

* The homopolymer exists in isotactic crystalline (CAS No. 25036-29-7) and atactic amorphous forms.

2.5   Other Olefins

Polyisobutylene becomes at high molecular weights a rather sluggish, unusable rubber. Polyisobutylene is used as an oil additive. The copolymer with 2 percent isoprene is the widely used butyl rubber.

Polymethylpentene is a predominantly isotactic polymer distinguished by unusual transparency, exceptionally low density (0.83), and high softening point (Vicat softening point 179°C). Limitations are high permeability and environmental stress cracking (14).

Ionomers are ionic polymers that contain metal or other inorganic counterions and are intended for film and plastic applications (4). Most commercial ionomers (as Surlyn®) are derived from ethylene copolymers. These copolymers contain anionic pendant groups introduced from an unsaturated carboxylic acid such as methacrylic acid. The copolymers are then treated with a metal derivative (typically of zinc or sodium) to make the carboxylic group appear to ionize reversibly with heat. The products can resemble thermosetting resins at ambient temperatures and linear thermoplastics at elevated temperatures (10).

Chlorosulfonated polyethylene and chlorinated polyethylene are discussed in Section 3. Section 5 mentions polymers containing ethylene and vinyl acetate units.

3   OLEFIN AND DIENE ELASTOMERS

3.1   Overview

The first identified elastomer, natural rubber, was described by Columbus as a ball that *bounced*. More practical early applications included primitive waterproof clothing and the rubber tires made for the carriage of Queen Victoria in 1846. Although the first specialty elastomers, polysulfides and polychloroprene, were commercialized in the 1930s, natural rubber was the major industry product until World War II when butadiene–styrene was established as a general purpose rubber and butadiene–acrylonitrile also became a major product. Today approximately two-thirds of all elastomers are used in tires and tubes (143). Many special purpose elastomers have been developed, particularly for high temperature service (approximately 130 to 150°C or above) during the last decade (144, 145).

All elastomers (see definition in Table 54.2) are composed of flexible macromolecular chains that "straighten out" under stress but "coil up" when the restraining force is released (7). For optimum properties as an elastomer, these chains must be anchored by cross-links as noted in Section 1.2; the cross-linking of elastomers is often described as vulcanization. Table 54.11 lists some basic properties typically associated with representative olefin and diene elastomers.

Table 54.11. Overview of Comparative Properties of Olefin and Diene Elastomers[a]

Property	ASTM Test Method	NR and IR Polyisoprene	SBR Styrene-Butadiene	EPDM Ethylene-Propylene + Diene	CR Neoprene	NBR Nitrile Rubber	FKM Vinylidene Fluoride-Hexafluoropropylene	CSM Chlorosulfonated Polyethylene	IIR Butyl	Perfluorinated Elastomer
Specific gravity (base material)	D 720	0.93	0.94	0.86	1.23	1.00	1.8	1.12	0.92	2.02
Glass transition temperature (°C)		-72	-60 to -65	-55	-45 to -50	-80	-25	-15 to -30	-71	-12
Maximum operating temperature (°C) (continuous service)		70	70	160-175	100	100-120	200-220	125	100	260-280
Brittleness temperature (°C)	D 796	-55	-60	-50 to -80	-45	-55	-40 to -45	-40 to -60	-60	-35 to -40
Hardness range[b] (Durometer A)	D 2240	30-90	40-90	30-90	40-95	40-95	50-90	40-95	40-75	70-90
Mooney viscosity, ML4 at 100°C	D 1646	60-100	40-70	40-70 (at 102°C)	40-60	45-60	50-80	30-95	40-80	Not measured
Tensile strength[c] (psi) 200% modulus at 24°C (psi)	D 412	>3000	<1000	<1000	>3000	<1000	1000-1200	3500-4000	1500	2000-2500
Tear strength	D 624	Very good	Fair	Fair-good	Good	Fair	Fair-good	Good	Good	Fair
Heat aging	D 573 or D 865	Poor-fair	Poor-fair	Very good	Fair-good	Good	Excellent	Good-very good	Good	Outstanding

Property	ASTM	−30 to −45	−25 to −45	−30 to −50	−20 to −40	−10 to −35	−10 to −30	−30 to −45	−20 to −40	−30 to −45
Low temperature serviceability (°C)		−30 to −45	−25 to −45	−30 to −50	−20 to −40	−10 to −35	−10 to −30	−30 to −45	−20 to −40	−30 to −45
Dielectric strength	D 149	Excellent	Excellent	Excellent	Very good	Poor	Very good	Very good	Excellent	Very good
Resistance to sunlight aging		Poor	Poor	Outstanding	Very good	Poor	Outstanding	Outstanding	Very good	Excellent
Resistance to gas permeability	D 1434	Good	Good	Good	Very good	Very good	Excellent	Very good	Excellent	Poor
Resistance to water	D 471	Very good	Good–very good	Excellent	Good	Good	Very good	Good	Very good	Excellent
Resistance to weak acids	D 471	Fair–good	Fair–good	Good–excellent	Excellent	Good	Good–excellent	Excellent	Excellent	Excellent
Resistance to strong acids	D 471	Fair–good	Fair–good	Fair–good	Good	Good	Excellent	Good	Very good	Excellent
Resistance to organic solvents	D 471	Poor–fair	Poor–fair	Poor–good	Poor–good	Fair–very good	Good	Poor–good	Poor–good	Excellent
Resistance to hydrocarbon fuels and oils	D 471	Poor	Poor	Poor	Good	Very good–excellent	Excellent	Good	Poor	Excellent

a Based on typical stocks or the middle of the commercial range of representative kinds of elastomers where a choice exists—as medium acrylonitrile NBR. The ratings shown are qualitative and comparative only. They should not be construed as recommendations. Specific compounding may be required for optimum performance. For additional data see Refs. 10, 11, 14, 16. Reference 11 gives various physical constants of polybutadiene as compiled by G. H. Stempel and of IR, SBR, butyl rubber, and CR as compiled by L. A. Wood. This table is supplied through the courtesy of D. H. Geschwind, E. I. du Pont de Nemours and Co., Polymer Products Dept., Wilmington, Del. 19898, and printed with permission.

b *Hardness*: "On the Shore A scale, 0 would be soft and 100 hard; as a comparison, a rubber band would be about 35 and rubber tire tread 70. Hard rubber (ebonite) is much more solid and is read on the Shore D scale, as a comparison, a hard-rubber pipe stem may read about 60 Shore D, and a bowling ball about 90 Shore D. Most products will fall between 40 and 90 durometer (unless otherwise specified, Shore A is assumed)" (16).

c *Tensile strength*: The value listed are for pure gum; carbon or black loaded polyisoprene and neoprene have similar values. Carbon loading can give values >2500 with CSM, and 1500–3000 with FKM. The listed effects of *water*, *acids*, and *solvents* are based on ordinary environmental temperatures.

3.1.1 Production and Processing

Production data are summarized in Table 54.2. Most rubber is sold raw or uncured in dry solid or liquid latex state. The basic steps in the manufacture of some types of dry synthetic rubber are polymerization, coagulation, washing, and drying. With a latex, the basic steps are polymerization, stabilization and, usually, concentration. Latex can be defined as a stable aqueous dispersion containing discrete polymer particles about 0.05 to 5 μm in diameter (146).

Emulsion polymerization systems contain water, monomer(s), initiator, and anionic or cationic surfactants. Solution polymerization with stereospecific catalysts involves the reaction of one or more monomers in an inert solvent; system conditions can be controlled to maximize a desired isomer arrangement in the polymer.

Antioxidants are generally added for shelf and processing stability. Unsaturation correlates with sensitivity to oxidation. Among synthetic rubbers, butadiene and isoprene polymers are much more sensitive to oxidation than ethylene–propylene polymers. Natural rubber contains some antioxidants from rubber trees.

Vulcanization is usually done with sulfur, sulfur-containing compounds, or peroxides, but it may also be accomplished with other compounds that yield free radicals at curing temperature or by radiation. Various supplementary materials such as cure accelerators, cure retarders, or reinforcing agents are commonly part of the compounding recipe. Vulcanization begins as the finished item is molded and stops when the desired state of cure is reached (148).

3.1.2 Analysis and Specifications

Measurements to determine the amounts of residual monomer have been of particular interest (31, 149, 150). Residual butadiene, acrylonitrile, and styrene can be determined to levels in solid rubber ≤ 1 ppm by headspace gas chromatography (150). Butadiene monomer in 0.1 g samples of latex can be measured by gas chromatography with a detection limit of 50 ppm (151). A gas chromatographic/mass spectrometric method to identify volatile materials released during simulated vulcanization at 160 to 200°C has been reported (152); the polymers tested were cis-polybutadiene, styrene–butadiene, and a blend of these two. Pyrolysis/mass spectrometry (samples ≤ 5 μg) has been used to distinguish between adhesives based on natural rubber, styrene–butadiene, or polychloroprene (153).

Residual acrylonitrile in nitrile rubber has been reported as varying from "non-detectable to something less than 100 ppm" (154). Latexes of this material are reported as containing 250 to 750 ppm acrylonitrile. The concentration of chloroprene monomer is described as <1 ppm in solid polychloroprene, but amounts as high as 5000 ppm have been reported in some samples of latex (149).

Accepted practice in raw rubber manufacturing calls for quality control of the polymer within various specifications (usually ± 10 percent). Standards for these measurements are generally formulated by the American Society for Testing and Materials (12) or product-oriented organizations such as the Society of Automotive Engineers (147). See also Section 1.3.

Elastomers meeting certain specifications are permitted in stated food-additive applications (92). Both natural rubber and polyisoprene are specifically identified among the elastomers permitted in rubber articles intended for repeated use. Natural rubber, natural latex, cis-1,4-polyisoprene, synthetic rubber, or rubber hydrochloride are variously permitted in adhesives, sealing gaskets, paper, paperboards, and coatings. Styrene–butadiene, ethylene–propylene copolymer (and certain diene-containing terpolymers), acrylonitrile–butadiene, and chloroprene polymers are similarly permitted in stated applications. The list of components permitted in rubber articles intended for repeated use also includes vinylidene fluoride–hexafluoropropylene copolymer with a minimum number average molecular weight of 70,000 and the tetrafluoroethylene terpolymer with a minimum number average molecular weight of 100,000. Chlorosulfonated polyethylene is also permitted in certain food contact or drinking water applications; specifications provide that the chlorine content and the sulfur content do not exceed 25 and 1.15 percent by weight, respectively, and that the molecular weight is in the range of 95,000 to 125,000. Permissible direct additive applications include the use of natural rubber, butadiene–styrene, and butyl rubber as masticatory substances in chewing gum base.

3.1.3 Toxicologic Potential

Residual monomer (or solvent) is usually a very minor industrial hygiene concern in dry solid polymer but can require evaluation with latex materials (see above).

Several recent reports (155–157) of ongoing or continued studies in the American and British rubber industries have described an excess of certain types of cancer. Cancer incidence was not linked to any specific chemical substances, but, rather, operations that involve exposure to multiple chemicals. The British study (157) also found a significant overall excess of cancer.

Industrial dermatitis from finished rubber products as a result of the various chemicals added during polymerization, curing, and processing is not at all uncommon (158–162). Mercaptobenzothiazole, tetramethylthiuram disulfide, N-isopropyl-N'-p-phenylenediamine (IPPD), and related compounds appear to be the most common offenders.

Table 54.12 indicates that elastomers degraded at high temperatures around

Table 54.12. Overview of Inhalation and Thermal Degradation Data on Representative Olefin and Diene Elastomers[a]

Polymer	TLV	Type of Study				Description of Material	Procedure	Observations/Conclusions	Reference
		Ambient Temperature	Elevated Temperature	Effluent Analysis	Species Exposed				
Polyisoprene	No specific TLV (see Section 1.4)		X	X	Mouse	Natural rubber	Rising and fixed temperature programs (Table 54.10, Ref. 66)	Time to incapacitation and death longer than median when compared to other polymers tested in rising temperature program series (see Table 54.8); fixed temperature program produced similar response in shorter time.[c] Significant yield of CO[d]	51, 66, 163[b]
			X	X		Natural rubber, synthetic rubber, and gutta percha	Degradation 287–400°C	Yield of 3–16% isoprene and dipentene, plus small amounts menthene	Cited in 105
			X	X			Degradation 450–800°C (natural rubber)	Dipentene main product at 450°C, optimum yield of isoprene in range 675–800°C	Cited in 105
Styrene–butadiene	As polyisoprene above		X	X	Mouse	25/75 SBR	Rising and fixed temperature programs (Table 54.10, Ref. 66)	As described for polyisoprene above	51, 66, 163[b]
			X	X			Degradation, 327–430°C	12% of products volatile at 25°C, 2% butadiene with other saturated and unsaturated hydrocarbons	Cited in 105
			X	X			Polymer burnt in electric oven to determine levels of pollutants	Causes "styrene pollution"	48
Ethylene–propylene	As polyisoprene above		X	X	Mouse	Ethylene–propylene–diene terpolymer	Fixed and rising temperature programs, (Table 54.10, Ref. 66)	Time to incapacitation and death as described for polyisoprene above and in Table 54.8. Effluent CO in some	51, 66, 163[b]

cases high enough to have caused death in observed time[d]

Polymer			Animal	Material	Test conditions	Results	Reference
Polychloroprene	X	X		As polyisoprene above	Fixed and rising temperature programs (Table 54.10, Ref. 66)	Time to incapacitation and death as described for polyisoprene above and in Table 54.8. Char residue in rising temperature program series from 10 to 39% for 4 samples (50)	50, 51, 66, 163[b]
	X		Mouse	Flexible foam (type used in upholstered furniture)		Nine different test conditions of polymer temperature and air flow correlated with time to incapacitation and death	164
	X		Mouse	Polymer with or without 5% zinc ferrocyanide (fire retardant)	Pyrolysis products generated by 20°C/min increase in temperature with 20 l/min air flow, 10-min exposure[e]	Decomposition products of polymer with zinc ferrocyanide found more toxic (factor >7) than those without this additive; both found more toxic than wood (Douglas fir)	165
	X	X		Uncured gum, cured rubber, cable insulation, and hose conduit (last two used in underground mining)	Thermal oxidative degradation under static and dynamic environments	First weight loss at about 300°C; at 370°C, up to 84% of the total chlorine evolved as hydrogen chloride. Additional products detected at elevated temperatures up to 600°C were sulfur dioxide, carbon disulfide, hydrogen sulfide, carbon monoxide, benzene, chloroprene, acetic acid, formic acid, formaldehyde, and various mercaptans	166
	X	X			Degradation, 377°C	Hydrogen chloride formed	Cited in 105
Acrylonitrile–butadiene	X	X	Mouse	As polyisoprene above	Fixed and rising temperature programs (Table 54.10, Ref. 66)	Time to incapacitation and death close to the median when compared to other polymers in rising tempera-	50, 51, 66, 163[b]

Table 54.12. (*Continued*)

Polymer	TLV	Ambient Temperature	Elevated Temperature	Effluent Analysis	Species Exposed	Description of Material	Procedure	Observations/Conclusions	Reference
								ture program series (see Table 54.8). Fixed temperature program produced response in much shorter time (incapacitation at 2 min, death at 4 min) than 5 other elastomers comparably tested; insufficient CO (~2400–5400 ppm) to be lethal (163[b]). Char residue in rising temperature program from 1.1 to 3.9% for 3 samples (50)	Cited in 105
			X	X		30/70 NBR	Degradation, 310–400°C	14% of products volatile at 25°C (saturated hydrocarbons)	
Vinylidene fluoride-based elastomer[f]	As polyisoprene above	Processing temperature range			Rat	Vinylidene fluoride–hexafluoropropylene copolymer	Polymer heated at 200°C or 248°C in presence of 3 l/min nitrogen; effluent diluted with oxygen, inhaled for 4 hr	Lethargy and labored breathing during exposure; some respiratory congestion and related signs intermittently during 2-week recovery period	167
			X	X	Rat	Vinylidene fluoride–hexafluoropropylene as copolymer, copolymer with additives, or terpolymer with tetrafluoroethylene; also polytetrafluoroethylene	Pyrolysis effluent mixed with 100 cm³/min air flow, inhaled for 30 min (some 5 min exposures)	Pyrolysates of the 3 elastomers at 550°C considered less toxic than that of polytetrafluoroethylene; at 625°C, increase in pyrolysis temperature of the 4 test materials to 800°C resulted in increased pyrolysate toxicity in all cases. CO and CO₂ considered below lethal levels. No correlation found between hydrolyzable fluoride and relative lethality of pyrolysates	168

Material			Species	Sample	Effects	References
	X		Rat	Vinylidene fluoride–hexafluoropropylene cured with 2 benzoyl peroxide recipes, a sulfur recipe, and radiation (5 passes)	Pyrolysis, effluent mixed with 2 l air/min and inhaled for periods ranging from 1 to 4 hr (or until death)	Effluents from the various products lethal at 300–375°C degradation but not at 250°C. Fatalities showed acute respiratory tract injury. Effluent from radiation-cured copolymer (unlike other 3 materials) reacted with glass of the pyrolysis chamber (consistent with likely formation of hydrogen fluoride) — 169, 170
	X		Rat	Unidentified	Pyrolysis products generated at 475°F (246°C)	Cloudy corneas in exposed rats; cloudiness attributed to reaction with free hydrogen fluoride — 171
Chlorosulfonated polyethylene	X	X	Mouse	As polyisoprene above	Fixed and rising temperature programs (Table 54.10, Ref. 66)	Time to incapacitation and death longer than median when compared to other polymers in rising temperature program series (see Table 54.8); fixed temperature program produced similar response in shorter time. Char residue in rising temperature program from 3 to 34% for 2 samples (50); effluent CO high enough in some cases to have caused death in observed time (163[b]) — 50, 51, 66, 163[b]
		X			Degradation at 300, 350, or 400°C, over 100 hr of test	High rate of decomposition ≥300°C; greatly accelerated above 350°C — 172

[a] Summarized as described in Table 54.10.

[b] This earlier reference gives identifications of various samples and gas concentrations determined in individual exposures.

[c] This effect is presumably related to more rapid generation of toxicants (66).

[d] Polyisoprene, styrene–butadiene, and ethylene–propylene–diene have been listed as "non-char-forming" by Hilado and Machado (51). The highest char yield in their nonchar group was 3.3%, and the yields for these three elastomers ranged from 0.0 to 0.6%. Appreciable yields of char may or may not develop with other elastomers; see data on acrylonitrile–butadiene, polychloroprene, and chlorosulfonated polyethylene in this table.

[e] Thirty-minute exposure also used in comparative testing but not with these polymers.

[f] See Section 3.7.3, concerning emissions evolved during press cure and postcure.

800°C can yield more toxic products than elastomers degraded at smoldering temperatures or gradually rising temperatures. This appears to be especially so with nitrile–butadiene.

3.1.4 Workplace Practice and Standards

At ambient temperatures the main concern with vulcanized polymer is possible inhalant effects from dust generated in grinding operations and dermatitis in "rubber-sensitive" individuals. Operations that may generate dust should, as a minimum, be controlled within recommended limits for nuisance dusts (27). Dermatitis can usually be controlled only by avoidance of either the specific materials that provoked the reaction or "rubber" in general. In some cases, avoiding occluded contact with the skin may result in elimination of dermatitis (173).

Unvulcanized rubber, depending on the particular polymer and type, may require special handling owing to residual monomer or solvent. Uncured chlorosulfonated polyethylene (Section 3.8) or other elastomers similarly prepared by dissolving an existing polymer generally contain residual solvent. With chlorosulfonated polyethylene, monitoring for carbon tetrachloride is desirable if bulk polymer is stored in warehouses with limited ventilation or if ventilation in the processing area is questionable. Ventilation adequate to protect against diffusion of the residual carbon tetrachloride should be adequate for any other vapors that may develop during normal curing operations.

Handling of latex requires consideration of the toxic potential of the monomer(s) and also of the particular operation(s) involved to determine if monitoring of the adjacent atmosphere is indicated. Standard operating procedures should be available where the opportunity for gross exposure is unavoidable, as in tank entry.

3.2 Polyisoprene

The CAS No. of polyisoprene is 9003-31-0; it is referred to as NR or IR.*

cis-1,4   trans-1,4

The major form of natural rubber (*Hevea*) consists primarily of the *cis* isomer.

---

* NR and IR, respectively, are preferred abbreviations for natural rubber and synthetic isoprene rubber (12). IIR designates isobutene–isoprene rubber.

Gutta-percha or *balata* (from the gutta tree or balata bush) is primarily *trans*-1,4-polyisoprene and has less commercial importance.[†] The 1,2 and 3,4 isomers may also be present in the polyisoprene chain.

3.2.1   Preparation

The latex of natural rubber is obtained from trees (*Hevea brasiliensis*); the actual monomer is isopentenyl pyrophosphate that has been formed by biosynthesis. Natural rubber contains low molecular weight impurities; small amounts of sugar, fatty acids, proteins and trace metals all play an important part in processing (174, 175). Synthetic polyisoprene (usually as the cis form) is prepared from isoprene in a solution system with a stereospecific catalyst (176, 177).

3.2.2   Molecular Weight, Properties, and Applications

Natural rubber comprises a range of polymers with varying molecular weight, estimated as $<100,000$ to 4×10^6 (14). The molecular weight of synthetic polyisoprene can be controlled during manufacture and is generally lower and considerably more uniform than that of natural rubber (176). Tables 54.4, 54.6, and 54.11 as well as Figures 54.1 and 54.4 give data on typical properties.

Both forms are widely used in tires. Natural rubber has been replaced by synthetics, particularly styrene–butadiene, as a general purpose rubber. High resilience, low heat buildup, and easy processing are particular advantages of natural rubber, which is often used in blends with synthetic polyisoprene and other elastomers. Natural rubber, both alone and in combination with neoprene, received a high rating for resistance to water, dimethyl sulfoxide, and some alcohols in a comparative test of glove materials (178); resistance to other solvents varied from good to poor. The high purity and high gum strength of synthetic polyisoprene make it particularly useful for food contact items, in the medical field, and in chemical synthesis.

3.2.3   Skin Irritation and Thermal Degradation Products

Splints of *trans*-1,4-polyisoprene (80 percent polymer) can be placed directly on the skin for up to several weeks, provided that periodic cleansing prevents occlusion dermatitis (179). The limited thermal degradation tests summarized in Table 54.12 reveal no unusual toxic characteristics from thermal decomposition products.

[†] Polymers composed primarily of the trans isomer have quite different properties that are more typically associated with thermoplastic polymers (14). Today the main uses of *trans*-polyisoprene are as golfball coverings, special purpose adhesives, and certain medical-dental applications (80).

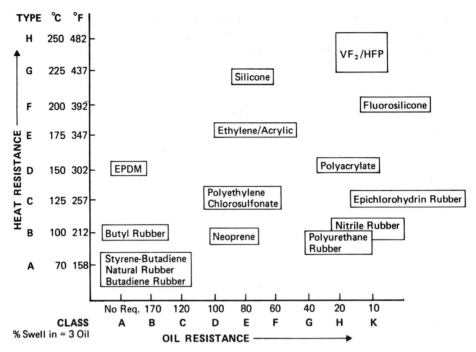

Figure 54.4   Heat and oil resistance of various elastomers. Courtesy of D. H. Geschwind, E. I. du Pont de Nemours and Company, Inc., Polymer Products Dept., Wilmington, Del.

3.3   Styrene–Butadiene

The CAS No. is 9003-55-8; styrene-butadiene is also called SBR,* butadiene–styrene, and GR-S, government rubber styrene.

$$\{(CH-CH_2)_x \; (CH_2-CH=CH-CH_2)_y\}_n$$

The basic copolymer is shown above. Styrene is commonly present at 23 to 29 percent; butadiene units (see Section 3.2) are a combination of cis, trans, and 1,2 (or vinyl) isomers, with trans often predominating.

3.3.1   Preparation

Styrene–butadiene rubber is formed by free radical polymerization of styrene and butadiene (180). A mercaptan such as dodecylmercaptan may be used as

---

* Preferred abbreviation for styrene–butadiene (12).

a chain transfer agent to control molecular weight. The original "hot" rubber polymerizations at about 50°C have been supplemented by "cold systems" at around 20 to 40°C (181), occasionally as low as −40°C (180) with more active initiators. The major differences between emulsion and solution styrene–butadiene appear to be in the linearity and molecular weight distribution of the polymer chains.

3.3.2 Molecular Weight, Properties, and Applications

The molecular weight for "hot" rubber is given (14) as 150,000 to 400,000 (viscosity average) or 30,000 to 100,000 (number average); for "cold" rubber the corresponding values are 280,000 or 110,000 to 260,000 (or 500,000 as weight average molecular weight). In general, mechanical properties are slightly inferior to those of natural rubber and variable according to composition (see directly below). "Cold" styrene–butadiene has improved abrasion resistance and better tensile strength. Tables 54.4, 54.6, and 54.11 as well as Figures 54.1 and 54.4 give comparative data on properties.

Copolymers with <20 percent styrene have low temperature applications in tires and coatings for wire and cable, whereas 43 to 48 percent styrene permits use in adhesives and sealants. Although increasingly challenged by ethylene–propylene (see Section 3.4), styrene–butadiene is currently the highest volume synthetic elastomer, with wide use in tires and many other rubber articles.

3.3.3 Thermal Degradation Products

The tests summarized in Table 54.12 suggest that thermal decomposition poses no unusual toxic hazards.

3.4 Ethylene–Propylene

Ethylene–propylene is referred to as EPM or EPDM.*
EPM (CAS No. 9010-79-1) has the structure

$$\{(CH_2 - CH_2)_{\overline{x}} \, \{CH_2 - \underset{\underset{CH_3}{|}}{CH})_y\}_n$$

where x and y vary from 10 to 20.
EPDM, available as Nordel® hydrocarbon rubber, can exist in many variations.

---

* EPM and EPDM, respectively, are preferred abbreviations for ethylene–propylene copolymer and for terpolymer containing ethylene and propylene in the backbone and a diene with the residual unsaturated portion of the diene in the side chain (12). EPDM is also used as a collective term for both the copolymer and terpolymer (143).

The terpolymer of ethylene, propylene, and 1,4-hexadiene (CAS No. 25038-37-3) has the structure

$$\text{---}[(CH_2\text{---}CH_2)_x\text{---}(CH_2\text{---}\underset{\underset{CH_3}{|}}{CH})_y\text{---}(CH_2\text{---}\underset{\underset{\underset{\underset{\underset{CH_3}{|}}{CH}}{\overset{\|}{CH}}}{\underset{\underset{CH}{|}}{CH_3}}}{CH})_z]_n$$

The main components are as in EPM above, with small amounts of hexadiene to provide curing sites.

3.4.1 Preparation

For copolymers, the ethylene and propylene monomers are copolymerized with Ziegler–Natta catalysts. The usual ethylene content is 55 to 75 weight percent. The saturated copolymers are vulcanized with organic peroxides. Terpolymers additionally contain several percent of a diene, such as 1,4-hexadiene, dicyclopentadiene, or ethylidenenorbornene, which permits sulfur vulcanization (14, 181, 182).

3.4.2 Molecular Weight, Properties, and Applications

The molecular weight has been placed at 100,000 to 200,000 (14). The outstanding feature of these rubbers is their inherent resistance to oxidation as a result of their essentially saturated molecular backbone. They are noted for excellent weather, heat aging, and electrical characteristics, as well as low permeability to gases. Table 54.11 and Figure 54.4 give comparative data on properties.

Applications include tires, automotive parts, as an oil additive, and as an impact modifier for polypropylene (143).

3.4.3 Thermal Degradation Products

The tests summarized in Table 54.12 indicate no unusual toxic hazard.

3.5 Polychloroprene

The CAS No. is 9010-98-4; polychloroprene is also referred to as CR,* neoprene,* and polychlorobutadiene.

* CR is a preferred abbreviation for chloroprene rubber (12). Neoprene is a generic term for polychloroprene.

Sulfur-modified polymers contain approximately 1 atom sulfur per 100 chloroprene units (14).

Commercial polychloroprene polymers are primarily trans-1,4 units with varying amounts of other isomers (14, 183).

3.5.1 Preparation

Polychloroprene is prepared by free radical, emulsion polymerization of 2-chloro-1,3-butadiene. The product is sold either as raw solid polymer (usually homopolymer) or as latex. More than 20 commercial grades of raw polymer were available in 1979 (183–186).

3.5.2 Molecular Weight, Properties, and Applications

The molecular weight of the sulfur-modified polymers has been given (14) as 20,000 to 950,000 (greatest frequency about 100,000) and the mercaptan-modified polymers as 180,000 to 200,000. Fire resistance is significantly greater than that of hydrocarbon rubbers although variable with changes in composition. Polychloroprene has good mechanical properties, good oil resistance, and good resistance to weather. Tables 54.4 and 54.11 and Figures 54.1 and 54.4 give comparative data on properties. (See Reference 187 concerning composition in relation to limiting oxygen index.)

Polychloroprene serves as a multipurpose industrial rubber. Solid polymers are used in wire and cable insulation, belts, tires, shoe heels and soles, adhesives, and as coatings in tanks and turbines. Latexes are used as adhesives, binders, coatings, foams, and sealants (183). Gloves made of polychloroprene have shown high resistance to water, alcohols, and other polar solvents, but little resistance to chlorinated and nonpolar solvents (178).

3.5.3 Oral Toxicity, Skin and Eye Irritation

The LD_{50} for a latex containing 46 percent polychloroprene was found to be >40,000 mg/kg; only one of the five rats dosed at this level died (188). Tests with several latex formulations conducted on occluded rabbit skin according to a Department of Transportation test protocol (189) showed that these formulations do not classify as corrosive (190). Dermatitis may result from other chemicals present in polychloroprene-based adhesives. An accelerator, ethyl butyl thiourea, was shown to be the cause of a severe foot rash in eight Oregon patients wearing polychloroprene soles (191). The dodecylmercaptan residue present in some polychloroprene-based adhesives used in a Spanish shoe industry has also been reported as the cause of shoe dermatitis (173).

Rabbit eye tests with several latex formulations (190, 192) showed mild reversible injury except for a formulation containing 4 percent of a quaternary ammonium soap. This material produced moderate damage that was still present to a residual extent at the final 14-day observation.

3.5.4 Carcinogenicity and Teratogenicity

The carcinogenic potential of the monomer has received close scrutiny (31). However, recent lifetime inhalation studies on rats and hamsters at levels as high as 50 ppm chloroprene have shown no convincing evidence of carcinogenicity (193). The 50 ppm level produced slight physiologic effects in both species but no response was observed in animals similarly exposed at the 10 ppm level. No evidence of carcinogenicity was observed (194) in an oral intubation test in which 150 mg/kg was given to pregnant rats on the seventeenth day of gestation and followed by weekly 100 mg/kg doses from weaning through the life-span of the progeny.

No teratogenic effect was found (193) in the offspring of mothers who inhaled chloroprene concentrations as high as 175 ppm during gestation. Some embryotoxicity was evident at maternally toxic levels (75 and 175 ppm).

3.5.5 Thermal Degradation Products

The tests summarized in Table 54.12 show no unusual toxic effects. The formation of hydrogen chloride accounts for most of the chlorine evolved.

3.5.6 Marine Degradation

Polyester coated with polychloroprene and nylon coated with polychloroprene both retained mechanical properties after immersion for 24 months in the sea off Japan (195).

3.6 Acrylonitrile–Butadiene

The CAS No. is 9003-18-3; it is also called nitrile rubber* and NBR*; originally it was called Perbunan or Buna N (see Section 3.1). The structure of the basic

---

* NBR is a preferred abbreviation for nitrile–butadiene rubber (12). The term "nitrile rubber" is generic and refers to any copolymers of an unsaturated nitrile and a diene; in practice these are usually acrylonitrile and butadiene, respectively.

copolymer is

$$+\mkern-6mu[CH_2-CH=CH-CH_2\mkern-2mu]_x \quad +\mkern-6mu(CH_2-\underset{\underset{CN}{|}}{CH}\mkern-2mu)_y]_n$$

Butadiene units may be cis-1,4, trans-1,4, or 1,2; definitions vary, but "low" acrylonitrile usually implies 18 to 20 percent acrylonitrile, "high" acrylonitrile about 40 percent acrylonitrile, and "very high" 45 to 50 percent acrylonitrile.

3.6.1 Preparation

Nitrile rubber is produced by polymerization of acrylonitrile and butadiene in varying proportions. The basic steps are essentially the emulsion polymerization techniques used for styrene–butadiene (196).

3.6.2 Molecular Weight, Properties, and Applications

Molecular weight determined by fractionation gives a range of 20,000 to 1,000,000 (14). The outstanding feature of nitrile rubber is its resistance to oil and chemicals. If maximum oil resistance is required, a high acrylonitrile content (see above) is indicated. Flexibility at very low temperatures is obtained with a lower acrylonitrile content; in turn, this gives some sacrifice in oil resistance (197). Low temperature flexibility for some applications can be modified by plasticizers or blends. Additional plus points are resistance to heat, abrasion, and water, also resistance to gas permeation. See Tables 54.4 and 54.11 and Figure 54.4 for comparative data on properties.

Applications include fuel lines, hoses, automobile parts, structural adhesives, oil resistant clothing or articles, gloves, and shoe soles and heels. Gloves made of nitrile rubber were found comparable to neoprene gloves in their resistance to organic solvents; the nitrile rubber gloves were superior in resistance to n-hexane and inferior in resistance to phenol (178).

3.6.3 Analysis

Free acrylonitrile monomer in NBR latex (198) has been determined by adding Na_2SO_3 to the double bond and then titrating the NaOH by-product with HCl (also see Section 3.1).

3.6.4 Thermal Degradation Products

The tests on mice summarized in Table 54.12 indicate that thermal degradation products generated at 800°C killed mice unusually rapidly. The rapid clinical action observed from the thermal degradation of this nitrile-containing polymer suggests the formation of hydrogen cyanide.

3.7 Vinylidene Fluoride-Based Elastomers

The most thoroughly studied and widely known polymer of this type is the copolymer with hexafluoropropylene, *vinylidene fluoride–hexafluoropropylene* (CAS No. 9011-17-0), also known as FPM,* FKM,* Viton® A fluoroelastomer, and Fluorel®.

$$\{ (CH_2CF_2)_x \quad (CF_2 - \overset{\overset{\displaystyle CF_3}{\displaystyle |}}{CF})_y \}_n$$

The ratio *x/y* or *x/y/z* (see directly below) varies; the copolymer usually contains 30 to 50 mole percent of hexafluoropropylene.

Variations include the terpolymer with tetrafluoroethylene: *vinylidene fluoride–hexafluoropropylene–tetrafluoroethylene* (CAS No. 25190-89-0), also known as FPM,* FKM,* and Viton® B fluoroelastomer.

$$\{ (CH_2CF_2)_x \quad (CF_2 - \overset{\overset{\displaystyle CF_3}{\displaystyle |}}{CF})_y \ (CF_2CF_2)_z \}_n$$

3.7.1 Preparation

Vinylidene fluoride-based elastomers have generally been prepared by free radical, emulsion polymerization of the monomers with organic or inorganic peroxide initiators (199). Low molecular weight fluid or semifluid polymers have been made with various chain transfer agents, special initiators, or dehydrofluorination oxidation methods (199).

3.7.2 Molecular Weight, Properties, and Applications

The molecular weight of some fluoropropene copolymers has been reported as 100,000 to 200,000 (200) and as low as 60,000 (14). Waxlike materials such as Viton® LM may have a molecular weight <5000, whereas fluoroalkoxy rubbers may have a molecular weight as high as 1,000,000 (200). The primary advantage of the fluororubbers is that even at relatively high temperatures (150 to 275°C for the Viton®–Fluorel® series) they are oil resistant, chemically inert, and retain their elastic properties. See Table 54.11 and Figure 54.4 for comparative data on properties.

---

* FPM is an abbreviation formerly recommended for vinylidene fluoride–hexafluoropropylene copolymer. FKM is now recommended (12) to indicate fluororubber of the polymethylene type having substituent fluorine and perfluoroalkyl or perfluoroalkoxy groups on the polymer chain. (See also Section 3.9; FFKM designates a separate subdivision for fully fluorinated elastomers.)

Applications in the aircraft, petroleum, and chemical industries include seals (O-rings, gaskets, extrusions for valves, pumps, and hoses) and protective coatings (paper, wood, rubber, leather, plastic, and fabric). The higher molecular weight copolymers have superior hot tear strength (14, 200). (See also Section 3.1.2.)

3.7.3 Analysis

Pelosi and co-workers (201) have reported quantitative determinations of volatile emissions evolved during press cure (typically 10 min at 193°C) and the usual 24-hr postcure. Stepwise analysis with several stocks (Viton® A and Viton® B) showed the greatest weight loss, 1.32 to 2.16 percent, during the first 12 hr of oven cure. During both curing operations, the emissions were primarily water and carbon dioxide, secondarily curing agent fragments; no fluorocarbon decomposition products were detected. Small amounts of hydrogen fluoride were evolved during press cure, and even smaller amounts during postcure.

3.7.4 Oral Toxicity, Skin and Eye Irritation, and Inhalation Toxicity at Elevated Temperatures

The oral LD_{50} for rats dosed intragastrically with a latex dispersion containing 60 to 65 percent terpolymer (vinylidene fluoride–hexafluoropropylene–tetrafluoroethylene) was >40,000 mg/kg (202). Rats fed the uncured copolymer crumb (vinylidene fluoride–hexafluoropropylene) for 2 weeks at 25 percent in the diet showed no clinical or nutritional signs of toxicity; large livers were seen at the end of the feeding period but there was some evidence of remission after a 14-day recovery period (203).

Uncured sheets and latex dispersions of both this copolymer and this terpolymer produced only minimal erythema when tested with 24-hr occluded contact on rabbit skin; eye tests with the dispersions showed only very mild, temporary conjunctival irritation (204). A subsequent skin test (205) was conducted to determine the irritancy potential of the copolymer when cured with 2-dodecyltetramethylguanidine (present at 1.33 percent in the compounded polymer). Rabbits showed erythema 2 days after a series of five 6-hr exposures. Reactions were somewhat more pronounced after four additional exposures but resolved completely within a 14-day recovery period.

Thermal degradation starts at 245 to 250°C. Table 54.12 summarizes pertinent data above and below this demarcation point. As might be expected with these fluorinated elastomers, the toxicity of the effluent depends strongly on the degradation temperature and the curing recipe.

3.8 Chlorosulfonated Polyethylene

This polymer is available as Hypalon® synthetic rubber and is also known as CSM.*

$$
\left[+CH_2CH_2CH_2 - \underset{\underset{Cl}{|}}{\overset{\overset{CH_3}{|}}{C}} - CH_2 - CH_2 \right)_x CH_2 - \underset{\underset{Cl}{\overset{|}{SO_2}}}{\overset{\overset{H}{|}}{C}} + \right]_n
$$

The substituent chlorine and SO_2Cl groups are randomly distributed along the polymer chain and frequently have an affinity for tertiary carbon atoms.

3.8.1 Preparation

This type of elastomer is made by dissolving polyethylene in carbon tetrachloride and then treating it with chlorine and sulfonyl chloride in the presence of a catalyst (181). After the desired degree of chlorosulfonation has been attained, the residual chlorine–sulfur dioxide mixture is stripped off, a stabilizer is added, and the commercial product is isolated as raw rubber in crumb or film form.

Chlorine content ranges from 20 to 43 percent, sulfur content from 1 to 2 percent (186, 206). Minimum stiffness is found with a chlorine content of near 30 percent when the base is low density polyethylene and about 35 percent when it is high density polyethylene. The sulfonyl chloride group is subject to hydrolysis and provides a curing site for reaction with metallic oxide curing agents.

3.8.2 Properties and Applications

Chlorosulfonated polyethylene is noted for toughness, resistance to abrasion, weatherability, and colorability. It is usually cured with heat in the presence of a divalent metal oxide (181). See Table 54.11 and Figure 54.4 for comparative data.

Major applications are hose, wire covering, or in sheet form for uses such as swimming pool liners or reservoir containers.

3.8.3 Oral Toxicity and Skin Contact

Rats fed 20 g/kg of a chlorosulfonated polyethylene in ground chow showed no clinical effects (207).

* CSM is a preferred abbreviation for chlorosulfonated polyethylene (12). No specific registry number for this structure is identified in the Ninth Collective Index of *Chemical Abstracts* (see Ethene, homopolymer, chlorosulfonated).

A limited skin sensitization test with two commercial solutions (U.S. and Japanese) produced one instance of sensitization; the causative agent was not identified (208). More extensive testing with four formulations containing 45 to 50 percent chlorosulfonated polyethylene, cured and uncured, produced no reactions, either irritation or sensitization (209). No specific instances of dermatitis attributable to chlorosulfonated polyethylene are known to have occurred in several decades of extensive commercial use.

3.8.4 Thermal Degradation Products

The tests summarized in Table 54.12 indicate no unusual toxic hazard.

3.8.5 Biodegradation

No fungal degradation was apparent in a test with chlorosulfonated polyethylene using ASTM Method D-1924-63 (125).

3.9 Other Elastomers

Butyl rubber, the copolymer of isobutylene with small amounts of isoprene, is a high volume elastomer (143) that has the advantages of low permeability to air and other inorganic gases plus good resistance to heat, aging, and weather. Butyl rubber also has good tear resistance and good dielectric properties. Chlorobutyl rubber consists of the butyl polymer with 1 to 2 percent chlorine to provide increased cure versatility and greater vulcanization flexibility with other elastomers.

Polybutadiene was developed as an alternate synthetic rubber that was much less expensive than synthetic polyisoprene. Emulsion polybutadiene is more readily processed than solution polybutadiene but today has been essentially replaced by solution polybutadiene where the microstructure can be controlled with stereospecific catalysts to produce desired isomer ratios. Solution polybutadiene has a relatively narrow molecular weight distribution that favors high abrasion resistance, high resilience, low temperature flexibility, resistance to heat degradation, but also imposes processing limitations. This relatively inexpensive elastomer is therefore used primarily in blends with styrene–butadiene, natural rubber, and as an elastomeric modifier for other polymers.

Polysulfide rubbers (Thiokol®) are noted for their excellent solvent resistance. As a group, they have relatively little high or low temperature flexibility.

Acrylic elastomers (found in the Hycar® series, as Cyanacryl®, Thiacril®, and more recently Vamac®) were developed to provide a saturated polymer that would be resistant to sulfur and oxygen. Their outstanding feature is their ability to withstand sulfur-modified oils at temperatures >150°C (196). They are less effective for use in water or water-soluble materials although ethylene/

acrylic copolymer (Vamac®) shows little effect from water or steam <100°C (210).

Epichlorohydrin elastomers (chlorine content of homopolymer 38%) and chlorinated polyethylene elastomers are special purpose materials that offer excellent resistance to aging. They are also flame resistant and oil resistant (14, 211). Thermal degradation products from chlorinated polyethylene appear to be similar to or slightly less in toxicity than those from polyethylene as judged by the time to death of exposed mice (95).

Perfluoroelastomers such as Kalrez® are suitable for long term service at 500 to 550°C and have excellent chemical resistance even at these elevated temperatures.

References previously cited (7, 10, 144, 145) provide an overview of established and developmental elastomers. Urethane and silicone elastomers are discussed in Sections 10 and 12, respectively. Plasticized polyvinyl chloride can be processed to serve as an elastomer in special cases such as flexible foam for electrical insulation; see Section 4.

4 VINYL HALIDES

4.1 Overview

The commercial polymers in this group contain chlorine atoms, fluorine atoms, or in a few cases both. In very diverse ways these halogens can be utilized to give vinyl polymers with increased resistance to water, oils, and solvents, plus other distinctive properties.

4.1.1 Polymers Containing Chlorine

The prototypes are polyvinyl chloride and polyvinylidene chloride. Polyvinyl chloride and its copolymers rank second or third in production/consumption volume among polymers in the United States and abroad (Table 54.2, also Reference 10). Their key point is low-cost versatility. Polyvinylidene chloride materials have an extremely regular, closely packed molecular structure that results in outstanding impermeability to water, oils, and gases (10, 15). Table 54.13 gives some general properties.

Production and Processing. Polyvinyl chloride-type polymers are available as pellets, powder, or in liquid forms that are processed into a great variety of moldings, extrusions, foams, films, coatings, and also fibers (particularly in Europe). Polyvinyl chloride is broadly described as rigid or plasticized, filled or unfilled. It is one of the least stable of commercial polymers; formulation into useful products requires stabilizers and other additives. Polyvinylidene chloride is processed mainly as films and coatings.

Analysis and Specifications. The intense scrutiny given to vinyl chloride has resulted in procedures for detecting it with parts per billion sensitivity (Section 4.2.3). Residual monomer in polyvinylidene chloride is also receiving increasing attention (Section 4.3.3).

Subject to specifications, polyvinyl chloride and polyvinylidene chloride are permitted in food-contact applications such as coatings and packaging. Additionally, polyvinyl chloride is permitted in cellophane and water-insoluble hydroxyethyl cellulose film, whereas polyvinylidene chloride is permitted in paper and paperboard and in coatings on cellophane and other packaging films (92).

Toxicologic Potential. When vinyl chloride was identified as a carcinogen (see Chapter 39, Volume 2B), the amount and possible migration of any residual monomer in polyvinyl chloride became a paramount toxicologic concern. Recent critical review by the International Agency for Research on Cancer (31) does not incriminate the polymer in this respect although available studies were considered inadequate with respect to both carcinogenicity and teratogenicity. Other than a bizarre case report (see Section 4.2.7), the human data cited in connection with the polymer appear to be complicated by probable excessive dust exposure, probable direct exposure to the monomer, or exposure to other materials used in processing.

Toxic additives that might leach into drinking water or food, or might be assimilated through medical use, are an additional concern with plasticized polyvinyl chloride, a rubberlike polymer typically containing very high levels of plasticizer. These qualities may permit the plasticizer and other adjuvants to migrate through the polymer at relatively high rates (24). In 1968, the use of an organotin stabilizer in polyvinyl chloride tubes intended for endotracheal use was associated with an increased incidence of inflammatory glottic reactions (213). Other investigators during this period confirmed the toxicity of organotin stabilizers in animal tests and identified the release of phthalate ester plasticizers from polyvinyl chloride into solutions and biologic media (Reference 33, citations generally dated 1960–1971). The additives in plasticized polyvinyl chloride as presently used in medical practice are subject to safety testing and surveillance (214). Accelerated experiments with simulating solvents and actual foods (including studies with radiolabeled plasticizer) have been used to develop mathematical models for predicting plasticizer migration from plasticized polyvinyl chloride to solid foods and "fatty" liquid foods (24).

Several reports from Europe describe the development of respiratory tract effects, ranging from functional changes to frank pneumoconiosis, in factory workers after prolonged and/or excessive exposure to polyvinyl chloride dust. Other reports describe effects of varying degree in experimental animals. In the United States, workplace exposure limited to the polymer is not known to

Table 54.13. Selected Properties Associated with Polymers of Polyvinyl Chloride and Polyvinylidene Chloride[a]

Properties	Polyvinyl Chloride				Polyvinylidene Chloride	Reference
	Homopolymer/Type Unspecified	Rigid	Flexible, Unfilled	Flexible, Filled		
Melting point (°C)	212–310				190–210	85
Glass transition temperature (°C)		75–105[b]	75–105[b]	75–105[b]	–18	82, 86, 212
Brittle point (°C)		<–40				14
Density		1.30–1.58	1.16–1.35	1.3–1.7	1.65–1.72	82, 212
Rockwell hardness		65–85 (Shore D)	50–100 (Shore A)	50–100	M50–M65	82
Refractive index, n_D		1.52–1.55			1.60–1.63	82, 87, 212
Soluble in	Tetrahydrofuran, cyclohexanone, dimethylformamide; solubility in various solvents may vary with molecular weight and % chlorination (Ref. 88)				Trichloroethane, pentachloroethane. Crystalline homopolymer dissolves in trichlorobenzene 170°C, amorphous polymer in tetrahydrofuran	14, 88

Swollen by	Aromatic and chlorinated hydrocarbons, nitroparaffins, aliphatic hydrocarbons (can leach plasticizers), oils	Oxygen-containing liquids (dioxane, cyclohexanone, dimethyl formamide)	14, 88
Nonsolvents/relatively unaffected by	Water, concentrated alkali, nonoxidizing acids, aliphatic hydrocarbons (can leach plasticizers), oils	Chloroform, vinylidene chloride, hydrocarbons, mineral oils, alcohols, phenols, certain organic solvents (see above)	14, 88
Decomposed by	Concentrated oxidizing acids (H_2SO_4, HNO_3, H_2CrO_4). Degraded by heat and light, requires stabilizers	Prolonged contact with ammonia and related compounds; attacked by chlorine (slowly by hot concentrated H_2SO_4). Copper, iron catalyze decomposition	14

[a] Data are summarized as described in Table 54.9.
[b] The 75–105° range is given in Ref. 82; Lee and Rutherford (86) cite references indicating the glass transition temperature as 81°C, rising with increasing syndiotactic content to 98°C.

have produced clinical complaints except in cases of "meat wrappers' asthma." This name, descriptive if imprecise, refers to clinical effects associated with fumes from thermal cutting and sealing of the polyvinyl chloride film generally used to package meat in retail markets. Particulate and plasticizer products have been identified as the responsible agents.

Although polyvinyl chloride per se does not readily support combustion, commercial formulations containing significant or large amounts of plasticizers or certain other additives may be flammable (38). Thermal decomposition of polyvinyl chloride yields hydrogen chloride as the main toxicant and also releases a variety of organic fragments. Decomposition can be rapid in fire environments. Bowes has estimated that some 70 percent of the possible hydrogen chloride may evolve in about 10 min at a temperature of 270°C in air (215). Much less extensive data on the thermal decomposition of polyvinylidene chloride suggest that hydrogen chloride would also generally be the major product.

Table 54.14 provides an entry to the literature on inhalation and thermal degradation with polyvinyl and polyvinylidene chloride. Table 54.15 summarizes the decomposition data that were obtained from a sample of polyvinyl chloride under varying conditions of oxygen supply, temperature, and heating rate.

Workplace Practices and Standards. Precaution should be taken to avoid dust exposure (see particularly Reference 216). Where opportunity for gross exposure exists, prudence may suggest periodic dust counts and medical surveillance. Knowledge of monomer levels in raw polymer would also appear advisable.

Bulk quantities of these vinyl halides should be stored away from flammable materials where possible. Malten and Zielhuis (251) consider that in a nonventilated room of 1000 ft^3 the thermal decomposition products from about 3 lb of plastic "may constitute a real toxic hazard."

In the event of fire involving significant quantities of vinyl chloride polymers, personnel entering the area should have respiratory protection* (see Table 54.14, References 244–248). Normal amounts of scrap (<10 lb) may often be incinerated with general plant waste, provided that there is no direct operator contact and adequate facilities are available to dilute the acid content of exhaust gases.

The likely toxicant contribution to a fire from commercial products containing polyvinyl chloride varies depending on type of use as well as quantity. Full-scale tests indicated that the use of a 3-mm thick polyvinyl chloride sheet wall lining in a corridor was found to be undesirable, but use of a decorative wall paper or cloth in a compartment did not significantly increase the risk beyond that presented by the basic wood fire (252). The actual concentration of evolved

---

* Eye protection may also be indicated under unusual circumstances; in practice, the use of full-face respirators would cover both contingencies.

Table 54.14. Overview of Inhalation and Thermal Degradation Data on Polyvinyl Chloride and Polyvinylidene Chloride[a]

Polymer	TLV	Type of Study			Species Exposed	Description of Material	Procedure	Observations/Conclusions	Reference
		Ambient Temperature	Elevated Temperature	Effluent Analysis					
Polyvinyl chloride (PVC)	No specific TLV[b] (see Section 1.4)	X			Human	PVC dust from drying, sacking, and blending operations (production plants in Porto Marghera, Italy)	Epidemiologic survey of 731 employees exposed to PVC dust and 431 exposed to monomer alone	PVC dust concentrations higher than 10 mg/m³ total dust in about 60% of samples; particles 1–5 μm in diameter = 4.31% of total dust weight. 20 cases of pneumoconiosis, all in group with PVC dust exposure. X-rays showed limited profusion; slight restrictive impairment of respiratory function in some cases	216a[c]
		X			Human	PVC dust from packing, drying, and miscellaneous operations (production plants at Hillhouse Works, England)	Epidemiologic survey of workforce sample of 818 men, including 528 with ≥5 years exposure in the more dusty jobs	Highest mean average respirable dust exposure over a shift was 2.88 mg/m³ (SD ± 1.84). Exposure to respirable PVC dust associated "with the presence of small rounded opacities on the chest radiograph and a decline in mean ventilatory capacity" but not associated with acute or chronic bronchitis, pneumonia, pleurisy, or asthma	216b[c]
		X			Human	PVC dust	Case report of 31-year-old man who worked unspecified length of time in factory where he shoveled PVC as fine powder	Pneumoconiosis attributed to PVC dust exposure. Biopsy revealed cyclohexanone-soluble material morphologically similar to dust collected from workplace, diffuse fibrosis of moderate degree	217[c]

Table 54.14. (Continued)

Polymer	TLV	Type of Study			Species Exposed	Description of Material	Procedure	Observations/Conclusions	Reference
		Ambient Temperature	Elevated Temperature	Effluent Analysis					
		X			Guinea pig, rat	PVC dust, "grain size varying predominantly around one micron"	Continuous 24-hr exposure in working environment for 2 to 7 months	Concentration near bagging operations given as 10,000 particles/cm^3. Animals showed fibrosis	218[c]
		X			Rat	PVC dust, 92% of particles <5 μm in diameter	1-hr exposures of rats for 1–12 months	PVC dust had "mechanical and toxic effect . . . inducing changes of the toxic pneumoconiosis type"	219a[c]
		X			Rat	PVC dust particles, 1.7 μm mass median diameter, from paste polymer containing detergent	6-hr exposures, 5 days/week for 15 weeks, at 10 mg/m^3	"Weak biological reactivity;" "small, randomly scattered, lung lesions" persistent through 15 weeks after cessation of exposure	219b[c]
			X		Human		Case histories of 3 meat wrappers	Dyspnea, coughing, wheezing, shortness of breath, positive function tests. All subjects cigarette smokers	220
			X	X			"Laboratory . . . test results showed that, even under artificially severe operating conditions, concentrations of hydrochloric acid or of total particulates in the operator's breathing zone did not reach the threshold limit value for occupational exposure. . . . By controlling the temperature of the cutting wire, it is possible to greatly reduce fume production"		221
			X		Human		NIOSH-sponsored survey of 5 South Dakota meat wrappers, 17 Kentucky meat wrappers	Throat irritation, headaches, sneezing, coughing, shortness of breath in varying degrees. Allergic history in 3 cases	222

X	Human		Questionnaire study of 145 Texas meat wrappers	Dry or sore throat; burning, itching, or tearing of eyes; stuffy or runny nose; coughing, chest soreness, wheezing, or shortness of breath. Meat wrappers who smoke presumably at greater risk. No symptoms in 17 of 145 wrappers working with mechanical blade machines rather than hot wires	223, 224
X	Human	X	Clinical studies of 24 meat wrappers and 8 meat cutters	Throat, eye, and nasal effects, cough, chest tightness, or pain, wheezing, shortness of breath. No acute changes in pulmonary function tests. HCl and bis(2-ethylhexyl adipate considered the primary decomposition products. Methods sensitive to 0.2 ppm vinyl chloride monomer showed none in emissions	225–227
X	Human		Case history of 1 meat cutter	Very minor effects from fumes of PVC soft wrap, intolerance to fumes of heat-activated price-label adhesive	228
X	Human		Case history of 1 meat cutter	Subject reported wheezing, tightness in chest; impairment confirmed by pulmonary function test. Phthalic anhydride identified among emissions but did not evoke response	229
X	Rat		Homopolymers, copolymer vinyl chloride/vinyl acetate 85:15, and formulations	Slow and rapid combustion conditions (Table 45.10, Ref. 109). Also exposure of rats 2–3 hr to pyrolysis products gen-	109, 230
				PVC among most toxic in series of polymers degraded under slow combustion conditions, but among least toxic when degraded under rapid combustion conditions. Death	

Table 54.14. (Continued)

Polymer	TLV	Ambient Temperature	Elevated Temperature	Effluent Analysis	Species Exposed	Description of Material	Procedure	Observations/Conclusions	Reference
							erated as temperature gradually increased to 600°C (Ref. 230)	from CO after exposure to pyrolysis products diluted with room air; when dilution made with O_2, rats survived exposure but showed lung irritation (attributed to effects from HCl) at necropsy 24 hours after exposure (Ref. 230)	
			X		Mouse	PVC and chlorinated PVC	Rising temperature program (Table 54.10, Ref. 66)	Times to incapacitation shorter than median when compared to products from some other polymers tested in rising temperature program. See Table 54.8	66
			X	X (Refs. 101, 232)	Mouse	Rigid board	Exposure 15 min, decomposition products generated at 350, 500, and 750°C (Refs. 101, 232). Also pyrolysis temperature gradually increased to 740°C during 15-min exposure (Ref. 231)	Cause of death considered HCl at 350°C decomposition, CO and HCl at 500 to 750°C. Eye damage ("eyeballs decolored and white") noted in first series (Ref. 231)	231, 232
			X		Mouse	Homopolymer, also plasticized formulation (46% homopolymer)	Based on sensory irritation tests, HCl considered "principal contributor to the toxicity of the thermal decomposition products of polyvinylchloride with other gases and/or aerosols evolved contributing only to small extent" (Ref. 233)		165, 233, 234
			X	X		Homopolymer as in Ref. 230,	Major products from thermal decomposition are HCl, CO, and CO_2. Essentially all available chlorine evolved		102, 235

Material	Conditions / Description			Reference
	as HCl when temperature increased from 25 to 280°C (first fraction). See Table 54.15			
also 1 copolymer and 3 formulations	Major products as directly above, plus "organic degradation products which result from the cyclization and/or scission of the dehydrochlorinated backbone" (Ref. 236)	X	X	236, 237
	Laser-induced decomposition — HCl, benzene, and toluene main products; monomer not detected	X	X	238
Commercially available PVC polymer and rigid sheet	Analysis with furnace temperatures 300–600°C indicated HCl and CO as the main toxic risks although some 75 organic materials (mainly aromatic and aliphatic hydrocarbons) detected. No oxygenated organic species (even with phosgene seeding to verify analytic capability)	X	X	239
	Summary of earlier literature pertaining to nature of combustion gases	X	X	240
"Pure" resin and 4 plasticized compositions used in mines	Oxidative degradation in dynamic system and under static conditions — HCl main product. Compositions contained phthalate ester plasticizers; some transformation into phthalic anhydride occurred under static conditions	X	X	241
	Degradation 200–300°C, 400°C, and 600°C in helium — At 200–300°C, quantitative yield of HCl at 400°C, saturated and unsaturated, aliphatic and aromatic hydrocarbons, with high yields of benzene and toluene. At 600°C, quantitative yield of HCl, remainder residue and hydrocarbons; benzene major volatile hydrocarbon product	X	X	Cited in 105
	Amount of smoke evolved from burning chlorinated PVC increases with degree of chlorination		X	242
	PVC produced "noticeably large amounts of soot," significantly more than cellulose		X	49

Table 54.14. (Continued)

| Polymer | TLV | Type of Study | | | | Description of Material | Procedure | Observations/Conclusions | Reference |
		Ambient Temperature	Elevated Temperature	Effluent Analysis	Species Exposed				
			X	X		PVC film, 36% chloride by weight	Burning of "torches" made from PVC interleaved with polyethylene	0–12% of combustible material recovered as soot and particles 0.03–0.11 μm; 19 mg HCl/g soot tightly bound	243
			X	X	Human		Review of hazard in fires; in a 1970 fire in a building electrical service vault 31 fire fighters were treated for what "appeared to be inhalation of hydrogen chloride." Subsequent tests indicated that the insulation was PVC "and that at 540°F, HCl, as well as CO and CO₂ were given off with no visible smoke. In conclusion, HCl generated in fires is probably not a serious threat to life, unless one is unable to move from its area of generation. The sensory warning given and the physically unbearable nature of a lethal concentration fortunately serve to warn of its presence"		244
			X		Human		Case reports describing a connection between clinical effects in fire fighters and burning PVC; specific identification of PVC products tenuous or absent		245–248
Polyvinylidene chloride (PVDC)	As polyvinyl chloride above		X	X	Rat, mouse, rabbit	Fiber	"Neither severe impairment nor death occurred in animals exposed to combustion products . . . of PVDC"		249
			X	X		PVDC (and related copolymers)	Degradation of PVDC per se "yields HCl as the only volatile product in more than trace quantities." If polymer was prepared in presence of O₂, rapid initial decomposition may also yield COCl₂, CH₃O, and CCL₄; slow decomposition typically yields HCl		250
			X	X			Degradation, 225–275°C	High yields of HCl	Cited in 105

[a] Summarized as described in Table 54.10.

[b] Yugoslavia has a total dust limit of 10 mg/m³ and respirable dust limit of 3 mg/m³ for "plastic material, dust (polyvinylchloride, amino and pheno-plastics" (28).

[c] Ref. 31 describes several additional reports pertaining to PVC dust, but monomer exposure might be a factor at least in the case of "polymerization workers" or "vinyl chloride workers." Refs. 216–219 as above discuss biologic changes associated with the polymer per se. The animal experiments described in Ref. 218 were apparently motivated by the case report of Ref. 217 although it is not clear if the same factory site was used.

hydrogen chloride from degradation by fire of vinyl floor covering can be influenced by the type of wall surfaces; experimental data have shown higher concentrations of hydrogen chloride in a room with a surface of nonabsorptive supergloss paint compared to one of unpainted asbestos cement (253).

4.1.2 Fluoropolymers

The attachment of one or more fluorine atoms to the straight chain carbon backbone of a polymer molecule generally gives high thermal and chemical stability, high density, low surface tension, and a low coefficient of friction. The unmodified homopolymers are among the most durable polymers known; their use in copolymers typically enhances useful product life. Table 54.16 gives some properties for several key types.

Production and Processing. The fluoropolymers are relatively low volume, specialty items (see polytetrafluoroethylene in Table 54.2). Polytetrafluoroethylene is available as powders, filled materials, and in dispersions. It cannot be processed by conventional molding techniques but requires press and sinter methods similar to those used in powder metallurgy. Other fluoropolymers can generally be processed by molding and extrusion techniques and are typically available as pellets, powders, dispersions, and/or films. Polyvinyl fluoride is currently offered only as film.

Relatively few additives are used with fluoropolymers. Properties can often be modified by the use of different polymer units to form copolymers or terpolymers.

Analysis and Specifications. The studies referenced in Table 54.17 identify various techniques of atmospheric analysis and methods for monitoring of urinary fluoride to corroborate possible excessive exposure to fluoropolymer dust. The International Agency for Research on Cancer identifies (31) several additional methods that may be useful to detect effluents generated at elevated temperatures.

Polytetrafluoroethylene resins have been differentiated on the basis of density, particle size, and physical properties. The performance of end use articles such as tape, film, capacitor hose, and wire may be defined by specifications of the Society of Automotive Engineers, by those of the Society of the Plastics Industry, or by military specifications (259). The use of polytetrafluoroethylene as a release agent in coatings and certain other food-contact applications is permitted under specifications of the Food and Drug Administration (92). Polyvinylidene fluoride resins may be similarly used as articles or components of articles intended for repeated use in contact with food according to prescribed conditions (92). Polyvinyl fluoride resins of specified type may be used as food

Table 54.15. Effects of Oxygen Supply, Temperature, and Heating Rate on Varying Combustion Products of Polyvinyl Chloride (Homopolymer; Molecular Weight, 109,000; Mesh Size, 80)[a,b]

Compound	Variation with O_2 Supply			Variation with Temperature					Variation with Heating Rate	
	Air (30 cm³/min)	Air (60 cm³/min)	Air (25 cm³/min) + O_2 (21 cm³/min)	25–280°C	280–350°C	350–430°C	430–510°C	510–580°C	3°C/min	50°C/min
CO_2	861.	619.	814.	—	9.7	181.	244.	237.	619.	397.
CO	357.	429.	401.	—	20.	46.	151.	181.	429.	269.
Methane	6.7	4.7	3.8	—	0.20	1.3	1.8	0.31	4.7	8.7
Ethylene	0.76	0.53	0.28	0.04	0.33	0.39	—	—	0.53	2.3
Ethane	2.6	2.1	1.7	—	0.12	0.94	0.41	—	2.1	3.5
Propylene	0.80	0.53	0.28	0.06	0.11	0.31	—	—	0.53	1.5
Propane	1.3	1.0	0.66	—	0.08	0.44	0.11	—	1.0	1.3
Vinyl chloride	0.51	0.59	0.66	0.04	0.25	0.17	0.02	—	0.59	0.64
1-Butene	0.25	0.18	0.06	0.02	0.04	0.08	—	—	0.18	0.67
Butane	0.53	0.31	0.15	—	0.03	0.20	0.02	—	0.31	0.69

Isopentane 0.02	0.01	—	—	0.005	0.001	—	0.02	0.02
1-Pentene 0.10	0.04	—	0.01	0.03	—	—	0.08	0.18
Pentane 0.26	0.11	—	0.01	0.08	0.01	—	0.20	0.29
Cyclopentene 0.07	0.03	—	0.02	0.01	—	—	0.05	0.19
Cyclopentane 0.08	0.03	—	0.01	0.02	—	—	0.07	0.11
1-Hexene 0.07	0.03	—	0.01	0.02	—	—	0.06	0.13
Hexane 0.16	0.09	—	0.01	0.05	0.01	—	0.14	0.20
Methylcyclopentane 0.06	0.03	—	—	0.02	—	—	0.05	0.08
Benzene 35.	32.	24.	6.6	0.35	0.16	—	31.	43.
Toluene 1.5	0.68	0.12	0.18	0.55	0.03	0.01	1.1	3.5
HCl 580 ± 5, "independent of air conditions and accounts for nearly all the chlorine atoms of the polymer" (235, p. 388)		"Production of HCl . . . roughly parallels that of benzene" (235, p. 388)			0.01	—		"The heating rate has no significant effect on the amount of HCl produced" (235, p. 389)

[a] Summarized from the data of E. A. Boettner, G. Ball, and B. Weiss, "Analysis of the Volatile Combustion Products of Vinyl Plastics," *J. Appl. Polym. Sci.*, **13**, 377–391 (1969).

[b] Combustion products are expressed in milligrams per gram of product.

Table 54.16. Selected Properties of Polyvinyl Fluoride, Polyvinylidene Fluoride, and Polytetrafluoroethylene[a]

	Polyvinyl Fluoride (PVF)	Polyvinylidene Fluoride (PVDF)	Polytetrafluoroethylene (PTFE)	Reference
Melting point (°C)	200–230	156–220 (138–148, Ref. 56)	327–342 (more commonly 327)	56, 82, 254
Glass transition temperature[b] (°C)	141 (70, Ref. 56)	~−60; ~40 (32, 50 on separate runs, Ref. 56)	−113, 127 (with remark "much data, some conflicting")	56, 86
Brittle point (°C)	≤80 (?)	−40	<−80	14
Maximum continuous service temperature (°C)	107		250	255, 256
Density	1.38–1.44	1.75–2.02	2.1–2.7	14, 82, 85
Rockwell hardness		D 80 (Shore)	D 50–55 (Shore)	82
Refractive index, n_D	1.46	1.42	1.35–1.43	14, 87
Soluble/swollen/plasticized in	Hot cyclohexanone, hot dimethylformamide (~150°C) hot dimethylacetamide, hot dimethyl sulfoxide	Cyclohexanone, dimethyl sulfoxide, dimethylacetamide; dissolves or swells in polar solvent such as dimethylformamide; highly oriented polymers insoluble	Perfluorokerosene at 350°C; hot "fluorinated oil" (as C_2F_{44}) has some effect	14, 87

Relatively unaffected by	Aliphatic, cycloaliphatic, aromatic hydrocarbons; resists concentrated acids and alkalis	Aliphatic, cycloaliphatic hydrocarbons, acetone, methyl isobutyl ketone; resists acids, alkalies, several solvents, but less inert than PTFE	Concentrated alkalies and acids, also common solvents, even when heated — 14, 87
Decomposed by/reacts with	Attacked by borate fume (such as found in certain welding fluxes)	Attacked by fuming H_2SO_4, strong amines	Molten alkali metals, alkali metal complexes (as sodium naphthalene in tetrahydrofuran; prolonged exposure to fluorine. Reacts at elevated temperatures with certain metals when both finely divided (as powdered resin mixed with powdered magnesium or aluminum >425°C)[c] — 14, 257, 258

[a] Data are summarized as described in Table 45.9.
[b] Each of these three fluorinated materials would appear to undergo more than one transition, as Roff and Scott (14) describe for polyetrafluoroethylene.
[c] This reaction has been utilized for military explosives.

Table 54.17. Overview of Inhalation and Thermal Degradation Toxicity of Various Fluoropolymers and Recommendations for Control

A. Overview of Experimental and Clinical Studies[a]

Polymer	TLV	Type of Study			Species Exposed	Description of Material	Procedure	Observations/Conclusions	Reference
		Ambient Temperature	Elevated Temperature	Effluent Analysis					
Polyvinyl fluoride (PVF)	As polytetrafluoroethylene below		X		Mouse		Rising temperature program (Table 54.10, Ref. 66)	Time to incapacitation and death longer than median when compared to products of some other polymers; see Table 54.8	66
			X	X			Degradation, 372–480°C	High yields of HF and products involatile at 25°C, little carbonization	Cited in 105
			X	X			Thermal decomposition of γ-ray irradiated polymer	HF eliminated as the first step, followed by evolution of unsaturated hydrocarbons as a consequence of main chain scission	260
Polyvinylidene fluoride (PVDF)	As polytetrafluoroethylene below		X		Mouse		Rising temperature program (Table 54.10, Ref. 66)	Time to incapacitation and death shorter than median when compared to products of some other polymers; see Table 54.8	66
			X				Test at 200, 300, or 350°C, up to 100 hr	Excellent resistance to degradation <200°C, rapid decomposition above this temperature	172
			X	X			Degradation, 400–530°C	35% HF, high yields of products involatile at 35°C, some carbonization	Cited in 105

Compound	Description				Species/Source	Comments	References
Polytetrafluoroethylene (PTFE)	Dust at ambient temperature, see Section 4.1.2. Oxidized decomposition products "can be quantitatively determined in air as fluoride to provide an index of exposure. No TLV is recommended pending determination of the toxicity of the products, but air concentrations should be minimal" (27)[b,c]	X			Human	Dust levels in plant industrial hygiene survey 0.4–5.5 mg/m³; local exhaust ventilation adequate most areas, housekeeping less than desirable. In 77 samples, soluble urinary fluoride level ranged from 0.098 to 2.19 liter (toxic level considered ≥3.0 mg/l) Plant dust from fabrication of PTFE	261, 262
		X	X	X		Criteria document providing single most complete guide to literature with 92 refs.; refs. on fluoropolymer decomposition dated 1955–1975. Also see Part B of this table for recommendations	263
		X	X	X	Rat (also bird and human, Ref. 267; also dog, Ref. 267)	Polymer heated <500°C yielded particulate with possible adsorbed toxicants; 500–650°C, carbonyl fluoride predominant; >650°C carbon tetrafluoride and carbon dioxide the major products. Rats exposed to PTFE products "containing hydrolyzable fluoride equivalent to 50 ppm" carbonyl fluoride for 1 hr showed 4 times normal excretion of urinary fluoride. Characterization of PTFE toxicant spectrum; see Figure 54.5	264 and refs. cited therein, 265–268

Table 54.17. *(Continued)*

A. Overview of Experimental and Clinical Studies[a]

| Polymer | TLV | Type of Study | | | Description of Material | Procedure | Observations/Conclusions | Reference |
		Ambient Temperature	Elevated Temperature	Effluent Analysis	Species Exposed			
			X	X	Rat		4-hr approximate lethal temperature of PTFE for rats confirmed (see **Ref. 268**) as 450°C. At this PTFE temperature exposed rats showed lung changes (including edema), collection grids revealed particles 0.02–0.04 μm in diameter; filtering fume through Millipore filter (pore size 0.2 μm) prevented clinical signs. No typical PTFE particles were evident when PTFE was heated to 425°C, nor were any lung changes evident by electron microscopy	269
			X	X	Mouse	Commercial (tubing 8 mm OD, 2 mm wall thickness, cut into small pieces)	Pyrolysis temperature increased 20°C/min; mice exposed for 10 or 30 min from first weight loss of sample	On a weight of sample basis, products of PTFE[d] considered more toxic than those of several other materials (polyvinyl chloride, wood, polyurethane, polyester, wire)	165, 234, 270
			X	X	Rat		PTFE fumes more toxic when polymer heated at 800°C than 625°C; see Table 54.12 for comparison to 3 fluorinated elastomers	168
			X		Human		Industrial hygiene survey of PTFE-fabricating plant revealed that 6 of 7 subjects with clinical effects smoked, nonsmoker worked at ovens. Urine samples from 1 subject who had been sintering granular polymer had 5 mg/l fluorine	271
			X		Human (also rabbit)		Survey in Russian factory found that cases of fever came from areas where thermal processing reached 480°C (rather than 360°C permissible) or where workers were smoking contaminated tobacco. Pyrogenic effect in rabbits given intravenous injections or dust inhalations correlated with type of fluoropolymer product administered. Onset of fluoroplast fever developing after working hours; causative agents generated during processing distinguished from those arising during smoking of contaminated tobacco	272

Material	Species	Cigarette smoking		Observations	Ref.
Cigarette smoking — Tetrafluoroethylene fluorocarbon telomer	Human (volunteers)	X		"The amount of fluorocarbon eliciting a typical polymer fume fever response was 0.40 mg in a single cigarette, and 9 of 10 smokers reported typical symptoms"	273
PTFE-impregnated tape	Human	X		Tape inadvertently wrapped around exhaust manifold of auxiliary power unit in C54 aircraft; illness in 39 of 40 persons aboard after several hours when this unit had been run extensively. Typical signs of polymer fume fever followed by general recovery within 24 hr	274
PTFE-coated frying pan	Human, bird	X		When pan containing water allowed to boil dry on unattended stove, patient had symptoms 60 min postexposure through 1 day. Patient's wife in room next to frypan experienced no symptoms although five pet cockatiels in this same room died (birds more sensitive than humans, see Ref. 264)	275
	Human	X		9 smokers in manufacturing plant had symptoms: all with chills and fever; some with chest pain, cough, and/or shortness of breath; X-ray in 1 subject thought to be consistent with "pulmonary edema (reversible). Seven nonsmokers had no symptoms"	276
			X	At 504–538°C, 95% monomer, 2–3% D_3F_6, no larger fragments (in vacuum). At 600–700°C (various pressures), either "pure monomer" or monomer, C_3F_6, and C_4F_8. At 1200°C, monomer yield drops, larger fragments appear (in vacuum)	Cited in 105
				See also listing for Ref. 277 directly below	
		X	X	Strongly acid pH when decomposition gases dissolved in water (procedure, Table 54.12)	48
		X		Oxidative pyrolysis of both PTFE and FEP yielded (a) carbonyl fluoride and (b) trifluoroacetyl fluoride as the main products; in humid air (a) formed HF and CO_2 and (b) formed trifluoroacetic acid and HF. Octafluoroisobutylene identified both under nitrogen and in oxidative pyrolysis with FEP, but only in nonoxidative pyrolysis with PTFE	277
Fluorinated ethylene propene (FEP) — See polytetrafluoroethylene above					
"Teflon" FEP wire coatings		X		Negligible degradation <300°C, rapid above this temperature. Test at 300, 400, or 454°C, up to 100 hr	172

Table 54.17. (Continued)

A. Overview of Experimental and Clinical Studies[a]

Polymer	TLV	Type of Study			Species Exposed	Description of Material	Procedure	Observations/Conclusions	Reference
		Ambient Temperature	Elevated Temperature	Effluent Analysis					
Polychlorotrifluoroethylene	See polytetrafluoroethylene above		X		Rat	¼ in. rod	No lethal concentration in 1-hr exposures at 375°C (despite pyrolysis of 43 g). LC$_{50}$ for 3 hr exposures at 400°C: 23.5 g/hr vs. 31.5 g/hr at 375°C. "Death may be due to inhalation of particles rather than toxic by-products"		278
			X	X			3-hr exposure, moist air, temperatures 160–325°C	Fluoride- and chloride-specific methods showed effluent from polymer heated >250°C	279
			X			"KEL-F"	Degradation, 347–418°C	25% of products volatile at 25°C (monomer with traces of C$_3$F$_5$Cl and C$_3$F$_4$Cl$_2$); 72% of larger chain fragments involatile at 25°C	Cited in 105
Perfluoro-alkoxy resin (PFA)	See polytetrafluoroethylene above		X			"Tefzel" nonreinforced	"Does not degrade below 250°C"		172

B. Recommendations for Control or Monitoring

Scope	Statement	Reference
General guidelines	*Permissible Environmental Limits:* "... NIOSH is not recommending an environmental limit for the decomposition products of fluorocarbon polymers. Instead, NIOSH recommends that exposure to the decomposition products of fluorocarbon polymers be controlled by the use of engineering and administrative controls and by strict adherence to work practices that will minimize worker contact with pyrolysis products of fluorocarbon polymers or with potentially pyrolyzable dust" (Ref. 263, pp. 86–87). See TLV, Part A of this table. *Local Exhaust Ventilation Systems:* "... require regular inspection and maintenance to ensure effective operation. Regularly scheduled inspections should include face-velocity measurements of the collecting hood and inspection of the air mover and collector. In addition to providing adequate ventilation, it may be appropriate, under certain circumstances, to isolate a process or system in an especially designed enclosure to minimize exposure. Other control measures may be intentionally redundant	262, 263, 280, 281

temperature monitoring systems and alarms to signal harmful thermal excursions" (Ref. 263, p. 76; see Refs. 280, 281).

Precautionary Measures: "The following precautionary measures should be followed in the use of [PTFE] or any synthetic resin.

1. Temperature Limitations . . . The upper limits of chemical stability should be ascertained from the manufacturer. Different types of fluorocarbon resins are available and apparently with different upper temperature limitations of use.
   Unless information from a competent source dictates otherwise, we recommend 500°F [260°C] as the limiting temperature except when mechanical ventilation is afforded for fume removal.

2. Handling of Formed [PTFE] . . . No precautions are necessary.

3. Ventilation . . . The Du Pont Company recommends ventilation when [PTFE] is subject to elevated temperatures on a somewhat continuous basis. . . . [If] one can keep his face from being directly in the fume rising directly from the process, the existing ventilation of such areas should be adequate, unless a few pounds of [PTFE are] burned.

4. Maintenance—Flame Cutting . . . Insofar as practical, other measures than burning or cutting should be used to disassemble equipment or elements containing [PTFE]. Burning off a pipe section close to a flange which may have a [PTFE] gasket, or burning any metal containing a [PTFE] element, (i.e., pump, etc.) should be done under conditions with mechanical ventilation to disperse the fumes, if there is 'he possibility of the [PTFE] element being burned . . . Burning in highly confined spaces, such as valve pits, should be conducted only with local exhaust ventilation to remove the fumes.

5. Smoking . . . should be prohibited in areas where [PTFE] is cut, machined, or processed where dust or chips are produced.

6. Chemical Laboratory Use [PTFE] . . . labware, used under conditions of direct applied heat, such as hot plates of flames, should be placed in fume hoods.

7. Disposal [PTFE] . . . scrap should not be disposed of by burning in incinerators, furnaces, etc., unless the fumes are vented to the outside and mechanical draft is provided. The disposal of [PTFE] scrap should preferably be done by burial in the ground when an appreciable quantity of scrap is involved.

8. Storage and Flammability . . . [PTFE] is non-flammable and will not propagate a flame but will decompose under heat with development of noxious and toxic gases and fumes. This is characteristic of many non-flammable materials. Therefore, the only storage restriction which need be applied is to prevent storage with flammable materials, such as oils, gases, solvents, etc. This type of restriction applies equally well to rubber belting, insulated wires and plastics. . . . Respiratory protective equipment used by fire fighting personnel, namely, the air supplied type, is adequate for entrance into burning areas where [PTFE] may be thermally decomposed. . . . Because of the development of acid fumes from the thermal decomposition of [PTFE], fire fighting personnel should be instructed to bathe and change clothes when the fire area involved considerable quantity of [PTFE], unless protective clothing was worn" (Ref. 262).

Ventilation rate for processing   258, 263, 282

1. Recommendation to reduce decomposition rates to a level of 1 ppm of decomposition products (Refs. 258, 263 p. 75):

Table 54.17. (*Continued*)

Scope	B. Recommendations for Control or Monitoring Statement			Reference
	Ventilation (ft³/min/lb resin)			
	Polymer Temperature (°C)	Polytetra-fluoroethylene	Fluorinated Ethylene–Propylene	
	230	0.50	0.86	
	260	1.3	2.1	
	290	3.5	21	
	320	10	42	
	340	33	180	
	370	73	650	
	400	180		
	425	325		
	2. "It is not recommended to extrapolate these data to other fluoropolymers" (282)			263
Medical/biologic evaluation	"… Studies on humans … and animals … showed that exposure to fluorocarbon polymer decomposition products produced respiratory irritation and pulmonary edema. A medical surveillance program should therefore give special attention to the respiratory tract. Medical attention should be provided for employees accidentially exposed to the decomposition products of fluorocarbon polymers" (p. 87)			
	"Urinary fluoride levels have been measured in humans and animals exposed to pyrolyzed fluorocarbon polymers or to their individual decomposition products. Abnormally high concentrations of fluoride in the urine of an employee who works with fluorinated polymers may be taken as signs of exposure to these compounds but cannot be used as reliable indices of the extent of exposure at present" (p.77)			
Determination of fluoride in human urine	"The response time used in this study was 3 minutes, which was found sufficient for the expected fluoride concentrations. … we studied … determinations of fluoride over the range 0–8 mg/l" (284)			283, 284

[a] Summarized as described in Table 54.10.

[b] The current OSHA standard for hydrogen fluoride and inorganic fluorides in the workplace air is 2.5 mg/m³, as fluoride.

[c] Australia and Belgium provide (28) statements similar to that of the ACGIH in the TLV list (27).

[d] Kidney damage reported from PTFE pyrolysis products but changes "difficult to study because of the closeness between the PTFE dose which initiates the visible and functional lesion (100 mg) and the point at which death occurs in a significant number of animals (LC$_{50}$ 165 mg)" (270). Barrow et al. (234) acknowledge "that PTFE which was rated the most hazardous in our tests has the very desirable property of being stable until 500°C was reached while all other materials tested released toxic components at much lower temperatures." (This 500°C temperature is probably not essentially different from the stability temperatures given earlier, since this method would likely have a lag in sample temperature.) The "potential fire load" of PTFE is typically low in real life situations; simply from a point of economics, few people would use PTFE as abundantly as wood if the latter would serve the intended purpose equally well!

4298

contact coatings for containers having a capacity of not less than 5 gal or in formulations for wall covering and ceiling tile in meat and poultry processing plants. See also Section 1.3.

Toxicologic Potential.   The main toxicologic concern with fluoropolymers is the unique phenomenon "polymer fume fever" that may develop from inhaling the products of fluoropolymers heated from below 200 to 375°C, depending on the type of fluoropolymer. With polytetrafluoroethylene, which is the most stable, the range in which this fever may develop is 315 to 375°C (285). This fever, colloquially called "the shakes," is particularly associated with the smoking of fluoropolymer-contaminated tobacco. The smoking of a cigarette contaminated with as little as 0.4 mg fluorocarbon may precipitate polymer fume fever (see Table 54.14, Reference 273). Occupational case histories indicate that exposure typically produces a nonfatal, influenza-like syndrome lasting 24 to 48 hr.

Laboratory experiments with animals indicate that somewhat higher temperatures yield toxic or lethal decomposition products. Perfluoroisobutylene is considered the most toxic product. However, the actual temperature range and decomposition conditions that yield perfluoroisobutylene appear to be limited.

Monomer residue has not been a concern with fluoropolymers. The International Agency for Research on Cancer (31) calls for long-term follow-up of patients who have received implants of polytetrafluoroethylene but notes that with typical commercial polytetrafluoroethylene the residue would be "very small" (see also Reference 29). Plasticizers are used infrequently if at all.

Highlights of the many varied features of inhalation and thermal degradation toxicity are summarized in Table 54.17. Figure 54.5 shows the toxicant spectrum of polytetrafluoroethylene.

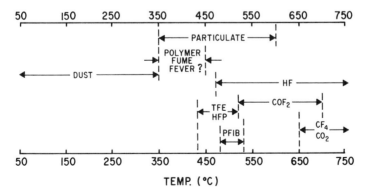

Figure 54.5   Inhalation toxicant spectrum of polytetrafluoroethylene as a function of temperature when heated in air. Reproduced with permission from R. S. Waritz (264).

Workplace Practice and Standards. Precaution should be taken to avoid excessive exposure, both directly and as a matter of housekeeping. A non-smoking rule should be enforced in all areas where fluorocarbon polymer dust may be present. Precautions should be taken against unwitting contamination of personal property with any dust particles that may later provide opportunity for the development of polymer fume fever. Du Pont has provided specific engineering control guidelines which, with additional recommendations, are quoted by the National Institute for Occupational Safety and Health (Table 54.17, Part B). The ventilation rates are based on the assumption that reducing the total concentration of decomposition products from these fluorocarbons resins to a level of 1 ppm offers an adequate safeguard.

Quantities of fluorocarbon resin >100 lb should be stored away from flammable materials. The points noted on the basis of more extensive experience with polymers containing chlorine (preceding subsection) may be applicable. Protective gear should be available for fire fighters in the event of a fire. Scrap (<10 lb) may often be incinerated with plant waste, provided that there is no direct operator contact and exhaust gases are adequately diluted.

4.2 Polyvinyl Chloride

The CAS No. is 9002-86-2; polyvinyl chloride (PVC) is marketed as Marvinol®, in the Vybak®, Geon®, and Koroseal® series; as Deckor® plastisols; and as Rhovyl®, Fibravyl®, Thermovyl®, and Isovyl® fibers (less common).

$$-\!\!\left[CH_2\!\!-\!\!CHCl\right]_n\!\!-$$

4.2.1 Preparation

Commercial polymerization of the vinyl chloride monomer typically involves free radical systems initiated by the thermal decomposition of peroxides and peroxydicarbonates. The four basic methods all use batch polymerization and are, in order of frequency, suspension, emulsion, mass or bulk, and solution (10, 213, 286–290). Suspension polymerization uses monomer-soluble initiators and water-soluble suspension agents such as gelatin, methyl cellulose, or polyvinyl alcohol. Trichloroethylene may be used as a modifier to control molecular weight. Polymerization proceeds under heat and agitation to the desired state; after monomer recovery, the water is removed and the resin is bagged for shipment.

Emulsion polymerization is similar except that emulsifying agents are used to maintain a colloidal dispersion. The resultant particle size is much smaller—about 0.5 to 2 μm compared to approximately 100 to 150 μm with the suspension process. Emulsion polymers can be produced with higher molecular weights and greater surface area/weight ratio (for plasticizing), but they have inferior water resistance and electrical properties owing to the presence of residual emulsifiers.

Bulk polymerization does not use water and yields the purest product. Liquefied vinyl chloride is polymerized with initiators such as benzoyl peroxide and azobisisobutyrronitrile. Unreacted monomer is withdrawn under pressure. Solution polymerization, a high quality process used to prepare copolymers, generally uses a solvent that dissolves the monomer but precipitates the polymer when it reaches a certain molecular weight.

Thermal stabilizers are invariably added; these include organometallic salts (tin, lead, barium, cadmium, calcium, and zinc). Antioxidants are generally used. Epoxides and phosphite chelators are secondary stabilizers used mainly in plasticized applications. Phthalate esters are the most widely used plasticizers; other types include aliphatic diacid esters, phosphate esters, and epoxidized oils. The quantity used varies greatly; very soft materials can contain plasticizer and resin in equal parts. Lubricants such as waxes, fatty acids, and metallic soaps of fatty acids facilitate melt flow and processing. Fillers (most commonly calcium carbonate) are used primarily to lower raw material cost. Pigments are used to impart color, opacity, and weatherability.

4.2.2 Molecular Weight, Properties, and Applications

Brydson (10) describes the number average molecular weight as generally 45,000 to 64,000, occasionally 40,000, and the weight average molecular weight as generally 100,000 to 200,000, occasionally 480,000. The properties of different types of polyvinyl chloride vary greatly. Rigid polyvinyl chloride is normally a hard, amber-colored material, but it can be rendered colorless and transparent, translucent, or opaque. Plasticized polyvinyl chloride can be made as flexible as required. See Tables 54.4, 54.6, and 54.16, plus Figure 54.1.

Approximately half the total production is used in construction materials such as pipe, conduit, flooring, siding, window trim, wire, and cable (291). Household products include furniture, phonograph records, shower curtains, garden hoses, and small appliance components. Other applications include footwear, medical tubing, credit cards, sporting goods, and vinyl car tops. Packaging applications use both rigid sheet and plasticized film.

4.2.3 Analysis

The International Agency for Research on Cancer has provided a comprehensive review of methods for the measurement of vinyl chloride in polyvinyl chloride,* in food or other contents of polyvinyl chloride containers, and also in the atmosphere (292). Two procedures can be mentioned briefly.

* Vinyl chloride tends to linger more in rigid and semirigid polyvinyl chloride in contrast to flexible film where the use of heat and plasticizer during manufacturing can dissipate monomer.

Steichen (150) has determined residual vinyl chloride monomer in a solution of dissolved polymer down to a level of 0.05 ppm. More recently Dennison and his associates at the Food and Drug Administration (293) have dissolved polyvinyl chloride in dimethylacetamide and sparged the solution with helium gas prior to headspace sampling and standard gas–solid chromatography with flame ionization detection. They report quantitative determination in the 1 ppb range. Other reports provide an entry to literature on the migration of the monomer and various other low molecular weight materials from polyvinyl chloride into food, water, food simulants, and miscellaneous items (24, 25, 32, 33, 213, 214, 294, and as cited in these references).

4.2.4 Oral and Skin Toxicity

Dogs fed 250 mg/kg of "PVC acrylic" sheeting for 5 days showed no adverse effects (295). A 2-year test with rats revealed no toxic effects from the feeding of a diet containing 1.5 or 12 percent of a copolymer of 95 percent vinyl chloride and 5 percent vinyl acetate (296). A Russian report describes blood changes and histologic effects in the liver and stomach from the feeding of an aqueous dispersion of a vinyl chloride–vinyl acetate copolymer that may also have contained other undetermined substances (297). Particles of polyvinyl chloride have been detected (298) in sediments of blood, bile, urine, and cerebrospinal fluid from animals that had been fed particles of 5 to 110 μm.

Skin reactions may develop from low molecular weight additives, organotin stabilizers (299), plasticizers, or other low molecular weight additives, but few have been reported (31).

4.2.5 In Vitro and Intraperitoneal Studies

Using an in vitro hemolysis technique, Richards et al. (300) found that one sample of polyvinyl chloride formed by standard processing procedures had 100 times the hemolytic effect of a second sample, and both samples contained "immeasurable amounts" of monomer (<1 ppm). The hemolytic potency of the more active sample could be reduced more than 60 percent with a single wash and even further by subsequent washes. They attributed this hemolytic potential to "the presence of a readily-soluble surface associated agent."

Styles and Wilson (301) measured the cytotoxicity of dusts from polyvinyl chloride (Corvic®), and several other polymers* as well as dusts from various minerals and chemicals. They also tested the tissue reaction to these materials

---

* Polyethylene (Alkathene®), polypropylene, and polyethylene terephthalate dusts were also placed in Group 1 (least cytotoxicity among three defined classes). Reactions of polyethylene and polypropylene appear to have been unremarkable; polyethylene is specifically stated to have produced no lesions after intraperitoneal administration. See Section 10.2.5 for additional discussion of this study.

by intraperitoneal injections in rats and found a good correlation between the cytotoxicity of these dusts to macrophages in culture and the degree of fibrosis in the animal experiments. Polyvinyl chloride (Corvic® without further description and also ball-milled) was placed in Group 1: less than 2 percent of peritoneal macrophages and less than 5 percent of alveolar macrophages were killed following phagocytosis of dust. At 1 month the ball-milled sample produced a macrophage-type lesions that contained refractile polymer; these lesions resolved by three months. Injections of the other sample produced no lesions after intraperitoneal administration.

4.2.6 Inhalation and Thermal Degradation Data.

Inhalation of prolonged and/or excessive levels of polyvinyl chloride dust has been associated abroad with impairment of respiratory function and pneumoconiosis. Fibrosis is also described in some instances that may involve exposure to monomer or where the available information on atmospheric concentrations suggests that they may have significantly exceeded concentrations permissible under accepted principles of industrial hygiene practice.

The numerous studies of thermal degradation products fall into two basic categories: (1) the "meat wrappers' asthma" associated with exposure to products emitted when hot wire or other heating elements are used on polyvinyl chloride wrap during the retail packaging of meat and (2) decomposition studies essentially comparable to those conducted with other polymers. Effects can vary from very minor to incapacitating. Cigarette smoking and an allergic history appear to be aggravating but not essential factors.

As noted in Section 3.1, hydrogen chloride is considered the most serious toxic threat in fires. Confined fires with high fuel loads of polyvinyl chloride, such as a fire in a vault with a high load of polyvinyl chloride-coated electrical wiring (244), may generate sufficient hydrogen chloride to cause irritation in fire fighters. Rapid combustion of relatively large amounts of polymer may yield either hydrogen chloride or lethal amounts of carbon monoxide, depending on the oxygen supply, as the primary hazard (230). Thermal decomposition also yields a variety of organic degradation products but only trace amounts, if any, of monomer or phosgene (see Table 54.15 and References 238, 239).* Polyvinyl chloride products appear to be relatively more toxic under smoldering rather than rapid combustion conditions. Burning polyvinyl chloride or polyvinyl chloride-filled materials may release considerable quantities of smoke; the smoke particles may contain adsorbed hydrogen chloride.

Table 54.14 surveys representative studies pertinent to the above discussion. Table 54.15 lists the specific degradation products of polyvinyl chloride detected

---

* Colardyn et al. (246) cite the O'Mara et al. study on vinyl chloride monomer (302) in connection with the production of "very small quantities of phosgene" from polyvinyl chloride.

by Boettner et al. under varying conditions. A brief report (123) suggests that the products from incineration of polyvinyl chloride can be particularly toxic to freshwater algae and rotifera, whereas marine algae and protozoa appear unaffected.

4.2.7 Carcinogenicity

Since vinyl chloride monomer is recognized as an occupational carcinogen* (see Chapter 39), polyvinyl chloride has received particular scrutiny. The International Agency for Research on Cancer (31) notes several studies that describe an elevated incidence of digestive system cancers, and possibly some other cancers, in workers involved in the fabrication of plastics; these data are considered insufficient to evaluate the carcinogenicity of polyvinyl chloride. What role the monomer may have played in the genesis of these cancers would appear difficult if not impossible to ascertain.

Casterline et al. (303) report an unusual case of buccal carcinoma in a 22-year-old man that they associate with polyvinyl chloride because since the age of eight the patient chewed materials used in his hobby of electronics. The rather uncommon lesion developed from one residual papule after an attack of acute aphthous stomatitis accompanied by "painful ulcers." The patient's intimate contact with unidentified material(s) presumably containing polyvinyl chloride may conceivably have elicited residual monomer; however, the authors support their statement about the "carcinogenic effects" of the polymer with three references on the activity of the monomer.

4.2.8 Biodegradation

Additives may facilitate microbial growth. Biodeterioration is apparently limited to the interface where microbial organisms can readily obtain water, inorganic nutrients, oxygen, and organic substrate (304). In mineral salts agar plate tests with *Pseudomonas aeruginosa, Candida lipolytica,* and a fungal mixture (ASTM D-1924-63), biodeterioration developed mostly around the perimeter of the samples, which contained 33 to 40 percent plasticizer and approximately 3 percent other additives. Measuring ammonia given off 48 hr after incubation of polyvinyl chloride samples with *Pseudomonas fluorescens* can serve as a rapid test for biodegradability; Sharpe and Woodrow (305) report an inverse correlation with weight loss of the film after 14 days. See also Reference 77a.

---

* Acroosteolysis and capillary abnormalities are also associated with occupational exposure to polymerization of the monomer but are not known to have been connected to the polymer per se.

4.3 Polyvinylidene Chloride and Copolymers (PVDC; Saran)

The term "Saran" is used generically to describe all copolymers of high vinylidene chloride content (31), usually in the form of film.* The homopolymer
$$-\!\!\left[CH_2CCl_2\right]_{\!n}-$$
known as *Saran A* (CAS No. 9002-85-1) has a softening point very close to decomposition temperature. Commercial products are based on copolymers with lower softening points, such as the following:

> *Saran B*, vinylidene chloride–vinyl chloride (CAS No. 9011-06-7)
> *Saran F*, vinylidene chloride–alkyl acrylate
> *Saran C*, vinylidene chloride–acrylonitrile (CAS No. 9010-76-8)

4.3.1 Preparation

The monomer 1,1-dichloroethylene is typically polymerized (or copolymerized) by free radical processes (10, 14, 306). Emulsion processes are used for latexes, higher molecular weight polymers, and where a relatively high concentration of additives is permissible; suspension polymerization is generally used for moldings and extrusions (31).

4.3.2 Molecular Weight, Properties, and Applications

Molecular weights described for the copolymers vary: approximately 20,000 to 50,000 (10), 10,000 to 100,000 for vinyl chloride copolymer (31), and 200,000 for emulsion copolymer (14). These copolymers are noted for toughness, flexibility, durability, clarity, chemical resistance, and water and gas impermeability. The high chlorine content gives fire resistance, but thermal stability is poor.

The vinyl chloride copolymers are used mainly as filaments and in food packaging film (10, 31). Other applications include latex coatings for packaging rods, pipes, and pipe liners.

4.3.3 Analysis

Vinylidene chloride monomer has been detected in "polyvinylidene chloride" by electron capture with gas chromatography or mass spectrometry (307).

Vinylidene chloride monomer that has eluted from polyvinylidene chloride film into edible oil remains unchanged after storage for 3 months at 50°C (308). However, monomer that has eluted into water decreases gradually until

---

* Saran fiber identifies a manufactured fiber in which the fiber-forming substance is any long chain synthetic polymer composed of at least 80 percent by weight of vinylidene chloride units.

practically no monomer can be found after 3 days when stored at 50°C and 3 months when stored at room temperature.

4.3.4 Oral and Skin Toxicity

The International Agency for Research on Cancer (31) cites an unpublished study from a 1954 publication indicating that rats and dogs fed a diet containing 5 percent of a vinylidene chloride–vinyl copolymer for 2 years showed no toxic effects. This agency also references a single report of human contact dermatitis.

4.3.5 Inhalation Toxicity and Thermal Degradation

Data summarized in Table 54.14 indicate no unusual toxic characteristics and indicate hydrogen chloride as the main decomposition product.

4.3.6 Carcinogenic Potential

Although results such as "limited evidence of carcinogenicity" (194) have been reported for the monomer, carcinogenic effect was not evident in three recent 2-year tests: maximum tolerated dose by gavage, given to rats and mice (309a); administration to rats, in drinking water at levels up to 200 ppm and also by inhalation at levels up to 75 ppm (309b).

4.4 Polyvinyl Fluoride

The CAS No. is 24981-14-4; polyvinyl fluoride (PVF), available as Tedlar® PVF film, has the structure

$$-[CH_2-CHF]_n-$$

4.4.1 Preparation

Vinyl fluoride monomer is polymerized in a free radical-initiated water suspension process. After filtration and drying, polymer may be cast into film using an organosol based on dimethylacetamide or an equivalent solvent system. Oriented polyvinyl fluoride film as sold for adhesive bonding may contain up to 0.5 percent of residual dimethylacetamide (310).

4.4.2 Molecular Weight, Properties, and Application

Molecular weight has been given as 60,000 to 180,000 (14). Polyvinyl fluoride is noted for inertness, flexibility, toughness, durability, and weather resistance (255, 310, 311). Comparative properties are shown in Table 54.16 and Figure 54.1.

Polyvinyl fluoride film is intended primarily for adhesive bonding to a variety of substrates. Construction surfaces, especially residential and commercial siding and aircraft interior surfaces, are the major application.

4.4.3 Skin Toxicity

Polyvinyl fluoride produced no skin reactions when tested for irritation and sensitization on 215 human subjects (312, 313).

4.4.4 Thermal Degradation Products

The main decomposition product is hydrogen fluoride. No unusual toxic effect was evident in a test where mice were exposed to pyrolysis products. See Table 54.17 for details.

4.5 Polyvinylidene Fluoride

The CAS No. is 24937-79-9; polyvinylidene fluoride (PVDF) is available as Kynar®. It has the structure

$$-\!\!\left[CH_2\!-\!CF_2\right]_{\overline{n}}$$

4.5.1 Preparation

The polymer can be polymerized under pressure from the monomer 1,1-difluoroethylene with free radical initiators such as peroxides (14).

4.5.2 Molecular Weight, Properties, and Applications

Commercial grades with molecular weights of approximately 300,000 and 600,000, respectively, have been reported (10). Properties are in many respects intermediate between the monofluorinated and perfluorinated analogues discussed elsewhere in this section. Wear resistance and creep resistance are typically greater than with perfluorinated polymer (see Sections 4.6 and 4.7) and weather resistance less than that of the monofluorinated analogue. See Table 54.16 for a comparative listing of properties.

Applications include electrical insulation, chemical process equipment (particularly pipe), and use as a base in durable finishes for metal siding.

4.5.3 Thermal Degradation Products

Thermal decomposition yields hydrogen fluoride among other products. Mice showed signs of incapacitation from the pyrolysis products of polyvinylidene fluoride somewhat sooner than mice similarly exposed to products from some other polymers tested in the same series. See Table 54.17 for details.

4.6 Polytetrafluoroethylene

The CAS No. is 9002-84-0; polytetrafluoroethylene (PTFE) is available commercially as Teflon® TFE fluorocarbon resin (series).* Its structure is

$$-\!\!\left[CF_2CF_2\right]_{\!\overline{n}}$$

4.6.1 Preparation

Polytetrafluoroethylene is generally made from tetrafluoroethylene gas by free radical polymerization, under pressure with oxygen, peroxides, or peroxydisulfates (10, 311; see also Reference 4). The "granular resins" have medium-size particles ranging from 30 to 600 µm. Colloidal aqueous dispersions, made by a different process, are concentrated to about 60 percent by weight of the polymer and have particles averaging about 0.2 µm. Coagulated dispersions with agglomerates averaging 450 µm are also available (259).

4.6.2 Molecular Weight, Properties, and Applications

The original commercial polymers of the granular type were in the number average molecular weight range of 400,000 to 10,000,000 (259); modern polymers (10) are more likely to be in the range 500,000 to 5,000,000. The outstanding characteristics of polytetrafluoroethylene are its nearly universal chemical inertness, high thermal stability, general insolubility (except for certain fluorinated solvents), lubricity, and useful mechanical characteristics over a wide temperature range. See Table 54.16 for comparative data.

The major applications are as components or linings for chemical process equipment, high temperature wire and cable insulation, molded electrical components, tape, and nonstick coatings (311). Applications for nonstick coatings include home cookware, tools, and food processing equipment. Polytetrafluoroethylene also has been used in a variety of medical applications (32).

Polytetrafluoroethylene fiber is the most chemical-resistant fiber known, also the most fire-resistant in oxygen-rich and high pressure atmospheres. Bleached fiber of Teflon® TFE was used for the inflight coveralls of the Apollo astronauts.

4.6.3 Oral Toxicity

No abnormalities were evident in a 7-month experiment where rats ingested a diet containing 0.5 percent of a mixture containing 21 percent polytetrafluoroethylene (Harris, as cited in Reference 1). Rats fed one of three types of polytetrafluoroethylene at a dietary level of 25 percent for 90 days "showed a

* Teflon® is a trademark for tetrafluoroethylene (TFE) fluorocarbon polymers but can also refer to fluorinated ethylene–propylene (FEP) resins (3).

shift in the number and distribution of white blood cells and, in one group fed unsintered Teflon® 6 TFE resin, an increase in the relative size of the liver, relative to body weight"; feeding did not produce any adverse effects on growth rate or behavior, or any microscopic evidence of tissue change (314).

This minimal response was most likely attributable to a surfactant used in polymerization. Subsequent 2- or 3-week tests at 25 or 10 percent dietary levels (315 and references cited therein) showed that polytetrafluoroethylene resins prepared with various volatile dispersing agents produced enlarged livers but the same resins heated to remove these volatiles did not produce enlarged livers. In one case feeding the surfactant per se did not produce the large livers observed under similar test conditions with feeding of the resins containing the surfactant; these data suggested that the "active agent" was produced in situ during the polymerization process (316). Several Russian reports (317–319) also describe effects consistent with the concept of low molecular weight additives leaching out under test conditions.

4.6.4 Inhalation and Thermal Degradation Toxicity

Grossly excessive exposure to polytetrafluoroethylene dust can elevate urinary fluoride. The key point is that polymer fume fever may result when even trace amounts of polymer heated in the range from 315 to 375°C are inhaled. The most common means of exposure is through the smoking of contaminated tobacco in any form.

Decomposition studies at higher temperatures reveal that the approximate lethal temperature for rats is 450°C. Tetrafluoroethylene monomer is first detected at 440°C, traces of perfluoroisobutylene become evident at 475°C, and carbonyl fluoride is the dominant oxidative pyrolysis product at 550°C. Above 650°C carbon tetrafluoride and carbon dioxide are the major products. Table 54.17 provides an entry into the copious literature.

4.6.5 Carcinogenic Potential/Implant Studies

Polytetrafluoroethylene fabricated in textile form has been effectively utilized for certain types of vascular grafts (32); thrombosis develops (320) when the prosthesis is used as a replacement for small arteries (diameter of 6 to 8 mm or less). Susceptibility of polytetrafluoroethylene to deformation under load makes it unsuitable for prostheses such as an acetabular cup that articulates with a femoral head prosthesis of stainless steel (321).

As with many other polymers (Section 1.4), implants in rodents have caused sarcomas at the site of implantation. Bischoff (29) speculated that "chemical factors were more important than the continuous surface in eliciting the carcinogenic response" in implants and suggested a "possible product difference." The International Agency for Research on Cancer noted (31) a single case of fibrosarcoma, without evidence of metastasis, 10½ years after implantation of a polyester–polytetrafluoroethylene arterial prosthesis (see Section 9.2). This case plus the studies on mice and rats provided insufficient evidence to assess the carcinogenic risk of exposure in humans. Long-term follow-up was recommended only in the case of patients with medical implants.

4.7   Other Vinyl Halides

The commercial polymers are mainly polymers that contain both chlorine and fluorine atoms, fluoroalkoxy materials, and also a variety of copolymers, terpolymers, tetrapolymers, etc. In common usage they may all be termed "fluorocarbon polymers" (263) or "fluoroplastics" (259, 311). Fluorinated ethylene–propylene, or "Teflon" FEP, is a fully fluorinated copolymer made from tetrafluoroethylene and hexafluoropropylene that has many of the desirable properties of polytetrafluoroethylene (311). Unlike polytetrafluoroethylene, fluorinated ethylene–propylene can be processed by conventional thermoplastic techniques such as extrusion; this advantage is gained at a sacrifice in maximum service temperature, which is 200°C. See Table 54.17 for comparative data on thermal degradation.

Other related polymers or copolymers with somewhat varied properties that may be desirable in particular applications include polychlorotrifluoroethylene, ethylene–chlorotrifluoroethylene copolymer, and ethylene–tetrafluoroethylene copolymer (313). Table 54.17 lists several pertinent decomposition studies.

Perfluoroalkoxy resins comprise a somewhat different class of melt processible fluoroplastics that are in many respects comparable to polytetrafluoroethylene (211). They offer improved mechanical properties over fluorinated ethylene–propylene at temperatures >150°C (311, 322). See Table 54.17.

5   POLYVINYL ACETATE, POLYVINYL ALCOHOL, AND DERIVATIVES

5.1   Overview

These vinyl esters are not typically used for molding or extrusion but have important special applications (5). Polyvinyl acetate, the most widely used vinyl ester, is noted for its adhesion to substrates and high cold flow. Polyvinyl acetate serves as the precursor for polyvinyl alcohol and, directly or indirectly, the polyvinyl acetals. Both polyvinyl acetate and polyvinyl alcohol are insoluble in many organic solvents but water sensitive. Polyvinyl acetate absorbs from 1 to 3 percent water, up to 8 percent on prolonged immersion (14). Polyvinyl alcohol absorbs 6 to 9 percent water when humidity-conditioned and can usually be dissolved completely in water above 90°C, but it can also be insolubilized by chemical treatment. Table 54.18 lists some comparative properties.

5.1.1 Production and Processing

United States manufacturers currently sell polyvinyl acetate in emulsion form and polyvinyl alcohol as granules. Polyvinyl alcohol is processed into films and formulated with other materials into emulsion intermediates. Both polymers are typically used in aqueous systems.

5.1.2 Analysis and Specifications

Gas chromatography can be used to detect vinyl acetate in aqueous latexes and in acrylic copolymers of vinyl acetate (citations in Reference 31). Reported sensitivity ranges from 10 to 200 ppm. See also Section 6.1.2 and 6.5.3.

Both polyvinyl acetate and polyvinyl alcohol meeting certain specifications are permitted in stated food-contact applications such as packaging, coatings, and adhesives. Ethylene–vinyl acetate copolymers and ethylene–vinyl acetate–vinyl alcohol terpolymers (CAS No. 26221-27-2) are similarly permitted in certain food-contact applications. Polyvinyl acetate with a minimum molecular weight of 2000 is permitted as a synthetic masticatory substance in chewing gum base (92).

5.1.3 Toxicologic Potential

Monomer residue has not been considered a problem in end use products. Latexes or solutions of polyvinyl acetate that are essentially intermediates may contain residual vinyl acetate, essential emulsifiers, or initiators (31). No detailed information is available on the amount of unreacted monomer in either polyvinyl acetate or polyvinyl alcohol resins (31).

Local sarcomas have been produced in rats with polyvinyl alcohol sponges, but implants of both polyvinyl alcohol and polyvinyl acetate in powder form did not produce tumors (see Section 1.4). The International Agency for Research on Cancer considered (31) that additional studies would be required prior to evaluation of carcinogenic potential.

Inhalation and combustion toxicity have not been considered problems (see Table 54.19). This may be attributed to polymer structure and degradation characteristics as well as the nature of ordinary intermediate and end use products.

5.1.4 Workplace Practices and Standards

With products formulated as solutions or emulsions, the potential for inhalation toxicity or skin reactions from residual monomer or additives may in some cases require evaluation. Recommended procedures for dealing with nuisance dusts (see Section 1.4) should provide adequate control for any dust hazards

Table 54.18. Selected Properties of Polyvinyl Acetate and Polyvinyl Alcohol[a]

Property	Polyvinyl Acetate	Polyvinyl Alcohol	Reference
Melting temperature (°C)		212–267 (syndiotactic)	85
Softening temperature (°C)	35–50[b]		323
Glass transition temperature (°C)	28–32	85°C	14, 86, 323
Brittle point (°C)		When suitably plasticized, flexible <40	14
Density	1.17–1.19	1.25–1.35 (depending on moisture content)	
Hardness (Shore units, at 20°C)	L 80–85		323
Refractive index, n_D	1.46–1.47	1.49–1.53 (depending on moisture content)	87, 323
Soluble in	Aromatic and chlorinated hydrocarbons (toluene, chloroform), ketones (acetone), lower alcohols with water. Incomplete solution in lower alcohols (anhydrous) and aqueous solvent mixture; slightly swollen (and whitened) by water alone	Water,[c] especially on warming to 70–80°C. Formamide, dimethyl sulfoxide (heat). Fibers can be prepared to dissolve at 44–50°C or 90°C (or insolubilized by treatment with formaldehyde)	6, 14, 88

Swollen/plasticized in	Plasticized by high boiling esters (dibutyl phthalate) and copolymerization; see also directly above	Moisture and water-soluble polyhydric alcohols, amides, and calcium chloride	14, 88
Nonsolvents/relatively unaffected by	Aliphatic hydrocarbons, most oils and fats, turpentine, ether; carbon disulfide	Oils, fats, almost all organic hydrocarbons and chlorinated hydrocarbons	14, 88
Decomposed by	Conc. alkalies (especially with heat) and conc. acids; decomposition accentuated in alcoholic solution	Conc. acids, especially oxidizing acids. Boiling with hydrogen peroxide lowers molecular weight and introduces acid groups	14

[a] Data are summarized as described in Table 54.9.

[b] Polyvinyl acetate softens to a viscous melt "from 70 to 190°C according to molecular weight" (14).

[c] The viscosity of aqueous solutions of polyvinyl alcohol varies with the grade, concentration, and temperature (see Ref. 324). Variations in pH may influence the degree of alcoholysis [N. Alkhalili and B. J. Meakin, "The Chemical Stability of Polyvinyl Alcohols," J. Pharm. Pharmacol., **29**, 35P (1977)].

Table 54.19. Overview of Inhalation and Thermal Degradation Data on Polyvinyl Acetate and Polyvinyl Alcohol[a]

Polymer	TLV	Ambient Temperature	Elevated Temperature	Effluent Analysis	Species Exposed	Procedure	Observations/Conclusions	Reference
							Type of Study	
Polyvinyl Acetate	No specific TLV (See Section 1.4)	X			Rat, dog	30 exposures, 4 hr daily, to 90 mg/m³ or 380 mg/m³, aerosol with mean particle diameter of 1.2 μm	No clinically significant changes[b]; no "potential inhalation hazard" indicated at these concentrations	325
			X			(a) Degradation at 213–235°C	Quantitative yields of acetic acid	Cited in 105
						(b) Degradation at 300°C	Small amounts of aromatics including benzene	
Polyvinyl alcohol	As polyvinyl acetate above	X	X		Rat (presumably)	15 inhalations of 45 mg/m³ dust; also exposure to products from thermoxidation at 170°C	Slightly depressed growth rate, biochemical change	326
			X	X		"Smoldering combustion" of vinylon at 300–400°C	Relatively large quantities of acrolein and formaldehyde; see Figure 54.3	103
			X	X		(a) Degradation at 240°C	Main products water and ethanol, also aldehydes and ketones	Cited in 105
						(b) Degradation at 250°C	Quantitative yields of water	

[a] Summarized as described in Table 54.10.

[b] A statistically significant increased liver to body weight ratio was observed in rats at the high dose level; body weight gain and histopathology were normal.

4314

from solid forms of polymer. Spillage of these polymers can produce slipping hazards; spills should be cleaned up immediately to prevent falls.

5.2 Polyvinyl Acetate

The CAS No. is 9003-20-7. Polyvinyl acetate, also known as PVA,* PVAC, and PVAc, is commercially available as Gelva®, Lemac®, and Vinac®, in the Bakelite® series.

$$
\left[CH_2 - \underset{\underset{\displaystyle O-CO-CH_3}{|}}{CH} \right]_n
$$

Commercial copolymers containing at least 60 percent vinyl acetate units are generally termed "polyvinyl acetate." The major copolymers are (31) as follows:

> Polyvinyl acetate–butyl acrylate (CAS No. 25067-01-0)
> Polyvinyl acetate–2-ethylhexyl acrylate (CAS No. 25067-02-1)
> Polyvinyl acetate–ethylene (CAS No. 24937-78-8)

Many other comonomers are also used; also, small amounts of vinyl acetate units may be polymerized with other monomers, such as vinyl chloride.

5.2.1 Preparation

Vinyl acetate can be easily polymerized by bulk, solution emulsion, or suspension procedures (10). The most common process involves emulsion polymerization of the monomer(s) in the presence of surfactants and a free radical initiator (10, 31 and references cited therein).

5.2.2 Molecular Weight, Properties, and Application

Molecular weight has been described as varying from 5000 to over 500,000 (14) and from 11,000 to 1,500,000 (as cited in Reference 31). Commercial polymers are atactic and therefore, if free from emulsifier, transparent (10). Polyvinyl acetate is noted for its high tensile strength and toughness. In general, copolymerization results in a particular hardening or softening effect desirable in the intended application. Properties vary with humidity. See Table 54.18 for comparative data.

Polyvinyl acetate, as a homopolymer and in copolymers, is widely used in emulsion paints; in adhesives for paper, wood, and textiles; and as sizing. It is also an intermediate in the preparation of other polymers (see Sections 5.3 and 5.4). Vinyl acetate polymerized with vinyl chloride yields a copolymer that in the molecular weight range of 10,000 to 28,000 is suitable for making fibers (6).

---

* PVA may also mean polyvinyl alcohol; PVAC and PVAc are preferable (14).

5.2.3 Oral, Skin, Inhalation, and Thermal Degradation Toxicity

Meager data indicate low toxicity. Rats and mice apparently survived a single oral administration of 25 g/kg polyvinyl acetate as well as 12-month administration at 250 mg/kg that produced some tissue damage (327). Animals fed various doses of several modifications of polyvinyl acetate showed effects ranging from increased "work capacity" to hepatocyte necrosis.

Weeks and Pope reported that the oral lethal dose of polyvinyl acetate formulated as an emulsion dust control material was >9.7 g/kg for rats (325). Skin irritation tests on rabbits with this emulsion or the base latex produced moderate to severe irritation on intact and abraded skin. (The irritation potential received the most emphasis in medical surveillance recommendations.) However, Hood (328) found no evidence of primary irritation or skin sensitization in any of 210 human subjects exposed to film made from two different polyvinyl acetate emulsions or cotton cloth impregnated with 40 percent polyvinyl acetate resin. Her subjects wore 1-in. squares of the test material for 6 days and then again for 1 day after a 10-day rest period.

Table 54.19 summarizes the limited available inhalation and thermal degradation data.

5.2.4 Carcinogenic Potential/Implant Studies

The International Agency for Research on Cancer reviewed (31) data on subcutaneous or intraperitoneal implantation of polyvinyl acetate powder in mice and rats. No local sarcomas were found.

5.2.5 Effects on Aquatic Life and Biodegradation

Polyvinyl acetate emulsions have been found nontoxic to pike, carp, daphnia, and other species of aquatic life at concentrations ranging from 0.3 to 80 mg/l (329–331).

Griffin and Mivetchi (332) report that polyvinyl acetate is biodegradable, and ethylene–vinyl acetate copolymer (see Section 5.4) shows the effect to a degree, depending on vinyl acetate content. Starch filler in polyvinyl acetate stimulates biodegradability.

5.3 Polyvinyl Alcohol

The CAS No. is 9002-89-5. Polyvinyl alcohol is also called PVA,* PVAL, and PVAl. Trade names include Elvanol®, Vinol®, and Vinylon® fiber.

$$(13) \qquad \left[CH_2 - \underset{\underset{OH}{|}}{CH} \right]_n$$

---

* PVA may also mean polyvinyl acetate; PVAL and PVAl are preferable (14).

5.3.1 Preparation

"Vinyl alcohol monomer" is the unstable enol form of acetaldehyde. Various indirect methods that involve alcoholysis (also called saponification or hydrolysis) of polyvinyl acetate are therefore used to prepare polyvinyl alcohol (4, 5, 10, 31 and references cited therein). Preparation can be carried out by dissolving polyvinyl acetate in methanol or ethanol with an alkaline or acid catalyst and heating to precipitate the polyvinyl alcohol from the solution.

5.3.2 Molecular Weight, Properties, and Applications

Molecular weight depends on and is approximately half that of the polyvinyl acetate from which the polyvinyl alcohol was derived. Properties vary depending on the degree of alcoholysis. Polyvinyl alcohol is amorphous to polycrystalline, depending on mechanical and heat treatment. When stretched as fibers the polymer becomes up to 60 percent crystalline (14).

Polyvinyl alcohol is noted for its resistance to oil, grease, and certain solvents. In a comparative test, solvents such as chloroform, methylene chloride, and methyl iodide readily penetrated five other glove materials but polyvinyl alcohol glove material showed <0.1 percent penetration after 0.5 hr exposure. However, water, dimethyl sulfoxide, dimethylformamide, and methanol penetrated polyvinyl alcohol >10 percent (178). See also Table 54.18.

The leading applications are in textiles, as warp sizing, as a finishing agent, and, primarily abroad, as water-soluble or insolubilized fibers (depending on treatment). Other applications include adhesives, as thickening and release agents, as a polymerization aid, in paper sizing, and also as packaging film where water solubility or grease resistance is desired. Plasticizing permits specialized molding/extrusion use (hoses, pipe, gloves, etc.).

5.3.3 Oral, Eye, and Inhalation Toxicity

Oral tests on animals date back to 1939, when Hueper reported (333) on a test with four rats conducted in connection with more extensive parenteral tests. The rats were fed a dietary level of 4 percent polyvinyl alcohol increased to 8 and then 29 percent at 2-week intervals; two were sacrificed at 4 weeks. The two surviving rats that received the higher dose level showed microscopic tissue changes in the liver, stomach, and sternum. Sections stained with iodine solution did not show the characteristic blue color indicative of polyvinyl alcohol deposits (see Section 5.3.4 below). Zaeva et al. (326) describe only minimal changes in the liver and myocardial cells of rats receiving 30 doses of 500 mg/kg. Yamatani and Ishikawa (334) found no polyvinyl alcohol in the urine or blood of rats dosed orally with 2 ml of 2 percent polyvinyl alcohol (molecular weight = 22,000).

In eye drops polyvinyl alcohol in soluble form increases viscosity; a 1.4 percent neutral solution of polymer with molecular weight >100,000 has been used successfully in tests on rabbit eyes and on human eyes (335, 336). Other reports suggest continuing interest in solutions of polyvinyl alcohol as a component of formulations for topical treatment of "dry eye" (337) and intraocular pressure (338).

The limited inhalation and thermal degradation data summarized in Table 54.19 indicate no unusual toxic characteristics.

5.3.4 Parenteral/Carcinogenic Potential/Mutagen Tests

Polyvinyl alcohol released into the circulation may cause serious dysfunction, particularly in the kidneys. This was first demonstrated by Hueper (333), who showed that polyvinyl alcohol injected subcutaneously into rats diffused throughout major body organs. Most of the kidneys showed degeneration or necrosis of the tubular epithelium, which was sometimes accompanied by other changes. During the 1960s Hall and Hall also conducted an extensive series of subcutaneous tests with various grades of polyvinyl alcohol. They confirmed that injections of polyvinyl alcohol may affect elements of the reticuloendothelial system and cause a nephrotic syndrome. Initial experiments (339, 340 and references cited therein) suggested a correlation with molecular weight; solutions of a medium-size molecule (average molecular weight = 133,000) appeared more injurious than those of a larger molecule (average molecular weight = 185,000) and those of a relatively small molecule (average molecular weight = 37,000). The last caused only a slight elevation of arterial pressure in a small percentage of rats and its inertness was presumed due to ready filtration of its small size through renal glomeruli. Additional work showed that the hypertensive syndrome found after injection of some but not all types of polyvinyl alcohol did not correlate with infiltration of the kidney; infiltration and varying degrees of anemia were considered the only common sequelae with various types of polymer (341).

The International Agency for Research on Cancer summarizes (31) various animal studies, some but not all of which produced tumors. All these studies were parenteral (see Section 1.4). Reference is also made to one case of hemangiopericytoma of the bladder in an occupationally exposed worker, which the authors "speculated" might be analogous to that of angiosarcoma of the liver and vinyl chloride. Polyvinyl alcohol (10 mg per petri plate, highest feasible concentration) was not mutagenic to any of five strains of *Salmonella typhimurium* (Ames test) when tested with or without activation (342).

5.3.5 Effects on Aquatic Life, Biodegradation, and Related Data

Bluegill sunfish exposed for 96 hr to 10,000 mg/l of polyvinyl alcohol showed no mortality or other evidence of physiologic response (343). Biodegradation of polyvinyl alcohol can be accomplished in properly designed and operated waste treatment systems where >90 percent of the polyvinyl alcohol in the influent may be removed by acclimated microorganisms (344, 345). The chemical oxygen demand rate coefficient was found to be $k = 1.2$ mg oxygen/mg polyvinyl alcohol/day in a domestic (municipal) activated waste system (344) and $k = 0.50$ in a textile waste treatment facility (345). Organisms degrading polyvinyl alcohol in mixed culture (346) have remained viable under hydrogen peroxide treatment conditions that were toxic to filamentous and nitrifying organisms. Product differences, particularly the number and location of residual acetate groups (79), may have to be considered. Ozonization (347) may lower polymer molecular weight and improve biodegradability. Additionally, see References 77a and 324.

5.4 Derivatives

Reacting polyvinyl alcohol or polyvinyl acetate with aldehydes yields polyvinyl acetals, of which polyvinyl butyral is the most important. This polymer is usually made from polyvinyl alcohol to give a very pure product that is extensively used in safety glass (10). Polyvinyl butyral has high adhesion to glass, toughness, light stability, clarity, and moisture insensitivity. Laminating several layers of glass and polyvinyl butyral film gives bulletproof glass. Polyvinyl butyral is also used in coatings for textiles and metals and in adhesives.

Polyvinyl formal is usually made directly from polyvinyl acetate. Polyvinyl formal is often used in conjunction with phenolic resins and in structural adhesives. Other applications include can coatings and wash primers (10).

Ethylene–vinyl acetate copolymers and the more exotic terpolymers (also described as ethylene–vinyl acetate–vinyl alcohol copolymers) are used with modifying resins, waxes, fillers, antioxidants, etc., to formulate hot melt adhesives that provide improved adhesion to substrates. Ninety-day feeding tests with male and female rats have been conducted at a dietary level of 5 or 10 percent ethylene–vinyl acetate copolymer (29 or 33 percent vinyl acetate by weight) without significant clinical, nutritional, or histopathologic effect. The male rats fed at the 10 percent dietary level developed a significant relative granulocytosis without increase in the total white cell count; similar changes were also observed in the control male rats fed 10 percent nonnutritive bulk (Alphacel® cellulose) in the diet (348, 349).

6 POLYSTYRENE AND ACRYLICS

6.1 Overview

The historical impetus for the use of polystyrene came after the end of World War II, when the availability of natural rubber decreased demand for styrene–butadiene elastomer and the accompanying production facilities (10). As indicated in Table 54.2, styrene-based plastics now rank third highest in production among the thermoplastics. "Styrene-based plastics" most commonly are polystyrene itself plus the derivatives containing butadiene, acrylonitrile, or both. The derivatives containing acrylonitrile are also called "acrylonitrile polymers" or "nitrile polymers."

Structurally the acrylic polymers include those containing repeating units of acrylonitrile, acrylic acid, acrylates, methacrylates, and all the various derivatives. "Acrylic plastics" may imply only polymers of acrylic or methacrylic acid ester (350), among which the prototype is polymethyl methacrylate. The demand for sheet polymethyl methacrylate dates from World War II when it was used for aircraft glazing (10).

Table 54.20 gives data on properties for typical polymers.

6.1.1 Production and Processing

Polystyrene is usually sold by the manufacturer in pellets. It is fabricated at process temperatures of 360 to 450°F (365). Depending on the particular conditions prevailing, melt processing of polymer may result in partial volatilization of unreacted monomer or other components. Molding formulations containing flame retardant chemical modifiers can release toxic and corrosive gases if material temperature exceeds 475°F (366). Postfabrication operations may include trimming, machining, drilling, buffing, and solvent bonding of parts (367).

Polyacrylonitrile is used primarily as fibers, commonly called "acrylic," that have been formulated with varying amounts of comonomer. Processing may involve blending with other fibers. Dyeing may be done either with the yarn or woven fabric. The recently commercialized "polymeric acrylonitrile" (Du Pont Types A and A-7) is a powder that can be hydrolyzed to form water-soluble polyacrylamides or cast as a plastic or film from solvent.

The main copolymer types derived from both styrene and acrylonitrile are (1) styrene–acrylonitrile copolymer resin and (2) acrylonitrile–butadiene–styrene (ABS resins), in which discrete butadiene particles are dispersed in a styrene–acrylonitrile copolymer matrix and then sold as pellets or powder. Temperatures described for acrylonitrile–butadiene–styrene processing are in the 380 to 525°F range (367).

Acrylate and methacrylate esters are generally available from the manufacturer in granules or powder. Dyes, pigments, plasticizers, or ultraviolet absorbers may be added during processing. Commercial processing of polymethyl methacrylate per se uses three intermediate types of approach (10): the melt state for injection molding and extrusion; sheets, rods, and tube that are machined or welded; and monomer–polymer dough, primarily for dentures. "Acrylic

Table 54.20. Characteristics of Polystyrene, Prototype Acrylic Polymers, and Derivatives[a]

A. Selected Properties of Polystyrene, Polyacrylonitrile (and its Fibers), and Polymethyl Methacrylate[a]

Property	Polystyrene	Polyacrylonitrile and Fibers	Polymethyl Methacrylate	Reference
Melting point (°C)		317–341 (acrylic fibers may show degradation with color formation ~105)	Isotactic: 160 Syndiotactic: >200	14, 82, 85, 351, 352
Glass transition temperature (°C)	80–100	87–104. Fibers show transition ~70 (Ref. 354)	Syndiotactic, atactic 105 Isotactic: 38	14, 82, 86, 351, 353–357
Softening point (°C)	~80	Acrylic fibers: 235–255 Modacrylic fibers: 150		14, 31
Density	1.04–1.127	Syndiotactic: 1.07–1.273 Isotactic: 1.538 Modacrylic fibers: 1.24–1.37	1.17–1.23	14, 31, 82, 85, 351, 355–358
Refractive index, n_D	1.57–1.60	1.52	1.49	14, 82, 87, 356
Moisture regain (fiber)		1.5%, commercial value		12[b]
Solvents	Cyclohexane >35°C, benzene, toluene, styrene, lower chlorinated aliphatic hydrocarbons, ethyl acetate. *Plasticized* by common ester and hydrocarbon plasticizers (which may alter mechanical properties), copolymerization. *Swollen* by oils, ketones, esters, some aliphatic hydrocarbons	Dimethylformamide, dimethylacetamide, dimethyl sulfoxide, succinonitrile, many other organic liquids; conc. acids (as 70% sulfuric); concentrated aqueous or solutions of soluble salts (as zinc chloride); quaternary ammonium salts. *Plasticized* by castor oil	Benzene,[c] toluene,[c] chloroform,[c] dioxane,[c] acetic acid,[c] ethyl acetate,[c] formic acid, nitroethane, ethanol/water. *Plasticized* by some esters, castor oil, copolymerization. *Swollen* by alcohols, phenols, ether, carbon tetrachloride.	14, 88
Nonsolvents/relatively unaffected by	Water, alcohols, alkalies, nonoxidizing acids (as acetic), phenol, acetone, some aliphatic hydrocarbons. (Stressed polystyrene may show cracking from lower alcohols, fats, paraffins, etc., that otherwise have little or no effect)	Acrylonitrile, hydrocarbons, chlorinated hydrocarbons, formamide, methylformamide. Acrylic fibers show "little effect" from weak inorganic acids, weak alkalies, and oxidizing agents; "negligible effect" from common organic solvents; "no effect" from "dry cleaning solvents" (358)	Methanol,[d] cyclohexane,[d] carbon tetrachloride, absolute ethanol, linseed oil	14, 88, 358

Table 54.20. (Continued)

A. Selected Properties of Polystyrene, Polyacrylonitrile (and its Fibers), and Polymethyl Methacrylate

Property	Polystyrene	Polyacrylonitrile and Fibers	Polymethyl Methacrylate	Reference
Decomposed by	Prolonged contact or boiling with oxidizing agents	Conc. alkalies, hot dilute alkalies, warm 50% sulfuric acid	Concentrated oxidizing acids, alcoholic alkalies	14
Serviceability	Deterioration from prolonged exposure to sunlight (or fluorescent lighting), which causes yellowing or fading, often with crazing of mechanical properties	Acrylic fibers in general: good to excellent resistance to sunlight aging, inferior abrasion resistance. Fibers of homopolymer unusually resistant to nuclear radiation (which can even improve strength)	Acrylic resins: fastness to light, resistance to weathering	6, 14, 358

B. Some Properties Associated With Styrene–Acrylonitrile Polymers and Acrylonitrile–Butadiene–Styrene Polymers

Property	Styrene–Acrylonitrile	Acrylonitrile–Butadiene–Styrene	Reference
Glass transition temperature (°C)	~80		56
Softening point (°C)	106–110	100–108	31, 359
Density	1.06–1.08	1.04–1.05 (Ref. 31); 1.02–1.07 depending on whether high, medium, or low impact (Ref. 360); minimum = 0.97, maximum = 1.122 (Ref. 361)	31, 359–361
Solvents/swelling	"Acetone, chloroform, dioxane, methyl ethyl ketone, pyridine, swells in benzene, ether, toluene" (359). "Soluble in ketones" (31). See directly below		
Nonsolvents/relatively unaffected by	"Carbon tetrachloride, ethyl alcohol, gasoline, lubricating oil" (359). "Insoluble in aromatic compounds, although swelling occurs; resistant to water, aqueous acids and alkalies, detergents and bleaches, gasoline and chlorinated hydrocarbons" (31). "These resins are hygroscopic and will absorb up to 0.6% moisture during storage" (362)	"Insoluble in alcohols, aliphatic hydrocarbons and mineral and vegetable oils . . . unaffected by water, inorganic salts and alkalis" (31)	31, 359, 362

C. Some Properties Associated with Polyacrylamide and Polyacrylic Acid

Property	Polyacrylamide	Polyacrylic Acid	Reference
Glass transition temperature (°C)	153	106	31, 363, 364
Density	1.302		363
Refractive index, n_D		1.53	87
Solvents/swelling	Water, morpholine	Water (deliquescent) (Ref. 31). Atactic: methanol, ethanol, ethylene glycol, methoxyethanol, dioxane, formamide, dimethylformamide, water, dilute alkali solutions; swells in dioxane at higher temperatures. Isotactic: dioxane/water (80:20)	31, 88, 363
Nonsolvents/relatively unaffected by	Hydrocarbons, alcohols, glycols, diethyl ether, tetrahydrofuran, esters, dimethylformamide, nitrobenzene	Ether, benzene, and cyclohexane (31). Atactic: aliphatic and aromatic hydrocarbons, esters, ketones. Isotactic: dioxane	31, 88, 363
Decomposed by		"Attacked by nitric and sulphuric acids, and by aldehydes, ketones, esters and chlorinated hydrocarbons"	31

[a] Summarized as described in Table 54.9.
[b] Standard D 1909-77.
[c] This material identified as solvent for polymethacrylates in general (88).
[d] This material identified as nonsolvent for polymethacrylates in general (88).

resins" derived from these esters are formulated into durable coatings that are applied by methods such as spraying, roller coating, or electrodeposition (16).

Table 54.2 lists production volumes for the major categories.

6.1.2 Analysis and Specifications

Acrylic fibers can be identified by traditional textile techniques (89, 90) or by sophisticated methods such as pyrolysis–mass spectrometry (372) and thermomechanical examination (373). The limited data available indicate that the amount of residual monomer may vary widely among different types and forms of plastics, latexes, and coatings. The International Agency for Research on Cancer (31) cites a number of methods for determining residual styrene, acrylonitrile, and methyl methacrylate in specific polymers. Free liquid or gaseous monomer can be determined in aqueous polymer latexes (or dispersions) by a gas chromatographic technique (368). Residual monomer analysis of coatings is included in a comprehensive general review categorized according to type of analytical technique (369). Related topics include monomer released to the atmosphere from latexes, paints, or coatings (31); migration of monomer from polymer packaging containers and dental or medical devices (370, 371); specifications for food-contact applications (92); and effluent products generated at elevated temperatures. Monomer content or its determination in specific polymers is also discussed in Sections 6.2 to 6.6.

6.1.3 Toxicologic Potential

Acrylic and modacrylic fibers have been occasionally connected with granuloma-like skin changes from embedded fibers or respiratory effects from the dust of synthetic fibers as discussed in Section 9.1.4.

The application of coatings, lacquers, or enamels containing acrylic or methacrylic esters or their derivatives frequently requires industrial hygiene evaluation. The problem typically arises from the carrier or supplemental ingredients in such formulations during their application; the end use coatings on the finished articles are essentially innocuous under conditions of normal use. Aqueous dispersions such as water-base acrylic paint present relatively little hazard in their application. However, the thinners in nonaqueous dispersions are generally organic solvents that are capable of causing toxic effects.

During the last several years residual acrylonitrile levels in various polymers derived from acrylonitrile have attracted comment. In 1977 the Food and Drug Administration banned (374) the use of soft drink bottles made from styrene–acrylonitrile copolymer with a residual acrylonitrile level of ≥ 3.3 ppm because the unreacted monomer might leach into the contents and become an indirect food additive (see Section 1.4, also Reference 375 concerning continuing discussion of possible migration of resin constituents). Subject to specifications, this copolymer is permitted in certain other food-related applications (92). Acrylonitrile levels (154) in acrylonitrile–butadiene–styrene and styrene–acrylonitrile resins have been reported as 30 to 50 ppm (30 to 50 mg/kg) and 15 ppm (15 mg/kg), respectively. Residual acrylonitrile levels in finished acrylic fibers "are usually well below 20 ppm" (376). Sampling of commercial acrylic and modacrylic fibers (154) has shown levels generally less than 1 ppm (1 mg/kg).

Residual styrene monomer has also been of interest, particularly in connection with the very extensive use of polystyrene in food packaging. However, a recent Canadian survey (377) showed the amount of monomer present in several types of food packaged in polystyrene containers to be ≤246 ppb, usually <50 ppb. Preliminary data with radio-labeled polystyrene indicate that some constituents of foods may penetrate into the polymer and enhance monomer migration (24).

Residual acrylate–methacrylate monomer has been detected in latexes and enamels as noted above, in various commercially prepared articles of polymethyl methacrylate or similar resins, and in monomer–polymer dough used for medical cement and dentures. The residual monomer in internally used medical cement has been associated with severe or even fatal reactions (31). Allergic reactions from residual monomer in commercially finished solid articles are uncommon except in the case of cold-cured dentures (160). However, heat-cured dentures that are incorrectly processed can cause "tissue sensitivity" (370).

Table 54.21 shows that polystyrene, polyacrylonitrile, polymethyl methacrylate, and their derivatives yield toxic products when thermally decomposed. The primary life hazards of burning polystyrene are carbon monoxide and, perhaps even more, dense smoke. Flame-retarded polystyrene also burns with sufficient heat and oxygen. "Acrylic fibers" generally contain small amounts (5 to 15 percent) of comonomer whereas "modacrylic fibers" contain 15 to 65 percent comonomer (see Section 6.1.1). Nonacrylic degradation products may be biologically significant depending on type and concentration.

In general, the most noteworthy thermal degradation product of polymers containing acrylonitrile units is hydrogen cyanide. Except for nitro compounds, organic materials containing nitrogen atoms commonly release hydrogen cyanide when heated in excess of 600°C (392). Evolution of hydrogen cyanide often starts at about 500°C and increases with rising temperature, although polyacrylonitrile may under nitrogen yield hydrogen cyanide as low as 300°C. Conditions of limited air supply—which often occur in smoldering combustion—promote formation of hydrogen cyanide. Oxidation reactions, particularly flaming combustion, may consume the hydrogen cyanide to a greater or lesser degree. Table 54.22 shows comparative data obtained at different temperatures with polyacrylonitrile, styrene–acrylonitrile, and acrylonitrile–butadiene–styrene under vacuum, whereas Figure 54.6 shows the effect of increasing air supply on the formation of hydrogen cyanide from polyacrylonitrile at 500, 650, and 800°C. Several decades ago tests conducted at the Underwriters Laboratories showed that "when these acrylic fiber fabrics burn in an excess of air (oxygen

Table 54.21. Overview of Inhalation and Thermal Degradation Data for Polystyrene and Some Acrylic Polymers[a]

			Type of Study					
Polymer	TLV	Ambient Temperature	Elevated Temperature	Effluent Analysis	Species Exposed	Description of Material	Procedure	Observations/Conclusions	Reference
Polystyrene (PS)	No specific TLV (see Section 1.4)	X			Human	Aerosols of 0.188, 0.557, or 1.305 μm particles	Comparison of respiratory rate, tidal volume and flow rates: "... the less the respiratory rate, the more the deposition," particularly with 1.305 micrometer particles. Aerosols of two smaller size particles showed "low deposition rates"	378
			X		Rat		DIN 53436 protocol (Table 54.10)	600°C degradation of polymer lethal, caused 74–80% carboxyhemoglobin (COHb); see Table 54.7. Low evolution of CO at 300–500°C	62
			X		Rat	Rigid expanded foam	Exposure 30 min, to products generated 300–600°C, German protocol DIN 4102	Exposures to products generated 300–600°C caused COHb levels ≤41%, rats survived	379, 380
			X		Rat		Slow and rapid combustion conditions (Table 54.10)	PS in the middle range of toxic effect when degraded under rapid combustion conditions and least toxic when degraded under slow combustion conditions	109
			X	X	Mouse		Fixed, rising, and incremental temperature programs (Table 54.10)	Time to death among the longest when series of polymer products tested; see Table 54.8. Influence of temperature and air flow evident in incremental temperature program; no toxic effects evident ≤400°C furnace temperature	66, 381

		Species	Material	Conditions	Comments	Ref.
X	X	Mouse	Rigid foam, commercial material	Fixed and rising temperature programs (Table 54.10, Ref. 66)	CO "an important if not the principal toxicant," levels in fixed temperature program in general the highest among the 6 cellular polymers tested	98
X	X	Rat	High impact PS containing 7% rubber; similar material, flame-retarded; foamed PS	Exposure 30 min, to pyrolysis products (maximum furnace temperature that does not result in autoignition) or to combustion products (continuous flame, ~25 to 150°C > pyrolysis)	"Exposure to the pyrolysis products of the three polystyrene formulations did not produce any fatalities at any dose level short of the explosive limits of the vapors in air; no exposures were attempted in such a range." Exposure to combustion products comparable in lethal effect to that observed with cellulose materials. "Heavy" smoke frequently observed	382
X	X	Rat		Exposure 100 min, to products generated with DIN 53436 protocol at unspecified temperature	PS products caused no deaths but produced quickest reaction in "swim" test with series of 10 polymers, suggesting strong incapacitant effect; average COHb, 3%	383[b]
X	X	Rat, mouse		Degradation products generated at 350°C, acute and subacute tests	Decrease in reduced glutathione immediately after exposure, followed by rebound effect	384
X	X		Powder		Polymer completely combusted at 450°C, major component of combustion in both gas and liquid phases was styrene monomer. Liquid-residue components also included benzene, toluene, benzaldehyde, phenol, diphenylethane, plus styrene and benzene derivatives	102 and as cited therein
X	X			As described in Table 54.10	In mg/g sample, at 50 or 100 l/hr air flow, respectively: 590 or 619 mg CO_2; 207 or 178 mg CO; 16 or 18 mg C_2H_4; 6 or 13 mg C_2H_4; 7 or 7 mg CH_4	104

Table 54.21. (Continued)

| Polymer | TLV | Type of Study | | | Species Exposed | Description of Material | Procedure | Observations/Conclusions | Reference |
		Ambient Temperature	Elevated Temperature	Effluent Analysis					
			X	X			Degradation in 300–400°C range yields 41% monomer, increased monomer with increased pressure of nitrogen. At 500–1200°C, C₁–C₆ fragments appear		Cited in 105
			X	X			Approximately 3–4× greater evolution of soot from PS than from polyethylene at high temperatures with limited air supply; less evolution of benzo[a]pyrene		49
			X	X		Number average molecular weight = 233,000	Pyrolysis at 425–1025°C: at 425°C, monomer content of pyrolyzate = 64%; evolution of monomer reaches maximum of 84% around 725°C and thereafter decreases; ethylbenzene, toluene, benzene, ethylene, and acetylene formed at higher temperatures		385
Polyacrylonitrile (PAN)	As polystyrene above	X			Human		Comparison of ventilatory symptoms and changes in ventilatory function in (1) 68 textile workers exposed to polyacrylonitrile fibers only (1.04 mg/m³). (2) 30 workers with previous exposure to cotton dust in same mill, and (3) 77 workers with previous exposure to hemp in another polyacrylonitrile fiber mill (0.42 mg/m³); prevalence of dyspnea significantly lower in workers exposed to synthetic fibers only; all groups showed mild reduction in ventilatory capacity, workers exposed in 1963 to hemp had practically same capacity in 1973		386
		X			Rat	Polymer, molecular weight = 25,000	Exposure 1 hr daily to 1.3 mg/l for 45 days	Mucous membrane irritation of respiratory tract, also albuminuria, changes in thyroid and blood counts	387
			X		Rat		As described above, this table	Products most lethal in series of 10 polymers tested; average COHb, 3%	383

	Animal	Material	Conditions	Results	Reference
X	Rat	Acrylic fabric	Exposure 60 min. Flame ignition plus supplemental conductive heat generated smoke in a 110-liter dual compartment exposure chamber	Fuel load of 8 g produced 50% mortality, compared to 7, 9, 10, and 11 g for nylon, newsprint, cotton, and wood, respectively. Flame ignition alone did not generate lethal smoke with acrylic, acetate, polyester, or wool but did with cotton	390
X	Rat, mouse, rabbit	Polyacrylonitrile and modacrylic fibers	Decomposition products obtained with 300-W heater, from 2 to 20 g lethal <18 min; high concentration of cyanide detected in blood of exposed animals; lack of additive effect with CO		249, 388, 389
X	Mouse	Fiber	Inhalation of decomposition products generated at 350, 500, or 750°C (as described Tables 54.10 and 54.14)	Products generated at 750°C lethal with toxicologically significant amounts of CO and HCN, particularly HCN. Diminished effects at 500°C, survival at 350°C	101, 232
X			Review: "ranking of nitrogen-containing materials based on the propensity of toxic gas generation can be variable because the quantity of HCN generated varies widely depending on the conditions of combustion"		391
X		Commercial polyacrylonitrile containing 24% nitrogen	Combustion pyrolysis in air and under nitrogen	Increased air flow results in decreased HCN; see Figure 54.6. Under nitrogen, evolution of HCN noted at 300°C, increases with temperature, and is proportional to nitrogen content at 900°C; see Figure 54.7	392
X			Pyrolysis under vacuum and with oxygen	Varying percentages of HCN and other gases evolved at 500–1000°C; see Table 54.22	107, 393, 394
X		Yarn	Samples decomposed at 400, 600, or 800°C under either nitrogen or in air: increased evolution of HCN with increased temperature, in both air and nitrogen; also 16 other nitriles. Earlier test with only 800°C degradation showed proportionately more HCN evolved from "acrylic fiber" than several other nitrogen containing polymers		395, 396

Table 54.21. (Continued)

Polymer	TLV	Ambient Temperature	Elevated Temperature	Effluent Analysis	Species Exposed	Description of Material	Procedure	Observations/Conclusions	Reference
			X	X			As described in Table 54.10	In mg/g sample, at 50 or 100 l/hr air flow, respectively: 630 or 556 mg CO_2; 132 or 108 mg CO; 59 or 56 mg HCN; 8 or 4 or 7 mg C_2H_2	104
			X	X		Woven fabric	HCN determined colorimetrically from degradation at 600, 800, and 1000°C		397
			X	X		Number average molecular weight = 13,000–43,000	Degradation in 280–450°C range: cyanogen, HCN, acrylonitrile, acetonitrile, vinyl acetonitrile, plus 11 minor products identified; ammonia detected only when traces of water present; black residue[a] (polynaphthyridine). Concise literature review		398
			X			Fibers, films, or powders (review)	Acrylic heated to ~150°C slowly develops intense colors: "the extent of color formation will depend on the environment of the acrylic segment in the chain: the presence or absence of oxygen; the type of co-monomers present; and the inevitable presence of impurities, polymerization defect, and molecular chain ends"		352
			X	X		Polymerization with azo initiator, dimethylformamide solution	Degradation 180–500°C: HCN the most important product "under air at temperatures below 250°C. The amount of HCN evolved is temperature dependent . . . cyclization competes with the evolution of HCN"		399
			X	X			As described in Table 54.12	Pyrolysis at 150–950°C showed slow gradual weight loss	48
			X	X		Acrylic fabrics	"Under conditions of oxygen deficiency, the combustion and decomposition products of the acrylic fiber fabrics include carbon dioxide and small amounts of carbon monoxide, hydrocyanic acid, ammonia, and oxides of nitrogen. When the acrylic fiber fabrics are subjected to thermal decomposition (pyrolysis) in at-		401

Polymer			Animal	Conditions	Observations	Cited in
	X	X			...mospheres of nitrogen (air absent), the volatile products include relatively large amounts of hydrocyanic acid, ammonia, hydrogen and methane, together with smaller amounts of unsaturated hydrocarbons	
Styrene–acrylonitrile (SAN) — As polystyrene above	X				Degradation <200°C: "colours through yellow, orange, red and black"	105
	X		Rat	As described in Table 54.10	Degradation products formed at 400 and 500°C lethal. Significant amounts of COHb at 500 and 600°C. See Table 54.7	62
	X		Rat	Slow and rapid combustion conditions as described in Table 54.10	SAN ranked among the more toxic polymers when tested under rapid combustion conditions, in the middle of the series of 10 polymers when tested under slow combustion conditions	109
	X			As described above, this table		107, 393, 394
	X			Pyrolysis under nitrogen at 900°C	Evolution of HCN approx. proportional to nitrogen content; see Figure 54.7	392
Acrylonitrile–butadiene–styrene (ABS) — As polystyrene above	X		Rat	Slow and rapid combustion conditions as described in Table 54.10	ABS ranked in the middle or slightly more toxic range in series of 10 polymers, under both types of testing conditions	109, 402
	X		Mouse	Fixed and rising temperature programs as described in 54.10	Time to incapacitation and death slightly shorter than the median among polymer products tested in this series; see Table 54.7	66
	X			As described above, this table		107, 393, 394
Fire-retardant ABS (402)				As described in Table 54.12	Most sample weight changes at 400–500°C; decomposition causes "styrene pollution"	48

Table 54.21. (Continued)

Polymer	TLV	Ambient Temperature	Elevated Temperature	Effluent Analysis	Species Exposed	Description of Material	Procedure	Observations/Conclusions	Reference
						Plastic sheet averaging 3 mm thick	Optical density test in NBS smoke chamber	Dense smoke, extremely so during flaming, also marked during smoldering	47
							As described above, this table		391
			X			Commercial sheet	As described in Table 54.17	Thermographs show appreciable weight loss starting area 250–300°C, <20% sample remaining at 500°C	55
						Waste pipe, 6-in. diameter		"More than 600 ft² of red oak, all burning at its maximum rate, would be required to produce the same rate of smoke release as the 8 ft of ABS pipe"	403
Polymethyl methacrylate (PMMA)	As polystyrene above	Industrial exposure			Human		Reference to reports, dated 1946–1962, which indicated some types of industrial work associated with fine, irritating dust or "sweetish vapours containing monomers"; no atmospheric concentrations stated		Cited in 251
			X		Mouse		Fixed and rising temperature programs, as described in Table 54.10	Time to incapacitation and death shorter than the median among polymers tested in this series; see Table 54.8	66
			X	X			As described in Table 54.10	PMMA yielded relatively minor amounts of fatty acids and acrolein, evolved formaldehyde at 700°C; see Figure 54.2	103
			X	X			As described in Table 54.10	In mg/g sample, at 50 or 100 l/hr air flow; respectively: 738 or 796 mg CO_2; 173 or 157 mg CO; 32 or 17 mg NH_3; 21 or 18 mg HCN; 20 or 16 mg CH_4; 13 or 10 mg C_2H_4; 4 or 9 mg C_2H_2	104

	X	X	At 425°C, pyrolyzate contains 99% monomer; monomer content of pyrolyzate decreases to 81% at 725°C and 20% at 1025°C as gaseous products increase correspondingly	385
	X	X	Degradation at 170–300°C yields 100% monomer; at 246–1200°C, monomer yield decreases as temperature is raised and a complex series of fragmented products form	Cited in 105
	X	X	PMMA pyrolyzed to monomer and formed large amounts of aliphatic aldehydes (130.2 ppm HCHO/0.1 g) As described in Table 54.12	48
	X	X	Products identified from pyrolysis in presence of air at 50–500°C were methyl methacrylate, methyl alcohol, acetone, n-C_4H_{10}, CO, CO_2, HCHO, propylene, isobutylene, acetic acid, H_2O. Depolymerization caused occasional explosions	404
	X	X	Simulated "worst possible" molding conditions (obvious resin degradation) yielded methyl methacrylate with much smaller amounts of ethyl acrylate and methanol	405
	X	X	As described in Table 54.10 In mg/g sample, at 50 or 100 l/hr air flow, respectively: 738 or 796 mg CO_2; 173 or 157 mg CO; 32 or 17 mg NH_3; 21 or 18 mg HCN; 20 or 16 mg CH_4; 13 or 10 mg C_2H_4; 4 or 9 mg C_2H_2	104
Polyacryl-amide	X	X	As polystyrene above	
	X	X	"Heating the polymer to 210°C in the absence of oxygen causes a weight loss because of loss of water. Continued heat (210–300°C) causes degradation of the amide group forming ammonia and water. As the temperature is increased to 500°C, a dark dry flake is formed containing only 40% of the original sample weight"	363

4333

Table 54.21. (*Continued*)

Polymer	TLV	Type of Study				Description of Material	Procedure	Observations/Conclusions	Reference
		Ambient Temperature	Elevated Temperature	Effluent Analysis	Species Exposed				
Polyacrylic acid	As polystyrene above							"When poly(acrylic acid) is heated in nitrogen at the rate of 10°C/min anhydride formation starts at 250–260°C and the polymer becomes insoluble. On further heating the anhydride decomposes. At about 400°C this decomposition becomes rapid"	364
"Cyanoacrylate glues"		X			Human			"These glues are known to be irritating to the eyes and respiratory tract when vaporized, but allergic reactions are considered [unlikely due to immediate bonding of acrylate to skin keratin]. . . . Normally water vapor in the atmosphere will polymerize any vaporized cyanoacrylate and remove the risk of irritation; but if there is not enough water vapor the cyanoacrylate vapor level increases in front of the operator doing close precision work"	406
"Vinylacrylic latex"			X	X	Rats	Gypsum wall board with 4 coats of latex paint	Exposure 15 min to products generated from material surface at 300°C with 100 l/min air flow: no significant difference in mortality from that observed in comparable tests with wallboard alone	407

[a] Summarized as described in Table 54.10.

[b] The polymers tested in this series were polystyrene, polyacrylonitrile fabric, pinewood, woolen fabric, laminated molded Bakelite, molded melamine resin, polyester fabric, and polyurethane foam; tests were conducted with strips measuring 1 m in length, 15 mm in breadth, and 2 mm in thickness. Except for polyacrylonitrile fabric, pinewood was considered the greatest acute hazard in fires. The decrease in swimming time with polystyrene combustion was attributed to the evolution of monomeric styrene.

Table 54.22. Percentage Composition of Gases Given Off Under Vacuum by PAN, SAN, and ABS at Different Temperatures[a]

Gas	Polyacrylonitrile						Styrene–Acrylonitrile						Acrylonitrile–Butadiene–Styrene					
	500°C	600°C	700°C	800°C	900°C	1000°C	500°C	600°C	700°C	800°C	900°C	1000°C	500°C	600°C	700°C	800°C	900°C	1000°C
HCN	15.7	12.6	15.0	31.5	30.9	35.9	15.2	15.2	16.8	16.7	12.0	9.0	6.8	8.3	9.7	14.3	11.6	7.9
H_2	28.6	31.5	32.0	30.3	40.6	42.6	24.1	27.8	24.3	34.0	56.3	70.9	17.2	19.4	24.6	31.0	52.9	68.2
CH_4	21.6	26.9	25.8	24.4	20.8	18.4	5.8	10.9	21.0	28.2	23.0	15.3	23.9	25.7	35.6	35.3	28.1	18.6
C_2H_6	4.4	4.3	2.5	0.1	Traces	Traces	0.9	0.8	0.8	0.6	0.2	Traces	10.0	5.5	2.7	0.9	0.2	Traces
C_3H_8	2.2	1.2	0.3	0.1	Traces	—	0.5	0.5	0.3	0.1	—	—	2.6	1.2	0.4	0.1	—	—
C_4H_{10}	0.2	0.1	Traces	—	—	—							0.7	0.2	Traces	—	Traces	0.1
C_2H_4	3.5	1.2	1.3	0.3	5.6	0.8	12.1	11.7	10.4	7.8	2.6	1.9	4.7	6.7	7.2	4.7	0.4	0.7
C_3H_6	3.9	2.9	3.8	1.5	0.6	0.2	3.8	5.5	3.2	0.5	0.1	Traces	9.4	9.3	4.3	0.6	0.4	0.1
C_4H_8	0.6	0.5	0.1	Traces	Traces	—	0.7	1.0	0.4	Traces	—	—	4.4	3.2	0.6	Traces	0.1	Traces
C_5H_{10}	0.4	0.2	—	—	—	—	0.3	0.3	Traces	—	—	—	2.0	1.7	0.1	—	0.1	—
C_2H_2	—	1.9	2.3	2.0	—	0.8	10.4	10.1	8.5	4.8	1.7	0.8	6.6	8.2	7.0	4.7	2.0	1.9
Butenes	0.4	0.5	0.3	0.3	0.1	—	0.3	0.6	0.6	0.2	—	—	0.2	0.3	0.7	2.6	2.2	1.8
Benzene	Traces	0.1	0.2	Traces	0.7	0.5	0.1	0.2	0.9	1.1	2.5	2.1	0.1	0.1	0.1	0.1	Traces	Traces
Toluene	—	—	—	—	—	Traces	—	Traces	0.1	Traces	Traces	—	1.0	0.9	0.2	Traces	0.1	Traces
Cyclopentene	0.7	0.5	0.2	0.1	—	—	0.1	0.2	0.2	Traces	—	—	0.2	0.5	0.4	0.2	0.1	Traces
CO_2	Traces	Traces	Traces	Traces	Traces	Traces	0.1	0.2	0.4	0.2	—	Traces	6.4	4.6	4.8	5.1	1.6	0.7
CO	—	—	—	—	—	0.8	4.7	1.0	—	—	—	Traces						
NH_3					0.7													
N_2	17.8	15.6	15.9	8.9	0.7	—	21.0	14.2	12.1	5.8	1.5	—						
Volume total (in l/kg)	90	128	264	435	512	524	1.9	8.5	63.2	208	422	616	17.6	53	139	302	434	695

[a] Summarized from the data of M. Chaigneau, "Analyse par spectrométrie de masse," *Analusis*, **5**(5), 223–227 (1977) and printed by permission of Masson Publisher, Paris. The styrene–acrylonitrile polymer is described as 21% acrylonitrile and 80% styrene. The acrylonitrile–butadiene–styrene polymer is described as 22% acrylonitrile, 19% butadiene, and 59% styrene (Ref. 394).

excess), the chief constituents of the gaseous products of combustion and thermal decomposition include carbon dioxide, together with small amounts of oxides of nitrogen" (60, 401).

The combustion toxicity of both styrene–acrylonitrile and acrylonitrile–butadiene–styrene obviously reflects their respective two or three monomer origin. Acrylonitrile–butadiene–styrene is noted for its capacity to produce dense smoke. At high temperatures in inert atmospheres, the amount of hydrogen cyanide evolved is likely to be a direct function of nitrogen content (see Figure 54.7).

Polymethyl methacrylate softens as it burns and depolymerizes with relatively low smoke formation by unzipping of the molecular chain. The pyrolyzate typically contains few high molecular weight products (38).

Some concern has recently been expressed about possible general effects of exposure to acrylamide (408) in polyacrylamide and the relative lack of data associated with exposure to polyacrylic acid (31). In many industrial situations where low concentrations of these polymers are present, significant exposure to residual monomer is unlikely. Partially polymerized acrylamide grout is a potential source of exposure to acrylamide both occupationally and from release into the environment. Polluted well water was connected with five cases of

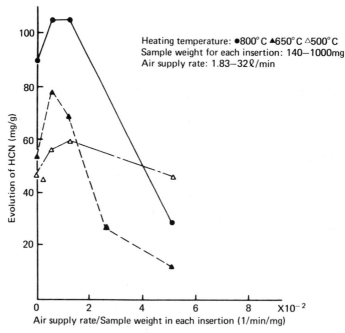

Figure 54.6 Evolution of HCN from polyacrylonitrile in air. From Morikawa (392). Reproduced by permission of *Journal of Combustion Toxicology*, Technomic Publishing Co., Inc., Westport, Conn.

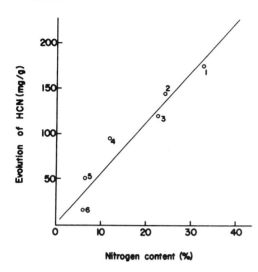

Figure 54.7 Relation between HCN evolution and nitrogen content of polymers at 900°C under nitrogen. Materials: 1, melamine resin; 2, polyacrylonitrile; 3, urea resin; 4, nylon 6; 5, styrene–acrylonitrile copolymer; 6, polyurethane flexible foam. From Morikawa (392). Reproduced by permission of *Journal of Combustion Toxicology*, Technomic Publishing Co., Inc., Westport, Conn.

acrylamide toxicity in a Japanese family in a 1975 report. Grout as a potential source of acrylamide toxicity would decrease in importance if the monomer is replaced with polyacrylamide to be cross-linked in situ, as recently proposed (discussion and citations in Reference 409).

6.1.4 Workplace Practices and Standards

Inhalation of fine dust, aerosol, or fume should be minimized. Solvent and/or residual monomer released during processing should be controlled by exhaust ventilation or other appropriate means. In the event of fire, precaution should be taken against inhalation of toxic gases and dense smoke. Combustion of polymers or copolymers of styrene may yield unusually dense smoke. "Smoldering combustion" in environments containing limited amounts of oxygen may be particularly hazardous with acrylic polymers. Disposal of water-soluble polymers may be accomplished by burial in sanitary landfill or by passage through water-treatment plants, depending on federal, state, or local regulations.

6.2 Polystyrene

The CAS No. is 9003-53-6. Polystyrene is also known as PS, Dylene®, Styron®, and Trycite®.

6.2.1 Preparation

Styrene can be made to yield high molecular weight polymer by all known methods of free radical initiation; specific properties of the polymer depend on the method of initiation (353). Various commercial processes use mass, suspension, solution, or emulsion methods, or combinations thereof (5, 10, 31, 363, 353, and citations therein). Mass or bulk techniques, which appear to be gaining in popularity, permit more efficient removal of residual impurities (365). Mass polymerization is usually initiated thermally; process temperatures typically range from 80 to 180°C.

Suspension polymerization is usually initiated with organic peroxides. Solution and suspension processes permit greater control of molecular weight but polymer clarity is impaired. Also, solvent recovery or drying is needed. Emulsion polymerization leaves a large soap/emulsifying agent residue that is permissible in latex but undesirable in other applications. All polymerization of styrene is exothermic.

6.2.2 Molecular Weight, Properties, and Applications

The molecular weight of commercial polymers has been reported as 50,000 to 200,000 (10) and 200,000 to 300,000 (14, 353).

Commercial products have been traditionally described as atactic (5), but syndiotactic segments may predominate (10). General purpose polystyrene is a clear, glassy, rigid, amorphous thermoplastic that has a characteristic metallic ring when dropped (10, 365). "Impact polystyrene" describes a product in which a rubber, most commonly polybutadiene, has been added to reduce brittleness (or increase toughness) but which keeps most of the characteristic properties of polystyrene (353, 365). High impact polystyrene and super impact polystyrene may contain 5 to 15 percent butadiene rubber. Foams may contain a blowing agent such as n-pentane or a fluorocarbon propellant. See Tables 54.4, 54.6, and 54.20, as well as Figure 54.1, for comparative data.

The major markets for polystyrene per se are packaging and housewares, including disposable dinnerware and containers for "fast food" and drugs. Specific applications include yogurt and cottage cheese containers, unit packages for jams and creams, hot and cold drink vending cups, and egg cartons. Both impact and general polystyrene are widely used in components of appliances, electronic and tape reels, furniture, and toys.

6.2.3 Analysis

Gas chromatography with headspace analysis over the dissolved polymer can be used to determine residual styrene at a level of 1 ppm (150). Migrant styrene (377, 410) in food packaged in polystyrene containers has been detected at levels of 3 to 246 ppb.

6.2.4 Oral Toxicity

Long-term feeding tests with polystyrene are known to have been conducted, but few details are available in the published literature. Rats fed a diet of 4 percent polystyrene-based plastic (Styron® 475 and Styron® 666) for an unspecified length of time showed no abnormal effects (411). A 1976 French trade report (412) mentions two earlier 2-year feeding tests with rats, neither of which revealed any toxic effects. The Huntingdon Research Centre conducted a study at a dietary level of 5 percent polystyrene with a sample that contained 120 ppm monomer; a BASF study used a 10 percent dietary level. Khamidullin et al. (413) report an 11-month test with "grade PS-Sp III suspension polystyrene,"* in which rats were given an aqueous extract that apparently contained up to 0.2 mg/l styrene monomer. No significant changes in behavior and general condition were observed; at histopathologic examination, "the spleen . . . contained considerable amounts of iron and hemosiderin, while the liver, kidneys and stomach showed a reactive proliferation of connective tissue." The implication of these changes is unclear.

6.2.5 Inhalation and Thermal Degradation

Under fire conditions, carbon monoxide and smoke are considered the primary life hazards. Effenberger's work (383) suggests that sustained decomposition of a sufficient fuel load at smoldering temperatures may release immobilizing amounts of styrene, but these conditions apparently did not develop in the rising temperature program of the Hilado team. See Table 54.21 for details.

6.2.6 Carcinogenic Potential/Implant Studies

Various implant studies have produced tumors that appear consistent with "solid state" carcinogenesis ascribable to physical rather than chemical factors (see Section 1.3). As summarized by the International Agency for Research on Cancer (31), "subcutaneous implantation of polystyrene discs, rods, spheres, or powder in rats induced local sarcomas, the incidences of which varied with the size and form of the implant." Additionally, a study with mice given intraperitoneal injections of polystyrene latex showed no carcinogenic effect (414a).

"No convincing evidence" for the carcinogenicity of styrene monomer was demonstrated in lifetime exposure of rats and mice dosed by gavage at levels as high as or exceeding the maximum tolerated dose (414b). Additional data plus some as yet unclarified indications of mutagenicity from the monomer in animals and also humans (who showed chromosome aberrations) have been discussed by the International Agency for Research on Cancer (31).

---

* Described as "containing 99.54% polystyrene, 0.05% zinc stearate and 0.15% residual monomer styrene."

6.2.7   Biodegradation

Biodegradation may occur in oligomers (415), but the polymer is considered highly resistant to biological decay (76, 124, 125, 416, 417). Canadian investigators (416) found degraded polystyrene more resistant to microbial attack than degraded polyethylene or degraded polypropylene. Radioactive tracer studies with a photodegraded polystyrene–vinyl ketone copolymer indicated a slow rate of biodegradation from bacteria. This slow rate was compared to "the known slow rates of biological degradation of lignin and other natural products containing aromatic residues."

6.3   Polyacrylonitrile

Polyacrylonitrile (CAS No. 25014-41-9) is the basic structure in *acrylic** and *modacrylic* fibers. Product names include PAN,* Orlon®, Acrilan®, Creslan®, and Zefran®.

$$\left[\text{--CH}_2\text{--}\underset{\underset{\text{CN}}{|}}{\text{CH}}\text{--}\right]_n$$

6.3.1   Preparation

Acrylonitrile monomer is generally copolymerized with other ingredients by free radical addition methods, primarily in suspension or solution with water or organic solvents (6, 14, 31, 418). As indicated above, the exact composition of "acrylic fibers" varies widely. Materials identified in patent literature as used with acrylonitrile include among others, vinyl acetate, vinyl chloride, styrene, isobutylene, acrylic esters, and acrylamide (6). A typical acrylic fiber may contain 85 to 91 percent acrylonitrile,* 7 to 8 percent of a comonomer such as methyl acrylate or methyl methacrylate, plus 2 to 5 percent other comonomers and additives such as dyeing aids, pigments, optical brighteners, stabilizers, and fire retardants (31). In modacrylic fibers, the nonacrylic moiety "consists of vinylidene chloride, vinyl chloride, vinyl bromide and other vinyl monomers. . . . The modacrylic fiber produced in the largest quantity is composed of 37% acrylonitrile, 40% vinylidene chloride, 20% isopropyl acrylamide and 3% methyl acrylate" (31). In 1979 marketed modacrylics were described as 25 to 60 percent comonomer (358).

---

* Acrylic fiber specifies "a manufactured fiber in which the fiber-forming substance is any long chain synthetic polymer composed of at least 85% by weight of acrylonitrile units." Modacrylic fiber specifies "a manufactured fiber in which the fiber-forming substance is any long chain synthetic polymer [except elastomeric substances and monohydric alcohol-acrylic acid esters] composed of less than 85 percent but at least 35 percent by weight of acrylonitrile units" (130). The term PAN is used as a short form of *polyacrylonitrile* and as the name for Bayer's modified filament fiber.

6.3.2   Molecular Weight, Properties, and Applications

The molecular weight for acrylic and modacrylic fibers has been described as 100,000 to 150,000 (31). "Continuous filament" contains molecules of different molecular weight; one calculation (weight basis) has shown about 45 percent of the molecules over 100,000, 33 percent from 50,000 to 100,000, and the remainder less than 50,000 (6).

Acrylic fibers are noted for their wool-like texture, hydrophobic properties, resistance to shrinkage and weathering, and versatility (dyeability, crimping capacity). Modacrylics have the additional property of flame resistance but have a much lower softening point than acrylics (31, 358). See Tables 54.4, 54.6, and 54.20, as well as Figure 54.1, for additional data.

Industrial applications of acrylic fibers are as filters for dry substances, in air pollution control, and in dye nets. Other pertinent uses are fiberfill batting, sandbags, and bagging. Industrial uses of modacrylic fibers include paint roller covers, filters for aqueous substances, and chemical-resistant twine. Both of these fiber types are used for felts, webbing, protective fabrics, carpets, upholstery, and draperies (31).

6.3.3   Analysis

Among the several methods for detecting residual acrylonitrile monomer (described in References 20, 31, and 369), the headspace gas chromatography of Steichen (150) would appear notable for sensitivity (to 0.5 ppm) and versatility. The U.S. Consumer Product Safety Commission (154) used high pressure liquid chromatography to determine residual acrylonitrile in acrylic fibers. See also Section 6.4.3.

6.3.4   Oral Toxicity, Skin and Eye Irritation

A 1962 Russian report (387) indicates that rats survived acute oral doses of 2 to 3 g polyacrylonitrile (molecular weight = 25,000 to 50,000) with little effect. Daily doses of 0.25 to 0.5 g/kg to rats for 6 months were reported to produce changes in the liver, kidneys, and thyroid that were reversible within the next 2 months. Skin and eye tests in rabbits produced no irritation.

General clinical experience in humans (160) indicates that acrylic fibers are not associated with allergic contact dermatitis. Representative experimental tests* have produced isolated reactions from a proposed surface modifier (421) or other additives (422–425), but reactions from commercial control materials were consistent with effects from occlusion.

---

* Methods have been described by Fleming (419) and more recently by Hood and Ivanova-Binova (420).

According to a 1974 report (426), none of eight samples of 100 percent acrylic fabric had a free formaldehyde content >30 ppm. This study involved 112 American made clothing samples of 39 types; 18 samples had a free formaldehyde content >750 ppm. (Formaldehyde was not associated with the intrinsic nature of any fabric but with finishing agents and, very briefly, "manufacturing techniques.")

6.3.5   Inhalation and Thermal Degradation

The one available animal inhalation study on polyacrylonitrile "dust" (387) describes respiratory irritation and some pathologic changes in rats exposed to concentrations of 1.3 mg/l of the same relatively low molecular weight polymer for oral and irritation tests. Workers (386) exposed to polyacrylonitrile fiber dust at mean concentrations of 0.42 or 1.04 mg/m^3 showed mild acute reductions of ventilatory capacity, but this response did not appear to be dose related. Several clinical reports that refer to acrylic and other synthetic fibers but lack data on exposure concentrations (426, 427), or are related more generally (428, 429), are discussed in Section 9.1.4.

Acrylic polymers are subject to thermal degradation at varying temperatures >150°C. They may evolve hydrogen cyanide, particularly under conditions of limited air supply and increasing temperature. In bioassay tests, degradation products of acrylonitrile polymers tend to be relatively more toxic than those of other polymers *under these experimental conditions*.

Table 54.21 presents a selective summary of studies on thermal degradation. Table 54.22 contrasts the degradation products of polyacrylonitrile with styrene–acrylonitrile and acrylonitrile–butadiene–styrene at different temperatures under vacuum.

6.3.6   Carcinogenic/Mutagenic Potential

Overall, no known data, directly or indirectly, implicate acrylic fibers or modacrylic fibers. (The issue is less clear-cut with modacrylic than with acrylic fibers, simply due to the more variable nature of their composition as discussed earlier.) Acrylonitrile monomer acts as a mutagen in predictive screening tests (*Salmonella typhimurium* and *Escherichia coli*); preliminary data indicate that it causes tumors in animals tests and suggest that extended exposure may elevate tumor incidence in humans (see Chapter 39, Volume 2B, and Reference 31). Given this state of knowledge, its practical significance for the polymer depends on the levels of residual acrylonitrile monomer (see Section 6.1) and how tightly the monomer is bound within the polymer matrix.

Bacterial mutagen tests are not recognized as definitive indications of carcinogenicity; however, the positive results from the monomer as cited above can be contrasted with the negative results from a test with polymerized acrylonitrile. The polymer, dissolved in dimethyl sulfoxide and tested at concentrations up to 10,000 μg polymer per plate (431), was not mutagenic to *Salmonella typhimurium* (strains TA 1535, TA 100, TA 1537, TA 1538, and TA 98, all tests performed with or without S·9 rat-liver homogenate activation).

6.3.7 Biodegradation

Acrylic fibers are considered resistant to attack by insects or microorganisms in the soil (358).

6.4 Styrene–Acrylonitrile and Acrylonitrile–Butadiene–Styrene

Styrene–acrylonitrile (CAS No. 9003-54-7) is also called SAN* and Lustran® (which is also ABS):

$$\left[\left[-CH-CH_2\right]_x \left[CH_2-CH\atop CN\right]_y\right]_n$$

where x may vary from 65 to 80 percent, and y from 20 to 35 percent.

Acrylonitrile–butadiene–styrene (CAS No. 9003-56-9) is available as ABS,* Lustran® (which is also SAN), Kralastic®, and in the Bakelite® series:

$$\left[\left[-CH-CH_2\right]_x \left[CH_2-CH\atop CN\right]_y + \left[CH_2-CH=CH-CH_2\right]_z\right]_n$$

where x, y, and z represent not a terpolymer[†] but a blend of SAN as above or its monomers with 5 to 30 percent polybutadiene or a butadiene-based elastomer.

6.4.1 Preparation

Styrene–acrylonitrile copolymers are manufactured from the acrylonitrile and styrene monomers by emulsion, suspension, and continuous mass processes

---

* See Reference 31 for a list of trade names. Some, such as Lustran®, may be used in product lines for both styrene–acrylonitrile and acrylonitrile–butadiene–styrene. Kralon® identifies rubber-modified polystyrene and acrylonitrile–butadiene–styrene, whereas Kralastic® denotes high impact acrylonitrile–butadiene–styrene (81).

† Acrylonitrile–butadiene–styrene plastics are somewhat confusingly described as both copolymers (on the basis of preparation) and terpolymers (on the basis of monomer composition). Morneau et al. (360) simply refer to "ABS polymers" or "ABS plastics" and state that ABS "is *not* a random terpolymer" (see References 4, 5, and 10 for discussion of random versus other types of copolymers).

(359). Additional typical ingredients may include water plus relatively small amounts of rosin soap, *tert*-dodecylmercaptan, and $K_2S_2O_8$ in an emulsion system. Water plus small amounts of dipentene, di-*tert*-butyl peroxide, acrylic acid–2–ethylhexyl acrylate (90:10) copolymer, and insoluble inorganic salts (most frequently calcium carbonate and calcium oxide) are typical additional ingredients in suspension polymerization. Commercial automated systems tend to be more complex.

In acrylonitrile–butadiene–styrene, styrene and acrylonitrile combine with butadiene in a complex blend. Emulsion and suspension processes have dominated manufacture (10, 360), but bulk polymerization has recently become important (360). The bulk process has the advantage of minimal waste water treatment and less energy consumed per kilogram of product; however, monomer conversion is less complete and devolatilization may be required to remove monomers. The emulsion process essentially involves three separate polymerizations: (1) formation of a polybutadiene substrate latex; (2) grafting of styrene and acrylonitrile onto this latex; and (3) formation of styrene–acrylonitrile copolymer. Suspension polymerization uses a very lightly cross-linked polybutadiene that is dissolved in the monomers; the process is basically that used to prepare suspension high impact polystyrene. The bulk process similarly starts with a butadiene polymer or copolymer that is dissolved in the other two monomers, after which the mixture is pumped into a bulk polymerizer. Initiators and other additives are used with all systems (see Reference 360 for typical recipes).

Other monomers may be used with the three basic monomers. For example, a styrene–butadiene rubber plus styrene–acrylonitrile–methyl methacrylate terpolymer yields a clear, impact-resistant product (uniform refractive index) rather than the usual translucent acrylonitrile–butadiene–styrene (360).

Washing styrene–acrylonitrile and acrylonitrile–butadiene–styrene in the final suspension polymerization stage with aqueous solutions of alkali metal sulfides has been reported (432) to be effective in reducing residual acrylonitrile content. Treating an acrylonitrile–butadiene–styrene suspension for 1 hr at 100 to 110°C with aqueous sodium sulfide ($NaHS/Na_2S$) reduced the monomer content in the polymer from 1.6 to <0.01 percent and in the waste water from 0.5 to 0.03 to 0.04 percent.

6.4.2 Molecular Weight, Properties, and Applications

The molecular weight of styrene–acrylonitrile has been reported as 100,000 to 400,000 (31); typical values are also given (359) as 30,000 to 150,000 and 150,000 to 400,000 for number average and weight average molecular weight, respectively. Styrene–acrylonitrile resins are transparent, rigid resins that are noted for their toughness and resistance to chemicals and solvents.

The presence of acrylonitrile gives a higher softening point, greater resistance to stress cracking, and enhanced impact strength that increases with acrylonitrile content and molecular weight. However, as acrylonitrile content increases, molding becomes more difficult and yellowness of the resin (impurities) may increase. Chemical properties are generally superior to polystyrene but weathering properties are inferior to those of polymethyl methacrylate (359, 362).

The molecular weight of acrylonitrile–butadiene–styrene copolymers has been given as 60,000–200,000 (31). The three-monomer system was developed to provide a polymer with less brittleness and greater versatility than styrene–acrylonitrile afforded (10). Today the production of the offspring is appreciably higher than that of the parent. The ABS plastics are now "a family of more than 15 different groups of engineering materials" (361). In general, emulsion processes yield materials of higher impact strength and suspension or bulk processes are preferred for materials with lower impact strength (360).

Table 54.20, Part B outlines some comparative properties for styrene–acrylonitrile and acrylonitrile–butadiene–styrene. See also Tables 54.4 and 54.6.

Styrene–acrylonitrile copolymers are widely used as components of consumer and industrial articles (359, 362). Typical items include automotive interior trim, brush handles, broom bristles, bottle sprayers, dentures, air conditioner parts, and filler bowls of washing machines.

In the U.S. acrylonitrile–butadiene–styrene market, pipe, automotive applications (structural parts, grilles, dashboards), and appliance parts account for more than half of sales. Among large appliances, which account for about three-quarters of appliance usage, the largest applications are in refrigerators: interior liners, vegetable pans, breaker strips contacted by the door gasket, door shelves, etc. Smaller markets include business machines, telephones, electrical and electronic equipment, snowmobile and boat parts, plus luggage and furniture (154). Food packaging is primarily for margarine tubs. In Western Europe, furniture is a significant market and pipe is not; instead, polyvinyl chloride is used for pipe (360).

The potential fire hazard of acrylonitrile–butadiene–styrene pipe in building plumbing systems and the optimum installation of connections was examined at the National Bureau of Standards. Tests were conducted in a vertical chase with pipe having a 2-hr fire resistance rating. The conclusion was that if certain specified installation requirements were met, the use of acrylonitrile–butadiene–styrene pipe as described in the chase installation would not "spread either fire or significant quantities of smoke from one floor to the next during a 2-hour fire exposure" (433).

6.4.3 Analysis

Interest in residual monomer has centered primarily on acrylonitrile and to a lesser extent on styrene. Detection of these monomers in the corresponding homopolymers is discussed earlier in this section. With acrylonitrile–butadiene–styrene, methods similar to those used in the impact polystyrene field as described by Moore and co-workers (434) may be employed for the detection of the "nonpolymeric materials present" (360).

The values for "residual SAN monomers" may be as high as 0.05 to 1.0 weight percent (359). However, the U.S. Consumer Product Safety Commission (154) listed residual acrylonitrile content as 15 ppm and 30 or 50 ppm in one styrene–acrylonitrile and two acrylonitrile–butadiene–styrene resins, respectively (data not available on several other commercial resins). (The Commission notes that although injection molding would have no effect, extrusion molding—the other primary means of processing these resins—would be expected to reduce monomer content by 20 percent. Thermoforming could reduce monomer content by 50 percent. Thermal evolution analysis of the volatiles collected by sweeping a typical acrylonitrile–butadiene–styrene resin with helium for 1.5 hr at ambient temperature yielded "10 ppb of styrene but negligible acrylonitrile" (435).

Although very few data are available on residual acrylonitrile in consumer products, "levels less than 50 ppm would generally be expected" (154). Extraction tests reported on two new refrigerator crisper pans showed 22 or 23 ppm residual acrylonitrile; extracted acrylonitrile was below the limit of detection (approximately 0.01 ppm, Food and Drug Administration test as cited in Reference 154).

6.4.4 Oral, Inhalation, and Thermal Degradation

Three-month feeding of a styrene–acrylonitrile/styrene–butadiene resin to rats and dogs at a 10 percent dietary level produced no clinical, nutritional, hematologic, urinary, biochemical, or pathologic evidence of toxicity (436). The resin had the following composition: 85 percent 80:20 acrylonitrile–styrene copolymer, 10 percent 80:20 butadiene–styrene rubber, and 5 percent 74:26 acrylonitrile–styrene graft on rubber.

Several combustion/pyrolysis tests suggest that effluents from both styrene–acrylonitrile and acrylonitrile–butadiene–styrene may be in the middle range of toxicity when compared to other polymers. Rats exposed according to a German standardized protocol showed significant amounts of carboxyhemoglobin from the degradation products of styrene–acrylonitrile generated at 500 or 600°C. Table 54.22 shows the relative percentages of HCN that styrene–acrylonitrile or acrylonitrile–butadiene–styrene can yield at 500–1000°C under vacuum. Table 54.21 provides a general summary; also see Figure 54.7.

6.4.5 Photodegradation and Weathering of Acrylonitrile–Butadiene–Styrene

Aging effects (437–439) that occur during processing and during aging by natural and artifical light may be explained by preferential brittling of the components derived from butadiene. Relatively small amounts of radiation in the range of 295 to 340 nm may be particularly destructive. Aging after external exposure of the unprotected polymer may occur within a few months to 1 year. Pigments, especially carbon black, can provide protection against ultraviolet light.

6.5 Polymethyl Methacrylate

This polymer (CAS No. 9011-14-7*) is also known as PMMA, Lucite® acrylic resin, Plexiglas®, and Perspex®. These trademarks specify the general type of product rather than pure PMMA; most commercial acrylate polymers are actually copolymers with acrylate.

$$\left[-CH_2-\underset{\underset{CH_3}{|}}{\overset{\overset{\displaystyle C-O-CH_3}{|}}{C}} - \right]_n$$

6.5.1 Preparation

Methyl methacrylate monomer can polymerize readily during storage unless an inhibitor, such as 0.1 percent hydroquinone, is added; the inhibitor is then removed prior to commercial polymerization (10). Free radical polymerization initiated by organic peroxides or azo initiators is usually conducted at temperatures up to about 100°C (10, 31, 357). To avoid the side development of methacrylate peroxides, polymerization is usually done in the absence of oxygen (either by bulk polymerization in a full cell or blanketing with inert gas). Bulk polymerization is extensively used to manufacture sheets and, to a lesser extent, rods and tubes. Suspension polymerization, which avoids the potentially serious exotherm problem found with bulk polymerization, gives a lower molecular weight polymer suitable for molding powder. Additives may include buffering agents, chain transfer agents, lubricants, and emulsifiers (10). Fire retardants such as halogen and antimony compounds have been used, but would appear to be relatively uncommon; if used, they may impair transparency or resistance to aging (38).

6.5.2 Molecular Weight, Properties, and Applications

The molecular weight would appear to be in the range of 50,000 to 60,000 for molding compounds and about 10^6 for acrylic sheet (10, 14). Polymethyl methacrylate is typically tougher than polystyrene but less tough than cellulose acetate or acrylonitrile–butadiene–styrene polymers (10). Polymethyl methac-

* CAS Registry numbers for the isotactic and syndiotactic forms are 25188-98-1 and 25188-97-0, respectively.

rylate is particularly noted for light transmission, plus shatter resistance and light weight as compared to glass. Polymethyl methacrylate cracks on impact but fragments are normally less sharp and jagged than those of glass; correspondingly, polymethyl methacrylate is inferior to glass in scratch resistance. Light transmission through unblemished sheet can exceed that of plate glass— 92 to 93 percent versus 89 percent (440). The major applications of polymethyl methacrylate include lighting fixtures, display signs, automotive lenses, sanitary ware, and glazing. Acrylic glazing is particularly useful in storm doors and storm windows.

Polymethyl methacrylate and its copolymers are widely used in molding resins, exterior and interior paints, enamels, polishes, adhesives, and miscellaneous coatings. Polymethyl methacrylate and other acrylic resins have been used for specified food contact items (92), dentures, and medical applications such as hard contact lenses, intraocular lenses, or bone cements (10, 31).

6.5.3 Analysis

The International Agency for Research on Cancer (31) lists a variety of references for determining residual methyl methacrylate in polymers and copolymers. Gas chromatography has been used for determinations in aqueous emulsions used in the textile industry (441) and in paints and varnishes (442). Methyl methacrylate and other volatile components have been determined (443) in polymeric formulations by vacuum pyrolysis and infrared spectrometry as part of a series of tests for evaluating candidate restorative and prosthetic materials.

Volatile monomer evolved at ambient temperature from a commercial sample of polymethyl methacrylate was measured by thermal evolution analysis (see Section 6.4.3). "After each equilibration, the pot was swept with helium to measure the monomer content. The monomer level was found to increase steadily to a constant value of 20 ppb of the resin after a two-week period. . . . This residual monomer did not reappear after the resin was placed in a vacuum oven at 80°C for one week. Thus, the monomer in the headspace seemed to be in equilibrium with its solution in the resin and not regenerated from the polymer" (435).

Migration of monomer from commercial acrylic bone cements into a tissue medium has been studied by Willert and associates (444). They report detecting concentrations of 0.7 to 5.1 weight percent in the fatty components of bone marrow.

6.5.4 Skin and Inhalation Toxicity, Thermal Degradation Data

Although the monomer is a recognized contact allergen, heat-polymerized polymethyl methacrylate appears essentially inert in this respect. The level of residual monomer in cold-cured dentures may be as much as 5 percent (445a) but can be reduced by heat curing to a level not detected by a person strongly sensitive to 0.1 percent methyl methacrylate (445b). The monomer is more commonly "a cause of hand dermatitis in dentists and dental mechanics than of denture sore mouth" (445c; see also References 160, 370, 445d). Sensitivity to added accelerators may also occur. Residual methyl methacrylate appears to be a weaker allergen than residual acrylates (see Section 6.7).

Mice inhaling decomposition products from polymethyl methacrylate showed effects somewhat sooner than mice comparably exposed to decomposition products from a series of other polymers. Elevated temperatures produce a classical unzipping of the polymer and can yield 99 percent monomer at 425°C; at higher temperatures monomer content decreases and aldehydes, particularly formaldehyde, may be evolved. Some manufacturing processes such as sawing or molding the polymer may, under certain circumstances, release fine dust or monomer-containing vapors (251); comparatively small amounts of methanol may also be released (405). The heating, lathing, molding, and grinding processes associated with the processing of polymethyl methacrylate lenses are capable of degrading the polymer sufficiently to cause a measurable, although not persistent, increase in monomer content of the polymer (371). See Table 54.21 for details.

6.5.5 Carcinogenic Potential/Implant Studies and Other Relevant Biologic Data

Experimental data on animals plus various types of human exposure all indicate no evidence of metastatic cancer but reveal that newly polymerized or implanted resin may not be entirely inert. The International Agency for Research on Cancer (31) called for experimental and epidemiologic studies on the basis of animal studies that essentially showed sarcomas at the site of implantation (see Section 1.4) and the evident human exposure arising from widespread industrial or medical use. Available data were considered inadequate to evaluate either the monomer or the polymer.

Two additional animal studies not cited in Reference 31 are relevant. Stinson (446) implanted polymethyl methacrylate particles ≤76 μm in the muscle and knee joint of guinea pigs. These particles produced "an extremely mild reaction," without evident erosion over a 3-year period. The author contrasts these results with his earlier results from disks in guinea pigs and also rats (when tumors did occur, as cited in Reference 31). Barvic (447) did an extensive study on 450 rats with both "small flat pieces of pure [poly]methyl-methacrylate and neutral specimens of [poly]hydrocolloidalmethacrylate." Perforated specimens caused no tumors and all observed neoplasms were induced by specimens prepared from a solid sheet. "Small specimens" rarely caused tumors, whereas "larger implants" produced a 12 to 30 percent incidence of tumors with a shorter latency (also see Reference 29).

One case report describes a chondrosarcoma intimately associated with the fibrous capsule surrounding polymethyl methacrylate spheres used as plombage

(448). The spheres had been implanted for compression of a tuberculous cavity 18 years prior to terminal hospital admission when the tumor was detected.

Tomatis (449) reports that mice given subcutaneous implants of ^{14}C and ^{3}H-labeled polymethyl methacrylate films showed increased radioactivity in the urine until the seventh to eighth week after implantation and sharply decreased radioactivity thereafter. Very low levels of radioactivity were detected in saline in which labeled implants were immersed.

Polymethyl methacrylate has been widely utilized as a cement in bone surgery, particularly for hip replacements in elderly patients and in malignant neoplastic fractures. Acute hypotension, cardiovascular collapse, and other misadventures have been associated with the use of acrylic cement in hip replacements, but the reported incidence of these events varies widely. Charnley's survey of some 3700 hip replacements (cited in Reference 450) lists four patients with cardiac arrest while Lipecz et al. (451) describe hypotension in 58 percent of 300 patients. Factors considered significant include operative technique, age and medical fitness of the patient, and release of the monomer from the implantation directly into the circulation. For an entry into the very extensive literature, see References 31, 443, and 450–455.

6.5.6 Biodegradation and Photodegradation

"Extremely low decomposition rates" were found when ^{14}C-labeled polymethyl methacrylate was exposed to fungi, soil invertebrates, and mixed microbial communities (417; see also Reference 456). Ultraviolet irradiation of polymethyl methacrylate (molecular weight = 190,000 or 240,000) may cause a rapid decrease in molecular weight (457). Commercial grades are extremely resistant to weathering as compared to other thermoplastics (10).

6.6 Polyacrylamide and Polyacrylic Acid

These two water-soluble homopolymers are widely used as flocculants and often occur together, since the capacity of polyacrylamides to ionize is a commercially useful property.

Polyacrylamide is available in many modifications, including various types of Separan® (see below) and Versenex®.

$$\left[\begin{matrix} & \overset{\displaystyle O}{\underset{\displaystyle |}{\overset{\displaystyle \|}{C}}}-NH_2 \\ -CH_2-CH- & \end{matrix} \right]_n$$

The nonionic homopolymer has CAS No. 9003-05-8. Polymers with <4.0 percent hydrolysis are usually considered nonionic (363). Examples of the structure of commercial polyacrylamide (458) are given below. In Separan® NP 10, considered essentially nonionic, x = approximately 19; in Separan® AP 30, considered anionic, x = approximately 3.

$$\left[\left(\begin{array}{c} CH_2-CH- \\ | \\ C=O \\ | \\ NH_2 \end{array}\right)_x -CH_2-CH- \begin{array}{c} \\ | \\ C=O \\ | \\ O^-Na^+ \end{array}\right]_n$$

Polyacrylic acid (CAS No. 9003-01-4) is available as various types of Acrysol® and Carbapol® (see Reference 31).

$$\left[CH_2-\overset{\overset{\displaystyle COOH}{|}}{CH}\right]_n$$

6.6.1 Preparation

Commercial polyacrylamide products are produced by copolymerization of acrylamide monomer or by postpolymerization reactions of the amide group (363); the first is usually preferred. Polymerization can take place in aqueous solution with free radical initiators (peroxides, persulfates, azo compounds). Alternately, polymerization can be accomplished in a solvent system containing water and alcohol or other organic solvents (363). Mixed solvent systems usually give a narrower molecular weight distribution; decreasing the level of organic solvent increases molecular weight.

Polyacrylic acid may be polymerized in aqueous solution at concentrations of 25 percent or less (364). Polymerization can be accomplished with a peroxydisulfate initiator at 90 to 100°C or a redox initiator (activated by trace of ferric iron) at 0 to 10°C. Polymerization of undiluted monomer is hazardous. Polymers may contain at least 5 percent moisture to facilitate re-solution (363). Polymerization can also be accomplished in nonaqueous solvents that dissolve the monomer but not the polymer.

6.6.2 Molecular Weight, Properties, and Applications of Polyacrylamide

The molecular weights for commercial polyacrylamides are typically very high. They range from at least 1,000,000 to 3,000,000 (data in Reference 458). Polyacrylamide acts as a polyelectrolyte and a flocculant.* It is very hygroscopic. The rate of moisture absorption increases with increasing ionic character of the derivative. The polymer is most commonly supplied as a solid (usually powder,

---

* Flocculation describes the aggregation of suspended solid particles in small clumps or wool-like tufts (3).

also beads) that is stable for up to 2 years (363). Alternate product forms include viscous aqueous solutions for dilution and emulsions in hydrocarbons. See Table 54.20 and Reference 324 for additional properties.

Polyacrylamides are extensively used in many types of liquid–solid separation. Applications include water clarification, waste treatment, and mining. Polyacrylamides are also used as additives and processing aids in the manufacture of paper and paperboard products, in textile sizing, and as binders for foundry sand. Anionic polymer solutions are increasingly important in enhanced oil recovery. High viscosity solution is pumped into an oil well to push the oil from rock or reservoir to another well (363).

6.6.3   Properties and Applications of Polyacrylic Acid

Polyacrylic acid, also noted for its flocculant action, is a weaker acid than its monomeric counterpart. The polyacid acts as an excellent buffer in the range of pH 4 to 6.4. Precise titration in water alone is difficult, but titration in 0.01 to 1 N solutions of neutral salts gives precise end points (364). Copolymers with only a few cross-links are gels; more cross-linking results in a brittle polymer (364). See Table 54.20 and Reference 324.

Polyacrylic acids have a wide range of very similar applications and have proved especially successful as warp sizes for nylon. The copolymers with divinylbenzene act as ion exchange resins. In drilling an oil well, "the addition of about 0.3 lb of a copolymer of acrylic acid and acrylamide (containing from 25% to about 70% acrylic acid and having a molecular weight of the order of 250,000) per barrel of mud decreases the water loss and prevents caking of the drilling mud" (364).

6.6.4   Analysis and Detection of Impurities

The level of residual acrylamide in domestic polyacrylamides may range between 0.05 and 0.75 percent, depending on the type of products. Approximately 1 ppm residual acrylonitrile may also be present (409). In one Japanese study (459) two monomer solutions for acrylamide soil-consolidating agents were found to contain 0.002 or 0.3 mg/l acrylonitrile, respectively. (The various components were intended to be mixed just prior to injection into the soil where polymerization occurred with the formation of a gel matrix.) After mixing as prescribed and polymerization with sand, concentrations of acrylonitrile were 117 and 4440 mg/m^3, respectively, in "the closed atmosphere over the gel layer."

Various methods of analyzing polyacrylamide, including detection of residual monomer, have been summarized (363). Liquid chromatographic methods permit detection at 10 ppm monomer in nonaqueous dispersions and 0.1 ppm in aqueous dispersions (459). Modifications permit determination within an hour and are suitable for use with water-in-oil emulsion copolymers that have cationic comonomers (461). Both acrylamide and acrylic acid have been determined in polyacrylamide by a German method (462) described as sensitive to 0.5 μg, with a standard deviation of ±2.5 percent. Another German report indicates that acrylic acid can be determined in polyacrylic acid at levels as low as 300 ppm (463).

Determination of trace amounts of polyacrylamide in water can be measured by spectrophotometry, with absorbance determined on the basis of organic nitrogen or after conversion to triiodide. See Reference 324 for additional data.

6.6.5 Animal Toxicity Tests, Human Exposure, and Thermal Degradation

Various mammalian toxicity tests with polyacrylamide polymers all indicate a very low order of toxicity. The single dose LD_{50} for rats by oral administration, and for rabbits by skin penetration, has been described as >8.2 g/kg; 20 percent solutions were reported as nonirritating to the rabbit eye (464). Only a very mild dermal inflammatory reaction developed in rabbits after 24 hr contact with pure moistened product. Cationic polyacrylamides may be somewhat more irritating (363). A brief Russian report states that "prolonged exposure," presumably of rats, to technical grade polyacrylamide at doses up to 70 mg/kg was not toxic, and histologic examination revealed no toxic effects (465).

McCollister and his colleagues (458) found that rats survived the maximum feasible oral dose, 4 g/kg, of a nonionic and anionic polyacrylamide, and conducted three separate 2-year feeding studies with rats. Two were done with nonionic product containing 0.02 or 0.08 percent residual acrylamide, respectively; in the third, anionic product containing 0.07 percent residual monomer was fed to both rats and dogs for 2 years. Dogs were also fed nonionic product (0.01 percent monomer) for 1 year. "The laboratory animals tolerated 5–10% in their total diet without effects other than those believed to be attributable indirectly to the large, hydrophilic, non-nutritive bulkiness of the materials. The unequivocal 'no ill-effect' levels were 1% in the diet of rats and 5–6% in the diet of dogs." Additional ^{14}C radioactive tracer studies with both types of product indicated that "negligible amounts of polymer, if any, pass through the walls of the intestinal tract of the rat."

Christofano et al. (466) administered "strongly cationic, high molecular weight, water-soluble . . . polyacrylamide resins" to rats and dogs. Tests included maximum single dose, 90-day feeding, and 2-year feeding with three-generation reproduction studies. Animals fed ≥1 percent dietary levels showed, in some cases, changes such as depressed weight gain and altered liver weight while levels ≤0.2 percent produced no effect.

Environmental studies have been conducted in polyacrylamide production plants of the Dow Chemical Company over a 5-year period (458). Airborne concentrations averaged approximately 1 mg/m^3 with almost all particles >50 μm in size. "It was postulated that about 5 mg/day might be ingested by human subjects working in such areas. . . . The physicians found no indication of any more pathology than one would expect to find in a similar group of men in the general population."

Polyacrylamides and polyacrylic acids may decompose at temperatures greater than 200 to 300°C. Thermal degradation of polyacrylamide in air at 700°C has yielded relatively small amounts of HCN and ammonia. See Table 54.21.

6.6.6   Aquatic Toxicity

Studies with the nonionic and anionic polyacrylamides discussed in Section 6.6.5 (458) also included several species of fish, such as Lake Emerald shiners, yellow perch, fathead minnows, rainbow trout, and blue gills. The anionic products at 2500 ppm caused 100 percent mortality: "the solution was so viscous that the fish had difficulty in swimming. . . . Fish were maintained in water containing 1000 ppm of these resins for 5 days and 100 ppm for 90 days without apparent adverse effect."

Two Japanese reports (467, 468) indicate that some types of polyacrylamide or their derivatives can be toxic to fish. The 48-hr median tolerance limit (TLm) values in one bioassay (467) appeared to vary from >80 to <1 ppm.

6.6.7   Biodegradation and Ozonization

Polyelectrolytes may act to restrict microbial growth by chelating essential metals or they may flocculate microbial cells. Certain polyacrylamides can actually serve as a nitrogen source in a chemically defined medium and stimulate bacterial growth (*Pseudomonas aeruginosa*), but they do not appear to be biodegradable in the usual sense of the word (469). The breakage of polymer chains by ozone has been reported to be strongly accelerated by ultraviolet radiation under acidic and neutral conditions (470). Formaldehyde was produced in the ozonization process.

6.7   Other Acrylics

An almost infinite number exist but the great majority appear to be variants of those already discussed. They all have in common a basic structural similarity to acrylic acid. These polymers are very widely used in paints, coatings, sealants, adhesives, and gels. The poly(2-hydroxyethyl methacrylate) hydrogels (471) have been used extensively for hydrophilic contact lenses ("soft lenses"). Applications such as blood-compatible prosthetic materials are limited by the thrombogenic capacity recently associated with radiation-grafted hydrogels (472).

The more common industrial hygiene concerns would seem to be dermatitis from "prepolymers," particularly from the light-sensitive acrylates, and atmospheric contamination during the processing of coatings and paints as discussed in Section 6.1. Many printing processes now use the light-sensitive acrylates that polymerize on exposure to ultraviolet light. Resins containing acrylic monomers tend to present a more serious dermatitis problem than those containing methacrylic monomers. An outbreak of dermatitis occurred in a French newspaper printing plant where photopolymer acrylic resins had been substituted for lead printing plates (473). Substitution of a resin containing tetraethylene glycol diacrylate with one containing tetraethylene glycol dimethacrylate eliminated this dermatitis. Review of additional case reports indicates (445c) that patch testing to methyl methacrylate has generally been negative in dermatitis associated with light-sensitive acrylates. In one case where acrylonitrile was used with methyl methacrylate in a plastic finger splint, acrylonitrile was found to be the sensitizer. The "instant adhesives" formulated from cyanoacrylates may cause respiratory tract irritation during processing (406, see Table 54.21) and a skin problem in end use: the fingers stick together (474a). These adhesive compositions have "pungent, unpleasant odors, and may be mildly lachrymatory" (474b; see also Reference 474c).

Fully polymerized resin, even if occurring as fine particulate, appears to cause relatively little problem. The National Institute for Occupational Safety and Health conducted an investigation of a textile plant where polyester fibers were saturated with an aqueous emulsion of ethyl acrylate polymer (pH of bulk solution = 3.9). The mean particle size of the settled dust of this polymer averaged 0.42 μm with 99 percent of all particles <4.5 μm. The mean breathing zone concentration of five samples was 4.10 mg/m^3. The dust was judged to be potentially more toxic than an inert dust because of its mildly acidic nature (pH following hydration = 5.7). However, discomfort was noted only by one worker, when overspray of the emulsion was prominent. "His discomfort was described as an itching of the skin when dust became lodged in facial creases, ears, and nose" (475).

Two acrylic resin paints used in the tin packaging industry yielded trimeric and hexameric methyl methacrylate plus "small quantities of methanol" at drying temperatures of 180 to 200°C. More than two-thirds of the waste gas condensates from the 26 paint binders of various types investigated in this German study appeared to be low molecular weight components from the binder rather than actual decomposition products (476).

7 CELLULOSICS AND OTHER POLYSACCHARIDES

7.1 Overview

In the broad sense, this section embraces all the natural polymers of the plant kingdom. Chemically, the major industrial polymers derived from living plants are the two carbohydrate polymers cellulose and starch. They are the most abundant polysaccharides produced in nature. In the form of various products such as wood, cotton, and grain they have long been an integral part of civilization. Today these polymers are being increasingly regarded as readily renewable industrial raw materials in an energy-hungry economy. This discussion emphasizes the traditional cellulosic products.

Cellulose and starch may be defined as specific polysaccharides or as plant-derived materials containing other ingredients, for example, lignin in the case of "wood cellulose." "Technical cellulose" may imply that portion of a plant cell wall derived exclusively from glucose and resembling cotton cellulose in its physical and chemical properties" (477).

Cellulose and starch are both essentially composed of units of D-glucose, $C_6H_{12}O_6$, but differ in the way these units are joined. In cellulose they are linked through carbons 1 and 4 by beta linkages (all substituents in the equatorial rather than axial plane as with starch). For a more detailed discussion of this point, see Stevens (4) and Bolker (477).

Natural cellulose is highly crystalline. The molecular weight varies from 300,000 to >1,000,000 (4). Wood, depending on the type, contains approximately 50 to 70 percent cellulose or cellulose-like material (478–480). Cotton and textile fibers of vegetable origin such as flax, hemp, and jute contain approximately 65 to 95 percent cellulose (4, 89, 481). Rayon is a regenerated cellulose prepared by dissolving natural cellulose and then precipitating it from solution, usually with some loss of crystallinity (4, 482). Cellulose derivatives of commercial importance are esters and ethers made by treatment of regenerated or solubilized cellulose.

The natural cellulosic products as a class are water-insoluble and often offer relatively little resistance to thermal or microbial degradation (38, 483). Synthetic cellulose derivatives vary in solubility, may be highly flammable, and generally offer more resistance to microbial attack. Table 54.23 gives some basic comparative properties.

7.1.1 Production and Processing

Production data for several cellulosic textiles are listed in Table 54.2. The natural products (wood, cotton, and other natural fibers) are frequently machined with relatively little cleaning or other preparation. Wood may be treated with preservatives and cotton may be blended with other fibers; both may be treated with fire retardants (38) as well as a host of other additives. Wood pulp and cotton linters (also rags, grasses, and plants) are processed (4, 477, 488, 489, 495) to form paper and cellulose. Cellulose may be made into rayon fibers, cellophane film, or cellulose esters and ethers, in the form of fibers, resins, coatings, or gums.

7.1.2 Analysis and Specifications

These products are so diverse that the references given here are primarily limited to reviews. References 89, 372, 490, and 491 may be helpful for textiles; Reference 23 for food additives; References 485–487 and 492 for cellulose esters and ethers; and Reference 480 for wood and wood products. Measurement of cotton dust is discussed in Section 7.1.4.

The pyrolysis of cellulose is reviewed with reference to its fine structure in Reference 493. Levoglucosan, the pyrolysis product that makes cellulose combustible, can be detected by the infrared method described in Reference 494. Reference 495 gives a comprehensive survey of the thermal behavior of textiles. See also Section 1.3 concerning analysis and Reference 92 for specifications relating to food contact.

7.1.3 Toxicologic Potential

At ambient temperatures the toxic characteristics of the cellulosic polymers are primarily attributable to a few natural constituents. Wood, a highly complex substance, appears to be the only polymer known to have caused cancer as a result of occupational exposure. Long-term exposures—frequently several decades—to presumably high, usually unmeasured, concentrations of wood dust are associated with an increased incidence of nasal cancer (496). The causative agent or agents have not been identified although certain aldehyde constituents or their quinone oxidation products may perhaps be suspect (see Section 7.2). The other main physiologic responses to wood are dermatitis and nonmalignant respiratory disease.

Inhalation of the dust from unwashed cotton, flax, and soft hemp frequently causes byssinosis (also called "brown lung"), a chronic respiratory disease that develops insidiously. Early symptoms are indistinguishable from chronic bronchitis or emphysema. Symptoms in the early stages are often precipitated by reexposure but with continued exposure over a period of years may become permanent and incapacitating. The causative agent is known to be water soluble and filterable and may be an aminopolysaccharide intimately associated with the cellulose stalk.

Regenerating cellulose to yield the products rayon and cellophane removes the natural impurities. These regenerated products are essentially inert unless new toxicants such as finishes and plasticizers are added in sufficient quantities to cause injury. Many cellulose derivatives would appear to be similarly inert. Large doses of some types of orally administered cellulose are regarded as nonnutritive bulk. Human studies with radiolabeled microcrystalline cellulose indicate that doses up to 30 g/day appear to be tolerated therapeutically as a bulk laxative. Microcrystalline cellulose was considered essentially nonabsorbable

Table 54.23. Some Typical Properties of Cellulosics[a]

A. Natural and Regenerated Forms of Cellulose

Property	Cellulose	Wood (~50–70% Cellulose)	Cotton (~90% Cellulose)	Viscose Rayon (Regenerated Cellulose Plus Finishes)	Reference
Ignition temperature[b] (°C)		360–540°C (spontaneous)	390–400	420	38, 484
Ignition temperature with pilot flame (°C)		225 (10 min) to 300 (1–2 min)			38
Charring temperature (°C)	Above 200	280			14, 38
Glass transition temperature (°C)	230–245; −30 to 160 ("conflicting data," Ref. 86)				86, 484
Density			~1.56, dry in He; ~1.60 in water		14, 38
		Most commercial U.S. species: 0.3–0.7 g/cm³			
Refractive Index, n_D	1.54		1.527–1.534	1.512–1.520	87, 484
Moisture regain			7–12%, commercial values for yarns	11%, commercial value	12[c]
Solvents/swelling/decomposition	CELLULOSE: No simple solvents. Soluble in aq. solutions of metal complex compounds, such as cuprammonium hydroxide (Schweizer's reagent) and cupriethylenediamine (cuene) . . . . ordinary regenerated cellulose dissolves in an aq. solution of a metal such as zinc (e.g., by aq. sodium zincate) and in ice-cold alkalis (e.g., 2N sodium hydroxide at −5°C); polynosic rayons are more resistant. . . . *Plasticised by water; polyhydric alcohols, e.g., glycerol. Relatively unaffected by many organic liquids, e.g., most types of hydrocarbons, chlorinated hydrocarbors, oils, esters, high alcohols. Swollen by water, and especially by moderately conc. alkales. . . . Decomposed by hot acids of sufficient strength and most oxidising agents. . . . In the presence of an alkaline medium, atmospheric oxygen causes some degradation . . .*				14

B. Cellulose Derivatives

Property	Cellulose Acetate	Cellulose Triacetate	Cellulose Nitrate	Carboxymethyl Cellulose[d]	Hydroxyethyl Cellulose	Ethyl Cellulose	Reference
Melting temperature (°C)	230–260 (with decomposition)	290–306	Flow temperature ~150. Plastics serviceable to 60[e]				14, 85, 485
Charring temperature (°C)				252			486
Browning temperature (°C)				227	205–210		

							Refs
Softening temperature (°C)	75–120	220–225	80–90 (Ref. 14); 155–220, Parr (Ref. 485)	135–140		152–162	14, 485, 486
Density	1.28–1.32 (sheeting). 122–1.34 (molding)	1.30	1.58–1.65 (cast film, Ref. 486). 1.35–1.5 (trinitrate; plastics, ~1.38, Ref. 14)	0.75 (bulk). 1.59 (film). 1.007 (2% solution)	0.6 (bulk). 1.34 (film; 50% relative humidity). 1.003 (2% solution)		14, 485–487
Refractive index	1.46–1.50	1.47–1.48	1.51	1.515 (film). 1.336 (2% solution)	1.51 (film). 1.336 (2% solution)	1.479	87
Moisture absorption	3% (depends on plasticizer)	1% (24 hr at 80% RH)				2% (24 hr at 80% RH)	14, 485, 486
Glass transition temperature (°C)	53–66 ("conflicting data")	49–478 ("Conflicting data; depend on acetate and water content and degree of crystallinity," Ref. 86). 180 (2nd-order transition, fibers, Ref. 14)				43	86
Solvents/nonsolvents/relatively unaffected by	Solubility varies with degree of substitution; water and 2-methoxyethanol can act as both solvents and nonsolvents		Soluble in methylene chloride, chloroform; acetone may act as solvent and nonsolvent (Ref. 88). Relatively unaffected by hydrocarbons, most oils and greases (Ref. 14)			Solubility varies with degree of substitution; generally, carboxymethyl cellulose and hydroxyethyl cellulose are soluble in water, ethyl cellulose insoluble; see Ref. 88	14, 88, 485, 486
Decomposed by	Less resistant than cellulose triacetate	Moderately concentrated acids, alkalies with pH >9.5					14
Biologic oxygen demand						Reported values after 5 days of incubation in the range of 7000–18,000 ppm depending on substitution of polymer; comparable value for cornstarch is over 800,000 ppm	486

[a] Summarized as described in Table 54.9.

[b] Ignition temperature depends on specimen size and composition, heating conditions, time, and test method. See Refs. 38, 39, and 54 for perspective and basic data.

[c] Standard D 1909-77, ASTM (see Ref. 12).

[d] Carboxymethyl cellulose is typically sold as the sodium salt (14).

[e] "Dry cellulose nitrate is extremely flammable; cellulose nitrates with high nitrogen content may explode when subjected to heat or sudden shock" (485).

and noncaloric. Any persorbed microcrystalline cellulose would be expected to be excreted in the urine or bile. The establishment of an acceptable daily intake in milligrams per kilogram of body weight was not deemed necessary by the World Health Organization.*

Dry, untreated cellulosic products are often highly flammable. In the general population harm is more likely to result from the flammability of these cellulosics or the means used to retard flammability than any other factor associated with their use. Untreated cellulosic materials exposed to smoldering flames readily generate serious or lethal amounts of carbon monoxide. Analysis of New York City autopsy records from a 2-year period (1966–1967) showed that 79 percent of fire victims who died during the first 12 hr had carboxyhemoglobin poisoning (497, also see Reference 36). Wood can generate lethal combustion products over a broad temperature range (99); the products generated in the vicinity of 450°C appear to be particularly toxic.

The overzealous use of unsuitable fire retardants may add a daily toxic stress to the entire exposed population to protect against a relatively infrequent disaster. A classic case in point is the use and then the withdrawal of tris-2,3-dibromopropyl phosphate (498) as a fire retardant for children's clothing. The hazard from its regular use was deemed to be more serious than the loss of its retardant properties in possible fire situations.

Table 54.24 gives an overview of inhalation and thermal degradation data pertaining to the major types of cellulose products. Figure 54.8 indicates that burning wood may cause respiratory rate depression (end point for respiratory irritation) at concentrations well below the lethal range. Figure 54.9 presents a comparison of the weight loss of cotton to that of several other fabrics at elevated temperatures. Table 54.25 gives a chronologic summary of studies concerned with the effects of cotton dust and their control.

7.1.4 Workplace Practices and Standards

In 1979 the American Conference of Governmental Industrial Hygienists proposed a change in the existing TLV of 5 mg/m^3 for nonallergenic wood dust. The 1981 adopted limit values (TLV-TWA) are 1 mg/m^3 for hardwood and 5 mg/m^3 for softwood. (For discussion of apropriate industrial hygiene measures, see Chapter 23, Volume 1). The maximum allowable concentration (MAK) in the German Democratic Republic (East Germany) is 10 mg/m^3 for "all types of domestic wood dust (pine, oak, fir and beech tree)" (28).

General high standards of personal hygiene are desirable to protect against dermatitis, bronchitis, and asthma from wood dust. Where opportunity for gross exposure exists, prudence would suggest periodic dust counts, identification of the principal species of woods involved, and epidemiologic monitoring.

---

* See Reference 23, *WHO Food Additive Ser.* 8, 1975, for general discussion of the acceptable daily intake and background literature.

Table 54.24. Overview of Inhalation and Thermal Degradation Data on Some Cellulosic Materials[a]

Polymer	TLV	Type of Study — Ambient Temperature	Type of Study — Elevated Temperature	Type of Study — Effluent Analysis	Species Exposed	Description of Material	Procedure	Observations/Conclusions	Reference
Wood	1 mg/m^3, hardwood; 5 mg/m^3, softwood (intended change 1979)	X			Human			Documentation for TLV: impairment of nasal mucociliary function may occur below 5 mg/m^3 and may be important in the development of nasal adenocarcinoma in furniture workers (exposure to hardwood)	496
		X			Human	Fine hardwood dust (beech, mahogany, afrormosia)	Follow-up of English furniture workers in High Wycombe area	"Adenocarcinoma of the ethmoid has been established as an industrial hazard in the woodworkers in the United Kingdom. . . . Prerequisites seem to be hardwoods and fine particles"	499
		X			Human		Epidemiologic appraisal of factors other than smoking in respiratory carcinogenesis	"The risk seems to extend to various trades exposed to dust in the furniture industry and to woodworkers outside the furniture industry"	500
		X			Human		Epidemiologic survey: "In 186 cases of nasal cancer diagnosed over the decade 1965–74, in a population of 2.0 million, 114 of 157 ectodermal tumors were found in men. Adenocarcinoma was found in 17 patients, two women and 15 men; 12 of these had a history of occupational exposure to wood dust in the furniture industry. The period of latency was from 28 to 57 years"	501
		X			Human	Dust from woodworking (more work with local hardwoods than exotic woods)	Review of 36 cases of ethmoid cancer at Rennes clinic, 1964–1974, also related literature: woodworkers considered a high-hazard population; malignancy generally develops after a very long interval, sometimes short	502

4361

Table 54.24. (*Continued*)

Polymer	TLV	Ambient Temperature	Elevated Temperature	Effluent Analysis	Species Exposed	Description of Material	Procedure	Observations/Conclusions	Reference
		X			Human			Adenocarcinoma of the nose associated with occupational exposure to organic dust (Ref. 503); incidence currently under study in Scandinavia (Ref. 504)	503, 504
		X			Human	Maple or pine dust		Cross-sectional survey of 1157 woodworkers; dry hardwood or softwood dust exposure correlated with reduced pulmonary flaw rates	505
		X			Human			"The coarse dust of sawmills may be a hazard, just as the fine dust of cabinet-makers' shops is"	506
		X			Human	Western red cedar		Review of asthmatic symptoms based on experience with 80 patients, related literature	507
		X			Human	Western red cedar		Survey of 49 men in factory showed 4 cases of rhinitis, 3 of asthma; asthma attributed to red cedar dust (visible as "faint haze," most particles <2 μm)	508
		X			Human	Maple		Case report of "maple-bark disease" (pneumonitis due to a fungus, *Coniosporium corticale*), which was associated with fine heavy black dust when bark removed. Biopsy revealed some fibrosis but X-ray study 2 weeks later interpreted as normal	509
			X	X	Human, rat, mouse			Mice showed 50% respiratory depression from 3-min exposure to 0.33 g/m³ of barely visible red oak smoke. Humans simultaneously exposed found this concentration irritating but tolerable, showed no incapacitation. LC$_{50}$s for rats exposed to pyrolysis or combustion products varied from 22 to 35 g/m³. Large-scale fire tests indicate elevated temperature and CO as limiting hazards, acrolein not considered significant	65, 99, 100
			X	X	Dog	1000 g of wood with 10 g of sawdust and 40 ml of kerosene		10- to 20-min exposure, head only, to wood smoke produced "50% mortality and pulmonary edema with increased sodium and water contents in the lungs." Comparable exposure to kerosene smoke produced neither death nor pulmonary edema. Analysis of smoke	510

		Species	Material	Method	Observations	Ref.
X		Rat	Pinewood, spruce, expanded cork, other types	DIN 53436 protocol as described in Table 54.10	for acrolein, formaldehyde, acetaldehyde, butyraldehyde; lung injury attributed to increased concentrations of aldehyde gases in wood smoke compared to kerosene smoke. Deaths occurred in some tests with 300°C decomposition, in all tests with ≥400°C decomposition; see Table 54.7	62
X	X	Rat	Spruce, cork, other types		German protocol DIN 53436, concept of "critical temperature range" as the range of pyrolysis temperatures above those not producing death and below those where death occurred. Critical temperature for spruce wood = 423–486°C. Ranking the hazards of 11 polymers by relative temperature showed spruce wood, cork, pinewood, plywood, and polyethylene all in the same 300–400°C range; others above or below this temperature range. Mortality from pyrolysis products of wood attributed to CO but with cork "other gases were responsible." Extensive review of published and unpublished data	63
X	X	Rat, rabbit	Douglas fir, mahogany		Toxicant production greater at 850°C (flaming conditions) than 400°C (smoldering conditions). Mahogany releases more CO and CO_2 but considered less toxic than fir, which toxicity presumably is due to release of aldehydes. Concept of *graded toxicants* that are frequently present in fire environment (as CO, CO_2, hypoxia, and heat) and *limiting toxicant* (as acrolein with wood)	511
X	X	Rat	Douglas fir	30-min exposure, "cup" furnace at several temperatures	Products generated at 450°C evoked clinical and lethal response at slightly lower atmospheric concentration (mg/l) than observed with products generated at 300 or 500°C; see Figure 54.8	72

Table 54.24. (Continued)

| Polymer | TLV | Type of Study | | | Species Exposed | Description of Material | Procedure | Observations/Conclusions | Reference |
		Ambient Temperature	Elevated Temperature	Effluent Analysis					
			X	X	Rat	Red oak	2-hr exposure to products generated at ~822°C, different amounts of various polymers burned. ALC_{50} = 22.7 mg/l (of 12 plastics, 5 had relative toxicity values that were comparable to red oak, 3 appeared less toxic, 4 more toxic).[b] CO considered primary toxicant, but small scale fire tests describing relative toxicities based on production of CO "can be misleading." CO data from large-scale tests presented for comparison	512
			X		Mouse	Douglas fir	As described in Table 54.12	Douglas fir tested as a reference material; various polymers considered more or less toxic	71, 165
			X	X	Rat (cannulated and uncannulated)	Douglas fir	20-min exposure to pyrolysis products generated by 5 W/cm² radiant heat flux	Compared to CO-exposed controls, smoke-exposed animals had less mortality but a greater reduction in respiratory rates, slower recovery of leg-flexion avoidance response, and slower desaturation of COHb; role of sensory irritation discussed	513
			X		Mouse	Douglas fir, also other hardwoods, softwoods	Rising and fixed temperature programs (Table 54.10, Ref. 66), also variations with 1 l/min air flow showed products of wood and cellulosic board approximately intermediate in toxicity among those of a series of polymers as judged by time to death; see Table 54.8. No particular difference observed between species of wood; relative toxicity appeared to increase with increasing pyrolysis temperature and introduction of air flow	66, 514–517

			Material	Description	Ref.
X	X	Mouse	Japanese cedar, untreated hardwood, fire-retardant treated plywood	"Wood-base materials treated to be fire-retardant or made difficult to burn by mixing the inorganic materials are fairly safe at around 350°C but at 500°C or higher, large quantities of CO are produced and there is a possibility of their being more hazardous than untreated materials" (232)	101, 231, 232
X	X		White pine, white oak, plywood, and Masonite	NBS Smoke chamber/gas chromatograph study of combustion gases showed major products to be CO, CO_2, and H_2O. White oak evolved 276 ppm/min CO and Masonite 208 ppm/min CO (high and low, respectively, of group studied). Methane, ethylene, propylene, methyl alcohol, acetaldehyde, and acetic acid also identified	518
X	X			Review: pyrolysis of cellulosic materials precedes ignition, weight loss slow up to 200°C. Wood lignins and hemicellulose undergo glassy transitions in this range; primary products 280–500°C range are CO, H_2, CH_4, CO_2, acetic and formic acids, ethanol, aldehydes, ketones, and tars	519
X	X		Untreated and variously impregnated samples of southern yellow pine	Decomposition in sealed tube (static study) showed that "creosote-treated pine . . . afforded a significantly lower quantity of CO than the untreated woods. . . ."	520
X	X			Review of pyrolysis in terms of major components: hemicellulose evolves more gases and less tar (no levoglucosan) than cellulose. Cellulose characteristically yields levoglucosan, which in turn pyrolyzes to water, formic acid, acetic acid, and phenols. Lignin is particularly productive of aromatic products. *Four distinct temperature zones in thermal decomposition of wood: >200°C, appearance of noncombustible gases; 200–280°C, much less water vapor, reactions endothermic; 280–500°C, active pyrolysis under exothermic conditions, leading to secondary reactions; >500°C, residue primarily charcoal, active site for secondary reactions*	521–523
X	X		Cottonwood, lignocellulose, others	Vacuum pyrolysis at 300–500°C yields tar containing mainly levoglucosan	524, 525

Table 54.24. (*Continued*)

Polymer	TLV	Type of Study — Ambient Temperature	Type of Study — Elevated Temperature	Type of Study — Effluent Analysis	Species Exposed	Description of Material	Procedure	Observations/Conclusions	Reference
				X			Review: "More smoke is produced under nonflaming combustion than under flaming combustion. . . . Results of tests with the NBS smoke chamber show that plywood treated with specific fire-retardant chemicals may give off more or less smoke than untreated wood depending on the chemicals employed and the conditions of burning"		526
			X			Red oak, white pine, veneer, plywood, hardboard, fiberboard, kraft paper	Fire performance determined according to Motor Vehicle Standard No. 302 (occupant compartments of motor vehicles)	"Only the 0.012-inch thick kraft paper burned at a rate in excess of the 4 inches per minute limitation of the standard. The other materials had zero or very low burn rates. Enamel and clear lacquer did not add any flammability by this test method to $\frac{1}{8}$" birch plywood	527
					Human		Documentation for TLV with critical review of proposed criteria for allowable dust concentrations: "The interpretation of the findings . . . would appear to be that there is no readily measurable limit for raw cotton dust that will completely eliminate Monday morning 'chest tightness' and reduction in 1-minute forced expiratory volumes"		496
Cotton	0.2 mg/m³ for raw cotton dust (fibers <15 μm long); *this value also OSHA standard*[a]	X							
			X		Rat		Slow and rapid combustion conditions as described in Table 54.10	Cotton the most toxic when degraded under rapid combustion conditions and among the less toxic of polymers when degraded under slow combustion conditions	109

(Table 54.25 summarizes representative studies pertaining to cotton dust disease and its control.)

4366

	Animal		Material	Conditions	Results	Cited in
X	Rat		Pyrolysis products, samples of equal weight ("1.2 g per 100 mm")		Deaths in 11 of 20 rats exposed to products generated at 250°C (effluent concentrations reaching 8000 ppm CO, 10 ppm HCN). No deaths in exposures to products generated at 200°C (effluent with 50 ppm CO, no HCN)	Cited in 63
X	Rat		Cotton flannel and twill	Exposure 2 hr, to products resulting from 3-min burn	"Untreated samples proved to be the least toxic. Treatment with flame retardants increased the toxicity of combustion products when burned under the conditions described"	528
X	Mouse		Cotton fiber and fabric (Ref. 530)	Time to incapacitation and death from products generated in rising temperature program (Table 54.10, Ref. 66) shorter than median among a series of fabrics comparably tested; see Table 54.8. Weight loss in air and nitrogen relative to several other fabrics shown in Figure 54.9	529–531
			Cotton batting	"Smoldering combustion" at 300–400°C	Cotton yielded moderate amounts of acrolein, formaldehyde, and fatty acids when compared to other polymers; see Figure 54.2	103
X		X			Aldehydes found in wood smoke (see above this table) also identified in similar or lesser concentrations from "burning" of clothing	497
X					Treatment, 0.5–5 min at 160–200°C; some effect on properties after 5-min heating	532
X			Cotton cellulose linters, purified		"Oxidative reactions are responsible for acceleration in the rates of weight loss and depolymerization of cellulose . . . on pyrolysis . . . below 300°C. . . . Above 300°C, the rate of pyrolysis is essentially the same in both air and nitrogen"	533
X		X	Cotton fabrics treated for flame retardation		Relative quantities of CO and CO_2 very heavily dependent on experimental conditions. HCN formation in 2 samples of flame-retarded cotton relatively slight in relation to fiber weight	534

Table 54.24. (Continued)

Polymer	TLV	Type of Study			Species Exposed	Description of Material	Procedure	Observations/Conclusions	Reference
		Ambient Temperature	Elevated Temperature	Effluent Analysis					
			X	X				"In the oxygen atmosphere the depolymerization of cotton begins already at about 200°C. At 250°C definite exothermic degradation reactions set in, and above 350°C rapid charring occurs"	535
			X	X		Cotton		At 400°C "only the cellulose fibers (cotton, viscose) develop CO in relatively significant quantities"	536
			X	X		Cotton, cellulose, and cellulose with various chemicals		Products determined by pyrolysis gas chromatography included CO, CO_2, H_2O, CH_4, C_2H_6, ethene, propane, propene, butene, acetaldehyde, furan, 2-methylfuran, methyl ethyl ketone, 2,5-dimethylfuran, 2,3-butanedione, methyl isopropenyl ketone (?), toluene, acetol, acetone, acetic acid, furfural, formic acid, 5-methyl-2-furfural, 2-acetylfuran, and furfuryl alcohol	537–539
			X	X				"Formaldehyde in appreciable concentrations is found in the fumes evolved by burning and smoldering cotton fabric under conditions of oxygen deficiency"	401
Flax, jute, hemp, and sisal	No specific TLV; see cotton above	X			Human	Soft hemp "cultivated and biologically retted in northeastern Yugoslavia"		Survey^c of cardroom and spinning room workers showed symptoms of byssinosis in 39% of 102 non-smoking female hemp workers. Tests indicated a "chronic effect of hemp dust on ventilatory function." Concise literature review	540
		X				Hemp		Review of earlier studies on lung mechanics in hemp workers, type of response, and relation to cotton dust	541
		X			Human	Flax		"No cases of byssinosis were found among workers in a flax plant which produces yarn by chemical degumming instead of biological retting. ... These findings support the view that the agent in flax which causes symptoms of byssinosis originates during biological retting of flax and is absent from unretted flax"	542

Substance	TLV	Species			Comments	Reference
Flax		Human	X		Factory survey, 176 linen workers, clinical examination and lung function tests showed workers with symptoms of byssinosis, 16% with chronic bronchitis. Symptoms and alteration in lung function related to duration of exposure. Factory airborne level >1 mg/m³	543
Sisal		Human	X		Survey in 6 Tanzanian factories. Workers in brushing exposed to sisal dust for 12 years had 48% incidence of byssinosis; smoking also a factor	544
Jute and hemp		Human	X		Working population simultaneously exposed to both dusts with atypical tightness of chest in 7%. Incidence of chronic bronchitis greater than in controls. Statistically significant reduction in 1-min forced expiration volume in all exposed workers at the end of shift	545
Rayon	No specific TLV; see Section 1.4	Mouse	X	X	Time to incapacitation and death longer from products generated in the rising temperature program (Table 54.10, Ref. 66) longer than the median among a series of fabrics comparably tested; see Table 54.8. Death attributed to CO (Ref. 547). Addition of an "organic phosphorus-sulfur based" flame retardant "appeared to increase char yield but had slight if any effect in reducing time to death"	66, 547
			X	X	"Aldehydes are present in very small amounts, if any, in the combustion and decomposition products of . . . acetate rayon" (see this ref. above, this table)	401
Cellulose acetate	As rayon above		X	X	Very thin films or bulk material degraded under vacuum between 230–320°C; main volatile product from films was acetic acid, some water also formed; tar from bulk degradation contained acetyl derivatives of D-glucose	549
			X	X	Change in sample weight at burning temperature <200°C (procedure, Table 54.12)	48
Cellulose nitrate	As rayon, above	Human	X	X	Review, long-term survivorship study: on May 15, 1929, "gases from the decomposition of an estimated 50,000 nitrocellulose x-ray films ignited and three explosions shook the Cleveland Clinic. . . . [The investigating committee] concluded that both CO and HCN contributed to causing the death of the 97 acutely fatal cases"	550

4369

Table 54.24. (Continued)

| Polymer | TLV | Type of Study | | | Species Exposed | Description of Material | Procedure | Observations/Conclusions | Reference |
		Ambient Temperature	Elevated Temperature	Effluent Analysis					
			X		Rat	Celluloid	DIN 53436 protocol as described in Table 54.10	200°C degradation products lethal, caused 60% COHb (data ≥300°C degradation, see Table 54.7)	67, 379
			X		Rat	"Nitrocellulose"	Comparison of toxicity of pyrolysis products generated from various polymers in standardized test; deaths from products generated at <200°C (only material so determined; PVC and several wood products were "<300")		Cited in 63
Ethyl cellulose	As rayon above		X	X				"Volatile products from ethyl cellulose include H_2O, CO, CO_2, C_2H_4, C_2H_6, C_2H_5OH, CH_3CHO, unsaturated aliphatic compounds, and furan derivatives." Products identified from degradation of thin films mainly water, ethanol, and acetaldehyde; products of bulk degradation separated into yellow-brown tar, liquid, and gas fractions (procedure, see above this table)	549

[a] Summarized as described in Table 54.10.
[b] "High impact polystyrene" and one of two samples of flame-retarded polycarbonate had the two lowest ALC_{50s}, 11.6 and 12.0 mg/l, respectively.
[c] Cosponsored by U.S. Public Health Service.
[d] See text for additional discussion of OSHA standard.

Dose–Response Curves
Douglas Fir

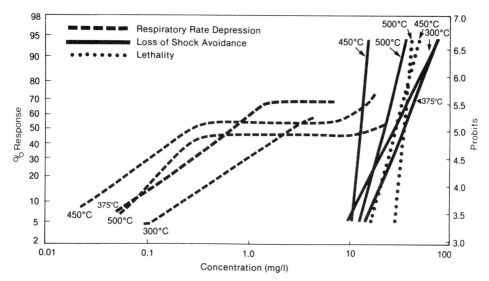

Figure 54.8 Dose-response curves for combustion atmospheres of Douglas fir: respiratory rate depression, loss of shock avoidance, and lethality as determined in rats with experiments conducted in three nonflaming modes (300, 375, and 450°C) and one flaming mode (500°C). Reproduced with permission from Burgess (72).

The permanent Occupational Safety and Health Administration standard on cotton dust (567) provides mandatory requirements for the control of employee exposure. The permissible exposure limit in yarn manufacturing is 0.2 mg/m³ of lintfree respirable cotton dust (unwashed cotton). This is a mean concentration averaged over an 8-hr period, "as measured by a vertical elutriator or a method of equivalent accuracy and precision." [Figure 54.10 shows the vertical elutriator cotton dust sampler (or its equivalent) that is specified in the OSHA standard; several suggested equivalent devices appear promising (573).] Similar limits are 0.750 mg/m³ in the slashing and weaving areas and 0.5 mg/m³ in other areas. Employees must receive written notification of exposure measurements and any corrective action required. The sampling program must include "at least one determination during each shift for each work area." The initial monitoring is to be repeated whenever the employer has any "reason to suspect an increase in employee exposure," or at least every 6 months. If feasible engineering and work practice controls are insufficient to reduce exposures to the applicable permissible exposure limit, the standard requires that the lowest feasible exposure limit should be supplemented by one of several types of respirator (type depends on whether the cotton dust concentration is 5, 10,

Table 54.25. Highlights in the Study of the Respiratory Effects of Cotton Dust, 1970–1979[a]

	Type of Study			
Clinical Data	Atmospheric Analysis	Means of Control	Scope	Observations/Conclusions	Reference
X			Clinical evaluation	Reliance on the forced expiration volume in 1 sec (FEV_1) as an objective basis for management: "Most individuals with Grade 1, 2, and 3 byssinosis have a moderate to marked decrease of [FEV_1] after six hours of dust exposure. However the evidence of no decrement in FEV_1 does not preclude the diagnosis of byssinosis in persons with symptoms" (551)	551, 552
X	X		Epidemiologic survey of 441 of 443 employees at cotton synthetic blend mill	Employees diagnosed as byssinotic included 20% in preparation areas, 20% in yarn processing areas, and 6% of all. Byssinotic risk determined by respirable dust levels, not total dust. Byssinosis and chronic bronchitis both influenced by cotton dust exposure and cigarette smoking	553
X	X		Evaluation for threshold limit value	TLV = 0.2 mg/m^3 of dust, fibers less than 15 µm in length: "The limitation on fiber length restricts the sample to those fibers that contain the active agent. The limit is intended to prevent Monday morning tightness in most of the workers so that the more susceptible may be detected and transferred out of the exposure before irreversible damage to health results[b]	496
X	X		Epidemiologic survey with development of dose response curves based on vertical elutriator sampling	Vertical elutriator sampling found a practical means of sampling; based on log-probit curves, "a reasonably safe level of lint free cotton dust is 0.1 mg/m^3, a level at which nearly 94% of the population exposed were found to have no symptoms of	554

4372

		Description	Findings	
	X	Respiratory screening program of 10,133 employees in 19 plants	byssinosis. A separate level of 0.75 mg/m^3 is for slashing and weaving areas. . . . Smokers had a significantly higher prevalence of byssinosis"	555
X	X	Comprehensive review of 116 references, recommendations for OSHA standard	Marked relationship between byssinosis and bronchitis, also between byssinosis and lowered pulmonary function; cigarette smoking increased incidence of bronchitis and lowered pulmonary function	
			"NIOSH cannot recommend an environmental limit of cotton that will prevent all adverse effects on workers' health. . . . Any permanent standard [should] also incorporate a program of medical monitoring and management, work practices, and administrative controls, as well as the lowest feasible environmental limit which has been indicated to be less than 0.2 mg lint-free cotton dust/cu m of air"	556
X	X	Comparison of various methods of cleaning or treating cotton to prevent byssinosis	Tests with 16 subjects for subjective symptoms, change in FEV$_1$, and five dust levels over 6 hr of exposure showed *washing eliminated detectable biologic effects* but left cotton difficult to process; steaming, particularly autoclaving, reduced byssinotic activity	557
X	X	Epidemiologic survey of 6,361 cotton textile employees in 14 plants	Majority of complaints of byssinosis among relatively small subset of employees located in high dust work areas of opening, picking, and carding. Byssinosis significantly associated with byssinosis in opening, picking and carding areas; 3% of population with subjective symptoms (history), 0.8% with both symptoms and objective sign ($\geq$10% decrement in FEV$_1$ during working day)	558
	X	Comparison of high volume sampler, multistage cascade impactor, and modified vertical elutriator	Dust concentrations of high volume sampler and modified vertical elutriator indistinguishable; cascade impactor gave approximately 40% lower value than that shown by high volume sampler	559

4373

Table 54.25. (Continued)

| Type of Study | | | | | |
Clinical Data	Atmospheric Analysis	Means of Control	Scope	Observations/Conclusions	Reference
	X		Evaluation of Lumsden-Lynch vertical elutriator	Detailed description of equipment, operating techniques, potential sources of errors and recommended sampling procedures; see Figure 54.10. 6-hr sampling more efficient at relatively low dust levels, 3-hr samples more efficient at medium and high dose levels	560, 561
			Chemical composition, related data	Composition of respirable dust affected more by growing location than variety; 81–95%, by weight, <15 μm in diameter. Differences in dust composition often greater from one plant to another than among processing stages within same plant	562, 563
			In vitro testing	Histamine present in cotton dust and plant parts, particularly bracts; the most potent extract released limited amount of histamine in swine lung	564, 565
			Literature review of suggested causative agents and means of control	An aminopolysaccharide in cotton dust considered a major cause of byssinosis. Bacteria and fungi not cause of byssinosis but studies cited to show relation of byssinosis to chemicals produced by bacteria or fungi. Water washing, at present a costly means of control, has definite potential (see directly below)	566
X	X	X	Proposed permanent OSHA standard providing mandatory requirements for the control of employee exposure to cotton dust (with certain provisions for variances)	Permissible exposure limits as summarized in Section 7.1. *Handling or processing of washed cotton is specifically exempted from provisions of this standard*	567

X	Reduction of dust levels by additive treatments	Hydrocarbon "mineral oil" and a formulation of hydrocarbon plus surfactant found effective as additives; when applied at levels below 0.4% prior to carding they can reduce dust levels in carding by 40% to 70%; considered a cost-effective means for controlling dust in the range 300 to 600 $\mu g/m^3$	568, 569
X	Sampling efficiency of vertical elutriator	Computer model developed to predict integral efficiencies for systematic assortment of size distributions of cotton dust	570
	Analysis of ash and protein contents of trash samples collected in model card room	"Trash samples from cotton grown under dry, arid conditions tend to have relatively high inorganic contents. Steaming tends to reduce the inorganic fraction of both the raw samples and their dried aqueous extracts but has little effect on the protein content"	571
	Comprehensive review	Discussion of research related to chemical composition of cotton dust and the components responsible for byssinosis; also of water washing, steam treatment, gas treatment, additives, and other possible means of controlling incidence of byssinosis	572
X	Investigation of short-term sampling devices as alternate to "bulky, noisy, heavy" vertical elutriator, which also requires 2–8 hr sampling time and 110-V current	Measurements by instruments that utilize either a quartz crystal microbalance (2 min sampling time) or the attenuation of beta radiation (12 min sampling time) can be correlated with measurements by the vertical elutriator for concentrations up to 1 mg/m^3. A third instrument that uses light scattering appears suitable for study of particle size distribution and concentration	573

[a] Only selected data are listed.
[b] The TLV value of 0.2 mg/m^3 is listed as an intended change in the 1972 edition of Ref. 27.

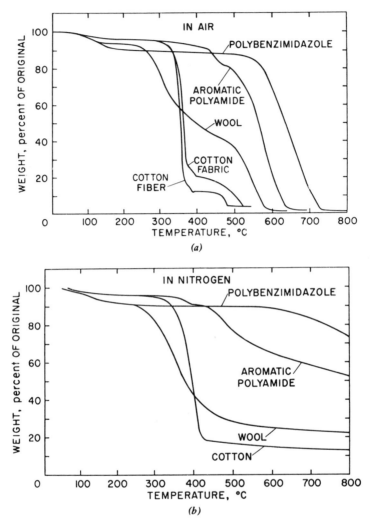

Figure 54.9 Thermogravimetric analysis of materials in (a) air and (b) nitrogen. From Hilado et al. (530). Reproduced by permission of *Journal of Combustion Toxicology*, Technomic Publishing Co., Inc., Westport, Conn.

100, or >100 times the applicable permissible exposure limit). The standard also stipulates programs for work practices, medical surveillance, and employee education and training.

The German MAK indices of 1.5 mg/m³ total dust roughly correspond to the index of 0.5 mg/m³ used in Great Britain (574). Values for several other countries are listed in Reference 28.

Dry wood and cotton should be handled as highly flammable materials.

Practices for handling regenerated cellulose and process waste depend primarily on the specific alkali, acids, or other materials involved. Rayon, cellophane, and cellulose derivatives may be handled without precaution or may require special precaution (generally due to flammability). See Sections 7.1.1 and 7.4 to 7.8, and Chapter 27, Volume 1.

7.2 Wood

7.2.1 Source and Composition

Commercial grades of wood come from many different types of living trees grown under varying conditions. The chief components of wood are cellulose from the cell walls, the related hemicelluloses, and lignin, a polyphenolic substance which acts as a cement. Cellulose and hemicelluloses account for some 50 and 10 to 30 percent, respectively, of wood substance. Lignin constitutes

Figure 54.10 Lumsden-Lynch vertical elutriator. From Neefus et al. (560). Reproduced by permission of the *American Industrial Hygiene Association Journal,* Akron, Ohio.

approximately 15 to 35 percent of wood and is generally higher in softwoods than in hardwoods. Miscellaneous constituents include other polysaccharides (starch, pectic substances, and water-soluble polysaccharides), other phenolic-type substances (tannins, lignins, coloring matter), terpenes, aliphatic acids, alcohols, proteins, and inorganic ash (480). "The elementary composition of most wood species used for construction is approximately 49 ± 1 percent carbon, 6.1 to 1.0 percent ash; the remainder is oxygen" (575).

Wood composition is also a function of anatomic structure, which varies widely between softwoods and hardwoods (477, 479, 480). The cells of hardwood are 0.7 to 2 mm long and spindle-shaped, whereas softwood has cells generally 3 to 4 mm long, with a larger lumen and frequent microscopic pits in the cell walls (477).

7.2.2 Properties and Applications

The structural variations in wood yield many grades with widely differing properties. Certain types of wood are particularly noted for appearance, texture, grain, insulation characteristics, or ease of fabrication, respectively. Dry untreated wood substance is porous and hygroscopic. Wood treated with inorganic fire retardants is usually more hygroscopic than untreated wood (526). See Tables 54.4, 54.6, and 54.23.

Wood was probably the first known fuel. Wood and wood-base products have for centuries been utilized for furniture, housing, and building construction. In the chemical industry, wood tanks are preferable to ordinary steel for the storage of cold, dilute acid. Plastic-lined or resin-impregnated wood tanks can combine structural advantages with resistance to chemicals. Wood tanks do not give a metallic taste to vinegar or sour food stored in them, nor do they yield traces of iron that can reduce the brilliance of many dyes.

Currently available fire-retardant treatments are concisely reviewed in Reference 38. Additionally, see Table 54.24.

7.2.3 Skin Toxicity

Woods and Calnan (576) give an exhaustive, indexed, review of the worldwide literature and discuss 83 cases of dermatitis seen over a 20-year period. Reactions to wood are not novel. Irritation of the nose and eyes from sawdust was described by Ramazzini in 1700. Wood, usually as sawdust or splinter, may affect the skin or mucous membranes by mechanical action or by chemical irritation and sensitization. Irritant chemicals may be found in the sap or latex, as well as in the bark or cracks of certain types of trees. Irritant reactions would appear to be more common among lumbermen and sensitization dermatitis more common among carpenters, polishers, and finishers; however, most of the dermatitis that occurs in the woodworking trade may be attributable to additives (preservatives, polishes, etc.) rather than the wood itself (576).

Suskind (577) found that although an extensive variety of woods may produce dermatitis, allergic reactions from handling hemlock, spruce, pine, and cedar are uncommon. Western red cedar would appear to be an exception (578; see also references cited therein). The strongly allergenic rosewood is one of several tropical woods that have been reported to give erythema multiforme-like eruptions (579).

The irritant and allergenic chemicals found in woods (rather than plants) would appear to be primarily quinones. The list of known or suspected causes of dermatitis also includes terpenes, furocoumarins, phenols, and stilbenes. Saponins irritate mucous membranes but do not penetrate unbroken skin (576).

In exceptional cases alkaloids or glycosides may be absorbed through skin abrasions. Toxic alkaloids from woods such as yew or oleander have been assimilated into food when these woods were used as containers, spoons, or spits.

Hausen (580) found that 2,6-dimethoxy-p-benzoquinone could be isolated from 21 different wood species, most of them timbers of commercial value. This chemical sensitized guinea pigs in a maximization test but did not elicit irritation reactions when tested at 10 percent in 40 humans. In earlier work this substance had produced a positive allergic skin reaction in a person with wood dermatitis. The quinone is the oxidation product of sinapaldehyde, a normal lignin constituent of many species. [Sinapaldehyde (CAS No. 4206-58-0) is a hydroxy-dimethoxy derivative of cinnamaldehyde.]

7.2.4 Inhalation Toxicity, Thermal Degradation, and Carcinogenic Potential

Inhalation of wood dust can produce effects ranging from bronchial asthma and allergic or sensitivity reactions to pneumonitis, fibrosis, bronchiectasis, and cancer. The incidence of adenocarcinoma among certain furniture workers in Great Britain during the decade 1956 to 1965 "was about 1000 times greater than that of the general male population" (496). Oak, beech, mahogany, and to a lesser extent Douglas fir, cherry, chestnut, walnut, and others were implicated. The known data are consistent with a long-continued exposure to a minor irritant. Schoental (581) has commented on the carcinogenic potential of sinapaldehyde derivatives in relation to the use of wood shavings as bedding for laboratory animals, "spontaneous" tumors of laboratory animals, and the occurrence of nasal tumors in workers exposed to high concentrations of wood dust.

The thermal degradation of such a complex and varied substance as wood obviously presents many ramifications. Burning wood can readily generate lethal quantities of carbon monoxide and changing quantities of many other dissimilar substances. The work of Zikria et al. (497) indicates that the non-CO moiety readily causes more subtle but nonetheless toxic reactions. The thermal degradation of wood—or perhaps even one type of wood such as Douglas fir— is sometimes taken as an arbitrary standard in toxicity tests or building codes. This may indirectly give "the false impression that wood is a relatively safe material in a fire situation" (582).

Many types of wood and wood products are treated with formaldehyde (583). Current investigation of the release of formaldehyde in urea–formaldehyde foam (see Section 11) can be expected to call attention to any excessive use of formaldehyde in construction material.

Table 54.24 outlines the scope of the enormous volume of literature. Section 7.1 discusses the TLV and suggested workplace practices.

7.2.5 Degradation at Ambient Temperature

Most varieties of wood are subject to decay from microorganisms or termites. Moisture content slightly above saturation promotes fungal growth. Species low in decay resistance include some types of pine, spruce, fir, hemlock, and birch, whereas cypress, redwood, cedar, and black walnut have decay-resistant heart-wood (479, 480). DeGroot and Esenther (584) give a comprehensive review of specific organisms that affect wood and approaches used to prevent its deterioration (see also References 78 and 585).

7.3 Cotton

7.3.1 Source and Composition

Cotton is the seed hair of the cotton plant. Cotton fibers contain from 88 to 96 percent cellulose (typically 94 percent), plus small amounts of protein, pectic substances, ash, wax, malic and other organic acids, sugars, and miscellaneous substances. Longer staple varieties can measure up to 5.7 cm and are fine, silky, and soft, whereas shorter staple varieties tend to be coarser with length ranging to less than 2.5 cm (481).

7.3.2 Properties and Applications

Cotton fibers are highly absorbent and can retain up to 50 percent water (89). Fabrics have good strength both dry and wet and can withstand relatively severe laundering. They are readily ignited. Cotton fabrics have a comparatively low heat of combustion but a high rate of heat generation and heat transfer (586). Untreated cotton clothing therefore presents a distinctly greater burn hazard than wool. Many cotton fabrics are now treated for fire retardance, rot resistance, or crease resistance; this treatment usually reduces absorbency. For additional data on properties, see Tables 54.4, 54.6, and 54.23.

Cotton has traditionally been the world's most important textile fiber. This position is now challenged by competition from synthetic fibers, but its popularity continues where absorbency is important. Cotton linters are used as raw materials for certain grades of cellulose acetate (see Section 7.6.1).

7.3.3 Skin and Inhalation Toxicity, Thermal Degradation

Purified cotton fibers are considered nonirritating. Cotton may be the fabric of choice when an absorbent material is required in treatment of dermatitis (160). Analyses of 24 clothing samples (426) showed free formaldehyde levels ranging from very low (≥ 1, <50 ppm) to relatively high (approximately 1400 to 1500 ppm) compared to those in other types of fabric. Some levels of this additive (see Section 6.3.4) may be above that necessary to induce allergic dermatitis.

Finished commercial products do not present an inhalation hazard. Unwashed cotton, as routinely handled in textile mills, is associated with byssinosis. Byssinosis describes the insidious, often progressive respiratory incapacitation that is characteristically associated with exposure to dusts of cotton, flax, and hemp. The term "byssinosis" is derived from the Greek word for flax. A frequent description of the early stages is an "atypical chest tightness," most noticeable on return to work (hence the nickname, "Monday morning fever"). Symptoms are relatively brief and intermittent in the early stages and then gradually extend through the week, always with an exacerbation on return to the work area after a weekend or similar absence. After a period of some 10 years, symptoms progress from temporary irritation or incapacitation to persistent asthma, often with cough, mucous expectoration, or bronchial irritation. In the advanced stage increasing shortness of breath on exertion causes strain on the right side of the heart that may ultimately cause cardiac failure. Postmortem examination does not reveal any features specific for byssinosis.

The rate at which byssinosis develops depends on two factors: (1) the amount of the biologically active constituent(s) inhaled and (2) the susceptibility of the individual. Individual susceptibility varies widely and is increased by smoking. A Duke University group has recently suggested that "byssinosis does not evolve into its more advanced stages among nonsmokers but does so among smokers" (587).

Schilling's suggested system (588, also as cited in Reference 556) for grading of reactions is as follows:

Grade ½	Occasional chest tightness on the first day of the working week
Grade 1 (or I)	Chest tightness and/or breathlessness on the first workday of the week
Grade 2 (or II)	Chest tightness and/or breathlessness extending beyond the first workday of the week

Grade 3 (or III)   Grade 2 symptoms accompanied by evidence of permanent incapacity from "diminished effort tolerance" and/or reduced ventilatory capacity

The biological active agent is known to be water soluble and has been tentatively identified as an aminopolysaccharide. The amount of active agent varies with the type of cotton and particularly the amount of the foreign matter or "trash" left by machine picking. Machine picking tends to increase the amount of trash collected with the harvested cotton fibers. Although the amount of active agent varies in cotton dust from cotton of different grades and from different processes, determination of dust levels in areas where raw cotton is handled offers the best available measure of risk. Several groups of investigators (see References 496, 556) have endeavored to correlate the probable incidence of byssinosis with lintfree or flyfree dust, which is the sum of respirable (<7 μm) and medium length fibers (<15 μm). Based upon these recommendations, most notably those of Merchant and his co-workers, the base line of 0.2 mg/m^3 was recommended as the TLV and then adopted with slightly higher limits for areas of lower risk, as the OSHA standard (see Section 7.1).

The high risk areas are cotton preparation: opening, cleaning, and carding. Ginning, depending on the type of operation, dissipates relatively little active agent although it, too, is under investigation (589). Activities after the carding operation offer less risk, hence the somewhat higher standard levels. The reduction in byssinogenic activity after cotton leaves the carding room is in agreement with a British study (590). A "remarkably low" incidence of byssinosis—0.1 percent—was found in workers at mills that used an antiquated dusty "willowing" process of reclaiming cotton from dirty waste. A plant survey of 60 operatives revealed three borderline cases (grade ½) that did not qualify for disability.

Inhalation of cotton dust may also result in other respiratory conditions (556), particularly bronchitis. Over half the workers studied by Chinn et al. (590) had symptoms of chronic bronchitis. An increased incidence of lung cancer would not be expected, particularly in absence of smoking, on the basis of recent studies cited by Heyden and Pratt (587).

As noted above, unwashed cotton is highly flammable. The toxicity of the combustion products is primarily that of carbon monoxide. Additives, however, frequently contain nitrogen or phosphorous compounds.

Table 54.24 presents an overview of inhalation and thermal degradation data. Table 54.25 summarizes work of the 1970s related to the incidence and control of byssinosis.

7.3.4 Carcinogenic Potential/Tissue Reactions

No documented reports of specific neoplasms attributed to cotton per se are known. Cotton sutures in human tissues examined up to 7 years after the original implantation (591) showed a reaction similar to silk, but generally more cellular.

7.3.5 Degradation at Ambient Temperature

Untreated cotton, essentially a natural, ordered form of cellulose, is susceptible to degradation by cellulose enzymes (89, 592). Warm, damp conditions promote fungal growth and rot. See Section 7.2.5.

7.4 Flax, Jute, and Other Natural Fibers of Vegetable Origin

Flax, jute, and hemp are all bast or stem fibers while sisal is a leaf fiber (89, 90). In the U.S. market they are low volume, specialty fibers. Excess exposure to dust during processing provokes respiratory reactions similar to those observed with cotton. See Table 54.24 for details.

7.5 Rayon and Cellophane

7.5.1 Source and Composition

All methods of preparation essentially depend upon solubilizing short-fibered forms of natural cellulose, reshaping it into long fibers or film by extrusion through a spinneret or slit aperture, then immediately converting the extruded product back into solid cellulose.

Rayon was first commercialized in the nineteenth century by the now discarded Chardonnet process that used highly flammable cellulose nitrate (see Section 7.7). Next came the cuprammonium process, still used to a limited extent to produce extremely fine, silklike filaments. Today the most widely used process is the xanthate or viscose process (4, 482). Generally, alkali cellulose is prepared by reacting wood pulp with excess sodium hydroxide (or other alkali), followed by aging to permit partial depolymerization. The alkali cellulose is reacted with carbon disulfide to form sodium xanthate, which is then dissolved in alkali and extruded into an acid bath that converts the filaments or film into cellulose. These filaments may be stretched, desulfurized, washed, dried, or otherwise finished.

Extruding the viscose through a slit into an acid bath yields cellophane. Cellophane can be plasticized by washing the product with glycerol, propylene glycol, or polyethylene glycol (310, 477).

Regenerated cellulose may also be prepared by saponification of cellulose acetate.

7.5.2 Properties and Applications

The specific properties of the many types of rayon vary widely (482). Rayon is readily blended with other fibers. Rayon can contribute pleasing texture and touch quality ("hand"), moisture absorbency, or strength. Cellophane is noted for clarity, crisp hand, and dimensional stability. Coatings of nitrocellulose or saran reduce moisture and oxygen permeability; polyethylene or ionomer coatings reduce heat loss. Both rayon and cellophane can be highly flammable. See Tables 54.4, 54.6, and 54.23.

Industrial uses of rayon include reinforcing cords for tires, belts, hoses, as well as the "disposable" market in nonwoven fabrics. It is also widely used in textiles, although less extensively than formerly. Cellophane films are widely used for sausage casings (478).

7.5.3 Skin Contact, Inhalation, and Thermal Degradation Data

Unprocessed rayon does not cause dermatitis (160). Commercial fabrics may contain free formaldehyde or formaldehyde resins. Analyses of 12 samples of 100 percent rayon clothing showed free formaldehyde levels ranging from 15 up to 3517 ppm (426; formaldehyde additive, see Section 6.3.4).

The toxicity of the combustion products is that of carbon monoxide plus products from additives. See Table 54.24.

7.5.4 Carcinogenic Potential/Implant Studies

Regenerated cellulose per se is not known to be associated with the development of neoplasms. Subcutaneous implants of cellophane characteristically produce tumors at the site of the injection (593–595 and references cited therein). The incidence of these tumors varies with the physical shape of the implant and is now generally recognized as consistent with solid state carcinogenesis (see Section 1.4). A study with surgical cotton or cotton linters (from which cellophane had been manufactured) showed no sarcomas (593).

7.6 Cellulose Esters

7.6.1 Cellophane Acetate and Triacetate

Cellulose acetate (CAS No. 9004-35-7), also known as secondary acetate, is a partially acetylated cellulose. Cellulose triacetate (CAS No. 9012-09-3), also known as primary acetate, is an almost completely acetylated cellulose.

Preparation. Cellulose triacetate plastic is customarily prepared by reacting cellulose* with acetic anhydride in the presence of a catalyst, often sulfuric acid (596). Fibers are most commonly prepared with the same reactants. Fibers can also be prepared by a solvent process with perchloric acid catalyst that dissolves

---

* Usually wood pulp, although acetate from cotton linters has better color and solution clarity.

the acetate as formed, or by a nonsolvent process that yields a product physically similar to the original cellulose (487). Acetic acid is the most common solvent. Cellulose acetate is prepared by hydrolyzing the triacetate to remove some of the acetyl groups.

Properties and Applications. Properties of the plastics vary with acetic acid content and molecular weight. Acetyl content <13 percent results in a relatively insoluble product whereas products with higher acetyl content may have a narrow range of solubility in certain solvents (485). The plastics are noted for their clarity. Acetate and triacetate fibers have a bright, lustrous appearance and lower strength and abrasion resistance compared to most other man-made fibers. They are weakened by prolonged exposure to elevated temperatures in air. See Tables 54.4, 54.6, and 54.23.

The fibers are used primarily in textiles and household applications where esthetic appeal is desirable. The plastic is used in photographic film, sheeting, lacquers, and for molded or extruded products.

Skin Toxicity and Thermal Degradation. Cellulose acetate and other cellulose esters may be used for plastic film attached to adhesive tape (597), eyeglasses, hearing aids, and other products that contact the skin for prolonged periods (598). Dermatitis from the final plastic product is uncommon but "may result from pressure, chemical irritation, or allergy" (598). Identified allergens include resorcinol monobenzoate, dyes for plastic colorant, and triphenyl phosphate. See also Fisher (160).

Thermal decomposition data indicate that cellulose acetate would be expected to yield carbon monoxide, water, acetic acid, and related materials, depending on specific conditions. See Table 54.24 for details.

7.6.2 Other Organic Esters

Cellulose acetate butyrate is noted for toughness, clarity, and ease of processing (485). It is extensively used in lacquers—for example, with acrylics on automobiles. Compared to cellulose acetate, polymers of cellulose acetate butyrate are typically of lower density, slightly softer, and may have slightly lower water absorption values (10). The properties of cellulose propionate and cellulose acetate propionate polymers ordinarily lie between those of the acetate and the acetate butyrate polymers (485), although properties of all four may be considerably modified by plasticizer, chain length, or degree of substitution (10).

7.6.3 Cellulose Nitrate

Cellulose nitrate is also called nitrocellulose (a misnomer).

Preparation. Cellulose nitrate is prepared from cellulose and nitric acid, usually in the presence of sulfuric acid (4). Completely nitrated cellulose contains 14.4 percent nitrogen. The three main types (485) contain approximately 12, 11.5, or 11 percent nitrogen, respectively. Celluloid is cellulose nitrate plasticized with camphor.

Properties and Applications. Cellulose nitrate is known for its extreme flammability, which limits its commercial use. Dry cellulose nitrate may explode when subjected to heat or sudden shock (485); cellulose nitrate is therefore generally shipped or handled wet with water or alcohol. Physical properties vary with the degree of nitration, which in turn, is usually directed towards the proposed end use. Cellulose nitrate with about 12 percent nitrogen is compatible with many synthetic resins, soluble in ketones and esters, and is used in lacquer coatings. Cellulose nitrate with about 11 percent nitrogen is more thermoplastic, soluble in ethyl or isopropyl alcohol, and is used widely in flexographic inks for paper and foil (485). See Tables 54.4, 54.6, and 54.23.

The term "celluloid" is now used as a general term for plasticized cellulose nitrate compositions, which are extremely flammable and have fair to poor chemical resistance. Celluloid is a reasonably tough thermoplastic that can be molded for balls with "bounce" and small articles requiring pleasant appearance (10).

Inhalation Toxicity and Thermal Degradation. Toxic combustion products, mainly carbon monoxide and oxides of nitrogen, are readily generated at relatively low temperatures (<200°C). See Table 54.24 for details.

Aquatic Toxicity, Biodegradation, and Disposal. No acutely toxic effects were observed among several species of fish, algae, or invertebrates exposed to nominal concentrations of 1000 mg/l except in the case of the green alga *Selenastrum capricornutum* (599). Based on the data obtained, a water quality criteria of 50 mg/l was proposed. A combined chemical–biologic treatment has been suggested for disposal (600).

7.7 Cellulose Ethers

Water-soluble cellulose ethers are known as gums. The main commercial water-insoluble ether is ethyl cellulose.

7.7.1 Cellulose Gums

The major products (486, 601, 602) are sodium carboxymethyl cellulose (Cellulose Gum® or CMC), methyl cellulose (Methocel®), hydroxypropyl methyl cellulose (also Methocel®), hydroxyethyl cellulose (Natrosol®), and hydroxypropyl cellulose (Klucel®).

Preparation. Carboxymethyl cellulose is prepared by reacting alkali cellulose with monochloroacetic acid, either as the acid or sodium salt. The other products are formed by the reaction of alkali cellulose with alkyl halides or alkylene oxides (486).

Properties and Applications. Carboxymethyl cellulose is an anionic cellulose with exceptionally high water-binding capacity (602). It is soluble in both hot and cold water. Carboxymethyl cellulose is used as a thickening, emulsifying, and bulk additive in food products, pharmaceuticals, detergents, and in sizing of paper and textiles.

Methyl cellulose and hydroxypropyl methyl cellulose are nonionic and can function as polymeric surfactants (486). They are both soluble in cold water (601). Methyl cellulose and its modifications have the unusual property that they are more soluble in cold than in hot water (486). This thermal gelation property is widely utilized in adhesives, where it provides a quick set and control of penetration into the substrate. Applications include pharmaceuticals, cosmetics, agricultural products, also cement and plaster formulations.

Hydroxyethyl cellulose is nonionic and soluble in both cold and hot water. Solution viscosity decreases with increasing temperature. Major applications include coatings and petroleum production; others are pharmaceuticals, agricultural products, textile processing, and construction (486). Hydroxypropyl cellulose is also nonionic but has a thermal gel point and other properties quite different from the hydroxyethyl derivative. It functions as a protective colloid in suspension polymerization, in paint removers, cosmetics, and pharmaceuticals, and also as a protective coating and glaze for food items.

Additional properties are given in Table 54.23 and in References 423 and 486.

Oral Toxicity. Data on the very low oral toxicity of all these products extend back for several decades. Hake and Rowe (492) review a number of tests with carboxymethyl cellulose, including a 6-month study with dogs, a 1-year test with guinea pigs, and a 25-month study with rats. Three humans fed 20 to 30 g/day showed a depression of protein digestion and an increase in fat digestion. A 1968 report (603) of 2-year feeding tests indicates that rats and mice showed little adverse response from dietary levels of 1 or 10 percent sodium carboxymethyl cellulose and also methyl ethyl cellulose. Rats showed an increase in food consumption and slight retardation of growth consistent with the non-nutritive value of the diet.

Methyl cellulose and hydroxypropyl methyl cellulose have also been considered essentially innocuous (492) by the oral route on the basis of extensive feeding tests extending back to at least 1942. More recently, McCollister and associates (604) found no evidence of toxicity in either rats or dogs fed these materials at concentrations up to 6 percent (methyl cellulose, 10 or 4000 cP or 10 percent (hydroxypropyl methyl cellulose, 10 cP) for 90 days. Two-year studies in rats with methyl cellulose (15, 400, or 4000 cP) at dietary levels of 1 or 5 percent were also without effect. Braun and his colleagues (605) found no significant absorption of a single dose of 500 mg/kg of methyl cellulose (3300 cP) or five daily doses. After administration of the single dose, the measured amount of ^{14}C activity in the feces was 102 percent. No accumulation of ^{14}C activity was detected in the body or in selected tissues after multiple dosing.

Hydroxyethyl cellulose has been administered to rats in single oral doses as high as 23,000 mg/kg (50 percent in corn oil) without observed toxic effects (606). Rats maintained for 2 years on diets containing ≤5 percent hydroxyethyl cellulose showed no adverse effects (Smyth et al., as cited in Reference 492).

Hydroxypropyl cellulose has been fed to rats for 90 days at dietary levels of 5, 1, or 0.2 percent. The only differences observed between any groups were an increase in food consumption and decrease in food utilization among rats fed both the hydroxyethyl cellulose and the unmodified cellulose control (607).

Bachmann et al. (608) recommend 0.5 percent methyl cellulose as a suspending medium of choice for oral administration of water-insoluble components in experimental studies. This dose was not observed to alter mitochondrial function or activity of mixed function oxidases, whereas these effects were observed at a higher dose level of methyl cellulose and with various levels of sodium carboxymethyl cellulose, gum arabic, or gum tragacanth.

Repeated ingestion of a short-chained methyl cellulose (Cologel®) was not tolerated in two patients with an erratic history of fluid retention and obesity (609).

Skin Reactions. The cellulose gums have for several decades been considered nonirritating and nonsensitizing. A number of negative tests on animals and humans are cited by Hake and Rowe (492) and Greminger (486). In a rare exception, eczema of the hands of 8 years' duration in a Japanese baker was attributed (610) to carboxymethyl cellulose after a positive closed patch test (2 percent in petrolatum).

Carcinogenic Potential/Parenteral Studies. Data cited above do not indicate any carcinogenic potential in normal use (see particularly References 492, 603, 604). Tumors have been observed at the site of injection in some animals given massive, repeated subcutaneous injections. Hematologic and biochemical changes are more characteristic; death in dogs given repeated intravenous injections of methyl cellulose was attributed to renal failure. Hypertension, anemia, and arterial lesions have been observed in rats given intraperitoneal or intravenous injections (citations in Reference 492).

Biodegradation. Cellulose gums are generally slowly biodegraded. Wirick (611) reviews some general principles and notes that the biologic oxygen demand for carboxymethyl cellulose is about 3 percent of the theoretical amount of oxygen required for complete oxidation during the normal 5-day test period. Acclimated bacteria may, however, cause some 55 to 75 percent biodegradation, depending on the percentage of saturated groups originally present in the two samples; approximately 90 percent of the unsaturated groups of both samples degraded during the standard test period. Reported 5-day values for two samples with different properties were 11,000 and 17,300 ppm, respectively, compared to a value of 800,000 ppm for starch (as cited in Reference 486). Testing of methyl cellulose (612) with activated sludge indicates that 96 percent can be dissipated in 20 days. Reported values for the biologic oxygen demand of two samples of hydroxyethyl cellulose are 7000 and 18,000 ppm, respectively, after 5 days of incubation (as cited in Reference 486).

7.7.2 Ethyl Cellulose

Ethyl cellulose resins have been used in foods and pharmaceuticals as diluents, binders, and fillers. They are insoluble in water and are most frequently used in solvent mixtures containing 60 to 80 percent aromatic hydrocarbon and 20 to 40 percent alcohol; this percentage of alcohol results in minimum viscosity. The ethoxyl content of ethyl cellulose affects thermal behavior. The melting point of approximately 210°C with an ethoxyl content of 44 percent may decrease to approximately 170°C with an ethoxyl content of 48 to 49 percent (486).

7.8 Starch

Obtained primarily from corn, starch is receiving increasing attention as a readily renewable raw material (324, 613). Most of the common starches contain both the linear polysaccharide amylose and the branched polysaccharide amylopectin. Amylose, which has an average molecular weight of about one-half million, resembles cellulose except that the glucose units are linked through alpha rather than beta linkages. Amylopectin may have a molecular weight of 10 million; branching through alpha 1,6 linkages occurs at about 1 in every 25 units (4, 613). Amylose forms strong flexible films and amylopectin is useful as a thickening agent. Starch can be mixed with other polymers as a filler, extender, or reinforcing agent, or incorporated into the structure to take advantage of biodegradability or other characteristics (see biologic oxygen demand above, this page). Starch has been recently grafted to acrylamide and sodium acrylate to make an absorbent polymer (starch-g-poly, SGP, "Super Slurper," Reference 614).

Amylose and amylopectin are regarded as food rather than as food additives.*
With amidated pectin, the acceptable daily intake was temporarily estimated*
in 1975 at 0 to 25 mg/kg body weight pending completion of additional tests
within the next 5 years. Modified starches prepared from natural starches have
been increasingly used in food products over the past several decades (615,
616, and references cited therein). During the 1970s the World Health
Organization evaluated a series of these products as acceptable, subject to
certain limitations.* The following synopsis provides an entry to recent toxicologic and related biochemical literature.

De Groot et al. (615) fed five chemically modified starches, unmodified corn
starch, and unmodified potato starch to groups of rats at dietary levels up to
30 percent in 2-year tests and at 10 percent in a three-generation test. The
modified starches were acetylated distarch phosphate, acetylated diamylopectin
phosphate, starch acetate, hydroxypropyl distarch glycerol, and phosphated
distarch phosphate. No "distinct effect of toxicological significance" was observed although "the males fed the 30% level of any of the modified starches
showed a slightly increased degree and incidence of focal hyperplasia of the
renal papillary and pelvic epithelium, accompanied by calcified patches in the
underlying tissue."

Truhaut and his associates (616) fed rats a cooked diet containing 62 percent
acetylated distarch adipate, acetylated distarch glycerol, or unmodified maize
starch. No relation between the incidence and type of renal lesions and the
administration of these modified starches was observed in this test, which
included special attention to renal lesions. No particularly important toxicologic effects were observed; slightly lower weights among the treated groups
correlated with significantly greater fat deposits in controls. Femur measurements showed no differences in skeletal growth.

Anderson et al. (617) conducted 25-day feeding tests with miniature pigs
starting at 3 days of age. The diets were formulated to provide 24 percent of
calories from thin-boiling waxy corn starch (100 percent amylopectin treated
with hydrochloric acid), phosphated distarch phosphate, distarch phosphate,
and hydroxypropylated distarch glycerol. Growth, blood chemistry, and composition of liver and carcass were similar in all groups.

The indiscriminate use of modified food starches during the first 6 months
of life has more recently been "discouraged until further information is
available" (618). Three principal concerns were identified: the bioavailability of

* See Reference 23, *WHO Food Additive Ser. 5*, 1974, concerning amylose and amylopectin and *Ser.
8*, 1975, concerning amidated pectin. The modified starches receiving a temporary "not limited"
evaluation pending additional study included starch acetate, hydroxypropyl starch, hydroxypropyl
distarch glycerol, distarch phosphate, phosphated distarch phosphate, acetylated distarch phosphate,
distarch glycerol, acetylated distarch glycerol, and acetylated distarch adipate (*Ser. 1*, 1972). Oxidized
starches and hydroxypropyl distarch phosphate subsequently were evaluated as "acceptable daily
intake . . . deemed not necessary" (*Ser. 6*, 1975).

the starch itself, the potential for gastrointestinal tract dysfunction, and the toxicologic implications of modifying chemical treatment.

Hydroxyethyl starch, which has been used as a colloidal plasma substitute, is known to be metabolized much more slowly than starch itself (619). Infusions in both normal volunteers (620) and in patients with renal disease (621) have resulted in elevated serum amylase; this hyperamylasemia was not associated with other clinical symptoms.

Starch granules have long been known to be absorbed intact into the blood or lymphatic system (622) from the intestinal tract (persorption). Use of starch as a surgical glove lubricant can result in postoperative granulomatous disease (623).

7.9 Other Polysaccharides

Many other polysaccharides and their derivatives are used as thickeners and stabilizers or are considered of potential economic importance (324, 624). Their use is not new; seaweeds and their extractives have been utilized as foods, medicinal agents, and fertilizers for some 5000 years (625). Agar, algin, and carrageenan, for example, are derived from seaweed, and gum arabic, guar gum, locust bean gum, and pectin come from plants or seeds. Xanthan gum, which is derived from the bacterium *Xanthomonas campestris,* is one of several biosynthetic gums (626).

Carrageenan is the generic name for calcium, sodium, potassium, and other salts of sulfated polysaccharides that are used extensively in food. Collins and associates (627) give a concise review of feeding tests extending back to 1959 and the use of carrageenan as a food additive. Pittman et al. (628) had found that carrageenan of high molecular weight was not absorbed and various feeding tests showed no adverse effects. However, dietary levels of 25 percent κ,λ-carrageenan were not tolerated. A three-generation reproduction and teratology test with rats fed 0.5, 1, 2.5, or 5 percent dietary levels of calcium κ,λ-carrageenan caused no effect on maternal weight gain, but dose-related and significant decreases in the weights of offspring were observed at weaning (627, 629). A later study (630) was conducted with calcium κ,λ-carrageenan, sodium κ,λ-carrageenan; and ι-carrageenan. The offspring of hamsters given doses ≤200 mg/kg on days 6 to 10 of gestation showed no fetotoxic or teratogenic effects.

Locust (or carob) bean gum, guar gum, and tara gum are galactomannans that provide no bioavailable calories or are at best a very limited source. In 1975 the World Health Organization estimated, on a temporary basis, that an acceptable daily intake of these gums did not have to be specified.* The

---

* See Reference 23, *WHO Food Additive Ser. 8,* 1975, for data on carob bean gum, guar gum, and tara gum; *Ser. 6,* 1975, for "xantham" gum (discussed below, this paragraph).

stipulation was made that additional specified studies should be conducted within a 5-year period. This organization also considered that xanthan (or xantham) gum caused no toxicologic effect in the rat at a level of 1000 mg/kg body weight and estimated the acceptable daily intake in man as 0 to 10 mg/kg body weight. Long-term tests were conducted with xanthan gum containing D-glucose, D-mannose, and D-glucuronic acid as the predominant hexose units (631).

8 POLYAMIDES AND POLYIMIDES

8.1 Overview

The synthetic polyamides and polyimides are all step-growth or polycondensation polymers. As a group, they are considered performance polymers, whereas the chain growth or addition polymers include the typical commodity polymers of polyethylene, polyvinyl chloride, and polystyrene as well as the high performance fluoropolymers. Polyamides are linked with the word nylon, the first major synthetic polyamide. Nylon was developed as a fiber in the 1930s and as a plastic in the 1940s. Polyamide polymers, however, basically include protein fibers such as wool and silk that have been commercially important for several millenia.

Nylon is a generic term for a synthetic aliphatic polyamide of well-defined structure and certain typical properties either as a fiber (6) or as a plastic (632). The name system reflects the chemical structure and preparation. Nylons 66, 610, and 612 are all prepared from a six-carbon diamine and a 6, 10, or 12-carbon dibasic acid, respectively. [The names can also be written in the style nylon 6/6, nylon 6.6, or nylon 6,6 to reflect the two-monomer origin; the simpler style of nylon 66—always "six six," or "six ten" for nylon 610—is usually preferred (633).] Nylons 6, 11, and 12 are prepared from an amino acid or derivative thereof with 6, 11, or 12 carbons, respectively. Nylon 66 was developed in the United States and nylon 6 was developed abroad (10), but both are now found worldwide. As a group the nylons are tough, strong, abrasion resistant, and resistant to alkalies, hydrocarbons, ketones, and esters.

Aromatic polyamides such as Nomex® were formerly called nylon (418), but aramid is now the official generic classification of the U.S. Federal Trade Commission and the International Standards Organization. Aramid denotes a long-chain synthetic polyamide fiber in which at least 85 percent of the amide linkages are attached directly to two aromatic rings, whereas nylon now indicates that less than 85 percent of the amide linkages are so attached (130). Aromatic polyamide fibers typically have many desirable properties of nylon fibers plus improved heat resistance and strength.

Polyimides are a completely synthetic class of polymers developed as a variation on polyamides to provide increased resistance to high temperature. Aromatic polyimides have exceptional heat resistance. Conventional tensile strength has been measured up to 500°C (634). Thermoplastic varieties, or those that become rubbery rather than melt at the glass transition temperature of approximately 310°C (635), retain high strength at almost 300°C.

Table 54.26 summarizes some typical properties. Figure 54.11 compares the performance of selected aramid and nylon fibers with other types in a standardized test.

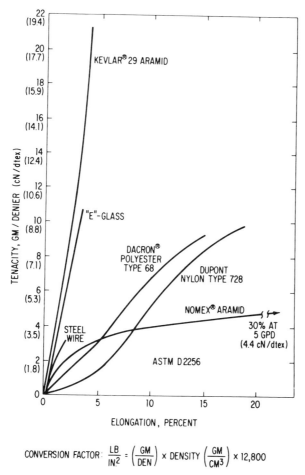

Figure 54.11 Stress-strain behavior of yarns, in textile units. From Wardle (642). Reproduced by permission of the *Journal of Coated Fabrics*, Technomic Publishing Co., Inc., Westport, Conn.

Table 54.26. Selected Properties of Representative Polyamide and Polyimide Polymers[a]

Property			A. Wool, Silk, and Nylon						Reference
	Wool	Silk	Nylon 66	Nylon 610	Nylon 612	Nylon 6	Nylon 11	Nylon 12	
Melting temperature (°C)			250–270	215–233	217	200–250	182–220	179	14, 85, 636
Glass transition temperature (°C)			~40 (conflicting data, most values ~40)	40 (variable data below 40)	46	40–87 (conflicting data, most values ~40–52)	43, 92 (affected by thermal history and relaxation effects, Ref. 86)	41	14, 86
Thermal deterioration	Discolors on prolonged drying >100°C	Brown on prolonged drying at 12°C	Fairly rapid oxidative degradation when heated >200°C in air; degradative discoloration may occur at 150°C (depending on type)			As nylon 66			14
Density	1.30–1.32	Raw silk, 1.36; weighted silk, ~1.6	1.20–1.25 (Ref. 85); 1.09, amorphous moldings; 1.13–1.45, crystalline moldings	1.152–1.189		1.220–1.25; 1.10, amorphous moldings; 1.12–1.14, crystalline moldings	1.04–1.05[b]	1.034–1.10	14, 85, 636
Refractive index, n_D			1.52–1.58	1.52–1.57		1.52–1.58	1.51–1.55		14
Water absorption (% (plastics))			8.5 ± 0.5 (moldings, 20–90°C, Ref. 636) ~1.5 (Ref. 14)	ca. 0.4		9.5 ± 0.5 (moldings, 20–90°C, Ref. 636); ~2.5 (can approach 100% on prolonged immersion, Ref. 14)	"Slightly less" than nylon 610 (can approach 2–3% on prolonged immersion)		14, 636
Water retention (% (fibers))			~25			15	3		14
Moisture regain (% (fibers))	13.6, recommended commercial value all yarns	11, commercial value	4.5, commercial value (Ref. 12); 3.9–4.2 (Ref. 14)	~1.7 (Ref. 14)		~15 (Ref. 14)	~1 (Ref. 14)		12,[c] 14
Solvents[d]	Hot conc. acids or alkalies (dissolution). Swells in water	Conc. aq. lithium thiocyanate (little or no decomposition); liquid ammonia (with ammonolysis). Swells in dilute alkalies, conc. organic acid (as acetic, formic)	*Room temperature:* trichloroethanol, phenol, cresols, formic acids, halogenated acetic acids, sulfuric acid, saturated acetic acids, sulfuric acid, saturated solutions of alcohol-soluble salts (as magnesium chloride in methanol). *At 120–180°C:* benzyl alco-	Chlorobenzene, diethyl succinate (79°C)		*m*-Cresol, chlorophenol, phenol, formic acid, acetic acid, trichloroacetic acid, sulfuric acid, phosphoric acid	Higher primary alcohols, DMF, DMSO		14, 88

Nonsolvents/relatively unaffected by	hol, diethylene glycol, acetic acid, formamide, DMSO	Water, organic liquids, liquids, dilute acid, alkalies	Hydrocarbons, chloroform, alcohols, ethers, ketones, esters	14, 88
Decomposed by	Concentrated acids and alkalies, especially when hot. Silk degraded by oxidizing conditions, especially hypochlorites	Hydrocarbons, aliphatic alcohols, chloroform, diethyl ether, aliphatic ketones and esters	Nylon in general: Concentrated mineral acids, hot dilute mineral acids, oxidizing agents, halogens (including chlorine-containing bleaching agents); hot concentrated alkales cause loss in weight	14

B. Several Aromatic Types

	Aromatic Polyamide		SP Polyimide	
	Poly-m-Phenylene-Isophthalamide	Poly-p-Phenylene-Terephthalamide	Poly(oxy-di-p-phenylene) pyromellitimide (Kapton®)	Reference
Melting temperature (°C)	350–400 (softening and extension)	500–560	None (begins to char >800°C)	85, 637, 638
Glass transition temperature (°C)	~275	~300–345		637
Decomposition temperature (°C)	440	>362 in air; >489 in nitrogen (differential scanning calorimetry); 590 (Ref. 637)		637, 639
Zero-strength temperature (°C)			815 (20 psi load for 5 sec)	640
Cut-through temperature (°C)			435 (1 mil); 525 (2.5 mil)	640
Density	1.14		1.42	85
Refractive index			1.78 (1 mil film, Becke Line)	640
Moisture absorption		2.9%, 5 days at 100% RH, 23°C (absorbs water reversibly)	1.3%, 50% RH; 2.9%, immersion for 24 hr	639, 640
Solvents			No known organic solvent[d]	640
Nonsolvents/relatively unaffected by			Some changes in mechanical properties after 1-year exposure to benzene, toluene, methanol, acetone at room temperature, also 100°C water for 4 days; no obvious change after 6 month exposure to 150°C transformer oil	640
Decomposed by			10% sodium hydroxide (5 days)	640

[a] Summarized as described in Table 54.9.
[b] The density of 1.04–1.05 for nylon 11 is from Ref. 14, which also gives values of 1.13–1.14 for nylon 6, 1.14 for nylon 66, 1.09 for nylon 610, and ~0.98 for nonfibrous polyamides.
[c] Standard D 1909-77.
[d] Solubility of nylon in hydrochloric acid at room temperature varies with type; proceeding from most soluble to least soluble the order is nylon 6, 66, 610, and 11. Nylon 6 is soluble in 4N HCl; nylon 11 is insoluble in 6N acid (14).
[e] Some derivatives have been structurally modified to provide solubility (641).

8.1   Production and Processing

Wool, silk, nylon, and aramid are processed as fibers. Nylon can also be readily processed as molded or extruded plastic or as film. In the United States "nylon" as a plastic usually refers to nylon 66, the first commercial product and still dominant in the U.S. market (632). Both nylon 66 and nylon 6 are used extensively for textiles in the United States (31). Nylon 6 is more prominent abroad. Polyimides are used as molded plastic, plastic film, and fiber. Table 54.2 gives amounts of production for some major categories.

8.1.2   Analysis and Specifications

Anton (643) discusses the identification of nylons by solubility tests, color tests, thermal methods, and infrared spectra. Many basic types of nylon can meet the requirements of the Food and Drug Administration for food-related contact in particular applications (92). See also Section 1.3 and directly below.

8.1.3   Toxicologic Potential

Available data concerning residual reactants or solvents in the synthetic polymers of this group are meager. Negligible amounts of residual reactants would be expected on a stoichiometric basis in nylons formed by polymerization of a nylon salt and the known aromatic polyamides or polyimides. In the case of nylon 6, residual caprolactam has been present in the polymer but little concern on this point appears to have developed (Reference 31 and citations therein). Residual solvent might be a concern in view of the powerful solvent systems required for polymerization and processing of the high melting aromatic polymers. Examples of such solvents for aromatic polyamides are dimethyl-acetamide, N-methylpyrrolidone, hexamethylphosphoramide, tetramethylurea, and mixtures of these solvents, which may be used with inorganic salts to increase solvating power (644). Subsequent processing—particularly as a textile—would generally remove most or all polymerization solvent.

More concern has been expressed concerning exposure to airborne particulates from these polymers than any low molecular weight constituents. Recent data indicate that prolonged exposure to an excess of at least some types of leather dust may be carcinogenic. One retrospective survey (645) showed a statistically significant excess of cancer among wool fiber preparers but not among any of four other groups of wool workers. Wool, silk, and nylon textiles have occasionally been connected with miscellaneous respiratory effects or skin reactions in humans. Skin reactions are relatively infrequent and, at least in the case of wool, often associated with an atopic history. Allergic dermatitis from currently manufactured, commercial nylon fabric would appear to be rare and associated with the dyed products. See Section 9.1 concerning several clinical reports that refer to inhalant or related reactions but lack specific exposure data.

Thermal degradation of polyamides and polyimides can yield toxic gases, particularly carbon monoxide, hydrogen cyanide, or ammonia. The temperatures at which these gaseous products are released can vary appreciably.

Table 54.27 provides an entry to the extensive literature pertaining to inhalation and thermal degradation effects. Figure 54.12 compares the evolution of ammonia from wool and nylon 6 with that of several other polymers when the materials were heated under nitrogen.

8.1.4 Workplace Practices and Standards

Inhalation of excessive dust or fume should be avoided. In the event of fire, precaution should be taken against inhalation of toxic gases and smoke.

8.2 Wool, Silk, and Leather

8.2.1 Wool

Source and Composition. Wool is shorn or collected from a variety of sheep and several types of goats. Wool fibers can be classified as fine, crossbred,

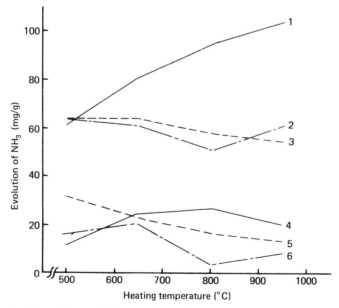

Figure 54.12 Evolution of ammonia from various polymers in nitrogen gas. Materials: 1, melamine resin; 2, urea resin; 3, wool; 4, nylon 6; 5, polyacrylonitrile; 6, polyurethane flexible foam. From Morikawa (392). Reproduced by permission of the *Journal of Combustion Toxicology*, Technomic Publishing Co., Inc., Westport, Conn.

Table 54.27. Overview of Inhalation and Thermal Degradation Data on Polyamide and Polyimide Polymers[a]

Polymer	TLV	Ambient Temperature	Elevated Temperature	Effluent Analysis	Species Exposed	Description of Material	Procedure	Observation/Conclusions	Reference
							Type of Study		
Wool	No specific TLV; see Section 1.4[b]	X			Human			Survey in 2 Yugoslavian textile mills: workers exposed to wool dusts >10 years showed significant differences in lung function (maximal expiratory flow curves, forced expiratory volume in 1 sec) and more signs of chronic respiratory effects, compared to workers exposed <10 years	646
		X			Human			Survey of 569 women and 478 men in Polish woolen mill: chronic bronchitis correlated with employment >10 years and increased dust levels (up to 24.3 mg/m^3 in weaving preparation), also found twice as often among smokers compared to nonsmokers	647
		X			Human			Survey of 800 Polish workmen: symptoms of chronic nonspecific pulmonary disease correlated with age (up to or over age 40), smoking or not smoking, and duration of employment (>15 years or ≤5 years)	648
		X			Human			Survey of 175 wool workers in India: incidence of chronic bronchitis among smokers and nonsmokers in workers nearly the same (18 vs. 14%); in control group incidence 25% among smokers, zero among nonsmokers	649
		X			Human			"In a series of 40 cases of occupational asthma . . . 19 cases were sensitive to sheep wool and sensitization had been acquired in the course of occupation in the woolen industries and in the clothing trade"	650
			X		Rat	Sheep's wool		Degradation products generated ≥300°C may be lethal, cause significant amount of COHb ≥400°C; see Table 54.7 (procedure, Table 54.10)	62, 379
			X		Rat			Mortality observed from degradation products generated at 300°C but not 250°C (procedure, Table 54.24)	63
			X		Rat	Worsted flannel		ALC$_{50}$ = 12.7 mg/l, products among the more toxic of polymers tested in this series (procedure, Table 54.24)	512

				Ref.
X	Rat, rabbit		Decomposition products from wool considered less toxic than those from polyacrylonitrile or silk; cause of death considered neither CO intoxication nor cyanide poisoning (procedure, essentially as described Table 54.21)	249, 388
X	Mouse		Time to death from products of 100% wool[c] shortest among a group of polymers so tested; see Table 54.8. Thermogravimetric analysis showed thermal stability greater than with cotton, generally less than with aromatic polyamide or benzimidazole; see Figure 54.9	66, 515, 530, 651–653
X	Mouse		Mortality observed from degradation products generated at 500°C but not at 350°C. Wool considered ~10 × more toxic than wood, less toxic than polyacrylonitrile (procedure, Table 54.10)	101, 654
X			Discussion of bioassays that "do not 'rank' wool as the fibre which produces the most toxic combustion products" and reply	655, 656
X		X	Evolution of HCN less than that observed with polyacrylonitrile or nylon. Wool yielded relatively high amounts of ammonia. See Figure 54.12 (procedure, tests under nitrogen, as described Table 54.21)	392
X	Untreated or treated with various flame retardants	X	Degradation up to 600°C: pyrograms in air of treated and untreated samples identical with those in helium at ≤450°C; at ≥500°C, pyrogram changes completely	657
X	Untreated or treated with nitrogen-free flame retardants	X	Thermal degradation in helium, 600–925°C: "HCN production depends on the chemical composition of the retardant and on the temperature of exposure"	658
X		X	Pyrolysis under vacuum yielded HCN, H_2, CH_4, CO_2, CO, H_2S, SCO, CS_2, CH_3SH; "combustion" with ≤8 l/hr air at 500–1000°C showed that increasing air flow lowered the temperature for relative high yields of HCN and H_2S	659
X	Fiber	X	COS, H_2S, C_2S, but not SO_2, evolved from 0.9 to 8.4 mg/g of wool under nitrogen or air flow, 600–800°C; SO_2 detected as principal gas under flaming conditions in O_2-enriched atmosphere	660

Table 54.27. (*Continued*)

Polymer	TLV	Type of Study			Species Exposed	Description of Material	Procedure	Observation/Conclusions	Reference
		Ambient Temperature	Elevated Temperature	Effluent Analysis					
			X	X			Degradation at 800°C in air: wool generally evolved slightly less HCN per weight of sample than acrylic fiber or nylon		396
			X	X		Flame-retarded fabric used in aircraft	Based on decomposition of wool and other materials in combustion tubes and under flaming conditions: gas yields were more reproducible for gases such as CO and HCN; "the utility of measuring NO_2, SO_2, and HCHO for the purpose of ranking interior materials is questionable . . ."		46
			X	X			Degradation of wool in "cup" furnace (**Ref.** 382)	Concentration profile of HCN for wool degraded at several temperatures	661
			X				Material heated to 500°C, 20°C/min: wool ranked behind nylon 6 and polyurethane but ahead of wood and polymethyl methacrylate in amount of smoke emitted (per gram of material/1 cm light path)		662
			X	X			Free burning: products listed = CO_2, CO, cyanide (HCN and RCN), trace "Chlorine–HCl." Smoldering: same products, also ammonia		663
			X	X			Within test limitations, results indicate that "the life hazards presented by the fumes from the acrylic fiber, silk, and woolen fabrics under fire conditions are similar"		401
			X				Manikin and sensor tests showed less transfer of heat than observed with cotton, acrylic, nylon, polyester, or polyester/cotton blend		586
Silk	No specific TLV; see Section 1.4	X			Human		Survey in >150 workers with occupational bronchial asthma and asthmoid bronchitis, who had been exposed to silk "at all stages of its obtaining and processing"; natural silk considered a strong occupational allergen		664

4400

Material	Species			Description	Reference
	Rat, rabbit	X		Decomposition products from silk considered comparable to those of polyacrylonitrile in toxicity, associated with a relatively high blood cyanide (procedure, essentially as described Table 54.21)	249, 388
	Mouse	X		Time to death from products of 100% silk longer than that observed with 100% wool but shorter than observed with most polymers in series tested; see Table 54.8 (procedure, Table 54.10)	66
	Mouse	X		Mortality observed from degradation products generated at 350°C (procedure, Table 54.10)	101
		X		Free burning: products listed = CO_2, CO, formaldehyde, cyanide (HCN and RCN), and ammonia. Smoldering: same products but amounts (g/g of sample) less for all products except ammonia	662
		X		See above this table under wool	401
Leather	Human		X	Study of boot and shoe workers in Northamptonshire, England: incidence rate for all types of tumors of nasal cavity and sinuses = 0.14 per thousand (press finishing) compared to an approximate rate of 0.01 per thousand in the general population. "The risk is virtually limited to the [workers] who have been exposed to dusty work in the preparation and finishing departments . . ." (667)	665–667
	As silk above		X		
	Rat	X		Degradation products formed at 300–600°C lethal, caused ≥54% COHb; see Table 54.7 (procedure, Table 54.10)	62
Nylon	Human		X	As summarized in Section 9.1.4	
	As silk above		X		
	Rat	X		Degradation products formed at 500–600°C lethal, caused 22–53% COHb; see Table 54.7 (procedure, Table 54.10)	62
Nylons 66 and 6	Rat	X		Fuel load of 7 g produced 50% mortality, compared to 8–11 g for several other materials; see Table 54.21	390
Nylon fabric					
Nylons 66, 6, 610, and 611	Guinea pig	X		Decomposition at 500°C in air current: all 4 materials produced HCN, ammonia, CH_4, C_2H_6, C_3H_2, and C_4H_{10}; vol. % of HCN greatest in effluent from nylon 6, vol. % of ammonia greatest in effluent from nylon 66. Guinea pigs exposed to gases of nylon 6 died when cyanide precipitate observed in effluent trap	668

Table 54.27. (*Continued*)

Polymer	TLV	Type of Study			Species Exposed	Description of Material	Procedure	Observation/Conclusions	Reference
		Ambient Temperature	Elevated Temperature	Effluent Analysis					
			X	X	Mouse	Nylon fiber/fabric (Ref. 66), polycaprolactam (Ref. 515)	Products of 100% nylon (all 3 types) among the least toxic as judged by time to death among a series of polymers so tested; see Table 54.8. CO considered an important toxicant in the pyrolysis of polycaprolactam (procedure, Table 54.10)	66, 515, 651
			X	X	Mouse	Nylon 66	Rapid deaths from products generated at 750°C attributed to HCN poisoning; no deaths in similar tests with 350 or 500°C decomposition (procedure, Table 54.10)	101
			X	X	Mouse	Nylon 66 (Ref. 232); polyamide (Ref. 653)	Products generated at 750 but not 350 or 500°C lethal; death attributed to CO and HCN. Products generated at 850°C considered about twice as toxic as those generated at 550°C (procedure, Table 54.14)	232, 653
			X	X	Mouse	"Polyamide" film, analyzed as nylons 6 and 66	Exposure 1 hr, film decomposed in airstream at 550°C. ALC = 79.6 g/polyamide per 1000 liters of air. Main contents of effluent = 2700 ppm CO, 700 ppm ammonia, small amount of propylene. Some mice died during exposure showed >55% COHb; some died at 1–2 days with pulmonary edema	669
			X	X		Nylon 6 with inorganic additives	Decomposition in air≥400°C: NaHCO₃, Na₂CO₃, Na borate additives decreased formation of HCN, and increased ammonia formation at 400°C	670
						Nylon 66	In mg/g sample, at 50 or 100 l/hr air flow respectively: 194 or 205 mg CO, 563 or 590 mg CO₂, 4 or 10 mg NH₃, 26 or 31 mg HCN, 39 or 40 mg CH₄, 82 or 94 mg C₂H₄, 7 or 15 mg C₂H₂ (procedure, Table 54.10)	104
						Commercial nylon 6 containing 12% nitrogen (Ref. 392)	Increased air flow results in decreased HCN evolved at temperatures 500–900°C. Evolution of HCN under nitrogen generally increases with temperature and is approximately proportional to nitrogen content at 900°C.	103, 392

		Material	Description	Reference
X	X	Nylon 6	Some ammonia, little formaldehyde evolved. See Figures 54.2, 54.7, and 54.12 (procedure, Table 54.21)	396
X	X		Degradation, 800°C in air: nylon generally evolved less HCN per weight of sample than acrylic fiber but more than wool	397
X	X		Significant quantities of ammonia evolved; amounts increased as decomposition increased from 400 to 1000°C. Other materials detected included alkanes, aldehydes, alkenes, alkylamines; HCN not listed (procedure, Table 54.21)	
X	X	Nylons 4, 7, 12, 66, 610 (also nylon 4,10, nylon 11,6, nylon 12,12)	Cyclic oligomers detected below 200°C; thermal decomposition above 350°C. Mass spectra recorded at 170°C and ~400°C are characteristic	672
X	X	Nylons 66 and 610	Degradation of nylon 66 under vacuum $\leq 1000°C$ yielded mostly NH_3, H_2O, CO, CO_2, cyclopentanone, and several hydrocarbons; similar procedure with nylon 610 showed H_2O, CO, CO_2, 1,5-hexadiene and other hydrocarbons as the major products, no ammonia detected	672
X	X	Nylon 66	Escape of volatile material from heating nylon above its melting point results in rapid gelation and color formation, even in the absence of oxygen	673
X	X	Nylon 66	HCN, CO, NH_3 evolved at 350°C in air, nitrogen oxides evolved at 600°C. Comparable temperatures for release of these gases in nitrogen identical or within 50°C (polyacrylonitrile yielded HCN at 250°C in air and nitrogen)	674
X	X	Nylon 66	Degradation 310–380°C: "H_2O, CO_2, cyclopentanone, traces of saturated and unsaturated hydrocarbons; purification from water and acid polymerization catalysts increases stability and decreases yield of CO_2"	Cited in 105
X	X	Nylon 6	Major products found during oxidative degradation to 1000°C were CO_2, H_2O, ε-caprolactam, methane, oligomers, also propene, propenenitrile, ethylene, acetonitrile, HCN; pyrolysis in helium gave high yield of oligomers, less ε-caprolactam, as main products	675

Table 54.27. (*Continued*)

Polymer	TLV	Ambient Temperature	Elevated Temperature	Effluent Analysis	Species Exposed	Description of Material	Procedure	Observation/Conclusions	Reference
								Type of Study	
			X	X		Nylons 6 and 66		Slightly more ammonia, slightly less cyanide and nitrogen oxide from Nylon 66 compared to Nylon 6	48
			X			Nylon		Molten droplets contained considerable energy but inhibited burning	586
Chlorinated and fluorinated polyamide			X	X	Mouse	Materials proposed for aircraft use	30-min exposure to products from pyrolysis at 700°C; LC_{50} values based on weight of sample charged	LC_{50} for fluorinated polyamide lowest in series of 7 polymers—approximately $\frac{1}{3}$ that of chlorinated polyamide, $\frac{1}{5}$ that of polyamide, and $\frac{1}{16}$ that of polyvinyl fluoride	676
Aromatic polyamide	As silk above	X		X	Rat	Poly(*p*-phenylene) terephthalamide (dust representative of workroom atmosphere)	4-hr acute exposure, also 4 hr/day, 5 days/week × 2 weeks; 150 and 130 mg/m³ in acute and repeated exposures, respectively[d]	Inactivity, shallow respiration during exposures; slow weight gain during repeated exposure series, normal weight gain in 2-week recovery period. Slight phagocytosis of foreign particles in lung found after 10th exposure, persistent to same degree 14 days later	677
			X		Mouse	Aromatic polyamide, Nomex® (Ref. 653)	Time to death longer than median among products of a group of polymers similarly tested; see Table 54.8 (procedure, Table 54.10). Thermogravimetric analysis showed less weight loss than with polybenzimidazole until ~450°C; see Figure 54.10	66, 530, 653
			X	X		Kevlar® [aramid fiber]	Fast pyrolysis of 2.8 mg by introduction into 650°C furnace yielded 63% CO, 35% benzene, 2% toluene, plus 0.6 mg residue. Slow pyrolysis at 40°C/min for 15 min yielded "at least 15 compounds and a series of nitrogen containing cyclic components [but] no ammonia or hydrogen cyanide was observed"	678

Material	Species				Comments	Ref
Two samples of aromatic polyamide, 1 identified as Nomex®		X	X		See under wool above, this table	46
"Polyaramide," as Kevlar® [aramid fiber] woven blanket and as a Derakane® 510 Resin System		X	X		Impingement under atmospheric O₂ conditions, also thermal radiation under atmospheric and low O_2 conditions	Polyaramide itself resistant to stress, few toxic combustion products formed. Addition of resin for structural strengthening "resulted in a rapidly deteriorating environment under all three stresses" — 679
Polyimide — SP Polymer [Kapton® polyimide film]	Rat	X	X	X — As silk above	Dust levels as high as 15 mg/l (nominal) produced signs of discomfort, inactivity and deep difficult respiration. Pyrolysis products produced death in 20–140 min at 450–500°C, did not produce death in 4-hr exposures ≤400°C. CO detected in effluent with pyrolysis temperature of 300°C. Rats dying after exposure to products generated at 450°C showed microscopic evidence of edema, congestion, and occasional hemorrhage in lungs, no effects in liver or kidneys; death considered due to either or both CO and lung irritant(s)	680
Polyimide flexible foam	Mouse	X	X		Products of polyimide foams generally associated with relatively short times to death in rising and fixed temperature programs (see Table 54.10, Ref. 66). With polyimide flexible foam, "the toxicants causing death were evolved at temperatures below 600°C; . . . the principal cause of death was not carbon monoxide, but probably hydrogen cyanide and/or nitrogen dioxide" (682)	98, 681
		X	X		Thermogravimetric analysis in air showed 1-step degradation starting at ~500°C, with complete combustion at ~650°C. CO, CO_2, ammonia, cyanide identified in effluent	37, 682
H-film		X	X		Degradation 300–510°C: thermal oxidation occurred at 426°C in air, degradation at 510°C in nitrogen. Vacuum stability of film exposed to air at elevated temperatures significantly less than that of film as received	683

4405

Table 54.27. (*Continued*)

Polymer	TLV	Type of Study			Species Exposed	Description of Material	Procedure	Observation/Conclusions	Reference
		Ambient Temperature	Elevated Temperature	Effluent Analysis					
			X	X		Cured polyimide resin ("Skybond 700")	Decomposition in vacuum using simultaneous mass spectral and differential thermal analysis	"The volatile decomposition products were H_2, CO, CO_2, H_2O, HCN, and minor quantities of benzonitrile, benzene, methane, and ammonia"	684
			X	X		Polypyromellitimide H-film	Decomposition in vacuum pyrolysis at 540°C yielded CO and CO_2 plus smaller or trace amounts of hydrogen, water, hydrogen cyanide, benzene, benzonitrile; comparison of data with those reported in Ref. 686		685
			X	X		Polypyromellitimide H-film	Data "suggest that the primary scission occurs at the imide bonds, most likely followed by a secondary cleavage resulting in the elimination of CO groups"		686

[a] Summarized as described in Table 54.10. ALC = approximate lethal concentration.

[b] Ref. 28 lists a MAC of 5 mg/m³ for wool in Czechoslovakia (not officially approved) and Yugoslovia. The German Democratic Republic (East Germany) has a MAC value of 5 mg/m³ for animal hair.

[c] Evolved gases were considered "particularly noxious" by laboratory personnel conducting tests (Ref. 653).

[d] A low but unspecified concentration of particules were <10 μm in diameter. The concentration for these exposures was calculated from an increase in filter weight (obtained by drawing a known volume of sample) per liter of air.

[e] Polymethacrylimide and polybismaleimide rigid foams were also tested.

4406

medium, long, or coarse (687). The outermost waxlike epicuticle of the fiber covers a flat scale cell layer that in turn covers the cortex, which constitutes more than 90 percent of the wool fiber (90). The cortex has two sections twisted spirally around one another and may enclose a central medulla that is either empty or has a loose cellular network.

The main structural material in wool is keratin, a protein. Some 18 to 20 different amino acid residues have been isolated, the principal ones being glutamic acid, cystine, leucine, serine, and arginine. The dithio groups in the cystine residues provide cross-linkages.

Molecular Weight, Properties, and Application. The molecular weight has been given as 60,000 minimum, with soluble fractions up to 80,000 (14). Wool is an unusually resilient fiber, a property derived primarily from the curled cross-linked molecular structure and to a lesser extent from the waviness of spiral core sections. This natural crimp makes possible a relatively strong yarn without knitting the fibers tightly. Wool fibers are poor conductors of heat but warmth is due primarily to the air trapped inside the fabric. Wool is among the less flammable textiles (586). Zirpro-treated* wool fabrics have been reported to provide particularly effective protection against conducted heat, as from molten aluminum splashes (688).

Table 54.26 gives additional data on properties.

Skin Reactions. Wool may be irritating to some people, particularly those with atopic dermatitis, and can act as a weak allergen. Inability to tolerate wool next to the skin was reported in >40 percent of 187 patients with atopic dermatitis (689). A skin test with untreated wool (690) by the Schwartz-Peck technique (420) showed one mild reaction after the initial 24-hr contact but none after 6-day contact or after the challenge test. In rare instances the sensitivity is such that skin contact produces urticaria (160).

Inhalation and Thermal Degradation Data. Wool dust appears to be a relatively minor irritant compared to cotton dust. When gross exposure to dust is common, sensitivity is more frequent (650). Inhalation of wool dust has been considered the dominant although not the only factor in many cases of atopic dermatitis (691).

Thermal degradation products are often relatively more toxic than are observed in comparable tests with other materials. However, decreased flammability may more than offset this toxicity in uncontrolled fire situations.

See Table 54.27 for specific references to the pertinent literature.

---

* Zirpro treatments involve the application of anionic metal complexes.

Biodegradation. The protein of wool provides an appetizing diet for the larvae of the clothes moth, black carpet beetle, and some other insects unless the fabric is protected by chemicals such as naphthalene or insecticides. Wool is susceptible to mildew, particularly if damp or slightly alkaline (90).

8.2.2 Silk

Source and Composition. Fibers of natural silk are derived mainly from the cocoons of the domesticated silkworm, *Bombyx mori*. Tussah silk is one of several "wild silks" produced by the genus *Antheraeae* (90).

Silk is derived from two continuous filaments of fibroin that are covered with silk gum. The gum acts as a size during weaving or knitting. Silk fibroin is a linear polycrystalline fiber containing at least 16 different kinds of amino acid residues. The principal residues are glycine, alanine, serine, and tyrosine. The polypeptide chains are fully extended with closely packed and aligned molecules.

Molecular Weight, Properties, and Applications. The molecular weight has been given as 84,000 minimum, with native fibroin probably 150,000 (14). Silk is a versatile fiber, strong, soft, and flexible (90). See Table 54.26 for comparative properties. The primary application of silk today is in luxury fabrics.

Skin Reactions, Inhalation, and Thermal Degradation. Skin allergy to finished silk is comparatively rare and may resemble atopic dermatitis (160). A brief case report (692) indicates that in a truly hypersensitive person urticaria may occur with even slight exposure.

The processing of natural silk may induce bronchial asthma and bronchitis. A Russian report (662) refers to natural silk as a "strong" occupational allergen on the basis of positive dermal and inhalation tests and clinical observations with more than 150 patients.

Thermal degradation of silk can produce carbon monoxide, also small amounts of cyanide and aldehydes, as summarized in Table 54.27.

Tissue Reaction. Silk sutures in humans may be absorbed after some years or become encapsulated with some cellular reaction (591). Faulborn et al. (693) found very little tissue reaction in experimental studies in rabbit corneas; these authors consider that mechanical irritation has been a factor in some earlier experimental studies.

8.2.3 Leather

Leather comes chiefly from the corium that lies below the epidermis of various animals. This fibrillar collagen is converted (tanned) by treatment with tannins, chromium salts, formaldehyde, or isocyanates to the heat-resistant product known as leather. Obviously dusty work in the processing of tanned leather for boots and shoes may be associated with an increased incidence of nasal cancer. Pyrolysis as low as 300°C generates large amounts of carbon monoxide as evidenced by high carboxyhemoglobin concentrations in exposed animals. For details see Table 54.27.

8.3 Nylon

The main products* in the American and European markets are nylon 66 and nylon 6. The commercially important polymers prepared from a diamine and a diacid are nylon 66, nylon 610, and nylon 612.

Nylon 66 [32131-17-2], polyhexamethylene adipamide:

$$H-[HN(CH_2)_6NH-CO(CH_2)_4CO]_n-OH$$

Nylon 610 [9008-66-6], polyhexamethylene sebacamide:

$$H-[HN(CH_2)_6NH-CO(CH_2)_8CO]_n-OH$$

Nylon 612 [24936-74-1], polyhexamethylene dodecaneamide:

$$H-[HN(CH_2)_6NH-CO(CH_2)_{10}CO]_n-OH$$

Nylon 6, nylon 11, and nylon 12 are the commercially important polymers prepared from an amino acid or derivative thereof.

Nylon 6 [25038-54-4], polycaprolactam:

$$H-[HN(CH_2)_5CO]_n-OH$$

Nylon 11 [25035-04-5], polyundecanoamide:

$$H-[HN(CH_2)_{10}CO]_n-OH$$

Nylon 12 [24937-16-4], polydodecaneamide:

$$H-[HN(CH_2)_{11}CO]_n-OH$$

8.3.1 Preparation of Nylons 66, 610, and 612

Nylon 66 can be manufactured (10, 632, 694) by mixing equimolar amounts of hexamethylene diamine and adipic acid in alcoholic or aqueous solution to form

---

* Trademarks frequently refer to a product line and are distinguished by code letters or numbers, such as Zytel® nylon resins for several chemically different polymers and Zytel® ST (Super Tough) nylon resin specifically. Other trademarks (81) include Tynex® (for nylon filaments), Vydyne® (for reinforced nylon and unreinforced resins), Rilsan (for nylon 11), Ultramid® (for nylons 6, 66, and 610), and Enkalon® (for nylon 6). See also glossaries in References 6 and 633.

hexamethylene diammonium adipate (nylon salt). This salt, which has a melting point approximately 190°C, precipitates out of solution and is then subjected to heat and pressure to condense it to a high molecular weight polyamide. Nylons 610 and 612 are prepared from hexamethylenediamine and either sebacic or dodecanedioic acid, respectively, by processes similar to those for nylon 66. Polymers of this type can also be prepared by interfacial polymerization (14). The polymer forms almost instantaneously when an aqueous solution of one reactant (as hexamethylenediamine) contacts a nonmiscible solution of the other reactant (as adipyl chloride dissolved in a chlorinated hydrocarbon).

8.3.2 Preparations of Nylons 6, 11, and 12

Nylons of one-monomer origin are commercially prepared either by self-condensation of an ω-amino acid or by opening of a lactam ring. Nylon 6 has been polymerized from ε-caprolactam by continuous and batch processes run at about 250°C. In a batch process, caprolactam, water that serves as catalyst, and a molecular weight regulator such as acetic acid may be reacted under a nitrogen blanket for about 12 hr. Low molecular weight materials can be removed by leaching and/or vacuum distillation to achieve desired physical properties (10). Alternatively, residual caprolactam may serve as an effective plasticizer and not be extracted after polymerization has reached equilibrium (632).

Casting in the mold with anionic polymerization has also been described (10).

Nylon 11 is polymerized from ω-aminoundecanoic acid, a derivative of castor oil, at temperatures slightly above 200°C. Nylon 12 is prepared from lauryl lactam at somewhat higher polymerization temperatures, above those used for polymerization of caprolactam. Both reactions are normally catalyzed. These polymerization reactions essentially go to completion and a washing step is not considered necessary (694).

8.3.3 Molecular Weight, Properties, and Applications of Above Nylons

For useful properties as polyamide fibers or plastics the number average molecular weight must be above 10,000 (14). Nylons are tough, strong, and have good abrasion resistance but are notch-sensitive (632). They have a low coefficient of friction and are suitable for use where lubricants are undesirable. Nylons show no appreciable change in aging properties at room temperature in indoor service or when protected from sunlight (10). When heat stabilized their continuous use temperature is about 250°F (632).

Nylons 610 and 612 are considered special purpose nylons. They absorb approximately one-half to one-fourth as much water as nylons 66 or 6, depending on humidity and other conditions of exposure (632). Although the melting point of nylon 610 is lower than that of nylon 66 (215°C and 264°C, respectively, in Reference 10), nylon 610 has greater thermal stability. Nylon 66 starts to gel (cross-link) after prolonged steam heating (695). Nylons 11 and 12 are also specialty nylons noted particularly for low water absorption. Tables 54.6 and 54.26 provide some comparative data. Nylons absorb a varying percentage of water; fibers are considered hydrophobic compared to cotton, whereas plastics are hygroscopic compared to many other synthetic polymers.

The automotive industry is the single largest user of nylon, in uses ranging from nylon tire cord to emission control canisters, fuel filter bowls, and door strikers. Speedometer and window-lift gears are molded from glass-reinforced nylon. Industrially, nylon is used in gears for electric drills, cams and slides, and bearing housings. Plumbing, electrical, personal, and household applications include devices such as pipe fittings, plugs, sockets, hairbrushes, toothbrushes, and vacuum cleaner brushes (632).

8.3.4 Other Types of Nylon

Many other types of nylon have been developed (4, 10, 633, 695, 696). Copolymers such as nylon 66/610/6, with an irregular structure that limits interchain bondings and crystallization, are soluble in alcohols and many other common polar solvents. Methyl methoxy nylons can be prepared by dissolving a nylon (as in 90 percent formic acid) and then treating it with formaldehyde and an alcohol in the presence of an acidic catalyst. Polymers with about 33 percent of the —NH— groups substituted are soluble in lower aliphatic alcohols and phenols and can absorb up to 21 percent moisture when immersed in water.

8.3.5 Oral Toxicity

Sherman and his associates (697) fed nylon 66 at 10 percent of the diet to male and female rats for 2 years and male dogs for 1 year. Figure 54.13 shows a slight retardation in the rate of weight gain in male rats compared to controls, but no differences were observed in tibiae lengths, hematologic or biochemical measurements, anatomic or histopathologic lesions, or incidences and types of tumors. Dogs showed no toxic effects and no tumor development. Similar tests with nylon 610 and two copolymers were devoid of any observed effect. The copolymers were composed, respectively, of (1) 36 percent nylon 66, 26 percent nylon 610, and 38 percent nylon 6 or (2) 50 percent each of nylon 66 and nylon 610.

The first three materials, and also nylon 6, had previously been fed to rats at a dietary level of 25 percent for 2 weeks. In the 2-week test average weight and food consumption was normal except in the group fed nylon 6; these animals showed a slower rate of weight gain, a lowered food consumption, and

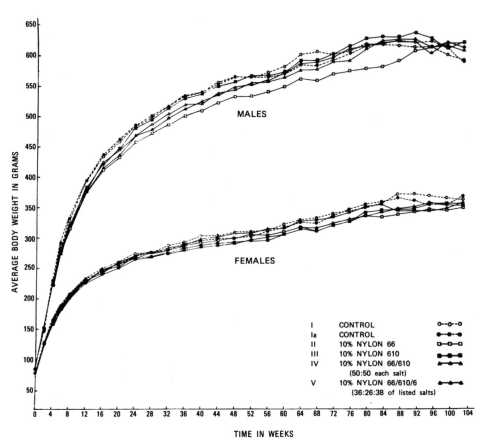

Figure 54.13 Growth response of male and female rats fed various nylon resins. Adapted with permission from H. Sherman et al. (697).

a decreased ability to utilize their food effectively. No anatomic injury attributable to any of these resins was observed after the 2-week feeding.

Nylon 612 was fed to rats and dogs of both sexes for 90 days at a dietary level of 10 percent without any clinical, nutritional, hematologic, urinary, biochemical, or pathologic evidence of toxicity (698).

8.3.6 Skin Reactions

Clinical dermatitis due to unprocessed nylon appears to be rare, although it was reported in 1955 and 1960 (citations in Reference 160). Nylon does not absorb perspiration readily and is lipophilic. Human volunteers exposed to unfinished nylon 66 fibers or fabrics by the Schwartz–Peck technique (420) occasionally show erythematous or papular reactions after 6 days' occluded contact. Reactions other than at 6 days are uncommon and are not typical of sensitization (699). Materials that contain stiff bristles sometimes produce temporary irritation. One instance of very mild erythema, at the initial 2-day reading only, was observed in one of 176 subjects exposed by the Schwartz-Peck technique to a sample of an untreated nylon carpet fiber (700). A recent modified Draize-Shelanski test (Reference 420, ten 24-hr applications plus 24-hr challenge after 2-week rest period) showed no reaction in 100 subjects exposed to a sample of knitted nylon (701).

Dermatitis associated with the dyes and finishes on nylon (419), sometimes called "nylon stocking dermatitis," is now rare because of improved dyeing technique (160). However, a 1979 case report describes a patient who showed stronger skin reactions to nylon blanks dyed with the component colors than to the dyes alone (702). The patient showed no reaction to undyed nylon.

8.3.7 Inhalation Toxicity and Thermal Degradation

Several reports from abroad describe inhalant reactions in humans in connection with occupational exposure (type of nylon unspecified). These data are discussed generally in Section 9.1.4.

The toxicity of the thermal degradation products of nylon is often attributed to carbon monoxide and/or hydrogen cyanide. Ammonia can be evolved in significant quantities—accounting for as much as half the nitrogen eliminated (695)—and its role in bioassay tests does not appear to have been defined. Concentrations of the pyrolysis/combustion products are generally not lethal until decomposition occurs, which varies from approximately 400 to 750°C, depending on test method. See Table 54.27 for details.

8.3.8 Carcinogenic Properties/Implant Studies

No known reports attribute any metastatic carcinogenic potential to nylon. The International Agency for Research on Cancer (31) considered that a definite assessment of the carcinogenicity of caprolactam and its polymer could not be made in view of the limited data available. Local sarcomas in rats have been reported after intraperitoneal implantation of nylon films, about 10 mm in diameter (Druckrey and Schmähl, as cited in Reference 31). A local tumorigenic response was also observed (594) when "nylon" was embedded as a flexible plain film, and to a lesser extent as a flexible perforated film, but not as a soft textile (7 of 26 rats, 2 of 31 rats, and 0 of 33 rats, respectively). The critical factor would most likely appear to be the physical form of the material. See also Sections 1.4 and 8.1.3.

Experimental tests in rabbit corneas have shown that nylon thread (Tübinger nylon) is well tolerated in microsurgical sutures (693). Human studies show that nylon sutures were well tolerated when examined at intervals from less than 1 year to more than 5 years after surgery (591).

Nylon has for some years been considered unsuitable as an endothesis (703). Fragmentations and crazing result in total unserviceability within a 1-year period (704).

8.3.9 Biodegradation

Nylon polymers per se are generally not considered susceptible to growth of microorganisms. Growth on nylon may occur if a supply of extraneous nutrients is available (as from lubricants, plasticizers, or contaminating organic matter). Forty-two day incubation of three strains of *Penicillium janthinellum* on agar containing strips of nylon 66 yielded an abundance of pink pigment in the agar and stains on the nylon strips (705). No significant changes in the tensile strength of nylon 66 occurred after samples were cultured on agar for 42 days with *P. janthinellum* and then buried in soil for 42 additional days.

8.4 Fatty Acid Polyamides

These materials, sometimes simply called "fatty polyamides," are obtained by prepolymerizing fatty acids to rather variable structure (10). These dimers are then treated with amines (706), typically ethylenediamine (10), to form viscous liquids or brittle resins of low molecular weight (2000 to 15,000). They are useful as hardeners or flexibilizers in epoxide resins, thixotropic paints, and adhesives.

8.5 Aromatic Polyamides

8.5.1 Source and Type

The current American products are most frequently aramid* fibers (Nomex®, Kevlar®, MPP-I, PABH-T or PABH-T X-500). The polymers are generally formed by treating aromatic diamines and aromatic diacid chlorides in a solution polymerization process with an amide-type solvent (644). Several types of aramid fibers have now been developed (644); the basic structures appear to be isophthalamide or terephthalamide with variations such as amide–hydrazide. The condensation product of *m*-phenylenediamine and isophthalic acid (10, 211) is

* Aramid refers to a polyamide fiber containing at least 85 percent of the specified aromatic amide structure (see Section 8.1).

The condensation product of *p*-phenylenediamine and terephthaloyl chloride (644, 707) is

$$\left[HN{-}\langle\bigcirc\rangle{-}NH{-}OC{-}\langle\bigcirc\rangle{-}CO \right]_n$$

The condensation products of *p*-aminobenzhydrazide and terephthaloyl chloride (211, 644) are

$$\left[\left(HN{-}\langle\bigcirc\rangle{-}CO{-}NHNH \right)_x OC{-}\langle\bigcirc\rangle{-}CO \right]_n$$

Other types of aromatic polyamide (211) include poly(quinazolinedione) fiber (AFT-2000) and transparent polyamide plastics that are mostly European in origin. The best known of the plastics is probably poly(trimethylhexamethylene terephthalamide). This polymer, known as Trogamid T®, is obtained by treating terephthalic acid with an isomeric mixture of trimethylhexamethylenediamines (10, 211, 708).

8.5.2 Molecular Weight, Properties, and Applications of Aramid Fibers

Preston (644) notes that the molecular weight "probably should be 60,000 or greater" for the best balance of tensile properties. The aramid fibers have been developed for (1) heat and flame resistance or (2) ultrahigh strength and high modulus. Polymers of the first type usually contain primarily *m*-oriented phenylene rings, whereas those of the second type contain primarily *p*-oriented phenylene rings (644). The basic isophthalamide and terephthalamide structures shown above have been modified with chlorine (some types of Durette®), fluorine, or phosphorus groups to provide particular properties. Aramid fibers characteristically burn with difficulty and produce a thick char. The limiting oxygen index of poly(*p*-phenylene terephthalamide) can be raised from approximately 28 to 30 up to 40 to 42 with the incorporation of 1 percent phosphorus (644). Table 54.26 and Figure 54.11 give some comparative data on properties.

The heat and flame resistance of the aramid fibers makes them suitable in such varied applications as filter bags for hot stack gases, insulation paper for electrical transformers, welder's clothing, and jump suits for forest fire fighters. Aramid pajamas and robes have been quite useful for certain nonambulatory patients. Aramid drapes are specified on some aircraft and all ships of the United States Navy. Aramid fibers of moderately high modulus (as Nomex®) are used in the reinforcement of fire hose and V-belts. Ultrahigh strength, high modulus fibers (as Kevlar®) are used in tire cord, V-belts, cables, body armor, and as reinforcement in aircraft and other vehicles of transportation.

8.5.3   Oral Toxicity and Skin Reactions

When administered by intragastric intubation to rats, the approximate lethal dose of poly(p-phenylene terephthalamide) was greater than 7500 mg/kg (709). This was considered the maximum feasible acute dose.

Skin tests conducted with two types of poly(m-phenylene isophthalamide) showed no sensitization reactions and no irritation reactions after 48-hr contact. Moderate erythema was observed after 6 days of occluded contact in one of 207 persons tested (710). (This isolated reaction was not considered sensitization and may well have been related to occlusion.) No reactions were observed in the panel of 200 people tested with another sample (711).

8.5.4   Inhalation Toxicity, Thermal Degradation, and Related Data

Intratracheal insufflation tests in rats have been conducted to assess the respiratory tract reaction to particles of two aromatic polyamides. A single 0.5 ml injection of a 5 percent suspension of poly(p-phenylene terephthalamide) in 0.9 percent saline (712) was given under mild ether anesthesia. Rats were sacrificed serially at 2, 7, and 49 days, and also at 3, 6, and 12 months; the last group of 10 rats was killed at 21 months. Respirable-sized dust particles (<10 μm) caused only the minimal lung tissue reaction seen with inert-type dusts. Large nonrespirable dust particles—up to 150 μm—produced mild foreign body granulomas. After 21 months the lung dust content, granulomatous reaction, and bronchiolar polypoid protrusions (considered a result of mechanical trauma) were decreased compared to those seen at 1 year.

Rats similarly given a dose of 0.25 ml of a 1 percent suspension of poly(m-phenylene isophthalamide) (paper dust) were sacrificed at 6 months, 1 year, and 2 years (713). The small-sized dust particles were mostly phagocytized within 2 years. Large dust particles—up to 100 μm in length and 20 μm in width—provoked giant cell reactions without any fibrosis or collagenization.

Rats inhaling the dust of poly(p-phenylene terephthalamide) at a maximum practicable concentration showed a depressed rate of weight gain, which was followed by normal gain when exposures were discontinued (see Table 54.27). The listed thermal degradation data for aromatic polyamides do not indicate that they release hydrogen cyanide or ammonia, but these eventualities should be anticipated if thermal decomposition occurs.

8.6   Polyimides

8.6.1   Structure, Synthesis, and Related Data

Although imides, characterized by the structure CO—NH—CO, can have an open chain arrangement, the imide unit in polymers is usually a cyclic, five-member ring. This ring may be fused to one or more cyclic or aromatic rings to maximize thermal resistance, inevitably with some loss in processability.

The commercial polycondensation methods of synthesis (4, 477, 714) involve the treatment of diamines with dianhydrides—most commonly pyromellitic dianhydride. Polyamide acid or polyamide is formed in the first stage and then ring closure with dehydration forms the polyimide. An example is the reaction product of pyromellitic dianhydride and 4,4′-diaminodiphenyl ether, known as poly(oxy-di-*p*-phenylene pyromellitimide) (CAS No. 25036-53-7) (Kapton® polyimide films, some types of Vespel® fabricated parts):

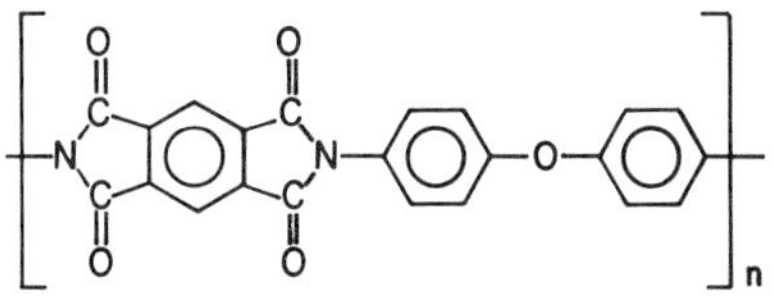

Polyimide films such as above that are sold for adhesive bonding and other types of processing may contain up to 1 percent by weight of residual dimethylacetamide solvent (638).

The term "polyimide" is also used as a category for other structurally similar polymers with greater or lesser thermal stability (4, 641). Substitution of trimellitic anhydride for pyromellitic dianhydride and treatment with an aromatic diamine yields polymers of the type known as polyamide–imide:

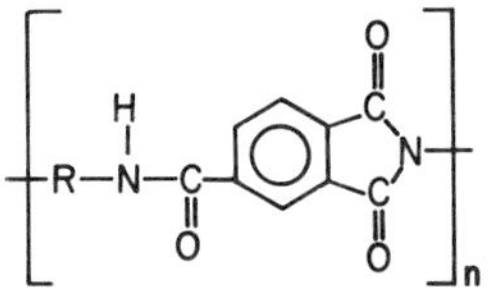

Some polyimides (Kinel®, Kerimid® laminate) can also be synthesized by addition polymerization of "short preimidized segments very similar in nature to those of condensation polyimides" (715). Many innovative methods have been described (4, 716).

8.6.2 Molecular Weight, Properties, and Application

The outstanding property of polyimide polymers is their thermal stability, which can be as high as 500 to 600°C. The "thermoset types" have no distinct softening point below their thermal degradation point (717). They can be formulated for good to excellent mechanical properties at elevated temperatures. Creep is typically almost nonexistent, even at high temperatures. Coefficients of thermal expansion are comparable to those of metals. See Table 54.26 for comparative data.

The first commercial applications of polyimides (10) were as wire enamels and insulating varnishes (Pyre-ML®). Polyimide film (as Kapton®) is used in electric motors where size is important and in wire insulation (aircraft wire cable, flat flexible cable, magnet wire). The qualities of the film permit the use of thinner insulation, which can result in greater horsepower with the same frame size. Polyimide resins can be prepared as molding formulations (Vespel®, Kinel®), which can then be machined into parts for aircraft engines, printed wiring boards used in computers, and bearing cages for high-precision gyroscopes (717). Polyimide can be spun as fiber or formulated as insulating foam (Skybond®) for use where operating temperatures are high, as in the sound deadening of jet engines (211).

8.6.3 Oral Toxicity and Skin Contact

Rats fed a diet containing 25 percent (weight/weight) of polyimide molding powder had slightly lower food consumption and a lower rate of weight gain during the 14-day feeding period as compared to control rats fed 25 percent Alphacel® cellulose. Gross pathologic examination showed chronic murine pneumonitis present to a slightly greater degree than in control animals, but no microscopic changes attributable to the test polymer were observed. In rats given a 14-day recovery period, food consumption and body weight returned to normal levels; pathologic examination was devoid of noteworthy changes. Microscopic examination of tissues showed no effects attributable to the test material. Skin tests on guinea pigs showed no allergic reaction and generally no irritation; mild transient erythema was occasionally observed in more susceptible animals (680).

8.6.4 Inhalation Toxicity and Thermal Degradation Data

Data summarized in Table 54.27 indicate that poly(oxy-di-p-phenylene) pyromellitimide dust has low acute inhalation toxicity. Temperatures $\geq 300°C$ can result in the evolution of carbon monoxide. Inhalation of the volatile products generated at temperatures $\geq 450°C$ can be lethal. Tests where mice were exposed to the pyrolysis products of a series of polymers showed that the products from imide-type polymers ranged among the more toxic as judged by time to death.

8.6.5 Aqueous Degradation

Immersion of polyimide film specimens in distilled water at 25 to 100°C for time periods lasting from one to several hundred hours (718) has been reported to cause a decrease in elongation to failure from 38 to approximately 5 percent. The reaction depended slightly on pH in the range 2.0 to 12.0.

9 POLYESTERS, POLYETHERS, AND RELATED POLYMERS

9.1 Overview

Commercial use of polyesters dates from the early twentieth century when alkyd resins were used in surface coatings (10). The polyesters are found today as fibers, films, laminating resins, molding resins, and engineering plastics. The high molecular weight polyethers are known primarily as engineering plastics,* as are the polysulfides and the polysulfones. Table 54.28 gives some basic properties.

9.1.1 Production and Processing

Production data for the general categories are listed in Table 54.2. Processing techniques vary widely and are discussed in the appropriate subsection.

9.1.2 Specifications and Test Methods

Clearance for certain food-contact applications with many of these materials has been obtained under Title 21 of the Code of Federal Regulations (92). Those specifically listed include polyethylene terephthalate, polybutylene terephthalate, polycarbonate, polyoxymethylene, polyoxyethylene derivatives, polyphenylene sulfide, and polysulfone. Specifications concerning mechanical and other types of performance depend on end use as summarized in Section 1.3.

9.1.3 Toxicologic Potential

Industrial hygiene concerns with these polymers include (1) styrene exposure during the fabrication of unsaturated polyester resins and in any subsequent release of styrene, (2) other volatile products generated at elevated temperatures or in fires, and (3) inhalation of dust or particulate generated in the manufacturing and processing of polyester fibers. The use of styrene in polyester resins that are fabricated into glass-reinforced plastics can provide the greatest intensity of exposure to styrene in workplace situations (734). Toxic vapors can also be released by solvent systems, particularly at elevated temperatures, and during fabrication of engineering resins. For example, the processing of polyoxymethylene in a poorly ventilated space may release biologically significant amounts of formaldehyde into the adjacent atmosphere.

Table 54.29 shows that the combustion products of the most concern are generally carbon monoxide and, in some cases, acetaldehyde. Sulfur-containing

---

* An exception is high molecular weight polyoxyethylene, a water-soluble packaging polymer.

Table 54.28. Selected Properties of Some Polyesters, Polyethers, and Related Polymers[a]

Property	A. Polyesters		Polycarbonate of Bisphenol A	Reference
	Polyethylene Terephthalate	Polybutylene Terephthalate		
Melting temperature (°C)	254–284; 256 (commercial PET); 271 (highly crystalline PET)	221–232	215–230	14, 85, 719, 720
Glass transition temperature (°C)	69 (Ref. 86). 67, amorphous; 81, crystalline; 125, crystalline and oriented (719)	17–80 ("conflicting data," Ref. 86); 22–43 (Ref. 10); unfilled, 50 (Ref. 721)	145–149	10, 86, 720, 721
Service temperature (°C)	−60 to +150	"Can be used for prolonged periods of time at 120–140°C" (722). Embrittlement by hydrolysis on long-term aging at 60–85°C under humid conditions (721)	135 maximum	14, 719–723
Density	1.33 (amorphous) 1.45 (crystalline)	1.31–1.32	1.20	10, 14, 723
Refractive index	Film: amorphous, 1.5760; crystalline and biaxially oriented, 1.64		1.585	14, 719, 720
Moisture absorption	0.55%, 24-hr immersion of commercial films; 0.8%, immersion in water at 25°C for 1 week	0.08%, 24-hr immersion at 23°C	0.2% in air at 60% relative humidity	10, 719, 720
Moisture regain (fiber)	0.4%, commercial value (Ref. 12) and "normal conditions" (Ref. 6)			6, 12[b]

4420

	Polyoxymethylene	Polyoxyethylene	Polyphenylene Oxide	Reference
Solvents	Crystalline: choral hydrate, phenol tetrachloro-ethane (1:1 vol.), nitrobenzene, DMSO (hot)	"Complex phenols" such as o-chlorophenol (724)	Methylene chloride and chloroform; less soluble in tetrachloroethane, trichloroethylene, dichloroethane, tetrahydrofuran, dioxane, cyclohexanone, dimethylformamide. Swells in benzene, chlorobenzene, acetone, ethyl acetate, carbon tetrachloride	88, 720, 724
Nonsolvents/relatively unaffected by	Crystalline: hydrocarbons, chlorinated hydrocarbons, aliphatic alcohols, ketones, carboxylic esters, ethers	"Very resistant to most chemicals" (724)	Hydrocarbons, styrene, carbon tetrachloride, acetone, lower esters	88, 724
Decomposes	Strong acids and bases, particularly when hot. Hydrolyzes slowly in water at elevated temperature (fibers with ~20% loss in strength after 1 week at 100°C, without measurable loss in strength after several weeks at 70°C)		Hot alcoholic alkalies, amines and other organic bases; surface attack by aqueous alkali. Hydrolyzes in water >60°C (can withstand relatively short exposures)	14, 725
B. Polyethers				
Melting point (°C)	175–200 (Copolymers, 167–170, Ref. 726)		298 (as poly-p-phenylene oxide); 261–272 (as poly(2,6-dimethyl-p-phenylene oxide)	85, 726, 727

Table 54.28. (Continued)

| | | B. Polyethers | | |
Property	Polyoxymethylene	Polyoxyethylene	Polyphenylene Oxide	Reference
Glass transition temperature (°C)			105–120 (Phenylene oxide-based resin, Ref. 82)	82
Density	1.40–1.42 for molded parts, range to 1.56 for special grades of resin (Ref. 727)		1.314 (as poly-2,6-dimethyl-p-phenylene oxide); 1.408 (as poly-p-phenylene oxide)	85, 727, 728
Refractive index	1.489–1.553	1.51–1.54		87, 726
Water absorption, 24 hr (%)	0.2–0.3		0.066%	727, 729
Solvents	At elevated temperature: benzyl alcohol, phenol, chlorophenols, aniline, formamide, DMF, γ-butyrolactone, bromobenzene, diphenyl ether	Benzene, chloroform, carbon tetrachloride, alcohols, cyclohexanone esters, DMF, water (cold), aqueous K_2SO_4 (0.45 M above 35°C); (swells in dioxane)	*Poly(oxy-2,6-dimethyl-1,4-phenylene)*c: amorphous, α-pinene (hot); crystalline, benzene, toluene, chloroform, chlorobenzene. *Poly(oxy-1,3-phenylene)*c: benzene, biphenyl, 3-pentanol, phenyl ether, pyridine, benzophenone, nitrobenzene, DMF, DMSO	88
Nonsolvents/relatively unaffected by	Aliphatic hydrocarbons, lower alcohols, diethyl ether, lower esters	Aliphatic hydrocarbons, ethers, water (hot)	*Poly(oxy-2,6-dimethyl-1,4-phenylene)*c: amorphous, α-pinene (cold), methanol, ethanol; crystalline, α-pinene (hot), methanol, ethanol, nitromethane. *Poly(oxy-1,3-phenylene)*c: methanol	88

C. Polyphenylene Sulfide and Polysulfones

Property	Polyphenylene Sulfide	Polysulfone of Bisphenol A	Polyethersulfone 200P	Reference
Melting point (°C)	~285–295	~190		85, 731
Glass transition temperature (°C)	97		230 (other polysulfones ranging to 315)	85, 732, 733
Density	1.440; 1.34, unfilled; 1.64, 40% glass (Ref. 731)	1.24		85, 731, 732
Refractive index		1.633		87
Water absorption	0.02, unfilled; 0.01, 40% glass			731
Solvents	Biphenyl, dimethyl-p-terphenyl, chloronaphthalene, some other solvents at elevated temperature (see Ref. 88)	Chlorinated hydrocarbons, dimethylformamide, N-methylpyrrolidone; swells in dimethyl sulfoxide		88, 731, 732
Nonsolvents/relatively unaffected by	At reflux temperature: toluene, pyridine, phenyl oxide, phenyl sulfide	Inorganic acids, alkalies, aliphatic alcohols		88, 732
Decomposes		Concentrated sulfuric acid (dissolves with degradation)	Alkalies with pH >9; acids with pH <4 (extended contact)	732; 730

[a] Summarized as described in Table 54.9.
[b] Standard D 1909-77.
[c] "Phenylene-oxide based resin" may soften or dissolve in certain halogenated or aromatic hydrocarbons. If an application requires such exposure, stressed samples should be tested under operating conditions (729).

4423

Table 54.29. Overview of Inhalation and Thermal Degradation Data for Some Polyesters, Polyethers, and Sulfur Polymers[a]

Polymer	TLV	Type of Study			Species Exposed	Description of Material	Procedure	Observations/Conclusions	Reference
		Ambient Temperature	Elevated Temperature	Effluent Analysis					
Polyethylene terephthalate[b]	No specific TLV; see Section 1.4		X	X	Rat	100% polyethylene terephthalate film, also film with different flame retardants	Exposure 1 hr, to pyrolysis products of 5–7.5 g film generated at 450, 550, or 650°C	No deaths when polymers pyrolyzed at 450°C; deaths observed with all samples pyrolyzed at 650°C, some samples pyrolyzed at 550°C. Higher pyrolysis temperatures associated with more rapid increase of CO	735
			X	X	Mouse	Thermal-resistant film used for oven cooking and sterilization; analyzed as polyethylene terephthalate	As described in Table 54.27	ALC = 30.5 g/polyester per 1000 liters of air; main contents of effluent = 3000 ppm CO, 1000 ppm acetaldehyde, small amount of benzene	669
			X	X			Acetaldehyde major product at 283 and 306°C; other products included CO, CO_2, H_2O, C_2H_4, 2-methyldioxolane, CH_4, C_6H_6		736
			X	X			Nature of thermal, hydrolytic, oxidative, and radiation-induced reactions; nongaseous thermal degradation products include cyclic oligomers (mainly the trimer)		737
			X	X			Major products of oxidative degradation to 1000°C included CO_2, oligomers, CO, H_2O, terephthalic acid, methane, acetaldehyde; pyrolysis in helium gave higher yield of oligomers		675
Polyester fiber or fabric[c]	As polyethylene terephthalate above	X			Human		As summarized in Section 9.1.4		
			X	X	Rat	Polyester fiber	Decomposition products obtained with 300-W heater from 10 g of sample were incapacitating <20 min		249

Material		Species		Observations	Ref.
Polyester	X	Mouse	X	Time to death from products of 100% polyester shorter than the median among a group of fabrics so tested; see Table 54.8	66, 531
Woven fabric	X		X	Ethylene, CO_2, CO, acetaldehyde, methane, other products evolved at 400°C; varying amounts generated at 600, 800, and 1000°C	397
Copolymers of polyethylene terephthalate containing 1–24% poly(diethylene glycol terephthalate)	X		X	Fibers prepared from 7 copolymers heat-aged at 121 and 204°C: decomposition mechanisms operative below melt temperatures, can rapidly destroy such copolymers	738
	X		X	Polyester considered relatively low in heat transfer capacity	586
100% polyester, also blends of polyester/cotton and 100% cotton	X		X	Depth of burn in manikin tests primarily related to fiber content—as polyester content increased, depth of burn decreased (tests without "clinging" fabric)	739
Polybutylene terephthalate	X	Rat	X	ALC_{50} = 15.3 and 22 mg/l, respectively, for polymer with flame retardant and as such (value for nonflame retarded sample within 95% confidence limits for red oak); see procedure, other data in Table 54.24	512
Polybutylene terephthalate as such and also flame retarded	X	Rat	X	Exposure 4 hr, to volatile products generated under nitrogen, sample temperature 255°C	740
3 Flame-retarded compositions	X	Rat	X	Rats exposed to one of three compositions showed no clinical signs during exposure, borderline enlarged liver incidence when killed 2 weeks later	740
Reinforced with 30% glass fibers; also with 10% decabromodiphenyl ether	X	Rat	X	No abnormalities, including organ weight or microscopic changes in liver, after 10 exposures in 2-week period (procedure as directly above)	741
As polyethylene terephthalate above	X		X	Degradation 240–280°C, in absence of oxygen: first-order mechanism of fission followed by evolution of butadiene; condensate included terephthalic acid and some unsaturated esters	742

4425

Table 54.29. (Continued)

Polymer	TLV	Type of Study			Species Exposed	Description of Material	Procedure	Reference
		Ambient Temperature	Elevated Temperature	Effluent Analysis				
Unsaturated polyester			X		Rat	Resin and fiberglass, with and without flame-retardant finish	Degradation products ≥400°C sometimes lethal, higher mortality ratios with flame-retardant sample; see Table 54.7	62
			X	X	Rat	Rigid foams based on unsaturated polyester resin with expanded glass pellets	No deaths according to DIN 53436 protocol when testing based on samples of equal volume, pyrolysis temperature up to 600°C; deaths at 350–400°C when comparison based on samples of equal weight, analyses indicated deaths not attributable to CO or HCN (procedure, Table 54.24)	63
			X		Mouse	Fiberglass-reinforced polyester	Respiratory depression occurred at lower concentration (mg/l) than with several other polymers comparably tested	743
			X				Propylene glycol based resin produced higher smoke levels than ethylene, diethylene, or neopentyl glycol based resin. Smoke density tests showed styrene-based monomers generally gave highest smoke emission, acrylate monomers lowest, and allyl monomers were generally in between	744
Polycarbonate[d]	As polyethylene teraphthalate above		X	X	Rat		Products evolved at 500°C not lethal to rats; at 600°C, all exposed rats died, deaths attributed to CO (procedure, Table 54.24)	63
			X	X	Rat		ALC_{50} = 43.1 mg/l for glass filled polycarbonate; "polycarbonate" and two of three flame-retarded samples had lower values comparable to red oak (procedure, Table 54.24)	512
			X	X	Mouse	Bisphenol A polycarbonate	Exposures with fixed furnace temperatures, with and without air flow, indicated that relative toxicity of different materials is "highly dependent on the test conditions used" (745); rising temperature program, see Table 54.8, CO likely principal toxicant	66, 94, 745, 746

Material	Sample	Species			Remarks	Ref.
			X	X	Degradation 300–389°C yielded CO_2 and bisphenol as main products, also phenol, 2-(4-hydroxyphenyl)-2-phenyl propane, CO, CH_4, and diphenyl carbonate	747
			X	X	Oxidative degradation at 100°C/min to 1000°C yielded 95% of the volatile products as CO_2, CO, and H_2O; other compounds found in greater than trace concentrations were CH_4, C_6H_5OH, and ethylene	675
Polyoxymethylene	As polyethylene terephthalate above	Rat	X'	X	6-hr exposure to 47 mg/l dust (nominal) caused emphysema and atelectasis evident at sacrifice 19 days after exposure. No evident effect from dust inhalation of 22 mg/l. Lethal atmospheres from 8 g samples of molding powder heated to 250°C for 2–3 hr or of film heated to 150°C for 4 hr	748
		Rat		X	Acute lethality, changes in blood chemistry, other data indicate that smoke toxicity greatest in the nonflaming (or pyrolysis) mode as compared to flaming combustion with or without external sources of heat; mode of combustion important consideration in evaluation of toxic hazard	749
		Mouse		X	Exposure at fixed furnace temperatures 400–800°C or with rising temperature program (see Table 54.10, Ref. 66), with or without air flow. Flash fires frequent. CO considered principal toxicant	95, 750
				X	100% monomer can be evolved at 222°C; polymers of different type vary in stability	751
Polyphenylene oxide-based resin (PPO)	As polyethylene terephthalate above	Rat		X	ALC_{50} = 11.1 mg/l for modified PPO, 23.5 mg/l for flame-retarded sample (procedure, Table 54.24; also see above this table)	512
		Rat		X	Some deaths observed in test with products generated from flame-retarded sample at 450°C	752
Polyphenylene oxide, modified		Mouse		X	Relatively short time to incapacitation in tests summarized in Table 54.8	66
Commercial samples		Rat		X	Volatile combustion products determined under varying conditions of temperature and air supply. CO and CO_2 major volatile products. Methane, toluene, ethylbenzene, styrene, benzene, saturated and unsaturated aliphatics through hexane, methanol, and acetaldehyde identified in minor or trace amounts	753

Table 54.29. (Continued)

Polymer	TLV	Type of Study			Species Exposed	Description of Material	Procedure	Observations/Conclusions	Reference
		Ambient Temperature	Elevated Temperature	Effluent Analysis					
			X	X				Thermal degradation under vacuum of unsubstituted and halogenated polyphenylene oxides: in general thermal stability decreased as degree of substitution increased, derivatives with chlorine groups relatively more stable than corresponding derivatives with bromine groups	754
Polyphenylene sulfide	As polyethylene terephthalate above		X	X	Mouse			As summarized in Table 54.8; CO evolved considered insufficient to have been the principal toxicant	66, 515
			X	X				Hydrogen sulfide the dominant volatile product from degradation under vacuum at 450°C; at 550–620°C evolution of H_2 becomes the dominant reaction	755
			X	X				Polyphenylene sulfide more thermally stable in inert and oxidizing atmospheres than fully fluorinated analog. Pyrolysis under vacuum at 275°C yields dibenzthiophene as principal volatile product	756
Polysulfone	As polyethylene terephthalate above		X	X	Mouse			As summarized in Table 54.8, also later work with forced air flow; CO evolved considered insufficient to have been principal toxicant	66, 515, 757
			X	X		Bisphenol A polysulfone		Degradation in helium or air atmospheres to >600°C showed most of the sulfur released as SO_2; relatively small amounts COS and H_2S also formed	102
			X	X		Polysulfone Phenoxy T		Degradation under vacuum ≤620°C released SO_2, CH_4, H_2, H_2O, CO, CO_2, C_6H_{14}, C_6H_5, $C_6H_5CH_3$. Elimination of SO_2 practically complete at 450°C	755

[a] Summarized as described in Table 54.10. ALC = approximate lethal concentration.

[b] Polyester fabric is listed separately below.

[c] "Polyester" as a commercial textile is generally modified polyethylene terephthalate or a copolymer thereof. Polyethylene terephthalate specifically identified as such is listed directly above.

[d] Identified specifically as poly[2,2-propanebis(4-phenyl carbonate)] in Refs. 675 and 747, presumed or stated commercial polymer in other references.

[e] "Inhaled polyformaldehyde particles" at a dose of "about 5 g to 1 rat" for repeated 30-min exposures were not associated with toxic effects; see Section 9.7.1, injection studies, concerning tissue response data from this study (814).

engineering resins typically have a high resistance to thermal deterioration but may yield hydrogen sulfide or sulfur dioxide if heated to decomposition temperatures.

Insufflation tests in rats (Section 9.2.1) indicate no particular hazard to humans from particles of commercial polyester fiber but suggest that sufficient inhalation of cyclic ethylene terephthalate trimer dust may produce a granulomatous tissue response. Both types of particles produced a nonspecific inflammatory tissue response when first introduced into the lungs (758). This initial response was transitory and was not followed by collagen formation or any development of fibrosis. However, small particles of crystalline cyclic ethylene terephthalate trimer dust produced more prominent foreign body giant cell reactions than small polyester fiber particles. The latter, formed during high speed processing by abrasive fracture of the surface of commercial fiber, produced a modest response consisting mostly of macrophages; the large unrespirable particles produced artifactual foreign body granulomas. See also Section 9.1.4.

9.1.4 Synthetic Fibers and Human Response

Prior to 1950 "textile dust" was essentially the dust of cotton and other natural fibers. Harmful health effects were particularly identified with the inhalation of the dusts of raw cotton, flax, and hemp in the workplace. Today the production of synthetic fibers is comparable to that of natural fibers (Table 54.2). No conclusive evidence links the manufacturing or processing of synthetic fibers with serious health effects.

Except for cotton and other byssinogenic dusts, relatively few epidemiologic studies that involve exposure to organic fibers have been conducted. Three studies that consider exposure to polyacrylonitrile fiber, rayon, or an unidentified synthetic fiber, respectively, do so in the context of a comparison to cotton. Valic and Zuskin (386) report a mild reduction of ventilatory capacity in 175 Yugoslav textile workers exposed to polyacrylonitrile fibers, including 30 with previous exposure to cotton and 77 with previous exposure to hemp; however, this response did not appear to be dose related. Tiller and Schilling (759) found an insignificant change in ventilatory capacity (less than 1 percent) among 26 English rayon workers, 13 of whom had been previously exposed to cotton. Another British report (760) describes a decreased prevalence of bronchitis among workers in two spinning mills using "man-made" fiber compared to workers in 14 cotton mills.

Several clinical reports record the symptoms or test results of affected textile workers but do little to clarify the origin of these observations. The significance of these reports outside the immediate environment is difficult to determine and quite likely limited. Pimentel et al. (427) describe seven patients in Portugal with a history of exposure to various textiles during the manufacturing process.

The seven patients were variously affected with asthma, extrinsic allergic alveolitis, chronic bronchitis with bronchiectasis, spontaneous pneumothorax, or pneumonia. All had worked with one or more synthetic fibers; five had also worked with wool and/or cotton. The terminal patient identified as Case 3 had been exposed to the dust of "wool, cotton, and synthetic fibres" over approximately a quarter century and had apparently continued to work with fibers for some 3 years after a hospital admission for progressive breathlessness upon exertion during the preceding 8 years. Many fibers were found in fibrotic tissue and some were identified as polyester. However, the mere presence of particles in the respiratory tract cannot be considered pathologic. The average person daily inhales some 20,000 liters of air laden with particles, many of which are deposited on the alveolar surface (761).

Bouhuys (429) points out that this report of case histories (427) suffers from lack of control biopsy samples, specific exposure data, and information on the nature and size of the populations at risk. A later report by Pimentel et al. (430) describes two patients with unusual "sarcoid-like granulomas" of the skin that are attributed to acrylic or nylon fibers, respectively; one of these patients also had respiratory tract lesions that were considered similar to those described in the first report. Unfortunately this second report does not remedy the deficiencies of the first report. Two Finnish investigators (428) cite the first Pimentel et al. study and also fail to give exposure data when reporting the results of several inhalation challenge tests in textile workers. Neither group provides specific evidence to support its implication of an immunogenic phenomenon. (By way of contrast, sensitization-type reactions have been associated with workplace exposure to cotton, also hemp, flax, and silk; see Sections 7 and 8.)

These clinical reports pay little or no attention to the dimensional characteristics of the inhaled fibers as distinct from the chemical nature of the polymer in question. Both size and shape are important factors in biologic responses to durable fibers—particularly carcinogenicity (see Section 1.4). Work with fibrous glass indicates that short fibers (≤8 μm in length) have negligible carcinogenic potential, but fine diameter fibers (≤1.5 μm) that are long (>8 μm) appear to increase in carcinogenic potential as their length increases (762). The synthetic organic fibers used in textiles are generally larger in diameter than this apparently critical 1.5 μm diameter (763).

9.1.5 Workplace Practices and Standards

No specific standards are known that pertain to ordinary industrial use of the finished polymeric products. Fabrication of the raw polymers or use of the prepolymers may release volatile materials and require appropriate industrial hygiene measures. Possible fire hazards should be considered when large amounts of these polymers are present in a given area, particularly with prepolymers of unsaturated polyester that contain appreciable amounts of flammable solvent.

In general, exposure to particles of polymeric fibers should be minimized by appropriate industrial hygiene measures for the control of exposure to insoluble organic dust. Particles of fiber dust should not be considered harmless but rather as particles that can—like particles of any dust—have biologically significant consequences if inhaled in gross concentrations. The presence of oligomers or other impurities may require evaluation in specific workplace situations.

The Textile Research Institute has provided a current review (763) on methods for collecting and measuring the particle sizes of respirable dust with special emphasis on dusts of importance to the fiber and textile industries.

9.2 Linear Terephthalate Polyesters

All high molecular weight polyester polymers of commercial significance as fibers or films are derived from dimethyl terephthalate (or terephthalic acid). The basic structure is polyethylene terephthalate (or its copolymers). Poly-1,4-cyclohexylenedimethylene terephthalate has also been used but to a much more limited extent.

Polybutylene terephthalate is known as an engineering resin. The two fiber-forming terephthalates can also be adapted for this purpose as homopolymers or copolymers (10).

9.2.1 Polyethylene Terephthalate and Polyester Fibers

Structure, Synthesis, and Processing of Polyester Fiber. As noted above, the basic structure in most polyester fiber is polyethylene terephthalate (CAS No. 25038-59-9):

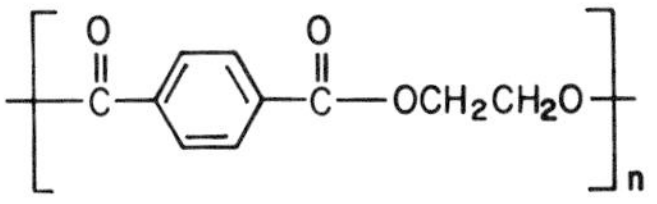

The structure of the end groups has been described as mainly hydroxyethyl ester with a small number of carboxyl end groups (6). The Federal Trade Commission defines a polyester fiber as "a manufactured fiber in which the fiber-forming substance is any long chain synthetic polymer composed of at least 85% by weight of an ester of a dihydric alcohol and terephthalic acid" (130).

Dacron® polyester fiber is composed primarily of polyethylene terephthalate, as are Fortrel®, Terylene®, and some types of Kodel®. The Kodel® tradename has also been used with fibers derived from polycyclohexylenedimethylene terephthalate (6). Other variations include terephthalate-substituted isophthalate copolymers that provide enhanced dyeability (418, 477).

Polymerization has two basic steps (6, 10, 418): (1) reaction of ethylene glycol with terephthalic acid (esterification) or dimethyl terephthalate (alcoholysis) to produce oligomeric hydroxyethyl terephthalate, ranging from dimer to pentamer; (2) polycondensation of this oligomeric mixture to the desired molecular weight and removal of the excess glycol and by-products.

Commercial polyester fibers are prepared by melt spinning. The extruded melt is formed into filamentary streams that are formed into spun filaments and then drawn to fibers, either in a separate drawing step or in combination with spinning (418).

Two impurities are "normally" present in polyester fibers (6). Ethylene glycol used in the synthesis may be converted to diethylene glycol and the corresponding ether group may be present to a slight degree (1 to 3 mole percent) in the polyester. Cyclic trimer is present to the extent of ≤ 1.5 percent in the polymer or in the fiber derived from it. Part of this trimer may be removed during dyeing or reprecipitate on the fiber.

"Snow" deposits that can develop on the surfaces of high speed friction twist texturing machines (764) are aggregates of irregular polymeric particles ("skin" particles*) rather loosely held together by finish oils. Analyses of these deposits average approximately half polymer, half finish. Generally the polymeric particles have the melting point and molecular weight characteristic of polyethylene terephthalate and should not be confused with cyclic trimer. "Snow" appears to be the result of local yarn heating caused by friction. Optimum control of yarn and process variables can reduce the level of snow generation.

Structure and Processing of Film.   The basic structure in most of these films is again polyethylene terephthalate (Mylar® polyester film, Melinex®). Polycyclohexylenedimethylene terephthalate (Kodar®) can also be used (3, 81). These films can be prepared by quenching (solidifying) extruded polymer to the amorphous state and then reheating and stretching the sheet approximately threefold in each direction at 80 to 100°C. Orienting the film and then annealing it under restraint at 180 to 210°C can raise the crystallinity to 40 to 42 percent (10).

Molecular Weight, Properties, and Applications.   The molecular weight of commercial polymers has been given as 15,000 to 20,000 number average and 20,000 to 30,000 weight average (14). Farrow and Hill (725) describe number average molecular weights in the range of 10,000 to 50,000, the latter figure representing fibers used particularly for industrial applications.

---

* The term describes the outermost layer of the individual fibers.

Fabrics made of polyester fibers are noted for their strength, wrinkle resistance, and resistance to moisture at ordinary temperatures. Weathering resistance is good and superior to that of the polyamides (14); resistance to sunlight is inferior to that of the acrylics. Polyester fibers blend well with cotton or wool and blends with cotton for clothing are easily the largest single end use (725). Industrial uses of polyester fibers include rubber reinforcing material, filter cloths, sieve cloths, and marine applications such as fishing nets or tarpaulins.

Polyethylene terephthalate film in its oriented crystalline state ranks among the strongest of the thermoplastics. Properties vary widely among the different types and subtypes but its basic advantages are its toughness, durability, excellent flex life, resistance to most organic solvents and mineral acids, very low moisture retention, and general absence of plasticizers (765). Applications include magnetic tape, X-ray and other photographic film, electrical insulation (metallized for capacitors), food packaging, and boil-in-bag food pouches. Polyethylene terephthalate can also be blow-molded to prepare bottles (10, 723).

Additional properties are listed in Table 54.28.

Oral Toxicity. Several 90-day feeding tests have been conducted. In a series of tests with films intended for food packaging, polyethylene terephthalate (766) and heat-treated polyethylene terephthalate (767, 768) were fed to rats and dogs. The dietary level was 10 percent in all cases. No clinical, nutritional, hematologic, urinary, biochemical, or pathologic evidence of toxicity was observed. Chloroform extracts of powdered resin produced no adverse effects when administered to rats in oral doses as high as 10 g/kg in an acute test and 400 mg/kg in a 90-day test (769).

Skin Contact. As with acrylic and nylon fabrics, clinical experience over the past several decades indicates that the basic polyethylene terephthalate fiber is essentially innocuous when applied to the skin. Tests conducted by the Schwartz-Peck procedure* with a 2-week rest period (770, 771) showed no sensitization but occasional irritation reactions consistent with occlusion. Fabric made from textured yarn (false twist) has also caused some slight reactions that appeared to be related to mechanical irritation from the rough edges formed during draw texturing and to increasing denier per filament (772). Several recent tests conducted by a modified Shelanski Repeated Insult Patch Test Procedure* (773, 774) or a modified Draize Repeated Insult Patch Test Procedure* (775) were entirely negative.

Like acrylic fabric, 100 percent polyester knit fabric is low in free formaldehyde content; 15 samples of American clothing contained ≤30 ppm free

---

* These procedures are described in Reference 420.

formaldehyde (426; formaldehyde not attributable to the intrinsic fiber as discussed in Section 6.3.4). Samples of polyester/cotton fabrics showed in some cases a relatively high level of free formaldehyde (see also Section 7.3.3).

Inhalation Toxicity, Thermal Degradation, and Related Data. Insufflation tests have been conducted with the cyclic ethylene terephthalate trimer dust (776) and also with the polyester "skin" particles (777) released from some types of filament yarns during high speed processing (see above). Under examination by a light microscope, the particles of trimer dust were refractile and varied in size from 1 to 7 µm; many particles were 1 to 3 µm. The dust was brilliantly birefringent under polarized light. Forty rats were given intratracheal injections of 0.25 ml of a 1 percent trimer dust suspension in 0.9 percent saline; three other groups of rats received injections of the same volume of a 10 percent suspension of quartz dust in saline, a 1 percent suspension of quartz dust in saline, or the saline diluent alone. Rats were killed serially in groups of five at 2, 7, 28, 91, 183, and 371 days after dosing. The remaining 10 rats of each group (or less if some died) were killed at 2 years after treatment.

The injected trimer dust was found to be scattered in the alveoli adjacent to the respiratory bronchioles. Rats that were killed 2 days after exposure showed extensive acute peribronchiolar pneumonia* from the dust accumulation. At 1 week after exposure the inflammatory exudate had disappeared and a small number of minute dust-laden granulomas developed. The granulomas were readily detected under polarized microscopic examination. Dust particles a few micrometers or less in diameter were directly surrounded by foreign body giant cells and lymphocytes. Some dust particles had been transported from the lung to the tracheal lymph nodes. The number of granulomas appeared to be somewhat less at 1 year and was further diminished at 2 years, although a few active foreign body granulomas were still evident at this time. No fibrogenic activity was evident in the granulomas or in the dust-laden macrophages of the tracheobronchial lymph nodes.

The pulmonary reactions from the quartz-treated groups were quite different. Two years after treatment at the 2.5 mg dose level the dust was almost entirely eliminated from the lungs and normal architecture was observed. At the 25 mg dose level quartz lesions were characterized by silicotic nodule formation with progressive collagenization.

The polyester fiber "skin" particles appeared irregularly constricted and sausage-like when examined under a light microscope and were brilliantly birefringent in polarized light (777). Size varied from 1 to 1000 µm in length and 1 to 40 µm in width. Forty rats given an intratracheal injection of 0.25 ml

---

* This reaction was much more intense than that observed after a 2-week series of 10 4-hr exposures to 0.4 mg/l trimer dust (mass median diameter = 5.9 ± 1.1 µm, histologically visible particles mostly <2 µm). The lungs of rats so inhaling trimer dust showed a nuisance dust cell reaction after the tenth exposure with significant reduction of dust-laden macrophage cells 2 weeks later (758).

of a 1 percent suspension of these particles were maintained on the sacrifice schedule described for the trimer dust. The long unrespirable particles were found in the small bronchi and terminal bronchioles whereas shorter fibers were trapped in the respiratory bronchioles and adjoining alveoli. On the second day after exposure, the inflammatory reaction consisted of bronchitis, bronchiolitis, and peribronchial pneumonia. At 1 week the inflammatory reaction had disappeared; the test material was retained in the foreign body giant cells or macrophages but no significant tissue reaction was observed. The intensity of the reaction progressively decreased at 1-month, 6-month, 1-year, and 2-year observations. At 2 years large unrespirable particles were retained in the terminal air passages while small respirable particles had been mostly removed by the lung clearing mechanism. No evidence of collagen formation, fibrosis, or significant alteration of the lung stromal architecture was observed.

Several clinical reports, generally of mixed etiology, are discussed in Section 9.1.4.

Combustion toxicity tests show no unusual hazard from the products evolved from polyester compared to those from other fabrics similarly tested. Thermal degradation at approximately 300°C yields primarily acetaldehyde whereas somewhat higher temperatures yield carbon monoxide. Details are summarized in Table 54.29.

Carcinogenic Potential/Cytotoxicity/Implant Studies. Many studies have been conducted with implants of polyethylene terephthalate, mostly in a form described as mesh or velour but also as sutures or powder. Polyethylene terephthalate has often been favored on the basis of minimal toxicological response, durability, and mechanical properties (778; see also 779–781). Sutures of this fiber have been preferred by some investigators (693) but not others (782). Although relatively inert, polyethylene terephthalate fiber is subject to slow degradation in body fluids (703, 783, 784).

Particulate polyethylene terephthalate showed little cytotoxicity when tested in rats (301, 785). Two instances have been reported where a prosthetic graft made from this polymer was associated with tumor development in humans (786, 787); pore size between the polymeric strands was identified as a determining factor in the second case. Vascular prostheses should be knitted with the largest possible pores to promote connective tissue organization and blood supply within the knitted structure of the prosthesis (784). A Russian report states that subcutaneous implants of polyethylene terephthalate "fibers" were resorbed in humans after an average interval of 30 years (788). See also Section 1.4.

Biodegradation. Human data as cited above indicate that polyethylene terephthalate is slowly biodegradable.

9.2.2   Polybutylene Terephthalate

The CAS No. is 24968-12-5; it is also called polytetramethylene terephthalate, PBT, and Valox®.

$$\left[\begin{array}{c}\overset{O}{\underset{\parallel}{C}}-\!\!\!\!\bigcirc\!\!\!\!-\overset{O}{\underset{\parallel}{C}}-OCH_2CH_2CH_2CH_2O\end{array}\right]_n$$

Synthesis, Molecular Weight, Properties, and Applications.   Polybutylene terephthalate can be made by the catalyzed condensation of 1,4-butanediol with either terephthalic acid or more frequently dimethyl terephthalate (724, 789). The low molecular weight polymers generally have number average molecular weights of 23,000 to 30,000 and weight average molecular weights of 36,000 to 50,000. High molecular weight polybutylene terephthalate resins, which have preferable mechanical properties, have number average molecular weights of 36,000 to 50,000 and weight average molecular weights of 60,000 to 90,000 (724).

Polybutylene terephthalate has a relatively low glass transition temperature (see Table 54.28) and crystallizes at a rapid rate; molding cycles are very short. These resins are noted for their low coefficients of friction and are resistant to abrasion. They can be readily glass-reinforced. Typical applications include exterior parts in the automotive and related fields, also connectors and fuse cases in the electrical and electronic industries. Polybutylene terephthalate materials are also used in appliances, pump housings, and impellers (724).

Oral Toxicity.   Polytetramethylene terephthalate with a weight average molecular weight of 45,000 to 85,000 has been fed at dietary levels up to 5 percent, to rats for 148 days, and to dogs for 90 days (790). At the 5 percent level both male and female dogs showed a somewhat enhanced food intake but otherwise no untoward effects were observed. Analysis of urine from rats and dogs fed the 5 percent dietary level revealed no evidence of free or combined terephthalic acid at the detection limit of 1 μg/ml; it was therefore concluded that <0.003 percent of the daily intake of polytetramethylene terephthalate was absorbed from the gastrointestinal tract and then eliminated via the urine. The investigators reasoned that if the 2.5 percent level of polymer intake was considered an acceptable no-effect level, and 1.8 mg extractable material from the polymer the likely maximum to migrate into the total food intake, the safety factor would be in excess of 10,000.

Inhalation Toxicity and Thermal Degradation Data.   Data summarized in Table 54.29 indicate no unusual inhalation toxicity of the pyrolysis/combustion products of polybutylene terephthalate per se. Flame-retarded samples may exhibit additional toxic properties; rats inhaling vapors from some but not all brominated samples heated to simulate processing temperatures developed slightly enlarged livers.

9.2.3 Elastomeric Polyester

This term has been used to describe a random copolymer of polybutylene terephthalate and polytetramethylene ether glycol (Hytrel®). This polyester elastomer has strength comparable to many thermoplastics plus a rubberlike extensibility (791).

9.3 Unsaturated Polyester Resins

The term "polyester resins" may include both these resins and the alkyds discussed in Section 9.5.

9.3.1 Structure, Synthesis, and Processing

Unsaturated polyester resins are made in several steps (10, 14, 38). A saturated dihydric alcohol is generally condensed with both a saturated and an unsaturated dicarboxylic acid. The alcohol used in the prepolymer is almost always a glycol (such as propylene, butylene, or diethylene glycols; ethylene glycol tends to give a crystalline final product). Polyhydric alcohols are sometimes used to provide strength and chemical resistance. Maleic acid (anhydride) and fumaric acid are the usual unsaturated acids although itaconic or mesaconic acid may be used to give flexibility. Phthalic anhydride is the most widely used saturated acid component.

This prepolymer (or first stage resin) is dissolved in a vinyl monomer, usually styrene, with appropriate inhibitors; the product may be supplied as a syrup containing 20 to 50 percent monomer (vinyl toluene, methyl methacrylate, diallyl phthalate, or other monomers can be used) (792). Glass fiber is added as reinforcement (alternatively, calcium carbonate, sisal, polyvinyl alcohol, other substances, or no reinforcement or filler is used). Curing is produced by free radical polymerization of styrene monomer and unsaturated acid residues. Benzoyl peroxide is frequently used for elevated temperature curing; methyl ethyl ketone peroxide or cyclohexane peroxide can be used with a cobalt accelerator at room temperature.

Products of high quality often require a relatively high percentage of unsaturation in the polymeric chain. This, however, can result in resins that during the curing process produce high exotherm temperatures, which in turn may result in an explosion (unless prevented by inhibitors). Styrene-containing unsaturated polyesters polymerize with time at ambient temperatures.

Maximum mechanical strength may not be attained until more than a week after curing. Unsaturated polyester may remain undercured—soft and in some cases tacky—if freely exposed to air during this period.

9.3.2 Molecular Weight, Properties, and Applications

The molecular weight for uncured resins has been given as 7000 to 40,000 (14). When used in polyester–glass laminates, the cured resins are noted for their good strength and rigidity, low density, toughness, and translucency. They can be formulated to be fire retardant and generally have superior heat resistance compared to most rigid thermoplastics available in sheet form.

Propylene and diethylene glycol are often used to influence the rigidity or flexibility of the resin. With phthalate–fumarate resins, increasing propylene glycol content increases hardness but may cause a reduction in tensile and flexural strength. Diethylene glycol can make the polyester more flexible and also more susceptible to water absorption.

Unsaturated polyester resins are used in automobiles to achieve lower weight and in marine applications, particularly pleasure boats. Other applications include tub and shower stalls, corrosion-resistant pipe, electric appliances, hand tools, and motor housings.

9.3.3 Skin Reactions

The finished, completely polymerized products are not considered dermatologic hazards (160). Exposure to the unsaturated polyester resin systems used in manufacturing was associated with outbreaks of dermatitis during the 1960s (793, 794), but this appears to have become less common during the 1970s (160, 795). However, dermatitis can readily become a major problem in plants with poor industrial hygiene practices (796). The incompletely hardened macromolecular resin was considered the main causative agent in 17 dermatitis cases cited in a 1962 Czech report (797).

The dermatitis is reported as mostly caused by primary irritation but occasionally by sensitizing agents (794, see also Reference 160). Reactions are eczematous and more frequent on the backs of the hands, wrists, and forearms. The suggested patch test concentration for the unsaturated polyester resin is 10 percent in acetone. Malten (795) discusses several earlier studies (including Reference 796), rates the sensitizing capacity of the polyester resin system low, and calls attention in this connection to a test with volunteers exposed to benzoyl peroxide that induced a 40 percent incidence of sensitization.

9.3.4 Inhalation and Thermal Degradation

As indicated earlier, styrene is frequently used as a cross-linking monomer in the preparation of unsaturated polyester products. Potential exposure to styrene during processing is generally considered to present the most serious inhalation hazard associated with polyester resin. Overall exposure to styrene monomer in 7 representative U.S. plants in the fiberglass plastic boat industry ranged from 2 to 183 ppm; mean exposures for the primary job categories ranged from 44 to 78 ppm (734a; see also References 31, 734b, 798). A Scandinavian report indicates that workers can be exposed "to a styrene concentration typically ranging from 20 to 300 ppm" (734c). A French study (799) conducted under model laboratory conditions indicates that a standard-type resin released the equivalent of 4.3 mg styrene/cm^2 surface in 4 hr, whereas a resin with reduced potential for evaporation of styrene yielded 0.7 mg/cm^2.

Fiberglass particles released during processing may be coated with a mixture of resins and finishes. Lim et al. (796) report finding minute amounts of trivalent and hexavalent chromium on fiberglass samples collected during walk-through plant surveys.

When exposed to fire, both conventional and fire-retarded formulations of unsaturated polyester resins typically yield copious amounts of smoke because the major decomposition product is usually styrene, which burns with a very smoky flame (38). Resins based on alkyl and particularly acrylate monomers may produce less smoke (744). Tests conducted by German investigators indicate that products evolved from pyrolysis in the 330 to 400°C range or above may be lethal to rats. Carbon monoxide can be released in biologically significant amounts but is not necessarily the cause of death. See Table 54.29 for additional details.

9.4 Alkyd Resins

Alkyd resins are polyesters in which the prepolymer characteristically contains unsaturated bonds in a fatty acid residue of a side chain. The prepolymer is typically formed from a dibasic acid such as phthalic acid, a dihydric alcohol such as ethylene glycol, plus a vegetable oil such as linseed oil, soybean oil, or tung oil that provides the fatty acid component. Properties are usually varied by modifying the type and quantity of the fatty acid (14, 38). The term "oil-free alkyds" describes products prepared without fatty acids. The fatty acids may be replaced by materials such as rosin, glycerol terephthalates, and water-soluble glycol esters (800).

Other substances that may be used as modifiers or additional ingredients include phenolic resins, epoxy resins, styrene, cobalt naphthenate, lead soaps, and fire retardants. Curing or cross-linking is accomplished by air oxidation of the unsaturated groups (14, 38).

Alkyd resins are very widely used in paints, lacquers, or other surface coatings. They provide durability, flexibility, gloss retention, and reasonable heat resistance. Alkyd resins may also be formulated as molding materials containing a mixture of polyester resins, cross-linking monomers, catalysts, fillers, and miscellaneous ingredients. An organic peroxide or other material that provides free radicals when heated serves to initiate the cross-linking of the polyester chains. A nonvolatile cross-linking monomer such as diallyl phthalate is typically used for curing. Applications include such items as distributor caps, ignition coil towers, other automotive parts, circuit breakers, switch parts, and transformer housing (801).

Alkyd resin binder paints may yield relatively high concentrations of phthalic acid compounds under industrial drying conditions. At a drying temperature of 160°C one alkyd resin yielded 0.8 g phthalic acid compounds (calculated as anhydride) per 100 g wet paint (37 mg/m^3 in first stack). An alkyd–melamine resin paint dried at 170°C yielded 1.7 g phthalic anhydride per 100 g wet paint but no compounds from the melamine component. However, at 130°C the yield of phthalic acid and maleic acid compounds from alkyd resins should be limited to trace quantities only (476).

Otherwise both uncured and cured alkyd resins appear to have attracted relatively little comment as a source of irritation or toxic effects. Fisher describes alkyd resin as a rare sensitizer in connection with its use in hypoallergenic nail enamel; alkyd resin is also listed among a series of resins that are used for antiwrinkle effect on clothing and "may cause dermatitis particularly when the resin is not completely cured" (160).

9.5 Allyl Polymers

The term "allyl resin" typically refers to unsaturated polyester cross-linked with allyl-type monomer rather than styrene. Such resins are notable for their retention of electrical properties under conditions of high temperature and high humidity. "Allyl molding compounds" may refer to nonpolymeric substances used in the preparation of thermoset moldings. Those in widest commercial use are the monomers and prepolymers of diallyl phthalate and diallyl isophthalate (802). The diethylene glycol bis(allyl carbonate) polymers (CR-39) are used for lenses and optical devices because of their light weight and resistance to impact, scratch, and abrasion.

CR 39 monomer is a colorless, slightly volatile liquid (molecular weight 274) prepared from diethylene glycol chloroformate and allyl alcohol (803). A peroxide catalyst (typically, benzoyl peroxide, isopropyl percarbonate, cyclohexyl percarbonate) is dissolved at about 3 percent by weight in the diallyl glycol carbonate and then the liquid is polymerized first to a gel and then to a fusible solid. The monomer is known to be an irritant. Skin contact during the polymerization process may result in a rapidly developing, irritant dermatitis with an incidence as high as 70 percent. Dermatitis frequently appears to result from direct contact with the liquid monomer. Some cases would appear to be traceable to residual monomer or other ingredient(s) in partially polymerized polymer since "adherence of the polymer to the moulds is often associated by the workers with the irritant nature of the resin" (803).

9.6 Polycarbonate

Trade names include Lexan®, Merlon®, and Makrolon®.*

9.6.1 Structure

Polycarbonates are essentially polyesters of carbonic acid. Although a number of derivatives have been developed (720), the only major commercial polycarbonate at present is the polycarbonate of bisphenol A (CAS No. 24936-68-3):

$$\left[\mathrm{O}\text{—}\underset{}{\bigcirc}\text{—}\overset{\underset{\displaystyle CH_3}{|}}{\underset{|}{\overset{\displaystyle CH_3}{C}}}\text{—}\underset{}{\bigcirc}\text{—}\mathrm{O}\text{—}\overset{\displaystyle O}{\underset{}{C}} \right]_n$$

9.6.2 Preparation

Commercial methods appear to favor the reaction of phosgene with bisphenol A in the presence of acid acceptors. The bisphenol A may be dissolved in pyridine (used as hydrohalide acceptor) or a mixture of pyridine and other solvent, or the phosgene may be reacted with pyridine to form a complex that is then dissolved in bisphenol A. Molecular weight can be controlled by a deficiency of one of the materials or a monofunctional reactant such as phenol (otherwise very high molecular weight polymers may be formed).

Polycarbonate can also be formed by transesterification, with diphenyl carbonate the preferred ester. This reaction is typically conducted in the range of 200°C under pressure; molecular weights above 30,000 are difficult to obtain without special equipment (4, 10).

9.6.3 Molecular Weight, Properties, and Applications

The molecular weight of commercial polymers can reflect intended processing and end use. Polycarbonate intended for injection molding can have a number average molecular weight range (10) of 23,000 to 32,000, whereas that intended for extrusion blow molding and solution casting into film can have similar

---

* In the United States, Lexan® and Merlon® are supplied as base resins and foamable grades. The largest use of sheet is as Lexan® and Tuffak® (804).

weight ranges of 32,000 to 39,000 and approximately 70,000 (or at least >50,000).

Polycarbonates are extremely tough and rigid, and have good electrical insulation characteristics (10, 38). They are partially crystalline and transparent with a pale yellow color (4, 10). Their resistance to creep or deformation under load is superior to that of acetal and polyamide plastics. They are notch-sensitive and have limited resistance to chemicals and ultraviolet light. Some mechanical properties may be impaired in use above 140°C. Autoclaving is probably less detrimental to tensile properties than either gamma irradiation or ethylene oxide sterilization,* although any of the three can probably be used satisfactorily once (805).

Polycarbonate and acrylic plastic used in equipment for physiologic experiments released sufficient carbon monoxide to interfere with the measurement of carbon monoxide production by individual rats (806). The amount of carbon monoxide liberated from the polycarbonate was approximately 10 times lower than that from the acrylic prior to deliberate exposure to carbon monoxide but 6.4 times higher after such exposure (5 percent CO in N_2 for 96 hr). The evolution of carbon monoxide was considered to be due to the slow removal of gas dissolved in the sheet plastic.

Table 54.28 lists additional data on properties.

Applications include medical equipment, beer pitchers, automotive lenses and trim extrusions, drapery fixtures, door and window components, furniture, and plumbing. Foamed polycarbonate is used in major automotive components such as a one-piece bus seat frame. Sheet glazing products that conform to specifications for safety glazing and burglar resistance may be used in schools, off-highway installations, and security facilities. Coated thin-gauge sheet has been used for protective eyewear and business machines (804). Registration of fast-neutron-induced recoil and (n, α) tracks in polycarbonate foils provides a sensitive means of dosimetry (807).

9.6.4 Processing

Bisphenol A polycarbonate can be readily processed in solution. Alternative techniques include thermoforming, machining, or bonding (720).

9.6.5 Oral Toxicity

Bisphenol A polycarbonate was considered physiologically inert when fed to rats at a level of 6 percent in the food (808; similar results are reported for nylon 6).

---

* Ethylene oxide can react with certain functional groups and its use requires adequate degassing of the polymer; gamma radiation doses above about 15 Mrad may produce polymer degradation.

9.6.6   Inhalation and Thermal Degradation

Tests summarized in 54.29 indicate no unusual effects from pyrolysis of polycarbonate compared to other polymers.

Thermal degradation yields carbon monoxide. See also mention of carbon monoxide above.

9.7   Polyethers

9.7.1   Polyoxymethylene

The CAS No. is 9002-81-7. Polyoxymethylene is also known as Delrin® acetal resin (homopolymer) and Celcon® (copolymer).

$$-[CH_2-O]_{\overline{n}}$$

Structure and Preparation.   At least four means of polymerization are known. Commercial polymers typically start with a very pure form of formaldehyde (10, 728) that can be produced from a low molecular weight polyformaldehyde (10). Esterification with acetic anhydride or other anhydrides can be conducted at approximately 130 to 200°C with sodium acetate as a catalyst. Copolymers can be made by copolymerization of formaldehyde with a cyclic ether as second monomer (10). The end groups of some polyoxymethylenes can be either acetate or methoxy; copolymers may contain terminal hydroxyethyl ethers for added stability (10).

Molecular Weight, Properties, and Applications.   The number average molecular weight of some commercial polyoxymethylenes has been reported in the range from 20,000 to 110,000 (10). The notable features of the acetal resins are their metal-like stiffness, resistance to fatigue, and resistance to organic solvents below 70°C. Coupled with low density (as compared to metal) and ease of fabrication, these qualities permit their use as replacements for metal in many applications. Acetal in contact with steel generally gives less friction than acetal with other metals such as aluminum, and acetal with nylon is usually preferable to either alone (10). Specific uses of acetal include gears, links in conveyor belts, blower wheels, cams, fan blades, valve stems, and shower heads.

See Table 54.28 for additional data.

Processing.   Acetal resins may be readily processed on conventional injection molding, blow molding, and extrusion equipment provided that overheating is avoided. Overheating can lead to the production of formaldehyde gas and a serious or even dangerous buildup of pressure. Gas pressure created by decomposition can rapidly become extremely high when processing machines are not properly vented. A shotgun-like reaction from a pellet hopper was reported to develop within a few minutes after failure of one of three heating zone circuits (809). Recommended control measures (810, 811) include (1) the use of vented feed screws and proper design of injection nozzle, (2) preventive maintenance and replacement of worn gaskets and valves, (3) automatic heat regulation and monitoring of excess temperature conditions by an audible alarm or warning device in direct view of the operator, (4) a standby water cooling device not dependent on the main electrical system, and (5) personnel instruction.

Oral Toxicity and Skin Contact.   Ninety-day feeding tests have been conducted with two different types of polyoxymethylene (96+ and 99 percent active ingredient, respectively, both with number average molecular weights of 25,000 to 30,000). Rats in groups of 40 and dogs in groups of four were fed a 10 percent dietary level of each test polymer or Alphacel® cellulose as a control (812). One or two rats died in each of the two test groups; one death was also observed in the corresponding control group. No deaths occurred among any of the dogs. No changes attributable to the test polymer were evident in behavior and appearance, body weight, food consumption, ophthalmoscopy, and hematologic, biochemical, or urinalysis studies. No gross or microscopic lesions or variations in organ weight attributable to treatment were observed in any of the animals.

Skin irritation and sensitization tests with 1-in. disks of polyoxymethylene extruded sheeting were conducted by the Schwartz-Peck procedure (420; 10-day rest period) on a panel of 212 volunteers (813). Some mild or moderate erythema consistent with occlusion was observed at the 6-day reading but otherwise no reactions were observed. The panel included one subject known to be sensitive to urea–formaldehyde who showed no reaction through 4 days after removal of the test polymer.

Inhalation Toxicity and Thermal Degradation Data.   A massive single 6-hr dust inhalation exposure (47 mg/l nominal concentration) caused emphysema and atelectasis in rats. Fumes evolved at temperatures as low as 150°C can, in sufficient quantity, be lethal to animals that are exposed for 4 hr. Both formaldehyde and carbon monoxide have been identified in the effluent evolved at elevated temperatures. See Table 54.29 for details, as well as the next paragraph.

Injection Studies.   Rats receiving intraperitoneal injections of an unstated concentration of a dust suspension showed granulomas with some scarring when examined 90 days later (814). Similar subcutaneous injections produced less reaction.

9.7.2 Polyoxyethylene*

Polyoxyethylene is available as Polyox® (high molecular weight).

$$-\!\!\!-\!\!\mbox{[CH}_2\!\!-\!\!\mbox{CH}_2\!\!-\!\!\mbox{O]}_n\!\!-$$

Type and Molecular Weight. The polyoxyethylenes such as Carbowax® range from liquids to waxy solids, with a molecular weight up to 10,000. Polyox® describes a material with a molecular weight described as $\geq 10^5$ (14) and ranging up to 4 million (815).

Properties and Applications. The Polyox® resins are water-soluble thermoplastic polymers that can be extruded as films. They can be used in packaging for food, as textile sizes, and for the reduction of hydrodynamic friction.

Oral Toxicity. Polyoxyethylene (Polyox® WSR-301, mean molecular weight of 4 million) was fed to rats for 90 days at dietary levels averaging 8.0 and 18.4 g/kg/day (815). These rats showed changes, rated minor, in the liver and kidney. Two-year tests with rats fed at dietary levels averaging up to 2.76 g/kg/day and dogs fed at dietary levels in the range of 0.6 g/kg/day showed no detectable effect. Testing with a ^{14}C-tagged sample revealed no significant absorption of the polymers per se from the gastrointestinal tract of the rat or dog. The content of glycols and polyglycols in the tagged sample was considered sufficient to account for all the radioactivity found in the urine (0.72 and 1.1 percent of the dose administered to 20 rats and 1 dog, respectively).

Skin Absorption. Rabbits were not injured by 20 ml of 5 percent aqueous solution applied to the clipped trunk (815).

Intravenous Toxicity. Solutions of 0.1 percent commercial material killed rats when first administered (815) at a dosage of 3 mg/kg (0.3 ml). Subsequent injection of this concentration after shearing the polymer in an Osterizer did not kill rats receiving 40 mg/kg (4 ml).

9.7.3 Polyphenylene Oxide-Based Resin

Resins are available as Noryl®.

Structure and Synthesis. Polyphenylene oxide-based resins have been prepared by oxidative coupling of 2,6-disubstituted phenols in the presence of

* The CAS registry number 25322-68-3 is used to identify polymers of both the Polyox® and Carbowax® types.

oxygen and a copper-complex catalyst (4, 14, 729). The poly(2,6-dimethyl-1,4-phenylene ether) was first introduced as PPO (or p.p.o.). The phenylene oxide-based resins may actually be polymer blends with polymers such as polystyrene or high impact polystyrene.

Properties and Applications.   The molecular weight of commercial materials has been given as 25,000 to 60,000 (10). These resins are rigid polymers with very good dimensional stability. They are noted for low water absorption and resistance to chemicals. Some halogenated polyphenylene oxides provide greater thermal stability. See also Table 54.28.

Applications include use in appliance and business machine housings, automotive dashboards, connectors, panels, grilles, and electrical connectors.

Thermal Degradation Data.   Tests summarized in Table 54.29 suggest that carbon monoxide is the main toxic hazard.

9.8 Sulfur Polymers

The main commercial polymers are polyphenylene sulfide and the polysulfones. Poly(p-phenylene sulfide) (CAS No. 53027-72-8) is also known as poly(thio-1,4-phenylene) and Ryton®.

The polysulfone of bisphenol A (CAS No. 25135-51-7) is available as Udel®.

Other polysulfones, such as Polyether sulfone 200P, Polyether sulfone 720P, and Astrel® 360, differ in the types of linkage between aromatic rings (and vary in their properties accordingly).

9.8.1   Synthesis, Properties, and Applications of Polyphenylene Sulfide

This polymer has been manufactured commercially by reacting p-dichloroben-zene and sodium sulfide, apparently in N-methylpyrrolidone (as cited in Reference 211). Heating under oxygen cures the product to form a fine white powder, apparently linear. The white, lightly crystalline polymer discolors on heating in air to form a brown product, presumably cross-linked. Polyphenylene sulfide is known (731) for thermal and chemical resistance, toughness, and flexibility. Additional properties are summarized in Table 54.28.

Polyphenylene sulfide molded parts can be used in submersible and centrifugal pumps, computers, and in telecommunications. Coatings of this polymer can be utilized for chemical resistance in industrial process operations (816).

9.8.2 Synthesis, Properties, and Applications of Polysulfone

For commercial products an aryl ether can be treated with sulfonyl chloride in the presence of catalysts to give sulfone groups (polyetherification), or sulfones can be treated with phenolates to give ether groups (polysulfonylation). The structures made by one route are usually not made by the other, although some can be made by either process. All polysulfones have excellent creep resistance. They are stable to oxygen and thermal degradation and are flame resistant. Their toughness is affected by the inclusion of bulky side groups in the polymer structure or deviations from the all-para orientation of groups linking the aromatic rings (211, 733). See Table 54.28 for additional data.

9.8.3 Inhalation and Thermal Degradation Data

Several experimental tests indicate that products evolved at elevated temperatures can be rapidly toxic to mice. Thermal degradation of polyphenylene sulfide may yield hydrogen sulfide; the sulfur of polysulfone is typically evolved as sulfur dioxide. See Table 54.29 for details.

10 POLYURETHANES

Polyurethanes are an extremely complex class of polymers (817) that are essentially ester–amide derivatives of carbonic acids (4). They were first developed commercially around 1937–1941 as nylonlike fibers (10, 14). Today they exist in such widely different forms as foams, elastomers, coatings, adhesives, and elastomeric fibers.

10.1 Overview

Polyurethanes, also called polyurethans, urethanes, or polycarbamates, contain significant numbers of urethane groups, regardless of what the rest of the molecule may be. A typical polyurethane may contain aliphatic and aromatic hydrocarbon residues, ester, ether, and amide linkages (818). Polyurethanes can be prepared from the reaction of bischloroformates with diamines or by other techniques but are most commonly made from the reaction of isocyanates with polyhydroxy compounds. These reactions characteristically involve either polycondensation or the addition of hydrogen across the carbon–nitrogen double bond of the isocyanate group (4). The term polyurethane is now generally understood to include all the complex polymers formed from isocyanates and polyols (10).

10.1.1 Production and Processing

Use of polyurethane is expanding worldwide and may exceed 7 billion lb in 1980, with the United States accounting for about 40 percent of world use (819). In 1979, 59 percent of the U.S. market was in flexible foam, 25 percent in rigid foam, and 16 percent in elastomeric and other types. Production data for polyether and polyester polyols are given in Table 54.2.

The hazards involved in processing vary with the type of product, but are basically isocyanate exposure that may exist in handling the prepolymer or uncured product.

10.1.2 Specifications and Analysis

Several recently published methods describing the determination of isocyanate in polyurethane paints (820), paint hardeners (821), prepolymers (822, 823), or adhesives (824) are available. A gel permeation chromatographic procedure (824) is reported as adaptable to products of different type and having a detection limit of about 0.1 percent monomer. Hexamethylene diisocyanate (821) can be detected in paint hardeners by high pressure liquid chromatography at levels as low as 0.01 percent. Additional citations are given in Reference 31.

Polyurethanes meeting stated requirements have been cleared for certain uses related to food contact (92). Section 1.3 provides a guide to general identification methods for the various types of polyurethane; see also Section 6.1.2 concerning identification of fibers and coatings.

10.1.3 Toxicologic Potential

As indicated in Table 54.30, the inhalation toxicity associated with the spraying or processing of polyurethane foam at ambient temperature is essentially that of the residual isocyanate groups. The most hazardous commercial isocyanate with respect to asthma and acute respiratory tract insufficiency is generally 2,4-toluene diisocyanate (TDI), for which the TLV is 0.02 ppm (1981 intended change, 0.005 ppm). Some of the aliphatic diisocyanates may be appreciably more irritating than TDI. Skin irritation and sensitization may occur from contact with both the diisocyanates and partially polymerized uncured resin. These effects are not associated with completely cured foam. However, several inhalation tests, an insufflation test, and cytotoxicity tests (see Table 54.30, also Sections 10.2.3 and 10.2.5) indicate that polyurethane foam dust may act as a tissue irritant or cause changes in respiratory function.

Seven of 17* polyurethanes were reported to have a greater "tumorigenic potential" than a polyethylene control in implant studies (859). In a subsequent amplifying test (860) with the most active sample, the data were "considered consistent with a mechanism of biological degradation of the polymer to yield an active carcinogenic metabolite"; the identity of this proposed metabolite is not apparent. Degradation by pyrolysis and autoclaving (Table 54.30, also Reference 861) of some polyurethane foams may yield aromatic amines that have been associated with carcinogenic effects.

Thermal decomposition of polyurethane in the absence of air yields hydrogen cyanide at about 800°C; at about 1000°C hydrogen cyanide accounts for about two-thirds of the available nitrogen. In air the evolution of hydrogen cyanide begins at about 500°C, but although it increases with temperature, the yield at 1000°C is only about half that obtained in the absence of oxygen. Large-scale tests with 135-kg loads of burning polyurethane foam showed a rapidly developing, very hot fire with high smoke levels and an early peak evolution of hydrogen cyanide, followed by a more sustained production of carbon monoxide (215). However, "the cyanide risk seems to be relatively low, in comparison with the risk from carbon monoxide, during fires involving domestic loads of these foams. . . . The hazard of isocyanate production during both industrial and domestic situations seems to be very much lower than might be expected from estimations from the thermal decomposition (laboratory) experiments" (846).

10.1.4   Workplace Practices and Standards

Work with uncured resins that may involve respiratory exposure or skin contact to isocyanates should be evaluated prior to use to assess the types and severity of hazard (see Chapter 23, Volume 1).

This discussion on spills of all kinds of partially polymerized polyurethane (862) is intended to illustrate the range of practices that may be advisable rather than serve as a didactic code. In the event of skin contact, immediately but gently remove excess polymer, then immediately flush with undiluted isopropyl alcohol. Flush with copious water. Repeat if necessary. This provides a simple yet effective means of cleansing intact skin. Dilutions of isopropyl alcohol such as 30 percent (160) may be advisable if the skin is irritated, but cleansing is much less effective and may need prolonged attention. *If eye contact occurs, irrigate eyes with water immediately and seek medical attention without delay.* Butyl rubber or comparable overshoes should be worn in areas susceptible to floor

---

* Based on results with male rats; in female rats 10 of 13 polyurethanes ranked higher than the control in "tumorigenic potential." The samples are not specifically identified as foams.

Table 54.30. Overview of Inhalation, Emission, and Thermal Degradation Data From Several Kinds of Polyurethane[a]

Polymer	TLV	Type of Study			Species Exposed	Description of Material	Procedure	Observations/Conclusions	Reference
		Ambient Temperature	Elevated Temperature	Effluent Analysis					
Polyurethane foam	No specific TLV; see Section 1.4	X		X	Human	Foam from 2-tank unit, 1 tank with polyol and catalyst, other with MDI	Spray application (majority of unreacted MDI in particulate, nearly all particles in respirable range) monitored in underground mine: exposure of ~1.3 ppm/min resulted in antibody response to MDI, exposure of ~0.9 ppm did not	825
					Human	MDI prepolymer	No "significant pulmonary symptomatology" evoked by exposure in work area where concentration of MDI ranged from 0.01 mg/m^3 to 0.06 mg/m^3	826
		X			Monkey	Rigid polyurethane foam representative of current use	Exposure 6 hr/day, 5 days/week, ×18 months, at 10 mg/m^3 (97.5% respirable, 0.74 μm geometric mean), with pulmonary function evaluated by plethysmography in anesthetized animals after 4, 14, or 18 months of exposure	Reduction in flow maximas at 75% vital capacity after 4 months, in end tidal forced expiratory volume in 0.5 sec after 14 months. Pathologic evaluation of test and control animals considered generally unremarkable; both test and some control animals showed particles in lungs	827
		X			Rat	Rigid polyurethane foam, freshly generated; based on sucrose polyether	Exposure 6 hr/day, 5 days/week × 12 weeks, 8.65 mg/m^3, maximal concentration with simulated use generator; 94% of particles <5 μm. Also comparison serves with TiO$_2$-exposed and air-exposed rats	No changes in appearance, behavior, body weight, or average survival time attributable to foam dust exposure through 140 weeks. High tumor rate in all groups without significant differences between groups, peritracheal and peribronchial lymphocytic infiltration within usual physiologic range	828

4450

		Animal	Material	Exposure conditions	Results	Ref.
X		Rat, hamster	Rigid polyurethane foam, standard commercial variety used for structural insulation	Exposure 6 hr/day, 5 days/week × 30 exposure days. Concentration of dust = 20 or 3.6 mg/m³; count median and mass median diameters of 0.5–0.7 and 1.8–4.0 μm, respectively	Squamous cell carcinoma in 1 rat dying 349 days after initial exposure to 20 mg/m³ and in 1 rat dying at 544 days after initial exposure to 3.6 mg/m³; 3 of 15 rat exposed at 20 mg/m³ with centrolobular emphysema at 256–365 days after initial exposure	829
X			Polyurethane foam generated from adduct (reaction product of excess TDI with polyol) and cross-linker (remaining formulation components)	Spray application out of doors, monitored by method sensitive to 0.001 mg TDI	TDI not detected more than 8 ft upwind from spray gun, but a few foam particles found as far as 200 ft downwind	830
X			Various types of polyurethane foam	Comprehensive review of studies where measurements were taken under factory conditions (also related studies)		831
X	X	Rat	Polyurethane foam (¼ in. sheet), also nylon coated on 1 side with polyurethane (30% polyurethane, 70% nylon canvas)	Exposure 60 min. Pyrolysis products of generated by cone heater, combustion products generated with match and heater	Pyrolysis products of both materials about equal in toxicity; cause of death considered acute asphyxia with polyurethane products, pulmonary edema with coated nylon products. No deaths when rats exposed to similar concentrations of combustion products of the 2 materials	832
X		Rat	Three elastomeric foams	6-hr exposure to pyrolysis products generated from 5-g samples at 250°C followed by 1 death due to pulmonary congestion and edema, 1 death at 7 days from pneumonia (perhaps coincidental) among total of 12 rats exposed. No deaths among 6 rats exposed to products generated at 560°C for 10 min	833

Table 54.30. (Continued)

| Polymer | TLV | Type of Study | | | Species Exposed | Description of Material | Procedure | Observations/Conclusions | Reference |
		Ambient Temperature	Elevated Temperature	Effluent Analysis					
			X	X	Rat	Commercial sheet stock	Exposure 30 min. Samples of increasing weight heated to 650–800°C	Both CO and HCN considered contributory to death	834
			X	X	Rat		Exposure 5–8 min. Products degraded in muffle furnace, rats restrained with head only in chamber	Some inhibition of cytochrome *c* oxidase in tissue, correlated with blood cyanide concentrations	835
			X	X	Rat	Ten flexible polyurethanes, both polyether and polyester based	"The order of toxicity has been shown to change significantly if one considered mortality versus avoidance response" (837). "Mutagenic activity is due mainly to . . . polynuclear aromatic hydrocarbons or heterocyclic components . . . but other components are also operative" (838)		836–838
			X	X	Rat	Rigid polyurethane foam with or without phosphorus-based flame retardant	20-min exposure to products from nonflaming combustion of foam without fire retardant caused elevated carboxyhemoglobin, no deaths; similar test with fire-retarded foam caused grand mal seizures and death. Toxic component identified as bicyclic phosphate generated from decomposition of flame retardant and trimethylolpropane polyol		839
			X	X	Rat		Confirmation of neurologic effects and increased toxicity from the nonflaming decomposition products of certain foams as described directly above. Absence of these effects in similar experiments with rigid polyurethane foams and polyester fabric samples containing a cyclic phosphonate derivative of trimethylolpropane considered to be related to the stability of the cyclic phosphonate flame retardant additive		840

			Species	Material	Test method	Effects	Cited in
X	X	X	Rat			"Combustion products [of a polyester-type polyurethane foam] included small amounts of carbon monoxide, hydrogen cyanide, and oxides of nitrogen. When the foamed polyurethane was heated in nitrogen, or a mixture of nitrogen and air, carbon monoxide was produced but much less oxides of nitrogen. Rats exposed [for 15 min to combustion or decomposition products evolved from samples ~0.5–1.3 g] showed no adverse effects"	Cited in 663
X			Mouse	Polyurethane rigid foams and polyurethane flexible foams (also polyisocyanurate foams)	Rising and fixed temperature programs as described in Table 54.10	Polyurethanes products among the more toxic of a series of polymer products as judged by decreasing time to death; see Table 54.8. CO considered an important or principal toxicant in tests with some but not all foams (98). Some samples of flexible polyurethane foam appeared to evolve less toxic products with age	66, 94, 98, 841
X			Mouse	Four polyether-based foams: flexible and rigid, with and without chlorine/phosphorus flame retardant	Sensory irritation test, which measures inhibition of respiration; experimental technique similar to that described in Table 54.17 (Ref. 165)	Visible smoke when furnace temperature reached 250–300°C. Flexible foams "more highly irritating" than rigid foams due to more complete combustion and more rapid evolution of decomposition products. Flame retardant did not significantly alter thermostability or degree of sensory irritation response	842
X				Six commercial samples: 4 foam samples of the type used for automobile seats and carpet backing, 2 samples in pellet form	As described in Table 54.10	"Combustion of polyurethanes may result in a wide variety of hydrocarbon and nitrogen-containing products. Carbon monoxide and cyanide are the only acutely toxic products" (682)	102, 682

4453

Table 54.30. (Continued)

Polymer	TLV	Type of Study				Description of Material	Procedure	Observations/Conclusions	Reference
		Ambient Temperature	Elevated Temperature	Effluent Analysis	Species Exposed				
Flexible polyester based and polyether based foams TDI (843–845), rigid foam prepared from MDI (843, 846)			X	X				"Pyrolysis of flexible polyurethane at about 300°C releases a polymeric yellow smoke which decomposes at temperatures above 800°C in nitrogen (500°C in air) to give a range of nitrogen-containing compounds, particularly hydrogen cyanide" (843). TDI produced especially at about 250–300°C in both air and nitrogen atmospheres, completely destroyed above about 700°C (844). HCN, acetonitrile, acrylonitrile, pyridine, and benzonitrile found during high temperature decomposition of both flexible and rigid foams (845, 846). See Figures 54.14 and 54.15	843–846
Four flexible polyurethane foams based on TDI or m-phenylene diisocyanate mixtures; 1 rigid polyurethane foam			X	X				Analysis of smokes from foams heated to 300–1000°C showed yields of main nitrogen-containing materials in 800–1000°C range; refs. to amine formation	847
Flexible polyurethane foam			X	X				Analysis of ~2 g sample heated in airstream to ≤470°C indicated that most of the isocyanate-active products appear to be oligomeric ureas having isocyanate end groups. "Yellow smoke" (843) considered polyurea; nitrogen oxide formation considered in cited ref.	848
Flexible foam based on 80:20 2,4/2,6 isomers of TDI			X	X				Pyrolysis in inert atmosphere 300–1000°C yielded TDI, 3-amino-4-methylphenyl isocyanate, 2,4-diaminotoluene,[b] other products	849
Four samples of all foam (presumably rigid)			X	X				Decomposition by stagnation burner to approximate mine conditions yielded CO_2, also CO, HCl, dichloroethane, and trichlorofluoromethane in some cases[b]	850

Material	Composition / type			Sample description	Observations	Ref.
Polyurethane varnish		X	X	Foam used in mines	"Urethane foam ignites easily and, while burning, produces black, acrid smoke"	851
	As for foam, above	X	X — Human	Commercial two-component systems	Materials that were pyrolyzed reformed the antecedent diisocyanates. Varnishes based on aromatic diisocyanates have higher yields than those based on aliphatic diisocyanates. Human "intoxication" mentioned	852
Polyurethane paint	As for foam, above	X	X	Commercial primer	Approximately 0.2% of deposited primer released as aromatic amine, principally toluenediamine, during heating or extraction; this event associated with unexpected yellowing of the topcoat when it cured	853
Polyurethane-covered wire	As for foam, above	X	X — Human	Wire used intermittently for winding particular type of coil	Soldering iron (300°C) released fumes into workplace atmosphere	Spasms of breathlessness and cough occurring sporadically in workers adjacent to coil winding area	854
Polyurethane films (degradation studies)		X	X	One aliphatic, 2 aromatic, and 1 segmented polyurethane (compositions identified)	Thermal oxidative degradation, emphasis on HCN evolution over range of temperatures (271–331°C, 450–480°C) and O_2 concentrations	HCN evolution during oxidative thermal degradation a relatively minor reaction compared with the many other concurrent reactions. Inhibition of HCN by copper or copper oxide during degradation due to catalytic oxidation of evolved HCN (856)	855, 856
		X	X	Polyurethane prepared from 1,4-butanediol and methylene bis(4-phenyl isocyanate)	Volatile degradation products start to evolve ~240°C. "Total reaction comprises a primary depoly-condensation process in which the two monomers are formed, followed by the subsequent reaction of these monomers to form the volatile products, tetrahydrofuran, dihydrofuran, carbon dioxide, water, butadiene, hydrogen cyanide, and carbon monoxide and residual carbodiimide and urea structures"	857

[a] Summarized as described in Table 54.10.

[b] Trichlorofluoromethane was known to be a blowing agent. Its thermal decomposition does not typically yield phosgene, which appears to have been a concern of these investigators.

[c] Formation of toluenediamine is viewed as a result of the decomposition of TDI. Toluenediamine was not detected during thermolysis of a foam derived from polymethylene polyphenylisocyanate (858).

contamination. Acid-suited* personnel should be swabbed down with decontamination solution by a helper *prior* to showering and removal of the suit. Contaminated personal clothing should be swabbed or immersed in decontaminated solution prior to laundering or else discarded. A mixture containing 20 parts isopropyl alcohol, 40 parts concentrated ammonia solution, 2 to 5 parts household type detergent, plus water to make 100 parts, provides an effective decontaminating solution† in areas of good ventilation. Ammonia–isocyanate mixtures can volatilize most unpleasantly in confined spaces. In areas of poor ventilation, a preparation such as Spartan Rinsable Degreaser WRD-160‡ should be worked into the spill and followed by 50/50 monethanolamine§–water. The Degreaser WRD-160 will also liquefy solidified polymer that should then be contained in dams of Auto-Dri‖ or equivalent absorbent before being neutralized for disposal.

Processing of polyurethanes at elevated temperatures should be conducted under exhaust ventilation or preferably in a sealed system. Bulk quantities of polyurethane foam should be stored separately from other flammable substances.

10.2 Polyurethane Foams

Polyurethane foams are typically made from the reaction of diisocyanates, polyester or polyether resins, a blowing agent, water, catalysts, and surfactants (863). They may also contain flame retardants, fillers, extenders, bacteriostats, and dyes. They can be classed as flexible, semirigid, or rigid. Flexible foams probably are the best known and most common type of polyurethane.

10.2.1 Synthesis, Properties, and Applications of Flexible Foams

Flexible foams are generally made (10, 864) by the one-shot process, the prepolymer process, or a combination (quasi-prepolymer or semiprepolymer). In the one-shot process all the ingredients are piped separately into the mixing head of the foam machine. The polyether one-shot process is the one most commonly used for flexible foam. The polyols are most frequently the triols derived from the polymerization of propylene oxide along with ethylene oxide, glycerol, trimethylolpropane, or 1,2,6-hexanetriol (or various combinations thereof, depending on the type of polymer desired). The isocyanate component is usually 80/20 toluene diisocyanate (ratio of the 2,4 isomer to the 2,6 isomer).

---

* Chem suits, which additionally offer the respiratory protection of an air supply, are recommended where the resin temperature exceeds 70°C or in the presence of unusually volatile material.
† This solution is flammable; an alternate nonflammable solution is aqueous ammonia about 10 percent by weight, plus a heavy-duty detergent. Neither is suitable for skin application.
‡ Spartan Rinsable Degreaser WRD-160, supplied by Spartan Chemical Co., 110 N. Westwood Ave., Toledo, Ohio 43607 (available in 55-gal drums).
§ Monoethanolamine itself requires care in handling.
‖ Auto-Dri, supplied by Engelhard Mineral and Chemical Division, Menlo Park, Edison, N.J. 08817.

The older prepolymer processes involve prereacting a portion of the polyol with an isocyanate to form a prepolymer. This isocyanate-terminated prepolymer is reasonably stable if kept sealed and dry. Mixing with a second blend of additional polyol, catalysts, water, surfactant, and blowing agent yields foam. The prepolymer processes have the advantage of less exotherm, less free isocyanate, and possibly superior mechanical properties, but the disadvantages of limited stability, viscosity in handling, and an extra process step (10, 863).

10.2.2 Synthesis, Properties, and Applications of Rigid Foams

Rigid foams are made by similar procedures to those described for flexible foams. However, the less volatile methane diphenyl diisocyanate (MDI) or polymethylene polyphenylisocyanate (PAPI®) is often substituted for TDI. A typical formulation (as cited in Reference 31) may contain sucrose or sorbitol based polyether polyols (100 parts by weight), polymethylene polyphenyl isocyanate (110 to 140 parts by weight), fluorocarbon 11 (25 to 35 parts by weight), flame retardant (0 to 15 parts), catalysts (1.5 to 3 parts), and surfactant (1.3 to 3 parts).

The density of rigid foam has been described as 0.017 to 0.66 g/cm^3, most frequently 0.025 to 0.058 g/cm^3. Maximum service temperatures can vary from at least 94 to 149°C, and the softening temperature from 128 to 145°C.

Rigid polyurethane foam is used primarily as an insulating material. In nonresidential construction the foam may be used as board stock, spray applied, or poured in place, for roofing insulation, cold storage, pipe covering, and mobile homes. It is widely used in refrigerator and freezer insulation as well as insulation for freight cars, fuel oil tanks, and natural gas pipelines. Other applications include flotation for marine salvage and trim or frames for furniture.

10.2.3 Dust Inhalation Toxicity and Related Data

Two earlier tests (829, 865) showed several rats with tumors, but subsequent, more extensive tests have not shown significant tumor induction. As summarized in Table 54.30, Laskin and his associates (829) exposed rats and hamsters to freshly generated polyurethane foam dust, 3.6 or 20 mg/m^3, 6 hr/day for 30 exposure days. One rat exposed at each concentration developed squamous cell carcinoma late in the life-span observation period but no neoplasms of this type were seen in the hamsters. Thyssen and co-workers (828) also exposed rats of the same strain 6 hr/day for 12 weeks to concentrations averaging 8.65 mg/m^3 of respirable dust from polyurethane foam. Control rats were exposed to titanium dioxide or to air alone. A high respiratory tumor rate was seen in all three of these exposed groups, with no significant differences among the groups, during more than 2 years of observation.

An intratracheal intubation test (865) in rats was conducted with approximately 5 mg samples from foam of the same composition used by Laskin et al. (829). The samples for intubation were either freshly prepared or from the identical block, with 94 percent of the particles 10 μm or less in diameter. The initial inflammatory response was followed by the development of areas of fibrosis, then nodular scars and perifocal emphysema; these latter changes appeared to be below the level that would cause functional impairment. Four rats showed a benign intrabronchial adenoma 18 months after intubation and one other rat developed a hyperplastic subpleural lesion without evident dust particles.

Monkeys inhaling 10 mg/m^3 6 hr/day for 18 months showed changes in pulmonary function that were considered suggestive of small airway obstructive impairment. No neoplasms were attributed to polyurethane foam exposure (827; see Table 54.30).

Intratracheal injection of 4 mg of fine dust (particles 1 to 15 μm in diameter) resulted in an "intermediate" degree of fibrosis 3 months after treatment (301, 785; see also Section 10.2.5).

10.2.4 Toxicity at Elevated Temperatures and Thermal Degradation Products

In general, polyurethanes start to undergo major decomposition somewhere between 150 and 300°C.* Carbon monoxide and hydrogen cyanide are probably the products of greatest concern. Intratracheal intubation of thermal degradation products can elevate blood cyanide in rats (866). Hydrogen cyanide is evolved in air as a secondary product at approximately 500°C and increases with rising temperature. Other nitrogen-containing decomposition products that have been reported include diaminotoluene (associated with degradation of TDI, Reference 849), and also acetonitrile, acrylonitrile, pyridine, and benzonitrile at higher temperatures. The initial decomposition processes of MDI-based rigid foams appear to be quite different from TDI-based flexible foams, but at temperatures in excess of 800°C the pattern of decomposition products is similar (867). Kimmerle found (868) that the decomposition temperature at which flexible urethane polyether-type foam was lethal to rats may be as low as 300°C whereas the corresponding temperature for rigid polyurethane foams and rigid isocyanurate foams sometimes exceeded 650°C. Table 54.30 provides amplifying data; see also Figures 54.14 and 54.15.

Gas temperatures in a polyurethane fire can reach 500 to 1000°C within 2 to 5 min, significantly more rapidly than observed with a comparable fire load of wood (843). Descriptions of uncontrolled fires involving polyurethane products

---

* Literature published during the 1970s indicates that commercial fire retardants are often without significant fire-retardant effects, although some approaches appear promising. However, phosphate fire retardants in trimethylol propane foams may produce materials with significant toxic effects during thermal decomposition (see Table 54.30).

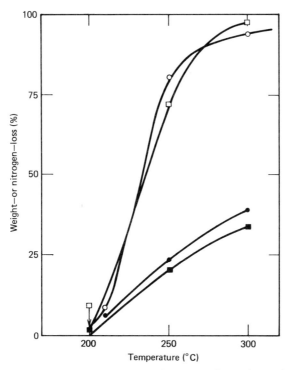

Figure 54.14 Comparisons between the weight—and nitrogen—losses during the thermal decomposition of flexible polyurethane foams between 200 and 300°C. O——O, Nitrogen loss of polyester foam; ●——●, weight loss of polyester foam; □——□, nitrogen-loss of polyester foam; ■——■, weight loss of polyether foam. Reproduced with permission from Woolley (845).

of an unspecified composition indicate that rapid fire spread and smoke can contribute significantly to the lethal potential of these fires (869).

10.2.5 Carcinogenic Potential/Cytotoxicity/Implant Studies

Overall the carcinogenic potential of the polymers tested in the United States and Germany appears to be low or negligible. Aromatic amines associated wth carcinogenic effect may develop during certain stages of thermal degradation. Formation of "active carcinogenic agents" from biodegradation of polyurethanes in implant studies has also been hypothesized (859; see also Reference 860 and Section 10.1.3). Other animal tests to determine the nature of tissue response and medical data do not suggest a carcinogenic effect, but rather a marked inflammatory effect.

In rats, the results of intraperitoneal tests with fine polyurethane foam dust (301, 785) at 3 months after treatment appear qualitatively similar to those in

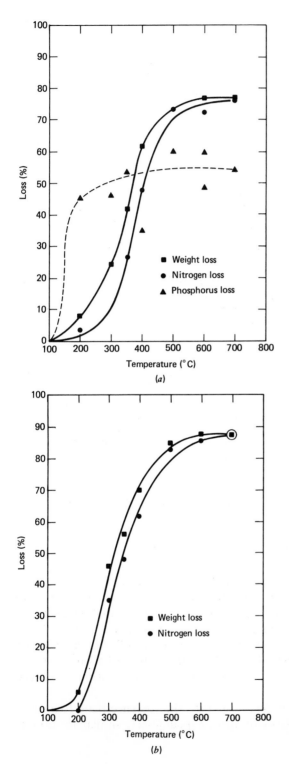

Figure 54.15 Decomposition data for four rigid polyurethane foams: (a) polyester; (b) polyether; (c) glycerol polyether; (d) similar to (c) but containing 12 percent tris(chloropropyl) phosphate flame retardant. From Woolley et al. (846). Reproduced by permission of the Building Research Establishment, Fire Research Station, Borehamwood, Hertfortshire, England; Crown copyright.

4460

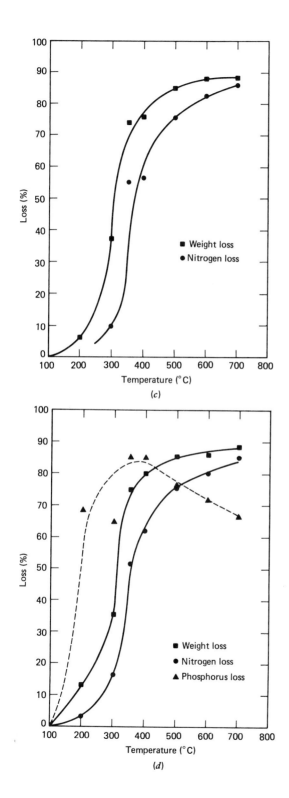

(c)

(d)

the intratracheal tests discussed in Section 10.2.3. Polyurethane caused fibrotic nodules with persistent granulomatous reactions. By way of comparison, asbestos and silica caused fibrotic nodules with collagen deposition. Cytotoxicity tests were also conducted with rat macrophage cells in culture. Polyurethane foam dust showed an intermediate or Group 2 reaction:* between 4 and 8 percent of peritoneal macrophages, and 10 to 40 percent of alveolar macrophages, were killed following phagocytosis of dust.

The local reaction to subcutaneous injections of polyurethane rigid foam has been reported to vary significantly in the mouse, rat, and rabbit (870). Inflammatory reactions appeared to be most marked in the mouse, less in the rat, and least in the rabbit.

In humans, marked foreign body reactions have been reported from breast implants of silicone coated with polyurethane foam (871). "Whorls of fibrous tissue" were sometimes present. Deterioration of the polyurethane foams began about 6 months after implantation and separation of the polyurethane coating from the silicone prosthesis took about 2 years.

The International Agency for Research on Cancer (31) cites additional studies on animals, as do Autian et al. (859, 860).

10.2.6   Microbial Biodegradation

Polyurethanes are unique among synthetic polymers in that the polymer itself provides a direct contribution toward the overall resistance or susceptibility to microbial growth of the formulated product. Polyester-based urethanes are characteristically less resistant than polyether-based urethanes, which can be quite resistant to microbial attack (872, 873).† Polyurethane based on polyester diol (no other specific detail) showed the only growth among 30 generic types of commercial "plastics" tested according to ASTM D-1924-63 (76).

10.3   Polyurethane Elastomers

Depending on composition, the versatile polyurethane elastomers can be processed by liquid casting systems, as vulcanizable millable rubbers, by thermoplastic techniques, or as spraying materials. The isocyanates that may be used include MDI, TDI, hexamethylene diisocyanate, naphthalene diisocyanate, dimethyl diphenyl diisocyanate, and methylene dicyclohexyl diisocyanate. The isocyanate is typically reacted with a hydroxyl terminated polymer and a low molecular weight diol, triol, or amine (often called chain extender). The hydroxyl-terminated polymer can be an adipic acid–ethylene glycol polyester,

---

* Several forms of silica also caused Group 2 reactions. Asbestos caused a Group 3 reaction: >10 percent of peritoneal macrophages and >60 percent of alveolar macrophages were killed following phagocytosis of dust. Group 1 reactions are discussed in Section 4.2.5.
† Reference 872 (preliminary report) specifically refers to rigid foams.

a glycol such as polybutylene glycol or polypropylene glycol, or a hydroxyl-terminated polybutadiene (874). Polyurethane elastomers have the special feature of segmented or block structure added to the basic characteristics of rubbery behavior. Figure 54.16 shows in simplified form the regions of "hard" segments derived from the isocyanate component and "soft" segments derived from the hydroxyl terminated polymer.

Polyurethane elastomers are noted for their exceptional abrasion resistance at moderate temperatures, high tensile and tear strength, and high hardness with mechanical strength. Figure 54.4 indicates superior resistance to oil at ambient temperatures. Polyester urethane elastomers show less swelling in oils and fuels than polyether elastomers but are more susceptible to deterioration by moisture at elevated temperatures. Polyurethane elastomers are used in solid tires for industrial trucks, potting and sealing of electronic components, shoe heels and soles, and elastic thread (874, 875). Specially formulated elastomers are considered promising for use in vascular repair and cardiovascular devices (876; see also Reference 871).

Toxicity data pertaining to both the cured and uncured elastomers per se are limited (see Section 10.1.3). Contact dermatitis in polyurethane molding appears to be relatively uncommon but has been reported in a plant using an aliphatic diisocyanate prepolymer with poor hygienic practices (877). "Polyurethane rubber" in shoes has been implicated as a cause of contact dermatitis (173), but this dermatitis may be attributable at least in part to polyurethane adhesives and coatings (see Section 10.4).

10.4 Polyurethane Coatings and Adhesives

Polyurethane coatings are typically tough, flexible, abrasion resistant, impact resistant, and fast curing, and adhere well to a variety of substrates. Solvent

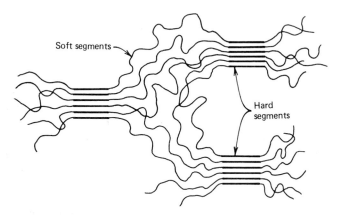

Figure 54.16 Segregation of hard segments in polyurethane elastomers (874). From *Rubber Technology*, second edition, Maurice Morton, ed., copyright © 1973 by Van Nostrand Reinhold Company. Reprinted by permission of the publisher.

resistance may vary from fair to excellent (878). They are often sensitive to extreme temperatures, humidity, and reactive contaminants. Isocyanate prepolymers are generally based on TDI but MDI, hexamethylene diisocyanate, or other aliphatic isocyanates may be used when clear or color-stable pigmented coatings are desired. Polyols may be polyesters or polyethers. The traditional types are defined by the American Society for Testing and Materials are the alkyd or oil urethanes (Type 1), the isocyanate prepolymers that cure with atmospheric moisture or by baking (Types 2 and 3, respectively), and the more cumbersome two-part prepolymer catalyst or prepolymer polyol systems (Types 4 and 5, respectively) that provide maximum desirable properties (864). Organic solvents may also be present (see Section 6.1.3).

The oil coatings are used in wood and machine finishes, marine enamels, and foundry core binders intended to hold sand. Applications of the other types include floor finishes, coatings for auto and machine parts, and finishes for nylon rainwear, leather, and rubber (864). Polyurethane adhesives are used in boots and shoes (10).

Toxicity data pertaining to these coatings and adhesives per se are limited (see Section 10.1.3). Cases of "intoxication" have occurred during processing; thermal decomposition at elevated temperatures frequently yields isocyanates (852) and may also yield aromatic amines (853).

Fully cured polyurethane coatings appear to be quite inert. Four thin coats of polyurethane applied to the skin contact surfaces of costume jewelry and nickel coins completely prevented dermatitis in 8 of 11 nickel-sensitive patients (879). Some polyurethane resins of the type associated with coatings and adhesives (Desmocoll®, Desmodur® materials in methylene chloride) have been reported as the cause of contact dermatitis from shoes (173). The urethane prepolymer coating used in the Letterflex® photoprepolymer mixture has also been connected with contact dermatitis (880); the causative agent in at least some of these cases appeared to be a polythiol cross-linking component, pentaerythritol tetrakis-3-mercaptopropionate.

10.5 Polyurethane Fibers

Fibers that have an extension at break in excess of 200 percent plus the property of rapid recovery when tension is released are classed as elastomers. Spandex, used as a generic term for polyurethane elastomeric fibers, is defined as a "manufactured fiber in which the fiber-forming substance is a long chain synthetic polymer comprised of at least 85 percent of a segmented polyurethane" (130). Lycra® spandex fiber and Vyrene® were the first products to be developed (6).*

* The first polyurethane fiber, Perlon® U, was not elastomeric but a nylonlike yarn spun in Germany during World War II. Unlike spandex, Perlon® U was a completely polyurethane fiber made by treating 1,4-butanediol with hexamethylene diisocyanate at 195°C (6).

10.5 Structure, Synthesis, Properties, and Applications

The spandex fibers are block polymers (see Figure 54.16). Relatively long "soft" segments of a very flexible nature, rubbery and noncrystalline, alternate in the molecular chain of the fiber with short "hard" segments that are usually cross-linked, crystalline, and polar. The soft segments, generally a polyester or a polyether, are easily deformed to give high extensions in contrast to the hard isocyanate segments that are not deformed. The use of preformed soft-segment polymer blocks results in a more regular spacing of tie points along the molecular chain than occurs in randomly vulcanized rubber (881).

Four stages of synthesis have been described (6). A low molecular weight linear polymer is made with terminal hydroxyl groups; this may be either polybutylene glycol or polyester made with excess glycols. This prepolymer is then treated with excess diisocyanate (either TDI or MDI) to give a polyurethane with terminal isocyanate groups. Third, water is added, just sufficient to convert some of the terminal isocyanate groups to amine groups. Heat then causes the amine and isocyanate groups to cure and form urea cross-linkages, which provide the snap-back property. Elastomeric fibers are commonly formed by dry spinning into a heated gas or by wet spinning into a dilute solvent bath (418).

The outstanding feature of these polyurethane fibers is their stretch and recovery. The stretch of spandex yarn before break may range from 520 to 610 percent, compared to 540 to 760 percent for rubber (see References 6 and 881). Spandex is relatively stronger and lighter than rubber and can be used uncovered, whereas rubber fiber is generally covered with rayon or nylon. Spandex has a melting point about 250°C, specific gravity about unity, and moisture regain about 0.3 percent. The fiber is soluble in boiling dimethylformamide and accepts dye readily. Spandex is mainly used in clothing. The technical importance of the yarns is considerably greater than may appear from total production figures since very small amounts can give textiles elastic properties (882).

10.5.2 Skin Reactions

During the mid-1960s a number of investigators here and abroad described allergic contact dermatitis with itching and erythema that was attributed to contact with spandex. This dermatitis typically developed under spandex panels in brassieres and in some cases resembled "classical rubber dermatitis" (883). In many cases sensitivity was clearly related to sensitivity to the 2-mercaptoben-zothiazole present as a trace contaminant in some, but not all, brands of spandex. However, three patients in the Netherlands all reacted to both *N*-phenyl-*N'*-isopropyl-*p*-phenylenediamine and zinc diethyl dithiocarbamate as well as "Wonderlastic" (884). All three had positive reactions to one or more additional substances—two reacted to both rubber cloth and 2-mercaptoben-zothiazole—but none reacted to "Lycra."*

Several human volunteer tests showed no evidence of chemical irritation or sensitization from commercial spandex materials without mercaptobenzothiazole. Panels of 200 subjects (885), 151 subjects (886), or 203 subjects (887) showed no reactions when tested according to the Schwartz-Peck technique† with fabrics containing three different types of "Lycra."

A rare case of vitiligo (888) observed after the wearing of an elasticized brassiere containing spandex may have involved a "depigmenting agent" in the yarn as well as friction and pressure. Depigmentation from contact with rubber containing antioxidants has been reported previously (160; see also Section 11.1.3).

10.5.3 Implant Studies/Carcinogenic Potential

No tumors were observed in Syrian hamsters at the site of a subcutaneous implant consisting of a tracheal graft containing an acetone-washed spandex fiber (889). Tracheal grafts containing these fibers showed a well-differentiated mucociliary epithelium 12 months after implantation. The grafts served as control implants in a test where other fibers coated with varying amounts of 2-methylcholanthrene had been inserted into the lumina of excised tracheas and then implanted into syngenetic recipients.

11 MISCELLANEOUS ORGANIC POLYMERS

11.1 Phenolics‡

11.1.1 Highlights

The phenolics were the first completely synthetic class of resins developed commercially and are still a major product line (Tables 54.1, 54.2). The term describes a resinous material produced by a phenol or a mixture of phenols with an aldehyde. Phenol itself, or sometimes cresols, are usually combined with formaldehyde, occasionally with other aldehydes such as furfural (10). The resin has sometimes been associated with dermatitis that appears to be

---

* Van Dijk (884) comments on the generic use of the term Lycra® for spandex, which probably has contributed to some of the confusion in the dermatologic literature.
† Reference 420, 2-week rest period.
‡ The CAS registry number for $(C_6H_6O.CH_2O)_n$ is 9003-35-4. The trademark Bakelite® was first used with these resins but now includes polyethylene, polypropylene, epoxy resins, polystyrene, vinyl resins, and other materials (3, 81). Other trademarks include Durez®, which also refers to other resins, and Genal®.

traceable to the phenolic monomer component, the resin itself, or less frequently formaldehyde. Inhalant effects from dust or fumes during processing have also been reported. No toxicologically significant emission of volatile substances is known to have been connected with the cured resin. Phenol-formaldehyde is "highly recalcifrant to biological decay" (417).

11.1.2 Synthesis, Molecular Weight, Properties, and Applications

The polymerization of phenol and formaldehyde readily becomes exothermic and may present an explosive hazard (10, 890). The reaction is generally conducted only to the intermediate condensate product (891, 892), which can be either a one-step *resol* (or resole) or a two-stage *novolak* (or novolac). Resols are prepared with an alkaline catalyst and sufficient formaldehyde (about 1.5 : 1 mole ratio with phenol) so that the resins can later cross-link and cure simply by heat. Their molecular weights are reported as 300 to 700 (892).

The novolaks are produced by the controlled reaction of phenol and formaldehyde in the presence of an acid catalyst to form a brittle resin. The amount of formaldehyde, about 0.75 to 0.90 mole formaldehyde/mole phenol, is insufficient to permit cross-linking; this last is accomplished by the addition of hexamethylene tetramine (or "hexa"). The novolaks may have molecular weights as high as 1200 to 1500 (892).

Both types of resin are available in granular, powdered, or liquid forms (see References 10, 12, 891, and 893). Wood flour is the general purpose filler; special purpose fillers include cotton flock, hydrated alumina, glass, mica, or other minerals. The molecular weight of a cured (or thermoset) phenolic resin may be in the range of several hundred thousand. The common property that forms the basis of the varied applications of phenolic resins is their capacity for being first a low molecular weight, fusible, soluble resin that can be easily handled and then subsequently polymerized to high molecular weight, strong, heat resistant products (891). They are also noted for surface hardness, chemical resistance, and useful electric properties. Comparing the two types, resols offer better resistance to cracking and essentially no ammonia emission (see Section 11.1.5), whereas novolaks offer greater molding latitude, better dimensional stability, and better long-term storage properties (893).

The cured resins are probably most widely known as electrical, automotive, and appliance components: for example, automotive distributor caps; water pumps; appliance bases, handles, and knobs; electrical switch gears, circuit breakers, and wiring devices. Some phenolic resins based on substituted phenols are permanently soluble and fusible. These resins are used in varnish coatings and as rubber tackifiers (891). Phenolic fibers (Kynol®) provide flame resistance and thermal insulation, but the upper temperature limit in air for long-term use appears to be around 150°C (894). Certain food-contact applications are permitted (92).

11.1.3 Analysis

Free phenol in phenol resole resins has been determined by gel permeation chromatography (895). See also Section 1.3.

11.1.4 Skin Contact

Foussereau et al. (897) associated allergy to *p-tert*-butyl phenol resins with their presence in polychloroprene adhesives. In Europe occupational allergy to these resins is reported as most common in the shoemaking industry and to a lesser extent in the automobile industry. Nonoccupational allergy may occur in connection with hypersensitivity to footwear or other items of clothing. These investigators consider that "formaldehyde plays only a very minor role in PTBP allergies." Both dermatitis (898) and vitiligo (899) have been reported in the United States; vitiligo would seem to be the more common complaint.

11.1.5 Inhalation of Dust and Fumes, Thermal Decomposition, and Related Data

Emissions from solid finished commercial articles appear to have presented relatively few if any problems due to the nature of the phenol–formaldehyde bond. Formaldehyde may be detected as described in Section 11.1.3 above. Ammonia is produced in the curing process of the "hexa"-cured novolaks and "over a period of time from the molded parts" (893). Resols contain virtually no "hexa" (and therefore are used in applications where ammonia emissions would be undesirable).

Several reports from abroad associate inhalation of "bakelite" dust with pulmonary changes in workers (900–902). In one case the dust concentration at a cutting machine is reported as 10.5 mg/m^3; however, the "comparatively large" size of the particles appeared to be related to the "nasal symptoms in 4 cases and history of nasal diseases in 4 cases" among 16 workers examined (900). "Slight pneumoconiotic opacities" have also been described (901). Extrinsic allergic alveolitis, which the author identifies as "caused by the inhalation of vegetable, animal, or chemical dust" (902), was reported in two patients with a history of occupational exposure to "bakelite" dust. Pulmonary changes have also been described in rats (903) and guinea pigs (900) exposed to unstated atmospheric concentrations of phenol–formaldehyde or "bakelite" dust.

In the United States, fumes and particulates from phenol–formaldehyde resins have been associated with effects on the respiratory tract. Long-term exposure to phenol–formaldehyde resin fumes was reported to cause airway obstruction as evidenced by pulmonary function tests (904). The resin fumes came from a production line where acrylic wool filters were dipped into a vat of liquid phenol–formaldehyde resin and then placed into a curing oven at 160°C. Levels of phenol found in breathing zone samples ranged from 7 to 10 mg/m³ (about 1.6 to 2.6 ppm). Comparable levels of formaldehyde varied from an atypical high of 16.3 mg/m³ to an estimated 0.5 to 1 mg/m³ (high about 14 ppm, estimated 0.4 to 0.8 ppm).

Workers in a tire manufacturing plant reported an excess of acute irritant symptoms and significant reductions in expiratory flow rates at low lung volumes (905). This plant used an adhesive system whereby resorcinol (as the phenol) and hexamethylenetetramine (as both a formaldehyde donor and catalyst) were added directly into the rubber stock during compounding. No significant association was found between changes in pulmonary function and environmental levels of resorcinol, formaldehyde, ammonia, or area sample particulates. Some associations were found between respirable particulate (personal samples) and decrements in lung function; the investigators suggest that the particulate may have carried an adsorbed agent deep into the lung. Although only small decreases in pulmonary function were found, the incidence of acute symptoms among the 40 current workers was high. These symptoms included eye irritation (98 percent), cough (72 percent), rash (35 percent), and phlegm (20 percent).

Phenolic resin paints can release appreciable quantities of phenol plus smaller amounts of ammonia and formaldehyde during industrial drying processes, according to a German study (476). Alcohols and phosphoric acid were also detected. In one particular case the emission of phenol was in the range of 700 g/hr (476).

Phenolic resins show an appreciable rate of decomposition at temperatures above 300°C, particularly under oxidative conditions (906–908). The primary products appear to be phenol and methyl phenols, plus carbon dioxide and/or carbon monoxide. Formaldehyde can be formed as a minor product under oxidative conditions at 400°C (908) or its yield may be "negligible" (907). At temperatures >600°C ring scission occurs; the amounts of phenol and methyl-substituted phenols decrease as benzene, toluene, benzaldehyde, and benzyl alcohol are formed.

11.2 Aminoplastics

The term aminoplastic describes a group of resinous polymers that are produced by the interaction of amines or amides with aldehydes. The two polymers of major commercial importance are urea–formaldehyde and melamine–formaldehyde. Other types include co-reacted polymers such as melamine–phenol–formaldehyde molding powders and benzoguanamine-based resins for coating applications.

The aminoplastics are generally prepared as liquid or dry resins of relatively low molecular weight that are then converted to insoluble and infusible end use items. Cure is ordinarily effected with heat or heat plus acidic catalysts and also without heat in the case of catalyzed urea resins (10, 909).

11.2.1   Urea–Formaldehyde

Highlights. These resins were first commercialized in the 1920s and were then developed for adhesives and resins in uses such as glues for wood, wet-strength paper applications, and crease-resistant finishes for clothing (see Tables 54.1 and 54.2). Complaints of the consumer public were primarily limited to odor and dermatitis that appear to have been relatively infrequent, considering its very extensive use. Complaints of both odor and some systemic effects, principally irritation of the respiratory tract, were more common in occupational exposure. Urea–formaldehyde foam for thermal insulation was first used extensively in Europe; extensive use in the United States came in connection with increased interest in residential energy conservation during the 1970s.

Formaldehyde is known to be emitted from the degassing of these foams, but presently available quantitative data do not permit an assessment of the contribution such foams are likely to make to ambient air quality in residences. The qualitative data—prolonged odor and also adverse health effects—reported as a consequence of installation of these foams have resulted in a series of investigations (910 and references cited therein) and banning of further installation of these foams in some areas (911). The current exposure limits to formaldehyde include a ceiling TLV of 2 ppm, a community air quality standard of 0.1 ppm recommended by the American Industrial Hygiene Association, and an indoor air standard of 0.1 ppm promulgated by the Netherlands (910). These standards were developed prior to any consideration of the possible carcinogenic effect of formaldehyde (912a).

Synthesis, Properties, and Applications of Molding Powders.   Urea–formaldehyde resins are usually prepared in a two-stage reaction. Urea and formaldehyde are first partially reacted under neutral or mildly alkaline conditions to form mono- and dimethylol ureas, the ratio of which varies depending on the urea–formaldehyde ratio. The resultant intermediate is then subjected (second stage) to acid conditions and heat to form a solution that converts to a gel; eventual evolution of water and formaldehyde yields a hard, colorless, transparent, infusible mass. The reaction is often arrested prior to gelation by changing to a slightly alkaline pH and removing some or all of the water, then continued when desired by changing again to an acid pH. The powdery condensation products are normally insoluble but can be dissolved in aqueous solutions of lithium bromide, lithium iodide, and magnesium perchlorate (912b).

Molding powders based on urea–formaldehyde usually contain a number of additional ingredients: filler, pigment, hardener, or accelerator, stabilizer, plasticizer, and lubricant. The filler is usually bleached wood pulp although wood flour may be used. The cellulose-filled materials are used in wiring devices, cosmetic closures, buttons, toilet seats, knobs and handles (909).

Synthesis, Properties, and Applications of Adhesives and Coatings. Urea and formaldehyde are heated to give dimethylol urea and other low molecular weight products that are acidified for further reaction and then stabilized with alkali. The resins are hardened by adding an aqueous solution of an acid donor (10) or mixed with fillers containing an acid catalyst for use in cold pressing (909).

Urea–formaldehyde glues are commonly used with wood, particularly particle board and plywood. Particle board is a mixture of sawdust or wood shavings held together with a urea–formaldehyde resin (10, 893). Cationic urea-formaldehyde coatings are used to improve wet strength, dry tensile strength, and bursting strength of paper (909). Certain types of urea–formaldehyde resins are permitted in food contact applications (92).

Composition, Properties, and Applications of Foams. Urea–formaldehyde-based foams can be generated from three types of formulations: aqueous solution, powder mixed on site in water, and concentrated solution diluted with water. The aqueous solution has been described as the most common in the United States (913). A typical foam formulation (909) may contain 54 percent urea–formaldehyde resin, 38 percent water, 7 percent phosphoric acid (75 percent), and 1 percent foaming agent (quaternary ammonium salt).

Urea–formaldehyde foam insulation burns with difficulty (10, 38, 913). The air-dried foam has been reported to weigh 0.75 lb/ft^3 (909), although a literature survey shows considerable variation in this parameter (913). The foam is reported as friable with negligible load-bearing characteristics, also hydrophobic (909) but capable of absorbing some water (913). In many respects data on properties are contradictory or not known to have been determined (913). Much of this variation is undoubtedly due to variation in formulation and application under conditions of relatively poor quality control. Shrinkage of the foams and resistance to temperature and humidity have been identified as problem areas in its major application of home insulation use (see section titled Highlights and section on emission of formaldehyde). Other reported applications of urea–formaldehyde foam also include its use in floral decorations, artificial snow in television, and firelighters made by saturating urea–formaldehyde foam with paraffin, and as an arrester bed to stop aircraft on runways (10).

Skin Reactions. Sensitivity to urea–formaldehyde resin does not necessarily involve sensitivity to formaldehyde. The issue was extensively explored during the 1960s (914–917) and has recently been reviewed (918). Hypersensitivity to formaldehyde may occur in some of these patients but does not seem to be a factor in their reactions to textiles finished with formaldehyde resins, nor do reactions to these fabrics correlate with the amount of free formaldehyde present therein. Dermatitis associated with industrial exposure to urea–formaldehyde molding powder was reported in the 1940–1950s (919, 920).

Emission of Formaldehyde at Ambient Temperature From Urea–Formaldehyde Resins. The emission of formaldehyde from urea–formaldehyde foam used in building construction received widespread public comment in the late 1970s (911). Earlier, a 1975 Danish report (921) described 23 dwellings where the concentration (range of 0.08 to 2.24 mg/m^3; mean 0.62 mg/m^3 or about 0.5 ppm) of formaldehyde was associated with its emission from chipboard (particle board, wood shavings held together with urea–formaldehyde glue). Specific data on the contribution of formaldehyde from these resins in relation to other contributing factors are generally lacking. Other factors may include cigarette smoke (supported by a measurement of 0.23 ppm formaldehyde from five cigarettes smoked in a 30 m^3 chamber, as cited in Reference 910) and combustion products released by cooking and burning fuels (no known measurements but undoubtedly a wide variation depending on type of residence and personal idiosyncrasy). Reference 918 gives a review of methods of analysis.

Most of the measurements in residential dwellings containing urea–formaldehyde foam appear to be less than 5 ppm, although levels as high as 10 ppm were reported in a Connecticut survey (910 and Sardinas et al. as cited therein). In a retrospective list (922), clinical symptoms commonly included eye and respiratory tract irritation; headaches, vomiting, and diarrhea were mentioned less frequently. Whether or not any other possible causes, or contributing causes of these symptoms, were considered is generally not evident.

Long et al. (923) attribute short-term emission of formaldehyde to free formaldehyde and associate longer-term emission with hydrolysis of the polymer. Increased temperature and decreased humidity appear to result in higher initial and lower subsequent concentrations of formaldehyde in the adjacent atmosphere. Varying intrinsic quality of the foam from use of inappropriate materials or failure to adhere to recommended instructions during installation can also affect emission of formaldehyde. Residual hot-water soluble, low molecular weight, acidic components affect the hydrolytic breakdown of urea–formaldehyde foams. An approximate sixfold reduction in the total quantity of formaldehyde emitted from a commercial product over a 30-day period (55°C, 90 percent relative humidity) was obtained when acidic components were removed by washing prior to testing (924).

According to a Japanese study (459), four commercial soil-consolidating agents contained formaldehyde up to 192 mg/ml and formic acid up to 21 mg/ml in the urea–formaldehyde condensate prior to gelation. After gelation of the resin with sand, formaldehyde up to 14,900 mg/m^3 and formic acid up to 96 mg/m^3 evaporated from the gel to air at equilibrium.

Laboratory tests conducted in France to determine the evolution of formaldehyde during a 90-min oxidative pyrolysis showed a steady increase with increasing temperature to aproximately 140°C (925). The emissions were (g/100 g) 0.19 at 5°C, 0.25 at 55°C, 0.58 at 88°C, 0.65 at 111°C, and 1.29 at 140°C.

The half-life for formaldehyde found in a commonly used Scandinavian particle board has been reported to be about 2 years with a ventilation rate of 0.3 changes/hr (926). The level of formaldehyde can be reduced to approximately half the original value by coating with a formaldehyde absorbent paint. Such painting may be less effective with urea–formaldehyde foam (922).

Toxicity of Thermal Degradation Products. Several brief reports (123, 927) indicate that pyrolysis products can be toxic to rats. Carbon monoxide, carbon dioxide, ammonia, and cyanide have been identified as (or in) effluent gases (102, 396, 682; see also Reference 928, Figures 54.7 and 54.12).

11.2.2 Melamine–Formaldehyde Resins

Highlights. Many features of these resins are very similar to those of the urea–formaldehyde resins. In general the melamine–formaldehyde resins offer more desirable properties than the urea–formaldehyde materials (10). Emissions of volatile substances have not been a problem in consumer use with these resins, which are cured at a higher temperature than urea–formaldehyde resins, particularly the foam. Melamine–formaldehyde has been cited as biodegradable (77a).

Synthesis, Properties, and Applications. Melamine–formaldehyde resins are synthesized by treating melamine with formaldehyde to yield a product with up to six methylol groups per molecule. The methylol content depends on the melamine–formaldehyde ratio and reaction conditions. Compared to urea–formaldehyde, melamine–formaldehyde moldings have low water absorption, better staining resistance to aqueous solutions, better electrical properties in damp weather, better heat resistance, and greater hardness. Like the phenolics and urea–formaldehyde resins, these resins may be sold as a low molecular weight product intended for further processing. These resins may be formulated (as Cymel®) for use as molding powders, laminates, adhesives, and for treatment of textiles. The molding powders and laminates are probably most familiar as dinnerware (Melmac®) and as laminates (Formica®). Dinnerware represents the largest single use. Subject to certain specifications, melamine–formaldehyde resins are permitted in food-contact applications such as coatings, components of packaging, and molded articles (92).

Analysis. A Taiwanese report (929) describes a formaldehyde content of 25 ppm in melamine bowls, compared to 20 to 42 ppm in urea resin in or coated on chopsticks. See Section 1.3.

Oral and Skin Toxicity. A melamine–formaldehyde resin sold as a low molecular weight, wet-strength additive was treated to simulate the conditions of cure that this resin generally meets during its intended use in papermaking (930). Accordingly a commercial batch of resin was dried as a thin layer in an oven for 3 days at 115°C. The resulting cured product was fed to rats, 40 males and 40 females per group, at dietary levels of 2.5, 5, and 10 percent for 2 years (931). Some decreases in mean weight gain were observed at the two higher levels; no effects were evident in hematologic studies or mortality patterns. Pathologic examination indicated no effects attributable to feeding of the resin except at the 10 percent level, where mean testis weight was significantly reduced. No evidence of reproductive toxicity was observed in a separate test where rats were fed a 10 percent dietary level of resin and mated three times.

The cured product was also fed to dogs for 2 years at dietary levels of 2.5 and 5 percent (two males and two females per group). Overall health and appearance were good, except for a slightly increased incidence of diarrhea compared to control dogs. Pathogenic examination showed that the testis weight was reduced in both males at the 5 percent level and one at the 2.5 percent level. Microscopic examination at the higher level showed depression of spermiogenesis and low-grade inflammation of the alimentary tract.

Overall, dermatitis appears to be less frequent than with urea—formaldehyde resin but may occur during processing. A 1963 Italian report (932) describes an incidence of 7 percent dermatitis in the production of phenolic, urea, and melamine resins; the relative importance of the melamine resins is unclear. A series of human patch tests with resin-treated paper and textiles indicates that these resins have "no significant sensitizing potential in their ordinary and usual uses in paper and textiles" (930).

Inhalation Toxicity and Thermal Degradation. Carbon monoxide, carbon dioxide, and formaldehyde were identified in the pyrolysis gases from melamine–formaldehyde and from all of a series of plastics tested (933). Anderson et al. (934, 935) suggest that low temperature degradations up to 350°C involve the loss of water and formaldehyde, the latter from reversible demethylolation reactions. At temperatures above about 375°C the resin structure breaks down to evolve products that include ammonia, hydrogen cyanide, carbon monoxide, and melamine. Weight loss with melamine–formaldehyde resins may increase, relatively, with decreasing melamine–formaldehyde ratio (936). See also Figures 54.7 and 54.12.

11.3 Furan Resins

Furan or furane resins may be prepared either from furfuryl alcohol or the reaction products of furfuryl alcohol with formaldehyde, furfural, urea–formaldehyde, or phenol–formaldehyde (937). Acids such as phosphoric acid, *p*-toluenesulfonic acid, and sulfuric acid are used for curing. The resulting polymers are not resistant to concentrated nitric acid and strong oxidizing acids but are generally resistant otherwise to strong alkalies, acids, and solvents including chlorinated solvents. They are used in the foundry industry as no-bake resins, in the manufacture of chemical-resistant tanks, laboratory tabletops, and mortars and grouts.

Asthmatic sensitization to the volatile products released by mixing the components of a furan-based binder system has been described in one case (938). Neither the resin (containing furfuryl alcohol, paraformaldehyde, and xylene) nor the catalyst (containing sulfuric acid, phosphoric acid, and butyl alcohol) produced a response, but the mixing operation "provoked late asthmatic responses and heightened nonspecific bronchial responsiveness to inhaled histamine."

Degradation mechanisms have been described as similar to those of phenol- and urea-based condensation polymers (939).

11.4 Polyvinylpyrrolidone

The CAS No. is 9003-39-8. Polyvinylpyrrolidone is also called PVP, Povidone,* Luviskol® (series), Kollidon®, and Peregal® ST (textile chemical series).

11.4.1 Preparation

The monomer N-vinylpyrrolidone is typically polymerized in bulk or in aqueous solution with heat and free radical catalysts, for example, hydrogen peroxide and ammonia (14, 80, 940).

11.4.2 Molecular Weight, Properties, and Applications

The molecular weight has been given as 10,000 to 700,000 (80). United States manufacture of povidone has included four basic viscosity grades (K-15, K-30, K-60, and K-90) with number average molecular weights of about 10,000, 40,000, 160,000, and 360,000, respectively. Polyvinylpyrrolidone is extremely soluble in water and also many organic solvents. Polyvinylpyrrolidone forms tight complexes with many substances; insoluble complexes formed by addition of polybasic acids such as polyacrylic acid or tannic acid can be reversed by neutralizing with base. Polyvinylpyrrolidone complexes with iodine retain antimicrobial effect with reduced toxicity and staining tendency (940).

---

* A cross-linked form is known as crospovidone (PVPP), Polyplasdone® XL, Kollidon® CL (pharmaceutical), and Polyclar® AT (food use).

Polyvinylpyrrolidone is useful as a thickening agent, emulsion stabilizer, and complexing agent in cosmetics and toiletries. In the textile and dye industries, the complexing and surfactant properties of polyvinylpyrrolidone can improve the dye receptivity of hydrophobic fibers such as polyacrylonitrile or polypropylene (940). Use as a plasma volume expander in the treatment of shock has been generally discontinued.

11.4.3 Analysis

Polyvinylpyrrolidone can be detected in foods, beverages, etc., at concentrations as low as 0.1 ppm ±5 percent by a chromatographic–colorimetric method (941). An infrared method permits detection of as little as 0.1 mg with ±1 percent accuracy in solutions as dilute as 0.1 percent (942). See Reference 31 for additional methods.

11.4.4 Oral Toxicity, Parenteral Toxicity, and Xenobiotica

Polyvinylpyrrolidone caused no adverse effects when fed to rats and dogs at dietary levels as high as 10 percent for 2 years (as reviewed in References 23,* 943, and 944; some viscosity average molecular weights given as 38,000 to 40,000). The only biologic effect attributed to oral administration is stool softening or diarrhea (944). Data obtained from a series of terminal cancer patients showed absorption of 0.013 to 0.048 percent of orally administered povidone, mean molecular weight 11,000 (Siber et al., as cited in Reference 944).

Parenterally administered povidone with a molecular weight <25,000 is readily excreted by the kidneys (945). At a mean molecular weight of 40,000 it may accumulate to some extent in kidney, lungs, liver, spleen, and lymph nodes (946). Daily subcutaneous injections administered over a span of years have been associated with the development of polyvinylpyrrolidone "storage disease" (947, 948). This disease is described as occurring only with polymer >50,000 in molecular weight (948).

For additional references to the extensive xenobiotic literature, see References 31, 944, and 949.

11.4.5 Inhalation Toxicity

Polyvinylpyrrolidone was first related to thesaurosis in beauticians in 1958 (see References 31 and 950). The nature of the factors involved has subsequently been disputed (951). Increased bronchovascular markings, but no thesaurosis, were noted in 11 of 227 subjects during a survey of beauticians. Hair spray

---

* *WHO Food Additive Ser. 8*, 1975, which gives a molecular weight of 38,000 for polymer tested in several 2-year tests and also cites other data.

exposure was considered the most likely origin of the pulmonary infiltrate present in five patients with major aerosol exposure. In 1977, 32 cases of thesaurosis had been identified, 10 of which were in beauticians. Discher (950) concluded that all these cases "seem to bear a remarkable resemblance to a recognized clinical condition, sarcoidosis, which existed prior to the use of PVP hairspray."

11.4.6 Carcinogenic Potential

In its review (31), the International Agency for Research on Cancer considered that available data did not permit an evaluation of the carcinogenicity of either the polymer or the monomer. Other than injection-site sarcomas, the tumors described by this agency occurred in parenteral tests that were reported some two decades ago, and are to some extent unclear. The lack of carcinogenic potential observed in several 2-year oral tests supports a current request for an estimate of the acceptable daily intake as 50 mg/kg in food additive use (944).

11.5 Polybenzimidazole

Polybenzimidazole is known as PBI.

The most common polymer, poly(2,2'-*m*-phenylene)-5,5'-bisbenzimidazole, can be prepared by heating tetraamino biphenyl and diphenyl isophthalate under nitrogen, typically about 200°C, to yield a low molecular weight, glassy, friable foam, which is generally ground and cured by heating to a typical temperature of 385°C. This resulting amorphous powder can be prepared as a heat-resistant fiber, by dry spinning from a solution of dimethylacetamide, as an adhesive, or as sheet film (952). Foam can be made by heating the low molecular weight, first stage product.

As a fiber, polybenzimidazole is noted for exceptional high temperature performance. The useful temperature limit is reported to be about 560°C (952).

Thermal decomposition of polybenzimidazole foams in an air environment can begin slowly at 375°C and become more rapid above 500°C (see also Figure 54.9). Studies conducted by different investigators in inert and air environments at temperatures up to 1000°C indicate that volatile products may include hydrogen, carbon monoxide, carbon dioxide, methane, water, hydrogen cyanide, ammonia, propene, acrylonitrile, phthalonitrile, benzene, aniline, and benzonitrile. Water may be the only significant product released under 550°C (953).

12 SILICONES

Commercial development of silicone resins started in 1931 with the objective of preparing polymers with properties intermediate between organic polymers and inorganic glasses (10). Silicone polymers have an alternating silicon–oxygen backbone described as siloxane that is somewhat similar to quartz and mica.

12.1 General Data

12.1.1 Structure and Synthesis

Silicones or silicon hydrides of increasing chain length up to about Si_6H_{14} are known. Above this length, however, the Si–Si chain becomes thermally unstable. Commercial silicone polymers contain the siloxane link, Si–O–Si.

  Polyorganosiloxanes are generally prepared by reacting chlorosilanes with water and then condensing this intermediate material to form a polymer. A broad variety of products are made from relatively few monomers (954). The linear polydimethylsiloxanes (or polymethylsiloxanes) may be represented by the generic formula:

$$H_3C-\underset{\underset{CH_3}{|}}{\overset{\overset{CH_3}{|}}{Si}}-O-\left[\underset{\underset{CH_3}{|}}{\overset{\overset{CH_3}{|}}{Si}}-O\right]_n\underset{\underset{CH_3}{|}}{\overset{\overset{CH_3}{|}}{Si}}-CH_3$$

where n may range from 0 to >10,000. In the cyclic oligomers the terminal end groups are lacking and n may be 3 to 8 with $n = 4$ and 5 predominating (955).

12.1.2 Toxicity Screening and Identification of Thermal Degradation Products

Hilado et al. (956) exposed mice to the pyrolysis gases evolved from samples of 11 silicone polymers—a transformer liquid, four elastomeric products, and six silicone resins. These polymers appeared to be among the least toxic, as judged by time to death or incapacitation, of approximately 300 materials tested in the rising temperature program (described in Table 54.10, Reference 66). Carbon monoxide was identified as an important toxicant.

  Programmed heating at 10°/min under vacuum causes the evolution of volatile products that are detectable at 343°C and reach a maximum at 443°C. Cyclic trimer, cyclic tetramer, and higher cyclics were detected. Benzene is also a significant product. Limited data indicate that oxygen accelerates production of volatile products and promotes cross-linking (957–959*).

* Reference 958 contains a brief discussion of poly(diphenylsiloxane) and methylphenylsiloxane polymers.

12.1.3 Compilations of Toxicity Data, Methods, and Specifications

The Food and Drug Administration has sponsored a monograph (960) pertaining to the safety of methylpolysiloxanes. This monograph summarizes various individual items found in the scientific literature from 1920 through the report date of September 1978.

Bischoff (29) has presented an extensive review of the literature pertaining to medical uses and carcinogenic potential through 1972. Liquid silicone has a tendency to migrate from an injection site and can cause granulomatous reactions.

Malignancies that vary with test conditions and species (tumors in mice but not in rats) have also been reported. Some of the medical applications of silicone polymers appear to have had a very high incidence of complications and some have been highly beneficial (see Section 12.4).

Analytic methods are listed in Reference 960. See Reference 92 for specifications pertaining to food contact.

12.2 Silicone Fluids and Resins

12.2.1 Properties and Applications

The most common liquid silicones are mostly polydimethylsiloxanes, which occur as oligomers up to polymers of high molecular weight (>500,000). Useful properties are generally retained over a wide range of temperature (-70 to 200°C). These silicones have prolonged stability at 150°C but oxidize at 250°C unless inhibited by antioxidants (10). They are water repellent and have excellent antiadhesive qualities. They also function as excellent heat transfer agents. Silicone fluids are used as cooling and dielectric fluids; in polishes, waxes, and coatings; and in packaging, paper, and textile treatment.

Silicone resins have very good heat resistance but are mechanically much weaker than the corresponding organic cross-linked polymers (10). High phenyl content resins are compatible with phenol–formaldehyde, urea–formaldehyde, melamine–formaldehyde, and oil-modified alkyd resins (not nonmodified alkyds). Silicone resins are used to prepare heat resistant glass-cloth laminates, particularly for electric motors, printed circuits, and transformers. Lesser uses include molding powders, water-repellent treatments, and release agents. The use of silicone resins as a heat-cured coating on baking pans aids the release of bread after baking.

12.2.2 Highlights of Toxicity Data

Early toxicologic studies were conducted by Rowe et al. (961, additional work cited in Reference 960). The initial series included acute and short-term tests with hexamethyldisiloxane* and dodecamethylpentasiloxane* (see Chapter 33, Volume 2A); two dimethylpolysiloxanes of grease-like consistency, one a sealing compound and the other an antifoam (Dow Corning Antifoam A); and three silicone resins, one a methylpolysiloxane and two methylphenylpolysiloxanes. Rats fed a dietary level of 0.3 percent Antifoam A for 2 years showed no significant toxic effect. Two other 2-year tests with rats fed a silicone oil emulsion or another antifoam agent, respectively, showed no direct evidence of toxic effect although the data were perhaps less conclusive (Gloxhuber and Hecht, as cited in Reference 960).

Cutler and co-workers (962) reported that mice given a lifetime diet containing 0.25 and 2.5 percent of a silicone antifoam agent from weaning showed no significant toxic effect. The antifoam agent contained 94 percent dimethylsiloxane silicone oil and a 6 percent finely divided silicon dioxide. Analysis of the whole animals revealed no silicone component in 10 mice that had been fed a diet containing 2.5 percent silicone for 75 weeks. Two other groups of mice received a single subcutaneous injection of 0.2 ml antifoam and 0.2 ml liquid paraffin, respectively, at weaning. The mice receiving the paraffin showed an increased incidence and earlier appearance of subcutaneous fibromas at the injection site in male mice, whereas the silicone-injected mice showed a greater incidence of cysts.

Several grades of polydimethylsiloxane caused no evident change when tested for reproductive and teratologic effects in rats and rabbits (963) and testicular effects in rabbits (964). Lack of effect on male reproductive function was also observed (965a) with polydimethylsiloxane (a Dow Corning® 200 fluid), a trimethyl end-blocked dimethylphenylmethylpolysiloxane (Dow Corning® 550 fluid), tris(trimethylsiloxy)phenylsilane (Dow Corning® 556 fluid), and trifluoropropylmethylpolysiloxane (Dow Corning® FS-1265 fluid). However, testicular effect was observed (964) in each of a series of 19 tests with rabbits given different batches of an equilibrated copolymer of mixed cyclopolysiloxanes.

Several polydimethylsiloxane fluids and formulations exhibited very low toxicity (965a) in tests with *Daphnia* water fleas, freshwater and marine fish, mallard ducks, bobwhite quail, and domestic chickens. Radioactive tracer studies ([^{14}C]polydimethylsiloxane) showed no detectable degradation by sewage microorganisms or bioaccumulation in bluegill sunfish. Silicon residues in eggs and tissue samples of white Leghorn chickens were below the detection limit of 2 to 4 ppm after the chickens had been maintained on diets containing 200 to 5000 ppm polydimethylsiloxane.

Dermal tests (rat, monkey, human) with a mixture of cyclic oligomers (specified as 70 percent tetramer, octamethylcyclotetrasiloxane, and 20 percent

---

* Hexamethyldisiloxane, dodecamethylpentasiloxane, and some other siloxanes are classed as Dow Corning 200 Fluids.

pentamer, heptamethylcyclopentasiloxane) showed no evidence of absorption (955). No increase in silicon levels was detected in expired air by atomic absorption analysis or in urine by emission spectroscopy.

The use of silicone gel in breast prostheses has been associated with delayed foreign body reactions (965b).

12.3 Silicone Elastomers

Dimethylsilicone rubbers are prepared primarily from the cyclic tetramer, octamethylcyclotetrasiloxane. The most outstanding property of the silicon elastomers is their service temperature, which can range from approximately -150 to $+600°F$ (966, 967).

Silicone rubber is widely used in aerospace, automotive, and electrical applications. Consumer uses include gaskets in household appliances, pharmaceutical stoppers, and weather coatings on walls and roofs (966). Silicone elastomer prostheses can provide restoration of joint function in crippled arthritic patients (968, 969) although sporadic complications have been reported (970, 971).

ACKNOWLEDGMENTS

I thank Seymour Yolles, Frank A. Bower, and Fran W. Lichtenberg for their critical review of the manuscript. Dr. Yolles also contributed to the organization of the data. Jeanne M. Skrotsky, Vickie L. McCall, and Lisa C. Pinder coordinated the typing and reference work. Appreciation is also due many of the Du Pont departmental liaison and library staff, particularly at Haskell and Textile Research Laboratories.

REFERENCES

1. R. Lefaux, *Practical Toxicology of Plastics*, English ed., CRC Press, Cleveland, Ohio, 1968.
2. H. F. Mark and S. Atlas, "Introduction to Polymer Science," in H. S. Kaufman and J. J. Falcetta, Eds., *Introduction to Polymer Science and Technology: An SPE Textbook*, Wiley, New York, 1977, pp. 1–23.
3. G. G. Hawley, *The Condensed Chemical Dictionary*, 9th ed., Van Nostrand Reinhold, New York, 1976.
4. M. P. Stevens, *Polymer Chemistry; An Introduction*, Addison-Wesley, Reading, Mass., 1975.
5. F. W. Billmeyer, Jr., *Textbook of Polymer Science*, 2nd ed., Wiley, New York, 1971.
6. R. W. Moncrieff, *Man Made Fibres*, 6th ed., Wiley, New York, 1975.
7. M. Morton, Ed., *Rubber Technology*, 2nd ed., Van Nostrand Reinhold, New York, 1973.

8. R. F. Boyer, "Transitions and Relaxations," in *Encyclopedia of Polymer Science and Technology*, 2nd ed., Suppl. 2, Wiley, New York, 1977, pp. 745–839.

9. G. T. Davis and R. K. Eby, "Glass Transition of Polyethylene: Volume Relaxation," *J. Appl. Phys.*, **44,** 4274–4281 (1973).

10. J. A. Brydson, *Plastics Materials*, 3rd ed., Whitefriars Press, London, 1975.

11. J. Brandrup and E. H. Immergut, Eds., *Polymer Handbook*, 2nd ed., Wiley, New York, 1975.

12. *1981 Annual Book of ASTM Standards*, American Society for Testing and Materials, Philadelphia, Pa. Updated annually. See D 1418-79a for nomenclature for rubber polymers.

13. J. L. Throne, *Plastics Process Engineering*, Marcel Dekker, New York, 1979.

14. W. J. Roff and J. R. Scott, with J. Pacitti, *Handbook of Common Polymers*, CRC Press, Cleveland, Ohio, 1971.

15. J. Haslam, H. A. Willis, and D. C. M. Squirrell, *Identification and Analysis of Plastics*, Heyden, London, 2nd ed., 1972, reprinted 1980.

16. C. A. Harper, Ed., *Handbook of Plastics and Elastomers*, McGraw-Hill, New York, 1975.

17. H. R. Allcock and F. W. Lampe, *Contemporary Polymer Chemistry*, Prentice-Hall, Englewood Cliffs, N.J., 1981.

18. *Guide to CAS ONLINE Commands*, Chemical Abstracts Service, Columbus, Ohio, 1981, p. 39. See bimonthly newsletter, *CAS ONLINE*, for current developments.

19. J. F. Rabek, *Experimental Methods in Polymer Chemistry*, Wiley, New York, 1980.

20. J. G. Cobler and C. D. Chow, "Analysis of High Polymers," *Anal. Chem.*, **51,** 287R–303R (1979).

21. C. W. Wadelin and M. C. Morris, "Rubber," *Anal. Chem.*, **51,** 303R–308R (1979).

22. R. E. Eckardt and R. Hindin, "The Health Hazards of Plastics," *J. Occup. Med.*, **15,** 808–819 (1973).

23. World Health Organization, *Toxicological Evaluation of Some Food Colours, Thickening Agents, and Certain Other Substances* (revised title), FAO Nutrition Meetings Report Series No. 55A, *WHO Food Additive Ser. No. 8*, Geneva, 1975 (also *WHO Tech. Rep. Ser. No. 576*). FAO is the Food and Agriculture Organization of the United Nations, Rome, which also issues this Nineteenth Report of the Joint FAO/WHO Expert Committee on Food Additives (subsequent reports issued by WHO in its Tech. Rep. Ser., also by FAO).

24. D. E. Till, R. C. Reid, A. C. Schwope, K. R. Sidman, J. R. Valentine, and R. H. Whelan, *A Study of Indirect Food Additive Migration* (Second Annual Technical Progress Report October 1, 1978 to September 30, 1979), Arthur D. Little, Inc., Project No. 81166, October 1979, Food and Drug Administration Contract Number 223-77-2360 (courtesy of FDA, Washington, D.C.).

25. J. T. Chudy and N. T. Crosby, "Some Observations on the Determination of Monomer Residues in Foods," *Food Cosmet. Toxicol.*, **15,** 547–551 (1977).

26. *Monsanto v. Kennedy et al.*, U.S. Court of Appeals for the District of Columbia Circuit; 613 F.2d, 947 (D.C. Cir. 1979).

27. *TLVs Threshold Limit Values for Chemical Substances in Workroom Air Adopted by ACGIH for 1980*, American Conference of Governmental Industrial Hygienists, Cincinnati, Ohio, 1980.

28. *Occupational Exposure Limits for Airborne Toxic Substances*, Occupational Safety and Health Series No. 37, International Labour Office, Geneva, 1977.

29. F. Bischoff, "Organic Polymer Biocompatibility and Toxicology," *Clin. Chem.*, **18,** 869–894 (1972).

30. R. D. Karp, K. H. Johnson, L. C. Buoen, H. K. G. Ghobrail, I. Brand, and K. G. Brand, "Tumorigenesis by Millipore Filters in Mice: Histology and Ultrastructure of Tissue Reactions as Related to Pore Size," *J. Natl. Cancer Inst.*, **51,** 1275–1285 (1973).

31. International Agency for Research on Cancer, *IARC Monographs on the Evaluation of the Carcinogenic Risk of Chemicals to Humans; Some Monomers, Plastics and Synthetic Elastomers and Acrolein*, Vol. 19, Lyon, France, 1979.

32. B. D. Halpern and W. Karo, "Medical Applications," *Encyclopedia of Polymer Science and Technology*, 2nd ed., Suppl. 2, Wiley, New York, 1977, pp. 368–403.

33. J. Autian, "Plastics," in L. J. Casarett and J. Doull, Eds., *Toxicology: The Basic Science of Poisons*, 2nd ed., Macmillan, New York, 1980, pp. 531–556.

34. "Polymers Put Cancer Drugs on Target," *Chem. Week*, **122**(17), 45, 48, 51 (April 1978).

35. S. Yolles, J. F. Morton, and B. Rosenberg, "Time-released Depot for Anticancer Agents. II," *Acta Pharm. Suec.*, **15**, 382–388 (1978).

36. J. B. Terrill, R. R. Montgomery, and C. F. Reinhardt, "Toxic Gases from Fires," *Science*, **200**, 1343–1347 (1978).

37. G. L. Ball and E. A. Boettner, "Toxic Gas Evolution from Burning Plastics," *Polym. Prepr., Am. Chem. Soc.*, **14**(2), 986–989 (1973).

38. National Materials Advisory Board, *Materials: State of the Art* (Fire Safety Aspects of Polymeric Materials, Vol. 1), Technomic, Westport, Conn., 1977.

39. National Materials Advisory Board, *Test Methods, Specifications, and Standards* (Fire Safety Aspects of Polymeric Materials, Vol. 2), Technomic, Westport, Conn., 1979.

40. R. G. Bauer, "Fire Retardant Polymers: A Review," *J. Fire Retard. Chem.*, **5**, 200–220 (1978).

41. K. Fischer, "Developing Flame Retarded Plastics: New Guidance Wanted," *Fire Mater.*, **3**, 167–170 (1979).

42. R. Fristrom, "Chemistry, Combustion and Flammability," *J. Fire Flammability*, **5**, 289–320 (1974).

43. R. Hindersinn, "Fire Retardancy," *Encyclopedia of Polymer Science and Technology*, 2nd. ed., Suppl. 2, Wiley, New York, 1977, pp. 271–339.

44. G. L. Nelson, "Smoke Evolution: Thermoplastics," *J. Fire Flammability*, **5**, 125–135 (1974).

45. A. F. Grand, "Defining the Smoke Density Hazard of Plastics," *J. Fire Flammability*, **7**, 217–235 (1976).

46. J. C. Spurgeon, L. C. Speitel, and R. E. Feher, "Oxidative Pyrolysis of Aircraft Interior Materials," *J. Fire Flammability*, **8**, 349–363 (1977).

47. G. Borsini and C. Cardinali, "Smoke Emission of Aircraft Seat Materials," *J. Fire Flammability*, **7**, 530–538 (1976).

48. M. van Grimbergen, G. Reybrouck, and H. van de Voorde, "Air Pollution Due to the Burning of Thermoplastics II," *Zentralbl. Bakteriol., Parasitenkd., Infektionskr. Hyg., Abt. 1: Orig., Reihe B*, **160**(2), 139–147 (1975).

49. T. Morikawa, "Evolution of Soot and Polycylic Aromatic Hydrocarbons in Combustion," *J. Combust. Toxicol.*, **5**, 349–360 (1978).

50. C. J. Hilado and C. J. Casey, "Pyrolysis of Polymeric Materials: I. Effect of Chemical Structure, Temperature, Heating Rate, and Air Flow on Char Yield and Toxicity," *J. Fire Flammability*, **10**, 140–167 (1979).

51. C. J. Hilado and A. M. Machado, "Effect of Char Yield and Specific Toxicants on Toxicity of Pyrolysis Gases from Synthetic Polymers," *Fire Technol.*, **15**, 51–62 (1979).

52. L. G. Imhof and K. C. Steuben, "Evaluation of the Smoke and Flammability Characteristics of Polymer Systems," *Polym. Eng. Sci.*, **13**, 146–152 (1973).

53. C. P. Fenimore and F. J. Martin, "Burning of Polymers," in L. A. Wall, Ed., *The Mechanisms of Pyrolysis, Oxidation, and Burning of Organic Materials*, NBS Special Publication 357, National Bureau of Standards, Washington, D.C., 1970, pp. 159–170.

54. C. J. Hilado, *Flammability Handbook for Plastics*, 2nd ed., Technomic Publishing Co., Westport, Conn., 1974.

55. D. A. Kourtides and J. A. Parker, "Thermochemical Characterization of Some Thermoplastic Materials," *J. Fire Flammability*, **8**, 59–94 (1977).

56. National Fire Protection Association, *Fire Protection Handbook*, 14th ed., Boston, Mass., 1976.

57. *The Hazards of Industrial Explosion from Dusts*, Conference organized by Oyez International Business Communications Ltd., London, May 1979; K. N. Palmer, Chairman, Fire Research Station.

58. National Fire Protection Association, *National Fire Codes*, Vol. 5, NFPA 654-1975, Boston, Mass., pp. 654-1 to 654-51.

59. M. M. Raftery, "Explosibility Tests for Industrial Dusts," *Fire Res. Tech. Pap.*, No. 21, 1975; *Chem. Abstr.* **88**, 140915u.

60. J. A. Zapp, "Toxic and Health Effects of Plastics and Resins," *Arch. Environ. Health*, **4**, 125–136 (1962).

61. Y. Alarie, C. K. Lin, and D. L. Geary, "Sensory Irritation Evoked by Plastic Decomposition Products," *Am. Ind. Hyg. Assoc. J.*, **35**, 654–661 (1974).

62. H. T. Hofmann and H. Sand, "Further Investigations into the Relative Toxicity of Decomposition Products Given Off from Smoldering Plastics," *J. Combust. Toxicol.*, **1**, 250–258 (1974).

63. G. Kimmerle, "Aspects and Methodology for the Evaluation of Toxicological Parameters During Fire Exposure," *J. Combust. Toxicol.*, **1**, 4–51 (1974).

64. C. Herpol and P. Vandervelde, "Calculation of a Toxicity Index for Materials Based on a Biological Evaluation Method," *Fire Mater.*, **2**, 7–10 (1978).

65. W. J. Potts and T. S. Lederer, "Some Limitations in the Use of the Sensory Irritation Method as an End-Point in Measurement of Smoke Toxicity," *J. Combust. Toxicol.*, **5**, 182–195 (1978).

66. C. J. Hilado, J. J. Cumming, and C. J. Casey, "Toxicity of Pyrolysis Gases from Natural and Synthetic Materials," *Fire Technol.*, **14**, 136–146 (1978).

67. C. J. Hilado, C. J. Casey, and J. E. Schneider, "Effect of Pyrolysis Temperature on Relative Toxicity of Some Plastics," *Fire Technol.*, **15**, 122–129 (1979).

68. R. C. Anderson and Y. C. Alarie, "An Attempt to Translate Toxicity of Polymer Thermal Decomposition Products into a Toxicological Hazard Index and Discussion on the Approaches Selected," *J. Combust. Toxicol.*, **5**, 476–484 (1979).

69. G. Kimmerle and F. K. Prager, "The Relative Toxicity of Pyrolysis Products. Part I. Plastics and Man-Made Fibers," *J. Combust. Toxicol.*, **7**, 42–53 (1980).

70. G. Kimmerle and F. K. Prager, "The Relative Toxicity of Pyrolysis Products. Part II. Polyisocyanate Based Foam Materials," *J. Combust. Toxicol.*, **7**, 54–68 (1980).

71. D. G. Farrar and W. A. Galster, "Biological End-Points for the Assessment of the Toxicity of Materials," *Fire Mater.*, **4**, 50–58 (1980).

72. B. A. Burgess, "A Profile Approach to the Evaluation of Combustion Toxicity," *Proceedings of the Tenth Conference on Environmental Toxicology*, Dayton, Ohio, November 1979, pp. 244–252 (AD A086341/5).

73. National Materials Advisory Board, *Smoke and Toxicity (Combustion Toxicology of Polymers) (Fire Safety Aspects of Polymeric Materials*, Vol. 3), Technomic, Westport, Conn., 1978.

74. W. D. Woolley and P. J. Fardell, "The Prediction of Combustion Products," *Fire Res.*, **1**, 11–21 (1977).

75. W. D. Woolley, S. A. Ames, and P. J. Fardell, "Chemical Aspects of Combustion Toxicology of Fires," *Fire Mater.*, **3**, 110–120 (1979).

76. J. E. Potts, R. A. Clendinning, W. B. Ackart, and W. D. Niegisch, "The Biodegradability of

Synthetic Polymers," in J. Guillet, Ed., *Polymers and Ecological Problems*, Vol. 3, Plenum Press, New York, 1973, pp. 61–79.

77. (a) J. Mills, "The Biodeterioration of Synthetic Polymers and Plasticizers," *CRC Crit. Rev. Environ. Control*, **4**, 341–351 (1974); (b) A. M. Kaplan, "Microbial Decomposition of Synthetic Polymeric Materials," in T. Hasegawa, Ed., *Proceedings of the First Intersectional Congress of AIMS*, Vol. 2, 1975, pp. 535–545.

78. F. J. Buono and G. A. Trautenberg, "Biocides," *Encyclopedia of Polymer Science and Technology*, 2nd ed., Suppl. 1, Wiley, New York, 1976, pp. 95–115.

79. J. J. Porter and E. H. Snider, "Long-Term Biodegradability of Textile Chemicals," *J. Water Pollut. Contr. Fed.*, **48**, 2198–2210 (1976).

80. M. Windholz, Ed., *The Merck Index*, 9th ed., Merck & Company, N.J., 1976.

81. J. B. Titus, *Trade Designations of Plastics and Related Materials*, (*Revised*), Plastec Note N9C, U.S. Army Materiel Development and Readiness Command (Plastics Technical Evaluation Center), Dover, N.J., 1978 (ADA 058395).

82. *Modern Plastics Encyclopedia*, 1977–78 ed., Vol. 54, No. 10A, McGraw-Hill, New York.

83. S. L. Aggarwal, "Physical Constants of Poly(ethylene)," in J. Brandrup and E. H. Immergut, Eds., *Polymer Handbook*, 2nd ed., Wiley, New York, 1975, V-13 to V-22.

84. S. L. Aggarwal, "Physical Constants of Poly(propylene)," in J. Brandrup and E. H. Immergut, Eds., *Polymer Handbook*, 2nd ed., Wiley, New York, 1975, V-23 to V-28.

85. R. L. Miller, "Crystallographic Data for Various Polymers," in J. Brandrup and E. H. Immergut, Eds., *Polymer Handbook*, 2nd ed., Wiley, New York, 1975, pp. III-1 to III-137.

86. W. A. Lee and R. A. Rutherford, "The Glass Transition Temperatures of Polymers," in J. Brandrup and E. H. Immergut, Eds., *Polymer Handbook*, 2nd ed., Wiley, New York, 1975, pp. III-139 to III-192.

87. L. Bohn, "Refractive Indices of Polymers," in J. Brandrup and E. H. Immergut, Eds., *Polymer Handbook*, 2nd ed., Wiley, New York, 1975, pp. III-241 to III-244.

88. O. Fuchs and H.-H. Suhr, "Solvents and Non-Solvents for Polymers," in J. Brandrup and E. H. Immergut, Eds., *Polymer Handbook*, 2nd ed., Wiley, New York, 1975, pp. IV-241 to IV-265.

89. J. E. Ford, Ed., *Fibre Data Summaries*, Shirley Institute, Manchester, England, 1966.

90. J. G. Cook, *Handbook of Textile Fibres*, 4th ed., Merrow Publishing, Herts, England, 1968.

91. H. F. Mark, S. M. Atlas, and E. Cernia, Eds., *Man-Made Fibers Science and Technology*, Wiley, New York, 1967.

92. *U.S. Code of Federal Regulations*, Supt. Doc., Washington, D.C., 21 CFR 170-199, April 1, 1981, also 21 CFR 500-599. Revised annually. *Guide* available from *Food Chemical News*.

93. J. Cooper, "*Plastic Containers for Pharmaceuticals-Testing and Control*," World Health Organization, Geneva, 1974.

94. C. J. Hilado and D. P. Brauer, "Concentration-Time Data in Toxicity Tests and Resulting Relationships," *J. Combust. Toxicol.*, **6**, 136–149 (1979).

95. C. J. Hilado and N. V. Huttlinger, "Toxicity of Pyrolysis Gases from Some Synthetic Polymers," *J. Combust. Toxicol.*, **5**, 361–369 (1978).

96. C. J. Hilado, N. V. Huttlinger, and D. P. Brauer, "Effect of Pyrolysis Temperature and Air Flow on Toxicity of Gases from a Polyethylene Polymer," *J. Combust. Toxicol.*, **5**, 465–475 (1978).

97. C. J. Hilado and D. P. Brauer, "Polyethylene: Effect of Pyrolysis Temperature and Air Flow on Toxicity of Pyrolysis Gases," *J. Combust. Toxicol.*, **6**, 63–68 (1979).

98. C. J. Hilado and A. M. Machado, "Toxicity of Pyrolysis Gases From Some Cellular Polymers," *J. Combust. Toxicol.*, **5**, 162–181 (1978).

99. W. J. Potts, T. S. Lederer, and J. F. Quast, "A Study of the Inhalation Toxicity of Smoke Produced Upon Pyrolysis and Combustion of Polyethylene Foams. Part I. Laboratory Studies," *J. Combust. Toxicol.*, **5**, 408–433 (1978).

100. R. L. Kuhn, W. J. Potts, and T. E. Waterman, "A Study of the Inhalation Toxicity of Smoke Produced Upon Pyrolysis and Combustion of Polyethylene Foams. Part II. Full Scale Fire Studies," *J. Combust. Toxicol.*, **5**, 434–464 (1978).

101. K. Kishitani and K. Nakamura, "Study on Toxicities of Combustion Products of Building Materials at Initial Stage of Fire," *J. Fac. Eng. Univ. Tokyo, Ser. B*, **34**, 295–313 (1977).

102. E. A. Boettner, G. L. Ball, and B. Weiss, *"Combustion Products from the Incineration of Plastics,"* U.S. Environmental Protection Agency, Cincinnati, Ohio, 1970 (PB 222001).

103. T. Morikawa, "Acrolein, Formaldehyde, and Volatile Fatty Acids from Smoldering Combustion," *J. Combust. Toxicol.*, **3**, 135–150 (1976).

104. T. Morimoto, K. Takeyama, and F. Konishi, "Composition of Gaseous Combustion Products of Polymers," *J. Appl. Polym. Sci.*, **20**, 1967–1976 (1976).

105. N. Grassie and A. Scotney, "Products of Thermal Degradation of Polymers," in J. Brandrup and E. H. Immergut, Eds., in *Polymer Handbook*, 2nd ed., Wiley, New York, 1975, pp. II-473 to II-479.

106. J. Michal, J. Mitera, and S. Tardon, "Toxicity of Thermal Degradation Products of Polyethylene and Polypropylene," *Fire Mater.*, **1**, 160–168 (1976).

107. M. Chaigneau, "Sur Les Gaz Dégagés Par La Pyrolyse De Diverses Matières Plastiques à 1000°C," *C. R. Acad. Sci. Ser. C.*, **278**, 109–111 (1974).

108. C. J. Hilado, J. E. Schneider, and D. P. Brauer, "Toxicity of Pyrolysis Gases from Polypropylene," *J. Combust. Toxicol.*, **6**, 109–116 (1979).

109. H. H. Cornish, K. J. Hahn, and M. L. Barth, "Experimental Toxicology of Pyrolysis and Combustion Hazards," *Environ. Health Perspect.*, **11**, 191–196 (1975).

110. J. R. Flesher, "Polyethylene," in *Modern Plastics Encyclopedia*, 1978–79 ed., Vol. 55, No. 10A, McGraw-Hill, New York, pp. 59, 60, 63, 64.

111. R. A. Botham, G. H. Jenkins, G. L. Ball, III, and I. O. Salyer, "Crosslinked HDPE Makes the Grade in Thermal Energy Storage," *Mod. Plast.*, **55**, 54–56, 58 (1978).

112. R. A. A. Hentschel, "Spunbonded Sheet Products," *Chem., Tech.*, **4**, 32–41 (1974).

113. W. O. Bracken, "UHMW Polyethylene," in *Modern Plastics Encyclopedia*, 1978–79 ed., Vol. 55, No. 10A, McGraw-Hill, New York, p. 64.

114. M. L. Westrick, P. Gross, and H. H. Schrenk, unpublished data, Industrial Hygiene Foundation of America, Pittsburgh, Pa., March 1956 (courtesy of the Society of the Plastics Industry, New York).

115. B. Y. Kalinin and L. P. Zimnitskaya, "Toxicology of Low-Pressure (High Density) Polyethylene," *Toksikol. Vysokomol. Mater. Khim. Syr'ya Ikh Sin., Gos. Nauch.-Issled. Inst. Polim. Plast. Mass.*, 21–40 (1966); as cited in *Chem. Abstr.* **66**, 103638.

116. C. S. Weil and P. E. Palm, unpublished data, Mellon Institute of Industrial Research, Pittsburgh, Pa., October 1960 (courtesy of the Society of the Plastics Industry, New York).

117. C. S. Weil and H. F. Smyth, Jr., unpublished data, Mellon Institute of Industrial Research, Pittsburgh, Pa., March 1966 (courtesy of the Society of the Plastics Industry, New York).

118. J. A. Boling, R. H. Grummer, and E. R. Hauser, "A Comparison of Plastic Dilution of Diets with Full and Limit-Fed Diets for Growing Swine," College of Agricultural and Life Sciences, Research Report 55, University of Wisconsin, Madison, Wis., 1-4 (April 1970).

119. C. J. Hilado, A. N. Solis, W. H. Marcussen, and A. Furst, "Effect of Temperature and Heating Rate on Apparent Lethal Concentrations of Pyrolysis Products," *J. Combust. Toxicol.*, **3**, 381–392 (1976).

120. R. W. Heimburg, "Environmental Effects of the Incineration of Plastics," *AIChE Symp. Ser.,* **68,** 21–27 (1972).

121. R. H. Rigdon, "Plastics and Carcinogenesis," *South. Med. J.,* **67,** 1459–1465 (1974).

122. J. Autian, "Film Carcinogenesis," in P. Bucalossi, U. Veronesi, and N. Cascinelli, Eds., *Chemical and Viral Oncogenesis,* American Elsevier, Vol. 2, New York, 1975, pp. 94–101.

123. B. F. Matlaga, L. P. Yasenchak, and T. N. Salthouse, "Tissue Response to Implanted Polymers: The Significance of Sample Shape," *J. Biomed. Mater. Res.,* **10,** 391–397 (1976).

124. J. E. Potts, R. A. Clendinning, and W. B. Ackart, "An Investigation of the Biodegradability of Packaging Plastics," Union Carbide Corporation, Bound Brook, N.J., 1972 (PB 213488; EPA-R2-72-046; sponsored by U.S. Environmental Protection Agency, Washington, D.C.).

125. J. E. Potts, R. A. Clendinning, and W. B. Ackart, "The Effect of Chemical Structure on the Biodegradability of Plastics," *Degradability Polym. Plast., [Prepr.] Conf.,* pp. 12-1 to 12-10 (1973); *Chem. Abstr.,* **85,** 130108g.

126. L. R. Spencer, M. Heskins, and J. E. Guillet, "Studies on the Biodegradability of Photodegraded Polymers: Identification of Bacterial Types," *Proc. Int. Biodegradation Symp.,* 3rd ed., 1976, pp. 753–763.

127. N. B. Nykvist, "Biodegradation of Low-Density Polyethylene," *Plast. Polym.,* **42,** 195–199 (1974).

128. D. S. Brackman, "Photodegradable Polyethylene," *Res. Discl.,* **156,** 5, 6 (1977); *Chem. Abstr.,* **86,** 172337g.

129. J. W. Tobias, L. J. Taylor, and S. J. Gaumer, "Controlled Environmental Deterioration of Plastics," U.S. Patent 4,051,306 (1977); *Chem. Abstr.,* **87,** 185521h.

130. U.S. Federal Trade Commission, *Rules and Regulations Under the Textile Fiber Products Identification Act,* Washington, D.C., effective March 3, 1960, as amended to November 1, 1974.

131. S. L. Cannon and W. O. Statton, "Springy Polypropylene: A Novel Elastomeric Biomaterial," *Int. J. Polym. Mater.,* **5,** 165–176 (1977).

132. J. S. Houston, "Polypropylene," in *Modern Plastics Encyclopedia,* 1978–79 ed., Vol. 55, No. 10A, McGraw-Hill, New York, pp. 74, 77–78.

133. B. Y. Kalinin, "Toxicity of Polypropylene," *Toksikol. Vysokomol. Mater. Khim. Syr'ka Ikh. Sin., Gos. Nauch-Issled. Inst. Polim. Plast. Mass.,* 55–63 (1966); *Chem. Abstr.,* **66,** 103640a.

134. F. K. Kinoshita, personal communication, Hercules, Inc., Wilmington, Del., July 1979 (tests conducted at Hercules 1958 to 1959).

135. P. J. Garvin, personal communication, Standard Oil Co., Chicago, Ill., August 1979 (tests conducted at Industrial Bio-Test Laboratories 1964 to 1966).

136. D. F. Edwards, unpublished data, E. I. du Pont de Nemours and Co., Inc., Haskell Laboratory, Wilmington, Del., Sept. 1975.

137. J. Vollmar and G. Ott-Heidelberg, "Experimentelle Geschwulstauslösung durch Kunststoffe aus chirurgischer Sicht," *Langenbecks Arch. Klin. Chir.,* **298,** 729–736 (1961).

138. T. C. Liebert, R. P. Chartoff, S. L. Cosgrove, and R. S. McCuskey, "Subcutaneous Implants of Polypropylene Filaments," *J. Biomed. Mater. Res.,* **10,** 939–951 (1976).

139. M. Reich and D. E. Hudgim, "Degradable Plastic," U.S. Patent 3,984,940, Oct. 1976; *Chem. Abstr.,* **86,** 15741s.

140. Y. Shimura, "Degraded Products in Weathered Polymer," *J. Appl. Polym. Sci.,* **22,** 1491–1507 (1978).

141. M. P. Schard, "Polybutylene," in *Modern Plastics Encyclopedia,* 1978–79 ed., Vol. 55, No. 10A, McGraw-Hill, New York, pp. 45.

142. G. Bornmann and A. Loeser, "Zur Toxikologie von Polybuten-(1)," *Arch. Toxikol.,* **23,** 240–244 (1968).

143. E. V. Anderson, "Downturn Looms for Synthetic Elastomers," *Chem. Eng. News,* **57**(10), 8–9 (1979).

144. H. E. Schroeder, "Recent Developments in Synthetic Elastomers," *Angew. Makromol. Chem.,* **16/17**, 1–25 (1971).

145. H. E. Schroeder, "Advances in High Temperature Elastomers," *Rubber Plast. News,* **8**, 21–22 (1978).

146. T. H. Rogers and K. C. Hecker. "Latex and Foam Rubber," in M. Morton, Ed., *Rubber Technology,* 2nd ed., Van Nostrand Reinhold, New York, 1973, pp. 459–495.

147. *SAE Handbook 1979,* Society of Automotive Engineers, Warrendale, Pa., pp. 12.07–12.23.

148. R. L. Bebb, "Chemistry of Rubber Processing and Disposal," *Environ. Health Perspect.,* **17**, 95–101 (1976).

149. A. Nutt, "Measurement of Some Potentially Hazardous Materials in the Atmosphere of Rubber Factories," *Environ. Health Perspect.,* **17**, 117–123 (1976).

150. R. J. Steichen, "Modified Solution Approach for the Gas Chromatographic Determination of Residual Monomers by Head-Space Analysis," *Anal. Chem.,* **48**, 1398–1402 (1976).

151. I. Li Giotti, M. Franzosi, and G. Bonomi (1972), as cited in C. W. Wadelin and M. C. Morris, "Rubber," *Anal. Chem.,* **45**, 333R–343R (1973).

152. S. M. Rappaport and D. A. Fraser, "Gas Chromatographic-Mass Spectrometric Identification of Volatiles Released from a Rubber Stock during Simulated Vulcanization," *Anal. Chem.,* **48**, 476–480 (1976).

153. J. C. Hughes, B. B. Wheals, and M. J. Whitehouse, "Pyrolysis Mass Spectrometry. A Technique of Forensic Potential," *Forensic Sci.,* **10**, 217–228 (1977).

154. U.S. Consumer Product Safety Commission, *Assessment of Acrylonitrile Contained in Consumer Products,* Final Report, Contract No. CPSC-C-77-0009, [Washington, D.C.], January 1978.

155. R. R. Monson and K. K. Nakano, "Mortality Among Rubber Workers," *Am. J. Epidemiol.,* **103**, 297–303 (1976).

156. D. Andjelkovich, J. Taulbee, M. Symons, and T. Williams, "Mortality of Rubber Workers with Reference to Work Experience," *J. Occup. Med.,* **19**, 397–405 (1977).

157. A. J. Fox and P. F. Collier, "A Survey of Occupational Cancer in the Rubber and Cable-making Industries: Analysis of Deaths Occurring in 1972–74," *Br. J. Ind. Med.,* **33**, 249–264 (1976).

158. E. L. Rhodes and J. Warner, "Contact Eczema. A Follow-Up Study," *Br. J. Dermatol.,* **78**, 640–644 (1966).

159. H. T. H. Wilson, "Rubber Dermatitis," *Br. J. Dermatol.,* **81**, 175–179 (1969).

160. A. A. Fisher, *Contact Dermatitis,* 2nd ed., Lea & Febiger, Philadelphia, Pa., 1973.

161. J. Roed-Petersen, N. Hjorth, W. P. Jordan, and M. Bourlas, "Postsorters' Rubber Fingerstall Dermatitis," *Contact Dermatitis,* **3**, 143–147 (1977).

162. J. Foussereau and C. Cavelier, "La N-isopropyl-N′-phénylparaphénylènediamine a-t-elle sa place dans la batterie standard d'allergènes? Importance de cet allergène dans l'intolérance au caoutchouc," *Dermatologica,* **155**, 164–167 (1977).

163. C. J. Hilado, K. L. Kosola, and A. N. Solis, "Effect of Heating Rate on Toxicity of Pyrolysis Gases from Some Elastomers," *J. Combust. Toxicol.,* **4**, 563–579 (1977).

164. C. J. Hilado and H. J. Cumming, "Polychloroprene Flexible Foam as a Reference Material," *J. Combust. Toxicol.,* **4**, 464–471 (1977).

165. R. C. Anderson and Y. C. Alarie, "Screening Procedure to Recognize "Supertoxic" Decomposition Products from Polymeric Materials Under Thermal Stress," *J. Combust. Toxicol.,* **5**, 54–63 (1978).

166. K. L. Paciorek, R. H. Kratzer, J. Kaufman, J. Nakahara, and A. M. Harstein, "Thermal Oxidative Decomposition Studies of Neoprene Compositions," *Am. Ind. Hyg. Assoc. J.*, **36**, 10–16 (1975).

167. G. T. Hall, unpublished data, E. I. du Pont de Nemours and Co., Haskell Laboratory, Wilmington, Del., April 1975.

168. V. L. Carter, Jr., D. A. Bafus, H. P. Warrington, and E. S. Harris, "The Acute Inhalation Toxicity in Rats from the Pyrolysis Products of Four Fluoropolymers," *Toxicol. Appl. Pharmacol.*, **30**, 369–376 (1974).

169. J. W. Williams, unpublished data, E. I. du Pont de Nemours and Co., Haskell Laboratory, Wilmington, Del., August 1957 (some data included in next reference).

170. J. W. Clayton, Jr., "The Toxicity of Fluorocarbons with Special Reference to Chemical Constitution," *J. Occup. Med.*, **4**, 262–273 (1962).

171. I. Einhorn, personal communication to R. S. Waritz, E. I. du Pont de Nemours and Co., Inc., Haskell Laboratory, Wilmington, Del., August 1971.

172. S. Barron, "An Investigation of the Effects of High Temperatures Upon Various Industrial Polymers," *J. Fire Flammability*, **7**, 387–400 (1976).

173. F. Grimalt and C. Romaguera, "New Resin Allergens in Shoe Contact Dermatitis," *Contact Dermatitis*, **1**, 169–174 (1975).

174. S. T. Semegen, "Natural Rubber," in M. Morton, Ed., *Rubber Technology*, 2nd ed., Van Nostrand Reinhold, New York, 1973, pp. 152–177.

175. E. M. Glymph, "Natural Rubber," in G. G. Winspear, Ed., *The Vanderbilt Rubber Handbook*, Vanderbilt Co., New York, 1968, pp. 3–33.

176. G. R. Himes, "Synthetic Polyisoprene," in M. Morton, Ed., *Rubber Technology*, 2nd ed., Van Nostrand Reinhold, New York, 1973, pp. 274–301.

177. J. Platner, "Synthetic Polyisoprene," in G. G. Winspear, Ed., *The Vanderbilt Rubber Handbook*, Vanderbilt Co., New York, 1968, pp. 81–88.

178. E. B. Sansone and Y. B. Tewari, "The Permeability of Laboratory Gloves to Selected Solvents," *Am. Ind. Hyg. Assoc. J.*, **39**, 169–174 (1978).

179. J. Peppiatt, "Trans-polyisoprene Sheeting—Its Use in Orthopedics and Other Medical Conditions," *Int. J. Polym. Mater.*, **5**, 251–269 (1977).

180. W. M. Saltman, "Styrene–Butadiene Rubbers," in M. Morton, Ed., *Rubber Technology*, 2nd ed., Van Nostrand Reinhold, New York, 1973, pp. 178–198.

181. R. S. Barrows, personal communication, E. I. du Pont de Nemours and Co., Inc., Elastomer Chemicals Dept., Wilmington, Del., May 1979.

182. E. L. Borg, "Ethylene/Propylene Rubber," in M. Morton, Ed., *Rubber Technology*, 2nd ed., Van Nostrand Reinhold, New York, 1973, pp. 220–248.

183. P. R. Johnson, "Neoprene," in *Encyclopedia of Chemical Technology*, 3rd ed., Vol. 8, Wiley, New York, 1979, pp. 515–534.

184. D. A. Horn, D. R. Tierney, and T. W. Hughes, "*Source Assessment: Polychloroprene State of the Art*," Research sponsored by the U.S. Environmental Protection Agency, Cincinnati, Ohio, December 1977 (PB 278777).

185. D. B. Forman, "The Neoprenes," in G. G. Winspear, Ed., *The Vanderbilt Rubber Handbook*, Vanderbilt Co., New York, 1968, pp. 119–129.

186. D. B. Forman, "Neoprene and Hypalon®," in M. Morton, Ed., *Rubber Technology*, 2nd ed., Van Nostrand Reinhold, New York, 1973, pp. 322–348.

187. P. R. Johnson, "A General Correlation of the Flammability of Natural and Synthetic Polymers," *J. Appl. Polym. Sci.*, **18**, 491–504 (1974).

188. O. L. Dashiell, unpublished data, E. I. du Pont de Nemours and Co., Haskell Laboratory, Wilmington, Del., September 1972.

189. *U.S. Code of Federal Regulations*, Supt. Doc., Washington, DC, 49 CFR 100-199 (revised annually; Ref. 190 conducted on basis of 1973 ed.).

190. D. F. Edwards, unpublished data, E. I. du Pont de Nemours and Co., Inc., Haskell Laboratory, Wilmington, Del., June 1975.

191. J. L. Roberts and J. M. Hanifin, "Athletic Shoe Dermatitis," *J. Am. Med. Assoc.*, **241**, 275–276 (1979).

192. K. M. Frank, unpublished data, E. I. du Pont de Nemours and Co., Inc., Haskell Laboratory, Wilmington, Del., October 1972.

193. H. J. Trochimowicz, personal communication, E. I. du Pont de Nemours and Co., Inc., Haskell Laboratory, Wilmington, Del., December 1981 (tests conducted at CIVO Laboratories, Netherlands, under the auspices of Du Pont, Denka, Bayer A.G., and British Petroleum; to be published).

194. V. Ponomarkov and L. Tomatis, "Long-Term Testing of Vinylidene Chloride and Chloroprene for Carcinogenicity in Rats," *Oncology*, **37**, 136–141 (1980).

195. Y. Isono, "Degradation and Fouling of Synthetic Fabrics During Immersion in the Sea," *Kasen Geppo*, **30**(6), 23–41 (1977); *Chem. Abstr.*, **87**, 86249u.

196. H. E. Minnerly, "Polyacrylic Rubber," in G. G. Winspear, Ed., *The Vanderbilt Rubber Handbook*, Vanderbilt Co., New York, 1968, pp. 156–173.

197. J. P. Morrill, "Nitrile and Polyacrylate Rubbers," in A. Morton, Ed., *Rubber Technology*, 2nd ed., Van Nostrand Reinhold, New York, 1973, pp. 302–321.

198. R. P. Taubinger, as cited in C. W. Wadelin and G. S. Trick, "Rubber," *Anal. Chem.*, **43**, 334R–344R (1971).

199. R. G. Arnold, A. L. Barney, and D. C. Thompson, "Fluoroelastomers," *Rubber Chem. Technol.*, **46**, 619–652 (1973).

200. D. A. Stivers, "Fluorocarbon Rubbers," in M. Morton, Ed., *Rubber Technology*, 2nd ed., Van Nostrand Reinhold, New York, 1973, pp. 407–439.

201. L. F. Pelosi, A. L. Moran, A. E. Burroughs, and T. L. Pugh, "The Volatile Products Evolved from Flurorelastomer Compounds During Curing," *Rubber Chem. Technol.*, **49**, 367–374 (1976).

202. O. L. Dashiell, unpublished data, E. I. du Pont de Nemours and Co., Inc., Haskell Laboratory, Wilmington, Del., August 1972.

203. H. Sherman, unpublished data, E. I. du Pont de Nemours and Co., Inc., Haskell Laboratory, Wilmington, Del., January 1968.

204. R. E. Reinke, unpublished data, E. I. du Pont de Nemours and Co., Inc., Haskell Laboratory, Wilmington, Del., September 1967.

205. M. E. McDonnell, unpublished data, E. I. du Pont de Nemours and Co., Inc., Haskell Laboratory, Wilmington, Del., February 1970.

206. P. R. Johnson, "Chlorosulfonated Polyethylene," in *Encyclopedia of Chemical Technology*, 3rd ed., Vol. 8, Wiley, New York, 1979, pp. 484–491.

207. O. L. Dashiell, unpublished data, E. I. du Pont de Nemours and Co., Inc., Haskell Laboratory, Wilmington, Del., September 1972.

208. M. E. McDonnell, unpublished data, E. I. du Pont de Nemours and Co., Inc., Haskell Laboratory, Wilmington, Del., April 1971.

209. Betro Laboratories, unpublished data, Philadelphia, Pa., November 1965 (work sponsored by Haskell Laboratory).

210. *Chemical Resistance of the Du Pont Elastomers*, E. I. du Pont de Nemours and Co., Inc. (Elastomer Chemicals Department Bulletin E-24722), Wilmington, Del., n.d.

211. H.-G. Elias, *New Commercial Polymers 1969–1975*, Gordon and Breach Science Publishers, New York, 1977.

212. E. A. Collins, C. A. Daniels, and C. E. Wilkes, "Physical Constants of Poly(Vinyl Chloride)," in J. Brandrup and E. H. Immergut, Eds., *Polymer Handbook*, 2nd ed., Wiley, New York, 1975.

213. W. L. Guess and J. B. Stetson, "Tissue Reactions to Organotin-Stabilized Polyvinyl Chloride (PVC) Catheters," *J. Am. Med. Assoc.*, **204**, 580–584 (1968).

214. (a) D. Pelling, M. Sharatt, and J. Hardy, "The Safety Testing of Medical Plastics. I. An Assessment of Methods, *Food Cosmet. Toxicol.*, **11**, 69–83 (1973); (b) W. L. Guess, "Safety Evaluation of Medical Plastics," *Clin. Toxicol.*, **12**, 77–95 (1978).

215. P. C. Bowes, "Smoke and Toxicity Hazards of Plastics in Fire," *Ann. Occup. Hyg.*, **17**, 143–157 (1974).

216. (a) G. Mastrangelo, M. Manno, G. Marcer, G. B. Bartolucci, C. Gemignani, G. Saladino, L. Simonato, and B. Saia, "Polyvinyl Chloride Pneumoconiosis: Epidemiological Study of Exposed Workers," *J. Occup. Med.*, **21**, 540–542 (1979); (b) C. A. Soutar, L. H. Copland, P. E. Thornley, J. F. Hurley, and J. Ottery, *An Epidemiological Study of Respiratory Disease in Workers Exposed to Polyvinylchloride Dust*, Institute of Occupational Medicine, Edinburgh, October 1979 (sponsored by Imperial Chemical Industries Ltd.; courtesy of The Society of the Plastics Industry). See also C. A. Soutar et al., *Thorax*, **35**, 644–652 (1980).

217. B. Szende, K. Lapis, A. Pinter, A Nemes, and E. Tarjan, "Pneumoconiosis Developing After Inhalation of Polyvinyl Chloride," *Orv. Hetil.*, **112**, 85–86 (1971).

218. N. Frongia, A. Spinazzola, and A. Bucarelli, "Experimental Pulmonary Lesions from Prolonged Inhalation of PVC Powders in the Working Environment," *Med. Lav.*, **65**, 1–12.

219. (a) J. Popow, "Effect of Poly(Vinyl Chloride) (PVC) Dust on the Respiratory System of the Rat," *Rocz. Akad. Med. Bialymstoku*, **24**, 5–48 (1969); *Chem. Abstr.*, **74**, 74463c; (b) R. J. Richards, F. A. Rose, T. D. Tetley, L. M. Cobb, and C. J. Hardy, "Effect in the Rat of Inhaling PVC at the Nuisance Dust Level (10 mg/m^3)," *Arch. Environ. Health*, **36**, 14–19 (1981).

220. W. N. Sokol, Y. Aelony, and G. N. Beall, "Meat-Wrapper's Asthma; a New Syndrome?," *J. Am. Med. Assoc.*, **226**, 639–641 (1973).

221. R. W. Van Houten, A. L. Cudworth, and C. H. Irvine, "Evaluation and Reduction of Air Contaminants Produced by Thermal Cutting and Sealing of PVC Packaging Film," *Am. Ind. Hyg. Assoc. J.*, **35**, 218–222 (1974).

222. P. L. Polakoff, N. L. Lapp, and R. Reger, "Polyvinyl Chloride Pyrolysis Products," *Arch. Environ. Health*, **30**, 269–271 (1975).

223. H. Falk and B. Portnoy, "Respiratory Tract Illness in Meat Wrappers," *J. Am. Med. Assoc.*, **235**, 915–917 (1976).

224. H. Falk and B. Portnoy, "Respiratory Tract Illness In Meat Wrappers," *J. Am. Med. Assoc.*, **238**, 1721–1722 (1977).

225. R. Vandervort and S. M. Brooks, *Health Hazard Evaluation/Toxicity Determination Report 74-24*, National Institute for Occupational Safety and Health, Cincinnati, Ohio, December 1975 (PB 249426).

226. R. Vandervort and S. M. Brooks, "Polyvinyl Chloride Film Thermal Decomposition Products as an Occupational Illness, 1. Environmental Exposures and Toxicology," *J. Occup. Med.*, **19**, 188–191 (1977).

227. S. M. Brooks and R. Vandervort, "Polyvinyl Chloride Film Thermal Decomposition Products as an Occupational Illness, 2. Clinical Studies," *J. Occup. Med.*, **19**, 192–196 (1977).

228. R. H. Andrasch and E. J. Bardana, "Thermoactivated Price-Label Fume Intolerance. A Cause of Meat-Wrapper's Asthma," *J. Am. Med. Assoc.*, **235**, 937 (1976).

229. S. A. Levy, J. Storey, and B. E. Phashko, "Meat Worker's Asthma," *J. Occup. Med.*, **20**, 116–117 (1978).

230. H. H. Cornish and E. L. Abar, "Toxicity of Pyrolysis Products of Vinyl Plastics," *Arch. Environ. Health,* **19,** 15–21 (1969).

231. K. Kishitani, "Study on Injurious Properties of Combustion Products of Building Materials at Initial Stage of Fire," *J. Fac. Eng., Univ. Tokyo, Ser. B,* **31,** 1–35 (1971).

232. K. Kishitani and K. Nakamura, "Toxicities of Combustion Products," *J. Fire Flammability,* **1,** 104–123 (1974).

233. C. S. Barrow, H. Lucia, and Y. C. Alarie, "A Comparison of the Acute Inhalation Toxicity of Hydrogen Chloride Versus the Thermal Decomposition Products of Polyvinylchloride," *J. Combust. Toxicol.,* **6,** 3–12 (1979).

234. C. S. Barrow, H. Lucia, M. F. Stock, and Y. Alarie, "Development of Methodologies to Assess the Relative Hazards from Thermal Decomposition Products of Polymeric Materials," *Am. Ind. Hyg. Assoc. J.,* **40,** 408–423 (1979).

235. E. A. Boettner, G. Ball, and B. Weiss, "Analysis of the Volatile Combustion Products of Vinyl Plastics," *J. Appl. Polym. Sci.,* **13,** 377–391 (1969).

236. M. M. O'Mara, "A Comparison of Combustion Products Obtained from Various Synthetic Polymers," *J. Fire Flammability,* **1,** 141–156 (1974).

237. M. M. O'Mara, "Combustion of PVC," *Pure Appl. Chem.,* **49,** 649–660 (1977).

238. R. M. Lum, "Laser-Induced Decomposition of PVC," *J. Appl. Polym. Sci.,* **20,** 1635–1649 (1976).

239. W. D. Woolley, "Decomposition Products of PVC for Studies of Fires," *Br. Polym. J.,* **3,** 186–192 (1971).

240. C. A. Clark, "The Burning Issue of PVC Disposal," *Soc. Plast. Eng. J.,* **28,** 30–34 (1972).

241. K. L. Paciorek, R. H. Kratzer, J. Kaufman, J. Nakahara, and A. M. Hartstein, "Oxidative Thermal Decomposition of Poly(Vinyl Chloride) Compositions," *J. Appl. Polym. Sci.,* **18,** 3723–3729 (1974).

242. E. J. Quinn, D. H. Ahlstrom, and S. A. Liebman, "Effect of Structure of Chlorinated Vinyl Polymers on Smoke Evolution," *Polym. Prepr., Am. Chem. Soc.,* **14,** 1022–1027 (1973).

243. J. P. Stone, R. N. Hazlett, J. E. Johnson, and H. W. Carhart, "The Transport of Hydrogen Chloride by Soot from Burning Polyvinyl Chloride," *J. Fire Flammability,* **4,** 42–51 (1973).

244. J. M. Ives, E. E. Hughes, and J. K. Taylor, *Toxic Atmospheres Associated with Real Fire Situations,* NBS Report 10807, National Bureau of Standards, Washington, D.C., February 1972.

245. R. F. Dyer and V. H. Esch, "Polyvinyl Chloride Toxicity in Fires. Hydrogen Chloride Toxicity in Fire Fighters," *J. Am. Med. Assoc.,* **235,** 393–397 (1976).

246. F. Colardyn, M. Van Der Straeten, H. Lamont, and T. Van Peteghem, "Acute Inhalation-Intoxication by Combustion of Polyvinylchloride," *Int. Arch. Occup. Environ. Health,* **38,** 121–127 (1976).

247. M. G. Genovesi, D. P. Tashkin, S. Chopra, M. Morgan, and C. McElroy, "Transient Hypoxemia in Firemen Following Inhalation of Smoke," *Chest,* **71,** 441–444 (1977).

248. D. P. Tashkin, M. G. Genovesi, S. Chopra, A. Coulson, and M. Simmons, "Respiratory Status of Los Angeles Firemen One-Month Follow-Up After Inhalation of Dense Smoke," *Chest,* **71,** 445 (1977).

249. K. Yamamoto, "Acute Toxicity of the Combustion Products from Various Kinds of Fibers," *Z. Rechtsmed.,* **76,** 11–26 (1975); *Chem. Abstr.,* **84,** 922m.

250. D. R. Roberts and A. L. Gatzke, "Thermal Decomposition of Poly(Vinylidene Chloride) Prepared in the Presence of Oxygen," *J. Polym. Sci.,* **16,** 1211–1219 (1978).

251. K. E. Malten and R. L. Zielhuis, *Industrial Toxicology and Dermatology in the Production and Processing of Plastics,* Elsevier (Elsevier Monographs on Toxic Agents), New York, 1964.

252. G. W. V. Stark and P. Field, *Toxic Gases and Smoke from Polyvinyl Chloride in Fires in the Fire*

Research Station Full-Scale Test Rig, Fire Research Station (Fire Research Note No. 1030), London, April 1975.

253. K. G. Martin and D. A. Powell, "Toxic Gases and Smoke Assessment Studies on Vinyl Floor Coverings with the Fire Propagation Test," *Fire Mater.,* **3,** 132–139 (1979).

254. C. A. Sperati, "Physical Constants of Poly(Tetrafluoroethylene)," in J. Brandrup and E. H. Immergut, Eds., 2nd ed., *Polymer Handbook,* Wiley, New York, 1975, pp. V-29 to V-36.

255. E. I. du Pont de Nemours and Company, Inc., *Du Pont Tedlar® PVF Film; Physical-Thermal Properties,* Plastics Products and Resins Dept. Bulletin TD-2 (E-01211), Wilmington, Del., October 1974.

256. E. I. du Pont de Nemours and Company, Inc., *Tetrafluoroethylene (TFE) Polymer,* Plastics Products and Resins Dept. Material Safety Data Sheet, Wilmington, Del., July 1976.

257. E. I. du Pont de Nemours and Company, Inc., *Du Pont Tedlar® PVF Film; Glazing and Solar Energy,* Plastics Products and Resins Dept. Bulletin TD-31 (E-17987), Wilmington, Del., n.d.

258. E. I. du Pont de Nemours and Company, Inc., *"Teflon" Fluorocarbon Resins—Safety in Handling and Use,* Plastics Products and Resins Dept. Brochure A-72531, Wilmington, Del., July 1970.

259. D. I. McCane, "Tetrafluoroethylene Polymers," *Encyclopedia of Polymer Science and Technology,* Vol. 13, Wiley, New York, 1970, pp. 623–671.

260. M. Hagiwara, G. Ellinghorst, and D. O. Hummel, "Thermal Decomposition of Gamma-ray Irradiated Poly(Vinyl Fluoride)," *Makromol. Chem.,* **178,** 2901–2912 (1977).

261. P. L. Polakoff, K. A. Busch, and M. T. Okawa, "Urinary Fluoride Levels in Polytetrafluoroethylene Fabricators," *Am. Ind. Hyg. Assoc. J.,* **35,** 99–106 (1974).

262. M. T. Okawa and P. L. Polakoff, "Occupational Health Case Reports—No. 7," *J. Occup. Med.,* **16,** 350–355 (1974).

263. National Institute for Occupational Safety and Health, *Occupational Exposure to Decomposition Products of Fluorocarbon Polymers,* DHEW (NIOSH) Publ. No. 77-193, Supt. Doc., Washington, D.C., 1977.

264. R. S. Waritz, "An Industrial Approach to Evaluation of Pyrolysis and Combustion Hazards," *Environ. Health Perspect.,* **11,** 197–202 (1975).

265. R. S. Waritz and B. K. Kwon, "The Inhalation Toxicity of Pyrolysis Products of Polytetrafluoroethylene Heated Below 500 Degrees Centigrade," *Am. Ind. Hyg. Assoc. J.* **29,** 19–26 (1968).

266. W. E. Coleman, L. D. Scheel, R. E. Kupel, and R. L. Larkin, "The Identification of Toxic Compounds in the Pyrolysis Products of Polytetrafluoroethylene (PTFE)," *Am. Ind. Hyg. Assoc. J.,* **29,** 33–40 (1968).

267. L. D. Scheel, W. C. Lane, and W. E. Coleman, "The Toxicity of Polytetrafluoroethylene Pyrolysis Products—Including Carbonyl Fluoride and a Reaction Product, Silicon Tetrafluoride," *Am. Ind. Hyg. Assoc. J.,* **29,** 41–48 (1968).

268. L. D. Scheel, L. McMillan, and F. C. Phipps, Biochemical Changes Associated with Toxic Exposures to Polytetrafluoroethylene Pyrolysis Products," *Am. Ind. Hyg. Assoc. J.,* **29,** 49–53 (1968).

269. K. P. Lee, J. A. Zapp, and J. W. Sarver, "Ultrastructural Alterations of Rat Lung Exposed to Pyrolysis Products of Polytetrafluoroethylene (PTFE, Teflon)," *Lab. Invest.,* **35,** 213–221 (1976).

270. H. L. Lucia, A. K. Burton, R. C. Anderson, M. F. Stock, and Y. C. Alarie, "Renal Damage in Mice Following Exposure to the Pyrolysis Products of Polytetrafluoroethylene," *J. Combust. Toxicol.,* **5,** 270–277 (1978).

271. R. J. Sherwood, "The Hazards of Fluon (Polytetrafluoroethylene)," *Trans. Assoc. Ind. Med. Off.,* **5,** 10–12 (1955).

272. O. A. Blagodarnaya, "Fluoroplastic Fever Caused by Inhalation of Products Secondary to

Thermal-Oxidation Destruction of the Fluoroplastic-4," *Gig. Tr. Prof. Zabol.,* **4**(5), 25–29 (1972); author's English abstract.

273. A. M. Kligman, unpublished data, as cited in J. W. Clayton, Jr., "Fluorocarbon Toxicity and Biological Action," *Fluorine Chem. Rev.,* **1,** 197–252 (1967).

274. J. B. Nuttall, R. J. Kelly, B. S. Smith, and C. K. Whiteside, "Inflight Toxic Reactions Resulting from Fluorocarbon Resin Pyrolysis," *Aerosp. Med.,* **35,** 676–683 (1964).

275. T. B. Blandford, P. J. Seamon, R. Hughes, M. Pattison, and M. P. Wilderspin, "A Case of Polytetrafluoroethylene Poisoning in Cockatiels Accompanied by Polymer Fume Fever in the Owner," *Vet. Rec.,* **96,** 175–176 (1975).

276. R. E. Brubaker, "Pulmonary Problems Associated with Use of Polytetrafluoroethylene, *J. Occup. Med.,* **19,** 693–695 (1977).

277. H. Arito and R. Soda, "Pyrolysis Products of Polytetrafluoroethylene and Polyfluoroethyle-nepropylene with Reference to Inhalation Toxicity," *Ann. Occup. Hyg.,* **20,** 247–255 (1977).

278. H. A. Birnbaum, L. D. Scheel, and W. E. Coleman, "The Toxicology of the Pyrolysis Products of Polychlorotrifluoroethylene," *Am. Ind. Hyg. Assoc. J.,* **29,** 61–65 (1968).

279. H. A. Watson, H. J. Stark, L. E. Sieffert, and L. B. Berger, *Decomposition Temperatures of Polytetrafluoroethylene and Polymonochlorotrifluoroethylene as Indicated by Halogen Liberation,* U.S. Department of the Interior, Bureau of Mines, Report of Investigations 4756, Pittsburgh, Pa., December 1950.

280. American Conference of Governmental Industrial Hygienists Committee on Industrial Ventilation, *Industrial Ventilation—A Manual of Recommended Practice,* 16th ed., Lansing, Mich., 1980.

281. American National Standards Institute Inc., *Fundamentals Governing the Design and Operation of Local Exhaust Systems,* ANSI Z9.2-1971, New York, 1971.

282. R. E. Putnam, personal communication, E. I. du Pont de Nemours and Company, Inc., Polymer Products Department, Parkersburg, W. Va., December, 1980.

283. M. S. Frant and J. W. Ross, "Electrode for Sensing Fluoride Ion Activity in Solution," *Science,* **154,** 1553–1555 (1966).

284. J. Tusl, "Direct Determination of Fluoride in Human Urine Using Fluoride Electrode," *Clin. Chim. Acta.,* **27,** 216–218 (1970).

285. J. A. Zapp, Jr., "Polyfluorines," *Encyclopaedia of Occupational Health and Safety,* Vol. II, McGraw-Hill, New York, 1972, pp. 1095–1096.

286. J. R. Mehall and K. L. Brenis, "Polyvinyl and Vinyl Copolymers," in *Modern Plastics Encyclopedia,* 1977–78 ed., Vol. 54, No. 10A, McGraw-Hill, New York, pp. 98, 99, 102.

287. M. J. R. Cantow, "Vinyl Polymers (Chloride)," in *Encyclopedia of Chemical Technology,* Vol. 21, Wiley, 1970, pp. 369–412.

288. W. D. Davis, "Rubber-Related Polymers II. Poly(Vinyl Chloride)," in M. Morton, Ed., *Rubber Technology,* 2nd ed., Van Nostrand Reinhold, New York, 1973, pp. 534–554.

289. G. F. Cohan, "Industrial Preparation of Poly(Vinyl Chloride)," *Environ. Health Perspect.,* **11,** 53–57 (1975).

290. Z. S. Kahn and T. W. Hughes, "Source Assessment: Polyvinyl Chloride," *Monsanto Research Corporation,* Dayton, Ohio, 1978 (PB 283395).

291. D. V. Porchey and H. J. Hall, "Polyvinyl and Vinyl Copolymers," in *Modern Plastics Encyclopedia,* 1978–79 ed., Vol. 55, No. 10A, McGraw-Hill, New York, pp. 97–102.

292. International Agency for Research on Cancer, *Environmental Carcinogens; Selected Methods of Analysis,* Vol. 2 (Methods for the Measurement of Vinyl Chloride in Poly(Vinyl Chloride), Air, Water and Foodstuffs), IARC Scientific Publications No. 22, Lyon, France, 1978.

293. J. L. Dennison, C. V. Breder, T. McNeal, R. C. Snyder, J. A. Roach, and J. A. Sphon, "Industrial Chemicals," *J. Assoc. Off. Anal. Chem.*, **61**(4), 813–819 (1978).

294. Scientific and Technical Assessment Report on Vinyl Chloride and Polyvinyl Chloride, U.S. Environmental Protection Agency, June 1975 (EPA-600/6-75-004).

295. W. S. Johnson and R. E. Schmidt, "Effects of Polyvinyl Chloride Ingestion by Dogs," *Am. J. Vet. Res.*, **38**, 1891–1892 (1977).

296. H. F. Smyth, Jr. and C. S. Weil, "Chronic Oral Toxicity to Rats of a Vinyl Chloride-Vinyl Acetate Copolymer," *Toxicol. Appl. Pharmacol.*, **9**, 501–504 (1966); study completed in 1947.

297. E. V. Ivanova and M. P. Shamina, "Toxicological Evaluation of an Aqueous Dispersion of a Vinyl Chloride–Vinyl Acetate Copolymer," *Aktual. Vopr. Lab. Prakt., Mater. Obl. Nauchno-Prakt. Konf. Vrachei-Laborantov, Biokhim., Bakteriol. Khim. Voronezh, Obl.*, **3**, 221–222 (1973); *Chem. Abstr.*, **83**, 109400v.

298. G. Volkheimer, "Hematogenous Dissemination of Ingested Polyvinyl Chloride Particles," *Ann. NY Acad. Sci.*, **246**, 164–171 (1975).

299. E. P. Zeisler, "Dermatitis from Elasti-Glass Garters and Wrist-Watch Straps," *J. Am. Med. Assoc.*, **114**, 2540–2542 (1940).

300. R. J. Richards, R. Desai, P. M. Hext, and F. A. Rose, "Biological Reactivity of PVC Dust," *Nature*, **256**, 664–665 (1975).

301. J. A. Styles and J. Wilson, "Comparison Between *in vitro* Toxicity of Polymer and Mineral Dusts and Their Fibrogenicity," *Ann. Occup. Hyg.*, **16**, 241–250 (1973).

302. M. M. O'Mara, L. B. Crider, and R. L. Daniel, "Combustion Products from Vinyl Chloride Monomer," *Am. Ind. Hyg. Assoc. J.*, **32**, 153–156 (1971).

303. C. L. Casterline, P. F. Casterline, and D. A. Jaques, "Squamous Cell Carcinoma of the Buccal Mucosa Associated with Chronic Oral Polyvinyl Chloride Exposure Report of a Case," *Cancer*, **39**, 1686–1688 (1977).

304. J. L. Osmon, R. E. Klausmeier, E. I. Jamison, "Rate-Limiting Factors in Biodeterioration of Plastics," *Biodeterior. Mater., Proc. Int. Biodeterior. Symp., 2nd*, 66–75 (1971).

305. A. N. Sharpe and M. N. Woodrow, "A Rapid Test for Biodegradability by Pseudomonas Organisms," *Biodeterior. Mater., Proc. Int. Biodeterior. Symp., 2nd*, 233–237 (1972).

306. R. A. Wessling, *Polyvinylidene Chloride*, Gordon and Breach Science Publishers, Midland, Mich., 1977.

307. T. J. Birkel, J. A. G. Roach, and J. A. Sphon, "Determination of Vinylidene Chloride in Saran Films by Electron Capture Gas-Solid Chromatography and Confirmation by Mass Spectrometry," *J. Assoc. Off. Anal. Chem.*, **60**, 1210–1213 (1977).

308. M. Ohta, S. Motegi, K. Ueda, and H. Tanaka, "Elution Study of Residual Vinylidene Chloride Monomer into Food Packaged in Polyvinylidene Chloride Film," *Bull. Jap. Soc. Sci. Fish.*, **43**(11), 1341–1350 (1977).

309. (a) U.S. National Toxicology Program, *Carcinogenesis Bioassay of Vinylidene Chloride*, draft report, January 1981, NTP-81-82 (DHHS 81-1784), subsequently reviewed February 1981; (b) M. McKenna, personal communication, Dow Chemical Company, Midland, Mich., December 1981 (tests conducted at Dow Chemical Company under the auspices of the Chemical Manufacturers Association).

310. P. J. Vanderhorst, personal communication, E. I. du Pont de Nemours and Company, Inc., Plastics Products and Resins Dept., Wilmington, Del., May 1979.

311. L. M. Schlanger and W. E. Titterton, "Fluoroplastics," in *Modern Plastics Encyclopedia*, 1978–79 ed., Vol. 55, No. 10A, McGraw-Hill, New York, pp. 15, 16, 25.

312. R. W. Morrow, unpublished data, E. I. du Pont de Nemours and Company, Inc., Haskell Laboratory, Wilmington, Del., October 1972.

313. C. H. Hine, unpublished data, Hine Laboratories, Inc., Research and Development, San Francisco, Calif., December 1972.

314. J. W. Clayton, Jr., "Fluorocarbon Toxicity and Biological Action," *Fluorine Chem. Rev.*, **1,** 197–252 (1967).

315. S. B. Fretz and H. Sherman, unpublished data, E. I. du Pont de Nemours and Company, Inc., Haskell Laboratory, Wilmington, Del., October 1968.

316. S. B. Fretz and H. Sherman, unpublished data, E. I. du Pont de Nemours and Company, Inc., Haskell Laboratory, Wilmington, Del., September 1969.

317. R. S. Khamidullin and G. A. Petrova, "Hygienic Evaluation of Fluoroplast-4 Films," *Gig. Vop. Proizvod. Primen. Polim. Mater.*, 242–248 (1969); *Chem. Abstr.*, **75,** 107796q.

318. R. S. Khamidullin and G. A. Petrova, "Experimental Hygienic Assessment of Teflon Film," *Gig. Sanit.*, **34,** 247–248 (1969).

319. R. S. Khamidullin, I. A. Petrova, G. Ya. Karaulova, and A. D. Promyslova, "Hygienic Assessment of a Fluoroplast 4 Film in Connection with its Use in the Food Industry," *Gig. Sanit.*, **1,** 38–41 (1977).

320. R. Preiss, "Properties and Toxicology of Polytetrafluoroethylene and Its Uses in Industry and Medicine," *Pharmazie*, **28,** 281–284 (1973).

321. R. A. Elson and J. Charnley, "The Direction and Resultant Force in Total Prosthetic Replacement of the Hip Joint," *Med. Biol. Eng.*, **6,** 19–27 (1968).

322. R. L. Johnson, A. B. Robertson, and E. C. Lupton, Jr., "Fluorinated Plastics," in *Encyclopedia of Polymer Science and Technology*, 2nd ed., Suppl. 1, Wiley, New York, 1976, pp. 260–267, 286–287.

323. M. K. Lindemann, "Physical Constants of Poly(Vinyl Acetate)," in J. Brandrup and E. H. Immergut, Eds., *Polymer Handbook*, 2nd ed., Wiley, New York, 1975, pp. V-51 to V-53.

324. R. L. Davidson, Ed., *Handbook of Water-Soluble Gums and Resins*, McGraw-Hill, New York, 1980.

325. M. H. Weeks and C. R. Pope, *Toxicological Evaluation of Polyvinyl Acetate (PVA) Emulsion Dust Control Material*, U.S. Army, Environmental Hygiene Agency, Aberdeen, Md., May 1973–March 1974 (AD 784603).

326. G. N. Zaeva, M. D. Babina, V. I. Fedorova, and V. A. Shchirskaya, "Toxicological Characteristics of Poly(Vinyl Alcohol), Polyethylene, and Polypropylene," *Toksikol. Novykh Prom. Khim. Veshchestv.*, **5,** 136–149 (1963); *Chem. Abstr.*, **61,** 6250h.

327. B. I. Shcherbak, L. Y. Broitman, T. B. Yatsenko, and S. I. Kolesnikov, "Toxicological Characteristics of Some Poly(Vinyl Acetate) Dispersions (PVAD)," *Uch. Zap. Mosk. Nauchno-Issled. Inst. Gig.*, **22,** 74–80 (1975); *Chem. Abstr.*, **86,** 66421v (also 66422w, 66423x).

328. D. B. Hood, unpublished data, E. I. du Pont de Nemours and Company, Inc., Haskell Laboratory, Wilmington, Del., January 1962.

329. V. A. Goreva, "Study of the Toxic Effect of Poly(Vinyl Acetate) Emulsion on Fish," *Tr. Sarat. Otd. GosNIORKh*, **13,** 89–92 (1975); *Chem. Abstr.*, **87,** 96940b.

330. G. V. Gurova, "Effect of Poly(Vinyl Acetate) Emulsion on Phytophilic Fish in Early Ontogenesis," *Tr. Sarat. Otd. GosNIORKh*, **13,** 92–95 (1975); *Chem. Abstr.*, **87,** 96941c.

331. A. E. Shapshal, N. I. Berlyakova, "Effect of Poly(Vinyl Acetate) Emulsion on Chironomus Dorsalis Meig. and *Daphnia magna* Straus," *Tr. Sarat. Otd. GosNIORKh*, **13,** 85–89 (1975); *Chem. Abstr.*, **87,** 96939h.

332. G. J. L. Griffin and H. Mivetchi, "Biodegradation of Ethylene/Vinyl Acetate Co-Polymers," in *Proc. Int. Biodegradation Symp.*, **3,** 807–817 (1975).

333. W. C. Hueper, "Organic Lesions Produced by Polyvinyl Alcohol in Rats and Rabbits," *Arch. Pathol.*, **28,** 510–531 (1939).

334. Y. Yamatani, S. Ishikawa, "Polyvinyl Alcohol as a Water-Soluble Marker. Part I. Absorption and Excretion of Polyvinyl Alcohol from the Gastro-Intestinal Tract of Adult Rat," *Agric. Biol. Chem.*, **32**(4), 474–478 (1968).

335. W. M. Grant, "Polyvinyl Alcohol," *Toxicology of the Eye*, 2nd ed., Charles C Thomas, Springfield, Ill., 1974, pp. 849–851.

336. N. Krishna and B. Mitchell, "Polyvinyl Alcohol as an Ophthalmic Vehicle," *Am. J. Ophthalmol.*, **59**, 860–864 (1965).

337. D. O. Shah and M. J. Sibley, "Treatment of Dry Eye," U.S. Patent 4,131,651 (Cl. 424-78; A 61K31/74), December 26, 1978, Appl. 844,555, October 25, 1977; *Chem. Abstr.*, **90**, 110015z.

338. V. I. Trautmann, "Poly(Vinyl Alcohol) as Vehicle for Eye Drops," *Dtsch. Gesundheitsw.*, **22**, 317–320 (1967).

339. C. E. Hall and O. Hall, "Polyvinyl Alcohol Nephrosis: Relationship of Degree of Polymerization to Pathophysiologic Effects," *Proc. Soc. Exp. Biol. Med.*, **112**, 86–91 (1963).

340. C. E. Hall and O. Hall, "Eclampsialike Syndrome in Rats Treated with Poly(Vinyl Alcohol) (PVA)," *Texas Dept. Biol. Med.*, **21**, 16–27 (1963).

341. C. E. Hall and O. Hall, "Polyvinyl Alcohol: Relationship of Physicochemical Properties to Hypertension and Other Pathophysiologic Sequelae," *Lab. Invest.*, **12**, 721–736 (1963).

342. M. E. Sippel, unpublished data, E. I. du Pont de Nemours and Company, Inc., Haskell Laboratory, Wilmington, Del., November 1977.

343. B. H. Sleight, unpublished data, Bionomics, Inc. Wareham, Mass., September 1971.

344. Q. D. Wheatley and F. C. Baines, "Biodegradation of Polyvinyl Alcohol in Wastewater," *Text. Chem. Colorist*, **8**, 23–28 (1976).

345. W. H. Hahn, E. L. Barnhart, and R. B. Meighan, "The Biodegradability of Synthetic Size Material Used in Textile Processing," *Proc. Ind. Waste Conf.*, **30**, 530–539 (1977).

346. J. P. Casey and D. G. Manly, "Polyvinyl Alcohol Biodegradation by Oxygenactivated Sludge," *Proc. Int. Biodegradation Symp. 3rd*, **33**, 819 (1976).

347. J. Suzuki, K. Hukushima, and S. Suzuki, "Effect of Ozone Treatment Upon Biodegradability of Water-Soluble Polymers," *Environ. Sci. Technol.*, **12**, 1180–1183 (1978).

348. H. Sherman and J. K. Wier, unpublished data, E. I. du Pont de Nemours and Company, Inc., Haskell Laboratory, Wilmington, Del., August 1961.

349. K. P. Lee, unpublished data, E. I. du Pont de Nemours and Company, Inc., Haskell Laboratory, Wilmington, Del., March 1971.

350. D. B. Bivens and R. B. Rector, "Acrylics," in *Modern Plastics Encyclopedia*, 1978–79 ed., Vol. 55, No. 10A, McGraw-Hill, New York, pp. 7–9.

351. J. F. Rudd, "Physical Constants of Poly(Styrene)," in J. Brandrup and E. H. Immergut, Eds., *Polymer Handbook*, 2nd ed., Wiley, New York, 1975, V-59 to V-62.

352. L. H. Peebles, Jr., "Acrylonitrile Polymers, Degradation," *Encyclopedia of Polymer Science and Technology*, 2nd ed., Suppl. 1, Wiley, New York, 1976, pp. 1–25.

353. R. F. Boyer, Ed., "Styrene Polymers," in *Encyclopedia of Polymer Science and Technology*, 2nd ed., Suppl. 2, Wiley, New York, 1977, pp. 128–447.

354. J. Banbaji, "Tensile Properties of Relaxed and Unrelaxed Pan Fibers," *J. Appl. Polym. Sci.*, **31**, 117–125 (1977).

355. W. Fester, "Physical Constants of Poly(Acrylonitrile)," in J. Brandrup and E. H. Immergut, Eds., *Polymer Handbook*, 2nd ed., Wiley, New York, 1975, V-37 to V-40.

356. W. Wunderlich, "Physical Constants of Poly(Methyl Methacrylate)," in J. Brandrup and E. H. Immergut, Eds., *Polymer Handbook*, 2nd ed., Wiley, New York, 1975, V-55 to V-57.

357. B. B. Kine and R. W. Novak, "Acrylic Ester Polymers," *Encyclopedia of Chemical Technology*, 3rd ed., Vol. 1, Wiley, New York, 1978, pp. 386–408.

358. P. H. Hobson and A. L. McPeters, "Acrylic and Modacrylic Fibers," *Encyclopedia of Chemical Technology*, 3rd ed., Vol. 1, Wiley, New York, 1978, pp. 355–386.

359. F. M. Peng, "Acrylonitrile Polymers (Survey and SAN)," *Encyclopedia of Chemical Technology*, 3rd ed., Vol. 1, Wiley, New York, 1978, pp. 427–442.

360. G. A. Morneau, W. A. Pavelich, and L. G. Roettger, "ABS Resins," *Encyclopedia of Chemical Technology*, 3rd ed., Vol. 1, Wiley, New York, 1978, pp. 442–456.

361. S. E. Ball, "ABS," in *Modern Plastics Encyclopedia*, 1978–79 ed., Vol. 55, No. 10A, McGraw-Hill, New York, pp. 4–5.

362. D. Tumminia, "Styrene–Acrylonitrile," in *Modern Plastics Encyclopedia*, 1978–79 ed., Vol. 55, No. 10A, McGraw-Hill, New York, pp. 108–110.

363. J. D. Morris and R. J. Penzenstadler, "Acrylamide Polymers," in *Encyclopedia of Chemical Technology*, 3rd ed., Vol. 1, Wiley, New York, 1978, pp. 312–330.

364. M. L. Miller, "Acrylic Acid Polymers," in *Encyclopedia of Polymer Science and Technology*, 3rd ed., Vol. 1, Wiley, New York, 1964, pp. 197–226.

365. S. R. Betso, "Polystyrene," in *Modern Plastics Encyclopedia*, 1978–79 ed., Vol. 55, No. 10A, McGraw-Hill, New York, pp. 78–88.

366. Dow Chemical Company, *Product Stewardship—Styrene Resins*, Midland, Mich., 1976.

367. Dow Chemical Company, *Styron Polystyrene Resins*, Midland, Mich., 1977.

368. M. Bollini, A. Seves, and B. Focher, "Determination of Free Monomers in Water Emulsions of Synthetic Polymers or Copolymers," *Ind. Carta*, **12**(7), 234–240, 1974; *Chem. Abstr.*, **81**, 121672b.

369. D. G. Anderson and J. T. Vandeberg, "Coatings," *Anal. Chem.*, **51**, 80R–90R (1979).

370. J. F. McCabe and R. M. Basker, "Tissue Sensitivity to Acrylic Resin," *Br. Dent. J.*, **140**, 347–350 (1976).

371. M. A. Galin, L. Turkish, and E. Chowchuvech, "Detection, Removal, and Effect of Unpolymerized Methylmethacrylate in Intraocular Lenses," *Am. J. Ophthalmol.*, **84**, 153–159 (1977). See also L. Turkish and M. A. Galin, *Arch. Ophthalmol.*, **98**, 120–121 (1980).

372. J. C. Hughes, B. B. Wheals, and M. J. Whitehouse, "Pyrolysis—Mass Spectrometry of Textile Fibres," *Analyst*, **103**, 482–491 (1978).

373. M. R. Martinelli, S. W. Mayer, and P. F. Jones, "Thermomechanical Examination of Fabric Composed of Synthetic Polymers," *J. Forensic Sci.*, **24**, 130–139 (1979).

374. "Monsanto Loses Plastic Bottle Fight," *Chem. Eng. News*, **55**(39), 6 (September 26, 1977).

375. "Risk Assessment Clearance of Acrylonitrile Possible, FDA Says," *Food Chem. News*, **23**(15), 3–6 (June 22, 1981).

376. U.S. Dept. of Labor, "Emergency Temporary Standard for Occupational Exposure to Acrylonitrile (Vinyl Cyanide); Notice of Hearing," *Fed. Reg.*, **43**, 2586–2621 (1978); for permanent standard, see 29 CFR 1910–1045 and *Fed. Reg.* update.

377. J. R. Withey and P. G. Collins, "Styrene Monomer in Foods; A Limited Canadian Survey," *Bull. Environ. Contam. Toxicol.*, **19**, 86–95 (1978).

378. H. Momotani, "The Respiratory Deposition of Polystyrene Aerosols in Man," *Jap. J. Hyg.*, **21**, 417–423 (1967); author's English abstract.

379. H. T. Hofmann and H. Oettel, "Comparative Toxicity of Thermal Decomposition Products," *Mod. Plast.*, **46**(10), 94, 98–100 (1969).

380. H. T. Hofmann and H. Oettel, "Relative Toxicity of Thermal Decomposition Products of Expanded Polystyrene," *J. Combust. Toxicol.*, **1**, 236–249 (1974).

381. C. J. Hilado, E. M. Olcomendy, and D. P. Brauer, "Effect of Pyrolysis Temperature and Air Flow on Toxicity of Gases from a Polystyrene Polymer," *J. Combust. Toxicol.*, **6**, 13–19 (1979).

382. W. J. Potts and T. S. Lederer, "A Method for Comparative Testing of Smoke Toxicity," *J. Combust. Toxicol.*, **4**, 114–162 (1977).

383. E. Effenberger, "Toxische Wirkungen der Verbrennungsprodukte von Kunststoffen," *Städtehygiene*, **23**(12), 275–280 (1972).

384. A. Zitting, P. Pfaffli, and H. Vainio, "Effects of Thermal Degradation Products of Polystyrene on Drug Biotransformation and Tissue Glutathione in Rat and Mouse," *Scand. J. Work Environ. Health*, **4**, 60–66 (1978).

385. F. A. Lehmann and G. M. Brauer, "Analysis of Pyrolyzates of Polystyrene and Poly(Methyl Methacrylate) by Gas Chromatography," *Anal. Chem.*, **33**, 673–676 (1961).

386. F. Valic and E. Zuskin, "Respiratory-Function Changes in Textile Workers Exposed to Synthetic Fibers," *Arch. Environ. Health*, **32**, 283–287 (1977).

387. G. V. Lomonova, "Toxic Properties of Acrylonitrile Polymer," *Gig. Tr. Prof. Zabol.*, **6**(6), 54–57 (1962).

388. K. Yamamoto and Y. Yamamoto, "On the Acute Toxicities of the Combustion Products of Various Fibers, with Special Reference to Blood Cyanide and PO_2 Values," *Z. Rechtsmed.*, **81**(3), 173–179 (1978).

389. K. Yamamoto, "Acute Combined Effects of Hydrocyanic Acid and Carbon Monoxide, with the Use of the Combustion Products from PAN (Polyacrylonitrile)—Gauze Mixtures," *Z. Rechtsmed.*, **78**(4), 303–311 (1976).

390. P. L. Wright and C. H. Adams, "Toxicity of Combustion Products from Burning Polymers: Development and Evaluation of Methods," *Environ. Health Perspect.*, **17**, 75–83 (1976).

391. Y. Tsuchiya, "Significance of Hydrogen Cyanide Generation in Fire Gas Toxicity," *J. Combust. Toxicol.*, **4**, 271–282 (1977).

392. T. Morikawa, "Evolution of Hydrogen Cyanide During Combustion and Pyrolysis," *J. Combust. Toxicol.*, **5**, 315–330 (1978).

393. M. Chaigneau and G. Le Moan, "Pyrolysis of Plastic Materials, VIII. Polyacrylonitrile and Copolymers," *Ann. Pharm. Fr.*, **32**, 485–490 (1974).

394. M. Chaigneau, "Mass Spectrometric Analysis of Compounds Formed by the Pyrolysis of Polyacrylonitrile and Copolymers," *Analasis*, **5**(5), 223–227 (1977).

395. Y. Tsuchiya and K. Sumi, "Thermal Decomposition Products of Polyacrylonitrile," *J. Appl. Polym. Sci.*, **21**, 975–980 (1977).

396. K. Sumi and Y. Tsuchiya, "Combustion Products of Polymeric Materials Containing Nitrogen in Their Chemical Structure," *J. Fire Flammability*, **4**, 15–22 (1973).

397. R. L. Schumacher and P. A. Breysse, "Combustion and Pyrolysis Products from Synthetic Textiles," *J. Combust. Toxicol.*, **3**, 393–424 (1976).

398. A. R. Monahan, "Thermal Degradation of Polyacrylonitrile in the Temperature Range 280–450°C," *J. Polym. Sci.*, **4**, 2391–2399 (1966).

399. D. Braun and R. Disselhoff, "Thermischer Abbau von Polyacrylnitril unter Luft," *Angew. Makromol. Chem.*, **74**, 225–248 (1978).

400. H. H. G. Jellinek and A. Das, "Hydrogen Cyanide Evolution During Thermal-Oxidative Degradation of Nylon 66 and Polyacrylonitrile," *J. Polym. Sci.*, **16**, 2715–2719 (1978).

401. R. E. Dufour, "Report on Textile Fiber, Miscellaneous Hazard 5148," Underwriters' Laboratories, Inc., Chicago, Ill., October 1950.

402. H. Cornish, "Toxicity of Thermal-Degradation Products of Plastics," in Armstrong Cork Company (Research and Development Center), *Symposium: Products of Combustion of (Plastics) Building Materials*, Lancaster, Pa., March 1973, pp. 30–33.

403. E. E. Smith, "Measuring Rate of Heat, Smoke and Toxic Gas Release," *Fire Technol.*, **8**, 237–245 (1972).

404. K. Thinius, E. Schroeder, and A. Gustke, "Analytical Chemistry of Plastics. XXI. Pyrolysis of Plastics in the Presence of Air," *Plaste Kautsch.*, **11**, 67–72 (1964); *Chem. Abstr.*, **62**, 9297c.

405. W. H. Martin, unpublished data, E. I. du Pont de Nemours and Company, Inc., Polymer Products Dept., Wilmington, Del., September 1980.

406. C. D. Calnan, "Cyanoacrylate Dermatitis," *Contact Dermatitis*, **5**, 165–167 (1979).

407. W. A. Skornik, R. S. Robinson, and D. P. Dressler, "Toxicity of Thermal Decomposition Products of Various Paints," *J. Combust. Toxicol.*, **3**, 71–85 (1976).

408. P. S. Spencer and H. H. Schaumburg, "Nervous System Degeneration Produced by Acrylamide Monomer," *Environ. Health Perspect.*, **11**, 129–133 (1975).

409. U.S. Environmental Protection Agency, Support Document Decision Not to Require Testing for Health Effects: Acrylamide, EPA-560/11-80-016, Washington, D.C., June 1980.

410. J. R. Withey, "Quantitative Analysis of Styrene Monomer in Polystyrene and Foods Including Some Preliminary Studies of the Uptake and Pharmacodynamics of the Monomer in Rats," *Environ. Health Perspect.*, **17**, 125–133 (1976).

411. I. Phillips and G. C. Marks, as cited in R. Lefaux, Ed., *Practical Toxicology of Plastics*, CRC Press, Cleveland, Ohio, 1968.

412. "Les Emballages en Polystyrène et la Santé du Consommateur," *Caoutch. Plast.*, **562** (July/August 1976).

413. R. S. Khamidullin, N. G. Fel'dman, and M. V. Vendilo, "Hygienic Assessment of Articles Made of PS-S Suspension Polystyrene," *Gig. Sanit.*, **33**, 331–334 (1968); in translation, *Hygiene and Sanitation*.

414. (a) J. H. Kennedy, D. W. Bailey, C. D. Armeniades, and B. S. Ruark, "Polymer Tumorigenesis; Relation of Polystyrene Conformational Change to Oncogenetic Activity," *Proceedings of the 26th Annual Conference on Engineering in Medicine and Biology,* Minneapolis, Minn., 1973; (b) U.S. National Cancer Institute, *Bioassay of Styrene for Possible Carcinogenicity,* NCI-CG-TR-185 (DHEW 79-1741), Washington D.C., 1979. See also NCI-CG-TR-170.

415. M. Sielicki, "Microbial Degradation of Styrene and Styrene Polymers," *Diss. Abstr. Int. B.*, **38**, 4076-B (1978).

416. (a) P. H. Jones and D. Prasad, M. Heskins, M. H. Morgan, and J. E. Guillet, "Biodegradability of Photodegraded Polymers," *Environ. Sci. Technol.*, **8**, 919–923 (1974); (b) J. E. Guillet, T. W. Regulski, and T. B. McAneney, "Biodegradability of Photodegraded Polymers," *Environ. Sci. Technol.*, **8**, 923–925 (1974).

417. D. L. Kaplan, R. Hartenstein, and J. Sutter, "Biodegradation of Polystyrene, Poly(methyl methacrylate), and Phenol Formaldehyde," *Appl. Environ. Microbiol.*, **38**, 551–553 (1979).

418. E. M. Hicks, R. A. Craig, E. L. Wittbecker, J. G. Lavin, N. A. Ednie, D. E. Howe, E. D. Williams, R. E. Seaman, N. C. Pierce, A. J. Ultee, and M. Couper, "The Production of Synthetic-Polymer Fibres," *Text. Progr.*, **3**, 1–113 (1971).

419. A. J. Fleming, "The Provocative Test for Assaying the Dermatitis Hazards of Dyes and Finishes Used on Nylon," *J. Invest. Dermatol.*, **10**, 281–291 (1948).

420. D. B. Hood and A. Ivanova-Binova, "The Skin as a Portal of Entry and the Effects of Chemicals on the Skin, Mucous Membranes and Eye," in *Principles and Methods for Evaluating the Toxicity of Chemicals. Part II,* Environmental Health Criteria Series 6, World Health Organization, Geneva, in press.

421. L. A. Wells, unpublished data, E. I. du Pont de Nemours and Company, Inc., Haskell Laboratory, Wilmington, Del., March 1968.

422. O. L. Dashiell, unpublished data, E. I. du Pont de Nemours and Company, Inc., Haskell Laboratory, Wilmington, Del., June 1974.

423. R. J. Neher, unpublished data, E. I. du Pont de Nemours and Company, Inc., Haskell Laboratory, Wilmington, Del., June 1965.

424. R. J. Neher, unpublished data, E. I. du Pont de Nemours and Company, Inc., Haskell Laboratory, Wilmington, Del., July 1964.

425. O. L. Dashiell, unpublished data, E. I. du Pont de Nemours and Company, Inc., Haskell Laboratory, Wilmington, Del., August 1974.

426. W. F. Schorr, E. Keran, and E. Poltka, "Formaldehyde Allergy," *Arch. Dermatol.*, **110**, 73–76 (1974).

427. J. C. Pimentel, R. Avila, and A. G. Lourenco, "Respiratory Disease Caused by Synthetic Fibres: A New Occupational Disease," *Thorax*, **30**, 204–219 (1975).

428. A. Muittari and T. Veneskoski, "Natural and Synthetic Fibers as Causes of Asthma and Rhinitis," *Ann. Allergy*, **41**, 48–50 (1978).

429. A. Bouhuys, "Fibers and Fibrosis," *Ann. Int. Med.*, **83**, 898–899 (1975).

430. J. C. Pimentel, "Sarcoid Granulomas of the Skin Produced by Acrylic and Nylon Fibres," *Br. J. Dermatol.*, **96**, 673–677 (1977).

431. A. Koops, unpublished data, E. I. du Pont de Nemours and Company, Inc., Haskell Laboratory, Wilmington, Del., April 1976.

432. S. A. Labofina, "Acrylonitrile–Styrene Copolymers," Belg. Patent 821,435 (Cl. Co8f), February 17, 1975, Appl. October 24, 1975; *Chem. Abstr.*, **83**, 180293t.

433. I. A. Benjamin and W. J. Parker, "Fire Spread Potential of ABS Plastic Plumbing," *Fire Technol.*, **8**, 104–119 (1972).

434. L. D. Moore, W. W. Moyer, and W. J. Frazer, "Molecular Structure Analysis of Graft Polymers," *Appl. Polym. Symp.*, **7**, 67–80 (1968).

435. J. Chiu and E. F. Palermo, "Polymer Characterization by Thermal Evolution Techniques," *Anal. Chim. Acta.*, **81**, 1–19 (1976).

436. H. Sherman, unpublished data, E. I. du Pont de Nemours and Company, Inc., Haskell Laboratory, Wilmington, Del., March 1973.

437. A. Casale, O. Salvatore, and G. Pizzigoni, "Measurement of Aging Effects of ABS Polymers," *Polym. Eng. Sci.*, **4**, 286–293 (1975).

438. E. Priebe and J. Stabenaw, "Effects in Aging of Impact-Resistant Modified Styrene–Acrylonitrile Polymers. 2. Relation Between Surface Damage and Impact Strength of ABS," *Kunststoffe*, **64**, 497–502 (1974); *Chem. Abstr.*, **82**, 44204s.

439. A. Davis and D. Gordon, "Rapid Assessment of Weathering Stability from Exposure of Polymer Films; II. The Effectiveness of Different Regions of the Solor Spectrum in Degrading an ABS Terpolymer," *J. Appl. Polym. Sci.*, **18**, 1173–1179 (1974).

440. E. I. du Pont de Nemours and Company, Inc., *Lucite*® *SAR*, Plastic Products and Resins Department, Wilmington, Del., n.d.

441. M. Bollini, A. Seves, and B. Focher, "Determination of Free Monomers in Aqueous Emulsions of Synthetic Polymers and Copolymers," *Textilia*, **51**(3), 25–28, 1975; *Chem. Abstr.*, **83**, 60039t.

442. N. A. Zaytseva, S. A. Tolstobrova, and D. T. Il'in, "Chromatographic Determination of Unreacted Monomers in Emulsions of Acrylic Copolymers, *Khim. Prom.*, **48**, 351–352 (1972); as translated in *Soc. Chem. Ind.*, **48**, 298–299 (1972).

443. E. O. Dillingham, N. Webb, W. H. Lawrence, and J. Autian, "Biological Evaluation of Polymers I. Poly(Methyl Methacrylate)," *J. Biomed. Mater. Res.*, **9**, 569–596 (1975).

444. H. G. Willert, H. A. Frech, and A. Bechtel, "Measurements of the Quantity of Monomer Leaching out of Acrylic Bone Cement into the Surrounding Tissues During the Process of Polymerization," *Am. Chem. Soc., Div. Org. Coat. Plast. Chem.,* **33,** 370–370g (1973).

445. (a) M. A. Guill and R. B. Odom, "Hearing Aid Dermatitis," *Arch. Dermatol.,* **114,** 1050–1051 (1978); (b) A. I. Fernström and G. Oquist, "Location of the Allergenic Monomer in Warm-Polymerized Acrylic Dentures," *Swed. Dent. J.,* **4,** 253–260 (Part II; see also Part I, pp. 241–252); (c) R. J. G. Rycroft, "Contact Dermatitis from Acrylic Compounds," *Br. J. Dermatol.,* **96,** 685–687 (1977); (d) S. Kaaber, H. Thulin, and E. Nielsen, "Skin Sensitivity to Denture Base Materials in the Burning Mouth Syndrome," *Contact Dermatitis,* **5,** 90–96 (1979).

446. N. E. Stinson, "Tissue Reaction Induced in Guinea-Pigs by Particulate Polymethylmethacrylate, Polythene and Nylon of the Same Size Range," *Br. J. Exp. Path.,* **46,** 135–146 (1965).

447. M. Barvic, "Reaction of the Body to the Presence of Acrylic Allografts and Possible Carcinogenic Effects of Such Grafts," *Acta. Univ. Carol. Med.,* **8,** 707–753 (1962).

448. J. R. Thompson and S. D. Entin, "Primary Extraskeletal Chondrosarcoma," *Cancer,* **23,** 936–939 (1969).

449. L. Tomatis, "Subcutaneous Carcinogenesis by [14]C- and ^{3}H-Labeled Poly(Methyl Methacrylate) Films," *Tumori,* **52,** 165–172 (1966).

450. "Acrylic Cement and the Cardiovascular System," *Lancet,* **1974-II,** 1002–1004.

451. J. Lipecz, C. Nemes, F. Baumann, and V. Csernohorszky, "Kreislaufkomplikationen bei Alloarthroplastiken des Hüftgelenkes," *Anaesthesist,* **23,** 382–388 (1974).

452. R. H. Ellis, "Hypotension and Methylmethacrylate Cement," *Br. Med. J.,* **1,** 236 (1973).

453. F. T. Schuh, S. M. Schuh, M. G. Viguera, and R. N. Terry, "Circulatory Changes Following Implantation of Methyl-methacrylate Bone Cement," *Anesthesiology,* **39,** 455–457 (1973).

454. G. Schlag, H.-J. Schliep, E. Dingeldein, A. Grieben, and W. Ringsdorf, "Does Methylmethacrylate Induce Cardiovascular Complications during Alloarthroplastic Surgery of the Hip Joint?" *Anaesthesist,* **25,** 60–67 (1976); authors' English abstract.

455. J. Kraft, "Polymethylmethacrylate—A Review," *J. Foot Surg.,* **16,** 66–68 (1977).

456. T. Awao, K. Komagate, I. Yoshimura, and K. Mitsugi, "Deterioration of Synthetic Resins by Fungi," *J. Ferment. Technol.,* **49,** 188–194 (1971).

457. M. Abouelezz and P. F. Waters, "Studies on Photodegradation of Poly(Methyl Methacrylate)," *National Bureau of Standards,* Department of Commerce, Washington, D.C., 1978 (PB 281828).

458. D. D. McCollister, C. L. Hake, S. E. Sadek, and V. K. Rowe, "Toxicologic Investigations of Polyacrylamides," *Toxicol. Appl. Pharmacol.,* **7,** 639–651 (1965).

459. Y. Matsumura and H. Arito, "Toxic Volatile Components of Organic Soil Consolidating Agents," *Ind. Health,* **13,** 135–149 (1975).

460. E. R. Husser, R. H. Stehl, D. R. Price, and R. A. DeLap, "Liquid Chromatographic Determination of Residual Acrylamide Monomer in Aqueous and Nonaqueous Dispersed Phase Polymeric Systems," *Anal. Chem.,* **49,** 154–157 (1977).

461. F. J. Ludwig and M. F. Besand, "High Performance Liquid Chromatographic Determination of Unreacted Acrylamide in Emulsion or Aqueous Homopolymers of Emulsion Copolymers," *Anal. Chem.,* **50,** 185–187 (1978).

462. G. Schmotzer, "Determination of Acrylamide and Acrylic Acids in Acrylamide Polymers," *Chromatographia,* **4,** 391–395 (1971); *Chem. Abstr.,* **76,** 25744c.

463. J. Brunn, K. Doerffel, H. Much, and G. Zimmerman, "Ultraviolet Photometric Determination of Residual Monomer Content in Technical Polymers of Acrylic Acids and Acrylic Acid Ethyl Esters," *Plaste Kautsch.,* **22,** 485–486 (1975); *Chem. Abstr.,* **83,** 115163m.

464. American Cyanamid Company, "Cyanamer P 26 Acrylamide Copolymer; Cyanamer P 250

Polyacrylamide," *American Cyanamid Brochure,* Process Chemicals Department, Wayne, N.J., July 1957.

465. N. A. Rakhmanina, "Toxicological Characteristics of Technical Grade Polyacrylamide," *Nauchn. Tr. Akad. Kommun. Khoz.,* **22,** 56–59 (1963); *Chem. Abstr.,* **61,** 15249d.

466. E. E. Christofano, J. P. Frawley, O. E. Fancher, and M. L. Keplinger, "The Toxicology of Modified Polyacrylamide Resin," *Toxicol. Appl. Pharmacol.,* **14,** 616 (1969).

467. Y. Matsuo, "Bioassay of the Acute Toxicity of Condensing Agents of High Molecular Weight," *Osakashi Suidokyoku Komubu Suishitsu Shikensho Chosa Hokoku Narabini Shiken Seiseki,* **24,** 7–11 (1972); *Chem. Abstr.,* **85,** 138141h.

468. N. Miyanaga, T. Horikawa, A. Tsubouchi, Y. Ueyama, and M. Shioya, "Toxicity of Polymer Flocculant in Water," *Mizu Shori Gijutsu,* **18**(4), 333–342 (1977); *Chem. Abstr.,* **88,** 65557y.

469. M. M. Grula and M. Huang, "Interactions of Polyacrylamides with Certain Soil Pseudomonads," *Dev. Ind. Microbiol.,* **22,** in press.

470. J. Suzuki, H. Harada, and S. Suzuki, "Ozone Treatment of Water-Soluble Polymers. V. Ultraviolet Irradiation Effects on the Ozonization of Polyacrylamide," *J. Appl. Polym.,* **24,** 999–1006 (1979).

471. B. D. Ratner and A. S. Hoffman, "Synthetic Hydrogels for Biomedical Applications," *Hydrogels Med. Relat. Appl. Symp.,* **31,** 1–36 (1975).

472. B. D. Ratner, A. S. Hoffman, S. R. Hanson, L. A. Harker, and J. D. Whiffen, "Blood-Compatibility-Water-Content Relationships for Radiation-Grafted Hydrogels," *J. Polym. Sci.,* **66,** 363–375 (1979).

473. J. Beurey, J.-M. Mougeolle, and M. Weber, "Accidents Cutanes Des Resines Acryliques Dans L'Imprimerie," *Ann. Dermatol. Syphiligr.,* **103,** 423–430 (1976).

474. (a) I. Picton-Robinson, "Danger of Instant Adhesives," *Br. Med. J.,* 581–582 (1977); (b) H. W. Coover, Jr., and J. M. McIntire, "2-Cyanoacrylic Ester Polymers," *Encyclopedia of Chemical Technology,* 3rd ed., Vol. 1, Wiley, New York, 1978, pp. 408–413; (c) R. F. Walker and R. Guiver, "Determination of Alkyl-2-Cyano-Acrylate Concentrations in Air," *Am. Ind. Hyg. Assoc. J.,* **42,** 559–565.

475. S. R. Cohen, A. A. Maier, and J. P. Flesch, "Occupational Health Case Report—No. 3, Ethyl Acrylate," *J. Occup. Med.,* **16,** 199–200 (1974).

476. H. Schulz and R. Gunther, "Paint Processing and Prevention of Air Pollution—Investigations Concerning Organic Substances from Paint Binders Appearing in the Drying Air During Paint Drying in Continuous Ovens," *Staub-Reinhalt. Luft,* **32**(12), 1–11 (1972) (English translation, U.S. Environmental Protection Agency).

477. H. I. Bolker, *Natural and Synthetic Polymers,* Marcel Dekker, New York, 1974.

478. A. F. Turbak, D. F. Durso, O. A. Battista, H. I. Bolker, and J. R. Colvin, "Cellulose," *Encyclopedia of Chemical Technology,* Vol. 5, 1979, pp. 70–88.

479. H. Tarkow, A. J. Baker, H. W. Eickner, W. E. Eslyn, G. J. Hajny, R. A. Hann, R. C. Koeppen, M. A. Millet, and W. E. Moore, "Wood," *Encyclopedia of Chemical Technology,* Vol. 22, 1970, pp. 358–387.

480. B. L. Browning, "Wood," *Encyclopedia of Polymer Science and Technology,* Vol. 15, Wiley, New York, 1971, pp. 1–40.

481. B. A. K. Andrews and I. V. de Gruy, "Cotton," *Encyclopedia of Chemical Technology,* Vol. 7, 3rd ed., 1979, pp. 176–195.

482. R. L. Mitchell and G. C. Daul, "Rayon," *Encyclopedia of Polymer Science and Technology,* Vol. 11, Wiley, New York, 1969, pp. 810–847.

483. J. Bevers and H. Verachtert, "Cellulose: Biodeterioration and Biodegradation," *Agricultura (Louvain),* **23**(3), 116 pp. (1975); *Chem. Abstr.,* **84,** 118407a.

484. E. Treiber, "Properties of Cellulose Materials," in J. Brandrup and E. H. Immergut, Eds., *Polymer Handbook,* 2nd ed., Wiley, New York, 1975, V-87 to V-121.

485. R. T. Bogan, C. M. Kuo, and R. J. Brewer, "Cellulose Derivatives, Esters," *Encyclopedia of Chemical Technology,* 3rd ed., Vol. 5, Wiley, New York, 1979, pp. 118–143.

486. G. K. Greminger, "Cellulose Derivates, Ethers," *Encyclopedia of Chemical Technology,* 3rd ed., Vol. 5, Wiley, New York, 1979, pp. 143–163.

487. G. A. Serad and J. R. Sanders, "Cellulose Acetate and Triacetate Fibers," *Encyclopedia of Chemical Technology,* 3rd ed., Vol. 5, Wiley, New York, 1979, pp. 89–117.

488. O. Kaipainen, "Cellulose and Derivatives," *Encyclopaedia of Occupational Health and Safety,* International Labour Office, Geneva, 1971 (McGraw-Hill Edition, New York, 1974), pp. 275–277.

489. W. L. Ball, "Paper, Paper Pulp Industry," *Encyclopaedia of Occupational Health and Safety,* International Labour Office, Geneva, 1972 (McGraw-Hill Edition, New York, 1974), pp. 997–1000.

490. W. Gunther, K. Koukoudimos, and F. Schlegelmilch, "Characterization of Textile Fibrous Materials by Pyrolysis and Capillary Gas Chromatography," *Melliand Textilberichte* (Engl. ed.), **8,** 490–491 (1979).

491. A. G. De Boos, "Chemical Testing and Analysis," *Text. Progr.,* **6**(3), 1–51 (1974).

492. C. L. Hake and V. K. Rowe, "Ethers," in F. A. Patty, Ed., *Industrial Hygiene and Toxicology,* 2nd ed., Vol. 2, Wiley, New York, 1963, pp. 1655–1718.

493. M. Lewin and A. Basch, "Fire Retardancy; Cellulose," *Encyclopedia of Polymer Science and Technology,* Suppl. Vol. 2, Wiley, New York, 1977, pp. 340–362.

494. A. Basch and M. Lewin, "The Infrared Determination of Leboglucosan Formed During the Pyrolysis of Cellulose," *J. Fire Flammability,* **4,** 92–98 (1973).

495. K. Slater, "The Thermal Behaviour of Textiles," *Text. Progr.,* **8**(3), 1–116 (1976).

496. *Documentation of the Threshold Limit Values,* 4th ed., American Conference of Governmental Industrial Hygienists, Cincinnati, Ohio, 1980.

497. B. A. Zikria, G. C. Weston, M. Chodoff, and J. M. Ferrer, "Smoke and Carbon Monoxide Poisoning in Fire Victims," *J. Trauma,* **12,** 641–645 (1972).

498. A. Blum and B. N. Ames, "Flame-Retardant Additives as Possible Cancer Hazards," *Science,* **195,** 17–23 (1977).

499. E. H. Hadfield and R. G. MacBeth, "Adenocarcinoma of Ethmoids in Furniture Workers," *Ann. Otol. Rhinol. Laryngol.,* **80,** 699–703 (1971).

500. J. F. Fraumeni, "Respiratory Carcinogenesis: An Epidemiologic Appraisal," *J. Natl. Cancer Inst.,* **55,** 1039–1046 (1975).

501. H. C. Andersen, I. Andersen, and J. Solgaard, "Nasal Cancers, Symptoms and Upper Airway Function in Woodworkers," *Br. J. Ind. Med.,* **34,** 201–207 (1977).

502. J.P. Curtes, E. Trotel, and J. Bourdiniere, "Adenocarcinoma of the Ethmoid in Wood Workers," *Arch. Mal. Prof. Med. Trav. Secur. Soc.,* **38,** 773–786 (1977).

503. U. Engzell, A. Englund, and P. Westerholm, "Nasal Cancer Associated with Occupational Exposure to Organic Dust," *Acta Oto-Laryngol.,* **86,** 437–442 (1978).

504. U. Engzell, "Occupational Etiology and Nasal Cancer. An Internordic Project," *Acta Oto-Laryngol.,* (Suppl.), **360,** 126–128 (1979).

505. L. W. Whitehead, T. Ashikaga, and P. Vacek, "Pulmonary Function Status of Workers Exposed to Hardwood or Pine Dust," *Am. Ind. Hyg. Assoc. J.,* **42,** 178–186 (1981). See also L. W. Whitehead et al., *Am. Ind. Hyg. Assoc. J.,* **42,** 461–467 (1981).

506. P. Ironside and J. Matthews, "Adenocarcinoma of the Nose and Paranasal Sinuses in Woodworkers in the State of Victoria, Australia," *Cancer,* **36,** 1115–1121 (1975).

507. M. Chan-Yeung and S. Grzybowski, "Occupational Asthma," *Can. Med. Assoc. J.,* **114,** 433–436 (1976).

508. B. Gandevia, "Ventilatory Capacity During Exposure to Western Red Cedar," *Arch. Environ. Health,* **20,** 59–63 (1970).

509. D. A. Emanuel, B. R. Lawton, and F. J. Wenzel, "Maple-Bark Disease. Pneumonitis Due to Coniosporium Corticale," *N. Engl. J. Med.,* **266,** 333–337 (1962).

510. B. A. Zikria, J. M. Ferrer, and H. F. Floch, "The Chemical Factors Contributing to Pulmonary Damage in 'Smoke Poisoning,'" *Surgery,* **71,** 704–709 (1972).

511. J.M. Jouany, R. Truhaut, and C. Boudene, "Compared Toxicities of Combustion Products of Two Woods: Douglas Fir and Mahogany," *Arch. Mal. Prof. Med. Trav. Secur. Soc.,* **38,** 751–772 (1977).

512. G. L. Nelson, E. J. Hixson, and E. P. Denine, "Combustion Product Toxicity Studies of Engineering Products," *J. Combust. Toxicol.,* **5,** 222–238 (1978).

513. S. C. Packham, R. B. Jeppsen, J. B. McCandless, T. L. Blank, and J. H. Petajan, "The Toxicological Contribution of Carbon Monoxide as a Component of Wood Smoke," *J. Combust. Toxicol.,* **5,** 11–24 (1978).

514. C. J. Hilado and L. A. Gall, "Relative Toxicity of Pyrolysis Products of Some Wood Samples," *J. Combust. Toxicol.,* **4,** 193–199 (1977).

515. C. J. Hilado and N. V. Huttlinger, "Concentration-Response Data on Toxicity of Pyrolysis Gases from Some Natural and Synthetic Polymers," *J. Combust. Toxicol.,* **5,** 196–213 (1978).

516. C. J. Hilado and D. P. Brauer, "Effect of Air Flow on Toxicity of Pyrolysis Gases From Wood in USF Toxicity Test," *J. Combust. Toxicol.,* **6,** 37–43 (1979).

517. C. J. Hilado, E. M. Olcomendy, and D. P. Brauer, "Effect of Pyrolysis Temperature and Air Flow on Toxicity of Gases from Douglas Fir in USF Toxicity Test," *J. Combust. Toxicol.,* **6,** 48–57 (1979).

518. M. M. O'Mara, "The Combustion Products from Synthetic and Natural Products, Part 1: Wood," *J. Fire Flammability,* **5,** 34–53 (1974).

519. J. R. Welker, "The Pyrolysis and Ignition of Cellulosic Materials: A Literature Review," *J. Fire Flammability,* **1,** 12–29 (1970).

520. K. L. Paciorek, R. H. Kratzer, J. Kaufman, J. Nakahara, and A. M. Hartstein, "Thermal Oxidative Degradation Studies of Woods," *J. Fire Flammability,* **5,** 243–254 (1974).

521. F. L. Browne, *Theories of the Combustion of Wood and Its Control,* U.S. Dept. of Agriculture, Forest Service, Madison, Wis., December 1958.

522. F. C. Beall and H. W. Eickner, *Thermal Degradation of Wood Components: A Review of the Literature,* U.S. Dept. of Agriculture, Forest Service Research Paper FPL 130, Madison, Wis., May 1979.

523. J. P. Wagner, "Survey of Toxic Species Evolved in the Pyrolysis and Combustion of Polymers," *Fire Res. Abstr. Rev.,* **14,** 1–23 (1972).

524. F. Shafizadeh and P. P. S. Chin, "Thermal Deterioration of Wood," *ACS Symposium Series, No. 43, Wood Technology: Chemical Aspects,* American Chemical Society, Washington, D.C., 1977, pp. 57–81.

525. F. Shafizadeh, R. H. Furneaux, T. G. Cochran, J. P. Scholl, and Y. Sakai, "Production of Levoglucosan and Glucose From Pyrolysis of Cellulosic Materials," *J. Appl. Polym. Sci.,* **23,** 3525–3539 (1979).

526. C. A. Holmes, "Effect of Fire-Retardant Treatments on Performance Properties of Wood," *Wood Technol.: Chem. Aspects,* **1977,** 82–106.

527. C. A. Holmes, "Flammability of Selected Wood Products Under Motor Vehicle Safety Standard," *J. Fire Flammability,* **4,** 156–164 (1973).

528. L. J. Nunez, G. W. Hung, and J. Autian, "Toxicity of Fabric Combustion Products," *J. Combust. Toxicol.,* **3,** 371–380 (1976).

529. C. J. Hilado, W. H. Marcussen, A. Furst, and H. A. Leon, "Effect of Species on Relative Toxicity of Pyrolysis Products," *J. Combust. Toxicol.,* **3,** 125–134 (1976).

530. C. J. Hilado, L. A. LaBossiere, H. A. Leon, D. A. Kourtides, J. A. Parker, and M. S. Hsu, "The Sensitivity of Relative Toxicity Rankings by the USF/NASA Test Method to Some Test Variables," *J. Combust. Toxicol.,* **3,** 211–236 (1976).

531. C. J. Hilado and H. J. Cumming, "Relative Toxicity of Pyrolysis Gases from Materials: Effects of Chemical Composition and Test Conditions," *Fire Mater.,* **2,** 68–79 (1978).

532. A. Hebeish, A. T. El-Aref, E. A. El-Alfi, and M. H. El-Rafie, "Effect of Short Thermal Treatment on Cotton Degradation," *J. Appl. Polym. Sci.,* **23,** 453–462 (1979).

533. F. Shafizadeh and A. G. W. Bradbury, "Thermal Degradation of Cellulose in Air and Nitrogen at Low Temperatures," *J. Appl. Polym. Sci.,* **23,** 1431–1442 (1979).

534. U. Einsele and I. Tarakcioglu, "Investigations into the Combustion Gases of Various Textile Fibers," *Melliand Textiber.* (Engl. ed.), **58,** 52–58 (1977).

535. B. Krahne, "Decomposition and Combustion Gases and Smoke Development of Textile Raw Materials and Their FR Modifications," *Melliand Textilber.* (Engl. ed.), **58,** 64–71 (1977).

536. R. Muller and P. Couchoud, "Analysis of the Pyrolysis and Combustion Gases of Textile Fabrics," *Melliand Textilber.* (Engl. ed.), **58,** 886–890 (1976).

537. P. D. Garn and C. L. Denson, "Pyrolysis Products from Cellulose Treated with Flame Retardants. Part I: Halogen-Containing Flame Retardants," *Text. Res. J.,* **47,** 485–491 (1977).

538. P. D. Garn and C. L. Denson, "Pyrolysis Products from Cellulose Treated with Flame Retardants. Part II: Nitrogen-Containing Flame Retardants," *Text. Res. J.,* **47,** 535–542 (1977).

539. P. D. Garn and C. L. Denson, "Pyrolysis Products from Cellulose Treated with Flame Retardants. Part III: Ancillary Studies," *Text. Res. J.,* **47,** 591–597 (1977).

540. F. Valic and E. Zuskin, "Effects of Hemp Dust Exposure on Nonsmoking Female Textile Workers," *Arch. Environ. Health,* **23,** 359–364 (1971).

541. A. Bouhuys, "Byssinosis; Airway Responses Caused by Inhalation of Textile Dusts," *Arch. Environ. Health,* **23,** 405–407 (1971).

542. A. Bouhuys, F. Hartogensis, and H. J. H. Korfage, "Byssinosis Prevalence and Flax Processing," *Br. J. Ind. Med.,* **20,** 320–323 (1963).

543. A. Rossi, G. Moro, L. Fabbri, F. Brighenti, C. Mapp, and E. De Rosa, "Byssinosis, Chronic Bronchitis and Pathology of Ventilatory Function in a Population of Workers Exposed to Flax Dust," *Med. Lav.,* **69,** 698–707 (1978).

544. K. Y. Mustafa, A. S. Lakha, M. H. Milla, and U. Dahoma, "Byssinosis, Respiratory Symptoms and Spirometric Lung Function Tests in Tanzanian Sisal Workers," *Br. J. Ind. Med.,* **35,** 123–128 (1978).

545. S. H. El Ghawabi, "Respiratory Function and Symptoms in Workers Exposed Simultaneously to Jute and Hemp," *Br. J. Ind. Med.,* **35,** 16–20 (1978).

546. E. Zuskin, F. Valic, and A. Bouhuys, "Byssinosis and Airway Responses Due to Exposure to Textile Dust," *Lung,* **154,** 17–24 (1976).

547. C. J. Hilado and D. P. Brauer, "Toxicity of Pyrolysis Gases From Viscose Fibers Containing a Flame Retardant," *J. Combust. Toxicol.,* **6,** 69–77 (1979).

548. A. Scotney, "The Thermal Degradation of Cellulose Triacetate—I; The Reaction Products," *Eur. Polym. J.,* **8,** 163–174 (1972).

549. W. P. Brown and C. F. H. Tipper, "The Pyrolysis of Cellulose Derivatives," *J. Appl. Polym. Sci.,* **22,** 1459–1468 (1978).

550. K. L. Gregory, V. F. Malinoski, and C. R. Sharp, "Cleveland Clinic Fire Survivorship Study, 1929–1965," *Arch. Environ. Health,* **18,** 508–515 (1969).

551. "Recommendations," in D. A. Fraser and M. C. Battigelli, Eds., *Transactions of the National*

Conference on Cotton Dust and Health (held on May 2, 1970), University of North Carolina, Charlotte, N.C., 1971, pp. viii–xi.

552. T. R. Harris, J. A. Merchant, K. H. Kilburn, J. D. Hamilton, "Byssinosis and Respiratory Diseases of Cotton Mill Workers," *J. Occup. Med.,* **14,** 199–206 (1972).

553. J. A. Merchant, K. H. Kilburn, W. M. O'Fallon, J. D. Hamilton, and J. C. Lumsden, "Byssinosis and Chronic Bronchitis Among Cotton Textile Workers," *Ann. Intern. Med.,* **76,** 423–433 (1972).

554. J. A. Merchant, J. C. Lumsden, K. H. Kilburn, W. M. O'Fallon, J. R. Ujda, V. H. Germino, J. D. Hamilton, "Dose Response Studies in Cotton Textile Workers," *J. Occup. Med.,* **15,** 222–230 (1973).

555. H. R. Imbus and M. W. Suh, "Byssinosis; A Study of 10,133 Textile Workers," *Arch. Environ. Health,* **26,** 183–191 (1973).

556. National Institute for Occupational Safety and Health, *Occupational Exposure to Cotton Dust,* HEW Publ. No. (NIOSH) 75-118, Supt. Doc., Washington, D.C., 1974.

557. J. A. Merchant, J. C. Lumsden, K. H. Kilburn, V. H. Germino, J. D. Hamilton, W. S. Lynn, H. Byrd and D. Baucom, "Preprocessing Cotton to Prevent Byssinosis," *Br. J. Ind. Med.,* **30,** 237–247 (1973).

558. C. F. Martin and J. E. Higgins, "Byssinosis and Other Respiratory Ailments," *J. Occup. Med.,* **18,** 455–462 (1976).

559. R. M. Bethea and P. R. Morey, "A Comparison of Cotton Dust Sampling Techniques," *Am. Ind. Hyg. Assoc. J.,* **37,** 647–654 (1976).

560. J. D. Neefus, J. C. Lumsden, and M. T. Jones, "Cotton Dust Sampling, II-Vertical Elutriation," *Am. Ind. Hyg. Assoc. J.,* **38,** 394–400 (1977).

561. M. W. Suh and J. D. Neefus, "Cotton Dust Sampling, III-Estimation of Statistical Errors with Vertical Elutriators," *Am. Ind. Hyg. Assoc. J.,* **38,** 387–390 (1977).

562. D. F. Brown, E. R. McCall, B. Piccolo, and V. W. Tripp, "Survey of Effects of Variety and Growing Location of Cotton on Cardroom Dust Composition," *Am. Ind. Hyg. Assoc. J.,* **38,** 107–115 (1977).

563. D. F. Brown, J. H. Wall, R. J. Berni, and V. W. Tripp, "Chemical Composition of Cotton-Processing Dusts," *Text. Res. J.,* **48,** 355–363 (1978).

564. M. C. Battigelli, P. L. Craven, J. J. Fischer, P. R. Morey, and P. E. Sasser, "The Role of Histamine in Byssinosis," *J. Environ. Sci. Health,* **A12,** 327–339 (1977).

565. M. C. Battigelli and J. J. Fischer, "The Biological Activity of Cotton Dust and Its Extracts," presented at the American Industrial Hygiene Association, Chicago, May 1979 (abstract #166).

566. Textile Research Institute, *Chemical Composition of Cotton Dust and Its Relation to Byssinosis; A Review of the Literature,* Report No. 1, Princeton, N.J., 1978.

567. Occupational Safety and Health Administration, "Cotton Dust," *Fed. Reg.,* **43,** 27350, 28473 (1978); *U.S. Code of Federal Regulations,* 29 CFR 1910.1043.

568. H. H. Perkins, Jr. and J. B. Cocke, "Cotton Dust Levels Reduced with Additives," *Text. Ind.,* **42,** 92–95 (1978).

569. H. H. Perkins, Jr. and J. B. Cocke, "Dust-Control Additives for Cotton and Cotton/Polyester Blends," *Text. Res. J.,* **49,** 131–136 (1979).

570. K. Q. Robert, Jr., "Cotton Dust Sampling Efficiency of the Vertical Elutriator," *Am. Ind. Hyg. Assoc. J.,* **40,** 535–542 (1979).

571. D. K. Mittal, R. D. Gilbert, R. E. Fornes, and S. P. Hersh, "Chemical Composition of Cotton Dusts, Part II: Analyses of Samples Collected in a Model Card Room," *Text. Res. J.,* **49,** 389–394 (1979).

572. T. F. Cooke, "Chemical Composition of Cotton Dust and Its Relation to Byssinosis: A Review of the Literature," *Text. Res. J.,* **49,** 398–404 (1979).

573. S. P. Hersh, R. E. Fornes, and M. Anand, "Short-Term Cotton Dust Sampling Utilizing Three Non-Gravimetric Methods," *Am. Ind. Hyg. Assoc. J.*, **40**, 578–587 (1979).

574. W. Stein and G. Dallmeyer, "Problems Involved in the Measurement of Textile Air Dusts," *Melliand Textilber.* (Engl. Ed.), **5**, 357–377 (1979).

575. L. E. Wise and E. C. Jahn, as cited in *Wood Chemistry*, 2nd ed., Reinhold, New York, 1944.

576. B. Woods and C. D. Calnan, "Toxic Woods," *Br. J. Dermatol.*, **95**, 1–97 (1976).

577. R. R. Suskind, "Dermatitis in the Forest Product Industries," *Arch. Environ. Health*, **15**, 322–326 (1967).

578. T. E. Barber and E. L. Husting, "Plant and Wood Hazards," in *Occupational Diseases*, U.S. Department of Health, Education, and Welfare, National Institute for Occupational Safety and Health, Washington, D.C., June 1977.

579. R. Holst, J. Kirby, and B. Magnusson, "Sensitization to Tropical Woods Giving Erythema Multiforme-Like Eruptions," *Contact Dermatitis*, **2**, 295–297 (1976).

580. B. M. Hausen, "Sensitizing Capacity of Naturally Occurring Quinones," *Contact Dermatitis*, **4**, 204–213 (1978).

581. R. Schoental, "Carcinogenicity of Wood Shavings," *Lab. Animals*, **7**, 47–49 (1973).

582. T. L. Junod, *Gaseous Emissions and Toxic Hazards Associated with Plastics in Fire Situations—A Literature Review*, National Aeronautics and Space Administration, Washington, D.C., 1976.

583. F. H. M. Nestler, *The Formaldehyde Problem in Wood-Based Products—An Annotated Bibliography*, United States Department of Agriculture, Forest Service, Madison, Wis., 1977.

584. R. C. DeGroot and G. R. Esenther, *Microbiological and Entomological Stresses on the Structural Use of Wood*, United States Department of Agriculture, Forest Service, Madison, Wis., May, 1978.

585. A. J. McQuire, "Preservation of Timber in the Sea," in E. B. G. Jones and S. K. Eltringham, Eds., *Marine Borers; Fungi and Fouling Organisms of Wood*, Organization for Economic Co-operation and Development, Paris, 1971, pp. 339–346.

586. J. R. Bercaw, "The Melt-Drip Phenomena of Apparel," *Fire Technol.*, **9**, 24–45 (1973).

587. S. Heyden and P. Pratt, "Exposure to Cotton Dust and Respiratory Disease; Textile Workers, 'Brown Lung,' and Lung Cancer," *J. Am. Med. Assoc.*, **244**, 1797–1798 (1980).

588. R. S. F. Schilling, E. C. Vigliani, B. Lammers, F. Valic, and J. C. Gilson, "A Report on a Conference on Byssinosis," *Proceedings of the Excerpta Medica*, Amsterdam, 1963, pp. 137–145.

589. R. A. Wesley and J. H. Wall, "Levels and Chemical Composition of Cotton Gin Dust," *Am. Ind. Hyg. Assoc. J.*, **39**, 962–969 (1978).

590. D. J. Chinn, F. F. Cinkotai, M. G. Lockwood and S. H. M. Logan, "Airborne Dust, Its Protease Content and Byssinosis in 'Willowing' Mills," *Ann. Occup. Hyg.*, **19**, 101–108 (1976).

591. R. W. Postlethwait, D. A. Willigan, and A. W. Ulin, "Human Tissue Reaction to Sutures," *Ann. Surg.*, **181**, 144–150 (1975).

592. K. Selby, "Mechanism of Biodegradation of Cellulose," in A. H. Walters and J. J. Elphick, Eds., *Biodeterioration of Materials*, Elsevier, New York, 1968, pp. 62–78.

593. B. S. Oppenheimer, E. T. Oppenheimer, and A. P. Stout, "Sarcomas Induced in Rodents by Imbedding Various Plastic Films," *Proc. Soc. Exp. Biol. Med.*, **79**, 366–369 (1952).

594. B. S. Oppenheimer, E. T. Oppenheimer, I. Danishefsky, A. P. Stout, and F. R. Eirich, "Further Studies of Polymers as Carcinogenic Agents in Animals," *Proc. Am. Assoc. Cancer Res.*, **2**, 333–340 (1955).

595. B. S. Oppenheimer, E. T. Oppenheimer, A. P. Stout, M. Willhite, and I. Danishefsky, "The Latent Period in Carcinogenesis by Plastics in Rats and Its Relation to the Presarcomatous Stage," *Cancer*, **11**, 204–213 (1958).

596. R. E. Scales, "Cellulosic," in *Modern Plastics Encyclopedia*, 1978–79 ed., Vol. 55, No. 10A, McGraw-Hill, New York, pp. 13–14.

597. W. P. Jordan and M. V. Dahl, "Contact Dermatitis from Cellulose Ester Plastics," *Arch. Dermatol.,* **105,** 880–885 (1972).

598. N. Hjorth, "Contact Dermatitis from Cellulose Acetate Film," *Berufsdermatosen,* **12,** 86–100 (1964).

599. R. E. Bentley, G. A. Leblanc, T. A. Hollister, and B. H. Sleight, III, *Laboratory Evaluation of the Toxicity of Nitrocellulose to Aquatic Organisms,* U.S. Army Medical Research and Development Command, Washington, D.C., 1977 (AD A037749).

600. T. M. Wendt and A. M. Kaplan, "A Chemical-Biological Treatment Process for Cellulose Nitrate Disposal," *J. Water Pollut. Contr. Fed.,* **48,** 660–668 (1976).

601. R. E. Klose and M. Glicksman, "Gums," in T. E. Furia, Ed., *Handbook of Food Additives,* 2nd ed., CRC Press, Cleveland, Ohio, 1972, pp. 295–359.

602. R. G. Rufe, "Cellulose Polymers in Cosmetics and Toiletries," *Cosmet. Perfum.,* **90,** 93, 94, 99, 100 (1975).

603. T. F. McElligott and E. W. Hurst, "Long-Term Feeding Studies of Methyl Ethyl Cellulose ('Edifas' A) and Sodium Carboxymethyl Cellulose ('Edifas' B) in Rats and Mice," *Food Cosmet. Toxicol.,* **6,** 449–460 (1968).

604. S. B. McCollister, R. J. Kociba, and D. D. McCollister, "Dietary Feeding Studies of Methylcellulose and Hydroxypropylmethylcellulose in Rats and Dogs," *Food Cosmet. Toxicol.,* **11,** 943–953 (1973).

605. W. H. Braun, J. C. Ramsey, and P. J. Gehring, "The Lack of Significant Absorption of Methylcellulose, Viscosity 3300cP, from the Gastrointestinal Tract Following Single and Multiple Oral Doses to the Rat," *Food Cosmet. Toxicol.,* **12,** 373–376 (1974).

606. Hercules Inc., *Natrosol® 250; Summary of Toxicological Investigations,* Coatings & Specialty Products Dept., Bulletin T-101B, Wilmington, Del., n.d.

607. Hercules Inc., *Klucel® Hydroxypropyl Cellulose—Summary of Toxicological Investigations,* Coatings & Specialty Products Dept. Bulletin T-122, Wilmington, Del., n.d.

608. E. Bachmann, E. Weber, M. Post, and G. Zbinden, "Biochemical Effects of Gum Arabic, Gum Tragacanth, Methylcellulose and Carboxymethylcellulose-Na in Rat Heart and Liver," *Pharmacology,* **17,** 39–49 (1978).

609. M. G. Crane, J. J. Harris, R. Herber, S. Shankel, and N. Specht, "Excessive Fluid Retention Related to Cellulose Ingestion: Studies on Two Patients," *Metabolism,* **18,** 945–960 (1969).

610. T. Hamada and S. Horiguchi, "Allergic Contact Dermatitis Due to Sodium Carboxymethyl Cellulose," *Contact Dermatitis,* **4,** 244 (1978).

611. M. G. Wirick, "Aerobic Biodegradation of Carboxymethylcellulose," *J. Water Pollut. Contr. Fed.,* **46,** 512–521 (1974).

612. F. A. Blanchard, I. T. Takahashi, and H. C. Alexander, "Biodegradability of [^{14}C] Methylcellulose by Activated Sludge," *Appl. Environ. Microbiol.,* **32,** 557–560 (1976).

613. W. M. Doane, "Starch: Renewable Raw Material for the Chemical Industry," *J. Coatings Technol.,* **50,** 88–98 (1978).

614. Henkel Corporation, "*SGP Absorbent Polymer,*" Minneapolis, Minn., 1978.

615. A. P. De Groot, H. P. Til, V. J. Feron, H. C. Dreef-van der Meulen, and M. I. Willems, "Two-Year Feeding and Multigeneration Studies in Rats on Five Chemically Modified Starches," *Food Cosmet. Toxicol.,* **12,** 651–663 (1974).

616. R. Truhaut, B. Coquet, X. Fouillet, D. Galland, D. Guyot, D. Long, and J. L. Rouaud, "Two-Year Oral Toxicity and Multigeneration Studies in Rats on Two Chemically Modified Maize Starches," *Food Cosmet. Toxicol.,* **17,** 11–17 (1979).

617. T. A. Anderson, L. J. Filer, Jr., S. J. Fomon, D. W. Andersen, R. L. Jensen and R. R. Rogers, "Effect of Waxy Corn Starch Modification on Growth, Serum Biochemical Values and Body Composition of Pitman-Moore Miniature Pigs," *Food Cosmet. Toxicol.,* **11,** 747–754 (1973).

618. E. Lebenthal, "Use of Modified Food Starches in Infant Nutrition," *Am. J. Dis. Child.*, **132**, 850–852 (1978).

619. A. J. Ryan, G. M. Holder, C. Mate, and G. K. Adkins, "The Metabolism and Excretion of Hydroxyethyl Starch in the Rat," *Xenobiotica*, **2**, 141–146 (1972).

620. J. M. Mishler, H. Borberg, P. M. Emerson, and R. Gross, "Hydroxyethyl Starch. An Agent for Hypovolaemic Shock Treatment II. Urinary Excretion in Normal Volunteers Following Three Consecutive Daily Infusions," *Br. J. Clin. Pharmacol.*, **4**, 591–595 (1977).

621. H. Kohler, W. Kirch, and H. J. Horstmann, "Hydroxyethyl Starch-Induced Macroamylasemia," *Int. J. Clin. Pharmacol.*, **15**, 428–431 (1977).

622. P. Grasso, "Persorption—A New Long-Term Problem?," *Food Cosmet. Toxicol.*, **14**, 497–498 (1976).

623. R. M. Jager, "Glove-Starch Granulomatous Disease," *J. Am. Med. Assoc.*, **235**, 2583–2584 (1976).

624. R. L. Whistler, Ed., *Industrial Gums*, 2nd ed., Academic Press, New York, 1973.

625. G. A. Towle, "Carrageenan," in R. L. Whistler, Ed., *Industrial Gums*, 2nd ed., Academic Press, New York, 1973, pp. 83–114.

626. W. H. McNeely and K. S. Kang, "Xanthan and Some Other Biosynthetic Gums," in R. L. Whistler, Ed., *Industrial Gums*, 2nd ed., Academic Press, New York, 1973, pp. 473–497.

627. T. F. X. Collins, T. N. Black, and J. H. Prew, "Long-Term Effects of Calcium Carrageenan in Rats-I. Effects on Reproduction," *Food Cosmet. Toxicol.*, **15**, 533–538 (1977).

628. K. A. Pittman, L. Golberg, and F. Coulston, "Carrageenan: The Effect of Molecular Weight and Polymer Type on its Uptake, Excretion and Degradation in Animals," *Food Cosmet. Toxicol.*, **14**, 85–93 (1976).

629. T. F. X. Collins, T. N. Black, and J. H. Prew, "Long-Term Effects of Calcium Carrageenan in Rats.-II. Effects on Foetal Development," *Food Cosmet. Toxicol.*, **15**, 539–545 (1977).

630. T. F. X. Collins, T. N. Black, and J. H. Prew, "Effects of Calcium and Sodium Carrageenans and iota-Carrageenan on Hamster Foetal Development," *Food Cosmet. Toxicol.*, **17**, 443–449 (1979).

631. G. Woodard, M. W. Woodard, W. H. McNeely, P. Kovacs, and M. T. I. Cronin, "Xanthan Gum: Safety Evaluation by Two-Year Feeding Studies in Rats and Dogs and a Three-Generation Reproduction Study in Rats," *Toxicol. Appl. Pharmacol.*, **24**, 30–36 (1973).

632. M. Meisters, "Nylon," in *Modern Plastics Encyclopedia*, 1978–79 ed., Vol. 55, No. 10A, McGraw-Hill, New York, pp. 28–30.

633. M. I. Kohan, Ed., *Nylon Plastics*, Wiley, New York, 1973.

634. C. E. Sroog, "Polyimides," *J. Polym. Sci., Part C*, **1967**(16), 1191–1209 (1967).

635. F. P. Recchia and W. J. Farrissey, Jr., "Polyimide; Thermoplastic Polyimide," in *Modern Plastics Encyclopedia*, 1978–79 ed., Vol. 55, No. 10A, McGraw-Hill, New York, pp. 66.

636. R. Pfluger, "Physical Constants of Poly(imino(1-oxohexamethylene) (Polyamide 6) and Poly(iminohexamethyleneiminoadipoyl) (Polyamide 66)," in J. Brandrup and E. H. Immergut, Eds., *Polymer Handbook*, 2nd ed., Wiley, New York, 1975.

637. J. R. Brown and B. C. Ennis, "Thermal Analysis of Nomex® and Kevlar® Fibers," *Text. Res. J.*, **47**, 62–66 (1977).

638. E. I. du Pont de Nemours & Co., Inc., *Kapton Polyimide Film*; Safety in Handling and Use, Polymer Products Dept., Bulletin E-36500, September 1980.

639. L. Penn and F. Larsen, "Physicochemical Properties of Kevlar 49 Fiber," *J. Appl. Polym. Sci.*, **23**, 59–73 (1979).

640. E. I. du Pont de Nemours & Co., Inc., *"Kapton" Polyimide Film-Type H Summary of Properties*, Plastic Products and Resins Dept., Bulletin H-1D, Wilmington, Del., n.d.

641. F. W. Harris and L. H. Lanier, "Structure-Solubility Relationships in Polyimides," in F. W. Harris and R. B. Seymour, Eds., *Structure-Solubility Relationships in Polymers,* Academic, New York, 1977, pp. 183–198.

642. M. W. Wardle, "High Performance Coated Fabrics of Kevlar® Aramid Fiber," *J. Coated Fabr.,* **7,** 3–23 (1977).

643. A. Anton, "Polyamides," *Encyclopedia of Industrial Chemical Analysis,* Vol. 17, Wiley, 1973, pp. 275–306.

644. J. Preston, "Aromatic Polyamide Fibers," *Encyclopedia of Polymer Science and Technology,* Vol. 2, Wiley, New York, 1977, pp. 84–112.

645. E. Moss and W. R. Lee, "Occurrence of Oral and Pharyngeal Cancers in Textile Workers," *Br. J. Ind. Med.,* **31,** 224–232 (1974).

646. E. Zuskin, F. Valic, and A. Bouhuys, "Effect of Wool Dust on Respiratory Function," *Am. Rev. Resp. Dis.,* **114,** 705–709 (1976).

647. K. Brysiewicz, H. Buluk, M. Cesarz-Fronczyk, J. Kordecka, B. Leszczynski, Z. Lukjan, and H. Sadokierska, "The Effect of Work in Dusty Surroundings on the Prevalence of Chronic Bronchitis Among the Workmen of Sierzan's Establishments of Wool Industry at Bialystok," *Gruzlica,* **38,** 657–660 (1970).

648. S. Jordeczka, S. Basa, and B. Basa, "The Prevalence of Chronic Non-Specific Bronchopulmonary Disease in Workmen of Wool Inudstry," *Gruzlica,* **38,** 643–650 (1970); authors' English summary.

649. K. C. Mathur and S. N. Misra, "Incidence of Pulmonary Disease Among Wool Workers," *Indian J. Chest Dis.,* **14,** 172–178 (1972).

650. H. H. Moll, "Occupational Asthma; With Reference To Wool Sensitivity," *Lancet,* **1933-I,** 1340–1342.

651. C. J. Hilado and C. R. Crane, "Comparison of Results with the USF/NASA and FAA/CAMI Toxicity Screening Test Methods," *J. Combust. Toxicol.,* **4,** 56–60 (1977).

652. C. J. Hilado and H. J. Cumming, "Standard Wool Fabric as a Reference Material," *J. Combust. Toxicol.,* **4,** 454–463 (1977).

653. C. J. Hilado and E. M. Olcomendy, "Toxicity of Pyrolysis Gases From Some Flame-Resistant Fabrics," *J. Combust. Toxicol.,* **7,** 69–72 (1980).

654. K. Kishitani and S. Yusa, "Study on Evaluation of Relative Toxicities of Combustion Products of Various Materials," *J. Fac. Eng. Univ. Tokyo,* **35,** 1–17 (1979).

655. L. Benisek, "Letter to the Editor," *J. Combust. Toxicol.,* **5,** 101–102 (1978).

656. C. Herpol and R. Minne, "Letter to the Editor," *J. Combust. Toxicol.,* **7,** 73 (1980).

657. P. E. Ingham, "The Pyrolysis of Wool and the Action of Flame Retardants," *J. Appl. Polym. Sci.,* **15,** 3025–3041 (1971).

658. E. Urbas and E. Kullik, "Pyrolysis Gas Chromatographic Analyses of Untreated and Flameproofed Wools," *Fire Mater.,* **2,** 25–26 (1978).

659. M. Chaigneau and G. Le Moan, "Analysis and Evolution of the Compounds Formed by the Pyrolysis and Combustion of Wool," *Analusis,* **4,** 28–33 (1976).

660. Y. Tsuchiya and J. G. Boulanger, "Carbonyl Sulphide in Fire Gases," *Fire Mater.,* **3,** 154–155 (1979).

661. M. Paabo, M. M. Birky, and S. E. Womble, "Analysis of Hydrogen Cyanide in Fire Environments," *J. Combust. Toxicol.,* **6,** 99–108 (1979).

662. A. J. Christopher, "Some Aspects of Smoke and Fume Evolution from Overheated Non-Metallic Materials," *J. Combust. Toxicol.,* **3,** 89–102 (1976).

663. R. E. Dufour, "Survey of Available Information on the Toxicity of the Combustion and

Thermal Decomposition Products of Certain Building Materials Under Fire Conditions," Underwriters' Laboratories, Inc.®, Bulletin of Research No. 53, Chicago, Ill., 1963.

664. V. P. Saakadze, "Natural Silk as an Occupational Allergen," *Gig. Tr. Prof. Zabol.*, **11,** 13–16 (1976); *Chem. Abstr.*, **86,** 126404d.

665. E. D. Acheson, R. H. Cowdell, and B. Jolles, "Nasal Cancer in the Northamptonshire Boot and Shoe Industry," *Br. Med. J.*, **1,** 385–393 (1970).

666. E. D. Acheson, R. H. Cowdell, and B. Jolles, "Nasal Cancer in the Shoe Industry," *Br. Med. J.*, **2,** 791 (1970).

667. E. D. Acheson, R. H. Cowdell, and E. Rang, "Adenocarcinoma of the Nasal Cavity and Sinuses in England and Wales," *Br. J. Ind. Med.*, **29,** 21–30 (1972).

668. M. Chaigneau and G. Le Moan, "Study of the Pyrolysis of Plastic Materials. VII. Polyamides," *Ann. Pharm. Fr.*, **31,** 495–501 (1973).

669. Y. Yoshida, K. Kono, A. Harada, S. Toyota, M. Watanabe, and K. Iwasaki, "Toxicity of Pyrolysis Products of Thermal-Resistant Plastics Including Polyamide and Polyester," *Jap. J. Hyg.*, **33,** 450–458 (1978).

670. T. Sugihara, "Effect of Some Inorganic Additives on the Formation of Gaseous Products from Heated Nylon 6," *Gifu Daigaku Kyoikugakubu Kenkyu Hokoku, Shizen Kagaku*, **6**(1), 107–114 (1977); *Chem. Abstr.*, **90,** 104566.

671. I. Luderwald and F. Merz, "Über den thermischen Abbau von Polyamiden der Nylon-Reihe," *Angew. Makromol. Chem.*, **74,** 165–185 (1978); from authors' English abstract.

672. I. J. Goldfarb and A. C. Meeks, *Thermal Decomposition of Polyamides*, Air Force Materials Laboratory, Wright-Patterson Air Force Base, Ohio, 1969 (AD 686076).

673. L. H. Peebles, Jr., and M. W. Huffman, "Thermal Degradation of Nylon 66," *J. Polym. Sci.*, **9,** 1807–1822 (1971).

674. B. Bott, J. G. Firth, and T. A. Jones, "Evolution of Toxic Gasses from Heated Plastics," *Br. Polym. J.*, **1,** 203–204 (1969).

675. N. Igarashi, "The Thermal Degradation of Nylon 6, Polyethylene Terephthalate and Polycarbonate Polymers," *Diss. Abstr. Int. B.*, **39,** 3458 (1979).

676. W. Young, C. J. Hilado, D. A. Kourtides, and J. A. Parker, "A Study of the Toxicology of Pyrolysis Gases from Synthetic Polymers," *J. Combust. Toxicol.*, **3,** 157–165 (1976).

677. R. M. Brown, unpublished data, E. I. du Pont de Nemours and Company, Inc., Haskell Laboratory, Wilmington, Del., April 1975.

678. M. L. Bazinet, C. DiPietro, and C. Merritt, Jr., *Thermal Degradation Studies of Fibers and Composite Base Materials*, U. S. Army Natick Research and Development Command, Natick, Mass., 1977 (AD A037604).

679. P. A. Tatem and F. W. Williams, "The Degradation Products from a Polyaramide Under Low and High Temperature Stresses," *J. Combust. Toxicol.*, **6,** 238–247 (1979).

680. R. S. Waritz and S. D. Morrison, unpublished data, E. I. du Pont de Nemours and Company, Inc., Haskell Laboratory, Wilmington, Del., February 1965.

681. C. J. Hilado, "The Practical Use of the USF Toxicity Screening Test Method," *J. Combust. Toxicol.*, **5,** 331–338 (1978).

682. G. Ball and E. A. Boettner, "Combustion Products of Nitrogen-Containing Polymers," *Am. Chem. Soc. Div. Org. Coatings Plast. Chem.*, **33,** 431–437 (1973).

683. C. Arnold, Jr. and L. K. Borgman, "Chemistry and Kinetics of Polyimide Degradation," *Ind. Eng. Chem. Prod. Res. Develop.*, **11,** 322–325 (1972).

684. T. H. Johnston and C. A. Gaulin, "Thermal Decomposition of Polyimides in Vacuum," *J. Macromol. Sci.-Chem.*, **A3,** 1161–1182 (1969).

685. J. F. Heacock and C. E. Berr, "Polyimides-New High Temperature Polymers: H-Film, A Polypyromellitimide Film," *SPE Trans.*, **5**, 105–110 (1965).

686. S. D. Bruck, "Thermal Degradation of an Aromatic Polypyromellitimide in Air and Vacuum II—The Effect of Impurities and the Nature of Degradation Products," *Polymer*, **6**, 49–61 (1965).

687. K. R. Makinson, "Wool," *Encyclopedia of Polymer Science And Technology*, Vol. 15, Wiley, New York, 1971, pp. 41–79.

688. L. Benisek, G. K. Edmondson, and W. A. Phillips, "Protective Clothing: Evaluation of Zirpro Wool and Other Fabrics," *Fire Mater.*, **3**, 156–166 (1979).

689. E. M. Hambly, L. Levia, and D. S. Wilkinson, "Wool Intolerance in Atopic Subjects," *Contact Dermatitis*, **4**, 240–241 (1978).

690. R. J. Neher, unpublished data, E. I. du Pont de Nemours and Company, Inc., Haskell Laboratory, Wilmington, Del., June 1962.

691. E. D. Osborne and P. F. Murray, "Atopic Dermatitis," *AMA Arch. Dermatol. Syphilol.*, **68**, 619–626 (1953).

692. E. Rudzki, "Contact Urticaria from Silk," *Contact Dermatitis*, **3**, 53 (1977).

693. J. Faulborn, G. Mackensen, K. Beyer, and J.-R. Moringlane, "Studies on the Tolerance of Silk, Nylon, Dacron, and Collagen Suture Material in the Cornea of the Rabbit," *Adv. Ophthalmol.*, **30**, 50–54 (1975).

694. E. C. Schule, "Polyamide Plastics," *Encyclopedia of Polymer Science and Technology*, Vol. 10, Wiley, New York, 1969, pp. 460–482.

695. W. Sweeny and J. Zimmerman, "Polyamides," *Encyclopedia of Polymer Science and Technology*, Vol. 10, Wiley, New York, 1969, pp. 483–597.

696. O. E. Snider and R. J. Richardson, "Polyamide Fibers," *Encyclopedia of Polymer Science and Technology*, Vol. 10, Wiley, New York, 1969, pp. 347–460.

697. H. Sherman, D. B. Hood, J. R. Barnes, and D. M. Gay, unpublished data, E. I. du Pont de Nemours and Company, Inc., Haskell Laboratory, Wilmington, Del., March 1959.

698. H. Sherman, unpublished data, E. I. du Pont de Nemours and Company, Inc., Haskell Laboratory, Wilmington, Del., April 1971.

699. R. W. Morrow, personal communication, E. I. du Pont de Nemours and Company, Inc., Haskell Laboratory, Wilmington, Del., February 1980 (tests conducted at Haskell Laboratory 1972–1976).

700. N. C. Goodman, unpublished data, E. I. du Pont de Nemours and Company, Inc., Haskell Laboratory, Wilmington, Del., June 1976.

701. J. F. Wilson and K. L. Gabriel, unpublished data, Biosearch, Inc., Philadelphia, Pa., October 1979 (sponsored by Haskell Laboratory).

702. S. A. Imbeau and C. E. Reed, "Nylon Stocking Dermatitis; An Unusual Example," *Contact Dermatitis*, **5**, 163–164 (1979).

703. R. I. Leininger, "Changes in Properties of Plastics During Implantation," in American Society for Testing and Materials, *Plastics in Surgical Implants*, ASTM Special Technical Publication No. 386, Philadelphia, Pa., 1965, pp. 71–76.

704. E. Roggendorf, "The Biostability of Silicone Rubbers, a Polyamide, and a Polyester," *J. Biomed. Mater. Res.*, **10**, 123–143 (1976).

705. M. R. Rogers and A. M. Kaplan, "Effects of *Penicillium Janthinellum* on Parachute Nylon—Is there Microbial Deterioration?" *Int. Biodeterior. Bull.*, **7**, 15–24 (1971).

706. D. E. Peerman, "Polyamides from Fatty Acids," *Encyclopedia of Polymer Science and Technology*, Vol. 10, Wiley, New York, 1969, pp. 597–615.

707. T. I. Bair, P. W. Morgan, and F. L. Killian, "Poly(1,4-phenyleneterephthalamides). Polymerization and Novel Liquid-Crystalline Solutions," *Macromolecules,* **10,** 1396–1400 (1977).

708. J. B. Titus, *New Plastics: Properties, Processing and Potential Uses,* U.S. Army Armament Research and Development Command, Dover, N.J., 1977 (AD A056990).

709. N. C. Goodman, unpublished data, E. I. du Pont de Nemours and Company, Inc., Haskell Laboratory, Wilmington, Del., November 1974.

710. J. W. McAlack, unpublished data, E. I. du Pont de Nemours and Company, Inc., Haskell Laboratory, Wilmington, Del., June 1973.

711. A. M. Kligman, unpublished data, Ivy Research Laboratories, Inc., Philadelphia, Pa., April 1977 (sponsored by Haskell Laboratory).

712. D. P. Kelly, unpublished data, E. I. du Pont de Nemours and Company, Inc., Haskell Laboratory, Wilmington, Del., September 1979.

713. J. B. Terrill, unpublished data, E. I. du Pont de Nemours and Company, Inc., Haskell Laboratory, Wilmington, Del., June 1977.

714. S. S. Hirsch and S. L. Kaplan, "Polyimides: Chemistry, Processing Properties," *Am. Chem. Soc. Div. Org. Coatings Plast. Chem.,* **34,** 162–169 (1974).

715. F. P. Darmory, "Processable Polyimides," in R. D. Deanin, Ed., *New Industrial Polymers,* ACS Symposium Series 4, American Chemical Society, Washington, D.C., 1972.

716. C. E. Sroog, "Polyimides," *J. Polym. Sci.: Macromol. Rev.,* **11,** 161–208 (1976).

717. F. P. Darmory and N. A. Sullo, "Thermoset Polyimide," in *Modern Plastics Encyclopedia,* 1978–79 ed., Vol. 55, No. 10A, McGraw-Hill, New York, pp. 66, 70.

718. R. DeIasi and J. Russell, "Aqueous Degradation of Polyimides," *J. Appl. Polym. Sci.,* **15,** 2965–2974 (1971).

719. E. L. Ringwald and E. L. Lawton, "Physical Constants of Poly(oxyethyleneoxyterephthaloyl) (Poly(ethylene Terephthalate))," in J. Brandrup and E. H. Immergut, Eds., Polymer Handbook, 2nd ed., Wiley, New York, 1975, V-71 to V-78.

720. L. Bottenbruch, "Polycarbonates," *Encyclopedia of Polymer Science and Technology,* Vol. 10, Wiley, New York, 1969, pp. 710–764.

721. P. G. Kelleher, G. H. Bebbington, D. R. Falcone, J. T. Ryan, and R. P. Wentz, "Thermal and Hydrolytic Stability of Poly(butylene Terephthalate)," *Soc. Plast. Eng., Tech. Pap.,* **25,** 527–531 (1979).

722. J. R. Caldwell, W. J. Jackson, Jr., and T. F. Gray, Jr., "Polyesters, Thermoplastic," *Encyclopedia of Polymer Science and Technology,* Vol. 1, Wiley, New York, 1976, pp. 444–467.

723. B. W. Pengilly, "Thermoplastic Polyester: PET," in *Modern Plastics Encyclopedia,* 1978–79 ed., Vol. 55, No. 10A, McGraw-Hill, New York, p. 50.

724. D. P. Wyman, "Thermoplastic Polyester: PBT," in *Modern Plastics Encyclopedia,* 1978–79 ed., Vol. 55, No. 10A, McGraw-Hill, New York, p. 49.

725. G. Farrow and E. S. Hill, "Polyester Fibers," *Encyclopedia of Polymer Science and Technology,* Vol. 11, Wiley, New York, 1969, pp. 1–41.

726. K. H. Burg and G. Sextro, "Physical Constants of Poly(oxymethylene)," in J. Brandrup and E. H. Immergut, Eds., *Polymer Handbook,* 2nd ed., Wiley, New York, 1975, V-63 to V-70.

727. E. I. du Pont de Nemours and Company, Inc., *Designing with Plastics,* Plastic Products and Resins Dept., Wilmington, Del., n.d.

728. K. J. Persak and L. M. Blair, "Acetal Resins," *Encyclopedia of Chemical Technology,* Vol. 1, Wiley, New York, 1978, pp. 112–123.

729. G. Bommi and P. LeBlanc, "Phenylene Oxide-Based Resin," in *Modern Plastics Encyclopedia,* 1978–79 ed., Vol. 55, No. 10A, McGraw-Hill, New York, pp. 36–38.

730. R. K. Johannes and J. Y. Dutour, "Acetal Homopolymer," in *Modern Plastics Encyclopedia,* 1978–79 ed., Vol. 55, No. 10A, McGraw-Hill, New York, pp. 6–7.

731. G. C. Bailey and H. W. Hill, Jr., "Polyphenylene Sulfide: A New Industrial Resin," *Am. Chem. Soc. Div. Org. Coatings Plast. Chem.,* **34,** 156–161 (1974).

732. R. N. Johnson, "Polysulfones," *Encyclopedia of Polymer Science and Technology,* Vol. 11, Wiley, New York, 1969, pp. 447–463.

733. V. J. Leslie, J. B. Rose, G. O. Rudkin, and J. Feltzin, "Polyethersulphone—A New High Temperature Engineering Thermoplastic," *Am. Chem. Soc. Div. Org. Coatings Plast. Chem.,* **34,** 142–155 (1974).

734. (a) M. S. Crandall, "Worker Exposure to Styrene Monomer in the Reinforced Plastic Boat-Making Industry," *Am. Ind. Hyg. Assoc. J.,* **42,** 499–502 (1981); (b) R. L. Schumacher, P. A. Breysse, W. R. Carlyon, R. P. Hibbard, and G. D. Kleinman, "Styrene Exposure in the Fiberglass Fabrication Industry in Washington State," *Am. Ind. Hyg. Assoc. J.,* **42,** 143–149 (1981); (c) A. Tossavainen, "Styrene Use and Occupational Exposure in the Plastics Industry," *Scand. J. Work Environ. Health,* **4,** 7–13 (1978).

735. J. B. Terrill, unpublished data, E. I. du Pont de Nemours and Company, Inc., Haskell Laboratory, Wilmington, Del., September 1975.

736. E. P. Goodings, "Thermal Degradation of Polyethylene Terephthalate," in Society of Chemical Industry, *High Temperature Resistance and Thermal Degradation of Polymers,* S. C. I. Monograph No. 13, MacMillan, New York, 1961, pp. 211–228.

737. L. H. Buxbaum, "The Degradation of Poly(ethylene Terephthalate)," *Angew. Chem. Int. Ed.,* **7,** 182–190 (1968).

738. W. L. Hergenrother, "Influence of Copolymeric Poly(diethylene Glycol) Terephthalate on the Thermal Stability of Poly(ethylene Terephthalate)," *J. Polym. Sci.,* **12,** 875–883 (1974).

739. W. T. Langstaff and L. C. Trent, "The Effect of Polyester Fiber Content on the Burn Injury Potential of Polyester/Cotton Blend Fabrics," *J. Consum. Prod. Flammability,* **7,** 26–39 (1980).

740. J. W. Sarver, unpublished data, E.I. du Pont de Nemours and Company, Inc., Haskell Laboratory, Wilmington, Del., October 1974.

741. O. L. Dashiell, unpublished data, E. I. du Pont de Nemours and Company, Inc., Haskell Laboratory, Wilmington, Del., June 1975.

742. V. Passalacqua, F. Pilati, V. Zamboni, B. Fortunato, and P. Manaresi, "Thermal Degradation of Poly(butylene Terephthalate)," *Polymer,* **17,** 1044–1048 (1976).

743. C. S. Barrow, Y. Alarie, and M. F. Stock, "Sensory Irritation and Incapacitation Evoked by Thermal Decomposition Products of Polymers and Comparisons with Known Sensory Irritants," *Arch. Environ. Health,* **33,** 79–88 (1978).

744. D. P. Miller, R. V. Petrella, and A. Manca, "An Evaluation of Some Factors Affecting the Smoke and Toxic Gas Emission from Burning Unsaturated Polyester Resins," *Am. Chem. Soc. Div. Org. Coatings Plast. Chem.,* **36,** 576–581 (1976).

745. C. J. Hilado, C. J. Casey, and J. E. Schneider, "Effect of Pyrolysis Temperature on Relative Toxicity of Some Plastics," *Fire Technol.,* **15,** 122–129 (1979).

746. C. J. Hilado and N. V. Huttlinger, "Concentration-Response Data on Toxicity of Pyrolysis Gases from Six Synthetic Polymers," *J. Combust. Toxicol.,* **5,** 81–101 (1978).

747. A. Davis and J. H. Golden, "Thermal Degradation of Polycarbonate," *J. Chem. Soc.,* **1,** 45–47 (1968).

748. J. W. Clayton and G. Limperos, unpublished data, E. I. du Pont de Nemours and Company, Inc., Haskell Laboratory, Wilmington, Del., August 1956–August 1957.

749. R. G. McKee, S. B. Martin, J. V. Dilley, and D. Palmer, "Effects of Combustion Mode on the

Smoke Toxicity of Polyoxymethylene (Delrin)," *Natl. SAMPE Tech. Conf.*, **11**, 582–592, 1979; *Chem. Abstr.*, **92**, 70553f.

750. C. J. Hilado, J. E. Schneider, and D. P. Brauer, "Toxicity of Pyrolysis Gases from Polyoxymethylene," *J. Combust. Toxicol.*, **6**, 30–36 (1979).

751. C. E. Schweitzer, R. N. MacDonald, and J. O. Punderson, "Thermally Stable High Molecular Weight Polyoxymethylenes," *J. Appl. Polym. Sci.*, **1**, 158–163 (1959).

752. R. S. Robinson, D. P. Dressler, D. L. Dugger, and P. Cukor, "Smoke Toxicity of Fire-Retardant Television Cabinets," *J. Combust. Toxicol.*, **4**, 435–453 (1977).

753. G. Ball, B. Weiss, and E. A. Boettner, "Analysis of the Volatile Combustion Products of Polyphenylene Oxide Plastics," *Am. Ind. Hyg. Assoc. J.*, **31**, 572–578 (1970).

754. J. M. Cox, B. A. Wright, and W. W. Wright, "Thermal Degradation of Poly(phenylene Oxides)," *J. Appl. Polym. Sci.*, **9**, 513–522 (1965).

755. G. F. L. Ehlers, K. R. Fisch, and W. R. Powell, "Thermal Degradation of Polymers with Phenylene Units in the Chain. II. Sulfur-Containing Polyarylenes," *J. Polym. Sci.*, **7**, 2955–2967 (1969).

756. N. S. J. Christopher, J. L. Cotter, G. J. Knight, and W. W. Wright, "Thermal Degradation of Poly(phenylene Sulfide) and Perfluoropoly(phenylene Sulfide)," *J. Appl. Polym. Sci.*, **12**, 863–870 (1968).

757. C. J. Hilado and E. M. Olcomendy, "Toxicity of Pyrolysis Gases from Polyether Sulfone," *J. Combust. Toxicol.*, **6**, 117–123 (1979).

758. K. P. Lee, personal communication, Haskell Laboratory, E. I. du Pont de Nemours and Company, Inc., Wilmington, Del., July 1980.

759. J. R. Tiller and R. S. F. Schilling, "Respiratory Function During the Day in Rayon Workers—A Study in Byssinosis," *Trans. Assoc. Ind. Med. Off.*, **7**, 161–162 (1958).

760. G. Berry, M. K. B. Molyneux, and J. B. L. Tombleson, "Relationships Between Dust Level and Byssinosis and Bronchitis in Lancashire Cotton Mills," *Br. J. Ind. Med.*, **31**, 18–27 (1974).

761. J. D. Brain, D. E. Knudson, S. P. Sorokin, and M. A. Davis, "Pulmonary Distribution of Particles Given by Intratracheal Instillation or by Aerosol Inhalation," *Environ. Res.*, **11**, 13–33 (1976).

762. M. F. Stanton, M. Layard, A. Tegeris, E. Miller, M. May, and E. Kent, "Carcinogenicity of Fibrous Glass: Pleural Response in the Rat in Relation to Fiber Dimension," *J. Natl. Cancer Inst.*, **58**, 587–597 (1977).

763. Textile Research Institute, *Methods of Measuring the Particle Sizes of Respirable Dusts; A Review of the Literature with Special Emphasis on Fibrous Dusts*, Report No. 5, Princeton, N.J., 1980.

764. R. C. Knowlton, N. C. Pierce, and P. Popper, "'Snow' Deposits in Friction Twisting," *Fiber Prod.*, **6**(6), 18–30 (1978).

765. J. M. Hawthorne and C. J. Heffelfinger, "Polyester Films," *Encyclopedia of Polymer Science and Technology*, Vol. 11, Wiley, New York, 1969, pp. 42–61.

766. H. Sherman, unpublished data, E. I. du Pont de Nemours and Company, Inc., Haskell Laboratory, Wilmington, Del., January 1973.

767. F. X. Wazeter and E. I. Goldenthal, unpublished data, International Research and Development Corp., Mattawan, Mich., June 1973 (sponsored by Haskell Laboratory).

768. F. X. Wazeter and E. I. Goldenthal, unpublished data, International Research and Development Corp., Mattawan, Mich., June 1973 (sponsored by Haskell Laboratory).

769. T. Otaka, K. Miyoshi, Y. Komai, and H. Nakayoshi, "Safety Evaluation of Polyethylene Terephthalate Resins. I. Studies on Acute and Subacute (3 months) Toxicities by Oral Administration in Rats," *Shokuhin Eiseigaku Zasshi*, **19**(5), 431–442 (1978); *Chem. Abstr.*, **91**, 14755r.

770. K. M. Frank, unpublished data, E. I. du Pont de Nemours and Company, Inc., Haskell Laboratory, Wilmington, Del., July 1972.

771. R. W. Morrow, unpublished data, E. I. du Pont de Nemours and Company, Inc., Haskell Laboratory, Wilmington, Del., November 1972.

772. J. W. McAlack and N. C. Goodman, unpublished data, E. I. du Pont de Nemours and Company, Inc., Haskell Laboratory, Wilmington, Del., December 1976.

773. M. V. Shelanski, unpublished data, Product Investigations, Inc., Conshohocken, Pa., October 1979 (sponsored by Haskell Laboratory).

774. M. V. Shelanski, unpublished data, Product Investigations, Inc., Conshohocken, Pa., December 1979 (sponsored by Haskell Laboratory).

775. J. F. Wilson and K. L. Gabriel, unpublished data, Biosearch, Inc., Philadelphia, Pa., October 1979 (sponsored by Haskell Laboratory).

776. J. B. Terrill and K. P. Lee, unpublished data, E. I. du Pont de Nemours and Company, Inc., Haskell Laboratory, Wilmington, Del., May 1977.

777. R. W. Hartgrove and K. P. Lee, unpublished data, E. I. du Pont de Nemours and Company, Inc., Haskell Laboratory, Wilmington, Del., July 1978.

778. R. N. King and D. J. Lyman, "Polymers in Contact with the Body," *Environ. Health Perspect.,* **11,** 71–74 (1975).

779. J. H. Harrison, D. S. Swanson, and A. F. Lincoln, "A Comparison of the Tissue Reactions to Plastic Materials," *Arch. Surg.,* **74,** 139–144 (1957).

780. H. Kus, K. Kawecki, and E. Szewczak, "Studies on Carcinogenesis and the Usefulness of Polyester Yarn in Alloplasty," *Arch. Immunol. Ther. Exp.,* **12,** 730–739 (1964).

781. H. C. Amstutz, W. F. Coulson, and E. David, "Reconstruction of the Canine Achilles and Patellar Tendons Using Dacron Mesh Silicone Prosthesis. I. Clinical and Biocompatibility Evaluation," *J. Biomed. Mater. Res.,* **10,** 47–59 (1976).

782. Y. M. Tardif, C. L. Schepens, and F. I. Tolentino, "Vitreous Surgery," *Arch. Ophthalmol.,* **95,** 229–234 (1977).

783. R. H. Hayward and F. L. Korompai, "Degeneration of Knitted Dacron Grafts," *Surgery,* **79,** 581–583 (1976).

784. H. Kulenkampff and G. Simonis, "Biological Tolerance of Vascular Protheses Made of Dacron® and Synthetic Suture Material," *Chirurg,* **47,** 189–192 (1976).

785. D. M. Conning, M. J. Hayes, J. A. Styles, and J. A. Nicholas, "Comparison Between *In Vitro* Toxicity of Dusts of Certain Polymers and Minerals and Their Fibrogenicity," in W. H. Walton, Ed., *Inhaled Particles III,* Vol. 1, Unwin Brothers Limited, Surrey, England, 1971, pp. 499–506.

786. W. A. Burnes, S. Kanhouwa, L. Tillman, N. Saini, and J. B. Herrmann, "Fibrosarcoma Occurring at the Site of a Plastic Vascular Graft," *Cancer,* **29,** 66–72 (1972).

787. T. X. O'Connell, H. J. Fee, and A. Golding, "Sarcoma Associated with Dacron Prosthetic Material," *J. Thorac. Cardiov. Surg.,* **72,** 94–96 (1976).

788. T. T. Daurova, O. S. Voronkova, S. D. Andreev, T. E. Rudakova, Y. V. Moiseev, L. L. Razumova, and G. E. Zaikov, "Kinetics of the Degradation of Poly(ethylene Terephthalate in Body Tissues," *Dokl. Akad. Nauk SSSR,* **231**(4), 919–920 (1976); *Chem. Abstr.,* **87,** 15639k.

789. W. F. H. Borman and M. Kramer, "Poly(butyleneterephthalate) Chemical and Physical Properties Affecting Its Processing and Usage," *Am. Chem. Soc. Div. Org. Coatings Plast. Chem.,* **34,** 77–85 (1974).

790. B. D. Astill and R. L. Raleigh, unpublished data, Health, Safety and Human Factors Laboratory, Rochester, N.Y., May 1974 (courtesy of the Society of the Plastics Industry, New York).

791. H. E. Schroeder, "Thermoplastic or Elastomer? A New Engineering Polymer," *Shell Polym., 3,* 70–73 (1979).

792. H. P. Cordts and J. A. Bauer, "Unsaturated Polyester," in *Modern Plastics Encyclopedia,* 1978–79 ed., Vol. 55, No. 10A, McGraw-Hill, New York, pp. 54–59.

793. L. B. Bourne and F. J. M. Milner, "Polyester Resin Hazards, *Br. J. Ind. Med., 20,* 100–109 (1963).

794. M. M. Key and D. P. Discher, "Polyester Resins—Their Dermatologic Aspects in Industry," *Cutis, 2,* 27–29 (1966).

795. K. E. Malten, "Occupational Dermatoses in the Processing of Plastics," *Trans. St. John's Hosp. Dermatol. Soc., 59,* 78–113 (1973).

796. J. Lim, J. L. Balzer, C. R. Wolf, and T. H. Milby, "Fiber Glass Reinforced Plastics," *Arch. Environ. Health, 20,* 540–544 (1970).

797. L. Jirasek, "Polyester Resins and Glass Laminates," *Prac. Lek., 14,* 120–124, 1962; *Chem. Abstr., 57,* 7560d.

798. J. S. McDermott, "Health and Safety in Reinforced Plastics Fabrication," in Society of Plastics Engineers, Inc., *Safety & Health with Plastics,* National Technical Conference, Denver, Colo., November 8–10, 1977, pp. 187–191.

799. H. L. Boiteau and F. Rossel-Renac, "Étude expérimentale de l'evaporation du styrène au cours de la polymérisation des résines polyesters insaturées," *Arch. Mal. Prof. Med. Trav. Secur. Soc., 39,* 52–59 (1977).

800. R. G. Mraz and R. P. Silver, "Alkyd Resins," *Encyclopedia of Polymer Science and Technology,* Vol. 1, Wiley, New York, 1964, pp. 663–734.

801. D. W. Lichtenberg, "Alkyd," in *Modern Plastics Encyclopedia,* 1978–79 ed., Vol. 55, No. 10A, McGraw-Hill, New York, pp. 10.

802. J. L. Thomas, "Allyl," in *Modern Plastics Encyclopedia,* 1978–79 ed., Vol. 55, No. 10A, McGraw-Hill, New York, pp. 9–10.

803. M. Lacroix, H. Burckel, J. Foussereau, E. Grosshans, C. Cavelier, J. C. Limasset, P. Ducos, D. Gradinski, and P. Duprat, "Irritant Dermatitis from Diallyglycol Carbonate Monomer in the Optical Industry," *Contact Dermatitis, 2,* 183–195 (1976).

804. R. O. Carhart, R. L. Stadterman, and R. O. L. Lynn, "Polycarbonate," in *Modern Plastics Encyclopedia,* 1978–79 ed., Vol. 55, No. 10A, McGraw-Hill, New York, p. 46.

805. R. E. Weyers, P. R. Blankenhorn, L. R. Stover, and D. E. Kline, "Effects of Sterilization Procedures on the Tensile Properties of Polycarbonate," *J. Appl. Polym. Sci., 22,* 2019–2024 (1978).

806. F. L. Rodkey, H. A. Collison, and R. R. Engel, "Release of Carbon Monoxide from Acrylic and Polycarbonate Plastics," *J. Appl. Physiol., 27,* 554–555 (1969).

807. M. Sohrabi and K. Z. Morgan, "A New Polycarbonate Fast Neutron Personnel Dosimeter," *Am. Ind. Hyg. Assoc. J., 39,* 438–447 (1978).

808. G. Bornmann and A. Loeser, "Monomer-Polymer," *Arzneim.-Forsch., 9*(9), 9–13 (1959); *Chem. Abstr., 53,* 11662f.

809. "Thermo Plastic Resin Decomposition," *Mich. Occup. Health, 16*(2), 8 (1970–71).

810. W. M. Cleary, "Thermoplastic Resins Decomposition," *Ind. Med., 39*(3), 34–36 (1970).

811. L. Beudouin, "Acetal Resins. Injection Molding: The Risks and Their Prevention," *Cah. Notes Doc., 85,* 545–552, 1976; *Chem. Abstr., 87,* 172178b.

812. F. X. Wazeter and E. I. Goldenthal, unpublished data, International Research and Development Corp., Mattawan, Mich., June 1973 (sponsored by Haskell Laboratory).

813. R. J. Neher, unpublished data, E. I. du Pont de Nemours and Company, Inc., Haskell Laboratory, Wilmington, Del., July 1964.

814. J. Kopecny, E. Cerny, and D. Ambroz, "Effect of Polyformaldehyde Dust in Rat Tissues," *Scr. Med.*, **41**, 405–409 (1968).

815. H. F. Smyth, Jr., C. S. Weil, M. D. Woodside, J. B. Knaak, L. J. Sullivan, and C. P. Carpenter, "Experimental Toxicity of a High Molecular Weight Poly(ethylene Oxide)," *Toxicol. Appl. Pharmacol.*, **16**, 442–445 (1970).

816. P. J. Boeke, "Polyphenylene Sulfide," in *Modern Plastics Encyclopedia*, 1978–79 ed., Vol. 55, No. 10A, McGraw-Hill, New York, pp. 73–74.

817. H. J. Fabris, "Thermal and Oxidative Stability of Urethanes," in K. C. Frisch and S. L. Reegen, Eds., *Advances in Urethane Science and Technology*, Vol. 6, Westport, Conn., 1978, pp. 173–196.

818. K. A. Pigott, "Polyurethanes," *Encyclopedia of Polymer Science and Technology*, Vol. 11, Wiley, New York, 1969, pp. 506–563.

819. "World Polyurethane Use Heads for 7 Billion Lb," *Chem. Eng. News*, **57**(42), 12–13 (1979).

820. E. Heuser, W. Reusche, K. Wrabetz, and R. Fauss, "Determination of Free Monomeric Tolylene Diisocyanates and Hexamethylene Diisocyanate in Polyurethane Paints," *Z. Anal. Chem.*, **257**(2), 119–125 (1971); *Chem. Abstr.*, **76**, 114874n.

821. G. B. Cox and K. Sugden, "The Determination of 1,6-Hexamethylene Diisocyanate in Paint Hardeners by High-Pressure Liquid Chromatography," *Anal. Chim. Acta*, **91**, 365–368 (1977).

822. P. McFadyen, "Determination of Free Toluene Diisocyanate in Polyurethane Prepolymers by High-Performance Liquid Chromatography," *J. Chromatogr.*, **123**, 468–473 (1976).

823. M. P. Potapova, V. I. Lushchik, T. A. Ermolaeva, and I. A. Pronina, "Gas Chromatographic Determination of Free Monomers of Hexamethylene and Tolylene Diisocyanates in Polyure-thane Prepolymers," *Plaste Kaut.*, **22**(12), 988–989 (1975); *Chem. Abstr.*, **84**, 60290p.

824. F. Spagnolo and W. M. Malone, "Quantitative Determination of Small Amounts of Toluene Diisocyanate Monomer in Urethane Adhesives by Gel Permeation Chromatography," *J. Chromatogr. Sci.*, **14**, 52–56 (1976).

825. R. B. Konzen, B. F. Craft, L. D. Scheel, and C. H. Gorski, "Human Response to Low Concentrations of *p,p*-Diphenylmethane Diisocyanate (MDI)," *Am. Ind. Hyg. Assoc. J.*, **27**, 121–127 (1966).

826. R. L. Ruhe and J. B. Lucas, *Health Hazard Evaluation/Toxicity Determination Report H.H.E. 73-156-205*, Goodyear Tire and Rubber Company, St. Mary's, Ohio, National Institute for Occupational Safety and Health, Cincinnati, Ohio, 1975 (PB 249385).

827. W. J. Moorman, J. B. Lal, R. E. Biagini, and W. D. Wagner, "Pulmonary Effects of Chronic Exposure to Polyurethane Foam (PUF)," *Toxicol. Appl. Pharmacol.*, in press.

828. J. Thyssen, G. Kimmerle, S. Dickhaus, E. Emminger, and U. Mohr, "Inhalation Studies with Polyurethane Foam Dust in Relation to Respiratory Tract Carcinogenesis," *J. Environ. Pathol. Toxicol.*, **1**, 501–508 (1978).

829. S. Laskin, R. T. Drew, V. P. Cappiello, and M. Kuschner, "Inhalation Studies with Freshly Generated Polyurethane Foam Dust," in T. T. Mercer, P. E. Morrow, and W. Stober, Eds., *Assessment of Airborne Particles*, Springfield, Ill., 1972, pp. 382–404.

830. J. E. Peterson, R. A. Copeland, and H. R. Hoyle, "Health Hazards of Spraying Polyurethane Foam Out-of-doors," *Am. Ind. Hyg. Assoc. J.*, **23**, 345–352 (1962).

831. National Institute for Occupational Safety and Health, *Occupational Exposure to Diisocyanates*, DHEW Publ. No. (NIOSH) 78-215, Supt. Doc., Washington, D.C., 1978.

832. H. N. MacFarland and K. J. Leong, "Hazards from the Thermodecomposition of Plastics," *Arch. Environ. Health*, **4**, 591–597 (1962).

833. J. A. Zapp, Jr., "Hazards of Isocyanates in Polyurethane Foam Plastic Production," *AMA Arch. Ind. Health*, **15**, 324–330 (1957).

834. K. Yamamoto and Y. Yamamoto, "Toxicity of Gases Released by Polyurethane Foams Subjected to Sufficiently High Temperature," *Jap. J. Legal. Med.*, **25**, 303–314 (1971).

835. W. C. Thomas and E. J. O'Flaherty, "Cytochrome c Oxidase Activity in Tissues of Rats Exposed to Polyurethane Pyrolysis Fumes," *Toxicol. Appl. Pharmacol.*, **49**, 463–472 (1979).

836. R. A. Parent, J. V. Dilley, S. B. Martin, and R. G. McKee, "Acute Toxicity in Fischer Rats of Smoke from Non-Flaming Combustion of Ten Flexible Polyurethane Foams," *J. Combust. Toxicol.*, **6**, 185–197 (1979).

837. R. A. Parent, G. H. Y. Lin, G. T. Pryor, S. B. Martin, R. G. McKee, and J. V. Dilley, "Behavioral Toxicity in Fischer Rats Exposed to Smoke from Non-Flaming Combustion of Ten Flexible Polyurethane Foams," *J. Combust. Toxicol.*, **6**, 215–227 (1979).

838. R. A. Parent, J. V. Dilley, and V. F. Simmon, "Mutagenic Activity of Smoke Condensates from the Non-Flaming Combustion of Ten Flexible Polyurethane Foams Using the Salmonella/ Microsome Assay," *J. Combust. Toxicol.*, **6**, 256–264 (1979).

839. J. H. Petajan, K. J. Voorhees, S. C. Packham, R. C. Baldwin, I. N. Einhorn, M. L. Grunnet, B. G. Dinger, and M. M. Birky, "Extreme Toxicity from Combustion Products of a Fire-Retarded Polyurethane Foam," *Science*, **187**, 742–744 (1974).

840. J. G. Keller, W. R. Herrera, and B. E. Johnston, "An Investigation of Potential Inhalation Toxicity of Smoke from Rigid Polyurethane Foams and Polyester Fabrics Containing Antiblaze® 19 Flame Retardant Additive," *J. Combust. Toxicol.*, **3**, 296–304 (1976).

841. C. J. Hilado and R. M. Murphy, "Effect of Foam Age on Toxicity of Pyrolysis Gases from Polyurethane Flexible Foams," *J. Combust. Toxicol.*, **5**, 239–247 (1978).

842. Y. C. Alarie, E. Wilson, T. Civic, J. H. Magill, J. M. Funt, C. Barrow, and J. Frohliger, "Sensory Irritation Evoked by Polyurethane Decomposition Products," *J. Fire Flammability/Combust. Toxicol.*, **2**, 139–150 (1975).

843. W. D. Woolley, "Toxic Products from Plastic Materials in Fires," *Plast. Polym.*, **41**, 280–286 (1973).

844. W. D. Woolley, "The Production of Free Tolylene Diisocyanate (TDI) from the Thermal Decomposition of Flexible Polyurethane Foams," *J. Fire Flammability/Combust. Toxicol.*, **1**, 259–267 (1974).

845. W. D. Woolley, "Nitrogen-containing Products from the Thermal Decomposition of Flexible Polyurethane Foams," *Br. Polym. J.*, **4**, 27–43 (1972).

846. W. D. Woolley, P. J. Fardell, and I. G. Buckland, *The Thermal Decomposition Product of Rigid Polyurethane Foams Under Laboratory Conditions*, Joint Fire Research Organization, Fire Research Station (Fire Research Note No. 1039), Borehamwood, Hertfordshire WD6, 2BL, England, August 1975.

847. J. Chambers and C. B. Reese, "The Thermal Decomposition of Some Polyurethane Foams," *Br. Polym. J.*, **8**, 48–53 (1976).

848. D. W. Skidmore and P. R. Sewell, "The Evolution of Toxic Gases from Heated Polymers—III; Isocyanate Products," *Eur. Polym. J.*, **11**, 139–142 (1975).

849. F. D. Hileman, K. J. Voorhees, and I. N. Einhorn, "Pyrolysis of a Flexible-Urethane Foam," National Academy of Sciences, *Physiological and Toxicological Aspects of Combustion Products* (International Symposium, March 18–20, 1974, University of Utah, Salt Lake City), Washington, D.C., 1976, pp. 226–244.

850. K. L. Paciorek, R. H. Kratzer, J. Kaufman, and A. M. Hartstein, "Oxidative Thermal Degradation of Selected Polymeric Compositions," *Am. Ind. Hyg. Assoc. J.*, **35**, 175–180 (1974).

851. D. W. Mitchell, E. M. Murphy, and J. Nagy, *Fire Hazard of Urethane Foam in Mines*, U.S. Dept. of the Interior, Bureau of Mines, Report of Investigations 6837, Washington, D.C., 1966.

852. V. J. Seemann and U. Wolcke, "Über die Bildung toxischer Isocyanatdämpfe bei der thermischen Zersetzung von Polyurethanlacken und ihren polyfunktionellen Härtern," *Zentralbl. Arbeitsmed. Arbeitsschutz*, **26**, 2–9 (1976); from authors' English abstract.

853. O. H. Bullitt, personal communication, E. I. du Pont de Nemours and Company, Inc., Fabrics & Finishes Dept., Wilmington, Del., January 1980.

854. D. P. G. Paisley, "Isocyanate Hazard from Wire Insulation: An Old Hazard in a New Guise," *Br. J. Ind. Med.*, **26**, 79–81 (1969).

855. H. H. G. Jellinek and K. Takada, "Toxic Gas Evolution from Polymers: Evolution of Hydrogen Cyanide from Linear Polyurethane," *J. Polym. Sci.*, **13**, 2709–2723 (1975).

856. H. H. G. Jellinek and K. Takada, "Toxic Gas Evolution from Polymers: Evolution of Hydrogen Cyanide from Polyurethane," *J. Polym. Sci.*, **15**, 2269–2288 (1977).

857. N. Grassie and M. Zulfiqar, "Thermal Degradation of the Polyurethane from 1,4-Butanediol and Methylene bis(4-Phenyl Isocyanate)," *J. Polym. Sci.*, **16**, 1563–1574 (1978).

858. K. J. Voorhees, F. D. Hileman, I. N. Einhorn, and J. H. Futrell, *An Investigation of the Thermolysis Mechanism of Model Urethanes*, presented at the Fourth International Cellular Plastics Conference of the Cellular Plastics Division, Society of the Plastics Industry, Montreal, Canada, November 1976 (Flammability Research Center, University of Utah FRC/UU-076, UTEC 76-238).

859. J. Autian, A. R. Singh, J. E. Turner, G. W. C. Hung, L. J. Nunez, and W. H. Lawrence, "Carcinogenesis from Polyurethans," *Cancer Res.*, **35**, 1591–1596 (1975).

860. J. Autian, A. R. Singh, J. E. Turner, G. W. C. Hung, L. J. Nunez, and W. H. Lawrence, "Carcinogenic Activity of a Chlorinated Polyether Polyurethan," *Cancer Res.*, **36**, 3973–3977 (1976).

861. T. D. Darby, H. J. Johnson, and S. J. Northrup, "An Evaluation of a Polyurethane for Use as a Medical Grade Plastic," *Toxicol. Appl. Pharmacol.*, **46**, 449–453 (1978).

862. B. R. Grant, personal communication, E. I. du Pont de Nemours and Company, Inc., Polymer Products Dept., Wilmington, Del., March 1980.

863. R. R. Beard, "Polyurethanes; Hazards & Control," *Mich. Occup. Health*, **16**(1), 2–7 (1970).

864. H. Ulrich, "Polyurethane," in *Modern Plastics Encyclopedia*, 1978–79 ed., Vol. 55, No. 10A, McGraw-Hill, New York, pp. 88, 90, 96, 97.

865. K. L. Stemmer, E. Bingham, and W. Barkley, "Pulmonary Response to Polurethane Dust," *Environ. Health Perspect.*, **11**, 109–113 (1975).

866. R. H. Bell, K. L. Stemmer, W. Barkley, and L. D. Hollingsworth, "Cyanide Toxicity from the Thermal Degradation of Rigid Polyurethane Foam," *Am. Ind. Hyg. Assoc. J.*, **40**, 757–762 (1979).

867. W. D. Woolley and M. M. Raftery, "Smoke and Toxicity Hazards of Plastics in Fires," *J. Hazard Mater.*, **1**, 215–222 (1975/76).

868. G. Kimmerle, "Toxicity of Combustion Products with Particular Reference to Polyurethane," *Ann. Occup. Hyg.*, **19**, 269–273 (1976).

869. A. E. Willey, "Fire Problems with Foam Plastics," in S. Steingiser, Ed., *Safety and Product Liability*, Westport, Conn., 1974, pp. 76–81.

870. R. H. Rigdon, "Local Reaction to Polyurethane—A Comparative Study in the Mouse, Rat, and Rabbit," *J. Biomed. Mater. Res.*, **7**, 79–93 (1973).

871. J. Šmahel, "Tissue Reactions to Breast Implants Coated with Polyurethane," *Plast. Reconstr. Surg.*, **61**, 80–85 (1978).

872. A. M. Kaplan, R. T. Darby, M. Greenberger, and M. R. Rogers, "Microbial Deterioration of Polyurethane Systems," in Society for Industrial Microbiology, *Dev. Ind. Microbiol.*, Vol. 9, American Institute of Biological Sciences, Washington, D.C., 1968, pp. 201–217.

873. R. T. Darby and A. M. Kaplan, "Fungal Susceptibility of Polyurethanes," *Appl. Microbiol.*, **16**, 900–905 (1968).

874. D. A. Meyer, "Urethane Elastomers," in M. Morton, Ed., *Rubber Technology*, Reinhold, New York, 1973, pp. 440–458.

875. D. A. Meyer, "Polyurethane Elastomers," in G. G. Winspear, Ed., *Rubber Handbook,* Vanderbilt, New York, 1968, pp. 208–220.

876. D. J. Lyman, W. J. Seare, Jr., D. Albo, Jr., S. Bergman, J. Lamb, L. C. Metcalf, and K. Richards, "Polyurethane Elastomers in Surgery," *Int. J. Polym. Mater.,* **5,** 211–229 (1977).

877. E. A. Emmett, "Allergic Contact Dermatitis in Polyurethane Plastic Moulders," *J. Occup. Med.,* **18,** 802–804 (1976).

878. J. A. Mock, "Plastic and Elastomeric Coatings," in C. A. Harper, Ed., *Handbook of Plastics and Elastomers,* McGraw-Hill, New York, 1975, pp. 9–1 to 9–51.

879. J. C. Moseley and H. J. Allen, Jr., "Polyurethane Coating in the Prevention of Nickel Dermatitis," *Arch. Dermatol.,* **103,** 58–60 (1971).

880. K. E. Malten, "Contact Sensitization to Letterflex Urethane Photoprepolymer Mixture Used In Printing," *Contact Dermatitis,* **3,** 115–121 (1977).

881. E. M. Hicks, Jr., A. J. Ultee, J. Drougas, "Spandex Elastic Fibers," *Science,* **147,** 373–379 (1965).

882. H. Oertel, "Structure, Modification Potential and Properties of Segmented Polyurethane Elastomer Filament Yarns," *Chemiefasern/Textilindustrie, 28/80,* E10–E12 (1978).

883. H. L. Joseph and H. I. Maibach, "Contact Dermatitis from Spandex Brassieres," *J. Am. Med. Assoc.,* **201,** 880–882 (1967).

884. E. Van Dijk, "Contact Dermatitis Due to Spandex," *Acta Dermato-Venereol.,* **48,** 589–591 (1968).

885. A. M. Kligman, unpublished data, Ivy Research Laboratories, Inc., Philadelphia, Pa., April 1977 (sponsored by Haskell Laboratory).

886. O. L. Dashiell, unpublished data, E. I. du Pont de Nemours and Company, Inc., Haskell Laboratory, Wilmington, Del., August 1974.

887. C. W. Colburn, unpublished data, E. I. du Pont de Nemours and Company, Inc., Haskell Laboratory, Wilmington, Del., January 1970.

888. S. S. Bleehen and P. Hall-Smith, "Brassiere Depigmentation: Light and Electron Microscope Studies," *Br. J. Dermatol.,* **83,** 157–160 (1970).

889. B. T. Mossman and J. E. Craighead, "Induction of Neoplasms in Hamster Tracheal Grafts with 3-Methylcholanthrene-coated Lycra Fibers," *Cancer Res.,* **38,** 3717–3722 (1978).

890. P. A. Waitkus and G. R. Griffiths, "Explosion Venting of Phenolic Reactors-Toward Understanding Optimum Explosion Vent Diameters," in Society of Plastics Engineers, Inc., *Safety & Health with Plastics,* National Technical Conference, Denver, Colo., November 8–10, 1977, pp. 181–186.

891. W. A. Keutgen, "Phenolic Resins," *Encyclopedia of Polymer Science and Technology,* Vol. 10, Wiley, New York, 1969, pp. 1–73.

892. W. R. Sorensen and T. W. Campbell, *Preparative Methods of Polymer Chemistry,* 2nd ed., Wiley, New York, 1968.

893. T. E. Steiner, "Phenolic," in *Modern Plastics Encyclopedia,* 1978–79 ed., Vol. 55, No. 10A, McGraw-Hill, New York, pp. 34–36.

894. J. Economy and L. Wohrer, "Phenolic Fibers," *Encyclopedia of Polymer Science and Technology,* Vol. 15, Wiley, New York, 1971, pp. 365–375.

895. M. Tsuge, T. Miyabayashi, and S. Tanaka, "Determination of Free Phenol in Phenol Resole Resin by Gel Permeation Chromatography," *Chem. Lett.,* **3,** 275–278 (1973).

896. M. Nagaoka and G. E. Myers, "Formaldehyde Emission: Methods of Measurement and Effects of Particleboard Variables," manuscript, Forest Products Laboratory, Madison, Wis., December 1979; to be submitted to *Forest Prod. J.*

897. J. Foussereau, C. Cavelier, and D. Selig, "Occupational Eczema from para-Tertiary-Butylphenol Formaldehyde Resins: A Review of the Sensitizing Resins," *Contact Dermatitis,* **2,** 254–258 (1976).

898. L. E. Gaul, "Absence of Formaldehyde Sensitivity in Phenol-Formaldehyde Resin Dermatitis," *J. Invest. Dermatol.*, **48**, 485–486 (1967).

899. A. A. Fisher, "Vitiligo Due to Contactants," *Cutis*, **17**, 431–448 (1976).

900. A. Abe and T. Ishikawa, "Studies on Pneumoconiosis Caused by Organic Dusts," *J. Sci. Labor*, **43**, 19–41 (1967); from authors' English abstract.

901. T. Sano, "Pathology and Pathogenesis of Organic Dust Pneumoconiosis," *J. Sci. Labor*, **43**, 3–18 (1967); from author's English abstract.

902. J. C. Pimentel, "A Granulomatous Lung Disease Produced by Bakelite," *Amer. Rev. Resp. Dis.*, **108**, 1303–1310 (1973).

903. O. M. Ratner and G. N. Tkachuk, "Toxicological-hygienic Study of the Dust of Glass-fiber-reinforced Plastic from Phenol-Formaldehyde Resin," *Gig. Tr. Prof. Zabol.*, **10**, 52–53 (1976); *Chem. Abstr.*, **86**, 12429k.

904. J. B. Schoenberg and C. A. Mitchell, "Airway Disease Caused by Phenolic (Phenol-Formaldehyde) Resin Exposure, *Arch. Environ. Health*, **30**, 574–577 (1975).

905. J. F. Gamble, A. J. McMichael, T. Williams, and M. Battigelli, "Respiratory Function and Symptoms: An Environmental-Epidemiological Study of Rubber Workers Exposed to a Phenol-Formaldehyde Type Resin," *Am. Ind. Hyg. Assoc. J.*, **37**, 499–513 (1976).

906. G. F. Heron, "Pyrolysis of a Phenol-Formaldehyde Polycondensate," in Society of Chemical Industry, *High Temperature Resistance and Thermal Degradation of Polymers*, S. C. I. Monograph No. 13, MacMillan, New York, 1961, pp. 475–498.

907. Y. Tsuchiya and K. Sumi, *Toxicity of Decomposition Products—Phenolic Resin*, Division of Building Research (Building Research Note No. 106), National Research Council of Canada, Ottawa, December 1975.

908. J. Q. Walker and M. A. Grayson, "Thermal Oxidative Degradation Characteristics of a Phenolic Resin," in Society of Plastics Engineers, Inc., *Safety & Health with Plastics*, National Technical Conference, Denver, Colo., November 8–10, 1977, pp. 171–173.

909. W. H. Fried, "Amino," in *Modern Plastics Encyclopedia*, 1978–79 ed., Vol. 55, No. 10A, McGraw-Hill, New York, pp. 10–13.

910. National Research Council, Assembly of Life Sciences, Board on Toxicology and Environmental Health Hazards, Committee on Toxicology, *Formaldehyde—An Assessment of Its Health Effects*, National Academy of Sciences, Washington, D.C., January 1980; prepared for the Consumer Product Safety Commission.

911. Consumer Product Safety Commission, "Public Hearing Concerning Safety and Health Problems That May Be Associated with Release of Formaldehyde Gas from Urea Formaldehyde (UF) Foam Insulation," *Fed. Reg.*, **44**, 69578–69583 (1979).

912. (a) National Research Council, Assembly of Life Sciences, *Formaldehyde and Other Aldehydes*, National Academy Press, Washington, D.C. (1981); (b) G. Widmer, "Amino Resins," *Encyclopedia of Polymer Science and Technology*, Vol. 2, Wiley, New York, 1965, pp. 1–94.

913. W. J. Rossiter, Jr., R. G. Mathey, D. M. Burch, and E. T. Pierce, *Urea–Formaldehyde Based Foam Insulations: An Assessment of Their Properties and Performance*, National Bureau of Standards, Washington, D.C., 1977.

914. A. A. Fisher, N. B. Kanof, and E. M. Biondi, "Free Formaldehyde in Textiles and Paper," *Arch. Dermatol.*, **86**, 753–756 (1962).

915. K. E. Malten, "Textile Finish Contact Hypersensitivity," *Arch. Dermatol.*, **89**, 215–221 (1964).

916. S. E. O'Quinn and C. B. Kennedy, "Contact Dermatitis Due to Formaldehyde in Clothing Textiles," *J. Am. Med. Assoc.*, **194**, 123–126 (1965).

917. H. Shellow and A. T. Altman, "Dermatitis from Formaldehyde Resin Textiles," *Arch. Dermatol.*, **94**, 799–801 (1966).

918. S. Hsiao and J. E. Villaume, *Occupational Health and Safety and Environmental Aspects of Urea-Formaldehyde Resins*, Science Information Services Department, Franklin Institute Research Laboratories, Philadelphia, PA, April 1978; supported by U.S. Army Medical Research and Development Command, Fort Detrick, Frederick, Md., (AD A054991/5ST).

919. D. K. Harris, "Health Problems in the Manufacture and Use of Plastics," *Br. J. Ind. Med.*, **10**, 255–268 (1953).

920. L. Schwartz, "Dermatitis from Synthetic Resins," *J. Invest. Dermatol.*, **6**, 239–255 (1945).

921. I. Andersen, G. R. Lundqvist, and L. Molhave, "Indoor Air Pollution Due to Chipboard Used as a Construction Material," *Atmos. Environ.*, **9**, 1121–1127 (1975).

922. U.S. Consumer Product Safety Commission, *Summary of In-Depth Investigations Urea Formaldehyde Foam Home Insulation*, (Table 1), [Washington, D.C.], July 1978.

923. K. R. Long, D. A. Pierson, S. T. Brennan, C. W. Frank, and R. A. Hahne, *Problems Associated with the Use of Urea-Formaldehyde Foam for Residential Insultation, Part I: The Effects of Temperature and Humidity on Formaldehyde Release from Urea-Formaldehyde Foam Insulation*, Oak Ridge National Laboratory, Oak Ridge, Tenn. (operated by Union Carbide Corporation for the Department of Energy), September 1979; ORNL/SUB-7559/I, Contract No. W-7504-eng-26.

924. G. G. Allan, J. Dutkiewicz, and E. J. Gilmartin, "Long-Term Stability of Urea-Formaldehyde Foam Insulation," *Environ. Sci. Technol.*, **14**, 1235–1240 (1980).

925. M. Chaigneau, G. Le Moan, and C. Agneray, "Study of the Pyrolysis of Plastic Materials. IX. Urea-Formaldehyde Resins," *Ann. Pharm. Fr.*, **36**, 551–554 (1978).

926. C. D. Hollowell, J. V. Berk, and G. W. Traynor, "Impact of Reduced Infiltration and Ventilation on Indoor Air Quality in Residential Buildings," *ASHRAE Trans.*, **85**, Part 1, 1979, preprint.

927. R. H. Moss, C. F. Jackson, and J. Seiberlich, "Toxicity of Carbon Monoxide and Hydrogen Cyanide Gas Mixtures. A Preliminary Report," *AMA Arch. Ind. Hyg. Occup. Med.*, **4**, 53–64, 1951.

928. K. Hiramatsu, "Mass-spectrometric Analysis of Pyrolysis Products of Melamine-Formaldehyde and Urea-Formaldehyde Resins," *Osaka Furitsu Kogyo-Shoreikan Hokoku*, **43**, 28–33 (1967); *Chem. Abstr.*, **69**, 28147b.

929. T. Kuo, J. Lai, and W. Lin, "Formaldehyde Content of Some Plastic Dinnerwares in Taiwan," *T'ai-wan I Hsueh Hui Tsa Chih*, **77**, 218–225; *Chem. Abstr.*, **89**, 71749z.

930. A. E. Sherr, personal communication, American Cyanamid Company, Toxicity Dept., Wayne, N.J., February 1980.

931. G. J. Levinskas, L. B. Vidone, and C. B. Shaffer, unpublished data, American Cyanamid Company, Toxicity Dept., Wayne, N.J., November 1960–February 1961.

932. G. Armeli and F. Azimonti, "Observations and Prevention of Occupational Diseases in the Production of Phenolic, Ureic, and Melamine Resins," *Med. Lav.*, **59**, 534–539 (1968).

933. E. Hagen and G. Friedrich, "Analytical Chemistry of Plastics. XXV. Analyses of Pyrolysis Gases of Some Molded Plastics," *Plaste Kautsch.*, **12**(4), 215–218 (1965); *Chem. Abstr.*, **63**, 13486h.

934. I. H. Anderson, M. Cawley, and W. Steedman, "Melamine-Formaldehyde Resins I. An Examination of Some Model Compound Systems," *Br. Polym. J.*, **1**(1), 24–28 (1969).

935. I. H. Anderson, M. Cawley, and W. Steedman, "Melamine-Formaldehyde Resins II. Thermal Degradation of Model Compounds and Resins," *Br. Polym. J.*, **3**(2), 86–92 (1971).

936. W. R. Moore and E. Donnelly, "Thermal Degradation of Melamine-Formaldehyde Resins," *J. Appl. Chem.*, **13**, 537–543 (1963).

937. J. Delmonte, "Furan," in *Modern Plastics Encyclopedia*, 1978–79 ed., Vol. 55, No. 10A, McGraw-Hill, New York, p. 26.

938. D. W. Cockcroft, G. Jones, S. M. Tarlo, A. Cartier, J. Dolovich, and F. E. Hargreave, "Asthma Caused by Occupational Exposure to a Furan-based Binder System," *J. Allergy Clin. Immunol.*, **65**, 168–169 (1980).

939. R. T. Conley, "Oxidative Mechanisms for the Degradation of Phenol-, Furan-, and Urea-based Condensation Polymers," *Proc. Battelle Symp. Therm. Stabil. Polym.; Chem. Abstr.*, **60**, 16049g.

940. D. H. Lorenz, "*n*-Vinyl Amide Polymers," *Encyclopedia of Polymer Science and Technology*, Vol. 14, Wiley, New York, 1971, pp. 239–251.

941. L. J. Frauenfelder, "Universal Chromatographic-Colorimetric Method for the Determination of Trace Amounts of Polyvinylpyrrolidone and Its Copolymers in Foods, Beverages, Laundry Products, and Cosmetics," *J. Assoc. Off. Anal. Chem.*, **57**, 796–800 (1974).

942. K. Ridgway and M. H. Rubinstein, "The Quantitative Analysis of Polyvinylpyrrolidone by Infrared Spectrophotometry," *J. Pharm. Pharmacol.*, **23**, 587–589 (1971).

943. L. W. Burnette, "A Review of the Physiological Properties of Polyvinylpyrrolidinone," *Pro. Sci. Sect. Toilet Goods Assoc.*, **1962**(38), 1–4.

944. J. F. Borzelleca and S. L. Schwartz, unpublished critique for acceptable daily intake, Medical College of Virginia, Richmond, Virginia; critique submitted to World Health Organization by L. Blecher, GAF Corporation, Wayne, N.J., December 1979. See also S. L. Schwartz, in press, *Yakuzaigaku*.

945. H. A. Ravin, A. M. Seligman, and J. Fine, "Polyvinyl Pyrrolidone as a Plasma Expander. Studies on Its Excretion, Distribution and Metabolism," *N. Engl. J. Med.*, **247**, 921–929 (1952).

946. R. K. Loeffler and J. Scudder, "Excretion and Distribution of Polyvinyl Pyrrolidone in Man; As Determined by Use of Radiocarbon as a Tracer," *Am. J. Clin. Pathol.*, **23**, 311–321 (1953).

947. E. Reske-Nielsen, M. Bojsen-Moller, M. Vetner, and J. C. Hansen, "Polyvinylpyrrolidone-storage Disease," *Acta Pathol. Microbiol. Scand., Sect. A*, **84**, 397–405 (1976).

948. M. Bojsen-Moller, E. Reske-Nielsen, M. Vetner, and J. C. Hansen, "The PVP-storage Disease" (from authors' English abstract), *Ugeskr. Laeg.*, **138**, 1017–1020 (1976).

949. W. Wessel, M. Schoog, and E. Winkler, "Poly(vinylpyrrolidinone) (PVP), its Diagnostic, Therapeutic, and Technical Application and Consequences Thereof," *Arzneim.-Forsch.*, **21**, 1468–1482 (1971).

950. D. P. Discher, "Inhalation of Hairspray Resin—Does it Cause Pulmonary Disease?," in Society of Plastics Engineers, Inc., *Safety & Health with Plastics*, National Technical Conference, Denver, Col., November 8–10, 1977, pp. 21–24.

951. J. M. Gowdy and M. J. Wagstaff, "Pulmonary Infiltration Due to Aerosol Thesaurosis," *Arch. Environ. Health*, **25**, 101–108 (1972).

952. R. H. Jackson, "PBI Fiber and Fabric—Properties and Performance," *Text. Res. J.*, **48**, 314–319 (1978).

953. I. N. Einhorn, D. A. Chatfield, and D. J. Wendel, *Thermochemistry of Polybenzimidazole Foams*, Paper Presented at the Third International Symposium on Analytical Pyrolysis, Amsterdam, September 1976 (Flammability Research Center, University of Utah, FRC/UU-065, UTEC 76-126, March 1976).

954. T. J. Gair and R. J. Thimineur, "Silicone," in *Modern Plastics Encyclopedia*, 1978–79 ed., Vol. 55, No. 10A, McGraw-Hill, New York, pp. 102–108.

955. J. C. Calandra, M. L. Keplinger, E. J. Hobbs, and L. J. Tyler, "Health and Environmental Aspects of Polydimethylsiloxane Fluids," *Polym. Prepr., Am. Chem. Soc., Div. Polym. Chem.*, **17**, 1–4 (1976).

956. C. J. Hilado, C. J. Casey, D. F. Christensen, and J. Lipowitz, "Toxicity of Pyrolysis Gases from Silicone Polymers," *J. Combust. Toxicol.*, **5**, 130–140 (1978).

957. N. Grassie and I. G. Macfarlane, "The Thermal Degradation of Polysiloxanes—I; Poly(Dimethylsiloxane)," *Eur. Polym. J.*, **14**, 875–884 (1978).

958. N. Grassie and I. G. Macfarlane, *Degradation Reactions in Silicone Polymers*, Department of Chemistry, Glasgow, Scotland, 1976 (AD A050366).

959. N. Grassie, I. G. Macfarlane, and K. F. Francey, "The Thermal Degradation of Polysiloxanes—II; Poly(methylphenylsiloxane)," *Eur. Polym. J.*, **15**, 415–422 (1979).

960. R. Dailey, *Methylpolysilicones*, Informatics, Inc., Rockville, MD; Prepared for Food and Drug Administration, Washington, D.C., Bureau of Foods, 1978 (PB 289396).

961. V. K. Rowe, H. C. Spencer, S. L. Bass, "Toxicological Studies on Certain Commercial Silicones," *J. Ind. Hyg. Toxicol.*, **30**, 332–352 (1948).

962. M. G. Cutler, A. J. Collings, I. S. Kiss, and M. Sharratt, "A Lifespan Study of a Polydimethylsiloxane in the Mouse," *Food Cosmet. Toxicol.*, **12**, 443–450 (1974).

963. G. L. Kennedy, Jr., M. L. Keplinger, J. C. Calandra, and E. J. Hobbs, "Reproductive, Teratologic, and Mutagenic Studies with Some Polydimethylsiloxanes," *J. Toxicol. Environ. Health*, **1**, 909–920 (1976).

964. E. J. Hobbs, O. E. Fancher, and J. C. Calandra, "Effect of Selected Organopolysiloxanes on Male Rat and Rabbit Reproductive Organs," *Toxicol. Appl. Pharmacol.*, **21**, 45–54 (1972).

965. (a) E. J. Hobbs, M. L. Keplinger, and J. C. Calandra, "Toxicity of Polydimethylsiloxanes in Certain Environmental Systems," *Environ. Res.*, **10**, 397–406 (1975); (b) R. J. Hausner, F. J. Schoen, M. A. Mendez-Fernandez, W. S. Henly, and R. C. Geis, "Migration of Silicone Gel to Axillary Lymph Nodes After Prosthetic Mammoplasty," *Arch. Pathol. Lab. Med.*, **105**, 371–372 (1981).

966. W. J. Bobear, "Silicone Rubber," in M. Morton, Ed., *Rubber Technology*, 2nd ed., Reinhold, New York, 1973, pp. 368–406.

967. M. M. Prisco, "Silicone Elastomers," in G. G. Winspear, Ed., *Rubber Handbook*, Vanderbilt, New York, 1968, pp. 174–188.

968. A. B. Swanson, letter concerning complications of silicone elastomer prostheses, *J. Am. Med. Assoc.*, **238**, 939 (1977).

969. R. M. Nalbandian, letter concerning complications of silicone elastomer prostheses, *J. Am. Med. Assoc.*, **238**, 939 (1977).

970. A. J. Christie, K. A. Weinberger, and M. Dietrich, "Silicone Lymphadenopathy and Synovitis; Complications of Silicone Elastomer Finger Joint Prostheses," *J. Am. Med. Assoc.*, **237**, 1463–1464 (1977).

971. A. J. Christie, K. A. Weinberger, and M. Dietrich, letter concerning complications of silicone elastomer prostheses, *J. Am. Med. Assoc.*, **238**, 939 (1977).

CHAPTER FIFTY-FIVE

Alcohols

V. K. ROWE, Sc.D. (Hon.), and
S. B. McCOLLISTER

1 INTRODUCTION

1.1 Sources, Uses, and Industrial Exposure

Some alcohols are made by fermentation but, in general, the alcohols of commercial significance are made synthetically. Since each synthesis route is different, the sources for each are discussed in the sections pertaining to the individual compounds. The alcohols have a variety of uses depending on their chemical and physical properties. Generally they are used as solvents, cosolvents, and chemical intermediates. Uses specific for individual materials are discussed in the relevant sections. Industrial exposure to the alcohols is very broad because of the diversity of their use. When data are available with respect to intensity of exposure and the consequences, they are discussed in the relevant sections.

1.2 Physical and Chemical Properties

Physical and chemical properties of individual compounds are given along with any other pertinent information available.

1.3 Determination in the Atmosphere

It was but a few years ago that analyses for alcohols in the atmosphere relied almost wholly upon bubbling air through some trapping solvent and then determining the concentration in that solvent by a physical or chemical method. In general most of these methods suffer from interferences by other materials, lack of specificity, sensitivity, and slowness. However, in some instances they may still be useful, particularly where more sophisticated equipment is not available. Hence references to some are given.

The current methodology involves adsorbing on charcoal or silica gel, desorbing with a suitable solvent or combination of solvents, and making the final analysis using chromatographic procedures coupled with flame ionization detectors or even mass spectrophotometers.

NIOSH has developed and published methods for many of the alcohols which are referenced in the relevant sections. Recently Mueller and Miller (1) and Langvardt and Melcher (2) have described methods that can be used in places where a variety of organic vapors may be present at the same time. Langvardt and Melcher (2) achieved exceptionally high recoveries of some low molecular weight aliphatic alcohols. They used a gas chromatographic procedure capable of measuring both polar and nonpolar compounds present simultaneously in the work environment at concentrations between $1/100$ and 1 times the threshold limit values. Airborne organics are collected on a single activated charcoal tube for periods of 3 to 6 hr and desorbed with a two-phase (water/carbon disulfide) desorption mixture. Carbon disulfide and water phases are analyzed separately on the same gas chromatographic column (Oronite NIW on Carbopack B), which is capable of resolving all the C_1 to C_5 saturated aliphatic alcohols. Recoveries for ethanol, 1- and 2-propanol, and 1-butanol ranged from 90 to 100 percent with coefficients of variation averaging ± 7 percent. The method is such that it is likely that it can be validated for other alcohols as well.

1.4 Physiologic Response

The physiologic effects of the alcohols vary greatly and are described in the sections dealing with the individual compounds.

2 METHANOL; Methyl Alcohol, Carbinol, Wood Spirits, Wood Alcohol, Columbian Spirits, Manhattan Spirits, CAS No. 67-56-1

$$CH_3OH$$

2.1 Source, Uses, and Industrial Exposure

Synthetic methyl alcohol made from carbon oxides and hydrogen has practically replaced that distilled from wood. Methyl alcohol is used extensively as a chemical intermediate, a denaturant for ethanol, and an industrial solvent. It is used in the lacquer industry and in the preparation of celluloid, films, plastics, textile soaps, wood stains, artificial leather, and nonshatterable glass. It is used in enamels, stains, dyes for straw hats, paint and varnish removers, cleaning and dewaxing preparations, embalming fluids, antifreeze mixtures, and as a fuel for internal combustion engines. It is also used as an intermediate and as an extracting medium in organic synthesis (3). Seventy-two occupations that offer exposure to methyl alcohol have been reported by the United States Department of Labor.

Industrial injuries or fatalities have been reported from the inhalation of high concentrations of methyl alcohol by persons engaged in varnishing beer vats (4), varnishing metal (5), varnishing the engine room of a submarine (6), shellacking lead pencils (7), shellacking hogsheads (8), coloring cloth and other articles in solutions of dyes in methyl alcohol (9–11), stiffening hats (12, 13), and manufacturing shoes (14).

As might be expected from the variety of uses and diversity of the operations involved, concentrations of methanol in the work environment have been found to vary greatly (7, 10, 16–19). Sterner (17), in a study of the manufacture of photographic film where the material is kept usually within a closed system, found concentrations ranging from 200 to several thousand ppm for short periods of time during the loading of mixers and the changing of filters. Nevertheless he has estimated that the daily average of the concentration to which the operators were exposed probably was between 400 and 500 ppm. Sterner believes these latter values would not ordinarily result in any serious effect or even moderate discomfort, since numbers of men known by him to have been exposed to such conditions while handling millions of gallons of this solvent failed to show any evidence of methyl alcohol intoxication.

2.2 Physical and Chemical Properties

Physical state	Colorless, volatile liquid
Molecular formula	CH_4O
Molecular weight	32.04
Boiling point	64.7°C (20)
Melting point	−97.8°C (20)
Specific gravity	0.7915 (20/4°C) (20)
Vapor pressure	160 mm Hg (30°C) (3)
Refractive index	1.3292 (20°C) (20)
Flammability	
Flash point	12°C (20)
Explosive limits	6.0 to 36.5 percent vol. in air (20)
Ignition temperature	470°C (20)
Percent in saturated air	21.05 (30°C)

Density of saturated air (air $= 1$)	1.02
Solubility	Miscible with water, alcohols, ketones, esters, halogenated hydrocarbons, and benzene
1 mg/l $\approx$	764 ppm at 25°C, 760 mm Hg
1 ppm $\approx$	1.31 mg/m³ at 25°C, 760 mm Hg

2.3 Determination in the Atmosphere

NIOSH methods Nos. S59 and P and CAM 247 (21) have been recommended. They involve drawing a known volume of air through silica gel to trap the organic vapors present (recommended sample is 5 liters at a rate of 0.2 l/min). The analyte is desorbed with water. The sample is separated by injection into a gas chromatograph equipped with a flame ionization detector and the area of the resulting peak is determined and compared with standards. Recently Mueller and Miller (1) and Langvardt and Melcher (2) have published methods of collection and analysis by gas chromatographic procedures that give excellent results and are applicable in situations where a variety of other materials may be present. The older chemical methods (22–27) would seem to have limited usefulness except where modern equipment is not available or where unusual conditions exist.

2.4 Physiologic Response

2.4.1 Summary

Methanol is low in acute oral toxicity as measured by lethality but sublethal doses can cause serious effects to the central nervous system, the liver, and particularly to the visual system. The *route* of exposure makes little difference; it is the dose and the resulting blood concentration that counts. Methanol is mildly irritating upon direct contact with the eyes and upon prolonged contact with the skin. Absorption of toxic amounts through the skin can occur but under reasonable conditions of exposure this would not seem to be very likely. Inhalation is the most likely route of exposure in industrial operations, and in spite of its wide use and high volatility, it is surprising that exposure by inhalation of methanol has not been associated with significant industrial illness. Ingestion, of course, has caused serious illnesses and deaths.

2.4.2 Effects in Animals

Single or Short-Term Exposure Data. *Single-Dose Oral.* Data on the single-dose oral toxicity of methanol to animals are given in Table 55.1.

**Table 55.1. Single-Dose Oral Toxicity Values
for Methanol in Animals**

Species	LD_{50} Values (g/kg)	Ref.
Rat	6.2	28
	9.1	29
	12.9	30
	13.0	31
Rabbit	14.4	37
Mouse	0.420^a	31
Monkey	$2-3^a$	32, 33
	7.0^a	34

^{a} MLD values.

Single-Dose Injection. When methanol was injected intravenously into mice, the LD_{50} was found to be 5.66 g/kg (54).

Eye Contact. Methanol is a mild eye irritant (39). Undiluted methanol caused moderate corneal opacity in three of six rabbits and conjunctival redness in six of six. This was a postive FHSLA eye irritation test (38). A 50 percent aqueous methanol solution caused minimal to no effects, and a 25 percent aqueous solution caused no adverse effects (38).

Skin Contact. The LD_{50} for rabbits is reported to be 20 ml/kg (40). Yant and his associates (41) drenched the entire body of unshaven dogs for several hours in such a manner as to eliminate any inhalation of the vapor. According to Schrenk (42), the unpublished data for these experiments reveal, from the amount of methyl alcohol in the blood, that so far as dogs are concerned the possibility of poisoning by this method was not great and certainly was very much less than from inhalation.

Inhalation. The effects upon animals of single exposures of methyl alcohol by inhalation are summarized in Table 55.2.

The minimum lethal concentration of methyl alcohol in the air breathed by different animals over short periods of time varies widely with the species and according to the investigator, as shown in Table 55.2.

The pathologic changes found in the tissues of animals exposed by inhalation to methyl alcohol are quite similar to those observed in animals following ingestion of this compound. In the eyes of the dog, Tyson and Schoenberg (50) found hyperemia of the choroid and edema of the ocular tissue with early signs of degeneration of the ganglionic cells of the retina and nerve fibers. Scott and his associates (51) also found that the vessels of the choroid of poisoned animals

Table 55.2. Results of Single Exposures of Animals to Vapors of Methanol

Animal	Concentration ppm	(mg/l)	Duration of Exposure (hr)	Signs of Intoxication	Outcome	Ref.
Cat	132,000	173.0	5–5.5	Narcosis	Died	43
	65,700	86.0	4.5	On side	50% died	43
	33,600	44.0	6	Incoordination	50% died	43
	18,300	24.0	6	None, but salivation	Survived	43
Mouse	72,600	95.0	54	Narcosis	Died	44
	72,600	95.0	28	Narcosis	Died	44
	54,000	70.7	54	Narcosis	Died	44
	48,000	62.8	24	Narcosis	Survived	44
	10,000	13.1	230	Ataxia	Survived	44
	152,800	200.0	94 min	Narcosis		45
	101,600	133.0	91 min	Narcosis		45
	91,700	120.0	95 min	Narcosis	Overall	45
	76,400	100.0	89 min	Narcosis	mortality	45
	61,100	80.0	134 min	Narcosis	45%	45
	45,800	60.0	153 min	Narcosis		45
	30,600	40.0	190 min	Narcosis		45
	173,000	227.0			Died	46
	139,000	182.0		Highest concentration endurable		46
Rat	60,000	78.5	2.5	Narcosis, convulsions		47
	31,600	41.4	18–20		Died	47
	22,500	29.5	8	Narcosis		47
	13,000	17.0	24	Prostration		47
	8,800	11.5	8	Lethargy		47
	4,800	6.3	8	None		47
	3,000	4.0	8	None		47
	50,000	65.4	1	Drowsiness	Survived	48
Dog	37,000	48.4	8	Prostration, incoordination		47
	13,700	17.9	4	None		47
	2,000	2.6	24	None		47

were markedly congested, the entire retina was edematous, and the ganglion cells were degenerated. Occasionally there were degenerative changes and fibrosis of the optic nerve. Although Weese (44) observed degenerative alterations in retinas of mice, he did not attribute them to the effects of methyl alcohol. Hemorrhage (50), edema (49, 51), congestion (51), and pneumonia (44, 52) were observed in the lungs of the various species that were exposed to vapors containing methyl alcohol. The livers and kidneys showed congestion (50), albuminous and fatty degeneration (52), and fatty infiltration (44, 49).

Cardiac dilatation (49) and myocardial degeneration (51, 52) were observed in the hearts of animals. Degenerative injuries of the central nervous system have been described by several investigators (49–52). Pinpoint hemorrhages and congestion of the gastric mucosa were believed by Tyson and Schoenberg (50) to be characteristic of poisoning from the inhalation of methyl alcohol.

Aspiration.   Aspiration of 0.2 ml methyl alcohol by rats resulted in death of only 1 of 10 treated animals (53).

Mode of Action.   Exposure of animals to concentrations of methyl alcohol in air may induce the following signs of intoxication: increased rate of respiration, a state of nervous depression followed by excitation, irritation of the mucous membranes, ataxia, partial paralysis, agony, prostration, deep narcosis, convulsions, decrease in rectal temperature, and loss of weight. Death following inhalation of lethal concentrations of methyl alcohol occurs from respiratory failure.

Repeated or Prolonged Exposure Data.   The narcotic effect of methyl alcohol is weaker than that of ethyl alcohol, but the toxic effect from accumulated doses of methyl alcohol, owing to slow elimination, is greater than that of ethyl alcohol. There is not a very wide difference between the concentration necessary to produce narcosis and that which is lethal. Temporary or permanent visual disturbances and blindness may result from repeated exposure to intermediate concentrations.

Repeated Oral.   Rats given 1 percent methanol in their drinking water daily for 6 months did not show physiologic values significantly different from those of untreated controls. A safety limit of 1 mg/l of drinking water was recommended (55).
  Repeated-oral doses of 3 to 6 g/kg methyl alcohol to rhesus monkeys for 3 to 20 weeks resulted in ultrastructural abnormalities of hepatocytes of a nature that indicated alteration of RNA metabolism of hepatic cells (56). Rats receiving oral doses of 10, 100, or 500 mg/kg/day methanol for 1 month showed liver changes characterized by focal proteinic degeneration of the hepatocytic cytoplasm, changes in the activity of some microsomal enzymes, and enlarged hepatic cells (57).

Repeated Inhalation.   Mice exposed to air containing 48,000 ppm for 3.5 to 4 hr/day up to a cumulative total of 24 hr were in a state of narcosis, but survived, whereas they succumbed in coma when correspondingly exposed for 54 hr to air containing 54,000 ppm (44). Sayers and his associates (59) exposed two dogs to 10,000 ppm of methanol for about 3 min in each of eight hourly periods per day for 100 consecutive days and found no noteworthy adverse effects attributable to the exposure. Sayers et al. (58) also exposed dogs 8 hr/day to 450 to 500 ppm of methyl alcohol vapors for 379 consecutive days. The animals showed no unusual behavior, impairment of vision, or loss of weight, and all survived. Ophthalmoscopic examinations disclosed no remarkable abnormalities. There were no significant changes in the formed elements or the chemical constituents of the blood of the animals, nor were there gross or microscopic abnormalities in their tissues at necropsy.

In rabbits exposed by inhalation to 61 mg/m^3 (46.6 ppm) methyl alcohol for 6 months (duration of exposure per day is not given), electron microscopic examination of the retina showed ultrastructural changes in the photoreceptor cells and Müller fibers (60).

Ubaidullaev (61) reports that male rats exposed continuously for 90 days to a concentration of 5.3 mg/m^3 (4 ppm) of methanol vapor exhibited changes in chronaxie ratio between antagonistic muscles, in whole blood cholinesterase activity, in urinary excretion of coproporphyrin, and in albumin–globulin ratio of the serum. Male rats exposed to 0.57 mg/m^3 (0.4 ppm) continuously for 90 days showed no changes (61).

Combined Repeated Inhalation and Repeated Oral.   Rats exposed to 0.022 mg methyl alcohol/l of air (16.8 ppm) 4 hr/day for 6 months and simultaneously administered orally 0.7 mg/kg of methanol daily showed changes in blood morphology, oxidation–reduction processes, and liver function. The effects of methyl alcohol given simultaneously by two routes seemed to be additive (62).

Absorption and Distribution.   In dogs exposed repeatedly to vapor concentrations of 450 or 500 ppm for 8 hr/day, methyl alcohol concentrations in the blood varied between 10 and 15 mg/100 ml at the end of an 8-hr exposure period, but occasional concentrations as high as 52 mg/100 ml were found (58). Median values of 6.5 and 14 mg methyl alcohol/100 ml blood were obtained for two dogs exposed to 10,000 ppm for about 3 min in each of eight periods/day at hourly intervals on 100 consecutive days (59).

The distribution of methyl alcohol within the tissues of dogs exposed to 4000 and 15,000 ppm in air over periods ranging from 12 hr to 5 days was found to be rapid (63). The quantities found in the various tissues were correlated to their water content; although the differences were not great, the highest concentrations were found in the blood, eye fluids, bile, and urine, and the lowest in the bone marrow and fatty tissue. Loewy and von der Heide (47) showed that the lipoid solubility of methyl alcohol is slight and that fat rats absorb less proportionally than the lean. The entire carcasses of rats exposed for 8 hr to 4500, 8500, or 22,500 ppm contained 0.65, 2.0, and 4.3 g of methanol/kg of body weight, respectively.

One to 7 mg of methyl alcohol/g of blood (100 to 700 mg per 100 ml) was found by Haggard and Greenberg (64) in the blood of rats following oral administration of 4 g methyl alcohol/kg of body weight.

Metabolism. Species differences exist with respect to the metabolic pathways for the metabolism of methyl alcohol. In the rat, the amino acid oxidase catalase system has the function of metabolizing methyl alcohol (65) with liver ADH having little or no importance. Methyl alcohol is also reported to have no reactivity with rabbit, guinea pig, or mouse liver ADH (35). It is noteworthy that Makar (66) found that methyl alcohol is oxidized in the monkey primarily by an ADH system. Potts (32) and Gilger and Potts (33) have shown in their studies that methanol poisoning in the monkey resembles that of human poisoning with acidosis and visual disturbances. According to them, the effect of methanol on nonprimates is due mostly to a narcotic effect similar to that of ethanol.

Sleeping time in rabbits given an oral dose of 2 ml/kg of various lower alcohols was shortest for methyl alcohol (36). It is generally agreed that the toxicity of methyl alcohol is due to its metabolites rather than methyl alcohol per se. There are differences of opinion as to whether the metabolite responsible for the toxicity is formaldehyde or formic acid. Although it has been established that formaldehyde can remain intact for but a short time because of its reaction with protein, Keeser (67) found it present for a short time in vitreous humor, spinal fluid, and abdominal fluids of rabbits poisoned by methyl alcohol. Keeser (68) also showed that methyl alcohol was oxidized in vitro by the freshly prepared vitreous humor of calves, but that hexamethylenetetramine could be formed if ammonium carbonate was present and the resulting amount of formaldehyde was materially decreased. Kini et al. (69) studied the effects of methanol and its metabolites on the retina of the rabbit and found formaldehyde to be the most toxic.

In the case for formic acid, following inhalation of vapors of methyl alcohol by dogs, Tyson and Schoenberg (50) found an increase in the electroconductivity of blood due to an increase in hydrogen ion content, which was substantiated by alkaline titration. The excretion of formic acid for several days following administration of methyl alcohol was reported by Pohl (70) and Hunt (71). Formic acid was excreted in the urine of rabbits during inhalation of methyl alcohol, according to Bachem (46). More recently, McMartin (72) showed that the formation of formic acid in the methanol-poisoned monkey was responsible for the development of severe metabolic acidosis. Ocular toxicity similar to that described for methanol poisoning in man was produced by intravenous administration of formate to monkeys (73).

Excretion. In rats that were given oral doses of 4 g/kg methanol, 70 percent of the methanol lost was eliminated in the expired air (64). The amount eliminated in unit time was determined by the concentration in the blood and the volume of the pulmonary ventilation. This was demonstrated by the exponential curve that resulted when the concentration of methyl alcohol in the blood was plotted against the time elapsing after the administration of methyl alcohol, under varying conditions of pulmonary ventilation. The rate of elimination of methyl alcohol from the blood was increased when pulmonary ventilation was increased by carbon dioxide or 2,4-dinitrophenol. Newman and Tainter (74) have shown that the increased elimination of methyl alcohol from the blood, following intramuscular injection of dinitrophenol, is due to pulmonary ventilation and not to oxidation, by the simple expedient of determining the rate of methyl alcohol elimination on the part of dogs so injected, under conditions of enforced rebreathing of the expired air, as compared with free breathing of ordinary atmosphere.

Following an oral dose of 2 ml/kg to rabbits, 5.3 percent was excreted in expired air and 7.8 percent in urine (36). Urinary excretion of methanol following an oral dose of 2 g/kg to rats was 1.81 percent of the dose administered (75). A number of other investigators have reported on the excretion of methanol; in a review of this information by Koivusalo (76), the percentages of the administered doses that were found in the expired air ranged from approximately 10 to 25 percent, and in the urine, from 3 to 19 percent.

A few milligrams of formate were found in the blood, muscle, kidney, and lung of a 7-kg animal by Pohl (70) on the day following the administration of 25 ml of methyl alcohol. These findings indicate there is very little storage of formate ion in the body. Bastrup (77) found rabbits that had been given a single oral dose of methyl alcohol (2 to 10 g/kg of body weight) excreted 0.1 to 1.1 percent of it as formate and 13 to 20 percent of it as methyl alcohol in the urine within 47 to 143 hr. Dogs given a single oral dose of methyl alcohol (1 to 2 g/kg) excreted in the urine 5 to 15 percent as formate and 5 to 8 percent as unchanged methyl alcohol. Lund (78, 79) likewise found that the dog excreted more formate than the rabbit after the ingestion of methanol. The rabbit excreted 10 percent of the dose (2.38 g/kg) unchanged in the urine within 95 hr but the urinary formate was only questionably above the normal amount. In the case of two dogs given the dosage of 1.70 and 1.97 g/kg, respectively, 6.0 and 8.7 percent were recovered in urine as methanol and 22.4 and 23.7 percent were found in the urine as formate within 100 hr.

Pharmacology. In anesthetized, mechanically ventilated dogs that received intravenous infusions of 20 percent methanol, blood alcohol concentrations of 130 to 200 mg/100 ml and higher progressively decreased stroke volume and cardiac output, systemic blood pressure, and flow through femoral and common carotid arteries, while total peripheral resistance increased (80).

Pharmacokinetics. The rate of elimination of methanol is much slower than that of ethanol (76). Following an oral dose of 2 g/kg to rats, methanol reached a higher blood level than ethanol and remained high for a longer period (75).

Following oral administration of 20 percent methanol to rabbits, blood levels reached a maximum at 2 hr and were still detectable after 40 hr; the concentration in expired air reached a maximum after 1 hr and was still detectable after 9 hr; urinary concentration peaked at 3 hr and was still detectable after 40 hr (35).

Saito (36) reported a maximum blood methanol concentration 6 hr following an oral dose of 2 ml/kg to rabbits; the alcohol was still detectable in the blood after 50 hr.

Mutagenicity. According to Simmon et al. (109), methanol was not mutagenic when tested on *Salmonella typhimurium* strains TA1535, TA1537, TA1538, TA98, and TA100 in activated and nonactivated systems.

2.4.3 Effects in Humans

Absorption/Excretion. Leaf and Zatman (81) have studied the metabolism of methanol under carefully controlled experimental conditions in man following ingestion and inhalation. Dosages of 71 to 84 mg/kg orally resulted in blood levels of 4.7 to 7.6 mg/100 ml of blood 2 to 3 hr afterward. The urine/blood concentration ratio was found to be relatively constant at about 1.30, similar to that found for ethanol in man by Haggard and Greenberg (64). This suggests that the urinary methanol concentration may be a reliable index of the concentration in body water during the excretory period.

Inhalation of from 500 to 1100 ppm methanol for periods of 3 to 4 hr gave urine concentrations of about 1 to 3 mg/100 ml of urine at the end of exposure; these concentrations seemed to be well correlated with time and exposure levels (81). The ingestion of 15 ml of ethanol at the same time as 4 ml of methanol gave a marked elevation in the peak concentration of methanol excretion in urine, although the time to return to a control level of methanol was unaltered. This indicated an inhibition of methanol oxidation by ethanol. An even more dramatic effect was noted when 10-ml doses of ethanol were given hourly (81). Elkins (82) has measured the concentration of methanol in the atmosphere of three plants and concurrently determined the concentration of methanol in the urine of the workers. The results were as follows:

Average Methanol in Air (ppm)	Methanol in Urine (mg %)
125	0.3, 0.6
440	0.35, 0.65, 0.9
2800	1.6, 4.4

The concentration of methanol in the blood was determined at the time of death of six men following the ingestion of methanol. Keeney and Mellinkoff (83) found the average concentration was 71.1 mg/100 ml of blood with the values ranging from 15.6 to 150. In five other fatal cases Lund (84) reported the following values:

	Methanol (mg %)	Formic Acid (mg %)
Blood	74–110	9–68
Urine	140–240	216–785
Liver	106	60–99

Lund (84) found no noteworthy increase in methanol or formic acid in the blood or urine of humans following the ingestion of a nonfatal estimated dose of 10 to 20 ml of methanol. However, 48 hr following the ingestion of nonfatal estimated doses of about 50 to 80 ml of methanol, formic acid could be demonstrated in the blood (2.6 to 7.6 mg percent), as well as an increased excretion in the urine (54 to 205 mg percent). Cooper (85) reported that blood methanol levels of 0.2 mg/ml had no effect on humans; blindness resulted with blood levels of 0.3 mg/ml, and death occurred at levels of 1.14 mg/ml.

Metabolism.   Methanol is oxidized by human liver ADH (35).

Pharmacokinetics.   In the human studies reported by Leaf and Zatman (81), in which doses of 71 to 84 mg/kg were administered orally, urine concentrations reached a peak rapidly in about an hour and declined exponentially reaching the blank or control value in 13 to 16 hr.

These excretion curves can be represented approximately by an equation

$$\log C - \log C_0 = kt$$

where C_0 and C are urinary concentrations at zero time and at t hr, respectively. The constant k was very similar in the three individuals studied (0.094 to 0.108). With these small dosages, the amount of the dose accounted for in the urine averaged only 0.7 percent of the total with amounts in a few expired air studies only slightly larger, indicating that metabolism (probably by oxidation) was accounting for the major part of the dose.

Mode of Action.   There are three stages in human methanol poisoning (86):

1.   Narcotic stage, similar to that produced by ethanol.
2.   Latent period (10 to 15 hr).
3.   Visual disturbances and CNS lesions—nausea, vomiting, dizziness, headache, failing eyesight, and deep respiration concomitant with the severe metabolic acidosis.

There is general agreement that the toxic symptoms exhibited during the third stage are due to metabolite(s) rather than to methanol per se (66). The generally accepted lethal dose in untreated methanol poisoning in humans is 0.8 to 1.5 g/kg (76).

Human Pathology.   Examination by Province et al. (87) of the tissues of five persons fatally poisoned by the ingestion of methyl alcohol disclosed the occurrence of catarrhal gastritis, acute enteritis, focal necrosis of the liver with infiltration by polymorphonuclear leukocytes, pulmonary edema, and early degeneration of neurons of the brain. Chew et al. (88) at necropsy of five other cases found cerebral edema, hypostatic pulmonary congestion, fatty infiltration of the liver, and passive congestion of all organs. On examining 11 similar cases Tonning (89) found superficial necrosis of the stomach accompanied by mucous distensions of epithelial cells, parenchymatous degeneration of the liver, engorgement of the pulmonary vessels, edema and hyperemia of the brain accompanied by occasional punctate hemorrhages and accumulation of brown pigment in the neurons, and irregular staining of the ganglion cells of the retina accompanied by eccentric nuclei, fraying, vacuolation, and autolysis. Necrosis of the pancreas of a woman succumbing from ingestion of methyl alcohol was the most characteristic finding of Branch (90). The fatal cases from ingestion of methanol that were studied by Keeney and Mellinkoff (83) revealed gastritis, pulmonary congestion, edema, and patchy necrosis, uniformly, with bronchial desquamation in two of six necropsies and bronchopneumonia in one case. Severe congestion of the glomerular tufts and cloudy swelling of the convoluted tubules, hepatic infiltration, congestion and parenchymal hemorrhage of the pancreas, and congestion of the spleen were also observed. Upon examination of the ganglion cells of the retina of humans dying of methanol, Roe (91) found severe degenerative cells. Roe attributes diminution in vision to acidosis, since he failed to obtain degenerative changes in the ganglion cells of rats and rabbits, which have a greater alkaline reserve than man.

Human Experience.   Bennett et al. (94) described the symptoms, laboratory and physical findings, pathologic alterations, and treatment following an epidemic of acute methyl alcohol poisoning (in Atlanta) that resulted from the ingestion of bootleg whiskey (mortality was 6.2 percent among 323 cases). These authors suggest that depression of carbon dioxide production is probably related to interference with controlling enzyme systems including succinic dehydrogenase. An account of a more recent outbreak of methanol poisoning has been reported by Kane et al. (95). Ocular disturbances and blindness in man have been reported by Campbell (96) and Woods (97) from repeated rubbing of the skin with methyl alcohol under conditions that did not prevent inhalation of the vapor. As may be seen, the available evidence as to the magnitude of the hazard from the percutaneous absorption is somewhat inadequate, but such as it is, it suggests that prolonged or frequently repeated exposure by this means should be avoided.

From the time of Buller and Wood (13) in 1904 until the present time, numerous cases of blindness and death due to the drinking of a few ounces of methyl alcohol have been reported. Articles by Jacobson and his associates (98), Kaplan and Levreault (99), and Voegtlin and Watts (100) record 24 deaths from drinking methyl alcohol. Twenty-three cases of poisoning following the ingestion of methanol by men in Korea were reported by Keeney and Mellinkoff (83). Six of these died but the others sustained no permanent injury, although they manifested nausea, epigastric pain, vomiting, headache, dizziness, delirium, varying degrees of transitory blindness, acidosis, and acetonuria. Hughes (101) reported the loss of vision of three workers following the ingestion of methanol.

General physical, ocular, and hematologic examinations by Greenberg and his associates (15) of 19 workers who had been repeatedly exposed to 22 to 25 ppm of methyl alcohol and 40 to 45 ppm of acetone revealed no significant abnormalities. Likewise, a survey by Yant and associates (41) of 36 men employed in the manufacture of methyl alcohol and of 24 drivers of trucks using methyl alcohol as an antifreeze disclosed no harmful effects.

Although the individual responses of man to methyl alcohol may vary considerably, industrial exposures are not very hazardous if concentrations are maintained within the upper limit of 200 ppm. Under varying conditions of severity and duration of exposure to the vapor of methyl alcohol the signs of intoxication may include irritation of all the mucous membranes, headache, roaring in the ears, tiredness, insomnia, nystagmus, trembling, vertigo, unsteady gait, dyspnea, nausea, vomiting, colic, constipation, dilated pupils, clouded vision, diplopia, blindness, itching of skin, eczema, and dermatitis (6, 11, 12, 16, 106).

Treatment. In acute cases, the use of intravenous injection of sodium bicarbonate or sodium lactate and glucose in physiologic saline has been recommended by Johnstone (102), Jacobson et al. (98), and Voegtlin (100). These investigators have also recommended the use of emetics, a high fluid intake, cardiac and respiratory stimulants, and oxygen or artificial respiration. Chew et al. (88) and Roe (103, 104) have also recommended the alkali treatment for methyl alcohol poisoning but disagree on the efficacy of the use of ethyl alcohol for poisoning from methyl alcohol. The studies of Leaf and Zatman (81) appear, however, to support the suggestion by Roe that ethanol may be a useful therapeutic agent in methanol poisoning.

Keeney and Mellinkoff (83) suggested that intravenous injection of glucose may be an adjuvant in the treatment of ketosis. Suprunov (105) has recommended the administration of vitamin C and thiamine in cases of poisoning by methyl alcohol, since he found a reduced amount of these vitamins in the tissues of rabbits following subcutaneous injections of sublethal amounts of methyl alcohol.

2.5 Hygienic Standard of Permissible Exposure

Both the ACGIH (107) and OSHA (108) have established industrial hygiene standards of 200 ppm or 260 mg/m^3. This same standard is in effect in Australia, Belgium, Finland, Federal Republic of Germany, Japan, Netherlands, Sweden, and Switzerland, and is considered appropriate. The standard is 250 mg/m^3 in Italy; 150 in Roumania; 100 in Czechoslovakia, German Democratic Republic, and Poland; 50 in Bulgaria, Hungary, and Yugoslavia; and 5 in the Soviet Union (110).

2.6 Odor and Warning Properties

Methyl alcohol does not have suitable warning odor or irritating properties except at high concentrations. Witte (43) found an initial salivation by cats when exposed to 18,300 ppm (24 mg/l). Methyl alcohol becomes unendurable in concentrations of 50,000 ppm (65 mg/l) (49). Scherberger et al. (92) state that a level of 2000 ppm is barely detectable by odor. May (93) reports that the odor of a concentration of 5900 ppm is barely perceptible.

3 ETHANOL; Ethyl Alcohol, Alcohol, Grain Alcohol, Methyl Carbinol, Ethyl Hydrate, Spirit of Wine, Cologne Spirits, CAS No. 64-17-5

$$C_2H_5OH$$

3.1 Source, Uses, and Industrial Exposure

Ethanol is manufactured by fermentation of starch, sugar, and other carbohydrates; from ethylene, acetylene sulfite waste liquors, and synthesis gas (CO + H); by hydrolysis of ethyl sulfate; and by oxidation of methane (111). Greater quantities of ethanol are synthesized from ethylene and sulfuric acid than are fermented from such agricultural products as molasses, starch, and cellulose (112, 113). Anhydrous ethanol is manufactured on a large scale by azeotropic distillation (112).

Ethanol is used in the manufacture of synthetic rubber, as an intermediate in the manufacture of certain chemicals and medicinals, as an antifreeze, as fuel, as a solvent or processing agent for various purposes including explosives, plastics, synthetic resins, nitrocellulose, lacquers, pharmaceuticals, cosmetics, adhesives, inks, and preservatives (114). Numerous denaturants for ethyl alcohol have been listed by Zangger (115). Mellan (114) lists denaturants authorized by the U.S. Treasury Department. The use of ethanol as a beverage and the problems related thereto are not considered herein except as they relate to industrial exposure. Because of the diversity of uses of ethanol, there is considerable opportunity for exposure, particularly to its vapors. Because of the strict regulations governing the accounting of its use, employers are careful to prevent ingestion and waste. The hazards from industrial exposure to ethyl alcohol are low.

3.2 Physical and Chemical Properties

Physical state	Clear, colorless, volatile liquid
Molecular formula	C_2H_6O
Molecular weight	46.07
Boiling point	78.5°C (111)
Solidifies below	−130°C
Melting point	−114.1°C (111)
Specific gravity	0.7904 (20°/20°C) (114)
Vapor pressure	50 mm Hg (25°C) (117)
Refractive index	1.361 (20°C) (111)
Flammability	
Flash point	9 to 11°C (111)
Explosive limits	3.28 and 18.95 percent by volume in air (118)
Autoignition temperature	793°F (119)
Percent in saturated air	6.58 (25°C)
Density of saturated air (air = 1)	1.04
Solubility	Miscible in all proportions with water and with most organic solvents
1 mg/l ≈	532 ppm at 25°C, 760 mm Hg
1 ppm ≈	1.88 mg/m^3 at 25°C, 760 mm Hg

3.3 Analytic Determination

3.3.1 Determination in the Atmosphere

NIOSH Method No. S56 (120) has been recommended. It involves drawing a known volume of air through charcoal to trap the organic vapors present (recommended sample is 1 liter at a rate of 0.2 l/min). The analyte is desorbed with carbon disulfide containing 1 percent 2-butanol. The sample is separated by injection into a gas chromatograph equipped with a flame ionization detector and the area of the resulting peak is determined and compared with standards. Recently Mueller and Miller (1) and Langvardt and Melcher (2) have published methods of collection and analysis by gas chromatographic procedures that give excellent results and are applicable in situations where a variety of other materials may be present. The older chemical methods (121–129) would seem to have limited usefulness except where modern equipment is not available or where unusual conditions exist.

3.3.2 Determination in Tissues and Blood

Most of the older methods for the determination of ethyl alcohol in tissues or blood are based upon measurement of the quantity of dichromate used up in oxidizing the alcohol (126, 130–134). More modern methods based on gas chromatographic procedures would seem adaptable to the analysis of tissues and body fluids (1, 2).

3.4 Physiologic Response

3.4.1 Summary

Ethanol is low in single-dose oral toxicity; it is appreciably irritating and even injurious to the eyes in undiluted form but much less so when diluted; it is not irritating to the skin nor is it absorbed through the skin in appreciable amounts. The effects from inhalation are not likely to be of a serious nature under reasonable conditions of handling and use. Prolonged repeated absorption of ethanol in excessive amounts can cause liver injury together with other adverse effects. The typical effects of excessive acute exposure are initial excitation followed by depression, and if death ensues, it is likely to result from respiratory failure. Some fetotoxicity has been observed in the offspring of female mice treated with rather large doses of ethanol during gestation. Some mutagenic changes of a transient nature have been observed in male but not female mice treated with rather large doses. Data on the carcinogenicity of ethanol are lacking.

3.4.2 Effects in Animals

Single or Short-Term Exposure Data. *Single-Dose Oral.* The single-dose oral data do not indicate that ethanol is particularly toxic to laboratory animals. The data are summarized in Table 55.3.

Eye Contact. Ethanol is capable of causing a response in the rabbit eye ranging from mild irritation to severe injury depending upon the concentration and amount used in testing (38, 40, 140, 141).

Skin Contact. Ethanol is not appreciably irritating to the intact or even abraded skin even upon repeated and prolonged contact (141–143). When it was given intradermally as a 95 or a 19 percent solution, severe erythema and some

Table 55.3. Single-Dose Oral Toxicity Values
for Ethanol in Animals

Species	LD_{50} Values (g/kg)	Ref.
Rat	13.7	135
	17.8[a]	136
	6.2[b]	136
	11.5[c]	136
Mouse	9.5	138
	8.3	139
Guinea pig	5.6	135
Rabbit	9.9	37
Rabbit	9.9[d]	137
	7.0[d]	35
Dog	5.5–6.6[e]	138

[a] Young adults.
[b] 14 days old.
[c] Older adults.
[d] MLD values.
[e] Lethal dose.

necrosis resulted (141). Ethanol penetrates the skin of animals, but not at a rate sufficient to induce serious effects. Deichmann (144) found values of 0.13 and 0.04 mg of alcohol, respectively, per 100 ml of blood, at 0.5- and 1-hr intervals following one application of 35 ml/kg of body weight upon the abdomen of a rabbit protected against inhalation of the vapor.

Inhalation.   Animals exposed to ethyl alcohol in air may manifest the following signs of intoxication: slight irritation of the mucous membranes, excitation followed by ataxia, drowsiness, prostration, narcosis, twitching, general paralysis, dyspnea, and occasionally death associated with respiratory failure. The effects of inhalation of ethyl alcohol in various concentrations in air by various animal species are given in Table 55.4.

Aspiration.   Gerarde and Ahlstrom (53) found that the aspiration of 0.5 ml of ethanol caused the death of 5 of the 10 rats treated.

Injection.   The data with respect to the single-dose intraperitoneal and intravenous injection toxicity for ethanol are given in Table 55.5.

Mode of Action—Signs.   Following single and 7-day repeated oral doses of 3 g/kg to dogs, the effects on renal function included elevated glomerular filtration rate, increased tubular reabsorption of sodium, and increased potassium

Table 55.4. Results of Single- or Short-Term Exposures of Animals to Vapors of Ethanol

Animal	Concentration ppm	mg/l	Duration of Exposure (hr)	Signs of Intoxication	Outcome	Ref.
Mouse	29,370	55.2	Short (?)		Died	46
	22,980	43.2	Short (?)	Tolerated	Survived	46
	31,900	70.0	0.33	Ataxia		145
	29,300	55.0	7.0	Narcosis	Died	145
	23,940	45.0	1.25	Narcosis		145
	13,300	25.0	1.33	Ataxia		145
	25,000	48.1	125 over several days	Narcosis	Survived	44
	16,700	31.4	24 over several days	Narcosis	Survived	44
	8,350	15.7	29 × 7.2	Ataxia	Survived	44
Guinea pig	45,000	84.6	3.75	Incoordination		116
	44,000	82.7	7.5	Deep narcosis		116
	50,170	94.3	10.2	Deep narcosis	Died	116
	19,260	36.2	3.75	None		116
	20,000	37.6	6.5	Incoordination		116
	21,900	41.2	9.8	Deep narcosis	Died	116
	9,080	17.1	5.25	None		116
	12,850	24.2	8.75	Incoordination		116
	13,300	25.0	24.0	Light narcosis		116
	6,400	12.0	8.0	None	Survived	116
Rat	32,000	60.1	8.0		Some died	146
	16,000	30.1	8.0		Some died	146
	16,000	30.1	8.0		Survived	40
	45,000	84.6	3.75	Deep narcosis		116
	44,000	82.7	6.5	Deep narcosis	Died	116
	19,260	36.2	2.0	Light narcosis		116
	21,960	41.2	9.8	Deep narcosis	Died	116
	18,200	34.2	1.0	Excitation		116
	18,200	34.2	1.75	Incoordination		116
	22,800	42.9	8.0	Deep narcosis		116
	22,100	41.5	15.0	Deep narcosis	Died	116
	10,750	20.2	0.5	None		116
	10,750	20.2	2.0	Incoordination		116
	12,400	23.3	8.5	Deep narcosis		116
	12,700	23.8	21.75	Deep narcosis	Died	116
	5,660	10.6	1.75	Incoordination		116
	6,400	12.3	12.0	Light narcosis	Survived	116
	3,260	6.1	6.0	None		116
	3,260	6.1	8.0	Drowsiness		116
	4,580	8.6	21.13	Ataxia	Survived	116

Table 55.5.   Single-Dose Injection Toxicity Values for Ethanol in Animals

Route	Species	LD_{50} Values (g/kg)	Ref.
Intraperitoneal	Rat	5.0	138
		6.0	40
	Mouse	3.2	139
		0.25[a]	148
Intravenous	Rat	1.8	139, 40
		4.2	149
	Mouse	2.2	139
		2.0	138
		2.8	54
	Guinea pig	2.4	149
	Dog	4.9	151
		5.4 [b]	138
	Cat	3.1 [b]	152
	Rabbit	7.4 [b]	150

[a] LD_o values.
[b] MLD values.

excretion. These results suggest that acute or chronic alcohol administration produces renal vasodilation (153). The main toxic effects of single large oral doses of alcohol are depression of the central nervous system and accumulation of triglycerides in the liver (154). Primary focus of acute ethanol toxicity is in the membranes of organ systems such as nervous tissue, kidney, heart, and red blood cells, which depend for their function on active cation transport and ATPase activity (155).

Cause of Death.   Dogs given intravenous doses of 3 to 10 g/kg died from respiratory insufficiency (156). Dogs given intravenous doses 1.25 g/kg/hr until death died of cardiac failure and circulatory depression (157). When dogs were given lethal oral doses of ethanol, deaths occurring in less than 12 hr after administration were associated primarily with respiratory failure. In the dogs that survived longer than 12 hr, death was preceded by a progressive fall in blood pressure, resulting in circulatory failure. Deaths in dogs following lethal intravenous doses of ethanol always appeared to be the result of respiratory failure (158).

Repeated or Prolonged Exposure Data.   *Repeated Oral.*   Rats given 3.26 *M* ethanol in drinking water for 12 weeks (equivalent to dose of 10.2 g/kg/day by end of study) exhibited a decrease in weight gain and fatty livers (159). Rats fed a diet in which 33 percent of caloric intake was ethanol for up to 14 weeks developed fatty livers; no changes were observed in myocardial ultrastructure (160). A monkey fed a fluid diet which provided 40 percent of total calories from ethanol for 3 months showed marked accumulation of triglycerides, cholesterol, and phospholipids in the serum and liver. Triglycerides and cholesterol ester were increased in the heart. Incorporation studies showed increased synthesis of triglycerides in heart muscle and liver. Fatty changes were observed in heart myocardium and liver (161).

Repeated Skin Contact. Daily applications for 187 days of 10 drops of 50 percent solution upon facial skin of rats produced no injury other than temporary irritation (162).

Repeated Inhalation. Weese (44) reports that reversible fatty infiltration of the liver occurs following repeated exposure to high concentrations. Mertens (163) exposed rabbits to air saturated with alcoholic vapors for periods ranging from 25 to 365 days and thereby induced cirrhosis of the liver as a common lesion. Petri (164) found hemorrhagic perivascular infiltrates in the tissues of animals subjected to high dosage. Smyth and Smyth (165) exposed guinea pigs for 4 hr/day, 6 days/week for 10½ weeks to vapor concentrations of 3000 ppm without any untoward effects. Rats, guinea pigs, rabbits, squirrel monkeys, and beagle dogs were exposed continuously to vapors containing 86 mg/m^3 ethanol for 90 days with no evidence of effects on behavior, mortality, hematologic values, and gross histopathology (166).

Teratology. Ethanol administered to pregnant rats at a dose of 5 ml/kg (3.95 g/kg) resulted in increased postimplantation death and retardation of fetal development (167). Maintenance on drinking water containing 30 percent ethanol by rats before and during pregnancy markedly decreased the size and number of offspring (168). Teratogenic effects were not observed in offspring of pregnant rats, rabbits, and mice given 15 percent ethanol in their drinking water during the period of major organogenesis. Minor skeletel variants considered to be due to retarded fetal growth were observed in mice and rats (169).

Carcinogenicity. No data.

Mutagenicity. Ethanol induced a dominant lethal mutation in male mice given three consecutive daily doses of 0.1 ml of 40 or 60 percent aqueous solutions of ethanol, equivalent to about 1 to 1.5 g/kg/dose. The effect was temporary and disappeared after 2 weeks (170). Females treated with a dose of 5 ml/kg in proestrus and estrus gave no indication of any dominant lethal effect, whereas known mutagens gave positive results (171). Mice that received 10 to 40 percent ethanol in their drinking water for 26 days did not show any damaging effect on chromosomes, as judged by examination of micronuclei in erythrocytes from bone marrow (172).

Absorption.   Loewy and von der Heide (116) found that a state of diminished excitability on the part of rats exposed to the vapors of alcohol was associated with the presence in their carcasses of concentrations of alcohol ranging from 0.16 to 0.27 g/kg of their total weight. Corresponding concentrations of 1 g/kg were associated with the induction of a state of narcosis, and concentrations ranging from 3.1 to 5.8 g/kg were found in the bodies of fatally poisoned animals. Guinea pigs, which are more resistant to the lethal effects of ethanol than rats, are severely intoxicated, but may survive in spite of the presence of concentrations of alcohol within the latter range.

Chickens exposed by Carpenter (173) to alcohol vapors for 2 to 29 hr usually had the highest concentration in the blood, although occasionally higher concentrations were found in the brain. The lowest concentrations of alcohol were found in the fat. According to Carpenter, when the concentration of alcohol in the blood of chickens was more than 2.5 g/kg, or when that in the whole body was more than 1.7 g/kg, the animals showed signs of abnormal behavior. Concentrations of 3.7 to 5.6 g/kg of body weight proved fatal. The concentration of alcohol in the blood of dogs was found to be about 60 percent higher than that in the body as a whole (174). This percentage is in excellent agreement with that of Carpenter. The concentration of alcohol in the blood of dogs breathing an undetermined concentration of alcohol rose from 0.8 g/kg of blood after 2 hr to 4.0 g/kg of blood after 6 hr.

The ratio of tissue/blood ethanol concentration found in cats is given below (recalculated on the basis of plasma–whole blood ratio of 1.12) (175):

Skeletal muscle	0.83
Liver	0.82
Kidney	0.86
Intestine	0.85
Spleen	0.90
Brain	0.88
Fat	0.18

Absorption of ethanol following oral doses is slower in rats than in guinea pigs. Blood levels for guinea pigs reached 6.8 mg/ml and for rats 2.1 mg/ml following oral doses of 6.4 g/kg. Intravenously injected ethanol was eliminated at the same rate in both species. Liver damage also occurred from lower doses in guinea pigs than in rats (176).

Metabolism.   Ethyl alcohol is oxidized to carbon dioxide and water, but small amounts of the unoxidized material remain in the blood and are excreted in the urine and expired air for several hours after exposure. Westerfield and associates (177) have shown the transitory presence of acetaldehyde in the oxidation of ethyl alcohol by the body. Rats and guinea pigs oxidized 66.5 to

98.9 percent of the absorbed alcohol when exposed for about 2 hr to air containing 27,600 to 41,400 ppm of ethyl alcohol (116).

Three enzyme systems can oxidize ethanol to acetaldehyde:

1. Alcohol dehydrogenase (178).
2. Catalase acting as peroxidase and coupled to a system supplying oxygenated water (178).
3. A microsomal ethanol oxidizing system (MEOS) requiring NADPH as coenzyme (179–181).

Lundsgaard (182) was the first to show conclusively that ethanol is primarily metabolized in the liver. About 20 percent of total ethanol metabolism takes place in other organs (183). Acetate produced from acetaldehyde in the liver is released into the blood and oxidized in peripheral tissues (184). Jacobsen (185) reported that the rate of oxidation of ethanol to acetaldehyde is much slower than the rate of oxidation of the aldehyde to acetic acid. Acetate from alcohol was proved by the use of ethyl alcohol labeled with deuterium (185). Acetic acid from alcohol joins the "acetylic pool" of normal metabolism. Although the enzyme system or systems that oxidize acetaldehyde to acetic acid (second step of oxidation of ethanol) have not been as well defined as those for the first step of oxidation of ethanol (ethanol to acetaldehyde), Jacobsen (185) postulates that this later step may involve the flavoproteins or the dehydrogenases. That the liver is the most important organ capable of oxidizing ethanol is shown by rate of disappearance of alcohol from the liver with concurrent measurement of oxygen consumption and the formation of carbon dioxide and acetic acid. Experiments on hepatectomized dogs revealed that very little, if any, ethanol was oxidized (185). Ethanol had no effect on oxygen uptake, decreased the carbon dioxide production, increased acid production, and decreased respiratory quotient of rat liver slices. Also, ethanol decreased pyruvate, resulting in an increased lactate/pyruvate ratio (186). Ethanol blocks the metabolism of some other alcohols by ADH via a competitive inhibition phenomenon (187), but the converse is also true, as discussed in later portions of this chapter.

Excretion. Following the ingestion by dogs of 3.3 g of alcohol per kilogram of body weight, Haggard and Greenberg (188) found that the concentration of alcohol in the urine in relation to that in the arterial blood corresponded closely to the relative solubility in vitro of alcohol in urine and blood (1.14 to 1). These observations were interpreted as suggesting that alcohol passes through the kidneys by simple diffusion. During the period of absorption from the gastroenteric tract, the concentration of ethyl alcohol found in a peripheral vein (femoral) was somewhat lower than that in the arterial blood. The same investigators found that during a 16-hr period following ingestion of alcohol by dogs, 2.1 to 4.3 percent of the total alcohol ingested was eliminated by the kidneys, the rate of elimination within this range being a function of the concurrent urinary volume. The ratio of the concentration of ethyl alcohol in the arterial blood to that in the alveolar air was found to be 1142 to 1 in correspondence with the distribution of alcohol in vitro in the media involved. The total amount eliminated by dogs in the expired air in a period of 8 hr following ingestion of 4 g/kg amounted to 4 percent of the total amount ingested.

After an oral dose of 2 ml/kg to rabbits, 0.80 percent of the dose was excreted in breath and 3.28 percent of the dose was excreted in urine (36). After an oral dose of 2 g/kg to rats, 2.02 percent of dose was excreted in urine (189).

Pharmacology. Sleeping time following an oral dose of 2 ml/kg to rabbits was approximately 2 hr. Of the various primary alcohols tested, only methanol had a shorter time (36). Newman and Lehman (190) demonstrated that dogs habituated to drinking alcohol showed better neuromuscular coordination under the influence of a given concentration of alcohol in the blood than did control dogs. Furthermore, since the brains of habituated rats contained a slightly higher concentration of alcohol than did those of control rats under corresponding condition of dosage, these investigators believe that acquired tolerance in animals is primarily an adaptation of the cells of the central nervous system.

Rats were subjected to an inclined plane test, in which the angle at which the rat lost its grip as a function of the inclination of the plane was compared before and after an oral dose of 0.043 mol/kg (about 2 g/kg) ethanol. Rats recovered very rapidly, within 2 hr following dosage (191). Rats were subjected to a moving belt test following oral administration of six different doses of ethanol, *n*-propanol, isopropanol, *n*-butanol, or *tert*-butanol. Almost identical dose-response curves were obtained after correction for thermodynamic activity. The procedure was repeated, giving ethanol to all rats. Rats showed complete cross-tolerance for all alcohols; all corrected dose-response curves shifted to the same degree. Authors concluded that the cross-tolerance is in the nervous system itself (192). Rats exposed to ethanol vapors at concentrations up to 24,000 ppm showed no effect on behavior after eight daily 4-hr exposures. At 32,000 ppm inhibitions of the avoidance excape conditioned response and the unconditioned response were obtained; this concentration produced overt symptoms of depression and ataxia (193).

Pharmacokinetics. Haggard and Greenberg (174) state that the rate of oxidation of ethyl alcohol is proportional to the amount in the body. Marshall and Fritz (194) employed a sensitive enzymic method for the determination of the concentration of ethanol in the plasma of dogs following the intravenous injection of ethanol. With very low concentrations of alcohol in the plasma (5 to 10 mg percent) the disappearance of ethanol followed an exponential curve, indicating that the rate of disappearance was proportional to the concentration present. The rate of disappearance at higher levels was greater than at lower levels but it was not determined whether this was due to an increased rate of oxidation because of the hourly biologic variation in metabolism. Aull et al. (195) observed variation of the rate of metabolism among rats given ethanol intraperitoneally in the dosage of 2.85 or 3.0 g/kg of body weight. Upon the basis that the rate of metabolism of ethanol is independent of the concentration, the investigators stated that alcohol disappeared at the rate of 270 mg/kg/hr when based on the declining concentration of alcohol in the blood with time, and disappeared at the rate of 293 mg/kg/hr when based on the difference between the dosage administered and the amount found in the carcass by analytic determination. These latter investigators state that the rat metabolizes ethanol more slowly than the mouse and more rapidly than the dog. In rats following an oral dose of 2 g/kg, blood levels peaked at 2 hr and dropped rapidly. Urinary excretion peaked after 1½ hr (87). After an oral dose of 20 percent ethanol (amount not specified) to rabbits (35), the concentration in blood peaked at 1 hr and then decreased; the concentration in the urine reached a maximum in 2 hr and the concentration in breath reached a maximum in 1 hr.

Mutagenicity—In Vitro. Ethanol was found to be not mutagenic in the *Salmonella*/microsome test (196).

3.4.3 Effects in Humans

Absorption. A study of the effects of inhalation of ethyl alcohol by man has been carried out by Lester and Greenberg (197). In this study, subjects were exposed to known concentrations of alcohol and observed for subjective and objective symptoms, ventilation rate, percent absorption, and blood levels of alcohol at varying times after exposure. These authors question the previous conclusions of Loewy and von der Heide (116) with regard to the conclusion that more than 0.1 percent in the air would be hazardous in workrooms. Since the subjects in the latter study were exposed for only 2 hr and since no blood levels were taken, it was difficult to interpret the symptoms that were produced at higher levels of exposure. Symptoms noted by Lester and Greenberg were as follows. At 10 to 20 mg/l (about 5000 to 10,000 ppm) there was some coughing and smarting of the eyes and nose, with disappearance of the symptoms within a few minutes. Although the atmosphere was not exactly confortable, it was felt that there would be no difficulty in tolerating these levels. At 30 mg/l (about 15,000 ppm) there was continuous lacrimation and coughing. Above 40 mg/l (about 20,000 ppm) it was impossible to tolerate the atmosphere even for a short period of time. The subjective symptoms increased when the rate of ventilation was increased by two or three times the resting level. Measurements were made of the percent absorption as related to changes in the ventilation rate from 7 to 25 l/min and at concentrations of acohol varying from 11 to 19 mg/l. The average absorption found was about 62 percent and seemed to be independent of concentration and rate of ventilation.

Studies on blood alcohol concentrations were made by exposing subjects to concentrations in the range of 13 to 16 mg/l for periods of 3 to 6 hr with ventilation rates varying from resting ventilation to as high as 25 l/min. The highest blood level obtained in any of the subjects was 47 mg alcohol/100 ml of blood after a 6-hr exposure to 16 mg/l (8500 ppm) of alcohol at a high ventilation rate of 22 l/min. When the ventilation rate was reduced to 15 l/min with an alcohol concentration in the air of 15 mg/l (8000 ppm), there was definitely a plateau reached in the blood level of about 7 to 8 mg/100 ml of blood. It was obvious that under these conditions the rate of intake of alcohol was equal to the rate of metabolism. They point out that, based on the results of these studies, a 70-kg man exposed to 1000 ppm of alcohol would have to breathe at a rate greater than 117 l/min in order to obtain any continuous rise in alcohol concentration. They conclude from this that since continuous hard work may involve a ventilation rate of the order of 30 l/min, a concentration of alcohol in the air not exceeding 7 mg/l (about 3500 ppm) seems to be a more realistic standard. Fassett (198) feels that his observations under practical conditions of exposure are in accord with Lester and Greenberg regarding irritant levels and lack of subjective symptoms at levels higher than 1000 ppm.

Determination of the content of ethyl alcohol in the blood, spinal fluid, urine, or even the breath is useful in determining the amount absorbed, when such data are considered in conjunction with accurate information on the time and duration of the exposure. Although there is considerable variation among individuals and in the same individual at different times, as well as some divergence in the data of various authorities, the clinical effects shown in Table 55.6 are commonly associated with the indicated concentrations in the blood (145).

Table 55.6. Effects of Various Concentrations of Ethanol in the Blood of Humans

Effect	Alcohol Concentration in Blood (%)
Beginning of uncertainty	0.06–0.08
Slow comprehension	0.10
Stupor	0.12–0.15
Drunkenness	0.16
Severe intoxication	0.2–0.4
Death	0.4–0.5

Table 55.7 shows the distribution of ethanol in human tissues obtained at autopsy of 93 corpses (199).

Metabolism. Ethanol can be oxidized by three enzyme systems in humans, similar to those in animals. According to Jacobsen (185), Antabuse (tetraethylthiuram disulfide) causes an increase in acetaldehyde concentration and a reduction in the formation of carbon dioxide. Casier and Polet (200) maintain that Antabuse prevents the oxidation of ethanol rather than causing the accumulation of acetaldehyde. These latter investigators propose that the disagreeable effects resulting from ethanol after treatment with this disulfide are not caused by the accumulation of acetaldehyde but rather by the combined effects of the alcohol and metabolites of the disulfide. Casier and Polet found no accumulation of radioactive acetaldehyde in mice treated beforehand with tetraethylthiuram disulfide, but instead they observed an increase in volatile reducing substances. Antabuse is known to chelate the zinc of ADH and the metals of the aldehyde-oxidizing enzymes (201). This compound can also inactivate dehydrogenases by formation of disulfide linkages with sulfhydryl groups essential for the activity of these enzymes (202). Thus the inhibitions of both the alcohol and aldehyde-metabolizing enzymes are the probable causes of the Antabuse–alcohol syndrome (203).

Excretion. The average rate of elimination of ethanol for man is 100 mg/kg/hr (204).

Pharmacology. Ethanol depresses the central nervous system over a wide range of doses (154). It is not clear whether ethanol per se produces chronic neurotoxic effects (205).

Pharmacokinetics. Jacobsen (185) suggested that ethyl alcohol can proceed to acetaldehyde by means of two different enzymes, namely, alcohol dehydrogenase and catalase, and that the shape of the disappearance curve for ethanol

Table 55.7. Ratio between the Concentration of Ethanol in Body Fluids and Tissues to that in Whole Blood

Fluid or Tissue	Ratio	Fluid or Tissue	Ratio
Blood	1	Kidney	0.92
Urine	1.33	Skeletal muscle	0.84
Cerebrospinal fluid	1.27	Spleen	0.89
Bile	1.10	Heart muscle	0.83
Testes	0.87	Brain	0.88
		Liver	0.64[a]

[a] Low figure attributed to postmortem enzymatic breakdown.

depends on the ratio between the effects of these two systems. If the former prevails, the curve is rectilinear and independent of the concentration, but if the latter takes prominence then the rate depends on the concentration.

Mode of Action. Ethanol affects the following organs/systems in the human body:

1. *Liver.* The production of fatty liver has been attributed to increased synthesis of fatty acids especially from the acetate formed from ethanol; increased transportation of fatty acids from peripheral fat depots; decreased transportation of lipids from liver; increased synthesis of triglycerides from free fatty acids and α-glycerophosphate; and decreased oxidation of fatty acids (206).

2. *Hematopoetic System.* Waller and Bonoehr (207), as a result of their observations of humans suffering from alcoholism, present most convincing arguments for their conclusions that alcoholism can result in hematologic changes up to and including pancytopenia. They give additional supportive references.

Petri (164) records that Fahr listed the following as the most commonly encountered lesions resulting from prolonged ingestion of toxic quantities of alcohol: fatty infiltration of the liver and heart muscle, chronic leptomeningitis, and chronic gastritis. Brezina (208), without much proof, accepts certain cardiac disturbances as effects of the inhalation of the vapors of alcohol.

Epidemiologic Data. Epidemiologic studies on alcoholics indicate that heavy drinking increases the risk of cancer of the larynx, esophagus, mouth, and pharynx (209). These increases were associated with smoking and cirrhosis of the liver.

Other Pertinent Observations. The well-known effects of chronic alcoholism from the excessive use of alcoholic beverages are not matters of concern in relation to occupational hazards, except to the extent that they may influence the effects of exposure to other substances and environmental conditions. In any case, they do not enter into the present discussion.

Although ethyl alcohol is relatively innocuous if proper ventilation is maintained, prolonged exposures to excessive concentrations may produce irritation of the mucous membrane, irritation of the upper respiratory tract, headache, nervousness, dizziness, tremors, fatigue, nausea, and narcosis (116, 210, 211). The effects of ethyl alcohol upon the power of concentration and alertness should be remembered in relation to the prevention of industrial accidents. Lehmann and Flury (145) are authority for the statement that intoxication has been seen among human beings subjected to inhalation of the vapors of hot alcohol.

In terms of symptomatology in relation to dosage, there is no doubt that a tolerance is acquired after repeated exposure to ethyl alcohol. However, no proof has been submitted of physiologic adaptation in man in terms of metabolic changes or of resistance to cellular injuries. Loewy and von der Heide (116) have shown that the symptoms are less severe and the time required to produce them is greater in subjects accustomed to alcohol than in those unaccustomed to it, Table 55.8.

Skin Irritation. A modification of the Draize test on men's forearms showed that a 21-day open test resulted in no irritation, whereas a 21-day occlusive test caused erythema and induration toward the end of the exposure period (143). Exposure of the skin of the forearm to ethanol 1 hr/day for 6 days resulted in injury only of the keratinized layer (212).

Skin Absorption. Bowers et al. (213) did not detect measurable amounts of alcohol in blood when both legs of a man were wrapped from the knee down with cotton dressing soaked in 95 percent alcohol.

Mutagenicity. In a test using human fibroblast cultures, an increase in chromatid breakage was considered to be consistent with a nonspecific cytotoxicity of ethanol rather than evidence of mutagenicity (214).

Table 55.8. Symptoms Induced in Humans by the Inhalation of Various Concentrations of Vapors of Ethanol (116)

Average Concentration		Duration of Exposure (min)	Symptoms
mg/l	ppm		
			A. Subject Unaccustomed to Alcohol
2.59	1380	39	None after 28 min; after 33 min, headache and slight numbness
6.28	3340	100	Sensation of warmth and coldness, nasal irritation, headache, and numbness
16.62	8840	64	Initial intolerable odor and difficulty in breathing, soon overcome, conjunctival and nasal irritation, feeling of warmth, headache, drowsiness, fatigue
			B. Subject Accustomed to Alcohol
9.45	5030	120	Slight headache after 20 min
11.50	6120	120	Odor intense, slight pressure in left temple
13.14	6990	109	Headache, conjunctival irritation, feeling of warmth, drowsiness, and fatigue

3.5 Hygienic Standard of Permissible Exposure

Both the ACGIH (215) and OSHA (108) have indicated that a standard of 1000 ppm or 1900 mg/m^3 would be appropriate, and we agree. This value is accepted in numerous countries with the following exceptions. The standard is 1500 mg/m^3 in Italy; 1000 mg/m^3 in Czechoslovakia, German Democratic Republic, Hungary, Poland, Roumania, and the Soviet Union; and 200 mg/m^3 in Bulgaria (110).

3.6 Odor and Warning Properties

Concentrations of 6000 to 9000 ppm have an intense odor and may be practically intolerable at first, but one becomes acclimated to them after a short time. Concentrations of this order of magnitude, however, should not be permitted. The minimum identifiable odor of ethyl alcohol at 350 ppm, as determined by Scherberger et al. (92), is well below 1000 ppm, the generally accepted hygienic standard.

May (93) found that a level of 50 ppm was barely detectable and that 100 ppm produced a distinct odor.

4 1-PROPANOL; *n*-Propanol, *n*-Propyl Alcohol, Propyl Alcohol, Optal, Osmosol Extra, CAS No. 71-23-8

$$CH_3CH_2CH_2OH$$

4.1 Source, Uses, and Industrial Exposure

n-Propyl alcohol is recovered as a by-product of methanol synthesis by high pressure, and in the propane–butane oxidation process (216).

n-Propyl alcohol, although not used as extensively as certain other alcohols, may be used as a solvent for vegetable oils; natural gums and resins such as soft copals, rosin, and shellac; certain synthetic resins; ethyl cellulose; and polyvinyl butyral. It is used in lacquers and dopes, cosmetics, dental lotions, cleaners, polishes, and pharmaceuticals (216, 217). Ill effects from industrial usage of *n*-propyl alcohol have not been reported.

4.2 Physical and Chemical Properties

Physical state	colorless, volatile liquid
Molecular formula	C_3H_8O
Molecular weight	60.09
Boiling point	97.3°C (217)

Freezing point	−126.2°C
Specific gravity	0.804 (20/4°C) (218)
Vapor pressure	20.8 mm Hg (25°C) (219)
Refractive index	1.385 (20°C) (217)
Flammability	
Flash point	59°F
Explosive limits in air	2 to 13 percent
Autoignition temperature	700°F
Percent in saturated air	2.7 (25°C)
Density of saturated air (air = 1)	1.028 (25°C)
Solubility	Miscible with water, alcohols, ethers, and most solvents
1 mg/l ≏	408 ppm at 25°C, 760 mm Hg
1 ppm ≏	2.45 mg/m³ at 25°C, 760 mm Hg

4.3 Determination in the Atmosphere

NIOSH Method No. S62 (120) has been recommended. It involves drawing a known volume of air (recommended sample is 10 liters at a rate of 0.2 l/min or less) through charcoal to trap the vapors present. The analyte is desorbed with carbon disulfide containing 1 percent 2-propanol. The sample is separated with a gas chromatograph and analyzed with a flame ionization detector. Recently Mueller and Miller (1) and Langvardt and Melcher (2) have published methods of collection and analysis by gas chromatographic procedures that give excellent results and are applicable in situations where a variety of other materials may be present. The older chemical methods, such as noted in Section 2.3, would seem to have limited usefulness except where modern equipment is not available or where unusual conditions exist.

4.4 Physiologic Response

4.4.1 Summary

n-Propanol is low in acute oral toxicity to animals, markedly irritating and injurious to the eyes of rabbits, and not irritating to the skin or absorbed through the skin in appreciable amounts. Although inhalation is a likely route of exposure, its odor and its capability to cause irritation of the mucous membranes tend to prevent exposure to acutely hazardous concentrations but may not deter excessive chronic exposure. Excessive acute exposure may cause varying degrees of mucous membrane irritation, ataxia, lethargy, prostration and narcosis. Prolonged treatment of rats by both the oral and the subcutaneous route has been reported to cause malignancies.

4.4.2   Effects in Animals

Single or Short-Term Exposure Data. *Single-Dose Oral.* Data on the single-dose oral toxicity of *n*-propanol are given in Table 55.9.

Eye Contact.   Instillation of 0.1 ml of 1-propanol into the conjunctival sac of rabbits produced marked to severe conjunctivitis, iritis, corneal opacities, and ulcerations. During a 14-day observation period, pannus formation and kerataconus developed in four of six exposed eyes (225).

Skin Contact.   1-Propanol is not appreciably irritating to the skin of rabbits even after prolonged contact (142, 220, 225) but it can be absorbed in significant amounts if confined to the skin. LD_{50} values of 5 ml/kg (4 g/kg) (220) and 6.7 g/kg (225) are reported for 24-hr exposures.

Inhalation.   Rats survived a 2-hr exposure to saturated vapor but two of six rats died after a 4-hr exposure at 4000 ppm (220). Exposure of rats to a minimum of 20,000 ppm for 1 hr resulted in no mortalities during a 14-day postexposure observation period (225).

Exposures of mice to 13,120 ppm (32.2 mg/l) for 160 min or 19,680 ppm (48.2 mg/l) for 120 min were fatal (44). Groups of mice (two in each) were exposed by Starrek (226) for decreasing lengths of time (480, 240, 135, 120, 90, and 60 min) to increasing concentrations of *n*-propyl alcohol in the atmosphere [3250 ppm (8 mg/l), 4100 ppm (10 mg/l), 8150 ppm (20 mg/l), 12,250 ppm (30 mg/l), 16,300 ppm (40 mg/l), and 24,500 ppm (60 mg/l)]. The length of time required for the appearance of ataxia, prostration, and deep narcosis was inversely proportional to the concentration to which the mice were exposed. Ataxia appeared in 10 to 14 min at 24,500 ppm and in 90 to 120 min at 3250 ppm. Prostration was evident in 19 to 23 min at the former concentration and in 165 to 180 min at the latter. Deep narcosis was manifest in 60 min at 24,500

Table 55.9.   Single-Dose Oral Toxicity Values for 1-Propanol in Animals

Species	LD_{50} Values (g/kg)	Ref.
Rat	1.87	220
	5.40	221
	5.35	222
	6.5	223, 228
Mouse	4.5	44
Rabbit	3.5 [a]	224
	2.82	37

[a] MLD values.

ppm and in 240 min at 4100 ppm. Only 1 of 12 mice that showed signs of intoxication died. Mice exposed for 480 min to 2050 ppm (5 mg/l) exhibited no signs of adverse effects.

Injection. When mice were given single subcutaneous injection of 1-propanol, the LD_{50} was found to be 3.23 g/kg (227); when the material was given intravenously, the LD_{50} was found to be 18.2 mmol or 1.09 g/kg (54).

Aspiration. Aspiration of 0.2 ml *n*-propanol caused mortality in nine of nine rats (53).

Mode of Action. Animals exposed to sufficient concentrations of the vapors of *n*-propyl alcohol may manifest the following signs of adverse effect: irritation of the mucous membranes, ataxia, lethargy, prostration, and narcosis. When rats were given 6.5 g/kg orally, the LD_{50} dose, they became comatose within a few minutes after treatment and deaths occurred between 2 and 18 hr posttreatment (228).

Repeated or Prolonged Exposure Data. *Repeated Oral.* No macroscopic liver lesions were observed and no deaths occurred when six rats were given oral doses of 2160 mg/kg/day *n*-propanol on each of four successive days (228).

Skin Absorption. Application of 38 ml/kg/day to the skin of rabbits for 30 days over a period of 6 weeks resulted in death of one-third of the animals treated (142).

Inhalation. Mice survived intermittent exposures for a total of 95 hr to 7874 ppm (19.3 mg/l) (44).

Carcinogenicity. 1-Propanol (reagent grade, twice redistilled) was administered by intubation in doses of 0.3 ml/kg twice a week to 18 Wistar rats aged 10 weeks at the beginning of the experiment. The average total dosage was 50 ml and the average survival time was 570 days. In addition to severe liver injury and hyperplasia of the hematopoietic parenchyma, five malignant tumors (two myeloid leukemias, two liver sarcomas, one liver cell carcinoma) and 10 benign tumors were observed. Three benign but no malignant tumors were found in the controls given saline (229). Ethanol, under similar test conditions did not cause any of these effects.

In a parallel study, 31 Wistar rats, aged 10 weeks at the beginning of experiment, were given subcutaneous injections of 0.06 ml/kg twice a week. The average total dosage was 6 ml and the average survival time was 666 days. Thirteen malignant tumors developed (five liver sarcomas, four myeloid leukemias, one local sarcoma, one kidney pelvis carcinoma, one bladder carcinoma, one uterine carcinoma) and seven benign tumors were identified. Severe liver injury and marked hematopoietic effects were noted in the treated rats. Control rats given saline had two benign but no malignant tumors (229).

Metabolism. Von Wartburg (230) reported that the rate of oxidation of *n*-propanol by ADH is approximately equal to that of ethanol. According to Teschke et al. (231), hepatic microsomes catalyzed the oxidation of propanol to its aldehyde in the rat. The reaction required oxygen and NADPH, was inhibited by carbon dioxide, and acted independently of catalase and alcohol dehydrogenase.

Excretion. Following an oral dose of 2 g/kg to rats, 0.13 percent of the dose was excreted in the urine (189). Saito (36) found 1.65 percent excreted in the expired air and 0.7 percent excreted in the urine following an oral dose of 2 ml/kg to rabbits. 1-Propanol inhibited the elimination of 2-propanol from rats' blood and from isolated perfused rat livers but 2-propanol did not inhibit elimination of 1-propanol (232).

Pharmacology. Rats were subjected to an inclined plane test in which the angle at which the rat lost its grip was compared before and after a nontoxic oral dose of one of several lower aliphatic alcohols. *n*-Propanol produced more severe and prolonged intoxication than did ethanol (191). In normothermic dogs given intravenous injections of *n*-propanol, with supported respiration and nitrous oxide anesthesia, 2-min cardiac standstill occurred at a blood level of 713 mg/ml (151). Blood pressure and mean pulse rate fell more rapidly after intravenous injection of 1-propanol than with ethanol, but less rapidly than with 1-butanol. Although the qualitative responses to those three alcohols are similar, a nearly exponential increase in their pharmacologic effects exists (151).

Pharmacokinetics. In rats given an oral dose of 2 g/kg, blood levels of 1-propanol were very low; they reached a maximum at 90 min and disappeared by 8 hr following the treatment; the disappearance rate was slower than that for ethanol (189). Saito (36) found that, following an oral dose of 2 ml/kg to rabbits, blood levels reached a maximum in 30 to 60 min and disappeared to a trace after 6 hr. *n*-Propyl alcohol was found in the blood of a dog for 275 min following the oral administration of 16.1 g of this alcohol. Acetone, derived from isopropyl alcohol, was found in the blood of a dog for 540 min following the oral administration of 15.8 g of isopropyl alcohol. Hence *n*-propyl alcohol is oxidized and eliminated faster than isopropyl alcohol (233). Basing his opinion upon the concentrations in the blood, the same investigator (233) concluded that *n*-propyl alcohol was oxidized and eliminated from the dog considerably more rapidly than was ethyl alcohol. Rabbits given *n*-propyl alcohol by the intravenous route also oxidized it more rapidly than ethyl alcohol (234).

Additional references dealing with such subjects as metabolism, pharmacology, enzyme activation, and inhibition may be of interest (241–253).

Mutagenicity—In Vitro. 1-Propanol inactivated *E. coli* CA 274 in a concentration-dependent manner and increased the reversion rate in these bacteria, predominantly inducing true reverse mutations but also some suppressor mutations (235).

Ames and mouse lymphoma tests were negative (236).

4.4.3 Effects in Humans

Skin Absorption. Blank (239), using abdominal autopsy skin in a diffusion chamber, found that 1-propanol penetrated the skin more rapidly from solution in nonpolar solvents such as isopropyl palmitate, olive oil, and mineral oil than from solution in saline.

Human Experience. Reports of adverse human experience from 1-propanol are rare in view of the extent of its use. In one case a positive patch test reaction to 1-propanol was elicited in a person sensitive to 2-propanol, suggesting that cross-sensitization may occur (240). No cases of adverse systemic effects were found.

4.5 Hygienic Standards of Permissible Exposure

Both the ACGIH (215) and OSHA (108) have indicated that 200 ppm or 490 mg/m^3 would be appropriate industrial hygiene standards. Most other countries agree, but Bulgaria, Poland, and Hungary have standards of 200 mg/m^3 and the Soviet Union has a standard of 10 mg/m^3 (110). Since significant hazards to health are associated with the synthesis of 1-propanol, they should be considered separately and not confused with the hazards of 1-propanol per se.

4.6 Odor and Warning Properties

Exposure to 400 ppm for 3 to 5 min produced mild irritation of eyes, nose, and throat (237). May (93) found the odor detection limit by sniff-testing to be 80 mg/m^3 or 30 ppm. Laing (238), by means of an automatic air dilution olfactometer, found the mean olfactory threshold to be 40 ppm for humans and 5 ppm for rats.

5 PROPANOL; Isopropyl Alcohol, Isopropanol, *sec*-Propyl Alcohol, Dimethyl Carbinol, Petrohol, CAS No. 67-63-0

$$CH_3CHOHCH_3$$

5.1 Source, Uses, and Industrial Exposure

Isopropyl alcohol is synthesized primarily from propylene, either by indirect hydration (strong-acid process) or by direct catalytic hydration (weak-acid process) (254). At present, the direct catalytic hydration technique has replaced the older indirect hydration technique in the United States. Isopropyl alcohol can be synthesized from acetone (255).

More than half of the isopropyl alcohol produced is used in the manufacture of acetone (254). Other uses are in liniments, skin lotions, cosmetics, permanent wave preparations, pharmaceuticals, and hair tonics; as a solvent in perfumes, oils, gums, and resins; and in extraction processes. It has gained widespread use as a rubbing alcohol and is an ingredient of antifreezes, liquid soaps and window cleaners.

NIOSH estimates that approximately 141,000 employees are potentially exposed to isopropyl alcohol in the United States (256). However, this alcohol does not constitute a significant industrial hazard.

5.2 Physical and Chemical Properties

Physical state	Colorless, volatile liquid
Molecular formula	C_3H_8O
Molecular weight	60.09
Boiling point	82.5°C (20)
Freezing point	−89.5°C
Melting point	−88.5°C (20)
Specific gravity	0.7874 (20/20°C) (257)
Vapor pressure	44 mm Hg (25°C) (257)
Refractive index	1.37723 (20°C) (20)
Flammability	
Flash point	11.7°C (closed cup) (20)
Explosive limits	2.5 to 11.80 percent (257)
Autoignition temperature	455.6°C (20)
Percent in saturated air	5.8 (25°C)
Density of saturated air (air = 1)	1.06
Solubility	Miscible with water, alcohol, ether, chloroform, and most organic solvents
1 mg/l ≎	408 ppm at 25°C, 760 mm Hg
1 ppm ≎	2.45 mg/m^3 at 25°C, 760 mm Hg

5.3 Determination in the Atmosphere

NIOSH Method No. S65 (258) has been recommended. It involves drawing a known volume of air through charcoal to trap the vapors present (recommended sample is 3 liters at a rate of 0.2 l/min). The analyte is desorbed with carbon disulfide containing 1 percent 2-butanol. The sample is separated by injection into a gas chromatograph equipped with a flame ionization detector; the area of the resulting peak is determined and compared with standards. Recently Mueller and Miller (1) and Langvardt and Metcher (2) have published methods for collection and analysis by chromatographic methods that give excellent results and are applicable in situations where a variety of other materials may be present. The older methods such as noted subsequently would seem to have limited usefulness except where modern equipment is not available. Isopropyl alcohol that has been collected in water may be oxidized to acetone by chromic acid, and the acetone subsequently measured iodimetrically (259). The use of sodium nitroprusside for determining the acetone seems feasible (260). After absorbing isopropyl alcohol on silica gel, Hahn (261) recovered it by steam distillation, converted it to isopropyl nitrite, and subsequently titrated the liberated nitrous acid according to the procedure of Knipping and Ponndorf (262).

5.4 Physiologic Properties

5.4.1 Summary

Isopropanol is low in acute toxicity to animals, appreciably irritating and even injurious to the eyes of rabbits, not appreciably irritating to the skin, and not absorbed through the skin in significant amounts, unless exposures are severe, prolonged, and repeated. Toxic amounts can result from severe exposure to vapors, but generally the odor and irritating effects should preclude exposure to acutely hazardous amounts of vapor. 2-Propanol is a depressant causing ataxia, prostration, narcosis, and even death, depending upon the severity of the exposure by any route. 2-Propanol in large amounts is fetotoxic but probably not teratogenic or carcinogenic. Reasonable care in handling and use should not present any serious industrial hazards to health.

5.4.2 Effects in Animals

Single or Short-Term Exposure Data. *Single-Dose Oral.* Data on the single-dose oral toxicity of 2-propanol are given in Table 55.10.

Eye Contact. Instillation of 0.1 ml of 70 percent 2-propanol resulted in significant irritation to the rabbit eye characterized by conjunctival redness, corneal opacity, and iritis of a temporary nature (265).

Skin Contact. 2-Propanol is not appreciably irritating to the intact or abraded skin of rabbits (142, 266). Also, it is not absorbed to any appreciable extent (142, 263) unless exposures are severe.

**Table 55.10.   Single-Dose Oral Toxicity Values for
2-Propanol in Animals**

Species	LD$_{50}$ Values (g/kg)	Ref.
Rats		
Young adults	4.7	28
Older adults	5.3	28
	5.8	263
Rabbit	7.9	37
	7.8	137
Dog	6.2	264

Inhalation.   Smyth found that rats survived when exposed for 4 hr to a concentration of 12,000 ppm but exposure for 8 hr to this concentration resulted in death in about one-half the group (267). Mice died if exposed to 12,800 ppm (31.4 mg/l) for 200 min or 19,200 ppm (47.1 mg/l) for 160 min (44). The length of time required for the development of ataxia, prostration, and deep narcosis on the part of mice exposed by Starrek (226) to vapors of isopropyl alcohol was inversely proportional to the concentration. Ataxia was manifest in 12 to 26 min at 24,000 ppm (60 mg/l), but it occurred with progressively decreasing rapidity at concentrations of 16,300 ppm (40 mg/l) 12,250 ppm (30 mg/l) 8150 ppm) (20 mg/l), and 4100 ppm (10 mg/l), until at 3250 ppm (8 mg/l) 180 to 195 min were required. Prostration appeared in 37 to 46 min at 24,500 ppm and in 340 to 350 min at 3250 ppm. The onset of deep narcosis ranged from 100 min at 24,500 ppm to 460 min at 3250 ppm. Only 1 of 12 mice exposed in these experiments succumbed. Mice exposed for 480 min to 2050 ppm (5 mg/l) gave no evidence of reaction.

Injection.   When mice were given single intravenous injections of 2-propanol, the LD$_{50}$ was found to be 31.0 mM or 1.86 g/kg (54).

Aspiration.   Deaths occurred in 6 of 10 rats following aspiration of 0.2 ml of 100 percent isopropyl alcohol, and in 1 of 10 rats following aspiration of 0.2 ml 70 percent concentration (53).

Mode of Action.   Isopropyl alcohol in large amounts is more toxic (264) and more narcotic (44, 268) than ethyl alcohol, but less so than *n*-propyl alcohol (152).

Animals subjected to vapor of isopropyl alcohol exhibit the following signs of intoxication: irritation of the mucous membranes, ataxia, prostration, deep narcosis, and death.

Repeated or Prolonged Exposure Data.   *Repeated Oral.*   Rats that ingested 0.5 to 10.0 percent isopropyl alcohol in their drinking water for 27 weeks showed decreased body weight gain (264). No gross or microscopic abnormalities of the brain, pituitary, lung, heart, liver, spleen, kidneys, or adrenals were observed.

Skin Contact. Slight erythema, dryness, and superficial desquamation resulted from prolonged and repeated exposure to the skin of rabbits; applications of 45 ml/kg/day for 30 days over a period of 6 weeks resulted in no deaths (142).

Inhalation. Rats were unaffected except for slight intoxication when exposed by Macht (152, 269) intermittently over a week (total number of hours of exposure not given) to air supposedly saturated with vapors of isopropyl alcohol. Mice subjected by Weese (44) to 10,900 ppm isopropyl alcohol in air (26.8 mg/ 1) for about 4 hr/day until they had accumulated 123 hr of exposure were narcotized but survived. Reversible fatty changes were observed in the liver.

Gorlova (311) and Baikov et al. (313) have reported on what appear to be the same studies in which three groups of 15 rats were exposed for 24 hr/day for 3 months to 20.5, 2.6, or 0.66 mg/m^3 of isopropanol vapor. These are equivalent to 8.4, 1.0, and 0.27 ppm, respectively. A fourth group served as controls. A variety of measurements were made. The positive findings were as follows: at 8.4 ppm, alterations in reflexes, enzyme activity, leukocyte fluorescence, BSP retention, total nucleic acids, coproporphyrin concentrations in the urine, and the morphology of the lung, liver, spleen, and central nervous system; at 1.0 ppm, lesser alterations in total nucleic acids, redox enzymes in the blood, and coproporphyrins in the urine; and at 0.27 ppm, no alterations. The significance of these reported findings is difficult to interpret because they are at variance with the other data available, particularly those involving organ morphology. Also, in view of the wide usage and extensive exposure to substantial amounts of isopropyl alcohol without apparent chronic effects, it is difficult to understand why more problems have not occurred if the material is as toxic as these data suggest. Verification of these findings is needed.

Carcinogenicity. Weil et al. (271) conducted studies in which mice were exposed by inhalation 5 days/week, 3 to 7 hr/day for 5 to 8 months to 3000 ppm* isopropyl alcohol vapors. Isopropyl oils from two manufacturing processes were similarly evaluated for carcinogenic potential. Isopropyl alcohol showed no tumorigenic activity, whereas increased tumor incidence was found in mice exposed to one of the oils. Weil (272) conducted a second series of experiments to further evaluate the tumorigenic potential of isopropyl oil; isopropyl alcohol was also studied in these investigations, which included subcutaneous injections and skin painting assays in mice. Again, isopropyl alcohol showed no tumorigenic potential. Results regarding the tumorigenicity of the oils were inconclusive.

---

* Publication (271) erroneously states exposures were to 0.0075 mg/m^3. This figure resulted from an error in converting 3000 ppm to mg/m^3. Correction by authors after personal communication with C. S. Weil.

Teratology. In a two-generation reproduction study in which male and female rats received 2.5 percent isopropyl alcohol in drinking water, the very early growth of the first generation offspring of treated rats was retarded (270), indicating a fetotoxic effect but not a teratologic effect. McLaughlin et al. (309) observed that isopropyl alcohol did not cause teratogenic effects when injected into fresh fertile chicken eggs before incubation.

Metabolism and Pharmacokinetics. The exact metabolism of isopropyl alcohol is not clearly defined. Part of it is oxidized to acetone (274), and some probably conjugates with glucuronic acid (275). However, these processes do not account for all of the administered isopropyl alcohol. Morris and Lightbody (282) gave 6 ml isopropyl alcohol per kilogram of body weight by mouth to each of six rabbbits, and found acetone in the urine of five of them during the following 72 hr, but none thereafter. Kemal (273) gave a progressively increasing daily dose of isopropyl alcohol (5 to 90 ml) by stomach tube to three dogs. From the thirteenth day on, when doses of 65 ml or more of isopropyl alcohol were given, 48 to 71 mg of acetone and 119 to 148 mg of isopropyl alcohol, per 100 ml of urine, were found in daily volumes of urine ranging from 1070 to 2250 ml. Acetone and isopropyl alcohol have been found in the expired air of animals following intake of isopropyl alcohol. In a period of 12 hr following the oral administration of 2.37 g of isopropyl alcohol to rabbits, 0.251 g of acetone (equivalent to 0.258 g of isopropyl alcohol) and 0.0281 g of isopropyl alcohol were found by Pohl (284) in the expelled air. These amounted to 10.9 and 1.2 percent, respectively, of that administered. The administration of adrenaline, histamine, or oxyphenylethylamine to dogs did not alter significantly the rate of the oxidation of isopropyl alcohol (284).

The distribution of isopropyl alcohol and acetone in the tissues of dogs 4 hr after the oral administration of 90 ml of isopropyl alcohol was determined by Kemal (273) using dogs that had been given progressively increasing amounts of the alcohol for the previous 59 days. In general, the concentrations of isopropyl alcohol found in the tissues and body fluids decreased in the following order: brain, urine, heart, kidney, and blood. The relationship of the concentrations of acetone in the tissues and urine was not as clearly defined as that for the alcohol. Except in the blood, where the value for acetone approached that for isopropyl alcohol, the concentration of isopropyl alcohol in the tissues and urine was about twice that for acetone.

Saito (36) studied the metabolic fate of several of the lower alcohols identified in various beverages. Signs of drunkenness appeared in rabbits in the order of *n*-butanol, *sec*-butanol, isobutanol, *n*-propanol, isopropanol, ethanol, *tert*-butanol, and methanol. Toxic responses showed the same descending order and were proportional to the number of carbon atoms. The more toxic alcohols showed a low concentration in the blood but remained for a long period. Primary alcohols did not form dehydrogenated or oxidized compounds while the secondary alcohols, isopropanol and *sec*-butanol, formed ketones. The ketone forming alcohols made the blood pH alkaline due to overrespiration. Oxidation reaction by alcohol dehydrogenase was observed strongly with the primary alcohols ethanol, *n*-propanol, *n*-butanol, and isobutanol, followed by the secondary alcohols isopropanol and *sec*-butanol. After oral administration of 2 ml/kg to rabbits of the ketone-forming alcohols, isopropanol and *sec*-butanol, about 6 percent of each was excreted in urine as the alcohol and about 25 percent as the ketone metabolite. In both cases the breath contained five to seven times more of the metabolites than did the urine. Gaillard and Derache (75) reported that 0.34 percent of an oral dose of 2 g/kg isopropyl alcohol to rats was excreted in the urine.

Nordmann et al. (310) treated groups of rats intraperitoneally with either pyrazole, an inhibitor of alcoholic dehydrogenase and catalase, or 3-amino-1,2,4-triazole, an inhibitor of catalase alone. These animals and untreated controls given saline or water were then given isopropanol intraperitoneally or orally and the blood concentrations of isopropanol and acetone were followed for 20 hr. Blood alcohol clearance was markedly reduced and acetone production was slowed in the pyrazole-treated animals as compared to the controls and the 3-amino-1,2,4-triazole treated animals. These data indicate that alcoholic dehydrogenase, but not catalase, plays an important role in the metabolism of isopropanol.

In the experiments of Lehman and associates (270) isopropyl alcohol was metabolized more slowly than was ethyl alcohol at two levels of concentration by cats, rabbits, and pigeons; but in dogs the rate of disappearance of isopropyl alcohol from the blood was more rapid than that of ethyl alcohol when the concentrations were high, and slower when they were low. The rate of decrease of isopropyl alcohol in the blood of dogs for a period of 4 to 24 hr following intravenous infusion of isopropyl alcohol (0.64 to 3.84 ml/kg) or oral administration (0.92 to 3.75 ml/kg) has been found to vary with the concentration of alcohol in the blood (268). Gaillard and Derache (283) determined the rate of metabolism of different alcohols in 18-hr fasted Wistar WAG 300 g rats given 2.0 g/kg of 20 percent aqueous solutions. In comparison with ethanol, isopropyl alcohol reached the same blood level as ethyl alcohol 2 hr after dosing but remained at a considerably higher level at 8 hr, roughly 25 mg/100 ml for ethyl versus 80 mg/100 ml for isopropyl alcohol.

Abshagen and Rietbrock (287) conducted studies in which isopropyl alcohol (1 g/kg) was injected intravenously and intraperitoneally in dogs and rats and the plasma levels of isopropyl alcohol and its metabolite, acetone, were determined by gas chromatography at intervals. The maximum concentration of acetone was 12.4 mM 9 hr after the injection of isopropyl alcohol in dogs and 12.0 mM 5 hr after the injection in rats. At 15 min postinjection the plasma concentration of isopropyl alcohol was about 22 mM in both species of animals. Both isopropyl alcohol and acetone are eliminated by simple exponential functions. The concentration of plasma acetone following isopropyl alcohol injection is a Bateman (288) function. The $t_{1/2}$ for the elimination of isopropyl alcohol by the dog and rat is 4 and 2 hr, respectively, and of acetone 11 and 5. The elimination of isopropyl alcohol has a pattern that differs from those of methanol and ethanol.

Pharmacology. The intoxicating effects of several alcohols were investigated by Wallgren (191) in a performance test. The angle of an inclined plane at which a rat slid down was recorded, and performance was measured before and after a nontoxic dose of a given alcohol; a dose of 2.6 g/kg of isopropyl alcohol was found to have about the same intoxicating effect as *n*-propanol and was 2.7 times as intoxicating as ethanol on a molar basis. The sleeping time in rabbits following an oral dose of 2 ml/kg isopropanol was longer than for ethanol and less than for *n*-propanol (36).

By measuring the electric current necessary to induce clonic convulsions in rabbits, rats, and cats, Chu et al. (285) found that isopropyl alcohol had anticonvulsant properties. These authors measured the increase in acetone in the blood and concluded that it paralleled the anticonvulsant action. However, Schaffarzick (286) measured the blood levels of both acetone and isopropyl alcohol in rats and rabbits, and concluded that this property was related to the isopropyl alcohol rather than the acetone.

A number of studies (276–281) have shown that isopropyl alcohol potentiates in animals the toxicity of carbon tetrachloride and other hepatotoxic chlorinated aliphatic hydrocarbons but not 1,1,1-trichloroethane, generally considered to be nonhepatotoxic (279).

5.4.3 Effects in Humans

Experimental Studies. Some controlled studies have been conducted on humans with isopropanol in order to learn more about its metabolism and toxicity. Although the total metabolic pathway of isopropanol is still obscure, acetone excretion in the expired air and urine following the absorption of isopropanol was established years ago (261, 293, 294) and confirmed in many cases of poisoning.

In a more recent study, Wills et al. (299) investigated the biochemical effects of daily ingestion of diluted isopropyl alcohol on three groups, each consisting of eight healthy men. The men in one group drank a daily dose of 2.6 mg/kg (0.003 ml/kg), whereas those in the second group drank a daily dose of 6.4 mg/kg (0.008 ml/kg). The third group was a control group who drank a placebo. The experiment was conducted for 6 weeks. During this time, various measurements were made on blood, serum, and urine on the first, third, and seventh day of each week. Serum cholesterol, acid and alkaline phosphatase, and glutamic-oxaloacetic transaminase activities were all normal. Retention of sulfobromophthalein in serum at the end of the experiment did not increase significantly in any group, suggesting that there had been no subacute liver damage. Ophthalmoscopic examinations at the end of the experiment showed no changes from examinations made before initiation of the experiment. Tachycardia occurred in one subject at the beginning of treatment at the low dosage level but disappeared on day three and did not recur. Symptoms of flushing and abdominal fullness were noted once during the study and more-than-normal constipation was noted once by one subject but consistently by another. Since all of these effects were in the group at the lowest dosage level, they may not have been treatment related. The authors conclude that no adverse effects resulted from the ingestion of isopropyl alcohol in daily doses of 2.6 and 6.4 mg/kg for 6 weeks.

Gorlova (311) reports that humans exposed to 2.5 and 3.0 mg/m^3 (1.0 and 1.2 ppm) of isopropanol suffered decreased ocular sensitivity to light whereas those exposed to 2.1 mg/m^3 (0.8 ppm) did not. The latent period of conditioned motor reflex to light was impaired by exposure to 1.0 ppm but not to 0.5 ppm. Likewise bioelectric activity of the cerebral cortex was changed by exposure to 0.5 ppm but not by exposure to 0.6 mg/m^3 (0.22 ppm). As a result Gorlova (311) recommends a community standard of 0.22 ppm. These data are difficult to interpret because the published record makes no mention of any data from control subjects, of variances in repetitive measurements, of duration of exposure, or of the reliability of the analytical method used to verify the concentrations.

Human Experience.   Reports of adverse effects on humans are relatively few in view of the widespread use of isopropyl alcohol. Although Nixon et al. (266) tested isopropyl alcohol on the skin of six human subjects and found its irritancy to be negligible, a few cases of dermal irritation and/or skin sensitization have been reported (296–298, 300, 301).

The use of 2-propanol as a sponge treatment for the control of fever has resulted in a number of intoxications and was likely the result of both absorption through the skin and inhalation (290–292, 302). Blood levels as high as 130 mg/100 ml were noted 95 min after admission for treatment (292). Unfortunately, there are no data with respect to amount used or duration of exposure. In all cases where high blood levels were noted, a comatose condition existed. Recovery occurred in all of these cases within 34 hr.

There have been a number of cases of poisoning due to the ingestion of isopropyl alcohol, particularly among alcoholics or victims of suicide (289, 303–307). The most common observation is a comatose condition. Pulmonary difficulty, nausea, vomiting, and headache accompanied by various degrees of central nervous system depression are typical. In the absence of shock, recovery usually occurred. King et al. (306) and Freireich et al. (307) have found extracorporeal hemodialysis to be effective in the treatment of severe cases of isopropanol poisoning.

Weil et al. (271) and Hueper (308) have reported on epidemiologic studies of employees of a manufacturing operation making isopropyl alcohol in which an excess of paranasal cancers occurred among workmen. The causative agent was *not* isopropyl alcohol but an intermediate, "isopropyl oil," and the two should not be confused.

Gorlova (311) reports on studies of human subjects living in the vicinity of a production plant making isopropanol using the sulfuric acid–propylene process. A variety of adverse effects was reported, such as two- to threefold increases in the incidence of complaints such as headache, weakness, colds, tickling of the throat, and stinging of the eyes over those in the control area. Children were thought to suffer a lag in physical development and to have low hemoglobin values. These effects seem to be attributed to exposure to concentrations of isopropanol ranging up to 3.0 mg/m^3 or 1.2 ppm, even though concentrations of sulfur dioxide ranged up to 0.54 ppm, sulfur trioxide up to 0.31 ppm, and other materials in unknown concentrations were acknowledged. That these persons suffered the effects cannot be disputed, but to attribute them to exposure to isopropyl alcohol seems unjustified under the conditions described.

In summary, reports of adverse effects in humans are few in view of the widespread use of isopropyl alcohol. Although there have been a few cases of dermal irritation and/or sensitization, and excluding the report of Gorlova (311) discussed above, no recorded cases of systemic poisonings due to industrial exposure have been found. Since a quantitative relationship between the dose of isopropyl alcohol and its metabolites has not been established, biologic monitoring cannot be used as a reliable index of exposure.

5.5 Hygienic Standards of Permissible Exposure

Both the ACGIH (215) and OSHA (108) have indicated that 400 ppm or 980 mg/m^3 would be an appropriate industrial hygiene standard. However, 400 ppm may cause some irritation of the eyes, nose, and throat, particularly in unacclimated persons. Australia, Belgium, Finland, Federal Republic of Germany, Japan, Netherlands, Switzerland, and Yugoslavia all agree on 400 ppm (980 mg/m^3). The standard in Czechoslovakia is 500 ppm, in the German Democratic Republic, 200 mg/m^3, and in Roumania, 400 mg/m^3 (110).

5.6 Odor and Warning Properties

Mild irritation of the eyes, nose, and throat was induced in human subjects exposed by Nelson and associates (237) for 3 to 5 min to 400 ppm of isopropyl alcohol. Although the effects of exposure to 800 ppm were not severe, most subjects found the atmosphere objectionable. From the viewpoint of comfort, these subjects found 200 ppm to be the highest concentration acceptable for an 8-hr exposure. Scherberger et al. (92) found the minimum concentration with identifiable odor of isopropyl alcohol to be 200 ppm. On the other hand, May (295) reports that 50 ppm is "definitely perceptible" and 40 ppm is the "smallest perceptible" concentration of isopropyl alcohol vapor. In this latter study a panel of eight men and eight women sniffed various concentrations made up in 5- to 10-liter bottles and determined using gas chromatography. Gorlova (311) reports that studies on 24 subjects showed the olfactory threshold to be 2.5 mg/m^3 (1.0 ppm). Since the accuracy of the colorimetric analytical procedure (312) employed is not know to us and since this value is so far below the others reported, its validity should be confirmed before it is accepted.

6 1-BUTANOL; *n*-Butyl Alcohol, Butyl Alcohol, Propyl Carbinol, Butyl Hydroxide, Butyric Alcohol, Hydroxybutane, CAS No. 71-36-3

$$CH_3CH_2CH_2CH_2OH$$

6.1 Source, Uses, and Industrial Exposure

1-Butanol is synthesized commercially from acetaldehyde or ethanol. Large quantities also are made by fermentation of carbohydrates (314, 315).

    n-Butyl alcohol is employed as a solvent for paints, lacquers (nitrocellulose), coatings, natural resins (congo, dammar, elemi, manila, sandrac, and shellac), gums, vegetable oils (castor and linseed oil), synthetic resins (ureas, alkyds, phenolics, ethyl cellulose, polyvinyl butyral), dyes, alkaloids, and camphor. It is used as an extractant in the manufacture of antibiotics, hormones, and vitamins. It is an intermediate in the manufacture of butyl acetate, dibutyl phthalate, and dibutyl sebacate (314, 315). The production, or in some cases, use of the following substances may offer exposure to *n*-butyl alcohol: artificial leather, butyl esters, rubber cement, dyes, fruit essences, lacquers, motion picture and photographic films, raincoats, perfumes, pyroxylin plastics, rayon, safety glass, shellac, varnish, and waterproofed cloth (314–319).

    Concentrations of 5 to 100 ppm of butyl alcohol in the air in six plants manufacturing raincoats and waterproofed cloths for sleeping pads have been reported by Tabershaw and associates (317). Over a period of 10 years Sterner et al. (319) conducted a study of men exposed to *n*-butyl alcohol in a baryta-coating operation in the manufacture of photographic paper. Initially the concentrations at the breathing zone exceeded 200 ppm but these were reduced to 100 ppm by lowering the concentration of butyl alcohol in the coating suspension and by the introduction of additional ventilation.

6.2 Physical and Chemical Properties

Physical state	Colorless, volatile liquid
Molecular formula	$C_4H_{10}O$
Molecular weight	74.12
Boiling point	117 to 118°C (20)
Melting point	−90°C (20)
Specific gravity	0.810 (20/4°C) (20)
Vapor pressure	6.5 mm Hg (25°C) (314)
Refractive index	1.3993 (20°C) (20)
Flammability	
Flash point	36 to 38°C (20)
Explosive limits	1.45 to 11.25 percent by volume
Ignition temperature	653°F (345°C) (320)
Percent in saturated air	0.86 (25°C)
Density of saturated air (air = 1)	1.01
Solubility	7.3 percent in water at 25°C (20, 36), miscible with alcohol, ether, and many other organic solvents
1 mg/l ⇌	330 ppm at 25°C, 760 mm Hg
1 ppm ⇌	3.03 mg/m³ at 25°C, 760 mm Hg

6.3 Determination in the Atmosphere

NIOSH Method No. S66 (321) has been recommended. It involves drawing a known volume of air through charcoal to trap the organic vapors present (recommended sample is 10 liters at a rate of 0.2 l/min). The analyte is desorbed with carbon disulfide containing 1 percent 2-propanol. The sample is separated by injection into a gas chromatograph equipped with a flame ionization detector and the area of the resulting peak is determined and compared with standards.

Recently Mueller and Miller (1) and Langvardt and Melcher (2) have published methods of collection and analysis by gas chromatographic procedures that give excellent results and are applicable in situations where a variety of other materials may be present. The older chemical methods (319, 322–324) would seem to have limited usefulness except where modern equipment is not available or where unusual conditions exist.

6.4 Physiologic Properties

6.4.1 Summary

Animal studies have shown that *n*-butyl alcohol is low in single-dose oral toxicity, markedly irritating and injurious to the eyes, somewhat irritating to the skin, and can be absorbed through the skin in toxic amounts. The primary effects from exposure to vapors for short periods are varying degrees of irritation of the mucous membranes and central nervous system depression depending on the intensity of exposure. Repeated intermittent inhalation exposures over a short period have not indicated unusual toxicity. Continuous exposure over a 4 month period, as reported by workers in the Soviet Union, indicate an unusually high toxicity that is difficult to rationalize in view of the balance of the data available. On the other hand, there are no other data adequate to permit a factual assessment of its significance. 1-Butanol has been found not to be mutagenic. The material is quite odorous to the unacclimated but olfactory senses readily become fatigued. Human experience has been remarkably free of serious complaints other than the transitory irritation of the mucous membranes and contact dermatitis that have occurred when exposures have been excessive.

6.4.2   Effects in Animals

Single or Short-Term Exposure Data.   *Single-Dose Oral.*   The LD_{50} values have been given as 4.36 (325) and 2.5 g/kg (326) for the rat and 3.4 (137) and 3.5 (37) g/kg for the rabbit. Saito (36) reports that 1.6 or 2.4 g/kg given orally to rabbits killed two of five and two of three respectively.

Single-Dose Injection.   When rats were given 1-butanol intraperitoneally the MLD was found to be 0.97 g/kg (329). When given intravenously to cats the MLD was found to be 0.24 g/kg (152); when given to mice, the LD_{50} was found to be 5.09 mmol/kg (0.38 g/kg) (54).

Eye Contact.   When tested in the rabbit eye, severe corneal irritation resulted from instillation of 0.005 ml undiluted *n*-butyl alcohol and from an excess of a 40 percent solution in propylene glycol. A 15 percent solution in propylene glycol caused minor corneal injury (40).

Skin Contact.   Single applications to the skin of rabbits show that 1-butanol is slightly to moderately irritating (142) and that it can be absorbed through the skin in toxic amounts. Skin absorption LD_{50} values for the rabbit have been given as follows: 5.3 (40) and 4.2 g/kg (390). DiVincenzo and Hamilton (327) applied 1-^{14}C-*n*-butanol to the skin of two male beagles and observed an absorption rate of 8.8 μg/min/cm^2. Application of 42 to 55 ml/kg/day for 1 to 4 consecutive days to rabbits resulted in 100 percent mortality. Repeated applications to rabbits of 20 ml/kg/day for 30 days over a period of 6 weeks produced no fatalities (142).

Inhalation.   Rats survived exposure to 8000 ppm for 4 hr (325). No evidence of intoxication was observed among mice exposed to 3300 ppm (10 mg/l) or 1650 ppm (5 mg/l) for 420 min (226). Exposures of mice to 6600 ppm produced giddiness after 1 hr, prostration after 1½ to 2 hr, deep narcosis with loss of reflexes after 3 hr and death of some animals (328).

Aspiration. Mortality of rats after aspiration of 0.2 ml of *n*-butanol was 9 of 10 (53). No reduction in the mortality ratio was obtained by dilution of *n*-butanol with water (85 percent *n*-butanol). The solubility limit is apparently responsible for the failure of water to decrease the aspiration hazard and toxicity of *n*-butyl alcohol.

Mode of Action. Animals exposed to *n*-butyl alcohol in the air may manifest the following signs of intoxication: restlessness, irritation of mucous membranes, ataxia, prostration, and narcosis. At high concentrations, *n*-butyl alcohol is more narcotic than is *n*-propyl alcohol (44, 226), but slightly less so than *sec*-butyl alcohol (226). Deaths from acute overexposure are believed due to respiratory failure.

Repeated or Prolonged Exposure Data. *Repeated Inhalation.* Guinea pigs exposed to 100 ppm, 4 hr/day, 6 days/week for 64 exposures showed a decrease in the red blood cell count, a relative and absolute lymphocytosis, some evidence of lung hemorrhage, albuminuria, early degenerative changes of the liver, and cortical and tubular degeneration in the kidneys (165). Mice subjected to 130 hr of total exposure (unstated number of hours per day for several days) to a concentration of 8000 ppm (24.3 mg/l of air) were narcotized repeatedly but gained in weight and survived. Reversible fatty changes were observed in the livers of the mice (44).

Rumyantsev et al. (330) gave a group of male rats and mice continuous 24-hr exposures for 4 months to concentrations of 0.0, 0.8, 6.6, and 40 mg/m³ (0.26, 2.2, and 13.2 ppm). They report numerous observations among which are the following: After 30 days of exposure both species showed decreases in hexabarbital sleeping time and in central nervous system subliminal impulses at the two top doses, and an increase in reflex activity at all three doses. At the end of the exposures there were no changes in resistance to radiation exposure (600R). In rats there was a decrease in work capacity, and in oxygen requirements at the two top levels, blood cholinesterase levels increased in a dose-related manner but the change reversed promptly after cessation of exposure, and an increase in thyroid activity occurred at all dose levels. Pathologic findings in the rats at the two top dose levels include dilation of vessels and diapedesis of erythrocytes, atelectasis and pulmonary edema, and necrotic changes in the parenchyma of the intestines. Some, but not all, of the rats at the 0.8 mg/m³ level showed the vascular effects noted above, but to a lesser extent. As a result of their observations, the authors concluded that a concentration of 0.1 mg/m³ (0.03 ppm) would be without effect when exposures are continuous and prolonged.

Absorption, Metabolism, Excretion. Following oral doses to rats of 450 mg/kg of 1-[14]C-*n*-butanol, the highest concentrations of radioactivity were in liver (3.88 percent) and blood (0.74 percent) at 8 hr. Overall distribution of radioactivity in other tissues was relatively low. The highest concentration of *n*-butanol in the plasma of the rats was 70.9 μg/ml at 1 hr (327). Following a 6-hr exposure of dogs to 50 ppm *n*-butanol, 55 percent of the inhaled vapor was absorbed through the lungs. The concentration of *n*-butanol in the blood was below the limit of detection both during and after the exposure (327). In dogs that were dosed orally with 200 mg/kg ethanol, then exposed for 6 hr to 50 ppm of *n*-butanol vapor, plasma concentrations of ethanol did not differ significantly from those of dogs dosed with ethanol alone (327). Since *n*-butanol has a very high reaction rate with ADH (37, 331), the lack of an effect may well be explained on the basis of dosage.

In rats given oral doses of 4.5, 45, or 450 mg/kg 1-[14]C-*n*-butanol, most of the radioactivity was excreted rapidly in the breath as $^{14}CO_2$. Less than 1 percent was excreted unchanged; 2.6 to 5.2 percent was excreted in the urine presumably as an *O*-glucuronide and as an *O*-sulfate. Overall excretion of radioactivity was similar regardless of dose (327). Following an oral dose of 2 g/kg to rats a very small amount (0.03 percent) of the administered dose was excreted in the urine (75). Excretion of *n*-butanol in breath and urine of rabbits following an oral dose of 2 ml/kg was less than 0.5 percent of the dose administered in each case (36).

Pharmacology. In an inclined plane test in which the angle at which a rat lost its grip as a function of the inclination of the plane was compared before and after a nontoxic dose of one of several aliphatic alcohols, *n*-butanol was the most intoxicating (191). Saito (36) also found *n*-butanol to be the most intoxicating as well as the most toxic of the lower alcohols. The effects of intravenous infusion of ethanol, *n*-propanol, and *n*-butanol in dogs were found to increase with increase in carbon number (151). The blood level of butanol associated with respiratory arrest was 84 mg percent. Decreases in mean arterial blood pressure and pulse rate were seen to follow a similar pattern, with butanol effecting the most significant decrease.

Pharmacokinetics. Oral administration of 2 g/kg of *n*-butanol to rats resulted in the appearance of this alcohol (19 mg/100 ml of blood) in blood some 15 min following dosing (75). The alcohol concentration reached a maximum of 51 mg percent after 2 hr and was reduced to 15 mg percent at 8 hr. Saito (36) found that blood levels reached a peak 30 to 60 min following an oral dose of 2 ml/kg to rabbits and became a trace after 10 hr. DiVincenzo and Hamilton (327) reported that in the rats given oral doses of ^{14}C-labeled *n*-butanol, the blood alcohol level reached a peak after 1 hr and disappeared rapidly; at 4 hr it was below the limit of detection. DiVincenzo and Hamilton (327) also found that when the dogs that were exposed to 50 ppm *n*-butanol vapor were removed from the exposure facility, the concentration of *n*-butanol in the expired air decreased precipitously and was below the limit of detection in 1 hr postexposure.

Mutagenicity—In vitro. *n*-Butanol was not found to be mutagenic in the Ames Salmonella/microsome test (196). *n*-Butanol inhibited the initiation of a new cycle of DNA replication in *E. coli* but permitted the completion of DNA replication initiated before addition of butanol to the medium (332).

6.4.3 Effects in Humans

Absorption. Åstrand et al. (333) exposed human subjects to 100 or 200 ppm (300 or 600 mg/m^3) for 2 hr to *n*-butyl alcohol vapor during rest and varying intensities of exercise (50, 100, or 150 W). The subjects at rest absorbed about 47 percent of the vapor inspired; those at light work, about 38 percent; those at medium work, about 40 percent; and those at heavy work, about 41 percent of the vapors inspired. Neither concentration nor duration of exposure influenced absorption significantly from a practical viewpoint. Alveolar concentrations corresponded to about 22 percent of that of the inspired air when the subjects were at rest, and increased to about 30 percent under work stress. The arterial blood concentration for those at rest exposed to 100 ppm was about 0.3 mg/kg, and 0.5 mg/kg for those exposed to 200 ppm. As the work load increased, the concentrations increased for those at the 100 ppm level from 0.6 to 0.9 to 1.3 mg/kg. For those at the 100 ppm level from 0.6 to 0.9 to 1.3 mg/kg. For those at the 200 ppm level, the arterial concentration plateaued at 1.1 mg/kg for all work levels. Venous concentrations varied, but were generally about 40 percent of the arterial level for exposure at 200 ppm and from 50 to 67 percent for exposure at the 100 ppm level.

DiVincenzo and Hamilton (327) have speculated regarding the absorption of *n*-butanol by inhalation versus that resulting from skin contact, and the possible total absorption when humans are exposed by both routes. Assuming the rate of absorption through the skin of *n*-butanol in man to be approximately that found in the dog, they calculated that if the hands of a man were immersed for 1 hr, the uptake of *n*-butanol by skin contact would be about 390 mg *n*-butanol. Exposure to 50 ppm *n*-butanol vapor for 1 hr leads to pulmonary uptake of about 91 mg [assuming a minute volume of 20 l/min and respiratory uptake of 50 percent (333)]. Thus direct contact of the hands with *n*-butanol for 1 hr may lead to absorption of four times the amount absorbed through the lungs from exposure to 50 ppm for 1 hr. The significance of skin contact periodically throughout the workday is apparent.

Epidemiologic Data. In the 10-year study of workers exposed to *n*-butyl alcohol, Sterner and associates (319) clinically determined and evaluated the hemoglobin, the erythrocyte count and cell volume, leukocyte count, differential leukocyte count, icterus index, and sedimentation rate. They conducted a complete urinanalysis and made serial chest X-rays. The eyes were examined by an ophthalmologist. Records of absence due to illness among the men exposed to *n*-butyl alcohol were compared to absence among all men in the entire plant. The tests for thymol–barbitone precipitation and cephalin–cholesterol flocculation were applied during the later part of the study. During the early portion of this study when the concentration was at 200 ppm or above, occasionally individuals developed increasing corneal inflammation associated with burning sensation, blurring of the vision, lacrimation, and photophobia, beginning at the middle of the week and growing more severe toward the end of the week. Slight to moderate corneal edema with injection and mild edema of the conjunctiva were reported by the ophthalmologist. In every instance the condition subsided over the weekend. At this period the mean erythrocyte count was slightly reduced. During the greater portion of this study when the concentration averaged 100 ppm, no systemic effects were observed and total absence, or absence due to illness, was no greater than among the total plant personnel. Only rare complaints of irritation or disagreeable odor were made when the concentration was at the 100 ppm level.

Experience. *n*-Butyl alcohol is potentially more toxic than any of the lower homologues, but the practical hazards associated with its industrial production and use (at ordinary temperatures) are appreciably diminished by its relatively low volatility. Exposure of human beings to vapors of this alcohol may induce the following symptoms: irritation of the nose, throat, and eyes; the formation of translucent vacuoles in the superficial layers of the cornea; headache; vertigo; and drowsiness (317, 335, 336). Contact dermatitis, involving the fingers and hands, also may occur. Hoch (324) reported that, in the manufacture of vitamins, *n*-butyl alcohol, as well as *sec*- and isobutyl alcohols, have been employed in variable quantities without giving rise to any evidence of systemic intoxication.

6.5 Hygienic Standards for Permissible Exposure

The ACGIH recommends a TLV of 50 ppm or 150 mg/m^3 (107) to be considered as a ceiling value, not an average value, and also warns about the significance of absorption through the skin. The authors support this standard, but believe the OSHA standard of 100 ppm or 300 mg/m^3 without any limitations is excessive (108). Japan, Italy, and Sweden have standards of 50 ppm; Yugoslavia, Poland, and Roumania, 66 ppm; and Australia, Belgium, Finland, Federal Republic of Germany, Netherlands, and Switzerland all have standards of 100 ppm. Czechoslovakia has a standard of 100 mg/m^3 (33 ppm) and Hungary and the Soviet Union have 10 mg/m^3 (3.3 ppm) (110).

6.6   Odor and Warning Properties.

n-Butanol has an odor similar to burnt fusel oil and can be detected at a level of 50 ppm. Following adaptation, the detection threshold increases to 10,000 ppm (334). Nelson et al. (237) reported mild irritation of the nose, throat, and eyes of subjects briefly exposed to 25 ppm and stated that exposure to 50 ppm was objectionable, because it produced pronounced irritation of the throat in all subjects and mild headaches in some instances. According to Tabershaw et al. (317), exposure in excess of 50 ppm is associated with irritation of the eyes. However, the more extensive 10-year study by Sterner and associates (319) revealed little or no irritation or complaints among workers when the average concentration was 100 ppm. Scherberger et al. (92) determined that the minimum concentration with identifiable odor of butyl alcohol was 15 ppm. May (93) reported that the odor detection limit for *n*-butanol was 11 ppm.

7   2-METHYL-1-PROPANOL; Isobutyl Alcohol, 2-Methyl-propanol-1, Isopropyl Carbinol, 1-Hydroxy-2-methylpropane, Fermentation Butyl Alcohol, CAS No. 78-83-1

$$(CH_3)_2CHCH_2OH$$

7.1   Source, Uses, Industrial Exposure

Isobutyl alcohol is obtained commercially as a by-product of high pressure methanol synthesis and by the oxidation of natural gas hydrocarbons (337).

Isobutyl alcohol is used to some extent in lacquers, paint removers, cleaners, and hydraulic fluids. It is employed in the manufacture of isobutyl esters, which serve as solvents, plasticizers, flavorings, and perfumes (338, 339).

7.2   Physical and Chemical Properties

Physical state	Colorless, volatile liquid
Molecular formula	$C_4H_{10}O$
Molecular weight	74.12
Boiling point	108°C (20)
Melting point	−108°C (20)
Specific gravity	0.8032 (20/20°C (338)
Vapor presure	12.2 mm Hg (25°C) (340)
Refractive index	1.3959 (20°C) (338)
Flammability	
Flash point	82°F (20)
Explosive limits	Lower 1.7 percent by volume, upper 10.9 percent at 212°F (119)
Ignition temperature	813°F (434°C) (320)

Percent in saturated air	1.61 (25°C)
Density of saturated air (air = 1)	1.03
Solubility:	9.5 to 10 percent in H_2O (20°C) (36, 338); miscible with alcohol, ether
1 mg/l $\approx$	330 ppm at 25°C, 760 mm Hg
1 ppm $\approx$	3.03 mg/m^3 at 25°C, 760 mm Hg

7.3 Determination in the Atmosphere

NIOSH Method No. S64 (341) has been recommended. It involves drawing a known volume of air through charcoal to trap the organic vapors present (recommended sample is 10 liters at a rate of 0.2 l/min). The analyte is desorbed with carbon disulfide containing 1 percent 2-butanol. The sample is separated by injection into a gas chromatograph equipped with a flame ionization detector and the area of the resulting peak is determined and compared with standards.

Recently Mueller and Miller (1) and Langvardt and Melcher (2) have published methods of collection and analysis by gas chromatographic procedures that give excellent results and are applicable in situations where a variety of other materials may be present. The older chemical methods such as that of Reference 342 would seem to have limited usefulness except where modern equipment is not available or where unusual conditions exist.

7.4 Physiologic Properties

7.4.1 Summary

Studies on animals have shown isobutyl alcohol to be low in single-dose oral toxicity, appreciably irritating to the eyes, not appreciably irritating to the skin, not readily absorbed through the skin in acutely toxic amounts, and low in acute inhalation toxicity. Systemically, isobutyl alcohol has a narcotic action. The results of long-term studies in which a small number of rats were given the material either orally or subcutaneously suggest that isobutyl alcohol may have tumorigenic and perhaps carcinogenic potential. A limited amount of documented human experience suggests that excessive exposure to isobutyl alcohol can cause irritation of the skin and mucous membranes and that at concentrations below 100 ppm these effects are not likely to occur.

7.4.2 Effects in Animals

Single or Short-Term Exposure Data. *Single-Dose Oral.* The following LD$_{50}$ values have been found: for the rat, 2.46 g/kg (220) and for the rabbit, 3.4 g/kg. (37).

Single-Dose Injection. When the material was given to mice intravenously, the LD_{50} was found to be 0.6 g/kg (54).

Eye Contact. One drop in a rabbit eye caused moderate to severe irritation but no permanent injury to the cornea (220, 343).

Skin Contact. Isobutyl alcohol is not irritating to the skin of the rabbit (220) and it is not absorbed readily in acutely toxic amounts, the LD_{50} being 4.24 g/kg (220).

Inhalation. Isobutyl alcohol is somewhat less lethal when inhaled in high concentrations than are normal or secondary butyl alcohol (44). Exposure of mice to concentrations of 10,600 ppm (32.2 mg/l) for 300 min, or 15,950 ppm (48.3 mg/l) for 250 min, resulted in fatalities (44). Smyth et al. (220) reported uniform survival among rats exposed for 2 hr to the saturated vapor (about 16,000 ppm) of isobutyl alcohol in air, but observed two deaths among a group of six rats when they were exposed for 4 hr to the concentration of 8000 ppm.

Mode of Action. Isobutyl alcohol is primarily a narcotic.

Repeated and Prolonged Exposure Data. *Inhalation.* Weese (44) subjected mice to a concentration of 2125 ppm (6.44 mg/l) for 223 hr in a series of intermittent exposures, each of which was of 9.25 hr duration, without any untoward effects. Mice were narcotized repeatedly in a series of intermittent exposures totaling 136 hr at 6400 ppm (19.3 mg/l), but survived.

Carcinogenicity. Nineteen Wistar rats, aged 10 weeks at the start of the experiment, were dosed with 0.2 ml isobutanol twice a week by oral intubation. The mean total dose was 29 ml and the average survival time was 495 days. The authors report that malignant tumors developed in three animals; one had a forestomach carcinoma and a liver cell carcinoma, another had a forestomach carcinoma and myelogenous leukemia, and the third, a myelogenous leukemia. A total of nine benign tumors were observed. No malignant tumors and three benign tumors were found in the controls (344).

Twenty-four Wistar rats were injected subcutaneously with 0.05 ml/kg isobutanol twice a week. The average total dose was 9 ml and mean survival time was 544 days. A total of eight malignant tumors developed, two forestomach carcinomas, two liver sarcomas, one spleen sarcoma, one mesothelioma, and two retroperitoneal sarcomas. A total of three benign tumors was recorded. Control rats had no malignant tumors and two benign tumors (344).

Metabolism and Excretion. Saito (36) observed that the reaction of isobutyl alcohol with ADH was about the same or perhaps slightly greater than that of ethanol. He also found that less than 0.5 percent of an oral dose of 2 ml/kg to rabbits was excreted in urine and breath. Gaillard and Derache (75) report that following an oral dose of 2 g/kg to rats, 0.27 percent was excreted in the urine.

Pharmacology. During a performance test in rats on an inclined plane, isobutyl alcohol was 3.6 times more intoxicating than ethanol on a molar basis, and was less intoxicating than *n*-butanol, *sec-*, or *tert*-butanol (191). The oral narcotic dose (ND_{50}) for rabbits, the dose which caused stupor and loss of voluntary movements in half the treated animals, is 19 mM or 1.4 g/kg (37).

Pharmacokinetics. The oral administration of 2 g/kg of a 20 percent aqueous solution of isobutanol to rats results in the initial appearance of the alcohol in the blood 15 min after dosing, achieving a peak concentration of 25 mg/100 ml blood at 90 min. This peak time is coincident with the presence of isobutanol in the urine (52 mg/100 ml urine) after which levels diminished to 27 mg/100 ml urine after 8 hr (17). Levels of isobutanol were barely detectable in the blood 8 hr after dosing. Saito (36) reported that following an oral dose of 2 ml/ kg to rabbits, blood levels of isobutanol reached a maximum in 30 to 60 min and disappeared to a trace after 6 hr.

Mutagenicity—In Vitro. Isobutyl alcohol inactivated *E. coli* in a concentration-dependent manner and increased the reversion rate in these bacteria (235).

7.4.3 Effect in Humans

Experience. Slight erythema without the formation of wheals was observed by Oettel (346) following the application of isobutyl alcohol to the skin of man. According to Schwartz and Tulipan (347), isobutyl alcohol may be a skin irritant. Irritation of the eyes and throat, formation of vacuoles in the superficial layers of the cornea, and loss of appetite and weight were reported among workers subjected to an undetermined, but apparently high, concentration of isobutyl alcohol and butyl acetate arising from the lacquering of cables under crowded conditions, ineffectual ventilation, and oppressive heat (89 to 107°F). Rectification of these conditions removed the symptoms (339). Fassett (198) has studied isobutyl alcohol under the same conditions of exposure as for *n*-butyl alcohol. No evidence of eye irritation was noted with repeated 8-hr exposures to levels on the order of 100 ppm. Animal data showed that it was very similar to *n*-butyl alcohol in its effect. It does, however, have a vapor pressure higher than *n*-butyl alcohol; therefore, under comparable conditions of use, higher concentrations of isobutyl alcohol are encountered.

7.5 Hygienic Standard for Permissible Exposure

The ACGIH recommends a TLV of 50 ppm or 150 mg/m^3 (348) and OSHA recommends 100 ppm or 300 mg/m^3 (108). The data base to support either value is inadequate. The standard in Belgium and Italy is 50 ppm; in Roumania and Yugoslavia, 66; and in Australia, Finland, Netherlands, and Switzerland, 100 ppm (110).

7.6 Odor and Warning Properties

May (93) reported that the limits of odor detection for isobutyl alcohol were 40 ppm or 120 mg/m^3. According to Salo (345), the odor threshold of isobutyl alcohol in a 9.4 percent grain spirit solution was 75 ppm.

8 2-BUTANOL; sec-Butyl Alcohol, Butylene Hydrate, 2-Hydroxybutane, Methyl Ethyl Carbinol, CAS No. 78-92-2

$$CH_3CH_2CH(OH)CH_3$$

8.1 Source, Uses, Industrial Exposure

2-Butanol is synthesized from butylene obtained in the cracking of petroleum. The butylene is reacted serially with sulfuric acid and then steam (349).

This secondary alcohol is employed to some extent as a solvent of lacquers, enamels, vegetable oils, gums, and natural resins. It finds usage in hydraulic brake fluids, industrial cleaning compounds, polishes, pain removers, and penetrating oils. It is employed in the preparation of ore-flotation agents, fruit essences, perfumes, and dye-stuffs. A large portion of 2-butanol is converted into methyl ethyl ketone (349, 350).

8.2 Physical and Chemical Properties

Physical state	Colorless, volatile liquid
Molecular formula	$C_4H_{10}O$
Molecular weight	74.12
Boiling point	99.5°C (20)
Freezing point	-144.7°C (20)
Specific gravity	0.8077 (20/20°C) (350)
Vapor pressure	23.9 mm Hg (30°C) (349)
Refractive index	1.3971 (20°C) (350)
Flammability	
Flash point	70°F
Explosive limits	1.7 to 9.8 percent at 212°F (119)
Ignition temperature	Air, 763°F (351); oxygen, 711°F (352)
Percent in saturated air	3.14 (30°C)
Density of saturated air (air = 1)	1.05

Solubility	12.5 percent in H_2O (36), miscible with alcohol and ether
1 mg/l ≈	330 ppm at 25°C, 760 mm Hg
1 ppm ≈	3.03 mg/m^3 at 25°C, 760 mm Hg

8.3 Determination in the Atmosphere

NIOSH Method No. S53 (353) has been recommended. It involves drawing a known volume of air through charcoal to trap the organic vapors present (recommended sample is 10 liters at a rate of 0.2 l/min). The analyte is desorbed with carbon disulfide containing 1 percent 2-propanol. The sample is separated by injection into a gas chromatograph equipped with a flame ionization detector and the area of the resulting peak is determined and compared with standards. Recently Mueller and Miller (1) and Langvardt and Melcher (2) have published methods of collection and analysis by gas chromatographic procedures that give excellent results and are applicable in situations where a variety of other materials may be present. The older chemical methods would seem to have limited usefulness except where modern equipment is not available or where unusual conditions exist.

8.4 Physiologic Properties

8.4.1 Summary

2-Butanol is low in acute oral toxicity for rats and rabbits, severely injurious to the eyes of rabbits, not irritating to the rabbit's skin, and low in acute inhalation toxicity for rats. It is a central nervous system depressant, being 4.4 times more effective than ethanol. It is absorbed readily from the gastrointestinal tract and not rapidly eliminated. Small amounts are excreted unchanged in the breath and urine. Substantial amounts are metabolized to methyl ethyl ketone and readily excreted in the breath and urine. No reports of adverse effects in humans were found.

8.4.2 Effects in Animals

Single-Dose Oral. The following LD_{50} values have been reported: for the rat, 6.5 g/kg, (9) and for the rabbit, 4.9 g/kg (37, 137).

Single-Dose Injection. The LD_{50} by the intravenous route is reported to be 0.8 g/kg (54).

Eye Contact. Smyth et al. (220) have reported that 2-butanol causes severe corneal injury when instilled directly into the eye of the rabbit.

Skin Contact.   Smyth et al. (220) did not find 2-butanol to be irritating to the skin of the rabbit.

Inhalation.   Five of six rats died following exposure for 4 hr to a concentration of 16,000 ppm (48.5 mg/l) 2-butanol (220). Concentrations of 10,670 ppm (32.3 mg/l) for 225 min and 16,000 ppm (48.5 mg/l) for 160 min were fatal for mice (44). Mice subjected repeatedly to a concentration of 5330 ppm (16.2 mg/l) 2-butanol for a total of 117 hr were narcotized, but survived (44).

Mode of Action.   2-Butanol exerts its principal toxic effect on the central nervous system. Ataxia, prostration, narcosis, and death occur progressively with increasing intensities of exposure.

Absorption, Metabolism, and Excretion.   Male rabbits were given 2 ml/kg orally and venous blood samples analyzed after 1, 2, 3, 4, 5, and 10 hr. The concentration peaked within an hour at about 1 mg/ml and disappeared to a trace after 10 hr. Unchanged *sec*-butyl alcohol was excreted to the extent of 3.3 percent of the dose in the breath and 2.6 percent in the urine. Methyl ethyl ketone, a metabolite, was detected in the blood and reached a maximum after 6 hr; it was excreted in amounts equivalent to 22.3 percent of the dose in the breath and 4.0 percent in the urine (36).

Pharmacology.   Rats were subjected to an inclined plane test, in which the angle at which the rat lost its grip as a function of the inclination of the plane was compared before and after a nontoxic oral dose of 0.0163 mol/kg (1.2 g/kg) *sec*-butyl alcohol. *sec*-Butyl alcohol showed 4.4 times the intoxicating effect relative to ethanol. No signs of recovery were evident within 2 hr following the dose of *sec*-butyl alcohol; a return to normal state was under way 7 hr after dosing (191).

   Groups of mice (two in each) were exposed by Starrek (226) for decreasing lengths of time (300, 190, 75, 60, 45, and 40 min) to increasing concentrations of *sec*-butyl alcohol in the air: 3300 ppm (10 mg/l), 6600 ppm (20 mg/l), 9900 ppm (30 mg/l), 13,200 ppm (40 mg/l), 16,500 ppm (50 mg/l), and 19,800 ppm (60 mg/l). As in the case of other alcohols studied by Starrek, the durations of exposure necessary to induce ataxia, prostration, or deep narcosis were inversely proportional to the concentration. Thus at 3330 ppm, ataxia, prostration, and narcosis became evident in 51 to 100 min, 120 to 180 min, and 300 min, respectively, whereas at 19,800 ppm these signs appeared in 7 to 8 min, 12 to 20 min, and 40 min, respectively. No deaths occurred among any of these 12 mice. No signs of intoxication were observed in mice exposed for 420 min to 1650 ppm (5 mg/l).

   Saito (36) observed that *sec*-butyl alcohol markedly increased the respiration rate of rabbits given 2 ml/kg orally. Based upon the effects of intraperitoneal injections into mice, Butler and Dickison (354) found the optically isomeric *sec*-butyl alcohols were equal in anesthetic activity.

8.4.3 Effects in Humans

No reports of case histories of systemic adverse effects in humans due to exposure to *sec*-butyl alcohol were found. However, it is reported that irritation of the eyes, nose, throat, headache, nausea, fatigue, and dizziness have been experienced from excessive exposure (355). It is also noted (355) that the material's malodorous and irritating properties are such that it is unlikely that voluntary exposure would be great enough to cause serious effects, and this may well be the reason for the lack of recorded cases of serious systemic effects.

8.5 Hygienic Standard for Permissible Exposure

Both the ACGIH (107) and OSHA (108) presently have standards of 150 ppm (450 mg/m^3). However, the ACGIH has indicated that it intends to lower its standard to 100 ppm. In the opinion of the authors, the revision seems justified. It also should be noted that there are no long-term data presently available to justify even this lower value for repeated exposure over a prolonged period of time. Australia, Belgium, Netherlands, and Switzerland all have 150 ppm as their standard, whereas the standard in the Federal Republic of Germany is 100 ppm, and in Italy, 250 mg/m^3 (82.5 ppm) (110).

9 2-METHYL-2-PROPANOL; *tert*-Butyl Alcohol, Trimethyl Carbinol, CAS No. 75-65-0

$$(CH_3)_3COH$$

9.1 Source, Uses, and Industrial Exposure

tert-Butyl alcohol is manufactured by catalytic hydration of isobutylene, and by reduction of *tert*-butylhydroperoxide (20). It is used for the removal of water from substances, in the manufacture of perfumes, flotation agents, flavors, cellulose esters, plastics, and lacquers, in paint removers, in the extraction of drugs, and as a solvent (20).

9.2 Physical and Chemical Properties

Physical state	Colorless, volatile liquid
Molecular formula	$C_4H_{10}O$
Molecular weight	74.12
Boiling point	82.4°C (356)

Melting point	25.6°C (20)
Specific gravity	0.783 (25/25°C) (356)
Vapor pressure	42.0 mm Hg (25°C) (357)
Refractive index	1.38468 (20°C) (20)
Flammability	
Flash point	52°F
Explosive limits	2.4 to 8.0 percent (119)
Ignition temperature	Air, 892°F; oxygen, 860°F (358)
Percent in saturated air	5.53 (25°C)
Density of saturated air (air = 1)	1.09
Solubility	Miscible with water, alcohols, esters, ethers, aromatic and aliphatic hydrocarbons
1 mg/l ≅	330 ppm at 25°C, 760 mm Hg
1 ppm ≅	3.03 mg/m^3 at 25°C, 760 mm Hg

9.3 Determination in the Atmosphere

NIOSH Method No. S63 (359) has been recommended. It involves drawing a known volume of air through charcoal to trap the organic vapors present (recommended sample is 10 liters at a rate of 0.2 l/min). The analyte is desorbed with carbon disulfide containing 1 percent 2-butanol. The sample is separated by injection into a gas chromatograph equipped with a flame ionization detector and the area of the resulting peak is determined and compared with standards.

Recently Mueller and Miller (1) and Langvardt and Melcher (2) have published methods of collection and analysis by gas chromatographic procedures that give excellent results and are applicable in situations where a variety of other materials may be present.

9.4 Physiologic Properties

9.4.1 Summary

tert-Butyl alcohol is low in single-dose oral toxicity, is not appreciably irritating to the skin, and its primary effect is that of a narcotic. Experience indicates that it is malodorous and is irritating to the mucous membrane in excessive concentrations.

9.4.2 Effect in Animals

Single-Dose. The following LD$_{50}$ values have been found: orally for the rat, 3.5 g/kg (360), and for the rabbit, 3.6 g/kg (37); intravenously for the mouse, 1.5 g/kg (54).

Skin Contact.   Prolonged contact with the rabbit's skin failed to cause irritation (355).

Mode of Action.   *tert*-Butyl alcohol, like the other butyl alcohols, has a narcotic action. It is stronger than the normal or isobutyl isomers in its narcotic effect upon animals (44).

Metabolism, Excretion, and Pharmacokinetics.   *tert*-Butyl alcohol is not a substrate for alcoholic dehydrogenase (36, 203, 331) and it is oxidized very slowly (75). It is excreted by the rat and the rabbit as the glycuronide (361, 362). Following an oral dose of 2 g/kg to rats, a maximum blood level of 124 mg percent was reached in 2 hr and decreased very slowly to 120 mg percent after 4 hr, and to 110 mg percent 8 hr after treatment; only about 1 percent of the dose was excreted in the urine (75). Saito (36) found *tert*-butyl alcohol in the blood of rabbits 70 hr after they were given 2 ml/kg orally.

Pharmacology.   In a performance test on an inclined plane, the angle at which the rat slides down was recorded, and performance after an oral dose was compared with pre-dose performance. A nontoxic dose of 0.0163 mol/kg of *tert*-butyl alcohol exhibited a strong intoxicating effect—4.8 times that of ethanol and exceeded only by *n*-butanol. Intoxication was of longest duration of all the alcohols tested; recovery was very slow with no improvement in performance 7 hr after administration (191). The oral narcotic dose (ND_{50}) for rabbits is 19 mM or 1.4 g/kg (37). Intoxication in rabbits (measured by sleeping time) following an oral dose of 2 ml/kg was of only slightly longer duration than for ethanol or methanol (36).

Mutagenicity—In Vitro.   *tert*-Butyl alcohol was not mutagenic in *Neurospora crassa* (363).

9.4.3   Effects in Humans

Experience.   Oettel (346) observed no reaction other than slight erythema and hyperemia following the application of *tert*-butyl alcohol to human skin. Schwartz and Tulipan (347) are authority for the statement that it may be a skin irritant. Irritation of the eye, nose, and throat, headache, nausea, fatigue, and dizziness are noted as symptoms of excessive exposure (355). No reports of systemic effect were found.

9.5   Hygienic Standard of Permissible Exposure

Both the ACGIH (348) and OSHA (108) presently have standards of 100 ppm or 300 mg/m^3 for *tert*-butyl alcohol. At the present time there are no data from long-term studies to justify these values. Hopefully, such data may come upon completion of the bioassay being conducted by the National Cancer Institute. Australia, Belgium, Federal Republic of Germany, Netherlands, and Switzerland all have 100 ppm as their standard. The German Democratic Republic and Yugoslavia use 66 ppm and Italy uses 82.5 ppm (110).

9.6 Odor and Warning Properties

The malodorous quality of *tert*-butyl alcohol together with its irritating effect upon the mucous membranes at excessive concentrations (355) may be adequate to prevent voluntary exposure to systemically toxic amounts of the material.

10 AMYL ALCOHOLS, GENERAL

10.1 Source, Uses, and Industrial Exposures

There are eight structural isomers of amyl alcohol. There are four primary amyl alcohols: 1-pentanol, 2-methyl-1-butanol, 3-methyl-1-butanol, and 2,2-dimethyl-1-propanol. The latter three of these are also known as isoamyl alcohols. There are three secondary amyl alcohols: 2-pentanol, 3-pentanol, and 3-methyl-2-butanol. There is one tertiary amyl alcohol, known as 2-methyl-2-butanol. Three of these, namely, 2-methyl-1-butanol, 2-pentanol, and 3-methyl-2-butanol, possess an asymmetric carbon atom; hence each may exist as two optical isomers in addition to the optically inactive form. 2-Methyl-1-butanol is formed from the fermentation of an optically active substance; it exists in commerce largely as the *d*-amyl alcohol (382). Commercial amyl alcohols are secured from different sources. Refined fusel oil or fermentation amyl alcohol is obtained from by-product fusel oil, which arises during the production of ethyl alcohol by fermentation. This consists largely of 3-methyl-1-butanol with minor amounts of 2-methyl-1-butanol and other alcohols (383). Fusel oil is essentially seven parts of 3-methyl-1-butanol and one part of the active form of 2-methyl-1-butanol (*d*-amyl alcohol) with small amounts of pyridine, furfurals, and esters (357). Amyl alcohols are also made synthetically from pentane by chlorination and hydrolysis. Synthetic amyl alcohol consists of about 70 percent of three of the primary alcohols (described above) and three other amyl alcohols. 3-Pentanol is available in high purity by synthetic methods, and 3-methyl-1-butanol is available as a pure product by separation from fusel oil (383). Pentasol is a mixture of several amyl alcohols (384). A commercial product containing about 80 percent of 2-pentanol and 20 percent of 3-pentanol is marketed as *sec*-amyl alcohol (384).

In the past, the handling and use of fusel oils have been the principal means of industrial exposure. Amyl alcohols are used in the manufacture of lacquers, chemicals, rubber, plastics, fruit essences, and explosives (357, 370, 372, 386). They are used as a component of paint stripper and hydraulic fluids and as an intermediate of ore-flotation agents and other amyl derivatives (383). 3-Methyl-1-butanol is used principally as an intermediate in the preparation of pharmaceuticals and isoamyl derivatives (383). The United States Department of Labor listed 20 occupations that offer exposure to amyl alcohol (372).

10.2 Physical and Chemical Properties

Except for 2,2-dimethyl-1-propanol, which is a crystalline solid, the amyl alcohols when purified are colorless liquids with a mild odor. They are solvents for camphor, alkaloids, natural and synthetic resins, iodine, phosphorus, sulfur, benzyl abietate, copal ester, ester gum, and so on (357). They are miscible with organic solvents but are only slightly soluble in water. The physical properties of the individual compounds are given along with the discussion of each.

$$1 \text{ mg/l} \simeq 278 \text{ ppm and } 1 \text{ ppm} \simeq 3.60 \text{ mg/m}^3 \text{ at } 25°C, 760 \text{ mm Hg}$$

10.3 Determination in the Atmosphere

In 1977 NIOSH (385) published Method S58 specifically recommended for isoamyl alcohol (3-methyl-1-butanol). The method involves absorption of vapors from a known volume of air passed through activated charcoal followed by desorption of the amyl alcohol from the charcoal with carbon disulfide containing 5 percent 2-propanol. Gas chromatography is then used to quantitate the isoamyl alcohol in the eluent. It is reasonable to believe that this method could be adapted for use with the other amyl alcohols. Recently Mueller and Miller (1) and Langvardt and Melcher (2) have published methods of collection and analysis by gas chromatographic procedures that give excellent results for a variety of organic materials and are applicable in situations where a variety of other materials also may be present. It is believed that these methods could be adapted to the determination of the amyl alcohols. Amyl alcohol may be determined by either of two nonspecific methods employed for the determination of fusel oil in distilled spirits: (a) oxidation by chromic acid or (b) measurement of the color produced from coupling with various aldehydes in the presence of sulfuric acid. Titration of valeric acid, distilled from a reaction mixture with sulfuric acid, and chromate was employed by Allen and Chattaway (364) and is given in the AOAC methods (365). Furfural (366), salicylaldehyde (367, 368), benzaldehyde, p-dimethylaminobenzaldehyde, and vanillin (368), in the presence of sulfuric acid, have been utilized for analysis of distilled spirits. Korenman (369) used furfural to determine the amount of amyl alcohol in the air. Quantities of the order of 0.6 to 2.0 mg in 3.5 liters of air were determined with an error of ±10 percent. These older chemical methods would seem to have limited usefulness except where modern equipment is not available or where unusual conditions exist.

10.4 Physiologic Response

10.4.1 Effects in Animals

Summary.* The amyl alcohols are more narcotic and toxic than the lower homologues. They are moderate to low in single-dose oral toxicity with LD_{50} values ranging from about 1 to 5.77 g/kg for various animal species. The primary alcohols are the least toxic, and the tertiary alcohol the most toxic. The *n*-amyl alcohol was found to be extremely toxic when aspirated; the others have not been studied. All are appreciably irritating to the eyes and capable of causing transient corneal opacity. All are somewhat irritating to the uncovered skin when exposures are repeated and severely irritating when confined to the skin. All can be absorbed through the skin of animals in toxic amounts when exposures are severe. High concentrations of vapor are irritating to the mucous membranes and typically cause central nervous system effects and death by respiratory failure. Short-term (7- and 90-day) repeated inhalation exposures have been conducted only on the tertiary amyl alcohol and indicate that the material is surprisingly low in systemic toxicity, especially in view of its slow metabolism and elimination. Only 3-methyl-1-butanol has been subjected to lifetime exposure studies. When it was given to rats, either in oral doses of 0.1 g/kg twice a week, or subcutaneously in doses of 0.04 g/kg once a week, it caused an excess of malignancies. The primary alcohols are rapidly metabolized largely via the ADH route, the secondary alcohols are slowly metabolized largely via the oxidative route, and the tertiary isomer is very slowly metabolized. Small amounts of the primary and secondary alcohols, and appreciable amounts of the tertiary isomer, are excreted unchanged in the respired air and urine.

10.4.2 Effects in Humans

Summary. Without regard to any specific isomer, most investigators have found that the inhalation of amyl alcohol vapors by man caused marked irritation of the eyes and respiratory tract, headache, and vertigo (106, 237, 370–375); dyspnea and cough (106, 374, 376); and nausea, vomiting, and diarrhea (371, 372, 106, 376, 377). Double vision, deafness, delirium, and occasionally fatal poisoning, preceded by severe nervous symptoms, have been attributed by Flury and Zernik (106) and by Eyquem (376) to the effects of the absorption of amyl alcohol. Coma, glycosuria, and sometimes methemoglobinemia are represented by Fuchter (378) and Underhill (375) as characteristic

---

* The basis for this summary is found in the discussion of the individual compounds that follows this general section.

of amyl alcohol intoxication. A few cases of industrial poisoning appear to have been caused by amyl alcohol, although in each reported instance some other solvent was present to becloud the issue. A neurasthenic brewery manager who inhaled the vapors (primary active amyl and isoamyl isomers) from fermentation vats exhibited psychic stimulation, insomnia, and chromatopsia (377). Eyquen (376) reported the following symptoms among workers engaged in producing smokeless powder: cough, irritation of the eyes, colic, diarrhea, vomiting, palpitation of the heart, nervous symptoms, headache, vertigo, disturbances of vision, forgetfulness, insomnia, somnolence, weakness, and one fatality. Although other alcohols and ether were employed, the signs increased as the use of amyl alcohol (probably fusel oil) increased. Two fatal cases from the use of a lacquer containing amyl alcohol and probably tetrachloroethane, for coating the inside surface of a tank, were reported by Zangger (379). A lacquerer exposed to amyl alcohol and amyl acetate had digestive symptoms and secondary anemia, according to Baader (380). An increase of urobilin in the urine of lacquer sprayers exposed to a solvent containing amyl and butyl alcohols, amyl and butyl acetates, and acetone was reported by Burger and Stockmann (381). Nonindustrial cases of poisoning from drinking fusel oil, characterized by coma, glycosuria, and methemoglobinuria, were seen by Fuchter (378).

11   1-PENTANOL; n-Amyl Alcohol; Pentyl Alcohol; n-Butyl Carbinol; CAS No. 71-41-0

$$CH_3(CH_2)_3CH_2OH$$

11.1   Sources, Uses, and Industrial Exposure

See general discussion of amyl alcohols, Section 10.1.

11.2   Physical and Chemical Properties

Physical state	Liquid
Molecular formula	$C_5H_{12}O$
Molecular weight	88.15
Boiling point	137.5°C (111)
Melting point	−79°C (111)
Specific gravity	0.8146 (20/4°C) (111)
Vapor pressure	1 mm Hg (13.6°C), 10 mm Hg (44.9°C) (119)
Refractive index	1.4103 (20°C) (111)
Flammability	
Flash point	100°F (closed cup) (111)
Explosive limits	1.2 to 10 percent at 212°F (119)
Ignition temperature	391°C (387)

Solubility:	2.7 g/100 ml water at 22°C (111), miscible with alcohol and ether
1 mg/l ≎	278 ppm at 25°C and 760 mm Hg
1 ppm ≎	3.60 mg/m^3 at 25°C and 760 mm Hg

11.3 Determination in the Atmosphere

See general discussion of amyl alcohols, Section 10.3.

11.4 Physiologic Response

11.4.1 Summary

1-Pentanol is low in acute toxicity when given orally and by various injection routes. It is markedly irritating to the eyes and to the skin when confined and can be absorbed through the skin in toxic amounts. Aerosolized material is appreciably toxic when inhaled for several hours. The primary effect is on the central nervous system and the target organs are the lung and kidney. The material is readily absorbed by various routes and readily metabolized; the metabolite, valeric acid, is slowly eliminated and is thought to be the cause of the "hangover" effect.

11.4.2 Effects in Animals

A number of investigations have been conducted on a mixture consisting of 74 percent 1-pentanol, 25 percent 2-methyl-1-butanol, and 1 percent 3-methyl butanol and not on pure 1-pentanol; those conducted on the mixture are identified here with an asterisk (*).

Single-Dose Oral.   The LD$_{50}$ values for rats are 3.03 (5) and 2.69* g/kg (388). Typical signs of toxicity were varying degrees of central nervous system depression and gastrointestinal irritation.

Single-Dose Injection.   The intravenous LD$_{50}$ for mice is 2.03 g/kg (54). Lendle (147) found that the toxicity of amyl alcohols following intraperitoneal injection to rats decreases in the order of 2-methyl-2-butanol (tertiary), 2-pentanol (secondary), and 1-pentanol (primary). Starrek (226) found *n*-amyl alcohol less toxic than 3-methyl-1-butanol (isoamyl) when administered subcutaneously.

Eye Contact.   The mixture*, when instilled into the rabbit eye, caused severe irritation characterized by conjunctival erythema, chemosis, discharge, iritis, and varying degrees of corneal injury (388).

Skin Contact.   The mixture* was severely irritating and injurious to the skin of rabbits upon confinement to the skin for 24 hr and was absorbed through the skin in toxic amounts (388). The percutaneous absorption LD_{50} values for the rabbit are >3.2* (388), 4.5 (389), and 2.0 g/kg (390). Signs of central nervous system depression were typical* (388).

Inhalation (388).   Groups of 10 mice, guinea pigs, and rats each were exposed for 6 hr to an atmosphere containing the aerosolized mixture* at a calculated concentration of 14,000 mg/m^3. Two rats and seven mice died during exposure; all others survived and recovered quickly from the signs seen during the exposure. During exposure, preconvulsive movements were noted in the mice and guinea pigs, whereas prostration was characteristic of the signs in rats. Histopathologic examination revealed the lung and kidney as the principal target organs. Appreciable lung edema was observed in the mice.

Aspiration.   Gerarde and Ahlstrom (53) reported instant death in 10 of 10 rats after aspiration of 0.2 ml *n*-amyl alcohol. They attribute death to cardiac and respiratory failure.

Absorption, Metabolism, and Excretion.   Haggard et al. (391), after repeated and diminishing intraperitoneal doses, found 76 mg percent of *n*-amyl alcohol in the jugular blood at the time of death of rats from respiratory failure. Rats given 1 g/kg (0.25 g at 15-min intervals) showed a peak blood concentration of about 21 mg percent 1 hr after dosing started and disappearance from the blood after 3½ hr. Only 0.88 and 0.29 percent were excreted in the expired air and urine, respectively. The disappearance times for the primary amyl alcohols ranged from 3½ to 9 hr; for the secondary alcohol, 13 to 16 hr; and for the tertiary alcohol, 50 hr.

Gaillard and Derache (189) gave rats an oral dose of 2 g/kg of *n*-amyl alcohol and found a peak blood concentration of 20 to 25 mg percent in 1 hr and essentially none after 4 hr. They did not detect any of the alcohol in the urine. Haggard et al. (391) showed the transitory presence of small amounts of valeraldehydes in the blood following the administration of the primary alcohols. In their studies, they noted irritation of the lungs and sedation of the animals that persisted for some time after the disappearance of the alcohol and the aldehyde from the blood. They postulated that the lung irritation may be due to the aldehyde, and the sedation to the valeric acid. Credence, but not proof of the hypothesis, was obtained when they observed similar signs after administration of the aldehyde and the acid. Further support comes from the observation that when *n*-amyl alcohol was given to partially (80 percent) hepatectomized rats, the blood levels rose to a peak of 100 mg percent as opposed to 23 mg percent in the normal animal, with marked prolongation of the disappearance curve. Also the observation of Von Wartburg (230) that *n*-amyl alcohol is a substrate for ADH and a potent competitor of ethanol oxidation further supports the role of the liver and the postulated metabolic pathway.

Mutagenicity—In Vitro.   *n*-Amyl alcohol inactivated *E. coli* in a concentration-dependent manner and increased the reversion rate (235). It was classed as a nonmutagen by Szybalski (392).

Skin Penetration—In Vitro.   Blank (239) studied the penetration of the homologous series of primary alcohols, methanol through octanol, through human abdominal skin, removed at autopsy within 12 hr of death. He found that solubility characteristics play an important part in determining the rate at which the alcohols penetrate skin. As the solubility of the alcohols in nonpolar liquids increases from methanol to octanol, the rate of penetration from an aqueous solution also increases. The alcohols penetrate more rapidly from vehicles in which they are less soluble. The rate of penetration of *n*-amyl alcohol from saline is about 20 times that of ethanol and about one-ninth that of octanol. The rate of penetration of *n*-amyl alcohol from a solution of olive oil does not continue to increase as the concentration becomes high; the rate becomes maximal at a given point and thereafter decreases as the concentration continues to rise.

11.4.3   Effects in Humans

See general discussion, Section 10.4.2.

11.5   Industrial Standard of Permissible Exposure

None is given by either ACGIH or OSHA. Bulgaria, German Democratic Republic, and Poland all have a standard of 100, Hungary has 200, and the Soviet Union, 10 mg/m^3 (110).

11.6   Odor and Warning Properties

May (93) reported that the lowest perceptible concentration of *n*-amyl alcohol was 10 ppm, and that 20 ppm was "definitely perceptible." Verschueren (386) describes the odor as sweet and pleasant.

12   ISOAMYL ALCOHOLS: 2-Methyl-1-butanol, 2-Methyl-1-butanol, and 2,2-Dimethyl-1-propanol

12.1 Source, Uses, and Industrial Exposure

This information is covered in the general discussion on amyl alcohols, Section 10.1.

12.2 Physical and Chemical Properties

These are given in Table 55.11.

12.3 Determination in the Atmosphere

The method described for amyl alcohols in general is applicable for these three materials. See Section 10.3.

12.4 Physiologic Response

12.4.1 Summary

2-Methyl-1-butanol and 3-methyl-1-butanol are low in single-dose oral toxicity. Both alcohols are severely irritating to the eyes, minimally irritating to the skin, and can be absorbed through the skin. Increased tumor production was reported in rats receiving repeated oral doses of 3-methyl-1-butanol. These two alcohols are rapidly metabolized to the corresponding aldehydes, then to the acids, with very low levels of unchanged alcohol excreted in the urine or expired air. No data regarding the physiologic activity of 2,2-dimethyl-1-propanol were found.

12.4.2 Effects in Animals

Single or Short-Term Exposure Data. *Single-Dose Oral.* The LD_{50} for 2-methyl-1-butanol in rats is reported to be 4.01 g/kg (40). The following LD_{50} values have been reported for 3-methyl-1-butanol: 5.77 g/kg (397) in rats, and 3.47 (137) and 3.44 g/kg (37) in rabbits. According to Munch and Schwartze (137), 3-methyl-1-butanol is less toxic upon oral administration to rabbits than 2-pentanol (a secondary alcohol), which is in turn less toxic than 2-methyl-2-butanol (a tertiary alcohol).

Injection. Starrek (226) found 3-methyl-1-butanol to be more toxic than 1-pentanol when administered subcutaneously. Single-dose intravenous injection of 3-methyl-1-butanol to mice gave an LD_{50} value of 233 mg/kg (54).

Eye Contact. Both 2-methyl-1-butanol (40) and 3-methyl-1-butanol (397) produced severe burns in the eyes of rabbits, with moderately severe burns in the eyes of rabbits, with moderately severe corneal necrosis from the instillation of

Table 55.11. Physical and Chemical Properties of Primary Amyl Alcohol Isomers, $C_5H_{12}O$

	2-Methylbutanol-1	3-Methylbutanol-1	2,2-Dimethylpropanol-1
Synonyms	Primary active amyl alcohol, sec-butyl carbinol, methyl ethyl carbinol, 2-methyl-butan-1-ol	Isoamyl alcohol, primary isobutyl carbinol, 2-methyl butan-4-ol, 3-methylbutan-1-ol	tert-Butyl carbinol, neopentyl alcohol
CAS No.	137-32-6	123-51-3	75-84-3
Structural formula	$C_2H_5CH(CH_3)CH_2OH$	$(CH_3)_2CHCH_2CH_2OH$	$(CH_3)_3CCH_2OH$
Physical and chemical properties			
Physical state	Liquid	Liquid	Crystalline solid
Molecular weight	88.15	88.15	88.15
Boiling point	128°C (393)	131.4°C (394)	114°C (393)
Melting point		−117.2°C (394)	52°C (395)
Specific gravity	0.816 (20/4°C) (393)	0.8129 (15/4°C) (394)	0.812 (395)
Vapor pressure	3.4 mm Hg (25°C) (357)	2.8 mm Hg (20°C) (394)	
Refractive index	1.4087 (25°C) (395)	1.4014 (20°C) (394)	
Flammability			
Flash point	122°F (open cup) (396)	114°F (closed cup) (387)	98°F (396)
Explosive limits		1.2 (5)–9.0% (396)	
Ignition temp.		343°C (387)	
Solubility	3.6% in H_2O; miscible with alcohol, ether	2.0% in H_2O (394); miscible with alcohol, ether	Sl. sol. H_2O; miscible with alcohol, ether

1 mg/l ≏ 278 ppm and 1 ppm ≏ 3.60 mg/m³ at 25°C and 760 mm Hg

4596

0.005 ml of undiluted material or an excess of a 15 percent solution in propylene glycol.

Skin Contact. Both of the above primary alcohols produced minimal irritation to the skin of the uncovered rabbit belly (397, 40). Dermal LD_{50} values are reported to be 2.89 g/kg for 2-methyl-1-butanol (40) and 3.24 g/kg for 3-methyl-1-butanol (397).

Inhalation. No deaths occurred in six rats exposed to concentrated vapors for 2-methyl-1-butanol. (40) and 3-methyl-1-butanol (397) for up to 8 hr.

Mode of Action. 3-Methyl-1-butanol has been shown to have a strong narcotic effect (37, 391).

Repeated Exposure Data. *Short-Term Repeated Oral.* Daily gavage doses of 0, 150, 500, or 1000 mg/kg/day of 3-methyl-1-butanol to rats for 17 weeks produced no effects on hematologic parameters, serum biochemical analyses, urinalyses, organ weights, or histopathologic findings. A slight decrease in body weight gain at the highest dose was shown to be due to reduced food intake (398).

Carcinogenicity. Fifteen Wistar rats received 0.1 ml/kg 3-methyl-1-butanol by gavage twice weekly for a total dose of 27 ml (229). The average survival time was 527 days. Four malignant tumors were reported: a myeloid leukemia, two liver cell carcinomas, and one antestomach carcinoma. No malignant tumors were found in the 25 control rats. Equal numbers of benign tumors were observed in treated and control groups (three in each). The above investigators also conducted subcutaneous studies, in which 0.04 ml/kg 3-methyl-1-butanol was injected into each of 24 Wistar rats once weekly for a total dose of 3.8 ml. A total of 10 malignant tumors were reported, occurring in the liver, spleen, antestomach, and retroperitoneal region. Five benign tumors were observed. The 25 control rats had no malignant and two benign tumors.

Absorption, Metabolism, Excretion, Pharmacokinetics Haggard et al. (391) reported small amounts of 2-methyl-1-butanol and 3-methyl-1-butanol excreted in the air or urine following intraperitoneal injections of 1 g/kg to rats:

	Percent Excreted		
	Expired Air	Urine	Total
2-Methyl-1-butanol	5.6	2.0	7.6
2-Methyl-1-butanol (levo form)	0.86	0.22	1.08
3-Methyl-1-butanol	0.97	0.27	1.24

Maximum blood alcohol concentrations ranged from 14 to 55 mg percent and disappeared from the blood in 4 to 9 hr. Gaillard and Derache (75) reported similar findings following an oral dose of 2 g/kg 3-methyl-1-butanol to rats. The alcohol was not detectable in the urine; a maximum blood level of 17 mg/100 ml was found after 1 hr, with only trace amounts detected after 4 hr. Gaillard and Derache (75) state that the low maximum blood levels could be a result of high rate of metabolism, or a fixation or solubilization in the peripheral tissues. Given the fact that 3-methyl-1-butanol has a high reaction rate with ADH, higher than does ethanol (331), rapid metabolism probably accounts for the low blood levels of these primary amyl alcohols. Haggard et al. (391) determined the minimum concentrations of 2-methyl-1-butanol, the levo form of 2-methyl-1-butanol, and 3-methyl-1-butanol in the blood of the jugular vein of rats necessary to cause death from respiratory failure. The respective concentrations were 110, 76, and 76 mg/100 ml of blood.

Haggard et al. (391) reported that in their studies on rats, 2-methyl-1-butanol and 3-methyl-1-butanol are metabolized to the corresponding aldehydes, and they showed the transitory presence of small amounts of valeraldehyde in the blood following administration of these alcohols. The aldehydes are then rapidly converted to the corresponding acids. These investigators also showed that the clinical effects, such as sedation, persisted long after the alcohol and aldehyde were undetectable in the blood, and they furnished suggestive evidence that valeric acid was the active agent, as described in the section on *n*-amyl alcohol. Guggenheim and Loffler (399) found that valeric acid was formed during perfusion of rabbit liver with 3-methyl-1-butanol. The conversion of alcohol to aldehyde depends on the action of the liver, since this conversion was largely inhibited in partially hepatectomized rats (391).

In studies on the metabolism of rat liver slices, Forsander (186) reported that 2-methyl-1-butanol and 3-methyl-1-butanol decreased carbon dioxide production and increased acid production, but had no effect on oxygen uptake. Ethanol showed similar influences on carbon dioxide and acid production, but to a greater extent. Both 2-methyl-1-butanol and 3-methyl-1-butanol increased the lactate/pyruvate ratio.

Pharmacology. Munch (37) showed that 3-methyl-1-butanol had a strong narcotic effect in rabbits following oral doses and that the ND_{50} dose was 8 mM or 0.7 g/kg. The therapeutic index (LD_{50}/ND_{50}) calculated from rabbit data for this material was relatively high, indicating a reasonable margin of safety.

12.4.3   Effects in Humans

See general discussion, Section 10.4.2.

12.5 Hygienic Standards of Permissible Exposure

Both the ACGIH (215) and OSHA (108) have set 100 ppm or 360 mg/m^3 3-methyl-1-butanol as a safe level for industrial exposure. Standards have not been set for 2-methyl-1-butanol or 2-2-dimethyl-1-propanol.

12.6 Odor and Warning Properties

Salo (345) reported that 3-methyl-1-butanol has an odor threshold of 7.0 ppm when determined by the triangular test.

13 SECONDARY AMYL ALCOHOLS; 2-Pentanol, 3-Pentanol, and 3-Methyl-2-butanol

13.1 Source, Uses, and Industrial Exposure

See general discussion, Section 10.1.

13.2 Physical and Chemical Properties

These data are given in Table 55.12.

13.3 Determination in the Atmosphere

See general discussion, Section 10.3.

13.4 Physiologic Response

13.4.1 Summary

The secondary amyl alcohols are low in single-dose toxicity. 3-Pentanol is appreciably irritating to the eyes, mildly irritating to the skin, and can be absorbed through the skin. All are metabolized to ketones and largely excreted as such via the expired air. All have a narcotic action. No indication that the materials present a serious hazard in humans was found but data adequate for hazard assessment are lacking.

13.4.2 Effects in Animals

Single-Dose Oral. The following values are given:

Rat LD_{50} = 1.87 g/kg (3-pentanol) (220)
Rabbit LD_{50} = 3.5 ml/kg (2-pentanol) (224)

Table 55.12. Physical and Chemical Properties of Secondary Amyl Alcohols

	2-Pentanol	3-Pentanol	3-Methyl-2-butanol
Synonyms	1-Methyl-1-butanol, *sec*-active amyl alcohol, methyl propyl carbinol	1-Ethyl-1-propanol, diethyl carbinol	2-Methyl-3-butanol, *sec*-isoamyl alcohol, methyl isopropyl carbinol
CAS No.	6032-29-7	584-02-1	598-75-4
Structural formula	$CH_3CH_2CH_2CH(OH)CH_3$	$CH_3CH_2CH_2CH(OH)CH_2CH_3$	$(CH_3)_2CHCH(OH)CH_3$
Physical and chemical properties			
Physical state	Liquid	Liquid	Liquid
Molecular formula	$C_5H_{12}O$	$C_5H_{12}O$	$C_5H_{12}O$
Molecular weight	88.15	88.15	88.15
Boiling point	119.3°C (111)	115.6°C (111)	113–114°C (111)
Freezing point	Glassy in liquid air	Glassy in liquid air	—
Specific gravity	0.8098 (20/4°C) (111)	0.8169 (20/20°C) (383)	0.818 (20/4°C) (400)
Vapor pressure	—	2 mm Hg (20°C) (383)	—
Refractive index	1.406 (20°C) (111)	1.4097 (20°C) (111)	1.4091 (20°C) (111)
Flammability			
Flash point	105°F (open cup) (111)	94°F (closed cup (119)	103°F (111)
Explosive limits	—	upper 9.0% (119)	1.2–9.0% (119)
Ignition temperature	343°C (387)	650–725°F (119)	657°F (119)
Solubility			
H_2O	16.6 g/100 ml at 20°C (111)	5.5 g/100 g at 30°C (111)	2.8 g/100 g at 30°C (111)
Alcohol, solvents	Miscible	Soluble in alcohol, ether	Miscible with alcohol, ether

1 mg/l ⇌ 278 ppm at 25°C, 760 mm Hg
1 ppm ⇌ 3.60 mg/m³ at 25°C, 760 mm Hg

Rabbit LD_{50} = 2.82 g/kg (isomer unspecified) (37)
Rat LD_{50} = 1.47 g/kg (isomer unspecified) (401)

The oral toxicity in rabbits of the three classes of amyl alcohol decreases in the following order: tertiary, secondary, primary (224).

Eye Contact.   Smyth et al. (220) found 3-pentanol to be moderately irritating to the eyes of rabbits.

Injection.   Lendle (329) found that the order of toxicity of amyl alcohols by the intraperitoneal route in rats was the same as that reported for oral administration in rabbits.

Skin Contact.   Smyth et al. (220) found 3-pentanol to be mildly irritating to the skin or the rabbit and to be absorbed through the skin to some extent, the LD_{50} being 2.5 ml/kg.

Absorption.   Concentrations of *sec*-amyl alcohol found in the blood of rats 1 hr after an intraperitoneal dose of 1 g/kg were 51 to 65 mg percent (391). Haggard et al. (391) determined the minimum concentrations of 2-pentanol, 3-pentanol, and 3-methyl-2-butanol in jugular vein blood at rats necessary to cause death from respiratory failure. The respective concentrations were 86, 87, and 90 mg/100 ml of blood.

Metabolism.   Haggard et al. (391) reported that secondary amyl alcohols are oxidized to the ketones, which were present in the blood in measurable quantities following administration of the alcohols to animals. The ketones were present about twice as long as the alcohols. Although there was a definite decrease in the conversion of secondary amyl alcohols to ketone in partially hepatectomized mice, the inhibition was not nearly as great as in the case of the primary alcohol (391).

Excretion.   Haggard et al. (391) reported on the excretion of *sec*-amyl alcohols following intraperitoneal administration to rats. They found appreciable amounts of the given dose excreted as ketones in the expired air, as shown in Table 55.13.
   Following oral administration of 2-pentanol to rabbits, small amounts of conjugated glucuronic acid were found in the urine (402, 403). Dogs excreted less glucuronic acid than did rabbits (402, 403).

Pharmacology.   The *sec*-amyl alcohols are reported to have the highest narcotic activity of the amyl alcohols, followed by tertiary, then primary (224). The narcotic dose 50 (ND_{50}) of *sec*-amyl alcohol (isomeric composition not given) in

Table 55.13. Pattern of Excretion of Secondary Amyl Alcohols by Rats after Intraperitoneal Injection

	Percent of Administered Dose				
	Expired Air		Urine		
Isomer	Alcohol	Ketone	Alcohol	Ketone	Total Excreted
2-Pentanol	5.4	36.8	1.2	2.0	45.4
3-Pentanol	0.3	50.0	0.1	4.6	55.0
3-Methyl-2-butanol	8.1	48.0	2.9	2.4	61.4

rabbits was found to be 5 mmol or 0.44 g/kg (37). This is defined as the dose producing stupor and loss of voluntary movements in one-half of the rabbits.

Pharmacokinetics. The rate of elimination of amyl alcohols from the body decreases in the following order: primary, secondary, and tertiary (391). Secondary amyl alcohols disappeared from the blood 13 to 16 hr following intraperitoneal administration of 1 g/kg to rats (391).

13.4.3 Effects in Humans

No effects upon the nerves of the skin of men, and no local wheal formation, erythema, or hyperemia were observed by Oettel (346) following the application of any *sec*-amyl alcohols. Schwartz and Tulipan (347) state that secondary amyl alcohols may be irritating to the skin.

For additional information, see general discussion, Section 10.4.2.

13.5 Industrial Standard of Permissible Exposure

Neither the ACGIH, OSHA, nor other countries have set a standard for safe exposure.

13.6 Odor and Warning Properties

No data are available.

14 2-METHYL-2-BUTANOL, *tert*-Amyl Alcohol, *tert*-Pentyl Alcohol, Dimethyl Ethyl Carbinol, *tert*-Pentanol, Amylene Hydrate, CAS No. 75-85-4

$$(CH_3)_2C(OH)CH_2CH_3$$

14.1 Source, Uses, and Industrial Exposure

See general discussion, Section 10.1.

14.2 Physical and Chemical Properties

Physical state	Volatile liquid
Molecular formula	$C_5H_{12}O$
Molecular weight	88.15
Boiling point	101.8°C (393)
Freezing point	−9.0°C (20)
Specific gravity	0.809 (20/4°C) (393)
Vapor pressure	10 mm Hg (17.2°C) (119)
Refractive index	1.406 (20°C) (393)
Flammability	
Flash point	67°F (closed cup) (20)
Ignition temperature	437°C
Solubility	8 g/100 ml water at 25°C (20). Miscible with alcohol, ether, benzene, chloroform, glycerol, oils (20)
1 mg/l ≈	278 ppm at 25°C, 760 mm Hg
1 ppm ≈	3.60 mg/m^3 at 25°C, 760 mm Hg

14.3 Determination of Presence

14.3.1 In the Atmosphere

See general discussion, Section 10.3.

14.3.2 In Plasma

Recently Langvardt and Braun (404) developed a method for the determination of *tert*-amyl alcohol in plasma with a detection limit of 0.1 μg/ml. The method requires as little as 50 μl of blood and about 15 min. It involves direct injection of heparinized plasma, separation by gas–liquid–solid chromatography, and quantitation via selective ion monitoring mass spectrophotometry.

14.4 Physiologic Response

14.4.1 Summary

tert-Amyl alcohol is moderate in acute oral toxicity, moderately irritating to the conjuctival membranes of the eye, and capable of causing corneal cloudiness.

It is mildly irritating to the skin, but if confined under an impervious covering, it can cause severe irritation including necrosis. It is not a skin sensitizer. Toxic amounts can be absorbed through the skin of rabbits from prolonged confinement, the LD_{50} being 1.7 g/kg. Repeated exposures to daily doses of 344 mg/kg to the open skin were well tolerated, causing only local skin effects, but daily doses of 3440 mg/kg caused systemic toxicity including some deaths. Single 6-hr inhalation exposure of rats to vapor concentrations ranging from 1100 to 15,800 ppm caused signs ranging from slight motor incoordination to deaths believed due to respiratory failure. Repeated exposures of dogs, rats, and mice over a 90-day period to concentrations up to 1000 ppm resulted in relatively minor adverse effects in the dogs, lesser effects in the rats and none in the mice. *tert*-Amyl alcohol is very slowly metabolized by all species tested and is largely excreted in the expired air. It is not mutagenic in *in vitro* test systems.

14.4.2   Effects in Animals

Single-Dose Oral.   The LD_{50} values are for rats, 1.0 g/kg (360) and between 1.0 and 2.0 g/kg (405) and for rabbits, 2.0 g/kg (37). Munch and Schwartze (137) give 2.0 g/kg as the lethal dose and note that for the rabbit, the oral toxicity decreases in the following order: tertiary, secondary, and primary.

Single-Dose Injection.   The intravenous LD_{50} for mice is 0.61 g/kg (54). Lendle (147) found that *tert*-amyl alcohol was more toxic than the secondary or primary amyl alcohol when given to rats intraperitoneally.

Eye Contact (405).   *tert*-Amyl alcohol is markedly irritating to the rabbit eye. When instilled undiluted it caused pain, moderate conjunctival irritation with swelling of the lids, moderate to severe corneal cloudiness, and moderate iritis. Healing was complete 14 days postexposure but not after 7 days.

Skin Irritation (405).   When *tert*-amyl alcohol was applied repeatedly to the uncovered skin of the rabbit it did not cause irritation but if the skin was covered with a cloth, redness and exfoliation resulted. Confinement of the material under an impervious covering resulted in severe necrosis.

Skin Sensitization (405).   When a group of 10 guinea pigs was tested with undiluted *tert*-amyl alcohol using a modified Maguire (407) method, none was sensitized, whereas all 10 of a positive control group were sensitized.

Skin Absorption—Single Dose (405).   When various amounts of *tert*-amyl alcohol were held in intimate contact with the intact belly skin of the rabbit under an impervious cuff for 24 hr, absorption of toxic amounts occurred; the LD_{50} was found to be 1720 mg/kg.

Skin Absorption—30-Day Exposure (405). Three groups of five male and five female rabbits were treated topically with 0 (water control), 344, or 3440 mg/kg/day of *tert*-amyl alcohol, 5 days/week for 4 weeks. Half of the dose was applied to the clipped backs of the rabbits twice a day. Numerous hematologic and clinical chemical parameters were evaluated during the course of the study, and extensive histopathologic examinations were conducted on animals that died during the course of the study and on those sacrificed at termination. No adverse effects were detected in those animals that received the low dose other than the expected local effects to the skin at the site of application. These included erythema, drying, thickening, cracking, and induration. The high dose (3440 mg/kg/day) caused similar but more extensive injury to the skin and caused serious systemic effects. Three of each sex became comatose and died or were sacrificed in a moribund condition after a few applications. Gross and histopathologic examination failed to reveal morphologic lesions that would account for the condition of these animals. With the exception of decreases in body size, adipose reserves, and glycogen of hepatocytes, all other parameters were within the normal range.

Inhalation—Single-Exposure (405). Groups of four rats each were exposed for 6 hr to nominal concentrations of 15,800, 5700, 3000, and 1100 ppm *tert*-amyl alcohol. At the top level, all rats exhibited motor incoordination within 90 min and all were dead within 3 hr. Those exposed to 5700 ppm were unconscious at the end of exposure and died within 24 hr. Deaths were believed to be due to respiratory failure. The rats exposed to 3000 ppm were also unconscious at the end of the exposure, lost weight for a few days and then recovered and survived. Those exposed to 1100 ppm exhibited slight motor incoordination at the end of the exposure, appeared normal 24 hr later, experienced no weight loss, and survived.

Inhalation—7-Day Exposure (405). Groups of eight rats each were exposed to approximately 0, 150, 500, and 1500 ppm of vapors of *tert*-amyl alcohol for 6 hr/day for seven consecutive days. Those animals at the top dose exhibited motor incoordination after the first two exposures but not after seven exposures, suggesting that some tolerance may have occurred. They appeared lethargic for 3 to 4 hr after exposure. There was an increase in both absolute and relative liver and kidney weights, and a decrease in blood glucose level; gross pathologic examination did not reveal any consistent changes attributable to exposure. It is clear from this experiment that 1500 ppm causes adverse effects but the data are too sparse to permit a conclusion that either 500 or 150 ppm is without some adverse effect.

Inhalation—90-Day Exposure (405). Four groups of 40 male and female rats and mice and four male beagle dogs were exposed to 0, 50, 225, or 1000 ppm of *tert*-amyl alcohol 6 hr/day, 5 days/week for a total of 59 to 60 exposures in 3 months. Impaired motor coordination was very apparent in the dogs during and for a short time after exposure to 1000 ppm, especially initially. The female rats also exhibited this effect but only after the first exposure. Tolerance developed in both species. Tearing, probably caused by irritation of the eyes, occurred in one dog exposed to 1000 ppm and in the rats at both 1000 and 225 ppm. Numerous hematologic and clinical chemistry parameters were measured during the course of the study. The only change noted was an increase in the serum alkaline phosphatase activity in the dogs at the top dosage level. All animals survived the entire study. Gross examination of tissues of all animals revealed only an enlargement of the liver in four out of four of the dogs and the male rats exposed to 1000 ppm. Detailed histopathologic examination was conducted on about 35 tissues from a sampling of both sexes of the rats and mice of the control and 1000 ppm dose level and on all of the dogs at all dosages. The tissues from the rats and mice at the two lower exposure levels were not examined because no changes were found at the high-dose level. Histologically, the livers of all the dogs appeared normal except for one out of four dogs in each exposure group having hepatocellular cytoplasmic inclusions. These are considered to be a response unique to the dog, minimal in degree and reversible in nature.

Absorption.   Haggard et al. (391) found a concentration of 12.5 mg percent of *tert*-amyl alcohol in the blood of rats 1 hr following an intraperitoneal dose of 1.0 kg. They also found a concentration of 191 mg percent in the jugular blood of rats at the time of death from respiratory failure. Concentrations in the plasma of rats about 0.6 hr after they had inhaled 150, 500, or 1500 ppm of *tert*-amyl alcohol for 6 hr were 9.6, 59.9, and 337 µg/ml, respectively (405).

Metabolism.   The metabolism of *tert*-amyl alcohol is very slow. According to Winer (331), the material is not oxidized by ADH, but Von Wartburg (203) reports some activity.

Excretion.   An appreciable amount of absorbed *tert*-amyl alcohol is excreted unchanged in the expired air but the amount varies from species to species. Dogs and cats exhale more unchanged amyl alcohol than do rabbits or rats. A dog weighing 11 kg, when injected subcutaneously with a dose of 0.1 ml of 2-methyl-2-butanol (*tert*-amyl alcohol) exhaled 65 percent of it within 5.75 hr. Another dog given a slightly larger dose of the same alcohol intravenously exhaled 52 percent within 6 hr (17). Cats also exhaled a large quantity unchanged (408). When rabbits were injected intravenously with two levels of dosage, one approximately the same as that given to dogs and the other half as much, they exhaled, in the first instance, 21 percent of the alcohol in unchanged form within 4 hr, and in the second, 22 percent within 3 hr. Over a period of 3 hr a dog eliminated 55 percent, whereas a rabbit expelled 17 percent. Haggard et al. (391) found that a rat given 1 g/kg of *tert*-amyl alcohol eliminated 26.4 percent unchanged in the expired air and 8.9 percent in the urine within 50 hr. No volatile metabolites were found in the blood (391). Following oral administration of 2-methyl-2-butanol to rabbits, small amounts of conjugated glucuronic acid were found in the urine (402, 403). Dogs excreted no glucuronic acid following administration of the tertiary alcohol (402, 403).

Pharmacokinetics.   The rate of elimination of amyl alcohols from the blood of rats decreases in the order of primary, secondary, and tertiary (391). *Tert*-amyl alcohol did not disappear from the blood of rats until 50 hr following an intraperitoneal dose of 1 g/kg (391). Rats were exposed to 150, 500, and 1500 ppm of *tert*-amyl alcohol and the absorption and elimination kinetics studied immediately after single exposures (405, 406). The plasma concentration time profile following the 150 ppm exposure is well described by either first-order or Michaelis–Menten elimination kinetics. The plasma concentration time profiles following the 500 and 1500 ppm exposures were significantly better described by Michaelis–Menten kinetics than by first-order elimination kinetics. The elimination of *tert*-amyl alcohol from the plasma following the 500- and 1500-ppm exposures clearly exhibited "exposure-dependent" or saturation kinetics.

Upon repeated exposure to 50 ppm, the material was cleared by rats ($t_{1/2}$ = 47 min) and dogs ($t_{1/2}$ = 69 min) in an apparent first-order manner. End exposure plasma concentrations after repeated exposure to 1000 ppm were ≈110 µg/ml in rats and mice and 480 µg/ml in dogs. Following repeated exposure to 1000 ppm the plasma clearance was more rapid in the mouse than in either the rat or the dog exposed to 50 ppm, the $t_{1/2}$ times being 29, 47, and 69 min, respectively. At 1000 ppm the clearance is apparently first order for the mouse and best expressed by Michaelis—Menten kinetics for the rat and dog. The data demonstrate that the kinetics of *tert*-amyl alcohol are highly species- and concentration-dependent. This is an important observation, because it suggests that toxicity may increase disproportionately at high exposure concentrations and that valid extrapolations depend on adequate knowledge of the kinetics (405, 406).

Pharmacology.   According to Munch and Schwartze (137), the narcotic action in rabbits of a series of amyl alcohols decreases in the order of secondary, tertiary, and primary. The narcotic dose 50 (ND_{50}) of *tert*-amyl alcohol in rabbits was found to be 8 mmol or 0.7 g/kg (11). From the increasing amounts of 2-methyl-2-butanol (tertiary) required to narcotize a dog in successive experiments, evidence of habituation (tolerance) was obtained (409). Similar observations upon a rabbit yielded no such evidence (409).

Mutagenicity. *tert*-Amyl alcohol was not found to be mutagenic when tested in both activated and nonactivated systems against *Salmonella typhimurium* tester strains TA1535, TA1537, and TA1538 and *Saccharomyces cerevisiae* (405).

14.4.3 Effects in Humans

A patient died following an enema of 35 g of *tert*-amyl alcohol (410). Anker (411) reported the recovery of a woman who intentionally drank about 27 g of *tert*-amyl alcohol. Recovery was preceded by coma, dyspnea, irregular pulse, and dilation, then contraction of the pupils. No effects upon the nerves of the skin of men, and no local wheal formation, erythema, or hyperemia were observed by Oettel (346) following the application of *tert*-amyl alcohol to the skin. Schwartz and Tulipan (347) reported that tertiary amyl alcohols may be irritating to the skin. For additional information, see general discussion, Section 10.4.2.

14.5 Hygienic Standard for Permissible Exposure

No standards for prolonged or repeated exposure to tertiary amyl alcohol have been recommended.

14.6 Odor and Warning Properties

According to Verschueren (386), 2-methyl-2-butanol (*tert*-amyl alcohol) has a sour odor, a threshold value of 8.2 mg/m^3 (2.3 ppm), an absolute perception limit of 0.04 ppm, and a 100 percent recognition level of 0.23 ppm.

15 1-HEXANOL; n-*Hexyl Alcohol, Amyl Carbinol, Pentyl Carbinol, 1-Hydroxyhexane, CAS No. 111-27-3*

$$CH_3(CH_2)_4CH_2OH$$

15.1 Source, Uses, and Industrial Exposure

1-Hexanol is found in several essential oils and aromas. It is prepared industrially by reducing ethyl caproate with sodium in absolute alcohol. It is used in the manufacture of antiseptics, hypnotics and fragrances (111, 414). Industrial exposure is limited.

15.2 Physical and Chemical Properties

Physical state   Liquid
Molecular formula $C_6H_{14}O$

Molecular weight	102.17
Boiling point	157°C (III)
Melting point	−51.6°C (111)
Specific gravity	0.8153 (25/4°C) (111)
Vapor pressure	0.98 mm Hg (20°C) (386)
Refractive index	1.4162 (25°C) (111)
Flash point	145°F (closed cup) (111)
Solubility	Slightly soluble in water and miscible with alcohol, ether, and benzene (111)
1 mg/1 ≈	239.3 ppm at 25°C and 760 mm Hg
1 ppm ≈	4.25 mg/m^3 at 25°C and 760 mm Hg

15.3 Determination in the Atmosphere

No specific methods have been described but it is suspected that NIOSH Method No. S60 (420) and those of Mueller and Miller (1) and Langvardt and Melcher (2) could be adapted (see Section 17.3).

15.4 Physiologic Response

15.4.1 Summary

1-Hexanol is low in single-dose oral toxicity, severely irritating and injurious to the eyes, moderately irritating when confined to the skin, slowly absorbed through the skin, and low in single-dose inhalation toxicity. It does not appear to be carcinogenic or to have tumor-promoting potential when applied to the skin of mice. No adverse human experience was noted. It was neither an irritant nor a sensitizer when tested on human subjects.

15.4.2 Effects in Animals

Single-Dose Oral. The following LD$_{50}$ values have been reported: for the rat, 4.87 (412) and 4.59 (325) and for the mouse, 4.0 g/kg (413).

Single-Dose Intravenous. The LD$_{50}$ dose for the mouse is 0.1 g/kg (54).

Eye Contact. 1-Hexanol was severely irritating and injurious when introduced directly into the eye of the rabbit (325). Electron microscope examination of the retina of the eyes of rabbits exposed by inhalation to vapors containing 118 mg/m^3 (28.3 ppm) hexyl alcohol for 6 months revealed untrastructural changes in the photoreceptor cells and Muller fibers (60). (The hexanol was not specifically identified as 1-hexanol and the number of hours per day, days per week of the exposure were not given.)

Skin Contact. 1-Hexanol applied full strength to intact or abraded rabbit skin for 24 hr under occlusion was moderately irritating (325, 414). It can be absorbed through the skin of rabbits in toxic amounts when exposure is prolonged; the following LD_{50} values are given: 2.53 (325, 390) and >5 g/kg (414).

Inhalation. The maximum period of survival of rats inhaling the "saturated" vapors of 1-hexanol was 8 hr (325).

Aspiration. Aspiration of 0.2 ml of hexyl alcohol by the rat caused respiratory arrest and instant death. The lungs showed small areas of focal pulmonary hemorrhage and edema (53).

Carcinogenicity. The tumor-promoting activity of 1-hexanol was studied on skin of mice treated with an initiating dose of 7,12-dimethylbenz(a)anthracene, then with 20 μl of a hexanol solution (20 g hexanol in 100 ml of cyclohexane) three times a week for 60 weeks. No skin tumors developed in 36 survivors (415).

Metabolism. 1-Hexanol has a high affinity for ADH, similar to amyl and n-octyl alcohol, and is a potent inhibitor of ethanol oxidation (230). 1-Hexanol is metabolized by direct conjugation with glucuronic acid and by oxidation to the carboxylic acid and eventually to carbon dixoide. In the rabbit, direct conjugation is a minor pathway and oxidation the major pathway (416).

15.4.3   Effects in Humans

No reports of adverse human experience due to 1-hexanol were found.

Skin Exposure. 1-Hexanol was tested as a 1 percent solution in petrolatum on human subjects and produced no irritation after a 48-hr closed-patch test (417). Using a maximization procedure described by Kligman (418) and Kligman and Epstein (419), Epstein (417) reported that a 1 percent solution of 1-hexanol in petrolatum caused no sensitization reactions in 21 volunteer test subjects.

15.5   Hygienic Standards for Permissible Exposure

No standards for permissible exposure have been given in the United States. No data adequate to establish an acceptable level are available. A MAC of 10 ppm has been established in the Soviet Union, 100 ppm in Yugoslavia, and 150 ppm average with a maximum of 250 ppm in Roumania (110). The basis for these standards is not known.

15.6 Odor and Warning Properties

According to Verschueren (386) the absolute perception concentration is 0.01 ppm and the recognition level is 0.09 ppm. The sensory odor threshold of 1-hexanol (in 9.4 percent w/w "grain spirits") was determined to be 5.2 ppm by the triangular test (345).

16 2-ETHYL-1-BUTANOL; 2-Ethylbutyl Alcohol, Isohexyl Alcohol, CAS No. 97-95-0

$$C_2H_5CH(C_2H_5)CH_2OH$$

16.1 Source, Uses, and Industrial Exposures

2-Ethylbutyl alcohol can be prepared commercially by the aldol condensation of acetaldehyde and butraldehyde (421). 2-Ethylbutyl alcohol is used as a solvent for printing inks, as a component in lacquers, and in the manufacture of surface active agents, synthetic lubricants, and lubricant additives (394). No data on the extent of exposure in industrial application was found.

16.2 Physical and Chemical Properties

Physical state	Colorless liquid
Molecular formula	$C_6H_{14}O$
Molecular weight	102.18
Specific gravity	0.8328 (394)
Refractive index	1.4208 (20°C) (394)
Melting point	$< -50°C$ (421)
Boiling point	149.4°C (394)
Vapor density	3.52 (air = 1)
Vapor pressure	1.80 mm Hg (20°C) (422)
Percent in "saturated" air	0.22 (20°C)
Density in "saturated" air	1.005 (20°C)
Solubility	0.43 percent in water (20°C) (394)
Flash point	135°F (57°C) (394)
1 mg/l ≈	239.3 ppm at 25°C and 760 mm Hg
1 ppm ≈	4.17 mg/m³ at 25°C, 760 mm Hg

16.3 Determination in the Atmosphere

No specific methods have been developed. Those noted under 4-methyl-2-pentanol likely are adaptable (see Section 17.3).

16.4 Physiologic Response

16.4.1   Effect on Animals

The toxicologic data obtained on animals for 2-ethylbutyl alcohol are given in Table 55.14.

Metabolism and Excretion.   Kamil et al. (275) have shown that rabbits excrete 40 percent of an ingested dose of 2-ethylbutyl alcohol as urinary glucuronide (diethyl acetyl glucuronide); that the 2-substituted primary alcohols are oxidized rapidly, resulting in very little conjugation with glucuronic acid if there is a methyl group at the 2 position (isobutanol and 2-methylbutanol), whereas the 2-ethyl substituted alcohols give rise to α-ethyl fatty acids (which are not readily oxidized in vivo) that are readily conjugated with glucuronic acid (2-ethylbutyl alcohol and 2-ethylhexyl alcohol). The fatty acid resulting in vivo from 2-ethylbutyl alcohol is not completely resistant, since a small amount of methyl n-propyl ketone is also excreted in rabbit urine following the ingestion of 2-ethylbutyl alcohol (275).

16.5 Hygienic Standards of Permissible Exposure

No standard was found. There are no reports of human experience. There are no data upon which a standard can be proposed.

Table 55.14.   Toxicologic Response of Animals to 2-Ethylbutyl Alcohol

Single-Dose	
Rat (orally)	LD_{50} 1.85 g/kg (220)
Rat (IP)	LD_{50} 0.80 g/kg (198)
Guinea pig (IP)	LD_{50} 0.45 g/kg (198)
Skin Absorption	
Rabbit (24-hr contact)	LD_{50} 1.26 ml/kg (220)
Guinea pig	LD_{50} <5.0 ml/kg (220)
Skin Irritation	
Rabbit	Capillary injection (220)
Guinea pig	Moderate irritation (198)
Inhalation	
Rat	Concentrated vapor, 8 hr, no deaths (220)

17 4-METHYL-2-PENTANOL; Methylamyl Alcohol, Methyl Isobutyl Carbinol, *sec*-Hexyl Alcohol, CAS No. 108-11-2

$$(CH_3)_2CHCH_2CHOHCH_3$$

17.1 Source, Uses, and Industrial Exposure

Methylamyl alcohol is prepared commercially by reduction of mesityl oxide (421) in acetic acid solution with platinum, or over reduced copper on asbestos at 120°C. It is used in brake fluids, as a frothing agent in ore flotation, and in the manufacture of ore flotation agents, lubricant additives, solvents, plasticizers, lacquers, and in organic synthesis (423, 424). No analysis of the extent of exposure in industrial applications was found.

17.2 Physical and Chemical Properties

Appearance	Colorless, stable liquid (425)
Molecular formula	$C_6H_{14}O$
Molecular weight	102.18 (426)
Melting point	−90°C (426)
Boiling point	131.8°C (425)
Specific gravity (20/20°C)	0.8079 (425, 426)
Refractive index	
(25°C)	1.4089 (425)
(20°C)	1.4113 (426)
Vapor density (air = 1)	3.52 (426)
Vapor pressure (20°C)	3.52 mm Hg (424)
Percent in "saturated air" (20°C)	0.46 (426)
Density of "saturated air" (air = 1)	1.01 (426)
Flash point	
TOC	131°F (427)
OC	105°F (425)
Solubility (20°C)	1.7 percent in H_2O; soluble in ethyl alcohol, hydrocarbons, most organic solvents (426)
1 mg/l ≈	239.3 ppm at 25°C and 760 mm Hg
1 ppm ≈	4.17 mg/m^3 at 25°C and 760 mm Hg

17.3 Determination in the Atmosphere

In 1977 NIOSH (420) published Method S60 specifically recommended for methyl isobutyl carbinol. The method involves absorption of vapors from a known volume of air onto charcoal, desorption with carbon disulfide containing 5 percent isopropanol, and quantitation of the eluent by gas chromatography.

Recently Mueller and Miller (1) and Langvardt and Melcher (2) published methods of collection and analysis by gas chromatographic procedures that give excellent results and are applicable in situations where a variety of other materials may be present. It is believed these methods could be adapted to the determination of 4-methyl-2-pentanol.

17.4 Physiologic Response

17.4.1 Summary

4-Methyl-2-pentanol is low in single-dose oral toxicity. It is appreciably irritating to the eyes and mildly irritating to the skin when exposures are short, but if exposures are repeated or the material is confined to the skin, appreciable irritation and absorption can occur. Inhalation of high concentrations is irritating to the mucous membranes, and central nervous system depression including anesthesia is reported. Deaths are due to respiratory failure following prolonged anesthesia. Irritation of the mucous membranes of humans begins at about 50 ppm.

17.4.2 Effects in Animals

Single-Dose Oral.   The LD_{50} for rats is reported to be 2.59 g/kg (325). McOmie and Anderson (428) gave mice 1.0 ml/kg and observed anesthesia in two of five and death in one of five; at doses of 1.5 and 2.0 ml/kg, all mice were anesthetized and four of five and five of five succumbed, respectively. Gastrointestinal irritation and mesenteric congestion were noted.

Eye Contact.   4-Methyl-2-pentanol is moderately irritating to the rabbit eye (325, 428) and if contact is sustained, corneal injury results (427).

Skin Contact.   Smyth et al. (325) found the material to be slightly irritating to the rabbit's skin and to be absorbed, the LD_{50} being 3.56 ml/kg. It is reported elsewhere (427, 428) that a 15-min exposure can cause erythema. Five repeated applications of 3 ml/kg caused "severe drying of the skin with some sloughing and cracking" but no systemic effects were noted (428).

Inhalation.   The results of inhalation studies on rats and mice are summarized in Table 55.15.

Autopsy of exposed mice revealed irritation of the respiratory tract and congestion of the lungs (428).

Table 55.15. Results of Single or Short-Term Exposures of Rats and Mice to Vapors of 4-Methyl-2-pentanol

Species	Vapor Concentration (mg/l)	Duration of Exposure (hr)	No. Deaths/No. Treated	Observations	Ref.
Rats	Saturated (4600 ppm or 20 mg/l)	2	0/6		325
Rats	2000 ppm	8	5/6		325
Mice	20	15	8/10	10/10 anesthetized	428
Mice	20	10	6/10	10/10 anesthetized	428
Mice	20	8.5	0/10	10/10 anesthetized	428
Mice	20	4	0/10	7/10 anesthetized	428
Mice	20	1	0/10	Somnolence	428
Mice	20	1 min	0/10	Irritation of respiratory passages	428
Mice	20	12 × 4	0/9	Preanesthesia irritation and mild anesthesia	427

17.4.3 Effects in Humans

Silverman et al. (429) exposed 12 humans of each sex to the vapors of 4-methyl-2-pentanol for 15 min. They found that 50 ppm caused eye irritation, more than 50 ppm was required to cause nasal and throat irritation, and 25 ppm was estimated to be the highest concentration acceptable for 8 hr. An exposure to 50 to 100 ppm for 5 min is expected to cause eye and nose irritation and some respiratory discomfort.

17.5 Hygienic Standard of Permissible Exposure

Both the ACGIH (107) and OSHA (108) have set standards of permissible exposure at 25 ppm for methyl isobutyl carbinol (4-methyl-2-pentanol). The ACGIH also calls attention to a hazard from absorption through the skin. Although these standards may be adequate for protection against irritation, there are no long-term data to support them. Surely, exposures that cause mucous membrane irritation should be avoided.

17.6 Odor and Warning Properties

The irritating properties of the vapors of 4-methyl-2-pentanol likely will prevent voluntary exposure to acutely toxic concentrations, but they may not be adequate to warn against excessive prolonged and/or repeated exposure to lower concentrations.

18 HEXANOL, Commercial Grade (Mixture Consisting of 44 Percent 1-Hexanol, 53 Percent Methyl Pentanols, and 3 Percent 2-Ethyl-1-butanol)

$$C_6H_{14}OH$$

18.1 Source, Uses, and Industrial Exposure

Commercial grade hexanol is manufactured by the oxo process in which pentenes are reacted with carbon monoxide and hydrogen using a cobalt catalyst. It is used as a solvent and as an intermediate in the production of plasticizers and other derivatives having a variety of uses. Industrial exposure may occur in handling and use (388).

18.2 Physical and Chemical Properties

The properties of 1-hexanol and 2-ethyl-1-butanol are given in the discussion of these individual compounds. The properties of the primary methyl pentanols are given in Table 55.16.

18.3 Determination in the Atmosphere

The methods discussed in Section 17.3 may be adaptable.

18.4 Physiologic Response

The only data on commercial grade hexanol are those from Scala and Burtis (388).

18.4.1 Single-Dose Oral

The LD_{50} for rats is 3.67 g/kg.

Table 55.16. Physical and Chemical Properties of Methyl-1-pentanols in Commercial Hexanol

	2-Methyl-1-Pentanol	3-Methyl-1-Pentanol	4-Methyl-1-Pentanol
Synonyms	Methyl *n*-amyl carbinol		Isoamyl carbinol, isohexyl alcohol
CAS No.	105-30-6	589-35-5	626-89-1
Molecular formula	$CH_3(CH_2)_2CH(CH_3)CH_2OH$	$C_2H_5CH(CH_3)CH_2CH_2OH$	$(CH_3)_2CH(CH_2)_2CH_2OH$
	$C_6H_{14}O$	$C_6H_{14}O$	$C_6H_{14}O$
Molecular weight	102.18	102.18	102.18
Boiling point	148°C (762 mm Hg) (395)	153°C (395)	152–3°C (395)
Specific gravity	0.8192 (25/4°C) (395)	0.822 (27/4°C) (395)	0.8205 (25/4°C) (395)
Refractive index	1.4230 (19°C) (395)	1.4182 (25°C) (395)	1.4177 (25°C) (395)
Flash point	135°F (open cup) (119)		

1 mg/l $\approx$ 239.3 ppm at 25°C and 760 mm Hg
1 ppm $\approx$ 4.25 mg/m^3 at 25°C and 760 mm Hg

18.4.2   Eye Contact

The material is severely irritating to the eyes of rabbits. The liquid caused persistent iritis, corneal opacity, and some corneal vascularization.

18.4.3   Skin Contact

Confinement of the material to the abdominal skin of rabbits for 24 hr resulted in appreciable irritation, including some necrosis. Sufficient amounts to cause signs of central nervous system depression, labored breathing, and ataxia can be absorbed through the skin; the LD_{50} was greater than 2.6 g/kg.

18.4.4   Inhalation

When mice, rats, and guinea pigs were exposed for 6 hr to essentially saturated vapors (1060 ppm), they showed moderate local irritation (reversible) of the upper respiratory tract, slight congestion of the lungs, and a questionable effect upon the central nervous system.

18.5   Hygienic Standard of Permissible Exposure

There are no data upon which a standard for prolonged and/or repeated exposure can be based. Certainly it would be prudent to avoid exposure which causes irritation of the mucous membranes. No reports of adverse human experience were found.

19   2-OCTANOL; Capryl Alcohol, sec-Caprylic Alcohol, Methyl Hexyl Carbinol, CAS No. 123-96-6

$$CH_3(CH_2)_5CH(OH)CH_3$$

19.1   Source, Uses, and Industrial Exposure

2-Octanol is produced commercially by heating a soap of castor oil with sodium hydroxide (421). The content of methyl hexyl ketone ranges from 2 percent maximum in the chemical grade to 14 percent maximum in the solvent grade (430). It is used in lacquers, enamels, and perfumes; as a solvent for fats and waxes; as an antifoaming agent; and as a chemical intermediate in the manufacture of plasticizers and disinfectant soaps (111, 421, 430).

19.2   Physical and Chemical Properties

Physical state   Colorless liquid
Molecular formula $C_8H_{18}O$

Molecular weight	130.22
Boiling point	178.5°C (111)
Freezing point	Solidifies at -38.6°C (111)
Specific gravity	0.8232 (20/20°C) (430)
Vapor pressure	1 mm Hg (32.8°C) (431); 40 mm Hg (98°C) (198)
Refractive index	1.42025 (20°C) (111)
Flash point	~140°F (111); 150°F solvent grade, 180°F chemical grade (430)
Percent in saturated air	0.132 (32.8°C)
Density of saturated air (air = 1)	1.001
Solubility	0.096 ml/100 ml of water and miscible with alcohol, ether, aromatic and aliphatic hydrocarbons (111)
1 mg/1 $\eqsim$	187.8 ppm at 25°C, 760 mm Hg
1 ppm $\eqsim$	5.32 mg/m^3 at 25°, 760 mm Hg

19.3 Determination in the Atmosphere

No specific methods have been developed. Those noted under 4-methyl-2-pentanol likely are adaptable (see Section 17.3).

19.4 Physiologic Response

19.4.1 Single-Dose Oral

2-Octanol is low in single-dose oral toxicity; the LD_{50} values for rats and mice respectively are >3.2 (198) and 4.0 g/kg (432).

19.4.2 Single-Dose Intravenous

The LD_{50} dose for the mouse is 0.51 mmol/kg (0.066 g/kg) (54).

19.4.3 Eye Contact

Application of the liquid material to the eye of the rabbit caused pronounced erythema (432).

19.4.4 Skin Contact

Fassett (5) has reported that 2-octanol is slightly irritating to skin of guinea pigs and that the LD_{50} by skin absorption is greater than 0.5 g/animal. When 2 ml/day of undiluted material was applied for 6 days to the shaven skin of rabbits, it caused erythema, inflammation, and cracking of the skin which healed in 10 to 12 days (432). Immersion of the tails of mice in the liquid for 3 hr caused some deaths due to absorption (432).

19.4.5 Inhalation

Rats were exposed 2 hr/day, 6 days/week for 4½ months to vapor concentrations ranging from 0.18 to 0.35 mg/l (33.8 to 56.3 ppm). Mild reversible central nervous system disorders (not described), depression of hemoglobin and erythrocyte numbers, and minor changes in the liver, kidneys, and myocardium were noted (432).

19.5 Hygienic Standard for Permissible Exposure

The data available are inadequate to justify a standard for prolonged and/or repeated exposure. However, the inhalation data available indicate that an acceptable level would be well below 34 ppm. No reports of adverse human effects were found.

20 2-ETHYL-1-HEXANOL; 2-Ethylhexanol, 1-Hexanol-2-ethyl, 2-Ethylhexyl Alcohol, CAS 104-76-7

$$CH_3(CH_2)_3CH(C_2H_5)CH_2OH$$

20.1 Source, Uses and Industrial Exposure

2-Ethylhexanol is synthesized on a commercial scale (421). It is commercially the most important of the higher alcohols and its principal use is as an intermediate in the manufacture of plasticizers. Another significant, but much smaller, use is in the manufacture of 2-ethylhexyl acrylate (430). Other minor uses are as a dispersing and wetting agent, as a solvent for gums and resins, a cosolvent for nitrocellulose, and in ceramics, paper coatings, rubber latex and textiles (394, 421, 430).

20.2 Physical and Chemical Properties

Physical state	Colorless liquid
Molecular formula	$C_8H_{18}O$
Molecular weight	130.22
Boiling point	184 to 185°C (20)
Melting point	< -76°C (421, 198)
Specific gravity	0.8344 (20/20°C) (20)

Vapor pressure	0.05 mm Hg (20°C) (20)
Refractive index	1.4300 (20°C) (20)
Flash point	81°C (20)
Percent in saturated air	0.0066 (20°C)
Density of saturated air (air = 1)	<1.001
Solubility	0.1 percent in water; soluble in ether, ethanol, and other organic solvents
1 mg/l ≈	187.8 ppm at 25°C, 760 mm Hg
1 ppm ≈	5.32 mg/m^3 at 25°C, 760 mm Hg

20.3 Determination in the Atmosphere

Recently Mueller and Miller (1) and Langvardt and Melcher (2) published methods of collection and analysis by gas chromatographic procedures that give excellent results and are applicable in situations where a variety of other materials may be present. Their methods would be expected to be adaptable for 2-ethylhexanol.

20.4 Physiologic Response

20.4.1 Summary

2-Ethylhexanol is low in acute oral toxicity to animals and appreciably irritating and injurious to the eyes of rabbits. Primary skin irritation potential is slight, and the material is not absorbed through the skin of rabbits in toxic amounts following single exposures. Acute inhalation exposures are irritating to animals, but are of a low order of toxicity. Repeated oral and dermal exposures to animals resulted in liver and kidney effects.

20.4.2 Effects in Animals

Single or Short-Term Exposure Data. *Single-Dose Oral.* LD$_{50}$ values in rats resulting from single oral doses have been reported as follows: 3.2 (433), 3.3 (434), 2.0 (397), and 3.7 g/kg (388). Scala and Burtis (388) reported evidence of gastrointestinal irritation at gross necropsy following single oral doses to rats. The LD$_{50}$ for mice is reported to fall between 3.2 and 6.4 g/kg (198).

Eye Contact. Instillation of one drop of the undiluted material into the eye of a rabbit produced reddening and swelling of the conjunctiva, tearing, and mucous secretion, but no corneal injury (434). The signs disappeared in 96 hr. Concentrations of 50 and 25 percent ethylhexanol in oil produced similar but less severe reactions, requiring 48 and 8 hr, respectively, to disappear. A 12.5

percent concentration had no observable effect. Scala and Burtis (388), in contrast to the findings of Schmidt et al. (434), reported severe eye irritation, with persistent, widespread corneal opacity, following application of 0.1 ml of undiluted 2-ethylhexyl alcohol to the eye of a rabbit; and Smyth et al. (397) reported a Grade 5 (severe injury) in rabbits. The reasons for these discrepancies are unknown.

Skin Contact. In primary skin irritation tests conducted by Smyth et al. (397), 2-ethylhexanol produced only mild irritation on the uncovered rabbit belly. Dermal tests in rabbits gave an LD_{50} of 2.0 g/kg following 24-hr exposure to the skin (397). Scala and Burtis (388) found that a 24-hr dermal application of doses up to 2.6 g/kg to rabbits resulted in no deaths nor any other signs of percutaneous toxicity. Moderate irritation of the skin was observed following the 24-hr occlusive exposure. Skin absorption in guinea pigs likewise appears to be minimal, with a reported LD_{50} >8.3 g/kg (198). Moderate irritation of the skin was reported in these animals.

Inhalation. Fassett (198) found 100 percent survival in rats exposed to 235 ppm for 6 hr, and Smyth et al. (397) reported that six of six rats survived single 8-hr exposures to concentrated vapors of 2-ethylhexanol. Scala and Burtis (388) also reported 100 percent survival in mice, rats, and guinea pigs exposed to nearly saturated atmospheres (227 ppm) of 2-ethylhexanol for 6 hr. Signs of central nervous system depression were observed, including labored respiration; local irritation involving the mucous membranes of the eyes, nose, throat, and respiratory passages was judged to be moderate in degree. Areas of slight hemorrhage were observed upon gross necropsy examination.

Injection. Reported LD_{50} values for rats following intraperitoneal injections of 2-ethylhexanol are 0.8 to 1.6 g/kg (198) and 0.65 g/kg (433). Values reported for mice are <0.4 (198) and 0.78 g/kg (433).

Mode of Action. The principal signs of effect following acute oral and inhalation exposures were central nervous system depression and labored respiration (388).

Repeated or Prolonged Exposure Data. ***Repeated Oral.*** Schmidt et al. (434) reported no significant cumulative effects after 16 daily doses of two-fifths of the LD_{50} to rats. Slightly decreased body weights were observed, and 1 of 10 rats died. Daily doses of one-fifth the LD_{50} produced no observable effects.

Male and female rats fed diets containing 0.01, 0.05, 0.25, or 1.25 percent 2-ethylhexanol for 90 days showed evidence of liver and kidney effects at the highest level (435). Increased absolute (g) and relative (g/100 g body weight) liver weights were found in both sexes fed 1.25 percent. The livers of these females appeared congested and/or swollen upon gross examination, and degenerative changes were grossly evident in the kidneys of male rats fed the highest level. Histopathologic examination revealed mild and reversible alterations in the livers and kidneys of these rats. Mortality, body weights, and food consumption were not affected at any concentration.

Repeated Skin Contact.　A dose of 2 ml/kg applied daily to the skin of 10 rabbits for 12 days produced slight reddening and formation of scabs after 10 days (434). Decreased body weights were observed after 9 and 10 days. Histopathologic examination revealed fatty degeneration in the liver, bronchial ectasia and inflammation in the lungs, edema in the heart, disturbed membrane permeability in the kidney, and interstitial edema and rarefied spermiogenesis in the testes.

Repeated Inhalation.　Rats exposed to 6, 10, or 110 mg/m^3 2-ethylhexanol for 4 to 6 months showed increased excitability. Slight dystrophic alterations were observed in the parenchymatous organs (436). (Unfortunately the duration of the daily exposures was not given.)

Excretion and Metabolism.　Kamil et al. (437) found that nearly 90 percent of a dose of 2-ethylhexanol when ingested by rabbits is excreted in the urine conjugated with glucuronic acid (α-ethylhexanonylglucuronide).

Albro (438) reported similar findings in rats, with 80 to 82 percent radioactivity excreted in the urine following a single oral dose of 2-ethyl(1-^{14}C)hexanol. Small amounts were excreted in respiratory carbon dioxide (6 to 7 percent) and feces (8 to 9 percent). The major urinary metabolite was 2-ethylhexanoic acid. Only about 3 percent of the ethylhexanol was excreted unchanged. 2-Ethylhexanol was found to be a competitive inhibitor of yeast alcohol dehydrogenase (ADH) but a good substrate for mammalian ADH.

Moody (439) reported a marked proliferation of liver peroxisomes associated with hepatomegaly in rats fed diets containing 2 percent of the test material for 3 weeks.

Pharmacology　2-Ethylhexanol, upon intravenous injection of doses ranging from 0.2×10^{-5} to 3.2×10^{-5} mol/kg, has been reported to induce hypotension in rabbits in a dose-related manner (440). Such an effect was not observed in rats or dogs (440).

20.4.3　Effects in Humans

Large quantities of 2-ethylhexanol have been used for a long time in industry without any apparent reports of injury.

20.5 Hygienic Standards of Permissible Exposure

Neither the ACGIH nor OSHA has set a standard for safe industrial exposure.

20.6 Odor and Warning Properties

No data are available.

21 ISOOCTANOL (Mixture Consisting Typically of 70 to 80 Percent Dimethyl-1-hexanols, 10 to 20 Percent Methyl-1-heptanols, and 5 to 10 Percent Other Homologous Primary Alcohols)

$$C_8H_{17}OH$$

21.1 Source, Uses, and Industrial Exposure

Isooctanol is manufactured by the oxo process in which appropriate olefins are reacted with carbon monoxide and hydrogen using a cobalt catalyst. It is used as a solvent and a chemical intermediate in the production of plasticizers and other products. Industrial exposure may occur in handling and use (388).

21.2 Physical and Chemical Properties

Physical state	Colorless liquid
Molecular formula	$C_8H_{18}O$
Molecular weight	130.22
1 mg/l ≎	187.8 ppm at 25°C and 760 mm Hg
1 ppm ≎	5.32 mg/m^3 at 25°C and 760 mm Hg

21.3 Determination in the Atmosphere

The methods discussed under 4-methyl-2-pentanol, Section 17.3, may be adaptable.

21.4 Physiologic Response

The only data on the isooctanols are those from Scala and Burtis (388).

21.4.1 Single-Dose Oral

The LD_{50} for rats is given as 1.48 g/kg.

21.4.2 Eye Contact

The material is severely irritating to the eyes of rabbits including persistent widespread corneal opacity which cleared within a week.

21.4.3 Skin Contact

Confinement of the isooctanol to the abdominal skin of rabbits for 24 hr resulted in appreciable irritation including some necrosis. Sufficient amounts were absorbed to cause central nervous system depression with labored breathing and ataxia. The LD_{50} was greater than 2.6 g/kg; one of four treated animals died at the 3.16 g/kg dose level.

21.4.4 Inhalation

Mice, rats, and guinea pigs were exposed for 6 hr to essentially saturated vapors (~200 ppm). Moderate local irritation of the mucous membranes of the upper respiratory tract but no signs of systemic effects were noted.

21.5 Hygienic Standard for Permissible Exposure

There are no data upon which to base a standard for prolonged and/or repeated exposure. Certainly it would be prudent to avoid exposure which causes irritation of the mucous membranes. No reports of adverse human experience were found.

22 NONYL ALCOHOLS

$$C_9H_{19}OH$$

There are numerous isomers and mixtures of isomers that are referred to an nonyl alcohol. They are dealt with as a group, with distinctions made where possible.

22.1 Source, Uses, and Industrial Exposure

1-Nonanol occurs in oil of orange. It can be made using the oxo process described below and by the reduction and cleavage of oleic acid ozonide (20). Most of the commercial nonyl alcohols are mixtures of branched chain alcohols made by the oxo process, which involves the addition of carbon monoxide and hydrogen to selected olefins using a cobalt catalyst. They are used as solvents, defoaming agents, and to a large extent in the preparation of lubricants and plasticizers (394). Industrial exposure is associated with contact in handling. Significant exposure to vapor is not likely unless the material is handled hot.

22.2 Physical and Chemical Properties

The available data on the materials used in toxicologic studies discussed are given in Table 55.17. Since different material was used by the different investigators, the composition of the material used is described.

Smyth et al. (441) used diisobutyl carbinol. Treon (142) and Kitzmiller (446) used a product rich in trimethylhexanol. Fassett (198) used nonyl alcohol containing 2 percent of 2-propylheptanol. Egorov and Andrianov (444), Egorov (445), and Gerarde and Ahlstrom (53) used 1-nonanol. Scala and Burtis (388) used an isononyl alcohol typically consisting of 75 to 85 percent of dimethyl-1-heptanols, 5 to 10 percent methyl-1-octanols, and 10 to 20 percent other homologous primary alcohols. Vendilo et al. (60) did not specify the isomeric composition of the alcohol they used.

22.3 Determination in the Atmosphere

No methods have been recommended. The chromatographic–mass spectrometric methods described by Mueller and Miller (1) and by Langvardt and Melcher (2) may be adaptable.

22.4 Physiologic Response

22.4.1 Effects in Animals

Single or Short-Term Exposure Data. The toxicologic data available on animals are given in Table 55.18.

Pathology. Examination of animals dying as a result of treatment with nonyl alcohol rich in trimethylhexanols revealed degenerative changes in the neurons in all parts of the brain and brain stem. Hepatocellular degeneration involved the liver cords and was pronounced in the central areas. Renal damage was manifested by severe degeneration and frequent necrosis of the epithelium of the proximal and loop tubules, as well as the epithelium of the glomeruli. There were slight degenerative changes in the myocardium (446).

Repeated or Prolonged Exposure Data. *Oral.* Administration to rabbits of 0.148 g/kg on each of 67 days over a period of 83 days resulted in normal growth, uniform survival, and no signs of intoxication (142).

Table 55.17. Physical and Chemical Properties of Some Nonyl Alcohols, $C_9H_{19}OH$

Structural formula	1-Nonanol (53) $CH_3(CH_2)_7CH_2OH$	3,5,5-Trimethylhexanol (142, 446) $(CH_3)_3CCH_2CH(CH_3)CH_2CH_2OH$	Diisobutyl Carbinol (441) $[(CH_3)_2CHCH_2]_2CHOH$	Nonyl Alcohol Rich in Trimethylhexanols (142, 446)	Nonyl Alcohol containing 2% 2-Propylheptanol (198)
Physical state	Liquid	Liquid	Liquid	Colorless liquid	Liquid
CAS No.	143-08-8	3452-97-9	108-82-7	—	—
Molecular weight	144.26	144.26	144.26	—	—
Boiling point (°C)	215 (5)	193–194 (442)	178.1 (394)	187–208	192
Melting point (°C)	—	−70 (119)	F.p. −65 (119)	—	−5
Specific gravity	0.8279 (20/4°C) (20)	0.8236 (25/4°C) (442)	0.809 (21/4°C) (443)	0.845 (15.6/15.6°C)	0.8274 (20/4°C)
Vapor pressure	—	—	0.21 mm (20°C) (394)	—	40 mm (129°C)
Refractive index	1.4338 (20°C) (20)	1.4300 (25°C) (442)	1.4229 (20°C) (443)	1.438	1.43347 (20°C)
Flammability					
Flash point	—	200°F (open cup) (119)	165°F (119)	—	—
Explosive limits	—	—	0.8–6.1% at 212°F (119)	—	—
Percent in saturated air	—	—	0.027	—	—
Density of saturated air (air = 1)	—	—	0.001	—	—
Solubility					
H_2O	Practically insoluble	Insoluble	0.06% (20°C) (394)	Insoluble	Insoluble
Alcohol, solvents	Miscible with alcohol and ether	—	Soluble in alcohol and ether	—	—

1 mg/l ⇌ 169.6 ppm at 25°C, 760 mm Hg
1 ppm ⇌ 5.90 m³ at 25°C, 760 mm Hg

Table 55.18. Toxicologic Responses of Animals to Single Exposures of Nonyl Alcohols

Species	Route	Dose	Responses	Ref.
Mouse	Oral	6.4–12.8 g/kg	LD_{50}	198
Rat		3.2–6.4 g/kg	LD_{50}	198
Rat		3.56 g/kg	LD_{50}	441
Rat		2.98 g/kg	LD_{50}	388
Mouse	Intraperitoneal injection	0.8–1.6 g/kg	LD_{50}	198
Rat		0.8–1.6 g/kg	LD_{50}	198
Rabbit	Intravenous injection	15–21 mg/kg	LD_{50}	142
Guinea pig	Percutaneous absorption	>10 ml/kg	LD_{50}	198
Rabbit		5.66 ml/kg	LD_{50}	441
Rabbit		<3.6 ml/kg	LD_{50}	142
Rabbit		2.96 g/kg	LD_{50}	445
Rabbit		>3.2 g/kg	LD_{50}	388
Rat	Inhalation	730 ppm, 6 hr	No deaths	198
Rat		215 ppm, 6 hr	No deaths	198
Rat		Sat'd vapor, 8 hr	No deaths	441
Rat		21.7 mg/l aerosol, 6 hr	2/10 deaths[a]	388
Guinea pig		21.7 mg/l aerosol, 6 hr	0/10 deaths[a]	388
Mouse		21.7 mg/l aerosol, 6 hr	1/10 deaths[a]	388
Mouse		5.5 mg/l (duration not given)	LC_{50}	444
Rat	Aspiration	0.2 ml	Instant death	53
Unspecified	Eye contact		Conjunctivitis and keratitis	444
Rabbit			Conjunctival irritation but no corneal effects noted with lower molecular weight alcohols	388

[a] In these studies (388) some local mucous membrane irritation, CNS depression, impaired respiration, and prostration were noted. Deaths occurred within 14 hr postexposure.

4628

Skin. Application of 5 ml (1.6 to 2.0 g/kg) of nonyl alcohol to the skin of rabbits for 1 hr/day on each of 50 days over a period of 75 days resulted in retarded growth and erythema of the treated skin but no mortality (142).

Inhalation. Rabbits were exposed by inhalation to vapor containing 84 mg/m^3 (14.2 ppm) nonyl alcohol for 6 months. Electron microscopy examination of the retina of the eye revealed ultrastructural changes in the photoreceptor cells and Müller fibers (60). In rabbits and rats exposed to concentrations of 0.8, 0.6, or 0.2 mg/l (136, 99, or 33 ppm) nonyl alcohol for 2 hr/day for 2 months, small amounts of deformed or degenerate glial elements diffusely scattered in the cerebral cortex and subcortex were observed (444).

22.4.2 Effects in Humans

No reports of adverse effects were found.

22.5 Hygienic Standards for Permissible Exposure

No standard has been recommended in the United States. A MAC value of 10 mg/m^3 (1.7 ppm) for 1-nonanol is in force in the Soviet Union, a MAC of 200 mg/m^3 (34 ppm) in Yugoslavia, and a time-weighted average of 150 mg/m^3 (25 ppm) and a MAC of 250 mg/m^3 (42 ppm) in Romania (110).

22.6 Odor and Warning Properties

None were found.

23 DECYL ALCOHOLS

$$C_{10}H_{21}OH$$

There are numerous isomers and mixtures referred to as decyl alcohol. They are dealt with as a group with distinctions made where possible.

23.1 Source, Uses, and Industrial Exposure

n-Decyl alcohol is a primary alcohol that is manufactured commercially by the catalytic reduction of coconut oil and of coconut oil fatty acids or their esters under pressure (421). Decyl (oxo) alcohol, which contains mixed isomers of decanol, is manufactured by the "oxo process." Surface active agents and detergents are manufactured by the sulfonation of *n*-decyl alcohol. *n*-Decyl alcohol is also used as an antifoaming agent and as an intermediate in the manufacture of perfumes (421). The decyl alcohols are used in the manufacture of plasticizers, synthetic lubricants, petroleum additives, herbicides, and solvents (111).

23.2 Physical and Chemical Properties

Physical and chemical properties are available only for 1-decanol, n-decyl alcohol, or nonyl carbinol, CAS No. 112-30-1.

Empirical formula	$C_{10}H_{22}O$
Molecular formula	$CH_3(CH_2)_8CH_2OH$
Physical state	Moderately viscous liquid
Molecular weight	158.28
Boiling point	232.9°C (111)
Freezing point	Solidifies at +6.4°C (111)
Specific gravity	0.8297 (20/4°C) (111)
Vapor pressure	1 mm Hg (69.5°C) (431)
Refractive index	1.43587 (20°C) (111)
Flash point	180°F (119)
Solubility	Insoluble in water; soluble in ether, alcohol, acetone, benzene, and glacial acetic acid
1 mg/1 $\backsimeq$	154.5 ppm at 25°C, 760 mm Hg
1 ppm $\backsimeq$	6.47 mg/m³ at 25°C, 760 mm Hg

Since different investigators used different decyl alcohols, the information available on the material used by each is given below.

Fassett (198) used a mixture of n-decyl and sec-decyl alcohols.

Smyth et al. (325) used mixed isomers (decyl oxo alcohol).

Scala and Burtis (388) used an isodecyl mixture (oxo) consisting typically of 95 percent by weight of trimethylheptanols and other homologous primary alcohols.

The other investigators (53, 415, 445, 447) used n-decyl alcohol.

23.3 Determination in the Atmosphere

No methods have been recommended. The chromatographic–mass spectrometric methods described by Mueller and Miller (1) and Langvardt and Melcher (2) may be adaptable.

23.4 Physiologic Response

23.4.1 Effects in Animals

Single or Short-Term Exposure Data. The toxicologic data available on animals are given in Table 55.19.

Table 55.19. Toxicologic Responses of Animals to Single Exposures of Decyl Alcohols

Species	Route	Dose	Responses	Ref.
Mouse	Oral	6.4–12.8 g/kg	LD$_{50}$	198
Rat		12.8–25.6 g/kg	LD$_{50}$	198
Rat		9.8 g/kg	LD$_{50}$	325
Rat		4.72 g/kg	LD$_{50}$	388
Mouse	Intraperitoneal injection	0.8–1.6 g/kg	LD$_{50}$	198
Rat		0.8–1.6 g/kg	LD$_{50}$	198
Guinea pig	Percutaneous absorption	>10 ml/kg	LD$_{50}$	198
Rabbit		3.56 ml/kg	LD$_{50}$	325
Rabbit		>2.6 ml/kg	LD$_{50}$	388
Rabbit		18.8 ml/kg	LD$_{50}$	445
Rat	Inhalation	906 ppm, 6 hr	No deaths	198
Rat		65 ppm, 6 hr	No deaths	198
Rat		Sat'd vapor, 8 hr	No deaths	325
Mouse		525 ppm (duration not given)	LC$_{50}$	447
Mouse, rat, guinea pig		95 ppm, 6 hr	Irritation of mucous membranes of eyes, nose, throat, and respiratory passages	388
Rabbit	Eye contact		Severely irritating to conjunctival membranes with corneal injury	388
			Corneal injury	447
Guinea pig	Skin contact		Moderately irritating	198
Rabbit		24 hr	Severely irritating	325
Rabbit		24 hr	Moderately irritating	388

4631

Repeated or Prolonged Exposure Data. *Inhalation.* Rats and rabbits were exposed 2 hr/day to 0.8, 0.6, or 0.2 mg/1 for 2 months. Serum cholinesterese activity was decreased (apparently at all levels). Local irritation was observed at 0.6 mg/1 (447).

Carcinogenicity. n-Decanol showed tumor-promoting activity when applied three times weekly for 60 weeks to the skin of mice that had previously received an initiating dose of dimethylbenzanthracene. Tumors occurred in 6 of 30 mice between 25 and 36 weeks of treatment, with two developing into squamous cell carcinomas. Severe cutaneous irritation was observed (415).

23.4.2   Effects in Humans

No reports of adverse effects were found.

23.5   Hygienic Standard for Permissible Exposure

No standard has been recommended in the United States. A MAC value of 10 mg/m^3 (1.5 ppm) for 1-decanol is in force in the Soviet Union, a MAC of 200 mg/m^3 (31 ppm) in Yugoslavia, and a TWA of 150 mg/m^3 (23 ppm) and a MAC of 250 mg/m^3 (39 ppm) in Romania (110).

23.6   Odor and Warning Properties

The threshold odor concentration in air is 6.3 ppb (386).

24   1-DODECANOL; Dodecyl Alcohol, Lauryl Alcohol, CAS No. 112-53-8

$$CH_3(CH_2)_{10}CH_2OH$$

24.1   Source, Uses, and Industrial Exposure

Lauryl alcohol was prepared commercially at one time by the reduction of ethyl laurate (from coconut oil) with sodium and absolute alcohol, but more recently has been manufactured by the catalytic hydrogenation of coconut oil fatty acids or their esters under high pressure (421).

Lauryl alcohol is used in the manufacture of detergents (421), and in the manufacture of sulfuric acid esters which are used as wetting agents (111).

24.2   Physical and Chemical Properties

Physical state	Crystalline solid
Molecular formula	$C_{12}H_{26}O$

Molecular weight	186.33
Boiling point	259°C (111)
Melting point	24°C (111)
Specific gravity	0.8309 (24/4°C) (421)
Vapor pressure	1 mm Hg (91°C) (448)
Flash point	260°F (119)
Solubility	Insoluble in water; soluble in alcohol and ether
1 mg/1 ≈	131.3 ppm at 25°C, 760 mm Hg
1 ppm ≈	7.62 mg/m³ at 25°C, 760 mm Hg

24.3 Physiologic Response

24.3.1 Effects in Animals

Single or Short-Term Exposure. *Single-Dose Oral.* The single-dose oral tox-icity of 1-dodecanol is very low. LD_{50} values found for the rat are >12.8 (198) and >36.0 ml/kg (142); and for the rabbit, >36 ml/kg (142).

Shaffer (449) reported that seven rats and seven rabbits that survived a dosage of either 24 or 36 ml of technical lauryl alcohol/kg of body weight demonstrated no significant gross or microscopic changes. One rat, which died 6 days after the oral administration of 36 ml/kg, exhibited fatty degeneration of the liver and confluent bronchopneumonia.

Single-Dose Injection. The LD_{50} when given by intraperitoneal injection to rats was between 0.8 and 1.6 g/kg (198).

Skin Contact. The material caused practically no irritation when applied to the skin of guinea pigs and the LD_{50} was greater than 10 ml/kg (198).

Aspiration. Aspiration of 0.2 ml of lauryl alcohol by rats caused the death of 9 of 10 animals, deaths being caused by pulmonary edema rather than cardiac arrest or respiratory failure as with the C_3 to C_{10} alcohols. Survival time was also longer than in C_3 to C_{10} alcohols; seven rats died between 7 and 30 min after dosing; two rats died 5 hr or longer after dosing. The lungs were dark red, resembling the aspiration lung injury caused by kerosene (53).

Repeated or Prolonged Exposure. *Carcinogenicity.* Lauryl alcohol demon-strated weak tumor-promoting activity when applied three times weekly for 60 weeks to the skin of mice that had previously received an initiating dose of dimethylbenzanthracene. Papillomas developed in 2 of 30 mice after 39 and 49 weeks of treatment. Severe cutaneous irritation was observed (415).

24.4 Hygienic Standards for Permissible Exposure

None are given nor does it seem that any are required.

24.5 Odor and Warning Properties

The threshold odor concentration is 7.1 ppb (386).

25 HEXADECANOLS; 1-Hexadecanol, Cetyl Alcohol, Hexadecyl Alcohol, Ethal, Ethol, Palmitic Alcohol, CAS No. 36653-82-4

$$CH_3(CH_2)_{14}CH_2OH$$

A branched form, 2,2-dialkyl-1-ethanol, hexadecyl alcohol, or hexadecanol, is also used:

$$(C_6H_{13})(C_8H_{15})CHCH_2OH$$

25.1 Source, Uses, and Industrial Exposure

Cetyl alcohol, which was first obtained from spermaceti, is prepared commercially by the catalytic reduction of fats containing palmitic acid and by the saponification of spermaceti wax. Cetyl alcohol is used in cosmetics, perfumes, lotions, toilet articles, and medicinals, and in manufacturing detergents (421).

The hexadecyl alcohol studied by Scala and Burtis (388) is prepared synthetically by the oxo process and is a mixture of 2,2-dialkyl-1-ethanols where the alkyl groups are typically methyl branched six- and eight-carbon chains. It is widely used in the cosmetic industry because of its excellent solubilizing properties for lanolin, hexachlorophene, and bromo acid dyes (451). The hexadecyl alcohol studied by Fassett (198) was a synthetic liquid C_{16} alcohol that may have contained some impurities.

There is extensive opportunity for exposure to both of these hexadecyl alcohols in their handling and particularly in their use in cosmetics.

25.2 Physical and Chemical Properties

	1-Hexadecanol	Hexadecyl Alcohol (388)
Physical state	White crystals	Liquid
Molecular formula	$C_{16}H_{34}O$	$C_{16}H_{34}O$
Molecular weight	242.43	242.43
Boiling point	344.2 (111)	B.p.$_{50}$ 195–205°C (450)

Melting point	49°C (111)	F.p. < − 60°C
Specific gravity	0.8176 (50/4°C) (198)	0.842 (20/20°C) (450)
Vapor pressure	1 mm Hg at 122.7°C (448)	
Refractive index	1.4283 at 78.9°C (421)	
Solubility	Both insoluble in water; soluble in alcohol, ether, and chloroform	
1 mg/1 ≈	100.9 ppm at 25°C, 760 mm Hg	
1 ppm ≈	9.91 mg/m^3 at 25°C, 760 mm Hg	

No physical and chemical properties are available on the material studied by Fassett (198).

25.3 Physiologic Response

25.3.1   Effects in Animals

Single or Short-Term Exposure Data. *Single-Dose Oral.* The following LD_{50} values have been reported: for the rat, 6.4 to 12.8 (198) and >8.42 g/kg (388); and for the mouse, 3.2 to 6.4 g/kg (198).

Single-Dose Intraperitoneal Injection. The LD_{50} for both the rat and the mouse lies between 1.6 and 3.2 g/kg (198).

Eye Contact. When 0.1 ml undiluted material was instilled into the conjunctival sac of rabbits, it produced slight irritation (388).

Skin Irritation. The material is but slightly irritating to the skin of the guinea pig (198) and of the rabbit (388).

Skin Absorption. LD_{50} values of <10 g/kg for the guinea pig (198) and >2.6 g/kg for the rabbit (388) have been reported.

Inhalation. The single-dose inhalation data available may be summarized as follows:

| Rat | 2.22 mg/l (calculated), 6 hr | All died within 2 days | (198) |

Rat	0.41 mg/l (calculated), 6 hr	All survived	(198)
Rats, mice, and guinea pigs	26 ppm for 6 hr	Slight local irritation. No deaths or adverse systemic effects were observed.	(388)
Rats and guinea pigs	9.6 mg/l (aerosol) for 10 min of every 30 min over a 4-hr period (80 min intermittent exposure)	No local or systemic effects were noted	(388)

Repeated or Prolonged Exposure Data. *Carcinogenicity.* The tumor-promoting activity of *n*-hexadecanol was studied by applying it three times weekly for 60 weeks to the skin of 40 mice that had previously received an initiating dose of dimethylbenzanthracene. One papillona appeared after 53 weeks of treatment. The authors concluded that hexadecanol probably possesses weak tumor-promoting activity (415).

26 BENZENEMETHANOL; Benzyl Alcohol, Phenyl Carbinol, Phenylmethanol, α-Hydroxytoluene, CAS No. 100-51-6

$$C_6H_5CH_2OH$$

26.1 Source, Uses, and Industrial Exposure

Commercially, benzyl alcohol is manufactured from benzyl chloride by refluxing with sodium or potassium carbonate (452). Benzyl alcohol has been used as a lacquer solvent and a plasticizer. It is employed in the manufacture of perfumes, pharmaceuticals, and dyestuffs (394, 452a). It is used as a preservative in injectable saline and other injectable drugs (456). It is useful as a solvent in cosmetics and inks. Its esters are utilized in soaps, perfumes, and flavors (452a).

26.2 Physical and Chemical Properties

Physical state	Colorless liquid
Molecular formula	C_7H_8O
Molecular weight	108.13
Boiling point	204.7°C (111)
Melting point	−15.19°C (111)
Specific gravity	1.0472 (20/20°) (394)
Vapor pressure	0.15 mm Hg (25°C) (452a)

Refractive index	1.54035 (20°C) (111)
Flash point	96°C (3); 213°F (closed cup), 220°F (open cup) (111)
Percent in saturated air (20°C)	0.02
Density of saturated air (air = 1)	1.0005
Solubility	Sparingly soluble in water, 4 percent (111); soluble in alcohol, ether, aromatic hydrocarbons
1 mg/1 ≈	226.1 ppm at 25°C, 760 mm Hg
1ppm ≈	4.42 mg/m^3 at 25°C, 760 mm Hg

26.3 Determination in the Atmosphere

No methods have been specified for the determination of benzyl alcohol in air but chromatographic and/or mass spectrophotometric methods should be applicable (1, 2). Where such modern equipment is not available, the methods of Callaway and Reznek (453) or Mohler and Hammerle (454) may be useful.

26.4 Physiologic Properties

26.4.1 Summary

Benzyl alcohol is low in single-dose oral toxicity, irritating and injurious to the eyes, irritating to the skin, particularly if confined, and can be absorbed in toxic amounts if exposure is severe. Inhalation of life-threatening amounts of vapor does not seem likely. When injected undiluted it is quite toxic but fairly large amounts of 0.9 percent solutions as used in pharmaceuticals are well tolerated. The material in toxic doses causes central nervous system depression and typical sequelae. It is readily metabolized to benzoic acid and excreted as such, or as hippuric acid. No long-term exposure data are available.

26.4.2 Effects in Animals

Single or Short-Term Exposure Data. *Single-Dose Oral.* LD$_{50}$ values of 3.1 g/kg (455) and 1.23 g/kg (326) have been reported for the rat and 1.58 g/kg (326) for the mouse. Approximate lethal dose (ALD) values of from 1.6 to 3.2 g/kg (198) and 2.25 g/kg (38) are reported. In the latter case (38), graded doses ranged from 670 to 5000 mg. The 5000 mg/kg dose produced prostration and death within 1 hr of treatment. Two animals receiving lethal doses of 3400 and 2250 mg/kg died within 18 hr. Sublethal doses of 1500 and 1000 mg/kg produced slight discomfort and transient weight loss. The animal receiving 670 mg/kg showed discomfort and weight loss for 1 week. Microscopic examination of tissues disclosed no pathologic changes. Lethal doses produced heavy breathing and peculiar gait, suggesting that the central nervous system was affected.

Short-Term Repeated Oral. Six male albino rats received 450 mg/kg benzyl alcohol as a 20 percent solution in peanut oil for 10 treatments. All animals showed an initial weight depression with no other clinical or pathologic signs of toxicity (38).

Single-Dose Injection—Intraperitoneal. The approximate LD_{50} for rats and guinea pigs given intraperitoneal injection is between 400 and 800 mg/kg (198).

Single-Dose Injection—Intravenous. Kimura et al. (456) gave rats, mice, and dogs intravenous injections of benzyl alcohol either as 0.9 percent solutions in saline or undiluted (94 percent material). They found the LD_{50} values for the 94 percent material to be for the mouse, <0.5 ml/kg; for the rat, between 0.05 and 0.08 ml/kg; and for the dog, between 0.050 and 0.055 ml/kg. Rats tolerated slow (1 ml/min) intravenous injection of 40 ml/kg (0.378 g/kg benzyl alcohol) of 0.9 percent benzyl alcohol in saline with no fatalities. When the 0.9 percent solution was injected rapidly (1 ml/5 sec), the LD_{50} was 33.4 ml/kg. Using the same treatment regimen, benzyl alcohol was found to be 20 to 30 times more toxic than ethanol. Zagradnik et al. (54) reported on intravenous LD_{50} in mice of 0.325 g/kg (3.00 mmol/kg).

Single-Dose Injection—Intraarterial. When 0.9 percent benzyl alcohol in saline or water was injected into the carotid artery of the anesthetized dog in amounts up to 40 ml, it caused minimal effects upon hematologic, respiratory, and cardiovascular parameters (456).

Single-Dose Injection—Subcutaneous. A comparison of the lethal doses of various alcohols when injected subcutaneously into mice indicated that benzyl alcohol is two and a half to three times as toxic as *n*-butyl or isopropyl alcohol (226).

Eye Contact. When benzyl alcohol was instilled into the rabbit eye it caused a severe response characterized by marked irritation of the conjunctive membranes and cloudiness of the cornea.

Skin Contact. The undiluted material was mildly irritating when applied in small amounts to the uncovered skin of rabbits (455). The undiluted material applied to depilated skin of guinea pigs for a period of 24 hr caused moderately strong primary irritation, and there was evidence of systemic symptoms with death from applications of less than 5 ml/kg. Systemic symptoms were still noted when it was applied under the same conditions as a 20 percent solution in acetone. It was not a skin sensitizer in guinea pigs (198).

Inhalation. Exposure of rats to saturated vapors (~200 ppm) resulted in an LC_0 of 2 hr, an LC_{33} of 4 hr, and an LC_{100} of 8 hr (455). No fatalities or symptoms were found in rats when exposed 6 hr to a calculated concentration of 61 ppm, nor were symptoms produced by exposures obtained by bubbling air through the liquid heated to 100 and 150°C (198). The approximate lethal concentration (ALC) 4 or 8 hr exposures was found to be >250 ppm (nominal). Gross and histopathology were negative (38). Subacute exposure of 216 to 270 ppm to six male rats for 4-hr periods produced no clinical or pathologic signs of toxicity (38).

Pharmacology. Doses of 0.2 ml/kg, or more, given to dogs by stomach tube induced emesis and defecation. This was apparently due to irritation of the gastric mucosa, since no such effects resulted from smaller doses given by this route or from the injection of larger doses (457). Diuresis was more pronounced in the rabbit than in the dog, after the administration of benzyl alcohol by various routes (458). The injection of 5 or 10 percent benzyl alcohol in oil of sweet almond in the region of the auditory meatus of cats caused temporary degeneration of the small facial nerves (459). Mice suffered respiratory stimulation, respiratory and muscular paralysis, convulsions, and narcosis following subcutaneous injection (226). Macht (461) also observed convulsions when lethal doses were injected into small animals. Single-oral doses to rats produced depression and coma within 10 to 15 min, and the animals were excitable for 3 to 4 days (326). Depression was observed in mice following single oral doses (326).

Kimura et al. (456) observed transient respiratory arrest in mice following intravenous injection of 50 ml/kg of a 0.9 percent solution of benzyl alcohol in saline. Intravenous injection of 1 ml/kg of 0.9 percent benzyl alcohol in saline to dogs and monkeys did not affect blood pressure, heart rate, respiration, electrocardiogram, or hematologic parameters. In earlier work, Macht (461) reported the occurrence of a decrease in the blood pressure of animals following various modes of injection. Unfortunately, experimental details are not available. Gruber (457) observed a decrease in the arterial blood pressure of rabbits, cats, and dogs following intravenous injection of benzyl alcohol, but failed to find such a decrease in the case of dogs following the oral administration of 0.1 to 1.0 ml/kg of body weight.

Deaths due to rapid intravenous injection of rats with 0.9 percent benzyl alcohol in saline were believed due to respiratory arrest (456). Intravenous injection of 94 percent benzyl alcohol into dogs caused dyspnea, diarrhea, ataxia, mydriasis, nystagmus, urination, respiratory arrest, collapse, and cardiac arrest. In some instances deaths were delayed; in these cases, death was due to pulmonary hemorrhage and edema (456). Macht (461) believed death from benzyl alcohol was caused by paralysis of the respiratory center, but Gruber (457) believed that cardiac paralysis might precede that of the respiratory center.

Anesthetic action of benzyl alcohol in the rabbit is independent of pH up to pH 9; at higher pH values, effect is increased (466). The blood sugar of fasting animals was increased somewhat by prolonged administration of benzyl alcohol (462). The administration of benzyl alcohol stimulated the rate of elimination of an injection pigment by the liver (462).

Pharmacokinetics. The plasma half-life of benzyl alcohol administered as a 2.5 percent solution in saline was found to be approximately 1.5 hr in dogs injected intravenously at doses of 52 and 105 mg/kg (456).

Metabolism/Excretion. Benzyl alcohol is oxidized by ADH. Winer (331) reports the activity of ADH on benzyl alcohol to be slightly less than on ethanol. Von Wartburg (230) reports benzyl alcohol to be approximately one-half as active as ethanol in reaction with ADH and to have affinity to ADH roughly equal to that of isopropyl and butyl alcohols. The animal organism readily oxidizes benzyl alcohol to benzoic acid, which, after conjugating with glycine, is rapidly eliminated as hippuric acid in the urine. Rabbits given 1 g of benzyl alcohol subcutaneously eliminated 300 to 400 mg of hippuric acid within the following 24 hr (463). Within 6 hr after the oral administration of 0.40 g benzyl alcohol/ kg of body weight, rabbits eliminated 65.7 percent of the dose as hippuric acid in the urine (464). Chief metabolic products are benzoic and hippuric acids. Metabolites identified in the urine of rabbits given an oral dose of 0.25 g/kg benzyl alcohol are given in the following tabulation: (465)

	Metabolite	Percent of Dose	Range (No. of Experiments)
	1. Mercapturic acid*	2	0–4 (326)
	2. Glycine conjugate*	74	71–76 (111)
	3. Copper reducing material (as glucuronic acid)	8	4–15 (111)
98%	4. Glucosiduronic acid	14	6–23 (111)
accounted	5. Ethereal sulfate	0	0 (452)
for	6. Ether-soluble acid	84	80–90 (111)

26.4.3   Effects in Humans

Metabolism/Excretion. Humans readily oxidize benzyl alcohol to benzoic acid, which, after conjugation with glycine, is rapidly eliminated as hippuric acid in the urine. Within 6 hr after taking 1.5 g of benzyl alcohol orally, human subjects eliminated 75 to 85 percent of the dose in the urine as hippuric acid (467).

---

* The difference between these ether-soluble acids (1 + 2) and total ether-soluble acid (6) probably represents the percent of dose excreted as unconjugated acid.

Experience. Seven cases of illness are associated with the use of a lacquer containing 5 percent benzene, 10 percent benzyl alcohol, acetone, denatured alcohol, butyl tartrate, and cellulose acetate (468). It was believed that benzene (which was present in the air in a poorly ventilated room to the extent of 0.3 mg/l) and benzyl alcohol were the chief causes of the violent headaches, vertigo, nausea, gastric pains, vomiting, diarrhea, and loss of weight. These signs of intoxication, which appeared after 1½ to 2 months of exposure, disappeared upon removal from the lacquer exposure. Local necrosis of tissue following the accidental injection of pure benzyl alcohol in preparation for a circumcision has been described by Macht (460).

26.5 Hygienic Standard for Permissible Exposure

No hygienic standards have been proposed. No long-term data are available to permit the establishment of an exposure level acceptable for prolonged and/or repeated exposure.

26.6 Odor and Warning Properties

Benzyl alcohol is reported to have a faint aromatic odor and a burning taste (111).

27 2-PHENYLETHANOL; β-Phenylethyl Alcohol, Phenethyl Alcohol, Benzyl Carbinol, β-Hydroxy Ethylbenzene, CAS No. 60-12-8

$$C_6H_5CH_2CH_2OH$$

27.1 Source, Uses, and Industrial Exposure

2-Phenylethyl alcohol is produced from benzene and ethylene oxide by the Friedel–Crafts reaction (452).

Along with citronellol and geraniol this alcohol forms the basis of rose-type perfumes. It is used to enhance other perfumes (452).

27.2 Physical and Chemical Properties

Physical state	Liquid
Molecular formula	$C_8H_{10}O$
Molecular weight	122.16
Boiling point	219.5–220°C (452)
Freezing point	Solidifies at −27°C (111)
Specific gravity	1.0235 (15/15°C) (198)
Refractive index	1.530 to 1.533 (20°C) (111)

Vapor pressure	1 mm Hg (58°C) (448); 10 mm Hg (100°C) (198)
Flash point	216°F (198)
Percent in saturated air	0.13 (58°C)
Solubility	2 percent in water (452); 1 part in 1 part of 50 percent alcohol (111); miscible with alcohol and ether
1 mg/l ≈	200.2 ppm at 25°C, 760 mm Hg
1 ppm ≈	5.00 mg/m³ at 25°C, 760 mm Hg

27.3 Determination in the Atmosphere

2-Phenylethyl alcohol may be identified as the *p*-nitrobenzoate (m.p. 106–107°C) and as β-phenylethyl-*p*-nitrobenzylphthalate (m.p. 84.3°C). It also forms styrene on treatment with alkali (452). The chromatographic and mass spectrometric methods of Mueller and Miller (1) and Langvardt and Melcher (2) may be adaptable.

27.4 Physiologic Response

27.4.1 Summary

2-Phenylethanol is moderate to low in single-dose oral toxicity, appreciably irritating to the eyes, slightly irritating to the skin of rabbits and guinea pigs but not to the skin of humans, and low in single-dose inhalation toxicity. The hazards of ordinary handling should be minimal.

27.4.2 Effects in Animals

Single-Dose Oral. The LD_{50} values are for the mouse, 0.8 to 1.5 g/kg (198); for the guinea pig, 0.4 to 0.8 g/kg (198); and for the rat, 1.79 (326) and 2.46 g/kg (469).

Single-Dose Injection. The LD_{50} values for intraperitoneal injection are for the mouse, 0.4 to 0.8 g/kg (198) and for the guinea pig, 0.2 to 0.4 g/kg (198).

Eye Contact. When instilled into the rabbit eye, 0.005 ml of undiluted material or 0.5 ml of 5 or 15 percent solutions in propylene glycol caused severe corneal irritation and iritis (469).

Skin Contact. The material was slightly irritating to the skin of guinea pigs (198) and rabbits (469). It was capable of penetrating the skin in toxic amounts; the LD_{50} given for guinea pigs is from 5 to 10 ml/kg (198) and for rabbits, 0.79 ml/kg (469).

Inhalation. An 8-hr exposure of rats to an atmosphere essentially saturated with vapors failed to cause any deaths (469).

Metabolism/Excretion. Phenylethyl alcohol is oxidized almost entirely to the corresponding acid (470). Metabolites excreted in urine following oral doses of 0.2 to 0.3 g/kg to rabbits were phenaceturic acid, the glycine conjugate of phenylacetic acid, 42 percent; glucosiduronic acid, 5 percent; and an ether-soluble acid, probably phenylacetic acid, 19 percent. Only 66 percent of the original dose could be accounted for (465). According to Winer (331), β-phenylethyl alcohol is a substrate for ADH, with an initial rate of oxidation nearly as high as that of allyl alcohol and appreciably greater than that of ethanol.

Mutagenicity—In Vitro. β-Phenylethyl alcohol has been found to obstruct genetic transformation in *B. subtilis* by rendering competent *B. subtilis* organisms unable to irreversibly incorporate transforming DNA (471). Yoshiyama et al. (472) have shown that β-phenylethyl alcohol stimulates or recovers the cell division ability in ultraviolet-irradiated cells of *E. coli* β by selectively suppressing the initiation of DNA replication.

27.4.3   Effects in Humans

Skin Contact. Katz (473) reported that a patch test using phenylethyl alcohol full strength for 24 hr produced no irritation in 20 human subjects. A skin sensitization test carried out on 25 volunteers at a concentration of 8 percent in petrolatum produced no sensitization reactions (474).

27.5   Hygienic Standard of Permissible Exposure

None have been suggested either in the United States or elsewhere. There are no data available that permit a recommendation.

27.6   Odor and Warning Properties

The sensory odor threshold of β-phenylethyl alcohol (in 9.4 percent w/w "grain spirits") was determined to be 7.5 ppm or 0.06 mM by the triangular test (345).

28   CYCLOHEXANOL; Cyclohexyl Alcohol, Hexalin, Hexahydrophenol, Hydrophenol, CAS No. 108-93-0

28.1 Source, Uses, and Industrial Exposure

Cyclohexanol may be prepared by (1) the catalytic hydrogenation of phenol at elevated temperatures and pressures or (2) catalytic oxidation of cyclohexane in the liquid phase at 100 to 250°C (475). Cyclohexanol is used as a solvent for oils, alkyd resins, shellac, alcohol-soluble phenolics, ethyl cellulose, basic chrome and acid dyes, metallic soaps, gums, and natural resins. It is used in fragrances, dry cleaning, textile cleaning, and laundry and household preparations. Cyclohexanol is used in leather lacquers, paint and varnish removers, polishes, rubber cements, and as an intermediate in the preparation of plasticizers and other chemicals (394, 475a). It is used in the manufacture of celluloid, finishing textiles, and insecticides (111). Because of its diversity of use, industrial exposure can be rather extensive.

28.2 Physical and Chemical Properties

Physical state	Colorless liquid; hygroscopic crystals
Molecular formula	$C_6H_{12}O$
Molecular weight	100.16
Boiling point	161°C (111)
Melting point	23 to 25°C (111)
Specific gravity	0.9493 (20/4°C) (394)
Vapor pressure	3.5 mm Hg (34°C) (476)
Refractive index	1.4656 (20°C) (394)
Flammability	
Flash point	68°C
Ignition temperature	572°F
Percent in saturated air	0.33 (30°C)
Density of saturated air (air = 1)	1.01
Solubility	3.6 percent soluble in H_2O at 20°C (394); miscible with most aliphatic, aromatic, hydrogenated, and chlorinated solvents
1 mg/l ≈	244.1 ppm at 25°C, 760 mm Hg
1 ppm ≈	4.10 mg/m³ at 25°C, 760 mm Hg

28.3 Determination in the Atmosphere

NIOSH Method No. S54 (477) has been recommended. It involves drawing a known volume of air through charcoal to trap the organic vapors present (recommended sample is 10 liters at a rate of 0.2 l/min). The analyte is desorbed with carbon disulfide containing 5 percent 2-propanol. The sample is separated by injection into a gas chromatograph equipped with a flame ionization detector and the area of the resulting peak is determined and compared with standards.

The methods of Mueller and Miller (1) and Langvardt and Melcher (2) may also be adaptable.

The concentration of cyclohexanol in the air may be determined colorimetrically by measuring the intensity of the straw color produced by the reaction with catechol and concentrated sulfuric acid (478). The error of analysis of an aqueous solution of cyclohexanol containing no other alcohol is ± 0.009 mg in the range of 0.05 to 0.25 mg.

28.4 Physiologic Response

28.4.1 Summary

Cyclohexanol is low in single-dose toxicity by the oral, skin, parenteral, or inhalation routes. It is markedly irritating to the eyes, only slightly irritating to the skin, and it can be absorbed through the skin in toxic amounts if exposures are severe. Excessive amounts of vapor are irritating to the mucous membranes and can cause varying degrees of narcosis, as well as degenerative changes in the vascular system and of the liver, kidney, heart, and brain. The material can interfere with reproduction but it is not mutagenic. Marginally excessive human exposure may result in conjunctival irritation and headache, and there is a report of nonspecific disturbances of the autonomic nervous system.

28.4.2 Effects in Animals

Single or Short-Term Exposure Data. *Single-Dose Oral.* An LD_{50} value for the rat is given as 2.06 g/kg (479) and an MLD value of 2.2 to 2.6 g/kg is given for the rabbit (480).

Single-Dose Injection. When given intravenously to the mouse the LD_{50} is 2.72 mmol/kg or 0.27 g/kg (54); when given intramuscularly, the LD_{50} is 1.0 g/kg (483).

Eye Contact. When the liquid was applied to the eyes of rabbits it caused moderately severe irritation and corneal injury (38, 482). The corneal injury was reversible (38).

Skin Contact. Cyclohexanol is slightly irritating to the rabbit's skin (482) and can be absorbed in toxic amounts from extensive exposure causing tremors, narcosis, hypothermia, and death (480).

Inhalation. Pohl (482) was unable to observe any effects when a dog was exposed to air saturated with cyclohexanol for 10 min/day on 7 successive days. Smyth et al. (481) reported that 8 hr was the maximum time for inhalation of concentrated cyclohexanol vapor by rats without deaths.

Mode of Action.   The oral administrations of lethal doses of cyclohexanol (2.6 g or more/kg of body weight) to rabbits caused severe vascular damage and extreme toxic effects with massive coagulation necrosis of the myocardium, lung, liver, kidney, and brain. Animals that survived somewhat smaller doses developed toxic degenerative lesions and vascular damage of much lesser degree (480). Unlike benzene, this compound, when absorbed by experimental animals over considerable periods of time, has shown no tendency to bring about injury to the cellular elements of the peripheral blood (478). Animal experiments have indicated that cyclohexanol is narcotic in high concentrations, with the possibility of damage to kidneys, liver, and blood vessels (119).

Repeated or Prolonged Exposure Data.   *Skin Contact.*   Only temporary ery-thema and superficial sloughing of the skin of rabbits occurred when an ointment consisting of potassium oleate and cyclohexanol, up to 15 percent by weight, was applied in 5-g portions for 1 hr/day over a period of 15 days (480). Necrosis, exudative ulceration, and thickening of the skin were observed following application of 10 ml cyclohexanol to the intact skin of a rabbit for 1 hr/day for 10 successive days (480). These 10-ml applications also induced narcosis, tremors, athetoid movements, and hypothermia, and the rabbit died the day after the tenth treatment (480). The pathologic changes observed following the cutaneous application of cyclohexanol were very similar, in general, to those caused by oral administration, and so afforded corroborative evidence of the absorption of the compound through the skin (480).

Inhalation.   Exposure of animals to sufficiently high concentrations of cyclo-hexanol in air for 6 hr/day, 5 days/week induced an intoxication characterized by conjunctival congestion and irritation, lacrimation, salivation, lethargy, incoordination, narcosis, and mild convulsions (478). Under suitable conditions this intoxication may terminate in death. Table 55.20 summarizes the response of animals that received repeated inhalation exposures to cyclohexanol vapors 6 hr/day, 5 days/week, for 5 to 11 weeks.

Toxic degenerative changes were found in the brain, heart, liver, and kidneys of rabbits exposed repeatedly to concentrations of cyclohexanol in air ranging from 997 to 1229 ppm (4.00 to 4.93 mg/l). Similar but less severe changes were seen in the myocardium, liver, and kidneys of rabbits exposed to vapors of cyclohexanol at the concentration of 272 ppm in air (1.09 mg/l). Rabbits exposed to a concentration of 145 ppm (0.58 mg/l) suffered little or no injury, slight degenerative changes being barely demonstrable in the liver and kidneys (478).

Rats were exposed to cyclohexanol vapor for 24 hr/day for 87 days, followed by a 6-day starvation period (20 percent of normal rations) (484). At a concentration of 0.61 mg/m^3 (0.149 ppm), changes were noted in the interre-lationship of the chronaxie of the extensor and flexor muscles (decreased after week 7), in cholinesterase activity (increased time needed for hydrolysis of

Table 55.20.   Results of Repeated Exposures of Animals to Vapors of Cyclohexanol (478)

Number of Animals Exposed	Concentration		Duration of Exposure[a] (hr)	Percentage Mortality	Signs of Intoxication
	mg/l	ppm			
4 rabbits	4.93	1229	150	50	Narcosis, lethargy, incoordination, conjunctival congestion and irritation, and salivation
4 rabbits	4.00	997	330	50	Narcosis, lethargy, conjunctival congestion and irritation, lacrimation, salivation, few convulsive movements
1 monkey	2.78	693	300	0	Lethargy, conjunctival irritation
4 rabbits	1.09	272	300	0	Slight conjunctival irritation
4 rabbits	0.58	145	300	0	None

[a] hr/day, 5 days/week.

acetylcholine), and in the ascorbic acid content of liver (increased by 30 percent by the end of month 2). No changes were noted in health and weight gain, reactive (sulfhydryl) groups of serum protein, or ascorbic acid content of the brain. At a concentration of 0.059 mg/m^3 (0.014 ppm) all properties studied remained normal. Confirmation of these observations is needed before significance is attached to them.

Reproduction.   An increase in mortality of the young during the first 21 days of life was found in the offspring of mice fed diets containing 1 percent cyclohexanol during gestation (485). In strain TB mice, the percent mortality in first and second generation treated offspring was 14.1 and 53.5 percent, respectively, compared to 11.9 percent for controls. The percent mortality of first generation offspring of treated strain MNRI mice was 43.1 percent, and for corresponding controls, 12.2 percent.

Carcinogenicity.   No information was found.

Mutagenicity.   Collin et al. (486) reported a few karyotypic changes in bone marrow cells of rats fed on a diet containing cyclohexanol. Cyclohexanol was not mutagenic in several strains of *Salmonella typhimurium* in concentrations of 500 µg/plate either in the presence or absence of a rat liver homogenate activation system. It did not significantly increase the spontaneous, or background, mutation frequency (38). Cattanach, in a review of the mutagenicity test data on cyclamates and their metabolites (487), concluded that none of the compounds can be considered mutagenic.

Metabolism/Excretion. Cyclohexanol is a substrate for ADH with a relative velocity or maximal turnover rate of 67 percent compared to ethanol, but has a stronger affinity for the enzyme and will competitively inhibit the oxidation of ethanol (230). Winer (331) showed that the initial rate of oxidation for cyclohexanol was 135 mol/l/min per mole of ADH, the same as for ethanol and *n*-octanol.

Cyclohexanol and its derivatives are generally not aromatized in vivo; large amounts of cyclohexanol may be excreted as urinary glucuronides (489), without prolonged retention in the organism (490). No urinary metabolites were detected by Bernhard (491) when a dose of 0.29 g/kg of cyclohexanol was given subcutaneously to dogs or by Weitzel (492), but Pohl (482) found glucuronic acid in the urine of a dog following oral administration. Following oral administration of cyclohexanol to rabbits, this compound is excreted in the urine in conjugation with sulfuric and glucuronic acids (480, 493). Similar results were obtained when animals were subjected to repeated inhalation of the compound (478). In general, the increased elimination of conjugated sulfates could be correlated with the increase in the concentration of cyclohexanol in the air (478). On the other hand, rabbits exposed to the lowest concentration used, 145 ppm cyclohexanol in the air, showed no increase in the conjugation of urinary sulfates but excreted five times the normal amount of glucuronic acid (478). When rabbits received 50 mg cyclohexanol as a single oral dose, the urine contained 25 percent of the unchanged alcohol and <1 percent of the glucuronide (494). Di Prisco (488) found a decrease of 1 to 2 percent in the reduced glutathione of the blood of rabbits subjected to the inhalation of vapors of cyclohexanol for 10 to 15 min upon alternate days over a period of 21 days.

Pharmacology. A transient decrease in the blood pressure of rabbits was observed by Sato (495) following the intravenous injection of this compound. Cyclohexanol decreased the arterial pressure in anesthetized cats and inhibited the activity of isolated frog heart (496). Toxic doses of cyclohexanol injected into mice and rats caused inhibition of motor activity, flaccidity, lateral position, clono-tonic spasms, and death; and cyclohexanol injected into mice intensified the sedative effect of phenobarbitone and sodium pentobarbitone and decreased the intensity of spasms caused by strychnine and Corazole (496).

28.4.3   Effects in Humans

Pharmacology. Of a group of 279 men and 174 women exposed daily to less than the permitted concentrations of cyclohexanol during caprolactam production, 114 individuals showed nonspecific disturbances of the autonomic nervous system during a 2-year period, as compared to 8 out of 100 individuals in a nonexposed control group (497).

Skin Contact. Cyclohexanol tested at a concentration of 4 percent in petrolatum produced no irritation after a 48-hr closed-patch test in human subjects (498).

Experience. Browning (490) has reported one case of suspected intoxication, characterized by vomiting, coated tongue, and slight tremors, in a worker engaged in spraying leather with a preparation that contained butyl acetate and cyclohexanol. The evidence in this case was not adequate to convict cyclohexanol as the offending agent. On the other hand, it is apparent that headache and conjunctival irritation have resulted from prolonged exposure to excessive concentrations.

28.5 Hygienic Standard of Permissible Exposure

The ACGIH (348) has recommended a TLV of 50 ppm or 200 mg/m^3 for repeated 8-hr exposures as has OSHA (108). This value is consistent throughout the world except in Japan, where 25 ppm is the standard (110).

28.6 Odor and Warning Properties

Irritation of the eyes, nose, and throat were induced in human subjects exposed by Nelson et al. (237) for 3 to 5 min to 100 ppm of cyclohexanol. The majority of these subjects believed that the highest permissible concentration for an 8-hr exposure, from the standpoint of comfort, should be less than 100 ppm. At the concentration of 272 ppm which produced irritation of the eyes and nose in rabbits, the odor of the compound was recognized at once by human subjects.

29 2-, 3-, or 4-METHYLCYCLOHEXANOL; Methylcyclohexanol, Hexahydrocresol, Methyl Hexalin, Hexahydromethyl Phenol, Methyl Adronal, Methyl Anol, Sextol, CAS No. 589-91-3 and CAS No. 25639-42-3

29.1 Source, Uses, and Industrial Exposure

Methylcyclohexanol is manufactured by the hydrogenation of *m*- and *p*-cresols (499). There are several structural isomers of this compound but the ortho, meta, and para isomers are most common and each may occur as cis and trans geometric isomers. No data are available on the other isomers. The commercial product consists essentially of a mixture of two cyclic secondary alcohols (meta and para isomers) (499). Methylcyclohexanol dissolves gums, oils, resins, and waxes. This cyclic alcohol serves as a solvent in lacquers, a blending agent in textile soaps, and an antioxidant in lubricants (394, 499).

29.2 Physical and Chemical Properties

Physical state	Colorless liquid
Molecular formula	$C_7H_{14}O$
Molecular weight	114.19
Boiling point	173 to 175.3°C (499)
Freezing point	$-50°C$ (499)
Specific gravity	0.913 (25/15.5°C) (500)
Vapor pressure	1.5 mm Hg (30°C) (500)
Refractive index	1.461 (20°C) (394)
Flash point	154°F (closed cup) (394)
Percent in saturated air	0.20 (30°C)
Density of saturated air (air = 1)	1.01
Solubility	3 to 4 percent soluble in water (20°C) (499); miscible with common solvents, plasticizers, and gum solutions
1 mg/l $\backsimeq$	214.2 ppm at 25°C, 760 mm Hg
1 ppm $\backsimeq$	4.67 mg/m³ at 25°C, 760 mm Hg

29.3 Determination in the Atmosphere

NIOSH Method No. S374 has been recommended (503). It involves drawing a known volume of air through charcoal to trap the vapors. The analyte is desorbed with methylene chloride. The sample is separated with a gas chromatograph equipped with a flame ionization detector and the area of the resulting peak is determined and compared with standards. The methods of Mueller and Miller (1) and Langvardt and Melcher (2) may be adaptable. The concentration of methylcyclohexanol in air also may be determined colorimetrically by measuring the intensity of the straw color produced by the reaction with catechol and sulfuric acid (478). The error of analysis of an aqueous solution of methylcyclohexanol containing no other alcohol is ±0.009 mg in the range of 0.05 to 0.25 mg.

29.4 Physiologic Properties

29.4.1 Summary

Methylcyclohexanol is low in single-dose oral toxicity, somewhat irritating to skin, and it can be absorbed through the skin in toxic amounts especially if exposures are severe and repeated. The material is not especially toxic when inhaled. Excessive exposure causes headache and irritation of the mucous membranes.

29.4.2 Effects in Animals

Single or Short-Term Exposure Data. *Single-Dose Oral.* An MLD of 1.75 to 2.0 g/kg is given for the rabbit (480).

Injection. On the basis of the comparative results of the intraperitoneal injection of the three common isomeric forms into mice, Fillipi (501) concluded that *o*-methylcyclohexanol is more toxic than the meta or para isomers.

Eye Contact. No information was found.

Skin Contact (480). The application of large doses of methylcyclohexanol upon the intact skin of rabbits induced fatal poisoning characterized by tremors, narcosis, and hypothermia. The application of 10 ml of methylcyclohexanol upon the intact skin of a rabbit for 1 hr on each of six successive days was fatal. The MLD was determined to be from 6.8 to 9.4 g/kg. At various stages of the experiment, weakness, tremors, deep anesthesia, and local petechiae, gross hemorrhage, and thickening of the skin were seen. The application upon the intact skin of rabbits of 5-g portions of a mixture of methylcyclohexanol (up to 15 percent by weight) in potassium oleate, for 1 hr/day over a period of 15 days, produced only temporary erythema and superficial sloughing of the skin.

Inhalation. According to Pohl (482) there were no signs of intoxication in a dog exposed for 10 min daily on seven consecutive days to air saturated with methylcyclohexanol.

Mode of Action. The oral administration of lethal doses of methylcyclohexanol to rabbits (2.0 or more/kg of body weight) induced severe acute toxic parenchymal and vascular changes in the heart, liver, and kidneys, and toxic vascular damage in the lungs. As a general rule, these lesions were accompanied by cerebral edema and congestion. Diffuse degenerative changes in the liver were the only histopathologic evidence of intoxication in the case of animals given sublethal oral doses of the compound (480). Comparable toxic lesions were found in the tissues of animals subjected to inhalation of the vapors in air and to percutaneous absorption of methylcyclohexanol. The severity of the toxic changes varied with the severity of the experimental conditions, assuming borderline or questionable significance (when compared with the incidental variations in the histologic pattern in the tissues of control animals) in the case of rabbits exposed to the least concentration of methylcyclohexanol in air (121 ppm, 0.56 mg/l) (478).'

Repeated or Prolonged Exposure Data. *Inhalation.* Rabbits subjected to 503 ppm methylcyclohexanol in air for 6 hr/day, 5 days/week over a total of 10 weeks developed salivation, conjunctival congestion and irritation, and lethargy (478). Corresponding conditions of exposure to lower concentrations (121 and 232 ppm) resulted in no observed effects. No evidence of specific or general change was found in the cellular elements of the peripheral blood of animals exposed repeatedly to any of these concentrations (478).

Metabolism/Excretion. Methylcyclohexanol is a substrate for ADH. Its initial rate of oxidation is 108 mol/l/min per mole of ADH, somewhat less than that of ethanol and cyclohexanol (331). Glucuronic acid has been found in the urine of a dog following the oral administration of methylcyclohexanol (482). In the case of rabbits, under similar conditions of administration, conjugation of the compound or a metabolite with both glucuronic and sulfuric acids has been demonstrated by analysis of the urine (480). Some conjugation product with sulfuric acid was found also in the urine of animals exposed to 232 and 503 ppm methylcyclohexanol in air (478). The rate of excretion of glucuronic acid in the urine of rabbits is correlated directly with the concentration of methyl-cyclohexanol in the air to which they have been subjected (478). Twice the normal quantity of glucuronic acid was found in the urine of rabbits subjected to 121 ppm methylcyclohexanol in the air (478).

29.4.3   Effects in Humans

Headache and irritation of the ocular and upper respiratory membranes may result from prolonged exposure to excessive concentrations of the vapor of methylcyclohexanol in air.

After examining several workers who had been exposed to a cellulose solvent containing methylcyclohexanol, Browning (502) concluded that a few of them had a slightly but significantly diminished total number of leukocytes in the peripheral bloodstream, and one had a slight relative lymphocytosis.

29.5   Hygienic Standard of Permissible Exposure

The ACGIH (348) has recommended a TLV of 50 ppm or 235 mg/m^3, the same as in numerous other nations (110). The OSHA level is 100 ppm (108), the same as the standard in Yugoslavia (110).

29.6   Odor and Warning Properties

Methylcyclohexanol vapor in air can be detected and recognized by its odor when present to the extent of 500 ppm, a concentration capable of causing upper respiratory irritation [from Patty, Volume 2, 1963 (426)].

30   2-FURANMETHANOL; Furfuryl Alcohol, 2-Furyl Carbinol, 2-Furancarbinol, 2-Hydroxymethylfuran, CAS No. 98-00-0

30.1   Source, Uses, and Industrial Exposure

Furfuryl alcohol is prepared by high pressure catalytic hydrogenation of furfural (504).

Furfuryl alcohol is a solvent for cellulose ethers and esters, ester gum, coumarone resins, and natural resins. It is used in the manufacture of dark-colored thermosetting resins and phenolic resins and as a solvent in the manufacture of abrasive wheels. It is employed in the textile industry as a solvent and dispersant for dyes (394, 504, 505). It is used in the manufacture of wetting agents (111).

Under ordinary usage in the industrial plant the use of furfuryl alcohol for about 20 years has resulted in no impairment of health. More recently furfuryl alcohol has been employed as a liquid propellant (506).

30.2   Physical and Chemical Properties

Physical state	Liquid
Molecular formula	$C_5H_6O_2$
Molecular weight	98.10
Boiling point	170°C (111)
Freezing point	-14.6°C (394, 504)
Specific gravity	1.1287 (20/4°C) (394)
Vapor pressure	1 mm Hg (31.8°C) (507)
Refractive index	1.484 (20°C) (394)
Flammability	
Flash point	167°F
Explosive limits	1.8 to 16.3 percent between 72 and 122°C (119)
Ignition temperature	736°F
Spontaneous ignition temperature	915°F (111)
Percent in saturated air (20°C)	0.13 (31.8°C)
Density of saturated air (air = 1)	1.003
Solubility	Miscible but unstable in water; miscible with most organic solvents but immiscible with petroleum hydrocarbons and most oils

| 1 mg/l $\approx$ | 249.4 ppm at 25°C and 760 mm Hg |
| 1 ppm $\approx$ | 4.01 mg/m^3 at 25°C and 760 mm Hg |

30.3   Determination in the Atmosphere

NIOSH Method No. S365 has been recommended (508). It involves drawing a known volume of air through a glass tube containing Porapak Q to trap furfuryl alcohol vapors (recommended sample size is 6 liters). The analyte is desorbed with acetone. The sample is analyzed by gas chromatography, and the area of the resulting peak is determined and compared with standards. The methods of Mueller and Miller (1) and Langvardt and Melcher (2) may be useful in specific situations. Comstock and Oberst (509) collected furfuryl alcohol from the air by passing it through glacial acetic acid. They then added 0.05 N pyridine bromine sulfate solution to the acetic acid. After reaction of the bromine with the furan moiety (1 hr in the dark), 5 percent KI was added and the solution was titrated with 0.1 N Na$_2$S$_2$O$_3$.

30.4   Physiologic Response

30.4.1   Summary

Furfuryl alcohol is moderately toxic when ingested and appreciably less toxic when injected. This is unusual and suggests that its instability in acid may result in more toxic breakdown products in the stomach. The material is markedly irritating and injurious to the eyes but not appreciably irritating to the skin when not confined to the skin. Prolonged exposure to vapors that can occur at room temperature may result in irritation of the mucous membranes but not lethal effects in the absence of pronounced signs or symptoms.

30.4.2   Effects in Animals

Single-Dose Oral.   LD$_{50}$ values for the rat of 0.275 (510) and 0.132 g/kg (511) have been reported.

Repeated-Dose Oral.   Thirty rats were given 40 mg/kg five times/week for 4 weeks. Three died during the course of treatment. Gross or microscopic pathology was not different from that of control animals (511).

Single-Dose Injection.   When given to rabbits the intravenous LD$_{50}$ was found to be 0.65 g/kg (510) and the subcutaneous LD to be 0.60 g/kg (513).

Eye Contact.   When furfuryl alcohol was instilled into the eyes of rabbits, one drop caused reversible irritation, and two drops caused severe irritation which required several weeks to heal (511).

Skin Contact.   When 0.1 ml/day was applied 5 days/week for 4 weeks to the rabbit ear, it caused only a very slight erythema (405).

Single-Dose Inhalation.   Several animal species have been exposed to various concentrations of vapor for various periods of time. The results are given in Table 55.21.

When rats were exposed to furfuryl alcohol in the atmosphere in the concentration of 700 ppm, the symptoms noted were initial excitement, followed by eye irritation and drowsiness; their eyes became red within 8 min (509). Slight eye irritation was noted in the monkey exposed to 260 ppm for 6 hr (511).

Repeated-Dose Inhalation.   No deaths occurred in rats or mice that received 30 or 15 respective 6-hr exposures to an average concentration of 19 ppm furfuryl alcohol (range 12 to 29 ppm) (509). These animals exhibited restlessness for the first 5 or 10 min and drowsiness during the remaining portion of the 6-hr periods of exposure. The only pathologic change observed was a moderate diffuse congestion in the respiratory tract of the rats but without any significant cellular changes.

Dogs exposed for 6 hr/day, 5 days/week for 4 weeks to 239 ppm showed no changes in behavior and no gross pathology. Microscopic examination showed minimal chronic inflammation of the bronchi (511). A monkey exposed for 6 hr/day for three consecutive days to 239 ppm showed no apparent irritation or toxicity (511).

Pharmacology.   Fine and Wills (515) suggested that furfuryl alcohol probably is distributed equally throughout most of the body and has little specific action

Table 55.21.   Results of Single Exposures of Animals to the Vapors of Furfuryl Alcohol

Species	LC_x	Duration of Exposure (hr)	Concentration (ppm)	Ref.
Mouse	0	6	700	512
	0	6	243	511
	0	6	597	511
Rats	8	6	47	511
	17	4	700	509
	25	8	700	509
	50	4	233	506
	100	6	243	511
Rabbits	0	6	416	511
Dogs	0	6	349	511
Monkey	0	6	260	511

on enzyme systems localized in special structures in the body. They observed that furfuryl alcohol (1) had a negative inotropic effect on the heart without a significant chronotropic one, (2) decreased the tonus and contractility of intestinal smooth muscle, and (3) depressed the central nervous system, producing changes in the electroencephalographic record similar to those produced by certain anesthetic agents. Fine and Wills (515) produced severe falls in blood pressure and temporary apnea when the intravenous dosage exceeded 0.500 to 0.600 g/kg of body weight. The central depressant action of lethal dosages of furfuryl alcohol was combated effectively with pentamethylenetetrazol, amphetamine, or ephedrine. Erdman (513) reported that small doses of furfuryl alcohol stimulated respiration in both man and rabbit. Larger doses were reported to depress respiration, to lower the body temperature, and to produce nausea, salivation, diarrhea, dizziness, and diuresis. Sensory nerves were paralyzed by dilute solutions of furfuryl alcohol (514). Death occurred from respiratory paralysis at intravenous doses from 0.8 to 1.4 g/kg (515).

Metabolism. Furfuryl alcohol is a substrate for ADH with an affinity to the enzyme approximately equal to ethanol and n-propanol but with a relative velocity or turnover rate of 51 percent compared to ethanol (230). Winer (331) found that furfuryl alcohol was capable of being oxidized by ADH. at an initial oxidation rate of 108 mol/l/min per mole of ADH, about half the rate of n-butyl alcohol, and less than that of ethyl alcohol.

30.4.3   Effects in Humans

Experience. Workers handling products containing furfuryl alcohol developed dermatitis. Other workers developed symptoms of respiratory irritation (516). A dose of 40 to 150 mg given orally to humans was without adverse effect (517).

30.5   Hygienic Standard of Permissible Exposure

The ACGIH lists a TLV of 5 ppm or 20 mg/m^3, but indicates it intends to revise this upward to 10 ppm or 40 mg/m^3 (348). The OSHA standard is 5 ppm or 20 mg/m^3 (108). Numerous other countries have standards of either 5 or 50 ppm (110).

30.6   Odor and Warning Properties

The mean detectable concentration of furfuryl alcohol is stated to be 8 ppm (506).

31 TETRAHYDRO-2-FURANMETHANOL; Tetrahydrofurfuryl Alcohol, Tetrahydro-2-furancarbinol, Tetrahydro-2-furylmethanol, CAS No. 97-99-4

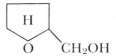

31.1 Source, Uses, and Industrial Exposure

Tetrahydrofurfuryl alcohol is prepared by liquid-phase hydrogenation of furfuryl alcohol over a nickel catalyst (504).

Tetrahydrofurfuryl alcohol is used in the preparation of esters, plasticizers, and as a chemical intermediate. It is a solvent for cellulose acetate, cellulose nitrate, ethyl cellulose, furfuryl alcohol, polymers, styrene, phenol–aldehyde resins, vinyl acetate (504), fats, waxes, and resins (111).

31.2 Physical and Chemical Properties

Physical state	Colorless liquid, hygroscopic
Molecular formula	$C_5H_{10}O_2$
Molecular weight	102.14
Boiling point	178°C (111)
Melting point	< -80°C (111)
Specific gravity	1.0495 (20/4°C) (198)
Vapor pressure	2.3 mm Hg (40°C) (385)
Refractive index	1.4520 (20°C) (111)
Flammability	
Flash point	183°F (111)
Explosive limits	1.5 to 9.7 percent by vol. (111)
Ignition temperature	540°F
Solubility	Miscible with water, alcohol, ether, acetone, chloroform, benzene
1 mg/l ≎	239.5 ppm at 25°C, 760 mm Hg
1 ppm ≎	4.18 mg/m³ at 25°C, 760 mm Hg

31.3 Determination in the Atmosphere

The methods of Mueller and Miller (1) and Langvardt and Melcher (2) may be adaptable.

31.4 Physiologic Response

31.4.1 Single-Dose Oral

The following LD_{50} values have been reported: for rats, 1.6 to 3.2 (198) and 2.50 g/kg (518); for guinea pigs, 0.8 to 1.6 (198) and 3.0 g/kg (518); and for the mouse, 2.3 g/kg (518).

31.4.2 Single-Dose Injection

When given intraperitoneally, the LD_{50} dose for rats and guinea pigs was found to be between 0.4 and 0.8 g/kg (198).

31.4.3 Skin Contact

When tested on the guinea pig the material was found to be moderately irritating, not to be a sensitizer, and to be absorbed through the skin, the LD_{50} being less than 5 ml/kg (198).

31.4.4 Inhalation

Fassett (198) reports that two of three rats exposed to a calculated concentration of 12,650 ppm died. Since this concentration is greater than the vapor pressure provides at room temperatures, the exposure atmosphere must have contained an appreciable amount of mist. None of three rats exposed to 655 ppm for 6 hr died; loss of coordination, prostration, and vasodilation of the ears and feet were noted.

31.5 Hygienic Standard of Permissible Exposure

No standards have been recommended and there are no data available upon which to set a standard.

31.6 Odor and Warning Properties

No data were found.

32 4-HYDROXY-4-METHYL-2-PENTANONE; Diacetone Alcohol, Pyranton, CAS No. 123-42-2

$$(CH_3)_2C(OH)CH_2COCH_3$$

32.1 Source, Uses, and Industrial Exposure

Diacetone alcohol is prepared by condensing acetone in the liquid phase in the presence of alkali and alkaline earth hydroxides (421). The technical grade contains up to 15 percent acetone (519). It is the toxic principal in *Stipa vaseyi*, commonly known as sleepy grass (526). Diacetone alcohol is a solvent for nitrocellulose, cellulose acetate, cellulose esters, epoxy resins, hydrocarbons, oils, fats, resins, gums, and dyes (394, 519). It is used in hydraulic fluids, metal-cleaning compounds, the manufacture of photographic film, making artificial silk and leather, and in coating compositions for paper and textiles (421, 519). It is used in some antifreeze solutions and as a preservative in pharmaceutical preparations (111). Unless neutral or slightly alkaline it will decompose to acetone (421). It is also used as a chemical intermediate.

32.2 Physical and Chemical Properties

Physical state	Clear liquid
Molecular formula	$C_6H_{12}O_2$
Molecular weight	116.16
Boiling point	167.9°C (111)
Freezing point	−42.8°C (394)
Specific gravity	0.9406 (20/20°C) (394)
Vapor pressure	0.97 mm Hg (20°C) (394); 1.5 mm Hg (25°C) (520)
Refractive index	1.4232 (20°C) (111)
Flammability	
Flash point	Pure diacetone alcohol 131°F (387); Reagent grade 151°F; commercial grade 48°F (closed cup), 55°F (open cup) (111)
Explosive limits	1.8 to 6.9 percent (119)
Autoignition temperature	1118°F (119)
Percent in saturated air (20°C)	0.13
Density of saturated air (air = 1)	1.005
Solubility	Miscible with water, alcohol, ether, and other solvents
1 mg/l ≑	216.5 ppm at 25°C, 760 mm Hg
1 ppm ≑	4.75 mg/m^3 at 25°C, 760 mm Hg

32.3 Determination in the Atmosphere

A method recommended by NIOSH (No. S55) involves the drawing of a known volume of air through charcoal to trap the organic vapors present (recommended sample is 10 liters at a rate of 0.2 l/min) (521). The analyte is desorbed with carbon disulfide containing 5 percent of 2-propanol. The sample is separated by injection into a gas chromatograph equipped with a flame ionization detector, and the area of the resulting peak is determined and compared with standards.

The methods of Mueller and Miller (1) and Langvardt and Melcher (2) may also be adaptable.

Diacetone alcohol may be differentiated qualitatively from acetone chemically, since with 2,4-dinitrophenylhydrazine, the former gives a red, and the latter a yellow, precipitate (522).

32.4 Physiologic Response

32.4.1 Summary

Diacetone alcohol is low in single-dose oral toxicity, would be expected to cause transient irritation of the eyes, is not appreciably irritating to the skin nor likely to be absorbed in toxic quantities, and is low in inhalation toxicity. It has a narcotic action. A daily intake of about 40 mg/kg/day over 4 to 6 weeks either orally or by inhalation was without adverse effect in rats. Grossly excessive vapor concentrations are very disagreeable if not voluntarily intolerable.

32.4.2 Effects in Animals

Single or Short-Term Exposure Data. *Single-Dose Oral.* The LD_{50} for rats is given as 4.0 g/kg (263).

Single-Dose Injection. When given intraperitoneally, the LD_{50} for mice is 933 mg/kg (524).

Skin Contact. Smyth and Carpenter (263) found that when applied to the skin of rabbits diacetone alcohol was essentially free of irritation and not likely to be absorbed in toxic quantities, the LD_{50} being 14.5 ml/kg.

Inhalation. Mice, rats, rabbits, and cats subjected for 1 to 3 hr to an atmosphere containing 2100 ppm (10 mg/l) of diacetone alcohol vapors manifested restlessness, symptoms of irritation, coryza, symptoms of excitation, then sleepiness (523). Kidney injury occurred in the rabbits. Smyth and Carpenter (263) found that 1500 ppm did not kill rats in 8 hr.

Mode of Action. Keith (525) observed a temporary decrease in the hemoglobin content and numbers of erythrocytes in the peripheral blood of rats for 1 to 4 days following the oral administration of a sublethal dose of diacetone alcohol.

Walton and his associates (522) observed that intravenous injections of diacetone alcohol into mice induced narcosis more rapidly than acetone, and that diacetone alcohol was about twice as toxic as acetone. Following intravenous, intramuscular, or oral administration to rabbits, diacetone alcohol depressed the respiration markedly, decreased the blood pressure, and induced narcosis leading to death from respiratory failure. The progressive decrease in the blood pressure of dogs injected repeatedly with diacetone alcohol, noted by these investigators, led them to believe that increased susceptibility had resulted from the repetition of the injections. Keith (525) described hepatic lesions characterized by vacuolization and granulation of the parenchymal cells, which reached the maximum stage in about 24 hr following the oral administration of a sublethal dose to rats.

Diacetone alcohol isolated from *Stipa vaseyi* (sleepy grass) produced respiratory and motor depression in rats in doses of 0.4 to 0.8 ml/kg (route not given), whereas doses of 2.0 ml/kg caused a mixed reaction. Doses of 4.0 to 8.0 ml/kg caused depression in mice and doses of 2.0 ml/kg caused a curious posture, the concave curving of the back (526).

Repeated or Prolonged Exposure Data. *Repeated Oral.* Smyth and Carpenter (263) administered diacetone alcohol in the drinking water of groups of rats for 30 days. The smallest daily dose that resulted in any evidence of micropathological alteration was 0.04 g/kg, whereas no effects were observed at the level of 0.01 g/kg.

Repeated Inhalation (527). Four groups of 12 male and 12 female rats were exposed to concentrations of 0, 232, 1035, and 4494 mg/m^3 (0, 50, 224, and 973 ppm, respectively) of diacetone alcohol 6 hr/day, 5 days/week for 6 weeks. The results are summarized in Table 55.22.

32.4.3 Effects in Humans

No report of adverse effects in humans was found.

32.5 Hygienic Standards for Permissible Exposure

The ACGIH (348) and OSHA (108) as well as the agencies of other governments (110), have set standards of 50 ppm or 240 mg/m^3 for repeated 8-hr exposures to diacetone alcohol.

32.6 Odor and Warning Properties

Silverman et al. (429) observed irritation of the eyes, nose, and throat in most human subjects exposed to 100 ppm for 15 min. Although some found the odor and taste unpleasant, this concentration was not intolerable.

Table 55.22.   Results of Repeated Exposures of Rats to the Vapors of Diacetone Alcohol

	Concentration in ppm (mg/m^3)			
Criteria	0	50 (232)	224 (1035)	973 (4494)
Signs	—	—	—	1
Body weight	—	—	—	2
Organ weight	—	—	3	3, 4
Urinanalysis	5	5	—	—
Hematology	—	—	—	—
Plasma clinical chemistry	—	—	—	6
Histopathology[a]	—	0	—	7

Key: — Studies done—no deviation from normal
    0 Studies not done
    1 Lethargy
    2 Slight decrease in body weight of females terminally
    3 Slight increase in liver weights
    4 Slight increase in kidney weights
    5 Glucose present
    6 Increase in LDH in females
    7 Eosinophilic droplets in proximal tubules of the kidneys

[a] Examination conducted on at least 28 different tissues from each sex.

Verschueren (386) gives the absolute threshold odor concentration as 0.28 ppm and the 100 percent recognition concentration as 1.7 ppm. She describes the odor as sweet, ranging from pleasant to unpleasant.

33   2-PROPEN-1-OL; Allyl Alcohol, 1-Propenol-3, Vinyl Carbinol, CAS No. 107-18-6

$$CH_2{=}CHCH_2OH$$

33.1   Source, Uses, and Industrial Exposure

Allyl alcohol is prepared (1) by high temperature chlorination of propylene yielding allyl chloride, which is subsequently hydrolyzed to the alcohol and (2) from glycerol by dehydration and reduction. Allyl alcohol is used as a flavoring agent as are many allyl esters. It is used in the preparation of allyl resin and plastics, in the preparation of pharmaceuticals, and in the chemical industry. Large amounts are used in the synthesis of glycerin. There are opportunities for exposure to allyl alcohol but these generally are confined to the industry and associated with transfers and maintenance of equipment.

33.2 Physical and Chemical Properties

Physical state	Colorless liquid
Molecular formula	C_3H_6O
Molecular weight	58.08
Boiling point	96 to 97°C (111)
Melting point	-50°C (111)
Specific gravity	0.8476 (25/4°C)
Vapor pressure	23.8 mm Hg (25°C)
Refractive index	1.41345 (20°C) (111)
Flammability	
Flash point	70°F (open cup); 75°F (closed cup) (111)
Explosive limits	2.5 and 18.0 percent (528)
Autoignition temperature	713°F (111)
Percent in saturated air	3.13 (25°C)
Density of saturated air (air $=$ 1)	1.031
Solubility	Miscible with water, alcohol, chloroform, ether, petroleum ether
1 mg/l $\approx$	422 ppm at 25°C, 760 mm Hg
1 ppm $\approx$	2.37 mg/m^3 at 25°C, 760 mm Hg

33.3 Determination in the Atmosphere

NIOSH Method No. S52 has been recommended (529). It involves drawing a known volume of air through charcoal to trap the organic vapors present (recommended sample is 10 liters at a rate of 0.2 l/min). The analyte is desorbed with carbon disulfide containing 5 percent 2-propanol. The sample is separated with a gas chromatograph equipped with a flame ionization detector and the area of the resulting peak is determined and compared with standards. The methods of Mueller and Miller (1) and Langvardt and Melcher (2) should be applicable.

Dunlap et al. (530) collected allyl alcohol in air by drawing it through water. To this aqueous sample was added 0.01 N bromine in acetic acid in the presence of a catalyst, mercuric acetate. Excess bromine was reduced by iodide and the iodine titrated with 0.01 N thiosulfate according to the method of Reid and Beddard (531).

33.4 Physiologic Response

33.4.1 Summary

Allyl alcohol is a rather toxic material. Its vapors, even in concentrations of a few parts per million, cause lacrimation and irritation of the eyes and irritation of the mucous membranes of the respiratory passages. Toxic concentrations of vapors can occur at room conditions. The warning properties are believed to be adequate to prevent voluntary exposure to acutely dangerous concentrations but inadequate to prevent excessive prolonged and/or repeated exposure. The liquid material is irritating to the skin and is readily absorbed through the skin. Absorption of small quantities causes pain deep in the tissues near the exposure site. Absorption of systemically toxic amounts can easily occur if exposure is extensive, prolonged, and/or repeated. The material is high in single-dose oral and single-dose inhalation toxicity, and lower than might be expected in repeated-dose toxicity by either route.

33.4.2 Effects in Animals

Single or Short-Term Exposure Data. *Single-Dose Oral.* LD_{50} values have been reported as follows: for the rat, 105 (young males) (530), 99 (for young adult males) (530), 64 (532), and 70 mg/kg (533); for the mouse, 96 mg/kg (530); and for the rabbit, 71 mg/kg (530). Jenner (533) noted that treated rats exhibited depression, colorless secretion from the eyes, a scrawny appearance for several days posttreatment, and that deaths occurred from 4 hr to 4 days after treatment.

Single-Dose Injection. The LD_{50} for mice given an intraperitoneal injection is reported to be 60 mg/kg (530); when given intravenously, the LD_{50} is 1.35 m mol/kg or about 80 mg/kg (54).

Eye Contact. Application of the liquid to the rabbit eye causes marked irritation and some transitory corneal opacity (263, 530).

Skin Contact. Application to the uncovered skin of rabbits causes very little local irritation (263, 532) but it is very rapidly absorbed through the skin in toxic and lethal amounts. LD_{50} values of 89 mg/kg (530) and 0.053 ml/kg (263) are given. It is also reported that contact for 1 hr with 0.1 ml killed both of two rabbits (543). Application of a pad wet with alcohol to the clipped bellies of two guinea pigs caused their death within 1½ hr. Lacrimation occurred almost immediately after application and increased in intensity until death; just before death there was a marked protrusion of the eyes (405).

Inhalation. The response of animals that received single or short-term repeated exposures to allyl alcohol by inhalation is summarized in Table 55.23.

Repeated or Prolonged Exposure Data. *Repeated-Dose Oral.* Daily doses of 25 mg/kg for 4 days to rats produced no deaths. Variable degrees of liver changes were observed macroscopically (228). In rats given daily oral doses of

Table 55.23. Results of Single or Short-Term Exposures of Animals to Vapors of Allyl Alcohol

Concentration (ppm)	Animal	Duration of exposure (hr)	Outcome	Ref.
29,300	Rat	0.5	100% lethal	537
14,500	Rat	1.25	100% lethal	537
14,500	Rat	0.5	Survived	537
14,500	Guinea pig	0.75	100% lethal	537
14,500	Guinea pig	0.25	Survived	537
5,800	Rat	2.0	100% lethal	537
5,800	Rat	0.5	Survived	537
2,900	Rat	3.0	100% lethal	537
2,900	Rat	1.0	Survived	537
2,900	Guinea pig	2.0	100% lethal	537
2,900	Guinea pig	1.0	Survived	537
1,000	Monkey	4	Death	534
1,000	Rabbit	4	100% lethal	534
1,000	Rat	4	100% lethal	534
1,060	Rat	1	LC_{50}	530
1,000	Rat	1	LC_{67}	263
500	Rat	1	Survived	146
290	Rat	8.0	100% lethal	537
290	Rat	3.0	Survived	537
290	Guinea pig	4.0	100% lethal	537
290	Guinea pig	1.0	Survived	537
250	Rat	4	Some deaths	535
165	Rat	4	LC_{50}	530
76	Rat	8	LC_{50}	530
200	Rat	2×7	100% lethal	534
200	Rabbit	18×7	100% lethal	534

14 or 28 mg/kg for 10 days, microscopic examination revealed engorgement of the heart, spleen, liver, and kidneys, swelling of the epithelium of the convoluted ducts of the kidney, and necrotic changes in the heart and liver (539). Daily dosage of 9.7 mg/kg given in the diet to rats over a period of 30 days resulted in reduced appetite, mortality, and microscopic lesions of the tissues. Daily dosage of 4 mg/kg given similarly had no effect (532). Allyl alcohol was administered to rats in drinking water for 90 days in amounts that provided doses ranging from approximately 0.15 to 70 mg/kg/day (530). Effects were noted at higher doses only, and included decreased body weight gain (42 and 70 mg/kg/day), well-localized areas of hepatic necrosis with regeneration (70

mg/kg/day), and increased relative liver and/or kidney weights (29, 42, and 70 mg/kg/day). Doses of 12 mg/kg/day produced no evidence of effects.

A similar study was conducted by Carpanini et al. (549). They maintained groups of 15 male and 15 female rats on drinking water containing either 0, 50, 100, 200, or 800 ppm of allyl alcohol for 15 weeks. No treatment-related effects were found upon hematologic examinations, urinanalyses, clinical chemistry of the serum, or histopathologic examination of the tissues. Water intake was reduced at all dose levels, and food intake and growth were depressed at the higher dose levels. Relative organ weights of the liver, kidneys, and/or spleen were increased in a dose-related way at all levels except the 50-ppm level. The authors judged the 50-ppm level, equivalent to 4.8 to 6.2 mg/kg/day of allyl alcohol, to be a no-adverse-effect level. These results are quite similar to those reported by Dunlap et al. (530) except with respect to the lack of histopathologic changes in the liver.

Rabbits that were given doses of 0.05 or 0.005 mg/kg/day allyl alcohol in drinking water for 8 months showed no morphologic changes in the internal organs. A dose of 2.5 mg/kg/day produced liver hemorrhage and necrosis, dystrophy of the epithelium of the convoluted tubules of the kidney, and diffuse reticular-endothelial hyperplasia in the pulp of the spleen (539). These data suggest that the rabbit is more sensitive to repeated exposure to allyl alcohol than the rat.

Repeated-Dose Inhalation. Dunlap et al. (530) and Torkelson et al. (540) reported on repeated inhalation exposures of animals to vapors containing allyl alcohol. The responses of the animals in these studies are summarized in Table 55.24.

Dunlap et al. (530) reported that retardation of growth appeared at 20 ppm, an increase in the weight of the lungs at 40 ppm, and an increase in the weight of the kidneys at 60 ppm. The vapor of allyl alcohol was toxic to rats when exposed on one occasion, but on repeated exposure the effects seemed to be those of repeated insult rather than those of accumulation in the tissue. For example, one exposure of 8 hr to the concentration of 76 ppm resulted in the death of one-half of the group of rats (LC_{50}), but the repeated exposure (7 hr/day) to the concentration of 60 ppm resulted in the death of only 1 of 10 rats. When rats were exposed to the vapor in the concentration of 100 ppm, their livers were slightly hemorrhagic; the lungs were pale and spotted, but the kidneys were normal. Microscopically, only slight congestion of the lungs and liver was observed. At 20 ppm and below, no gross or microscopic lesions were found (530).

On the other hand, Torkelson et al. (540) reported mild reversible changes in the liver (dilation of the sinusoids, cloudy swelling, and necrosis) and kidneys (necrosis of the epithelium of the convoluted tubules and proliferation of the interstitial tissue) of experimental animals subjected repeatedly to 7 ppm. No

Table 55.24. Results of Repeated Exposures of Animals to the Vapors of Allyl Alcohol

Concentration (ppm)	Animals	No. of 7-Hr Exposures[a]	Outcome	Signs of Intoxication	Ref.
150	Rat	10	100% lethal	Gasping, depression, nasal discharge, eye irritation, 1 case of corneal opacity	530
100	Rat	55	6/10 deaths during first 46 exposures	As above, but less intense	530
60	Rat	60	1/10 deaths	Eye irritation, persistent, but gasping only during first few exposures	530
40	Rat	60	Survived	Irritation during first few exposures	530
20	Rat	60	Survived	None	530
7	Rat Guinea pig Rabbit	28	Survived	None	540
5	Rat	60	Survived	None	530
2	Rat Guinea pig Rabbit Dog	134	Survived	None	540

[a] Animals exposed 7 hr/day, 5 days/week.

effects were found on growth, behavior, mortality, or final body and organ weights in animals exposed to 7 ppm. These investigators found no histologic alterations in the tissues of animals exposed repeatedly to allyl alcohol in the concentration of 2 ppm nor any significant deviations from normal in a variety of parameters measured.

Teratology. No conventional teratologic data on allyl alcohol are available for mammals. McLaughlin et al. (544) studied the effect on chicken embryos in fertile eggs. They found it not to be teratogenic. They did observe that high doses, which caused severe mortality of the embryos, also caused corneal opacity, a finding not observed at lower doses.

Mode of Action. Exposure to vapor concentrations causing toxic effects within hours causes lacrimation, irritation of the respiratory membranes, drowsiness, unsteadiness, and a weak narcotic effect. However, if unconsciousness does develop, death usually follows in a short time (537). Histologic examination of tissues from animals that died or were killed soon after exposure revealed significant changes in the lungs and kidneys and lesser changes in the liver (537). Regardless of the route of administration, signs of toxicity were similar for the various species. These included apathy, excitability, lacrimation, tremors, convulsions, diarrhea, coma, pulmonary and visceral congestion, and varying degrees of liver injury (530, 539).

Kodama and Hine (538) state that the primary cause of death from allyl alcohol is probably cardiovascular failure, which probably follows depression of the respiratory and vasomotor centers and a decrease in effective blood pressure.

Metabolism, Pharmacology, Pharmacokinetics. Allyl alcohol is apparently oxidized readily since within a few minutes after the intravenous injection of rats with the dosage of 30 mg/kg the vena caval blood contained an average concentration of about 24 μg/ml; within 15 min the concentration was about 4 μg/ml; and within an hour the alcohol had almost disappeared from the blood. During constant intravenous infusion the allyl alcohol disappeared at the rate of about 23 mg/hr (538). During the period of 15 to 120 min after the administration of a single oral dose of allyl alcohol (120 mg/kg) to rats, the mean concentration of this alcohol in the portal vein was between 9 and 15 μg/ml (538). Severe inhibition of oxygen uptake was demonstrated in liver slices of rats given an oral administration of 120 mg allyl alcohol/kg, but only slight inhibition occurred at the dosage of 60 mg/kg (538). The kidney slices from the above rats manifested an increase in oxygen uptake but the oxygen uptake of the cerebral cortex was not altered. Both phenoxybenzamine and hexamethonium, when given to rats prior to the oral administration of 120 mg/kg of allyl alcohol, afforded partial protection to the liver against (1) inhibition of oxygen uptake by the liver and (2) increase in water content. Phenoxybenzamine more than doubled the mean time to death for rats given allyl alcohol orally, but it did not alter mortality. These results indicate that the hepatic injury from alcohol is not the primary cause of death.

Blyuger et al. (536) investigated the development of liver necrosis induced by allyl alcohol. Doses of 1 ml/100 g body weight were administered orally to rats. At intervals up to 48 hr following the dose, the activities of several organelle-specific enzymes were determined and parallel electron microscopy studies of the liver were conducted. The results indicated that allyl alcohol induces early changes in all types of organelles, primarily the mitochondria and lysosomes, thereby causing the cells to die and foci of necrosis to develop.

Kim (545) gave rabbits an oral dose of 1 ml/kg and found that it lowered the plasma fibrinogen level and the thromboplastin-generating activity but increased the level of bilirubin and the transaminase and alkaline phosphatase activity.

Berg (546) found that 0.04 ml of allyl alcohol given orally to the rat caused changes in the fatty acid pattern in the liver. The phospholipid fraction was least affected and the neutral fat and cholesterol were changed significantly. The total saturated and unsaturated fatty acids were increased in all lipid fractions although some individual acids were decreased. These changes occurred before morphologically visible changes developed.

Reid (547), in his studies of the mechanism of allyl alcohol-induced hepatic necrosis, gave rats intraperitoneally 0.05 ml/kg of ^{14}C-labeled allyl alcohol 2 hr after pretreating them with either saline; pyrazole, an alcoholic dehydrogenase inhibitor; or phenobarbitone, an inducer of microsomal enzymes. The animals were sacrificed 8 and 24 hr later. In the animals pretreated with saline, periportal necrosis occurred in the liver but not in the lung or kidney. Autoradiographic study showed that radioactivity equivalent to about 120 mmol of allyl alcohol per gram of protein was bound in the periportal area after 8 hr and about half as much after 24 hr. Pyrazole pretreatment reduced by 80 percent the amount bound and prevented necrosis. Phenobarbitone had no effect on the binding or the intensity of the necrosis. The author concludes that a metabolite, such as acrolein, formed at the site of the necrosis, is the necrotizing agent.

Kaye and Young (550) and Kaye (551) found that allyl alcohol and allyl esters of weak acids were metabolized in the rat and found in the urine as 3-hydroxypropyl mercapturic acid, 3-hydroxypropyl-L-cysteine, N-acetyl-S-(3-hydroxypropyl)-L-cysteine, and some acrolein. Esters of strong acids were metabolized not only to the 3-hydroxypropyl derivatives but to some allyl mercapturic acid as well.

Serafini-Cessi (548) incubated rat liver preparations with allyl alcohol with and without the addition of NAD^+ or $NADP^+$. In the presence of NAD^+, a cofactor for liver alcohol dehydrogenase, acrolein was formed and identified. The presence of $NADP^+$ did not affect the formation of acrolein. Conversion was prevented or markedly reduced when inhibitors of alcoholic dehydrogenase were added to the system. This study strongly supports the hypothesis that allyl alcohol is converted to acrolein by alcoholic dehydrogenase in the periportal region of the liver and that this metabolite is the ultimate cause of the necrosis.

According to Racker (541) and Winer (331), allyl alcohol is an excellent substrate for ADH. The initial rate of oxidation is 192 mol/l/min per mole of ADH as compared to 135 for ethanol and 215 for n-butanol (331). Liver dehydrogenases (in vitro) were inhibited only by excessively high concentrations ($2 \times 10^{-1} M$ to $1 M$), which are unlikely to be obtained in vivo (538). A rapid drop in the carotid blood pressure, a definite reduction in respiratory rate and amplitude, and an increase in hemoconcentration (hematocrit), but no increase in the concentration of histamine in the plasma, followed the intravenous injection of 40 mg/kg into a dog (538). Oral doses of 0.005 and 0.05 mg/kg/day for 6 months had no effect on higher nervous activity in rats, whereas doses of

2.5 mg/kg/day weakened the process of stimulation and strengthened the inhibition reaction in rats (542).

Mutagenicity. No data were found.

Effects in Humans

Mode of Action. Unlike the vapors of the saturated lower aliphatic alcohols, the vapor of allyl alcohol does not possess particular narcotic properties but rather is an irritant to the mucous membranes and to the lungs.

Epidemiologic Data. No evidence of liver damage or disturbed kidney function was noted among a group of employees working for 10 years with allyl alcohol (530).

Experience. In contrast to the slight irritation produced on the skin of animals, Dunlap et al. (530) reported that skin irritation is quite common with this compound and that absorption through the skin leads to deep pain, probably due to muscle spasm. Intravenous injection of, or infiltrating the site with, calcium gluconate has been found to relieve the pain. Eye contamination from the liquid may cause a severe chemical burn (530). Torkelson et al. (540) corroborate delayed eye irritation and muscular spasm resulting from the skin contact with allyl alcohol in industry.

Dunlap et al. (530) reported that when the air was moderately contaminated with allyl alcohol, men frequently manifested lacrimation, retrobulbar pain, photophobia, and some blurring of vision. They reported that no permanent or irreversible severe scarring of the cornea or loss of corneal substances had been observed following exposure to these vapors. Smyth (146) reported that the vapor of allyl alcohol temporarily blinded one man through delayed corneal necrosis.

33.5 Hygienic Standard of Permissible Exposure

The ACGIH (348) has recommended a TLV of 2 ppm (5 mg/m^3) with a caution against skin contact. OSHA (108) also has the same standard as do numerous other countries throughout the world (110).

33.6 Odor and Warning Properties

Allyl alcohol in high concentrations has a pungent odor and is a potent lacrimator and irritant; in low concentrations the odor is alcoholic and not unpleasant (386).

Threshold, odor	<0.78 ppm (530)
Eye irritation, slight	6.25 ppm (530)
Eye irritation, moderate	25.0 ppm (530)
Threshold, nasal irritation	<0.78 ppm (530)
Nasal irritation, moderate	12.5 ppm (530)
Threshold, pulmonary discomfort	>25.0 ppm (530)
Threshold, CNS effects	>25.0 ppm (530)

Five out of 10 human volunteers reported a definite odor but no irritation when exposed to 2 ppm of allyl alcohol for 1 to 3 min (540).

34 2-PROPYN-I-OL; Propargyl Alcohol, Ethynol Carbinol, Acetylenyl Carbinol, Propiolic Alcohol, CAS No. 107-19-7

$$HC\equiv CCH_2OH$$

34.1 Source, Uses, and Industrial Exposure

Propargyl alcohol can be prepared by heating β-bromoallyl alcohol with concentrated potassium hydroxide, from formaldehyde and acetylide, from epichlorohydrin and sodium, or from acetylene and formaldehyde (111). It is used as an inhibitor of corrosion of steel by hydrochloric acid, as a stabilizer in certain chlorinated hydrocarbon formulations, as a soil fumigant and as a chemical intermediate in organic chemical syntheses.

34.2 Physical and Chemical Properties

Physical state	Moderately volatile, clear to slightly straw-colored liquid
Molecular formula	C_3H_4O
Molecular weight	56.06
Boiling point	114 to 115°C (111)
Melting point	−48 to − 52°C (111)
Specific gravity	0.9715 (20/4°C) (111)
Vapor pressure	11.6 mm Hg (20°C) (119)
Refractive index	1.43064 (20°C) (111)
Flash point	33°C (111)
Solubility	Miscible with water, benzene, chloroform, 1,2-dichloroethane, ethanol, ether, acetone, dioxane, tetrahydrofuran, pyridine; Immiscible with aliphatic hydrocarbons
1 mg/l ≏	437 ppm at 25°C and 760 mm Hg
1 ppm ≏	2.29 mg/m^3 at 25°C and 760 mm Hg

34.3 Determination in the Atmosphere

The methods of Mueller and Miller (1) and Langvardt and Melcher (2) may be adaptable.

34.4 Physiologic Response

34.4.1 Summary

Propargyl alcohol is highly toxic when ingested, damaging to the eyes, not appreciably irritating to the skin, but rapidly absorbed through the skin in amounts dangerous to life. Atmospheric concentrations readily attainable under room conditions are dangerous to life even from exposures of short duration. The limited repeated exposure data available suggest that the chronicity of the material is low.

34.4.2 Effects in Animals

Single or Short-Term Exposure Data. *Single-Dose Oral.* The oral toxicity of propargyl alcohol is high. The LD_{50} values reported for the rat are 20 to 50 (405) and 70 mg/kg (111, 554); for the mouse 50 mg/kg (552, 555); and for the guinea pig, 60 mg/kg (554).

Eye Contact. When instilled into the conjunctival sac of the rabbit eye the undiluted material causes marked pain, marked irritation, and corneal injury judged to be permanent; a 10 percent aqueous solution causes very slight pain and irritation which clears in a few days; and a 1 percent aqueous solution is not irritating (405).

Skin Contact. The undiluted material causes hyperemia, edema, and some superficial necrosis and is rapidly absorbed through the skin of rabbits in lethal amounts, the LD_{50} being about 16 mg/kg. A 10 percent aqueous solution causes mild hyperemia and edema, and lethal amounts can be absorbed if exposure is extensive or prolonged. A 1 percent aqueous solution is without apparent adverse effect (405). The material is reported not to be a skin sensitizer (554).

Inhalation. A single exposure of 6 min to an essentially saturated atmosphere at room temperature caused the death of two of three rats, and exposures of 12 min or longer were fatal to all exposed animals (405). Stasenkova and Kochetkova report that a 2-hr exposure to 2 mg/l (874 ppm) is lethal to mice (555).

Repeated or Prolonged Exposure Data. *Skin Absorption.* The application of propargyl alcohol to young adult rabbits in daily doses of 1, 3, or 10 mg/kg/day over a 63-day period, and 20 mg/kg/day over a 28-day period, caused no systemic effects as evidenced by studies of weight gain, hematology, blood chemistry, and histopathologic examination of the tissues (553).

Inhalation. Groups of 12 male and 12 female rats were exposed 7 hr/day, 5 days/week on 59 days over a period of 89 days to 80 ppm of propargyl alcohol. Similar groups of rats served as controls. After the first exposure, the animals appeared to have sore eyes and to be lethargic but thereafter no such signs were observed. The primary target organs were the liver and kidney; a few other deviations of questionable significance were noted. Liver weights in the males and both the liver and kidney weights in the females were increased. Histopathologic examination of various organs showed degenerative changes in the livers and kidneys, with the females showing the most injury (405).

34.4.3   Effects in Humans

No reports of adverse effects were found.

34.5   Hygienic Standard of Permissible Exposure

The ACGIH (348) has recommended a TLV of 1 ppm or 2 mg/m^3, as has OSHA (108). This value is generally accepted throughout the world except for the Federal Republic of Germany, which recommends 2 ppm (110).

34.6   Odor and Warning Properties

The material is reported to have a geranium-like odor (554) but no data on odor threshold were found. It is doubtful if the material has warning properties adequate to prevent dangerous exposure.

35   1-HEXYN-3-OL; Hexynol, CAS No. 105-31-7

$$CH_3(CH_2)_2CHOHCCH$$

35.1   Source, Uses, and Industrial Exposure

1-Hexyn-3-ol, like the other acetylenic alcohols, is prepared commercially by the catalytic addition of an acetylene moiety across the carbonyl of an aldehyde or ketone. Hexynol is used as a corrosion inhibitor for steel in the presence of mineral acids. It is used in the acidizing of oil wells, the pickling of mild steel, in metal cleaning operations involving acid systems, and in electroplating (556). Industrial exposure usually occurs from direct contact with the liquid or solutions containing it.

35.2 Physical and Chemical Properties

Physical state	Pale yellow liquid (556)
Molecular formula	$C_6H_{10}O$ (556)
Molecular weight	98.14 (556)
Boiling point	142°F (556)
Specific gravity	0.882 (20/20°C) (556)
Vapor pressure	13 mm Hg (25°C) (556)
Flash point	Approx. 125°F (TOC) (557)
Solubility	Soluble in water to 3.8 percent, miscible with most hydrocarbons, chlorinated solvents, alcohols, ketones and glycols (556)
1 mg/l ≈	249 ppm at 25°C and 760 mm Hg
1 ppm ≈	4.01 mg/m³ at 25°C and 760 mm Hg

35.3 Determination in the Atmosphere

The methods of Mueller and Miller (1) and Langvardt and Melcher (2) may be adaptable.

35.4 Physiologic Response

35.4.1 Summary

Hexynol is one of a few compounds that are more toxic when applied to the skin than when swallowed. It is moderately irritating to both the eyes and skin. Inhalation of concentrations of vapor that can occur at room temperature for a matter of an hour or less can be dangerous to life. A death of a human is believed due to prolonged contact with the skin.

35.4.2 Effects in Animals

Single-Dose Oral. The LD_{50} for rats is reported to be between 88 and 176 mg/kg (557) and between 126 and 256 mg/kg (405). Liver and kidney injury was observed among the survivors at autopsy several days postexposure.

Eye Contact. When instilled into the conjunctival sac of the rabbit eye, the undiluted material caused primary irritation characterized by conjunctival irritation, corneal injury, and iritis; recovery was not complete within a week (405, 557).

Skin Contact. The material is slightly to moderately irritating to the skin of rabbits upon prolonged contact but far more importantly, the material is rapidly absorbed in lethal amounts through the intact skin. The LD_{50} values from 24-hr exposure have been found to be between 15.8 and 126 (405) and <200 mg/kg (557). A dose of 252 mg/kg applied to 81 cm^2 of skin of a rabbit for 1 hr and then washed off caused death (405).

Inhalation. Rats were exposed to an atmosphere essentially saturated with vapor at room temperature. When exposure was for 12 min, all four animals survived, but when exposure was for 30 or 60 min all died. The lung, liver, and kidney were target organs (405). A 1-hr LC_{50} of >20 mg/l is reported (557).

Effects in Humans

One death is presumed to have resulted in 1961 from prolonged skin contact with hexynol. A workman spilled the material on his trousers, did not remove them, and delayed cleansing himself. Death occurred from kidney failure within 24 hr (405).

35.5 Hygienic Standards of Permissible Exposure

None has been established. Care must be taken to prevent skin contact.

35.6 Odor and Warning Properties

The odor is described as musty and slightly piercing (557).

36 2-CHLOROETHANOL, β-Chloroethyl Alcohol, Ethylene Chlorohydrin, Glycol Chlorohydrin, CAS No. 107-07-3

$$ClCH_2CH_2OH$$

36.1 Source, Uses, and Industrial Exposure

β-Chloroethyl alcohol may be produced by passing chlorine and ethylene simultaneously into water (558).

β-Chloroethyl alcohol is used to produce ethylene glycol and ethylene oxide (558). Huntress (559) states that β-chloroethyl alcohol is employed for the separation of butadiene from hydrocarbon mixtures, in dewaxing and removing naphthenes from mineral oil, in the refining of rosin, in the extraction of pine lignin, and as a solvent for cellulose acetate, cellulose ethers, and various resins. Ambrose (560) reports that it is an effective agent in hastening the early sprouting of potatoes and has been proposed for treating seeds for the inhibition of biologic activity. It is used in the manufacture of insecticides (111).

Several deaths have been reported by Koelsch (561), Middleton (562), Cavallazzi (563), Dierker and Brown (564), and Goldblatt and Chiesman (565), from the manufacture or other industrial exposure to β-chloroethyl alcohol by inhalation and/or percutaneous contact. In the fatal case reported by Dierker and Brown (564), the deceased had been using β-chloroethyl alcohol for 2 hr to clean trays upon which rubber strips were stored. Analysis of the air revealed a concentration of about 1 mg/l (304 ppm). Percutaneous absorption was a significant factor in the death reported by Middleton (562). Nine cases of nonfatal intoxication during the manufacture of β-chloroethyl alcohol were reported by Goldblatt and Chiesman (565). The average concentration of β-chloroethyl alcohol in the plant at the time of these nonfatal cases was 18 ppm. Smyth and Carpenter (566) warn that rubber gloves offer little protection, since dangerous amounts of β-chloroethyl alcohol or its aqueous solution rapidly penetrate through rubber.

36.2 Physical and Chemical Properties

Physical state	Colorless liquid
Molecular formula	C_2H_5ClO
Molecular weight	80.52
Boiling point	128 to 130°C (111)
Freezing point	Solidifies at −67°C (111)
Specific gravity	1.2045 (20/20°C) (558)
Vapor pressure	4.9 mm Hg (20°C) (558); 10 mm Hg (30.3°C) (448)
Refractive index	1.4419 (20°C) (111)
Flammability	
Flash point	135°F (closed cup) (558)
Explosive limits	4.9 and 15.9 percent (394)
	When β-chloroethyl alcohol is heated to 184°C, it decomposes into ethylene chloride and acetaldehyde; when heated with water to 100°C, β-chloroethyl alcohol decomposes into glycol and aldehyde (561)
Percent in saturated air	0.644 (20°C)
Density of saturated air (air = 1)	1.011 (20°C)
Solubility	Miscible with water, alcohol (111)

1 mg/l $\approx$ 303.8 ppm at 25°C, 760 mm Hg
1 ppm $\approx$ 3.29 mg/m^3 at 25°C, 760 mm Hg

36.3 Determination in the Atmosphere

NIOSH recommends method No. S103 (567) in which a known volume of air is drawn through charcoal to trap the organic vapors present (recommended sample is 20 liters at a rate of 0.2 l/min). The analyte is desorbed with 5 percent isopropyl alcohol in carbon disulfide. The sample is separated by injection into a gas chromatograph equipped with a flame ionization detector, and the area of the resulting peak is determined and compared with standards.

Recently Mueller and Miller (1) and Langvardt and Melcher (2) published methods of collection and analysis by gas chromatographic procedures that give excellent results and are applicable in situations where a variety of other materials may be present. Their methods would be expected to be adaptable for 2-chloroethanol.

Uhrig (568) describes a method for the determination of large quantities of this compound. An aqueous solution of this chlorinated alcohol is refluxed with potassium hydroxide. In the presence of nitric acid, excess standardized silver nitrate is titrated with ammonium thiocyanate, using ferric ammonium sulfate as indicator. This older method would seem to have limited usefulness except where modern equipment is not available or where unusual conditions exist.

36.4 Physiologic Response

36.4.1 Summary

2-Chloroethanol is moderately toxic upon single-dose oral administration. The undiluted material or high concentrations are extremely irritating and potentially damaging to the eye. Undiluted 2-chloroethanol produces little if any reaction upon contact with rabbit skin and it is not a sensitizer of guinea pig skin. Intradermal applications showed a high degree of toxicity potential. 2-Chloroethanol is highly irritating to mucous membranes. Toxic amounts are readily absorbed through the skin, and skin contact should be prevented. It is highly toxic when inhaled even for a short time at concentrations readily attainable under room conditions. Repeated administration by several routes at doses of one-fifth the LD_{50} or LC_{50} showed a lack of cumulative toxicity, indicating rapid detoxification. Detoxification is believed to occur through the conjugation of the metabolite, chloroacetaldehyde, with glutathione, and if the reserves of glutathione are not depleted by excessive doses, toxicity is limited. Both in vitro and in vivo studies have in some, but not all, cases indicated mutagenic potential for 2-chloroethanol. The limited data available are inadequate to permit an assessment of potential carcinogenicity. The material is pharmacologically active.

36.4.2   Effects in Animals

Single or Short-Term Exposure Data.   *Single-Dose Oral.*   The following oral LD_{50} values have been reported for 2-chloroethanol: for the rat, 95 (31), 72 (569), and 71 mg/kg (570, 571); for the mouse, 81 (570) and 91 mg/kg (571); and for the guinea pig, 110 mg/kg (31).

Short-Term Repeated Oral.   Rats receiving daily doses equivalent to one-fifth of the LD_{50} for 20 days showed no evidence of "marked cumulative" toxicity (571).

Eye Contact.   Lawrence et al. (570) reported that undiluted or 20 to 80 percent solutions of 2-chloroethanol were extremely irritating to the eye of a rabbit, with possible corneal damage resulting. Observations made following application of lower concentrations to the eye were as follows: 2.5 to 10 percent solutions—iritis with significant edema; 1.25 percent—conjunctival irritation and edema; less than 1 percent—no irritation. McDonald et al. (572) reported that 1 percent and lower concentrations administered topically were nontoxic and nonirritating, and intraocular administration of 0.5 percent and lower solutions showed no toxic effects. Observations at higher concentrations (up to 20 percent) included conjunctival redness, chemosis, discharge, flare, iritis, pannus, corneal opacity (transient—topical, and nontransient—intraocular), lens capsule rupture, and opaque lens. Guess (141) reported that no visible effects were observed following application of a 10 percent solution in the eye of a rabbit. The undiluted material caused transient clouding of the cornea, iritis, and excessive swelling and redness of the conjunctiva. The latter effect required 1 week for complete recovery. Proportionately less severe effects were noted on the iris and conjunctiva following instillation of 50 and 20 percent solutions, with no evidence of corneal damage.

Skin and Mucous Membrane Irritation.   Undiluted 2-chloroethanol is not significantly irritating to the skin of rabbits (141, 570). However, 2-chloroethanol produced severe edema and erythema when applied undiluted to the penile tissue of rabbits (141). Similar but less severe effects resulted from a 10 percent solution, and mild transient signs of irritation were observed following contact with 1 percent. These reactions with penile membranes (and with conjunctival membranes noted earlier) demonstrate the highly irritating property of 2-chloroethanol to mucosal tissue.

Single-Dose Injection.   When injected intradermally, the undiluted material or a 10 percent aqueous solution were both severely irritating and injurious to local tissue (141, 570). A 5 percent aqueous solution caused only slight irritation (570), a 2 percent solution barely perceptible irritation (141), and a 1 percent solution practically no irritation (570). The intradermal applications revealed a high degree of toxicity (141).

When the material is given intraperitoneally, the LD_{50} values reported are 56 (569) and 64 mg/kg (570) for rats, 98 mg/kg for mice, 85 mg/kg for rabbits, and 86 mg/kg for guinea pigs (570). When given subcutaneously, a LD_{50} of 72 mg/kg was determined for rats (575). Guess (141) gave intramuscular injections to rabbits and reported necrosis following administration of undiluted and 50 percent solutions of 2-chloroethanol. Solutions of 10 and 2 percent produced slight irritation, a 1 percent solution resulted in questionable blanching, and no response was observed following injection of a 0.2 percent solution.

Skin Absorption. Toxic amounts of 2-chloroethanol can be readily absorbed through intact skin of rabbits without irritation to the skin. The dermal LD_{50} for rabbits is reported to be 67.8 mg/kg (570). According to Ambrose (560), rabbits died following a few repeated applications of the undiluted compound. Semenova et al. (571) reported a dermal LD_{50} of 84 mg/kg for rats, and Ambrose (560) also reported that undiluted and aqueous solutions of 2-chloroethanol were lethal when applied to the skin of rats. Goldblatt (569) reported that application of 0.03 to 0.09 ml of this material to the skin of mice resulted in fatalities. Wahlberg and Boman (592) reported that 0.1 ml of undiluted material applied to the skin of guinea pigs killed all of them within 24 hr, a 0.1 ml of 35 percent aqueous solution killed about half of them, and 0.1 ml of a 10 percent aqueous solution killed 1 of 20. Smyth and Carpenter (566) applied 2-chloroethanol to a poultice that was maintained in contact with the skin of guinea pigs. They reported an LD_{50} of 84 mg/kg when the undiluted chemical was applied, with the majority of the deaths among the guinea pigs occurring within 24 hr. When contact was limited to 2 hr, the LD_{50} was about 361 mg/kg. When 2-chloroethanol was applied as a 10 percent solution in water, the LD_{50}, in terms of the chemical, was about 1.373 g/kg (566).

Skin Sensitization. Lawrence et al. (573) conducted a guinea pig maximization test on 2-chloroethanol and found no evidence of skin sensitization.

Inhalation. Table 55.25 gives the experimental data available in relation to single or short-term repeated exposures of animals to air containing 2-chloroethanol.

Ambrose (560) exposed rats to air that was bubbled through aqueous solutions of β-chloroethyl alcohol maintained at 40°C. Rats exposed for 1 hr to air passed through 12.5, 25, or 50 percent aqueous β-chloroethyl alcohol died 1 or 2 hr after the exposure. Rats that were exposed, during three periods each of 1-hr duration over a total period of less than 2 days, to air bubbled through a 6.25 percent aqueous solution of this alcohol died followed the third exposure. A total of four exposure periods to air bubbled through a 3.13 percent solution,

Table 55.25. Result of Single or Short-Term Exposures of Animals to the Vapors of 2-Chloroethanol

Animal	Dose (mg/l)	(ppm)	Duration of Exposure (hr)	Outcome	Ref.
Guinea pig	18.0	5468	0.25	Death	574
Mouse	7.0	2430	2.0	Death	569
Guinea pig	5.0	1544	0.9	Survived	569
Mouse	4.5	1367	0.5	Death	569
Rat	4.0	1215	0.5	Death	569
Mouse	4.0	1215	0.25	LC_{67}	569
Guinea pig	3.6	1094	1.0	Death	574
Rat	3.4	1033	$0.25 \times 3, 6,$ or 11	Death	569
Guinea pig	3.0	911	1.8	Death	569
Mouse	3.0	911	1.0	Death	569
Guinea pig	3.0	911	0.5	Survived	569
Rat	3.0	911	0.25	Survived	569
Cat	2.5	760	4×3	Death	574
Mouse	1.2	365	2	LC_{17}	564
Mouse	1.0	304	2	Survived	569
Mouse	0.38	115	?	LC_{50}	571
Rat	0.29	88	?	LC_{50}	571
Rat	0.11	33	4	$\cong LC_{50}$	535

each of 2 hr duration on two consecutive days, resulted in morbidity, depression, paralysis, and mortality among rats.

Lawrence (570) reported an LT_{50} of 13.3 min for mice exposed to air that had been bubbled through liquid 2-chloroethanol at a rate of 1 l/min.

Upon microscopic examination of the liver, kidneys, and lungs of a mouse that died following inhalation of 2-chloroethanol, Dierker and Brown (564) reported edema, capillary engorgement, and interstitial hemorrhages of all these organs. Goldblatt (569) believed that the kidneys were the earliest focus of stress, since following exposure to the vapor there were large numbers of hemorrhages, mainly at junctional areas between the cortex and medulla, and there was a complete disintegration of the cells of the convoluted tubules. Congestion, formation of pigment, and fatty degeneration were observed in the liver. Pulmonary hemorrhages and collapse were noted.

Semenova et al. (571) reported no "marked cumulative" effects in rats receiving 20 daily inhalation exposures equivalent to one-fifth the LC_{50} or 60 mg/m^3 (18 ppm).

Mode of Action. Inhalation of 2-chloroethanol by animals resulted in nasal irritation, incoordination, convulsions, prostration, and respiratory failure.

True narcosis was absent (569). Death following intravenous injections to rabbits was believed to be due to failure of the respiratory center of the medulla; direct cardiotoxic or neuromuscular effect was also thought possible (573).

Repeated or Prolonged Exposure Data. *Oral.* Rats maintained on diets containing 2-chloroethanol in concentrations ranging from 0.01 to 0.08 percent for at least 220 days grew normally, but those fed on diets containing 0.12 to 0.24 percent had retarded growth. The tissues of the rats at all levels were without histopathologic alteration (560).

Oser et al. (576) conducted 90-day oral toxicity studies on 2-chloroethanol in rats, monkeys, and dogs. Rats showed no effect following gavage doses of 45 mg/kg/day. Increased mortality and decreased body weights were observed at doses of 67.5 mg/kg/day. Monkeys failed to gain weight at any of the doses administered by syringe (30, 45, or 62.5 mg/kg/day), but showed no other effects. Ingestion of 2-chloroethanol in the diet by dogs produced severe emesis; the highest dose that could be tolerated was approximately 18 to 20 mg/kg/day. The treated dogs failed to gain weight, and some variations in hemoglobin and packed cell volume were observed. Gross and histopathologic examination did not reveal any evidence of toxicity in any species. The tolerance that these animals showed for relatively high oral doses of 2-chloroethanol given repeatedly (greater than one-half of the LD_{50} in the case of rats) indicates lack of cumulative toxicity of this material. Johnson (577) reported a similar absence of chronic toxic effects in rats that ingested 2-chloroethanol in drinking water at doses of 0, 4, 8, or 16 mg/kg/day for up to 2 years.

Repeated Inhalation. Semenova et al. (571) exposed rats for 4 hr daily for 4 months to 10 or 1 mg/m^3 (3.1 or 0.31 ppm) 2-chloroethanol. Decreased body weights and alterations in various biochemical measurements were observed at the higher concentration and remained unchanged 2 weeks after exposure was terminated; histopathologic examination revealed necrosis and decreased glycogen in the hepatic parenchyma of the liver and irritation of pulmonary parenchyma. Rats exposed to 1 mg/m^3 showed a few biochemical alterations which returned to normal 2 weeks after exposure ceased; histopathology was negative for these rats except for signs of activation of Kupffer's cells. Both concentrations caused changes in nervous system function as evidenced by means of chronaximetry and the summation-threshold index.

Repeated Injection. Lawrence et al. conducted studies on the effect of repeated intraperitoneal injections of 2-chloroethanol (573). Rats were injected three times weekly for 3 months with doses of 32, 12.8, or 6.4 mg/kg with no effect except increased mortality at 32 mg/kg. Rats that were injected for 30 days with a dose of 12.8 mg/kg/day showed increased mortality and marked decrease in body weight, but no histopathologic changes in the major organs examined. No effects were observed at the 6.4 mg/kg/day dose level. The authors concluded that 2-chloroethanol is rapidly detoxified.

Carcinogenicity. In the chronic study of Johnson (577), using small numbers of animals (six/dose), no evidence of carcinogenicity was observed in rats ingesting doses of 4, 8, or 16 mg/kg/day 2-chloroethanol in drinking water for up to 2 years. Mason et al. (575) gave subcutaneous injections twice weekly to rats at doses of 0.3, 1.0, 3, or 10 mg/kg for 1 year, followed by a 6-month observation period. No evidence of carcinogenicity or toxicity was observed.

Mutagenicity—In Vitro. The mutagenicity of 2-chloroethanol has been investigated in a number of in vitro studies. The material is reported to induce mutations of the base-substitution type in *S. typhimurium* (581–585); inhibit growth of DNA polymerase-deficient bacteria (581); and increase the number of histamine revertants with or without postmitocondrial mouse liver fractions (583, 584). McCann et al. (586) reported weak mutagenic activity of 2-chloroethanol when tested against *S. typhimurium* TA100 without microsomal activation, stronger activity with activation, but little, if any activity when tested against TA1535. Pfeiffer and Dunkelberg (592a) also found 2-chloroethanol to be weakly mutagenic to both *S. typhimurium* strains TA1535 and TA100. Contrary to the findings of McCann et al. (586), they found strain TA1535 to be more sensitive than TA100. When tested against strains TA1537 and TA98, no activity was found (592a). Stolzenberg and Hine (590) also showed weak activity both with and without activation when tested against *S. typhimurium* TA100 at a dose of 100μmol/plate but no activity at a dose of 10 or 1 μmol/plate. 2-Chloroethanol showed no mutagenic activity when tested against two strains of yeast (587).

Mutagenicity—In Vivo. 2-Chloroethanol was evaluated for mutagenic potential in rats that received inhalation exposures of 1 or 10 mg/m³, 4 hr daily for up to 4 months (578). Bone marrow was examined for chromosomal aberrations after 1, 15, 60, and 120 days of exposure, and after 15 days recovery following 120 days of exposure. Both concentrations significantly increased the incidence of cells with structural chromosome breakages and retarded the mitotic process. The number of abnormal cells increased with time for up to 2 months, but a decline in the number of abnormal cells from previous levels was observed upon examination after 4 months of exposure. The authors proposed the appearance of an antimutagenic defense mechanism. The incidence of abnormal cells was further decreased 2 weeks after withdrawal from exposure. Epstein et al. (589) conducted a dominant lethal assay on mice with what was reported to be δ-chloroethanol, and found it to be negative. (Since δ-chloroethanol is unknown, it seems likely the material tested was 2-chloroethanol.)

Metabolism. Johnson (579, 580) has conducted studies on the metabolism of 2-chloroethanol and found that it is converted to *S*-carboxymethylglutathione by rat liver in vivo and in vitro (580). He concluded that the apparent innocuity of repeated oral doses in his chronic study in rats (577) was due to absorption of much of the dose through the portal system and detoxification by this enzymic conversion to *S*-carboxymethylglutathione in the liver. An oral dose of 2-chloroethanol (0.68 mmol/kg) to rats resulted in reduction of liver glutathione levels to approximately 17 percent of control values and kidney glutathione levels to 60 percent of control values (579). Ethanol administered orally along with 2-chloroethanol reduced the acute toxicity of the latter and lessened its early effect on rat liver glutathione (580). The inhibition of these effects of 2-chloroethanol by ethanol is believed by the author to be due to ethanol competing for an enzyme dehydrogenating 2-chloroethanol. Johnson (580) also reported that 2-chloroethanol is metabolized to chloracetaldehyde in vivo and suggested that the latter may be the toxic agent.

Pharmacology. Goldblatt (569) reported that intravenous injection (but not inhalation) of 2-chloroethanol induced a decrease in blood pressure and inhibition of respiration in cats, but that vagal action and cardiovascular reflexes were not altered. This compound is an inhibitor of the perfused frog's heart. 2-Chloroethanol inhibited both the tone and rhythm of the smooth muscle. In contact with the nerve it induced complete nerve block, which is reversible (569).

More recently, Lawrence et al. (573) reported similar pharmacologic activity of 2-chloroethanol in rabbits. Intravenous administration of doses up to 1577 mg/kg induced a prompt drop in diastolic pressure, probably caused by vasodilation since systolic pressure, heart rate, and the electrocardiogram pattern were unchanged. Later, the systolic pressure dropped in some animals. Doses of 606.5 mg/kg and more caused a dose-related drop in both diastolic and systolic pressures as well as in respiration rate and depth, until death. Lawrence et al. (573) also found that the sleeping time of mice given 50 mg/kg of phenobarbital was prolonged by the intraperitoneal administration of 6.4 and 12.8 mg/kg of 2-chloroethanol. The intravenous injection of 2-chloroethanol impaired and blocked nerve-muscular function.

36.4.3 Effects in Humans

In fatalities among humans exposed to 2-chloroethanol the following signs of intoxication were usually reported: nausea, vomiting, incoordination of the legs, vertigo, weakness, weak irregular pulse, and respiratory failure (562, 564, 565). Vomiting of bile, profuse perspiration, headache, visual disturbance, decreased blood pressure, hematuria, and spastic contracture of the hands were reported by Goldblatt and Chiesman (565). These authors stated that fatty degeneration of the liver and edema, collapse, and extravasation of the lungs were insufficient to account for death. Unfortunately, kidney sections were unavailable. Microscopic examination of the tissues by Dierker and Brown (564) revealed edema, swelling, and vacuolation of the hepatic cells. Necrosis and engorgement of the hepatic blood vessels with interstitial hemorrhages were also observed. The tubules of kidneys showed cloudy swelling; there was intense engorgement of the tubules and some parenchymal cells in the tubules.

In nonfatal cases, Goldblatt and Chiesman (565) reported nausea, epigastric pain, repeated vomiting, occasionally with bile, signs of circulatory shock, headaches, confusion, vertigo, incoordination, slight albuminuria, polyuria, cough, and erythema of the skin.

A case of acute poisoning is reported in which a 24-year-old subject suffered central nervous system, respiratory, and cardiovascular disturbances, resulting in hypoxia and dysfunction of organs and collapse (591). Dystrophic and necrobiotic changes in the heart, and liver and kidney damage were reported. Death occurred in less than 12 hr following ingestion of approximately 2 ml of 2-chloroethanol by a 2-year-old child (588). Vomiting, cyanosis, and respiratory difficulty occurred rapidly after ingestion.

Lawrence et al. (570) estimate that by extrapolation of the dermal LD_{50} for rabbits to man, slightly more than a teaspoonful of 2-chloroethanol could be lethal to a 70-kg man by skin contact, if the material were not washed off immediately.

36.5 Hygienic Standard of Permissible Exposure

The ACGIH (348) has set 1 ppm or 3 mg/m^3 as a safe level for industrial exposure, with the designation that this is a ceiling value, not a time-weighted average, and also a caution against skin contact. The OSHA standard is 5 ppm or 16 mg/m^3 (108). The standard in the Soviet Union is 0.5 mg/m^3 (0.15 ppm), in Switzerland 1 ppm, and in numerous other countries, 5 ppm or 16 mg/m^3 (110).

36.6 Odor and Warning Properties

No data were found regarding these properties.

37 MISCELLANEOUS ALCOHOLS

Limited data are available on a number of saturated alcohols of lesser industrial importance. These data are given in Table 55.26. Data with respect to the unsaturated alcohols are given in Table 55.27. All these alcohols have uses as

Table 55.26. Physical, Chemical, and Toxicologic Data on Miscellaneous Saturated Alcohols

	1-Chloro-2-Propanol	2-Chloro-1-Proponal	2,3-Dichloro-1-Propanol	2,2-Dimethyl-1-Butanol	2-Methyl-1-Pentanol	4-Methyl-1-Pentanol	2-Heptanol
Common name	Propylene chlorohydrin	Propylene chlorohydrin	Dichloropropanol	Hexyl alcohol	Hexyl alcohol	Hexyl alcohol	Heptyl alcohol
CAS No.	127-00-4	78-89-7	616-23-9	1185-33-7	105-30-6	626-89-1	543-49-7
Structural formula	$CH_2ClCHOHCH_3$	$CH_3CHClCH_2OH$	$CH_2ClCHClCH_2OH$	$C_2H_5C(CH_3)_2CH_2OH$	$CH_3(CH_2)_2CH(CH_3)CH_2OH$	$CH_3CH(CH_3)(CH_2)_2CH_2OH$	$CH_3(CH_2)_4CHOHCH_3$
Molecular formula	C_3H_7ClO	C_3H_7ClO	$C_3H_6Cl_2O$	$C_6H_{14}O$	$C_6H_{14}O$	$C_6H_{14}O$	$C_7H_{16}O$
Molecular weight	94.5	94.5	129	102.2	102.2	102.2	116.2
1 mg/l $\Leftrightarrow$ x ppm at 25°C and 760 mm Hg; x =	258	258	189.5	236.9	236.9	236.9	210.4
1 ppm $\Leftrightarrow$ y mg/m^3 at 25°C and 760 mm Hg; y =	3.90	3.90	5.28	4.22	4.22	4.22	4.75
Toxicity data							
Single-dose oral, rats, LD$_{50}$	0.1–0.3 g/kg	0.22 ml/kg	0.09 g/kg	2.33 ml/kg	1.41 g/kg	6.50 ml/kg	2.58 g/kg
Eye irritant[a]	Yes, marked	8	5	7	8	5	9
Skin irritant[b]	No	1	2	2	1	2	3
Skin absorption, rabbit, LD$_{50}$ (ml/kg)	~0.5 g/kg	0.48	0.2	1.77	3.56	3.97	1.78
Inhalation, rat, sat'd vapor. Max. exposure with no deaths (hr)	[c]	0.25	—	—	8	8	8
Reference	405	397	263	481	220	481	220

Table 55.26. (Continued)

	3-Heptanol	2,3-Dimethyl-1-pentanol	Methyl Heptanol (Mixed Primary Isomers)	2-Ethyl-4-methyl-1-pentanol	2,2,4-Trimethyl-1-pentanol	Octanol—Mixed Primary Isomers, 75% 2-Ethyl-hexanol, 25% 2-Ethyl-4-methylpentanol
Common name	Heptyl alcohol	Heptyl alcohol	Octyl alcohol	Octyl alcohol	Octyl alcohol	
CAS No.	3913-02-8	10143-23-4		106-67-2	123-44-4	—
Structural formula	$CH_3(CH_2)_3CHOHC_2H_5$	$C_2H_5CH(CH_3)CH(CH_3)$-CH_2OH	$C_7H_{15}CH_2OH$	$C_2H_5CH(CH_3)CH_2CH$-$(C_2H_5)OH$	$CH_3CH(CH_3)CH_2C$-$(CH_3)_2CH_2OH$	
Molecular formula	$C_7H_{16}O$	$C_7H_{16}O$	$C_8H_{18}O$	$C_8H_{18}O$	$C_8H_{18}O$	$C_8H_{18}O$
Molecular weight	116.2	116.2	130.2	130.2	130.2	130.2
1 mg/l ⇌ x ppm at 25°C and 760 mm Hg; x =	210.4	210.4	187.8	187.8	187.8	187.8
1 ppm ⇌ y mg/m³ at 25°C and 760 mm Hg; y =	4.75	4.75	5.32	5.32	5.32	5.32
Toxicity data						
Single-dose oral, rats, LD50	1.87 g/kg	2.38 ml/kg	5.16 ml/kg	4.29 ml/kg	3.73 ml/kg	5.19 ml/kg
Eye irritant[a]	5	9	5	5	7	5
Skin irritant[b]	3	2	2	3	2	3
Skin absorption, rabbit, LD50 (ml/kg)	4.36	2.5	2.52	>5.0	6.30	5.66
Inhalation, rat, sat'd vapor. max. exposure with no deaths (hr)	4	8	4	8	8	8
Reference	325	481	481	481	481	435

	Cyclohexyl-1-ethanol	2,6-Dimethyl-4-heptanol	Diisobutyl Carbinol	2-Phenyl-2-propanol	3,3,5-Trimethyl-1-cyclohexanol	2-Propyl-1-heptanol
Common name	Cyclohexyl-1-ethanol	Nonyl alcohol	Nonyl alcohol	Phenylpropanol	Trimethylcyclohexanol	Decanol
CAS No.	4442-79-9	108-82-7	108-82-7	617-94-7	116-02-9	100042-59-8
Structural formula	⬡(H)$(CH_2)_2OH$	$CH_3CH(CH_3)(CH_2)_3CH$-$(CH_3)CH_2OH$	$(CH_3CH(CH_3)(CH_2)_2)_2$-$CHOH$	$C_6H_5C(CH_3)_2OH$	$(CH_3)_3$—⬡(H)—OH	$CH_3(CH_2)_4CH(C_3H_7)CH_2$ OH
Molecular formula	$C_8H_{16}O$	$C_9H_{20}O$	$C_9H_{20}O$	$C_9H_{12}O$	$C_9H_{18}O$	$C_{10}H_{22}O$
Molecular weight	128.2	144.2	144.2	136.2	142.2	158.3
1 mg/l ⇌ x ppm at 25°C and 760 mm Hg; x =	190.7	169.6	169.6	179.5	172.0	154.4
1 ppm ⇌ y mg/m³ at 25°C and 760 mm Hg; y =	5.24	5.90	5.90	5.57	5.82	6.47

Toxicity data (continued)

Single-dose oral, rats, LD_{50}	0.94 g/kg	3.16 g/kg	3.56 g/kg	1.07 ml/kg	3.25 g/kg	6.73 ml/kg
Eye irritant[a]	—	2	2	5	9	2
Skin irritant[b]	No	2	1	3	5	2
Skin absorption, rabbit, LD_{50} (ml/kg)	1.22 g/kg	>10.0	5.66	1.0	2.8	>10.0
Inhalation, rat, sat'd vapor. Max. exposure with no deaths (hr)	—	8	8	8	—	8
Reference	593	441	441	435	441	481

	2-Butyl-1-octanol	2,6,8-Trimethyl-4-nonanol	Tridecanol (Mixed Primary Isomers)	Tetradecanol (Mixed Primary Isomers)	2,8-Dimethyl-6-isobutyl-4-nonanol	Heptadecanol (Mixed Primary Isomers)
Common names	Dodecanol	Dodecanol	Tridecanol	Tetradecanol		Heptadecanol
CAS No.	3913-02-8	123-17-1	26248-42-0	27196-00-5		52788-44-5
Structural formula	$CH_3(CH_2)_5CH(C_4H_9)CH_2OH$	$(CH_3)_2CHCH_2CHOHCH_2\text{-}CH(CH_3)CH_2CH(CH_3)_2$	$C_{12}H_{25}CH_2OH$	$C_{13}H_{27}CH_2OH$	$(CH_3)_2CHCH_2CHOHCH_2\text{-}CH(C_4H_9)CH_2CH(CH_3)_2$	$C_{16}H_{33}CH_2OH$
Molecular formula	$C_{12}H_{26}O$	$C_{12}H_{26}O$	$C_{13}H_{28}O$	$C_{14}H_{30}O$	$C_{15}H_{32}O$	$C_{17}H_{36}O$
Molecular weight	186.3	186.3	200.3	214.3	228.4	256.5
1 mg/l ⇌ x ppm at 25°C and 760 mm Hg; x =	131.2	131.2	122.1	114.1	107	95.3
1 ppm ⇌ y mg/m^3 at 25°C and 760 mm Hg; y =	7.62	7.62	8.19	8.76	9.35	10.49
Toxicity data						
Single-dose oral, rats, LD_{50}	12.9 g/kg	17.0 g/kg	17.2 ml/kg	32.5 ml/kg	16.3 g/kg	51.6 ml/kg
Eye irritant[a]	1	1	2	1	1	1
Skin irritant[b]	3	3	4	2	3	3
Skin absorption, rabbit, LD_{50} (ml/kg)	>20	11.2	7.07	7.13	—	16.8
Inhalation, rat, sat'd vapor. Max. exposure with no deaths (hr)	8	8	8	8	8	8
Reference	325	220	481	397	220	481

[a] Scale of 1–10; 1 = no observed reaction; 10 = most severe.

[b] Grade 1—no reaction from undiluted material; 2—trace of capillary injection; 3—strong capillary injection; 4—slight erythema; 5—strong erythema, edema, or slight necrosis; 6 and 7—necrosis from undiluted material.

[c] A few repeated 6-hr exposures to 1000 ppm 1-chloro-2-propanol caused lethargy, some deaths, and lung and liver injury in rats; 15 exposures to 250 ppm caused lethargy, irregular weight gains, and congestion and perivascular edema in the lungs; 15 exposures to 100 ppm caused some similar alterations in the lung; and fourteen 6-hr exposures to 30 ppm caused no apparent adverse effects (594).

Table 55.27. Physical, Chemical, and Toxicologic Data on Miscellaneous Alcohols

	3-Methyl-1-Buten-3-ol, or 2-Methyl-3-Buten-2-ol	3-Butyn-2-ol	3-Methyl-1-Butyn-3-ol, or 2-Methyl-3-Butyn-2-ol	3-Methyl-1-Pentyn-3-OL	4-Ethyl-1-Octyn-3-Ol	1-Ethynyl-1-Cyclohexanol
Common name	Methyl butenol	Butynol	Methyl butynol	Methyl pentynol	Ethyl octynol	Ethynyl cyclohexanol
CAS No.	115-18-4	2028-63-9	115-9-5	77-75-8	5877-42-9	78-27-3
Structural formula	$(CH_3)_2COHCHCH_2$	$CH_3CHOHCCH$	$(CH_3)_2COHCCH$	$C_2H_5C(CH_3)OHCCH$	$CH_3(CH_2)_3CH(C_2H_5)-CHOHCCH$	$C_6H_{10}OHCCH$
Molecular formula	$C_5H_{10}O$	C_4H_6O	C_5H_8O	$C_6H_{10}O$	$C_{10}H_{18}O$	$C_8H_{12}O$
Molecular weight	86.1	70.1	84.1	98.2	154.2	124.2
Boiling point (°C)	97	107–109	103.6	121.4	197.2	180
Freezing point (°C)	−30.5		3	−30.6 (602)	—	30–31
Specific gravity (20/20°C)	0.825		0.867	0.872	0.873 (598)	0.97 (599)
Vapor Pressure (mm Hg)	51 at 25°C		14 at 20°C	6.5 at 20°C	30 at 25°C (598)	15 at 25°C (599)
Flash Point (°F)	64 (T.O.C.)		77 (T.O.C.)	80 (T.C.C.)	76 (T.C.C.)	150 (T.C.C.) (599)
Solubility in water (wt % at 20°C)	13.1		Miscible	9.9 (602)	Insoluble	2.4
1 mg/l ≏ x ppm at 25°C and 760 mm Hg; x =	284	349	291	249	158.6	196.7
1 ppm ≏ y mg/m³ at 25°C and 760 mm Hg; y =	3.52	2.86	3.44	4.02	6.30	5.08
Toxicity data						
Single-dose oral, rats, LD50 (g/kg)	1.4	0.034	1.9	0.89 Liver & kidney injury (405)	2.1	0.60
Eye irritant	Yes	Severely damaging	Yes. Corneal damage (405) 10% in H2O-Not irritating (405)	Yes Severe (405)	—	Yes—corrosive
Skin irritant	No	—	No	No Slight when confined (405)	—	—
Skin sensitizer	Possibly	—	Possibly	No	Possibly	—
Skin absorption, rabbit, LD50 (g/kg)	>1.0	between 0.032 and 0.063	>1.0 >5.0 (405)	Possibly >1.0	0.2–1.0	1.0
Inhalation, rat, 1-hr LC50 (mg/l)	>20	Sat'd vapor-6 min. exp. 100% lethal	>20^a	>20^b	>20	No deaths—sat'd vapor for 8 hr
Reference unless indicated otherwise	595	405	596	601	597	600

[a] A single 4-hr exposure of rats to 3000 ppm of methyl butynol caused death whereas a 7-hr exposure of rats to 2000 ppm did not; both exposures caused liver and kidney injury. A single 7-hr exposure to 1000 ppm did not cause grossly apparent injury. Rats that received 81 7-hr exposures in 115 days to 76 ppm exhibited no adverse effects (405).

[b] A single 7-hr or a single 4-hr exposure of rats to 4600 ppm of methyl pentynol caused anesthesia, considerable injury to the lungs, liver, and kidney, and deaths. A 2-hr exposure did not cause deaths but did cause kidney injury and weight loss. A few deaths occurred after 7- and 4-hr exposures to 2000 ppm whereas all survived a 7-hr exposure to 1000 ppm but suffered kidney injury. Groups of rats tolerated 67 7-hr exposures in 98 days to 100 ppm without adverse effects (405).

solvents and chemical intermediates, but the acetylenic alcohols also have rather wide application as corrosion inhibitors for steel in acid systems such as oil well acidizing, mild steel pickling, acid cleaning systems, and in some plating applications. The acetylenic alcohols are prepared commercially by the catalytic addition of an acetylenic moiety across the carbonyl of an aldehyde or ketone. Specific methods for determination of these alcohols in the atmosphere have not been described, but it is suggested that the methods of Mueller and Miller (1) and Langvardt and Melcher (2) may be adaptable.

REFERENCES

1. F. X. Mueller and J. A. Miller, *Am. Ind. Hyg. Assoc. J.*, **40**, 380 (1979).

2. P. W. Langvardt and R. G. Melcher, *Am. Ind. Hyg. Assoc. J.*, **40**, 1006 (1979).

3. I. Mellan, *Industrial Solvents*, Reinhold, New York, p. 202, (1939).

4. C. A. Wood, *J. Am. Med. Assoc.*, **59**, 1962 (1912).

5. E. Browning, *Med. Res. Council Ind. Health Res. Board, Rep. No. 80*, H. M. Stationery Office, London, (1937).

6. S. L. Zeigler, *J. Am. Med. Assoc.*, **77**, 1160 (1921).

7. *N.Y. State Dept. Labor Bull. No. 86* (1917).

8. A. B. Hale, *J. Am. Med. Assoc.*, **37**, 1447, 1450 (1901).

9. A. Hamilton, *Industrial Poisons in the United States*, Macmillan, New York, (1925), p. 427.

10. *Natl. Res. Council Can. Bull. No. 15*, 20 (1930).

11. J. M. Robinson, *J. Am. Med. Assoc.*, **70**, 148 (1918).

12. C. Baskerville, "Wood Alcohol: A Report on Chemistry, Technology, and Pharmacology of and the Legislation Pertaining to Methyl Alcohol," Appendix VI, Vol. 2, Rept., N.Y. State Factory Investigating Commission, 1913, p. 917.

13. F. Buller and C. A. Wood, *J. Am. Med. Assoc.*, **43**, 1117 (1904).

14. E. R. Hayhurst, *Occupational Survey of Ohio*, 1915.

15. L. Greenberg, M. R. Mayers, L. J. Goldwater, and W. J. Burke, *J. Ind. Hyg. Toxicol.*, **20**, 148 (1938).

16. A. Loewy, *Vierteljahrsschr. Gerichtl. Med.*, **48**, Suppl., 93 (1914).

17. J. H. Sterner, personal communication, as reported by J. F. Treon in F. A. Patty, Ed., *Industrial Hygiene and Toxicology*, Vol. 2, Wiley, New York, 1962.

18. A. E. Goss and G. H. Vance, *Ind. Hyg. Newsl.*, **8**(9), 15 (1948).

19. R. G. McAllister, *Am. Ind. Hyg. Assoc. Q.*, **15**, 26 (1954).

20. M. Windholz, Ed., *The Merck Index*, 9th ed., Merck and Co., Inc., Rahway, N.J., (1976).

21. U.S. Dept. of Health, Education and Welfare, *NIOSH Manual of Analytical Methods*, Vol. 1, 2nd ed., 1977, P and CAM 247, and Vol. 2, 2nd ed., 1977, S59.

22. E. Deniges, *Compt. Rend.*, **150**, 832 (1910).

23. R. M. Chapin, *Ind. Eng. Chem.*, **13**, 543 (1921).

24. E. Elvove, *Ind. Eng. Chem.*, **9**, 295 (1917).

25. L. O. Wright, *Ind. Eng. Chem.*, **19**, 750 (1927).

26. C. M. Jephcott, *Analyst*, **60**, 588 (1935).

27. C. F. Ackerbauer and R. J. Lebowich, *J. Lab. Clin. Med.*, **28**, 372 (1942).

28. E. T. Kimura, D. M. Ebert, and P. W. Dodge, *Toxicol. Appl. Pharmacol.*, **19**(4), 699 (1971).

29. H. Welch and G. G. Slocum, *J. Lab. Chem. Med.*, **28**, 1440 (1943), from *Water Quality Criteria*, 1963.

30. W. B. Deichmann, *J. Ind. Hyg. Toxicol.*, **30**, 373 (1948).

31. H. F. Smyth, Jr., J. Seaton, and L. Fisher, *J. Ind. Hyg. Toxicol.*, **23**, 259 (1941).

32. A. M. Potts, *Am. J. Ophthalmol*, **39**, 86 (1955). Cited by M. Koivusalo, "Methanol," *Int. Encycl. Pharmacol. Ther.*, **20**, 465 (1970).

33. A. P. Gilger and A. M. Potts, *Am. J. Ophthalmol.* **39**, 63 (1955). Cited by M. Koivusalo, "Methanol," *Int. Encycl. Pharmacol. Ther.*, **20**, 465 (1970).

34. J. R. Cooper and P. Felig, *Toxicol. Appl. Pharmacol.*, **3**, 202 (1961).

35. I. Takeda, *Nichidai Igaku Zasshi*, **31**(6), 518 (1972).

36. M. Saito, *Nichidai Igaku Zasshi*, **34**(8–9), 569 (1975).

37. J. C. Munch, *Ind. Med. Surg.*, **41**(4), 31 (1972).

38. E. I. duPont De Nemours Inc., Haskell Laboratory, unpublished data.

39. C. P. Carpenter and H. F. Smyth, *Am. J. Ophthalmol.*, **29**, 1363 (1946); from NIOSH, *Occupational Exposure to Methyl Alcohol*, March 1976.

40. Carnegie-Mellon Institute of Research, unpublished data.

41. W. P. Yant, H. H. Schrenk, and R. R. Sayers, *Ind. Eng. Chem.*, **23**, 551 (1931).

42. H. H. Schrenk, personal communication.

43. R. Witte, Dissertation, Wurzburg, 1931.

44. H. Weese, *Arch. Exp. Pathol. Pharmakol.*, **135**, 118 (1928).

45. L. M. Mashbitz, R. M. Sklianskaya, and F. M. Urieva, *J. Ind. Hyg. Toxicol.*, **18**, 117 (1936).

46. C. Bachem, *Arch. Exp. Pathol. Pharmakol.*, **122**, 69 (1927).

47. A. Loewy and R. von der Heide, *Biochem. Z.*, **65**, 230 (1914).

48. R. Muller, (cited in Loewy and von der Heide), *Z. Angew. Chem.*, **23**, 351 (1910).

49. K. B. Lehmann and F. Flury, *Toxikologie und Hygiene der technischen Losungsmittel*, Springer, Berlin, 1938, p. 149.

50. H. H. Tyson and M. J. Schoenberg, *J. Am. Med. Assoc.*, **63**, 915 (1914).

51. E. Scott, M. K. Helz, and C. P. McCord, *Am. J. Clin. Pathol.*, **3**, 311 (1933).

52. A. A. Eisenberg, *Am. J. Public Health*, **7**, 765 (1917).

53. H. W. Gerarde and D. B. Ahlstrom, *Arch. Environ. Health*, **13**(4), 457 (1966).

54. R. Zagradnik, M. Chvapil, J. Vostal, and J. Teisinger, *Farmakol. Toksikol.*, **25**, 618 (1962).

55. H. P. Kloecking, *Fortschr. Wasserchem. Granzgeb.*, **14**, 189 (1972).

56. J. H. Garcia and J. P. VanZandt, *Proc., Electron Microsc. Soc. Am.*, **27**, 360 (1969).

57. B. K. Skirko, A. M. Ivanitskii, R. A. Pilenitsyna, N. N. Pyatnitskii, I. S. Zilova, V. M. Zhminchenko, N. V. Kleimenova, and N. P. Sugonyaeva, *Vopr. Pitan.*, **5**, 70 (1976).

58. R. R. Sayers, W. P. Yant, H. H. Schrenk, J. Chornyak, S. J. Pearce, F. A. Patty, and J. G. Linn, *J. Ind. Hyg. Toxicol.*, **26**, 255 (1944).

59. R. R. Sayers, W. P. Yant, H. H. Schrenk, J. Chornyak, S. J. Pearce, F. A. Patty, and J. G. Linn, *U.S. Bur. Mines Rep. Invest. No. 3617*, (1942).

60. M. U. Vendilo, Y. L. Egorov, and N. G. Fel'dman, *Gig. Tr. Prof. Zabol.*, **15**(1), 17 (1971).

61. R. Ubaidullaev, *Gig. Sanit.*, **31**(4), 8 (1966).

62. S. M. Pavlenko, *Gig. Sanit.*, **37**(1), 40 (1972).

63. W. P. Yant and H. H. Schrenk, *J. Ind. Hyg. Toxicol.*, **19**, 337 (1937).

64. H. W. Haggard and L. A. Greenberg, *J. Pharmacol.*, **66**, 479 (1939).

65. D. Gaillard and R. Derache, *C. R. Soc. Biol.*, **157**, 2097 (1963). Cited by R. Derache, "Toxicology, Pharmacology, and Metabolism of Higher Alcohols.," *Int. Encycl. Pharmacol. Ther.*, **20**, 507 (1970).

66. A. B. Makar, *Methanol Metabolism in the Monkey*, Univ. Microfilms (Ann Arbor, Mich.). Order No. 66-9029, 97 pp. 1966.

67. E. Keeser, *Dtsch. Med. Wochenschr.*, **57**, 398 (1931).

68. E. Keeser, *Arch. Exp. Pathol. Pharmakol.*, **160**, 687 (1931).

69. M. M. Kini, D. W. King, Jr., and J. R. Cooper, *J. Neurochem.*, **9**, 119 (1962).

70. J. Pohl, *Arch. Exp. Pathol. Pharmakol.*, **31**, 281 (1893).

71. R. Hunt, *Bull. Johns Hopkins Hosp.*, **13**, 213 (1902).

72. K. E. McMartin, *Diss. Abstr. Int. B.*, **38**(7), 3154 (1978).

73. G. Martin-Amat, K. E. McMartin, M. S. Hayreh, S. S. Hayreh, and T. R. Tephly, *Toxicol. Appl. Pharmacol.*, **45**, 201 (1978).

74. H. W. Newman and M. L. Tainter, *J. Pharmacol.*, **57**, 67 (1936).

75. D. Gaillard and R. Derache, *Trav. Soc. Pharm. Montp.*, **25**(1), 51 (1965).

76. M. Koivusalo, *Int. Encycl. Pharmacol. Ther.*, **20**, 465 (1970).

77. J. T. Bastrup, *Acta Pharmacol.*, **3**, 303 (1947).

78. A. Lund, *Acta Pharmacol.*, **4**, 99 (1948).

79. A. Lund, *Acta Pharmacol.*, **4**, 108 (1948).

80. A. DeFelice, W. Wilson, and J. Ambre, *Toxicol. Appl. Pharmacol.*, **38**, 631 (1976).

81. G. Leaf and L. J. Zatman, *Br. J. Ind. Med.*, **9**, 19 (1952).

82. H. B. Elkins, personal communication.

83. A. H. Keeney and S. M. Mellinkoff, *Ann. Int. Med.*, **34**, 331 (1951).

84. A. Lund, *Acta Pharmakol.*, **4**, 205 (1948).

85. J. R. Cooper, *Biochem. J.*, **82**, 164 (1962). Cited by I. Takeda, *Nichidai Igaku Zasshi*, **31**(6), 518 (1972).

86. A. P. Gilger, A. M. Potts, and I. S. Farkas, *Am. J. Ophthalmol.*, **42**, 244 (1956). Cited by M. Koivusalo, *Int. Encycl. Pharmacol. Ther.* **20**, 465 (1970).

87. W. D. Province, R. A. Kritzler, and F. P. Calhoun, *Bull. U.S. Army Med. Dept.*, **5**, 114 (1946).

88. W. B. Chew, E. H. Berger, O. A. Brines, and M. J. Capron, *J. Am. Med. Assoc.*, **130**, 61 (1946).

89. D. J. Tonning, *Nova Scotia Med. Bull.*, **24**, 1 (1945).

90. A. Branch, *Can. Med. Assoc. J.*, **51**, 428 (1944).

91. O. Roe, *Acta Ophthalmol.*, **26**, 169 (1948).

92. R. F. Scherberger, G. P. Happ, F. A. Miller, and D. W. Fassett, *Am. Ind. Hyg. Assoc. J.*, **19**, 494 (1958).

93. J. May, *Staub*, **26**(9), 385 (1966).

94. I. L. Bennett, F. H. Casey, G. L. Mitchell, and M. N. Cooper, *Medicine*, **32**, 431 (1935).

95. R. L. Kane, W. Talbert, J. Harlan, G. Sizemore, and S. Cataland, *Arch. Environ. Health*, **17**, 119 (1968).

96. J. A. Campbell, *J. Ophthalmol. Otol. Laryngol.*, **21**, 756 (1915).

97. H. Woods, *J. Am. Med. Assoc.*, **60,** 1762 (1913).

98. B. M. Jacobson, H. K. Russell, J. J. Grimm, and E. C. Fox, *U.S. Naval Med. Bull.*, **44,** 1099 (1945).

99. A. Kaplan and G. V. Levreault, *U.S. Naval Med. Bull.*, **44,** 1107 (1945).

100. W. L. Voegtlin and C. E. Watts, *U.S. Naval Med. Bull.*, **41,** 1715 (1943).

101. J. P. Hughes, *J. Am. Med. Assoc.*, **156,** 234 (1954).

102. R. T. Johnstone, *Occupational Diseases*, p. 169. Saunders, Philadelphia, 1941.

103. O. Roe, *Acta Med. Scand.*, **126,** 182, 253 (1946); *Chem. Abstr.*, **41,** 2805 (1947).

104. O. Roe, *Q. J. Studies Alc.*, **11,** 107 (1950).

105. A. T. Suprunov, *Farmakol. Toksikol.*, **9,** 49 (1946); *Chem. Abstr.*, **41,** 3221 (1947).

106. F. Flury and F. Zernik, *Schadliche Gase*, Springer, Berlin (1931).

107. American Conference of Governmental Industrial Hygienists, *Threshold Limit Values for Chemical Substances in Workroom Air*, 1979.

108. U.S. Government, 29 CFR 1910.1000, July 1, 1978.

109. V. F. Simmon, K. Kauhanen, and R. G. Tardiff in D. Scott, B. A. Bridges, and F. N. Sobels, Eds., *Progress in Genetic Toxicology*, Vol. 2, Elsevier/North-Holland, 1977, p. 249.

110. International Labour Office, Geneva, *Occupational Exposure Limits for Airborne Toxic Substances*, Occupational Safety and Health Series No. 37 (1977).

111. *The Merck Index*, 9th Ed., M. Windholz, Ed., Merck and Co., Inc., Rahway, N.J., 1976.

112. I. Mellan, *Industrial Solvents*, Reinhold, New York, 1950, p. 454.

113. A. K. Doolittle, *The Technology of Solvents and Plasticizers*, Wiley, New York. (1954), p. 617.

114. I. Mellan, *Industrial Solvents*, Reinhold, New York, (1950), p. 460.

115. H. Zangger, *Arch. Gewerbepathol Gewerbehyg.*, **2,** 205 (1931).

116. A. Loewy and R. von der Heide, *Biochem. Z.*, **86,** 125 (1918).

117. *International Critical Tables of Numerical Data, Physics, Chemistry, and Technology*, Vol. III, McGraw-Hill, New York, 1928, p. 27.

118. G. W. Jones, *Chem. Rev.*, **22,** 1 (1938).

119. N. I. Sax, *Dangerous Properties of Industrial Materials*, 3rd ed., Van Nostrand-Reinhold, New York, (1968).

120. U.S. Dept. of Health, Education and Welfare, *NIOSH Manual of Analytical Methods*, Vol. 2, 2nd ed., No. S56, (1977).

121. M. B. Jacobs, *Analytical Chemistry of Industrial Poisons, Hazards, and Solvents*, Interscience, New York, (1941), p. 482.

122. H. W. Haggard and L. A. Greenberg, *J. Pharmacol.*, **52,** 137 (1934).

123. M. Nicloux, *C. R. Soc. Biol.*, **3,** 841 (1896).

124. A. O. Gettler and A. Tiber, *Arch. Pathol. Lab. Med.*, **3,** 218 (1927).

125. W. D. McNally, *Toxicology*, Industrial Medicine, Chicago, (1937), p. 648.

126. R. N. Harger, *J. Lab. Clin. Med.*, **20,** 746 (1935).

127. Z. Hepter, *Nahr. Genussm.*, **26,** 342 (1913).

128. C. H. Werkman and O. L. Osburn, *Ind. Eng. Chem. Anal. Ed.*, **3,** 387 (1931).

129. R. D. Stanley, *J. Assoc. Off. Agr. Chem.*, **22,** 594 (1939).

130. A. S. Chaikelis and R. D. Floersheim, *Am. J. Clin. Pathol. Tech. Suppl.* **10,** 180 (1946), *Biol. Abstr.* **21,** 5367 (1947).

131. J. Rochat, *Helv. Chim. Acta*, **29,** 819 (1946), *Analyst*, **72,** 450 (1947).

132. W. D. McNally and H. M. Coleman, *J. Lab. Clin. Med.*, **29**, 429 (1944); *Chem. Abstr.*, **38**, 5858 (1944).

133. R. Gingras and R. Gaudry, *Laval Med.*, **9**, 661 (1944); *Chem. Abstr.*, **39**, 954 (1945).

134. R. J. Henry, C. F. Kirkwood, S. Berkman, R. D. Housewright, and J. J. Henry, *J. Lab. Clin. Med.*, **33**, 241 (1948).

135. H. F. Smyth, Jr., *J. Ind. Hyg. Toxicol.*, **23**, 253 (1941).

136. E. T. Kimura, D. M. Ebert, and P. W. Dodge, *Toxicol. Appl. Pharmacol.*, **19**, 699 (1971).

137. J. C. Munch and E. W. Schwartze, *J. Lab. Clin. Med.*, **10**, 985 (1925).

138. W. S. Specter, Ed., *Handbook of Toxicology*, Vol. I, *Acute Toxicities*, Saunders, Philadelphia and London, 1956, p. 128. Cited by H. Maling, *Int. Encycl. Pharmacol. Ther.*, **20**, 277 (1970).

139. W. Bartsch, G. Sponer, K. Dietmann, and G. Fuchs, *Arzneim. Forsch.*, **26**(8), 1581 (1976).

140. E. I. du Pont de Nemours, Inc., Haskell Laboratories, unpublished data, 1974.

141. W. L. Guess, *Toxicol. Appl. Pharmacol.*, **16**, 382 (1970).

142. J. F. Treon, unpublished data.

143. L. Phillips, M. Steinberg, H. Maibach, and W. Akers, *Toxicol. Appl. Pharmacol.*, **21**, 369 (1972).

144. W. Deichmann, personal communication.

145. K. B. Lehmann and F. Flury, *Toxikologie und Hygiene der technischen Losungsmittel*, Springer, Berlin, 1938, p. 152.

146. H. F. Smyth, Jr., *Am. Ind. Hyg. Assoc. Q.*, **17**, 129 (1956).

147. L. Lendle, *Arch. Exp. Pathol. Pharmakol.*, **132**, 214 (1928).

148. E. G. Worthley and C. D. Schott, *Lloydia*, **29**(2), 123 (1966).

149. J. Tremolieres and R. Lowy, *Actual. Pharm.* **17**, 119 (1964). Cited by H. Maling, *Int. Encycl. Pharmacol. Ther.*, **20**, 277 (1970).

150. A. J. Lehman and H. W. Newman, *J. Pharmacol. Exp. Ther.*, **61**, 103 (1937).

151. D. C. MacGregor, E. Schonbaum, and W. G. Bigelow, *Can. J. Physiol. Pharmacol.*, **42**(6), 689 (1964).

152. D. I. Macht, *J. Pharmacol. Exp. Ther.*, **16**, 1 (1920).

153. W. Q. Sargent, J. R. Simpson, and J. D. Beard, *J. Pharm. Exp. Ther.*, **188**(2), 461 (1974).

154. H. M. Maling, *Int. Encycl. Pharmacol. Ther.*, **20**, 277 (1970).

155. H. Kalant and Y. Israel, in R. P. Maichel, Ed., *Biochemical Factors in Alcoholism*, Pergamon Press, Oxford, 1967, p. 25. Cited by S. W. French, "Acute and Chronic Toxicity of Alcohol," Chapter 14 in B. Kissin, Ed., *The Biology of Alcoholism*, Vol. 1, Plenum Press, New York 1971, pp. 437–511.

156. T. Polaczek-Kornecki, T. Zelazny, F. Walczak, S. Dendura, and E. Szpak, *Anaesthesist*, **21**(6) (1972).

157. J. C. Garriott, A. B. Richards, F. W. Hughes, and R. B. Forney, *J. Forensic Sci.*, **12**, 8 (1967). Cited by S. W. French, "Acute and Chronic Toxicity of Alcohol," Chapter 14 in B. Kissin, Ed., *The Biology of Alcoholism*, Vol. 1, Plenum Press, New York, 1971, pp. 437–511.

158. G. I. Klingman and H. B. Haag, *Q. J. Studies Alc.*, **19**, 203 (1958). Cited by S. W. French, "Acute and Chronic Toxicity of Alcohol," Chapter 14 in B. Kissin, Ed., *The Biology of Alcoholism*, Vol. 1, Plenum Press, New York, 1971, pp. 437–511.

159. L. Kager and J. Ericsson, *Acta Pathol. Microbiol. Scand.*, Sect. A, **82**, 534 (1974).

160. J. L. Hall and D. T. Rowlands, *Am. J. Pathol.*, **60**(2), 153 (1970).

161. S. C. Vasdev, R. N. Chakravarti, D. Subrahmanyam, A. C. Jain, and P. L. Wahi, *Cardiovasc. Res.*, **9**, 134 (1975).

162. L. L. Boughton, *J. Am. Pharm. Assoc. Sci. Ed.*, **33**, 111 (1944).

163. H. Mertens, *Arch. Int. Pharmacodyn.*, **2**, 127 (1896).

164. E. Petri, in F. Henke and O. Lubarsch, eds., *Handbuch der Speziellen Pathologischen Anatomie und Histologie*, Vol. X, Springer, Berlin, 1930, p. 276.

165. H. F. Smyth and H. F. Smyth, Jr., *J. Ind. Hyg.*, **10**, 261 (1928).

166. R. A. Coon, R. A. Jones, L. J. Jenkins, Jr., and J. Siegel, *Toxicol. Appl. Pharmacol.*, **16**, 646 (1970).

167. A. M. Skosyreva, *Akush. Ginekol.*, **4**, 15 (1973).

168. W. J. Tze and M. Lee, *Nature*, **257**, 479 (1975).

169. B. A. Schwetz, F. A. Smith, and R. E. Staples, *Teratology*, **18**(3), 385 (1978).

170. F. M. Badr and R. S. Badr, *Nature*, **253**, 134 (1975).

171. L. Machemer and D. Lorke, *Mutation Res.*, **29**, 209 (1975).

172. R. C. Chaubey, B. R. Kavi, P. S. Chauhan, and K. Sundaram, *Mutation Res.*, **43**, 441 (1977).

173. T. M. Carpenter, *J. Pharmacol.*, **37**, 217 (1929).

174. H. W. Haggard and L. A. Greenberg, *J. Pharmacol.*, **52**, 167 (1934).

175. M. G. Eggleton, *J. Physiol.* (*London*), **98**, 228 (1940). Cited by H. Wallgren, *Int. Encycl. Pharmacol. Ther.*, **20**, 161 (1967).

176. O. Strubelt, C. P. Siegers, and H. Breining, *Arch. Toxicol.*, **32**, 83 (1974).

177. W. W. Westerfield, E. Stotz, and R. L. Berg, *J. Biol. Chem.*, **149**, 237 (1943).

178. R. Derache, *Int. Encycl. Pharmacol. Ther.*, **20**, 507 (1970).

179. W. H. Orme-Johnson and D. M. Ziegler, *Biochem. Biophys. Res. Commun.*, **21**, 78 (1965). Cited by J. Von Wartburg, "The Metabolism of Alcohol in Normals and Alcoholics: Enzymes," Chapter 2 in B. Kissin, Ed., *The Biology of Alcoholism*, Vol. 1, Plenum, New York, 1971, pp. 63–102.

180. C. S. Lieber and L. M. DeCarli, *Science*, **162**, 917 (1968). Cited by J. Von Wartburg, "The Metabolism of Alcohol in Normals and Alcoholics: Enzymes," Chapter 2 in B. Kissin, Ed., *The Biology of Alcoholism*, Vol. 1, Plenum, New York, 1971, pp. 63–102.

181. M. K. Roach, W. N. Reese, Jr., and P. J. Creaven, *Biochem. Biophys. Res. Commun.*, **36**, 596 (1969). Cited by J. Von Wartburg, "The Metabolism of Alcohol in Normals and Alcoholics: Enzymes," Chapter 2 in B. Kissin, Ed., *The Biology of Alcoholism*, Vol. 1, PLenum, New York, 1971, pp. 63–102.

182. E. Lundsgaard, *C. R. Trav. Lab. Carlsbg.*, **22**, 333 (1938). Cited by F. Lundquist, "Enzymatic Pathways of Ethanol Metabolism," *Int. Encycl. Pharm. Ther.*, **20**, 95–116 (1967).

183. J. A. Larsen, *Scand. J. Clin. Lab. Invest.*, **11**, 340 (1959). Cited by F. Lundquist, "Enzymatic Pathways of Ethanol Metabolism," *Int. Encycl. Pharm. Ther.*, **20**, 95–116 (1967).

184. O. Forsander, N. Räihä, and H. Suomalainen, *Z. Physiol. Chem.*, **318**, 1 (1960). Cited by J. Von Wartburg, "The Metabolism of Alcohol in Normals and Alcoholics: Enzymes," Chapter 2 in B. Kissin, Ed., *The Biology of Alcoholism*, Vol. 1, Plenum, New York, 1971, pp. 63–102.

185. E. Jacobsen, *Pharmacol. Rev.*, **4**, 107 (1952).

186. O. A. Forsander, *Biochem. J.*, **105**(1), 93 (1967).

187. P. Gervais, *Presse Med.*, **74**(24), 1253 (1966).

188. H. W. Haggard and L. A. Greenberg, *J. Pharmacol.*, **52**, 150 (1934).

189. D. Gaillard and R. Derache, *Trav. Soc. Pharm. Montp.*, **25**(1), 51 (1965).

190. H. W. Newman and A. J. Lehman, *J. Pharmacol.*, **62**, 301 (1938).

191. H. Wallgren, *Acta Pharmacol. Toxicol.*, **16**, 217 (1960).

192. A. E. LeBlanc and H. Kalant, *Toxicol. Appl. Pharmacol.*, **32**, 123 (1975).

193. M. E. Goldberg, C. Haun, and H. F. Smyth, Jr., *Toxicol. Appl. Pharmacol.*, **4,** 148 (1962).

194. E. K. Marshall, Jr., and W. F. Fritz, *J. Pharmacol. Exp. Ther.*, **109,** 431 (1953).

195. J. S. Aull, Jr., W. J. Roberts, Jr., and F. W. Kinard, *Am. J. Physiol.*, **186,** 380 (1956).

196. J. McCann, E. Choi, E. Yamasaki, and B. Ames, *Proc. Natl. Acad. Sci. U.S.*, **72**(12), 5135 (1975).

197. D. Lester and L. A. Greenberg, *Q. J. Studies Alc.*, **12,** 167 (1951).

198. D. W. Fassett, personal communication, in F. A. Patty, Ed., *Industrial Hygiene and Toxicology*, Vol. 2, Wiley-Interscience, New York, (1963).

199. W. Van Hecke, H. Handovsky, and F. Thomas, *Ann. Med. Leg.*, **3,** 291 (1951). Cited by H. Wallgren, *Int. Encycl. Pharm. Ther.*, **20,** 161–188 (1967).

200. H. Casier and H. Polet, "Radioisotopes in Scientific Research," *Proc. First (UNESCO) International Conference*, Pergamon, New York, 1958, p. 481.

201. D. E. Drum, J. H. Harrison, T. K. Li, J. L. Bethune, and B. L. Vallee, *Proc. Natl. Acad. Sci.*, **57,** 1434 (1967). Cited by J. Von Wartburg, "The Metabolism of Alcohol in Normals and Alcoholics: Enzymes," Chapter 2 in B. Kissin, Ed., *The Biology of Alcoholism*, Vol. 1, Plenum, New York, 1971, pp. 63–102.

202. A. H. Neims, D. S. Coffey, and L. Hellerman, *J. Biol. Chem.*, **241,** 3036, 5941 (1966). Cited by J. Von Wartburg, "The Metabolism of Alcohol in Normals and Alcoholics: Enzymes," Chapter 2 in B. Kissin, Ed., *The Biology of Alcoholism*, Vol. 1, Plenum, New York, 1971, pp. 63–102.

203. J. Von Wartburg, "The Metabolism of Alcohol in Normals and Alcoholics: Enzymes," Chapter 2 in B. Kissin, Ed., *The Biology of Alcoholism*, Vol. 1, Plenum, New York, 1971, pp. 63–102.

204. H. Wallgren, *Int. Encycl. Pharm. Ther.*, **20,** 161 (1967).

205. G. Freund, *Ann. Rev. Pharmacol.*, **13,** 217 (1973).

206. O. Forsander, *Int. Encycl. Pharm. Ther.*, **20,** 117 (1967).

207. H. D. Waller and H. C. Bonoehr, *Therapiewoche*, **26**(50), 8457, 8461 (1976).

208. E. Brezina, *Internationale Uersicht uber Gewerbekrankheiten*, Springer, Berlin, 1929, p. 83.

209. A. B. Lowenfels, Ann. N.Y. Acad. Sci., **252,** 366 (1975).

210. F. Koelsch, *Zentr. Gewerbehyg. Unfallverhut*, **9,** 203 (1921).

211. E. Roth, in R. Abel, *Handbuch der prakt. Hygiene*, Vol. 2, Gustav Fischer, Jena, 1913, p. 232.

212. K. Malten, D. Spruit, H. Boemaars, and M. de Keizer, *Berufsdermatosen*, **16**(3), 135 (1968).

213. R. V. Bowers, W. D. Burlason, and J. F. Blades, *Q. J. Stud. Alcohol*, **3,** 31 (1942). Cited by H. Wallgren, *Int. Encycl. Pharm. Ther.*, **20,** 161 (1967).

214. L. F. Meisner and S. L. Inhorn, *Acta Cytologica*, **16**(1), 41 (1972).

215. American Conference of Governmental Industrial Hygienists, *Threshold Limit Values for Chemical Substances in Workroom Air, 1978*, Cincinnati, Ohio.

216. I. Mellan, *Industrial Solvents*, Reinhold, New York, 1950, pp. 466, 467.

217. A. K. Doolittle, *The Technology of Solvents and Plasticizers*, Wiley, New York, 1954, p. 627.

218. Beilstein, *Handbuch de Organischen Chemie*, Vol. I, 4th ed., Springer, Berlin, 1918, p. 350.

219. I. Mellan, *Industrial Solvents*, Reinhold, New York, 1939, p. 215.

220. H. F. Smyth, Jr., C. P. Carpenter, C. S. Weil, and U. C. Pozzani, *Arch. Ind. Hyg. Occup. Med.*, **10,** 61 (1954).

221. Industrial Hygiene Foundation of America, Chemistry and Toxicology Series, *Bulletin*, **6,** 1 (1967).

222. L. I. Golovinskaya, *Sud. Med. Ekspert*, **19**(2), 33, (1976).

223. P. M. Jenner, E. C. Hagan, J. M. Taylor, E. L. Cook, and O. G. Fitzhugh, *Food Cosmet. Toxicol.*, **2**(3), 327 (1964).

224. J. C. Munch and E. W. Schwartze, *J. Lab Clin. Med.* **10**, 985 (1925).

225. Industrial Hygiene Foundation of America, Inc. (Mellon Inst.) "Range-Finding Toxicity Tests for Normal Propanol for Celanese Corporation," June–July, 1962.

226. E. Starrek, dissertation, Wurzburg, 1938.

227. E. I. Goldenthal, *Toxicol. Appl. Pharmacol.*, **18**, 185 (1971).

228. J. Taylor, P. Jenner, and W. Jones, *Toxicol. Appl. Pharmacol.*, **6**(4), 378 (1964).

229. W. Gibel, Kh. Lohs, and G. P. Wildner, *Arch. Geschwulstforsch.*, **45**(1), 19 (1975).

230. J. Von Wartburg, "The Metabolism of Alcohol in Normals and Alcoholics: Enzymes," Chapter 2 in B. Kissin, Eds., *The Biology of Alcoholism*, Vol. 1, Plenum, New York, 1971, pp. 63–91.

231. R. Teschke, Y. Hasumura, and C. S. Lieber, *Biochem. Biophys. Res. Commun.*, **60**(2), 851 (1974).

232. U. Abshagen and N. Reitbrock, *Arch. Pharmakol.*, **265**(5), 411 (1970).

233. M. Neymark, *Skand. Arch. Physiol.*, **78**, 242 (1938).

234. S. M. Berggren, *Skand. Arch. Physiol.*, **78**, 249 (1938).

235. H. Hilscher, E. Geissler, Kh. Lohs, and W. Gibel, *Acta Biol. Med. Ger.*, **23**(6), 843 (1969).

236. Celanese Corp., unpublished data.

237. K. W. Nelson, J. F. Ege, M. Ross, L. E. Woodman, and L. Silverman, *J. Ind. Hyg. Toxicol.*, **25**, 282 (1943).

238. O. Laing, *Chem. Senses Flavor*, **1**, 257 (1975).

239. I. H. Blank, *J. Invest. Dermatol.*, **43**, 415 (1964).

240. E. Ludwig and B. M. Hausen, *Contact Dermatitis*, **3**(5), 240 (1977).

241. N. W. Weisbrodt, *Proc. Soc. Exp. Biol. Med.*, **142**(2), 450, (1973).

242. F. Beauge, M. Clement, J. Nordmann, and R. Nordmann, *Biochimie (Paris)*, **56**(8), 1157 (1974).

243. S. G. Hedlund and K. H. Kiessling, *Acta Pharmacol. Toxicol.*, **27**(5), 381 (1969).

244. K. Takagi and Y. Gomi, *Yakugaku Zasshi*, **86**(6), 479 (1966).

245. H. P. Ammon, F. Heim, C. J. Estler, G. Fickeis, and M. Wagner, *Biochem. Pharmacol.*, **16**(8), 1533 (1967).

246. D. C. Villenueve, G. J. Mulkins, H. L. Trenholm, K. A. McCully, and W. P. McKinley, *J. Agric. Food Chem.*, **17**, 101 (1969).

247. M. A. D'Agostino, K. M. Lowry, and G. F. Kalf, *Arch. Biochem. Biophys.*, **166**(2), 400 (1975).

248. A. Thore and H. Baltscheffsky, *Acta Chem. Scand.*, **19**(7), 1591 (1965); *Acta Chem. Scand.*, **19**(8), 1975, (1965).

249. P. W. Gage, *J. Pharmacol. Exp. Ther.*, **150**(2), 236 (1965).

250. P. Seeman, M. Chau, M. Goldberg, T. Sauks, and L. Sax, *Biochim. Biophys. Acta*, **225**(2), 185 (1971).

251. P. Seeman, *Biochem. Pharmacol.*, **15**(10), 1632 (1966).

252. S. N. Listvinova and N. S. Gryaznova, *Antibiotiki (Moscow)*, **14**(9), 808 (1969).

253. W. Tuganowski, K. Sowa, and M. Tendera, *Acta Physiol. Pol.* **22**(3), 373 (1971).

254. Anonymous, "Isopropyl Alcohol—Salient Statistics," in *Chemical Economics Handbook*, Stanford Research Institute, Menlo Park, Calif., 1972, pp. 668.5030A, 668.5030B, 668.5060. Cited in "Criteria for a Recommended Standard, Occupational Exposure to Isopropyl Alcohol," U.S. NTIS, PB Rep., Issue PB-273873 (1976).

255. E. J. Wickson, "Isopropyl Alcohol," in *Kirk-Othmer Encyclopedia of Chemical Technology.*, Vol. 16, 2nd ed., Wiley, New York, 1968, pp. 564–578. Cited in "Criteria for a Recommended Standard, Occupational Exposure to Isopropyl Alcohol, U.S. NTIS, PB Rep., Issue PB-273873 (1976).

256. Anonymous, "Criteria for a Recommended Standard Occupational Exposure to Isopropyl Alcohol," U.S. NTIS, PB Rep., Issue PB-273873 (1976).

257. I. Mellan, *Industrial Solvents*, Reinhold, New York, 1950, pp. 467–482.

258. U.S. Dept. of Health, Education and Welfare, *NIOSH Manual of Analytical Methods*, Vol. 2, 2nd ed., 1977, S65.

259. M. B. Jacobs, *The Analytical Chemistry of Industrial Poisons, Hazards, and Solvents*, Interscience, New York, 1941, p. 487.

260. G. Kleyer, *Pharm. Ztg.*, **72**, 1262 (1927).

261. E. Hahn, *Biochem. Z.*, **292**, 148 (1937).

262. H. W. Knipping and W. Ponndorf, *Z. Physiol. Chem.*, **160**, 25 (1926).

263. H. F. Smyth, Jr. and C. P. Carpenter, *J. Ind. Hyg. Toxicol.*, **30**, 63 (1948).

264. A. J. Lehman and H. F. Chase, *J. Lab. Clin. Med.*, **29**, 561 (1944). Cited in "Criteria for a Recommended Standard, Occupational Exposure to Isopropyl Alcohol," U.S. NTIS, PB Rep., Issue PB-273873 (1976).

265. F. N. Marzulli and D. I. Ruggles, *J. Assoc. Off. Anal. Chem.*, **56**, 905 (1973). Cited in "Criteria for a Recommended Standard, Occupational Exposure to Isopropyl Alcohol," U.S. NTIS, PB Rep., Issue PB-273873 (1976).

266. G. A. Nixon, C. A. Tyson, and W. C. Wertz, *Toxicol. Appl. Pharmacol.*, **31**, 481 (1975). Cited in "Criteria for a Recommended Standard, Occupational Exposure to Isopropyl Alcohol," U.S. NTIS, PB Rep., Issue PB-273873 (1976).

267. H. F. Smyth, Jr., *Am. Ind. Hyg. Assoc. J.*, Philadelphia, April 25, 1956.

268. A. J. Lehman, H. Schwerma, and E. J. Richards, *Pharmacol.*, **82**, 196 (1944).

269. D. I. Macht, *Arch. Int. Pharmacodyn.*, **26**, 285 (1922).

270. A. J. Lehman, H. Schwerma, and E. J. Richards, *Pharmacol. Exp. Ther.*, **85**, 61 (1945). Cited in "Criteria for a Recommended Standard, Occupational Exposure to Isopropyl Alcohol," U.S. NTIS, PB Rep., Issue PB-273873 (1976).

271. C. S. Weil, H. F. Smyth, Jr., and T. W. Nale, *Arch. Ind. Hyg. Occup. Med.*, **5**, 535–47 (1952). Cited in "Criteria for a Recommended Standard, Occupational Exposure to Isopropyl Alcohol," U.S. NTIS, PB Rep., Issue PB-273873 (1976).

272. C. S. Weil, Sept. 1975. Cited in "Criteria for a Recommended Standard, Occupational Exposure to Isopropyl Alcohol," U.S. NTIS, PB Rep., Issue PB-273873 (1976).

273. H. Z. Kemal, *Physiol. Chem.*, **246**, 59 (1937).

274. F. W. Ellis, *J. Pharmacol. Exp. Ther.*, **105**, 427 (1952). Cited in "Criteria for a Recommended Standard, Occupational Exposure to Isopropyl Alcohol," U.S. NTIS, PB Rep., Issue PB-273873 (1976).

275. I. A. Kamil, J. N. Smith, and R. T. Williams, *Biochem. J.*, **53**, 129 (1953). Cited in "Criteria for a Recommended Standard. Occupational Exposure to Isopropyl Alcohol," U.S. NTIS, PB Rep., Issue PB-273873 (1976).

276. H. H. Cornish and J. Adefuin, *Arch. Environ. Health*, **14**, 447 (1967). Cited in "Criteria for a Recommended Standard, Occupational Exposure to Isopropyl Alcohol." U.S. NTIS, PB Rep., Issue PB-273873 (1976).

277. G. J. Traiger and G. S. Plaa, *Toxicol. Appl. Pharmacol.*, **20**, 105 (1971). Cited in "Criteria for a Recommended Standard, Occupational Exposure to Isopropyl Alcohol," U.S. NTIS, PB Rep., Issue PB-273873 (1976).

278. G. J. Traiger and G. S. Plaa, *Arch. Int. Pharmacodyn. Ther.*, **202**, 102 (1973). Cited in "Criteria for a Recommended Standard, Occupational Exposure to Isopropyl Alcohol," U.S. NTIS, PB Rep., Issue PB-273873 (1976).

279. G. J. Traiger and G. S. Plaa, *Arch. Environ. Health*, **28,** 276 (1974). Cited in "Criteria for a Recommended Standard, Occupational Exposure to Isopropyl Alcohol," U.S. NTIS, PB Rep., Issue PB-273873 (1976).

280. G. J. Traiger and G. S. Plaa, *Can J. Physiol. Pharmacol.*, **51,** 291 (1973). Cited in "Criteria for a Recommended Standard, Occupational Exposure to Isopropyl Alcohol," U.S. NTIS, PB Rep., Issue PB-273873 (1976).

281. G. J. Traiger and G. S. Plaa, *J. Pharmacol. Exp. Ther.*, **183,** 481 (1972). Cited in "Criteria for a Recommended Standard, Occupational Exposure to Isopropyl Alcohol," U.S. NTIS, PB Rep., Issue PB-273873 (1976).

282. H. J. Morris and H. D. Lightbody, *J. Ind. Hyg. Toxicol.*, **20,** 428 (1938).

283. D. Gaillard and R. Derache, *C. R. Soc. Biol.*, **158,** 1605 (1964).

284. J. Pohl, *Biochem. Z.*, **127,** 66 (1921).

285. N. Chu, R. L. Driver, and P. J. Hanzlik, *J. Pharmacol.*, **92,** 291 (1948).

286. R. W. Schaffarzick, *Proc. Soc. Exp. Biol. Med.*, **74,** 211 (1950).

287. U. Abshagen, and N. Rietbrock, *Arch. Pharmakol.*, **264**(2), 110 (1969).

288. H. Bateman, *Proc. Cambridge Phil. Soc.*, **15,** 423 (1910).

289. W. M. McCord, P. K. Schwitzer, and H. H. Brill, Jr., *Southern Med. J.*, **41,** 639 (1948).

290. R. F. Garrison, *J. Am. Med. Assoc.*, **152,** 317 (1953). Cited in "Criteria for a Recommended Standard, Occupational Exposure to Isopropyl Alcohol," U.S. NTIS, PB Rep., Issue PB-273873 (1976).

291. S. W. McFadden and J. E. Haddow, *Pediatrics*, **43,** 622 (1969). Cited in "Criteria for a Recommended Standard, Occupational Exposure to Isopropyl Alcohol," U.S. NTIS, PB Rep., Issue PB-273873 (1976).

292. E. H. Senz and D. L. Goldfarb, *J. Pediatr.*, **53,** 3223 (1958). Cited in "Criteria for a Recommended Standard, Occupational Exposure to Isopropyl Alcohol," U.S. NTIS, PB Rep., Issue PB-273873 (1976).

293. H. Kemal, *Biochem. Z.*, **187,** 461 (1927).

294. H. C. Fuller and O. B. Hunter, *J. Lab. Clin. Med.*, **12,** 326 (1927).

295. J. May, *Staub*, **26**(9), 385 (1966).

296. C. Wasilewski, Jr., *Arch. Dermatol.*, **98,** 502 (1968). Cited in "Criteria for a Recommended Standard, Occupational Exposure to Isopropyl Alcohol," U.S. NTIS, PB Rep., Issue PB-273873 (1976).

297. A. McInnes, *Br. Med. J.*, **1,** 357 (1973). Cited in "Criteria for a Recommended Standard, Occupational Exposure to Isopropyl Alcohol," U.S. NTIS, PB Rep., Issue PB 273873 (1976).

298. D. R. Richardson, C. M. Caravati, Jr., and P. E. Weary, *Cutis*, **5,** 1115 (1969). Cited in "Criteria for a Recommended Standard, Occupational Exposure to Isopropyl Alcohol," U.S. NTIS, PB Rep., Issue PB-273873 (1976).

299. J. Wills, E. M. Jameson, and F. Coulston, *Toxicol. Appl. Pharmacol.*, **15**(3), 560 (1969).

300. S. Fregert, O. Groth, N. Hjorth, B. Magnusson, H. Rorsman, and P. Ovrum, *Acta Dermatol. Vener.*, **49,** 493 (1969).

301. S. Fregert, O. Groth, B. Gruvberger, B. Magnusson, H. Mobarken, and H. Rorsman, *Acta Dermatol. Vener.*, **51,** 271 (1971).

302. J. R. Wise, *N. Engl. J. Med.*, **280,** 840 (1969).

303. L. Adelson, *Am. J. Clin. Pathol.*, **38,** 144 (1962).

304. L. Juncos and J. T. Laguchi, *J. Am. Med. Assoc.*, **204,** 732 (1968).

305. M. A. Chapin, *J. Maine Med. Assoc.*, **40,** 288 (1949).

306. L. H. King, Jr., K. P. Bradley, and D. L. Shires, Jr., *J. Am. Med. Assoc.*, **211**, 1855 (1970).

307. A. W. Freireich, T. J. Cinque, G. Xanthaky, and D. Landau, *N. Engl. J. Med.*, **277**, 699 (1967).

308. W. C. Hueper, *Occupational and Environmental Cancers of the Respiratory System*, Springer-Verlag, New York, 1966.

309. J. McLaughlin, Jr., J. P. Marliac, M. J. Verrett, M. K. Mutchler, and O. G. Fitzhugh, *Am. Ind. Hyg. Assoc. J.*, **25**, 282 (1964).

310. R. Nordmann, C. Ribiere, H. Rouach, G. Beauge, Y. Giudicelli, and J. Nordmann, *Life Sci.*, **13**, 919 (1973).

311. O. E. Gorlova, *Gig. Sanit.*, **35**, 9 (1970).

312. I. A. Selina, in *Predel'no Dopustimye Konts. Atm. Zagryazn.*, **8**, 162, (1964).

313. B. K. Baikov, O. E. Gorlova, M. I. Gusev, Y. V. Novkov, T. V. Yudiva, and A. N. Sergeev, *Gig. Sanit.*, **4**, 6 (1974).

314. I. Mellan, *Industrial Solvents*, Reinhold, New York, 1950, pp. 482–488.

315. A. K. Doolittle, *The Technology of Solvents and Plasticizers*, Wiley, New York, 1954, pp. 640–642.

316. D. G. Cogan and W. M. Grant, *Arch. Ophthalmol.*, **33**, 106 (1945).

317. I. R. Tabershaw, J. P. Fahy, and J. B. Skinner, *J. Ind. Hyg. Toxicol.*, **26**, 328 (1944).

318. U.S. Dept. Labor, *Bur. Labor Stat. Bull. No. 41*, 38 (1942).

319. J. H. Sterner, H. C. Crouch, H. F. Brockmyre, and M. Cusack, *Am. Ind. Hyg. Assoc. Q.*, **10**, 53 (1949).

320. N. Thompson, *J. Ind. Eng. Chem.*, **21**, 134 (1929).

321. U.S. Dept. of Health, Education and Welfare, *NIOSH Manual of Analytical Methods*, Vol. 2, 2nd ed., 1977, S66.

322. L. M. Christensen and E. I. Fulmer, *Ind. Eng. Chem. Anal. Ed.*, **7**, 180 (1935).

323. J. B. Ficklen, *Manual of Industrial Health Hazards*, Service to Industry, West Hartford, Conn., 1940, p. 50.

324. S. M. Hoch, personal communication.

325. H. F. Smyth, C. P. Carpenter, and C. S. Weil, *AMA Arch. Ind. Hyg. Occup. Med.*, **4**, 119 (1951).

326. P. M. Jenner, E. C. Hagan, J. M. Taylor, E. L. Cook, and O. G. Fitzhugh, *Food Cosmet. Toxicol.*, **2**(3), 327 (1964).

327. G. D. DiVincenzo and M. L. Hamilton, *Toxicol. Appl. Pharmacol.*, **48**, 317 (1979).

328. E. Starrek, "Uber die Wirkung einiger Alkohole, Glykole und Ester." Cited by Lehman and Flury, in *Toxicology and Hygiene of Industrial Solvents*, Williams and Wilkins, Baltimore, 1943.

329. L. Lendle, *Arch. Exp. Pathol. Pharmakol.*, **132**, 214 (1928).

330. A. P. Rumyantsev, N. A. Ostroumova, S. A. Astapova, Z. R. Kustova, I. Ya. Lobanova, L. V. Tiunova, V. V. Chernikova, and P. A. Kolesnikov, *Gig. Sanit.*, **11**, 12 (1976).

331. A. D. Winer, *Acta. Chem. Scand.*, **12**, 1695 (1958). Cited by R. Derache, *Int. Encycl. Pharmacol. Ther.*, **20**, 507 (1970).

332. Y. Yoshiyama, K. Nagai, H. Some, and G. Tamura, *Agric. Biol. Chem.*, **37**(6), 1317 (1973).

333. I. Astrand, P. Ovrum, T. Lindquist, and M. Hultengren, *Scand. J. Work Environ. Health*, **3**, 165 (1976).

334. R. W. Moncreiff, *Manuf. Chem. Aerosol. News*, **38**(11), 57 (1967).

335. F. Sander, Inaugural Dissertation, Koln, 1933.

336. E. Kruger, *Arch. Gewerbepathol. Gewerbehyg.*, **3**, 798 (1932).

337. I. Mellan, *Industrial Solvents*, Reinhold, New York, 1950, p. 488.

338. A. K. Doolittle, *The Technology of Solvents and Plasticizers*, Wiley, New York, 1954, p. 647.

339. D. Steinkoff, *Zentr. Arbeitsmed. Arbeitsschutz,* **2,** 13 (1952).

340. I. Mellan, *Industrial Solvents,* Reinhold, New York, 1939, pp. 226–236.

341. U.S. Dept. of Health, Education and Welfare, *NIOSH Manual of Analytical Methods,* Vol. 2, 2nd ed., 1977, S64.

342. H. H. Weber and W. Koch, *Chem. Ztg.,* **57,** 73 (1933).

343. C. P. Carpenter and H. F. Smith, *Am. J. Ophthalmol.,* **29,** 1363 (1946).

344. W. Gibel, Kh. Lohs, and G. P. Wildner, *Arch. Geschwulstforsch.,* **45**(1), 19 (1975).

345. P. Salo, *J. Food Sci.,* **35**(1), 95 (1970).

346. H. Oettel, *Arch. Exp. Pathol. Pharmcol.,* **183,** 641 (1936).

347. L. Schwartz and L. Tulipan, *Occupational Diseases of the Skin,* Lea & Febiger, Philadelphia, 1939.

348. American Conference of Governmental Industrial Hygienists, *Threshold Limit Values for Chemical Substances in Workroom Air, 1980,* Cincinnati, Ohio.

349. I. Mellan, *Industrial Solvents,* Reinhold, New York, 1950, pp. 488–493.

350. A. K. Doolittle, *The Technology of Solvents and Plasticizers,* Wiley, New York, 1954, pp. 644, 645.

351. M. G. Zabetakis, A. L. Furno, and G. W. Jones, *Ind. Eng. Chem.,* **46,** 2173 (1954).

352. G. S. Scott, G. W. Jones, and F. E. Scott, *Anal. Chem.,* **20,** 238 (1948).

353. U.S. Dept. of Health, Education and Welfare, *NIOSH Manual of Analytical Methods,* Vol. 2, 2nd ed., 1977, 353.

354. T. C. Butler and H. L. Dickison, *J. Pharmacol.,* **69,** 225 (1940).

355. Shell Chemical Corporation, Safety Data Sheet S.C., 57–97.

356. I. Mellan, *Industrial Solvents,* Reinhold, New York, 1950, pp. 493–495.

357. I. Mellan, *Industrial Solvents,* Reinhold, New York, 1939, p. 236.

358. W. J. Huff, *U.S. Bur. Mines Rep. Invest. No. 3669* (1942).

359. U.S. Dept. of Health, Education and Welfare, *NIOSH Manual of Analytical Methods,* Vol. 2, 2nd ed., 1977, S63.

360. R. W. Schaffarzick, *Science,* **116,** 663 (1952).

361. R. Derache, *Int. Encycl. Pharmacol. Ther.,* **20,** 507 (1970).

362. I. A. Kamil, J. N. Smith, and R. T. Williams, *Biochem. J.,* **49,** xxxviii (1951). Cited by H. Wallgren, *Acta Pharmacol. Toxicol.,* **16,** 217 (1960).

363. F. H. Dickey, G. H. Cleland, and C. Lotz, *Proc. Natl. Acad. Sci. U.S.,* **35,** 581 (1949).

364. A. H. Allen and W. Chattaway, *Analyst,* **16,** 102 (1891).

365. *Official and Tentative Methods of Analysis of the Association of Official Agricultural Chemists,* 5th ed., Association of Official Agricultural Chemists, Washington, D.C., 1940, p. 175.

366. H. P. Basset, *Ind. Eng. Chem.,* **2,** 389 (1910).

367. A. Komarowsky, *Chem. Ztg.,* **27,** 1086 (1903).

368. W. B. D. Penniman, D. C. Smith, and E. I. Lawshe, *Ind. Eng. Chem. Anal. Ed.,* **9,** 91 (1937).

369. I. M. Korenman, *Arch. Hyg.,* **109,** 108 (1932).

370. L. Carozzi, *Occupation and Health,* International Labor Office, Geneva, 1930, p. 115.

371. E. B. Ley and F. J. Vintinner, *The Toxicology and Prevention of Industrial Disease,* 3rd ed., U.S. War Dept. Eighth Service Command, 1944, p. 13.

372. *U.S. Dept. Labor Bur. Labor Stat. Bull. No. 41,* 33 (1942).

373. *Natl. Res. Council Can. Bull. No. 15,* 21 (1930).

374. L. Resnick, *Eye Hazards in Industry,* Columbia University Press, New York, 1941, p. 251.

375. F. P. Underhill, *Toxicology,* 2nd ed., Blakiston, Philadelphia, 1928, p. 212.

376. Eyquem, *Ann. Hyg. Publ. Med. Leg.*, **3**, 71 (1905).

377. L. Lewin, *Gifte und Vergiftungen*, Stilke, Berlin, 1929, p. 407.

378. T. B. Fuchter, *Am. Med*, **2**, 210 (1901).

379. H. Zangger, *Arch. Gewerbepathol. Gewerbehyg.*, **4**, 117 (1933).

380. E. Baader, *Verhandl. Deut. Ges. Inn. Med.*, **45**, 318, (1933).

381. G. E. C. Burger and B. H. Stockmann, *Zentr. Gewerbehyg. Unfallverhuet.*, **9**, 29 (1932).

382. F. C. Whitmore, *Organic Chemistry*, Van Nostrand, New York, 1937, p. 125.

383. A. K. Doolittle, *The Technology of Solvents and Plasticizers*, Wiley, New York, 1954, pp. 653–655.

384. L. L. Fieser and M. Fieser, *Organic Chemistry*, Heath, Boston, 1944, p. 124.

385. U.S. Department of Health, Education and Welfare, *NIOSH Manual of Analytical Methods*, Vol. 2, 2nd ed., 1977, S-58.

386. K. Verschueren, *Handbook of Environmental Data on Organic Chemicals*, Van Nostrand Reinhold, New York, 1977.

387. Committee on Flammable Liquids, *Fire Hazard Properties*, Nat'l Fire Protec. Assoc., Boston, 1941, pp. 6, 17.

388. R. A. Scala and E. G. Burtis, *Am. Ind. Hyg. Assoc. J.*, **34**(11), 493 (1973).

389. J. A. Monick, *Alcohols: Their Chemistry, Properties and Manufacture*, Reinhold, 1968.

390. Y. L. Egorov, *Toksikol. Gig. Prod. Neftekhim*, **98**, 102 (1972).

391. H. W. Haggard, D. P. Miller, and L. A. Greenberg, *J. Ind. Hyg. Toxicol.*, **27**, 1 (1945).

392. W. Szybalski, *Ann. N.Y. Acad. Sci.*, **76**, 475 (1958).

393. *International Critical Tables of Numerical Data, Physics, Chemistry, and Technology*, Vol. I, McGraw-Hill, New York, 1926, p. 193.

394. A. K. Doolittle, *The Technology of Solvents and Plasticizers*, Wiley, New York, 1954.

395. R. C. Weast, Ed., *Handbook of Chemistry and Physics*, 45th ed., CRC Press, Cleveland, Ohio, 1964.

396. *Fire Protection Guide on Hazardous Materials*, National Fire Protection Association, Boston, 1973.

397. H. F. Smyth, Jr., C. P. Carpenter, C. S. Weil, U. C. Pozzani, J. A. Striegel, and J. S. Nycum, *Am. Ind. Hyg. Assoc. J.*, **30**(5), 470 (1969).

398. F. M. B. Carpanini, I. F. Gaunt, I. S. Kiss, P. Grasso, and S. D. Gangolli, *Food Cosmet. Toxicol.*, **11**(5), 713 (1973).

399. M. Guggenheim and W. Loffler, *Biochem. Z.*, **72**, 325 (1916).

400. *Handbook of Chemistry and Physics*, The Chemical Rubber Co., Cleveland, Ohio, 1964.

401. J. V. Marhold, *Sbornik Vysledku Toxixologickeho Vysetreni Latek A Pripravku*, Institut Pro Vychovu Vedoucicn Pracovniku Chemickeho Prumyclu Prah, 1972, p. 36. Cited in *Registry of Toxic Effects of Chemical Substances, 1977*, DHEW Publ. No. (NIOSH) 78-104-B.

402. O. Neubauer, *Arch. Exp. Pathol. Pharmakol.*, **46**, 133 (1901).

403. H. Thierfelder and J. V. Mering, *Z. Physiol. Chem.*, **9**, 511 (1885).

404. P. W. Langvardt and W. H. Braun, *J. Anal. Toxicol.*, **2**, 83, (1978).

405. The Dow Chemical Company, unpublished data.

406. R. J. Nolan, P. W. Langvardt, and G. E. Blau, *Toxicol. Appl. Pharmacol.*, **48**, A164 (1979) (Abstract No. 329).

407. H. C. Maguire, *J. Soc. Cosmet. Chem.*, **24**, 151 (1973).

408. J. Pohl, *Arch. Exp. Pathol. Pharmakol. Suppl.* (1908), 427.

409. J. Biberfeld, *Biochem. Z.*, **92**, 198 (1918).

410. W. Jacobi and E. Speer, *Therap. Halbmonatsh.*, **34,** 445 (1920). Cited by L. Lewin, *Gifte und Vergifungen*, Stilke, Berlin, 1929, p. 40.

411. M. Anker, *Therap. Monatsh.*, **6,** 623 (1892).

412. E. Bar and F. Griepentrog, *Med. Ernahr*, **8,** 244 (1967). Cited by D. L. Opdyke, *J. Food Cosmet. Toxicol.*, **13,** Suppl., 695 (1975).

413. G. N. Zaeva and V. I. Fedorova, *Toksikol. Novykh Prom. Khim. Veshchestv.*, **5,** 51 (1963). Cited by D. L. Opdyke, *J. Food Cosmet. Toxicol.*, **13,** Suppl., 695 (1975).

414. O. M. Moreno, Report to RIFM, Aug. 26, 1974. Cited by D. L. Opdyke, *J. Food Cosmet. Toxicol.*, **13,** Suppl., 695 (1975).

415. J. Sice, *Toxicol. Appl. Pharmacol.*, **9**(1), 70 (1966).

416. R. T. Williams, *Detoxication Mechanisms. The Metabolism and Detoxication of Drugs, Toxic Substances and Other Compounds*, 2nd ed., Chapman & Hall Ltd., London, 1959.

417. W. L. Epstein, Report to RIFM, August 27, 1974. Cited by D. L Opdyke, *J. Food Cosmet. Toxicol.*, **13,** Suppl., 695 (1975).

418. A. M. Kligman, *J. Invest. Dermatol.*, **47,** 393 (1966).

419. A. M. Kligman and W. Epstein, *Contact Dermatitis*, **1,** 231 (1975).

420. U.S. Dept. of Health, Education and Welfare, *NIOSH Manual of Analytical Methods*, Vol. 2, 2nd ed., 1977, S60.

421. R. E. Kirk and D. F. Othmer, Eds., *Encyclopedia of Chemical Technology*, Vol. 1, Interscience, New York, 1947.

422. I. Mellan, *Industrial Solvents*, Reinhold, New York, 1950, pp. 507–509.

423. A. K. Doolittle, *The Technology of Solvents and Plasticizers*, Wiley, New York, 1954, pp. 400–401, 661–664.

424. I. Mellan, *Industrial Solvents*, Reinhold, New York, 1950, pp. 503–507.

425. *Condensed Chemical Dictionary*, 8th ed., 1971.

426. F. A. Patty, *Industrial Hygiene & Toxicology*, 2nd ed., Vol. II, p. 1459, Interscience, New York, 1963.

427. Shell Chemical Corporation, *Ind. Hyg. Bull.* (1957).

428. W. A. McOmie and H. H. Anderson, *U. Cal. Publ. Pharmacol.*, **2,** 217 (1949).

429. L. Silverman, H. F. Schulte, and M. W. First, *J. Ind. Hyg. Toxicol.*, **28,** 262 (1946).

430. I. Mellan, *Industrial Solvents*, Reinhold, New York, 1950, p. 512.

431. D. R. Stull, *Ind. Eng. Chem.*, **39,** 517 (1947).

432. G. N. Zaeva and V. J. Fedorova in *Toxicol. New Ind. Chem. Substances, Moscow*, **5,** 51 (1963). Cited by P. Schmidt, R. Gohlke, and R. Rothe, *Z. Ges. Hyg. Grenzgeb.*, **19,** 485 (1973).

433. H. Hodge, *Proc. Exptl. Biol. Med.*, **53,** 20 (1943).

434. P. Schmidt, R. Gohlke, and R. Rothe, *Z. Ges. Hyg. Grenzgeb.*, **19**(7), 485 (1973).

435. Union Carbide Corp., unpublished data.

436. O. N. Mashkina, *Mater. Kanf. Fiziol. Biokhim. Farmakol. Uchast. Prakt.*, 168 (1966); *Chem. Abstr.* **67,** 57066.

437. I. A. Kamil, J. N. Smith, and R. T. Williams, *Biochem. J.*, **53,** 137 (1953).

438. P. W. Albro, *Xenobiotica*, **5**(10), 625 (1975).

439. D. E. Moody, *J. Cell Biol.*, **1976,** 70/211 (No. 1088).

440. K. Hollenbach, P. Schmidt, and D. Stremmel, *Z. Ges. Hyg.*, **18,** 418 (1972).

441. H. F. Smyth, Jr., C. P. Carpenter, and C. S. Weil, *J. Ind. Hyg. Toxicol.*, **31,** 60 (1949).

442. I. Heilbron, *Dictionary of Organic Compounds*, Vol. 4, Oxford University Press, New York, 1953, p. 606.

443. I. Heilbron, *Dictionary of Organic Compounds*, Vol. 2, Oxford University Press, New York, 1953, p. 262.

444. Yu. L. Egorov and L. A. Andrianov, *Uch. Zap., Mosk. Nauchn.-Issled. Inst. Gig.* (9), 47 (1961); *Chem. Abstr.*, **61**, 3601 (1964); **60**, 12576d (1964).

445. Yu. L. Egorov, *Toksikol. Gig. Prod. Neftekhim., Neftekhim. Proizvod. Vses. Knof. (Dokl.) 2nd*, 98 (1972); *Chem. Abstr.*, **80**, 91721b (1974).

446. K. V. Kitzmiller, personal communication.

447. Yu. L. Egorov and L. A. Andrianov, *Uch. Zap., Mosk. Nauchn.-Issled. Inst. Gig.*, (9), 47 (1961); *Chem. Abstr.*, **60**, 12576d (1964).

448. T. E. Jordan, *Vapor Pressure of Organic Compounds*, Interscience, New York, London, 1954.

449. F. E. Shaffer, personal communication.

450. W. W. Edman and W. H. Lowden, *Drug Cosmet. Ind.*, **93**, 631 (Nov. 1963); cited by Merck.

451. M. Coopersmith and A. J. Rutkowski, *Drug Cosmet. Ind.*, **96**, 630 (1965).

452. R. E. Kirk and D. F. Othmer, Eds., *Encyclopedia of Chemical Technology*, Vol. 2, Interscience, New York, 1948.

452a. I. Mellan, *Industrial Solvents*, Reinhold, New York, 1950, pp. 86, 519.

453. J. Callaway, Jr. and S. Reznek, *J. Assoc. Offic. Agric. Chem.*, **16**, 285 (1933).

454. H. Mohler and W. Hammerle, *Z. Anal. Chem.*, **122**, 202 (1941); *Chem. Abstr.*, **36**, 4970 (1942).

455. H. F. Smyth, Jr., C. P. Carpenter, and C. S. Weil, *Arch. Ind. Hyg. Occup. Med.*, **4**, 119 (1951); H. F. Smyth, Jr., personal communication.

456. E. T. Kimura, T. D. Darby, R. A. Krause, and H. D. Brondyk, *Toxicol. Appl. Pharmacol.*, **18**(1), 60 (1971).

457. C. M. Gruber, *J. Lab. Clin. Med.*, **9**, 15, 92 (1923).

458. C. M. Gruber, *J. Lab. Clin. Med.*, **10**, 284 (1924).

459. D. Duncan and W. H. Jarvis, *Anesthesiology*, **4**, 465 (1943).

460. D. I. Macht, *J. Pharmacol. Proc.*, **13**, 509 (1919).

461. D. I. Macht, *J. Pharmacol.* **11**, 263, 419 (1918).

462. I. Hosino, *Zikken Syokakibyogaku (Exp. Gastorenterol.)*, **15**, 117 (1940); *Jap. J. Med. Sci. II, Biochem.*, **4**(4), Abstr. (in English), 104 (1941).

463. J. A. Stekol, *J. Biol. Chem.*, **128**, 199 (1939).

464. S. L. Diack and H. B. Lewis, *J. Biol. Chem.*, **77**, 89 (1928).

465. H. G. Bray, S. P. James, and W. V. Thorpe, *Biochem. J.*, **70**, 570 (1958).

466. J. M. Ritchie, B. Ritchie, and P. Greengard, *J. Pharmacol. Exp. Ther.*, **150**(1), 152 (1965).

467. I. Snapper, A. Grunbaum, and S. Sturkop, *Biochem. Z.*, **155**, 163 (1925).

468. R. deGaulejac and P. Dervillee, *Ann. Med. Leg. Criminol. Police Sci.*, **18**, 146 (1938).

469. C. P. Carpenter, C. S. Weil, and H. F. Smyth, Jr., *Toxicol. Appl. Pharm.*, **28**, 313 (1974).

470. R. T. Williams, *Detoxication Mechanisms. The Metabolism and Detoxication of Drugs. Toxic Substances and Other Organic Compounds*, 2nd ed., Chapman and Hall Ltd., London 1959, p. 318. Cited by D. L. Opdyke, *J. Food Cosmet. Toxicol.*, **13**, Suppl., 903 (1975).

471. J. E. Urban and O. Wyss, *J. Gen. Microbiol.*, **56**, 69 (1969).

472. Y. Yoshiyama, K. Nagai, K. Arima, and G. Tamura, *Agric. Biol. Chem.* **37**(3), 527 (1973), cited by Y. Yoshiyama, K. Nagai, H. Some, and G. Tamura, *Agric. Biol. Chem.*, **37**(6), 1317 (1973).

473. A. Katz, *Spice Mill*, **69**, (July), 46 (1946). Cited by D. L. Opdyke, *J. Food Cosmet. Toxicol.*, **13**, Suppl, 903 (1975).

474. N. Greif, *Am. Perfumer Cosmet.* **82** (June), 54 (1967). Cited by D. L. Opdyke, *J. Food Cosmet. Toxicol.*, **13**, Suppl., 903 (1975).

475. R. E. Kirk and D. F. Othmer Eds., *Encyclopedia of Chemical Technology*, Vol. 4, Interscience, New York, 1949, p. 769.

475a. I. Mellan, *Industrial Solvents*, Reinhold, New York, 1950, pp. 519–521.

476. G. S. Gardner and J. E. Brewer, *Ind. Eng. Chem.*, **29**, 179 (1937).

477. U.S. Dept. of Health, Education, and Welfare, *NIOSH Manual of Analytical Methods*, Vol. 2, 2nd ed., 1977, S54.

478. J. F. Treon, W. E. Crutchfield, Jr., K. V. Kitzmiller, *J. Ind. Hyg. Toxicol.*, **25**, 323 (1943).

479. F. Bar and F. Griepentrog, *Med. Ernahr*, **8**, 244 (1967). Cited by D. L. Opdyke, *J. Food Cosmet. Toxicol.*, **13**, Suppl., 777 (1975).

480. J. F. Treon, W. E. Crutchfield, Jr., and K. V. Kitzmiller, *J. Ind. Hyg. Toxicol.*, **25**, 199 (1943).

481. H. F. Smyth, Jr., C. P. Carpenter, C. S. Weil, U. C. Pozzani and J. A. Striegel, *Am. Ind. Hyg. Assoc. J.*, **23**, 95 (1962).

482. J. Pohl, *Zentr. Gewerbehyg. Unfallverhuet.*, **12**, 91 (1925).

483. B. J. Northover and J. Verghese, *J. Sci. Ind. Res.*, **21C**, 342 (1962). Cited by D. L. Opdyke, *J. Food Cosmet. Toxicol.*, **13**, 777 (1975).

484. A. A. Dobrinskii, *Biologicheskoe Deistvie i Gigienicheskoe Znachenie Atmosfernykh Zagryaznenii*, p. 119. Cited from *Chem. Abstr.* **65**, 14324 (1966) by D. L. Opdyke, *J. Food Cosmet. Toxicol.*, **13**, Suppl., 777 (1975).

485. E. Gondry, *J. Eur. Toxicol.*, **5**(4), 227 (1973).

486. J. P. Collin, *Diabete*, **19**, 215 (1971). Cited by B. M. Cattanach, *Mutat. Res.*, **39**, 1 (1976).

487. B. M. Cattanach, *Mutat. Res.*, **39**, 1 (1976).

488. L. DiPrisco, *Minerva Med.*, **2**, 423 (1932).

489. R. T. Williams, *Detoxication Mechanisms, The Metabolism and Detoxication of Drugs, Toxic Substances and Other Organic Compounds*, 2nd ed., Wiley, New York, 1959, pp. 114, 116–117.

490. E. Browning., *Toxicity and Metabolism of Industrial Solvents*, Elsevier, London, 1965,. p. 385.

491. K. Bernhard, *Z. Physiol. Chem.*, **248**, 256 (1937).

492. G. Weitzel, *Z. Physiol. Chem.*, **285**, 58 (1950). Cited by D. L. Opdyke, *J. Food Cosmet. Toxicol.*, **13**, Suppl. 777 (1975).

493. Y. Sasaki, *Acta Schol. Med. Univ. Imp. Kioto*, **1**, 413 (1917).

494. H. Ichibagase, S. Kojima, K. Inoue, and A. Suenaga, *Chem. Pharm. Bull.*, Tokyo, **20**, 175 (1972). Cited by D. L. Opdyke, *J. Food Cosmet. Toxicol.*, **13**, Suppl. 777 (1975).

495. K. Sato, *Jap. J. Med. Sci. IV Pharmacol. Trans. Abstr.*, **3**(1), (1928).

496. N. Ya. Myasoedova, *Toksikol. Gig. Prod. Neftekhim. Neftekhim Proizvod.*, 76 (1968). Cited from *Chem. Abstr.*, **74**, 123228 (1971) by D. L. Opdyke, *J. Food Cosmet. Toxicol.*, **13**, Suppl. 777 (1975).

497. Yu. N. Pestrii, *Gig/Tr. Prof. Zabol.*, **14**, 37 (1970). Cited by D. L. Opdyke, *J. Food Cosmet. Toxicol.*, **13**, Suppl., 777 (1975).

498. W. L. Epstein, report to RIFM June 14 (1974). Cited by D. L. Opdyke, *J. Food Cosmet. Toxicol.*, **13**, Suppl., 777 (1975).

499. I. Mellan, *Industrial Solvents*, Reinhold, New York, 1950, pp. 521–522.

500. The Barrett Division of Allied Chemical and Dye Corp., personal communication.

501. E. Fillipi, *Arch. Farmacol. Sper.*, **18**, 178 (1914).

502. E. Browning, *Toxicity of Industrial Organic Solvents*, H. M. Stationery Office, London, 1953, pp. 239–240.

503. U.S. Department of Health, Education and Welfare, *NIOSH Manual of Analytical Methods*, Vol. 4, 2nd ed., 1978, S374.

504. R. E. Kirk and D. F. Othmer, Eds., *Encyclopedia of Chemical Technology*, Vol. 6, Interscience, New York, 1951, pp. 997, 1002–1003.

505. I. Mellan, *Industrial Solvents*, Reinhold, New York, 1950, p. 523.

506. K. H. Jacobson, W. E. Rinehart, H. J. Wheelwright, Jr., M. A. Ross, J. L. Papin, R. C. Daly, E. A. Greene and W. A. Groff, *Am. Ind. Hyg. Assoc. J.*, **19,** 91 (1958).

507. D. R. Stull, *Ind. Eng. Chem.*, **36,** 517 (1947).

508. U.S. Dept. of Health, Education and Welfare, *NIOSH Manual of Analytical Methods*, Vol. 4, 2nd ed., 1978, S365.

509. C. C. Comstock and F. W. Oberst, *Chem. Corps. Med. Labs. Res. Rep. No. 139* (Oct. 1952).

510. J. Gajewski and W. Alsdorf, *Fed. Proc.*, **8,** 294 (1949).

511. The Quaker Oats, Co., unpublished data, Dec. 26, 1957.

512. NDRC, Univ. of Chicago, unpublished report, Feb. 21, 1942. Cited by C. C. Comstock and F. W. Oberst, *Chem. Corps. Med. Labs. Res. Rep. No. 139* (Oct. 1952).

513. E. Erdmann, *Arch. Exp. Pathol. Pharmacol.*, **48,** 233 (1902). Cited by E. H. Fine and J. H. Wills, *Arch. Ind. Hyg. Occup. Med.*, **1,** 625 (1950).

514. M. J. Okubo, Pharm. Soc. Jap., **539,** 39 (1937). Cited by E. H. Fine and J. H. Wills, *Arch. Ind. Hyg. Occup. Med.*, **1,** 625 (1950).

515. E. H. Fine and J. H. Wills, *Arch. Ind. Hyg. Occup. Med.*, **1,** 625 (1950).

516. E. J. Mastromatteo, *Occup. Med.*, **7**(10), 502 (1965).

517. G. Joachimoglu and N. Klissunis, *Prakt. (Akad. Athenon)*, **7,** 39 (1932). In German. *Chem. Abstr.* **28,** 5257 (1934).

518. *Hyg. Sanit.*, **32,** 273 (1967). Cited in E. J. Fairchild, Ed., *Registry of Toxic Effects of Chemical Substances*, Vol. II, DHEW Publication No. (NIOSH) 78-104-B, 1977.

519. I. Mellan, *Industrial Solvents*, Reinhold, New York, 1950, pp. 609–610.

520. G. S. Gardner, *Ind. Eng. Chem.*, **32,** 226 (1940).

521. U.S. Dept. of Health, Education and Welfare, *NIOSH Manual of Analytical Methods*, Vol. 2, 2nd ed., 1977, S55.

522. D. C. Walton, E. F. Kehr, and A. S. Lovenhart, *J. Pharmacol.*, **33,** 175 (1928).

523. Unpublished work of E. Gross. Cited by K. B. Lehmann and F. Flury, *Toxikologie and Hygiene der Technischen Losungsmittel*, Springer, Berlin, 1938, p. 248. Translation by E. King and H. F. Smyth, Jr., p. 209 (1941).

524. Shell Chemical Co., unpublished report (1961). Cited in E. J. Fairchild, Ed., *Registry of Toxic Effects of Chemical Substances* Vol. II, DHEW Publication No. (NIOSH) 78-104-B, 1977.

525. H. M. Keith, *Arch. Pathol.*, **13,** 707 (1932).

526. W. Epstein, K. Gerber, and R. Karler, "The Hypnotic Constituent of *Stipa Vaseyi*, Sleepy Grass," *Experientia*, **20**(7), 390 (1964).

527. Shell Research Limited, London, 1979.

528. T. R. Torkelson, M. A. Wolf, F. Oyen, and V. K. Rowe, *Am. Ind. Hyg. Assoc. J.*, **20,** 224 (1959).

529. U.S. Dept. of Health, Education and Welfare, *NIOSH Manual of Analytical Methods*, Vol. 2, 2nd ed., 1977, S52.

530. M. K. Dunlap, J. K. Kodama, J. S. Wellington, H. H. Anderson, and C. H. Hine, *A.M.A. Arch. Ind. Health*, **18,** 303 (1958).

531. V. W. Reid and J. W. Beddard, *Analyst*, **79**, 456 (1954).

532. H. F. Smyth, Jr., C. P. Carpenter, C. S. Weil, *Arch. Ind. Hyg. Occup. Med.*, **4**, 199 (1951).

533. P. M. Jenner, E. C. Hagan, J. M. Taylor, E. L. Cook and O. G. Fitzhugh, *Food Cosmet. Toxicol.*, **2**(3), 327 (1964).

534. C. P. McCord, *J. Am. Med. Assoc.*, **98**, 2269 (1932).

535. C. P. Carpenter, H. F. Smyth, Jr., and U. C. Pozzani, *J. Ind. Hyg. Toxicol.*, **31**, 343 (1949).

536. A. Blyuger, A. Majore, M. A. Bolotova, T. Stroze, and V. Zalcmane, *Byull. Eksp. Biol. Med.*, **77**(3), 109 (1974).

537. E. M. Adams, H. C. Spencer, and D. D. Irish, *J. Ind. Hyg. Toxicol.*, **22**(2), 79 (1940).

538. J. K. Kodama and C. H. Hine, *J. Pharmacol. Exp. Ther.*, **124**, 97 (1958).

539. K. S. Al'meev and V. E. Karmazin, *Faktory Vnesh, Sredy Ikh Znachenie Zdoroy'ya Naseleniya* (1), 31 (1969).

540. T. R. Torkelson, M. A. Wolf, F. Oyen, and V. K. Rowe, *Am. Ind. Hyg. Assoc. J.*, **20**, 224 (1959).

541. Racker, *Methods in Enzymology*, Vol. 1, Academic Press, Inc., New York, 1955, p. 502. Cited by M. Legator and D. Racusen, *J. Bacteriol.*, **77**, 120 (1959).

542. V. E. Karmazin, *Faktory Vnesh. Sredy Ikh Znachenie Zdorov'ya Naseleniya* (1), 35 (1969).

543. Shell Chemical Corporation, Industrial Hygiene Bulletin No. SD57-77.

544. J. McLaughlin, Jr., J. P. Marliac, M. J. Varrett, M. K. Mutchler, and O. G. Fitzhugh, *Am. Ind. Hyg. Assoc. J.*, **25**, 282 (1964).

545. D. H. Kim, *K'at'ollik Taehak Uihakpu Nonmunjip*, **18**, 37 (1970); *Chem. Abstr.*, **74**, 123224m (1971).

546. G. Berg, U. Troll, H. Breining, and O. Strubelt, *Med. Exp.*, **18**(2), 113 (1968). In German. *Chem. Abstr.*, **71**, 55849j (1969).

547. W. D. Reid, *Experientia*, **28**, 1058 (1972). Information Bulletin, *BIBRA*, **12**, 245, (1973) abstr. no. 2576.

548. F. Serafini-Cessi, *Biochem. J.*, **128**, 1103 (1972). *Food Cosmet. Toxicol.*, **11**(2), 325 (1973) abstr. no. 2488.

549. F. M. B. Carpanini, I. F. Gaunt, J. Hardy, S. D. Gangolli, K. R. Butterworth, and A. G. Lloyd, *Toxicology*, **9**, 29 (1978).

550. C. M. Kaye and L. Young, *Biochem. J.*, **127**, 97 (1972).

551. C. M. Kaye, *Biochem. J.*, **134**, 1093 (1973).

552. *Toksikologiya Novykh Promyshlennykh Khimischeskikh Veshchestv.* (Akademiya Meditsinskikh Nauk S.S.R., Moscos, U.S.S.R.), **8**, 97 (1966). Cited by *Registry of Toxic Effects of Chemical Substances* Vol. II, E. J. Fairchild, Ed., DHEW Publication No. (NIOSH) 78-104-B, 1977.

553. General Aniline and Film Corporation, unpublished data.

554. General Aniline and Film Corporation, Antara Chemicals Division, Bulletin 1M-7-61, AP-51.

555. K. P. Stasenkova and T. A. Kochetkova, *Toksikol. Novykh. Prom. Khim. Veshchestv.*, **8**, 97 (1966). In Russian. *Chem. Abstr.*, **67**, 89293b (1967).

556. Air Products and Chemicals, Inc., product bulletin on *Hexynol*. No. 120-042.16, 1980.

557. Air Products and Chemicals, Inc., Material Safety Data Sheet, *Hexynol*, 1979.

558. R. E. Kirk and D. F. Othmer, Eds., *Encyclopedia of Chemical Technology*, Vol. 3, Interscience, New York, 1949.

559. E. T. Huntress, *The Preparation, Properties, Chemical Behavior, and Identification of Organic Chlorine Compounds*, Wiley, New York, 1948, p. 705.

560. A. M. Ambrose, *Arch. Ind. Hyg. Occup. Med.*, **2**, 591 (1950).

561. F. Koelsch, *Zbl. Gewerbehyg.*, **14,** 312 (1927). Cited by E. Browning, *Toxicity of Industrial Organic Solvents*, Chemical Publishing Co., New York, 1953, p. 249.

562. E. L. Middleton, *J. Ind. Hyg. Toxicol.*, **12,** 265 (1930).

563. D. Cavallazzi, Samml. *Vergiftungsfallen*, **12,** 79 A-910 (1942). Cited by E. Browning, *Toxicity of Industrial Organic Solvents*, Chemical Publishing Co., New York, 1953, p. 249.

564. H. Dierker and P. Brown, *J. Ind. Hyg. Toxicol.*, **26,** 277 (1944).

565. M. W. Goldblatt and W. E. Chiesman, *Br. J. Ind. Med.*, **1,** 207 (1944).

566. H. F. Smyth, Jr., and C. P. Carpenter, *J. Ind. Hyg. Toxicol.*, **27,** 93 (1945).

567. *NIOSH Manual of Analytical Methods*, Vol. II, 2nd ed., USD HEW Publication No. 77-157-B, April 1977.

568. K. Uhrig, *Ind. Eng. Chem. Anal. Ed.*, **18,** 469 (1946).

569. M. W. Goldblatt, *Br. J. Ind. Med.*, **1,** 213 (1944).

570. W. H. Lawrence, J. E. Turner, and J. Autian, *J. Pharm. Sci.*, **60**(4), 568 (1971).

571. V. N. Semenova, S. S. Kazanina, and B. Y. Ekshtat, *Hyg. Sanit.*, **36**(6), 376 (1971).

572. T. McDonald, M. Roberts, and A. Borgman, *Toxicol. Appl. Pharmacol.*, **21,** 143 (1972).

573. W. H. Lawrence, K. Itoh, J. E. Turner, and J. Autian, *J. Pharm. Sci.*, **60**(8), 1163 (1971).

574. F. Koelsch, *Zbl. Gewerbehyg. N. F.*, **4,** 312 (1927). Cited by K. B. Lehmann and F. Flury, *Toxicology and Hygiene of Industrial Solvents*, Transl. by E. King and H. F. Smyth, Jr., Williams & Wilkins, Baltimore, 1948, p. 215.

575. M. M. Mason, C. C. Cate, and J. Baker, *Clin. Toxicol.*, **4**(2), 185 (1971).

576. B. Oser, K. Morgareidge, G. Cox, and S. Carson, *Food Cosmet. Toxicol.*, **13,** 313 (1975).

577. M. K. Johnson, *Food Cosmet. Toxicol.*, **5,** 449 (1967).

578. G. K. Isakova, B. Y. Ekshtat, and Y. Y. Kerkis, *Hyg. Sanit.*, **36**(11), 178, (1971).

579. M. K. Johnson, *Biochem. Pharm.*, **14**(9), 1383 (1965).

580. M. K. Johnson, *Biochem. Pharm.*, **16**(1), 185 (1967).

581. H. S. Rosenkranz and T. J. Wlodkowski, *J. Agric. Food Chem.*, **22**(3), 407 (1974).

582. S. Rosenkranz, H. S. Carr, and H. S. Rosenkranz, *Mutat. Res.*, **26,** 367 (1974).

583. C. Malaveille, H. Bartsch, A. Barbin, A. M. Camus, and R. Montesano, *Biochem. Biophys. Res. Comm.*, **63**(2), 363 (1975).

584. H. Bartsch, C. Malaveille, and R. Montesano, *Int. J. Cancer*, **15,** 429 (1975).

585. U. Rannug, R. Gothe, and C. A. Wachtmeister, *Chem-Biol. Interactions*, **12,** 251 (1976).

586. J. McCann, V. Simmon, D. Streitwieser, and B. Ames, *Proc. Natl. Acad. Sci. U.S.*, **72**(8), 3190 (1975).

587. N. Loprieno, R. Barale, S. Baroncelli, H. Bartsch, G. Bronzetti, A. Cammellini, D. Corsi, D. Frezza, R. Nieri, C. Leporini, D. Rossellini, and A. M. Rossi, *Cancer Res.*, **36,** 253 (1977).

588. V. Miller, R. J. Dobbs, and S. Jacobs, *Arch. Dis. Childhood*, **45**(242), 589 (1970).

589. S. S. Epstein, E. Arnold, J. Andrea, W. Bass, and Y. Bishop, *Toxicol. Appl. Pharmacol.*, **23,** 288 (1972).

590. S. J. Stolzenberg and C. H. Hine, *Environ. Mutagenesis*, **2,** 59 (1980).

591. A. O. Saitanov and A. M. Kononova, *Gig. Tr. Prof. Zabol.*, **2,** 49 (1976). Abstract from *Chem. Abstr.*, **84,** 145550Y (1976).

592. J. E. Wahlberg, and A. Boman, *Chem. Abstr., Dermatologica*, **156**(5), 299 (1978).

592a. E. H. Pfeiffer and H. Dunkelberg, *Food Cosmet. Toxicol.*, **18,** 115 (1980).

593. D. L. Opdyke, *J. Food Cosmet. Toxicol.*, **13,** Suppl., 785 (1975).

594. J. C. Gage, *Br. J. Ind. Med.,* **27,** 1 (1970).

595. Air Products and Chemicals, Inc., Material Safety Data Sheet, *Methyl Butenol,* (1979).

596. Air Products and Chemicals, Inc., Material Safety Data Sheet, *Methyl Butynol,* (1980).

597. Air Products and Chemicals, Inc., Material Safety Data Sheet, *Ethyl Octynol,* (1980).

598. Air Products and Chemicals, Inc., Product Bulletin No. 120-042.15, *Ethyl Octynol,* (1980).

599. Air Products and Chemicals, Inc., Product Bulletin No. 120-042.17, *Ethynyl Cyclohexanol,* 1980.

600. Air Products and Chemicals, Inc., Material Safety Data Sheet, *Ethynyl Cyclohexane,* 1979.

601. Air Products and Chemicals, Inc., Material Safety Data Sheet, *Methyl Pentynol,* 1980.

602. Air Products and Chemicals, Inc., Product Bulletin No. 120-042.18, *Methyl Pentynol,* 1980.

CHAPTER FIFTY-SIX

Ketones

W. J. KRASAVAGE, J. L. O'DONOGHUE,
V.M.D., Ph.D., and G. D. DIVINCENZO,
Ph.D.

1 INTRODUCTION

1.1 General Comments

A ketone is an organic compound containing a carbonyl group (C=O) attached to two carbon atoms and can be represented by the general formula

$$R—\overset{\displaystyle O}{\overset{\displaystyle \|}{C}}—R'.$$

Several billion pounds of ketones are produced annually for industrial use in the United States. Those with the highest production volumes include acetone, methyl ethyl ketone, methyl isobutyl ketone, cyclohexanone, 4-hydroxy-4-methyl-2-pentanone, isophorone, mesityl oxide, and acetophenone.

Ketones are used extensively in industry and commerce because of the ease of production, low manufacturing costs, excellent solvent properties, and desirable physical properties such as low viscosity, moderate vapor pressure, low to moderate boiling points, high evaporation rates, and a wide range of miscibility with other liquids. The low molecular weight aliphatic ketones are miscible with water and organic solvents whereas the high molecular weight aliphatic and aromatic ketones are generally immiscible with water. Most ketones are chemically stable. The exceptions are mesityl oxide, which can form peroxides, and methyl isopropenyl ketone, which polymerizes. Most ketones are generally of low flammability.

Common methods used to manufacture ketones include aliphatic hydrocarbon oxidation, alcohol dehydration with subsequent oxidation, dehydrogenation of phenol, alkyl aromatic hydrocarbon oxidation, and condensation reactions. Ketones are commonly used in industry as solvents, extractants, chemical intermediates, and to a lesser extent, flavor and fragrance ingredients. Recently, the use of branch chain ketones has fallen into disfavor owing to their propensity to contribute to the formation of photochemical smog.

1.2 Occupational Exposures

In an occupational setting the primary routes of exposure to ketones are inhalation and skin contact. Ingestion is rare. Since most ketones have a high vapor pressure at room temperature, exposure by inhalation in the workplace is likely to occur. The principal hazard associated with exposure to ketone vapors is irritation of the eyes, nose, and throat. Many ketones have excellent warning properties and can be easily detected by the olfactory sense. Accidental overexposure should be relatively rare provided the warning properties are not ignored and olfactory fatigue does not occur. The classic symptoms produced by an overexposure to ketones include, progressively, irritation of the eyes, nose, and throat, headache, nausea, vertigo, incoordination, central nervous system depression, narcosis, and cardiorespiratory failure. Recovery is usually rapid and without residual toxic effects. In the case of accidental spills, personnel should wear protective clothing including respiratory protection. Contaminated clothing should be removed promptly and the exposed areas of the body should be thoroughly flushed with water. Many ketones are absorbed through the skin; therefore caution should be exercised to avoid repeated or prolonged skin contact. The vapors produced by accidental spills may present a fire or explosion hazard.

1.3 Toxic Effects

Although the relative toxicity of most ketones is low and the effects of acute exposures are well recognized, the effects of chronic exposure are less well understood. In some cases, metabolic studies have helped to elucidate the toxic effects of several ketones. Generally, when ketones are absorbed into the bloodstream, they may be eliminated unchanged in the expired air, reduced to secondary alcohols, or oxidized to hydroxyketones, diketones, and carbon dioxide by a variety of metabolic pathways. Recent studies indicate that carbonyl reduction, α and ω-1 oxidation, decarboxylation, and transamination play important roles in the metabolism of aliphatic ketones. Aromatic ketones and ketones such as cyclohexanone and isophorone may undergo oxidative metabolism by dehydrogenation, ring hydroxylation, or substituent group oxidation. In addition, aromatic and aliphatic ketones may be conjugated with glucuronic acid, sulfuric acid, or glutathione prior to excretion in the urine. Glucuronic and sulfuric acid conjugation usually occur after a ketone is reduced to a secondary alcohol or oxidized to a carboxylic acid. Of the various conjugation mechanisms that occur, glucuronic acid conjugation appears to be the predominant pathway.

1.4 Structure/Activity Relationships

Alkanes, primary and secondary alcohols, carboxylic acids, glycols, diketones, epoxides, hydroxy acids, and ketones are metabolically related in many biologic systems. Thus a knowledge of the structure/activity relationships of these various compounds adds to our understanding of their individual and/or combined toxicities. Much of our present knowledge about the structure/activity relationships of ketones has surfaced in the past five years as the result of an occupationally related outbreak of neurotoxicity. Since this incident, the emphasis on ketone toxicity has been directed primarily toward neurotoxicity. It must be realized, however, that these same neurotoxic ketones produce toxic effects other than neurotoxicity.

Table 56.1 lists ketones and related compounds that have been examined for neurotoxicity. Those indicated as positive are substances that showed a specific anatomic and morphologic type of nerve degeneration characterized by large multifocal axonal swellings, often referred to as "giant axonal" neuropathy. These swellings are filled with masses of disorganized neurofilaments and other organelles. Myelin damage also occurs but is generally considered to be a secondary effect. Clinical symptomatology in man includes bilaterally symmetrical paresthesia, best described as a "pins and needles" feeling, and muscle weakness, primarily in the legs and arms.

The metabolic interrelationships of some of these neurotoxins are shown in Figure 56.1. These findings have led to the theory that neurotoxicity is related to a common metabolic pathway leading to the formation of a γ-diketone, which is the toxic metabolite that produces the neuropathy. Except for 2,5-heptanedione and 3,6-octanedione, all metabolic interconversions are oxidation of the ω-1 carbon(s), first to an alcohol or diol, then to a γ-diketone. When the ω carbon is oxidized in preference to the ω-1 carbon, as when n-hexane is converted to 1-hexanol, no γ-diketone is formed. In the case of n-heptane, where ω-1 oxidation may occur, the ketone formed would be a δ-diketone such as 2,6-heptanedione, which is not neurotoxic. The neurotoxicity of 5-nonanone appears to involve two metabolic pathways, one to 2,5-nonanedione, a mechanism similar to that of the other compounds shown in Figure 56.1, and the other to methyl n-butyl ketone via a series of oxidative and decarboxylative pathways (Figure 56.2).

Initially, studies of n-hexane and methyl n-butyl ketone neurotoxicity revealed that the γ-diketone 2,5-hexanedione was a neurotoxin. Subsequently, a series

Table 56.1. Neurotoxicity of Ketones and Related Substances

Chemical	Structure	Neuro-toxicity[a]	Ref.
	Three-Carbon Structure		
Acetone	$CH_3-\overset{\displaystyle O}{\overset{\|}{C}}-CH_3$	−	237
	Four-Carbon Structures		
Methyl ethyl ketone	$CH_3-\overset{\displaystyle O}{\overset{\|}{C}}-CH_2-CH_3$	−	256, 236
1,4-Butanediol	$HOCH_2-CH_2-CH_2-CH_2OH$	−	237
	Five-Carbon Structures		
Methyl n-propyl ketone	$CH_3-\overset{\displaystyle O}{\overset{\|}{C}}-CH_2-CH_2-CH_3$	−	256
Diethyl ketone	$CH_3-CH_2-\overset{\displaystyle O}{\overset{\|}{C}}-CH_2-CH_3$	−	256
2,4-Pentanedione	$CH_3-\overset{\displaystyle O}{\overset{\|}{C}}-CH_2-\overset{\displaystyle O}{\overset{\|}{C}}-CH_3$	[b]	256
	Six-Carbon Structures		
n-Hexane	$CH_3-CH_2-CH_2-CH_2-CH_2-CH_3$	+	131
Practical grade hexanes	Mixed hexanes	+	131
1-Hexanol	$HOCH_2-CH_2-CH_2-CH_2-CH_2-CH_3$	−	185, 131
2-Hexanol	$CH_3-\overset{\displaystyle OH}{\overset{\|}{C}}HCH_2-CH_2-CH_2-CH_3$	+	185, 131
6-Amino-1-hexanol	$HOCH_2-CH_2-CH_2-CH_2-CH_2-CH_2NH_2$	−	256
Methyl n-butyl ketone	$CH_3-\overset{\displaystyle O}{\overset{\|}{C}}-CH_2-CH_2-CH_2-CH_3$	+	121, 131, 157, 235, 236, 251
Methyl isobutyl ketone	$CH_3-\overset{\displaystyle O}{\overset{\|}{C}}-CH_2-\overset{\displaystyle CH_3}{\overset{\|}{C}}H-CH_3$	−	235, 236, 256
2,5-Hexanediol	$CH_3-\overset{\displaystyle OH}{\overset{\|}{C}}H-CH_2-CH_2-\overset{\displaystyle OH}{\overset{\|}{C}}H-CH_3$	+	131, 237
1,6-Hexanediol	$HOCH_2-CH_2-CH_2-CH_2-CH_2-CH_2OH$	−	237

Table 56.1. (*Continued*)

Chemical	Structure	Neuro-toxicity[a]	Ref.
Six-Carbon Structure (*Continued*)			
5-Hydroxy-2-hexanone	$CH_3-\overset{\overset{\displaystyle O}{\|}}{C}-CH_2-CH_2-\overset{\overset{\displaystyle OH}{\|}}{CH}-CH_3$	+	131
2,3-Hexanedione	$CH_3-\overset{\overset{\displaystyle O}{\|}}{C}-\overset{\overset{\displaystyle O}{\|}}{C}-CH_2-CH_2-CH_3$	−	174, 175, 237, 256
2,4-Hexanedione	$CH_3-\overset{\overset{\displaystyle O}{\|}}{C}-CH_2-\overset{\overset{\displaystyle O}{\|}}{C}-CH_2-CH_3$	−	174, 175, 237, 256
2,5-Hexanedione	$CH_3-\overset{\overset{\displaystyle O}{\|}}{C}-CH_2-CH_2-\overset{\overset{\displaystyle O}{\|}}{C}-CH_3$	+	131, 173, 174, 237, 256
Seven-Carbon Structures			
n-Heptane	$CH_3-CH_2-CH_2-CH_2-CH_2-CH_2-CH_3$	−	256
Methyl *n*-amyl ketone	$CH_3-\overset{\overset{\displaystyle O}{\|}}{C}-CH_2-CH_2-CH_2-CH_2-CH_3$	−	116
Methyl isoamyl ketone	$CH_3-\overset{\overset{\displaystyle O}{\|}}{C}-CH_2-CH_2-\overset{\overset{\displaystyle CH_3}{\|}}{CH}-CH_3$	−	256
Ethyl *n*-butyl ketone	$CH_3-CH_2-\overset{\overset{\displaystyle O}{\|}}{C}-CH_2-CH_2-CH_2-CH_3$	+[c]	121 256
Di-*n*-propyl ketone	$CH_3-CH_2-CH_2-\overset{\overset{\displaystyle O}{\|}}{C}-CH_2-CH_2-CH_3$	−	256
2,5-Heptanedione	$CH_3-\overset{\overset{\displaystyle O}{\|}}{C}-CH_2-CH_2-\overset{\overset{\displaystyle O}{\|}}{C}-CH_2-CH_3$	+	256, 174
2,6-Heptanedione	$CH_3-\overset{\overset{\displaystyle O}{\|}}{C}-CH_2-CH_2-CH_2-\overset{\overset{\displaystyle O}{\|}}{C}-CH_3$	−	256, 174
3,5-Heptanedione	$CH_3-CH_2-\overset{\overset{\displaystyle O}{\|}}{C}-CH_2-\overset{\overset{\displaystyle O}{\|}}{C}-CH_2-CH_3$	−	237
Eight-Carbon Structures			
3,6-Octanedione	$CH_3-CH_2-\overset{\overset{\displaystyle O}{\|}}{C}-CH_2-CH_2-\overset{\overset{\displaystyle O}{\|}}{C}-CH_2-CH_3$	+	256, 174

Table 56.1. (Continued)

Chemical	Structure	Neuro-toxicity [a]	Ref.
	Nine-Carbon Structures		
5-Nonanone	$\begin{matrix} & & & & \text{O} \\ & & & & \parallel \\ \text{CH}_3\text{—CH}_2\text{—CH}_2\text{—CH}_2\text{—C—CH}_2\text{—CH}_2\text{—CH}_2\text{—CH}_3 \end{matrix}$	+	256
5-Methyl-2-octanone	$\begin{matrix} \text{O} & & & \text{CH}_3 \\ \parallel & & & \mid \\ \text{CH}_3\text{—C—CH}_2\text{—CH}_2\text{—CH—CH}_2\text{—CH}_2\text{—CH}_3 \end{matrix}$	d	256
Diisobutyl ketone	$\begin{matrix} \text{CH}_3 & & \text{O} & & \text{CH}_3 \\ \mid & & \parallel & & \mid \\ \text{CH}_3\text{—CH—CH}_2\text{—C—CH}_2\text{—CH—CH}_3 \end{matrix}$	−	256
	Eleven-Carbon Structures		
Diisoamyl ketone	$\begin{matrix} \text{CH}_3 & & \text{O} & & \text{CH}_3 \\ \mid & & \parallel & & \mid \\ \text{CH}_3\text{—CH—CH}_2\text{—CH}_2\text{—C—CH}_2\text{—CH}_2\text{—CH—CH}_3 \end{matrix}$	−	256

[a] − Indicates that the material was tested experimentally and found not to be neurotoxic; + indicates the material may produce giant axonal neuropathy.

[b] 2,4-Pentanedione produces a clinically, anatomically, and morphologically distinct neurotoxic response which differs from that produced by the other ketones.

[c] Ethyl *n*-butyl ketone is metabolized to 2,5-heptanedione which is neurotoxic.

[d] Commercial samples of 5-methyl-2-octanone may contain 5-nonanone, which is neurotoxic. 5-methyl-2-octanone enhances 5-nonanone neurotoxicity.

of diketones were examined for their ability to produce "giant axonal" neuropathy in rats. Table 56.2 lists these compounds and further emphasizes the necessity of the γ-diketone spacing for the production of neuropathy. These data also suggest that, as chain length increases, the neurotoxicity of the diketone decreases, possibly owing to steric hindrance. However, chain length may not be as important for compounds, such as 5-nonanone, which is metabolized to methyl *n*-butylketone and 2,5-hexanedione. When the neurotoxic potential of aliphatic ketones and related substances is being considered, it is important to understand the necessity of the γ-diketone structure and secondary modifying factors such as chain length.

1.5 Summary

A summary of the toxicologic properties of the ketones included in the following monographs is presented in Table 56.3. During the past 20 years a significant

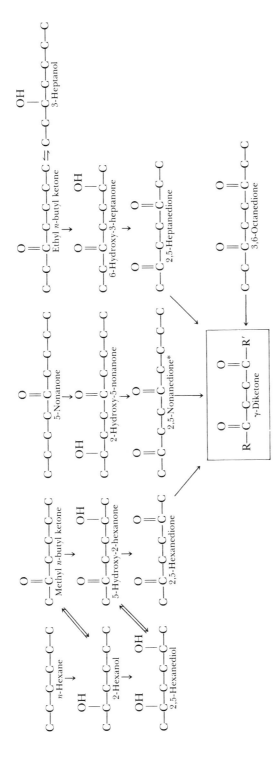

* Further oxidative and decarboxylative pathways lead to the formation of methyl *n*-butyl ketone and 2,5-hexanedione (see Figure 56.2).

Figure 56.1 Relationships of alkanes, alcohols, and ketones that produce "giant axonal" neuropathy. Hydrogen atoms are included only when present as hydroxylions.

4715

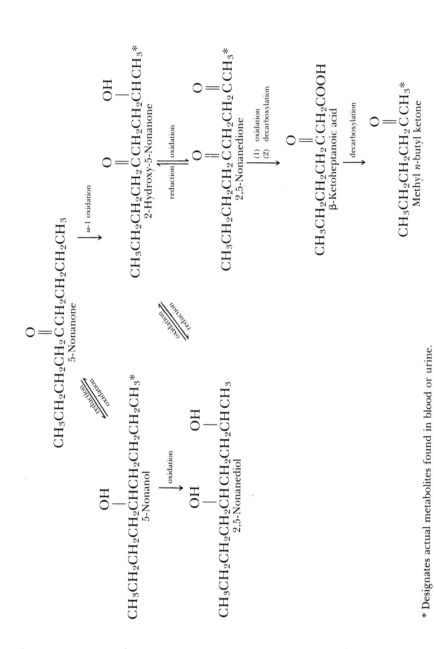

Figure 56.2 Metabolism of 5-nonanone.

* Designates actual metabolites found in blood or urine.

Table 56.2. Structure/Activity Relationships of Diketones Tested

Diketone	Structure	Ketone Spacing	Giant Axonal Neuropathy
2,4-Pentanedione	CH_3—C—CH_2—C—CH_3 (O, O)	β	$-$[a]
2,3-Hexanedione	CH_3—C—C—CH_2—CH_2—CH_3 (O O)	α	$-$
2,4-Hexanedione	CH_3—C—CH_2—C—CH_2—CH_3 (O, O)	β	$-$
2,5-Hexanedione	CH_3—C—CH_2—CH_2—C—CH_3 (O, O)	γ	$+$
3,5-Heptanedione	CH_3—CH_2—C—CH_2—C—CH_2—CH_3 (O, O)	β	$-$
2,5-Heptanedione	CH_3—C—CH_2—CH_2—C—CH_2—CH_3 (O, O)	γ	$+$
2,6-Heptanedione	CH_3—C—CH_2—CH_2—CH_2—C—CH_3 (O, O)	δ	$-$
3,6-Octanedione	CH_3—CH_2—C—CH_2—CH_2—C—CH_2—CH_3 (O, O)	γ	$+$

[a] 2,4-Pentanedione produces central nervous system damage which is clinically, anatomically, and morphologically distinguishable from "giant" axonal neuropathy.

amount of data has been accumulated on the biologic and toxicologic effects of ketones in experimental animals and man. With the exception of studies conducted after 1973, which have shown that several ketones produce a toxic polyneuropathy, these findings support the existing concepts of the relatively innocuous biologic effects of most ketones. The most widely and extensively used ketones such as acetone, methyl ethyl ketone, methyl isobutyl ketone, and cyclohexanone appear to be the least toxic.

Generally, there is a paucity of data on the chronic effects of ketones, particularly in the areas of teratogenicity, mutagenicity, carcinogenicity, and reproduction. More information is needed, not only in these areas but also for chronic, low level exposures to ketones and the effects of various mixtures of

Table 56.3. Toxicologic Properties of Ketones

Compound	Approximate Oral Rat LD$_{50}$ (ml/kg)	Lowest Reported Lethal Air Concn. Rat (ppm/hr)	Skin Irritation[a]	Ocular Injury[a]	TLV[b]	Toxic Effects
Acetone	8–11	16,000/4	SI	M	1000 (750)	Narcosis, cataracts, hepatic enzyme inducer
Methyl ethyl ketone	3–7	4,000/2	M	M	200	Narcosis, hepatic enzyme inducer, n-hexane and methyl n-butyl ketone interaction, fetotoxic
3-Butyn-2-one	0.01	10/4	SV	SV	—	Lacrimator, lung, kidney, and liver toxicity
Methyl n-propyl ketone	5	>2,000/4	SI	M	200	Narcosis
3-Methyl-2-butanone	4–7	5,700/4	SI	SI	—	Narcosis
3-Pentyn-2-one	0.1	Sat'd./0.1	SV	SV	—	Brain, liver, and kidney toxicity
Methyl isopropenyl ketone	0.2	125/4	M	SV	—	Lacrimator, narcosis, severe respiratory irritant
Methyl n-butyl ketone	3	8,000/4	SI	SI	5	Narcosis, nerve, testes, white blood cell toxicity
Methyl isobutyl ketone	5–6	4,000/6	SI	SI	50	Narcosis, liver, lung, spleen, and kidney toxicity
Mesityl oxide	1	500/8	SI	M	15	Narcosis, liver, lung, spleen, and kidney toxicity

Compound						Toxicity
4-Hydroxy-4-methyl-2-pentanone	4	>Sat'd/8	SI	M	50	Narcosis, liver and kidney toxicity
Methyl n-amyl ketone	2	4,000/4	M	SI	50	Narcosis, liver and kidney toxicity
Methyl isoamyl ketone	4	3,813/6	SI	SI	100 (50)	Narcosis, liver and kidney toxicity
Ethyl n-butyl ketone	3	4,000/4	SI	SI	50	Narcosis, possible neurotoxicity, white blood cell depression
Di-n-propyl ketone	4	2,670/6	SI	SI	50	Narcosis, liver and kidney toxicity
2-Octanone	11	—	M	SI	—	Narcosis
3-Octanone	>6	—	M	—	—	
5-Methyl-3-heptanone	4	3,484/4	SI	M	25	Narcosis
5-Nonanone	>2	—	—	—	—	Neurotoxicity
Diisobutyl ketone	7	Sat'd/8	SI	SI	25	Narcosis, liver, kidney, and cardiovascular toxicity
Trimethyl nonanone	10	>Sat'd/4	SI	SI	—	
2,4-Pentanedione	1	1,000/2	M	SV	—	Narcosis, brain and thymus toxicity
2,5-Hexanedione	1.5–3	Sat'd/1	SI	SV	—	Neurotoxic, testes and blood cell toxicity
Cyclohexanone	2	2,000/4	M	SI	25	Narcosis, cataracts, lymphoid changes
Methyl cyclohexanones	1–3	2,800/4	M	M	50 (S)	Narcosis
Acetophenone	1–3	>Sat'd/8	M	M	—	Narcosis
Isophorone	2->3	1,840/4	SI	SV	5	Lung and kidney toxicity

a SI—slight, M—moderate, SV—severe.

b TLV of American Conference of Governmental Industrial Hygienists (1981); numbers in parentheses indicate proposed changes; (S) indicates a skin notation applicable to TLV.

ketones, particularly their ability to act synergistically to produce toxic effects. Despite the information gaps that remain, ketones appear to present a low degree of hazard in the workplace. Most of the ketones have good warning properties at low concentrations and available data indicate that adverse effects such as neuropathy and renal and hepatic changes have threshold doses that are well above the levels expected to be encountered in the workplace.

2   SPECIFIC COMPOUNDS

2.1   Acetone; 2-Propanone, Dimethyl Ketone

$$CH_3—\overset{\overset{\displaystyle O}{\|}}{C}—CH_3$$

2.1.1   Physical-Chemical Properties

Acetone is a clear, colorless, volatile liquid with a characteristic odor. The odor threshold is about 20 ppm (265). Acetone is highly flammable and poses a fire hazard. Additional physical-chemical properties are summarized in Table 56.4.

2.1.2   Biologic Effects

Acetone is considered to be one of the least toxic solvents used in industry. However, exposure to high concentrations can produce central nervous system depression and narcosis. Prolonged or repeated skin contact may defat the skin and produce dermatitis. Liquid acetone is moderately irritating to the eyes, whereas the vapors produced only slight eye irritation at or below 1000 ppm.

Animal Studies.   *Single Dose.*   Various investigators have reported oral LD_{50}s of 8.5, 9.1, and 10.7 ml/kg for rats (227, 258, 123). The oral LD_{50} ranged from 4000 to 8000 mg/kg for mice (256). An oral LD_{50} of 5.3 g/kg was reported for rabbits (239). Intravenous administration of 4 ml/kg and 6 to 8 ml/kg was lethal to rats and rabbits, respectively (270). The minimum lethal oral dose for dogs was 8 g/kg (4). The minimum lethal intraperitoneal dose was 500 mg/kg for rats (205). The acute intraperitoneal LD_{50} for mice was 1297 mg/kg (257). An intramuscular dose of 5 ml/kg produced central nervous system depression in rabbits (270). The minimum lethal subcutaneous dose was 5 g/kg for dogs and guinea pigs (4, 71).

No evidence of teratogenicity was found when 39 or 78 mg of acetone was injected into the yolk sacs of fertile chick eggs prior to incubation (154). DiPaolo et al. (46) added 0.2 percent acetone to the growth medium of cultured Syrian hamster embryonic cells and detected no evidence of cellular transformation. Acetone was not mutagenic in the Salmonella/microsome test (153).

Repeated Dose.   Van Duuren et al. (263) reported that acetone did not produce tumors when applied to the skin of mice three times a week for 1 year.

Eye.   Carpenter and Smyth (25) reported that acetone produced moderate corneal injury to rabbit eyes. Larson et al. (136) reported mild ocular edema. Small multiple doses of acetone administered percutaneously (0.5 ml) or subcutaneously (0.05 ml) over a period of 3 to 8 weeks produced cataracts in guinea pigs. The eyes were examined at 60 or 90 days after the initial application, then at 30-day intervals up to 6 months. Two of the guinea pigs exposed by skin contact for 3 weeks developed cataracts by the third month. Subcutaneous administration produced cataracts in 7 of 16 guinea pigs. No cataracts were seen in control animals. Microscopic examination revealed extensive lens damage with eosinophilic deposits in subcapsular areas. The lens damage was considered to be reversible in some cases. In a subsequent study conducted similarly, acetone produced cataracts in guinea pigs, but not in rabbits (196).

Skin.   Acetone applied to the skin of rabbits (395 mg) produced mild skin irritation. The dermal LD_{50} was 20 g/kg (258).

Inhalation.   The minimum lethal concentration was 16,000 ppm for rats exposed to acetone vapors for 4 hr (227) and 46,000 ppm for mice exposed for 1 hr (71). Lethal concentrations for mice, rats, and guinea pigs were 46,000, 42,000, and 40,000 ppm, respectively (97, 71, 232). Guinea pigs exposed to 50,000 ppm showed immediate signs of irritation. Slight irritation was seen at 20,000 ppm and no irritation was noted at 10,000 ppm (232). Lethal concentrations reported for rats, guinea pigs, and cats were 126,000 ppm (2 hr), 50,000 ppm (¾ hr), and 21,000 ppm (3 hr), respectively (265). The anesthetic concentration of acetone was 99 mg/l for male and female mice (214). Rats exposed to 3000, 6000, 12,000, and 16,000 ppm of acetone 4 hr/day for 10 days showed some behavioral changes, particularly at the high levels, for example, the inability to climb a pole within 2 sec of receiving a stimulus. Tolerance developed after additional exposures (88). Rats exposed to 19,000 ppm of acetone 3 hr/day, 5 days/week for 8 weeks, and sacrificed at 2, 4, 8, and 10 weeks of exposure showed no evidence of toxic effects (22).

Table 56.4. Physical-Chemical Properties of Ketones

Compound	Molecular Formula	Mol. Wt.	Boiling Point (°C)	Melting Point (°C)	Specific Gravity[a]	Refractive index (20°C)	Vapor Pressure (mm Hg)[b]
Acetone	C_3H_6O	58.1	56.2	−95.4	0.791	1.3588	180 (20)
Methyl ethyl ketone	C_4H_8O	72.1	79.6	−86.6	0.807 (20/4)	1.3788	77.5 (20)
3-Butyn-2-one	C_4H_4O	68.1	83.0		0.879	1.4024	
Methyl n-propyl ketone	$C_5H_{10}O$	86.1	102.2	−77.8	0.809 (20/4)	1.3895	16.0
3-Methyl-2-butanone	C_5H_7O	83.1	93.0	−92.0	0.803 (20/0)	1.3879	
3-Pentyn-2-one	C_5H_6O	82.1	133.0	−28.7	0.910	1.141	
Methyl isopropenyl ketone	C_5H_8O	84.1	97.7	−53.7	0.855	1.4220	42.0
Methyl n-butyl ketone	$C_6H_{12}O$	100.2	127.5	−59.6	0.821	1.4007	3.8
Methyl isobutyl ketone	$C_6H_{12}O$	100.2	115.8	−83.5	0.802	1.3959	7.5
Mesityl oxide	$C_6H_{10}O$	98.2	129.6	−46.4	0.857	1.4440	9.5
4-Hydroxy-4-methyl-2-pentanone	$C_6H_{12}O$	116.2	169.2	−44.0	0.941	1.4242	1.2
Methyl n-amyl ketone	$C_7H_{14}O$	114.2	150.6	−26.9	0.817	1.4073	2.1
Methyl isoamyl ketone	$C_7H_{14}O$	114.2	144.0	−73.9	0.813	1.4062	4.5 (20)
Ethyl n-butyl ketone	$C_7H_{14}O$	114.2	147.6	−39.0	0.816	1.3994	1.4
Di-n-propyl ketone	$C_7H_{14}O$	114.2	143.7	−32.6	0.817 (20/4)	1.4069	5.5 (20)
2-Octanone	$C_8H_{16}O$	128.2	172.9	−16.0	0.819 (20/4)	1.4151	1.2
3-Octanone	$C_8H_{16}O$	128.2	167.0		0.822	1.4150	2.0
5-Methyl-3-heptanone	$C_8H_{16}O$	128.2	160.5	−56.7	0.850 (0/4)	1.4160	2.0
5-Nonanone	$C_9H_{18}O$	142.2	188.4	−4.8	0.822	1.4195	
Diisobutyl ketone	$C_9H_{18}O$	142.2	168.1	−46.0	0.809	1.4210	2.4
Trimethyl nonanone	$C_{12}H_{24}O$	184.3	207–228		0.817	1.4273	
2,4-Pentanedione	$C_5H_8O_2$	100.2	140.0	−23.2	0.976	1.4494	7.0
2,5-Hexanedione	$C_6H_{10}O_2$	114.2	194.0	−5.5	0.973	1.4230	1.6
Cyclohexanone	$C_6H_{10}O$	98.2	155.6	−45.0	0.948 (20/4)	1.4507	5.2
2-Methylcyclohexanone	$C_7H_{12}O$	112.2	165.1	−19.0	0.925	1.4440 (25)	
3-Methylcyclohexanone	$C_7H_{12}O$	112.2	170.0	−74.0	0.920 (20/4)	1.4449	
4-Methylcyclohexanone	$C_7H_{12}O$	112.2	171.3	−40.6	0.916 (20/4)	1.4451	10 (54)
Acetophenone	C_8H_8O	120.2	201.0	19.0	1.026 (20/4)	1.5363	0.33 (20)
Isophorone	$C_9H_{14}O$	138.2	215.2	−8.0	0.923		0.26 (20)

[a] Specific gravity is at 20/20°C unless otherwise noted.
[b] Vapor pressure is at 25°C unless otherwise noted.
[c] Closed cup unless otherwise noted. Figures in parentheses are °C.
[d] S = readily soluble, Sl = slightly soluble, I = insoluble.
[e] At 25°C, 760 mm Hg.

Vapor Density (Air = 1)	Max. Vapor Concn. (ppm @ 20°C)	Flash Point[c] °F (°C)	Flammability (%)		Solubility[d]			Conversion Factors[e]	
			Lower Limit	Upper Limit	Water	Alcohol	Ether	1 ppm ⇔ mg/m³	1 mg/l ⇔ ppm
2.0	23,700	0 (18)	2.2	13.0	S	S	S	2.37	422
2.4	102,000	24 (−4)	1.8	12.0	S	S	S	2.94	340
		90 (32)	1.3	8.8				2.78	360
3.0	21,000	45 (7)	1.6	8.2	S	S	S	3.52	284
								3.40	294
								3.35	298
	54,300	49 (9)			S	S	S	3.44	291
3.5	4,900	73 (23)	1.2	8.0	S	S	S	4.10	244
3.5	9,700	64 (18)	1.4	7.6	S	S	S	4.10	244
3.4	12,300	90 (32)	1.3	8.8	Sl	S	S	4.02	249
4.0	1,550	136 (58)	1.8	6.9	S	S	S	4.74	211
3.9	2,700	102 (39)	1.1	7.9	Sl	S	S	4.66	214
3.9	5,900	110 (43)	1.1	8.2	Sl	S	S	4.66	214
3.9	1,800	98 (37)	1.4	8.8	I	S	S	4.66	214
	7,200	120 (49)			I	S	S	4.66	214
4.4	1,550	155 (68)			S	S	S	5.24	191
	2,600	135 (57)			I	S	S	5.24	191
	2,600	110 (43)			I	S	S	5.24	191
					I	S	S	5.81	172
4.9	3,100	140 (60)	0.8	7.1	Sl	S	S	5.81	172
		195 (91) [O.C.]			I	S	S	7.54	133
3.4	9,055	105 (41)	2.4	11.6	S	S	S	4.10	244
	2,100	185 (85)			S	S	S	4.66	214
3.4	6,700	111 (44)	1.2	8.1	Sl	S	S	4.02	249
3.9		118 (48)			I	S	S	4.58	218
3.9		118 (48)			I	S	S	4.58	218
3.9	11,800	118 (48)	1.1	7.4	I	S	S	4.58	218
4.2	430	180 (82)			S	S	S	4.95	204
4.7	340	184 (84)	0.8	6.5	Sl	S	S	5.65	177

Human Studies. Human experience confirms the relatively low toxicity of acetone. Nelson et al. (168) reported that unacclimatized volunteers exposed to 500 ppm acetone experienced eye and nasal irritation. Lower concentrations were without effect. DiVincenzo et al. (52) reported that the concentration of acetone in the expired air and blood of volunteers exposed to 100 or 500 ppm acetone was directly proportional to the level of exposure. No effects were seen at these concentrations. However, acetone was detected by the olfactory senses at 500 ppm. Approximately 75 percent of the inhaled vapor was absorbed into the bloodstream. The half-life for the elimination of acetone in expired air was about 3 hr. Mild exercise during exposure produced increased pulmonary uptake of acetone which resulted in substantially higher concentrations of acetone in the expired air during the postexposure period. Parmeggiani and Sassi (183) exposed workers to 833 ppm acetone vapor for 3 hr twice daily with a 1-hr break between exposures. They found levels of 190 μg/l of acetone in the expired air at the end of the day. Sixteen hours later the concentration of acetone in expired air had decreased to 32 μg/l. The concentration of acetone returned to background values over the weekend. These results suggest that repeated exposure to high concentrations of acetone may lead to accumulation during the work week. Matsushita et al. (151) exposed groups of five students to 0, 100, 250, 500, or 1000 ppm acetone vapor for 6 hr. Irritation of the eyes, nose, and throat was noted at 500 and 1000 ppm. The odor of acetone was detected at 100 ppm and acclimatization occurred rapidly.

Raleigh and McGee (192) surveyed workers exposed for 8 hr/day to an average atmospheric concentration of 1006 ppm. Eye irritation was transient and generally occurred at atmospheric concentrations greater than 1000 ppm. There was no indication of central nervous system effects. It was concluded that 1000 ppm produced no untoward effects with the exception of slight, transient irritation of the eyes, nose, and throat.

Folland et al. (72) reported on an outbreak of illness when 14 workers in an isopropyl alcohol packaging plant were accidently exposed to carbon tetrachloride. Four people developed renal failure or hepatitis. Several workers had elevated levels of acetone in their expired air. The authors suggested that since acetone is a metabolite of isopropyl alcohol, the acetone formed may have adversely predisposed the workers to carbon tetrachloride toxicity.

A worker overcome by acetone vapor while cleaning a storage tank developed slight intoxication, headache, lassitude, drowsiness, loss of appetite, nausea, and vomiting. Recovery occurred after removal from the exposure site. However, the blood concentration of acetone was still elevated the following day (201). Harris and Jackson (102) reported a case of acute acetone poisoning in a 10-year-old boy. A solvent mixture containing 90 percent acetone and 9 percent pentane was used to set a cast for a broken leg. Eight hours after the cast was in place, the boy became restless and complained of a headache. Four hours later he vomited and collapsed. After removal of the cast the boy became comatose. The skin covered by the cast appeared normal. Recovery was complete after 4 days.

Eight workers employed in the production of cellulose acetate fibers were exposed to 307 to 918 ppm acetone. Somnolence, occasional dizziness, insomnia, eye and throat irritation, headache, and a feeling of inebriation were reported (183).

Workers exposed to 1000 ppm acetone for 3 hr/day for 7 to 15 years complained of respiratory tract irritation, dizziness, and loss of strength (267). Elevated concentrations of acetone were detected in expired air and urine during the work week. These levels returned to normal after the weekend. Gitelson et al. (87) described a case history of a 42-year-old man who ingested 200 ml of acetone. One hour later he was stuporous and showed signs of shallow respiration but had a normal pulse, body temperature, blood pressure, and tendon reflexes. Irritation of the throat was present and acetone was detected in the expired air. He became comatose and was admitted to the hospital. Supportive therapy was initiated and he regained consciousness 12 hr later. Subsequently hyperglycemia of an unknown cause was diagnosed. The author suggested that the hyperglycemia may have been related to the ingestion of acetone.

Eight men cleaning an enclosed pit were exposed to 12,000 ppm acetone and 50 ppm 1,1,1-trichloroethane and showed signs of throat and eye irritation, limb weakness, headache, dizziness, lightheadedness, and a feeling of general malaise (197). Urinary levels of acetone collected 90 min after exposure were elevated. The authors concluded that the symptoms reported were due to acetone intoxication.

Mirchev (159) described the onset of hepatorenal lesions in two men and two women acutely exposed to acetone. One person had inhaled acetone vapors whereas the others had ingested acetone. Clinical manifestation of liver injury was observed in all four workers and renal lesions were detected in two. No other reports of hepatorenal toxicity due to acetone were found.

Fifty-seven of 100 workers exposed to acetone vapors had high blood catalase activity compared to controls, while 13 workers had lower levels (38). These findings indicate that acetone exposure may be associated with a period of catalase activation followed by inhibition. Similar results have been reported for rats acutely exposed to 100 to 189 g/m^3 acetone (38).

Cesaro and Pinerolo (31) described studies in which eight men were enclosed in a sealed box for 20 to 30 min with their heads protruding through a hole at the top. The chamber was saturated with acetone vapor. Acetone was also sprinkled on the skin during exposure. Total ketone bodies in the blood after exposure were unchanged compared to preexposure levels. It was concluded that acetone had not been extensively absorbed through the skin. In a similar study, volunteers were placed in a chamber saturated with acetone vapor and acetone saturated pads were placed on the skin for 30 min (183). After 90 min of exposure the concentrations of acetone in the blood and urine were substantially elevated over preexposure values.

Application of acetone to human skin produced cellular damage in the stratum corneum and stratum spinosum as evidenced by marked intracellular edema of keratinized cells and vacuolization of spinous cells (143, 144). Restoration of normal ultrastructural patterns was seen 72 hr after exposure was terminated. Fukaburi et al. (74) applied acetone to the skin (12.5 cm^2) for 2 hr/day for four consecutive days. Elevated concentrations of acetone were found in blood (5 to 12 μg/ml), alveolar air (5 to 12 μg/ml), and urine (8 to 14 μg/ml). Skin penetration of acetone was rapid and peak concentrations in venous blood occurred at the end of each application. The absorption of acetone increased directly with the frequency and the extent of the exposure. After cessation of the exposure, the concentrations of acetone in blood, alveolar air, and urine decreased rapidly and returned to preexposure values by the next morning. Lupulescu and Birmingham (142) taped small glass tubes containing 1 ml acetone to the forearms of volunteers for 90 min. Biopsies (4 mm) were taken and examined by light and electron microscopy. Mild edema and hyperemia occurred at the application site. Electron microscopy revealed a reduction and desquamation of horny layers, with intercellular edema and the presence of vacuoles surrounding nuclei of the cells of the stratum spinosum.

Toxicokinetics.   Acetone is a naturally occurring constituent of blood and urine. It is readily absorbed by all routes of administration and its high solubility in water ensures widespread distribution in the tissues. Large doses of acetone are predominantly excreted unchanged in expired air whereas small doses (up to 7 mg/kg) are largely oxidized to carbon dioxide (190, 203, 274). Kagan (118) reported that 71 to 77 percent of inhaled acetone vapor was absorbed by the pulmonary route. The amount eliminated in the expired air during exposure was equivalent to the dead space volume of the respiratory tract (23 to 29 percent). Four subjects exposed to 600 ppm of acetone absorbed 83 percent of the dose. Urinary excretion of acetone increased after exposure and was proportional to the exposure concentration (241). Egle (60) reported that the pulmonary retention of acetone in the dog was about 52 percent.

Radio-labeled carbon was found in cholesterol, hepatic glycogen, various amino acids, and in carcass protein of rats dose with [^{14}C]-acetone (164, 190, 204). Acetone may be converted to 1,2-propanediol which enters the glycolytic pathway and possibly the one carbon pool (158, 198, 199, 203). More recently, acetone has been shown to be converted to lactate in mice. The rate-limiting step appears to be the conversion of acetone to a hydroxylated intermediate. Rats and mice exposed to 30 mg/l of acetone, and rabbits and guinea pigs exposed to 72 mg/l for 2 hr, had increased levels of acetone, acetoacetic acid, and β-hydroxybutyric acid in the blood and urine immediately after exposure and 24 hr later (33).

Many investigators have shown that acetone induces the hepatic microsomal mixed function oxidase system. Acetone enhanced the rate of microsomal aniline *p*-hydroxylase activity (12, 13, 32, 67, 108, 125, 189, 217, 218, 261). Inhaled acetone enhanced ethoxycoumarin *O*-demethylase activity and cytochrome P-450 content of the liver in rats but had no effect on NADPH cytochrome P-450 reductase activity (262). Acetone produced a Type I binding spectrum in the 9000 × g supernatant fraction of rat liver and may inhibit microsomal enzyme reactions involving Type I binding substrates and increase the metabolism of Type II binding substrates by increasing the availability of NADPH (75).

Pretreatment with acetone for 6 days (one-tenth the LD_{50}) potentiated acute ethanol toxicity in rats. The combined treatment with acetone and ethanol for 6 days potentiated acetone depression of total liver microsomal protein and cytochromes P-450 and b5, prolonged the pharmacologic effect of hexobarbital, and decreased plasma hexobarbital concentrations in rats recovering from narcosis (169). Plaa et al. (186) and Traiger and Plaa (248) demonstrated that acetone pretreatment potentiates chlorinated hydrocarbon toxicity.

It has been demonstrated that isopropanol is oxidized extensively to acetone in vivo by laboratory animals (172). Kohli et al. (129) showed that acetone protected animals against electroshock or isonicotinic acid hydrazide-induced convulsions. Acetone increased serum alanine and α-ketoglutarate transaminase activities and enhanced the hepatotoxicity of 1,1-dichloroethylene (200 ppm) in rats (112). Acetone was reported to inhibit nonspecific cholinesterase activity in various tissues and to stimulate catalase activity (44, 99, 220). It has been suggested that acetone is a potent bactericidal agent, and is of value for routine disinfection of surgical instruments (53, 54). It may also possess antiviral activity (82).

2.1.3 Industrial Exposure

Accidental overexposure to acetone is rare in industry. Millions of man hours of occupational exposure has led to few untoward effects. Exposure to less than 1000 ppm acetone vapor produces only slight eye, nose, and throat irritation. Atmospheric concentrations in excess of 10,000 ppm are likely to produce central nervous system depression and narcotic effects.

2.1.4 Hygienic Standards

The American Conference of Governmental Industrial Hygienists has established 1000 ppm (with an "intended" change to 750 ppm) as a threshold limit value for acetone (243). The National Institute of Occupational Safety and Health has recommended that the permissible exposure level be established at 250 ppm, based on the subjective evidence of eye and nasal irritation (171).

2.2 Methyl Ethyl Ketone; 2-Butanone, MEK

$$\underset{CH_3-C-CH_2-CH_3}{\overset{\displaystyle O}{\overset{\displaystyle \|}{}}}$$

2.2.1 Physical-Chemical Properties

Methyl ethyl ketone is a volatile, water-soluble, colorless liquid with a characteristic odor resembling that of acetone. Its odor threshold has been reported to be about 2 ppm (265), although a substantially lower odor threshold of 0.25 ppm has been reported (161). Methyl ethyl ketone is highly flammable and may pose a fire hazard. Additional physical-chemical properties are summarized in Table 56.4.

2.2.2 Biologic Effects

Methyl ethyl ketone has a low toxicity following both acute and chronic exposures. High atmospheric concentrations of methyl ethyl ketone are irritating to the eyes, nose, and throat, and prolonged exposure may produce central nervous system depression and narcosis. No untoward effects have been reported for chronic exposure to low concentrations. Prolonged skin contact may defat the skin and produce dermatitis. If splashed in the eyes it may produce painful irritation and corneal injury. Methyl ethyl ketone may enhance the neurotoxicity of n-hexane and methyl n-butyl ketone (9).

Animal Studies. *Single Dose.* The oral LD_{50} was reported to be 3.3, 3.4, and 3.6 g/kg and 6.86 ml/kg for rats (123, 227, 257). The intraperitoneal LD_{50} was 616 mg/kg for mice (180). Guinea pigs dosed intraperitoneally with 1500 or 2000 mg/kg had elevated serum ornithine carbamyl transferase activity 24 hr after treatment. One animal died at the 2000 mg/kg dose, and 750 mg/kg produced no effects (48). Methyl ethyl ketone produced narcosis when given intravenously to mice (0.15 ml) (41). Terhaar et al. (256) reported intraperitoneal LD_{50}s of 1.27 and 0.73 ml/kg for rats and guinea pigs, respectively.

Repeated Dose. Cats injected subcutaneously with 150 mg/kg methyl ethyl ketone twice daily, 5 days/week for up to 8.5 months failed to develop nervous system damage (236).

Beagle dogs receiving similar treatments as the cats also did not develop neuropathologic changes (256). Rats injected intraperitoneally for 35 weeks with 100 to 200 mg/kg of methyl ethyl ketone, a 9/1 mixture of methyl ethyl ketone and methyl isobutyl ketone, or a 9/1 mixture of methyl ethyl ketone and methyl n-butyl ketone showed no clinical or neuropathologic changes (256).

Eye. Methyl ethyl ketone is irritating to the eyes and has caused corneal edema in laboratory animals (136). Exposure to 100,000 ppm for 20 min produced transient corneal opacity in experimental animals (184).

Skin. Methyl ethyl ketone is absorbed through intact skin. The acute dermal LD_{50} was reported to be 8 and 10 ml/kg for rabbits. Skin contact for 24 hr produced moderate skin irritation, and dermatitis is likely to occur after repeated exposures (227, 257). Methyl ethyl ketone applied to intact or abraded rabbit skin for 24 hr under occlusion was moderately irritating (126). Methyl ethyl ketone in combination with methyl n-butyl ketone was absorbed through the skin of dogs (51). Other investigators used methyl ethyl ketone as a vehicle to enhance skin penetration (277). Methyl ethyl ketone, methyl ethyl ketone/methyl isobutyl ketone (9/1), or methyl ethyl ketone/methyl n-butyl ketone (9/1), with or without DMSO, dropped on the backs of guinea pigs daily (2 ml) for 31 weeks, produced only mild desquamation (256).

Inhalation. Rats were exposed to 2000 ppm of methyl ethyl ketone vapor for 2 hr without apparent toxic effects. Exposure to 4000 ppm for the same duration killed four of six animals (221). Exposure to 8000 ppm killed three of six rats in 8 hr, while 10,000 ppm for 1 hr produced irritation of the eyes and nose. Exposure for a few minutes to 100,000 ppm caused no deaths in guinea pigs (184). Specht et al. (233) found methyl ethyl ketone to be less irritating than other ketones in guinea pigs exposed to 25,000 ppm for 10 min. Guinea pigs exposed to 3300 ppm of methyl ethyl ketone for 13.5 hr showed no signs of intoxication (257). Exposure to 10,000 ppm for a similar period produced irritation of the eyes and nose within 4 min and narcosis between 4 and 5 hr. Exposure to 33,000 ppm for 200 min produced narcosis and death. Exposure to 100,000 ppm for 45 to 55 min produced narcosis after 10 min (257).

Rats exposed to 2150 ppm methyl ethyl ketone vapor for 6 weeks or 4740 ppm for 4 weeks showed no signs of neurotoxicity (43). Saida et al. (202) exposed rats 24 hr/day, 7 days/week to methyl ethyl ketone vapors (1125 ppm) and to a combination of methyl ethyl ketone and methyl n-butyl ketone (1125 and 225 ppm). Methyl ethyl ketone alone produced no signs of peripheral neuropathy after 5 months. Rats exposed to the combination of methyl ethyl ketone and methyl n-butyl ketone developed paralysis at 25 days, and exposure to 225 ppm methyl n-butyl ketone alone produced paralysis after 66 days. These results indicate that methyl ethyl ketone shortened the latency period for the onset of methyl n-butyl ketone-induced neuropathy. Duckett et al. (55) reported histologic changes indicating a neuropathy in rats exposed to a mixture of 2000 ppm methyl ethyl ketone and 200 ppm methyl n-butyl ketone 8 hr/day, 5 days/week for 6 weeks.

Rats were exposed to either 10,000 ppm n-hexane, a combination of 1100 ppm methyl ethyl ketone and 8900 ppm n-hexane, or 6,000 ppm methyl ethyl ketone. Enhancement of n-hexane neurotoxicity and shortened onset of morphologic and clinical signs of neuropathy were seen in the animals exposed to the combination of methyl ethyl ketone and n-hexane. Methyl ethyl ketone alone did not produce neuropathy. The findings suggest that commercial solvent mixtures containing both methyl ethyl ketone and n-hexane may produce neuropathy (8).

Groups of pregnant rats were exposed to methyl ethyl ketone 7 hr/day on the sixth through the fifteenth day of gestation. Concentrations of 1000 or 3000 ppm produced no maternal toxicity. The total number of litters with fetuses with skeletal alterations was increased at the lower concentration. The high concentration (3000 ppm) increased the incidence of soft tissue alterations and sternebral variations. It was concluded that methyl ethyl ketone was embryotoxic, fetotoxic, and potentially teratogenic to rats (210). The same investigators conducted a second study to determine the repeatability of these effects. The high dose (3000 ppm) produced slight maternal toxicity. The slight fetotoxicity seen in the initial study was seen in this study; however, no embryo toxicity or teratogenicity were seen (260).

Geller et al. (78) and Garcia et al. (76) reported that baboons exposed to 20 to 40 ppm of methyl ethyl ketone showed no impairment of discriminatory behavior. Exposure to 100 ppm produced an apparent delayed behavioral response time. Repeated exposure to methyl ethyl ketone for 7 days produced no behaviorial changes.

Human Studies.   Nelson et al. (168) determined sensory thresholds in 10 volunteers exposed for 3 to 5 min to methyl ethyl ketone. Exposure to 25 ppm for 5 min was without untoward effects. Exposure to 100 ppm gave an identifiable odor and produced mild irritation of the nose and throat. Exposure to 200 ppm for 15 min produced complaints of irritation of the nose and throat and a strong odor. Methyl ethyl ketone was objectionable at 300 ppm. At concentrations greater than 300 ppm, irritation of the eyes, nose, and throat and nausea were seen, particularly in unacclimatized subjects. The threshold for eye and nose irritation was about 200 ppm for 50 percent of the unacclimatized workers (257). Elkins (61) reported that workers exposed to 300 to 500 ppm complained of headache, irritation, and nausea. Exposures to 700 ppm led to no permanent untoward effects. Momentary exposure to 33,000 and 100,000 ppm produced intolerable irritation of the eyes, nose, and throat (184).

Exposure to vapor concentrations of 90 to 270 ppm methyl ethyl ketone for 4 hr shortened time estimations in men and increased the variation in time estimation tests in women (167). Workers exposed via inhalation (300 to 600 ppm) and skin contact complained of numbness in the upper extremities (219). Polyneuritis was reported in one worker exposed to methyl ethyl ketone and tetrahydrofuran polyester glue (266).

Gilchrist et al. (85), Billmaier et al. (20), and Allen et al. (5, 6) reported an outbreak of peripheral neuropathy in a coated fabrics plant in which workers were exposed to methyl ethyl ketone and methyl *n*-butyl ketone. Subsequent studies showed that methyl *n*-butyl ketone produced peripheral neuropathy in experimental animals, and that methyl ethyl ketone may have interacted with methyl *n*-butyl ketone to shorten the onset of neuropathy (7–9, 55, 236).

Altenkirch et al. (7) reported that people who had sniffed a commercially available glue thinner developed "glue sniffers' neuropathy." This polyneuropathy was nearly identical to that found after exposure to methyl *n*-butyl ketone. The polyneuropathies occurred in close time relation to the denaturation of a thinner with methyl ethyl ketone. The thinner did not contain methyl *n*-butyl ketone but did contain *n*-hexane, which has been shown to produce a neuropathy identical to that produced by methyl *n*-butyl ketone. Apparently, methyl ethyl ketone shortened the onset of the neurotoxicity produced by *n*-hexane (9).

Munies and Wurster (166) added methyl ethyl ketone to an absorption cell affixed to the palmar surface of the forearm of volunteers. Shortly after skin contact was initiated, methyl ethyl ketone was detected in the expired air. The concentration of methyl ethyl ketone reached a steady state in 2 to 3 hr. When the skin was hydrated a more rapid uptake occurred. DiVincenzo et al. (51) reported that a combination of methyl ethyl ketone and methyl *n*-butyl ketone was absorbed through human skin. Exposure of the forearm to methyl ethyl ketone for 1 hr/day for six consecutive days produced damage to the horny layer of the epithelium (148). Munies (165) and Munies and Wurster (166) demonstrated that methyl ethyl ketone can defat and partially dehydrate the stratum corneum of human skin without producing irritation or inflammation. Methyl ethyl ketone did not produce skin sensitization when tested in man (126, 128).

Workers exposed to methyl ethyl ketone by inhalation (300 to 600 ppm) and by skin contact developed dermatitis and a numbness in the extremities (219). Retrobulbular neuritis occurred in one worker exposed to methyl ethyl ketone. Complaints of headache, mild vertigo, and diminished vision were noted. Methanol and formaldehyde were detected in the blood 10 hr after exposure. The authors suggested that the worker suffered optic nerve toxicity induced by methanol, which may have been related to the metabolism of methyl ethyl ketone (19). Vaider et al. (266) suggested that methyl ethyl ketone may have produced peripheral neuropathy in a 55-year-old worker exposed to several organic solvents. However, the cause of neuropathy was not well documented. Methyl ethyl ketone has also been implicated in occupational polyneuropathies in shoe workers and in Swedish steel workers. However, direct evidence implicating methyl ethyl ketone is lacking (58, 66).

Toxicokinetics. Methyl ethyl ketone is readily absorbed by all routes of exposure. It is eliminated unchanged in expired air and its metabolites as conjugates of glucuronic acid are eliminated in the urine (41, 274). In humans, about 75 percent of inhaled methyl ethyl ketone was absorbed by the pulmonary route (241). In dogs, 30 to 33 percent of the dose (0.3 to 0.5 g/kg) was excreted unchanged in the expired air (274).

DiVincenzo et al. (49) dosed guinea pigs intraperitoneally with 450 mg/kg methyl ethyl ketone and characterized the metabolites in the serum. Methyl ethyl ketone was reduced to 2-butanol and oxidized by ω-1 oxidation to 3 hydroxy-2-butanone. The latter compound was reduced to 2,3-butanediol. The oxidative product 2,3-butanedione was not detected. The half-life of methyl ethyl ketone in serum was 270 min. The serum elimination times for methyl ethyl ketone and its metabolites were 12 and 16 hr, respectively. Methyl ethyl ketone may also undergo α oxidation with ultimate conversion to carbon dioxide (256). A similar metabolic pathway for methyl ethyl ketone has been reported by Dietz and Traiger (45).

Hetland et al. (105) and Couri et al. (35) reported that hepatic microsomal enzyme activities were significantly enhanced in rats exposed to methyl ethyl ketone vapor. Hexobarbital sleeping times were reduced following exposure to methyl ethyl ketone or a combination of methyl ethyl ketone and methyl n-butyl ketone. Aniline hydroxylase activity was increased in liver microsomal preparations from rats exposed continuously or intermittently (7 hr/day) for 7 days to a combination of methyl ethyl ketone and methyl n-butyl ketone. Continuous exposure increased aminopyrine demethylase, p-nitrobenzoate reductase, and neoprontosil reductase activites.

Traiger and Bruckner (246) showed that methyl ethyl ketone pretreatment in rats markedly enhanced the hepatotoxic response to carbon tetrachloride as measured by increases in serum enzyme activities. Dietz and Traiger (45) suggested that metabolites of methyl ethyl ketone, that is, 3-hydroxy-2-butanone and 2,3-butanediol, may augment the necrogenic effects of carbon tetrachloride in methyl ethyl ketone pretreated animals. Oral administration of 1.87 ml/kg methyl ethyl ketone or 2.2 ml/kg 2-butanol stimulated the mixed function oxidase system in the rat. Animals killed 16, 28, and 48 hr after dosing had an increase in acetanilide hydroxylase activity. Aminopyrine n-demethylase hydroxylase activity was significantly elevated in the animals treated with methyl ethyl ketone. Microscopic examination of hepatocytes revealed an increase in the smooth endoplasmic reticulum. These findings suggest that the enhanced carbon tetrachloride hepatotoxicity may be related to methyl ethyl ketone stimulation of the hepatic mixed function oxidase system (247).

2.2.3   Industrial Exposure

Methyl ethyl ketone is slightly more irritating to the mucous membranes and conjunctiva than acetone. Central nervous system depression can occur at high atmospheric concentrations but good warning properties should prevent over exposure. The principal routes of exposure are by inhalation and skin contact. Prolonged contact with the skin may produce dermatitis. Methyl ethyl ketone may interact with methyl *n*-butyl ketone or *n*-hexane to shorten the onset of neuropathy and may predispose the liver to injury from hepatotoxins.

2.2.4 Hygienic Standards

The American Conference of Governmental Industrial Hygienists has established a threshold limit value of 200 ppm for methyl ethyl ketone (243).

2.3 3-Butyn-2-one; Methyl Ethynyl Ketone, Acetylacetylene, 1-Butyn-3-one, Butynone, Acetylethyne

$$\underset{\text{CH}_3-\text{C}-\text{C}\equiv\text{CH}}{\overset{\displaystyle\overset{\text{O}}{\|}}{}}$$

2.3.1 Physical-Chemical Properties

3-Butyn-2-one is a clear, colorless liquid which is chemically reactive and has a penetrating odor. The autoignition temperature for 3-butyn-2-one is 274°C (256). Additional physical-chemical properties are summarized in Table 56.4.

2.3.2 Biologic Effects

3-Butyn-2-one is an extremely toxic material. It is a strong lacrimator, and both the liquid and vapor states are very irritating to the eyes, skin, and mucous membranes. It is readily absorbed percutaneously and is highly toxic both dermally and orally.

Animal Studies. *Single Dose.* The oral LD_{50} of 3-butyn-2-one was between 6.3 and 12.6 mg/kg for rats. Clinical signs included tremors, diarrhea, and depression (254).

Eye. Application of the undiluted material to rabbit eyes produced pain and severe damage. Lethal amounts may be absorbed through the conjunctival membranes. When the material was diluted to a 1 percent solution with propylene glycol, severe irritation was still evident (254).

Skin. Undiluted and 10 percent solutions of 3-butyn-2-one were severe skin irritants and were lethal when applied for a few hours. One percent solutions were also strong irritants particularly to abraded skin. When applied as a 10 percent solution in propylene glycol under an impervious cuff, the LD_{50} was 40 to 50 mg/kg with deaths occurring within hours of application (254).

Inhalation.   Rats exposed to 10 ppm of 3-butyn-2-one vapors died between 4 and 7 hr of exposure. At 25 ppm, deaths occurred between 0.5 and 1 hr of exposure. At 50 ppm, deaths occurred between 12 min and 2 hr of exposure. At 100 ppm, all deaths occurred between 12 min and 1 hr of exposure. At 200 ppm, all deaths occurred between 3 min and 30 min of exposure. At atmospheric saturation, all rats died within 6 min (254).

The vapors produced immediate, severe eye and respiratory irritation at all concentrations studied. Lung congestion was observed in all test animals and liver and kidney damage was apparent in animals receiving longer exposures (254).

Human Studies.   Limited human experience suggests that man is at least as sensitive to 3-butyn-2-one as experimental animals. 3-Butyn-2-one is a strong lacrimator.

2.3.3   Industrial Exposure

Industrial exposures to 3-butyn-2-one may occur by inhalation, skin, or eye exposures. High oral toxicity, the ease of skin and eye absorption, and the highly irritating nature of vapors and solutions demand extreme caution.

2.3.4   Hygienic Standards

Data to suggest a safe level are not available but prudence would suggest avoiding all exposures by the use of impermeable gloves, protective clothing, and an appropriate respirator.

2.4   Methyl n-Propyl Ketone; Methyl Propyl Ketone, 2-Pentanone, MPK

$$CH_3-\overset{\displaystyle O}{\overset{\displaystyle \|}{C}}-CH_2-CH_2-CH_3$$

2.4.1   Physical-Chemical Properties

Methyl n-propyl ketone is a clear, colorless liquid with a powerful, ethereal, fruity odor. The odor threshold concentration for methyl n-propyl ketone has been reported to be 8 ppm (265). The autoignition point for this compound is

505°C. Its moderate flammability indicates a possible fire hazard. Additional physical-chemical properties are summarized in Table 56.4.

2.4.2 Biologic Effects

Methyl n-propyl ketone is moderately toxic orally and can produce slight to moderate irritation of the skin and eyes. Repeated doses show no evidence of cummulative toxicity.

Animal Studies. *Single Dose.* The oral LD_{50} for rats has been calculated to be 3730 mg/kg by Smyth et al. (227). Unpublished data indicated the oral LD_{50} for both rats and mice was between 1600 and 3200 mg/kg (256). The intraperitoneal LD_{50} for methyl n-propyl ketone was 800 mg/kg for rats and 1600 mg/kg for mice (256).

Giroux et al. (86) injected mice intraperitoneally and found that doses of 250 to 500 mg/kg produced agitation and hyperesthesia. Doses of 2000 mg/kg produced agitation, loss of balance, staggering, paresis, loss of sensitivity, and deaths in 20 percent of the group. A dose of 2500 mg/kg caused 100 percent mortality.

Haggard et al. (98) reported respiratory failure and death in rats after an intraperitoneal dose of 2530 mg/kg. The plasma concentration of methyl n-propyl ketone was 156 mg/dl.

Repeated Dose. Methyl n-propyl ketone was administered in the drinking water to rats at concentrations of 0.25, 0.5, and 1.0 percent. The mean daily doses were calculated to be 144 mg/kg for 10 months and 250 mg/kg and 454 mg/kg for 13 months. Adverse effects due to 1.0 percent methyl n-propyl ketone were limited to a slight reduction (9 percent) in weight gain compared to controls after 67 days of exposure. Clinical signs, organ weights, and histology were normal. No neuropathologic changes were noted in the central or peripheral nervous systems. (256)

Eye. When the material was placed in the conjunctival sac of rabbit eyes, slight to moderate irritation was observed (256).

Skin. Single doses of methyl n-propyl ketone applied to the backs of guinea pigs under an impervious cuff produced only slight irritation (256). Doses up to 20 ml/kg on the skin of guinea pigs did not produce clinical abnormalities, although percutaneous absorption appeared to occur. Repeated application for 10 days to the backs of guinea pigs resulted in slight to moderate erythema and slight edema (256). The dermal LD_{50} of methyl n-propyl ketone for rats has been reported to be 8 ml/kg (227).

Inhalation. Smyth et al. (227) studied the inhalation toxicity of methyl *n*-propyl ketone and reported that none of six rats died within a 2-week period following a 30-min exposure to a saturated vapor (227). The LC_{50} for rats receiving a 4-hr exposure was greater than 2000 ppm.

Yant et al. (275) reported that exposures to 30,000 to 50,000 ppm of methyl *n*-propyl ketone for 30 to 60 min produced lethality in guinea pigs. Guinea pigs exposed to 5000 ppm for 1 hr, 2000 ppm for 8 hr, or 1500 ppm for several hours showed slight or no abnormal clinical signs (275). Specht et al. (233) reported that methyl *n*-propyl ketone produced narcosis, depression of cardiac and respiratory rates, and loss of cranial nerve reflexes. Generalized vascular congestion was noted at levels of 2500 to 40,000 ppm.

Five rats were exposed to methyl *n*-propyl ketone at a calculated level of 305 ppm to determine its neurotoxic potential. The exposures were given during two 16-hr and two 20-hr periods on four consecutive days per week over approximately a 17.5 week period for a total of 1240 hr of exposure. A broad histopathologic screen of tissue taken from these animals was normal except for the possibility of slight hepatocyte enlargement. Central and peripheral nerves embedded in epoxy resins did not reveal any neurologic damage (256).

Human Studies. Yant et al. (275) reported that exposure to 1500 ppm of methyl *n*-propyl ketone in the air was associated with a strong odor and caused irritation to the nasal passages and eyes.

Toxicokinetics. Schwartz administered methyl *n*-propyl ketone orally to dogs (358 to 381 mg/kg) and found 21.2 to 27.1 percent of the dose in the expired air within 24 hr of administration (209). Neubreuer (170) reported that metabolites of methyl *n*-propyl ketone were excreted in the urine as a glucuronide conjugate.

2.4.3 Industrial Exposure

The low degree of toxicity of methyl *n*-propyl ketone would indicate its relative safety provided skin and eye protection are available. Inhalation should not be hazardous if adequate ventilation is provided. The greatest hazard associated with its use is probably its moderate flammability.

2.4.4 Hygienic Standards

The threshold limit value for methyl *n*-propyl ketone established by the American Conference of Governmental Industrial Hygienists is 200 ppm (243). This level is low enough to prevent depression of the nervous system.

2.5 3-Methyl-2-Butanone; Methyl Isopropyl Ketone, Methyl Butanone, 2-Acetyl Propane, 2-Methyl-3-Butanone

$$CH_3-\overset{\overset{\displaystyle O}{\|}}{C}-\overset{\overset{\displaystyle CH_3}{|}}{CH}-CH_3$$

2.5.1 Physical-Chemical Properties

3-Methyl-2-butanone is a colorless, flammable liquid with an acetone-like odor. Additional physical-chemical properties are summarized in Table 56.4.

2.5.2 Biologic Effects

3-Methyl-2-butanone is slightly toxic by the oral and inhalation routes. Contact with the eyes produces slight irritation, and skin contact may produce slight to moderate irritation.

Animal Studies. *Single Dose.* Carpenter et al. (28) reported the oral LD_{50} for 3-methyl-2-butanone to be 5.66 ml/kg in rats. In other studies, the oral LD_{50} for male rats and mice was 3200 mg/kg (256). Clinical signs observed were weakness, prostration, and ataxia. Haggard et al. (98) reported an oral LD_{50} of 4.1 g/kg for rats. The intraperitoneal LD_{50} was reported as 800 mg/kg for male rats and between 200 and 400 mg/kg for male mice (256).

Eye. Carpenter et al. (28) reported slight ocular injury when 3-methyl-2-butanone was tested in the rabbit eye.

Skin. Slight irritation was observed when 3-methyl-2-butanone was applied to the uncovered abdomen of rabbits. Moderate irritation was observed when the material was applied undiluted to intact and abraded rabbit skin for 24 hr under an occlusive dressing (98). The acute dermal LD_{50} for rabbits was reported to be 6.35 ml/kg and >5 g/kg (28, 98).

3-Methyl-2-butanone applied to the abdomens of guinea pigs at doses of 5 to 20 ml/kg under an occlusive dressing for 24 hr produced moderate skin irritation as evidenced by edema and necrosis at 24 hr and residual eschars and scarring after 2 weeks (256).

Inhalation. Exposure to an airborne concentration of 5700 ppm for 4 hr resulted in the death of one of six rats (28).

Human Studies. Skin sensitization tests on 25 volunteers, using a 10 percent concentration of 3-methyl-2-butanone, did not produce sensitization or irritation (98).

Toxicokinetics. When 1000 mg/kg 3-methyl-2-butanone was administered to rats, 38 to 54 percent was eliminated unchanged in 25 to 35 hr; primarily in the expired air (98). Williams (274) reported that 3-methyl-2-butanone increases glucuronic acid output in urine.

Toxic Mechanisms. 3-Methyl-2-butanone appears to be biologically active primarily by its irritant properties and by the induction of narcosis.

2.5.3 Industrial Exposure

The main hazards associated with 3-methyl-2-butanone are slight eye irritation, slight to moderate skin irritation, inhalation of its vapors, and its flammability. Eye and skin protection plus adequate ventilation would be expected to prevent serious injury.

2.5.4 Hygienic Standards

No threshold limit value has been established for 3-methyl-2-butanone.

2.6   3-Pentyn-2-one

$$CH_3-\overset{\displaystyle O}{\overset{\displaystyle \|}{C}}-C\equiv C-CH_3$$

2.6.1 Physical-Chemical Properties

3-Pentyn-2-one is a water-clear, chemically reactive liquid with a pronounced unpleasant odor. Additional physical-chemical properties are summarized in Table 56.4.

2.6.2 Biologic Effects

3-Pentyn-2-one is moderately toxic orally, extremely irritating to the skin and eyes, and appears to be readily absorbed and is highly toxic percutaneously. Since inhalation of the vapors and skin contact are the primary routes of exposure, this compound presents a definite hazard if handled without proper protective equipment.

Animal Studies. *Single Dose.* Range-finding studies indicated an oral LD_{50} of 3-pentyn-2-one between 63 and 128 mg/kg for rats (254). Clinical signs of toxicity were tremors and convulsions; death occurred within 2 days. Necropsy of survivors revealed severe liver and kidney damage and marked irritation of the stomach mucosa.

Eye. When applied undiluted or as a 10 percent solution, 3-pentyn-2-one produced severe eye irritation; thus eye contact may cause impairment of vision (254).

Skin. The undiluted compound or a 10 percent solution was severely irritating to the skin. When held in contact with the skin of rabbits (1 to 2 ml) 3-pentyn-2-one produced death within a few hours. The LD_{50} was between 6 and 12 mg/kg using the Draize technique (254). Thus this material appears to be more toxic via the dermal route than orally.

Inhalation. Three rats exposed for 6 min to a saturated atmosphere of the vapor died over a 5-day period. Immediate irritation of eyes and nose was evident (254).

Human Studies. Human experience with 3-pentyn-2-one is limited but suggests that man is at least as sensitive to its effects as experimental animals.

2.6.3 Industrial Exposure

The limited range-finding studies reported indicate that this compound is moderately toxic orally and is highly toxic dermally. It is very irritating to the skin, eyes, and mucous membranes. Precautions should be taken to avoid all possible contact. Protective clothing and respiratory equipment should be worn when handling 3-pentyn-2-one.

2.6.4 Hygienic Standards

No threshold limit value has been established for 3-pentyn-2-one.

2.7 Methyl Isopropenyl Ketone; 2-Methyl-1-butene-3-one, 3-Methyl-3-butene-2-one, Isopropenyl Methyl Ketone, Methyl Butenone

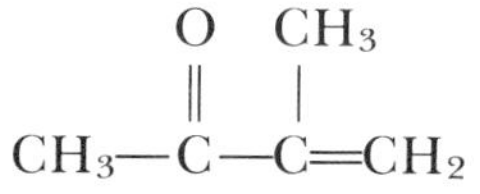

$$CH_3-\overset{\overset{\displaystyle O}{\|}}{C}-\overset{\overset{\displaystyle CH_3}{|}}{C}=CH_2$$

2.7.1 Physical-Chemical Properties

Methyl isopropenyl ketone is a chemically reactive, clear, colorless liquid with a pungent odor and moderate flammability. A threshold odor concentration of 0.291 ppm has been reported (265). It is stored at subzero temperatures to prevent polymerization. Additional physical-chemical properties are summarized in Table 56.4.

2.7.2 Biologic Effects

Methyl isopropenyl ketone is highly toxic by the oral, percutaneous, and respiratory routes. Topical exposure to the skin may be moderately irritating while topical or vapor exposure to the eye may be severely irritating.

Animal Studies. *Single Dose.* The oral LD_{50} for methyl isopropenyl ketone was reported to be 180 mg/kg for rats (225). The oral LD_{50} for guinea pigs was reported to be between 60 and 250 mg/kg, indicating a moderate to high degree of toxicity (254).

Eye. When tested by application to rabbit eyes, methyl isopropenyl ketone was found to cause severe irritation and damage. Vapors of methyl isopropenyl ketone may also be very irritating.

Skin. Methyl isopropenyl ketone was a moderate irritant to rabbit skin particularly when applied under a wrap. The percutaneous LD_{50} has been calculated to be 230 mg/kg for the rabbit (225).

Inhalation. Smyth et al. (225) reported that a 4-hr exposure to 125 ppm killed five of six rats, while a saturated atmosphere for 2 min killed all test animals. Rats survived a 90-min exposure to nominal vapor concentrations of 524 ppm, but exhibited marked irritation of ocular and nasal tissues (254). At a nominal vapor concentration of 1455 ppm for 30 min, methyl isopropenyl ketone was very irritating and deaths occurred in minutes. Animals dying during these exposures were cyanotic and died in convulsion. Delayed deaths were related to irritation of the respiratory systems.

Seven-hour vapor exposures of rats, guinea pigs, and rabbits for 20 exposures in 28 days at a concentration of 30 ppm resulted in marked ocular and nasal irritation in all species, death and weight loss in rats, and decreased growth in guinea pigs (254). In rats, the lungs were severely affected and unspecified changes were found in the kidneys and spleen. Leukocytosis was also observed. Organs of the other species were not significantly affected. One hundred 7-hr exposures of 15 ppm administered to rats, guinea pigs, and rabbits over 140-day period produced ocular and nasal irritation in all animals. Also, mortality, leukocytosis, and slight renal damage characterized by increased kidney weight and slight tubular damage occurred in the rats (254).

Human Studies. Short-term exposures to concentrations of methyl isopropenyl ketone ranging from 0.291 to 14.5 ppm have been reported (254). There was a questionable odor but no irritation at 0.291 ppm. At 0.77 ppm, there was an immediate definite odor and the vapors were evident to the eyes after a few minutes. At 1.45 and 2.91 ppm, eye irritation was apparent. At 14.5 ppm, there was an immediate strong odor with eye irritation apparent after 2 min. Ocular irritation rapidly increased in severity with time.

2.7.3 Industrial Exposures

Methyl isopropenyl ketone is a highly toxic, highly irritating liquid or vapor. Industrial exposures are likely to be by inhalation, skin and eye contact. Its strong irritating and lacrimatory properties may not prevent toxicity because these effects are often not immediately sensed.

2.7.4 Hygienic Standards

A maximum concentration of 0.3 ppm for repeated exposures should be low enough to prevent local and systemic toxicity. Engineering features in areas where this material is handled should include good local ventilation. Protective equipment, including occlusive eye protection to prevent eye damage by vapors, should be available when methyl isopropenyl ketone is used.

2.8 Methyl *n*-Butyl Ketone; 2-Hexanone, Hexan-2-one, Methyl Butyl Ketone, MBK, MnBK, Propylacetone

$$CH_3-\overset{\overset{\displaystyle O}{\|}}{C}-CH_2-CH_2-CH_2-CH_3$$

2.8.1 Physical-Chemical Properties

Methyl *n*-butyl ketone is a water-clear liquid with an odor resembling acetone but more pungent. It has a relatively low cool flame ignition temperature of 188°C, and an autoignition temperature of 424°C. Additional physical-chemical data are summarized in Table 56.4.

2.8.2 Biologic Effects

The acute oral toxicity of methyl *n*-butyl ketone is low. Topical damage to the skin and eyes is also low but transient eye irritation can occur. Percutaneous absorption may occur and with concomitant inhalation exposure may contribute to methyl n-butyl ketone toxicity. High vapor concentrations of methyl n-butyl ketone may produce ocular and respiratory irritation followed by central nervous system depression and narcosis. Chronic inhalation produces toxic manifestations different from those produced by acute exposure. These include degenerative axonal changes primarily in the peripheral nerves and long spinal cord tracts, atrophy of testicular germinal cell epithelium, and depression of circulating white blood cells. The neurologic and hematologic changes occur at low dose levels, whereas high dose levels are required to produce germinal cell damage. The principal hazards in industry are by repeated inhalation and skin absorption.

Animal Studies. *Single Dose.* Smyth et al. (226) reported an oral LD_{50} for methyl n-butyl ketone of 2590 mg/kg for rats. The oral LD_{Lo} for mice was 1000 mg/kg. The intraperitoneal LD_{50} for rats and guinea pigs was 1142 and 1140 ml/kg, respectively (256).

Repeated Dose. Since 1974, a series of reports have been published describing neurotoxicity after repeated exposure to methyl n-butyl ketone. Spencer and Schaumburg (236) injected cats subcutaneously with 150 mg/kg methyl n-butyl ketone or a 9/1 mixture of methyl ethyl ketone and methyl n-butyl ketone twice a day, 5 days/week for 8.5 months. Methyl n-butyl ketone alone produced clinical signs of peripheral neuropathy after 8 to 10 weeks of exposure. Weakness was first evident in the hind limbs and later in the fore limbs. Pathologic changes included axonal damage with secondary myelin changes in the peripheral and central nervous systems. The long spinal tracts in the central nervous systems were particularly affected. Neuropathologic changes but not clinical neurotoxicity were produed by the 9/1 mixture of methyl ethyl ketone and methyl n-butyl ketone. Beagle dogs receiving treatments similar to the cats had a similar clinical course and developed similar neuropathologic lesions, except that the 9/1 mixture of methyl ethyl ketone and methyl n-butyl ketone did not produce evidence of neurotoxicity (256).

Groups of 10 rats were administered methyl n-butyl ketone in their drinking water at concentrations of 1.0 percent (560 mg/kg), 0.5 percent (266 mg/kg), or 0.25 percent (143 mg/kg) daily for 10 to 13 months (132).

Methyl n-butyl ketone produced a dose-dependent reduction in body weight gain which was present by the second week in the 1.0 and 0.5 percent groups and by the third week in the 0.25 percent group. Clinical neurotoxicity consisting of hind limb weakness was evident after 6 weeks in the 1.0 percent animals and after 10 weeks in the 0.5 percent animals. The 0.25 percent group was clinically normal. Axonal degeneration was found in the cerebellum, brainstem, spinal cord, peripheral nerves, and skeletal muscle of all groups.

Misumi et al. (160) reported that rats given 415 mg/kg/day methyl n-butyl ketone 5 days/week for several weeks had decreased motor and sensory nerve conduction times, and decreased motor and sensory nerve action potential amplitudes that correlated with pathologic findings. Eben et al. (59) reported no overt signs of neurotoxicity when doses of 400 mg/kg/day methyl n-butyl ketone were administered by stomach tube over a 40-week period. Reduced body weight gain was present after the second week of treatment, and transient hind limb weakness was observed between the 17th and 28th weeks. Rats injected intraperitoneally for 35 weeeks with 100 to 200 mg/kg of methyl n-butyl ketone, or 9/1 mixture of methyl ethyl ketone and methyl n-butyl ketone, showed no clinical or neuropathologic changes (256). Guinea pigs given 0.25 or 0.1 percent methyl n-butyl ketone in their drinking water had an increased body weight gain, whereas studies with other species have consistently shown decreased body weight gain (3). Locomotor activity was depressed and pupillary responses were impaired. Johnson et al. (117) found that 68 mg/kg methyl n-butyl ketone given by gavage produced a reduction in bar pressing response rate which they attributed to mild narcosis.

The relative neurotoxicity of methyl n-butyl ketone, n-hexane, practical grade hexane, and their metabolites was studied by Krasavage et al. (131). They found that methyl n-butyl ketone (6.6 mmol/kg or 660 mg/kg for 90 days) was less neurotoxic than its metabolites 2,5-hexanediol, 5-hydroxy-2-hexanone, and 2,5-hexanedione, but more neurotoxic than 2-hexanol, n-hexane, and practical grade hexane. In addition to neurotoxicity, testicular germinal cell atrophy was evident. Methyl n-butyl ketone can produce body weight depression and neurotoxicity after repeated doses as low as 143 mg/kg. Testicular atrophy is seen at doses between 560 and 660 mg/kg for 90 days.

Eye. Smyth et al. (226) reported that methyl n-butyl ketone caused mild eye irritation and minor transient corneal injury.

Skin. Tests on rabbit skin indicated that methyl n-butyl ketone produced no significant irritation. The dermal LD_{50} was 5.99 ml/kg (226). Repeated skin contact may be irritating due to the ability of methyl n-butyl ketone to defat the skin resulting in dermatitis.

Absorption through human and dog skin has been quantified by DiVincenzo et al. (51). Two volunteers exposed to radio-labeled methyl n-butyl ketone had percutaneous absorption rates of 4.8 and 8.0 μg/min/cm^2 of exposed skin. Skin exposure to a 9/1 mixture of methyl ethyl ketone and methyl n-butyl ketone resulted in absorption rates of 4.2 and 5.6 μg/min/cm^2 for methyl n-butyl ketone. When exposed under similar conditions, dogs absorbed 23.1 μg/min/cm^2 of methyl n-butyl ketone and 45 μg/min/cm^2 of the mixture; thus dog skin was more permeable to methyl n-butyl ketone than human skin. There was a sharp increase in methyl n-butyl ketone absorption through dog skin after 20 min of exposure, indicating that permeability had been altered by prolonged contact with the solvent. Based on these findings it was calculated that about twice as much methyl n-butyl ketone would be absorbed during a 1-hr dermal exposure than would be absorbed by inhalation of 25 ppm of methyl n-butyl ketone vapor for 1 hr. Therefore, repeated dermal exposure to methyl n-butyl ketone may contribute to its toxicity. Application of 2 ml methyl n-butyl ketone or a 9/1 mixture of methyl ethyl ketone/methyl n-butyl ketone, with or without dimethyl sulfoxide, to the backs of guinea pigs twice a day, 5 days/week for 31 weeks did not produce clinical or histologic evidence of toxic neuropathy. The only adverse effect noted was desquamation due to chronic defatting of the skin (256).

Inhalation. Smyth et al. (226) reported that more than 30 min of exposure to a saturated atmosphere of methyl n-butyl ketone was required to kill rats. Exposure to 8000 ppm for 4 hr resulted in the death of all exposed rats whereas 4000 ppm for 4 hr did not kill any of the animals. Schrenk et al. (208) exposed guinea pigs to a maximum of 1000 ppm for several hours and produced slight or no abnormal clinical signs. A maximum of 1500 ppm for several hours had no serious effect; and a maximum of 3000 ppm for 60 min did not cause serious disturbances. Exposures to 4000 to 6000 ppm were potentially lethal after several hours, 10,000 to 20,000 ppm was potentially lethal in 30 to 60 min, and >20,000 ppm killed guinea pigs within a few minutes. The acute effects of methyl n-butyl ketone exposure were eye and nasal irritation followed by narcosis and death. Pathologic examination revealed vascular congestion in major organs. Specht et al. (233) reported that acute inhalation exposure of guinea pigs to methyl n-butyl ketone produced depression of body temperature, respiratory and heart rate, loss of corneal, auditory, and equilibrium reflexes, narcosis, coma, and death.

Spencer et al. (235) reported that rats exposed to 1300 ppm methyl n-butyl ketone 6 hr/day, 5 days/week showed slight narcosis by 4 hr of exposure, incoordination after 5.5 hr, decreased weight gain, slow progressive weight loss beginning after the 73rd exposure, and symmetrical hind limb weakness between the third and fourth months of exposure. Pathologic changes included "giant" axonal swellings, paranodal myelin retraction or loss, and axonal degeneration in peripheral nerves, spinal cord, medulla oblongata, and cerebellum. Electron microscopy showed masses of neurofilaments, numerous mitochondria, and sparse neurotubules in the swellings of myelinated and unmyelinated axons. Mendell et al. (157) described an identical neuropathy in chickens, rats, and cats exposed to 100 to 200 ppm, 24 hr/day, 7 days a week. Clinical signs of weakness were observed in chickens at 4 to 5 weeks and in cats at 5 to 8 weeks. Electrodiagnostic findings and pathologic changes comparable to those described in humans exposed to methyl n-butyl ketone were found in experimental animals.

Duckett et al. (55) described a neuropathy in rats exposed to 200 ppm methyl *n*-butyl ketone or 200 ppm methyl *n*-butyl ketone/2000 ppm methyl ethyl ketone, 8 hr/day, 5 days/week for 6 weeks. These findings are difficult to integrate with the other studies of methyl *n*-butyl ketone neurotoxicity because the method used to determine the concentration of methyl *n*-butyl ketone was not specified. The rats were weak after each exposure, which is an unlikely effect from an exposure to only 200 ppm, and the pathologic changes were not characteristic of methyl *n*-butyl ketone. Similar difficulties were also encountered when trying to evaluate a report of subclinical neuropathologic changes in rats exposed to 50 ppm methyl *n*-butyl ketone (56).

Johnson et al. (117) reported that exposure of rats and monkeys to 1000 ppm methyl *n*-butyl ketone for 6 hr/day, 5 days/week for 25 weeks produced decreased body weight, clinical signs of neurotoxicity, and abnormal electrodiagnostic parameters in the peripheral nerves and visual system. Rats exposed for 29 weeks and monkeys exposed for 41 weeks to 100 ppm of methyl *n*-butyl ketone had nerve conduction velocities and evoked muscle action potentials which differed significantly from the controls. Recovery of nerve conduction velocities occurred in monkeys 6 months after cessation of exposure to 1000 ppm and 2 months after exposure to 100 ppm. Recovery from clinical neuropathy has also been observed in dogs (256). Chronic inhalation of methyl *n*-butyl ketone at 100 and 330 ppm for 6 hr/day, 5 days/week for 72 weeks (rats) or 2 years (cats) produced subclinical neuropathy in cats at 330 ppm but not at 100 ppm, and a dose-dependent reduction of body weight gain in rats after 4 months (330 ppm) and 9 months (100 ppm) (256). Other effects due to methyl *n*-butyl ketone exposure include testicular germinal cell atrophy (131) and a reduction in total circulating white blood cells (121).

Enhancement of methyl *n*-butyl ketone neurotoxicity by methyl ethyl ketone has been reported by Saida et al. (202). They compared continuous exposures of 225 ppm methyl *n*-butyl ketone, 1125 ppm methyl ethyl ketone, and 225 ppm methyl *n*-butyl ketone plus 1125 ppm methyl ethyl ketone in rats, and demonstrated that the addition of high levels of methyl ethyl ketone to a neurotoxic dose of methyl *n*-butyl ketone shortened the onset of paralysis from 66 to 25 days. It would be expected that exposure of animals to a subneurotoxic dose of methyl *n*-butyl ketone plus a high dose of methyl ethyl ketone might also produce neurotoxicity, as has been reported for *n*-hexane and methyl ethyl ketone (9).

Human Studies.   Schrenk et al. (208) reported that volunteers exposed for several minutes to 1000 ppm methyl *n*-butyl ketone experienced moderate eye and nasal irritation.

Gillchrist et al. (85), Billmaier et al.(20), and Allen et al. (6) reported an outbreak of toxic neuropathy which resulted when methyl *n*-butyl ketone was substituted for methyl isobutyl ketone in a mixture with methyl ethyl ketone. The clinical course in affected workers included distal symmetrical sensory changes and weakness, most severely affecting the legs, and an unexplained weight loss. The amount of time from the introduction of the ketone change to the onset of neuropathy was approximately 7 months. Although ambient solvent levels were measured, it is questionable whether they reflect the actual exposure levels. Similar case reports on painters (147, 202), cabinet finishers (39), and a screen cleaner (273) are in the literature. In each case, exposure was probably by both the inhalation and the dermal routes. DiVincenzo et al. (51) studied the respiratory, gastrointestinal, and percutaneous absorption of radiolabeled methyl n-butyl ketone in man. The ketone was readily absorbed by all three routes. Exposure to 10 or 50 ppm for 7.5 hr or 100 ppm for 4 hr resulted in the absorption of between 75 and 92 percent of the vapor. Excretion was primarily as carbon dioxide. Slow excretion (65 percent of an oral dose in 8 days) probably allows accumulation of neurotoxic metabolites in the body, even with low exposures.

Toxicokinetics. Metabolism of hydrocarbons and methyl ketones is a common capability of many forms of life and has many similarities from microbes to man (49, 50, 51, 107, 124, 141, 149). Since 1974 much work has been done on the metabolic interrelationships of n-hexane, methyl n-butyl ketone, and their neurotoxic metabolites, particularly 2,5-hexanedione (2, 36, 37, 49, 50, 51, 59, 131, 185). This work clearly indicates that the toxicity of n-hexane and methyl n-butyl ketone are linked through their oxidation to secondary alcohols (2-hexanol, 2,5-hexanediol, and 5-hydroxy-2-hexanone) and to the γ-diketone, 2,5-hexanedione. Uptake of the parent compounds is rapid through pulmonary, gastrointestinal, and percutaneous tissues, as is their metabolic conversion to neurotoxic metabolites. The principal neurotoxic metabolite, 2,5-hexanedione, is slowly excreted from the body, giving rise to its possible accumulation in target organs. Methyl n-butyl ketone is eliminated from the body primarily as carbon dioxide in expired air, and as 2-hexanol, 5-hydroxy-2-hexanone, 2,5-hexanedione, 2,5-dimethylfuran, γ-valerolactone, norleucine, and urea in the urine (3, 36, 50, 51).

Simultaneous exposure to methyl n-butyl ketone and other chemicals can alter the neurotoxicity observed. Methyl ethyl ketone, in particular, can accelerate the onset of neurotoxicity (202). Enhancement may be due to the induction of the hepatic mixed function oxidase system. Inhalation of methyl ethyl ketone/methyl n-butyl ketone shortened hexobarbital sleeping time and increased in vitro production of p-aminophenol, formaldehyde, p-aminobenzoate, and sulfanilamide by rat hepatocyte microsomes (35). On the other hand, Abdel-Rahman et al. (2) reported that phenobarbital, also a microsomal inducer, protects against methyl n-butyl ketone neurotoxicity.

Toxic Mechanisms. The mechanism of acute methyl n-butyl ketone toxicity is by progressive depression of the central nervous system resulting in coma and cardiorespiratory failure, but the mechanisms involved in its chronic toxicity are unclear. Though studies in experimental animals have shown that 2,5-hexanedione is the active metabolite in methyl n-butyl ketone neurotoxicity and that the γ spacing of the carbonyl groups is essential to produce neurotoxic effects (174), the mechanisms for the interaction of the γ diketone and the target tissue are still unclear. Interference with energy metabolism (200), membrane structure (37), and vitamins or cofactors (30, 207) have been proposed. The interference with cell proliferation observed in tissue cultures (211) relates directly with the effects of methyl n-butyl ketone on the testis, although the mechanism of this action is not known.

2.8.3 Industrial Exposure

Industrial exposure by the oral route is unlikely in the normal work environment. The greatest hazards of methyl n-butyl ketone are by topical (eye and skin) and inhalation routes. Contact with the eyes should be avoided by wearing appropriate eye protection. Skin contact should be avoided to prevent dermatitis due to the defatting action of methyl n-butyl ketone and to prevent percutaneous absorption. Inhalation exposure, particularly repeated exposures, should be kept to a minimum.

2.8.4 Hygienic Standards

The American Conference of Governmental Industrial Hygienists (243) has established a threshold limit value of 5 ppm (20 mg/m^3) for methyl n-butyl ketone to prevent subclinical neuropathy.

2.9 Methyl Isobutyl Ketone; 4-Methyl-2-pentanone, Isopropyl Acetone, Hexone, MIBK

$$CH_3-\overset{\displaystyle O}{\overset{\displaystyle \|}{C}}-CH_2-\underset{\displaystyle \underset{\displaystyle CH_3}{|}}{CH}-CH_3$$

2.9.1 Physical-Chemical Properties

Methyl isobutyl ketone is a clear liquid with a sweet, sharp odor. The odor threshold concentration has been reported to be 0.10 ppm (265). Its flammability is such that it may present a fire hazard. Additional physical-chemical properties are summarized in Table 56.4.

2.9.2 Biologic Effects

Methyl isobutyl ketone has a low degree of oral toxicity. Skin exposures should produce little irritation unless the compound is held in close contact. Repeated skin contact may produce a dermatitis due to its defatting properties. Repeated doses, by various routes, indicate that methyl isobutyl ketone is not neurotoxic. Vapors of methyl isobutyl ketone inhaled at high concentrations can produce narcosis and death. However, the low odor threshold and its irritant effects should preclude overexposure.

Animal Studies. *Single Dose.* Smyth et al. (225) reported an oral LD_{50} of 2.08 g/kg for rats when methyl isobutyl ketone was administered as a 20 percent emulsion in Tergitol 7 surfactant. Smyth (255) also reported an LD_{50} of 5.7 ml/kg for rats when the compound was given undiluted. Oral LD_{50}s between 1600 and 3200 mg/kg for both rats and guinea pigs, and intraperitoneal LD_{50}s between 400 and 800 mg/kg for rats and between 800 and 1600 mg/kg for guinea pigs, have been reported (256). Terhaar et al. (256) determined the intraperitoneal LD_{50} for rats and guinea pigs to be 1.14 and 0.92 ml/kg, respectively. When methyl isobutyl ketone was administered in combination with methyl ethyl ketone (9/1 MEK/MIBK), the intraperitoneal LD_{50}s for these species were essentially unchanged, 1.64 ml/kg for the rat and 1.07 ml/kg for the guinea pig (256). Other investigators (276) have reported a 24-hr oral LD_{50} for methyl isobutyl ketone of 1.9 g/kg for mice. Batyrova (18) conducted acute oral studies in mice and rats and found the lethal dose to be 2.85 and 4.6 g/kg, respectively.

Repeated Doses. The neurotoxic potential of methyl isobutyl ketone has been studied in rats, cats, and dogs (256). Rats were given intraperitoneal injections of methyl isobutyl ketone or a mixture of methyl ethyl ketone and methyl isobutyl ketone (9/1 by volume) five times weekly for 35 weeks. The dose levels of 10, 30, and 100 mg/kg were increased to 20, 60, and 200 mg/kg after 2 weeks of treatment. Except for body weight suppression after 3 to 4 weeks of treatment, the only other effect noted was transient anesthesia during the first month of treatment in the 200 mg/kg animals. No toxic neuropathy was noted. Electromyographic examination of dogs administered subcutaneous doses of methyl isobutyl ketone (300 mg/kg) daily for 11 months revealed no evidence of neurotoxicity (256). Spencer and Schaumburg (236) also treated cats subcutaneously with 150 mg/kg methyl isobutyl ketone or methyl ethyl ketone/methyl isobutyl ketone (9/1) twice daily, five times a week for up to 8.5 months. They found no detectable nervous system damage attributable to methyl isobutyl ketone. Beagle dogs receiving similar treatments as the cats also did not show neurotoxic changes due to methyl isobutyl ketone or the combination of methyl ethyl ketone and methyl isobutyl ketone (9/1 by volume) (256).

Eye. Undiluted methyl isobutyl ketone (0.1 ml) produced some irritation within 10 min when instilled in the rabbit eye. Inflammation and swelling occurred in 8 hr, and inflammation, swelling, and exudate were present at 24 hr (257).

Skin. A single application of methyl isobutyl ketone to the skin of rabbits produced only transient erythema, but daily applications of 10 ml for 7 days caused drying and flaking of the skin (257). Undiluted methyl isobutyl ketone (5 and 10 ml) held in contact with the depilated skin of guinea pigs under an occlusive wrap for 24 hr produced slight irritation with no clinical evidence of absorption (256). Other studies (258) showed that 500 mg of methyl isobutyl ketone produced moderate irritation of rabbit skin after 24 hr. Methyl isobutyl ketone or methyl ethyl ketone/methyl isobutyl ketone (9/1), with or without dimethyl sulfoxide, dropped on the backs of guinea pigs in amounts up to 2 ml twice daily for 31 weeks produced only desquamation with no clinical or histologic evidence of toxic neuropathy (256).

Inhalation. Exposure to 19,500 ppm of methyl isobutyl ketone produced anesthesia in 7 of 10 mice within 30 min. When the animals were removed to fresh air, four recovered immediately and three were awake within 5 min. Concentrations above 20,000 ppm produced anesthesia within 30 min with subsequent death of most of the animals. Gross examination at necropsy revealed congestion of the lungs (257). Daily exposures to 20,000 ppm for 20 min for 15 days killed 6 of 10 mice. Smyth et al. (225) reported that 15 min in a saturated atmosphere of methyl isobutyl ketone was the minimum time necessary to kill the first animal. Six rats survived a 4-hr exposure to 2000 ppm methyl isobutyl ketone, but 4000 ppm killed all six animals in 4 hr. Specht (231) and Specht et al. (233) exposed guinea pigs to concentrations of 1000, 16,800, and 28,000 ppm methyl isobutyl ketone. The 1000-ppm level caused little or no irritation of the eyes and nose of the animals; however, the operator did experience eye and nose irritation. Guinea pigs showed a decreased respiratory rate during the first 6 hr of exposure which the authors attributed to a low-grade narcosis. The 16,800 ppm level caused immediate signs of eye and nose irritation followed by salivation, lacrimation, ataxia, and death. Nine of 10 guinea pigs died within 6 hr of exposure. The highest concentration used (28,000 ppm) killed 50 percent of the animals within 45 min. Only a few guinea pigs survived 60 min of exposure. Fatty livers and congestion of the brain, lungs, and spleen were noted. The heart and kidneys were not affected.

Exposures of 21,000 ppm killed all rats within 53 min, whereas exposures of

4000 ppm for 6 hr caused loss of coordination and prostration. All animals recovered and showed no adverse signs (256).

Spencer et al. (235) reported on groups of male rats exposed to 1300 ppm methyl n-butyl ketone for 4 months or 1500 ppm methyl isobutyl ketone for 5 months. Methyl n-butyl ketone produced a toxic distal axonopathy whereas methyl isobutyl ketone produced some minimal distal axonal changes. However, this may have been due to the 3 percent methyl n-butyl ketone present as a contaminant in the methyl isobutyl ketone or, more likely, was a compression neuropathy due to the type of cages used (236). Batyrova (18) showed that 20 to 30 ppm of methyl isobutyl ketone 4 hr/day for 4.5 months caused disturbances in the conditioned reflexes of rats and interfered with the detoxifying function of the liver; the eosinophil count was also elevated. Geller et al. (80) reported that rats exposed to 25 ppm methyl isobutyl ketone showed a minimal statistical increase in pressor lever response, but the discriminatory behavior of baboons was not impaired by exposures of 20 to 40 ppm. Geller et al. (79) reported delayed behavioral response times in baboons exposed to 50 ppm of methyl isobutyl ketone alone, but no alteration of response was seen when methyl isobutyl ketone was combined with methyl ethyl ketone (100 ppm). Garcia et al. (76) also reported subtle behavior alterations in rats exposed to low levels of methyl isobutyl ketone. Vernot et al. (264) and MacEwen et al. (145) reported increased kidney weights and kidney/body weight ratios in rats exposed to 100 ppm methyl isobutyl ketone continuously for 2 weeks. Kidney and liver weights and the organ/body weight ratios were also increased after exposure to 200 ppm for 2 weeks and to 100 ppm for 90 days. Similar effects were not seen in dogs, monkeys, or mice after the 2-week exposures. MacKenzie (146) described kidney damage at 2 weeks after 100 ppm methyl isobutyl ketone as hyaline droplet tubular nephrosis. This damage was reversible, even after 90 days of exposure, when the rats were removed from the exposure environment.

Human Studies.   Elkins (61) reported that one group of workers exposed to 100 ppm developed headache and nausea, whereas another group complained only of respiratory tract irritation. Tolerance seemed to develop during the work week but was lost over the weekend. Most of these effects were not noted at 20 ppm. Silverman et al. (216) reported that 12 persons exposed to methyl isobutyl ketone for 15-min periods indicated that 100 ppm was the highest concentration they considered satisfactory for an 8-hr exposure period, and that 200 ppm had an objectionable odor and definitely caused eye irritation. Others (257) have reported that 5-min exposures of unconditioned volunteers revealed that the odor threshold for all subjects was <100 ppm; 200 to 400 ppm produced eye irritation in 50 percent of the subjects, and 400 ppm caused nasal irritation in 50 percent. Armeli et al. (16) reported on 19 workers exposed to concentrations of 80 to 500 ppm MIBK and other chemicals for 30 min/day. Common complaints were weakness, loss of appetite, headache, eye and throat irritation, nausea, vomiting, and diarrhea. When examined 5 years later after improved ventilation reduced exposure levels to 50–105 ppm, these signs were greatly reduced. May (152) conducted studies which showed that the first perceptible concentration of methyl isobutyl ketone detected by a test subject was 8 ppm, whereas 15 ppm was reported as being definitely detectable. Wayne and Orcutt (271) directed concentrations of 5 or 20 ppm methyl isobutyl ketone at subjects wearing eye masks, and found that methyl isobutyl ketone was somewhat more irritating to the eyes than several other solvents tested.

Toxicokinetics. DiVincenzo et al. (49) have shown that methyl isobutyl ketone is metabolized in guinea pigs by ω-1 oxidation to the corresponding hydroxy ketone, 4-hydroxy-4-methyl-2-pentanone, and by carbonyl reduction to the secondary alcohol, 4-methyl-2-pentanol. Zlatkis and Liebick (278) showed the presence of methyl isobutyl ketone and other substituted 2-pentanones in 24-hr urine samples of unexposed men and women.

Toxic Mechanisms. Smyth (221) has stated that the most important manifestation of methyl isobutyl ketone toxicity was its narcotic effect. In addition to effects on the central nervous system, exposure to methyl isobutyl ketone may also affect the cardiovascular system. It is moderately irritating to the eyes, nose, and mucous membranes (216, 231, 271). Renal and hepatic toxicity have been reported by several investigators (18, 145, 146, 264).

2.9.3 Industrial Exposure

The most likely exposures to methyl isobutyl ketone in the workplace is by inhalation of the vapors and by skin and eye contact. The principal health hazard is inhalation. Levels 20 times the recommended TLV (i.e., 1000 ppm) produce central nervous system depression and narcosis. Exposure to 100 ppm appears to present a problem only because of the objectionable odor, whereas exposure to 200 to 400 ppm is irritating to the eyes, nose, and throat. However, reported industrial exposures show that levels from 100 to 500 ppm also produce gastrointestinal effects such as nausea, vomiting, loss of appetite, and diarrhea. Methyl isobutyl ketone should be handled where adequate ventilation is available. Skin contact should be avoided because the defatting property of this ketone may produce dermatitis. Eye contact may produce painful irritation, but if it does occur, immediate flushing of the eyes with water should be palliative. If exposure to high concentrations of methyl isobutyl ketone is likely to occur, proper respiratory protection and protective clothing should be used.

2.9.4 Hygienic Standards

The American Conference of Governmental Industrial Hygienists (243) has established a threshold limit value for methyl isobutyl ketone of 50 ppm (205 mg/m^3).

2.10 Mesityl Oxide; Methyl Isobutenyl Ketone, 4-Methyl-3-pentene-2-one

$$\begin{array}{ccc} O & & CH_3 \\ \parallel & & | \\ CH_3-C-CH & = & C-CH_3 \end{array}$$

2.10.1 Physical-Chemical Properties

Mesityl oxide is a colorless, oily liquid with a peppermint-like odor, which may darken upon standing. Its odor threshold is 0.017 ppm (265). Mesityl oxide is flammable and may present a moderate fire hazard. Additional physical-chemical properties are summarized in Table 56.4.

2.10.2 Biologic Effects

Mesityl oxide can produce marked irritation and transient corneal injury to the eye. Occasional skin contact may produce some irritation; prolonged contact may produce dermatitis and, if the dose is high enough, systemic injury. High concentrations of the vapors produce narcosis and may injure the lungs, liver, and kidneys. However, its penetrating odor and eye and nose irritation at low levels should prevent overexposure.

Animal Studies. *Single Dose.* The acute oral LD_{50} has been reported to be 1120 mg/kg for the rat (258). The oral LD_{50} was 1000 mg/kg for the rabbit (257). The single-dose intraperitoneal LD_{50} was 354 mg/kg for mice (257). The lowest doses to produce effects when given subcutaneously were 840 and 1400 mg/kg for the rabbit and the frog, respectively (15).

Repeated Dose. Ito (109) reported growth inhibition in rats given repeated subcutaneous doses of 0.2 ml/kg, three times a week for 30 days.

Eye. Undiluted mesityl oxide (43 mg) is intensely irritating and produces corneal damage when in contact with the rabbit eye (25).

Skin. When applied to the uncovered skin of rabbits, 430 mg of mesityl oxide produced only mild irritation. The acute dermal LD_{50} was reported to be 5.99 ml/kg for the rabbit (258). Doses of 0.5 ml (20,000 mg/kg) dropped on the skin of mice killed all (10/10) animals in 3 to 9 hr. Marked irritation occurred within a few minutes and the animals were ataxic and narcotized within 15 min. Applications of 0.1 ml to the backs of mice produced local irritation and excitement within 5 min. One of 10 animals died in 12 hr; the others recovered (103). Because of its defatting properties, repeated contact may cause dermatitis

(257). The 14-day dermal LD_{50} for guinea pigs has been reported to be 1.9 g/kg. However, exposure was over a 4-day period on a poultice which may have absorbed some of the compound, thus underestimating the LD_{50}. Gross pathology included mottled liver, pale spleen, and kidneys, and congestion of the stomach and intestines (255).

Inhalation. Vapor concentrations of 2300, 5000, and 10,000 ppm for up to 8 hr produced a dose-dependent narcotic response in guinea pigs. Clinical signs of toxicity were respiratory irritation, reduction of body temperature, respiratory and heart rates, loss of reflexes, coma, and death at all dose levels (233). Smyth et al. (229) found that 12,000 ppm mesityl oxide killed rats and guinea pigs after 1 hr of exposure. Eight-hour exposures to 2500, 1000, and 500 ppm killed 100, 68, and 30 percent of the test animals, respectively. Carpenter et al. (26) subjected groups of six male or female rats to acute 4-hr exposures of increasing concentrations of mesityl oxide until two to four of the animals died within a 14-day observation period. They found that a concentration of 1000 ppm was necessary to produce this mortality. Hart et al. (103) exposed mice to concentrations of 0.6 to 2.4 percent mesityl oxide (v/v) in air. Clinical signs of toxicity were ocular and nasal irritation, labored breathing convulsions, narcosis, vasodilation, cyanosis, and death. The time of death was concentration dependent, and ranged from 23 to 135 min. Rabbits exposed to a 1.3 percent concentration for 30 or 90 min exhibited only eye and nose irritation. Rats exposed to 300 ppm for 2 hr/day for 30 days showed leucocytosis and hypertrophy of the liver, kidney, and spleen. Exposure of rabbits to 25 ppm 4 hr/day for 189 days produced anemia and leucopenia (109).

Hart et al. (103) also exposed mice and rabbits repeatedly to 1.3 percent mesityl oxide. Five daily exposures of 15 min produced no deaths in mice, whereas 11 exposures killed 3/10 animals. Increasing the exposure time to 30 min/day killed all 10 mice within 6 days. The rabbits exposed to 1.3 percent mesityl oxide for 30 min/day showed only eye and nose irritation after 15 days, whereas exposures of 60 min duration produced spastic paralysis in 10 days and death within the following 7 to 11 days. Smyth et al. (229) exposed 10 male rats and 10 guinea pigs of both sexes to mesityl oxide at concentrations of 50, 100, 250, and 500 ppm for 8 hr/day, 5 days/week for 6 weeks. No deaths occurred in the 50, 100, or 250 ppm exposed groups. The 500 ppm group was terminated after 10 days due to high mortality (65 percent). Pathologic changes were seen at all concentrations except 50 ppm, and included poor growth, congestion of the liver, dilated Bowman's capsules and swollen convoluted tubular epithelium in the kidney, and often congestion of the lung. The frequency and intensity of these changes were dose dependent.

Human Studies. Silverman et al. (216) reported that 25 ppm mesityl oxide produced eye irritation in humans, and 50 ppm also produced nasal irritation, an objectionable odor, and a bad taste. Other studies using unconditioned subjects revealed that, when exposed for 5 min, half of the people could detect the odor of mesityl oxide at 12 ppm and all detected it at 25 ppm. Eye irritation was evidenced at 50 to 100 ppm. Half of the test subjects experienced nasal irritation and pulmonary discomfort at 25 ppm (257). Ito (109) reported the possibility of anemia and leucopenia in workers due to mesityl oxide exposure.

Toxic Mechanisms.   Mesityl oxide, like many other ketones, can produce narcosis. Sublethal concentrations of the vapors produce congestion, primarily in the kidneys. The liver and lung are affected to a lesser degree. Death is generally attributed to its narcotic action.

2.10.3 Industrial Exposure

No reports are available regarding human exposure resulting from the industrial handling of mesityl oxide. Brief skin exposure may produce irritation, and prolonged or repeated contact may result in dermatitis. Systemic injury may occur due to percutaneous absorption. Mesityl oxide has a distinct odor and its irritant properties at low concentrations should prevent overexposure.

2.10.4 Hygienic Standards

The American Conference of Governmental Industrial Hygienists (243) has adopted a threshold limit value of 15 ppm for mesityl oxide. The National Institute of Occupational Safety and Health (171) has recommended that the concentration of mesityl oxide not exceed 10 ppm in the workplace.

2.11   4-Hydroxy-4-methyl-2-pentanone; Diacetone Alcohol, 2-Methyl-2-pentanol-4-one

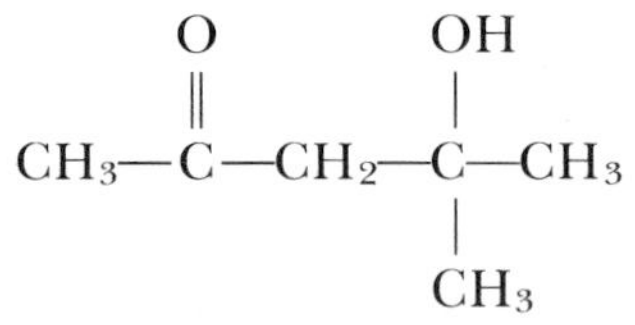

$$
\begin{array}{ccc}
\text{O} & & \text{OH} \\
\parallel & & | \\
\text{CH}_3\text{—C—CH}_2\text{—C—CH}_3 \\
& & | \\
& & \text{CH}_3
\end{array}
$$

2.11.1   Physical-Chemical Properties

4-Hydroxy-4-methyl-2-pentanone is a colorless, flammable liquid which becomes yellow with age. It has an odor threshold of about 0.28 ppm (265). Additional physical-chemical properties are summarized in Table 56.4.

2.11.2   Biologic Effects

4-Hydroxy-4-methyl-2-pentanone has a low degree of oral toxicity. It is only slightly irritating to the skin but prolonged contact may defat the skin and cause dermatitis. Eye contact may produce moderate to severe irritation. 4-Hydroxy-4-methyl-2-pentanone is slightly toxic by skin absorption. High concentrations of the vapor produce narcosis and systemic injury. Eye, nose, and throat irritation occur at levels well below those required to produce systemic injury.

Animal Studies.   *Single Dose.*   The oral LD_{50} of 4-hydroxy-4-methyl-2-pentanone has been reported to be 4.0 g/kg for rats (223). Other investigators (257) reported that 2 ml/kg given orally produced transient liver damage in rats, and 4 ml/kg killed the animals. Single doses of 2.4 to 4.0 ml/kg produced narcosis in rabbits, and 5 ml/kg killed the animals (70). The intraperitoneal LD_{50} was reported to be 933 mg/kg for mice (257). Intravenous injections of 1.0 to 1.5 ml/kg produced narcosis in rabbits and 3.25 mg/kg was the lowest dose to produce lethality. The lowest intramuscular dose that produced lethality in rabbits was between 3 and 4 ml/kg (70).

Repeated Doses.   Renal effects were seen in rats ingesting 40 mg/kg/day of 4-hydroxy-4-methyl-2-pentanone in the drinking water for 30 days. No effects were seen in rats ingesting 10 mg/kg/day (223). Oral administration of 2 ml daily for 12 days produced narcosis, kidney damage, and death in three of four rabbits (70). Repeated subcutaneous injections of 0.08 ml of 4-hydroxy-4-methyl-2-pentanone produced narcosis followed by recovery in rats (70).

Eye.   Undiluted 4-hydroxy-4-methyl-2-pentanone has been reported to produce moderate eye irritation and transient corneal damage in the rabbit (25).

Skin.   At most, only mild skin irritation was produced by application of undiluted 4-hydroxy-4-methyl-2-pentanone to the skin of rabbits (257, 258). The acute dermal LD_{50} was 14.5 ml/kg for rabbits (223).

Inhalation.   Rats survived 8-hr exposures to saturated atmospheres of 4-hydroxy-4-methyl-2-pentanone (25). Mice, rats, rabbits, and cats exposed to 2100 ppm for 1 to 3 hr exhibited restlessness, irritation of the mucous membranes, excitement, and later somnolence. Kidney injury was observed (268).

Human Studies.   Controlled studies in humans revealed that 100 ppm 4-hydroxy-4-methyl-2-pentanone produced eye and throat irritation (216). This level also produced a bad taste and an objectionable odor. Nasal irritation occurred only at levels >100 ppm. Other investigators (257) reported that humans exposed to 400 ppm for 15 min experienced pulmonary discomfort and eye, nose, and throat irritation. A level of 100 ppm produced the same results seen by Silverman et al. (216) except that nasal irritation was also noted.

Toxic Mechanisms. The primary systemic effect of 4-hydroxy-4-methyl-2-pentanone is narcosis. High concentrations produce irritation and pulmonary discomfort. Death may result from respiratory failure due to the depression of the respiratory center. Additional effects are renal and hepatic injury and decreased blood pressure.

2.11.3   Industrial Exposure

Occupational exposure to 4-hydroxy-4-methyl-2-pentanone is most likely to be by inhalation and skin contact. It presents a low degree of hazard if good work practices are observed. Appropriate protective clothing and eye protection should be available. The occurrence of eye, nose, and throat irritation and an objectionable odor at low concentrations should protect against overexposure to 4-hydroxy-4-methyl-2-pentanone.

2.11.4   Hygienic Standards

The American Conference of Governmental Industrial Hygienists (10) has established a threshold limit value of 50 ppm (240 mg/m^3) for diacetone alcohol. This level was set on the basis of the eye, nose, and respiratory irritation that occurred in humans exposed to 100 ppm.

2.12   Methyl *n*-Amyl Ketone; 2-Heptanone, Methyl Pentyl Ketone

$$CH_3-\overset{\overset{\textstyle O}{\|}}{C}-CH_2-CH_2-CH_2-CH_2-CH_3$$

2.12.1   Physical-Chemical Properties

Methyl *n*-amyl ketone is a liquid of low volatility with a penetrating fruity odor. It has an odor threshold of about 0.020 ppm (265). Additional physical-chemical properties are summarized in Table 56.4.

2.12.2 Biologic Effects

Single acute and subchronic toxicity studies and repeated inhalation studies indicate that methyl n-amyl ketone has a low degree of toxicity. Evidence is available which shows that it is not a neurotoxin. Although high concentrations are capable of producing serious narcotic effects, the penetrating odor and eye and nose irritation seen at these levels should be good warning signs.

Animal Studies. *Single Dose.* The acute oral LD_{50} was 1670 mg/kg for rats (226) and 730 mg/kg for mice (238). Other investigators have reported the acute oral LD_{50} for rats as approximately 1600 mg/kg and the intraperitoneal LD_{50} as 800 mg/kg. Mice survived oral doses of 1600 mg/kg; the intraperitoneal LD_{50} was between 400 and 800 mg/kg (256).

Repeated Doses. Gaunt et al. (77) gave doses of 20, 100, and 500 mg/kg methyl n-amyl ketone by gavage in oil to male and female rats for 13 weeks. No untoward effects were seen in the animals given 20 mg/kg. Liver weights were increased in both sexes, and kidney weights were increased in the males only after 100 and 500 mg/kg of methyl n-amyl ketone. No statistically significant effects were found on body weight gain, feed consumption, hematology, histopathology, or renal concentration. Rats given 0.5 percent methyl n-amyl ketone in their drinking water for 12 weeks failed to develop clinical or histologic signs of peripheral neuropathy (237). Anger et al. (14) exposed rats to methyl n-amyl ketone intraperitoneally and by inhalation. The rats were tested for behavioral changes on a multiple fixed ratio–fixed interval schedule of reinforcements. Little or no change was seen in the fixed interval response after 18 mg/kg given intraperitoneally; moderate decreases were seen after 37 and 74 mg/kg; and near-cessation of response was found after 175 mg/kg. Inhalation exposures <1500 ppm did not produce behavioral changes, but exposures greater than 1575 ppm produced effects similar to those seen after intraperitoneal administration of 175 mg/kg.

Eye. Smyth et al. (227) reported mild eye irritation in rabbits.

Skin. Smyth et al. (227) reported that methyl n-amyl ketone applied uncovered to the abdomen of rabbits produced moderate skin irritation. The acute dermal LD_{50} was 12.6 ml/kg. Methyl n-amyl ketone applied undiluted to the intact or abraded skin of rabbits under an occlusive wrap for 24 hr was not irritating (139). The undiluted compound in quantities from 5 to 20 ml/kg when held in contact with the depilated skin of guinea pigs under an occlusive wrap for 24 hr produced slight to moderate irritation. No deaths occurred and there was no evidence of percutaneous absorption (256).

Inhalation. Single exposures of guinea pigs to concentrations of 1500 ppm produced irritation of the mucous membranes; 2000 ppm was strongly narcotic; and 4800 ppm produced narcosis and death in 4 to 8 hr of exposure (233). Rats survived 4-hr exposures to 2000 ppm, whereas 4000 ppm killed all six animals (227). A concentration of 5100 ppm killed all three rats exposed within 4 hr, and 4100 ppm killed all three rats within 6 hr of exposure. All rats survived 6-hr exposures to concentrations of 830 to 2000 ppm. Clinical signs of toxicity at all concentrations were piloerection, vasodilation, hyperpnea, ataxia, prostration, and dyspnea (256). Johnson et al. (116) measured the motor nerve conduction velocity and the evoked muscle action potential of rats and monkeys exposed by inhalation to 131 and 1025 ppm methyl *n*-amyl ketone 6 hr/day, 5 days/week for 9 months. No neurologic deficits were noted at the highest level tested. Tison et al. (244) reported that exposure to 650 ppm of methyl *n*-amyl ketone did not produce peripheral neuropathy in rats or chickens.

Human Studies. Skin sensitization studies were done on 26 human volunteers. Methyl *n*-amyl ketone was tested at a concentration of 4 percent in petrolatum and produced no positive reactions (64).

Toxicokinetics. Unpublished data (256) reveal that methyl *n*-amyl ketone undergoes carbonyl reduction to a secondary alcohol and ω-1 oxidation to a hydroxyketone which is further oxidized to 2,6-heptanedione. 2,6-Heptanedione has been tested and shown not to be neurotoxic in rats (256).

2.12.3 Industrial Exposure

The principal route of exposure in the occupational setting is by inhalation. Skin and eye contact may also occur. Its low odor threshold, its penetrating odor, and its irritant properties at levels below those that produce serious effects should preclude injury due to this ketone.

2.12.4 Hygienic Standards

The American Conference of Governmental Industrial Hygienists has established a threshold limit value of 50 ppm (243) for methyl *n*-amyl ketone.

2.13 Methyl Isoamyl Ketone; 5-Methyl-2-hexanone, 2-Methyl-5-hexanone,
Methyl Isopentyl Ketone, MIAK

$$CH_3-\overset{\overset{\displaystyle O}{\|}}{C}-CH_2-CH_2-\overset{\overset{\displaystyle CH_3}{|}}{CH}-CH_3$$

2.13.1 Physical-Chemical Properties

Methyl isoamyl ketone is a clear, colorless liquid with a sharp but pleasant, sweet odor. The odor threshold level for methyl isoamyl ketone is .012 ppm (265). Its autoignition temperature is 425°C. Additional physical-chemical properties are summarized in Table 56.4.

2.13.2 Biologic Effects

Methyl isoamyl ketone has a low degree of toxicity. Repeated exposure to large amounts of this ketone may lead to hepatic toxicity.

Animal Studies. *Single Dose.* Methyl isoamyl ketone has an oral LD_{50} of 3200 mg/kg for rats and the LD_{50} is between 3200 and 6400 mg/kg for mice. The intraperitoneal LD_{50} is between 400 and 800 mg/kg for rats and approximately 800 mg/kg for mice (256).

Repeated Dose. Groups of three rats were administered 1000, 2000, or 4000 mg/kg of methyl isoamyl ketone by gavage, 5 days/week for 3 weeks. At 4000 mg/kg, all rats died within 1.5 hr of dosing due to central nervous system depression and subsequent cardiorespiratory failure; pathologic lesions were predominantly due to vascular congestion. At 2000 mg/kg, all rats survived, and effects noted were reduced weight gain, reduced feed consumption, chronic irritation of the gastric mucosa, hepatocyte hypertrophy, and renal hyaline droplet formation. At 1000 mg/kg the only effect noted was evidence of minor chronic gastric irritation (256). Doses of 2000 mg/kg methyl isoamyl ketone were given by gavage 5 days/week for 13 weeks to a group of eight rats to determine its neurotoxic potential. Feed consumption was reduced only the first week of exposure, but this produced an early body weight gain depression that was not fully recovered by the end of the study. Hematologic determinations were comparable to controls, but serum clinical chemistries showed slight elevations of hepatic enzymes. Absolute and relative liver and adrenal gland weights and relative kidney weights were increased. Histopathologic changes included chronic gastric irritation, individual hepatocyte degeneration, diffuse hepatocyte hypertrophy, and microfoci of hepatocyte hyperplasia (256).

Eye.   Methyl isoamyl ketone produced slight eye irritation when dropped into the conjunctival sacs of six rabbits' eyes, three of which were washed with water promptly after instillation of the material (256).

Skin.   When undiluted methyl isoamyl ketone (5 to 20 ml/kg) was held in contact with the abdomen of guinea pigs under an occlusive wrap for 24 hr, slight irritation developed. Percutaneous toxicity may have occurred at 20 ml/ kg since weight gain was reduced (256). Smyth et al. (227) have reported a dermal LD_{50} of 10 ml/kg for rabbits. Repeated daily applications to the backs of guinea pigs by an open rub-on technique resulted in exacerbation of the irritative response. After 10 applications, all animals developed cracked eschars (256). Slight skin sensitization was observed in one of five guinea pigs immunized with methyl isoamyl ketone and Freund's complete adjuvant (256).

Inhalation.   Single, 6-hr exposures of rats to 802 ppm methyl isoamyl ketone produced no effects. As the concentration reached 1603 ppm, a decreased response to noise was observed. A concentration of 3207 ppm produced eye irritation and a decreased respiratory rate, narcosis, and the death of one of four rats. Four rats exposed to 5678 ppm died within 2 hr. The 6-hr LC_{50} was calculated to be 3813 ppm for rats (256). Smyth et al. (227) reported that the lowest dose to produce an effect was 2000 ppm for 4 hr of exposure. Five rats exposed to 400 ppm, 5 days/week for 12 exposures showed no effects on body weights, organ weights, hematology, clinical chemistries, and gross and histo-pathology (256).

Human Studies.   No significant adverse effects to methyl isoamyl ketone exposure have been reported.

2.13.3   Industrial Exposure

Exposure to methyl isoamyl ketone in the workplace is most likely to occur by skin contact, eye contact, or inhalation, particularly at elevated temperatures. Repeated skin exposure may result in contact dermatitis of both the allergic and nonallergic types. Good industrial hygiene practices should include skin and eye protection, and adequate ventilation. Good warning properties of methyl isoamyl ketone should prevent overexposure.

2.13.4   Hygienic Standards

The American Conference of Governmental Industrial Hygienists has established a threshold limit value of 100 ppm for methyl isoamyl ketone based on its eye irritation and its chemical structure similarities to methyl isobutyl ketone (10). However, an "intended" change to 50 ppm has been suggested by ACGIH (243).

2.14 Ethyl *n*-Butyl Ketone; 3-Heptanone

$$CH_3-CH_2-\overset{\displaystyle O}{\overset{\displaystyle \|}{C}}-CH_2-CH_2-CH_2-CH_3$$

2.14.1   Physical-Chemical Properties

Ethyl *n*-butyl ketone is a clear liquid, with a strong fruity odor. Additional physical-chemical properties are summarized in Table 56.4.

2.14.2   Biologic Effects

Ethyl *n*-butyl ketone is generally recognized as safe (GRAS, 1965) and is approved by the FDA for use as an artificial flavoring substance in foods. Its toxicity is low by the oral and percutaneous routes. It produces only mild irritation when in contact with the skin and eyes, but repeated or prolonged skin exposure may result in dermatitis. Ethyl *n*-butyl ketone is metabolized by rats to 2,5-heptanedione which, like its six-carbon counterpart, 2,5-hexanedione, has been shown to be neurotoxic. However, repeated inhalation exposures to high concentrations of ethyl *n*-butyl ketone failed to produce serum levels of 2,5-heptanedione high enough to produce neuropathy in rats. The low volatility of this ketone along with good industrial hygiene practices should prevent the occurrence of serious toxic effects.

Animal Studies.   *Single Dose.*   The acute oral LD_{50} of ethyl *n*-butyl ketone was reported to be 2.76 g/kg for rats (224).

Repeated Doses.   Homan et al. (106) reported that ethyl *n*-butyl ketone did not produce discernible neurotoxic effects when administered in the drinking water of rats for 120 days.

Eye.   Ethyl *n*-butyl ketone caused only slight irritation of rabbit eyes (22).

Skin.   Smyth et al. (224) reported that ethyl *n*-butyl ketone produced mild skin irritation in rabbits. More recent studies by Moreno (163) indicated that ethyl *n*-butyl ketone applied undiluted to the intact or abraded skin of rabbits under an occlusive wrap for 24 hr was moderately irritating. The dermal LD_{50} was greater than 20 ml/kg for rabbits (224). Ethyl *n*-butyl ketone was tested for its potential to produce contact dermatitis in the guinea pig and was found to be negative (213).

Inhalation.   Smyth et al. (224) reported that rats survived 4-hr exposures to 2000 ppm of ethyl *n*-butyl ketone, but all rats exposed for 4 hr to 4000 ppm died.

A single, 6-hr exposure to 5800 ppm ethyl *n*-butyl ketone killed all three rats exposed. Six-hour exposures to 420 and 3000 ppm did not produce death. Ataxia, prostration, and narcosis were seen at the higher levels (256). Rats and guinea pigs survived 6-hr daily exposures to 400 to 500 ppm for 10 days. Clinical signs included piloerection, vasodilation, and lacrimation. No pathologic lesions were found in the lungs, livers or kidneys (256).

Katz et al. (121) studied the neurotoxic potential of ethyl *n*-butyl ketone in rats. Exposures to 700 ppm ethyl *n*-butyl ketone vapors for 24 weeks produced no clinical signs of systemic toxicity or neurotoxicity. No significant effects were seen on body weight gain, hematology, clinical chemistries, and histopathology except for a significant depression in total white blood cells. It was postulated that the lack of neurotoxicity may have been due to the low concentration of the known neurotoxic metabolite, 2,5-heptanedione, found in the serum of these animals.

Human Studies.   Ethyl *n*-butyl ketone (4 percent in petrolatum) produced no irritation to human skin after 48 hr under an occlusive patch. This same concentration used in maximization tests on 25 volunteers failed to produce sensitization (65).

Toxicokinetics.   Analysis of the sera of rats exposed to 700 ppm ethyl *n*-butyl ketone revealed the presence of 6-hydroxy-3-heptanone and 2,5-heptanedione (121). These compounds are the seven-carbon atom counterparts of the neurotoxic metabolites of methyl *n*-butyl ketone, 5-hydroxy-2-hexanone, and 2,5-hexanedione. These data suggest that ethyl *n*-butyl ketone follows a metabolic pathway similar to that of methyl *n*-butyl ketone (see Figure 56.1). Ethyl *n*-butyl ketone was identified in urine of unexposed humans (278).

2.14.3   Industrial Exposure

Exposure in the industrial setting would be principally by inhalation and by skin and eye contact. Appropriate protective equipment and good industrial hygiene practices should preclude injury to the skin and eyes, and the low volatility of ethyl *n*-butyl ketone should prevent injury due to inhalation of the vapors.

2.14.4   Hygienic Standards

A threshold limit value of 50 ppm has been established by the American Conference of Governmental Industrial Hygienists (243). This level should be low enough to prevent injurious effects due to inhalation of ethyl *n*-butyl ketone vapors.

2.15 Di-*n*-propyl Ketone; Dipropyl Ketone, 4-Heptanone, Propyl Ketone, Butyrone, Heptan-4-one

$$CH_3—CH_2—CH_2—\overset{\overset{\displaystyle O}{\|}}{C}—CH_2—CH_2—CH_3$$

2.15.1 Physical-Chemical Properties

Di-*n*-propyl ketone is a stable, colorless, liquid with a pleasant but penetrating odor and a burning taste. Additional physical-chemical properties are summarized in Table 56.4.

2.15.2 Biologic Effects

Di-*n*-propyl ketone has a relatively low degree of toxicity. Repeated exposures to high concentrations may lead to hepatic enlargement and reduced blood glucose activity.

Animal Studies. *Single Dose.* The LD_{50} of di-*n*-propyl ketone was found to be 3.73 ml/kg for rats (28) and >3200 mg/kg for mice (256).

Repeated Doses. Di-*n*-propyl ketone administered undiluted by gavage to eight rats at 2000 mg/kg 5 days/week produced severe central nervous system depression and reduced weight gain. One rat died after a week of treatment due to cardiorespiratory failure. Lowering the dose to 1000 mg/kg resulted in improved weight gain and the absence of any deleterious clinical effects over a 12-week period. Hematologic determinations and serum clinical chemistries, except for a reduction in glucose, were unaffected. Relative liver and kidney weights were increased, and histologically there was hepatocyte hypertrophy. Repeated contact with the stomach resulted in hyperkeratosis and evidence of chronic irritation of the nonglandular gastric epithelium (256).

Eye. When di-*n*-propyl ketone was applied to the eyes of rabbits, only slight irritation was observed. (256).

Skin. Single applications of di-*n*-propyl ketone to the skin of guinea pigs under an occlusive wrap produced slight irritation. Repeated applications of 0.5 ml for a 10-day period produced slight erythema (256). Although absorption through the skin may occur, doses of 5, 10, and 20 ml/kg were not lethal when applied under an occlusive wrap for 24 hr (256). The dermal LD_{50} has been reported to be 5.66 ml/kg for rabbits (28).

Inhalation. The LC_{50} of a single 6-hr exposure of rats to di-*n*-propyl ketone was 2690 ppm. A concentration of 400 ppm di-*n*-propyl ketone decreased respiration and 825 ppm produced depression. Narcosis occurred at 1600 ppm, and 3220 ppm killed three of four rats. All four rats exposed to a saturated atmosphere (4970 ppm) died (256). Exposure to 2000 ppm for 4 hr did not kill any of the rats, whereas 4000 ppm for 4 hr killed all six animals exposed (28). Six-hour exposures to 1200 ppm of di-*n*-propyl ketone 5 days/week for 2 weeks resulted in a slightly decreased response to stimulation during exposure. Marginal liver enlargement, but no changes in hematology, clinical chemistries, or pathology were seen (256).

Human Studies.   No published reports of exposures to humans were found.

2.15.3   Industrial Exposure

Industrial exposures to di-*n*-propyl ketone are expected to be primarily by inhalation, and by skin and eye contact. Single topical exposures may be slightly irritating but repeated exposures may exacerbate the irritation. Oral exposures are not expected to be hazardous unless large quantities of di-*n*-propyl ketone are consumed. Repeated exposure to high levels may lead to liver enlargement and lower blood glucose.

2.15.4   Hygienic Standards

The American Conference of Governmental Industrial Hygienists have established a threshold limit value of 50 ppm for di-*n*-propyl ketone (243).

2.16   2-Octanone; Methyl *n*-Hexyl Ketone, Hexyl Methyl Ketone

$$CH_3-\overset{\overset{\displaystyle O}{\|}}{C}-CH_2-CH_2-CH_2-CH_2-CH_2-CH_3$$

2.16.1   Physical-Chemical Properties

2-Octanone is a colorless liquid with a pleasant fruitlike odor resembling the essence of apple. 2-Octanone is low in volatility and has an odor threshold of 248 ppm (265). There is little fire or explosion hazard with the use of 2-octanone in an industrial setting. Additional physical-chemical properties are summarized in Table 56.4.

2.16.2   Biologic Effects

2-Octanone is relatively low in toxicity. Direct skin contact may cause defatting and irritation of the skin. Inhalation may produce mild symptoms of eye, nose, and throat irritation at low concentrations, and may cause narcosis at high concentrations.

Animal Studies   *Single Dose.* The acute oral LD_{50} of 2-octanone was greater than 3.2 g/kg for rats and mice (256). Abbasov et al. (1) reported LD_{50}s of 3.1 g/kg for mice and 9.2 g/kg for rats. Hall and Carlson (101) reported an oral LD_{50} of 1.6 g/kg for mice. An oral LD_{50} greater than 5 g/kg was reported for rats (178). Intraperitoneal LD_{50}s were between 800 and 1600 mg/kg for both rats and mice (256). 2-Octanone was negative in the Ames salmonella microsome bacterial assay (155).

Eye.   2-Octanone produced slight irritation when applied undiluted to rabbit eyes (256).

Skin.   2-Octanone applied to the skin of guinea pigs produced slight to moderate skin irritation. Guinea pigs lost weight during a 2-week period following occluded skin contact, suggesting that 2-octanone was absorbed through the skin (256). When 500 mg of 2-octanone was applied to the skin of rabbits or rodents for 24 hr, slight irritation was noted (178). 2-Octanone did not produce skin sensitization in guinea pigs (213). The acute dermal LD_{50} was greater than 5 g/kg for rats (178).

Inhalation.   Rats exposed to calculated vapor concentrations of 8.9 mg/l for 6 hr exhibited signs of mild eye irritation (256). Guinea pigs exposed to saturated atmospheres of 2-octanone (~1300 ppm) developed immediate signs of eye and nasal irritation. Exposure for 10 hr produced evidence of central nervous system depression, and after 12 hr the animals were comatose (233).

Human Studies.   2-Octanone did not produce skin sensitization in human volunteers (126, 128).

Toxic Mechanisms.   Studies conducted in rats and mice indicate that 2-octanone significantly lowers serum cholesterol in both species, and triglyceride and glycerol levels in rats. Rats given 20 mg/kg of 2-octanone per day for 16 days showed significantly elevated serum lipase activity (24, 101).

2.16.3   Industrial Exposure

Occupational exposure is expected to be by inhalation and skin contact. 2-Octanone is relatively innocuous and does not pose a hazard in the workplace when good industrial hygiene practices are followed.

2.16.4 Hygienic Standards

A threshold limit value has not been established for 2-octanone.

2.17 3-Octanone; Ethyl *n*-Amyl Ketone, Ethyl Pentyl Ketone, Ethyl Amyl Ketone, EAK

$$CH_3-CH_2-\overset{\displaystyle O}{\overset{\displaystyle \|}{C}}-CH_2-CH_2-CH_2-CH_2-CH_3$$

2.17.1 Physical-Chemical Properties

3-Octanone is a clear liquid with a penetrating, but pleasant fruity odor. It is of low volatility and should not present a fire hazard unless used at high temperatures. Additional physical-chemical properties are summarized in Table 56.4.

2.17.2 Biologic Effects

3-Octanone should present only a low degree of hazard in the occupational setting. It has a low acute oral toxicity. Skin contact may cause moderate irritation and repeated skin contact may produce dermatitis due to its defatting action. No data are available on the subchronic or chronic effects of 3-octanone or on the effects of inhaling its vapors.

Animal Studies. *Single Dose.* An acute oral LD_{50} of greater than 5 g/kg has been reported for rats (215). The intraperitoneal LD_{50} was 406 mg/kg for mice (257).

Eye. No reports of eye irritation studies in experimental animals were found.

Skin. 3-Octanone applied full strength to the intact or abraded skin of rabbits under an occlusive wrap for 24 hr produced moderate irritation. The acute dermal LD_{50} was reported to be greater than 5 g/kg (215). Using a modified Draize procedure, Sharp (213) found that 3-octanone was not a sensitizer for guinea pigs.

Human Studies. Kligman (127) tested 3-octanone at 2.0 percent in petrolatum on 25 volunteers. It produced no skin irritation after 48 hr under an occlusive patch and a maximization test revealed no sensitization reactions.

2.17.3 Industrial Exposure

The principal route of exposure in the workplace is by inhalation of the vapors. Occasional skin contact should present no hazard. However, repeated contact may result in a dermatitis. The low volatility of 3-octanone and its penetrating odor should minimize the occurrence of overexposures.

2.17.4 Hygienic Standards

No threshold limit value has been established for 3-octanone.

2.18 5-Methyl-3-Heptanone; Ethyl sec-Amyl Ketone, Ethyl Isoamyl Ketone, Methyl Heptanone

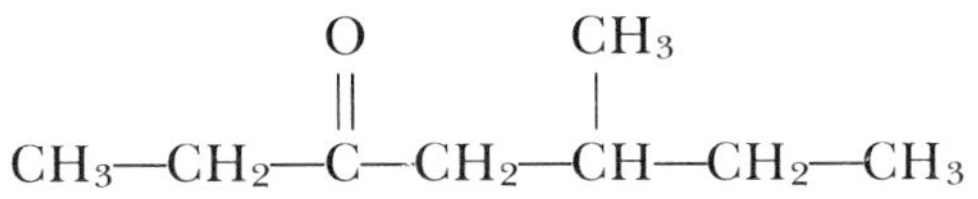

$$CH_3{-}CH_2{-}\overset{\displaystyle O}{\overset{\displaystyle \|}{C}}{-}CH_2{-}\overset{\displaystyle CH_3}{\overset{\displaystyle |}{CH}}{-}CH_2{-}CH_3$$

2.18.1 Physical-Chemical Properties

5-Methyl-3-heptanone is a colorless liquid of low volatility. It has an agreeable penetrating odor which resembles the essence of apricots and peaches. Additional physical-chemical properties are summarized in Table 56.4. The threshold odor concentration has been reported to be 6 and <5ppm (257,265).

2.18.2 Biologic Effects

The systemic toxicity of 5-methyl-3-heptanone is relatively low. Exposure to the vapors of 5-methyl-3-heptanone may produce irritation of the eyes, nose, and throat. 5-Methyl-3-heptanone can be detected readily at low atmospheric concentrations. Inhalation exposures to high concentrations may produce headache, nausea, anesthesia, narcosis, and death. Repeated skin exposure may produce dermatitis due to its defatting action.

**Animal Studies.   *Single Dose.*   **The oral LD_{50} was 3.5, 3.8, and 2.5 g/kg for rats, mice, and guinea pigs, respectively (257).

Eye.   5-Methyl-3-heptanone can produce eye irritation with some transient corneal damage in laboratory animals (257).

Skin.   5-Methyl-3-heptanone is mildly irritating to the skin and prolonged repeated exposures may cause defatting and dermatitis (257).

Inhalation.   Exposure to 3000 ppm 5-methyl-3-heptanone for 4 hr was lethal to three of six mice. Similar exposures conducted in rats were not lethal but obvious signs of irritation of the respiratory tract and eyes were noted. Exposure of rats and mice to 3484 ppm for 4 and 8 hr, respectively, was lethal to some of the animals. Exposure to 5888 ppm for 4 hr killed four of six rats (257).

Human Studies.   The odor of 5-methyl-3-heptanone can be detected at 5 ppm by most individuals. Exposure to 25 ppm produced a strong odor and mild irritation of the nasal passages. Exposures to 50 and 100 ppm caused moderate irritation of the eyes, nose, and throat, as well as headache and nausea (257).

2.18.3   Industrial Exposure

Any occupational exposures are likely to occur by inhalation, and skin and eye contact. Excellent warning properties at low vapor concentrations should prevent overexposure.

2.18.4   Hygienic Standards

The American Conference of Governmental Industrial Hygienists (243) has established a threshold limit value of 25 ppm for 5-methyl-3-heptanone.

2.19   5-Nonanone; Di-*n*-Butyl Ketone, Dibutyl Ketone

$$\text{CH}_3-\text{CH}_2-\text{CH}_2-\text{CH}_2-\overset{\displaystyle \overset{\text{O}}{\|}}{\text{C}}-\text{CH}_2-\text{CH}_2-\text{CH}_2-\text{CH}_3$$

2.19.1   Physical-Chemical Properties

5-Nonanone is a stable, colorless to light yellow liquid. Additional physical-chemical properties are summarized in Table 56.4.

2.19.2   Biologic Effects

5-Nonanone is moderately toxic orally. Repeated exposure may produce neurotoxicity which may be enhanced by concomitant exposure to other ketones.

Animal Studies.   *Single Dose.*   The oral LD_{50} of 5-nonanone was >2000 mg/kg for rats (256). The intravenous LD_{50} has been reported to be 1379 mg/kg for mice (47).

Repeated Dose. Repeated administration of 5-nonanone by gavage at doses of 1000 and 2000 mg/kg, 5 days/week for 13 and 4 weeks, respectively, produced a neuropathy in rats which was indistinguishable from that produced by methyl *n*-butyl ketone. An oral dose of 233 mg/kg 5-nonanone administered to rats 5 days/week for 13 weeks produced a minor subclinical neuropathy. When the same amount of 5-nonanone (233 mg/kg) was present as an 11 percent contaminant in a commercial lot of 5-methyl-2-octanone, neuropathy developed in less than 90 days. The neuropathy following 5-nonanone administration results in axonal damage in the peripheral nerves, spinal cord, brainstem, and cerebellum (256). Testicular germinal cell atrophy may be a possible effect resulting from repeated exposure to this solvent.

Human Studies. No adverse effects to humans due to 5-nonanone have been reported.

Toxicokinetics. 5-Nonanone is absorbed from the gastrointestinal tract of the rat. Metabolic transformations include ω-1 oxidation to 2,5-nonanedione with subsequent oxidative and decarboxylative steps to produce methyl *n*-butyl ketone and 2,5-hexanedione (Figure 56.2). In addition, 5-nonanone is oxidized to carbon dioxide (38 percent of the dose) (256). Radioactivity excreted in the urine accounted for approximately 50 percent of the dose after 72 hr (256). The half-life of 5-nonanone in the blood of rats dosed with 500 mg/kg was 1.6 hr (256). No unchanged 5-nonanone was detected in the urine, but both methyl *n*-butyl ketone and 2,5-hexanedione were present (256).

Toxic Mechanisms. The toxicity of 5-nonanone appears to be directly related to its metabolic conversion to 2,5-hexanedione; thus its mechanism of action is expected to be similar to γ-diketones.

2.19.3 Industrial Exposure

Any exposure in the workplace is expected to be by inhalation and by skin or eye contact. Repeated skin exposures could result in dermatitis and the accumulation of neurotoxic metabolites. Although exposures to 5-nonanone are not expected to occur frequently, 5-nonanone may be found as a contaminant in other solvents which may potentiate its toxic effects.

2.19.4 Hygienic Standards

No threshold limit value has been established for 5-nonanone.

2.20   Diisobutyl Ketone; 2,6-Dimethyl-4-heptanone, Isobutyl ketone, Isovalerone, DiBK

$$\underset{\displaystyle CH_3-CH-CH_2-C-CH_2-CH-CH_3}{\overset{\displaystyle \underset{|}{CH_3}\qquad\quad \underset{\|}{O}\qquad\quad \underset{|}{CH_3}}{}}$$

2.20.1   Physical-Chemical Properties

Diisobutyl ketone is a colorless stable liquid with a mild ketone odor. The threshold odor concentration has been reported to be <0.11 ppm (265). Flammability is low. Additional physical-chemical properties are presented in Table 56.4.

2.20.2   Biologic Effects

Experimental animal data and limited controlled human studies indicate that diisobutyl ketone presents a low degree of hazard in the industrial setting. Oral toxicity is low, and eye and skin contact may produce slight to moderate irritation. Concentrations of the vapor well below those which cause irritation and narcosis have an objectionable odor and cause discomfort, and thus may serve as warnings to preclude overexposure.

Animal Studies.   *Single Dose.* Smyth et al. (224) reported an oral LD_{50} of 5800 mg/kg for rats. Single oral doses as high as 3200 mg/kg failed to produce death in rats (256). An oral LD_{50} of 1416 mg/kg has been reported for mice (257). Intraperitoneal doses as high as 1600 mg/kg caused no deaths in rats (256).

Repeated Doses. Daily doses of 4000 mg/kg given by gavage killed all rats after two or three treatments (256). Deaths were due to severe depression of the central nervous system, hepatotoxicity, and dehydration. Pulmonary congestion and edema and renal toxicity were contributing factors. Doses of 2000 and 1000 mg/kg administered daily for 3 weeks failed to kill any of the animals. The 2000 mg/kg dose produced central nervous system depression initially, but a tolerance developed after several doses. Body weight gain and feed consumption were not affected. Hematologic and serum clinical chemistry determinations were comparable to the control values. Absolute and relative liver and kidney weights were slightly increased. Histologically, only minor changes were seen in the livers and kidneys (256).
   Diisobutyl ketone was tested at 2000 mg/kg in a 90-day gavage study (256). Abnormalities noted were reduced glucose levels, increased absolute and relative liver and adrenal gland weights, increased relative kidney weights, and de-

creased absolute brain and heart weights. Compound-related histopathology included minor changes in the stomach, liver, and kidneys. No neurotoxicity was evident.

Eye. Diisobutyl ketone instilled undiluted into the rabbit eye produced essentially no irritation (25, 224, 256).

Skin. Application of the undiluted liquid to covered and uncovered guinea pig skin caused moderate irritation (256). Ten milligrams per kilogram held in contact with the skin of rabbits for 24 hr under an occlusive wrap, and 500 mg/kg applied to the uncovered skin of rabbits, produced only mild irritation (224, 258). Smyth et al. (224) have reported that diisobutyl ketone is a mild skin irritant with a dermal LD_{50} of greater than 20 ml/kg for the rabbit.

Inhalation. McOmie and Anderson (156) reported no deaths of rats or guinea pigs exposed for 7.5 to 16 hr to saturated vapors of diisobutyl ketone. A single 6-hr exposure of rats to 1500 ppm caused ataxia and drowsiness but no deaths (256). Smyth et al. (224) reported a higher degree of toxicity for rats; an 8-hr exposure to essentially a saturated atmosphere (2000 ppm) caused the deaths of five of six animals. Exposure for 4 hr was reported to be the shortest time that produced a toxic effect in rats exposed to 2000 ppm (26). McOmie and Anderson (156) reported that mice survived twelve 3-hr exposures to a saturated atmosphere. Carpenter et al. (27) exposed guinea pigs and rats to concentrations of 125 to 1650 ppm for thirty 7-hr exposures. The 125 ppm concentration was a no-effect level for both species. The 250 ppm concentration increased liver and kidney weights of female rats, but decreased the liver weight of the male guinea pigs. Rats exposed to intermediate concentrations of 530 and 920 ppm had increased liver and kidney weights. Increased mortality and liver and kidney injury were seen at 1650 ppm.

Human Studies. Silverman et al. (216) studied the sensory response of humans to diisobutyl ketone vapors and reported that concentrations higher than 25 ppm had an objectionable odor and caused eye irritation. No nose and throat irritation was observed at 50 ppm. Carpenter et al. (27) found that a 3-hr exposure to 50 and 100 ppm caused slight eye, nose, and throat irritation in three volunteers. The 100 ppm level was judged to be "unsatisfactory" whereas 50 ppm was "satisfactory."

2.20.3 Industrial Exposure

In the industrial setting, exposure to diisobutyl ketone may occur by inhalation of the vapors and by eye and skin contact. However, the principal hazard to health would be inhalation of the vapors. Precautions should be taken to provide proper ventilation. Protective clothing should be available in case of emergency.

2.20.4 Hygienic Standards

A threshold limit value of 25 ppm has been established based on repeated inhalation in animals and human sensory response studies (10). This level is below those that produce an objectionable odor and eye, nose, and throat irritation, and should preclude adverse effects.

2.21 Trimethyl Nonanone; 2,6,8-Trimethyl-4-nonanone

$$
\begin{array}{ccccc}
CH_3 & O & CH_3 & CH_3 \\
| & \| & | & | \\
CH_3\!-\!CH\!-\!CH_2\!-\!C\!-\!CH_2\!-\!CH\!-\!CH_2\!-\!CH\!-\!CH_3
\end{array}
$$

2.21.1 Physical-Chemical Properties

Trimethyl nonanone is a colorless liquid with a pleasant fruity odor. Additional physical-chemical properties are summarized in Table 56.4.

2.21.2 Biologic Effects

Trimethyl nonanone is slightly toxic orally and is a mild dermal and ocular irritant.

Animal Studies. *Single Dose.* The LD_{50} of trimethyl nonanone was reported to be 8.47 g/kg for rats (225).

Eye. Trimethyl nonanone was a mild irritant when applied undiluted to rabbit eyes (225).

Skin. Ten milligrams applied to the skin of rabbits for 24 hr under a cover and 500 mg/kg applied uncovered produced mild skin irritation (225). Repeated application may produce irritation by defatting the skin. The dermal LD_{50} was 11 ml/kg for rabbits (225).

Inhalation. Saturated vapor exposures of 4 hr caused no deaths in rats (225).

Human Studies. No published reports of exposures to humans were found.

2.21.3 Industrial Exposures

Industrial exposures are expected to be primarily by inhalation and skin and eye contact. Single topical exposures may be slightly irritating but repeated exposures may exacerbate the irritation. Oral exposures should not be hazardous unless large volumes of trimethyl nonanone are swallowed. Trimethyl nonanone has a relatively high flash point and should not present a fire hazard.

2.21.4 Hygienic Standards

No threshold limit value has been established for trimethyl nonanone.

2.22 2,4-Pentanedione; Acetylacetone, Acetoacetone, Diacetyl Methane, Acetyl 2-Propanone, Pentanedione

$$\underset{\text{CH}_3-\text{C}-\text{CH}_2-\text{C}-\text{CH}_3}{\overset{\overset{\text{O}}{\|}\qquad\overset{\text{O}}{\|}}{}}$$

2.22.1 Physical-Chemical Properties

2,4-Pentanedione is a clear liquid with an unpleasant rancid odor. The odor threshold level is 0.010 ppm (265). It is a moderate fire hazard. Additional physical-chemical properties are summarized in Table 56.4.

2.22.2 Biologic Effects

High doses of 2,4-pentanedione produce dyspnea, severe central nervous system depression, and death in experimental animals. Similar clinical effects are produced by lower repeated doses, except that some animals survive and develop a central nervous system disorder characterized by an irreversible cerebellar syndrome. Thymic necrosis and atrophy accompany the central nervous system damage.

Animal Studies. *Single Dose.* Smyth and Carpenter (224) reported the oral LD_{50} of 2,4-pentanedione to be 1000 mg/kg for rats. Others reported an approximate oral LD_{50} of 800 mg/kg for rats and 951 mg/kg for mice (256). The intraperitoneal LD_{50} was 750 mg/kg for mice.

Repeated Dose. Groups of five rats were administered 100, 500, or 1000 mg/kg 2,4-pentanedione by gavage daily over a 1 to 15 day period (256). The 1000 mg/kg dose produced rapid onset of dyspnea and depression followed by prostration and deaths within 1 hr. The rats given 500 mg/kg reacted similarly, except that tremors and ataxia were observed and deaths were delayed. At 100 mg/kg, slight depression was observed in all rats. Examination at necropsy revealed no compound related changes in the 1000 mg/kg animals. In the 500 mg/kg group, one of five rats had a distended urinary bladder, congested lungs, and clouding of the cornea. Histologically, all animals in the 500 mg/kg group showed thymic necrosis, hepatocyte swelling, and hepatic congestion. No compound-related changes were found in the 100 mg/kg animals. An additional group of five male rats was administered 100 mg/kg 2,4-pentanedione 10 times over a 14-day period. No adverse effects were detected with regard to clinical signs, weight gain, feed consumption, hematology, serum clinical chemistries, absolute and relative liver and kidney weights, gross pathology, and histopathology.

Repeated administration of 250 mg/kg 2,4-pentanedione by gavage to male rats twice a day produced deaths on the second and fourth days (256). Signs of increased respiratory rate and volume and severe depression were observed.

Two rats added to the study received three to seven doses of 150 mg/kg twice daily over a 3 to 16 day period. One rat developed a head tilt and increased muscle tone after three doses. Bowel and bladder function and induced flexor reflex responses were normal. After four doses all rats were weak and debilitated. Following a 3-day rest period, the dose was lowered to 100 mg/kg twice a day. Three rats were added to the test group so that five animals were on test. Except for depression in two rats and increased muscle tone and abnormal positioning of the limbs in one of these animals, the other animals were unremarkable. Subsequently, the dose was raised to 125 mg/kg twice a day. After seven doses, one rat died, and after 16 doses, one rat developed quadriparesis. This animal partially recovered 5 days later. This rat showed slight weakness, ataxia, tremors, and a shuffling gait for the next 38 days. After 43 doses one rat showed weakness, tremors, and an alternating rolling motion of the head. The remaining rat died after 44 doses over a 61-day period. Significant changes seen at autopsy included poor general condition, gastric hemorrhage, gastric ulceration, and a healed gastric ulcer. Histologic changes were variable owing to the alterations in dosage and the length of time between the onset of clinical signs and death. In the stomach, both acute and chronic inflammatory changes were evident on the mucosal surface. In the thymus, acute changes were seen as massive necrosis of cortical lymphocytes and chronically as thymic atrophy. In the brain, acute changes were characterized by perivascular edema, hemorrhage into the Virchow-Robbin spaces and endothelial cell swelling, all primarily localized in the brainstem and cerebellum. Chronic central nervous system lesions were bilateral, symmetrical areas of malacia and gliosis centered on the cerebellar peduncles, olivary nuclei, and lower brainstem.

To exclude the possibility that these changes were peculiar to the rat, groups of two male New Zealand rabbits were administered 250, 500, and 1000 mg/kg of 2,4-pentanedione once a day for 5 days/week by gavage (256). At 1000 mg/kg, both rabbits died within 24 hr after a single dose. At 500 mg/kg, one rabbit died on the ninth day of the study, and the second rabbit died on the twelfth day following severe central nervous system depression. One rabbit receiving 250 mg/kg survived 14 days. At autopsy, rabbits at the 1000 mg/kg level showed congestion of the brain, lungs, and thymus. Gross changes in one rabbit at 500 mg/kg included hemorrhage in the brain, an atrophic thymus, pulmonary congestion, and gastric mucosal hemorrhages. Gross changes in rabbits at 250 mg/kg were absent. Histologically, both rabbits at the 1000 mg/kg level showed congestion and hemorrhage in the thymus and in one of the rabbits the thymus was very small. Both rabbits in the 500 mg/kg group showed congestion of the brain. Hemorrhage into the spinal cord was observed in one rabbit, and the one rabbit in which the thymus and stomach were examined showed hemorrhage into the mediastinal fat, marked thymic atrophy with heavy macrophage infiltration, and gastric mucosal hemorrhage. Animals in the 250 mg/kg group showed no compound-related histologic lesions.

Eye.   Instillation of 2,4-pentanedione into rabbit eyes produced severe irritation (256). Smyth and Carpenter (222) reported that 2,4-pentanedione was as irritating to the eye as acetone.

Skin.   2,4-Pentanedione applied to the depilated abdomens of guinea pigs under an occlusive wrap for 24 hr produced moderate skin irritation. Four guinea pigs receiving 20 ml/kg under a wrap died within 24 to 72 hr of application (256). Other studies (258) indicate that 2,4-pentanedione is a mild primary skin irritant with a dermal LD_{50} of 5 g/kg for rabbits. Ten applications of 2,4-pentanedione to the uncovered backs of guinea pigs did not exacerbate the primary irritation response (256). In a standardized skin sensitization test, one of four guinea pigs showed a weak sensitization reaction (256).

Inhalation.   Smyth and Carpenter (222) reported that 30 min was the maximum time that rats could be exposed to a saturated atmosphere of 2,4-pentanedione before death occurred. The same level was lethal to all animals in 1 hr (258). One thousand parts per million killed two of six rats in 2 hr and the other four animals in 4 hr. All animals were anesthetized by the test atmosphere (258).

Human Studies.   2,4-Pentanedione (2 to 14 ppm) has been reported to produce nausea and headaches in several persons, and may have been associated with the occurrence of hives in one worker (256).

Toxic Mechanisms.   Little information about the in vivo mechanisms of action of 2,4-pentanedione exists, but in vitro studies indicate that 2,4-pentanedione may inactivate enzyme activity in several ways. The first involves its metal chelating properties. 2,4-Pentanedione binds iron in peroxidase and thus prevents the oxidation of human serum proteins (81). Iron chelation may also be involved in the degradation of heme in cytochrome P-450. 2,4-Pentanedione suppresses cytochrome P-450 without lowering other hepatic microsomal enzymes (110). Enzymatic inactivation can also occur through reactions involving arginine and lysine. Reaction with the arginine of proteins leads to the formation of stable pyrimidines whereas reaction with lysine residuals lead to enamines (83). Of the two amine reactions, that of arginine is the slower and the more irreversible one (83). In correlating the in vitro activity of 2,4-pentanedione with its toxicity following repeated administration to rats and rabbits, it appears that the neurologic and lymphoid damage may be due to the inactivation of B vitamins or their coenzymes. Neuropathologic lesions due to 2,4-pentanedione closely resemble those produced by thiamine deficiency in experimental animals (256). Antilymphoid effects of folate antagonists are well known. In vitro, 2,4-pentanedione inactivates dihydrofolate reductase activity, which is a target of anti-folate drugs for cancer chemotherapy (182). It also inactivates the pyridoxal phosphate and pyridoxamine phosphate forms of porcine heart aspartate aminotransferase. The similarity between the morphologic damage produced by 2,4-pentanedione and acute vitamin B deficiency diseases and the known ability of 2,4-pentanedione to inactivate the lysyl residues of vitamin B coenzymes suggest that the toxicity of 2,4-pentanedione is due to its ability to produce deficiencies of thiamine, folic acid, pyridoxine, or their combinations.

2.22.3 Industrial Exposure

Exposure to 2,4-pentanedione may produce toxicity by the oral, dermal, or inhalation routes. Although ingestion would not appear to be a common route of exposure, accidental ingestion could be lethal or could produce irreversible neurologic impairment. Single high level exposures dermally or by inhalation should be avoided. Repeated exposures to low concentrations may be cumulative.

2.22.4 Hygienic Standards

No threshold limit value has been established for 2,4-pentanedione. Available data indicate that the no-effect level for neurologic damage in animals is probably below a cumulative dose of 100 mg/kg/day. The no-effect level for lymphoid damage is between 100 and 250 mg/kg/day in experimental animals.

2.23 2,5-Hexanedione; Acetonyl Acetone, α,β-Diacetyl Ethane, 2,5-Diketohexane, 1,2-Diacetylethane, Diacetonyl

$$\underset{\displaystyle CH_3-C-CH_2-CH_2-C-CH_3}{\overset{\displaystyle \overset{O}{\|}\qquad\qquad\overset{O}{\|}}{}}$$

2.23.1 Physical-Chemical Properties

2,5-Hexanedione is a clear liquid of low volatility with a sweet aromatic odor. It is a relatively reactive chemical and has an autoignition temperature of 493°C. Additional physical-chemical properties are summarized in Table 56.4.

2.23.2 Biologic Effects

Acute exposures to 2,5-hexanedione pose a low hazard, but chronic exposures may produce severe neurologic damage and testicular germinal cell damage. There is also the possibility of effects on the formed elements of the blood after prolonged exposures. 2,5-Hexanedione may produce moderate to marked eye irritation. Repeated skin contact may produce immunologically mediated skin sensitization and dermatitis. Acute toxicity by inhalation of 2,5-hexanedione is unlikely because its irritant properties should preclude significant exposures. Repeated skin and inhalation exposure may lead to neurologic damage.

Animal Studies. *Single Dose.* Smyth and Carpenter (222) reported an approximate oral LD_{50} of 2.7 g/kg for rats. Other studies (256) report the approximate oral LD_{50} to be 1600 mg/kg for male rats and to be between 1600 and 3200 mg/kg for male mice. When 2,5-hexanedione was administered intraperitoneally, the LD_{50} was approximately 800 mg/kg for rats and 1600 mg/ kg for mice (256).

Repeated Dose. Spencer and Schaumburg (234) administered 2,5-hexanedione subcutaneously to six rats, 5 days/week for 19 to 23 weeks at a dose of 97 to 388 mg/kg/day. The mean dose was 340 mg/kg/day. Axonal degeneration, similar to the "giant" axonal degeneration associated with methyl *n*-butyl ketone, was found in the peripheral nerves and medulla oblongata. O'Donoghue et al. (173) exposed groups of 10 male rats to 0.25, 0.5, and 1.0 percent (v/v) of 2,5-hexanedione in the drinking water for up to 165 days. Mean calculated doses based on water consumption and body weight were 240, 450, and 750 mg/kg/ day, respectively. Seven of 10 rats ingesting the 1.0 percent solution died between 14 and 43 days. All survivors showed clinical neurologic defects in the hind limbs by the 46th day of exposure. Hind limb weakness was evident in the 0.5 percent group at 52 days, and in the 0.25 percent group at 101 days. Neuropathologic changes identical to those produced by methyl *n*-butyl ketone were seen in the cerebellum, medulla oblongata, spinal cord, and peripheral nerves. Additional toxic responses included a dose-dependent body weight loss or reduced body weight gain, testicular germinal cell degeneration in all test groups, and hypocellularity of the bone marrow in the 1.0 and 0.5 percent groups. Intraperitoneal injections to rats 5 days/week up to 7.5 months produced axonal damage in the hind limb nerves and the medulla oblongata in 1.5 months. The morphologic damage in the peripheral nerve progressed with time but not to an extent which produced clinical signs of neurotoxicity. Testicular atrophy was not observed.

Four cats were exposed to 0.5 percent 2,5-hexanedione in the drinking water for up to 136 days to examine the effects on the visual system (206). Hind limb weakness was evident in 60 to 75 days; followed by quadriparesis. Electron microscopy revealed widespread axonal degeneration in the optic tracts, mammillary bodies, lateral geniculate nucleus, and superior colliculus, indicating that chronic exposure to relatively high levels of 2,5-hexanedione can result in degeneration in the visual pathways.

Dogs exposed to approximately 110 mg/kg 2,5-hexanedione developed a neurotoxic syndrome similar to that seen in cats and rats (133). Powell et al. (188) emphasized that Schwann cell damage may occur *pari passu* with axonal damage. Misumi et al. (160) demonstrated that changes in nerve conduction velocity coincide with neuropathologic abnormalities.

Krasavage et al. (131), in a comparative study with 2,5-hexanedione and other metabolites of *n*-hexane and methyl *n*-butyl ketone, showed that the ability of *n*-hexane and methyl *n*-butyl ketone to produce neurotoxicity was related to their quantitative conversion to 2,5-hexanedione. 2,5-Hexanedione was 3.3 times as neurotoxic as methyl *n*-butyl ketone and 38 times as neurotoxic as *n*-hexane.

Compared to other γ diketones, such as 2,5-heptanedione and 3,6-octanedione, 2,5-hexanedione produced neurotoxicity at lower doses (174).

Eye.   When placed on the cornea of rabbits, 2,5-hexanedione produced moderate to marked eye irritation with transient corneal injury (25).

Skin.   Skin irritation was slight when 5 to 20 ml/kg was applied to the abdomens of guinea pigs under an occlusive wrap for 24 hr. Systemic toxicity due to percutaneous absorption was not evident but the compound had a tanning effect and stained the skin (256). 2,5-Hexanedione stains human skin a light orange-brown. Repeated application to the uncovered, shaved backs of guinea pigs was not irritating (256). In a standardized skin sensitization test, four of five guinea pigs developed a weak positive skin sensitization reaction (256). Smyth and Carpenter (222) reported a dermal LD_{50} of 6.6 ml/kg for guinea pigs.

Inhalation.   Smyth and Carpenter (222) exposed rats to a saturated (1700 ppm calculated) atmosphere for a maximum of 1 hr and did not produce deaths. Specht et al. (233) stated that guinea pigs exposed to a saturated atmosphere at 25°C (400 ppm) showed a slight decrease in respiratory rate and slight nasal and ocular irritation. Five male rats exposed to 650 ppm 2,5-hexanedione for four 16-hr periods and three 20-hr periods over 10 days developed orange-brown staining of the hair and lost body weight. Limited histopathologic examination showed testicular germinal cell degeneration but no neuropathologic changes (256).

Human Studies.   No reports of adverse effects in humans due to the handling of 2,5-hexanedione were found. The discoloration of the skin produced by 2,5-hexanedione may serve as an indicator that skin contact has occurred.

Toxicokinetics. The toxicokinetics of 2,5-hexanedione have been studied primarily in relation to the metabolism of *n*-hexane, methyl *n*-butyl ketone, and their metabolites. In each of these cases the effects produced by the parent compound depend on the amount of 2,5-hexanedione produced when they are metabolized (131). DiVincenzo et al. (49) showed that guinea pigs given an intraperitoneal injection of 450 mg/kg 2,5-hexanedione converted a fraction of the dose to 5-hydroxy-2-hexanone and that both compounds were cleared from the serum in 16 hr. The serum half-life of 2,5-hexanedione was 100 min. Rats given 0.5 percent 2,5-hexanedione in the drinking water for 49 days had a mean 2,5-hexanedione serum level of 17.4 μg/ml (0 to 74.05 μg/ml). A level of 4.49 μg/ml 5-hydroxy-2-hexaneone was detected in the serum of one animal. A concentration of 0.25 percent 2,5-hexanedione in the drinking water did not produce detectable serum levels (≥ 0.1 μg/ml) of 2,5-hexanedione, 5-hydroxy-2-hexanone, or 2,5-hexanediol in the rats (173). The chronic toxicity of 2,5-hexanedione is at least partially due to its prolonged retention in the body; small doses accumulate and produce adverse effects.

Diketones without a γ spacing between carbonyl groups do not produce the same type of neurotoxicity as 2,5-hexanedione (174). Available data suggest that among aliphatic dicarbonyls a γ-diketone is necessary for neurotoxicity and that among γ-diketones, a six-carbon chain length is more neurotoxic than a seven- or eight-carbon chain length (174).

Toxic Mechanisms. Local effects of 2,5-hexanedione are probably related to its ability to interact with surface proteins as it does in commercial tanning processes. Reaction with epidermal proteins by cross-linking or by pyrrole formation may be the reason skin sensitization was observed in guinea pigs. Some effort has been expended to understand the neurotoxic mechanisms of 2,5-hexanedione but no unifying concept of neurotoxic action has yet been accepted. Sabri et al. (200) hypothesized that 2,5-hexanedione inhibited enzymes necessary for energy production and the maintenance of axonal integrity by its interaction with sulfhydryl groups of neural fructose-6-phosphate kinase, resulting in axonal degeneration. Couri and Nachtman (37) hypothesized that 2,5-hexanedione may damage myelin by producing shifts of the myelin lipid bilayer.

2.23.3 Industrial Exposure

Any toxicity due to 2,5-hexanedione is expected to be from eye and skin contact and from inhalation. Topically, 2,5-hexanedione may be a slight skin irritant and a moderate eye irritant. Acute inhalation hazard would be slight because of its good warning properties. Chronic exposure to 2,5-hexanedione by either the skin or inhalation routes could result in severe impairment of the central and peripheral nervous systems, the testes, and the circulating elements of the blood. Skin and eye contact should be prevented and inhalation exposures minimized.

2.23.4   Hygienic Standards

No threshold limit value has been established for 2,5-hexanedione.

2.24   Cyclohexanone; Pimelic Ketone, Ketohexamethylene, Hexanon, Hytrol O, Nadone, Pimelin Ketone, Sextone, Anone

2.24.1   Physical-Chemical Properties

Cyclohexanone is a clear to pale yellow liquid with an odor similar to that of acetone. The odor threshold is about 0.12 ppm (265). Additional physical-chemical properties are summarized in Table 56.4.

2.24.2   Biologic Effects

Cyclohexanone is slightly toxic and may be irritating to the skin, eyes, and nasal passages. It may produce transient corneal injury. Irritation is common at atmospheric concentrations above 50 ppm. Exposure to high concentrations can be expected to produce central nervous system depression.

Animal Studies.   *Single Dose.* The oral LD_{50} was reported to be 1.62 ml/kg for rats (228). Deichmann and LeBlanc (42) reported an oral LD_{50} of 1.84 g/kg for rats. The minimum lethal dose was 1.3 g/kg for mice (111). Treon et al. (249) reported a minimum lethal dose between 1.6 and 1.9 g/kg for rabbits. Gupta et al. (95) reported oral LD_{50}s of 2.07 and 2.11 ml/kg for male and female mice, respectively. The oral LD_{50} was 1.8 g/kg for both sexes of rats (95). Dogs dosed intravenously with 630 mg/kg died within 60 min of dosing (29). Mice injected intraperitoneally with 0.5 ml of cyclohexanone showed signs of excitation, paresis of the hind quarters, hypothermia, and death (68). The intraperitoneal LD_{50} was 1.23 and 1.13 g/kg for mice and rats, respectively, and 0.93 and 1.54 g/kg for guinea pigs and rabbits, respectively (95). The subcutaneous LD_{50} was 2.17 g/kg for rats (42).

Repeated Doses. Cyclohexanone (1.0 percent) fed in the diet to mice reduced growth rate in the first generation offspring. Growth of the second generation pups was normal (92). Cyclohexanone also increased the mortality rate of the offspring during the first 21 days of life. Cyclohexanone had no effect on the fertility of female mice given 50 mg/kg (100). Griggs et al. (93) and Weller and

Griggs (272) exposed embryonated chicken eggs to cyclohexanone vapors for 3, 6, or 12 hr prior to or after 96 hr of incubation. Embryos examined after 13 days of incubation showed reduced body weight gain but no gross morphologic changes. Newly hatched chicks appeared normal initially, but subsequently showed neuromotor changes of increasing severity. The authors suggested that cyclohexanone produced functional rather than morphologic deficits in the chick.

Collin (34) reported that achromatic regions, breaks, and deletions were seen in karyograms of human leukocytes which had been cultured on media containing 10^{-2}, 10^{-3}, or 10^{-4} M cyclohexanone. Pretreatment with cyclohexanone did not change the recessive lethal mutation frequency in *Drosophila Melanogaster* after γ- and X-irradiation (90, 91).

Eye. Cyclohexanone placed in the eyes of rabbits caused moderate irritation and some corneal injury (228). Rengstorff et al. (194, 195) administered small multiple doses of cyclohexanone or combinations of cyclohexanone and acetone percutaneously (0.5 ml) or subcutaneously (0.05 ml) to albino guinea pigs for a period of 8 weeks. Lens changes occurred as early as 8 weeks and as late as 6 months. The changes consisted of focal or extensive subcapsular vacuolated areas extending from the periphery toward the center of the lens. Cataracts developed in 29 of 120 animals given either cyclohexanone or acetone.

Skin. Cyclohexanone is moderately irritating when held in contact with guinea pig skin for 24 hr and mildly irritating when applied to the skin of rabbits (95, 119, 358). The acute dermal LD_{50} for rabbits was 1.0 ml/kg (228). Treon et al. (249) reported a minimum lethal dose of 10.2 to 23.0 g/kg for rabbits.

Inhalation. The minimum lethal concentration of cyclohexanone in rats exposed for 4 hr was 2000 ppm (228). Guinea pigs exposed to 4000 ppm for 6 hr showed signs of lacrimation, salivation, decreased body temperature, decreased heart rate, opacity of the cornea, and central nervous system depression (233). Mice, guinea pigs, and cats exposed to 3800 ppm of cyclohexanone showed signs of central nervous system depression (70). Monkeys and rabbits exposed repeatedly to 190 ppm 6 hr/day for 50 days showed slight kidney and liver injury. Vapor exposures to 309 and 773 ppm produced eye irritation and salivation. An atmospheric concentration of 3082 ppm produced light narcosis and central nervous system effects (249). Rabbits exposed to 500 ppm for 6 hr/day for 80 to 110 days developed lymphocytosis, followed by lymphopenia and a reduction in lung ventilation and in the rate of oxygen consumption (269). Exposure of mice to about 18.6 to 19.2 mg/l of cyclohexanone resulted in a mean time to death (LT_{50}) of about 100 min (95).

Human Studies. Nelson et al. (168) exposed volunteers of both sexes for 3 to 5 min to a range of atmospheric concentrations of cyclohexanone. The lowest objectionable concentration was judged to be 25 ppm. Eye, nose, and throat irritation was noted at 50 ppm. Exposure to 75 ppm produced marked irritation of the eyes, nose, and throat. Voronin (269) reported irritation of the mucous membranes at 125 ppm.

Toxicokinetics. The principal metabolic pathway of cyclohexanone is reduction to cyclohexanol and subsequent conjugation with glucuronic acid (113). Treon et al. (249) reported that rabbits exposed by inhalation to cyclohexanone excreted increased quantities of sulfate and glucuronic acid conjugates in the urine. There was a decrease in the ratio of inorganic sulfates to total sulfates. About 45 to 50 percent of the dose was excreted in the form of glucuronic acid conjugates. Rats and rabbits dosed orally with cyclohexanone excreted trace amounts of hydroxycyclohexyl mercapturic acid and *cis*-2-hydroxycyclohexyl mercapturic acid in the urine. Rabbits dosed orally with cyclohexanone excreted cyclohexyl glucuronide as the predominant urinary metabolite (62). Cyclohexanone did not produce significant changes in the phenobarbital-induced sleeping time of male mice (95). Boyland and Chasseaud (21) reported that cyclohexanone treatment depressed glutathione levels in the livers of female rats.

2.24.3 Industrial Exposure

Occupational exposure is likely to occur by inhalation and skin and eye contact. Excellent warning properties should reduce the likelihood of overexposure.

2.24.4 Hygienic Standards

The American Conference of Governmental Industrial Hygienists (243) has established a threshold limit value of 25 ppm for cyclohexanone.

2.25 Methylcyclohexanones: 2-Methylcyclohexanone, *o*-Methylcyclohexanone; 3-Methylcyclohexanone, *m*-Methylcyclohexanone; 4-Methylcyclohexanone, *p*-Methylcyclohexanone

Ortho   Meta   Para
Methylcyclohexanones

2.25.1 Physical-Chemical Properties

The isomers of methylcyclohexanone are colorless liquids of low volatility and have an acetone-like odor. Methylcyclohexanones are chemically stable, but may darken when exposed to light. Although methylcyclohexanones are available as pure compounds, they are usually found as isomeric mixtures. Methylcyclohexanones are flammable at temperatures commonly encountered in the workplace and may pose a fire hazard. Additional physical-chemical properties are summarized in Table 56.4.

2.25.2 Biologic Effects

The isomers of methylcyclohexanone do not present a high degree of health hazard in the workplace. They are moderately toxic to laboratory animals and may cause moderate skin irritation. Low concentrations are characterized by a strong persistent odor and by irritation to the eyes, nose, and throat. High concentrations produce central nervous system depression and narcosis. Lethal concentrations of methylcyclohexanone vapors are not usually reached at the temperatures commonly encountered in the workplace.

Animal Studies. *Single Dose.* The oral LD_{50} of 2-methylcyclohexanone is 2.14 g/kg for rats (228). Treon et al. (249) reported that 1.0 to 1.2 g/kg was a minimum lethal dose for rabbits. 4-Methylcyclohexanone had an oral LD_{50} of 800 to 1600 mg/kg for rats and 1600 to 3200 mg/kg for mice (256). Intravenously, the lethal dose for dogs ranged from 270 to 370 mg/kg depending upon the isomer studied (29). The intraperitoneal LD_{50} in mice was 200 mg/kg (228).

Eye. 2-Methylcyclohexanone produced moderate irritation with some corneal damage to rabbit eyes (228).

Skin. 2-Methylcyclohexanone was moderately irritating when held in contact with the skin of guinea pigs for 24 hr. The minimum lethal dose by skin application was from 4.9 to 7.2 g/kg for rabbits (249). Smyth et al. (228) reported a dermal LD_{50} of 1.77 g/kg for rabbits.

Inhalation. Exposures to 2800 ppm 2-methylcyclohexanone vapor for 4 hr produced lethality in three of six rats (228). Monkeys and rabbits exposed to 50 ppm of methylcyclohexanone for 6 hr/day for 50 days showed lethargy, salivation, lacrimation, and eye irritation. The "no observed effect" level was 182 ppm (249). Flury and Klimmer (70) exposed mice, guinea pigs, and rats to 3500 ppm methylcyclohexanone vapor for 30 min and found irritation of the mucous membranes and central nervous system effects. Rabbits and cats became somnolent after a 1-hr exposure to 2500 ppm. A concentration of 450 ppm produced irritation in mice.

Toxicokinetics. Methylcyclohexanones are reduced to methylcyclohexanols and excreted in the urine as sulfuric and glucuronic acid conjugates (62, 242, 249).

2.25.3 Industrial Exposure

Occupational exposure is likely to occur by inhalation and eye and skin contact. The methylcyclohexanones have good warning properties and are considered to be relatively nonhazardous in the workplace provided good hygienic standards are followed.

2.25.4 Hygienic Standards

The threshold limit value for 2-methycyclohexanone is 50 ppm with a skin notation (243).

2.26 Acetophenone; Acetylbenzene, Methyl Phenyl Ketone, 1-Phenylethanone, Benzoyl Methide, Hypnone

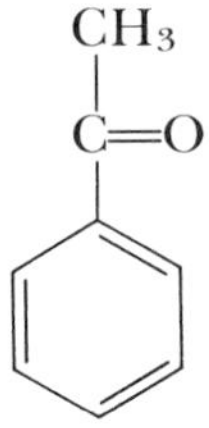

$$CH_3$$
$$|$$
$$C{=}O$$

2.26.1 Physical-Chemical Properties

Acetophenone is a colorless liquid with an orange blossom or jasmine-like odor. Depending on the ambient temperature, acetophenone may be a solid or a liquid. The odor threshold is about 0.30 ppm (265). Additional physical-chemical properties are summarized in Table 56.4.

2.26.2 Biologic Effects

Acetophenone has a low to moderate degree of toxicity. The principal hazard of acetophenone is irritation of the skin and eyes. Eye contact may produce transient corneal injury. The low volatility of acetophenone makes significant exposure by the pulmonary route unlikely. Ingestion has produced anesthetic effects.

Animal Studies. *Single Dose.* The oral LD_{50} was reported to be between 0.9 and 3.2 g/kg for rats (114, 222, 223). Central nervous system depression was observed in mice given intraperitoneal doses of 0.4 to 0.5 g/kg acetophenone. The intraperitoneal LD_{50} was 1.07 g/kg for mice (191).

Repeated Dose.   No adverse effects were noted in rats fed 102 mg/kg/day for 30 days (223). Rats fed 10,000 ppm acetophenone in the diet for 17 weeks showed no signs of toxic effects (96). Application of 0.48 g/kg of acetophenone to the skin of pregnant rats on days 10 through 15 of pregnancy did not cause any change in the gestation period, size of litter, weight of the offspring, time for appearance of teeth or hair, opening of the eyes, or appearance of reflexes (134).

Eye.   Undiluted acetophenone (771 mg) applied to the eyes of rabbits produced moderate irritation and transient corneal injury (223).

Skin.   The dermal LD_{50} was greater than 20 ml/kg for guinea pigs (222). When held in contact with rabbit skin for 24 hr, acetophenone produced irritation described as a mild burn (120). When the tails of mice were immersed in acetophenone for 4 hr all the animals died, indicating that acetophenone is readily absorbed percutaneously (134).

Inhalation.   All rats exposed to a saturated atmosphere of acetophenone for 8 hr survived (222).

Human Studies.   No skin sensitization was noted when 2 percent acetophenone in petrolatum was tested on humans (176).

Toxicokinetics.   Acetophenone is converted to methyl phenylcarbonal and benzoic acid. The benzoic acid is conjugated with glycine and excreted in the urine as hippuric acid (274). Minor metabolites identified in rabbit urine include o-, m-, p-hydroxyacetophenone and mandelic acid (122, 240). Acetophenone apparently neither induces its own metabolism (94) nor is its metabolism affected by phenobarbital pretreatment (138).

2.26.3   Industrial Exposure

Occupational exposure to acetophenone may occur by inhalation and eye and skin contact. Its low volatility and good warning properties should reduce the likelihood of overexposure.

2.26.4   Hygienic Standards

No threshold limit value has been established for acetophenone.

2.27 Isophorone; 3,5,5-Trimethyl-2-cyclohexenone, 3,5,5-Trimethyl-2-cyclohexen-1-one, 1,1,3-Trimethyl-3-cyclohexen-5-one, Trimethylcyclohexenone, Isoacetophenone, α-Isophorone

2.27.1   Physical-Chemical Properties

Isophorone is a high boiling, colorless liquid with low volatility and a sharp peppermint-like odor. The odor threshold has been reported to be 0.20 ppm (265). Additional physical-chemical properties are summarized in Table 56.4.

2.27.2   Biologic Effects

The most common route of exposure to isophorone is by inhalation. The sharp odor of isophorone may induce olfactory fatigue. Its vapors are also irritating to the eyes and mucous membranes. Prolonged skin contact may cause irritation.

Animal Studies.   *Single Dose.*   The oral LD_{50} of isophorone was 2.0 and 2.37 g/kg for mice and rats, respectively (23). Other studies (256) reported oral LD_{50}s of >3200 mg/kg for rats and mice. The intraperitoneal LD_{50} was 400 mg/kg for mice and 400 to 800 mg/kg for rats (256).

Eye.   Isophorone applied to the eyes of rabbits produced opacity of the cornea, inflammation of the eyelids and conjunctiva, and a purulent discharge (252). Moderate corneal injury and ocular burns have been reported by Carpenter and Smyth (25) and Bukhalovskii and Shugaev (23).

Skin.   Isophorone had a weak irritant action on rabbit (252) and guinea pig skin (23). The dermal LD_{50} was reported to be 1500 mg/kg for rabbits (258). When isophorone (5 to 20 mg/kg) was held in contact with guinea pig skin for 24 hr under an occlusive covering, moderate skin irritation was noted (256).

Inhalation.   The minimum lethal concentration for rats exposed for 4 hr was 1840 ppm (230). Rats and guinea pigs exposed to 100 to 500 ppm isophorone 8 hr/day, 5 days/week for 6 weeks had decreased weight gain. Exposure to 500 ppm produced chronic conjunctivitis and pulmonary inflammation. Kidney damage occurred at the higher concentrations. No effects were noted after exposure to 25 ppm (229).

Human Studies.   Isophorone vapor (25 ppm) produced irritation to the eyes, nose, and throats of unacclimatized volunteers (216). Workers exposed to 5 to 8 ppm isophorone complained of fatigue and malaise. No complaints were reported after exposure to 1 to 4 ppm (171). Workers exposed to 40, 85, 200, and 400 ppm experienced eye, nose, and throat irritation. Some complained of nausea, headache, dizziness, faintness, inebriation, and a feeling of suffocation at the higher concentrations.

Toxicokinetics.   Rabbits dosed orally with 1 g/kg isophorone excreted 5,5-dimethylcyclohexen-3-one, 5,5-dimethylcyclohexene-1-carboxylic acid, and glucuronic acid conjugates in the urine (251). These studies demonstrate that isophorone undergoes methyl group oxidation and subsequent conjugation with glucuronic acid. Rabbits partially eliminated isophorone unchanged in the expired air and urine (57, 253). The remainder was oxidized to a carboxylic acid and reduced to isophorol, which was eliminated in the urine as a glucuronide conjugate.

2.27.3   Industrial Exposure

Isophorone is not considered to be hazardous in the workplace owing to its low volatility. However, in the case of an accidental spill, caution should be exercised. Workers exposed to relatively low vapor concentrations of isophorone are subject to irritation of the eyes, nose, and throat. Central nervous system effects may occur also.

2.27.4   Hygienic Standards

The American Conference of Governmental Industrial Hygienists (243) has established a threshold limit value of 5 ppm for isophorone. The National Institute of Occupational Safety and Health (171) has recommended a time-weighted average exposure of 4 ppm. These standards were based on subjective sensory responses of volunteers (216).

REFERENCES

1. B. M. Abbasov, K. G. Zeinaloba, and I. A. Safarova, "Toxicology of Some Ketones," *Tr. Azerb. Nauchno-Issled. Inst. Gig. Pr. Prof. Zabol.*, **11**, 145–148 (1977).

2. M. S. Abdel-Rahman, L. B. Hetland, and D. Couri, "Toxicity and Metabolism of Methyl *n*-Butyl Ketone," *J. Am. Ind. Hyg. Assoc.*, **37**, 95–102 (1976).

3. M. S. Abdel-Rahman, J. J. Saladin, C. E. Bohman, and D. Couri, "The Effect of 2-Hexanone and 2-Hexanone Metabolites on Pupillmotor Activity and Growth," *Am. Ind. Hyg. Assoc. J.*, **39**, 94–99 (1978).

4. P. Albertoni, *Arch. Exp. Pathol. Pharmakol.*, **18**, 219, (1884).

5. N. Allen, J. R. Mendell, D. J. Billmaier, and R. E. Fontaine, "Outbreak of Previously

Undescribed Toxic Polyneuropathy Due to Industrial Solvent," *Trans. Am. Neurol. Assoc.*, **99**, 74–79 (1974).

6. N. Allen, J. R. Mendell, D. J. Billmaier, R. E. Fontaine, and J. O'Neill, "Toxic Polyneuropathy Due to Methyl *n*-butyl Ketone," *Arch. Neurol.*, **32**, 209–218 (1975).

7. H. Altenkirch, J. Mager, G. Stoltenburg, and J. Helmbrecht, "Toxic Polyneuropathies after Sniffing a Glue Thinner, *J. Neurol.*, **214**, 137–152 (1977).

8. H. Altenkirch, G. Stoltenburg, and H. M. Wagner, "Experimental Studies on Hydrocarbon Neuropathies Induced by Methyl Ethyl Ketone (MEK)," *J. Neurol.*, **219**, 159–170 (1978).

9. H. Altenkirch, G. Stoltenburg-Didinger, and H. M Wagner, "Experimental Data on the Neurotoxicity of Methyl Ethyl Ketone (MEK)," *Experientia*, **35**, 503–504 (1979).

10. American Conference of Governmental Industrial Hygienists, *Documentation of Threshold Limit Values for Substances in Workroom Air*, Cincinnati, Ohio, 1971.

11. M. W. Anders, "Acetone Enhancement of Microsomal Aniline Hydroxylase Activity," *Arch. Biochem. Biophys.*, **126**, 269–275 (1969).

12. M. W. Anders, "Stimulatory Effect of Acetone on the Activity of Microsomal Aniline Para-hydroxylase," in, J. R. Gillette, Ed., *Proceedings of the Symposium on Microsomes and Drug Oxidation*, Academic Press, New York, 1968, pp. 533–540.

13. M. W. Anders, "Stimulation *in vitro* of Microsomal Aniline Hydroxylation by 2,2'-bipyridine," *Biochem. Pharmacol.*, **18**, 2561–2565 (1969).

14. W. K. Anger, M. K. Jordan, and D. W. Lynch, "Effect of Inhalation Exposure and Intraperitonal Injections of Methyl n-amyl Ketone on Multiple Fixed Ratio, Fixed-Interval Response Rates in Rats," *Toxicol. Appl. Pharmacol.*, **49**, 407–416 (1979).

15. *Arch. Exp. Pathol. Pharmakol.*, **56**, 346 (1906).

16. G. Armeli, F. Linari, and G. Martorano, "Rilievi Clinici ed Ematochimici in Operai Esposti All'azione di un Chetone Superiore (MIBK) Ripetuti a Distanza di 5 anni," *Lav. Um.*, **20**, 418–423 (1968).

17. D. M. Aviado, "Physiological and Biomedical Response to Specific Group of Inhalants: Concluding Remarks," *Fed. Proc.*, **37**, 2508–2509 (1978).

18. T. F. Batyrova, "Materials for Substantiation of Maximal Permissible Concentration of Methyl Isobutyl Ketone in the Atmosphere of a Work Zone," *Gig. Tr. Prof. Zabol.*, **17**(11), 52–53 (1973).

19. E. F. Berg, "Retrobulbar Neuritis: A Case Report of Presumed Solvent Toxicity, *Ann. Ophthalmol.*, **3**, 1351–1353 (1971).

20. D. J. Billmaier, H. T. Yee, N. Allen, R. Craft, N. Williams, S. Epstein, and R. Fontaine, "Peripheral Neuropathy in a Coated Fabrics Plant," *J. Occup. Med.*, **16**, 665–671 (1974).

21. E. Boyland and L. F. Chasseaud, "Effect of Some Carbonyl Compounds on Rat Liver Glutathione Levels," *Biochem. Pharmacol.*, **19**, 1526–1528 (1970).

22. J. V. Bruckner and R. G. Peterson, "Effect of Repeated Exposure of Mice and Rats to Concentrated Toluene and Acetone Vapors," *Toxicol. Appl. Pharmacol.*, **45**, 359 (1978).

23. A. A. Bukhalovskii and V. V. Shugaev, "Toxicity and Hygienic Standardization of Isophorone, Dihydroisophorone and Dimethylphenylcarbinol," *Promst. Sint. Kauch.*, **2**, 4–5 (1976).

24. G. L. Carlson, I. H. Hall, and C. Piantadosi, "Cycloalkanones. Part 7: Hypocholesterolemic Activity of Aliphatic Compounds Related to 2,8-dibenzylcyclooctanone, *J. Med. Chem.*, **18**, 1024–1026 (1975).

25. C. P. Carpenter and H. F. Smyth, Jr., "Chemical Burns of the Rabbit Cornea," *Am. J. Ophthalmol.*, **29**, 1363–1372 (1946).

26. C. P. Carpenter, H. F. Smyth, Jr., and U. C. Pozzani, "The Assay of Acute Vapor Toxicity and

the Grading and Interpretation of Results on 96 Chemical Compounds," *J. Ind. Hyg. Toxicol.*, **31**, 343–346 (1949).

27. C. P. Carpenter, U. C. Pozzani, and C. S. Weil, "Toxicity and Hazard of Diisobutyl Ketone Vapors," *Arch. Ind. Hyg. Occup. Med.*, **8**, 377 (1953).

28. C. P. Carpenter, C. S. Weil, and H. F. Smyth, Jr., "Range Finding Toxicity Data: List VIII," *Toxicol. Appl. Pharmacol.*, **28**, 313–319 (1974).

29. F. Caujolle, P. Couturier, G. Roux, and Y. Gase, "Toxicite de la Cyclohexanone et de Quelques Cétones Homologues," *Compt. Rend.*, **236**, 633 (1953).

30. J. B. Cavanagh, "The 'Dying-Back Process': A Common Denominator in Many Naturally Occurring and Toxic Neuropathies," *Arch. Pathol. Lab. Med.*, **103**, 659–664 (1979).

31. A. N. Cesaro and A. Pinerolo, "Percutaneous Absorption of Acetone," *Med. Lav.*, **38**, 384–387 (1947).

32. H. Clark and G. Powis, "Effect of Acetone Administered *in vivo* Upon Hepatic Microsomal Drug Metabolizing Activity in the Rat," *Biochem. Pharmacol.*, **23**, 1015–1019 (1974).

33. T. L. Coleman, "Acetone Metabolism in Mice; Increased Activity in Mice Heterozygous for Obesity Genes," *Proc. Natl. Acad. Sci. U.S.*, 290–293 (1980).

34. J. T. Collin, "Cytogenetic Effect of Cyclamate, Cyclohexanone and Cyclohexanol," *Diabetes* **19**, 215–221 (1971).

35. D. Couri, L. B. Hetland, M. S. Abdel-Rahman, and H. Weiss, "The Influence of Inhaled Ketone Solvent Vapors on Hepatic Microsomal Biotransformation Activities," *Toxicol. Appl. Pharmacol.*, **41**, 285–289 (1977).

36. D. Couri, M. S. Abdel-Rahman, and L. B. Hetland, "Biotransformation of n-Hexane and Methyl n-Butyl Ketone in Guinea Pigs and Mice," *Am. Ind. Hyg. Assoc. J.*, **39**, 295–300 (1978).

37. D. Couri and J. P. Nachtman, "Biochemical and Biophysical Studies of 2,5-Hexanedione Neuropathy," *Neurotoxicology*, **1**, 269–283 (1979).

38. A. Csontos, A. Dienes, and Z. Szabo, "Alterations in Blood Catalase Activity in Persons Working with Organic Solvents," *Rem. Med. Rev.*, **12**, 6–10 (1968).

39. J. G. Davenport, D. F. Farrell, and S. M. Sumi, "'Giant Axonal Neuropathy' caused by Industrial Chemicals: Neurofilamentous Axonal Masses in Man," *Neurology*, **26**, 919–923 (1976).

40. G. W. Dawson, A. L. Jennings, E. Rider, and D. Drozdowski, "The Acute Toxicity of 47 Industrial Chemicals to Fresh and Salt Water Fishes," *J. Hazardous Mater.*, **1**, 303–318 (1977).

41. S. G. DeCastiglia, S. J. Cembal, A. H. Fraga deSuarez, J. O. Nicolini, M. G. Nato, and A. E. A. Mitta, "Quality Control of Compounds Labeled with Technetium-99M," *Argent. Com. Nac. Energ. At. CNEA-327* (1972).

42. W. B. Deichmann and T. V. Leblanc, "Determination of the Approximate Lethal Dose with About Six Animals," *J. Ind. Hyg. Toxicol.*, **25**, 415–417 (1943).

43. P. V. DeJesus, D. F. Pleasure, A. K. Asbury, M. J. Brown, and C. M. Paradise, "Effects of Methyl Butyl Ketone on Peripheral Nerves and its Mechanism of Action," Contract No. TENN CDC 99-76-16, New Haven, Conn., Yale University and West Haven, Conn. Veterans Administration Hospital (1977).

44. S. Dienes, K. Tofalvi, and A. Csontos, "Modification of Blood Catalase Following Acetone Intoxication. I. Acute Experimental Intoxication," *Rev. Med.*, **11**, 146–148 (1965).

45. F. K. Dietz and C. J. Traiger, "Potentiation of CCl_4 Hepatotoxicity in Rats by a Metabolite of 2-Butanone: 2,3-butanediol," *Toxicology*, **14**, 209–215 (1979).

46. J. A. DiPaolo, P. Donovan, and R. Nelson, "Quantitative Studies of *in vitro* Transformation by Chemical Carcinogens," *J. Natl. Cancer Int.*, **42**, 867–874 (1969).

47. T. DiPaolo, "Molecular Connectivity in Quantitative Structure-Activity Relationship Study of Anesthetic and Toxic Activity of Aliphatic Hydrocarbons, Ethers, and Ketones," *J. Pharm. Sci.,* **67,** 566–568 (1978).

48. G. D. DiVincenzo and W. J. Krasavage, "Serum Ornithine Carbamyl Transferase as a Liver Response Test for Exposure to Organic Solvents," *Am. Ind. Hyg. Assoc. J.,* **35,** 21–29 (1974).

49. G. D. DiVincenzo, C. J. Kaplan, and J. Dedinas, "Characterization of the Metabolites of Methyl n-Butyl Ketone, Methyl Isobutyl Ketone and Methyl Ethyl Ketone in Guinea Pig Serum and Their Clearance," *Toxicol. Appl. Pharmacol.,* **36,** 511–522 (1976).

50. G. D. DiVincenzo, C. J. Kaplan, and J. Dedinas, "Metabolic Fate and Disposition of ^{14}C-Labeled Methyl n-Butyl Ketone in the Rat," *Toxicol. Appl. Pharmacol.,* **41,** 547–560 (1977).

51. G. D. DiVincenzo, M. L. Hamilton, C. J. Kaplan, W. J. Krasavage, and J. L. O'Donoghue, "Studies on the Respiratory Uptake and Excretion and the Skin Absorption of Methyl n-Butyl Ketone in Humans and Dogs," *Toxicol. Appl. Pharmacol.* **44,** 593–604 (1978).

52. G. D. DiVincenzo, F. J. Yanno, and B. D. Astill, "Exposure of Man and Dog to Low Concentrations of Acetone Vapor," *Am. Ind. Hyg. Assoc. J.,* **34,** 329–336 (1973).

53. R. C. Drew, "Concentrated Acetone for the Sterilization of Ophthalmic Instruments," *Rev. Bras. Ostal.,* **35,** 279–283 (1976).

54. R. C. Drew, "Acetone Sterilization in Ophthalmic Surgery," *Ann. Ophthalmol.,* **9,** 781–784 (1977).

55. S. Duckett, N. Williams, and S. Frances, "Peripheral Neuropathy Associated with Inhalation of Methyl n-Butyl Ketone," *Experientia,* **30,** 1283–1284 (1974).

56. S. Duckett, L. J. Streletz, R. A. Chambers, M. Auroux, and P. Galle, "50 ppm MnBK Subclinical Neuropathy in Rats," *Experientia,* **35,** 1365–1366 (1979).

57. H. Dutertre-Catella, N. Phu Lich, D. Quoc Quan, and R. Truhaut, "Metabolic Transformations of the Trimethyl-3,5,5-Cyclohexene-2-One(Isophorone)," *Toxicol. Eur. Res.,* 209–216 (1978).

58. F. M. Dyro, "Methyl Ethyl Ketone Polyneuropathy in Shoe Factory Workers, *Clin. Toxicol.,* **13,** 1371–1376 (1978).

59. A. Eben, W. Flucke, F. Mihail, J. Thyssen, and G. Kimmerle, "Toxicological and Metabolic Studies of Methyl n-Butyl Ketone, 2,5-Hexanedione, and 2,5-Hexanediol in Male Rats," *Ecotoxicol. Environ. Saf.,* **3,** 204–217 (1979).

60. J. L. Egle, Jr., "Retention of Inhaled Acetone and Ammonia in the Dog," *Am. Ind. Hyg. Assoc. J.,* **34,** 533–539 (1973).

61. H. B. Elkins, "The Chemistry of Industrial Toxicology," 2nd ed., Wiley, New York, 1959.

62. T. H. Elliott, D. V. Parke, and R. T. Williams, "Studies in the Detoxication 79. The Metabolism of Cyclo[^{14}C]Hexane and its Derivatives," *Biochem. J.,* **72,** 193–200 (1959).

63. T. H. Elliott, E. T. Jacob, and R. C. Tao, "*In vitro* and *in vivo* Metabolism of Optically Active Methyl Cyclohexanols and Methyl Cyclohexanones," *J. Pharm. Pharmacol.,* **21,** 561–572 (1969).

64. W. L. Epstein, *Report to Research Institute of Fragrance Materials, 16 April, 1974.* Cited in G. L. J. Opdyke, *Food Cosmet. Toxicol.,* **16,** 731 (1978).

65. W. L. Epstein, *Report to Research Institute of Fragrance Materials, July 1976,* cited in G. L. J. Opdyke, *Food Cosmet Toxicol.* **16,** 731, (1978).

66. J. Fagius and B. Gronqvist, "Function of Peripheral Nerves and Signs of Polyneuropathy in Solvent Exposed Workers at a Swedish Steel Works," *Acta Neurol. Scand.,* **57,** 305–316 (1978).

67. T. A. Farquharson and B. H. Stock, "Modified Response of Hepatic Microsomes from Rats Administered Polar Aprotic Solvents," *Aust. J. Pharm. Sci.,* **4,** 111–115 (1975).

68. E. Fillipe, "Physiological Action and Behavior of Some Derivatives of Benzene Compared to Those of Cyclohexane," *Arch. Farmacol. Spe.,* **18,** 178–193 (1941).

69. R. R. Fisher and E. Jacobs, "Resolution and Reconstitution of *Rhodospirillium Rubrum* Pyridine Dinucleotide Transhydrogenase: Chemical Modification with N-Ethylmalemide and 2,4-Pentanedione," *Biochemistry*, **18**, 4315–4322 (1979).

70. F. Flury and O. Klimmer, *Toxicology and Hygiene of Industrial Solvents*, K. B. Lehman and F. Flury, Eds., Williams and Wilkins Co., Baltimore, Md. 1938.

71. F. Flury and W. Wirth, *Arch. Gewerbepathol. Gewerbehyg.*, **5,** 1 (1934).

72. D. F. Folland, W. Schaffner, H. E. Ginn, O. B. Crofford, and D. R. McMurray, "Carbon Tetrachloride Toxicity Potentiated by Isopropyl Alcohol. Investigation of an Industrial Outbreak," *J. Am. Med. Assoc.*, **236**, 1853–1856 (1976).

73. I. Fridovich, "A Study of the Interaction of Acetoacetic Decarboxylase with Several Inhibitors," *J. Biol. Chem.*, **243**, 1043–1051 (1968).

74. S. Fukabori, K. Nakaaki, and O. Taga, "Cutaneous Absorption of Acetone," *Rodo Kagaku*, **55,** 525–532 (1979).

75. R. L. Furner, E. D. Neville, K. F. Talarico, and D. B. Feller, "A Common Modality of Action of Simulated Space Stresses on the Oxidative Metabolism of Ethyl Morphine, Aniline and p-Nitroanisole by Male Rat Liver," *Toxicol. Appl. Pharmacol.*, **21**, 569–581 (1972).

76. C. R. Garcia, I. Geller, and H. L. Kaplan, "Effects of Ketones on Lever Pressing Behvaior of Rats," *Proc. West. Pharmacol. Soc.*, **21**, 433–438 (1978).

77. I. F. Gaunt, F. M. B. Carpanini, M. G. Wright, P. Grasso, and S. D. Gangolli, "Short Term Toxicity of Methyl Amyl Ketone in Rats," *Food Cosmet. Toxicol.*, **10**, 625–626 (1972).

78. I. Geller, R. L. Martinez, R. J. Hartmann, and H. L. Kaplan, "Effects of Ketones on a Match to Sample Task in the Baboon," *Proc. Wes. Pharmacol. Soc.*, **21**, 439–442 (1978).

79. I. Geller, E. Gause, H. Kaplan, and R. J. Hartmann, "Effects of Acetone, Methyl Ethyl Ketone, and Methyl Isobutyl Ketone on a Match to Sample Task in the Baboon," *Pharmacol. Biochem. Behav.*, **11**, 401–406 (1979).

80. I. Geller, J. R., Rowlands, and H. L. Kaplan, "Effects of Ketones on Operant Behavior of Laboratory Animals, DHEW Publ. Adm- 79–779, *Voluntary Inhalation of Industrial Solvents, 1978*, pp. 363–376.

81. A. Germent, "Inhibition of Oxidation by Peroxidase of Human Serum Proteins," *Mol. Biol. Rep.*, **3**, 283–287 (1977).

82. Y. Ghendon and G. Samoilova, "Antiviral Effect of Acetone," *J. Gen. Virol.*, **3,** 271–273 (1968).

83. H. F. Gilbert, III and M. H. O'Leary, "Modification of Arginine and Lysine in Proteins with 2,4-Pentanedione," *Biochemistry*, **14**, 5194–5198 (1975).

84. H. F. Gilbert, III and M. H. O'Leary, "Reversible Modification of Amino Groups in Aspartate Aminotransferase," *Biochem. Biophys. Acta*, **483**, 79–89 (1977).

85. M. A. Gilchrist, W. E. Hunt, N. Allen, H. T. Yee, D. J. Billmaier, "Toxic Peripheral Neuropathy," *Morb. Mort. Wk. Rep.* **23**, 9–10 (1974).

86. J. Giroux, R. Granger, and P. Monnier, "Comparative Study of 2-Pentanone and 2-Ethyl Butanone in the Mouse," *Trav. Soc. Pharm. Montp.*, **14**, 342–346 (1954).

87. S. Gitelson, A. Werczberger, and J. B. Herman, "Coma and Hyperglycemia Following Drinking of Acetone," *Diabetes*, **15,** 810–811 (1966).

88. M. E. Golberg, H. E. Johnson, U. C. Pozzani, and H. F. Smyth, Jr., "Effect of Repeated Inhalation of Vapors of Industrial Solvents on Animal Behvaior," *Am. Ind. Hyg. Assoc. J.*, **25,** 369–375 (1964).

89. J. J. Gomer, *Hyg. Infektrouskrankh*, **130,** 680 (1960).

90. R. I. Goncharova, "Genetic Activity of Some Cyclohexane Derivatives," *Genet. Tsitol.*, 137–142 (1970).

91. R. I. Goncharova and E. A. Laryutina, "Effect of Some Cyclohexane Derivatives on the Frequency of Lethal Mutations Induced by γ and X-Ray Irradiation in Drosophila," *Genet. Tsitol.,* 142–148 (1970).

92. E. Gondry, "Research on the Toxicity of Cyclohexylamine, Cyclohexanone and Cyclohexanol and Metabolites of Cyclamate," *J. Eur. Toxicol.,* **5,** 227–238 (1973).

93. J. H. Griggs, E. M. Weller, T. A. Polmisano, and W. Niedermeier, "The Effect of Noxious Vapors on Embryonic Chick Development," *Ala. J. Med. Sci.,* **8,** 342–345 (1971).

94. I. Gruebner, W. Klinger, and H. Ankermann, "Various Substances and Substance Classes with Inducer Properties II. *Arch. Int. Pharmacodyn. Ther.,* **196,** 288–297 (1972).

95. P. K. Gupta, W. H. Lawrence, J. E. Turner, and J. Autian, "Toxicological Aspects of Cyclohexanone," *Toxicol. Appl. Pharmacol.,* **49,** 525–533 (1979).

96. E. C. Hagan, W. H. Hansen, O. G. Fitzhugh, P. M. Jenner, W. I. Jones, J. M. Taylor, E. L. Long, A. A. Nelson, and J. B. Brouwer, "Food Flavourings and Compounds of Related Structure. II. Subacute and Chronic Toxicity," *Food Cosmet. Toxicol.,* **5,** 141 (1967).

97. H. W. Haggard, L. A. Greenburg, and J. M. Turner "The Physiological Principles Governing the Action of Acetone Together with Determination of Toxicity," *J. Ind. Hyg. Toxicol.,* **26,** 133 (1944).

98. H. W. Haggard, D. P. Miller, and L. A. Greenberg, "The Amyl Alcohols and Their Ketones: Their Metabolic Fates and Comparative Toxicities," *J. Ind. Hyg. Toxicol.,* **27,** 1–14 (1945).

99. Y. Hagiwara, "Benzoyl Cholinesterase II. Activation of Nonspecific Cholinesterase by Polyhydric Alcohols," *Jap. J. Pharmacol.,* **9,** 137–149 (1960).

100. I. H. Hall, G. L. Carlson, G. S. Abernethy, and C. Piantadosi, "Cycloalkanones No. 4 Antifertility Activity," *J. Med. Chem.,* **17,** 1253–1257 (1974).

101. I. H. Hall and G. L. Carlson, "Cycloalkanones. 9. Comparison of Analogs which Inhibit Cholesterol and Fatty Acid Synthesis. *J. Med. Chem.,* **19,** 1257–1261 (1976).

102. L. C. Harris and R. H. Jackson, "Acute Acetone Poisoning Caused by Setting Fluid for Immobilizing Casts. *Br. Med. J.,* **2,** 1024–1026 (1952).

103. E. R. Hart, J. N. Shick, and C. D. Leake, "The Toxicity of Mesityl Oxide," Univ. Calif. Berkeley, *Publ. Pharmacol.,* **1**(13), 161–173 (1939).

104. S. Hayman and R. F. Colman, "Reaction of Essential Lysyl Residues of Pig Heart Diphosphopyridine Nucleotide Dependent Isocitrate Dehydrogenase with 2,4-Pentanedione," *Biochemistry,* **16,** 998–1005 (1977).

105. L. B. Hetland, D. Couri, and M. S. Abdel-Rahman, "The Influence of Inhaled Ketone Solvent Vapors on Hepatic Microsomal Biotransformation Activities," *Toxicol. Appl. Pharmacol.,* **37,** 111 (1976).

106. E. R. Homan and R. R. Maronpot, "Neurotoxic Evaluation of Some Aliphatic Ketones," *Toxicol. Appl. Pharmacol.,* **45,** 312 (1978).

107. C. T. Hou, R. Patel, A. I. Laskin, N. Barnabe, and I. Marczak, "Microbial Oxidation of Gaseous Hydrocarbons: Production of Methyl Ketones from Their Corresponding Secondary Alcohols by Methane and Methanol-Grown Microbes," *Appl. Environ. Microbiol.,* **38,** 135–142 (1979).

108. M. Ikeda, H. Ohtsuji, and T. Imamura, "Comparative Studies on Aniline Hydroxylation and p-Nitrotoulene Hydroxylation by the Liver," *Jap. J. Pharmacol.,* **22,** 479–491 (1972).

109. Itos., "Industrial Toxicological Studies of Mesityl Oxide," *Yokohama Igaku Publ.* **20**(3), 253–265 (1969).

110. K. M. Ivanetich, S. Lucas, J. A. Marsh, M. R. Zinman, I. D. Katz, and J. J. Bradshaw, "Organic Compounds: Their Interaction with and Degradation of Hepatic Microsomal Drug-Metabolizing Enzymes *in vitro,*" *Drug Metab. Dispos.,* **6,** 218–225 (1978).

111. C. Jacobi, *Hayashi* and Szubinski *Arch. Exp. Pathol. Pharmacol.,* **50,** 199 (1903).

112. R. J. Jaeger, R. B. Conolly, E. S. Reynolds, and S. D. Murphy, "Biochemical Toxicology of Unsaturated Halogenated Monomers," *Environ. Health Perspect.*, **11**, 121–128 (1975).

113. S. P. James and R. H. Waring, "The Metabolism of Alicyclic Ketones in the Rabbit and Rat," *Xenobiotica*, **1**, 573–580 (1971).

114. P. M. Jenner, E. C. Hagan, J. M. Taylor, E. L. Cook, and O. G. Fitzhugh, "Food Flavorings and Compounds of Related Structure. I. Acute Oral Toxicity," *Food Cosmet. Toxicol.*, **2**, 327 (1964).

115. B. L. Johnson, J. V. Setzer, T. R. Lewis, and W. K. Anger, "Effects of Methyl *n*-Butyl Ketone on Behavior and the Nervous System," *Am. Ind. Hyg. Assoc. J.*, **38**, 567–579 (1972).

116. B. L. Johnson, J. V. Setzer, T. R. Lewis, and R. W. Hornung, "An Electrodiagnostic Study of the Neurotoxicity of Methyl n-Amyl Ketone," *Am. Ind. Hyg. Assoc. J.*, **39**, 866–872 (1978).

117. B. L. Johnson, W. K. Anger, J. W. Setzer, D. W. Lynch, and T. R. Lewis, "Neurobehavioral Effects of Methyl n-Butyl Ketone and Methyl n-Amyl Ketone in Rats and Monkeys: A summary of NIOSH Investigators," *J. Environ. Pathol. Toxicol.*, **2**, 123–133 (1979).

118. E. Kagan, *Arch. Hyg.*, **94**, 41 (1924).

119. P. T. Kan, M. A. Simetskii, and V. I. Il'yashchenko, "Effect of Organic Acaricide Solvents on the Skin and Ocular Conjunctiva of Rabbits," *Tr. Vses. Nauch.-Issled. Inst. Vet. Sanit.*, **39**, 369–372 (1971).

120. A. E. Katz, *Spice Mill*, **69**, 46 (1946).

121. G. V. Katz, J. L. O'Dohoghue, G. D. DiVincenzo, and C. J. Terhaar, "Comparative Neurotoxicity and Metabolism of Ethyl n-Butyl Ketone and Methyl n-Butyl Ketone in Rats," *Toxicol. Appl. Pharmacol.*, **52**, 153–158 (1980).

122. M. Kiese and W. Lenk, "Hydroxyacetophenones: Urinary Metabolites of Ethyl Benzene and Acetophenone in the Rabbit," *Xenobiotica* **4**, 337–343 (1974).

123. E. T. Kimura, D. M. Ebert, and P. W. Dodge, "Acute Toxicity and Limits of Solvent Residue for Sixteen Organic Solvents," *Toxicol. Appl. Pharmacol.*, **19**, 699–704 (1971).

124. R. D. King and G. H. Clegg, "The Metabolism of Fatty Acids, Methyl Ketones and Secondary Alcohols by *Penicillium roqueforti* in Blue Cheese Slurries," *J. Sci. Food Agric.*, **30**, 197–202 (1979).

125. M. Kitada, T. Kamataki, and H. Kitagawa, "Enhancement *in vivo* of Drug Oxidations Following Administration of Benzphetamine, Acetone, Metyrapone and Dimethylsulfoxide," *Jap. J. Pharmacol.*, **28**, 213–221 (1978).

126. A. M. Kligman, "The Identification of Contact Allergens by Human Assay III. The Maximization Test, A Procedure for Screening and Rating Contact Sensitizers," *J. Invest. Dermatol.*, **47**, 393 (1966).

127. A. M. Kligman, "Report to Research Institute for Fragrance Materials, January 1972, in D. L. J. Opdyke, "Fragrance Raw Materials, Ethyl Amyl Ketone," *Food Cosmet. Toxicol.*, **12**, 715 (1972).

128. A. M. Kligman and W. Epstein, "Updating the Maximization test for Identifying Contact Allergens," *Contact Dermatitis*, **1**, 231–239 (1975).

129. R. P. Kohli, K. Kishor, P. R. Dua, and R. C. Saxena, "Anti-Convulsant Activity of Some Carbonyl Containing Compounds," *Indian J. Med. Res.*, **55**, 1221–1225 (1967).

130. P. A. Kolesnikov, L. A. Tiunov, E. A. Zhercin, S. M. Faier, and I. S. Kolosova, "Combined Action of Acetone and X-Rays," *Sarmakol. Toksikol.*, **37**, 446–450 (1974).

131. W. J. Krasavage, J. L. O'Donoghue, G. D. DiVincenzo, and C. J. Terhaar, "The Relative Neurotoxicity of Methyl n-Butyl Ketone and n-Hexane and Their Metabolites," *Toxicol. Appl. Pharmacol.*, **52**, 433–441 (1980).

132. W. J. Krasavage, J. L. O'Donoghue, and C. J. Terhaar, "Oral Chronic Toxicity of Methyl n-Propyl Ketone, Methyl n-Butyl Ketone and Hexane in rats," *Toxicol. Appl. Pharmacol.* (Part 2), A205 (1979).

133. G. Krinke, H. H. Schaumburg, P. S. Spencer, P. Thomann, and R. Hess, "Clinoquinol and 2,5-Hexanedione Induced Different Types of Distal Axonopathy in the Dog," *Acta Neuropathol.* (Berl.), **47**, 213–221 (1979).

134. Z. Y. Lagno and G. Z. Bakhtizina, "The Skin Resorptive Action of Acetophenone," *Tr. Ufim. Nauch-Issled. Inst. Gig. Prof. Zabol.*, **5**, 90–94 (1969).

135. S. S. Lande, P. R. Durkin, D. H. Christopher, P. H. Howard, and J. Saxena, "Investigation of Selected Potential Environmental Contaminants: Ketonic Solvents," U.S. NTIS Public Report PB #252970, 1976.

136. T. S. Larson, J. K. Finnegan, and H. D. Haag, "Observations on the Effect of the Edema-Producing Potency of Acids, Aldehydes, Ketones and Alcohols," *J. Pharmacol. Exp. Ther.*, **116**, 119 (1956).

137. N. W. Lazarew, A. J. Brussilowskaja, and J. N. Lawrow, "Quantitative Untersuchingen Über die Resorption Einiger Organisher Gifte Durch die Haut ins Blut," *Arch. Gewerbepathol. Gewerbehyg.*, **2**, 641 (1931).

138. K. C. Leibman, "Reduction of Ketones in Liver Cytosol, *Xenobiotica*, **1**, 97–104 (1971).

139. I. Levinstein, Report to RIFM, in *Monographs on Fragrance Raw Materials*, D. L. J. Opdyke, "Methyl *n*-Amyl Ketone," *Food Cosmet. Toxicol.*, **13**, 847–848 (1975).

140. L. A. Linyuchepa, L. A. Tiunov, and T. S. Kolosova, "Effect of Acetone on Level of Ketone Bodies in Blood and Urine of Experimental Animals of Different Species," *Farmakol. Toksikol.*, **32**, 465–467 (1969).

141. H. B. Lukins and J. W. Foster, "Methyl Ketone Metabolism in Hydrocarbon-Utilizing Microbacteria," *J. Bacteriol.*, **85**, 1074–1086 (1963).

142. A. P. Lupulescu and D. J. Birmingham, "The Effect of Protective Agent Against Lipid Solvent Induced Damages. Ultrastructural and Scanning Electronmicroscopical Study of Human Epidermis," *Arch. Environ. Health*, **31**, 33–36 (1976).

143. A. P. Lupulescu, D. J. Birmingham, and H. Pinkus, "Electron Microscopic Study of Human Epidermis After Acetone and Kerosene Administration," *J. Invest. Dermatol.*, **60**, 33–45 (1973).

144. A. P. Lupulescu, H. Pinkus, and D. J. Birmingham, "Effect of Acetone and Kerosene on Skin Ultrastructure," *Proc. Electron Microsc. Soc. Am.*, **30**, 92–93 (1972).

145. J. D. MacEwen, E. H. Vernot, and C. C. Haun, "Effect of 90-day Continuous Exposure to Methylisobutyl Ketone on Dogs, Monkeys, and Rats," U.S. Nat. Tech. Info. Serv., AD Rep; ISS #730291, 1971.

146. W. F. MacKenzie, "Pathological Lesions Caused by Methyl Isobutyl Ketones," U.S. Nat. Tech. Info. Serv., AD RED; ISS #751444, 1971, pp. 311–322.

147. J. S. Mallov, "Methyl n-Butyl Ketone Neuropathy Among Spray Painters," *J. Am. Med. Assoc.*, **235**, 1445–1457 (1976).

148. K. E. Malten, D. Spruit, H. G. M. Boemaars, and M. J. M. DeKeizer, "Horny Layer Injury by Solvents," *Berufsdermatosen*, **16**, 135–147 (1968).

149. A. J. Markovetz, "Intermediates from the Microbial Oxidation of Aliphatic Hydrocarbons," *J. Am. Oil Chem. Soc.*, **55**, 430–434 (1978).

150. L. Martis, T. Tolhurst, M. T. Koeferl, T. R. Miller, and T. D. Darby, "Disposition Kinetics of Cyclohexanone in Beagle Dogs," *Toxicol. Appl. Pharmacol.*, **55**, 545–553 (1980).

151. T. Matsushita, T. Yoshea, A. Yoshimune, T. Inoue, F. Yamaka, and H. Suzuki, "Experimental Studies for Determining the Maximum Permissible Concentration of Acetone—1. Biologic Reactions in One Day Exposure to Acetone," *Jap. J. Ind. Health*, **11**, 477–485 (1969).

152. J. May, "Odor Thresholds of Solvents for Evaluating Solvent Odors in Air," *Staub*, **26**, 385 (1966).

153. J. McCann, E. Choi, E. Yamasaki, and B. N. Ames, "Detection of Carcinogens as Mutagens in the Salmonella/Microsome Test—Assay of 300 Chemicals," *Proc. Natl. Acad. Sci. U.S.*, **72**, 5135–5139 (1975).

154. J. McLaughlin, Jr., J. -P. Marliac, M. J. Verrett, M. K. Mutchler, and O. G. Fitzhugh, "Toxicity of 14 Volatile Chemicals as Measured by the Chick Embryo Method," *Am. Ind. Hyg. Assoc. J.*, **25**, 282–284 (1964).

155. R. E. McMahon, J. C. Cline, and C. Z. Thompson, "Assay of 855 Test Chemicals in 10 Tester Strains Using a New Modification of the Ames Test for Bacterial Mutagens," *Cancer Res.*, **39**, 682–693 (1979).

156. W. A. McOmie, and H. H. Anderson, "Comparative Toxicologic Effects of Some Isobutyl Carbinols and Ketones," *Univ. Calif. Publ. Pharmacol.*, **2**, 217 (1949).

157. J. R. Mendell, K. Saida, M. F. Ganansia, D. B. Jackson, H. Weiss, R. W. Gardier, C. Chrisman, N. Allen, D. Couri, J. O'Neill, B. Marks, and L. Hetland, "Toxic Polyneuropathy Produced by Methyl *n*-Butyl Ketone," *Science*, **185**, 787–789 (1974).

158. O. H. Miller, C. G. Huggins, and K. Arai, "Studies on the Metabolism of 1,2-Propandiol-1-Phosphate," *J. Biol. Chem.*, **202**, 263–271 (1953).

159. H. Mirchev, "Hepatorenal Lesions in Acute Acetone Poisoning," *Vutr. Voles*, **17**, 89–92 (1978).

160. J. Misumi, M. Kawakami, T. Hitoshi, and S. Nomura, "Effects of n-Hexane, Methyl n-Butyl Ketone and 2,5-Hexanedione on the Conduction Velocity of Motor and Sensory Nerve Fibers in Rats' Tail," *Sangyo Igaku*, **21**, 180–181 (1979).

161. B. M. Mokhitov and A. A. Zimbekov, "Hygienic Evaluation of Methyl Ethyl Ketone as an Atmospheric Pollutant," *Vopr. Gig. Tr. Profzabol.*, 232–234 (1972).

162. G. L. J., Opdyke, "Monographs on Fragrance Raw Materials," Special Issue IV, *Food Cosmet. Toxicol.*, **16**, 819–820 (1978).

163. O. M. Moreno, *Report to Research Institute of Fragrance Materials, August 1976*. Cited in G. L. J. Opdyke, *Food Cosmet. Toxicol.*, **16**, 731 (1978).

164. G. A. Mourkides, D. C. Hobbs, and R. E. Koeppe, "The Metabolism of Acetone-2-C^{14} by Intact Rats," *J. Biol. Chem.*, **234**, 27–30 (1959).

165. R. Munies, "Investigation of Some Factors Influencing the Percutaneous Absorption of Methyl Ethyl Ketone," University of Wisconsin, *Dissertation Abstracts International*, **26**, 5384 (1965).

166. R. Munies and D. E. Wurster, "Investigation of Some Factors Influencing Percutaneous Absorption. III. Absorption of Methyl Ethyl Ketone," *J. Pharmacol. Sci.*, **54**, 1281–1284 (1965).

167. K. Nakaaki, "Effects of Exposure to Organic Solvent Vapor in Human Subjects," *Rodo Kagaku*, **50**, 89–96 (1974).

168. K. W. Nelson, J. F. Ege, Jr., N. Ross, L. E. Woodman, and L. Silverman, "Sensory Response to Industrial Solvent Vapors," *J. Ind. Hyg. Toxicol.*, **25**, 282–285 (1943).

169. P. Nenov, M. Spasovski, M. and A. Bainova, "Changes of Some Microsomal Liver Enzymes in Rats Exposed to the Combined Effect of Acetone and Alcohol," *Khig. Zdraveopaz.*, **20**, 325–332 (1977).

170. O. Neubruer, "Concerning the Paring of Glucuronic Acid with Materials of the Aliphatic Series," *Arch. Exp. Pathol. Pharmacol.*, **46**, 133–154 (1901).

171. NIOSH, "Criteria for a Recommended Standard for Occupational Exposure to Ketones," U.S. Dept. of Health, Education and Welfare, National Institute for Occupational Safety and Health DHEW (NIOSH) Publication No. 78–173, 1978.

172. R. Nordmann, C. Ribiere, H. Rouach, F. Beauge, Y. Giudicelli and H. Nordman, "Metabolic Pathways Involved in the Oxidation of Isopropanol into Acetone by the Intact Rat," *Life Sci.*, **13**, 919–932 (1973).

173. J. L. O'Donoghue, W. J. Krasavage, and C. J. Terhaar, "Toxic Effects of 2,5-Hexanedione," *Toxicol. Appl. Pharmacol.*, **45**, 269 (1978).

174. J. L. O'Donoghue and W. J. Krasavage, "Hexacarbon Neuropathy: A γ-Diketone Neuropathy?" *J. Neuropathol. Exp. Neurol.*, **38**, 333 (1979).

175. J. L. O'Donoghue and W. J. Krasavage, "The Structure-Activity Relationship of Aliphatic Diketones and Their Potential Neurotoxicity," *Toxicol. Appl. Pharmacol.* (Part 2), **48**, A55 (1979).

176. D. L. J. Opdyke, "Monographs on Fragrance Raw Materials," *Food Cosmet. Toxicol.*, **11**, 99–100 (1973).

177. D. L. J. Opdyke, "Fragrance Raw Materials. Ethyl Amyl Ketone," *Food Cosmet. Toxicol.*, **12**, 715 (1974).

178. D. L. J. Opdyke, "Monographs on Fragrance Raw Materials. Methyl Hexyl Ketone," *Food Cosmet. Toxicol.*, **13**, 861 (1975).

179. D. L. J. Opdyke, "Monographs on Fragrance Raw Materials. Methyl *n*-Amyl Ketone," *Food Cosmet. Toxicol.*, **13**, 847–848 (1975).

180. D. L. J. Opdyke, "Fragrance Raw Materials Monographs: Methyl Ethyl Ketone," *Food Cosmet. Toxicol.*, **15**, 627–632 (1977).

181. D. L. J. Opdyke, "Monographs on Fragrance Raw Materials. Special Issue IV," *Food Cosmet. Toxicol.*, **16**, 637 (1978).

182. H. B. Otwell, K. L. Cipollo, and R. B. Dunlap, "Modification of Lysl Residues of Dihydrofolate Reductase with 2,4-Pentanedione," *Biochem. Biophys. Acta*, **568**, 297–306 (1979).

183. L. Parmeggiani and C. Sassi, "Occupational Poisoning with Acetone—Clinical Disturbances, Investigations in Workrooms and Physiopathological Research," *Med. Lav.*, **45**, 431–468 (1954).

184. F. A. Patty, H. H. Schrenk, and W. P. Yant, *Butanone U.S. Public Health Rep.*, **50**, 1217 (1935).

185. L. Perbellini, D. DeGrandis, F. Semenzato, N. Rizzuto, and A. Simonati, "An Experimental Study on the Neurotoxicity of n-Hexane Metabolites: Hexanol-1 and hexanol-2," *Toxicol. Appl. Pharmacol.*, **46**, 421–427 (1978).

186. G. L. Plaa, G. J. Traiger, G. K. Hanasono, and H. Witschi, *Proc. Int. Symp. Alcohol Drug Res.*, 225–244, J. M. Khawna, Y. Israel, and H. Kalant, Eds., Alcohol and Drug Addiction Research Foundation, Toronto, Canada, 1975.

187. V. Postolache, L. Safta, B. Cuparencu, and L. Steiner, "Animal Studies of the Mutal Effect of Acetone in Certain Tranquilizing Agents—Chlorodiazepoxide and Chlorpromazine," *Rev. Roum. Morphol. Physiol.*, **20**, 53–55 (1974).

188. H. C. Powell, T. Koch, R. Garrett, and P. W. Lampert, "Schwann Cell Abnormalities in 2,5-Hexanedione Neuropathy," *J. Neurocytol.*, **7**, 517–528 (1978).

189. G. Powis and A. R. Boobis, "The Effect of Pretreating Rats with 3-Methycholanthrene upon the Enhancement of Microsomal Aniline Hydroxylation by Acetone and Other Agents," *Biochem. Pharmacol.*, **24**, 424–426 (1975).

190. T. D. Price and D. Rittenberg, "The Metabolism of Acetone. I. Gross Aspects of Catabolism and Excretion," *J. Biol. Chem.*, **185**, 449 (1950).

191. A. Quevauviller, "Toxicite, Pouvoir Hypnotique de l'Acetophenone et des Thienylcetones" *C. R. Soc. Biol.*, **140**, 367–369 (1946).

192. R. L. Raleigh and W. A. McGee, "Effects of Short High Concentration Exposures to Acetone as Determined by Observation in the Work Area," *J. Occup. Med.*, **14**, 607–610 (1972).

193. A. Ramu, J. Rosenbaum, and T. F. Blaschke, "Disposition of Acetone Following Acute Intoxication," *West. J. Med.*, **29**, 429–432 (1978).

194. R. H. Rengstorff, J. P. Petrali, and V. M. Sim, "Cataracts Induced in Guinea Pigs by Acetone, Cyclohexanone and Dimethyl Sulfoxide," *U.S. Nat. Tech. Inform. Serv.*, AD Rep. Iss. No. 730902, 1971.

195. R. H. Rengstorff, J. P. Petrali, and V. M. Sim, "Cataracts Induced in Guinea Pigs by Acetone, Cyclohexanone, and Dimethyl Sulfoxide," *Am. J. Optom.*, **49**, 308–319 (1972).

196. R. H. Rengstorff, J. P. Petrali, and V. M. Sim, "Attempts to Induce Cataracts in Rabbits by Cutaneous Application of Acetone," *Am. J. Optom. Physiol. Opt.*, **53**, 41–42 (1976).

197. D. F. Ross, "Acute Acetone Intoxication Involving 8 Male Workers," *Ann. Occup. Hyg.*, **16**, 73–75 (1973).

198. J. A. Ruddick, "Toxicology, Metabolism and Biochemistry of 1,2-Propanediol," *Toxicol. Appl. Pharmacol.*, **21**, 102–111 (1972).

199. H. Rudney, "The Metabolism of 1,2-Propanediol," *Arch. Biochem.*, **29**, 231–232 (1950).

200. M. I. Sabri, K. Ederle, C. E. Holdsworth, and P. S. Spencer, "Studies on the Biochemical Basis of Distal Axonopathies II. Specific Inhibition of Fructose-6-Phosphate Kinase by 2,5-Hexanedione and Methyl n-Butyl Ketone," *Neurotoxicology*, **1**, 285–297 (1979).

201. G. Sack, "Ein Fall von Gewerblicher Azeton Vergiftung," *Arch. Gewerbepathol. Gewerbehyg.*, **10**, 80–86 (1940).

202. K. Saida, J. R. Mendell, and H. F. Weiss, "Peripheral Nerve Changes Induced by Methyl n-Butyl Ketone and Potentiation by Methyl Ethyl Ketone," *J. Neuropathol. Exp. Neurol.*, **35**, 207–225 (1976).

203. W. Sakami and J. M. Lasaye, "Formation of Formate and Labile Methyl Groups from Acetone in the Intact Rat," *J. Biol. Chem.*, **187**, 369–378 (1950).

204. W. Sakami and H. Rudney, *Brookhaven Symp. Biol.*, **5**, 1976–1979 (1952).

205. D. M. Sanderson, "A Note on Glycerol Formal as a Solvent in Toxicity Testing," *J. Pharm. Pharmacol.*, **11**, 150–156 (1959).

206. H. H. Schaumburg and P. S. Spencer, "Environmental Hydrocarbons Produced Degeneration in Cat Hypothalamus and Optic Tract," *Science*, **199**, 199–200 (1978).

207. R. Schoental and J. B. Cavanagh, "Mechanisms Involved in the "Dyingback" Process—An Hypothesis Implicating Co-Enzymes," *Neuropathol. Appl. Neurobiol.*, **3**, 145–157 (1977).

208. H. H. Schrenk, W. P. Yant, and F. A. Patty, *Hexanone U.S. Public Health Rep.*, **51**, 624 (1936).

209. L. Schwarz, "Uber die Oxydation des Acetones und Homologer Ketone der Fett Saurerethe," *Arch. Expl. Pathol. Pharmakol.*, **40**, 168 (1898).

210. B. A. Schwetz, B. K. Leong, and P. J. Gehring, "Embryo and Feto Toxicity of Inhaled Carbon Tetrachloride, 1,1-Dichloroethane and Methyl Ethyl Ketone in Rats," *Toxicol. Appl. Pharmacol.*, **28**, 452–464 (1974).

211. D. J. Selkoe, L. Luckenbill-Edds, and M. L. Shelanski, "Effects of Neurotoxic Industrial Solvents on Cultured Neuroblastoma Cells: Methyl n-Butyl Ketone, n-Hexane, and Derivatives," *J. Neuropathol. Exp. Neurol.*, **37**, 768–789 (1978).

212. E. S. Shalaby, M. El Danafory, A. A. E. Massoud, and S. M. Hathous, "Toxic Effects of Fat Solvents Used in Paints on Liver, Blood, and Lung," *J. Egypt Med. Assoc.*, **56**, 340–347 (1973).

213. D. W. Sharp, "The Sensitization Potential of Some Perfume Ingredients Tested Using a Modified Draize Procedure," *Toxicology*, **9**, 261–271 (1978).

214. B. Shekhtman and G. S. Allev, "Differences in the Action of Hydrocarbon in Relation to their Physicochemical Properties," *Azerb. Med. Zh.*, **51**, 58–64 (1974).

215. M. V. Shelanski, *Report to Research Institute for Fragrance Materials, November 1973*. Cited in D. L. J. Opdyke, "Fragrance Raw Materials, Ethyl Amyl Ketone," *Food Cosmet. Toxicol.*, **12**, 715 (1974).

216. L. Silverman, H. F. Schulte, and M. W. First, "Further Studies on Sensory Response to Certain Industrial Solvent Vapors," *J. Ind. Hyg. Toxicol.*, **28**, 262–266 (1946).

217. I. G. Sipes, D. Stripp, G. Krishna, H. N. Maling, and J. R. Gillette, "Enhanced Hepatic

Microsomal Activity by Pretreatment of Rats with Acetone or Isopropanol," *Proc. Soc. Exp. Biol. Med.,* **142,** 237–240 (1973).

218. I. G. Sipes, M. L. Slocumb, and G. Holtzman, "A Stimulation of Microsomal Dimethyl Nitrosamine-*n*-Demethylase by Pretreatment of Mice with Acetone," *Chem.-Biol. Interact.,* **21,** 155–166 (1978).

219. A. R. Smith and N. R. Mayers, "Poisoning and Fire Hazards of Butanone and Acetone," *Ind. Bull. N.Y. State Dept. of Labor,* **23,** 174 (1944).

220. Y. S. Smusin, "Certain Problems in Toxico-Dynamics and Post Mortem Diagnosis of Poisoning with Substances Having an Anticholinesterase Action," *Fb. Tr. IV-oi Vses. Konf. Sodebn. Medikov. Riga,* 443–445 (1962).

221. H. F. Smyth, Jr., "Hygienic Standards for Daily Inhalation, *Amer. Ind. Hyg. Assoc. Q.,* **17,** 129–185 (1956).

222. H. F. Smyth, Jr. and C. P. Carpenter, "The Place of the Range Finding Test in the Industrial Toxicology Laboratory," *J. Ind. Hyg. Toxicol.,* **26,** 269 (1944).

223. H. F. Smyth, Jr. and C. P. Carpenter, "Further Experience with the Range Finding Test in the Industrial Toxicology Laboratory," *J. Ind. Hyg. Toxicol.,* **30,** 63–68 (1948).

224. H. F. Smyth, Jr., C. P. Carpenter, and C. S. Weil, "Range Finding Toxicity Data, List III," *J. Ind. Hyg. Toxicol.,* **31,** 60–62 (1949).

225. H. F. Smyth, Jr., C. P. Carpenter, and C. S. Weil, "Range Finding Toxicity Data: List IV," *Arch. Ind. Hyg. Occup. Med.,* **4,** 119–122 (1951).

226. H. F. Smyth, Jr., C. P. Carpenter, C. S. Weil, and U. C. Pozzani, "Range Finding Toxicity Data: List V," *Arch. Ind. Hyg. Occup. Med.,* **10,** 61–68 (1954).

227. H. F. Smyth, Jr., C. P. Carpenter, C. S. Weil, U. C. Pozzani, and J. A. Striegel, "Range Finding Toxicity Data: List VI," *Am. Ind. Hyg. Assoc. J.,* **23,** 95–107 (1962).

228. H. F. Smyth, Jr., C. P. Carpenter, C. S. Weil, U. C. Pozzani, J. A. Striegel, and J. S. Nycum, "Range Finding Toxicity Data VII," *Am. Ind. Hyg. Ass. J.,* **30,** 470–476 (1969).

229. H. F. Smyth, Jr., J. Seaton, and L. Fischer, "Response of Guinea Pigs and Rats to Repeated Inhalation of Vapors of Mesityl oxide and Isophorone," *J. Ind. Hyg. Toxicol.,* **24,** 46–50 (1942).

230. H. F. Smyth, Jr. and J. Seaton, "Acute Response of Guinea Pigs and Rats to Inhalation of the Vapors of Isophorone," *J. Ind. Hyg. Toxicol.,* **22,** 477–483 (1940).

231. H. Specht, "Acute Response of Guinea Pigs to Inhalation of Methyl Isobutyl Ketone," *U.S. Public Health Rep.,* **53,** 292 (1938).

232. H. Specht, J. W. Miller, and P. J. Valaer, "Acute Response of Guinea Pigs to the Inhalation of Dimethyl Ketone (Acetone) Vapor in Air," *Public Health Rep.,* **54,** 944–954 (1939).

233. H. Specht, J. W. Miller, P. J. Valaer, and R. R. Sayers, "Acute Response of Guinea Pigs to the Inhalation of Ketone Vapors," U.S. Public Health Service, NIH Bull., No. 176, Div. Ind. Hyg., 1940, pp. 1–66.

234. P. S. Spencer and H. H. Schaumburg, "Experimental Neuropathy Produced by 2,5-Hexane-dione—A Major Metabolite of the Neurotoxic Solvent Methyl n-Butyl Ketone," *J. Neurol. Neurosurg. Psychol.,* **38,** 771–775 (1975).

235. P. S. Spencer, H. H. Schaumburg, R. L. Raleigh, and C. J. Terhaar, "Nervous System Degeneration Produced by the Industrial Solvent Methyl *n*-Butyl Ketone," *Arch. Neurol.,* **32,** 219–222 (1975).

236. P. S. Spencer and H. H. Schaumburg, "Feline Nervous System Response to Chronic Intoxication with Commercial Grades of Methyl n-Butyl Ketone, Methyl Iso-Butyl Ketone, and Methyl Ethyl Ketone," *Toxicol. Appl. Pharmacol.,* **37,** 301–311 (1976).

237. P. S. Spencer, M. C. Bischoff, and H. H. Schaumburg, "On the Specific Molecular Configuration

of Neurotoxic Aliphatic Hexacarbon Compounds Causing Central-Peripheral Distal Axono-pathy," *Toxicol. Appl. Pharmacol.,* **44,** 17–28 (1978).

238. H. Srepel and B. Akacic, "Testing the Antihelmintic Effectiveness of Volatile Oils from Plants of the Genus *Ruta,*" *Acta Pharm. Jugosl.,* **12,** 79–87 (1962).

239. P. G. Stecher, Ed., *Merck Index,* 8th ed., 1968.

240. H. R. Sullivan, W. M. Miller, and R. E. McMahon, "Reaction Pathways of *in vivo* Stereoselective Conversion of Ethyl Benzene to Mandelic Acid," *Xenobiotica,* **6,** 49–54 (1976).

241. O. Tada, K. Nakaaki, and S. Fukabori, "Experimental Study on Acetone and Methyl Ethyl Ketone Concentrations in Urine and Expired Air after Exposure to Those Vapors," *Rodo Kagaku,* **48,** 305–331 (1972).

242. C. C. Tao and T. H. Elliott, "The Metabolism of [^{14}C] Methyl Cyclohexane," *Biochem. J.,* **84,** 38–39 (1962).

243. *Threshold Limit Values for Chemical Substances and Physical Agents in the Workroom Environment with Intended Changes for 1981,* American Conference of Governmental Industrial Hygienists, Cincinnati, Ohio.

244. J. H. Tison, L. D. Prockup, and E. D. Means, in S. S. Lande, P. R. Durkin, D. H. Christopher, P. H. Howard, and J. Saxena, "Investigation of Selected Potential Environmental Contaminants; Ketonic Solvents," U.S. NTIS Public Report PB #252970, 1976.

245. G. J. Traiger, G. L. Plaa, G. K. Hanasono, and H. Witschi, "Effect of Alcohols on Various Forms of Chemical Induced Liver Injury," *Alcohol Liver Pathol.* [Proc. Int. Symp. Alcohol Drug RES], 225–244 (1973).

246. G. J. Traiger and J. V. Bruckner, "The Participation of 2-Butanone in 2-Butanol Induced Potentiation of Carbon Tetrachloride Hepatotoxicity," *J. Pharmacol. Exp. Ther.,* **196,** 493–500 (1976).

247. G. J. Traiger, J. V. Bruckner, and P. H. Cooke, "Effect of 2-Butanol and 2-Butanone on Rat Hepatic Ultra Structure and Microsomal Drug Metabolizing Enzyme Activity, *Toxicol. Appl. Pharmacol.,* **33,** 132 (1975).

248. G. J. Traiger and G. L. Plaa, "Chlorinated Hydrocarbon Toxicity. Potentiation by Isopropyl Alcohol and Acetone," *Arch. Environ. Health,* **28,** 276–278 (1974).

249. J. F. Treon, W. E. Crutchfield, Jr., and K. V. Kitzmiller, "The Physiological Response of Rabbits to Cyclohexane, Methyl Cyclohexane and Certain Derivatives of These Compounds. I. Oral Administration and Cutaneous Application," *J. Ind. Hyg. Toxicol.,* **25,** 199–214 (1943).

250. J. F. Treon, W. E. Crutchfield, Jr., and K. V. Kitzmiller, "The Physiological Response of Animals to Cyclohexane, Methyl Cyclohexane and Certain Derivatives of These Compounds. II. Inhalation," *J. Ind. Hyg. Toxicol.,* **25,** 323–347 (1943).

251. R. Truhaut, H. Dutertre-Catella, and P. Nguyen, "Metabolic Study of an Industrial Solvent, Isophorone, in the Rabbit," *C. R. Acad. Sci. Ser. D,* **271,** 1333–1336 (1970).

252. R. Truhaut, H. Dutertre-Catella, P. Nguyen, "Study of the Toxicity of an Industrial Solvent, Isophorone. Irritating Capacity with Regard to the Skin and Mucous Membranes," *J. Eur. Toxicol.,* **5,** 31–37 (1972).

253. R. Truhaut, P. L. Nguyen, J. L. Cluet, and H. Dutertre-Catella, "Metabolic Transformations of 3,5,5-Trimethylcyclohexanone (Dihydroisophorone). New Metabolic Pathway Dismutation," *C. R. Acad. Sci. Ser. D.,* **276,** 2223–2228 (1973).

254. Unpublished data, Biochemical Research Laboratory, The Dow Chemical Co., Midland, Mich.

255. Unpublished data; Chemical Hygiene Fellowship, Mellon Institute, Pittsburgh, Pa.

256. Unpublished data, Health, Safety and Human Factors Laboratory, Eastman Kodak Co., Rochester, N.Y.

4800　　　　　W. J. KRASAVAGE, J. L. O'DONOGHUE, and G. D. DIVINCENZO

257. Unpublished data, Shell Chemical Company, Houston, Texas.

258. Unpublished data, Med. & Tox. Dept., Union Carbide Corp., New York, N.Y.

259. U.S. Dept. of Labor, Occupational, Safety, and Health Standards (29 CFR 1910.100). Code of Fed. Reg., Rev. Ed. Table Z-1, 1976.

260. M. M. Deacon, M. D. Pliny, J. A. John, B. A. Schwetz, F. J. Murray, H. O. Yakel, and R. A. Kuna,"Embryo- and Fetotoxicity of Inhaled Methyl Ethyl Ketone in Rats," *Toxicol. Appl. Pharmacol.*, **59**, 620–622 (1981)

261. H. Vainio and O. Hanninen, "Enhancement of Aniline p-Hydroxylation by Acetone in Rat Liver Microsomes," *Xenobiotica*, **2**, 259–267 (1972).

262. H. Vainio and A. Zitting, "Interaction of Styrene and Acetone with Drug Biotransformation Enzymes in Rat Liver," *Scand. J. Work Environ. Health*, **4**, Suppl. 2, 47–52 (1978).

263. B. L. Van Duuren, A. Sizak, and S. Melchiomne, "Cigarette Smoke Carcinogenesis—Importance of Tumor Promoters," *J. Natl. Cancer Inst.*, **47**, 235–240 (1971).

264. E. H. Vernot, J. D. MacEwen, and E. S. Harris, "Continuous Exposure of Animals to Methyl Isobutyl Ketone Vapors," U.S. Natl. Tech. Info. Ser., AD Rep.; ISS #751443, 1971, p. 11.

265. K. Verschueren, *Handbook of Environmental Data on Organic Chemicals*, Van Nostrand Reinhold, New York, 1977.

266. F. Viader, B. Lechevalier, and T. Morin, "Toxic polyneuritis in a Plastics Worker—Possible Implications of Methyl Ethyl Ketone," *Nouv. Tresse Med.*, **4**, 1813–1814 (1975).

267. E. C. Viglini and N. Zurlo, "Experiences of Aquanics del Lauor with Some Maximum Concentrations of Poisons of Industry at the Place of Work (MAK)," *Arch. Gewerbepathol. Gewerbehyg.*, **13**, 528–534 (1955).

268. W. F. von Oettingen, "Aliphatic Alcohols," *Public Health Bull. No.* **281**, 138 (1943).

269. A. P. Voronin, "Chronic Intoxication of Animals with Cyclohexanone Vapor," *Tr. Nauch. Sessii Leningr. Nauch.-Issledo- Inst. Gig. Tr. Profzabol. Leningr.*, 261–264 (1956).

270. D. C. Walton, E. F. Kehr, and A. S. Loevenhart, "A Comparison of the Pharmacological Action of Diacetone Alcohol and Acetone," *J. Pharmacol. Exp. Ther.*, **33**, 175–183 (1928).

271. L. G. Wayne and H. H. Orcutt, "The Relative Potentials of Common Organic Solvents as Percursors of Eye-Irritants in Urban Atmospheres," *J. Occup. Med.*, **2**, 383 (1960).

272. E. M. Weller and J. H. Griggs, "Hazards of Embryonic Exposures to Chemical Vapors," *Anat. Rec.*, **184**, 561 (1976).

273. C. W. Wickersham, III and E. J. Fredricks, "Toxic Polyneuropathy Secondary to Methyl *n*-Butyl Ketone," *Conn. Med.*, **40**, 311–312, (1976).

274. R. T. Williams, "Detoxication Mechanisms," in *The Metabolism and Detoxication of Drugs, Toxic Substances and Other Organic Compounds*, 2nd. ed., Chapman and Hall, London, 1959, p. 96.

275. W. P. Yant, F. A. Patty, and H. H. Schrenk, *Pentanone U.S. Public Health Rep.*, **51**, 392 (1936).

276. S. Zakhari, P. Levy, M. Liebowitz, and D. M. Aviado, "Acute Oral, Intraperitoneal, and Inhalation Toxicity of Methyl Isobutyl Ketone in the Mouse," in *Isopropanol and Ketones in the Environment*, L. Goldberg, Ed., CRC Press, Cleveland, 1977.

277. J. Ziegenmeyer, N. Reuter, and F. Meyer, "Local Anesthesia after Percutaneous Application, Part II," *Arch. Int. Pharmacodyn. Ther.*, **224**, 338–350 (1976).

278. A. Zlatkis and H. M. Liebich, "Profile of Volatile Metabolites in Human Urine," *Clin. Chem.*, **17**, 592 (1971).

CHAPTER FIFTY-SEVEN

Organic Phosphates

ROBERT J. WEIR, Ph.D., and
LLOYD W. HAZLETON, Ph.D.

1 GENERAL CONSIDERATIONS

The organic phosphates are most widely recognized in their usage as insecticides, and the bulk of this chapter is devoted to discussion of chemicals intended for this use. Their biocidal properties present appreciable toxicologic problems from the standpoint of manufacture and use. Organic phosphates are also used as gasoline additives, hydraulic fluids, cotton defoliants, fire retardants, plastic components, growth regulators, and industrial intermediates, where their highly toxic effect is neither desirable nor always apparent. In dealing with this group of chemicals, it is clear that the toxicity of the organic phosphate pesticide is not a universal characteristic of the organic phosphates as a class. One must only consider those that occur naturally in the body, such as the phospholipids, phosphonucleotides, and phosphoproteins, to observe that some of the class members are relatively nontoxic.

The importance of this class of chemicals as insecticides rests in their high biocidal activity and their short life as residues. The loss from the market of DDT (1) and other chlorinated hydrocarbon insecticides such as Aldrin, Dieldrin, Endrin, Heptachlor, and Chlordane because of their persistence and toxicity has enhanced the use of the organic phosphate insecticides.

The nomenclature of the organic phosphates frequently is confusing. Trade names, generic names, and manufacturers' experimental designations add to the confusion. For the purpose of this chapter, class names follow those outlined by Negherbon (2). The individual examples are listed as trade names or generic names depending on common usage, but in all cases both are given if used. The chemical name is also given in the listing of individual compounds.

1.1 Symptoms in Animals

The universal signs of intoxication of the organic phosphates that are insecticidal appear to result from the inhibition of the cholinesterases (these esterases hydrolyze acetylcholine, butyrylcholine, benzoylcholine, acetyl-β-methylcholine, etc., depending on the species). It is important to mention that this is not an exclusive function of the organic phosphate insecticides, for it is shared with the class of insecticides known as carbamates and typified by Sevin®.

Following excessive exposure, the signs of toxicity reflect stimulation of the autonomic and central nervous systems, resulting from inhibition of acetylcholinesterase and consequent accumulation of acetylcholine. Prolongation and intensification of the acetylcholine action results in two degrees of response, depending on dosage and specific action of the inhibitor. The initial action is on smooth muscles, cardiac muscle, and exocrine glands and, in general, is comparable to stimulation of the postganglionic parasympathetic nerves. This phase of action results in the early signs of toxicity resembling those of muscarine and hence is referred to as the muscarinic action of acetylcholine. The action, and hence the signs, can be counteracted by atropine. The most common early signs are intestinal cramps, tightness in the chest, blurred vision, headache, diarrhea, decrease in blood pressure, and salivation.

The second stage of intoxication results from stimulation of the peripheral motor system and of all autonomic ganglia. Experimentally, these actions can be counteracted by curare and ganglionic blocking agents and in other respects resemble the classical action of nicotine; hence they are referred to as the nicotinic action of acetylcholine. The complexity of toxic action during the second stage includes neuromuscular and ganglionic blockade; thus curare therapy would be contraindicated. Ultimately the toxic manifestations of poisoning are referable to stimulation and/or paralysis of the somatic, autonomic, and central nervous systems. A more detailed understanding of this complex action and the influence of adequate atropinization may be obtained from Goodman and Gilman (3). A basic review in terms of health problems has been published by Hazleton (4). The role of cholinesterase activity is reviewed by Wills (5).

The signs of toxicity outlined above do not apply to animals that receive small doses over a long period of time. In this case, the correlation of signs of toxicity and inhibition of the cholinesterase(s) is not clear-cut. With many organic phosphates, inhibition in rats may be so complete in plasma and red blood cells that the cholinesterase activity is immeasurable with present techniques, and the brain activity may be markedly inhibited, although the animals appear normal in all respects. With other members of the organic phosphate class, chronic exposure in rats produces marked inhibition of plasma and red cell cholinesterase and moderate inhibition of brain activity, and results in diarrhea and tremors as the only toxic signs. These examples point out the variability of signs of toxicity, which depend on the toxicant, vehicle, route of administration, dosage, species studied, specific enzyme system on which the toxicant acts (true or pseudocholinesterase), metabolic conversion products, reversibility of the inhibition, etc. These variables and their ultimate discussion are beyond the scope of this book and only those factors that have bearing on industrial hygiene are covered here.

1.2 Gross Pathology in Animals

The greatest bulk of the organic phosphate insecticides do not produce morphologic alterations in animals. Some of the more recent additions to the group are chlorinated and are suspected of being capable of producing liver and kidney damage similar to that produced by the halogenated hydrocarbons. There is almost no literature to support the supposition. Even the organic phosphates that have been fed to rodents at maximum tolerated doses according to NCI carcinogenesis protocol are not carcinogens (6–12).

It has long been known that certain organic phosphates are capable of producing a delayed paralysis in animals and man; this effect is the result of a syndrome resembling "jake leg" or "ginger jake" paralysis, which has been described for Jamaica ginger poisoning. Early investigators described the associated histopathology as demyelination of the peripheral nerves; this terminology has been widely used, as is evidenced by the following reports. Paralytic effects of TOCP (tri-o-cresyl phosphate) have been described by Smith et al. (13, 14). More recent reports on this phenomenon after administration of TOCP are reported by Durham et al. (15), Barnes and Denz (16), Hine et al. (17), and Frawley et al. (18). Durham et al. (15) also studied Chlorthion® (O,O-dimethyl-O-3-chloro-4-nitrophenyl thionophosphate), DDVP (dimethyl 2,2-dochlorovinyl phosphate), Systox [O,O-diethyl-O-2-(ethylmercapto)ethyl thionophosphate], Diazinon® [O,O-diethyl-O-(2-isopropyl-6-methyl-4-pyrimidyl) phosphorothioate], OMPA (octamethyl pyrophosphoramide), EPN (ethyl-p-nitrophenyl benzenethionophosphonate), malathion [O,O-dimethyl S-(1,2-dicarboethoxyethyl) dithiophosphate], and Isopestox® [bis(monoisopropylamino) fluorophosphine oxide]. Of these, only Isopestox® was found to produce the demyelination syndrome. Later, Frawley et al. (18) showed that EPN also produced demyelination, contrary to the findings of Durham et al (15); Barnes and Denz (16) and Austin and Davies (19) demonstrated that DFP (diisopropyl fluorophosphate) also produced demyelination in animals. The symptomatic observations have been confirmed with histologic evidence in all cases (DFP, EPN, TOCP, and Isopestox®).

Later investigations revealed that the causative lesion for the delayed paralysis syndrome was not demyelination, but rather a degeneration of the axons in the spinal cord and peripheral nerves. A complete review of this subject, with references, is available as the proceedings of a conference sponsored by the

U.S. Environmental Protection Agency (20). An even newer review of this subject can be found in the extensive work on neurotoxicity by Spencer and Schaumberg (21). There are several important aspects to be considered in regard to this syndrome as a feature of organic phosphate poisoning: it is always delayed, never acute; it is not due to cholinesterase inhibition; it is structure specific and not caused by all organic phosphates; there is considerable species susceptibility variation; it may follow a single massive dose (if cholinergic effects are antidoted) or may result from the cumulative action of repeated small exposures; there is no known effective treatment; mild cases may slowly return to normal while in severe cases the paralysis is usually permanent; and finally, there are well-documented cases of delayed neuropathy due to organic phosphates in man.

1.3   Exposure in Man

The most common, and most important, route of industrial exposure to the organic phosphates is by accidental spillage on the skin. Most of the materials later discussed in detail are rapidly absorbed through the skin. Percutaneous absorption frequently is unnoticed since dermal irritation rarely occurs, unless the solvent systems of the formulated materials possess this irritative property.

The second most frequent exposure route is through the respiratory tract. Intoxication may occur with some of the more highly toxic members of the group, such as TEPP (tetraethyl pyrophosphate) and Phosdrin® (alpha isomer of 2-carbomethoxy-1-methylvinyl dimethyl phosphate), but in general it is agreed that exposure is due to particulate matter rather than to vapor (4). The organic phosphates as a group have extremely low vapor pressures; despite this, Kay and co-workers (22) give analytical values for parathion (vapor pressure 3.78×10^{-5} mm Hg) in the air of treated orchards for up to 3 weeks after application. Summerford et al. (23) correlated blood cholinesterase level with symptomatology in plant personnel, orchard workers, and other groups during and after a spray season. Inhalation exposure was not distinguished from dermal exposure. Fatal or near-fatal illnesses have resulted from brief, massive exposure to parathion due to gross carelessness rather than repeated exposure.

Oral exposure is rarely a problem except for accidental ingestion by children and in the case of suicide. The more toxic members of the organic phosphate group may be an ingestion problem in manufacturing and spraying operations if good personal hygiene practices are not followed.

The organic phosphate insecticides share the biologic action of inhibiting cholinesterase(s). Although this is generally recognized not to be the sole toxic action, it does provide the toxicologist and industrial hygienist with an excellent tool for the measurement of exposure of animals or workers to the toxicant. This measurement of exposure serves as a warning of impending toxicity and is useful in prophylactic programs. Beyond this its reliability, either diagnostic or prognostic, is of little value.

As noted above, the delayed neuropathy caused by organic phosphates is not due to inhibition of cholinesterase, and hence the assay of that enzyme level is not an indicator of risk. A "neurotoxic esterase" (NTE) has been described (24); this enzyme is believed to be responsible for the delayed neurotoxicity or paralytic effect displayed by some organophosphates.

For practical reasons, the most widely used method for determination of cholinesterase inhibition appears to be that of Michel (25) or some modification of it. The manometric method of Ammon (26) is also reliable. The colorimetric method of Metcalf (27) and the modified method for whole blood described by Fleisher and Pope (28) are also useful, and the field kits such as described by Edson (29) are based on this method. More recently, many investigators have been using the colorimetric method of Ellman (30) because it can be used in enzyme-substrate complexes where the action is rapidly reversible. This is more important with carbamates but can be true with organic phosphates. When the cause of intoxication is unknown, this is an extremely important consideration. A kit has been developed by Boehringer Mannheim for this method.

Using the electrometric technique, Wolfsie and Winter (31, 32) have evaluated the plasma and red blood cell cholinesterase levels of men and women who were not exposed to organic phosphate insecticides (ingestion of residues from treated crops could not be eliminated). Mean values were as follows:

| Red blood cell | 0.67 to 0.86 Δ pH units/hr |
| Plasma | 0.70 to 0.97 Δ pH units/hr |

In the prophylactic program, both red blood cell and plasma values should be obtained. Measurement should be made frequently and with regularity. The workers' previous values should be available for inspection and comparison. In addition to knowledge of the specific action of the compound, judgment and experience are essential to adequate interpretation of results. Marked or severe depression of either the plasma or red blood cell value may be considered strong evidence of exposure, whether accompanied by gross signs or not. The red cell level is more significant, since it represents the true neurohormone esterase level for humans, acetylcholinesterase. Plasma enzyme inhibition is less specific but may be important as a diagnostic aid in acute exposure since it usually, but not always, responds more rapidly and at lower dosage. Whole blood analysis is least reliable since it may give a composite effect and may mask the individual level of either the plasma or cells.

In the event exposure occurs, the symptoms in man are qualitatively similar to those described for animals and include headache, vertigo, blurred vision, lacrimation, salivation, sweating, muscular weakness and ataxia, dyspnea, diarrhea, abdominal cramps, vomiting, coma, pulmonary edema, and death.

Although the acute signs in man may at times resemble paralysis, this is not to be confused with the possibility of delayed neuropathy and paralysis which would occur only after eight or more days following acute exposure. Commercial organic phosphate pesticides and other chemicals have generally been screened for neurotoxic action and do not present a hazard under ordinary use conditions. In cases of severe exposure which require intensive control of the cholinergic signs and symptoms, it would be well to carefully follow the patient for some time after the acute phase.

1.4 Treatment

The onset of symptoms is rapid and maximum effects may develop within a few hours. It is thus important that medical care be obtained without delay. Since the early symptoms of headache, malaise, and so on, are easily confused with other diseases, it is important that workers exposed to the organic phosphates be instructed to report any such indications.

Adequate atropinization is essential to relieve the muscarinic effects, and to provide central respiratory stimulant action. An average adult may require 12 to 24 mg total dose of atropine intravenously during the first 24 hr. Since this is far in excess of the usual therapeutic dose, the physician unacquainted with the mutually antagonistic action of this drug and the organic phosphate may be hesitant to employ such large doses. A general rule is that atropine should be administered until visible effects of atropinization are observed. Since, as pointed out above, the muscarinic effects are only a part of the action produced by heavy exposure, it is essential that the patient be treated symptomatically with artificial respiration, postural drainage, warmth, etc. Prognosis depends largely upon the exposure, type of compound, and adequacy of treatment. Care should be exercised until the patient is obviously free of any signs of toxicity. A detailed survey of this subject is presented by Gordon and Frye (33) and has been more recently updated by Clyne and Shaffer (34).

Since the toxicity of the organic phosphates is due to the inhibition of cholinesterase, the reactivation of these enzymes would offer great promise as a therapeutic measure. This action is apparently achieved by pyridine-2-aldoxime methiodide (PAM, PAM-2, 2-PAM, P-2-AM), diacetyl monoxime (DAM), and other oximes. Namba and Hiraki (35) report both experimental and clinical investigations of PAM. In cases of parathion poisoning, they recommended intravenous doses of 1 g of PAM or more, if indicated.

Grob and Johns (36) give a step-by-step outline of combined therapy for organic phosphate intoxication. After removal from exposure, a patent airway and artificial respiration should be established. The therapeutic regimen includes atropine, 2 to 4 mg intravenously, repeated frequently until muscarinic symptoms disappear. PAM or DAM is recommended in doses of 2000 mg intravenously.

The industrial hygiene physician should contact the local poison control center if there is a likelihood of organic phosphate intoxication. The most recent advice on the use of oximes can thus be obtained. They are, however, in no way a substitute for atropine.

2 SPECIFIC COMPOUNDS

These compounds are representative of the organic phosphates used to the greatest degree, but this list is not all-inclusive. Information on methods for determination in the atmosphere has come from methods for crop residues which emphasize the need for cleanup. Since air samples do not usually have pigments and other interfering substances, the cleanup steps in these methods can be eliminated. If the method employs gas or high pressure liquid chromatography, or similar modern analytical technique, there is all the more reason to eliminate cleanup.

2.1 Abate®; O,O,O',O'-Tetramethyl O,O'-Thiodi-p-phenylene Phosphorothiate, Temephos, Biothion®

2.1.1 Source, Uses, and Industrial Exposures

The major use of temephos is in the public health field for control of mosquito larvae and a number of adult flies and biting midges. The product's low toxicity to birds, fish, and other beneficial species is an advantage.

2.1.2 Physical and Chemical Properties (37)

Physical state	Yellow to brown, viscous liquid (technical)
Purity	90% (technical)
Boiling point	Decomposes at 120 to 125°C
Melting point	10 to 15°C
Refractive index, $n_D^{25°C} =$	1.586 to 1.588
Vapor pressure	7.17×10^{-8} mm Hg at 25°C
Solubility	Soluble in acetonitrile, ethylene dichloride, toluene; insoluble in hexane, methyl cyclohexane, water

2.1.3   Determination in the Atmosphere

A method is provided by Pasarela and Orloski (38) that may be suitable for determination in the atmosphere.

2.1.4   Physiologic Response

Acute.   The acute oral LD_{50} of Abate® (technical) is 2030 and 2300 mg/kg in male and female rats, respectively (34). The acute dermal LD_{50} is 1930 and 970 mg/kg, respectively, for male and female rabbits. It is not an irritant either in the eye or on the skin of rabbits. Rats showed no toxic signs when exposed to 1 hr of saturated vapor in an inhalation (whole body) exposure.

Subchronic and Chronic.   Male rats fed dietary levels of 250, 500, and 1000 ppm Abate® for 30 days resulted in significant death at 1000 ppm; growth and food consumption were reduced at 500 ppm. Tremors were apparent at all levels. Cholinesterase was depressed at all levels. There was no gross pathology at necropsy.

In 90-day dietary feeding studies male and female rats survived dosages of 350 ppm. A dietary level of 2 ppm produced no effects in either sex. At 6 and 18 ppm red cell cholinesterase inhibition was observed (only males were affected at the lower dose).

In a second 3-month study in rats at dietary levels of 0, 6, 18, or 54 ppm there was no cholinesterase depression at 6 ppm. Although 18 ppm produced slight red cell cholinesterase depression, the red cell effect returned to normal levels in a group of rats fed 18 ppm and then placed on a control diet for 2 weeks. At 54 ppm, brain, red cell, and plasma activity were moderately depressed.

Dogs were fed dietary levels of Abate® at 2, 6, and 18 ppm and a fourth level of 700 ppm to assure no pathologic tissue changes would occur. No deaths occurred at any level. There were no changes in weight or food consumption. Signs of cholinergic stimulation occurred at 700 ppm but lower doses were without signs. The high dose was accordingly reduced to 500 ppm where red cell and plasma levels were almost completely inhibited, without toxic signs. The lower dose groups showed no anticholinesterase activity, and no histopathology was observed at any level.

In a 21-day repeated dermal study in rats at 12 and 60 mg/kg there were no observed effects.

Neurotoxicity studies in hens fed 920 ppm Abate® (approximately one-fourth the single dose LD_{50}) as a daily dose for 30 days resulted in no neurologic pathology. This test is designated to determine potential for peripheral neuropathy.

Three-generation reproduction studies at 25 or 125 ppm in albino rats resulted in no effect on fertility, gestation, reproduction, or lactation. No terata associated with Abate® were produced in either dosage group.

2.1.5   Hygienic Standard of Permissible Exposure

The American Conference of Governmental Industrial Hygienists (39) has adopted a time-weighted average (TWA) of 10 mg/m^3 for Abate® and a tentative value for a short-term exposure limit (STEL) of 20 mg/m^3.

2.2 Acephate; O,S-Dimethyl Acetylphosphoramidothioate, Orthene®

2.2.1   Source, Uses, and Industrial Exposure (40, 41)

Acephate has moderate persistence, with residual systemic activity of 10 to 15 days, and is used to control insects and aphids in ornamentals, where it has a reasonably broad spectrum. It is also cleared for use on beans, cotton, head lettuce, celery, soybeans, and bell peppers. It controls parasites of cattle, goats, hogs, horses, poultry, and sheep, where tolerances have been set for milk, eggs, fat, and meat.

2.2.2   Physical and Chemical Properties (40)

Physical state	White solid (technical)
Specific Gravity	1.35
Volatility	Low, 1.7×10^{-6} mm Hg (24°C)
Melting point	82 to 89°C (technical)
Stability	Relatively stable; store in cool place
Solubility	Good in water (approximately 65%); relatively low in organic solvents (less than 5% in aromatic solvents; over 10% in acetone and alcohol)

2.2.3   Determination in the Atmosphere

The method used (42) is for multiple pesticides, but this should not be a problem with air analysis because no cleanup is necessary; the method can be applied as is to the trapped acephate (43).

2.2.4   Physiologic Response (40)

Acute.   The technical material has an acute oral LD$_{50}$ in rats of 945 mg/kg (female) and 866 mg/kg (male), and 361 mg/kg in mice. The minimum effective oral dose for emesis in the dog is 215 mg/kg and the MLD (minimum lethal dose) in the dog is 681 mg/kg. Eye exposure in the rabbit results in slight conjunctival irritation which clears in a week. The acute dermal LD_{50} in the rabbit is greater than 2000 mg/kg body weight. Dermal irritation studies in the rabbit and sensitization studies in the guinea pig are without effect. Four-hour acute inhalation studies with the rat confirmed the low volatility of acephate, for no mortality, morbidity, or change in cholinesterase values occurred.

Subacute and Chronic. In 90-day subacute toxicity studies using the rat, 300 ppm revealed no change in weight gain, food consumption, survival, blood and urine values, gross and microscopic pathology, or organ weights and ratios.

In 90-day rat and dog subacute studies, feeding at doses of 10 ppm, no reduction of cholinesterase activity was observed, and this dose is considered to be the no-effect dose in both species.

In teratologic studies in rats and mice, no effect was noted. Mutagenic effects were absent in a dominant lethal study.

Chronic 2-year dog and rat studies conducted at 100 ppm showed no gross or histologic changes. In the dog, cholinesterase was slightly depressed at 100 ppm, but in the rat 30 ppm caused slight to moderate cholinesterase inhibition. Growth was slightly depressed in the rat at 100 ppm.

Atropine sulfate is antidotal.

2.2.5 Hygienic Standard of Permissible Exposure (39)

The American Conference of Governmental Industrial Hygienists has not yet set a threshold limit value for acephate.

2.3 Azinphos-methyl; O,O-Dimethyl S-[[4-Oxo-1,2,3-benzotriazin 3(4H)-yl] methyl] Phosphorodithioate, Guthion®, Bay 17147, R1582

2.3.1 Source, Uses, and Industrial Exposure (41)

Azinphos-methyl or Guthion® is a broad-spectrum insecticide for which EPA has established tolerances on a number of crops such as grasses, fruit, vegetables, milk, meats, and meat by-products and fibers. The tolerances range from 0.1 to 10 ppm depending on usage and crop.

2.3.2 Physical and Chemical Properties (44)

Physical state	Brown waxy solid (technical)
Molecular weight	317

Melting point	73°C
Solubility	Soluble in water to about 29 ppm at 25°C; soluble in most organic solvents except aliphatics
Stability	Decomposes at elevated temperature with gas evolution. Subject to hydrolysis
Refractive index, $n_D^{76°C}$	1.6115

2.3.3 Determination in the Atmosphere (42)

Azinphos-methyl may be analyzed by collecting a sample in a series of scrubbers containing a suitable solvent. The method as described by Thompson (42) may then be applied.

2.3.4 Physiologic Response (44)

Acute. The acute oral LD_{50} in the rat is 13 to 16.4 mg/kg for the technical material. The dermal LD_{50} is 200 mg/kg.

Atropine sulfate in large therapeutic doses is antidotal. Repeat as necessary to the point of tolerance. 2-PAM and Torogonin® (Merck) are also antidotal and may be administered in conjunction with atropine.

2.3.5 Hygienic Standard of Permissible Exposure

The American Conference of Governmental Industrial Hygienists (39) has established a time-weighted average (TWA) exposure level of 0.2 mg/m³ for skin.

2.4 Chlorthion®; O,O-Dimethyl O-3-Chloro-4-nitrophenyl Thionophosphate, Bayer 22/190

2.4.1 Source, Uses, and Industrial Exposures

Chlorthion® controls a wide range of agricultural and household insects, including houseflies, mosquitoes, and roaches. Chlorthion® is no longer in common usage.

2.4.2 Physical and Chemical Properties (45)

Physical state	Yellowish brown liquid
Molecular weight	295.5
Density	1.433 (20°C)

Boiling point	112°C (0.04 mm Hg)
Vapor pressure	7.0×10^{-6} mm Hg (30°C)
Refractive index, $n_D^{20°C}$	1.5680
Solubility	1:25,000 in water; readily miscible with benzene, toluene, alcohols, ethers, oils; unstable in alakaline solutions
Flash point and flammable limits	Not flammable at normal temperature; thermal decomposition point about 150°C

2.4.3 Determination in the Atmosphere

See parathion, Section 2.17 (pure material for color standards available from Chemagro Corporation).

2.4.4 Physiologic Response

Acute. The acute oral LD_{50} in rats is 1500 mg/kg, and the intraperitoneal LD_{50} is 750 mg/kg (46). The dermal application of approximately 1400 mg/kg to rabbits produced no signs of toxicity.

Subacute and Chronic. Daily administration of Chlorthion® in the diet of groups of rats at 50, 100, and 200 mg/kg/day resulted in cholinesterase inhibition at all levels. The two lower dosages produced 50 percent cholinesterase inhibition. Forty percent mortality resulted at 100 mg/kg/day over the 60-day feeding period. A dosage of 200 mg/kg/day produced 75 percent cholinesterase inhibition and was not tolerated for more than 5 to 10 days. The 50 mg/kg/day dosage was tolerated without mortality for 60 days (2, 46).

Chlorthion® is an active cholinesterase inhibitor in vitro. Low mammalian toxicity is ascribed to slow absorption (2). As a typical member of the organic phosphate group, no pathology has been described in animals receiving repeated high dosages.

2.4.5 Hygienic Standard of Permissible Exposure

No threshold limit values have been established for Chlorthion® (39).

2.5 DDVP; Dimethyl 2,2-Dichlorovinyl Phosphate, Vapona®

2.5.1   Source, Uses, and Industrial Exposures

DDVP is used almost exclusively as a quick, knockdown agent for the control of houseflies. Its action depends on its fumigant action as it is a poor stomach and contact poison.

2.5.2   Physical and Chemical Properties (45, 47)

Physical state	Oily liquid
Molecular weight	221.0
Density	1.415 (25°C)
Boiling point	84°C (1 mm Hg)
Vapor pressure	0.01 mm Hg (30°C)
Refractive index, $n_D^{25°C}$	1.451
Solubility	Miscible with alcohol and most nonpolar solvents; 1.0 percent in water and 0.5 percent in glycerin at room temperature
Flash point and flammable limits	Practically nonflammable

2.5.3   Determination in the Atmosphere

A micro method, which could be adapted to air analysis, has been published by Giang et al. (48).

2.5.4   Physiologic Response

Acute.   The acute oral LD_{50} is 80 mg/kg in male rats and 55 mg/kg in female rats (49). The acute dermal LD_{50} is 107 mg/kg in male rats and 75 mg/kg in female rats.

Subacute and Chronic.   Edson (50) has evaluated the cumulative effects of DDVP by administering to rats repeated one-half LD_{50} doses in rapid succession over a short period of time. The reversible nature of the cholinesterase inhibition produced is unique among organic phosphates.

2.5.5   Hygienic Standard of Permissible Exposure (39)

A threshold limit has been adopted for DDVP. The threshold limit value (TLV) is 0.1 ppm or 1.0 mg/m^3. The short-term exposure level (STEL) is 0.3 ppm or 3 mg/m^3.

2.6 Delnav; 70% of 2,3-*p*-Dioxanedithiol *S,S*-bis(*O,O*-diethyl Phosphorodithioate) and 30% related compound, Dioxathion, Hercules AC528, Navadel®

2.6.1 Source, Uses, and Industrial Exposures

Delnav is an insecticide-miticide used on citrus, stone fruit, grapes, and walnuts. It is also used for control of ticks, lice, and horn fly on cattle, goats, hogs, horses, and sheep when sprayed or dipped.

2.6.2 Physical and Chemical Properties

Physical State	Nonvolatile tan liquid
Melting point	$-20°C$
Solubility	Insoluble in water; soluble in aromatic hydrocarbons, ethers, esters, ketones
Refractive index, $n_D^{20°C}$	1.5420

2.6.3 Determination in the Atmosphere

A method is provided in the compendium by Thompson and Watts (42) that is suitable for determination in the atmosphere.

2.6.4 Physiologic Response

Acute. The acute oral LD_{50} of delnav has been investigated for a number of species including rats, dogs, and mice; the oral LD_{50} ranged between 43 mg/kg and 176 mg/kg (51). The acute oral toxicity of 15 other pharmacologically related pesticides has also been studied for evidence of potentiation. Delnav does not potentiate the acute toxicity of the other pesticides. The percutaneous LD_{50} in rats and rabbits ranges between 63 and 235 mg/kg. The 1-hr LC_{50} for mice and rat by inhalation is 340 and 398 µg/l, respectively. Instillation in the eye of rabbits of 0.1 ml concentrated delnav, as well as 5 and 25 percent solutions in corn oil, produced no corneal damage and only mild transient conjunctivitis.

Subacute and Chronic. Subacute exposure of chickens to delnav did not produce myelin degeneration under conditions that are positive for known degenerators.

Morphologic changes could not be produced in rats at 90-day dietary levels as high as 100 ppm delnav. Rats fed delnav in the diet at various doses for 90 days showed no plasma, brain, or erythrocyte inhibition at 1.0 and 3.0 ppm. Dogs administered delnav orally for 90 days showed no cholinesterase depression at doses of 0.075 mg/kg and below. Human volunteers have tolerated daily oral doses of delnav for 5 weeks at 0.075 mg/kg without plasma, erythrocyte cholinesterase depression, or any other toxic effect.

2.6.5 Hygienic Standard of Permissible Exposure

A threshold limit value has been established for delnav by the American Conference of Governmental Industrial Hygienists (39). The time-weighted average (TWA) for occupational exposure is 0.2 mg/m^3 for skin.

2.7 Diazinon®; O,O-Diethyl O-(2-Isopropyl-6-methyl-4-pyrimidyl) Phosphorothioate, G-24480

2.7.1 Source, Uses, and Industrial Exposures

Diazinon® is a broad-spectrum insecticide and acaricide. It has also received much use in the control of cockroaches, particularly those resistant to chlorinated hydrocarbon pesticides.

2.7.2 Physical and Chemical Properties (45)

Physical state	The pure material is a colorless liquid; the technical material is a pale to dark brown liquid
Molecular weight	340.4
Density	1.116 to 1.118 (25°C)
Boiling point	83 to 84°C (0.002 mm Hg)
Vapor pressure	1.4 × 10^{-4} mm Hg (20°C)
Refractive index, $n_D^{20°C}$	1.4978 to 1.4981
Solubility	0.004 percent in water at room temperature; miscible with alcohol, xylene, acetone, petroleum oils
Flash point and flammable limits	Practically nonflammable

2.7.3 Determination in the Atmosphere

A method of analysis of Diazinon® has been described by Harris (52) and can be adapted to air analysis.

2.7.4 Physiologic Response

Acute. The acute oral LD_{50} of 95 percent technical Diazinon® in rats is 100 to 150 mg/kg. The LD_{50} of a 23 percent wettable powder is 264.5 mg/kg on the basis of active ingredient (53, 54). This discrepancy was later explained by Gysin and Margot (55) as resulting from the formation of a number of possible isomerization and decomposition products resulting from the technical form. Presumably the responsible form is monothiono-TEPP (56). Formulation prevented the degradation.

Subacute and Chronic. Male and female rats receiving 100 and 1000 ppm (weight) technical Diazinon® in the diet for 4 weeks showed no gross signs of intoxication, alteration of growth, or gross pathology at autopsy. Red blood cells and brain cholinesterase levels were significantly inhibited at both dosages. Plasma activity at both dosages was comparable to the control.

Rats received 10, 100, and 1000 ppm (weight) active Diazinon® as a wettable powder in the diet for 72 weeks with no apparent gross signs of toxicity. Dogs received orally various doses of active Diazinon® as a wettable powder for 46 weeks. No pathology, gross or microscopic, was observed at the lowest dosage (4.6 mg/kg/day) in 2 weeks. After 12 weeks cholinesterase inhibition was complete at the lowest dosage. At a dosage of 9.3 mg/kg/day for 5 weeks signs of toxicity and complete cholinesterase inhibition were observed. Withdrawal of Diazinon® at the highest dosage resulted in reversal of signs and regeneration of cholinesterase activity to normal limits after 2 weeks.

2.7.5 Hygienic Standard of Permissible Exposure

A threshold limit has been adopted for Diazinon®; on skin the adopted TWA value is 0.1 mg/m^3 and the STEL is 0.3 mg/m^3 (39).

2.8 Dimethoate; O,O-Dimethyl S-(N-Methylcarbamylmethyl) Phosphorodithioate, Cygon®, Rogor®

$$[CH_3O]_2 \overset{\overset{\text{S}}{\|}}{P}-S-CH_2CONHCH_3$$

2.8.1 Source, Uses, and Industrial Exposures

Dimethoate is a very low toxicity pesticide which can be used on fruits, vegetables, cereals, coffee, cotton, olives, rice, tea, tobacco, etc. (57). The EPA has set tolerances of 2.0 ppm in many cases.

2.8.2 Physical and Chemical Properties

Molecular weight	229.3
Color and state	White crystalline solid
Flash point	130 to 132°C
Melting point	43.5 to 45.8°C
Refractive index	1.5377 at 50°C
Specific gravity	1.28 at 20°C
Solubility	Soluble in most organic solvents; soluble in water (7 percent at 80°C); very slightly soluble in aliphatic hydrocarbons
Stability	Unstable upon heating and at low pH; decomposes above 170°C

2.8.3 Determination in the Atmosphere

The residues in the atmosphere may be captured by drawing a known quantity of air through several scrubbers containing a useful solvent. The entrapped dimethoate may be analyzed by the method of Chilwell and Beecham (58) without cleanup.

2.8.4 Physiologic Response (34)

Acute. The acute oral LD_{50} of technical dimethoate was 280 mg/kg for the male and 240 mg/kg for the female rat. In the mouse these figures are 510 and 220 mg/kg for the male and female, respectively. The dermal LD_{50} for the male rat and guinea pig is >800 mg/kg. Dimethoate is not an irritant in the eyes or on the skin of rabbits.

Subacute and Chronic. Dietary levels of 2, 8, and 32 ppm were tolerated by rats for 90 days without toxic signs or effects on cholinesterase activity. Diets containing 50 ppm dimethoate for 33 days produced decreased plasma, red cell, and brain cholinesterase activity.

Repeated feeding of dimethoate to dogs for 13 weeks at 2, 10, and 50 ppm produced no effect at 2 ppm, slight red cell cholinesterase depression at 10 ppm and further suppression at 50 ppm. Dogs fed 1500 ppm or more for 13 weeks were without effect with regard to gross or microscopic pathology.

Hens fed 130 ppm dimethoate for 4 weeks showed no effect on the neurologic system.

Three-generation reproduction studies in mice at 5, 15, or 50 ppm were carried out. There were no alterations of fertility, gestation, viability, lactation, tissue pathology, or fetus morphology that could be related to the dimethoate treatment.

2.8.5 Hygienic Standard of Permissible Exposure

No threshold limit value has been established for dimethoate by the American Conference of Governmental Industrial Hygienists (39).

2.9 Dipterex®; O,O-Dimethyl 2,2,2-Trichloro-1-hydroxyethyl Phosphonate, Bayer L 13/59

2.9.1 Source, Uses, and Industrial Exposures

Dipterex has been used to control houseflies. It has shown promise in agriculture against lepidopterous and dipterous insects and mites.

2.9.2 Physical and Chemical Properties (45)

Physical state	White to pale yellow crystalline solid
Molecular weight	275.5
Density	1.73 (20°C)
Melting point	78 to 80°C
Boiling point	120°C (0.4 mm Hg)
Vapor pressure	Volatile
Refractive index, $n_D^{20°C}$	1.3439 (10 percent aqueous solution)
Solubility	Soluble in water to 13 to 15 percent at 25°C; soluble in alcohols, benzene, toluene, chloroform; slowly unstable in water; decomposition speeded by heat or alkali; decomposition product DDVP
Flash point and flammable limit	Practically nonflammable

2.9.3 Determination in the Atmosphere

A method suitable for air analysis of Dipterex® has been published by Giang et al. (59).

2.9.4 Physiologic Response

Acute. The acute oral and intraperitoneal LD_{50} in rats has been reported by DuBois (60) as 450 and 225 mg/kg, respectively. The dermal LD_{50} is 72000 mg/kg.

Subacute and Chronic. Daily intraperitoneal dosages of 100 mg/kg (more than one-fourth of the LD_{50} dose) produced 40 percent mortality of treated rats in 60 days. Duration of toxic action was brief; complete recovery occurred in a few hours after dosing.

2.9.5 Hygienic Standard of Permissible Exposure

No threshold limit or values have been set for Dipterex® (39).

2.10 EPN; Ethyl p-Nitrophenyl Benzenethionophosphonate

2.10.1 Source, Uses, and Industrial Exposures

EPN has been used as an insecticide and acaricide. It has shown a broad spectrum of activity against mites and insects. Present use in agriculture is limited.

2.10.2 Physical and Chemical Properties (45)

Molecular weight	323.3
Density	1.268 (25°C)
Melting point	36°C
Vapor pressure	3.0×10^{-4} mm Hg (100°C)
Refractive index, $n_D^{30°C}$	1.5978
Solubility	Practically insoluble in water; soluble in most of the common organic solvents; stable at ordinary temperatures and in neutral and acid media

2.10.3   Determination in the Atmosphere

The method of Averell and Norris (61) for parathion is applicable to EPN. This method has been modified by Gunther and Blinn (62) and more recently by Wilson et al. (63).

2.10.4   Physiologic Response (64)

Acute.   The acute oral LD_{50} for pure EPN is 42 and 14 mg/kg for male and female rats, respectively; for the technical materials, values of 28 to 33 and 7 to 13 mg/kg are given by Hodge et al. (64).

Subacute and Chronic.   In 2-year chronic feeding studies in rats, doses of 150 ppm for males and 75 ppm for females produced no effect. Doses of 2.0 mg/kg/day were administered to male and female dogs for 1 year without effect. EPN has been found to potentiate the effects of malathion (details under malathion, Section 2.14).

2.10.5   Hygienic Standard of Permissible Exposure

The threshold limit for EPN as a dust, fume, or mist is 0.5 mg/m^3 (skin) (39).

2.11   Fonofos; *O*-Ethyl *S*-Phenyl Ethylphosphonodithioate, Dyfonate®

2.11.1   Source, Uses, and Industrial Exposures (41)

Fonofos is an organic phosphate with negligible tolerances (0.1 ppm) set for sorghum, soybean, spearmint, peanuts, peas, peppermint, corn, and beans (as both the crop and its hay), as well as for asparagus, sugar beets, strawberries, and vegetables (fruit, leafy, root crop, seed, and pod).

2.11.2   Physical and Chemical Properties (44)

Physical state	liquid
Flash point	>200°F (Tagliabue closed cup)
Specific gravity	1.154 at 20/20°C
Solubility	13 ppm in water at 20°C
Vapor pressure	0.21 μm Hg at 25°C

| Boiling point | 100°C at 0.3 mm Hg |
| Chemical reactivity and stability | No known violent chemical reactions have been reported; normally stable at ambient conditions; does not react with water. |

2.11.3 Determination in the Atmosphere

A known volume of air can be trapped in a series of scrubbers containing a suitable solvent. The method for fonofos (65) may then be applied without further cleanup.

2.11.4 Physiologic Responses (66)

Acute. The acute oral LD_{50} of fonofos in rats ranges between 4 and 43 mg/kg with signs of toxicity of organic phosphate esters. The acute dermal LD_{50} in rabbits is 32 to 261 mg/kg with marked signs produced from single doses of 21.5 or 46.4 mg/kg. The acute dermal LD_{50} in rats is 147 mg/kg and in the guinea pig, 278 mg/kg. By the standard Draize skin irritation technique, fonofos is not irritating, but in one of three such tests fonofos was lethal. Application to the eye of rabbits also produced death.

The acute LC_{50} in rats was 0.46 and 0.9 mg/l for 4- and 1-hr exposures, respectively.

Subacute and Chronic. In a 90-day dog study, daily ingestion of 0.4 to 6.0 mg/kg fonofos resulted in plasma, red cell, and brain cholinesterase inhibition and moderate toxic signs. In 2-year dog studies with daily ingestion of 0.4 to 6.0 mg/kg, in addition to cholinesterase inhibition there was increased liver weight, congestion of the small intestine, decreased weight gain, soft stools, alopecia, increased nasal, salivary, and lacrimal secretions, nervous behavior, tremors, increased serum alkaline phosphatase, and altered liver morphology. No effects were seen at a dosage of 0.2 mg/kg/day.

In a 90-day rat study, daily ingestion of 100 ppm resulted in moderate plasma, red cell, and brain cholinesterase inhibition. No compound-related effects were seen at 10 and 31.6 ppm. In a 2-year chronic rat study, 31.6 and 100 ppm technical fonofos resulted in inhibition of plasma and red cell cholinesterase, nervous behavior, and tremors. No effects were noted at 10 ppm.

In a three-generation reproduction study in rats, doses of 1 to 3 mg/kg/day technical fonofos resulted in no adverse effects.

In a 46-day ingestion study in chickens, doses of 2 to 20 mg/kg daily produced no suggestion of delayed neurotoxicity.

2.11.5 Hygienic Standard of Permissible Exposure (39)

The American Conference of Governmental Industrial Hygienists has established a TWA (time-weighted average for 40-hr week) of 0.1 mg/m^3.

2.12 Imidan®; N-(Mercaptomethyl)phthalimide-S-(O,O-dimethyl Phosphorodithioate), Prolate® (Animal Health)

2.12.1 Source, Uses, and Industrial Exposures (41)

Imidan® is cleared for use as an insecticide on grapes, kiwi and citrus fruit, nuts, peaches, plums, and potatoes, as well as for parasites of goats, hogs, horses, and sheep.

2.12.2 Physical and Chemical Properties (44)

Physical state	White to off-white crystalline solid
Water solubility	25 ppm at 20°C
Vapor pressure	6×10^{-2} μm Hg at 25°C
Melting point	67 to 70°C
Reactivity	No known violent reactions have been reported
Stability	Stable under normal storage conditions

2.12.3 Determination in the Atmosphere

A known volume of air is collected in a series of scrubbers containing a suitable solvent. Imidan® is the oxygen analogue of Phosmet® and can be analyzed by the Phosmet® method (67). In order to be sure all of the Phosmet® is converted to Imidan® it may be oxidized with dilute bromine water. No cleanup is required.

2.12.4 Physiologic Response (68)

Acute. The acute oral LD$_{50}$ in rats is in the range of 147 to 316 mg/kg following ingestion of technical Imidan®. Signs of toxicity include tremors, ataxia, salivation, cyanosis, diarrhea, excessive urination, and death. The acute oral LD_{50} in mice is in the range of 23 to 43 mg/kg with similar signs. The acute dermal LD_{50} in rabbits is greater than 4640 mg/kg and no signs of toxicity were observed at this dose. In dermal and eye irritation studies following the Draize technique using the rabbit, Imidan® was mildly irritating to rabbit skin and eye.

Subacute and Chronic. Daily ingestion of 20, 40, and 400 ppm technical Imidan® by dogs for 2 years resulted in decreased red cell and plasma cholinesterase activity and lacrimation. One dog in six dosed at 400 ppm also showed hyperactivity, salivation, hyperemia of mouth, mucoid feces, and mortality. The same doses were used in a 2-year rat feeding study and no effect was observed at 20 and 40 ppm. Plasma and red cell cholinesterase activity was depressed at 400 ppm; a reduction in weight gain and liver cell vacuolation also occurred.

Daily ingestion of 40 ppm Imidan® in a three-generation reproduction study or 80 ppm in a two-generation reproduction study resulted in no changes in reproduction, body weight gain, general observations, or survival.

Repeated dermal application of 10, 30, or 60 mg/kg 5 days a week for 3 weeks prior to mating and for 3 weeks after mating in rabbits resulted in reduced plasma and red blood cell cholinesterase activity but reproduction and teratologic parameters were unaffected.

Daily ingestion of 100, 316, and 1000 ppm Imidan® technical by chickens for 6 weeks did not result in neurotoxicity.

2.12.5   Hygienic Standard of Permissible Exposure

The American Conference of Governmental Industrial Hygienists has not established a TLV for Imidan® (39).

2.13   Isopestox®; Bis(monoisopropylamino)fluorophosphine Oxide, Mipafox®

2.13.1   Source, Uses, and Industrial Exposures

Isopestox® was introduced as an effective systemic insecticide and acaricide. After being placed on the market, it was found responsible for near fatalities and paralysis of several workers in England, which resulted in its being withdrawn from commerce. It is included here for its historical value as one of the first organic phosphates that caused organic phosphate neurotoxicity.

2.13.2   Physical and Chemical Properties (45)

Physical state   Crystalline solid
Molecular weight   182.2
Density   1.2 (25°C)
Melting point   60 to 65°C
Boiling point   125 to 126°C (2 mm Hg)
Vapor pressure   0.001 mm Hg (5°C)
Solubility   Soluble in water and polar organic solvents; slightly soluble in petroleum oils

2.13.3   Determination in the Atmosphere

Presumably the general method of Giang and Hall (69) for enzymic determination of organic phosphorous insecticides could be applied to measure Isopestox® air contamination.

2.13.4   Physiologic Response

Acute.   The acute oral LD_{50} in various species has been reported to range from 25 to 100 mg/kg (45, 70, 71). Near lethal doses produce severe neurotoxic signs.

Subacute and Chronic.   In chickens, rabbits, and humans, flaccid paralysis has been demonstrated and appears to result from "demyelinization" of nerves, as described for TOCP (tri-o-cresyl phosphate) (71–73). In chickens, a single dose of 1.0 mg/kg orally produced the response in 10 to 14 days. Like other organic phosphates, Isopestox® reduces the activity of cholinesterase(s), both in vitro and in vivo. The signs of toxicity, except for the demyelination syndrome, are typical of the class, as indicated in the discussion of neurotoxicity, above.

2.13.5   Hygienic Standard of Permissible Exposure

No threshold limit or pesticide tolerance has been set for Isopestox® (39).

2.14   Malathion; O,O-Dimethyl S-[1,2-Dicarboethoxyethyl] Dithiophosphate, Experimental Insecticide No. 4049 (American Cyanamid Co.)

2.14.1   Source, Uses, and Industrial Exposures

Malathion is regarded as the least toxic of the class and is generally considered as a wide spectrum insecticide for fruits, vegetables, and ornamental plants.

2.14.2   Physical and Chemical Properties (45, 74)

Physical state	Clear amber liquid, dependent on purity
Molecular weight	330
Density	1.2315 at 25°C
Melting point	2.85°C
Boiling point	156 to 157°C (0.7 mm Hg)
Vapor pressure	4.0×10^{-5} mm Hg (30°C)
Refractive index, $n_D^{25°C}$	1.4985
Solubility	145 ppm (weight) in water; miscible with alcohols, ethers, vegetable oils
Flash point and flammable limits	Not flammable at normal temperature; thermal decomposition above the boiling point

2.14.3   Determination in the Atmosphere

The colorimetric method of Norris et al. (75) is applicable to analysis for malathion in air.

2.14.4   Physiologic Response

Acute.   The acute LD_{50} for various species and routes has been compiled by Golz and Shaffer (76). The acute oral LD_{50} has been reported to vary widely, but it is generally agreed to range around 1400 mg/kg for female rats when administered in vegetable oils. Acute vapor inhalation exposure does not appear to be too great a problem owing to the low vapor pressure and low inherent toxicity of malathion. Single dermal application of 2460 to 6150 mg/kg (90 percent grade) to rabbits produces some toxicity.

Subacute and Chronic.   Subacute and chronic exposures of rats (76, 77) indicate that an oral dosage of 100 ppm (weight) can be tolerated without effect on cholinesterase(s) activity. Although 5000 ppm (weight) has slight effect on survival, food consumption, and growth, some rats have survived 20,000 ppm (weight) in the diet for 2 years. Frawley et al (78, 79) demonstrated that the simultaneous administration of malathion and EPN resulted in mortality greater than expected from the administration of the components alone (potentiation). Slight potentiation was also found in subacute studies of these materials when cholinesterase activity was studied as the criterion of evaluation. DuBois and Coon (80) found that EPN interfered with the enzymic hydrolysis of malathion and thus produced the potentiating effect by disrupting the detoxication mechanism of malathion.

2.14.5   Hygienic Standard of Permissible Exposure

The threshold limit value (skin) for malathion as a dust, fume, or mist is 10 mg/m^3 of air (39).

2.15   Methyl Parathion; O,O-Dimethyl O-p-Nitrophenyl Phosphorothioate, Metacide®

$$O_2N\!\!\diagup\!\!\diagdown\!\!-O-\overset{\overset{\text{S}}{\|}}{P}\diagup\!\!\overset{OCH_3}{\diagdown_{OCH_3}}$$

2.15.1   Source, Uses, and Industrial Exposures

Methyl parathion is closely related to parathion in its chemistry and toxicology. It controls aphids, boll weevils, and mites especially well, although its spectrum for control of insects is nearly as broad as parathion.

2.15.2   Physical and Chemical Properties (45)

Physical state	White crystalline solid in pure form; brown liquid crystallizing at 29°C as the technical material
Molecular weight	263.3
Density	1.358 (20°C)
Melting point	35 to 36°C (pure)
Vapor pressure	0.5 mm Hg (109°C)
Refractive index, $n_D^{35°C}$	1.5515
Solubility	50 ppm (weight) in water at 25°C; soluble in most aromatic solvents; slightly soluble in paraffin hydrocarbons
Flash point and flammable limits	Not flammable at normal temperatures

2.15.3   Determination in the Atmosphere

The method of Averell and Norris (61) for parathion is applicable to methyl parathion. This method has been modified by Gunther and Blinn (62) and more recently by Wilson et al (63). The method is applicable to the analysis of air for the presence of methyl parathion.

2.15.4 Physiologic Response

Acute. The acute oral LD_{50} for methyl parathion in the rat ranges between 9 and 25 mg/kg (45), depending on the purity of the material studied. Methyl parathion is toxic by all routes but has been described as especially hazardous via the eye.

Subacute and Chronic. DuBois and Coon (80) state the methyl parathion is approximately as toxic as parathion to rats but is a much less potent cholinesterase inhibitor. This material, like parathion, is a poor cholinesterase inhibitor in vitro but is converted in the mammalian liver to the toxic form, (81) which is the oxygen analogue.

2.15.5 Hygienic Standard of Permissible Exposure

A threshold limit in air has been established at 0.2 mg/m^3 (39).

2.16 Naled; 1,2-Dibromo-2,2-dichloroethyl Dimethyl Phosphate, Dibrom®

2.16.1 Source, Uses, and Industrial Exposures (41)

Naled has been cleared for use as an insecticide and a miticide on nuts, seed, root, and leaf vegetables; cotton; peppers, eggplant, tomatoes; citrus fruit, grapes, hops, squash, melons, mushrooms, stone fruit, pumpkin, rice, strawberries, and grasses; and the meat, fat, and by-products of cattle, goats, hogs, horses, poultry, and sheep. In this use it combines fast knockdown and broad spectrum with short residual life of deposits, plus a low order of mammalian toxicity.

2.16.2 Physical and Chemical Properties (45, 82)

Physical state	Pure—white solid; technical—light straw-color liquid
Melting point	27°C (pure)
Molecular weight	381
Specific gravity	1.97 at 20°C/20°C (technical)
Volatility	Low
Vapor pressure	0.0002 mm Hg at 20°C
Boiling point	110°C at 0.5 mm Hg
Stability	Hydrolyzes in water or aliphatic solvents, highly soluble in aromatic solvents

2.16.3 Determination in the Atmosphere (83)

A known volume of air can be collected in a suitable solvent in a series of scrubbers. The referenced method can be applied without cleanup.

2.16.4 Physiologic Response

Acute. The acute oral LD_{50} in rats is 430 mg/kg for purified naled. The acute dermal LD_{50} in rabbits is 1100 mg/kg for purified naled.

Subacute and Chronic. Technical naled was fed in the diet of rats at 100 ppm for 12 weeks without effect. In chronic 2-year studies, technical naled was fed in the diet of rats at 100 ppm. No toxic effects were observed. In a 2-year chronic dog study, naled was administered at 7.5 mg/kg/day without effect.

Rats and guinea pigs were exposed to the vapor of naled at 0.5 mg/ft^3 on a 6 hr/day, 5 day/week basis for 5 weeks. No toxic effects were observed.

2.16.5 Hygienic Standard of Permissible Exposure (39)

The American Conference of Governmental Industrial Hygienists has published a threshold limit value (TLV) for Dibrom® in air as 3 mg/kg for the TWA and as 6 mg/kg for the STEL.

2.17 Parathion; O,O-Diethyl O-p-Nitrophenyl Phosphorothioate, E-605, Compound 3422

2.17.1 Source, Uses, and Industrial Exposures (41)

Parathion is one of the best known of the class and, despite relatively high toxicity, it has received extensive use in agriculture as a broad spectrum insecticide. It is cleared for use on fruits, vegetables, grasses, and nuts.

2.17.2 Physical and Chemical Properties (45)

Physical state	Technical grade clear, medium to dark brown liquid
Molecular weight	291.27
Density	1.265 (25°C)
Melting point	6.1°C

Boiling point   157 to 162°C (0.6 mm Hg); 375°C (760
                                     mm Hg)

Vapor pressure   0.00003, 0.00066, and 0.0028 mm Hg at
                                     24.0, 54.5, and 70.7°C, respectively

Refractive index, $n_D^{20°C}$ 1.53668

Solubility   24 ppm (weight) in water at 25°C; miscible
                                     with most organic solvents

Flash point and flammable limits   Flash point at 120 to 160°C until flam-
                                     mable impurities of technical material
                                     are removed; boils with decomposition
                                     at 215°C and residue supports combus-
                                     tion when temperature reaches 221°C

2.17.3   Determination in the Atmosphere

The methods of Averell and Norris (61), Gunther and Blinn (62), and, more recently, Wilson et al. (63) may be applied to analysis of parathion in air. A method of analysis in biologic material has been reported by Hazleton and Holland.

2.17.4   Physiologic Response

Acute.   According to Hazleton and Holland (84), the oral LD_{50} for parathion is 3.5 and 12.5 mg/kg for female and male rats, respectively. Essential aspects of pharmacology and acute toxicology and antidotes for parathion appeared in 1948 in reports by DuBois et al. (85), Hagan and Woodard (86), and Hazleton and Godfrey (87).

Subacute and Chronic.   Parathion was not found to be stored in tissues of rats in subacute feeding studies (84). No effect was noted in growth, food consumption, gross or microscopic morphology, or survival when 100 ppm (weight) parathion was fed to male rats for a period of 2 years. Cholinesterase inhibition was observed at lower levels. Parathion was found by Gardocki and Hazleton (88) to be excreted in the urine as p-nitrophenol. This is a useful method, together with cholinesterase determinations, for measurement of exposure to parathion. The toxicity to humans can be evaluated from the studies of Grob et al. (89) and acute human intoxication incidences; a picture similar to that observed in laboratory animals is apparent.

2.17.5   Hygienic Standard of Permissible Exposure

The threshold limit value (skin) for parathion as a dust, fume, or mist is 0.1 mg/m^3 air (39). Tentative values have been set for short-term exposure levels (STEL) at 0.3 mg/m^3 (39).

2.18   Phosdrin®; Alpha Isomer of 2-Carbomethoxy-1-methylvinyl Dimethyl Phosphate, OS2046, Mevinphos®

$$CH_3O \\ \underset{CH_3O}{\overset{\displaystyle \| }{>}} P-O-\underset{\underset{}{}}{\overset{CH_3}{C}}=CHCOOCH_3$$

2.18.1   Source, Uses, and Industrial Exposures

Phosdrin® is useful in the control of aphids, mites, thrips, and lepidopterous larvae on a wide variety of crops (90).

2.18.2   Physical and Chemical Properties (91)

Physical state   Light yellow to orange liquid
Molecular weight   224.1
Density   1.23 (20°C)
Boiling point   106 to 107.5°C (1.0 mm Hg)
Vapor pressure   0.0029 mm Hg (21°C)
Refractive index, $n_D^{25°C}$ 1.4493
Solubility   Miscible with water, alcohols, ketones, chlorinated and aromatic hydrocarbons; slightly soluble in aliphatic hydrocarbons
Flash point   175°F (Tag open cup)

2.18.3   Determination in the Atmosphere

Standard micro and macro methods for determination of Phosdrin® are available from the Shell Corporation (91). An older method of analysis for the determination of Phosdrin® was reported by Zweig at the 136th Meeting of the American Chemical Society, October, 1956. The method is based on incubation of the water-soluble Phosdrin® with a standard amount of enzyme acetylcholinesterase, addition of a standard amount of acetylcholine, and the measurement of unhydrolyzed acetylcholine remaining by means of a color reaction with alkaline hydroxylamine and ferric chloride.

2.18.4   Physiologic Response

Acute.   The acute oral toxicity of Phosdrin® for rats is 6.0 to 7.0 mg/kg. The dermal LD_{50} for rabbits is approximately 34 mg/kg. The LC_{50} in inhalation studies of 1-hr duration is approximately 14 ppm for female rats.

Subacute and Chronic.   In chronic studies the minimal lethal dose for rats has been established as between 100 and 200 ppm in the diet (weight). Lower doses affect tissue cholinesterase levels. Dosages below 5.0 ppm have no effect on cholinesterase activity (92).

2.18.5   Hygienic Standard of Permissible Exposure

A threshold limit (skin) has been set for Phosdrin® insecticide of 0.01 ppm or 0.1 mg/m^3 (39).

2.19   Phosphamidon; 2-Chloro-2-(diethylcarbamoyl)-1-methylvinyl Dimethyl Phosphate, Dimecron®

2.19.1   Source, Uses, and Industrial Exposures (41)

Phosphamidon has been cleared by EPA for use as an insecticide on pome and citrus fruit, cole crops, melons, cucurbits, tomatoes and bell peppers, potatoes, sugar cane, walnuts, and cotton. Tolerances range from 0.1 to 1.0 ppm.

2.19.2   Physical and Chemical Properties (93)

Physical state	Colorless and odorless liquid. 89% phosphamidon; 3% related compounds
Specific gravity	1.2 at 20°C
Molecular weight	299.5
Volatility	Low
Boiling point	160°C at 1.5 mm Hg
Stability	Stable under ordinary storage conditions
Solubility	Miscible in all proportions with water, alcohol, and many other organic solvents

2.19.3   Determination in the Atmosphere (94)

Phosphamidon may be analyzed in the atmosphere by first collecting air in a suitable solvent in a series of scrubbers. The indicated method can then be applied without cleanup.

2.19.4   Physiologic Response (93)

Acute.   The acute oral toxicity of phosphamidon technical (89% phosphamidon and 3% related compounds) in the albino rat is 28.3 mg/kg. The acute dermal LD$_{50}$ in rabbits is 267 mg/kg.

Subacute and Chronic. In subacute feeding studies in dogs, phosphamidon was tolerated at a dose of 5 mg/kg for 90 days without effect. In subacute inhalation studies, dogs, guinea pigs, and rats were exposed to an atmosphere containing 0.125 mg/l phosphamidon for 6 hr/day, 5 days/week for 90 days without any marked toxic effects.

2.19.5   Hygienic Standard of Permissible Exposure

No TLV has been established for phosphamidon by the American Conference of Governmental Industrial Hygienists (39).

2.20   Ronnel; O,O-Dimethyl O-[2,4,5-Trichlorophenyl] Phosphorothioate, ET-57, Korlan®

2.20.1   Source, Uses, and Industrial Exposures (41)

Ronnel is a systemic organic phosphate that has been shown to be highly effective in the control of insects affecting cattle as well as plants. As a systemic livestock pest control agent, it is effective against cattle grub, and occasional reports indicate utility against sheep keds, sheep nasal botfly, chicken lice and mites, dog fleas, and ticks.

2.20.2   Physical and Chemical Properties (95, 96)

Physical state   White to light tan crystalline solid
Molecular weight   321.56
Melting point   40.97°C
Vapor pressure   0.0008 mm Hg (25°C)
Solubility   Insoluble in water; soluble in most organic solvents

2.20.3   Determination in the Atmosphere

A method is presented in the compendium of Thompson and Watts (42) that can be used for determination of ronnel in the atomsphere.

2.20.4   Physiologic Response

Acute. Ronnel has a low toxicity to warm-blooded animals. The acute oral LD_{50} of ronnel in rats is 3000 mg/kg. The LD_{50} in rabbits is around 1000 mg/kg, and in chickens between 4000 and 5000 mg/kg.

Subacute and Chronic.   Ronnel is also a weak inhibitor of cholinesterase. It affects the pseudoesterase of the plasma predominantly, rather than the true acetylcholinesterase of the red blood cells, upon both single and repeated oral doses (97).

2.20.5   Hygienic Standard of Permissible Exposure

A threshold limit has been adopted as a time-weighted average of 10 mg/m^3 for ronnel (39).

2.21   Ruelene®; 4-*tert*-Butyl-2-chlorophenyl Methyl Methylphosphoramidate; Crufomate

2.21.1   Source, Uses, and Industrial Exposures (41)

Ruelene® has a broad spectrum of use as an anthelmintic in cattle, sheep, and goats, and in the systemic control of grubs, lice, and horn flies on cattle by oral or topical application.

2.21.2   Physical and Chemical Properties (44)

Physical state	White crystalline solid
Density	1.1618 (70/4°C)
Melting point	58.7°C (technical grade 92%)
Vapor pressure	0.01 mm Hg at 117°C
Solubility	Soluble in benzene, carbon tetrachloride, ethyl ether, methanol, cyclohexane; practically insoluble in water

2.21.3   Determination in the Atmosphere (42)

A method is described by Markus and Puma (94) that is suitable for Ruelene® determination in the atmosphere.

2.21.4   Physiologic Response (98)

Acute.   The acute (single oral dose) toxicity of Ruelene® has been studied in rabbits, rats, guinea pigs, and dogs. The LD$_{50}$ values ranged from 490 to more than 1000 mg/kg. Single oral doses of Ruelene® were administered to rats together with each of the 14 then-commercial cholinesterase-inhibiting insecticides. No combination produced greater than additive ratios of expected to observed LD_{50} values except Ruelene® with malathion, where the ratio was 5:3. Application of Ruelene® to the eye of rabbits as undiluted or 10 percent solutions resulted in corneal cloudiness and conjunctival irritation, along with slight pain. Ruelene® produced slight erythema when applied undiluted or as a 10 percent solution to the abraded or intact skin of the rabbit.

Subacute and Chronic. In 90-day dietary studies in rats at doses of 30 to 1000 ppm (weight) blood and/or brain cholinesterase was decreased 40 to 60 percent as compared to control animals. No other adverse effect was observed at any dose. In dogs a 75-day feeding study was performed with blood cholinesterase inhibition evident at 250 ppm (weight) together with slight liver morphologic changes. Lower levels, 40 or 125 ppm, produced neither cholinesterase nor morphologic changes.

In a 2-year dietary study in rats, cholinesterase activity was not depressed in either sex regardless of tissue at 40 ppm. Higher levels produced slight cholinesterase depression in various tissues depending on sex (females more sensitive) and dose. At 100 ppm there was no morphologic change. After 2 years at 1000 ppm, in addition to marked cholinesterase depression, there was retarded growth, atrophy of the muscles of the hind limbs, and slight degeneration of the sciatic nerves. At 12, 18, and 24 months 1000 ppm produced reduced testes weight to about one-half, reflecting degeneration and atrophy of the seminiferous tubules.

In a 2-year dietary study in dogs, 2000 ppm caused marked effects by 4 weeks, including decreased appetite with loss of weight, decreased activity, slow awkward gait, stiffness of the hindquarters, loss of flexor and extensor reflexes, and swaying of the hindquarters. In five of the dogs the hind feet were bent over at the torsal joints. One dog died. Because of the severity of the signs, the remaining seven dogs were placed on control diet until day 94 or 95. At termination at 95 days, three of these dogs showed slight weakness in the hindquarters whereas the other four dogs had recovered completely.

Histopathology on these dogs with emphasis on the central nervous system showed no changes. A moderate but significant decrease in cholinesterase activity at 200 ppm was the only effect after 2 years. Male and female dogs receiving 20 ppm Ruelene® for 2 years showed no evidence of adverse effect.

No effect was observed in multigeneration reproduction studies in the rat at feeding dose levels up to 500 ppm Ruelene®.

No information has been found on the mutagenic evaluation of Ruelene®.

2.21.5 Hygienic Standard of Permissible Exposure (39)

A threshold limit value has been established for Ruelene® by the American Conference of Governmental Industrial Hygienists. The time-weighted average

(TWA) for occupational exposure is 5 mg/m^3 and the short-term exposure limit (STEL) is 20 mg/m^3.

2.22 Schradan; Octamethyl Pyrophosphoramide, OMPA, Bis[bis(dimethylamino)] Phosphonous Anhydride

2.22.1   Source, Uses, and Industrial Exposures

Schradan is a systemic insecticide that is included here for historical value. It is the first member of the class used as an insecticide. It is no longer used.

2.22.2   Physical and Chemical Properties (45)

Physical state	Viscous, dark brown liquid
Molecular weight	286.3
Density	1.1343 (25°C)
Melting point	Below −10°C
Boiling point	135 to 137°C (1 mm Hg)
Vapor pressure	0.0003 mm Hg (25°C)
Refractive index, $n_D^{25°C}$	1.4612
Solubility	Miscible with water; soluble in ethanol, acetone, chloroform, benzene; insoluble in heptane and petroleum ether

2.22.3   Determination in the Atmosphere

A method suitable for the analysis of schradan in air has been reported by Hartley et al. (99).

2.22.4   Physiologic Response

Acute.   In acute studies, the LD$_{50}$ has been reported by Lehman (100) to be 13.5 mg/kg orally in rats. DuBois et al. (101) reported that oral LD$_{50}$ in rats to be 10 mg/kg and the acute intraperitoneal LD$_{50}$ to be 8.0 mg/kg. Frawley et al. (102) listed the oral LD$_{50}$ as 35.5 mg/kg for female rats and 13.5 mg/kg for male rats, thus indicating some sex variation. Reports from these and other investigators indicate that there is very little, if any, species variation to the acute response.

Subacute and Chronic. The chronic toxicity in rats has been studied by Barnes and Denz (103). At 50 ppm (weight) in the diet for 1 year, male rats showed toxic signs, growth suppression, and marked cholinesterase depression without producing tissue pathology. At dosages of 10 and 50 ppm (weight) whole blood cholinesterase activity was reduced but brain activity was unaffected. Further studies in male rats at 1.0 ppm (weight) showed that the acetylcholinesterase (true cholinesterase) in the red cell was unaffected. A dosage of 0.3 ppm produced no effect. This was confirmed by Edson et al. (104), who fed 0.25 ppm (weight) to rats without effect. Studies conducted on humans indicate that a level of 0.6 ppm (weight) was without effect in six male and six female subjects. One subject was administered 2.4 ppm (weight) in the total diet, which produced marked plasma and red cell cholinesterase depression (104).

2.22.5 Hygienic Standard of Permissible Exposure

No threshold limit has been set for schradan (39).

2.23 Systox®; *O,O*-Diethyl *O*-[2-(Ethylmercapto)ethyl] Thionophosphate, Dematon®

2.23.1 Source, Uses, and Industrial Exposures (45)

Systox® is a systemic insecticide, which also has contact action. It is particularly useful for the control of sucking insects such as mites and aphids.

2.23.2 Physical and Chemical Properties (45)

Physical state	Pale yellow to light brown liquid
Molecular weight	258
Density	1.1183 (20°C)
Boiling point	134°C (2 mm Hg)
Vapor pressure	0.001 mm Hg (33°C)
Refractive index, $n_D^{20°C}$	1.4875
Solubility	0.01 percent in water; miscible with most organic solvents

2.23.3 Determination in the Atmosphere

The method of Gardner and Heath (105) presumably can be applied to the analysis of Systox® in air.

2.23.4   Physiologic Response

At this time it is not practical to give the LD_{50} values for Systox® without further qualification, because it is a mixture of isomers. This is further complicated by an apparent literature discrepancy as to the structure of the isomers. The isomer (formula above) is referred to as Systox® by Deichmann (106) and Martin and Miles (107), whereas Barnes and Denz (103) apply this term to the mixture and designate the above structure as the P=S isomer. They designate the second isomer as P=O, an exchange of S and O positions, whereas Deichmann (106) designates the second isomer as Iso Systox and replaces the O with S, in effect producing a dithio compound.

Based on the above terminology, the Barnes and Denz (103) data for the mixture are as follows: the oral LD_{50} for rats is 4.0 mg/kg for females and 10 mg/kg for males.

2.23.5   Hygienic Standard of Permissible Exposure (39)

The threshold limit value adopted for Systox® (skin) is 0.01 ppm or 0.1 mg/m³ for TWA and 0.03 ppm or 0.3 mg/m³ for short-term exposure limit (STEL).

2.24   TEPP; Tetraethyl Pyrophosphate

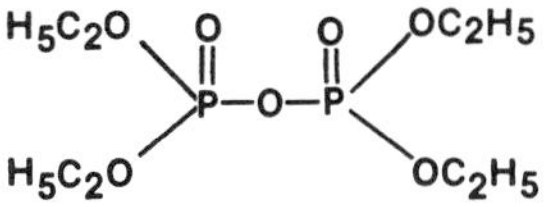

2.24.1   Source, Uses, and Industrial Exposures

TEPP is a relatively insoluble, nonsystemic insecticide used in the control of some aphids, spider mites, mealybugs, leafhoppers, and thrips. It is particularly useful close to harvest because of its rapid degradation and absence of residue in a short period after application.

2.24.2   Physical and Chemical Properties (45)

Physical state	Amber liquid
Molecular weight	290.2
Density	1.1810 (25°C)
Boiling point	104 to 110°C (0.08 mm Hg)
Vapor pressure	Volatile
Refractive index, $n_D^{25°C}$	1.4170 to 1.4180
Solubility	Miscible with water resulting in rapid hydrolysis (half-life 6.8 hr at 25°C, pH 7.0); miscible with

most organic solvents except kerosene of low
aromatic content

2.24.3 Determination in the Atmosphere

Methods for analysis of TEPP have been accepted by the Association of Official
Agricultural Chemists (108) and presumably could be adapted to air analysis.

2.24.4 Physiologic Response

Acute. TEPP is highly toxic by ingestion, skin absorption, and by way of the
eye. The LD_{50} for all species in general is 50 mg/kg or less, according to Harris
(109). He also reports general acute toxicity by inhalation: The LC_{50} is 200
ppm (volume) following a 1-hr exposure. In dermal studies, Harris gives the
LD_{50} for animals in general as 200 mg/kg following a 24-hr exposure.

Subacute and Chronic. Chronic exposure to sublethal amounts lowers the
blood cholinesterase level. Some experience has been gained with the use of
TEPP as a therapeutic agent in myasthenia gravis. Single dosages of 5.0 mg or
3.6 mg daily for 2 days, or 7.2 mg every 3 hr orally for 3 to 5 doses, produce
symptoms in normal subjects, as did somewhat larger doses in myasthenia gravis
patients. Symptoms appeared wtihin 30 min after final dose and were typical
of organic phosphate poisoning. The estimated lethal dose in man is 20 mg
intramuscularly or 100 mg orally.

2.24.5 Hygienic Standard of Permissible Exposure (39)

The adopted threshold limit value (skin) for TEPP as a dust, fume, or mist is
0.004 ppm or 0.05 mg/m^3 time-weighted average.

2.25 Trithion®; *O,O*-Diethyl *S*-(*p*-Chlorophenyl)Thiomethyl Phosphorodithioate,
R-1303

2.25.1 Source, Uses, and Industrial Exposures (45)

Trithion® is a nonsystemic insecticide, miticide, and ovicide with a relatively
long residual action.

2.25.2   Physical and Chemical Properties (110)

Physical state	Light amber liquid
Molecular weight	342.9
Density	1.265 to 1.285 (25°C)
Vapor pressure	Very low
Refractive index, $n_D^{25°C}$	1.590 to 1.597
Solubility	Not appreciably soluble in water; miscible with vegetable oils and most organic solvents

2.25.3   Determination in the Atmosphere

Air may be analyzed for Trithion® by the method described by Patchett (111).

2.25.4   Physiologic Response

Acute.   The acute oral LD_{50} of Trithion® for male albino rats is 17.2 to 28 mg/kg.

Subacute and Chronic.   Exposure of rats and dogs to a near saturated vapor (in air) of Trithion® for 4 weeks resulted in plasma cholinesterase depression in dogs, but there was no reduction in enzyme activity in rats. The animals appeared normal throughout the exposure. It would appear that Trithion® is not volatile enough to be a hazard by the inhalation route. Subacute studies in rats and dogs have been performed by Weir and Fogleman (112). Dosages of 100 ppm by weight in the diet of rats produced tremors and reduced body weight gains. Cholinesterase activity was markedly depressed, but no gross or microscopic morphologic changes occurred. Dosages of 5.0 ppm (weight) produced no effect. Dosages of 1.0 ppm (weight) produced plasma and red cell cholinesterase depression without overt signs or pathology.

2.25.5   Hygienic Standard of Permissible Exposure

No threshold limit has been established for Trithion® (39).

REFERENCES

1. *Fed. Reg.*, **37,** 13369 (July 1972).
2. W. O. Negherbon, *Handbook of Toxicology*, Vol. III, Saunders, Philadelphia, 1959.
3. L. S. Goodman and A. Gilman, *The Pharmacological Basis of Therapeutics*, 4th ed., Macmillan, New York, 1970.

4. L. W. Hazleton, *J. Agric. Food Chem.,* **3,** 312 (1955).

5. J. H. Wills, *Crit. Rev. Toxicol.,* **1,** 153 (1972).

6. Anon., *National Cancer Institute Carcinogenesis Technical Report Series, NCI-CG-TR-10* (1977).

7. *Ibid., NCI-CG-TR-24* (1978).

8. *Ibid., NCI-CG-TR-33* (1978).

Ibid., NCI-CG-TR-69 (1978).

10. *Ibid., NCI-CG-TR-70* (1979).

11. *Ibid., NCI-CG-TR-192* (1979).

12. *Ibid., NCI-CG-TR-16* (1979).

13. M. I. Smith, E. Ehrve, and W. H. Frazier, *Public Health Rep. U.S.,* **45,** 2509 (1930).

14. M. I. Smith and R. D. Lillie, *Am. Med. Assoc. Arch. Neurol. Psychiat,* **26,** 976 (1931).

15. W. F. Durham, T. B. Gaines, and W. J. Hayes, Jr., *Am. Med. Assoc. Arch. Ind. Health,* **13,** 326 (1956).

16. J. M. Barnes and F. A. Denz, *J. Pathol. Bacteriol.,* **65,** 587 (1953).

17. C. H. Hine, E. F. Gutenburg, M. M. Coursey, K. Seligman, and R. M. Gross, *J. Pharmacol. Exp. Therap.,* **113,** 28 (1955).

18. J. P. Frawley, R. E. Zwickey, and H. N. Fugat, *Fed. Proc.,* **15,** 424 (1956).

19. L. Austin and D. R. Davies, *Br. J. Pharmacol.,* **9,** 145 (1954).

20. R. Baron, Ed., *Pesticide Induced Delayed Neurotoxicity, Proceedings of a Conference Sponsored by the Environmental Protection Agency, 600/1-76-025,* July 1975.

21. P. S. Spencer and H. H. Schaumberg, Eds., *Experimental and Clinical Neurotoxicology,* Williams & Wilkins, Baltimore/London, 1980.

22. K. Kay, L. Monkman, J. P. Windish, T. Doherty, J. Park, and C. Racicat, *Arch. Ind. Hyg. Occup. Med.,* **6,** 252 (1952).

23. W. T. Summerford, W. J. Hayes, Jr., J. M. Johnston, K. Walker, and J. Spillane, *Arch. Ind. Hyg. Occup. Med.,* **7,** 383 (1953).

24. M. K. Johnson, "Mechanism of Action of Neurotoxic Organophosphorous Esters," in R. Baron, Ed., *Pesticide Induced Delayed Neurotoxicity, Proceedings of a Conference Sponsored by the EPA, 600/1-76-025,* July 1975, p. 51.

25. H. O. Michel, *J. Lab. Clin. Med.,* **34,** 1564 (1949).

26. R. Ammon, *Arch. ges. Physiol. Pflüger's,* **233,** 57 (1933).

27. R. L. Metcalf, *J. Econ. Entomol.,* **44,** 883 (1951).

28. J. H. Fleisher and E. J. Pope, *Arch. Ind. Hyg. Occup. Med.,* **9,** 323 (1954).

29. E. F. Edson, *World Crops,* **1,** (August 1958).

30. G. L. Ellman et al., *Biochem. Pharmacol.,* **7,** 88 (1961).

31. J. H. Wolfsie and G. D. Winter, *Arch. Ind. Hyg. Occup. Med.,* **6,** 43 (1952).

32. J. H. Wolfsie and G. D. Winter, *Arch. Ind. Hyg. Occup. Med.,* **9,** 396 (1954).

33. A. S. Gordon and C. W. Frye, *J. Am. Med. Assoc.,* **159,** 1181 (1955).

34. R. M. Clyne and C. B. Shaffer, *Toxicological Information, Cyanamid Organic Pesticides,* 3rd ed., American Cyanamid Company, Princeton, N.J., 1975.

35. T. Namba and K. Hiraki, *J. Am. Med. Assoc.,* **166,** 1834 (1955).

36. D. Grob and R. J. Johns, *J. Am. Med. Assoc.,* **166,** 1855 (1955).

37. *Abate®, Product Bulletin,* American Cyanamid Company, Princeton, N.J., 1980.

38. N. R. Pasarela and E. J. Orloski, "Abate® Insecticide," J. Sherma and G. Zweig, Eds., *Analytical*

Methods for Pesticides and Plant Growth Regulators, Vol. 8, Thin-layer and Liquid Chromatography, Pesticides of International Importance, Academic Press, New York, 1973.

39. Anon., *Threshold Limit Values for Chemical Substances and Physical Agents in the Workroom Environment,* American Conference of Governmental Industrial Hygienists, Cincinnati, Ohio, 1978.

40. Anon., "Orthene® Insecticide," *Technical Information, Exp. Data Sheet,* Chevron Chemical Co., Research Labs, Richmond, Calif., 1976.

41. Anon., *Tolerances for Pesticides In or On Agricultural Commodities,* EPA, Washington, D.C., July 9, 1980.

42. J. F. Thompson and R. R. Watts, *Analytical Reference Standards and Supplementary Data for Pesticides and other Organic Compounds, EPA-600/9-78-0012,* U.S. EPA, Health Effects Research Lab, ORD, Research Triangle Park, N.C., 1978.

43. M. A. Luke, J. E. Froberg, and H. T. Masumoto, *J. Offic. Anal. Chem.,* **58,** 1020 (1975).

44. Anon., *Farm Chemicals Handbook,* Meister Publishing Co., Willough, Ohio, 1975.

45. Anon., *Pesticide Official Publication and Condensed Data on Pesticide Chemicals,* Association of American Pest Control Officials, College Park, Md., 1955. Data in the text have been gathered from unpublished sources and manufacturers' data sheets and may vary from these basic references.

46. K. P. DuBois, J. Doull, J. Deroin, and O. K. Cummings, *Arch. Ind. Hyg. Occup. Med.,* **8,** 350 (1953).

47. Anon., *Pesticide Official Publication and Condensed Data on Pesticide Chemicals,* Association of American Pest Control Officials, College Park, Md., 1957.

48. P. A. Giang, F. T. Smith, and S. A. Hall, *J. Agric. Food Chem.,* **4,** 621 (1956).

49. A. M. Mattson, J. T. Spillane, and G. W. Pearce, *J. Agric. Food Chem.,* **3,** 319 (1955).

50. E. F. Edson, personal communication.

51. J. P. Frawley, R. Weir, T. Tusing, K. P. DuBois, and J. C. Calandra, *Toxical Appl. Pharmacol,* **2**(5), 605–624 (1963).

52. H. J. Harris, *A Tentative Ultraviolet Method for Analysis of Diazinon® in Spray Residues,* Geigy Chemical Corp., Yonkers, N.Y., 1953.

53. R. B. Bruce, *J. Agric. Food Chem.,* **3,** 1017 (1955).

54. R. B. Bruce, *Fed. Proc.,* **13** (1954).

55. H. Gysin and A. Margot, *J. Agric. Food Chem.,* **6,** 900 (1958).

56. H. Gysin, personal communication.

57. Anon., *Cygon®, Dimethoate Systemic Insecticide,* Cyanamid International Technical Information, American Cyanamid Co., Wayne, N.J., 1967.

58. E. D. Chilwell and P. I. Beecham, *J. Agric. Food Chem.,* **13,** 178 (1964).

59. P. A. Giang, W. F. Barthel, and S. A. Hall, *J. Agric. Food Chem.,* **2,** 1281 (1954).

60. K. P. DuBois and C. J. Cotter, *Arch. Ind. Hyg. Occup. Med.,* **11,** 53 (1955).

61. P. R. Averell and M. V. Norris, *Anal. Chem.,* **20,** 753 (1948).

62. F. A. Gunther and R. C. Blinn, *Adv. Chem. Ser.,* **1,** 75 (1950).

63. C. W. Wilson, R. Baier, D. Genung, and J. Mullowney, *Anal. Chem.,* **23,** 1487 (1951).

64. H. C. Hodge, E. A. Maynard, L. Hurwitz, V. DiStefano, W. L. Downs, C. K. Jones, and H. J. Blanchet, Jr., *J. Pharmacol. Exp. Ther.,* **112,** 29 (1954).

65. M. C. Bowman and M. Beroza, *J. Assoc. Offic. Anal. Chem.,* **54,** 1086 (1971).

66. Anon., *Prod. Safety Info. Dyfonate® Technical,* Stauffer Chemical Co., Agricultural Chem. Div., Westport, Conn., 1978.

67. B. D. Ripley, R. J. Wilkinson, and A. S. Y. Chau, *J. Assoc. Offic. Anal. Chem.*, **57,** 1033 (1974).

68. Anon., *Prod. Safety Info., Imidan®*, Stauffer Chemical Co., Agricultural Chem. Div., Westport, Conn., 1978.

69. P. A. Giang and S. A. Hall, *Anal. Chem.*, **23,** 1830 (1951).

70. H. Martin, *Guide to the Chemicals Used in Crop Protection*, 2nd ed., University of Western Ontario, 1955.

71. W. E. Ripper, *Ct. Rd. IIIme Congr. Int. Phytopharm.*, Paris, 1952.

72. D. R. Davies, *J. Pharm. Pharmacol.*, **6,** 1 (1954).

73. D. R. Davies, *Proc. Roy. Soc. Med.*, **45,** 570 (1952).

74. Anon., *Malathion Concentrate (Technical Bulletin)*, Cyanamid Intl. (American Cyanamid Co.), Wayne, N.J., 1966.

75. M. V. Norris, W. A. Vail, and P. R. Averell, *J. Agric. Food Chem.*, **2,** 570 (1954).

76. H. H. Golz and C. B. Shaffer, *Malathion, Summary of Pharmacology and Toxicology*, Central Medical Dept., American Cyanamid Co., New York, revised January 1955.

77. L. W. Hazleton and E. G. Holland, *Arch. Ind. Hyg. Occup. Med.*, **8,** 399 (1953).

78. J. P. Frawley, E. C. Hagan, O. G. Fitzhugh, H. N. Fuyat, and W. I. Jones, *J. Pharmacol. Exp. Ther.*, **119,** 147 (1957).

79. J. P. Frawley, H. N. Fuyat, E. C. Hagan, J. R. Blake, and O. G. Fitzhugh, *J. Pharmacol. Exp. Ther.*, **121,** 96 (1957).

80. K. P. DuBois and J. M. Coon, *Arch. Ind. Hyg. Occup. Med.*, **6,** 9 (1952).

81. R. L. Metcalf, *Organic Insecticides,* Interscience Publishers, New York/London, 1955.

82. Anon., *Technical Info., Ortho Dibrom®*, Chevron Chemical Co., Richmond, Calif., 1970.

83. W. P. McKinley, *J. Assoc. Offic. Anal. Chem.*, **48,** 748 (1965).

84. L. W. Hazleton and E. G. Holland, *Advances Chem. Ser.*, **1,** 31 (1950).

85. K. P. DuBois, J. Doull, and J. M. Coon, *Fed. Proc.*, **7,** 216 (1948).

86. E. C. Hagan and G. Woodard, *Fed. Proc.*, **7,** 224 (1948).

87. L. W. Hazleton and E. Godfrey, *Fed. Proc.*, **7,** 226 (1948).

88. J. F. Gardocki and L. W. Hazleton, *J. Am. Pharm. Assoc. Sci. Ed.*, **40,** 491 (1951).

89. D. Grob, W. L. Garlick, and A. M. Harvey, *Bull. Johns Hopkins Hosp.*, **87,** 106 (1950).

90. Anon., *Pesticide Official Publication, 1957 Supplement*, Association of American Pest Control Officials, College Park, Md.

91. Anon., *Bull. SC: 59-37, Summary of Basic Data for Phosdrin® Insecticide*, Shell Chemical Corp., New York.

92. J. K. Kodama, C. H. Hine, and M. S. Morse, *Arch. Ind. Hyg. Occup. Med.*, **9,** 54 (1954).

93. Anon., *Tech. Info. Bulletin: Phosphamidon*, Chevron Chemical Co., Richmond, Calif., 1970.

94. J. R. Markus and B. Puma, *Phosphamidon, Pesticide Analysis Manual*, Vol. 2, *Pesticide Regulation,* Sec. 120.239, Food & Drug Administration, August 1977.

95. Anon., *Korlan Information Sheet*, The Dow Chemical Co., May 1, 1958.

96. Anon., *Trolene (Dow-ET-57) The First Animal Insecticide, Agric. Chem. Dept. Inf. Bull. No. 108*, Dow Chemical Co., 1957.

97. D. C. McCollister, F. Oyen, and V. K. Rowe, *J. Agric. Food Chem.*, **7,** 689 (1959).

98. D. D. McCollister, K. J. Olsen, V. K. Rowe, O. E. Paynter, R. J. Weir, and W. H. Dietrich, *Food Cosmet. Toxicol.*, **6,** 185–198 (1968).

99. G. S. Hartley, D. F. Heath, J. M. Hulme, D. W. Pound, and M. Whittaker, *J. Sci. Food Agric.*, **2,** 303 (1951).

100. A. J. Lehman, *Assoc. Food Drug Offic. U.S. Q. Bull.,* **15,** 122 (1951).

101. K. P. DuBois, J. Doull, and J. M. Coon, *J. Pharmacol. Exp. Ther.,* **99,** 376 (1950).

102. J. P. Frawley, E. C. Hagan, and O. G. Fitzhugh, *J. Pharmacol. Exp. Ther.,* **105,** 156 (1952).

103. J. M. Barnes and F. A. Denz, *Br. J. Ind. Med.,* **11,** 11 (1954).

104. E. P. Edson, K. P. Fellowes, and F. MacL. Carey, *Med. Dept. Rep.,* Fisons Pest Control Ltd., England, 1954.

105. K. Gardner and D. F. Heath, *Anal. Chem.,* **25,** 1849 (1953).

106. W. B. Deichmann, *Fed. Proc.,* **13,** 346 (1954).

107. H. Martin and J. R. W. Miles, *Guide to the Chemicals Used in Crop Protection,* Science Service, Dominion of Canada Dept. of Agriculture, London, Ontario, 1952.

108. W. Horwitz, *Official Methods of Analysis of the Association of Official Agricultural Chemists,* 8th ed., Association of Official Agricultural Chemists, Washington, D.C., 1955, pp. 86–87.

109. J. S. Harris, *Agric. Chem.,* **2,** 27 (1947).

110. Anon., *Prod. Safety Info., Trithion® Technical,* Stauffer Chemical Co., Westport, Conn., 1979.

111. G. G. Patchett, *Determination of R-1303 Spray Residues in Oranges, Lemons and Alfalfa,* Stauffer Chemical Co., Richmond, Calif., 1956.

112. R. J. Weir and R. W. Fogleman, unpublished data.

CHAPTER FIFTY-EIGHT

Cyanides and Nitriles

ROLF HARTUNG, Ph.D.

1 GENERAL CONSIDERATIONS

Although cyanides are among the most acutely toxic of all industrial chemicals and are produced in large quantities, used in many different applications, they cause few serious accidents or deaths. This is partly because the word *cyanide* is synonymous with a highly poisonous substance, and a certain amount of care in handling is thereby insured. The good record is also due in no small part to the provision by manufacturers of adequate precautions and first-aid treatment.

The annual production of cyanides in the United States is now more than 318 million kg and is increasing (1).

Within limits the acute effects of poisoning by cyanide, and those effects of nitriles that are related to the release of cyanide, can be controlled by appropriate first-aid measures. Therefore, it is essential that all personnel working with processes involving cyanides or nitriles be specially trained so that they are fully aware of the hazards and follow faithfully all rules laid down for safe handling. It is also essential that special training be given in the specific first-aid measures, and that adequate specific antidotes be available for first aid and for use by physicians.

1.1 Symptoms Produced by Cyanides and Nitriles

For purposes of the toxicologist, cyanides and nitriles can be thought of as belonging to several classes: (1) hydrogen cyanide, cyanogen, simple salts of hydrogen cyanide that dissociate readily to release CN^- ions (such as sodium, potassium, calcium, and ammonium cyanide); (2) halogenated compounds such as cyanogen chloride or bromide; (3) simple and complex salts of hydrogen cyanide that do not dissociate readily to release CN$^-$ ions (such as cobalt cyanide trihydrate, cupric and cuprous cyanide, silver cyanide, ferricyanide, and ferrocyanide salts); (4) cyanide glycosides produced by plants (such as amygdalin and linamarin); and (5) nitriles, such as acetonitrile (methyl cyanide), acrylonitrile, and isobutyronitrile.

Hydrogen cyanide itself and its simple soluble salts, such as those in group 1, are among the most rapidly acting of all known poisons. A few breaths of higher concentrations of hydrogen cyanide vapor or the ingestion of amounts as low as 50 to 100 mg sodium or potassium cyanide may be followed by almost instantaneous collapse and cessation of respiration. At much lower dosages, the earliest symptoms may be simply those of weakness, headache, confusion, and occasionally nausea and vomiting. The respiratory rate and depth usually increase at the beginning, and at later stages become slow and gasping. Blood pressure is usually normal, especially in the mild or moderately severe cases, although the pulse rate is usually more rapid than normal. It is characteristic that the heartbeat may continue for some time even after respirations have ceased. If cyanosis is present, it usually indicates that respiration either has ceased or has been very inadequate for a few minutes.

The halogenated materials, cyanogen bromide and cyanogen chloride, are also highly toxic and possess some of the same properties as hydrogen cyanide and its soluble salts. However, at low concentrations, these materials behave more like the highly irritating vesicant gases, and cause severe lacrimatory effects and both acute and delayed pulmonary irritation and pulmonary edema.

Many nitriles such as acrylonitrile, isobutyronitrile, and propionitrile can cause the same general symptoms as hydrogen cyanide, but the onset of symptoms is apt to be slower and seems to be related to the ease with which cyanide is metabolically released from the compound. They are also apt to be more active as primary irritants on the skin or eye, and are frequently absorbed rapidly and completely through the intact skin. Skin absorption, however, is not absent with hydrogen cyanide or even its soluble salts, and this may be a prominent factor in preventing recovery unless all the material is removed from the skin, and the contaminated clothing removed.

Chronic exposure to cyanides and some nitriles, notably acetonitrile, can interfere with iodine uptake by the thyroid gland, which can lead to thyroid enlargement. This effect seems to be related to the concentrations and duration of excessive blood levels of thiocyanate ions resulting from the metabolism of cyanide ions (2, 3). Acrylonitrile may be carcinogenic. Details for these effects are provided in the appropriate sections where individual components are discussed.

A number of other related materials, such as cyanamide, calcium cyanamide, cyanates, isocyanates, isonitriles, thiocyanates, ferri- and ferrocyanides, and cyanoacetates, do not have all the typical toxic properties of cyanides and nitriles and may act by different mechanisms. Most of these (a notable exception being certain isocyanates) have a somewhat lower order of toxicity, although there is great variability in their activity.

1.2 Mode of Action and Metabolism of Cyanides

Cyanide readily forms relatively stable complexes with a number of biologically active metal ions. The most important of these reactions appears to be the interaction of cyanide with Fe^{3+} in cytochrome oxidase, which results in a 50 percent inhibition of the enzymatic activity of cytochrome oxidase at a 10^{-8} M concentration. Dixon and Webb (4) point out that cyanide can inhibit many other enzymes containing for the most part iron or copper. They list 42 enzyme reactions that can be inhibited by cyanides; however, cytochrome oxidase seems to be the most sensitive among these. The inhibition of cytochrome oxidase prevents the oxidation of reduced cytochrome C, thus stopping the utilization of molecular oxygen by cells. Since cytochrome oxidase occupies a central role in the utilization of oxygen in practically all cells, its inhibition rapidly leads to loss of cellular functions and then to cell death.

Cyanide does not combine appreciably with either the oxidized or reduced form of hemoglobin in the blood, although it combines with the 2 percent or so of methemoglobin (in which the iron is in the Fe^{3+} form) normally present.

The most specific pathologic finding in acute cases is the bright red color of venous blood. This is striking, visible evidence of the inability of the tissue cells to utilize oxygen, as a result of which the venous blood is only about one volume percent lower in oxygen content than arterial blood—in contrast to the usual arterial-venous difference of 4 to 5 volumes percent. As recovery takes place, the arterial-venous oxygen difference returns to normal.

Cyanide also reacts rapidly with methemoglobin, which contains Fe^{3+}. Albaum et al. (5) demonstrated in vitro that both cytochrome oxidase and methemoglobin can compete reversibly for cyanide, so that the addition of methemoglobin to a cyanide-inhibited cytochrome oxidase solution could partially restore the activity of the cytochrome oxidase. This phenomenon is an important aspect of the first-aid measures used for acute cyanide poisoning, which are discussed in greater detail in a subsequent section.

The reaction of cyanide with cytochrome oxidase is rapid. Schubert and Brill (6) found that the liver cytochrome oxidase was maximally inhibited 5 to 10 min after an intraperitoneal injection. In mice the cytochrome oxidase activity returned to normal after 5 to 20 min, but required up to 1 hr, or even more, in rats or gerbils. McNamara (7) estimated that the rate of metabolism of intravenously injected hydrogen cyanide in man was about 0.017 mg/kg/min.

If the concentration of cyanide ion is not so great as to cause death, then it is released from its combination with the ferric iron of cytochrome oxidase or methemoglobin, converted to thiocyanate ion (SCN^-) and excreted in the urine.

Lang (8), and Himwhich and Saunders (9), reported that the enzyme rhodanese, a sulfur transferase enzyme, is able to convert cyanide to thiocyanate. If the thiocyanate is not readily excreted in the urine, then it may be partially reconverted to cyanide by thiocyanate oxidase (10):

$$\text{Thiosulfate} + \text{cyanide} \xrightleftharpoons[\text{thiocyanate oxidase}]{\text{rhodanese}} \text{sulfite} + \text{thiocyanate}$$

The toxicity of thiocyanate is significantly less than that of cyanide, but chronically elevated levels of blood thiocyanate can inhibit the uptake of iodine by the thyroid gland and reduce thereby the formation of thyroxine (11).

Some free HCN is excreted unchanged in breath, saliva, sweat, and urine (12), and minor metabolic pathways include its oxidation to formate and its conjugation with cysteine to form 2-iminothiazolidine-4-carboxylic acid (13).

Low concentrations of cyanide are frequently found in normal human blood. Thus Feldstein and Klendshoj (14) found 0 to 10.7 μg CN^-/100 ml in 10 blood plasma samples. After exposure, plasma cyanide levels in man tend to return to normal levels within 4 to 8 hr after cessation of exposure (12, 14), suggesting a plasma half-life of 20 min to 1 hr after nonlethal exposures. Plasma thiocyanate levels are better indicators of exposure than plasma cyanide levels. Smokers tend to have higher plasma thiocyanate levels (2.1 to 2.9 mg thiocyanate/100 ml) (15), and smokers without known cyanide exposures also tend to have higher thiocyanate concentrations in their urine (4.4 mg/l for smokers and 0.17 mg/l for nonsmokers) (16).

1.3   Treatment of Poisoning by Cyanides and Nitriles

Although specific and effective antidotes for poisonous substances are very rare, partially effective antidotes are available for some cyanides and nitriles. The effectiveness of the antidotes derives from two basic facts: (2) methemoglobin binds well with any free cyanide ion. Therefore, if additional methemoglobin can be formed in the blood, this traps circulating cyanide ions that are either being continually absorbed from the stomach or through the skin or possibly returning to the bloodstream from the tissues. Since the formation of 10 or 20 percent methemoglobin usually involves no great risk, this can provide a large amount of cyanide-binding substance. (2) Since the amounts of thiosulfate formed in the body are relatively limited, the introduction of excess thiosulfate ions appears to increase the activity of the rhodanese enzyme and thus increase the rate of conversion of cyanide to the less toxic thiocyanate.

The combined use of methemoglobin-forming agents plus thiosulfate is capable of protecting experimental animals against at least 20 lethal doses of cyanide. A detailed account of this therapy can be found in reviews by Chen et al. (17), Wolfsie and Shaffer (18), and Wolfsie (19). In practice the antidotes are administered with the combined use of artificial respiration (when necessary;

however, mouth to mouth resuscitation is usually inadvisable, and the use of mechanical resuscitators is preferable) and the simultaneous inhalation of amyl nitrite vapor from ampuls crushed in a handkerchief and held by an assistant close to the nose of the victim. Several ampuls may be used in the course of the first half hour. This procedure alone may suffice for some of the milder cases, provided the sources of absorption are also removed (e.g., residual material on skin).

In more serious cases, if there is no response to the above, it may be necessary to produce methemoglobin in greater amounts and to use thiosulfate. This can be done by intravenous injection of 0.3 g sodium nitrite (10 ml of a 3 percent solution at a rate of 2.5 to 5 ml/min), followed at once by 12.5 g of sodium thiosulfate intravenously (50 ml of a 25 percent solution at the same rate as sodium nitrite). The sodium nitrite and thiosulfate therapy should be repeated in an hour, at half the original dose, if symptoms recur or persist. In milder cases, where the patient is conscious and not having much respiratory difficulty, recovery may occur without any specific therapy.

In order to use effectively such first-aid and medical therapy, it is necessary that all personnel and all physicians and nurses be thoroughly familiar with the toxic effects of cyanide and with the specific first-aid therapy. Because of the rapidity of onset, it is necessary to have first-aid kits in convenient locations in all areas where these materials are to be used. Specific training of employees in artificial respiration and in the use of amyl nitrite ampuls is extremely important.

1.4   Chronic Poisoning from Cyanide Exposure

Hardy et al. (15) and Wolfsie and Shaffer (18) discuss the possibility of chronic poisoning from cyanide exposure. Hardy suggests that in some individuals the thiocyanate excretion may be inadequate, and that increased thiocyanate levels may produce either goiters or symptoms of thiocyanate intoxication, or both.

The Hardy study has been questioned, however, because one of her cases came from an area of known endemic goiter. More recently, El Ghawabi et al. (2) reported finding 20 cases of mild to moderate thyroid enlargement among 36 male electroplating workers who had been exposed for up to at least 15 years to an average of 6.4 to 10.4 ppm cyanide in their breathing zones. Alterations in ^{131}I uptake suggested abnormalities in iodine uptake during exposure. The occurrence of thyroid abnormalities after long cyanide exposures has been only rarely reported. It is of interest in this connection that a long-term study in rats of the chronic toxicity of foods containing hydrogen cyanide at levels up to 300 ppm produced no typical evidence of chronic toxicity even though definite increases in thiocyanate concentrations could be found in the tissues of these animals (20). It is also interesting that no changes were found in the thyroid even though rats are sensitive to goitrogenic agents.

1.5 Fire and Explosion Hazards

Some materials discussed herewith are flammable or, in some cases, explosive. The hazard is increased by the release of hydrogen cyanide under the influence of heat, moisture, or acid. Fires involving nitriles or materials capable of generating hydrogen cyanide are always potentially very hazardous.

2 CYANIDES

2.1 Hydrogen Cyanide; (HCN) Prussic Acid, Hydrocyanic Acid

2.1.1 Source

Although hydrogen cyanide can be prepared by treating cyanide salts with dilute sulfuric acid, it is now manufactured largely by the reaction of ammonia, air, and methane in the presence of a platinum catalyst. In 1975 about 700 million lb of cyanide were produced. About 52 percent was used for acrylonitrile production, 18 percent for methyl methacrylate production, and 7 percent for sodium cyanide production (1).

2.1.2 Uses and Industrial Exposures

Hydrogen cyanide has wide usage, which may involve many different types of exposure. The chief uses are in fumigation of ships, buildings, orchards, and various foods; in electroplating; in mining; in the production of various resin monomers such as acrylates, methacrylates, and hexamethylenediamine; and in the production of other nitriles. It also has many uses as a chemical intermediate, and may be generated in such operations as blast furnaces, gas works, and coke ovens.

2.1.3 Physical and Chemical Properties

Physical state	Colorless liquid with characteristic odor
Molecular weight	27.03
Melting point	$-13.2°C$
Boiling point	$25.7°C$
Refractive index	1.2619 (20°C)
Vapor density	0.94 (air = 1)
Vapor pressure	807.23 mm Hg (27.22°C)
Percent in "saturated" air	100 (25.7°C)
Solubility	Soluble in alcohol, ether; miscible with water
Flash point	$-17.8°C$ (closed cup)
1 mg/m^3 $\approx$	0.9 ppm at 25°C, 760 mm Hg

2.1.4 Determination in the Atmosphere

Methods have been described in the NIOSH criteria document (21). Detector tubes, operating on the length of stain principle, are useful, but results should be interpreted with caution in the presence of reactive materials such as hydrogen sulfide, styrene, nitric acid, hydrochloric acid, and various organic vapors.

2.1.5 Physiologic Response

The typical symptoms and mode of action have been described previously. Relatively little gross or microscopic pathology can be seen following fatal inhalation of hydrogen cyanide. Though there may be scattered hemorrhages and scattered congestion, these are probably the result of anoxia. Venous blood may appear a brighter red color than normal. Hydrogen cyanide vapor is absorbed extremely rapidly through the respiratory tract; the liquid and possibly the concentrated vapor are absorbed directly through the intact skin.

There is relatively close agreement between the lethal concentrations in various species and in man. Of the usual experimental animals, the dog is most sensitive. The responses to various concentrations of hydrogen cyanide in animals and in man are listed in Tables 58.1 and 58.2.

Table 58.1. Physiologic Response to Various Concentrations of Hydrogen Cyanide in Air—Animals (22, 23)

| Animal | Concentration | | Response |
	mg/l	ppm	
Mouse	1.45	1300	Fatal after 1 to 2 min
Mouse	0.12	110	Fatal after $\frac{3}{4}$ hr exposure
Mouse	0.05	45	Fatal after $2\frac{1}{2}$ to 4 hr exposure
Cat	0.350	315	Quickly fatal
Cat	0.20	180	Fatal
Cat	0.14	125	Markedly toxic in 6 to 7 min
Dog	0.350	315	Quickly fatal
Dog	0.0125	115	Fatal
Dog	0.1	90	May be tolerated for hours; death after exposure
Dog	0.07–0.04	65–35	Vomiting, convulsions, recovery; may be fatal
Dog	0.035	30	May be tolerated
Guinea pig	0.035	315	Fatal
Guinea pig	0.23	200	Tolerated $1\frac{1}{2}$ hrs without symptoms
Rabbit	0.350	315	Fatal
Rabbit	0.13	120	No marked toxic symptoms
Monkey	0.14	125	Distinctly toxic after 12 min
Rat	0.12	110	Fatal after $1\frac{1}{2}$ hr exposure

Table 58.2.   Physiologic Response to Various Concentrations of Hydrogen Cyanide in Air—Man (22, 23)

Response	Concentration	
	mg/l	ppm
Immediately fatal	0.3	270
Estd. human LC_{50} after 10 min (7)	0.61	546
Fatal after 10 min	0.2	181
Fatal after 30 min	0.15	135
Fatal after $\frac{1}{2}$–1 hr or later, or dangerous to life	0.12–015	110–135
Tolerated for $\frac{1}{2}$–1 hr without immediate or late effects	0.05–0.06	45–54
Slight symptoms after several hours	0.02–0.04	18–36

2.1.6   Precautions in Handling

Detailed precautions for handling in industry may be obtained from the NIOSH criteria document on cyanide and cyanide salts (21). It is essential that all personnel be adequately trained in recognition of odors of hydrogen cyanide and in the application of proper emergency first-aid measures. All sources of vapor or liquid exposure should be carefully studied and adequate local exhaust ventilation be provided.

First-aid kits should be properly and conveniently located; and all employees in potentially hazardous areas should be under constant supervision in case of emergency.

Because hydrogen cyanide is highly toxic to all species living in water, special attention should be given to the possibility of water pollution (24).

2.1.7   Permissible Exposure Limit

The current permissible exposure limit (PEL) for hydrogen cyanide is 10 ppm (11 mg/m^3) as a time-weighted average (25). The concentration "immediately dangerous to life or health" (IDLH) representing a level from which one could die within 30 min is 50 ppm (25). NIOSH has proposed a reduction to 5 mg/m^3 as a 10-min ceiling concentration (21). The 1980 ACGIH TLV has a "skin" notation.

2.1.8   Flammability

Hydrogen cyanide is definitely flammable and burns in air with a bluish flame. The flammable limits are from 5.6 to 40 percent by volume in air.

2.1.9   Odor and Warning Properties

Hydrogen cyanide has a characteristic odor, which can be recognized by trained individuals at 2 to 5 ppm (26). The sense of smell is, however, easily fatigued; and there is wide individual variation in the minimum odor threshold.

2.2   Sodium Cyanide

NaCN

2.2.1   Source

Sodium cyanide can be prepared by heating sodium amide with carbon, by melting sodium chloride and calcium cyanamide together in an electric furnace, or by direct reaction of hydrogen cyanide with caustic soda to form sodium cyanide (27).

2.2.2   Uses and Industrial Exposures

Some of the more important uses of sodium cyanide are in the extraction of gold and silver from ores, the heat-treating of metals, electroplating, various organic reactions, and the manufacture of adiponitrile (27).

2.2.3   Physical and Chemical Properties

Physical state	White crystalline solid, deliquescent
Molecular weight	49.02
Melting point	564°C
Boiling point	1496°C
Vapor pressure	1.0 mm Hg (817°C), 10.0 mm Hg (983°C) (these temperatures may be encountered in metal-treating processes using cyanide)
Solubility	Readily soluble in water; slightly soluble in alcohol

2.2.4   Physiologic Responses

Sodium cyanide produces all the typical symptoms of other sources of cyanide ion. It can produce acute symptoms by inhalation and by skin absorption as well as by ingestion. The fatal dosage by oral ingestion varies considerably, depending on whether or not food is present in the stomach, etc. It is probably on the order of 1 to 2 mg/kg in man, as it is in a variety of experimental animals (28).

The symptoms and therapy are the same as those described previously in the introduction of this chapter. It is important to remove all remaining dust or solutions containing sodium cyanide from the skin in the event of acute exposures, since cyanide salts appear to be readily absorbed through the intact skin (21).

2.2.5 Permissible Exposure Limits

The permissible exposure limit currently suggested by NIOSH/OSHA for cyanide dust is 5 mg/m^3 expressed as CN (25).

2.2.6 Odor

The solid may have a light odor of hydrogen cyanide, especially if moisture is present.

2.3 Potassium Cyanide

KCN

2.3.1 Source

Potassium cyanide is produced by methods similar to those for sodium cyanide.

2.3.2 Uses and Industrial Exposures

Similar to sodium cyanide.

2.3.3 Physical and Chemical Properties

Physical state	White crystalline solid, deliquescent
Molecular weight	65.11
Melting point	636°C
Specific gravity	1.560
Solubility	Readily soluble in water—at 25°C, 1000 g dissolves 716 g KCN; slightly soluble in alcohol—at 10.5°C, 100 g dissolves 0.875 g KCN

2.3.4 Physiologic Response

The responses to excessive exposures by potassium cyanide are similar to those reported for sodium cyanide. In plating operations the exposures may involve both of these salts including other cyanide salts, depending on the metal that is being plated. Mucous membrane irritation and skin irritation have been reported in plating operations by Barsky (29). Tovo (30) cited a case history in which a fish poacher died apparently due to the percutaneous absorption of potassium cyanide which was intended as a fish poison. Streicher (31) has studied the effect of temperature on the toxicity of potassium cyanide in mice. The LD_{50} following oral administration at 23 to 25°C was 6.02 ± 3.3 mg/kg. When the mice were kept at temperatures of 40°C, the LD_{50} was 2.86 ± 1.6 mg/kg, indicating that toxicity increased with a decrease in environmental temperature.

Liebowitz and Schwartz (32) discuss the recovery of a patient with no specific therapy after a suicidal ingestion of an unusually large amount of potassium cyanide (3 to 5 g).

2.3.5 Odor

Similar to HCN.

2.4 Calcium Cyanide; Calcyanide, Cyanogas

$$Ca(CN)_2$$

2.4.1 Source

Calcium cyanide may be prepared by fusing calcium cyanamide ($CaCN_2$) with sodium chloride to give a crude mixture of calcium cyanide and sodium cyanide.

2.4.2 Uses and Industrial Exposures

Calcium cyanide is used as a fumigant and pesticide.

2.4.3 Physical and Chemical Properties

Physical state	Amorphous white powder
Molecular weight	92.12
Solubility	Readily soluble in water (with gradual liberation of HCN); soluble in alcohol

2.4.4 Physiologic Response

The physiologic properties are similar to those for other cyanide salts.

2.4.5 Permissible Exposure Limit

The current permissible exposure limit as suggested by NIOSH/OSHA for cyanide dusts is 5 mg/m^3 expressed as CN (25).

2.4.6   Odor and Warning Properties

The solid may have a rather definite odor of hydrogen cyanide.

2.5   Calcium Cyanamide; Calcium Carbimide, Cyanamid

$$CaCN_2$$

2.5.1   Source

The process of making calcium cyanamide involves three raw materials—coke, coal, and limestone—plus nitrogen. The limestone (calcium carbonate) is burned with coal to produce calcium oxide. The calcium oxide is then allowed to react with amorphous carbon in the furnace at about 2000°C with the formation of calcium carbide (CaC_2). Finely powdered calcium carbide is heated at about 1000°C in an electric furnace into which pure nitrogen is passed. It is then removed and uncombined calcium carbide removed by leaching (27).

2.5.2   Uses and Industrial Exposures

Calcium cyanamide has its major use as a fertilizer. However, it has a number of other uses such as an herbicide and a defoliant for cotton plants. It is finding increased use as a chemical intermediate. For example, it is being used to produce dicyandiamide, which in turn can be polymerized to form the widely used resin monomer, melamine. The conversion to calcium cyanide and hence into a variety of other uses is also important commercially.

2.5.3   Physical and Chemical Properties

Physical state	White crystalline solid
Molecular weight	80.11
Melting point	1300°C (sublimes >1150)
Specific gravity	2.3
Solubility	Decomposes in water, liberating ammonia

2.5.4   Physiologic Response

The principal exposures to calcium cyanamide dust (other than during manufacturing processes) result from its application as a fertilizer. The character of the toxic effect seems to be principally that of a transient vasomotor disturbance of the upper portion of the body. Irritation of the exposed mucous membranes and skin can occur, but this is probably related to the caustic content. It is thought that in the presence of body fluids calcium cyanamide reacts with carbon dioxide to form calcium carbonate and cyanamide ($CH_2 N_2$). Cyanamide is not converted to cyanide and its method of action is unknown. Glaubach (33) suggests that cyanamide may react with the sulfur groups of glutathione and thus influence enzymatic oxidation–reduction processes. Apparently, there is a wide variation in the sensitivity to the vasomotor effect and some evidence that it is increased by the simultaneous intake of alcohol (23). The possible Antabuse-like effect of cyanamide has been discussed by Hald et al. (34). Reports from older literature (23) of polyneuritis following acute exposures do not appear to be confirmed. DeLarrard and Lazarini (35) discuss five cases of poisoning in farmers characterized by dermatitis, vasomotor changes, and dyspnea.

2.5.5   Threshold Limit Value

$$\begin{array}{ll} \text{TWA} & 0.5 \text{ mg/m}^3 \text{ (1980 ACGIH)} \\ \text{STEL} & 1.0 \text{ mg/m}^3 \text{ (1980 ACGIH)} \end{array}$$

2.6   Cyanogen; Dicyan, Dicyanogen

$$N{=}CC{=}N$$

2.6.1   Source

Cyanogen can be prepared by slowly dropping potassium cyanide solution into copper sulfate solution or by heating mercury cyanide (37).

2.6.2   Uses and Industrial Exposures

Cyanogen has been used as a fumigant and may be encountered in situations in which there is heating of nitrogen-containing carbon bonds and in blast-furnace gases, etc.

2.6.3   Physical and Chemical Properties

Physical state	Colorless gas
Molecular weight	52.04
Melting point	$-34.4°C$
Boiling point	$-27.17°C$
Specific gravity	1.8064 (air = 1)
Solubility	Soluble in water (450 cm^3/100 ml water at 20°C); ethyl alcohol (2300 cm^3/100 ml alcohol at 20°C); ethyl ether (500 cm^3/100 ml ether at 20°C)

1 mg/l ≎ 469.6 ppm and 1 ppm ≎ 2.127 mg/m^3 at 25°C, 760 mm Hg

2.6.4   Determination in the Atmosphere

Cyanogen may be determined in the presence of hydrogen cyanide by first scrubbing out the cyanide with silver nitrate solution and then estimating the cyanogen by the ferrocyanide or thiocyanate methods (38).

2.6.5   Physiologic Response

The effect of cyanogen is similar in nature to that of other cyanides. It is thought to be converted in the body partly to hydrogen cyanide and partly to cyanic acid (HOCN). It appears to be somewhat more irritating than hydrogen cyanide; quantitatively, it appears to be less potent in a variety of species (see Table 58.3).

The effects on man are similar to those on animals, although it appears to be more irritating than hydrogen cyanide. Quantitative data on symptoms at various exposure levels appear to be lacking. Therapy and precautions are the same as for hydrogen cyanide.

2.6.6   Flammability

Flammable within the range of 6.60 to 42.60 percent by volume in air. Burns with peach-blossom red flame.

2.6.7   Odor

Pungent odor.

Table 58.3.   Toxicity of Cyanogen in Air for Various Animal Species (23)

Animanl	Concentration		Length of Time (hr)	Response
	mg/l	ppm		
Mouse	0.5	235	15 min	Recovered
Mouse	5.5	2,600	12 min	Fatal
Mouse	31.5	15,000	1 min	Fatal
Rat	0.59	350	1	LC_{50} (36)
Rabbit	0.21	100	4	Practically no effect
Rabbit	0.42	200	4	Slight symptoms
Rabbit	0.63	300	3.5	Severe symptoms; delayed death
Rabbit	0.84	400	1.8	Fatal
Cat	0.1	50	4	Severe symptoms but recovered
Cat	0.21	100	2–3	Fatal
Cat	0.42	200	$\frac{1}{2}$	Fatal
Cat	4.26	2,000	13 min	Fatal

2.6.8 Threshold Limit Value

TWA 10 ppm (1980 ACGIH)

2.7 Cyanogen Chloride; Chlorine Cyanide, Chlorocyanogen

CNCl

2.7.1 Source

Cyanogen chloride is produced by the action of chlorine on moist sodium cyanide suspended in carbon tetrachloride and kept cooled to $-3°C$, followed by distillation (37).

2.7.2 Uses and Industrial Exposures

Cyanogen chloride is used in organic synthesis (37) and as a warning agent in fumigant gases.

2.7.3 Physical and Chemical Properties

Physical state	Colorless liquid or gas
Molecular weight	61.48
Melting point	$-6°C$
Boiling point	13.8°C
Density of liquid	1.218 (4/4°C)
Vapor density	2 (air = 1)
Vapor pressure	1000 mm Hg at 2°C
Solubility	Soluble in water (2500 cm^3/100 ml H_2O at 20°C); ethyl alcohol (10,000 cm^3/100 ml alcohol at 20°C); ether (5000 cm^3/100 ml ether at 20°C). Dissolves readily, soluble in all organic solvents. Tends to form polymers upon storage

1 mg/l ≈ 398 ppm and 1 ppm ≈ 2.51 mg/m^3 at 25°C, 760 mm Hg

2.7.4 Determination in the Atmosphere

A colorimetric method is described by Jacobs (38).

2.7.5 Physiologic Response

Cyanogen chloride possesses the same general type of toxicity and mode of action as hydrogen cyanide, but is much more irritating even in very low concentrations. It can cause a marked irritation of the respiratory tract, with a hemorrhagic exudate of the bronchi and trachea, and pulmonary edema. Because of the high degree of irritant properties, it is improbable that anyone would voluntarily remain in areas with a high enough concentration to exert a typical nitrile effect. Tables 58.4 and 58.5 indicate the relationship of concentration to symptoms produced in various animal species and in man.

If the patient is conscious, first-aid and medical treatment should generally be directed toward the relief of any pulmonary symptoms. The patient should immediately be put to bed with the head slightly elevated and a medical examination carried out as quickly as possible. Oxygen should be administered if there is any dyspnea or evidence of pulmonary edema. If the patient has been trapped in an area so that the exposure was prolonged, it is possible that both cyanide effects and pulmonary edema may develop.

This situation has been studied experimentally by Jandorf and Bodansky (39). Dogs were exposed in pairs to concentrations of cyanogen chloride varying from 2.3 to 3.6 mg/l for periods of 1 to 2 min. One member of each pair received artificial respiration and inhalation of amyl nitrite (0.3 cm^3 administered by a nose cone) beginning 45 to 90 sec after the end of exposure. This was continued until the animals resumed spontaneous respiration or until a fatality occurred. At these high levels of exposure, 8 percent of the untreated animals recovered, whereas 77 percent of the amyl nitrite-treated animals survived. When the exposures were increased to concentrations ranging from 3.1 to 5.9 mg/l for 1.5 to 2.5 min, there was no benefit from the amyl nitrite therapy. Similar results were obtained with mice. Therefore, in cases with symptoms of both the nitrile-type effects and pulmonary edema, the combined therapy of oxygen plus amyl nitrite inhalations and artificial respiration seems to be indicated.

Table 58.4. Effects of Cyanogen Chloride Inhalation on Various Animal Species (23)

| | Concentration | | Length of Time | |
Animal	mg/l	ppm	(min)	Response
Mouse	0.2	80	5	Tolerated by some animals
Mouse	0.3	120	3.5	Fatal to some animals
Mouse	1.0	400	3	Fatal
Rabbit	3.0	1200	2	Fatal
Cat	0.1	40	18	Delayed fatalities after 9 days
Cat	0.3	120	3.5	Fatal
Cat	1.0	400	1	Fatal
Dog	0.05	20	20	Recovered
Dog	0.12	48	6 hr	Fatal
Dog	0.3	120	8	Severe injury, recovered
Dog	0.8	320	7.5	Fatal
Goat	2.5	1000	3	Fatal after 70 hr

Table 58.5. Effects of Varying Concentrations of Cyanogen Chloride in Air on Man (23, 26)

Concentration		Response
mg/l	ppm	
0.4	159	Fatal after 10 min
0.12	48	Fatal after 30 min
0.05	20	Intolerable concentration, 1-min exposure
0.005	2	Intolerable concentration, 10-min exposure
0.0025	1	Lowest irritant concentration, 10-min exposure

Effects of cyanogen chloride on experimental animals have been studied by Aldridge and Evans (40). These authors also point out the combined effect of pulmonary edema and the interference of the cellular metabolism by the cyanide ion.

The metabolism of cyanogen chloride has been studied by Aldridge (41). Cyanogen chloride is apparently converted to cyanide ion in vivo by a reaction with hemoglobin and glutathione, which eventually liberates the CN ion.

2.7.6 Threshold Limit Value

The 1980 ACGIH TLV lists a 0.3 ppm ceiling.

2.7.7 Odor and Warning Properties

Cyanogen chloride has a pungent odor detectable at 2.5 mg/m^3 (1 ppm) (26).

2.8 Cyanogen Bromide; Bromine Cyanide, Bromocyanogen

CNBr

2.8.1 Source

Cyanogen bromide may be prepared by either the action of bromine on potassium cyanide or the interaction of sodium bromide, sodium cyanide, sodium chlorate, and sulfuric acid (37).

2.8.2 Uses and Industrial Exposures

Cyanogen bromide is used in organic synthesis, as a fumigant and pesticide, and in gold-extraction processes. It has also been used in connection with cellulose technology.

2.8.3 Physical and Chemical Properties

Physical state	Colorless crystals (needles or cubes)
Molecular weight	105.93
Specific gravity	2.015 (20/4°C)
Melting point	52°C
Boiling point	61.6°C
Vapor density	3.62 (air = 1)
Vapor pressure	92.0 mm Hg (20 °C)
Density of "saturated" air	1.32 (20°C)
Percent in "saturated" air	12.1 percent (20°C)
Solubility	Soluble in water with hydrolysis, and in alcohol and ether

1 mg/l ≈ 230.9 ppm and 1 ppm ≈ 4.33 mg/m^3 at 25°C, 760 mm Hg

2.8.4 Determination in the Atmosphere

A colorimetric method is described by Jacobs (38).

2.8.5 Physiologic Response

Cyanogen bromide appears to be similar to cyanogen chloride in its effect. The sytemic toxicity may be greater and the irritant properties somewhat less than cyanogen chloride. Tables 58.6 and 58.7 indicate the response to various concentrations of cyanogen bromide in animals and man.

2.8.6 Threshold Limit Value

No official limit suggested; should certainly be less than 0.5 ppm.

2.8.7 Odor and Warning Properties

Has a penetrating odor and a bitter taste.

Table 58.6. Response of Animals to Various Concentrations of Cyanogen Bromide in Air (23)

Concentraton		Response	
mg/l	ppm	Mice	Cats
1	230	Fatal	Fatal
0.3	70	Paralysis after 3-min exposure	Paralysis after 3-min exposure
0.15–0.05	35–12	—	Severe injury; fatal on prolonged inhalation

Table 58.7. Response to Man to Various Concentrations of Cyanogen Bromide in Air (23, 26)

Concentration		Response
mg/l	ppm	
0.4	92	Fatal after 10 min
0.085	20	Intolerable concentration, 1-min exposure
0.035	8	Intolerable concentration, 10-min exposure
0.006	1.4	Lowest irritant concentration, 10-min exposure

2.9 Dimethyl Cyanamide

$$(CH_3)_2NCN$$

2.9.1 Physical and Chemical Properties

Physical state	Colorless liquid
Molecular weight	70.10
Melting point	$-41.0°C$
Boiling point	162 to 164°C
Density	0.8768 (30°C)
Vapor density	2.41
Vapor pressure	40 mm Hg (80°C)
Flash point	71°C (closed cup)

2.9.2 Physiologic Response

Fassett (42) found the oral LD_{50} in rats and guinea pigs to be 50 to 100 mg/kg. The symptoms were weakness, ataxia, gasping respirations, and unconsciousness. It was readily absorbed through the guinea pig skin (LD_{50} less than 5 ml/kg) with little skin irritation. It was not a severe eye irritant to the rabbit. No skin sensitization was produced in the guinea pig.

The compound may be hazardous, especially by skin absorption.

3 NITRILES

The nitriles are readily absorbed by all routes. Many of them display toxicologic effects that appear to be related to cyanide toxicity. However, not all nitriles dissociate readily to produce cyanide; thus various additional toxicologic effects can be noted for specific nitriles. The cyanohydrins (43), malononitrile, succinonitrile, and adiponitrile liberate cyanide, which then appears as elevated thiocyanate in the urine (11). The release of cyanide from some nitriles, such as acetonitrile, may be relatively slow (11). The toxicity of the individual nitriles is sufficiently different to discourage considering them collectively as "organic cyanides."

3.1 Acrylonitrile; Vinyl Cyanide, Propenenitrile

$$CH_2{=}CHCN$$

3.1 Source

Acrylonitrile is synthesized by the reaction of propylene with ammonia and oxygen in the presence of a catalyst (44).

3.1.2 Uses and Industrial Exposures

Acrylonitrile is one of the major industrial intermediates. In 1976 approximately 680 million kg was produced in the United States (44). A considerable proportion of this was used in the production of acrylic and modacrylic fibers, acrylonitrile–butadiene–styrene (ABS), and styrene–acrylonitrile (SAN) plastics.

3.1.3 Physical and Chemical Properties (45–47)

Physical state	Clear, colorless, volatile liquid (some technical grades slightly yellowish)
Molecular weight	53.06
Specific gravity	0.8060 (20/4°C)
Freezing point	−83.5°C
Boiling point	77.5–77.9°C
Refractive index	1.3911 (20°C)
Vapor density	1.9 (air = 1)
Vapor pressure	110 to 115 mm Hg (25°C)
Percent in "saturated" air	14.5 (25°C)
Density of "saturated air"	1.13 (25°C) (air = 1)
Solubility	In water, 7.3 percent by weight; soluble in all common organic solvents; forms azeotropes with water–benzene; soluble in isopropyl alcohol
Flash point	0 ± 2.5°C (39)

1 mg/l $\approx$ 460.5 ppm and 1 ppm $\approx$ 2.168 mg/m^3 at 25°C, 760 mm Hg

3.1.4 Determination in the Atmosphere

Collection of samples on activated charcoal or in methanol, followed by gas chromatographic analysis using an FID detector, has been found suitable.

3.1.5 Physiologic Response

Acrylonitrile can be readily absorbed by mouth, through intact skin, or by inhalation. It has long been known to possess a high degree of toxicity and to possess some of the characteristics of poisoning by the cyanide ion. Some typical values for which various toxic effects occur are listed in Table 58.8 (48).

Brieger et al. (49) have reviewed the literature regarding human exposures and have evaluated the effects of acrylonitrile in dogs, rats, and monkeys. They reported that the level of cyanide ion in blood appeared to be correlated with the degree of poisoning, and that the symptoms were similar to those due to cyanide. However, metabolic studies show that the degree of metabolic conversion of acrylonitrile to cyanide is relatively low. Thus Gut et al. (50) showed 20 percent conversion to cyanide after oral administration of acrylonitrile to Wistar rats, albino mice, or Chinese hamsters; 2 to 4 percent conversion after intramuscular or subcutaneous administration, and only 1 percent after intravenous administration. They also found acrylonitrile to be strongly bound to red blood cells. In vitro acrylonitrile has been found to conjugate readily with glutathione (51).

Also, the findings of mucous membrane irritation with hyperemia, lung edema, alveolar thickening, and hemosiderosis of the spleen in rats after subchronic inhalation exposure to 56 ppm acrylonitrile reported by Dudley et al. (48) are not compatible with cyanide-like effects, and are more likely to direct effects of acrylonitrile itself. Hashimoto and Kanai (52) have suggested

Table 58.8. Physiologic Response to Various Concentrations of Acrylonitrile in Air—Animals

Animal	Concentration mg/l	Concentration ppm	Response
Rat	1.38	636	Fatal after 4-hr exposure
Rat	0.28	129	Slight transitory effect
Rat	0.21	97	Slight transitory effects
Rabbit	0.56	258	Fatal during or after exposure
Rabbit	0.29	133	Marked transitory effects
Rabbit	0.21	97	Slight transitory effects
Cat	0.60	276	Markedly toxic
Cat	0.33	152	Markedly toxic, sometimes fatal
Guinea pig	1.25	576	4 hr LC_{50}
Guinea pig	0.58	267	Slight transitory effect
Dog	0.24	110	Fatal to three fourths of the dogs
Dog	0.213	98	Convulsions and coma; no death
Dog	0.12	55	Transitory paralysis; 1 dog died
Dog	0.063	29	Very slight effects

that the major mechanism of toxicity is due to the cyanoethylation of important sulfhydryl group containing enzymes.

Murray et al. (53) found that daily oral doses of 56 mg/kg/day given by gavage on days 6 to 15 to Sprague-Dawley rats produced significant maternal toxicity, fetal malformation, and embryotoxicity. No statistically significant effects were seen at doses of 10 or 25 mg/kg/day.

Milvy and Wolff (54) reported that acrylonitrile was mutagenic to *Salmonella typhimurium* strains TA 1535, TA 1538, and TA 1978 when activated with mouse liver homogenate; however, no dose-response relationships were detected. McMahon et al. (55) found a positive mutagenic response to acrylonitrile in both *E. Coli* and *S. typhimurium*, using a variant of the Ames assay.

Quast et al. (56) found increased cases of glial cell tumors (astrocytomas) in male and female Sprague-Dawley rats at and above 35 ppm acrylonitrile in the drinking water. Zymbal gland carcinomas and squamous cell carcinomas of the gastrointestinal tract were also noted. Maltoni et al. (57) exposed rats to acrylonitrile vapors at 5, 10, 20, and 40 ppm 4 hr/day, 5 days/week for 12 months and observed them for the remainder of their life-spans. He found slight increases in tumors in mammary glands of males and females, in the forestomach of males, and in the skin of females.

Human epidemiologic studies of workers exposed to acrylonitrile for long periods suggest an increased cancer risk. A study by O'Berg (58) showed a statistically significant increase in total cancer cases and especially in respiratory cancers. The study, however, lacked an analysis of smoking history. Another study (59) also showed slightly elevated cancer rates of the respiratory tract, genitourinary tract, and Hodgkin's disease, but not in overall cancer deaths for another group of workers who had been exposed to acrylonitrile.

3.1.6   Permissible Exposure Limit

The present exposure limit for acrylonitrile is 2 ppm.

3.1.7   Flammability

Acrylonitrile forms explosive mixtures with air in the range of 3.05 to 17.0 ± 0.5 percent by volume. The ignition temperature is 481°C. It is considered a fire and explosion hazard (60).

3.1.8   Odor and Warning Properties

Acrylonitrile has a characteristic unpleasant odor somewhat resembling that of pyridine. It can be detected with an odor threshold of 8 to 40 mg/m^3, but the sense of smell fatigues rapidly and is unreliable as an index of exposure.

3.2   2-Methyl-2-propenenitrile; Methacrylonitrile, Isopropenylnitrile

$$CH_2=C(CH_3)CN$$

3.2.1   Source

Methacrylonitrile can be derived from isobutyraldehyde.

3.2.2   Uses and Industrial Exposures

It is used as a monomer.

3.2.3   Physical and Chemical Properties

Physical state	Colorless liquid
Molecular weight	67.09
Melting point	$-35.8°C$
Boiling point	$90.3°C$
Specific gravity	0.8001 (20/4°C)
Vapor density	2.31 (air = 1)
Vapor pressure	40 mm Hg (12.8°C); 65 mm Hg (25°C); 100 mm Hg (32.8°C)
Density of saturated air	1.17 (air = 1)
Solubility	In water, 2.5 percent (20°C)
Flash point	13°C (open cup)

$$1 \text{ ppm} \approx 2.74 \text{ mg/m}^3 \text{ and } 1 \text{ mg/l} \approx 365 \text{ ppm}$$

3.2.4   Physiologic Response

McOmie (61) has made a comparative study of methacrylonitrile and acrylonitrile. The approximate LD_{50} for mice exposed 1 hr was 630 ppm (1700 mg/m³); for 4 hr it was about 400 ppm. None out of six mice was killed by an 8-hr exposure to 75 ppm. The animals showed respiratory paralysis and convulsions.

Methacrylonitrile was found to penetrate the rabbit skin readily, causing fatalities in doses of 2 to 4 ml/kg. One of the rabbits was treated with 20 mg/kg of sodium nitrite intravenously and was revived, indicating a typical nitrile effect.

Fassett (42) found the oral LD_{50} in mice to be 20 to 25 mg/kg and in rats 25 to 50 mg/kg. Symptoms were those of weakness, tremors, cyanosis, and convulsions. There was no damage to the rabbit cornea. It was absorbed readily through guinea pig skin with no skin irritation. It was not a skin sensitizer in this species. Inhalation of 9880 ppm for 2 hr killed three out of three rats. It should be handled the same as other toxic nitriles.

3.2.5 Odor and Warning Properties

Methacrylonitrile has a slight odor resembling cyanide.

3.2.6 Permissible Exposure Limit

TWA	1.0 ppm (1980 ACGIH)
STEL (skin)	2.0 ppm

3.3 Acetonitrile; Methyl Cyanide, Ethanenitrile

$$CH_3CN$$

3.3.1 Source

Acetonitrile is prepared by heating acetamide with glacial acetic acid (37).

3.3.2 Uses and Industrial Exposures

Acetonitrile is an important solvent. It is used as an intermediate in the synthesis of acetophenone, 1-naphthaleneacetic acid, and thiamine (43).

3.3.3 Physical and Chemical Properties

Physical state	Colorless liquid
Molecular weight	41.05
Specific gravity	0.7768 (25/4°C)
Melting point	-43 ± 2°C
Boiling point	81.6°C (760 mm Hg)
Refractive index	1.34596 (16.5°C)
Vapor pressure	87 mm Hg (24°C)
Percent in "saturated" air	9.6
Vapor density	1.42 (air = 1)
Density of "saturated" air	1.04 (air = 1)
Solubility	Infinitely soluble in water; readily miscible with alcohol, ether, acetone, chloroform, carbon tetrachloride, ethylene chloride
Flash point	48°C (Cleveland open cup)

1 mg/l $\approx$ 595.3 ppm and 1 ppm $\approx$ 1.68 mg/m^3 at 25°C 760 mm Hg

3.3.4 Determination in the Atmosphere

Acetonitrile can be analyzed by gas chromatography after trapping in a suitable solvent or activated charcoal.

3.3.5   Physiologic Response

The acute toxicity of acetonitrile in animals is summarized in Table 58.9.

The skin and eye irritation resulting from exposures to acetonitrile have been reported to be similar to those produced by acetone (62). In the acute oral studies in rats, the major symptoms observed were labored breathing, ataxia, cyanosis, and coma (64).

The LC_{50} for a single, 8-hr inhalation in male rats is 7500 ppm, with females being somewhat less sensitive. The rabbit and guinea pig are somewhat more sensitive. In the case of dogs, no fatalities occurred up to and including 8000 ppm for a 4-hr exposure; deaths occurred at levels of 16,000 and 32,000 ppm. Symptoms in animals appear to be those of prostration, followed by convulsive seizures. Autopsy findings indicate pulmonary hemorrhage and vascular congestion. At the lower dosage levels, the deaths always appeared to be delayed (65).

Repeated inhalation studies have also been made on a variety of species. Rats exposed 7 hr/day to acetonitrile vapor for a period of 90 days showed no specific effects at 166 or 330 ppm. At 665 ppm, a pulmonary inflammatory change and minor changes in the kidney and liver were noted in some animals. No mention was made of any effect on the thyroid (65).

Although these animals excreted some thiocyanate, apparently the amount was not proportional to the acetonitrile inhaled. Dogs and monkeys were exposed to acetonitrile vapor for 7 hr/day, 3 days/week, for 91 days. The mean concentration was approximately 350 ppm. The symptoms produced were not remarkable and only some minor variations in weight, hematocrit, and hemoglobin were reported. At autopsy some cerebral hemorrhage was noted in the monkeys and some evidence of focal emphysema and proliferation of alveolar septa in the lung. Rather marked pigment-bearing macrophages were consistently noted in monkeys. A similar picture in the lungs was noted in dogs. Small amounts of cyanide and thiocyanate ion were present, but the significance of this seems somewhat uncertain at the lower levels of exposure (65).

Human volunteers were studied at levels of 40, 80, and 160 ppm of acetonitrile vapor for periods of 4 hr. No specific subjective responses were noted. There

Table 58.9.

Route of Administration	Species	LD_{50} or LC_{50}	Remarks	Ref.
Oral	Rat	3.8 g/kg		62
Oral	Rat	2.46 g/kg		63
Oral	Rat	0.16 to 3.5 g/kg	Young more sensitive	64
Dermal	Rabbit	3.9 g/kg		62
Inhalation	Rat	16,000 ppm	4 hr	65
Inhalation	Guinea pig	5,655 ppm	4 hr	65
Inhalation	Rabbit	2,828 ppm	4 hr	65

was no consistent change in the blood cyanide level or urinary thiocyanate. From these various studies, the authors conclude that, at least in the case of dogs, the fatal concentrations were associated with the formation of cyanide in vitro, but that in some instances it may be that direct action of acetonitrile was responsible. They point out the important fact that determination of blood cyanide or urinary thiocyanate should not be relied on as evidence for brief inhalations of lower concentrations of acetonitrile vapor (65).

Acetonitrile has been shown to affect the thyroid. Matine et al. (66) showed that a progressive bilateral exophthalmos could be produced in rabbits with a daily intramuscular injection of 0.05 ml acetonitrile, and that this reaction could be inhibited by feeding of fresh vegetables. The degree of exophthalmos was related to the thyroid hyperplasia, and could be prevented by prior administration of iodine.

In 1955, Grabois (67) described a fatality and several cases of accidental poisoning in workers exposed to acetonitrile vapor. A comprehensive discussion of this incident has been given by Amdur (68).

The fatality occurred in a 23-year-old man who had been engaged for 2 days in hand-painting the interior of a tank with a resin containing 30 to 40 percent acetonitrile as well as other substances, such as diethylenetriamine and a mercaptan. Acetonitrile was the major volatile component. About 4 hr after leaving the job, the subject complained of chest pain, vomited, and had a massive hematemesis, followed by convulsions. About 9 hr later, he was admitted to the hospital in a comatose state, with an ashen-gray color and irregular and infrequent respirations; he expired about an hour after admission with convulsive seizures and marked rigidity of the neck. A postmortem examination disclosed only generalized vascular congestion. Examination of the blood and various organs showed high levels of cyanide ion (μg percent: blood, 796; urine, 215; kidney, 204; spleen, 318; lungs, 128; liver, 0).

Two additional cases were hospitalized with severe symptoms consisting of nausea and vomiting, respiratory depression, extreme weakness, and a semicomatose state. Both these men were treated with oxygen, fluids, and whole blood intravenously, as well as with ascorbic acid and sodium thiosulfate. Amyl nitrite was not used. One of the cases developed a transient weakness of the flexor muscles of the arms and wrists, and both developed urinary frequency, associated in one instance with albuminuria and in the other with the passage of a small oxalate-type urinary calculus. These cases showed also elevated blood cyanide levels and somewhat increased serum thiocyanate levels. All other exposed workers were evaluated; increased blood cyanide and thiocyanate values were found occasionally, and the symptoms described previously were found in lesser degree. It is interesting that none of these individuals developed any enlargement of the thyroid or alteration in thyroid function.

Dequidt et al. (69, 70) reported the fatal exposure of a photographic laboratory worker to acetonitrile. After a massive exposure to acetonitrile he left work, ate his evening meal, and began experiencing gastric distress and nausea about 4 hr after the exposure. He vomited during the night. When he was found on the next morning he was sweating profusely and was alternately crying out sharply and lapsing into a comatose state. Other symptoms were hypersalivation, conjunctivitis, very low urine output, low blood pressure, and albumin in the urine and cerebrospinal fluid. He experienced a cardiac and respiratory arrest from which he was resuscitated by cardiac massage and an intracardiac injection of adrenaline. In spite of continued treatment, he died 6 days later.

A possible explanation of the delayed onset of symptoms might be that acetonitrile is metabolized more slowly to cyanide than other similar nitriles, and that the final clinical picture is a result of effects of the intact molecule combined with the effects of gradually released cyanide ions.

In view of the report of severe intoxication from exposure to high concentrations of acetonitrile vapor, it seems important that the same protective measures should be applied as in the case of other nitriles, especially education of personnel and proper ventilation and protective measures. Certainly in the event of expected high concentrations such as those inside tanks, precautions should include supplied-air respirators and complete skin protection.

33.6   Permissible Exposure Limits

The present PEL is 40 ppm (25). A recent NIOSH criteria document (43) has proposed lowering that value to 20 ppm.

3.3.7   Odor and Warning Properties

Acetonitrile has an ethereal odor with a burning, sweetish taste. Minimum odor threshold level is not known; it is probable that the odor sensation is rapidly fatigued, which would make it an unreliable index of exposure.

3.4   Propionitrile; Ethyl Cyanide, Propanenitrile

$$C_2H_5CN$$

3.4.1   Source

Propionitrile is made by heating barium–ethyl sulfate and potassium cyanide, with subsequent distillation (37).

3.4.2   Uses and Industrial Exposures

Propionitrile is used as a solvent and in organic synthesis (37).

3.4.3 Physical and Chemical Properties

Physical state	Colorless liquid
Molecular weight	55.08
Specific gravity	0.7770 (25/4°C)
Boiling point	97.1 to 97.4°C
Melting point	-98 ± 6°C
Refractive index	1.3659 (24°C)
Vapor density	1.9 (air = 1)
Vapor pressure	40 mm Hg (22°C)
Density of "saturated" air	1.05 (22°C) (air = 1)
Solubility	Soluble in water up to 12 percent at 40°C; infinitely soluble in alcohol; slightly soluble in ether

3.4.4 Physiologic Response

Propionitrile has a high degree of toxicity and is thought to produce its action by fairly rapid metabolism to the cyanide ion. The fate of the other portion of the molecule is somewhat uncertain (13). Smyth et al. (71) indicates that the oral single dose LD_{50} in the rat is about 39 mg/kg. The LD_{50} by skin absorption in the rabbit was 0.21 ml/kg. A 2-min inhalation of saturated vapor killed all rats. A 4-hr exposure to 500 ppm gave a mortality of two out of six rats. Application to the skin and eyes of rabbits did not result in severe damage. Fassett (42) found that the approximate oral LD_{50} in the rat was 50 to 100 mg/kg and in the guinea pig 25 to 50 mg/kg. Intraperitoneally, the values for the rat were 25 to 50 mg/kg and for the guinea pig 10 to 25 mg/kg. The material was only slightly irritating to the skin, and had an LD_{50} of less than 5 ml/kg by skin in the guinea pig. Exposure of $1\frac{1}{2}$ hr to 9500 ppm killed all rats.

The subcutaneous LD_{50} in the guinea pig was found by Ghiringhelli (72) to be 18 mg/kg, and 70 percent of the dose was accounted for as thiocyanate. This material is obviously highly toxic and requires the same care in handling as other toxic nitriles.

Numerous authors, including Szabo and Selye (73) and Szabo and Reynolds (74) have demonstrated that propionitrile is a potent inducer of duodenal ulcers when injected subcutaneously in the rat.

3.4.5 Permissible Exposure Limits

The proposed permissible exposure limit to propionitrile is 6 ppm (43).

3.5 *n*-Butyronitrile; Butanenitrile, *n*-Propyl Cyanide

$$CH_3CH_3CH_2CN$$

3.5.1 Physical and Chemical Properties

Physical state	Colorless liquid
Molecular weight	69.11
Refractive index	1.3816 (24°C)
Density	0.796 (15°C)
Melting point	−112.6°C
Boiling point	116 to 117°C (760 mm Hg)
Vapor pressure	10 mm Hg (15.4°C); 40 mm Hg (38.4°C)
Vapor density	2.4 (air = 1)
Density of "saturated" air	1.07 (38.4°C) (air = 1)
Solubility	Slightly soluble in water; soluble in alcohol

3.5.2 Physiologic Response

Fassett (42) found the oral LD_{50} in rats to be 50 to 100 mg/kg and intraperitoneally, less than 50 mg/kg. The symptoms were weakness, tremors, vasodilatation, labored respiration, and terminal convulsions—similar to other active nitriles.

Similar symptoms were produced in mice. The LD_{50} by skin contact in the guinea pig was 0.1 to 0.5 ml/kg. Skin and eye irritation were slight. Inhalation of vapor readily produced fatalities in rats with symptoms of nitrile toxicity.

n-Butyronitrile is considered a highly hazardous material and full precautions should be used to prevent skin contact or inhalation of the vapor.

The first-aid and medical therapy should be the same as for hydrogen cyanide.

3.5.3 Permissible Exposure Unit

A recent NIOSH criteria document (43) has proposed a PEL of 8 ppm for n-butyronitrile.

3.6 Isobutyronitrile; 2-Methyl Propanenitrile, Isopropylcyanide

$$(CH_3)_2CHCN$$

3.6.1 Source

Isobutyronitrile can be derived from isobutyraldehyde.

3.6.2 Uses and Industrial Exposures

Isobutyronitrile is used in organic synthesis and as a gasoline additive.

3.6.3 Physical and Chemical Properties

Physical state	Colorless liquid
Molecular weight	69.11
Density	0.773
Melting point	$-75°C$
Boiling point	$107°C$
Vapor density	2.4 (air = 1)
Solubility	Slightly soluble in water; soluble in alcohol and ether

3.6.4 Physiologic Response

Fassett (42) found the oral LD_{50} in rats to be 50 to 100 mg/kg and in mice, 5 to 10 mg/kg. The symptoms were those of weakness, vasodilatation, tremors, and convulsions, similar to those caused by other nitriles. Thiocyanate was present in the urine. The LD_{50} by skin contact in the guinea pig was less than 5 ml/kg with only slight irritation noticed.

Vapor inhalation at a calculated concentration of 5500 ppm for about 1 hr killed all rats with similar symptoms to those seen after oral dosages.

Tsurumi and Kawada (75) reported depressed body weights and slight changes in some organ weights in male and female rats receiving orally 200 mg/kg isobutyronitrile daily for 14 days. Rats receiving daily doses of 38.6 mg/kg intraperitoneally for 14 days also demonstrated parenchymatous degeneration in the liver.

Theiss and Hey (76) reported a case history in which a 44-year-old man became unconscious while filling a tank with isobutyronitrile. He exhibited tonic-clonic movements of the arms, dilated pupils, weak pulse, shallow and gasping breathing, and cyanosis. After successive treatment with 1 mg norepinephrine intravenously, amyl nitrite, sodium nitrite, and sodium thiosulfate, followed by intravenous lobeline and phenobarbital, the patient's condition improved rapidly. Four hours after the exposure the patient was fully conscious, but complained of a headache during the following days. He was discharged 14 days after hospital admission.

3.6.5 Permissible Exposure Limits

A recent NIOSH criteria document has proposed a PEL of 8 ppm (43).

3.7 3-Hydroxypropionitrile; Ethylene Cyanohydrin, Hydracrylonitrile, 3-Hydroxypropanenitrile, Glycol Cyanohydrin

$$HOCH_2CH_2CN$$

3.7.1   Source

3-Hydroxypropionitrile can be prepared by reacting ethylene oxide with hydrogen cyanide or by reacting ethylene chlorohydrin with sodium cyanide.

3.7.2   Uses and Industrial Exposures

Its major use is in the synthesis of acrylonitrile.

3.7.3   Physical and Chemical Properties

Physical state	Colorless or straw-colored liquid
Molecular weight	71.08
Density of liquid	1.059 (0/4°C)
Melting point	−46°C
Boiling point	221°C, 724 mm Hg
Refractive index	1.4241 (25°C)
Vapor pressure	0.08 mm Hg (25°C); 1 mm Hg (58.7°C); 10 mm Hg (102°C)
Vapor density	2.45 (air = 1)
Solubility	Soluble in water and alcohol
Flash point	<27°C

3.7.4   Physiologic Response

Smyth (77) found the oral LD_{50} in rats to be 10 g/kg. Saturated vapor inhalation for 8 hr produced no effect. Sunderman and Kincaid (78) report that Hamblin found the minimum lethal dose in rabbits to be 0.9 to 1.4 g/kg. The oral LD_{50} in mice was 1.8 g/kg. Single applications to the skin caused moderate local irritation but no toxicity up to 3.8 g/kg in the rabbit. In 15 repeated applications to the rabbit skin, there was no injury. 3-Hydroxypropionitrile was applied to guinea pig skin (0.5 ml/guinea pig) on a gauze pad 1-in. square. No effect was noted in 24 hr. Rats and guinea pigs were exposed to vapor in an 8-liter chamber. Dry air at a rate of 0.9 l/min was passed through 250 ml in a 5-in. sintered glass tube. No effect was produced in rats or guinea pigs by a 1-hr exposure (78).

Fassett (42) found that the oral LD_{50} in the rat was between 3200 and 6400 mg/kg, with about the same values intraperitoneally. Little evidence of skin irritation was noted, and there was no significant skin absorption.

This material seems to be of a very low order of toxicity compared to some nitriles. Apparently, when the hydroxy group is in the beta position relative to the nitrile group, the compound is not readily hydrolyzed in the body to release cyanide. When the hydroxyl group is in the alpha position adjacent to the CN group, the extreme toxicity of nitriles is retained (77, 78). In view of the very low vapor pressure and the lack of significant toxicity in animals and the lack of reports of human injury, the customary training of personnel and handling precautions would seem sufficient.

3.8 Lactonitrile; 2-Hydroxypropanenitrile, Acetaldehyde Cyanohydrin, Ethylidene Cyanohydrin

3.8.1 Physical and Chemical Properties

Physical state	Colorless or straw-colored liquid
Molecular weight	71.08
Melting point	$-40°C$
Boiling point	103°C, 50 mm Hg
Refractive index	1.4058 (18.4°C)
Specific gravity	0.992 (0/18.4°C)
Vapor density	2.45
Vapor pressure	10 mm Hg (74°C)
Flash point	76.7°C (170°F)
Solubility	Readily miscible with water, acetone, alcohol, and other organic solvents

3.8.2 Physiologic Response

Lactonitrile is reported (79) to be an extremely toxic compound by oral administration and skin or eye contact. The acute oral LD_{50} (species not mentioned) was 21 mg/kg with deaths occurring as low as 10 mg/kg. As little as 0.05 ml of the undiluted compound applied to the eye was fatal to all animals within a period of 5 min. The LD_{50} by skin application was less than 1 ml/kg; all deaths occurred within a period of 1 hr.

It is unknown whether or not lactonitrile produces its effects by virtue of hydrolysis to give the cyanide ion, or whether it acts as an intact molecule. Methods for determination in blood have been reported (80, 72). Extreme care is necessary in handling this material, with particular attention to the education of personnel and availability of prompt first-aid and medical treatment.

3.9 2-Methyllactonitrile; Acetone Cyanohydrin, Hydroxyisobutyronitrile

3.9.1 Uses

Intermediate in resin synthesis.

3.9.2 Physical and Chemical Properties

Physical state	Liquid
Molecular weight	85.11
Melting point	$-20°C$
Boiling point 95°C	(760 mm Hg)
Density	0.932
Vapor density	2.93
Vapor pressure	0.8 mm Hg (20°C), 23.0 mm Hg (82°C)
Solubility	Soluble in water, alcohol, ether, acetone, benzene

3.9.3 Physiologic Response

Acetone cyanohydrin is readily absorbed by all routes, and it may be largely metabolized to yield free cyanide. The dermal LD_{50} in rats is 140 mg/kg (78). Motoc et al. (81) administered 5 mg acetone cyanohydrin per rat twice a week for 3 to 8 months. They found fatty changes in the liver, hepatic necrosis, and kidney lesions. In a series of intermittent static inhalation exposures (nominal concentration: 10.2 mg/l; duration of each individual exposure unknown), Motoc et al. (81) reported desquamation of the bronchial epithelium leading eventually to superficial bronchial ulcerations. Kidney lesions were also evident in the inhalation experiment.

Sunderman and Kincaid (78) also reported two human fatalities after acetone cyanohydrin exposure. In one of those cases a worker was splashed with an unknown quantity of acetone cyanohydrin when a tank overflowed. After 3 hr he complained of nausea and was examined at a hospital but returned to work on the advice of a physician. At work he became nauseated again, lost consciousness, and became convulsive. He died 6½ hr after the initial exposure.

Thiess and Hey (76) described another human case of acetone cyanohydrin poisoning, mostly by skin absorption. Nausea, vomiting, loss of consciousness, and tonic-clonic convulsions were evident. Treatment with sodium nitrate and sodium thiosulfate was effective.

3.9.4 Permissible Exposure Limit

A recent NIOSH criteria document suggests an exposure ceiling value of 1 ppm for acetone cyanohydrin (43).

3.10 Glycolonitrile

$$HOCH_2CN$$

3.10.1 Uses

Glycolonitrile is used as an organic intermediate in the synthesis of bactericides and fungicides, hydantoin, chloroacetonitrile, glycine, α-aminonitriles and α-ethylpiperazines. It is also used as a barrier resin additive.

3.10.2 Physical and Chemical Properties

Physical state	(Anhydrous glycolonitrile) colorless, odorless oil with sweetish taste
Molecular weight	57.05 (theoretical)
Boiling point	183°C (slight decomposition)
Density	1.104 (19°C)
Vapor density	1.96
Vapor pressure	1 mm Hg (63°C); 14 mm Hg (102°C)
Solubility	Soluble in water, ethanol, ether

3.10.3 Physiologic Response

The oral LD_{50} in mice has been reported as 10 mg/kg (82). The oral LD_{50} in rats is 16 mg/kg (83). The data for the dermal toxicity of glycolonitrile vary greatly. Thus Wolfsie (82) reported a dermal LD_{50} of 105 to 130 mg/kg in rabbits, whereas Smyth et al. (83) reported a dermal LD_{50} of 5 mg/kg in the same species. An inhalation exposure of mice, rats, and guinea pigs to 27 ppm glycolonitrile for 8 hr resulted in 6/7 deaths in mice, 2/7 deaths in rats, and 0/7 deaths in guinea pigs during the exposure period. During the next 18 hr the single surviving mouse and four more rats died. All guinea pigs survived (82). The ingestion of up to 62 mg/kg/day glycolonitrile in the diet of male rats, or the ingestion of up to 92 mg/kg/day by female rats for 13 weeks produced no observable effects (82).

A human dermal exposure to 70 percent glycolonitrile (82) (possibly accompanied by an inhalation exposure) resulted in complaints of headache, dizziness, "rubbery legs," and unsteady gait. The exposed worker vomited several times, was pale, and appeared bewildered; he spoke irrationally and became unresponsive. The pulse was rapid and irregular. His condition improved upon treatment with amyl nitrite, oxygen, and sodium thiosulfate. He returned to work the next day, but complained of weakness and nausea for five more days, and of congestion of the pharyngeal mucosa for a longer period.

3.10.4   Permissible Exposure Limits

A recent NIOSH criteria document (43) suggests a ceiling limit of 2 ppm for glycolonitrile.

3.11   **Succinonitrile;** Butanedinitrile, Ethylene Cyanide

$$CNCH_2CH_2CN$$

3.11.1   Physical and Chemical Properties

Physical state	Colorless, waxy solid
Molecular weight	80.09
Melting point	57 to 57.5°C
Boiling point	265 to 267°C (760 mm Hg)
Density	0.9867
Vapor density	2.8 (air = 1)
Vapor pressure	6 mm Hg (125°C)
Solubility	Soluble in water (12.8 g/100 ml of H_2O); soluble in alcohol, benzene, ether
Flash point	270°C

3.11.2   Physiologic Response

The acute toxicity appears somewhat lower than materials such as propionitrile or butyronitrile. The oral LD_{50} in rats is 450 mg/kg. The effects on the skin of rabbits of a 95 percent water solution were those of mild irritation. Continued contact of the solution with rabbit skin for 18 hr produced fatalities, indicating a probable hazard by skin absorption. A 24-hr exposure of mice to vapor from a 95 percent solution caused no symptoms (84). Adequate precautions should be taken against skin or eye contact. The inhalation hazard is uncertain. Contessa and Santi (85) determined that rats and rabbits converted about 60 percent of the succinonitrile to cyanide.

3.11.3   Permissible Exposure limit

A recent NIOSH criteria document (43) suggested a time-weighted limit of 6 ppm for succinonitrile.

3.12   Adiponitrile; Tetramethylene Dicyanide, Adipyl Dinitrile

$$CN(CH_2)_4CN$$

3.12.1   Source and Uses

Adiponitrile is derived from butadiene and used as an intermediate for hexamethylenediamine in nylon manufacturing (86).

3.12.2   Physical and Chemical Properties

Physical state	Colorless liquid
Molecular weight	108.14
Melting point	1–3°C
Boiling point	295°C (100 mm Hg)
Density	0.965 (20/4°C)
Vapor density	3.73
Vapor pressure	2 mm Hg (119°C)
Solubility	Slightly soluble in water; soluble in alcohol, chloroform
Flash point	199.4°C (open cup)

3.12.3   Physiologic Response

Ghiringhelli (72) reports a subcutaneous LD_{50} in the guinea pig of about 50 mg/kg, and that it is hydrolyzed to hydrogen cyanide in the body, giving rise to SCN in the urine. Seventy-nine percent of the dose was eliminated as SCN in the urine. In exposed guinea pigs thiosulfate was a more effective treatment than nitrites. No effect was seen on the blood of guinea pigs from repeated doses (3 to 30 mg/kg subcutaneously, 6 days/week for 40 to 70 days). Skin penetration was suggested by the increase in SCN in the urine of guinea pigs after application to depilated skin. Greater quantities were absorbed when the skin was abraded. The oral LD_{50} of adiponitrile in rats was reported to be 300 mg/kg (43).

Svirbely and Floyd (87) undertook an extensive evaluation of adiponitrile in rats and dogs. A 2-year drinking water exposure of Wistar rats at 0.5, 5.0, and 50 ppm adiponitrile produced significant adrenal degeneration in female rats at all three concentrations and at 50 ppm in male rats. All body weights and organ weight ratios of spleen, liver, and kidney were within normal ranges.

Exposures of pregnant Sprague-Dawley rats at 10, 100, and 500 ppm in drinking water did not change fertility, gestation, or viability of offspring. Exposure of mongrel dogs to approximately 10, 100, 500, and 1000 ppm adiponitrile in the diet resulted in greatly decreased food intake and vomiting at 1000 ppm. No hematologic abnormalities were found. Kidney and liver function were normal at 500 ppm and below.

A case history of a human exposure by Ghiringhelli (72) reports the effects of drinking "a few milliliters" of adiponitrile by an 18-year-old male. About 20 min after ingestion he experienced tightness in the chest, headache, weakness with difficulty in standing, and vertigo. He became cyanotic, respirations were rapid, and he had low blood pressure and tachycardia. The pupils were dilated and barely reacted to light. He exhibited mental confusion and tonic-clonic contractions of limbs and facial muscles. His stomach was pumped out without effect on symptoms. Intravenous treatment with sodium thiosulfate and glucose resulted in rapid recovery which lasted for 4 hr, after which the patient relapsed into the previous state, possibly with greater severity for 2 hr. After another course of treatment with sodium thiosulfate and glucose, the patient recovered slowly and completely.

Zeller et al. (88) reported that human skin exposures to adiponitrile result in skin irritation and inflammation, and cite one case in which adiponitrile caused massive destruction of the skin on one foot.

3.12.4 Odor and Warning Properties

There are none.

3.12.5 Permissible Exposure Limits

A recent NIOSH criteria document has proposed a limit of 4 ppm for adiponitrile as a time-weighted average (43).

3.13 3-Dimethylaminopropionitrile

$$(CH_3)_2NCH_2CH_2CN$$

3.13.1 Physical and Chemical Properties

Physical state	Colorless, mobile fluid
Molecular weight	98.15
Melting point	$-44.2°C$
Boiling point	172°C
Density	0.8617 (30°C)
Vapor density	3.4 (air = 1)

Vapor pressure	10 mm Hg (57°C)
Density of "saturated" air	1.03 (air = 1)
Solubility	Miscible with water, alcohol, other solvents
Flash point	64°C (closed cup)

3.13.2 Physiologic Response

Preliminary data (89) indicate a low order of acute toxicity in mice and rats. The oral LD_{50} in mice is 1.5 g/kg. The oral LD_{50} in rats is 2600 mg/kg and the dermal LD_{50} in rabbits is 1410 mg/kg (36). The vapor was thought to be hazardous, and care in handling was suggested.

3.14 3-Isopropylaminopropionitrile

$$(CH_3)_2CH_2NHCH_2CH_2CN$$

3.14.1 Physical and Chemical Properties

Physical state	Liquid
Molecular weight	112.18
Melting point	$<-20°C$
Boiling point	87°C, 17 mm Hg
Density	0.864 (25°C)
Vapor density	3.9 (air = 1)
Vapor pressure	2 mm Hg (60°C)
Solubility	Miscible with water and other solvents
Flash point	$>41°C$ (open cup)

3.14.2 Physiologic Response

Similar to 3-dimethylaminopropionitrile (89). The oral LD_{50} in mice is 2175 mg/kg.

3.15 3-Methoxypropionitrile

$$CH_3OCH_2CH_2CN$$

3.15.1 Physical and Chemical Properties

Physical state	Colorless liquid
Molecular weight	85.1
Melting point	$-62.9°C$
Boiling point	160°C

Density	0.9299 (30°C)
Vapor density	2.9 (air = 1)
Vapor pressure	10 mm Hg (55°C)
Solubility	In water, 33.5 g/100 g; miscible with alcohol, toluene, other solvents
Flash point	65°C (closed cup)

3.15.2 Physiologic Response

Similar to 3-dimethylaminopropionitrile (89). The oral LD_{50} in mice is 3.2 g/kg.

3.16 3-Isopropoxypropionitrile

$$(CH_3)_2CH_2OCH_2CH_2CN$$

3.16.1 Physical and Chemical Properties

Physical state	Liquid
Molecular weight	113.16
Melting point	−67°C
Boiling point	177°C
Density	0.883 (25°C)
Vapor density	3.9 (air = 1)
Soluble	In water, 6.4 g/100 g water; miscible with acetone, benzene, other solvents

3.16.2 Physiologic Response

Similar to 3-dimethylaminopropionitrile (89). The oral LD_{50} in mice is 4450 mg/kg 3-Isopropoxypropionitrile is a skin irritant in rabbits.

3.17 3-Chloropropionitrile

$$ClCH_2CH_2CN$$

3.17.1 Physical and Chemical Properties

Physical state	Colorless liquid
Molecular weight	89.53
Melting point	−51°C
Boiling point	132°C, 200 mm Hg
Density	1.1363 (25°C)
Vapor density	3.1 (air = 1)

Vapor pressure	5 mm Hg (46°C)
Soluble	In water, 4.5 g/100 g at 25°C; miscible with acetone, carbon tetrachloride, benzene, other solvents
Flash point	76°C (Closed cup)

3.17.2 Physiologic Response

3-Chloropropionitrile is reported to be highly toxic (84). The oral LD_{50} in mice is 9 mg/kg, and in rats, 100 mg/kg. Symptoms are those of deep anesthesia with no demonstrable pathology. Exposure to the vapor of 0.01 ml in a 1-liter beaker killed all mice in 18 hr. It is probably absorbed through the intact skin. The mechanism of action, however, appears unknown. The marked increase in toxicity associated with the 3-chloro in contrast to the 3-hydroxy substitution, and the atypical symptoms suggest a different mode of action. It is also of interest that the substitution of a methyl group (as in n-butyronitrile) results in the retention of the typical symptoms and potency of an active nitrile.

Substitution of an amino group (see 3-aminopropionitrile, Section 3.18) causes an even more extraordinary change in response, namely, that of an alteration of growth of mesodermal tissues at low levels in the diet. If the other hydrogen of the amino group is replaced by a second propionitrile (see 3-3'-iminodipropionitrile, Section 3.19) group, the effect changes to one of marked central nervous system damage.

The remarkable variety of toxicologic effects produced by this series of compounds indicates that they should be handled with caution and all exposed persons closely followed medically.

3.18 3-Aminopropionitrile; BAPN

$$NH_2CH_2CH_2CN$$

3.18.1 Physical and Chemical properties

Physical state	The free base is a liquid; the hydrochloride is a crystalline solid
Molecular weight	70
Boiling point	79 to 81°C, 16 mm Hg
Refractive index	1.4396
Vapor pressure	2 mm Hg (38 to 40°C)

3.18.2 Physiologic Response

3-Aminopropionitrile has been studied extensively since the isolation of its glutamyl derivative as the causative factor in the toxic effect of sweet peas

(90–92). The disease produced by ingestion of large quantities of sweet peas in man is known as lathyrism and is characterized by paralysis of the legs, and other central nervous system symptoms. In young rats, and various avian species, it produces severe skeletal deformities and aneurysms, leading to rupture of the aorta. The effective doses to turkey poults may be as low as 0.01 percent in the diet (93).

In the rat, somewhat higher concentrations may be necessary (0.1 to 0.2 percent) (94). The mechanism of the effect is unknown, but it is thought to be by some action on growth of certain mesodermal tissues. It is not due to one of its major metabolites, cyanoacetic acid (95), and both the free amino group and the cyano group seem essential for activity. It is not produced if the amino group is in the alpha position, or if placed in the gamma position in butyronitrile. On the contrary, aminoacetonitrile appears fully potent.

Some other related compounds found not to produce growth effects were propionitrile, potassium cyanide, bis(3-cyanoethyl)amine (3,3'-iminodipropion-itrile), ethylene cyanohydrin, 3-methylaminopropionitrile, 3-dimethylamino-propionitrile, and trimethylenediamine (91, 92).

3.19 3,3'-Iminodipropionitrile; Bis(3-cyanoethyl)amine

$$HN(CH_2CH_2CN)_2$$

3.19.1 Physical and Chemical Properties

Physical state	Colorless liquid
Molecular weight	123.2
Melting point	$-5.50°C$
Boiling point	173°C, 10 mm Hg
Density	1.0165 (30°C)
Vapor density	4.2 (air = 1)
Vapor pressure	1 mm Hg (140°C)
Solubility	Soluble in water, ethanol, acetone, benzene
Flash point	>80°C

3.19.2 Physiologic Response

The LD_{50} is greater than 3000 mg/kg when 3,3'-iminodipropionitrile is given orally to mice. The oral LD_{50} in rats is 2700 mg/kg (36). Central nervous system damage was apparent in 3 days, and persisted for prolonged periods. The same symptoms were noted after skin application. Damage to the lens of the eye was noted after oral dosage, but not after skin contact. The inhalation hazard is unknown (89). Injection of 1 to 2 g/kg in rats, mice, birds, and fish was followed in 2 to 10 days by a great increase in motor activity, changes in behavioral patterns, backward walking, and head twitching, similar to results caused by lysergic acid diethylamide, except for the delay in onset and permanence of symptoms. Marked histologic damage was found in the brain (96).

3.20 Malononitrile; Malonic Dinitrile, Methylene Cyanide, Propanedinitrile, Cyanoacetonitrile

$$CH_2(CN)_2$$

3.20.1 Uses

Malononitrile is used as a lubricating oil additive. It is used in the synthesis of thiamine, pteridine-type anticancer agents, acrylic fibers, and dyes.

3.20.2 Physical and Chemical Properties

Physical state	Colorless solid
Molecular weight	66.06
Melting point	30 to 31°C
Boiling point	218 to 220°C (760 mm Hg)
Density	1.049 (34°C)
Solubility	Soluble in water (13 g/100 ml of water); soluble in alcohol, ether, benzene

3.20.3 Physiologic Response

Stern et al. (97) found that 14 mg/kg subcutaneously in rats produced severe symptoms of dyspnea, cyanosis, and convulsions, and was a nearly fatal dose. Studies of tissue homogenates exposed to malononitrile showed that cyanide and thiocyanate are produced, along with an inhibition of respiration, and an increase of aerobic glycolysis resembling the action of cyanide.

The oral LD_{50} in rats is 61 mg/kg; the oral LD_{50} in mice is 19 mg/kg (36).

Fifty percent of mice and rats exposed for 2 hr to 200 to 300 mg/m^3 malononitrile vapor died (98). The white mice developed signs of restlessness. Following exposure, the respiratory rate first increased and then decreased, accompanied by lethargy. The mice became cyanotic and movements became incoordinated and were followed by tremors and convulsions leading to death in some animals. Oral administration of malononitrile near the LD_{50} produced moderate destruction of the mucosa of the stomach in mice, and a general hyperemia of all organs. Panov (98) also found evidence that malononitrile is absorbed through the intact tail skin of mice. In an inhalation exposure of rats to 36 mg/m^3 for 2 hr/day for 35 days, Panov (99) found no mortalities. Body weights were within normal ranges, but the lung weight increased. There was a slight decrease in the blood hemoglobin concentration accompanied by an increase in reticulocytes.

Van Breemen and Hiraoka (100) reported that 6 to 8 mg/kg malononitrile (route unknown) in rats produced extensive electron-microscopic changes in spinal ganglia. Hicks (101) reported lesions in the corpus streatum accompanied by a proliferation of microglia and oligodendroglia.

In the late 1940s malononitrile was used experimentally in the treatment of schizophrenia and depression (102). It was thought that malononitrile might stimulate the formation of proteins and polynucleotides in nerve tissue and thereby restore normal function. Patients were given an intravenous infusion of 5 percent malononitrile for 10 to 69 min. The total dose during each treatment ranged from 1 to 6 mg/kg. Treatments were given two to three times per week, with at least 1-day intervals. Ten to twenty minutes after the beginning of the infusion, all patients experienced tachycardia. In addition, redness, nausea, vomiting, headache, shivering, muscle spasms, and numbness were reported with varying frequency. Two patients experienced convulsions, and one case of cardiac collapse was encountered.

3.20.4 Permissible Exposure Limits

A recent NIOSH criteria document (43) has proposed a limit of 3 ppm malononitrile as a time-weighted average.

3.21 Cyanoacetic Acid; Malonic Mononitrile, Cyanoethanoic Acid

$$CNCH_2COOH$$

3.21.1 Source

Reaction of sodium chloroacetate and potassium cyanide.

3.21.2 Uses and Industrial Exposures

As a chemical intermediate.

3.21.3 Physical and Chemical Properties

Physical state	White crystals
Molecular weight	85.06
Melting point	66°C
Boiling point	108°C, 15 mm Hg
Solubility	Soluble in water, alcohol

3.21.4 Physiologic Response

Although no studies were found concerning industrial hazards, cyanoacetic acid has been studied with reference to its possible role in the production of the symptoms of lathyrism by 3-aminopropionitrile (95). Injection of ^{14}C-labeled 3-aminopropionitrile in rats showed that 25 to 30 percent could be recovered as cyanoacetic acid. In order to evaluate this metabolite, rats were given drinking water containing 200 mg cyanoacetic acid/100 ml daily for 7 weeks. No toxic effects of any sort were noted, indicating that cyanoacetic acid is not responsible for the skeletal deformities, etc., produced by feeding 3-aminopropionitrile. The intraperitoneal LD_{50} in mice is 200 mg/kg (36).

3.22 **2-Cyanoacetamide;** Propionamide Nitrile, Nitrilomalonamide
$$CNCH_2CONH_2$$

3.22.1 Physical and Chemical Properties

Physical state	White powder
Molecular weight	84.08
Melting point	119°C
Boiling point	Decomposes
Solubility	Soluble in water—15 g/100 g water; soluble in ethanol— 2 g/100 g of ethanol

3.22.2 Physiologic Response

Fassett (42) noted that the oral LD_{50} in rats was greater than 3200 mg/kg and greater than 800 mg/kg intraperitoneally. Contact with the skin of guinea pigs caused slight irritation with no evidence of toxic symptoms by skin absorption. Valdecasas (103) states that it has very low toxicity.

3.23 **Methyl Cyanoacetate;** Cyanoacetic Acid Methyl Ester

$$CH_3OOCCH_2CN$$

3.23.1 Physical and Chemical Properties

Physical state	Liquid
Molecular weight	99.09
Melting point	-22.5°C
Boiling point	203°C
Specific gravity	1.123 (15/4°C)
Vapor density	3.4 (air = 1)
Solubility	Insoluble in water; soluble in alcohol, ether

3.23.2   Physiologic Response

Fassett (42) found that oral LD_{50} in the guinea pig to be 400 to 800 mg/kg, and the same value intraperitoneally. Some toxic effects following skin contact were noted. Although there are no reports of injury to humans handling the material, care should be used to avoid skin contact and inhalation of vapor, especially heated vapor.

3.24   Ethyl Cyanoacetate; Cyanoacetic Acid Ethyl Ester

$$CH_2(CN)COOC_2H_5$$

3.24.1   Physical and Chemical Properties

Physical state	Colorless liquid
Molecular weight	113.12
Melting point	$-22.5°C$
Boiling point	205 to 208°C
Specific gravity	1.0560 (25/4°C)
Vapor pressure	1 mm Hg (68°C)
Solubility	Slightly soluble in water; soluble in alcohol, ether

3.24.2   Physiologic Response

Fassett (42) found the oral LD_{50} in rats to be greater than 400 and less than 3200 mg/kg. The LD_{50} by skin contact in the guinea pig was greater than 5 ml/kg. No skin irritation was noted, although some effects were probably produced by skin absorption. Ghiringhelli (72) obtained a subcutaneous LD_{50} of about 1100 mg/kg in the guinea pig.

3.24.3   Odor

Ethyl cyanoacetate has a mild, pleasant odor.

3.25   Methylcyanoformate; Cyanomethyl Carbonate, Methyl Cyanomethanoate

$$CNCOOCH_3$$

3.25.1   Physical and Chemical Properties

Physical state	Colorless liquid
Molecular weight	85.03
Boiling point	97°C
Density	1.08

3.25.2   Physiologic Response

Methylcyanoformate is said to act like hydrogen cyanide (23) but to be more active at lower concentrations. A dog recovered from a 10- to 20-min exposure to 29 ppm (0.1 mg/l). Cats were severely affected and developed pulmonary damage by short exposures to 3 to 18 ppm. Mice succumbed to 15-min exposures of 86 ppm (0.3 mg/l).

3.26   Ethylcyanoformate; Cyanoethyl Carbonate, Ethyl Cyanomethanoate

$$CNCOOC_2H_5$$

3.26.1   Physical and Chemical Properties

Physical state	Colorless liquid
Molecular weight	99.05
Boiling point	116°C
Density	1.013 (20/4°C)
Solubility	Insoluble in water, soluble in alcohol

3.26.2   Physiologic Response

Similar to, but slightly less potent than, methylcyanoformate (23).

3.27   Methylisocyanide; Methylcarbylamine, Isocyanic Acid Methyl Ester

$$CH_3NC$$

3.27.1   Physical and Chemical Properties

Physical state	Colorless liquid
Molecular weight	41.05
Melting point	−45°C
Boiling point	59.6°C
Density	0.756 (4°C)
Solubility	Soluble in water—10 g/100 ml at 15°C; soluble in alcohol

3.27.2   Physiologic Response

Methylisocyanide vapor is highly irritating to humans at 2 ppm. The oral LD_{50} in rats is 71 mg/kg. The inhalation LC_{50} in rats for 4 hr is 5 ppm. The dermal LD_{50} in rabbits is 220 mg/kg (36).

3.27.3   Odor

Methylisocyanide has a strong odor.

3.28   Cyanuric Chloride; Trichloro-s-Triazine, Tricyanogen Chloride

3.28.1   Uses and Industrial Exposures

Cyanuric chloride is used as a chemical intermediate.

3.28.2   Physical and Chemical Properties

Physical state	Colorless crystal
Molecular weight	184.4
Melting point	145.8°C
Boiling point	190°C
Solubility	Slightly soluble in water; soluble in alcohol

3.28.3   Physiologic Response

Cyanuric chloride is a lacrimator and respiratory irritant similar to cyanogen chloride. The oral LD_{50} in mice was 1000 mg/kg and in rats, 425 mg/kg (84). Deaths were delayed with evidence of corrosive damage to the intestinal tract. Repeated application to the skin increased injury. A repeated oral dose of 37 mg/kg daily for 5 weeks caused no injury in rabbits.

Fassett (42) noted similar effects with an oral LD_{50} in the mouse of 400 to 800 mg/kg, but less than 10 mg/kg intraperitoneally in this species. In the guinea pig, slight initial skin irritation was noted, which later developed a hard eschar. Eye damage was severe.

Cyanuric chloride should be handled with full precautions against skin or eye contact and inhalation of dust avoided.

3.28.4   Odor

Cyanuric chloride has a pungent odor.

3.29 Bromobenzyl Cyanide; Bromobenzylnitrile, Bromophenylacetonitrile

$$C_6H_5CHBrCN$$

3.29.1 Physical and Chemical Properties

Molecular weight	182.03
Melting point	25°C
Boiling point	225°C
Density	1.47 (solid)
Vapor density	6.6 (air $=$ 1)
Vapor pressure	0.01 mm Hg (20°C)

3.29.2 Physiologic Response

Bromobenzyl cyanide is a highly potent lacrimator. The CN group is probably released and converted to SCN (13). Like some other potent lacrimators, it probably acts by a progressive reaction with SH groups (104).

Prentiss (26) gives the physiologic effects of various levels in air as follows:

Lowest detectable level	0.09 mg/m^3
Lowest irritant concentration	0.15 mg/m^3
Intolerable concentration	0.8 mg/m^3 (10 min)
Lethal concentration	900.0 mg/m^3 (30 min); 3500.0 mg/m^3 (10 min)

3.30 Toluene-2,4-diisocyanate; Tolylene 2,4-Diisocyanate, TDI

3.30.1 Uses and Industrial Exposures

TDI is used in the manufacture of polyurethane foams, foam-type insulation, etc.

3.30.2 Physical and Chemical Properties

Physical state	White liquid
Molecular weight	174

Boiling point   250°C
Specific gravity   1.21 (28°C)
Vapor density   6.0 (air = 1)
Vapor pressure   1 mm Hg (80°C)
Solubility   Insoluble in water; soluble in acetone, ethyl acetate, tol-
                         uene, kerosene

$$1 \text{ ppm} \backsim 7.12 \text{ mg/m}^3 \text{ and } 1 \text{ mg/l} \backsim 140.5 \text{ ppm}$$

3.30.3   Physiologic Response

Since the introduction of this material in the manufacture of synthetic foams, there have been a number of reports of severe immunotoxicologic reactions in man, usually beginning after a latent period with repeated exposures, and characterized by an acute asthmalike reaction (105). Direct pulmonary irritant effects are noted in animals (106), and the specific picture of a sensitization in animals has been demonstrated.

Merewether (107) points out that the isocyanates are very reactive substances and are known to react with various groupings in proteins, and thus should be capable of forming antigens. The reaction is probably with free amino groups. In the case of phenyl isocyanate, the reaction leads to the formation of a phenylhydantoic acid (108). There seems little doubt that the pulmonary reaction is at least partly based on some type of delayed or sensitization-type reaction.

The oral LD_{50} in rats is 6170 mg/kg and the 6-hr inhalation LC_{50} in rats is 600 ppm (36).

Because pulmonary effects can be produced in man at very low levels in air, TDI should be used only in areas with adequate general and local ventilation or with air-supplied respirator equipment. Skin contact should be avoided. Persons with chronic respiratory disease or respiratory allergies should not be exposed.

Other isocyanates may have irritant properties on the eyes or respiratory tract.

3.30.4   Permissible Exposure Limits

The PEL is 0.02 ppm (0.14 mg/m^3) as a ceiling value (25). This may not protect persons having previous specific sensitization. A recent NIOSH criteria document (109) has proposed 0.02 ppm as a 20-min ceiling, and 0.005 ppm as the time-weighted average.

3.30.5   Odor

The odor threshold is 0.4 ppm in about half of the subjects.

3.31 Sodium Dicyanamide

$$NaN(CN)_2$$

3.31.1 Uses and Industrial Exposures

Sodium dicyanamide is used as a chemical intermediate.

3.31.2 Physical and Chemical Properties

Physical state   Colorless crystals
Molecular weight   89.04
Melting point   315°C (decomposes)
Solubility   Soluble in water—26.5 g/100 g of water at 30°C; soluble
                          in methanol—4.2 g/100 g of methanol at 30°C

3.31.3 Physiologic Response

The oral LD_{50} in mice is about 1000 mg/kg and the intraperitoneal LD_{50}, 610 mg/kg. It is not absorbed in significant amounts through the intact skin of rabbits, although it apparently penetrates the abraded skin of this species (89).

3.31.4 Flammability

Heavy metal salts may explode on heating (89).

3.32 Dicyandiamide; Cyanoguanidine

$$NH$$
$$\parallel$$
$$NH_2CNHCN$$

3.32.1 Physical and Chemical Properties

Physical state   Crystalline solid
Molecular weight   84.08
Melting point   209 to 211°C
Boiling point   Decomposes
Specific gravity   1.40 (14°C)
Solubility   Soluble in water—23 g/100 g of water; slightly soluble in
                          alcohol, ether

3.32.2   Physiologic Response

Hald et al. (34) found the oral LD_{50} in mice to be greater than 4 g/kg when given with alcohol, and greater than 3 g/kg in rabbits.

3.33   Sodium Cyanate

NaOCN

3.33.1   Physical and Chemical Properties

Physical state	Colorless solid
Molecular weight	65.9
Specific gravity	1.937 (20°C)
Solubility	Soluble in water

3.33.2   Physiologic Response

Birch and Schutz (110) noted that the LD_{50} in rats intramuscularly was 310 mg/kg. Lower doses caused drowsiness. Larger doses caused drowsiness with intermittent clonic convulsions, terminating in tonic convulsions. Loss of weight and apathy was caused by repeated intramuscular doses of 50 to 100 mg/kg in rats and rabbits.

Increased urinary output and diarrhea were also present.

No details of metabolism were found, but presumably the toxic effect is produced by the OCN ion, and not by breakdown products. Care should be used to avoid inhalation of dust and prolonged or repeated skin contact.

3.34   Potassium Cyanate

KOCN

3.34.1   Uses and Industrial Exposures

Potassium cyanate is used as a chemical intermediate and as a weed killer.

3.34.2   Physical and Chemical Properties

Physical state	White solid
Molecular weight	81.1
Melting point	315°C
Specific gravity	2.056 (20°C)
Solubility	Decomposes in hot water; soluble in alcohol

3.34.3   Physiologic Response

The LD_{50} in rats and mice by oral doses is about 1000 mg/kg. Dogs given 400 mg/kg intraperitoneally show severe or fatal symptoms (vomiting, defecation, urination, lacrimation, salivation, rapid respiration, tremors, and convulsions) (84).

Birch and Schutz (110) have described somewhat similar symptoms with the sodium salt (see sodium cyanate, Section 3.33).

The degree of hazard appears less than with cyanides and some nitriles, but care should be used to avoid inhalation of dust and prolonged and repeated skin contact.

3.35   Potassium Ferricyanide

$$K_2Fe(CN)_6$$

3.35.1   Source

Oxidation of ferrocyanide yields potassium ferricyanide.

3.35.2   Uses and Industrial Exposures

It finds use as a chemical reagent and in metallurgy, photography, and pigments.

3.35.3   Physical and Chemical Properties

Physical state	Red solid
Molecular weight	298.97
Specific gravity	1.8109
Solubility	Soluble in water

3.35.4   Physiologic Response

It is only slightly toxic (see potassium ferrocyanide, Section 3.36), and is converted rapidly to ferrocyanide (111).

3.36   Potassium Ferrocyanide

$$K_4Fe(CN)_6 \cdot 3H_2O$$

3.36.1   Uses and Industrial Exposures

Potassium Ferrocyanide is used as a chemical reagent, in metallurgy, and in graphic arts.

3.36.2 Physical and Chemical Properties

Physical state	Lemon-yellow solid
Molecular weight	422.39
Melting point	Loses water at 60°C
Specific gravity	1.85 (17°C)
Solubility	Soluble in water

3.36.3 Physiologic Response

It appears only slightly toxic (42). Fassett found the oral LD_{50} in rats to be 1600 to 3200 mg/kg. The handling hazard is slight. No dermatitis was observed in workers handling ferro- or ferricyanide over a number of years. Dogs tolerated 35 cm^3/kg of a 7.5 percent solution intravenously (2626 mg/kg) of crystalline ferrocyanide. It is rapidly excreted by glomerular filtration, similarly to creatinine (105).

Poisoning from oral ingestion seems to have been questionable.

3.37 Nitroprusside; Sodium Nitroferricyanide

$$Na_2Fe(NO)(CN)_5 \cdot H_2O$$

3.37.1 Uses and Industrial Exposures

Nitroprusside used as an analytical reagent; it also has been tried in hypertension.

3.37.2 Physical and Chemical Properties

Physical state	Red crystals
Specific gravity	1.72
Solubility	40 g/100 ml; soluble in alcohol

3.37.3 Physiologic Response

Nitroprusside is said to be decomposed in vivo to liberate cyanide. Five milligrams by kilogram by mouth produces a fall in blood pressure similar to nitrites. It is of interest that methemoglobin is not formed. There is evidence that the cyanide liberated is converted to SCN, as in the case of other nitriles (112).

REFERENCES

1. L. E. Towill, J. S. Drury, B. L. Whitfield, E. B. Lewis, E. L. Galyan, and A. S. Hammons, *Reviews of the Environmental Effects of Pollutants: V. Cyanide*, Oak Ridge Natl. Lab., Rep. No. ORNL/EIS-81, 1978.

2. S. H. El Ghawabi, M. A. Gaafar, A. A. El-Saharti, S. H. Ahmed, K. K. Malash, and R. Fanes, *Br. J. Ind. Med.*, **32**, 215 (1975).

3. S. H. Wollman, *Am. J. Physiol.*, **186**, 453 (1956).

4. M. Dixon and E. C. Webb, *Enzymes*, Academic Press, New York, 1958.

5. H. G. Albaum, J. Tepperman, and O. Bodansky, *J. Biol. Chem.*, **163**, 641 (1964).

6. J. Schubert and W. A. Brill, *J. Pharmacol. Exp. Ther.*, **162**, 352 (168).

7. B. P. McNamara, *Estimates of the Toxicity of Hydrocyanic Acid Vapors in Man*, Edgewood Arsenal Tech. Rep. EN-TR-76023, 1976.

8. K. Lang, *Biochem. Z.*, **259**, 243 (1933).

9. W. A. Himwhich and J. P. Saunders, *Am. J. Physiol.*, **153**, 348 (1948).

10. F. Goldstein and F. Reiders, *Am. J. Physiol.*, **173**, 287 (1953).

11. J. L. Wood, *Chemistry and Biochemistry of Thiocyanic Acid and its Derivatives*, A. A. Newman, Ed., Academic Press, New York, 1975, pp. 156–221.

12. M. Ansell and F. A. S. Lewis, *J. Forensic Med.*, **17**, 148 (1970).

13. R. T. Williams, *Detoxication Mechanisms*, Chapman & Hall, London, 1959.

14. M. Feldstein and N. C. Klendshoj, *J. Lab. Clin. Med.*, **44**, 166 (1954).

15. H. L. Hardy, W. McK. Jeffries, M. M. Wasserman, and W. R. Wadell, *N. Engl. J. Med.*, **242**, 968 (1950).

16. B. Radojicic, *Ark. Hyg. Rada*, **24**, 227 (1973).

17. K. K. Chen, C. L. Rose, and G. H. A. Clowes, *J. Indiana Med. Assoc.*, **37**, 344 (1944).

18. J. H. Wolfsie and B. C. Shaffer, *J. Occup. Med.*, **1**, 281 (1959).

19. J. H. Wolfsie, *Arch. Ind. Hyg. Occup. Med.*, **4**, 1 (1951).

20. J. W. Howard and R. F. Hanzal, *J. Agric. Food Chem.*, **3**, 325 (1955).

21. NIOSH, *Occupational Exposure to Hydrogen Cyanide and Cyanide Salts*, 1975.

22. H. C. Dudley, T. R. Sweeney, and J. W. Miller, *J. Ind. Hyg. Toxicol.* **24**, 255 (1942).

23. F. Flury and F. Zernik, *Schadliche Gase*, Springer, Berlin, 1931.

24. P. Doudoroff, *Toxicity to Fish of Cyanides and Related Components*, Ecol. Res. Ser., EPA 600/3-76-038, 1976.

25. F. W. Mackison, R. S. Stricoff, and L. J. Partridge, Jr., *NIOSH/OSHA Pocket Guide to Chemical Hazards*, DHEW (NIOSH) Pub. No. 78-210, 1978.

26. A. M. Prentiss, *Chemicals in War*, McGraw-Hill, New York, 1937.

27. R. N. Shreve, *The Chemical Process Industries*, McGraw-Hill, New York, 1956.

28. R. Gosselin et al., *Clinical Toxicology of Commercial Products*, 4th ed., Williams & Wilkins, Baltimore, 1976.

29. M. H. Barsky, *N.Y. State J. Med.*, **37**, 1031 (1937).

30. S. Tovo, *Minerva Med.*, **75**, 158 (1955).

31. E. Streicher, *Proc. Soc. Exp. Biol. Med.*, **76**, 536 (1951).

32. D. Liebowitz and H. Schwartz, *Am. J. Pathol.*, **18**, 965 (1950).

33. S. Glaubach, *Arch. Exp. Pathol. Pharmakol.*, **117**, 247 (1926).

34. J. Hald, E. Jacobson, and V. Larson, *Acta Pharmacol. Toxicol.*, **8**, 329 (1952).

35. J. DeLarrad and H. J. Lazarini, *Arch. Maladies Prof. Med. Trav. Sec. Soc.*, **15**, 282 (1954).

36. NIOSH, *Registry of Toxic Effects of Chemical Substances*, 1977.

37. F. M. Turner, *The Condensed Chemical Dictionary*, 4th ed., Reinhold, New York, 1950.

38. M. B. Jacobs, *Analytical Chemistry of Industrial Poisons, Hazards and Solvents*, 2nd ed., Interscience, New York, 1949.

39. B. J. Jandorf and O. Bodansky, *J. Ind. Hyg. Toxicol.*, **28**, 125 (1946).

40. W. N. Aldridge and C. L. Evans, *Q. J. Exp. Physiol.*, **33**, 241 (1946).

41. W. N. Aldridge, *Biochem. J.*, **48**, 271 (1951).

42. D. W. Fassett, unpublished data, Eastman Kodak Co., Rochester, N.Y.

43. NIOSH, *Criteria for a Recommended Standard . . . Occupational Exposure to Nitriles*, 1978.

44. J. F. Finklea, *Am. Ind. Hyg. Assoc. J.*, **38**, 417 (1977).

45. E. R. Blout and H. Mark, *Monomers*, Interscience, New York, London, 1951.

46. H. C. Dudley and P. A. Neal, *J. Ind. Hyg. Toxicol.*, **24**, 27 (1942).

47. R. C. Weast, *CRC Handbook of Chemistry and Physics*, CRC Press, Boca Raton, F., 1980.

48. H. E. Dudley, T. R. Sweeney, and J. W. Miller, *J. Ind. Hyg. Toxicol.*, **24**, 255 (1942).

49. H. L. Brieger, F. Rieders, and W. A. Hodes, *Arch. Ind. Hyg. Occup. Med.*, **6**, 128 (1952).

50. I. Gut, J. Nerudova, J. Kopecky, and V. Halecek, *Arch. Toxicol.*, **33**, 151 (1975).

51. E. Boyland and L. F. Chasseand, *Biochem. J.*, **104**, 95 (1967).

52. K. Hashimoto and R. Kanai, *Ind. Health (Kawasaki, Japan)*, **3**, 30 (1965).

53. F. J. Murray, B. A. Schwetz, K. D. Nitschke, J. A. John, J. M. Norris, and P. J. Gehring, *Food Cosmet. Toxicol.*, **16**, 547 (1978).

54. P. Milvy and M. Wolff, *Mutation Res.*, **48**, 271 (1977).

55. R. E. McMahon, J. C. Cline, and C. Z. Thompson, *Cancer Res.*, **39**, 682 (1979).

56. J. F. Quast, C. E. Wade, C. G. Humiston, R. M. Carreon, E. A. Hermann, C. N. Park, and B. A. Schwetz, "A Two-year Toxicity and Oncogenicity Study with Acrylonitrile Incorporated in the Drinking Water of Rats," Chemical Manufacturers Association Report, 1980.

57. C. Maltoni, A. Ciliberti, and V. DiMaio, *Med. Lav.*, **68**, 401 (1977).

58. M. T. O'Berg, *J. Occup. Med.*, **22**, 245 (1980).

59. *Fed. Reg.*, **43**, 45762 (1978).

60. American Industrial Hygiene Association, "Hygienic Guide, Acrylonitrile," *Am. Ind. Hyg. Assoc. Q.*, **18**, 78 (1957).

61. W. A. McOmie, *J. Ind. Hyg. Toxicol.*, **31**, 113 (1949).

62. H. F. Smyth and C. P. Carpenter, *J. Ind. Hyg. Toxicol.*, **30**, 63 (1948).

63. Union Carbide Corp., *Toxicology Studies—Acetonitrile*, Union Carbide Corp., Ind. Med. & Toxicol. Dept., New York, 1970.

64. E. T. Kimura, D. M. Ebert, P. W. Dodge, *Toxicol. Appl. Pharmacol.*, **19**, 699 (1971).

65. U. C. Pozzani, C. P. Carpenter, P. E. Palm, C. S. Weil, and J. H. Nair, *J. Occup. Med.*, **1**, 634 (1949).

66. D. Matine, S. H. Rosen, and A. Cipra, *Proc. Soc. Exp. Biol. Med.*, **30**, 649 (1933).

67. B. Grabois, *N.Y. State Dept. Labor Monthly Rev., Div. Ind. Hyg.*, **34**, 1 (1955).

68. M. L. Amdur, *J. Occup. Med.*, **1**, 625 (1949).

69. J. Dequidt, D. Furon, and J. M. Haguenoer, *Bull. Soc. Pharm. Lillie*, **4**, 143 (1972).

70. J. Dequidt, D. Furon, F. Wattel, J. M. Haguenoer, P. Scherpereel, B. Gosselein, and A. Ginestet, *Eur. J. Toxicol.*, **7**, 91 (1974).

71. H. F. Smyth, C. P. Carpenter, and C. S. Weil, *Arch. Ind. Hyg. Occup. Med.*, **4**, 119 (1951).

72. G. L. Ghiringhelli, *Med. Lav.*, **46**, 221, 229 (1955); **47**, 192 (1956); **49**, 683 (1958).

73. S. Szabo and H. Selye, *Arch. Pathol.*, **93**, 390 (1972).

74. S. Szabo and E. S. Reynolds, *Environ. Health Perspect.*, **11**, 135 (1975).

75. K. Tsurumi and K. Kawada, *Gifu Ika Daigaku Kiyo*, **18**, 655 (1971).

76. A. M. Thiess and W. Hey, *Arch. Toxikol.*, **24**, 271 (1969).

77. H. F. Smyth, Jr., *J. Ind. Hyg. Toxicol.,* **26,** 269 (1944).
78. F. W. Sunderman and J. F. Kincaid, *Arch. Ind. Hyg. Occup. Med.,* **8,** 371 (1953).
79. *Am. Cyanamid Co. New Products Bull.,* revised ed., Collective Vol. I, New York, 1952.
80. R. B. Bruce, J. W. Howard, and R. F. Hanzal, *Anal. Chem.,* **27,** 1346 (1955).
81. F. Motoc, S. Constantinescu, G. Filipescu, M. Dobre, E. Bichir, and G. Pambuccian, *Arch. Mal, Prof. Med. Trav. Sec. Soc.,* **32,** 653 (1971).
82. J. H. Wolfsie, *J. Occup. Med.,* **2,** 588 (1960).
83. H. F. Smyth, Jr., C. P. Carpenter, C. S. Weil, U. C. Pozzani, and J. A. Striegel, *Am. Ind. Hyg. Assoc. J.,* **23,** 95 (1962).
84. *Am. Cyanamid Co. New Products Bull.,* revised ed., Collective Vol. 1, New York, 1952.
85. A. R. Contessa and R. Santi, *Biochem. Pharmacol.,* **22,** 827 (1973).
86. W. N. Aldridge, *Analyst,* **69.** 262 (1944).
87. J. L. Svirbely and E. P. Floyd, *Toxicologic Studies of Acrylonitirle, Adiponitrile and β,β-Oxydipro-prionitrile—III Chronic Studies,* USDHEW, Robert A. Taft Sanitary Engineering Center, 1964.
88. H. V. Zeller, H. T. Hofmann, A. M. Thiess, and W. Hey, *Zentralbl. Arbeitsmed. Arbeitsschutz,* **19,** 225 (1969).
89. *Am. Cyanamid Co. New Products Bull.,* Collective Vol. II, New York, 1952.
90. E. D. Schilling, *Fed. Proc.,* **13,** 290 (1954).
91. T. E. Backhuber, J. J. Lalich, D. M. Angevine, E. D. Schilling, and F. M. Strong, *Proc. Soc. Exp. Biol. Med.,* **89,** 294 (1955).
92. S. Wawzonek, I. V. Ponseti, R. S. Shepard, and L. E. Wiedenmans, *Science,* **121,** 63 (1955).
93. B. D. Barnett, H. R. Bird, J. J. Lalich, and F. M. Strong, *Proc. Soc. Exp. Biol. Med.,* **94,** 67 (1957).
94. W. Dasler, *Proc. Soc. Exp. Biol. Med.,* **85,** 485 (1954).
95. J. J. Lalich, *Science,* **128,** 206 (1958).
96. H. A. Hartman and H. F. Stich, *Fed. Proc.,* **16,** 358 (1957).
97. J. Stern, C. Weil, M. Malherbe, and R. H. Green, *Biochem. J.,* **52,** 114 (1952).
98. I. K. Panov, *J. Eur. Toxicol.,* **2,** 292 (1969).
99. I. K. Panov, *J. Eur. Toxicol.,* **3,** 58 (1970).
100. V. L. Van Breemen and J. Hiraoka, *Am. Zool.,* **1,** 473 (1961).
101. S. P. Hicks, *Arch. Pathol.,* **50,** 545 (1950).
102. H. Hyden and H. Hartelius, *Acta Psychiatr. Neurol. Suppl.,* **48,** 1 (1948).
103. F. G. Valdecasas, *Arch. Inst. Farmacol. Exp. Madrid,* **5,** 64 (1953); *Chem. Abstr.,* **48,** 13084e (1954).
104. M. Dixon, *Biochemical Society Symposia No. 2,* Cambridge University Press, Mass., 1948.
105. American Industrial Hygiene Association, "Hygienic Guide, Toluene 2,4-Diisocyanate," *Am. Ind. Hyg. Assoc. Q.,* **18,** 370 (1957).
106. J. A. Zapp, Jr., *Arch. Ind. Health,* **15,** 324 (1957).
107. E. R. A. Merewether, *Industrial Medicine and Hygiene,* Vol. III, Butterworth, London, 1956.
108. J. S. Fruton, *General Biochemistry,* 2nd ed., Wiley, New York, 1959.
109. NIOSH, *Criteria for a Recommended Standard, Occupational Exposure to Toluene-2,4-diisocyanate,* 1978.
110. K. M. Birch and F. Schutz, *Br. J. Pharmacol.,* **1,** 186 (1946).
111. R. W. Berliner, *Am. J. Physiol.,* **160,** 325 (1950).
112. T. Sollmann, *A Manual of Pharmacology,* Saunders, Philadelphia, 1957.

CHAPTER FIFTY-NINE

Aliphatic Carboxylic Acids

DEREK GUEST, Ph.D., GARY V. KATZ,
Ph.D., and BERNARD D. ASTILL, Ph.D.

1 GENERAL CONSIDERATIONS

Organic acids constitute a very wide range of chemicals, and perform a very wide range of industrial functions. Many are naturally occurring, many serve an important function in nutrition, and many are intermediates in biochemical processes. This chapter is limited to aliphatic carboxylic acids, which are comprised of the saturated and unsaturated aliphatic mono- and dicarboxylic acids. Because of the wide range of uses, the differing circumstances of isolation, and frequently the complexity of structure, many organic acids are described by a variety of names. In this chapter the name used in the text is usually the trivial or most common name, followed by alternative names. The most recent systematic or Chemical Abstracts name is given in the tables describing physicochemical properties. The commercial and industrial importance of this class of compounds is indicated by recent production figures; thus acetic acid has a production volume in excess of 2.5 billion lb/year, propionic acid in excess of 84 million lb/year, acrylic acid in excess of 284 million lb/year, and 2-hexenoic acid in excess of 16 million lb/year. Large production volumes are also encountered with esters and salts of organic acids; thus stearic acid in the form of its salts and esters is used in excess of 100 million lb/year.

In general, serious physiological concerns are not associated with the acids discussed in this chapter; rather do such concerns arise with individual acids on an idiosyncratic basis. This behavior is attested to by the many members of this class that find use as food additives, flavoring agents, and stabilizers or serve

4901

as food materials. The primary concern is usually of an acute nature arising from the primary irritant effect, particularly of the short chain acids. As the molecular weight increases and the water solubility decreases, the irritating capacity in general decreases. In addition, sensitization is quite rare with the aliphatic carboxylic acids.

2 SATURATED MONOCARBOXYLIC ACIDS

2.1 Introduction

2.1.1 Industrial Applications

Saturated, aliphatic monocarboxylic acids are used in a variety of applications. Many are employed in the production of synthetic fiber materials, resins, plastics, and dyestuffs. A number of the acids or their esters are important chemical intermediates or solvents and are used in cosmetics or food applications.

2.1.2 Health Effects

The major physiologic effect associated with the monocarboxylic acids is that of primary irritation of the eye, skin, or mucous membranes. The degree of irritation is largely influenced by the strength of the acid (dissociation), its water solubility, and its ability to penetrate the skin. The dissociation constants and aqueous solubilities are given in Table 59.1. With the more volatile acids, the partial vapor pressure influences the potential for exposure through vapor contact.

The short chain acids such as formic, acetic, and propionic acids are relatively strong acids and can produce burns similar to those of the mineral acids. Higher molecular weight acids, such as lauric and stearic acid, are not irritants. This is probably due mostly to poor skin penetrability and low water solubility, since their dissociation constants are similar to those of many of the shorter chain acids.

2.1.3 Metabolic Fate

Saturated, straight chain monocarboxylic acids are incorporated into normal intermediary metabolism and are broken down by the β-oxidation pathway. This produces acetate and an acid with two fewer carbons than the original acid. The process is repeated until the end product is acetate, propionate, or butyrate. Acetate and butyrate are utilized for energy via the citric acid cycle, or converted to acetoacetate and subsequently other ketone bodies. Ketone bodies may be oxidized or excreted in the urine, depending on the nutritional state of the organism. Propionate originates from odd chain acids and is converted to carbohydrate and other intermediary metabolites.

In some cases medium chain acids are partly oxidized by ω-oxidation. This produces dicarboxylic acids, which may be broken down by β-oxidation from either end. ω-Oxidation does not normally occur with straight chain acids having more than 12 carbon atoms. It may occur when the capacity for β-oxidation is either exceeded because of a large dose or blocked because of substitution in the α- or β-position on the molecule.

In general, α-ethyl-substituted acids are not readily metabolized and are eliminated primarily by conjugation with glucuronic acid and excretion in the urine. Some dealkylation also occurs.

Medium and long chain fatty acids may be metabolized via chain elongation. This occurs by the addition of two-carbon (acetate) units to the carboxyl group of the original acid. Short chain acids, such as butyric, caproic, and caprylic acids, may be converted to long chain fatty acids, but this occurs largely by cleaving into two-carbon fragments, which are used as the building units.

2.2 Formic Acid

The formula is HCOOH; formic acid is also known as methanoic, formylic, and hydrogen carboxylic acid.

2.2.1 Major Uses

It is used as a reducing agent in wool dyeing and decalcifying, in tanning, dehairing, and plumping hides, for latex coagulation and regenerating old rubber, in the electroplating industries, and as an animal feed additive, food preservative, flavor adjunct, and brewing antiseptic.

2.2.2 Acute Studies

A detailed review of the health aspects of formic acid, sodium formate, and ethyl formate has been published (34). The acute toxicity of formic acid is summarized in Table 59.2.

The major hazard associated with formic acid is that of severe damage to skin, eye, or mucosal surfaces, simila to the effects of other relatively strong acids. Atmospheric concentrations as low as 32 mg/l are corrosive to skin and mucosal membranes (3). Rabbits tolerated 20 mg/kg ethyl formate dermally, but this dose caused severe corneal burning when applied to the eyes (16).

In guinea pigs formic acid vapor (0.3 to 42 ppm) for 1 hr is a more potent irritant than formaldehyde (35). Rabbits tolerated a 300-mg/kg subcutaneous administration without adverse effect (36). No toxic or teratogenic effects were noted after injection into fertilized chicken eggs of up to 20 mg formic acid (1).

Table 59.1. Physical Properties of Saturated Aliphatic Monocarboxylic Acids

Acid	CAS Acid Name	CAS Reg. No.	Mol Wt	M.P. (°C)	B.P. (°C)	Solubility[a] H₂O	Alcohol	Other	Specific Gravity	Acid Dissociation (pKₐ)	Vapor Pressure (mm) (°C)
Formic	Methanoic	64-18-6	46.03	8.4	100.8	∞	∞	Ether (∞) Acetone (v) Benzene (s)	1.220 (20/4)	3.75	35 (20)
Acetic	Ethanoic	64-19-7	60.05	16.7	118.0	∞	∞	Ether (∞) Acetone (∞) Benzene (∞)	1.049 (20/4)	4.76	11.4 (20)
Propionic	Propanoic	79-09-4	74.08	−20.8	141.1	∞	∞	Ether (∞) Chloroform (∞)	0.993 (20/4)	4.87	3.3 (27.6)
Butyric	Butanoic	107-92-6	88.12	−7.9	163.5	∞	∞	Ether (∞) Chloroform (∞)	0.958 (20/4)	4.82	0.8 (20)
Isobutyric	2-Methylpropanoic	79-31-2	88.12	−47	154.7	20%	∞	Ether (∞) Chloroform (∞)	0.949 (20/4)	4.86	1 (14.70)
Valeric	Pentanoic	109-52-4	102.15	−34/−58	186.4	3.3%	∞	Ether (∞) Chloroform (∞)	0.942 (20/4)	4.84	1 (42)
Isovaleric	3-Methylbutanoic	503-74-2	102.15	−29.3	176.5	4.2%	∞	Ether (∞) Chloroform (∞)	0.925 (20/4)	4.78	1 (34)
Caproic	Hexanoic	142-62-1	116.18	−5	205	1.1%	s	Ether (s)	0.927 (20/4)	4.87	1 (72)
Isocaproic	4-Methylpentanoic	646-07-1	116.18	−33.0	201	sl s	s	Ether (s)	0.923 (20/4)	4.84	
2-Methylvaleric	2-Methylpentanoic	97-61-0	116.18		193.5	0.6%	s	Ether (s)	0.927 (16/14)		0.02 (20)

Common name	IUPAC name	CAS No.	Mol. wt.	m.p.	b.p.	Water solubility		Other solvents	Density		
2-Ethylbutyric	2-Ethylbutanoic	88-09-5	116.18	−31.8	194	Sl s	S	Ether (s)	0.924 (20/4)	4.73	0.08 (20)
	Heptanoic	111-14-8	130.21	−7.5	223	0.2% (15°)	S	Ether (s)	0.920 (20/14)	4.88	1 (78)
Caprylic	Octanoic	124-07-2	144.23	16.7	240	V sl s	S	Acetone (s) Ether (s) Chloroform (s)	0.910 (20/4)	4.90	1 (78)
2-Ethylhexanoic	2-Ethylhexanoic	149-57-5	144.23		228 (755 mm)	S (hot)	Sl s	Ether (s)	0.902 (25/4)	4.95	0.03 (20)
	Nonanoic	112-05-0	158.24	13	254	Ins	S	Ether (s) Chloroform (s)	0.906 (20/4)		1 (108)
Capric	Decanoic	334-48-5	172.27	31.6	268	Ins	S	Methanol (s) Benzene (s) Chloroform (s)	0.888 (35/4)		1 (128)
Undecylic	Undecanoic	112-37-8	186.29	30	284	Ins	S	Ether (s)	0.891 (30/4)		1 (101)
Lauric	Dodecanoic	143-07-7	200.36	44	225 (100 mm)	Ins	S	Ether (s) Methanol (s) Benzene (s)	0.868 (50/4)		1 (121)
Myristic	Tetradecanoic	544-63-8	228.40	54.2	250.5 (100 mm)	Ins	S	Ether (s) Chloroform (s) Benzene (s)	0.844 (80/4)		1 (142)
Palmitic	Hexadecanoic	57-10-3	256.43	64	267 (100 mm)	Ins	S	Ether (∞) Acetone (s) Benzene (s)	0.836 (91/4)		1 (154)
Stearic	Octadecanoic	57-11-4	284.47	69.4	291 (110 mm)	0.03% (25°)	S	Ether (s) Acetone (s) Chloroform (s)	0.839 (80/4)		1 (174)

ᵃ ∞, soluble in all proportions; v, very soluble; s, soluble; sl s, slightly soluble; v sl s, very slightly soluble; ins, insoluble.

Table 59.2. Acute Toxicity of Saturated Monocarboxylic Acids

Acid	Species	Route	LD_{50} (mg/kg)	Ref.
Formic	Mouse	Oral	1076	1
	Mouse	Iv	142	1
	Mouse	Ip	940	2
	Rat	Oral	1830	2
	Rabbit	Oral	>4000 (LD_{Lo})	3
	Rabbit	Iv	239 (LD_{Lo})	4
Acetic	Mouse	Oral	4960	5
	Mouse	Iv	525	6
	Mouse	Inhaln.	5620 ppm (LC_{50}/1 hr)	7
	Rat	Oral	3310	5
	Rat	Oral	3530	8
	Rat	Inhaln.	16000 ppm (LC_{Lo}/4 hr)	8
	Rabbit	Oral	1200 (LD_{Lo})	9
	Rabbit	Skin	1060	10
	Rabbit	Sc	1200 (LD_{Lo})	9
Propionic	Mouse	Oral	5100[a]	11
	Mouse	Iv	625	6
	Rat	Oral	4260	12
	Rat	Oral	2600	13
	Rat	Oral	5160[b]	11
	Rabbit	Skin	496	12
Butyric	Mouse	Oral	500 (LD_{Lo})	14
	Mouse	Iv	800	6
	Mouse	Ip	3180	15
	Mouse	Sc	3180	15
	Rat	Oral	8790	16
	Rat	Oral	2940	8
	Rabbit	Oral	3600 (LD_{Lo})	14
	Rabbit	Skin	6083	16
	Rabbit	Skin	530	17
Isobutyric	Rat	Oral	280	12
	Rabbit	Oral	8000	18
	Rabbit	Skin	500	12
Valeric	Mouse	Iv	1290	6
	Mouse	Sc	3590	6
	Rat	Oral	1055	12
	Rat	Oral	1844	19
	Rabbit	Skin	660	12
	Rabbit	Skin	290	19
Isovaleric	Mouse	Iv	1120	6
	Rat	Oral	2000	20
	Rat	Oral	<3200	21
	Rabbit	Skin	3560	22
	Rabbit	Skin	310	20
Caproic	Mouse	Ip	3180	15
	Mouse	Sc	3180	15

Table 59.2. (Continued)

Acid	Species	Route	LD$_{50}$ (mg/kg)	Ref.
	Mouse	Iv	1725	6
	Rat	Oral	5970	16
	Rat	Oral	3000 (LD$_{Lo}$)	23
	Guinea pig	Skin	5000 (LD$_{Lo}$)	23
	Rabbit	Skin	630	16
Isocaproic	Rat	Oral	2050[c]	24
	Rat	Oral	>3200	21
	Rabbit	Skin	1050[c]	24
2-Methylvaleric	Rat	Oral	1890	16
	Rat	Oral	1600–3200	21
2-Ethylbutyric	Rat[d]	Oral	2033	16
	Rabbit	Skin	480	16
Heptanoic	Mouse	Oral	6400	25
	Mouse	Iv	1200	6
	Rat	Oral	7000	26
Caprylic	Mouse	Iv	600	6
	Rat	Oral	1280	12
	Rat	Oral	10080	27
	Rabbit	Skin	650	12
	Rabbit	Skin	>5000	28
2-Ethylhexanoic	Rat	Oral	3000	23
	Guinea pig	Skin	5690	23
	Rabbit	Skin	1260	29
Nonanoic	Mouse	Iv	224	6
	Rat	Oral	3200[e] (LD$_{Lo}$)	21
	Rabbit	Skin	>5000	30
Capric	Mouse	Iv	129	6
	Rat	Oral	3320[c]	12
	Rabbit	Skin	1575[c]	12
	Rabbit	Skin	>5000	31
Undecylic	Mouse	Iv	140	6
Lauric	Mouse	Iv	131	6
	Rat	Oral	12000	26
Myristic	Mouse	Iv	43	6
Stearic	Mouse	Iv	23	6
	Mouse	Iv	56	32
	Rat	Iv	22	6
	Rat	Oral	>5000	33
	Rabbit	Skin	>5000	33

[a] Sodium salt.

[b] Calcium salt.

[c] Experiments using mixed isomers of unspecified composition.

[d] Female.

[e] 10% corn oil solution.

Von Oettingen has reported that doses of 0.46 to 1.25 mg/kg intravenously to rabbits caused central nervous system depression, vasoconstriction, and diuresis; larger doses (approximately 4 g/kg) produced convulsions and death in rabbits and methemoglobinemia in dogs (3). In sheep formic acid given orally (150 mg/kg) was without adverse effect, except for some indication of anorexia (37).

Formic acid has been reported to be mutagenic in *E. coli* and in *Drosophila* germ cells, but did not affect DNA transformation in *B. subtilis* at concentrations up to 0.46 percent (38–40).

2.2.3   Repeated Exposures

In young rats administered formic acid in the diet (0.5 or 1.0 percent) or drinking water (0.5 or 1.0 percent) for 6 weeks the body weight gain and the size of most organs were reduced (2). Rats receiving 8 to 360 mg/kg formic acid in drinking water for 2 to 27 weeks showed no adverse effects other than a reduced rate of body weight gain (and feed intake) at the highest dose level. No adverse effects were observed when calcium formate (0.2 percent) was included in the drinking water of male and female adult rats and their offspring for three successive generations (1). No histologic changes were noted when formic acid (8 percent in water) was painted on the ears of Swiss mice for as many as 50 days (41).

2.2.4   Physiologic Effects in Humans

Formic acid is a natural constituent of some fruits, nuts, and dairy products. Signs and symptoms from accidental or intentional overdoses (50 g or more) are salivation, vomiting, a burning sensation in the mouth and pharynx, and severe pain. Circulatory collapse may follow, causing death. Ingestion of, or skin contact with, smaller quantities of formic acid may produce ulceration of membranes. Contact with eyes may cause permanent scarring of the cornea. Dilute solutions (e.g., 10 percent) appear to be noncorrosive (42). Workers exposed to formic acid in a textile plant reportedly complained of nausea from concentrations around 15 ppm (43).

2.2.5   Biochemistry

Formate is a normal constituent of intermediary metabolism and a precursor of many amino acids and purines which may be incorporated into nucleic acids, proteins, lipids, and carbohydrates. Some formic acid may be excreted unchanged, the amount depending on the species, dose, and route of administration (3).

Formate is metabolized in the rat primarily via the one-carbon pool, but in some circumstances the catalase–peroxidative pathway may serve as an alternative route of oxidation (44). Oxidation occurs in a variety of organs and tissues, including liver, lung, and erythrocytes, the end products being carbon dioxide and water (1, 3, 45, 46).

2.2.6 Regulations and Standards

Formic acid is classified as a primary irritant. Damage to the tissue can be severe, particularly after ingestion (47). The threshold limit value (TLV) adopted for formic acid in workroom air by the American Conference of Governmental Industrial Hygienists (ACGIH), and the standard time-weighted average (TWA) concentration issued by the Occupational Safety and Health Administration (OSHA) is 5 ppm (9 mg/m^3) (48, 49). Formic acid has been affirmed as generally recognized as safe as an indirect food substance in paper and paperboard (21 CFR 186.1316), a synthetic flavoring substance or adjuvant (21 CFR 172.515), and a preservative in hay crop silage (21 CFR 573.480).

2.3 Acetic Acid

The formula is CH$_3$COOH; acetic acid is also known as ethanoic, ethylic, and methanecarboxylic acid.

2.3.1 Major Uses

The primary uses are in the manufacture of synthetic fiber materials, cellulose acetate, acetate rayon, and in plastics and rubber. Acetic acid is used in printing and dyeing, food preserving, and the manufacture of vitamins, antibiotics, hormones, and organic and photographic chemicals.

2.3.2 Acute Studies

The acute toxicity of acetic acid is summarized in Table 59.2. A review of the biologic effects of acetic acid has been published (50). The physiologic effects of strong solutions (10 to 20 percent) of acetic acid may be due partly to corrosive action, in common with other relatively strong acids (51). This is particularly the case with direct application to the gastric mucosa of cat and rat, which leads to corrosion of the gastrointestinal tract. In mice inhalation of >1000 ppm acetic acid rapidly produced symptoms of irritation of the conjuctiva and upper respiratory tract; postmortem examination revealed hepatic swelling and congestion of the viscera (7). Acetic acid was a mild irritant to guinea pigs and rabbits when applied dermally (20 mg/24 hr) (52). Application of a larger quantity of acetic acid (0.5 ml, 525 mg) showed no corrosive effects in rabbits after 4 hr (22) but produced a severe irritation with necrosis after 24 hr (8).

Nixon et al. (53) found no effect in guinea pigs or rabbits after application of 10 percent acetic acid to intact or abraded skin patches.

Acetic acid (sodium salt) elicited no mutagenic response in *Salmonella typhimurium* or *Saccharomyces cerevisiae*, with or without liver preparations from mouse, rat, or monkey (54). Pregnant rabbits administered apple cider vinegar (1.6 g/kg/day) showed no increased fetal abnormalities or mortality compared to sham-treated controls (55). No teratogenic effects on developing chicken embryos were observed after injection of sodium acetate (100 mg/kg) into the yolk or air cell of eggs after 96 hr incubation. The LD_{50} of sodium acetate after injection into the yolk of unincubated eggs was 91.5 mg, whereas 200 mg/kg injected into the air cell was not toxic (55).

2.3.3 Repeated Exposures

Rats receiving acetic acid in their drinking water (up to 0.5 percent) for 2 to 4 months (daily doses up to 390 mg/kg) were found to lose body weight (apparently due to anorexia) at the highest dose, but no such effects were observed up to concentrations equivalent to 195 mg/kg daily; no fatalities occurred in any dose group (56). In rats fed acetic acid (4.5 g/kg/day) in the diet for 30 days, gastric lesions occurred in some animals, whereas others revealed slight forestomach wall thickening or inflammatory changes (57).

Male rats given sodium acetate orally (350 mg/kg, three times weekly for 63 days followed by 140 mg/kg, three times weekly for 72 days) showed no histologic evidence of tumors (58).

2.3.4 Physiologic Effects in Humans

Acetates are common constituents of animal and plant tissues and are formed during the metabolism of food substances. Typical concentrations of acetic acid occurring naturally in foods are 700 to 1200 mg/kg in wines, up to 860 mg/kg in aged cheeses, and 2.8 mg/kg in fresh orange juice (55). Acetic acid is present in vinegar at 3 to 6 percent and, as such, is extensively used in food preservation. Estimated possible average daily intakes (based on food intake concentrations) for persons more than 2 years old were given as 2.1 g/day for acetic acid and 0.23 g/day for sodium acetate (55).

According to Smyth (59), human exposure (8 hr) to acetic acid at 10 ppm could produce some irritation of eyes, nose, and throat; at 100 ppm marked lung irritation might result and possible damage to lungs, eyes, and skin. These predictions were based on animal experiments and industrial exposure. Ghiringhelli and DiFabio (7) reported conjuctival irritation, upper respiratory tract irritation, and hyperkeratotic dermatitis in 12 workers exposed for two or more years to an estimated mean acetic acid airborne concentration of 0.125 mg/l.

The clinical picture of oral acetic acid poisoning is similar to that described for formic acid, characterized by severe pain in the mouth and throughout the digestive tract; vomiting, respiratory, and circulatory distress may follow and some patients die in a coma (40).

2.3.5 Biochemistry

There are no reports of cumulative toxicity for acetic acid, presumably because of its incorporation into intermediary metabolism. Radiolabel from parenterally administered sodium [1-^{14}C]acetate is incorporated into cholesterol in the liver and brain of immature rats and guinea pigs, but this effect was not observed in mature animals (60). Intravenous administration of the compound results in distribution of radioactivity into all major lipid fractions of the brain of rats (61). Acetates serve as precursors for a large variety of intermediary compounds. When dogs were administered large doses (1 to 2 g/kg, intraperitoneally or subcutaneously) of sodium acetate, only small amounts appeared in the urine, which is evidence of the rapid utilization of acetic acid (62).

Acetic acid is absorbed from the gastrointestinal tract and through the lungs (50). It is readily metabolized by most tissues and may give rise to the production of ketone bodies as intermediates (50). In vitro experiments have demonstrated that acetate is incorporated into phospholipids, neutral lipids, steroids, sterols, and saturated and unsaturated fatty acids in a variety of human and animal tissue preparations (55). Hevesy (63) indicated that metabolism of [^{14}C]acetate in mice resulted in radiolabel associated with the protein fractions of plasma and most major tissues.

2.3.6 Regulations and Standards

Acetic acid is a strong irritant. Prolonged skin contact with glacial acetic acid may result in tissue destruction. Vapor concentrations of about 1000 ppm cause marked irritation of eyes, nose, and upper respiratory tract and can not be tolerated for more than 3 min (50).

The OSHA standard for airborne acetic acid is listed as a TWA value of 10 ppm (48) and the ACGIH-adopted (49) TLV is 10 ppm. Acetic acid has been affirmed as generally recognized as safe as a multipurpose food additive (21 CFR 182.1005), as a substance migrating to food from cotton and cotton fabrics used in dry-food packaging (21 CFR 182.70), as a substance migrating to food from paper and paperboard products (21 CFR 182.90), and as a general purpose food additive for animal feed (21 CFR 582.1005).

2.4 Propionic Acid

The formula is CH_3CH_2COOH; propionic acid is also known as propanoic, methylacetic, and ethanecarboxylic acid.

2.4.1 Major Uses

Propionic acid is used as an esterifying agent, in the production of calcium and sodium propionates which are used as antimicrobial agents, as an intermediate for cellulose propionate thermoplastics, and in the synthesis of herbicides, pharmaceuticals, flavorings, and perfumes.

2.4.2 Acute Studies

A review of the health aspects of propionic acid and calcium and sodium propionates has been published (11). The acute toxicity of propionic acid is summarized in Table 59.2. Local damage may occur to skin, eye, or mucosal surfaces on contact with concentrated solutions of propionic acid (21). Smyth et al. (12) reported that 10 mg/24 hr produced tissue necrosis in the rabbit irritation test, but the same quantity of propionic acid had little effect as a 10 percent solution in acetone (47). Severe corneal damage in rabbits resulted from the application of 990 µg (64) or an excess of a 5 percent aqueous solution (unspecified volume) of propionic acid (12).

In vitro mutagenicity assays with propionic acid, using *Salmonella typhimurium* or *Saccharomyces cerevisiae*, were negative with or without a mammalian liver preparation (65).

2.4.3 Repeated Exposure

Rats fed calcium or sodium propionate at 1 percent in the diet for 4 weeks (about 750 mg of propionic acid kg/day) followed by 3 percent for 3 weeks showed no changes in weight gain compared with controls (11). Male and female rats were fed a bread diet that contained 5 percent sodium propionate (approximately 4 g/kg/day) for 1 year with no adverse effects (66, 67). No effect on maternal or fetal survival and no increase in the number of fetal abnormalities were seen when calcium propionate was fed to pregnant mice and rats (up to 300 mg/kg/day for 10 days), hamsters (up to 400 mg/kg/day for 5 days), or rabbits (up to 400 mg/kg/day for 13 days) (11).

2.4.4 Physiologic Effects in Humans

Propionic acid is a normal intermediary metabolite formed in the oxidation of odd-number carbon fatty acids and from the side chain of cholesterol. In man, propionic acid represents up to 4 percent of the normal, total plasma fatty acid and is utilized by most organs and tissues (13). Propionic acid is a constituent of a variety of foods, including dairy products, and may constitute as much as 1 percent of Swiss cheese (11). In addition, propionic acid and its calcium and sodium salts are added to foods as antimicrobials. The estimated average daily intake of calcium propionate for persons over 2 years of age was given as 45 mg and for sodium propionate as 33 mg (11). No cumulative effects are known from industrial exposures (21). Bässler (68) has reported that an adult male receiving 6000 mg sodium propionate daily was found to have faintly alkaline urine but showed no other effects. Solutions of sodium propionate at concentrations up to 15 percent had no irritating effect in man and have been used in the treatment of external infections of the eyes (69).

2.4.5 Biochemistry

The metabolic fate of propionic acid has been reviewed (13). Propionic acid is rapidly absorbed through the gastrointestinal tract, and even after large doses, no significant amounts are excreted in the urine (70, 71).

Propionate is formed as the terminal three-carbon fragment (as propionyl-coenzyme A) in the oxidation of odd-number carbon fatty acids and from oxidation of the side chain of cholesterol (13). Radioactivity from [^{14}C]propionate administered to fasted rats may appear in glycogen, glucose, citric acid cycle intermediates, amino acids, and proteins. The route of metabolism of propionic acid involves interaction with coenzyme A, carboxylation to form methylmalonyl-coenzyme A, and conversion to succinic acid, which enters the citric acid cycle. Propionic acid may be oxidized without forming ketone bodies (21, 71), and in contrast to acetic acid, is incorporated into carbohydrate as well as lipid (13).

2.4.6 Regulations and Standards

Propionic acid is a moderate irritant to human skin (61). The ACGIH has adopted a TLV value of 10 ppm (49). Propionic acid is generally recognized as safe for use in food as a chemical preservative (21 CFR 182.3081), as a substance migrating to food from paper and paperboard products (21 CFR 182.90), and as a chemical preservative in animal feed (21 CFR 582.3081).

2.5 Butyric Acid

The formula is $CH_3(CH_2)_2COOH$; butyric acid is also known as butanoic, ethylacetic, and 1-propanecarboxylic acid.

2.5.1 Major Uses

The major uses are in the manufacture of esters for artificial flavorings, as a food additive, in the manufacture of varnishes, and in decalcifying hides.

2.5.2   Acute Studies

The acute toxicity of butyric acid is summarized in Table 59.2. The irritant action of butyric is less than that of propionic acid (71), although it is a moderately strong irritant in the guinea pig (21). In the closed, rabbit skin patch irritation test, 10 mg butyric acid produced a severe reaction in 24 hr (16), but in the open test, 500 mg elicited only a moderate response (17). Severe corneal burns were produced in rabbits with an excess of a 5 percent solution of butyric acid (unspecified volume) (16).

Large intravenous doses of butyric acid (sodium salt) cause temporary central nervous system depression in rabbits (1.6 g/kg) and dogs (0.86 g/kg) (71); similar effects were produced with subcutaneous doses in cats. Small doses of butyric acid have no effect on the cardiovascular system. Large doses impair cardiac function and depress blood pressure; the similarity to the clinical symptoms of diabetic coma suggests that ketone bodies, β-hydroxybutyrate and acetoacetate, may be responsible for these effects (71).

Prolonged exposure (of unspecified duration) of mice, rats, and rabbits to an atmospheric concentration of 0.1 to 0.2 mg/l butyric acid caused a massive increase in circulating lymphocytes and neutrophils, attributed to the irritant nature of the compound (72). Lethal vapor concentrations for mice, rats, and rabbits could not be reached by inhalation (72). After a 90-min exposure to a butyric acid aerosol (40 mg/l), rabbits displayed increased lethargy and dyspnea. Pathologic examination showed signs of bronchial and capillary dilation and emphysema (73). Smyth et al. (8) reported no lethalities when rats were exposed for 8 hr to air saturated with butyric acid vapor.

2.5.3   Repeated Exposures

No information on prolonged administration of butyric acid is available. No cumulative effects are known, presumably since butyric acid is readily metabolized via normal intermediary pathways (see Section 2.5.5).

2.5.4   Physiologic Effects in Humans

Butyric acid can act as a mild irritant in man (74). The effects are similar to those of propionic acid (21). According to Oettel (75), application to intact human skin elicits a moderate burning only after 52 min and erythema is hardly noticeable. Slight epidermal scaling may follow within 24 hr. Butyric acid (as an ester) is present in butter to a level of 4 to 5 percent (74) and has also been detected in essential oils of a number of herbs and spices.

2.5.5   Biochemistry

Butyric acid is readily absorbed from the gastrointestinal tract and rapidly metabolized by the liver (71). In rats a considerable portion of the butyric acid is metabolized to acetic acid. Butyric acid metabolism gives rise to ketone bodies, β-hydroxybutyrate, acetoacetate, acetone, and acetic acid, which may be excreted in the urine or incorporated into normal processes of fat metabolism (21, 71).

2.5.6   Regulations and Standards

Butyric acid is a mild, local irritant which causes readily reversible changes that disappear after exposure is terminated (47). Butyric acid is generally recognized as safe for use as a synthetic flavoring substance (21 CFR 182.60) and for use in animal feed (21 CFR 582.60).

2.6   Isobutyric Acid

The formula is $(CH_3)_2CHCOOH$; isobutyric acid is also known as 2-methyl-propanoic, dimethylacetic, and α-methylpropionic acid.

2.6.1   Major Uses

Isobutyric acid is primarily used as a food and animal feed fungistat and in the production of esters for solvents, flavoring, and perfume bases. It is also used in varnishes and disinfecting agents and for deliming hides and tanning.

2.6.2   Acute Studies

The acute toxicity of isobutyric acid is summarized in Table 59.2. In the open rabbit-skin irritation test isobutyric acid (0.01 ml, 10 mg) caused some necrosis in 24 hr (12). Severe corneal burning resulted from application of a 5 percent aqueous solution of the acid (unspecified volume) to rabbit eyes (12). No fatalities occurred when rats inhaled air saturated with isobutyric acid vapor for 8 hr (12).

2.6.3   Repeated Exposures

No reports of repeated exposures are available.

2.6.4   Physiologic Effects in Humans

Isobutyric acid is a naturally occurring component of food (cheese, butter, milk protein, vinegar, and beer) and feedstuffs, and is produced during the intermediary hepatic (76) and microbial (77) metabolism of valine. Isobutyric acid has been identified as a normal constituent of the fermentative organs of a number of species and is present in human feces, presumably due to action of intestinal microflora (78). It has also been detected in human blood (79) and saliva (80). No reports of studies with isobutyric acid in man were available.

2.6.5 Biochemistry

Isobutyric acid differs from butyric acid in that its metabolism does not produce ketone bodies (71). Large doses (5 to 6 g/kg) do not cause coma in rabbits, in contrast to butyric acid (71). Isobutyric acid is metabolized to propionic acid which, in turn, is converted to succinic acid (see Section 2.4.5) and ultimately to glucose and glycogen (78). Rats fed high doses of isobutyric acid excrete increased amounts of 2-methylmalonic acid in the urine (an intermediate in the conversion of propionic acid to succinic acid) (81). Metabolism in dairy cattle is apparently rapid, since there was no carryover of isobutyric acid into milk in animals fed 170 mg/kg/day for 10 days, and no increase in the peripheral blood concentrations of isobutyric acid (78). Male and female rats administered [1-^{14}C]isobutyric acid by oral gavage (4 to 400 mg/kg) rapidly eliminated 70 to 80 percent of the dose as $^{14}CO_2$ within 4 hr, and 90 to 96 percent in 48 hr; approximately 3 to 4 percent of the dose was present in the urine as [^{14}C]urea, also formed from $^{14}CO_2$ (78).

2.6.6 Regulations and Standards

Isobutyric acid is generally recognized as safe for use as a synthetic flavoring substance (21 CFR 172.515) and has exemption from the requirement of a tolerance on grasses and grains following its use as a fungistat (40 CFR 180.1028).

2.7 Valeric Acid

The formula is $CH_3(CH_2)_3COOH$; valeric acid is also known as pentanoic, propylacetic, and 1-butanecarboxylic acid.

2.7.1 Major Uses

Valeric acid is used in flavorings and as an intermediate in the manufacture of perfumes and pharmaceutical products.

2.7.2 Acute Studies

The acute toxicity of valeric acid is summarized in Table 59.2. In rabbits, intravenous administration of valeric acid (0.7 and 1.35 g/kg as 10 percent solutions) causes a moderate central nervous system depression. Up to 1 g/kg in cats and 0.5 g/kg in dogs causes somnolence followed by diuresis, vomiting, and defecation, with rapid recovery (71). In contrast to butyric acid oral administration of valeric acid did not result in coma in rabbits (82). According to Fassett (21), valeric acid is a strong skin irritant in the undiluted form. Using mixed isomers of valeric acid (unspecified composition), Smyth et al. (12, 13) demonstrated some tissue necrosis with 10 mg of material in the 24-hr open, rabbit skin irritation assay. They reported severe corneal injury in rabbits with an unspecified volume of a 1 to 5 percent solution of the mixture. No fatalities occurred when rats were exposed for 8 hr to air saturated with valeric acid vapor (mixed isomers) (12, 13).

2.7.3 Repeated Exposures

No cumulative effects have been reported. Valeric acid is a normal constituent of the fermentative organs of ruminant animals, from which it is absorbed into the blood (83). Valeric acid is formed by rumen microorganisms during the metabolism of proline, leucine, isoleucine, norleucine, and several intermediates of carbohydrate metabolism (84–86).

2.7.4 Physiologic Effects in Humans

There are no reports of industrial or experimental exposure of humans to this compound. Valeric acid is present in a variety of plants, fruits, and dairy and meat products (86) and has been detected in human saliva (80) and feces (87), probably because of the microbial metabolism of dietary constituents.

2.7.5 Biochemistry

Valeric acid is rapidly metabolized in rat liver to acetate and propionate (88), giving rise both to glycogen and ketone bodies.

2.7.6 Regulations and Standards

The TLV in the Soviet Union is 1.2 ppm (89). Valeric acid is generally recognized as safe (GRAS) for use as a synthetic flavoring substance (21 CFR 172.515) and has exemption from tolerance for use as an odorant in pesticides (40 CFR 180.1001). A specified mixture of aliphatic acids, including valeric acid, is GRAS as a secondary food additive when used in lye peeling of fruits and vegetables (21 CFR 173.315).

2.8 Isovaleric Acid

The formula is $(CH_3)_2CHCH_2COOH$; isovaleric acid is also known as 3-methylbutanoic, isopropylacetic, and isopentanoic acid.

2.8.1   Major Uses

Isovaleric acid is used as a flavoring ingredient in nonalcoholic beverages and in foods, as a fragrance ingredient in perfumes, and in the manufacture of pharmaceutical products.

2.8.2   Acute Studies

The acute toxicity of isovaleric acid is summarized in Table 59.2. Isovaleric acid is a mild irritant that produces a moderate response in the 24-hr open, rabbit skin irritation test with 0.5 ml of the compound and a mild irritation of the cornea when 940 μg was applied to rabbit eyes (20). Vernot et al. (22) found no skin corrosion when 0.5 ml isovaleric acid was applied as a closed patch to rabbits for 4 hr. The undiluted acid is a strong skin irritant in the guinea pig (21).

2.8.3   Repeated Exposures

According to Fassett (21), no cumulative effects are known. The maximum no-effect dietary level of isovaleric acid in rats was 5 percent (90).

2.8.4   Physiologic Effects in Humans

No reports of industrial exposure or human studies are available. Isovaleric acid occurs naturally in many plants, including tobacco, valeriana, and the essential oils of cypress, citronella, geranium, hops, laurel leaves, and American peppermint. Isovaleric acid is an intermediate in leucine metabolism.

2.8.5   Biochemistry

In contrast to valeric acid, isovaleric acid is ketogenic (71). It is readily absorbed from the gastrointestinal tract in man (70) and is metabolized by the liver to give two- and three-carbon fragments (91). The isopropyl fragment of isovaleric acid is readily converted to acetoacetate (92) and is an efficient source of carbon atoms for fatty acid and cholesterol synthesis (91, 93, 94).

Isovaleric acid is an intermediate in the metabolism of the amino acid leucine and has been identified in the fermentative organs of ruminant animals (78, 95). High blood concentrations of isovaleric acid occur in patients with the clinical disorder "isovaleric acidemia." This is a genetic defect of leucine metabolism in which the enzyme isovaleryl-coenzyme A dehydrogenase is inhibited or absent. The disease is characterized by episodic acidosis, slight mental retardation, and an unpleasant body odor (94).

2.8.6   Regulations and Standards

Isovaleric acid is generally recognized as safe for use as a synthetic flavoring substance (21 CFR 172.515).

2.9   Caproic Acid

The formula is $CH_3(CH_2)_4COOH$; caproic acid is also known as hexanoic, *n*-hexoic, and 2-butylacetic acid.

2.9.1   Major Uses

Caproic acid is used in the synthesis of esters for artificial flavorings, resins, driers, and pharmaceuticals, and as an intermediate in the manufacture of hexyl compounds.

2.9.2   Acute Studies

The acute toxicity of caproic acid is summarized in Table 59.2. Caproic acid is a skin and eye irritant in rabbits. According to Smyth et al. (16), some necrosis occurred when 10 mg of the compound was applied to open skin patches of rabbits for 24 hr; an unspecified volume of a 15 percent solution of caproic acid produced severe burns of rabbit cornea (16, 96). Rats exposed to air saturated with caproic acid for 8 hr suffered no fatalities (16).

2.9.3   Repeated Exposures

Rats fed caproic acid for 3 weeks at 2, 4, or 8 percent in the diet showed no changes in liver size, liver to body weight ratio, serum lipids, or peroxisome-related enzymes (97).

2.9.4   Physiologic Effects in Humans

No reports were available.

2.9.5   Biochemistry

Caproic acid is rapidly metabolized by β-oxidation in mitochondria (98). Rittenberg et al. (99) reported that caproic acid is not stored or converted to higher fatty acids. Oxidation occurs in the mitochondria of liver, kidney, and heart, but only hepatic mitochondria produce ketone bodies (100, 101).

2.9.6   Regulations and Standards

The TLV in the Soviet Union is 1 ppm (89). Caproic acid is generally recognized as safe (GRAS) for use as a synthetic flavoring substance (21 CFR 172.515). A specified mixture of aliphatic acids, including caproic acid, is GRAS for use in lye peeling of fruits and vegetables (21 CFR 173.315).

2.10   Isocaproic Acid

The formula is $(CH_3)_2CH(CH_2)_2COOH$; isocaproic acid is also known as 4-methylpentanoic, isohexanoic, and isobutylacetic acid.

2.10.1   Major Uses

Isocaproic acid is an intermediate in the manufacture of plasticizers, pharmaceuticals, and perfumes.

2.10.2   Acute Studies

The acute toxicity of isocaproic acid is summarized in Table 59.2. Isocaproic acid (0.5 ml) produced a moderate irritation to rabbit skin (open) and a mild reaction when 930 µg was applied to the rabbit cornea (24). Application of the undiluted compound to guinea pig skin and rabbit eye caused severe damage (21). The symptoms in rats after toxic oral doses were weakness, vasodilation, respiratory distress, and hematuria (21).

2.10.3   Repeated Exposures

No information is available. The compound is present in the fermentative organs of ruminant species.

2.10.4   Physiologic Effects in Humans

No data are available. The acid is a constituent of many edible plants and dairy products.

2.10.5   Biochemistry

Isocaproic acid is metabolized by β-oxidation, yielding isobutyric acid (102, 103) which in turn is converted to propionic acid (103). Isocaproic acid is therefore both glycogenic and ketogenic.

2.10.6   Regulations and Standards

Isocaproic acid is deemed to be generally recognized as safe by the Flavor and Extract Manufacturers' Association (FEMA).

2.11   2-Methylvaleric Acid

The formula is $CH_3(CH_2)_2CH(CH_3)COOH$; the acid is also known as 2-methylpentanoic, α-methylvaleric, and methylpropylacetic acid.

2.11.1   Major Uses

2-Methylvaleric acid is used in the production of plasticizers, vinyl stabilizers, metallic salts, and alkyd resins.

2.11.2   Acute Studies

The acute toxicity of 2-methylvaleric acid is summarized in Table 59.2. The compound caused only a mild irritation in rabbit skin (500 mg) (24) but elicited a severe reaction in guinea pigs (unspecified quantity of undiluted acid) (21). Application of a 5-percent solution to rabbit eyes (unspecified volume) caused severe necrosis (16). An 8-hr exposure of rats to air saturated with 2-methylvaleric acid caused no fatalities (16).

2.11.3   Repeated Exposures

No data are available.

2.11.4   Physiologic Effects in Humans

No data are available. 2-Methylvaleric acid is a constituent of many plants, including tea and tobacco.

2.11.5   Biochemistry

The presence of an α-methyl group tends to block β-oxidation (104). Since α-substituted acids have, in effect, two β-substituents, oxidation may proceed via either. Oxidation and removal of the methyl group would yield valeric acid, whereas the alternative scheme would produce two three-carbon (propionic acid) molecules. The latter route of metabolism appears to predominate because no ketone bodies (from valeric acid metabolism) were produced following 2-methylvaleric acid administration to rabbits (104).

2.11.6   Regulations and Standards

2-Methylvaleric acid is generally recognized as safe for use as a synthetic flavoring substance (21 CFR 172.515).

2.12   2-Ethylbutyric Acid

The formula is $(C_2H_5)_2CHCOOH$; 2-ethylbutyric acid is also known as 2-ethylbutanoic, diethylacetic, and 3-pentanecarboxylic acid.

2.12.1   Major Uses

The acid finds use in ester formation and as an intermediate for drugs, dyestuffs, chemicals, and flavorings.

2.12.2   Acute Studies

2-Ethylbutyric acid was a mild irritant (10 mg/24 hr) to rabbit skin but caused severe burns when applied as a 5-percent solution (unspecified volume) to rabbit eyes (16). No deaths resulted from an 8-hr exposure of rats to air saturated with 2-ethylbutyric acid vapor (16).

2.12.3   Repeated Exposures

The maximum no-effect level in a 90-day feeding study in rats was 0.6 percent (90).

2.12.4   Physiologic Effects in Humans

No data are available.

2.12.5   Biochemistry

The metabolic fate of 2-ethylbutyric acid involves several pathways. In general, α-ethyl-substituted acids are not readily metabolized (104). The predominant route of disposition may be by glucuronidation accounting for 20 to 50 percent of an oral or subcutaneous 1-g dose to rabbits or a 100-mg dose to rats (105). Some dealkylation has been reported in human diabetics and in perfused rat livers (104). This produces butyric acid and ketone bodies. β-Oxidation of 2-ethylbutyric acid, followed by decarboxylation to give methyl propyl ketone, has been reported after subcutaneous administration to dogs (106).

2.12.6   Regulations and Standards

2-Ethylbutyric acid is generally recognized as safe for use as a synthetic flavoring substance (21 CFR 172.515).

2.13   Heptanoic Acid

The formula is $CH_3(CH_2)_5COOH$; heptanoic acid is also known as enanthic, *n*-heptylic, and 1-hexanecarboxylic acid.

2.13.1   Major Uses

The major uses are organic synthesis and the production of special lubricants for aircraft and brake fluids.

2.13.2   Acute Studies

The acute toxicity of heptanoic acid is summarized in Table 59.2.

2.13.3   Repeated Exposures

According to Fassett (21) mice receiving 125 mg/kg/day by intraperitoneal injection died within 2 to 4 days after dosing began. This contrasts with the high intravenous LD_{50} value reported by Orö and Wretlind (6) (Table 59.2).

2.13.4   Physiologic Effects in Humans

No reports are available.

2.13.5   Biochemistry

Metabolism of heptanoic acid, which occurs by β-oxidation, follows the same pattern as other odd carbon number, straight chain, aliphatic acids. This results in the appearance of ketone bodies. Metabolism of the remaining propionic acid moiety results in glucose and glycogen formation (71, 104, 107, 108).

2.13.6   Regulations and Standards

Heptanoic acid is a primary irritant in concentrated solutions (21). A specified mixture of aliphatic acids, including heptanoic acid, is generally recognized as safe (GRAS) as a secondary, direct additive for use in the lye peeling of fruits and vegetables (21 CFR 173.315). Heptanoic acid is deemed GRAS by the FEMA.

2.14 Caprylic Acid

The formula is $CH_3(CH_2)_6COOH$; this acid is also known as octanoic, n-octylic and 1-heptanecarboxylic acid.

2.14.1 Major Uses

The chief uses include as an intermediate for dyes, drugs, perfumes, and flavorings and in the manufacture of foods, dental compositions, antiseptics, fungicides, and plasticizers.

2.14.2 Acute Studies

The acute toxicity of caprylic acid is summarized in Table 59.2. The acid caused slight necrosis when applied to rabbit skin (10 mg/24 hr) and severe corneal injury as a 5-percent solution (unspecified volume) to rabbit eyes (12). The maximum no-effect time for inhalation of concentrated caprylic acid vapor for rats was 4 hr (12).

Caprylic acid (10 mM, 3 hr) caused some mitotic abnormalities in eggs of the palmate newt (109), but was not mutagenic in *Salmonella typhimurium* or *Saccharomyces cerevisiae* with or without liver preparations from mouse, rat, or monkey (110).

2.14.3 Repeated Exposures

Feeding of 1 to 5 percent caprylic acid to dogs (for an unspecified period) caused diarrhea (21).

2.14.4 Physiologic Effects in Humans

Caprylic acid is present in many plants, fruits, and nuts. It produces relatively mild irritation to skin and mucous membranes (111). A 1-percent petrolatum solution caused no irritation after a 48-hr occluded-patch test in humans and produced no sensitization reactions (28).

2.14.5 Biochemistry

Caprylic acid administered to rats is readily metabolized by the liver and other tissues, forming carbon dioxide and two-carbon fragments which are incorporated into long-chain fatty acids (112–114). Ketone bodies detectable in vitro indicate that oxidation to acetate occurs. Other water-soluble products are also formed (71, 112, 113). Metabolism of caprylic acid has been reported in kidney, muscle, heart, brain, and adipose tissue (112, 115).

2.14.6   Regulations and Standards

Caprylic acid vapors are irritating and can cause coughing (47). Caprylic acid is affirmed as generally recognized as safe for use as a direct (21 CFR 184.1025) and indirect (21 CFR 186.1025) food substance. It was approved by the Council of Europe (1974) for food use (28).

2.15   2-Ethylhexanoic Acid

The formula is $CH_3(CH_2)_3CH(C_2H_5)COOH$; 2-ethylhexanoic acid is also known as 2-ethylcaproic, 2-ethylhexoic, and butylethylacetic acid.

2.15.1   Major Uses

The metallic salts are used as driers for paints and varnishes and as gelling agents for hydrocarbons.

2.15.2   Acute Studies

2-Ethylhexanoic acid caused mild skin irritation, but severe corneal irritation in rabbits (23). Other acute data are summarized in Table 59.2.

2.15.3   Repeated Exposures

Male rats fed 2 percent 2-ethylhexanoic acid in the diet for 3 weeks displayed no changes in body weight gain compared to controls, but showed hepatomegaly. The liver-to-body weight ratio increased by 50 percent (97). Serum triglyceride concentrations were reduced but no increases in mortality were reported (see also Section 2.15.5).

2.15.4   Physiologic Effects in Humans

A single case of corneal injury, with prompt healing, has been reported (21).

2.15.5   Biochemistry

Rats fed 2-ethylhexanoic acid (2 percent) for 3 weeks showed decreased serum triglycerides, hepatomegaly, and a large increase in hepatic peroxisomes (97). It has not been established whether the proliferation of these microbodies is responsible for the depressed serum lipid concentrations, but similar effects have been reported with a variety of hypolipidemic agents (116–119). 2-Ethylhexanoic acid has a marked effect on lipid metabolism, in vivo and in vitro, causing inhibition of triglyceride biosynthesis in intestinal mucosa. This leads to changes in absorption of fatty acids and cholesterol (120–122).

2.15.6   Regulations and Standards

2-Ethylhexanoic acid has exemption from the requirement of a tolerance as a pesticide residue when used as a cosolvent and defoamer (40 CFR 180.1001).

2.16   Nonanoic Acid

The formula is $CH_3(CH_2)_7COOH$; nonanoic acid is also known as pelargonic, pelargic, and 1-octanecarboxylic acid.

2.16.1   Major Uses

The primary uses are in organic synthesis and the manufacture of lacquers, plastics, pharmaceuticals, synthetic odors and flavorings, gasoline additives, flotation agents, lubricants, and vinyl plasticizers.

2.16.2   Acute Studies

The acute toxicity of nonanoic acid is summarized in Table 59.2. Nonanoic acid (500 mg/24 hr) was a moderate irritant to rabbit skin but produced a severe reaction in guinea pigs (21, 30). A severe irritation was produced by application of 91 mg of the acid to rabbit eyes (123). No symptoms of toxicity could be produced by inhalation of concentrated vapors in rats (21).

2.16.3   Repeated Exposures

Nonanoic acid, fed for 4 weeks at 4.17 percent in the diet, depressed the rate of growth only in vitamin B_{12}-deficient rats (124). Five percent in the diet was fairly well utilized by growing chicks (125).

2.16.4   Physiologic Effects in Humans

A 12-percent solution in petrolatum produced no irritation in humans after a 48-hr closed patch test (30). Higher concentrations (0.5 or 1.0 M in propanol) caused irritation when applied under occlusive patches. No sensitization reactions were produced in 25 volunteers after patch testing with nonanoic acid (12 percent in petrolatum) (30).

2.16.5   Biochemistry

Nonanoic acid is metabolized by the liver to produce ketone bodies (108, 126). Metabolism occurs via β-oxidation, and no evidence was found in rats of chain elongation or tissue storage of the acid (107). Metabolism of the terminal propionic acid residue results in increased glucose and glycogen synthesis (104). According to Bach et al. (126), infusion of 20 percent trinonanoate into dogs increased urinary ketone body levels by approximately 700 percent and elevated the lactate/pyruvate ratio in plasma.

2.16.6 Regulations and Standards

Nonanoic is generally recognized as safe for use as a synthetic flavoring substance (21 CFR 172.515) and was approved by the Council of Europe (1974) for food use (30).

2.17 Capric Acid

The formula is $CH_3(CH_2)_8COOH$; capric acid is also known as decanoic, decylic, and 1-nonanecarboxylic acid.

2.17.1 Major Uses

Capric acid is chiefly used in the production of esters for perfumes and fruit flavorings, as a base for wetting agents, and in the production of intermediates, plasticizers, resins, and food additives.

2.17.2 Acute Studies

The acute toxicity of capric acid is summarized in Table 59.2. The compound was a moderate to severe irritant when applied undiluted for 24 hr to intact or abraded rabbit skin in the occluded patch test (31). According to Smyth et al. (12), capric acid (mixed isomers) produces severe corneal burns when applied as a 5 percent solution to rabbit eyes, and is moderately irritating to rabbit skin in the open patch test. No deaths occurred in rats exposed for 8 hr to concentrated capric acid vapor (12).

2.17.3 Repeated Exposures

No gastric lesions were evident in rats fed capric acid (10 percent in diet) for 150 days (127). Capric acid administered daily (37 mg/kg) to pregnant rabbits increased sensitivity to oxytocin-induced labor (31).

2.17.4 Physiologic Effects in Humans

Capric acid produced no irritation when applied to human skin as a 1-percent solution in petrolatum for 48 hr in a closed-patch test (31). At higher concentrations (up to 1.0 M in propanol), the compound produced signs of irritation within 8 days in occlusive patch tests in human volunteers (128). No sensitization reactions were seen.

2.17.5   Biochemistry

Capric acid is metabolized by the β-oxidative pathway, giving rise to ketone bodies in rats (129), rabbits (130), dogs (71), piglets (131), and goats (132). ω-Oxidation, leading to the excretion of sebacic acid (104), and chain elongation reactions (130–132) have been observed. Metabolism of capric acid is rapid; in humans given [1-^{14}C]decanoic acid orally, about 52 percent of the radioactivity was recovered within 2.5 to 4 hr (133).

2.17.6   Regulations and Standards

Capric acid is generally recognized as safe for use as a food additive (21 CFR 182.60) and was approved by the Council of Europe (1974) for food use (31).

2.18   Undecylic Acid

The formula is $CH_3(CH_2)_9COOH$; undecylic acid is also known as undecanoic, hendecanoic, and 1-decanecarboxylic acid.

2.18.1   Major Uses

The chief use is in chemical syntheses.

2.18.2   Acute Studies

Little information is available on this compound (see Table 59.2).

2.18.3   Repeated Exposures

No information is available.

2.18.4   Physiologic Effects in Humans

Saturated fatty acids (C_3 to C_{18}) were applied daily under occlusive patch tests to human skin until detectable erythema appeared. The most irritating acids were C_8 to C_{12} (128). Small amounts of these acids are present in castor oil.

2.18.5   Biochemistry

Undecylic acid is metabolized by β-oxidation, analogously to other odd carbon number fatty acids, and the terminal propionic acid moiety is glycogenic (104).

Other routes of metabolism have been reported, including ω-oxidation to nonanedicarboxylic acid and reductive cleavage to give sebacic acid (104). In the latter case desaturation to 10-undecenoic acid is presumed to occur; this acid is known to cleave at the double bond to form sebacic acid.

2.18.6 Regulations and Standards

Undecanoic acid is deemed to be generally recognized as safe by FEMA at an average maximum level of 2 ppm in baked goods (134).

2.19 Lauric Acid

The formula is $CH_3(CH_2)_{10}COOH$; lauric acid is also known as dodecanoic, duodecylic, and 1-undecanecarboxylic acid.

2.19.1 Major Uses

The major uses are in soaps, detergents, wetting agents, alkyd resins, cosmetics, insecticides, and food additives.

2.19.2 Acute Studies

Lauric acid is not considered a skin irritant or sensitizer (21). Infusion of sodium laurate into guinea pigs caused platelet aggregation in arterial blood (135). Other acute data are given in Table 59.2.

2.19.3 Repeated Exposures

Fitzhugh et al. (136) fed lauric acid (10 percent of the diet) to male rats for 18 weeks and reported no adverse effects on weight gain, organ weights, or pathology and no mortality. No effects from lauric acid glycerides were found after 2 years of feeding at 25 percent in the diet of rats (21). Lauric acid was not carcinogenic (137). According to Holsti (138), lauric acid (as a 20-percent solution in chloroform) promoted papilloma formation in mice initiated with 9,10-dimethyl-1,2-benzanthracene. No histologically malignant tumors were found.

2.19.4 Physiologic Effects in Humans

There are no available reports of effects in man. Lauric acid is a constituent of many vegetable fats, mostly in the form of glycerides.

2.19.5   Biochemistry

Lauric acid may be metabolized in several ways. Chain elongation to myristic, palmitic, and stearic acids is known to occur, apparently by condensation of the methyl function of acetate with the carboxyl group of the fatty acid (104). The acid is rapidly absorbed from the blood in rats and is oxidized or incorporated into tissue lipids, predominantly neutral lipids rather than phospholipids (139). Das et al. (140) have demonstrated that ω-oxidation of lauric acid is accomplished by rat liver microsomes in vitro and a similar reaction occurs in renal cortex microsomes (141). This metabolic route may not be significant in vivo; Kam et al. (142) reported that the contribution of ω-oxidation to fatty acid oxidation in the rat and monkey is very small. In general, it is believed that ω-oxidation does not occur significantly in straight chain aliphatic acids with more than 12 carbons (104). Lauric acid is ketogenic, being metabolized by β-oxidation, and does not give rise to glycogen synthesis (143). In vitro conversion to a monounsaturated fatty acid was negligible in rat liver homogenates (144).

2.19.6   Regulations and Standards

Lauric acid is generally recognized as safe for use as a direct food additive (21 CFR 172.860).

2.20   Myristic Acid

The formula is $CH_3(CH_2)_{12}COOH$; myristic acid is also known as tetradecanoic acid, crodacid, and 1-tridecanecarboxylic acid.

2.20.1   Major Uses

The chief uses are in soaps and perfumes, in the synthesis of esters for flavorings and perfumes, and   a component of food-grade additives.

2.20.2   Acute Studies

Little information is available on the acute toxicity of myristic acid (see Table 59.2).

2.20.3   Repeated Exposures

Studies on the carcinogenic potential of myristic acid were negative (137).

2.20.4   Physiologic Effects in Humans

Myristic acid was a moderate irritant when applied to human skin (75 mg total over 3 days) (145).

2.20.5   Biochemistry

The metabolism of myristic acid is similar to that of lauric acid. In the rat, the rates of oxidation of the higher fatty acids were inversely related to chain length; thus lauric acid metabolism occurred more rapidly than myristic acid metabolism, which in turn was more rapid than stearic acid metabolism (139). In addition to metabolism by β-oxidation, myristic acid has been shown to undergo chain elongation to palmitic and stearic acids, desaturation to myristoleic acid and incorporation into hepatic neutral lipids (and to a lesser extent, phospholipids) (104, 139, 146). ω-Oxidation is believed not to occur in rat liver with myristic acid or longer chain, unsubstituted fatty acids (104) but occurs in vitro with myristic acid in pig renal cortex microsomes (141).

2.20.6   Regulations and Standards

Myristic acid is generally recognized as safe as a food additive (21 CFR 172.860) and has exemption from tolerance for use as a diluent for pesticides (40 CFR 180.1001).

2.21   Palmitic Acid

The formula is $CH_3(CH_2)_{14}COOH$; palmitic acid is also known as hexadecanoic, cetylic, and 1-pentadecanecarboxylic acid.

2.21.1   Major Uses

The major uses are in the manufacture of soaps, lubricating oils, and waterproofing materials and in the synthesis of metal palmitates and food-grade additives.

2.21.2   Acute Studies

Little information is available on the acute toxic effects of palmitic acid (see Table 59.2).

2.21.3   Repeated Exposures

Studies on the carcinogenic potential of palmitic acid were negative (137). When diets containing 5 to 40 percent palmitic acid were fed as the monoglyceride for 3 weeks to weanling mice depression of growth was observed at all dietary levels except the lowest. High mortality resulted at dietary levels of 20 to 40 percent, with 100 percent of mortality at the 40-percent level (147).

2.21.4 Physiologic Effects in Humans

Palmitic acid was a mild irritant when applied to human skin (75 mg total over 3 days) (145). It is a naturally occurring fatty acid component of animal fats and vegetable oils and fats.

2.21.5 Biochemistry

Palmitic acid is rapidly metabolized, primarily by β-oxidation. In addition to oxidative breakdown, palmitic acid undergoes a variety of interconversion reactions in the liver and intestinal mucosa to stearic, oleic, palmitoleic, and myristic acids (148, 149). ω-Oxidation, prior to β-oxidation, may account for 5 to 10 percent of hepatic metabolism of palmitic acid in the starved rat (150). After oxidation or conversion to other long chain fatty acids the carbon skeleton of palmitic acid is esterified or returned to the plasma, depending upon the nutritional state of the organism (151). Esterification produces mainly neutral fats rather than phospholipids (152).

2.21.6 Regulations and Standards

Palmitic acid is generally recognized as safe for use as a direct food additive (21 CFR 182.60) and is exempt from tolerance requirements under 21 CFR 180.1001 for use as diluent (adjuvant) for pesticide chemicals.

2.22 Stearic Acid

The formula is $CH_3(CH_2)_{16}COOH$; stearic acid is also known as octadecanoic, cetylacetic, and 1-heptadecanecarboxylic acid.

2.22.1 Major Uses

The primary uses are in pharmaceuticals, cosmetics, metal stearates (as polymer stabilizers), plastics, lubricants, coatings, food packaging, rubber softeners, baked goods, and confectioneries.

2.22.2 Acute Studies

The acute toxicity of stearic acid is summarized in Table 59.2. Single intraperitoneal doses of stearic acid in mice, ranging from 1.4 to 44 μmol (approximately 15 to 500 mg/kg), caused no fatalities, but at the highest dose caused a loss of body weight (153). In cats low doses of stearic acid produced elevated pulmonary but decreased systemic blood pressure. Doses greater than 5 mg caused apnea, a fall in blood pressure, and convulsions leading to death (6).

Intravenous infusion of large doses of stearic acid were thrombogenic in rats, rabbits, and dogs, causing blood platelet aggregation and acute heart failure (33).

In the occluded patch test undiluted stearic acid was a moderate irritant to intact or abraded rabbit skin after 24 hr (33).

2.22.3 Repeated Exposures

When diets containing 5 to 50 percent stearic acid (as the monoglyceride) were fed to weanling mice for 3 weeks depression of weight gain was seen above the 10-percent dietary level. Mortality occurred only with the 50-percent diet (147). The effects were less noticeable in adult mice. Rats fed 5 percent stearic acid as part of a high-fat diet for 6 weeks, or 6 percent stearic acid for 9 weeks, showed a decreased blood clotting time and hyperlipemia (33). Stearic acid is one of the least effective fatty acids in producing hyperlipemia, but the most potent in diminishing blood clotting time (33). No significant pathologic lesions were observed in rats fed 3000 ppm stearic acid orally for about 30 weeks, but anorexia, increased mortality, and greater incidence of pulmonary infection were observed (154).

When stearic acid was administered subcutaneously to female mice once weekly for 25 weeks (total dose 1.3 to 130 mg) no incidence of sarcoma at the injection site was observed (155). Subcutaneous sarcomas were previously observed by the same authors in one strain of mice at the lowest dose level (156). This finding was unexpected and unexplained.

2.22.4 Physiologic Effects in Humans

Stearic acid is a naturally occurring fatty acid component of tallow, animal fats and oils, and some vegetable oils. Stearic acid produced no skin irritation in man when applied as 7-percent solution in petrolatum for 48 hr in a closed-patch test (33). A 1.0 M solution in propanol produced no irritation when applied daily to skin for 10 days. Some hardening and mild erythema occurred when 0.1 M stearic acid in olive oil was injected intradermally (128). No sensitization reactions were observed with 7 percent stearic acid in petrolatum applied to human subjects (33).

2.22.5 Biochemistry

Stearic acid metabolism via β-oxidation, ω-oxidation, and (ω-1)oxidation has been demonstrated in rat liver (157–159). Removal of a single acetate moiety can occur to produce palmitic acid, and both this and stearic acid may be desaturated producing oleic and palmitoleic acids, respectively (160). After [^{14}C]stearic acid was injected into rats about 50 percent of the liver ^{14}C was recovered as oleic acid, indicating that extensive desaturation occurs (149). Desaturation occurs only to a small extent extrahepatically (149) but has been detected in adipose tissue and in cells of mammary tissue (149, 161). Human liver biopsy samples have oxidative desaturation capacity for stearic acid (162). Stearic acid is also incorporated into phospholipids, di- and triglycerides, cholesterol, cholesterol esters, and other sterol esters (149, 163–165).

Long chain, saturated fatty acids are less readily absorbed than unsaturated or short chain acids (32). Stearic acid is the most poorly absorbed of the common fatty acids. Its absorption in sheep is facilitated by the release of oleic and linoleic acids in response to bile and pancreatic secretions (166). Stearic acid absorption in chicks was increased from 14 to 49 percent by the addition of oleic acid to the diet and decreased to 2 percent when palmitic acid was added (167).

2.22.6   Regulations and Standards

Stearic acid is generally recognized as safe as a food additive (21 CFR 172.860) and a gum base (21 CFR 172.615). The Council of Europe (1974) has approved its use as an artificial flavoring substance (33).

3   SATURATED POLYCARBOXYLIC ACIDS

3.1   Introduction

3.1.1   Industrial Applications

Saturated, aliphatic polycarboxylic acids are used in similar applications to the monocarboxylic acids.

3.1.2   Health Effects

The physiologic effects of the saturated, aliphatic polycarboxylic acids are similar to those described for the monocarboxylic acids.

The acid dissociation constants and water solubilities of the aliphatic dicarboxylic acids are presented in Table 59.3. The short chain acids, in particular oxalic and malonic acids, are relatively strong acids, whereas the longer chain acids (e.g., pimelic and sebacic) are not irritants. The presence of the second carboxyl function in oxalic acid greatly increases its acidity, compared with formic acid. The fact that oxalic acid is a solid and of lower water solubility may partly counteract the effect of the increased ionization.

3.1.3 Metabolic Fate

The dicarboxylic acids appear to be less extensively metabolized than the monocarboxylic acids. An exception to this is succinate, which is rapidly metabolized by incorporation into intermediary metabolic pathways to produce glucose and glycogen.

Appreciable quantities of many of the dicarboxylic acids are excreted in the urine, the amount depending on the dose and route of administration. Some metabolism of the longer chain acids does occur by β-oxidation, producing a dicarboxylic acid having two fewer carbons than the original. This process may continue as far as succinate. Oxalic acid is excreted unchanged and malonic acid is only partly degraded, apparently by decarboxylation to acetate.

3.2 Oxalic Acid

The formula is HOOC·COOH; oxalic acid is also known as ethanedioic acid, ethanedionic acid and oxalsäure (German).

3.2.1 Major Uses

The major uses are in textile finishing, stripping and cleaning, calico printing and dyeing, paint, varnish, and rust removal, metal and equipment cleaning, wood cleaning, dye manufacture, chemical intermediates, and in the paper, ceramics, photographic, and rubber industries.

3.2.2 Acute Studies

The acute toxicity of oxalic acid is summarized in Table 59.4. Oxalic acid (500 mg/24 hr) was a moderate irritant to rabbit skin but produced a severe reaction when applied (250 μg) to rabbit eyes (179). Oxalic acid behaves similarly to other strong acids, producing severe local burns of eyes, mucous membranes, and skin (49). Intravenous administration of 5 mg/kg to dogs resulted in a moderate, transient hypotension, whereas doses of 42 mg/kg were lethal (183). The first signs of acute toxicity may be anorexia, salivation, and nasal discharge with progressive weakness, respiratory distress, and collapse. A dilute solution of oxalic acid (5 percent) and malonic acid (5 percent) is used extensively in veterinary practice as a hemostatic agent (74).

3.2.3 Repeated Exposures

The toxic effects of oxalic acid on repeated exposure arise from its ability to immobilize blood and tissue calcium. Calcium oxalate is insoluble at physiologic

Table 59.3. Physical Properties of Saturated Aliphatic Polycarboxylic Acids

Acid	CAS Acid Name	CAS Reg. No.	Mol Wt	M.P. (°C)	B.P. (°C)	Solubility[a]			Specific Gravity	Acid Dissociation		Vapor Pressure (mm) (°C)
						H_2O	Alcohol	Other		pK_{a1}	pK_{a2}	
Oxalic	Ethanedioic	144-62-7	90.04	189.5	157 (sublimes)	8.3%	V	Ether (sl s) Chloroform (ins) Benzene (ins)	1.90 (17/4)	1.46	4.40	0.54 (105)
Malonic	Propanedioic	141-82-2	104.07	135.6 (decomp.)	Dec. 140	V	V	Methanol (v) Ether (s) Pyridine (s)	1.63 (16/4)	2.80	5.85	
Succinic	Butanedioic	110-15-6	118.10	189.0	235 (dec.)	Sl s (v, hot)	S	Ether (s) Acetone (s) Benzene (ins)	1.57 (25/4)	4.17	5.64	0.03 (47)
Malic (DL)	Hydroxybutanedioic	6915-15-7	134.10	131		S	S	Methanol (s) Ether (sl s) Benzene (ins)	1.60 (20/4)	3.40	5.05	
Thiomalic	Mercaptobutanedioic	70-49-5	150.15	154		S	S	Ether (sl s) Acetone (s) Benzene (ins)				
Tartaric (L)	2,3-dihydroxy-butanedioic	87-69-4	150.10	171		V	V	Ether (sl s) Benzene (ins)	1.76 (20/4)	2.93	4.23	
Adipic	Hexanedioic	124-04-9	146.16	153	338	Sl s	V	Ether (s) Acetone (s) Benzene (ins)	1.36 (25/4)	4.43	5.52	1 (160)
Citric	2-Hydroxy-1,2,3-propanetricarboxylic	77-92-9	192.14	153		V	V	Ether (s) Chloroform (ins) Ethyl acetate (s)	1.66 (18/4) (anhyd.)	3.08	4.75	
Pimelic	Heptanedioic	111-16-0	160.19	106	272 (100 mm, subl.)	S	S	Ether (s) Benzene (ins)	1.33 (15/4)	4.47	5.42	
Suberic	Octanedioic	505-48-6	174.22	144	300 (subl.)	Sl s	S	Ether (sl s) Chloroform (ins) Methanol (s)	1.27 (25/4)			
Azelaic	Nonanedioic	123-99-9	188.25	106.5	287 (100 mm)	Sl s	V	Ether (sl s) Benzene (sl s)	1.03 (20/4)	4.54	5.52	1 (178)
Sebacic	Decanedioic	111-20-6	202.28	134.5	295 (100 mm)	Sl s	V	Ether (v) Benzene (ins) Acetone (s)	1.27 (20/4)	4.55	5.52	

[a] ∞, soluble in all proportions; v, very soluble; s, soluble; sl s, slightly soluble; v sl s, very slightly soluble; ins, insoluble.

Table 59.4. Acute Toxicity of Saturated Polycarboxylic Acids

Acid	Species	Route	LD$_{50}$ (mg/kg)	Ref.
Oxalic	Rat (male)	Oral	475[a]	22
	Rat (female)	Oral	375[a]	22
	Rabbit	Skin	20000[a] (not lethal)	22
	Cat	Sc	112 (LD$_{Lo}$)	168
	Dog	Oral	1000 (LD$_{Lo}$)	168
Malonic	Mouse	Oral	4000	25
	Mouse	Ip	300	169
	Mouse	Ip	Approx. 1500	21
	Rat	Oral	1310	170
	Rat	Ip	Approx. 1500	21
Succinic	Rat	Oral	2260	171
Malic	Mouse	Oral	1600–3200	171
	Mouse	Ip	50–100	171
	Rat	Oral	>3200	171
	Rat	Ip	100–200	171
Thiomalic	Mouse	Ip	500	172
	Rat	Oral	800–1600[b]	21
	Rabbit	Iv	>1000	173
	Guinea pig	Skin	>2000	21
Tartaric	Mouse	Iv	485	174
	Mouse	Oral	4360[c]	175
	Rabbit	Oral	5000 (LD$_{Lo}$)	176
	Rabbit	Oral	5290[d]	175
	Dog	Oral	5000 (LD$_{Lo}$)	177
Adipic	Mouse	Iv	680	174
	Mouse	Oral	1900	174
	Mouse	Ip	275	178
	Rat	Oral	3600 (LD$_{Lo}$)	179
Citric	Mouse	Iv	42	180
	Mouse	Oral	5040	181
	Mouse	Ip	961	180
	Mouse	Sc	2700	181
	Rat	Oral	11700	181
	Rat	Ip	883	180
	Rat	Sc	5500	181
	Rabbit	Iv	330	180
	Rabbit	Oral	7000 (LD$_{Lo}$)	176
Pimelic	Mouse	Oral	4800	25
	Rat	Oral	>3200	21
	Rat	Oral	7000	182
Sebacic	Mouse	Oral	6000	25
	Mouse	Ip	500	169

[a] As a 5% aqueous solution.
[b] As a 10% solution.
[c] Sodium salt.
[d] 3/7.

pH values (184) and may accumulate in renal tubules and in brain tissue. The resulting hypocalcemia produces severe disturbances in the actions of cardiac and nervous tissues. Feeding rats with dietary potassium oxalate caused retardation of growth and bone formation. These effects could be prevented by supplementing the diet with calcium, phosphorus, and vitamin D (185). Male and female rats fed 2.5 or 5.0 percent oxalic acid in the diet for 70 days showed restricted growth rates. At the 5.0-percent level, the organ weights of several visceral and endocrine tissues were reduced, but the organ/body weight ratios were increased. Estrous cycles in female rats were disrupted (186).

Oxalic acid (6 or 12 g/day) was fed to ewes from breeding to lambing. Although by crossing the placental barrier it did cause deposition of oxalate crystals in the kidneys of most lambs, they did not abort (187). Shetland ponies fed dietary oxalic acid (1 percent) demonstrated an increased fecal excretion of calcium and decreased calcium absorption; there was no compensatory decrease in urinary calcium excretion (188).

Oxalic acid toxicity may occur in animals grazing on plants that contain high levels of the compound. If the intake is sufficient, the capacity of the rumen microorganisms to metabolize oxalic acid may be exceeded, and acute and/or chronic effects occur (184). These toxic effects may be related to hypocalcemia, damage to the vascular system of the gut, or to renal insufficiency caused by crystal deposition (184).

3.2.4 Physiologic Effects in Humans

Solutions of 5 to 10 percent oxalic acid are irritating to the skin after prolonged exposure (49). There are numerous reports of acute poisoning from ingestion of oxalate-containing plants, especially rhubarb (leaves) and sorrel grass. There has been some dispute as to whether the oxalic acid content of rhubarb leaves is sufficient to explain these cases. This issue is dealt with in some detail in a comprehensive review of oxalates in food (184).

Ingestion of sufficient oxalic acid (about 5 g) can be fatal in man (49). It can result in corrosive damage to the gastrointestinal tract, shock, convulsions, and renal damage. When spinach was provided in the diet of humans, equivalent to 0.6 or 0.7 g oxalic acid daily for 3 to 4 weeks, there were no obvious alterations in calcium balance (189, 190). According to Fassett (184), a high intake of oxalate-containing food and a simultaneous low calcium and vitamin D intake over a prolonged period would be required for chronic toxic effects to be observed.

3.2.5 Biochemistry

Oxalic acid is poorly absorbed from the gut (1 to 6 percent) in the presence of food, but up to 50 percent of a small intake may be absorbed and excreted in the urine under fasting conditions (184). A significant proportion of endogenous urinary oxalic acid is derived from ascorbic acid and glycine metabolism. In a number of species oxalic acid is excreted unchanged in the urine and feces. Williams (46) has reported that oxalic acid, except for its combination with calcium, is metabolically inert. Some 40 percent of a low dose in rats is eliminated unchanged in the urine and large amounts are found in bone, probably as the calcium salt (191, 192). Bacterial metabolism occurs, particularly in ruminants, and about 1 percent of small doses may be oxidized to carbon dioxide (184, 191, 192).

3.2.6   Regulations and Standards

The ACGIH-adopted (1971) TLV for oxalic acid is 1 mg/m^3 (49) and the OSHA standard in air is a TWA value of 1 mg/m^3 (48). Oxalic acid is cleared for use as a polymerization catalyst aid (21 CFR 177.1010) and as a catalyst for phenolic resins (21 CFR 177.2410).

3.3   Malonic Acid

The formula is $CH_2(COOH)_2$; malonic acid is also known as propanedioic acid, carboxyacetic acid and dicarboxymethane.

3.3.1   Major Uses

Malonic acid is used as an intermediate in the manufacture of barbiturates and other pharmaceuticals.

3.3.2   Acute Studies

The acute toxicity of malonic acid is summarized in Table 59.4. Malonic acid is a relatively strong acid which can damage skin and mucous membranes (21). It is considered to be a strong irritant (47, 74), although it produced only a mild effect on rabbit skin (500 mg/24 hr) (170). Severe damage to rabbit eyes resulted from application of 100 mg of malonic acid (170).

3.3.3   Repeated Exposures

No data are available.

3.3.4   Physiologic Effects in Humans

Although malonic acid is an inhibitor of succinate dehydrogenase (see Section 3.3.5), no reports of human health effects are available (21).

3.3.5   Biochemistry

Malonic acid is metabolized to a greater extent than oxalic acid (104) but appreciable amounts are excreted unchanged. It is a competitive inhibitor in vitro of succinate dehydrogenase, which catalyzes the conversion of succinic acid to fumaric acid. Inhibition may also occur in vivo because rats administered malonic acid excrete succinic acid, citric acid, and α-ketoglutaric acid (194). Malonic acid may be metabolized by decarboxylation to acetate, which is metabolized to succinic acid via the citric acid cycle (195). The relative contribution of these two mechanisms to succinic acid formation from malonic acid remains to be determined.

3.3.6   Regulations and Standards

No data are available.

3.4   Succinic Acid

The formula is $HOOC \cdot (CH_2)_2 COOH$; succinic acid is also known as butanedioic, 1,2-ethanedicarboxylic, and ethylenesuccinic acid.

3.4.1   Major Uses

Succinic acid finds use in the manufacture of lacquers, dyes, esters for perfumes, and medicinal compounds. Succinic acid is also used in organic syntheses and photography and as a sequestrant, buffer, and neutralization agent in foods.

3.4.2   Acute Studies

Succinic acid is a moderate local irritant (47), producing slight skin irritation and strong eye irritation in rats (171). Application of a 15-percent solution of succinic acid produced severe damage in rabbit eyes (96). The symptoms of acute toxicity in rats are weakness and diarrhea (171). Other data are given in Table 59.4.

3.4.3   Repeated Exposures

No toxic effects were seen when succinic acid was fed to rats for 11 days at 0.1 or 1.0 percent in the diet (97 or 935 mg/kg/day intake) (171). Large intravenous doses of sodium succinate produced vomiting and diarrhea in cats, but these effects were nonspecific and were obtained with similar doses of sodium bicarbonate (196). Repeated administration of small oral doses of sodium or magnesium succinate produced no signs of systemic toxicity (196).

3.4.4   Physiologic Effects in Humans

No reports of human toxicity are available and no cases of injury in industrial use have been reported (21). Succinic acid occurs normally in human urine (1.9 to 8.8 mg/l) (46).

3.4.5   Biochemistry

Succinic acid is a normal intermediary metabolite and a constituent of the citric acid cycle. It is readily metabolized when administered to animals (197) but may be partly excreted unchanged in the urine if large doses are fed (46).

3.4.6   Regulations and Standards

Succinic acid is generally recognized as safe (GRAS) for use as a flavor enhancer and pH control agent in foods (21 CFR 184.1091). Disodium succinate is deemed to be GRAS by the FEMA (134).

3.5   Malic Acid

The formula is $HOOC \cdot CH(OH)CH_2COOH$; malic acid is also known as hydroxybutanedioic and hydroxysuccinic acid.

3.5.1   Major Uses

Malic acid is used in medicine, wine manufacture, preparation of esters and salts, as a chelating agent, and as a food acidulant and flavoring. The L form is found in fruits and plants. Commercial malic acid is a racemic mixture (DL-malic acid).

3.5.2   Acute Studies

Malic acid is a fairly strong acid and when undiluted can cause irritation of skin and mucous membranes (21). The symptoms of acute poisoning in rats and mice are weakness, retraction of the abdomen, respiratory distress, and cyanosis. It was a strong irritant in guinea pigs (21) and moderately irritating to rabbit skin (500 mg/24 hr) (179). Application of 750 μg malic acid caused severe eye irritation in rabbits (179). Other data are given in Table 59.4.

3.5.3   Repeated Exposures

No cumulative effects have been reported (21), presumably because of the incorporation of malic acid into intermediary metabolism.

3.5.4 Physiologic Effects in Humans

No accounts of toxicity in industrial use have been reported (21). Malic acid is a constituent of the human diet because of its presence in wine and in many edible plants and fruits (especially apples).

3.5.5 Biochemistry

Malic acid is an intermediate in the citric acid cycle. It is formed from fumaric acid and is oxidized to oxaloacetic acid. It is also metabolized to pyruvic acid by malic enzyme which is present in many biologic systems, including bacteria and plants (198). L-Malic and DL-malic acid are both rapidly metabolized in the rat. Orally or intraperitoneally administered L- or DL-malic acid was extensively eliminated as carbon dioxide (83 to 92 percent). No differences between the two forms were found in the rates (90 to 95 percent in 24 hr) or routes of excretion (199).

3.5.6 Regulations and Standards

Malic acid is affirmed as generally recognized as safe for use in food as a flavoring agent, adjuvant, and pH control agent (21 CFR 184.1069) and as an acidulant in the treatment of wine (21 CFR 240.1051).

3.6 Thiomalic Acid

The formula is $HOOC \cdot CH(SH)CH_2COOH$; thiomalic acid is also known as mercaptobutanedioic and mercaptosuccinic acid.

3.6.1 Major Uses

Thiomalic acid is used in biochemical research as a chemical intermediate, a rust inhibitor, and an additive in the rubber industry.

3.6.2 Acute Studies

The acute toxicity of thiomalic acid is summarized in Table 59.4. The symptoms of toxicity in rats were weakness, retraction of abdomen, depressed respiration, and cyanosis (21). Application of 2000 mg/kg (the dermal LD_{50}) to guinea pig skin for 24 hr caused severe skin damage (21).

3.6.3 Repeated Exposures

No reports are available.

3.6.4   Physiologic Effects in Humans

Allergic dermatitis in humans has been reported (47). No other data are available.

3.6.5   Biochemistry

Thiomalic acid has been used as an antidote for heavy metal poisoning (21). Under aerobic conditions and alkaline pH, it is oxidized by microorganisms to liberate hydrogen sulfide (200). It has some hypoglycemic activity in alloxan-diabetic rabbits (201).

3.6.6   Regulations and Standards

No data are available.

3.7   Tartaric Acid

The formula is HOOCCH(OH)CH(OH)COOH; tartaric acid is also known as 2,3-dihydroxybutanedioic, threaric, and 2,3-dihydroxysuccinic acid. There are three isomeric forms: L-tartaric, D-tartaric, and *meso*-tartaric acid. The natural form is the L isomer; small amounts of D-tartaric and the DL-racemic mixture also occur naturally (74). Commercial tartaric acid is primarily the L-form [or (+) tartaric acid].

3.7.1   Major Uses

Tartaric acid finds use in soft drinks, baking and confectionery, tanning, photography, ceramics, coloring metals, and silvering mirrors. Esters are used in lacquers and textiles.

3.7.2   Acute Studies

The acute toxicity of tartaric acid is summarized in Table 59.4. It is a relatively strong acid which can cause local irritation (21). Intravenous administration of large doses (0.2 to 0.3 g) in rats or rabbits caused renal damage (13).

3.7.3   Repeated Exposures

Packman et al. (202) fed sodium tartrate to rabbits at a concentration of 7.7 percent (equivalent to 5 percent free acid) for 150 days. No gross or histopathologic changes or differences in survival or growth rate were observed, compared to controls. Rabbits fed 1150 mg/kg disodium tartrate daily for 17 days suffered no fatalities; 3680 mg/kg daily for 19 days killed three out of six animals (175). Male and female rats fed diets containing 0.1 to 1.2 percent tartaric acid for 2 years showed no changes in growth rate (for the first year), mortality, gross pathology, or histopathology (203).

3.7.4 Physiologic Effects in Humans

Sodium tartrate has been prescribed as a laxative with daily doses of up to 20 g (13). Daily doses of 10 g sodium tartrate to 26 patients, for an average of 11.8 doses, gave laxative responses in 66 percent of the subjects. Occasional nausea, vomiting, and abdominal cramps were observed as side effects in 2 percent of the subjects.

3.7.5 Biochemistry

Species variations have been observed in the metabolic fate of tartaric acid. Rats eliminated about 73 percent of an oral dose unchanged in the urine. Small doses of tartrates to rabbit, dog, and rat are almost quantitatively excreted unchanged in the urine, whether administered orally or parenterally (205). About 20 percent of a small oral dose in man is excreted in the urine (205, 206) with no unchanged tartrate in the feces (205). After intramuscular injection, man excretes tartaric acid almost quantitatively in the urine within 10 hr (206). The differences observed in urinary excretion of tartaric acid between the oral and parenteral routes therefore seem to be due to the metabolism of the acid in the gastrointestinal tract.

Only a small proportion of a dose of tartaric acid is metabolized by mammalian tissues. Rats injected with 400 mg/kg sodium [^{14}C]-L-tartrate eliminated 7.5 percent of the radioactivity as $^{14}CO_2$ and <1.0 percent in the feces. Some 82 percent of the radioactive dose appeared in the urine within 48 hr (207). The major site of tartaric acid metabolism appears to be the intestinal microflora, since 67 percent of an intracecal injection of sodium [^{14}C]-L-tartrate was expired as $^{14}CO_2$, whereas only 2 percent of this dose appeared in the urine (208). Several genera of bacteria have been shown to metabolize tartaric acid via the metabolic sequence tartaric acid → dihydrofumarate → oxaloacetate → glycerate (209).

The metabolic fates of radioactively labeled L- and DL-tartaric acids in rats may differ (210). After seven successive oral doses of 2.73 mg/kg, rats given the racemic DL form showed differences in intrarenal distribution of radioactivity and a greater retention of material compared with rats given the L form. These findings may be attributable to the precipitation of the less soluble calcium DL-tartrate in the tubules, the more soluble L-tartrate being removed more rapidly. Plasma radioactivity also declined more rapidly after dosing with the L isomer than with the DL form.

3.7.6   Regulations and Standards

Tartaric acid is generally recognized as safe for use as a general-purpose additive for food (21 CFR 182.1099) and animal feed (21 CFR 582.1099), a sequestrant in food (21 CFR 182.6099) and animal feed (21 CFR 582.6099), a substance migrating from cotton fabrics used in dry-food packaging (21 CFR 182.170), and a material authorized for treatment of wine (21 CFR 240.1051).

3.8   Adipic Acid

The formula is $HOOC(CH_2)_4COOH$; adipic acid is also known as hexanedioic, 1,4-butanedicarboxylic, and adipinic acid.

3.8.1   Major Uses

Adipic acid is used in artificial resins, nylon, polyurethane foams, esters for plasticizers and lubricants, and adhesives. It is used as a neutralizer and flavoring agent in foods and in baking powders as an alternative to tartaric acid.

3.8.2   Acute Studies

Adipic acid is slightly toxic on acute exposure (21) but produces severe eye irritation in rabbits (20 mg/24 hr) (179). The available data on acute toxicity are summarized in Table 59.4.

3.8.3   Repeated Exposures

Male and female rats exposed to adipic acid as an aerosol dust (126 µg/l, 6 hr daily for 15 days) showed no signs of toxicity. Blood parameters at sacrifice and organ pathology at autopsy were normal (211). The teratogenic potential of adipic acid has been studied. Oral administration of up to 263 mg/kg adipic acid to pregnant mice from day 6 to day 15 of gestation had no discernible influence on nidation, maternal or fetal survival, or fetal abnormalities (212).

3.8.4   Physiologic Effects in Humans

No information was available and no cases of human injury in industrial handling were reported (21).

3.8.5   Biochemistry

In the rat adipic acid is absorbed and metabolized by normal intermediary metabolic processes. In fasted rats fed [^{14}C]adipic acid the urine contained radioactively labeled adipic acid, urea, glutamate, lactate, β-ketoadipate, and citrate (213). Adipic acid appears to be metabolized via β-oxidation to acetate, thence to other normal intermediates. According to Weitzel (214), adipic acid is also metabolized to some degree in man; unchanged acid may appear in the urine, depending on the dose ingested.

3.8.6 Regulations and Standards

Adipic acid is generally recognized as safe as a buffer and neutralizing agent (21 CFR 182.1009), a synthetic flavoring agent (21 CFR 172.515), a component in adhesives (21 CFR 175.105), a component of paper and paperboard in contact with aqueous and fatty foods (21 CFR 176.170), and a component of a variety of resins, coatings, and polymeric materials (21 CFR 175.300, 21 CFR 175.380, 21 CFR 175.390, 21 CFR 177.1210, 21 CFR 175.320, 21 CFR 177.1200, 21 CFR 177.1680, 21 CFR 177.2420).

3.9 Citric Acid

The formula is $HOOCCH_2(HOOC)C(OH)CH_2COOH$; citric acid is also known as 2-hydroxy-1,2,3-propanetricarboxylic and β-hydroxytricarballylic acid.

3.9.1 Major Uses

Citric acid finds use as an acidulant in foods, beverages, confectionery, effervescent salts, and pharmaceutical preparations. It is used in the manufacture of alkyd resins and plastics, in metal polishes, as a mordant, in electroplating, in anticoagulant citrate solutions such as citrate dextrose solution, to facilitate abscission of fruit in harvesting, and in cultured dairy products.

3.9.2 Acute Studies

The acute toxicity of citric acid is summarized in Table 59.4. Citric acid is a moderately strong acid with some irritant and allergenic properties (47). Application of 500 mg citric acid to rabbit skin produced a moderate irritation in 24 hr, whereas 750 μg causes severe effects in the rabbit eye (179).

3.9.3 Repeated Exposures

Citric acid (sodium salt) was incorporated into the diet of rabbits at 7.7 percent (equivalent to 5 percent free acid) for 150 days without producing any gross or histopathologic changes, or differences in growth and survival (202). Diets containing 1.2 percent citric acid had no harmful effects on the growth of two successive generations of rats over a 90-week period. No effect on reproduction, blood characteristics, pathology, or calcium was observed, although a slight increase in dental attrition was reported (215). In dogs a daily oral dose of 1,380 mg/kg for 112 to 120 days produced no evidence of renal damage (216).

Wright and Hughes (217) studied the effect of citric acid on the survival time of immature and sexually mature male mice, and on the survival time and reproductive capacity of rats and mice. Citric acid (5 percent in the diet) did not depress food intake but caused a loss in body weight gain and reduced survival time in mice, with a slightly greater influence on mature animals. No effect was detected on the litter size or survival up to weaning of young in mice or rats. The effects on body weight gain and survival time may have resulted from the chelating ability of citric acid, which could impair absorption of calcium and iron. In a previous report, a high dietary intake of citric acid was without effect on growth rate unless the animals were on a low-calcium diet, in which case a reduced body weight was observed (218).

3.9.4 Physiologic Effects in Humans

Citric acid is an intermediate in normal metabolism and occurs in many foods. Citric acid is generally considered to be largely innocuous, although some hypocalcemic effects were reported during transfusion of large volumes of citrated blood (21). Man's total daily consumption of citric acid from natural sources and from food additive sources may exceed 500 mg/kg (218). Frequent or excessive intake of citric acid may cause erosion of teeth and local irritation. This effect occurs also with lemon juice, which contains about 7 percent citric acid and has a pH <3 (13).

Potassium citrate, up to 10 g daily, has been used as a potassium supplement; the potassium and sodium salts have been used, in similar dosages, as mild diuretics in man (13).

3.9.5 Biochemistry

Citric acid is a normal metabolite and an intermediate in cellular oxidative metabolism. Citric acid and citrate salts occur naturally in many foods. It is formed in the mitochondrion after condensation of acetate with oxaloacetate. The six-carbon acid is then successively degraded to a series of four-carbon acids, effectively accomplishing the oxidation of acetate in the cell.

3.9.6 Regulations and Standards

Citric acid is generally recognized as safe for use as a general purpose food (21 CFR 182.1033) and animal feed (21 CFR 582.1033) additive, a sequestrant in food (21 CFR 182.6033) and animal feed (21 CFR 582.6033), and a material authorized for treatment of wine (21 CFR 240.1051); it is also exempt from tolerance requirements when used as an adjuvant for pesticide chemicals applied to animals (21 CFR 582.99) or to raw agricultural commodities (21 CFR 182.99). Production of citric acid is authorized using yeast fermentation (*Candida lypolytica*) (21 CFR 173.165) or fermentation of *Aspergillus niger* (21 CFR 173.280).

3.10 Pimelic Acid

The formula is $HOOC(CH_2)_5COOH$; pimelic acid is also known as heptanedioic, 1,5-pentanedicarboxylic, and heptane-1,7-dioic acid.

3.10.1 Major Uses

The major uses are in plasticizers and polymers.

3.10.2 Acute Effects

The acute toxicity is summarized in Table 59.4. Pimelic acid is considered to have low acute toxicity and is neither absorbed through nor an irritant to guinea pig skin (21).

3.10.3 Repeated Exposures

No data are available.

3.10.4 Physiologic Effects in Humans

No data are available.

3.10.5 Biochemistry

Pimelic acid is largely excreted unchanged in man (46) and dog (104), the extent varying with the dose. The mechanism of oxidation of the metabolized portion of pimelic acid is not clear. There is evidence that β-oxidation may occur with the dicarboxylic acids, resulting in the formation of dicarboxylic acids having two fewer carbon atoms than the parent acid (104). Further evidence comes from the reports that pimelic acid has been identified as a metabolite of azelaic acid in *Pseudomonas* (219) and *Micrococcus* (220). Pimelic acid is known to be an intermediate in the biosynthesis of biotin in microrganisms (221, 222).

3.10.6 Regulations and Standards

No data are available.

3.11   Azelaic Acid

The formula is $HOOC(CH_2)_7COOH$; azelaic acid is also known as nonanedioic acid.

3.11.1   Major Uses

Azelaic acid is used in lacquers, alkyd resins, plasticizers, adhesives, polyamides, urethane elastomers, and organic syntheses.

3.11.2   Acute Effects

Azelaic acid appears to have low acute toxicity. In the rabbit 500 mg azelaic acid produced only a mild skin irritation in 24 hr, and 3 mg was only a mild eye irritant (123).

3.11.3   Repeated Exposures

No data are available.

3.11.4   Physiologic Effects in Humans

No data are available.

3.11.5   Biochemistry

Azelaic acid is largely excreted unchanged in man (46) and dog (104). The mechanism of metabolism of this dicarboxylic acid does not appear to have been fully elucidated. Bloor (223) considered it improbable that direct β-oxidation of the odd carbon-numbered dicarboxylic acids occurred. Recent evidence indicates, however, that azelaic acid is activated by a mitochondrial, ATP-dependent enzyme system, which is similar to that responsible for activation of palmitic acid (224), and that it is quickly degraded and used again in stearic and palmitic acid biosynthesis (225). After 8 hr, 6 percent of the radioactivity of a tracer dose of [^{14}C]azelaic acid to rats was recovered as $^{14}CO_2$ (225). These reports provide strong evidence that metabolism of azelaic acid occurs via β-oxidation.

3.11.6   Regulations and Standards

Azelaic acid is generally recognized as safe for use in adhesives (21 CFR 175.105), resinous and polymer coatings (21 CFR 175.300), and resinous and polymeric coatings for polyolefin films (21 CFR 175.320).

3.12   Sebacic Acid

The formula is $HOOC(CH_2)_8COOH$; sebacic acid is also known as decanedioic and sebacylic acid.

3.12.1   Major Uses

Sebacic acid is used in alkyd and polyester resins, polyurethanes, fibers, paint products, polyester rubbers, low-temperature lubricants and hydraulic fluids, candles, and perfumes.

3.12.2   Acute Effects

The acute toxicity is summarized in Table 59.4.

3.12.3   Repeated Exposures

No data are available.

3.12.4   Physiologic Effects in Humans

No data are available.

3.12.5   Biochemistry

Metabolism of sebacic acid appears to follow a similar pattern to that of azelaic acid. Succinic and adipic acids have been identified as metabolites of sebacic acid (104) and sebacic acid is a substrate for the ATP-dependent activating enzyme discussed in Section 3.11.5. A large proportion of a dose of sebacic acid is excreted unchanged in the urine of men (46) and dogs (46, 104).

3.12.6   Regulations and Standards

Sebacic acid is generally recognized as safe for use in adhesives (21 CFR 175.105), as a component of paper and paperboard in contact with aqueous and fatty foods (21 CFR 176.170), in closures with sealing gaskets for food containers (21 CFR 177.1210), and in a variety of resinous and polymeric coatings (21 CFR 175.300, 21 CFR 175.380, 21 CFR 175.390, 21 CFR 175.320, 21 CFR 177.2420).

4 UNSATURATED MONOCARBOXYLIC ACIDS

4.1 Introduction

4.1.1 Industrial Applications

This group of acids includes the monomers acrylic and methacrylic acids, which play an important role in a variety of polymeric materials. Other interesting uses include sorbic acid as a chemical preservative, undecylenic acid as a popular fungistat in man, and linoleic and oleic acids as important soap and food constituents.

4.1.2 Health Effects

Physicochemical properties of this group of acids are presented in Table 59.5. In general, the health effects are characterized by low acute toxicity, coupled with a marked capacity for irritation in the lower molecular weight members of the series. Reports indicate that acrylic, crotonic, and methacrylic acids may possess some sensitization potential. The widespread use of acrylic, crotonic, methacrylic, and oleic acids has led to interest in oncogenic potential, presumably because of the presence of the double bond linkage. Although there have been reports of site sarcomas following a repeated subcutaneous injection with sorbic acid, no data to date indicate that members of this series possess any systemic oncogenic potential.

4.2 Propiolic Acid

The formula is CH≡CCOOH; propiolic acid is also known as 2-propynoic, acetylenecarboxylic, and propargylic acid.

4.2.1 Major Uses

Propiolic acid is an intermediate in monohalogenated acetylene synthesis (226) and a corrosion inhibitor for steel (227, 228).

4.2.2 Acute Studies

The undiluted acid was a moderately strong skin irritant when held in contact with the depilated guinea pig abdomen under an occlusive wrap for 24 hr (229). Additional acute studies are presented in Table 59.6. No other toxicity information on propiolic acid was available.

Table 59.5. Physical Properties of Aliphatic Unsaturated Monocarboxylic Acids

Acid	CAS Acid Name	CAS Reg. No.	Mol Wt	M.P. (°C)	B.P. (°C)	Solubility[a] H2O	Alcohol	Other	Specific Gravity	Vapor Pressure mm Hg	°C
Propiolic	2-Propynoic	471-25-0	70.05	18	144 (dec)	∞	∞	Ether (∞) CHCl₃ (∞) Benzene (∞)	1.138 (20/4)	10.5 50	54–55 70–75
Acrylic	2-Propenoic	79-10-7	72.06	13	141.6	∞	∞	Ether (∞) Acetone (sl s) Benzene (sl s)	1.0511 (20/4)	3.2 10 15	20 39 48.5
Crotonic	(E)-2-Butenoic	107-93-7	86.09	71.5–71.7	185	V	V	Ether (sl s) Acetone (sl s)	1.018 (15/4)	0.19	20
Methacrylic	2-Methyl-2-propenoic	79-41-4	86.09	16	162–163 (757 mm)	Sl s	∞	Acetone (∞)	1.0153 (20/4)	0.65 1.0 1.4	20 25 30
Pentenoic	4-Pentenoic	591-80-0	100.13	−22.5	188–189	Sl s	∞	Acetone (∞)	0.9809 (20/4)	20	93
Hexenoic	2-Hexenoic	1191-04-4	114.15	94				Benzene (sl s)		14	183
Sorbic	(E,E)-2,4-Hexadienoic	110-44-1	112.14	134.5	228 (dec)	Sl s (hot)	Sl s	Ether (v)	1.204 (19/4)		
Heptenoic	2-Heptenoic	18999-28-5	128.9								
Undecylenic	10-Undecenoic	112-38-9	184.28	24.5	275	Ins	Sl s	Ether (sl s) CHCl₃ (sl s) Benzene (sl s)		10 15	160 165
Linolenic	9,12,15-Octadecatrienoic	1955-33-5	278.44	−11.3		Ins	Sl s	Ether (sl s) Benzene (sl s)	0.9164 (20/4)	0.05 17	125 230
Linoleic	9,12-Octadecadienoic	2197-37-7	280.46	−5		Ins	∞	Ether (∞) Acetone (∞) CHCl₃ (∞) Benzene (∞)	0.9022 (20/4)	16	229
Elaidic	(E)-9-Octadecenoic	112-79-8	282.47	45		Ins	Sl s	Ether (sl s) CHCl₃ (sl s) Benzene (sl s)	0.8734 (41/4)	100	288
Oleic	(Z)-9-Octadecenoic	112-80-1	282.47	13.2		Ins	∞	Ether (∞) Acetone (∞) CHCl₃ (∞) Benzene (∞)	0.895 (20/4)	10 100	225 286
Ricinoleic	(Z)-12-Hydroxy-9-octadecenoic	141-22-0	298.47	5.5	(Dec.)	Ins	Sl s	Ether (sl s)	0.950 (21.4/4)	10	226
Arachidonic	(Z,Z,Z,Z)-5,8,11,14-Eicosatetraenoic	506-32-1	304.48	−49.5		Ins	Sl s	Ether (sl s) Acetone (sl s)			

[a] ∞, Soluble in all proportions; v, very soluble; sl s, slightly soluble; ins, insoluble.

Table 59.6. Acute Toxicity of Unsaturated Monocarboxylic Acids

Acid	Species	Route	LD_{50} (mg/kg)	Ref.
Propiolic	Mouse	Oral	100–200	229
	Mouse	Ip	25–50	229
	Rat	Oral	100–200	229
	Rat	Ip	25–50	229
	Guinea pig	Skin	0.1–1.0 (ml/kg)	229
Acrylic	Mouse	Oral	2400	236
	Mouse	Ip	128 (LD_{Lo})	237
	Rat	Oral	340–3200	232, 234, 235
	Rat	Ip	22–24	233, 235
	Rat	Inhaln.	3.6 g/m^3 (LC_{50}/ 4 hr)	235
	Rat	Inhaln.	19 g/m^3 (LC_{Lo}/5 hr)	211
	Rat	Inhaln.	Satd. atm. (LC_{50}/3.5 hr)	238
	Rat	Inhaln.	11.5 g/m^3 (no mort./4 hr)	238
	Rat	Inhaln.	Satd. atm. (no mort./8 hr)	239
	Rabbit	Skin	280	230
Crotonic	Mouse	Oral	400–4800	236, 243
	Mouse	Ip	25–50	243
	Mouse	Sc	3590	15
	Rat	Oral	400–1000	23, 243
	Rat	Ip	25–100	236, 243
	Guinea pig	Ip	60	244
	Guinea pig	Skin	200	243
	Rabbit	Skin	600	23
Methacrylic	Mouse	Oral	1600	246
	Mouse	Ip	48	248
	Rat	Oral	2260, 9400	246, 247
	Guinea pig	Skin	1–5 ml/kg	246
Pentenoic	Mouse	Oral	610	27
	Mouse	Ip	315	15
	Mouse	Sc	315	15
	Rat	Oral	470	27
Hexenoic	Mouse	Ip	1840	15
	Mouse	Sc	1840	15
Sorbic	Mouse	Oral	3200–6400	246, 275
	Mouse	Ip	1600–3200	15, 275
	Mouse	Sc	2820	15
	Rat	Oral	3200–7360	246, 275, 276
	Rat	Ip	800–1600	275
Heptenoic	Mouse	Ip	1600	15
	Mouse	Sc	1600	15
Undecylenic	Mouse	Oral	>3200	246, 296
	Mouse	Ip	960	296

Table 59.6. (*Continued*)

Acid	Species	Route	LD$_{50}$ (mg/kg)	Ref.
	Rat	Oral	>2500	246, 295
	Guinea pig	Skin	0.05–0.24	246
Linolenic	Mouse	Oral	>3200	246
	Rat	Oral	>3200	246
	Guinea pig	Skin	>20 (ml/kg)	246
Linoleic	Mouse	Oral	>3200	310
	Rat	Oral	>3200	310
	Guinea pig	Skin	>20 (ml/kg)	310
Elaidic	Mouse	Ip	512 (LD$_{Lo}$)	318
	Mouse	Iv	100	319
Oleic	Rat	Oral	64 (ml/kg)	310
Arachidonic	Mouse	Iv	100 (LD$_{Lo}$)	345
	Rat	Iv	100 (LD$_{Lo}$)	345
	Rabbit	Iv	1 (LD$_{Lo}$)	345

4.3 Acrylic Acid

The formula is $CH_2{=}CHCOOH$; acrylic acid is also known as 2-propenoic, acroleic, and vinylformic acid.

4.3.1 Major Uses

Acrylic acid is used in the manufacture of plastics, molding powder for signs, construction units, decorative emblems and insignias, polymer solutions for coatings applications, emulsion polymers, paint formulations, leather finishings, and paper coatings (226).

4.3.2 Acute Studies

Application of 500 mg to the uncovered skin of rabbits produced severe irritation (230). A standardized test for skin sensitization potential, in which footpad injections were used in guinea pigs, followed by a topical challenge to the back 1 week later, resulted in low-to-moderate activity in three of five animals (231). Application of 1 mg to conjunctival sacs of rabbit eyes produced severe irritation (230, 232).

A single dose of 14 mg/kg injected intraperitoneally into pregnant rats was teratogenic; doses as low as 2 mg/kg produced resorptions, gross and skeletal abnormalities, and a decrease in fetal birth weight (233).

Lethal doses or concentrations following administration by oral, intraperitoneal, skin, and inhalation routes are presented in Table 59.6.

4.3.3 Repeated Exposures

The International Agency for Research on Cancer Working Group (IARC) reviewed the literature but found no data to indicate that acrylic acid was carcinogenic (240).

Four groups of five female rats were injected once intraperitoneally with 0, 2.5, 4.7, or 8 mg/kg on days 5, 10, and 15 of pregnancy. Significant increases in the number of gross abnormalities occurred in the offspring of those given the two highest dose levels and skeletal abnormalities were significantly increased in pups of those given the highest dose level. Embryotoxicity also occurred in those animals given the highest dose level (233).

A group of eight rats (four male, four female) given four 6-hr exposures by inhalation to 1500 ppm exhibited nasal discharge, lethargy, and weight loss. Histologic examination revealed some kidney congestion. In a follow-up study two groups of eight rats each (four male, four female) received twenty 6-hr exposures of 80 or 300 ppm. Clinical changes seen in the former were nasal irritation, lethargy, and retarded weight gain. No toxic signs were observed at 80 ppm. No compound-related lesions were detected on gross examination at either exposure concentration (211). Exposure of rats by inhalation to 700 mg/m^3, 4 hr/day for 5 weeks produced strong local irritation resulting in nonreversible changes in skin and eyes, reduced body weight, increased reticulocyte counts, and reduced urine concentrating capacity of kidney. Injury to gastric mucosa and inflammation of upper respiratory tract was seen on histopathologic examination (235).

4.3.4 Regulations and Standards

Acrylic acid and certain esters may be used safely as a component of the uncoated or coated food-contact surface of paper and paperboard intended for use in producing, manufacturing, packing, processing, preparing, treating, packaging, transporting, or holding dry food (21 CFR 176.180). The ACGIH has adopted a TLV of 10 ppm (30 mg/m^3) in workroom air (241).

4.4 Crotonic acid

The formula is $CH_3CH{=}CHCOOH$; crotonic acid is also known as (E)-2-butenoic and 3-methylacrylic acid.

4.4.1 Major Uses

Crotonic acid is used in the synthesis of resins, polymers, plasticizers, and drugs (242), as a softening agents for synthetic rubber, and in medicinal chemicals (74).

4.4.2 Acute Studies

The compound was a strong irritant to rabbit skin, 10 mg/24 hr (23) and to guinea pig skin, 100 mg/kg/24 hr (243).

A standardized test using footpad injections in guinea pigs followed by a topical challenge to the back 1 week later resulted in one out of five animals responding positively (243).

Solutions are irritating and damaging to eyes when tested in rabbits (23).

Lethal doses associated with oral, intraperitoneal, and subcutaneous administration are presented in Table 59.6.

4.4.3 Regulations and Standards

The compound is identified as a corrosive material by the Department of Transportation (245). It is lawful for use as a component of the food-contacting surface of paper and paperboard in contact with all foods (21 CFR 176.180).

4.5 Methacrylic Acid

The formula is $CH_2{=}C(CH_3)COOH$; methacrylic acid is also known as 2-methyl-2-propenoic and 2-methylenepropionic acid.

4.5.1 Major Uses

Methacrylic acid is used primarily in the manufacture of methacrylic resins and plastics (74).

4.5.2 Acute Studies

Irritation was severe when 1, 5, or 10 ml were applied to a depilated guinea pig abdomen for 24 hr under an occlusive wrap (246). Eye irritation, based on skin irritation, is assumed to be severe. Lethal doses following oral, intraperitoneal, and skin application are presented in Table 59.6.

4.5.3 Repeated Exposures

Doses of 1000, 100, or 50 mg/kg/day by gavage for 5 days in rats resulted in sufficient reduced feed intake and weight loss to terminate the study (319). In a follow-up study doses of 10 and 5 mg/kg/day for 10 treatments produced no effect on feed intake, weight gain, hematology, serum clinical chemistries, and gross pathology. Histopathologic examination at both dose levels revealed some slight to moderate alveolar hemorrhage and lipid granuloma in the lungs and moderate to severe granularity of liver cytoplasm. The effects were less marked at the lower dose level. The lesions may be due to the corrosive action of the compound (246).

Intravenous injection of methacrylic acid and certain esters in anesthetized dogs increased the respiratory rate, decreased the heart rate, and produced electrocardiogram changes.

Daily application to the clipped backs of guinea pigs by rub-on for 10 days produced necroses (246). Methacrylic acid did not sensitize any of five guinea pigs tested by topical challenge to the back 1 week after a footpad injection (246).

Four rats (two male, two female) received five 5-hr inhalation exposures to 1300 ppm. Compound-related clinical signs were nose and eye irritation and weight loss. Blood and urine tests and the results of gross examination were normal (211). Rats received twenty 6 hr inhalation exposures to 300 ppm of methacrylic acid. Clinical signs and gross pathology were normal. Slight renal congestion was observed on histopathologic examination (211). Chronic inhalation exposures of rats to 116 ppm, 8 hr daily for 6 months produced significantly lowered adiposity and intestinal transit performance (250). Methacrylic acid was reviewed by the IARC Working Group. No monograph was prepared because no adequate data were available (251).

Methacrylic acid and 12 methacrylate esters were dissolved in Locke's solution. All compounds reduced the heart rate and force of contraction of isolated perfused rabbit heart at concentrations of 1:1000, 1:10,000, and 1:100,000. Most, but not all, of the compounds reduced coronary flow rate (248). These compounds also produced inhibition of spontaneous contractions of isolated guinea pig ileum and antagonized the stimulant actions of acetylcholine and barium chloride (252).

4.5.4   Physiologic Effects in Humans

No correlation was found between changes in methacrylic acid concentrations and changes in blood pressure following orthopedic surgery in which methyl-methacrylate cements that undergo hydrolysis to methacrylic acid (253) were used. Circulating levels of methacrylic acid were 0 to 15 μg/l. In 37 cases studied, acrylic acid resins produced a contact sensitization in a small number of workers. Methacrylic acid appeared to be the sensitizing agent (254).

4.5.5   Biochemistry

Up to 88 percent of a single dose of ^{14}C-labeled methylmethacrylate in rats was expired as carbon dioxide in 10 days (65 percent in 2 hr), irrespective of the route of administration. Small amounts of labeled methylmalonic, succinic, hydroxybutyric, and unlabeled 2-formylpropionic acid in the urine, as well as the formation of labeled normal physiologic metabolites, suggest that the metabolic pathway of methylmethacrylate involves the intermediary metabolism, probably via the mitochondria (255).

4.5.6 Regulations and Standards

Methacrylic acid and its methyl, ethyl, propyl, and butyl esters may be used as components of the food contacting surface of paper and paperboard in contact with all foods (21 CFR 176.180).

4.6 Pentenoic Acid

The formula is $CH_2{=}CH(CH_2)_2COOH$; pentenoic acid is also known as 4-pentenoic and allylacetic acid.

4.6.1 Acute Studies

A single intraperitoneal dose (200 mg/kg) in rats elevated mean ammonia levels in plasma approximately fourfold (256) and caused hypoglycemia in fed and fasted rats (256–261). Mice receiving 15 mg/kg intraperitoneally (262) showed hypoglycemia, whereas infusion with pentenoic acid altered electrolyte and glucose balance in dogs (263, 264). LD_{50}s following single oral, intraperitoneal, or subcutaneous administrations are presented in Table 59.6.

4.6.2 Repeated Exposure

Enlarged livers with extensive fatty degeneration and increased blood urea nitrogen were observed in rats that received 50 mg/kg intraperitoneally every 4 hr for 10 doses, followed by a single dose of 200 mg/kg (256, 265). Five mg/kg/day intraperitoneally for 3 days decreased triglycerides and free fatty acids in the liver and plasma of rats (266).

4.6.3 Biochemistry

A concentration of 1 to 0.1 mM 4-pentenoic acid strongly inhibited oxygen uptake, mitochondrial biochemical synthesis, and enzyme reactions in rat hepatocytes (267–272). The major metabolic actions of the acid in dogs were the inhibition of long chain fatty acid oxidation and increased renal calcium excretion (273).

4.7 Hexenoic Acid

The formula is $CH_3(CH_2)_2CH=CHCOOH$; hexenoic acid is also known as 2-hexenoic acid.

4.7.1 Major Use

The primary use of hexenoic acid is its fungicidal activity (274).

4.7.2 Acute Studies

Acute information is presented in Table 59.6. No other toxicity information was available.

4.8 Sorbic Acid

The formula is $CH_3CH=CHCH=CHCOOH$; sorbic acid is also known as (E,E)-2,4-hexadienoic acid.

4.8.1 Major Uses

Sorbic acid is used as a fungicide, a food preservative, in copolymerization, in upgrading of drying oils, as a cold rubber additive, and as an intermediate for plasticizers and lubricants (242).

4.8.2 Acute Studies

It was not a primary irritant or sensitizer when applied in 0.1 M concentrations to guinea pig skin (275). Lethality studies following administration of single doses are presented in Table 59.6.

4.8.3 Repeated Exposure

Wistar rats (six males) given subcutaneous injections of sorbic acid, 2 mg/0.5 ml in arachis oil, twice weekly for 65 weeks developed local sarcomas (5/6). The first tumor was observed at 82 weeks (277). Similar findings were also observed in follow-up studies (278, 279). However, the continuous administration of sorbic acid in drinking water (10 mg/100 ml) of rats for 64 weeks did not produce tumors (277). Tumors were also not observed in rats (16 male, 8 female) on diets that contained 8 percent sorbic acid for 20 weeks (280) or in mice after 40 mg/kg/day for 17 months (281).

No adverse effects were noted in rats fed sorbic acid at dietary levels of 1, 2, 4, and 8 percent for 90 days (282). Similarly, there were no adverse findings when sorbic acid was fed to puppies at a 4-percent dietary level for 90 days (282). No effects were observed in rats and dogs when sorbic acid was incorporated in the diets at the 5 percent level on a moisture free basis (283), but 30 daily doses of about 50 mg/kg in the feed of rats did cause some reduced weight gain (283).

Sorbic acid at dietary levels of 0, 1, 5, or 10 percent for 80 weeks with male and female mice and 0, 1, 5, or 10 percent for 2 years with male and female rats caused no increase in the number of deaths or the incidence of spontaneous histologic lesions, including tumors (284, 285). There was no adverse effect on the blood or internal organs of rats, guinea pigs, rabbits, and dogs after prolonged feeding at 1 to 500 times the amounts used in foods (286).

4.8.4 Physiologic Effects in Humans

Application of 150 mg to skin for 1 hr in man produced severe irritation (287). Ointments that contained sorbic acid may have caused itching of the face; three subjects may have been sensitized (287). There are other reports of skin sensitization (288) but sorbic acid was felt to be neither a primary irritant or a sensitizing agent for humans (276). An allergic response to sorbic acid was reported in 0.8 percent of eczema patients tested over a 3-year period (289). A case report of contact sensitivity is found in the literature (290).

4.8.5 Biochemistry

Metabolism in rats is identical to that of normally occurring fatty acids (291). Under normal conditions of intake sorbic acid is almost completely oxidized to carbon dioxide and water (283). Traces (0.1 percent of dose) may be converted by oxidation to *trans,trans*-muconic acid (292).

4.8.6 Regulations and Standards

Sorbic acid is generally recognized as safe for use as a chemical preservative (21 CFR 182.3089).

4.9 Heptenoic Acid

The formula is $CH_3(CH_2)_3CH=CHCOOH$; heptenoic acid is also known as 2-heptenoic acid.

4.9.1 Acute Studies

Acute information relating to heptenoic acid lethality in the mouse is presented in Table 59.6. No other toxicity information was available.

4.10 Undecylenic Acid

The formula is $CH_2{=}CH(CH_2)_8COOH$; undecylenic acid is also known as 10-undecenoic and 10-hendecenoic acid.

4.10.1 Major Uses

Undecylenic acid is used in perfumes, flavoring, medicinals, and plastics and as a modifying agent (plasticizer, lubricant additive) (242). In diluted form it is used as a fungistat in man.

4.10.2 Acute Studies

Skin irritation was moderate when the compound was applied to the depilated guinea pig abdomen under an occlusive wrap for 24 hr (246). Skin irritation was severe in rabbits following application of 500 mg to uncovered skin for 24 hr (293). In a standardized test in which guinea pigs were challenged topically on the back 1 week after an initial footpad injection, none of the animals responded positively (246). Eye irritation was mild in one study with rabbits (294), but based on skin tests the compound should be presumed to be an irritant. Lethal doses that follow administration by the oral, intraperitoneal, and skin routes are presented in Table 59.6.

4.10.3 Repeated Exposures

No signs of toxicity were observed in rats after the daily intake of 400 mg/kg in diet for periods up to 6 to 9 months (297). Concentrations of 5 to 25 g/kg in the diet markedly reduced rat growth (296). Rats dosed by gavage with 1000 mg/kg/day died after three to eight doses. Changes included moderate decreases in weight gain, minimal feed intake, dark urine, unsteady gait, difficulty in breathing, and enlarged livers. Histopathologic examination revealed changes in liver and kidneys (246). A dose of 100 mg/kg/day by gavage for 13 days had no effect on feed intake, weight gain, hematologic and serum clinical chemistry determinations, absolute and relative liver and kidney weights, and gross pathology. Histopathology showed minimal changes (246).

Daily applications to the clipped backs of guinea pigs for 10 days by rub-on moderately exacerbated the initial irritative response (246).

Few effects were found in studies with rats receiving subchronic inhalation exposures of saturated atmospheres at room temperatures, but respiratory distress was evident during exposures to aerosols or to saturated vapor atmospheres generated at 100°C (294).

4.10.4   Physiologic Effects in Humans

Daily oral doses of 6 to 14 g produced gastrointestinal disturbances, headache, fever, dizziness, urticaria, folliculitis, and conjunctivitis (298). Undecylenic acid was irritating to man when applied in a 21-day continuous closed patch test to humans at concentrations of 10, 20, and 40 percent (299). The acid was nonirritating at concentrations of 10, 20, 40, and 60 percent over a 21-day period when open testing was performed (299). There was no irritation after a 48-hr closed patch test with 4 percent of undecylenic acid (293). Undecylenic acid in a powder vehicle was tested in 104 patients with confirmed tinea pedis and appeared to be a safe and effective agent in the treatment of cutaneous fungal infections (300).

4.10.5   Regulations and Standards

Undecylenic acid was recommended for generally recognized as safe status by the Flavoring Extract Manufacturers Association (301). The Council of Europe included undecylenic acid at a level of 0.5 ppm in a list of artificial flavoring substances that may be added to foodstuffs without hazard to health (302).

4.11   Linolenic Acid

The formula is $CH_3CH_2CH{=}CHCH_2CH{=}CHCH_2CH{=}CH(CH_2)_7COOH$; linolenic acid is also known as 9,12,15-octadecatrienoic acid.

4.11.1   Major Uses

Linolenic acid is used in medicine, biochemical research, and drying oils (242).

4.11.2   Acute Studies

Skin irritation was slight when the acid was applied to the depilated abdomens of guinea pigs under an occlusive wrap for 24 hr. Lethal doses following administration orally and to the skin are presented in Table 59.6.

4.11.3   Physiologic Effects in Humans

The daily oral administration of 14 ml of linseed oil, which supplied 27 μmol linolenic acid/day to atherosclerotic women for 8 weeks with alternating 8- to 10-week control periods, decreased plasma fibrinogen levels and tended to increase platelet adhesiveness (303, 304). Linolenic acid stabilized human erythrocytes against hypotonic hemolysis (305). Lipids were isolated from the blood of six patients suffering from schizophrenia and from six age-matched healthy controls. A significantly higher proportion of linolenic acid was found in schizophrenics than in controls (306).

4.11.4 Biochemistry

Increased linolenic acid reduces platelet prostaglandin synthesis (307) and may inhibit heart microsomal enzymes involved in linoleate-arachidonate conversion in rats (308, 309). Linolenic acid, however, is an essential fatty acid and a normal constituent of the diet. Because it is handled as a food, untoward effects as a result of dietary intake at normal levels are not expected.

4.12 Linoleic Acid

The formula is $CH_3(CH_2)_4CH=CHCH_2CH=CH(CH_2)_7COOH$; linoleic acid is also known as 9,12-octadecadienoic and 9,12-linoleic acid.

4.12.1 Major Uses

Linoleic acid is used in soaps, special driers for protective coatings, emulsifying agents, medicine, feeds, biochemical research, dietary supplement, and margarine (242).

4.12.2 Acute Studies

Skin irritation was slight when the compound was applied to the depilated abdomens of guinea pigs under an occlusive wrap for 24 hr (310). Toxicity studies relating to lethal doses are presented in Table 59.6.

4.12.3 Repeated Exposures

Increased dietary levels of linoleic acid produced damage to erythrocyte membranes of dogs (311), platelet membranes of rabbits (312), and changes in rat platelet activity (311, 313). Linoleic acid was administered intraperitoneally in doses of 40 and 200 mg/kg every second day for 4 weeks to rats fed a fat-free diet. Both doses resulted in decreased microsomal activity, which may affect metabolic activity (314, 315).

4.12.4 Physiologic Effects in Humans

The fat in the normal diet of 19 apparently well men was partly replaced by linoleic acid. This produced changes in many platelet function tests which are ascribed to decreased platelet activation. Twenty controls maintained on a normal diet showed no such changes. It was concluded that normal people have a degree of platelet activation which may be decreased by linoleic acid. This may be relevant to the benefits attributed to a diet containing polyunsaturated fats (316).

4.12.5 Biochemistry

The addition of linoleic acid to baby hamster kidney cells reduced the cell-to-substrate adhesion, caused morphologic changes, and altered the cellular growth properties. The data indicate that the effects are probably due to actual changes in the surface membrane lipids (317). Linoleic acid, an essential dietary unsaturated fatty acid in mammals, is converted in a series of biosynthetic elongation and desaturation reactions to eicosadienoic and arachidonic acids. The biosynthetic pathways may be found in general texts on biochemistry.

4.12.6 Regulations and Standards

Linoleic acid is generally recognized as safe as a nutrient and/or dietary supplement (21 CFR 182.5065).

4.13 Elaidic Acid

The formula is $CH_3(CH_2)_7CH\!=\!CH(CH_2)_7COOH$; elaidic acid is also known as (E)-9-octadecenoic and *tran*-soleic acid.

4.13.1 Major Uses

Elaidic acid is used in medicinal research and as an analytic standard in chromatography (242).

4.13.2 Acute Studies

Toxicity studies relating to lethality are presented in Table 59.6. No other acute toxicity information was available.

4.13.3 Repeated Exposures

Elaidic acid was administered intraperitoneally at doses of 40 and 200 mg/kg every second day for 4 weeks to rats fed a fat-free diet. A slight reduction of weight gain was observed. The higher dose decreased microsomal activity, which may compromise metabolic activity (314).

4.13.4   Biochemistry

In rat liver, intraperitoneally injected, radioactively labeled elaidic acid was preferentially incorporated into the free fatty acid fraction of phospholipids. The turnover kinetics of 5-*cis*-9-*trans*-octadecadienoic acid, a metabolite of elaidic acid, as well as its distribution in liver and plasma, suggested that desaturation of elaidic acid occurred in its nonesterified fatty acid form (320).

4.14   Oleic Acid

The formula is $CH_3(CH_2)_7CH{=}CH(CH_2)_7COOH$; oleic acid is also known as (Z)-9-octadecenoic acid.

4.14.1   Major Uses

Oleic acid is used as a soap base, in the manufacture of oleates, ointments, cosmetics, polishing compounds, lubricants, and surface-coatings, and as a food additive (242).

4.14.2   Acute Studies

The acid was slightly irritating to rabbit skin and eye (310). Acute lethal doses for the rat are presented in Table 59.6.

4.14.3   Repeated Exposures

Dogs received weekly injections of 0.09 g/kg of oleic acid over a period of 1 to 3 months and responded with a variety of pulmonary changes. Early changes were thromboses and cellular necrosis. These changes were followed by a repair stage with proliferation of type 2 cells and fibrotic foci in subpleural areas. A later change was pulmonary fibrosis. The extent of lesions was related to the number of oleic acid injections (321).

The effect of oleic acid on insulin secretion was studied in the isolated perfused rat pancreas. In the absence of glucose a continuous infusion of oleic acid (1500 μ*M*) induced a biphasic insulin release. The results suggest that high concentrations of oleic acid stimulate insulin release from the isolated perfused rat pancreas and modulate the insulin response to arginine or glucose (322).

Studies on tumorigenic potential include subcutaneous injection into rabbits (323, 324), guinea pigs (323), rats (325), and mice (313, 323, 326), skin painting in mice (138, 327–330), oral administration in rats (331), and intraperitoneal injection in rats (325). No tumorigenic activity was found.

4.14.4   Physiologic Effects in Humans

Oleic acid in human blood reversibly altered the shape of erythrocytes, led to the reduction of viscosity of the blood in vitro, and reduced the erythrocyte sedimentation rate (332, 333).

Neutrophils from healthy volunteers were isolated and incubated with albumin-bound oleic acid. Standard in vitro function tests including phagocytosis, bactericidal activity, and chemotaxis were performed after the incubation. Oleic acid caused no changes in bactericidal activity and only moderate decreases in phagocytosis and chemotaxis at high concentrations (334).

4.14.5   Biochemistry

Oleic acid is one of the most abundant fatty acids in nature. The normal metabolic pathway of palmitic and stearic acids in mammals produces oleic acid. Oleic acid, on a series of elongation and desaturation steps, may be converted into longer chain eicosatrienoic and nervonic acid. The biosynthesis of this fatty acid may be found in many general biochemistry texts.

4.14.6   Regulations and Standards

Oleic acid is generally recognized as safe for use in paper and paperboard products that come in contact with food (21CFR 182.90).

4.15   Ricinoleic Acid

The formula is $CH_3(CH_2)_5CHOHCH_2CH{=}CH(CH_2)_7COOH$; ricinoleic acid is also known as (Z)-12-hydroxy-9-octadecenoic, ricinic, and castor oil acid.

4.15.1   Major Uses

Ricinoleic acid is used in soaps, Turkey red oils, textile finishing, ricinoleate salts, and 12-hydroxystearic acid synthesis (242). It is the active principle of castor oil.

4.15.2   Acute Studies

No acute toxicity information was available.

4.15.3   Repeated Exposures

Four rabbits given intermittent subcutaneous injections of 3120 mg/kg for 52 weeks developed subcutaneous neoplasms (335).

The tumorigenic potential of ricinoleic acid injected intravaginally in groups of 20 mice was studied. No tumors attributable to ricinoleic acid treatment were observed in three studies after 18 to 20 months (336–338). Tests with rabbits and guinea pigs suggest that ricinoleic acid may alter the net fluid transport and mucosal permeability of the colon. These structural and functional alterations may contribute to the intraluminal accumulation of fluid and catharsis that can result from administration of ricinoleic acid (339, 340).

4.15.4 Physiologic Effects in Humans

The lowest dose at which mortality occurred following an accidental ingestion was 5000 mg/kg (341). Perfusion studies were performed in healthy volunteers to test the hypothesis that net fluid secretion induced by fatty acids is accompanied by a parallel reduction in solute transport. Ricinoleic acid provoked a marked net secretion of fluid and concomitantly inhibited the absorption of all solutes tested, including glucose, xylose, L-leucine, L-lysine, folic acid, and 2-monoolein. The mechanism may be related to mucosal damage and altered mucosal permeability (342).

4.15.5 Biochemistry

Ricinoleic acid is a hydroxy-substituted long-chain fatty acid present in animal lipids. Perfusion of the colon with ricinoleic acid produces fluid and electrolyte accumulation. The mechanism is unknown, but 0.5 mM ricinoleate produced significant increases in electrical potential differences and cyclic AMP across isolated rat colonic mucosa. This suggests that hydroxy-fatty-acid-induced fluid and electrolyte accumulation is driven by an active ion secretory process (343).

Ricinoleic acid (1 ml) was administered by gastric intubation to thoracic duct-cannulated rats and lymph was collected over a 48-hr period. Ricinoleic acid was incorporated maximally into triglycerides, diglycerides, and monoglycerides in 30 hr and into free fatty acids in 24 hr. Ricinoleic acid was not found in phospholipids or cholesterol esters of lymph lipids (344).

4.16 Arachidonic Acid

The formula is $CH_3(CH_2)_4CH=CHCH_2CH=CHCH_2CH=CHCH_2CH=CH(CH_2)_3COOH$; arachidonic acid is also known as (Z,Z,Z,Z)-5,8,11,14-eicosatetraenoic acid.

4.16.1 Major Use

Arachidonic acid is mainly used in biochemical research.

4.16.2 Acute Studies

Lethal doses following single administrations are presented in Table 59.6. A 50 μg/kg intravenous dose stimulated erythropoiesis in mice (346). In rats the compound causes hypotension by peripheral dilation (347). Intravenous administrations reduce systemic arterial pressure, increase pulmonary vascular pressure (348), and cause platelet thrombi formation in the lung microvasculature of rabbits (349) and also have a marked effect on pulmonary vessels and airways in dogs (350). However, no major circulatory, lysosomal, or platelet effects were observed in dogs dosed with 150 μg/kg/min (351). Intradermal injection in rats causes increased vascular permeability and edema (352). A solution of 0.8 to 5 percent arachidonic acid injected subconjunctivally into rabbit eyes causes increased intraocular pressure (353).

In vitro studies suggest that arachidonic acid decreases rat heart muscle and endothelioid cell viability (354) and may modulate thyroid responsiveness in dogs (355).

4.16.3 Repeated Exposure

Repeated administration up to 1000 mg/kg intraperitoneally in mice decreased microsomal activity (356).

4.16.4 Physiologic Effects in Humans

Arachidonic acid produces cell lysis in cultures of human platelets exposed to concentrations as low as 0.5 mM (357) and modulates human neutrophil response in culture (358).

4.16.5 Biochemistry

Arachidonic acid is incorporated into phospholipids in rat liver (359, 360), rabbit alveolar macrophages (361), and guinea pig lung (362). It activates guanylate cyclase in guinea pig lung (363) and is metabolized by rat spleen microsomal cyclooxygenase and a lipoxygenase into prostaglandin, thromboxane, and shorter chain acids (364). Arachidonic acid is readily synthesized by man from linoleic acid. The pathway, which requires a series of biosynthetic elongation and desaturation steps, may be found in general textbooks of biochemistry.

5 UNSATURATED POLYCARBOXYLIC ACIDS

The unsaturated dicarboxylic acids find a number of industrial uses, such as monomers for resin manufacture, edible preservatives, mordants in dyeing, and synthetic intermediates. Several of them occur as normal constituents of the intermediary metabolism, including fumaric, aconitic, and itaconic acids. The principal physiologic effect is generally that of primary irritation. Table 59.7 presents the physical properties of unsaturated polycarboxylic acids.

5.1   Maleic Acid

The formula is $HOOCCH=CHCOOH$; maleic acid is also known as (Z)-2-butenedioic and *cis*-1,2-ethylenedicarboxylic acid.

5.1.1   Major Uses

Maleic acid is used chiefly in the manufacture of artificial resins, in dyeing and finishing wool, cotton, and silk, in salts of antihistamines (74), and as a preservative for oils and fats (242).

5.1.2   Acute Studies

Administration of 400 mg/kg by intraperitoneal injection markedly decreased blood sugar levels in fasted rats, causing a simultaneous increase in plasma free fatty acids and acetoacetate (365). Application of 500 mg for 24 hr produced mild skin irritation in the rabbit. In guinea pigs irritation was moderate (366). Doses of 100 mg (366) or 1 percent for 2 min (367) when administered to the rabbit eye caused moderate to severe eye irritation. A 1-hr inhalation exposure of six rats to 0.72 mg/l, produced generalized inactivity, hyperpnea, and sedation within 15 min of exposure. There was no mortality and there were no significant findings on gross examination (366). Lethal doses following single administrations are presented in Table 59.8.

5.1.3   Repeated Exposures

In chronic feeding studies with rats some toxic effects were noted at concentrations as low as 0.5 percent. The differences from controls were not marked and the pathology was nonspecific (203). Daily doses of 0.5 to 2.0 mg/rat injected subcutaneously in sesame oil for 60 days were tolerated, whereas 5 to 10 mg/rat produced death, retarded growth, and patchy hair distribution (203).

5.1.4   Physiologic Effects in Humans

Maleic acid produces marked irritation of the skin and mucous membranes. Severe effects, particularly in the eye, can result from aqueous concentrations as low as 5 percent. There are no reports of cumulative toxic effects in man (369).

Table 59.7. Physical Properties of Aliphatic Unsaturated Polycarboxylic Acids

Acid	CAS Acid Name	CAS Reg. No.	Mol Wt	M.P. (°C)	B.P. (°C)	Solubility[a] H2O	Alcohol	Other	Specific Gravity	Vapor Pressure mm Hg	°C
Maleic	(Z)-2-Butenedioic	110-16-7	116.07	139–140	135°C (dec)	V	V	Ether (sl s) Acetone (v) CHCl3 (insol) Benzene (insol)	1.590 (20/4)		
Fumaric	(E)-2-Butenedioic	110-17-8	116.07	300–302 (sealed tube)	290	V sl	Sl s	Ether (v sl) Acetone (v sl) CHCl3 (v sl)	1.635 (20/4)	1.7	165
Mesaconic	(E)-2-Methyl-2-butenedioic	498-24-8	130.1	204.5	Subl.	V sl	V	Ether (sl s) CHCl3 (v sl) Benzene (v sl)	1.466 (20/4)		
Citraconic	(Z)-2-Methyl-2-butenedioic	498-23-7	130.1	93–93.8 (dec)		V		Ether (v sl) CHCl3 (insol) Benzene (v sl)	1.617		
Itaconic	Methylenebutanedioic	97-65-4	130.1	175	(Dec)	Sl s	Sl s	Ether (v sl) Acetone (sl s) CHCl3 (v sl) Benzene (v sl)	1.632		
Aconitic	1-Propene-1,2,3-tricarboxylic	499-12-7	174.11	130	198–205 (dec)	Sl s		Ether (v sl)			

[a] s, soluble in all proportions; v, very soluble; sl s, slightly soluble; v sl, very slightly soluble; ins, insoluble.

Table 59.8. Acute Toxicity of Unsaturated Polycarboxylic Acids

Acid	Species	Route	LD_{50} (mg/kg)	Ref.
Maleic	Mouse	Oral	2,400	368
	Rat	Oral	708	366
	Rat	Inhaln.	>0.72 g/m^3 (LC_{50}/1 hr)	366
	Guinea pig	Skin	>1,000	366
	Rabbit	Skin	1,560	366
Fumaric	Mouse	Ip	200	380
	Rat	Oral	10,700	22
	Rat	Ip	587 (LD_{Lo})	381
	Rabbit	Oral	5,000 (LD_{Lo})	379
Mesaconic	Mouse	Ip	500 (LD_{Lo})	378
Citraconic	Mouse	Oral	2,260	27
	Rat	Oral	1,320	27
	Guinea pig	Oral	1,350	27
Aconitic	Mouse	Iv	180	394

5.1.5 Biochemistry

Maleic acid produces in rats a condition analogous to the human Fanconi syndrome, characterized by increased urinary elimination of glucose, amino acids, and other biochemicals, and resulting from an impaired tubular reabsorption of these materials (370–374). Ultrastructural changes occur at concentrations as low as 1.5 mM maleic acid (372). In dogs the intravenous administration of maleic acid (50 mg/kg) increased sodium, potassium, and phosphate excretion (375). Maleic acid may enhance anaphylactic histamine release from guinea pig lung by its metabolic utilization in the tricarboxylic acid cycle (376).

5.1.6 Regulations and Standards

Maleic acid may be used in adhesives and coatings for packaging, transporting, or holding food (21CFR 175.105, 300) (21CFR 177.1200).

5.2 Fumaric Acid

The formula is HOOCCH=CHCOOH; fumaric acid is also known as (E)-2-butenedioic and trans-1,2-ethylenedicarboxylic acid.

5.2.1 Major Uses

Fumaric acid is a substitute for tartaric acid in beverages and baking powders, an antioxidant, and a mordant in dyeing and printing inks; it is also used in the manufacture of polyhydric alcohols and synthetic resins (74).

5.2.2   Acute Studies

Fumaric acid injected intraperitoneally (10 mg/kg) in rats causes hepatotoxicity, tremors, and hypothermia, and at 100 mg/kg decreases motor activity and causes diuresis (377). In rabbits 500 mg produced only mild skin irritation in 24 hr, whereas 100 mg produced a severe effect in eyes (378). Lethal doses following single administrations are presented in Table 59.8.

5.2.3   Repeated Exposures

Sodium fumarate was fed to rabbits at 6.9 percent (equivalent to 5 percent free acid) in the diet for 150 days. No gross or histopathologic changes or differences in growth or survival were noted (202). Fumaric acid in the diet (1.5 percent) for 2 years produced an increased mortality and some testicular atrophy. No major visceral damage or histopathologic changes were seen and only minor differences between control and treated animals were noted. At 0.5 and 1.0 percent, fumaric acid had no measurable effect on rats (203). Chronic toxicity studies of young rats with dietary 0.1 or 1.0 percent fumaric acid or 1.38 percent sodium fumarate resulted in no significant variations from control animals in growth, blood hemoglobin concentration, red or white blood cell counts, or bone ash or histopathology of liver, kidney, spleen, or stomach. In guinea pigs 1.0 percent fumaric acid had no effect on growth, reproduction, or lactation (381).

5.2.4   Physiologic Effects in Humans

Fumaric acid is a mild irritant on human skin and mucous membranes. Humans can tolerate 500 mg/day for a year without ill effect (369). Fumaric acid is metabolized in human erythrocytes by conversion to oxalacetic acid, which is then decarboxylated (383).

5.2.5   Biochemistry

Fumaric acid (50 mg/kg) has an antiulcer action probably based on its ability to inhibit gastric juice secretion and to dilate the stomach muscle (382).

   Fumaric acid is a normal metabolite and constituent of the citric acid cycle. It is formed in the cell by the oxidation of succinic acid and hydrated under the influence of the enzyme fumarase to form maleic acid. Fumaric acid also acts as a link between the citric acid cycle and the urea cycle in its condensation with arginine, producing argininosuccinic acid. For more details of these reactions the reader is referred to general biochemistry texts.

5.2.6 Regulations and Standards

Fumaric acid is deemed to be generally recognized as safe by the Flavoring Extract Manufacturers' Association and has a wide variety of lawful uses in food (21CFR 172.350) and food-packaging materials (21CFR 176.170, 180).

5.3 Mesaconic Acid

The formula is $HOOCCH=C(CH_3)COOH$; mesaconic acid is also known as (E)-2-methyl-2-butenedioic and methylfumaric acid.

5.3.1 Major Uses

Mesaconic acid is used in electropolishing stainless steel (384) and as a laminant of polyethylene (385).

5.3.2 Acute Studies

Lethal doses following single administrations are presented in Table 59.8. No other toxicity information was available.

5.4 Citraconic Acid

The formula is $HOOCCH=C(CH_3)COOH$; citraconic acid is also known as (Z)-2-methyl-2-butenedioic and methylmaleic acid.

5.4.1 Major Uses

Citraconic acid is used in synthetic or copolymer oils (387), the treatment of metals for adhesion of polyolefins (388), and thermoplastic materials (389).

5.4.2 Acute Studies

Lethal doses following single administrations are presented in Table 59.8. No other toxicity information was available.

5.5 Itaconic Acid

The formula is $CH_2=C(COOH)CH_2COOH$; itaconic acid is also known as methylenebutanedioic and methylenesuccinic acid.

5.5.1 Major Uses

The major uses of itaconic acid are in copolymerizations, resins, plasticizers, and lube oil additives and as an intermediate (242).

5.5.2 Acute Studies

Single oral doses of 0.5 g/kg caused vomiting and diarrhea in cats, probably due to irritation in the gastrointestinal tract, since doses up to 1 g/kg failed to produced signs of systemic toxicity. Oral doses of 5 g/kg proved fatal with severe gastrointestinal disturbances, convulsions, and prostration (390). Rabbits given a 2000-mg oral dose showed a five-fold increase in urinary excretion of succinic acid in 24 hr (391).

5.5.3 Repeated Exposures

Daily oral administrations to cats of 100 mg/kg for 14 weeks were without effect on nutritional status, hematology, liver and kidney function, electrocardiogram, and histopathology of liver and kidney (390).

When rats were maintained on diets that contained 1 or 2 percent of itaconic acid for 210 days there were no significant effects on feed consumption, gross pathology, or histopathology of tissues and organs. Growth rate, however, was reduced (391).

5.5.4 Biochemistry

Itaconic acid is a component of the normal intermediary metabolism. Its metabolism in mammalian and bacterial systems has been demonstrated (391). Its inhibition of succinic dehydrogenase in vitro may explain the increased urinary excretion of succinic acid by rabbits and the decreased growth rate in rats following large doses (392).

5.5.5 Regulations and Standards

Itaconic acid may be used lawfully as a component of the food-contacting surface of paper and paperboard (21CFR 176.180).

5.6 Aconitic Acid

The formula is $HOOCCH=C(COOH)CH_2COOH$; aconitic acid is also known as 1-propene-1,2,3-tricarboxylic and *cis*-aconitic acid.

5.6.1   Major Uses

The major uses are in the manufacture of itaconic acid, as a plasticizer for buna rubber and plastics (74), in the preparation of wetting agents, and as an antioxidant (242).

5.6.2   Acute Studies

Perfusion of frog hearts with aconitic acid (1 mM) produced a slight inhibition of the force and rate of myocardial contraction and decreased cardiac output (393).

Aconitic acid decreases the carcinogenic action of 3,4-benzopyrene (BP) up to complete inhibition of tumor development in 4- to 5-week-old mice when injected subcutaneously. Doses of 30 mg decreased the BP tumor incidence from 97 to 25 percent (395, 396). Acute lethality data are presented in Table 59.8.

5.6.3   Biochemistry

Aconitic acid is a product of the citric acid cycle. Its reversible conversion from citric acid involves only a dehydration reaction catalyzed by one enzyme, aconitase. Although citric acid is favored, a small amount of aconitic acid is present at equilibrium and in respiring tissues. For greater detail the reader is referred to general texts on biochemistry.

5.6.4   Regulations and Standards

Aconitic acid is a direct food substance affirmed generally recognized as safe (21CFR 184.1007).

REFERENCES

1. G. Malorny, Z. Ernaehrungswiss, **9**, 332–339 (1969).
2. A. Sporn, V. Marin, and C. Schöbesch, Igenia (Bucharest), **11**, 507–515 (1962), cited in Ref. 34.
3. W. F. Von Oettingen, Arch. Ind. Health, **20**, 517–531 (1959).
4. U. Sammartino, Arch. Farmacol. Speri, **56**, 364–371 (1933), cited in Ref. 34.
5. G. Woodard, S. W. Lang, K. W. Nelson, and H. O. Calvery, J. Ind. Hyg. Toxicol., **23**, 78–82 (1941).
6. L. Orö and A. Wretlind, Acta Pharmacol. Toxicol., **18**, 141–152 (1961).
7. L. Ghiringhelli and A. DiFabio, Med. Lav., **48**, 559–565 (1957).
8. H. F. Smyth, C. P. Carpenter, and C. S. Weil, Arch. Ind. Hyg. Occup. Med, **4**, 119–122 (1951).

9. *C. R. Seances Soc. Biol.*, **83,** 136 (1920), cited in *Registry of Toxic Effects of Chemical Substances*, NIOSH ed., 1978.

10. Union Carbide Data Sheet, 8/7/63, cited in *Registry of Toxic Effects of Chemical Substances*, NIOSH ed., 1978.

11. "Evaluation of the Health Aspects of Propionic Acid, Calcium Propionate, Sodium Propionate, Dilauryl Thiodipropionate and Thiodipropionic Acid as Food Ingredients," Life Sci. Res. Office, Fed. Am. Soc. Exp. Biol., Bethesda, Md., 1979, NTIS No. PB80-104599.

12. H. F. Smyth, C. P. Carpenter, C. S. Weil, U. C. Pozzani, and J. A. Striegel, *Am. Ind. Hyg. Assoc. J.*, **23,** 95–107 (1962).

13. WHO Food Additive Series, No. 5, "Toxicological Evaluation of Some Food Additives Including Anticaking Agents Antimicrobials, Antioxidants, Emulsifiers and Thickening Agents," Geneva, 1974.

14. *Toksikol. Nov. Promys. Khim. Veshchestv.*, **4,** 19 (1962) cited in *Registry of Toxic Effects of Chemical Substances*, NIOSH ed., 1978.

15. *J. Pharm. Pharmacol.*, **21,** 85 (1969), cited in *Registry of Toxic Effects of Chemical Substances*, NIOSH ed., 1978.

16. H. F. Smyth, C. P. Carpenter, C. S. Weil, and U. C. Pozzani, *Arch. Ind. Hyg. Occup. Med.*, **10,** 61–68 (1954).

17. Union Carbide Data Sheet, 4/10/68, cited in *Registry of Toxic Effects of Chemical Substances*, NIOSH ed., 1978.

18. R. Ehrmann, *Z. Klin Med.*, **72,** 500 (1911), cited in Ref. 71.

19. H. F. Smyth, C. P. Carpenter, C. S. Weil, U. C. Pozzani, J. A. Striegel, and J. S. Nycum, *Am. Ind. Hyg. Assoc. J.*, **30,** 470–476 (1969).

20. Union Carbide Data Sheet, 1/31/72, cited in *Registry of Toxic Effects of Chemical Substances*, NIOSH ed., 1978.

21. D. W. Fassett, in *Industrial Hygiene and Toxicology*, 2nd ed, Vol. II, D. W. Fassett and D. D. Irish, Eds., Wiley-Interscience, New York, 1963.

22. E. H. Vernot, J. D. MacEwen, C. C. Haun, and E. R. Kinkead, *Toxicol. Appl. Pharmacol.*, **42,** 417–423 (1977).

23. H. F. Smyth and C. P. Carpenter, *J. Ind. Hyg. Toxicol.*, **26,** 269–273 (1944).

24. Union Carbide Data Sheet, 2/4/59, cited in *Registry of Toxic Effects of Chemical Substances*, NIOSH ed., 1978.

25. *Biochem. J.*, **34,** 1196 (1940), cited in *Registry of Toxic Effects of Chemical Substances*, NIOSH ed., 1978.

26. Food and Drug Research Labs., paper 123 (1976), cited in *Registry of Toxic Effects of Chemical Substances*, NIOSH ed., 1978.

27. P. M. Jenner, E. C. Hagan, J. M. Taylor, E. L. Cook, and O. G. Fitzhugh, *Food Cosmet. Toxicol.*, **2,** 327–343 (1964).

28. D. L. J. Opdyke, "Monographs on Fragrance Raw Materials, Caprylic Acid," *Food Cosmet. Toxicol.*, **19,** 237–245 (1981).

29. Union Carbide Data Sheet, 11/4/71, cited in *Registry of Toxic Effects of Chemical Substances*, NIOSH ed., 1978.

30. D. L. J. Opdyke, "Monographs on Fragrance Raw Materials: Pelargonic Acid," *Food Cosmet. Toxicol.*, **16,** 839–841 (1978).

31. D. L. J. Opdyke, "Monographs on Fragrance Raw Materials: Capric Acid," *Food Cosmet. Toxicol.*, **17,** 735–742 (1979).

32. U.S. Army Armament Res. Devpt. Commd., Chem. Syst. Lab., NX #0482, cited in *Registry of Toxic Effects of Chemical Substances*, NIOSH ed., 1978.

33. D. L. J. Opdyke, "Monographs on Fragrance Raw Materials: Stearic Acid," *Food Cosmet. Toxicol.*, **17**, 383–388 (1979).

34. "Evaluation of the Health Aspects of Formic Acid, Sodium Formate and Ethyl Formate as Food Ingredients," Life Sci. Res. Office, Fed. Am. Soc. Exp. Biol., Bethesda, Md., 1976, NTIS No. PB 266-82.

35. M. O. Amdur, *Int. J. Air Pollut.*, **3**, 201–220 (1960).

36. A. Lund, *Acta Pharmacol. Toxicol.*, **4**, 99–107 (1948).

37. H. Neumark, *J. Agric. Sci.*, **69**, 297–303 (1967).

38. E. B. Freese, J. Gerson, H. Taber, J. Rhaese, and E. Freese, *Mutation Res.*, **4**, 517–531 (1967).

39. M. Demerec, G. Bertani, and J. Flint, *Am. Nat.*, **85**, 119–136 (1951), cited in Ref. 34.

40. B. F. A. Stumm-Tegethoff, *Theor. Appl. Genetics*, **39**, 330–334 (1969), cited in Ref. 34.

41. J. V. Frei and P. Stephens, *Br. J. Cancer*, **22**, 83–92 (1968).

42. K. E. von Muhlendahl, U. Oberdisse, and E. G. Krienke, *Arch. Toxicol.*, **39**, 299–314 (1978).

43. J. P. Fahy and H. B. Elkins, unpublished data (1954), cited by Ref. 49.

44. M. Palese and T. R. Tephly, *J. Toxicol. Environ. Health*, **1**, 13–24 (1975).

45. G. Malorny, *Z. Ernaehrungswiss.*, **9**, 340–348 (1969).

46. R. T. Williams, Ed., *Detoxication Mechanisms*, Wiley, New York, 1959.

47. N. I. Sax, Ed., *Dangerous Properties of Industrial Materials*, Van Nostrand Reinhold, New York, 1975.

48. *Fed. Regist.*, **39**, 23540 (1974).

49. *Documentation of the Threshold Limit Values of Substances in Workroom Air*, Vol. 3, American Conference of Governmental Industrial Hygienists, 1971, p. 119.

50. W. F. von Oettingen, *Arch. Ind. Health*, **21**, 28–65 (1960).

51. S. Okabe, J. L. Roth, and C. F. Pfeiffer, *Experientia*, **27**, 146–148 (1972).

52. R. L. Roudabush, C. J. Terhaar, D. W. Fassett, and S. P. Dziuba, *Toxicol. Appl. Pharmacol.*, **7**, 559–565 (1965).

53. G. A. Nixon, C. A. Tyson, and W. C. Wertz, *Toxicol. Appl. Pharmacol.*, **31**, 481–490 (1975).

54. "Mutagenic Evaluation of Compound FDA 75-3, Sodium Acetate ·3H$_2$O NF Granular," Litton Bionetics, Inc., Kensington, Md., 1975, NTIS No. PB-254 514/3ST.

55. "Evaluation of the Health Aspects of Acetic Acid, Sodium Acetate and Sodium Diacetate as Food Ingredients," Life Sci. Res. Office, Fed. Am. Soc. Exp. Biol., Bethesda, Md., 1977, NTIS No. PB-274 670.

56. T. Sollmann, *J. Pharmacol. Exp. Ther.*, **16**, 463–474 (1921).

57. K. Mori, *Gann*, **43**, 433–466 (1952), cited in Ref. 55.

58. A. U. Pardoe, *Br. J. Pharmacol.*, **7**, 349–357 (1952).

59. H. F. Smyth, *Am. Ind. Hyg. Assoc. Q.*, **17**, 129–185 (1956).

60. H. J. Nicholas and B. E. Thomas, *Brain*, **84**, 320–328 (1961).

61. G. A. Dhopeshwarkar, C. Subramanian, and J. F. Mead, *Biochim. Biophys. Acta*, **248**, 41–47 (1971).

62. H. J. Deuel and A. T. Milhorat, *J. Biol. Chem.*, **78**, 299–309 (1928).

63. G. Hevesy, *Nature* (London), **16**, 1007 (1949), cited in Ref. 50.

64. Union Carbide Data Sheet, 3/24/70, cited in *Registry of Toxic Effects of Chemical Substances*, NIOSH ed., 1978.

65. "Mutagenic Evaluation of Compound FDA 71-36, Calcium Propionate," Litton Bionetics, Inc., Kensington, Md., 1976, NTIS No. PB 245-439.

66. W. D. Graham, H. Teed, and H. C. Grice, *J. Pharm. Pharmacol.*, **6**, 534–545 (1954).

67. W. D. Graham and H. C. Grice, *J. Pharm. Pharmacol.*, **7**, 126–134 (1955).

68. K. H. Bässler, *Z. Lebensm. Untersuch. Forsch.*, **110**, 28 (1959), cited in Ref. 13.

69. F. H. Theodore, *J. Am. Med. Assoc.*, **143**, 226–228 (1950).

70. A. M. Dawson, C. D. Holdsworth, and J. Webb, *Proc. Soc. Exp. Biol. Med.*, **117**, 97–100 (1964); *Chem. Abstr.*, **65**, 16545c.

71. W. F. von Oettingen, *Arch. Ind. Health*, **21**, 100–113 (1960).

72. K. P. Stasenkova and T. A. Kochetkova, *Toksikol. Nov. Prom. Khim. Veshchestv.*, **4**, 19–28 (1962); *Chem. Abstr.*, **58**, 9543f.

73. S. L. Danishevskii and B. I. Monastyrskaya, *Tr. Leningr. Sanit.-Gig. Med. Inst.*, **62**, 78–84 (1960); *Chem. Abstr.*, **57**, 3743i.

74. *The Merck Index*, 9th ed., M. Windholz, Ed., Merck & Co., Inc., Rahway, NJ, 1976.

75. H. Oettel, *Arch. Exp. Pathol. Pharmakol.*, **183**, 641 (1936), cited in Ref. 71.

76. D. S. Kinnary, Y. Takeda, and D. M. Greenberg, *J. Biol. Chem.*, **212**, 385–396 (1955).

77. M. Puukka, S. Laakso, and V. Nurmikko, *Acta Chem. Scand.*, **27**, 720–722 (1973).

78. G. D. DiVincenzo and M. L. Hamilton, *Toxicol. Appl. Pharmacol.*, **47**, 609–612 (1979).

79. V. Mahadevan and L. Zieve, *J. Lipid, Res.*, **10**, 338–341 (1969).

80. G. Guggenheim, L. Ertlinger, and R. Muhlemann, *Pathol. Microbiol.*, **28**, 77–83 (1965); *Chem. Abstr.*, **62**, 8152h.

81. K. Thomas and Kh. Stalder, *J. Physiol. Chem.*, **313**, 22–29 (1958).

82. A. Loewy and R. Ehrmann, *Z. Klin. Med.*, **72**, 502 (1911), cited in Ref. 71.

83. E. F. Annison and R. J. Pennington, *Biochem. J.*, **57**, 685–692 (1954).

84. H. E. Amos, C. O. Little, and G. E. Mitchell, *J. Agric. Food Chem.*, **19**, 112–115 (1971); *Chem. Abstr.*, **74**, 84262a.

85. W. Ritter and H. Hänni, *Pathol. Microbiol.*, **23**, 669–680 (1960); *Chem. Abstr.*, **55**, 6598a.

86. T. Nakae and J. Elliot, *J. Dairy Sci.*, **48**, 287–292 (1965); *Chem. Abstr.*, **62**, 15093.

87. R. Rubenstein, A. V. Howard, and A. M. Wrang, *Clin. Sci.*, **37**, 549–564 (1969).

88. I. Siegel and V. Lorber, *J. Biol. Chem.*, **189**, 571–576 (1951).

89. K. Verschueren, Ed., *Handbook of Environmental Data on Organic Chemicals*, Van Nostrand Reinhold, New York, 1977.

90. J. E. Amoore, M. R. Gumbmann, A. N. Booth, and D. H. Gould, *Chem. Senses Flavor (Engl.)*, **3**, 307–317 (1978).

91. I. Zabin and K. Bloch, *J. Biol. Chem.*, **185**, 117–129 (1950).

92. G. W. E. Plout and H. A. Lardy, *J. Biol. Chem.*, **192**, 435–445 (1951).

93. H. Kodama, T. Azumi, T. Shimomura, and Y. Fujii, *Acta Med. Okayama*, **20**, 107–113 (1966); *Chem. Abstr.*, **66**, 44789f.

94. I. Zabin and K. Bloch, *J. Biol. Chem.*, **185**, 131–138 (1950).

95. S. J. Henning and F. J. R. Hird, *Brit. J. Nutr.*, **24**, 145–155 (1970).

96. C. P. Carpenter and H. F. Smyth, *Am. J. Ophthalmol.*, **29**, 1363–1372 (1946).

97. D. E. Moody and J. K. Reddy, *Toxicol. Appl. Pharmacol.*, **45**, 497–504 (1978).

98. F. J. R. Hird and M. J. Weidemann, *Biochem. J.*, **98**, 378–388 (1966).

99. D. Rittenberg, R. Schoenheimer, and E. A. Evans, *J. Biol. Chem.*, **120**, 503 (1937), cited in Ref. 71.

100. C. Bode and M. Klingenberg, *Biochem. Z.*, **341**, 271–299 (1965); *Chem. Abstr.*, **62**, 10693c.

101. F. J. R. Hird and R. H. Symons, *Biochem. J.*, **84**, 212–216 (1962).

102. A. L. Grafflin and D. E. Green, *J. Biol. Chem.*, **176**, 95–115 (1948).

103. W. A. Atchley, *J. Biol. Chem.*, **176**, 123–131 (1948).

104. H. J. Deuel, Ed., *The Lipids, Their Chemistry and Biochemistry*, Vol. III, Interscience, New York, 1957.

105. D. D. Dziewiatowski, A. Ventakaraman, and H. B. Lewis, *J. Biol. Chem.*, **178**, 169–177 (1949).

106. L. Blum and M. Koppel, *Chem. Ber.*, **44**, 3576–3578 (1911), cited in Ref. 104.

107. A. M. Nervi, O. Mercuri, A. Marzzi, and R. R. Brenner, *Acta Physiol. Lat. Am.*, **16**, 357–365 (1966).

108. H. A. Krebs and R. Hems, *Biochem. J.*, **119**, 525–533 (1970).

109. P. Sentein and H. Vannereau, *Chromosoma*, **40**, 1–41 (1973); *Chem. Abstr.*, **79**, 49517u.

110. "Mutagenic Evaluation of Compound FDA 75-38.000124-07-2, Caprylic Acid 98%," Litton Bionetics, Inc., Kensington, Md., 1976, MTIS No. PB 257 872.

111. *Encylopaedia of Occupational Health and Safety*, Int. Labour Office, Geneva, 1971.

112. R. Scheig and G. Klatskin, *J. Am. Oil Chem. Soc.*, **45**, 31–33 (1968).

113. N. J. Greenberger, J. J. Franks, and K. J. Isselbacher, *Proc. Soc. Exp. Biol. Med.*, **120**, 468–472 (1965).

114. S. L. Kirscher and R. Harris, *J. Nutr.*, **73**, 397–402 (1961).

115. R. Scheig in *Medium Chain Triglycerides Trans., 1967*, J. R. Senior, Ed., University of Pennsylvania Press, Philadelphia, 1968.

116. M. K. Reddy, P. F. Hollenberg, and J. K. Reddy, *Biochem. J.*, **188**, 731–740 (1980).

117. S. J. Morton and R. J. Rubin, *Abstr. 19th Ann. Meet. Soc. Toxicol.*, No. 214, A72 (1980).

118. J. K. Reddy and M. S. Rao, *J. Natl. Cancer Inst.*, **59**, 1645–1650 (1977).

119. J. K. Reddy, D. L. Azarnoff, and C. E. Hignite, *Nature*, **283**, 397–398 (1980).

120. S. A. Hyun, G. V. Vahouny, and C. R. Treadwell, *Biochim. Biophys. Acta*, **137**, 306–314 (1967).

121. G. V. Vahouny, J. Nelson, and C. R. Treadwell, *Proc. Soc. Exp. Biol. Med.*, **128**, 495–500 (1968).

122. T. H. Chung, G. V. Vahouny, and C. R. Treadwell, *J. Atheroscler. Res.*, **10**, 217–227 (1969).

123. Emery Industries, Inc., Data Sheet S3B, (1964), cited in *Registry of Toxic Effects of Chemical Substances*, NIOSH ed., 1978.

124. L. P. Dryden and A. M. Hartman, *J. Nutr.*, **101**, 589–592 (1971).

125. M. Yoshida, H. Morimoto, and R. Oda, *Agric. Biol. Chem.*, **34**, 1301 (1970), cited in Ref. 30.

126. A. Bach, D. Guisard, P. Metais, and G. Debry, *Nutr. Metab.*, **14**, 203–209 (1972).

127. K. Mori, *Jap. J. Cancer Res.*, **44**, 421 (1953), cited in Ref. 31.

128. M. A. Stillman, H. I. Maibach, and A. R. Shalita, *Contact Dermatitis*, **1**, 65–69 (1975).

129. F. Wada and M. Usami, *Biochim. Biophys. Acta*, **487**, 261–268 (1977).

130. J. D. Bu'Lock and G. N. Smith, *Biochem. J.*, **96**, 495–499 (1965).

131. S. Molnar and H. Mohme, *Z. Tierphysiol. Tierernaehr. Futtermittelkd.*, **25**, 229 (1969), cited in Ref. 31.

132. P. E. Swenson and P. S. Dimick, *J. Dairy Sci.*, **57**, 290–295 (1974).

133. L. Forsgren, *Ark. Kemi*, **30**, 355 (1969), cited in Ref. 31.

134. "The Food Chemical News Guide," Food Chemical News, Inc., Washington, D.C., 1981.

135. D. Raffenbeul and G. Zbinden, *Acta Haemat.* (Basel), **57**, 87–95 (1977).

136. O. G. Fitzhugh, P. J. Schouboe, and A. A. Nelson, *Toxicol. Appl. Pharmacol.*, **2**, 59–67 (1960).

137. P. Shubik and J. L. Hartwell, *U.S. Public Health Serv. Publ.* **149** (Suppl. 1) (1957), cited in Ref. 21.

138. P. Holsti, *Acta Pathol. Microbiol. Scand.*, **46**, 51–58 (1959).

139. G. Göransson, *Acta Physiol. Scand.*, **64**, 383–386 (1965).

140. M. L. Das, S. Orrenius, and L. Ernster, *Eur. J. Biochem.*, **4**, 519–523 (1968).

141. K. Ichihara, E. Kusunose, and M. Kusunose, *Biochim. Biophys. Acta*, **202**, 560–562 (1970).

142. W. Kam, K. Kumaran, and B. R. Landau, *J. Lipid Res.*, **19**, 591–600 (1978).

143. J. S. Butts, H. Blunden, W. Goodwin, and H. J. Deuel, *J. Biol. Chem.*, **117**, 131–133 (1937).

144. M. Nakagawa and M. Uchiyama, *J. Biochem.* (Tokyo), **66**, 95–97 (1969).

145. V. A. Drill and P. Lazar, Eds., *Cutaneous Toxicity*, Academic, New York, 1977.

146. G. A. Rao and S. Abraham, *Lipids*, **9**, 269–271 (1974).

147. S. B. Tove, *J. Nutr.*, **84**, 237–243 (1964).

148. P. Boucrot and J. Clement, *Arch. Sci. Physiol.*, **19**, 181–196 (1965).

149. J. Elovson, *Biochim. Biophys. Acta*, **106**, 291–303 (1965).

150. I. Björkhem, *J. Lipid Res.*, **19**, 585–590 (1978).

151. G. Göransson and T. Olivecrona, *Acta Physiol. Scand.*, **62**, 224–239 (1964).

152. T. Olivecrona, *Acta Physiol. Scand.*, **54**, 295–305 (1962).

153. M. C. Hardegree and R. Kirschstein, *Ann. Allergy*, **26**, 259 (1968), cited in Ref. 33.

154. W. B. Deichmann, J. L. Radomski, W. E. MacDonald, R. L. Kascht, and R. L. Erdmann, *Arch. Ind. Health*, **18**, 483–487 (1958).

155. B. L. Van Duuren, C. Katz, M. B. Shimkin, D. Swern, and R. Weider, *Cancer Res.*, **32**, 880–881 (1972).

156. D. Swern, R. Wieder, M. McDonough, D. R. Meranze, and M. B. Shimkin, *Cancer Res.*, **30**, 1037–1046 (1970).

157. G. J. Antony and B. R. Landau, *J. Lipid Res.*, **9**, 267–269 (1968).

158. P. Bjorntorp, *J. Biol. Chem.*, **243**, 2130–2133 (1968).

159. I. Björkhem and H. Danielsson, *Eur. J. Biochem.*, **17**, 450–459 (1970).

160. J. F. Mead, *Am. J. Clin. Nutr.*, **6**, 652–655 (1958).

161. J. E. Kinsella, *J. Dairy Sci.*, **53**, 1757–1765 (1970).

162. I. N. de Gomez Dumm and R. R. Brenner, *Lipids*, **10**, 315–317 (1975).

163. D. L. Trout, E. H. Estes, H. L. Hilderman, and G. D. Long, *Am. J. Physiol.*, **203**, 1024–1028 (1962).

164. R. H. Coots, *J. Lipid Res.*, **5**, 468–472 (1964).

165. K. W. J. Wahle, *Comp. Biochem. Physiol.*, **48**, 87–105 (1974).

166. W. M. F. Leat, *Biochem. J.*, **94**, 21P (1965).

167. R. J. Young and R. L. Garrett, *J. Nutr.*, **81**, 321–329 (1963).

168. *Abdernalden's Handbuch der Biologischern Arbeitsmethoden*, **4**, 1377 (1935), cited in *Registry of Toxic Effects of Chemical Substances*, NIOSH ed., 1978.

169. National Technical Information Service, NTIS No. AD 277-689, cited in *Registry of Toxic Effects of Chemical Substances*, NIOSH ed., 1978.

170. Industrial Bio-Test Labs., Inc., BioFax Data Sheets 22 - March, 1971, cited in *Registry of Toxic Effects of Chemical Substances*, NIOSH ed., 1978.

171. Unpublished information, Health, Safety and Human Factors Laboratory, Eastman Kodak Co., Rochester, N.Y., 1981.

172. National Technical Information Service, NTIS No. AD 691-490, cited in *Registry of Toxic Effects of Chemical Substances*, NIOSH ed., 1978.

173. J. F. Danielli, M. Danielli, J. B. Fraser, P. D. Mitchell, L. N. Owen, and G. Shaw, *Biochem. J.*, **41**, 325–328 (1947).

174. H. J. Horn, E. G. Holland, and J. W. Hazelton, *J. Agric. Food Chem.*, **5**, 759–761 (1957).

175. A. Locke, R. B. Locke, H. Schlesinger, and H. Carr, *J. Am. Pharm. Assoc.*, **31**, 12–14 (1942).

176. *Ind. Eng. Chem.*, **15**, 628 (1923), cited by *Registry of Toxic Effects of Chemical Substances*, NIOSH ed., 1978.

177. T. L. Sourkes and T. Koppanyi, *J. Am. Pharm. Assoc.*, **39**, 275–276 (1950).

178. A. R. Singh, W. H. Lawrence, and J. Autian, *Toxicol. Appl. Pharmacol.*, **32**, 566–576 (1975).

179. "Sbornik Vysledku Toxixologickeho Vysetreni Latek A Pripravku," J. V. Marhold, Ed., Institut Pro Vychovu Vedoucicn Pracovniku Chemickeho Prumyclu Praha, Czechoslovakia (1972), cited in *Registry of Toxic Effects of Chemical Substances*, NIOSH ed., 1978.

180. C. M. Gruber and W. A. Halbeisen, *J. Pharmacol. Exp. Ther.*, **94**, 65–67 (1948).

181. *Takeda Kenkyusho Ho* (J. Takeda Res. Labs.), **30**, 25 (1971), cited in *Registry of Toxic Effects of Chemical Substances*, NIOSH ed., 1978.

182. W. B. Deichmann and H. W. Gerarde, Eds., *Toxicology of Drugs and Chemicals*, 4th ed., Academic, New York, 1969.

183. P. P. Singh, L. K. Kothari, and H. S. Sharma, *Asian Med. J.*, **16**, 287–293 (1973); *Chem. Abstr.*, **80**, 530z.

184. D. W. Fassett, "Oxalates," Chapter 16 in *Toxicants Occurring Naturally in Foods*, National Academy of Science, Washington, D.C., 1973.

185. C. G. MacKenzie and E. V. McCollum, *Am. J. Hyg.*, **25**, 1–10 (1937); *Chem. Abstr.*, **31**, 2655^9.

186. M. Goldman, G. J. Doering, and R. G. Nelson, *Res. Commun. Chem. Pathol. Pharmacol.*, **18**, 369–372 (1977).

187. B. Schiefer, M. P. Hewitt, and J. D. Milligan, *Zentralbl. Veterinaermed.*, *Reihe A*, **23**, 226–233 (1976); *Chem. Abstr.*, **85**, 88146r.

188. J. A. Swartzman, H. F. Hintz, and H. F. Schryver, *Am. J. Vet. Res.*, **39**, 1621–1623 (1978).

189. F. A. Johnston, T. J. McMillan, and G. D. Falconer, *J. Am. Diet. Assoc.*, **28**, 933–938 (1952), cited in Ref. 184.

190. P. Bonner, F. C. Hummel, M. F. Bates, J. Horton, H. A. Hunscher, and I. G. Macy, *J. Pediatr.*, **12**, 188 (1933), cited in Ref. 184.

191. S. Weinhouse and B. Friedmann, *J. Biol. Chem.*, **191**, 707–717 (1951).

192. C. O. Curtin and C. G. King, *J. Biol. Chem.*, **216**, 539–548 (1955).

193. E. K. Shirley and K. Schmidt-Nielsen, *J. Nutr.*, **91**, 496–502 (1967).

194. H. A. Krebs, E. Salvin, and W. A. Johnson, *Biochem. J.*, **32**, 113–117 (1938).

195. J. S. Lee and N. Lifson, *J. Biol. Chem.*, **193**, 253–263 (1951).

196. V. L. Friend and H. Gold, *J. Am. Pharm. Assoc.*, **36**, 50–56 (1947).

197. S. Rous, L. Luethi, M. J. Burlet, and P. Favarger, *Biochim. Biophys. Acta*, **152**, 462–471 (1968).

198. J. S. Fruton and S. Simmonds, Eds., *General Biochemistry*, 2nd ed., Wiley, New York, 1963.

199. J. W. Daniel, *Food Cosmet. Toxicol.*, **7**, 103–106 (1969).

200. V. L. von Reisen, *J. Bacteriol.*, **85**, 248–249 (1963).

201. T. Chiba, *Yakugaku Zasshi*, **89**, 1138–1143 (169); *Chem. Abstr.*, **71**, 111,089w.

202. E. W. Packman, D. D. Abbott, and J. W. E. Harrison, *Toxicol. Appl. Pharmacol.*, **5**, 163–167 (1963).

203. O. G. Fitzhugh and A. A. Nelson, *J. Am. Pharm. Assoc.,* **36,** 217–219 (1947).

204. J. Gry and J. C. Larson, *Arch. Toxikol.,* **36,** Suppl. I, 351–353 (1978).

205. F. P. Underhill, C. S. Leonard, E. G. Gross, and T. C. Jaleski, *J. Pharmacol.,* **43,** 359–380 (1931).

206. P. Finkle, *J. Biol. Chem.,* **100,** 349–355 (1933).

207. L. F. Chasseaud, W. H. Down, and D. Kirkpatrick, *Experientia,* **33,** 998–1003 (1977).

208. V. S. Chadwick, A. Vince, M. Killingley, and O. M. Wrong, *Clin. Sci. Mol. Med.,* **54,** 273–282 (1978).

209. L. D. Kohn and W. B. Jakoby, *Biochem. Biophys. Res. Commun.,* **22,** 33–37 (1966).

210. J. D. Lewis, *Acta Pharmacol. Toxicol.,* **41,** Suppl. I, 144–145 (1977).

211. J. C. Gage, *Br. J. Ind. Med.,* **27,** 1–18 (1970).

212. "Teratogenic Evaluation of FDA-71-50 (Adipic Acid)," Food and Drug Res. Labs., Inc., East Orange, N.J., 1972, NTIS No. PB 221 802.

213. I. I. Rusoff, R. R. Baldwin, F. J. Domingues, C. Monder, W. J. Ohan, and R. Thiessen, *Toxicol. Appl. Pharmacol.,* **2,** 316–330 (1960).

214. G. Weitzel, *Chem. Zentr.* II, 556 (1942); *Chem. Abstr.,* **37,** 4453.

215. S. L. Bonting and B. C. Jansen, *Voeding,* **17,** 137 (1956), cited in Ref. 13.

216. S. Krop, H. Gold, and C. A. Paterno, *J. Am. Pharm. Assoc.,* **34,** 86–89 (1945).

217. E. Wright and R. E. Hughes, *Nutr. Rep. Int.,* **13,** 563–566 (1976).

218. E. Wright and R. E. Hughes, *Food Cosmet. Toxicol.,* **14,** 561–564 (1976).

219. L. Janota-Bassilik and L. D. Wright, *Nature,* **204,** 501–502 (1964).

220. K. Ogata, T. Tochikura, M. Osugi, and S. Iwahara, *Agric. Biol. Chem.,* **30,** 176–180 (1966); *Chem. Abstr.,* **64,** 14615e.

221. R. E. Eakin and E. A. Eakin, *Science,* **96,** 187–188 (1942).

222. M. A. Eisenberg, *Biochem. J.,* **98,** 15C–17C (1966).

223. W. R. Bloor, Ed., *Biochemistry of the Fatty Acids and their Compounds, the Lipids,* Reinhold, New York, 1943.

224. J. E. Pettersen and M. Aas, *Biochim. Biophys. Acta,* **326,** 305–313 (1973).

225. N. Dousset and L. Douste-Blazy, *Biochimie,* **55,** 1279–1285 (1973); *Chem. Abstr.,* **80,** 106,272g.

226. *Encyclopedia of Chemical Technology,* 2nd ed., Interscience, New York, 1963.

227. *Chem. Abstr.,* **52,** 5263c (1958).

228. *Chem. Abstr.,* **66,** 13352d (1967).

229. Unpublished data, Health, Safety, and Human Factors Laboratory, Eastman Kodak Company, Rochester, N.Y., 1972.

230. Unpublished data, Union Carbide Data Sheet, 1965.

231. Unpublished data, Health, Safety and Human Factors Laboratory, Eastman Kodak Co., Rochester, N.Y., 1975.

232. M. L. Miller, in *Encylopedia of Polymer Science and Technology,* N. M. Bikales, Ed., Wiley-Interscience, New York, Vol. 1, 1964, pp. 197–226.

233. A. R. Singh, W. H. Lawrence, and J. Autian, *J. Dent. Res.,* **51,** 1632–1638 (1972).

234. C. P. Carpenter, C. S. Weil, and H. F. Smyth, Jr., *Toxicol. Appl. Pharmacol.,* **28,** 313–319 (1974).

235. J. Majka, K. Knobloch, and J. Stetkiewicz, *Med. Pr.,* **25,** 427–435 (1974).

236. E. Boyland, *Biochem. J.,* **34,** 1196–1201 (1940).

237. National Research Council-Chemical Biological Coordination Center, *Summary Tables of Biological Tests,* **3,** 51 (1951).

238. Union Carbide Corporation, N.Y. Ind. Med. Toxicol. Dept., 1977.

239. H. F. Smyth, Jr., C. P. Carpenter, C. S. Weil, O. C. Pozzani, and J. A. Stiegel, *Am. Ind. Hyg. Assoc. J.,* **23,** 95–107 (1962).

240. International Agency for Research on Cancer, *Monographs on the Evaluation of the Carcinogenic Risk of Chemicals to Humans,* **19,** 47–52 (1979).

241. American Conference of Governmental Industrial Hygienists, Inc., *Documentation of Threshold Limit Values for Substances in Workroom Air,* 4th ed., Cincinnati, Ohio, 1980.

242. *The Condensed Chemical Dictionary,* 9th ed., G. G. Hawley, Ed., Van Nostrand Reinhold, New York, 1977.

243. Unpublished data, Health, Safety and Human Factors Laboratory, Eastman Kodak Co., Rochester, N.Y., 1965.

244. *Toxicology of Drugs and Chemicals,* W. B. Deichmann, and H. W. Gerarde, Eds., Academic, New York, 1969.

245. *Fed. Regist.,* **41,** 57018 (1976).

246. Unpublished data, Health, Safety and Human Factors Laboratory, Eastman Kodak Co., Rochester, N.Y. (1980).

247. *Handbook of Toxicology,* W. S. Spector, Ed., Vol. 1, Saunders, Philadelphia, 1956, pp. 198–199.

248. G. N. Mir, W. H. Lawrence, and J. Autian, *J. Pharm. Sci.,* **62,** 778–82. (1973)

249. G. N. Mir, W. H. Lawrence, and J. Autian, *J. Pharm. Sci.,* **63,** 376–81 (1974).

250. M. F. Tansy, F. M. Kendall, S. Benhayem, F. J. Hohenleitner, W. E. Landiu, and M. Gold, *Environm. Res.,* **11,** 66–77 (1976).

251. International Agency for Research on Cancer, *Monograph on the Evaluation of the Carcinogenic Risk of Chemicals to Humans,* **19,** 497 (1979).

252. G. N. Mir, W. H. Lawrence, and J. Autian, *J. Pharm. Sci.,* **62,** 1258–1261 (1973).

253. D. H. G. Crout, J. A. Corkill, M. L. James, and R. S. M. Ling, *Clin. Orthop. Relat. Res.,* **141,** 90–95 (1979).

254. K. Jansen, *Arbeitsmed. Sozialmed. Praventivmed.,* **9,** 206–207 (1974).

255. H. Bratt and D. E. Hathway, *Br. J. Cancer,* **36,** 1014–1019 (1977).

256. A. M. Glasgow and H. P. Chase, *Biochem. Biophys. Res. Commun.,* **62,** 362–366 (1975).

257. A. E. Senior and H. S. A. Sherratt, *Biochem. J.,* **108,** 46P–47P (1968).

258. A. E. Senior and H. S. A. Sherratt, *Biochem. J.,* **104,** 56P (1967).

259. A. E. Senior and H. S. A. Sherratt, *Biochem. J.,* **100,** 71P–72P (1966).

260. A. E. Senior and H. S. A. Sherratt, *Biochem. J.,* **110,** 499–509 (1968).

261. A. E. Senior and H. S. A. Sherratt, *J. Pharm. Pharmacol.,* **21,** 85–92 (1969).

262. C. Corredor, K. Brendel, and R. Bressler, *Proc. Natl. Acad. Sci.,* **58,** 2299–2306 (1967).

263. M. Hohenegger, H. Brechtelsbauer, U. Finsterer, and P. Prucksunand, *Pfluegers Arch.,* **351,** 231–240 (1974).

264. J. G. Kleinman, J. Mandelbaum, and M. L. Levin, *Am. J. Physiol.,* **224,** 95–101 (1973).

265. A. M. Glasgow and H. P. Chase, *Pediatr. Res.,* **9,** 133–138 (1975).

266. T. Fujita and M. Yasuda, *Jap. J. Pharmacol.,* **27,** 73–76 (1975).

267. M. B. Wilson, *West Indian Med. J.,* **22,** 178–182 (1974).

268. A. M. Glasgow, and H. P. Chase, *Biochem. J.,* **156,** 301–307 (1976).

269. H. Osmundsen, D. Billington, and H. S. A. Sherratt, *Biochem. Soc., Trans.,* **3,** 331–333 (1975).

270. N. B. Ruderman, C. J. Toews, C. Lowy, I. Vreeland, and E. Shafrir, *Am. J. Physiol.,* **219,** 51–57 (1970).

271. J. R. Williamson, M. H. Fukami, M. J. Peterson, S. G. Rostand, and R. Scholz, *Biochem. Biophys. Res. Commun.*, **36,** 407–413 (1969).

272. K. Aoyagi, M. Mori, and M. Titibani, *Biochim. Biophys. Acta,* **587,** 515–521 (1979).

273. H. H. Yeok, L. E. Rice, A. Maggio, and M. L. Levin, *Am. J. Physiol.*, **231,** 216–221 (1976).

274. H. Gershon, M. W. McNeil, R. Parmegiani, P. K. Godfrey, and J. M. Baricko, *Antimicrob. Agents Chemother.*, **4,** 435–438 (1973).

275. Unpublished data, Health, Safety and Human Factors Laboratory, Eastman Kodak Co., Rochester, N.Y., 1955.

276. H. F. Smyth, Jr., and C. P. Carpenter, *J. Ind. Hyg. Toxicol.,* **30,** 63–70 (1948).

277. F. Dickens, H. H. Jones, and H. B. Waynforth, *Br. J. Cancer,* **20,** 134–144 (1966).

278. F. Dickens, H. H. Jones, and H. B. Waynforth, *Br. J. Cancer,* **22,** 762–768 (1968).

279. F. Dickens, H. H. Jones, and H. B. Waynforth, *Br. Emp. Canc. Camp.*, **46,** 108 (1968).

280. T. T. Miyaji, *J. Exp. Med.,* **103,** 331–369 (1971).

281. A. J. Shtenberg, and A. D. Ignat'ev, *Food Cosmet. Toxicol.,* **8,** 369–380 (1970).

282. D. F. Chichester, and F. W. Tanner, Chapter 3, in *Handbook of Food Additives*, T. E. Furia, Ed., CRC Press, Cleveland, Ohio, 1972, pp. 115–184.

283. W. H. Gardner, Chapter 5, in *Handbook of Food Additives*, T. E. Furia, Ed., CRC Press, Cleveland, Ohio, 1972, pp. 225–270.

284. R. J. Hendy, J. Hardy, I. F. Gaunt, I. S. Kiss, and K. R. Butterworth, *Food Cosmet. Toxicol.,* **14,** 381–386 (1976).

285. I. F. Gaunt, K. R. Butterworth, J. Hardy, and S. D. Gangolli, *Food Cosmet. Toxicol.,* **13,** 31–45 (1975).

286. I. P. Barchenko, and S. G. Vasiliu, *Vopr. Ratsion. Pitan,* **6,** 21–25 (1970).

287. L. E. Fryklof, *J. Pharm. Pharmacol.,* **10,** 719–720 (1958).

288. F. N. Marzulli and H. I. Marbach, *J. Soc. Cosmet. Chem.,* **24,** 399–421 (1973).

289. M. Hannuksela, M. Kousa, and V. Pirila, *Contact Dermatitis,* **2,** 105–110 (1976).

290. E. M. Saihan and R. R. Harman, *Br. J. Dermatol.,* **99,** 583–584 (1978).

291. H. J. Deuel, Jr., R. Alfin-Slater, C. S. Weil, and H. F. Smyth, *Food Res.,* **19,** 1–12 (1954).

292. D. V. Parke, Ed., *The Biochemistry of Foreign Compounds,* Permagon, New York, 1968, p. 141.

293. D. L. J. Opdyke, *Food Cosmet. Toxicol.,* **16,** 883–884 (1978).

294. M. H. Weeks, C. R. Pope, and J. A. Macko, Army Environmental Hygiene Agency, Aberdeen Proving Ground, Md., Report No. USAEHA-51-087-73/76 (1974).

295. *Pesticide Index* E. H. Frear, Ed., Vol. 4, College Science Publications, State College, Pa., 1969, p. 386.

296. G. W. Newell, A. D. Petretti, and L. Reiner, *J. Invest. Dermatol.,* **13,** 145–149 (1949).

297. R. Tislow, S. Margolin, E. J. Foley, and S. W. Lee, *J. Pharm. Exp. Therap.,* **98,** 31–32 (1950).

298. *Clinical Toxicology of Commercial Products, Acute Poisoning,* 3rd ed., Gleason et al., Eds. Williams and Wilkins, Baltimore, 1969.

299. L. Phillips, II, M. Steinberg, H. J. Maibach, and W. A. Akers, *Toxicol. Appl. Pharmacol.,* **21,** 369–382 (1972).

300. E. B. Smith, R. F. Powell, J. L. Graham, and J. A. Ulrich, *Int. J. Dermatol.,* **16,** 52–56 (1977).

301. Flavoring Extract Manufacturer's Association Survey of Flavoring Ingredient Usage Levels, No. 3247, *Food Technol. Champaign,* **24,** 25 (1970).

302. Council of Europe, Natural Flavoring Substances, Their Sources, and Added Artificial Flavoring Substances. Partial Agreement in the Social and Public Health Field. List 1, No. 689, Strasbourg 1974, p. 271.

303. T. Geill and R. Dybkaer, *Scand. J. Clin. Lab.*, **23,** 255–258 (1969).

304. P. A. Owren, *Ann. Intern. Med.*, **63,** 167–184 (1965).

305. A. Raz and A. Lione, *Biochim. Biophys. Acta*, **311,** 222–229 (1973).

306. F. O. Obi and E. A. Nwanze, *J. Neurol. Sci.*, **43,** 447–454 (1979).

307. D. H. Hwang and A. E. Carroll, *Am. J. Clin. Nutr.*, **33,** 590–597 (1980).

308. P. Dewailly, A. Nouvelot, G. Sezille, J. C. Fruchart, and J. Jaillard, *Lipids*, **13,** 301–304 (1978).

309. J. S. McCutcheon, T. Umermura, M. Bhatnager, and B. L. Walker, *Lipids*, **11,** 545–552 (1976).

310. Unpublished data, Health, Safety and Human Factors Laboratory, Eastman Kodak Co., Rochester, N.Y., 1979.

311. M. Chegnard, J. Lefort, and B. B. Vargaftig, *Prostaglandins*, **14,** 909–927 (1977).

312. V. M. Andreolie, *Eur. J. Pharmacol.*, **7,** 314–318 (1969).

313. L. McGregor and S. Renaud, *Thromb. Res.*, **12,** 921–927 (1978).

314. E. Hietanen, O. Hanninen, M. Laitinen, and M. Lang, *Enzyme*, **23,** 127–134 (1978).

315. M. Lang, *Gen. Pharmacol.*, **7,** 415–419 (1976).

316. J. R. O'Brien, M. D. Etherington, and S. Jamieson, *Lancet*, **2,** 995–996 (1976).

317. R. L. Hoover, R. D. Lynch, and M. J. Karnvosky, *Cell*, **12,** 295–300 (1977).

318. National Research Council, Chemical Biological Coordination Center, *Summary Tables of Biological Tests*, **2,** 188 (1950).

319. U.S. Army Armament Research and Development Command, Chemical Systems Laboratory, NIOSH Exchange Chemicals, NX #00371.

320. N. Munsch and M. Pascaud, *Bull. Soc. Chim. Biol.*, **51,** 1575–1590 (1970).

321. C. M. Derks and D. Jacobovitz-Derks, *Am. J. Pathol.*, **87,** 143–158 (1977).

322. J. E. Campillo, A. S. Luyckx, M. D. Torres, and P. J. Lefebvre, *Diabetologia*, **16,** 267–273 (1979).

323. C. P. White, *J. Pathol. Bacteriol.*, **14,** 450–462 (1910).

324. C. P. White, *Proc. Pathol. Soc. G. B. Ireland*, **14,** 145 (1910).

325. H. Burrows and J. W. Cook, *Am. J. Cancer*, **27,** 267–278 (1936).

326. I. Hieger, *Br. J. Cancer*, **16,** 716–721 (1962).

327. I. Hieger, *Biochem. J.*, **24,** 505–511 (1930).

328. E. L. Kennaway and I. Hieger, *Br. Med. J.*, **1,** 1044–1046 (1930).

329. C. C. Twort and J. D. Fulton, *J. Pathol. Bacteriol.*, **33,** 119–143 (1930).

330. K. Setala, L. Merenmies, L. Stjernvall, Y. Aho, and P. Kajanne, *J. Natl. Cancer Inst.*, **23,** 925–951 (1959).

331. K. Takahashi, *Sci. Papers Inst. Phys. Chem. Res.* (Tokyo), **51,** 103–132 (1926).

332. A. M. Ehrly, *Fette, Seifen, Anstrichm.*, **71,** 557–559 (1969).

333. A. M. Ehrly, F. Gramlick, and H. H. Mueller, *Klin. Wochenschr.*, **43,** 943–945 (1965).

334. H. P. Hawley and G. B. Gordon, *Lab. Invest.*, **34,** 216–222 (1976).

335. M. L. Chevrel-Bodin, and M. Cormier, *C. R. Seances Soc. Biol. Fil.*, **137,** 760–762 (1943).

336. E. Boyland and F. J. C. Roe, *Br. Emp. Canc. Camp.*, **42** (Part 2), 22–23 (1964).

337. E. Boyland, R. T. Charles, and N. F. C. Gowing, *Br. J. Cancer*, **15,** 252–256 (1961).

338. E. Boyland, F. J. C. Roe, and B. C. V. Mitchley, *Brit. J. Cancer*, **20,** 184–189 (1966).

339. I. K. M. Morton, G. E. Rose, S. H. Saverymutter, and J. R. Wood, *Br. J. Pharmacol.*, **56,** 350 (1976).

340. T. S. Gaginella, V. A. Chadwick, J. C. Debongnie, J. C. Lewis, and S. G. Phillips, *Gastroenterology*, **73,** 95–101 (1977).

341. *Clinical Toxicology of Commercial Products—Acute Poisoning*, 3rd ed., Gleason et al., Eds. Williams and Wilkins, Baltimore, 1969.

342. H. V. Ammon, P. J. Thomas, and S. F. Phillips, *Gut*, **18**, 805–813 (1977).

343. L. C. Racusen and H. J. Binder, *J. Clin. Invest.*, **63**, 743–749 (1979).

344. M. K. G. Rao, N. Reisser, and E. G. Perkin, *Proc. Soc. Exp. Biol. Med.*, **131**, 1369–1372 (1969).

345. C. Kohler, W. Wooding, and L. Ellenbogen, *Thrombosis Res.*, **9**, 67 (1976).

346. J. E. Foley, D. M. Gross, P. K. Nelson, and J. W. Fisher, *J. Pharmacol. Exp. Ther.*, **207**, 402–409 (1978).

347. J. Damas and J. Troquet, *Arch. Int. Physiol. Biochem.*, **86**, 1147–1151 (1978).

348. P. L. Lee and V. S. Murthy, *Clin. Exp. Hypertens.*, **1**, 685–701 (1979).

349. M. J. Silver, W. Hoch, J. J. Kocsia, C. M. Ingerman, and J. B. Smith, *Science*, **183**, 1085–1087 (1974).

350. E. W. Spannhake, R. J. Lemen, M. J. Wegmann, A. L. Hyman, and P. J. Kadowitz, *J. Appl. Physiol.*, **44**, 397–405 (1978).

351. G. A. Bridenbaugh, J. T. Flynn, and A. M. Lefer, *Am. J. Physiol.*, **231**, 112–119 (1976).

352. J. P. Giroud and J. Timsit, *Therapie*, **6**, 761–770 (1976).

353. P. Conquet, B. Plazonnet, and J. C. LeDouarec, *Invest. Ophthalmol.*, **14**, 772–775 (1975).

354. D. G. Wenzel and T. W. Hale, *Toxicology*, **11**, 119–125 (1978).

355. J. M. Boeynaems, J. Van Sande, C. Decoster, and J. E. Dumont, *Prostaglandins*, **19**, 537–550 (1980).

356. D. Pessayre, P. Mazel, V. Descatoire, E. Rogier, G. Feldmann, and J. P. Benhamou, *Xenobiotica*, **9**, 301–310 (1979).

357. C. Ts'ao and C. M. Holly, *Prostaglandins*, **5**, 775–784 (1979).

358. J. T. O'Flaherty, H. S. Showell, E. L. Becker, and P. A. Ward, *Inflammation*, **3**, 431–436 (1979).

359. R. H. Coots, *J. Lipid Res.*, **6**, 494–497 (1965).

360. G. Goransson, *Acta Physiol. Scand.*, **64**, 1–5 (1965).

361. S. Sahu and W. S. Lynn, *Inflammation*, **2**, 191–197 (1977).

362. F. Al-Ubaidi and Y. S. Bakhle, *J. Physiol.*, **295**, 445–455 (1979).

363. D. Y. Gruetter and L. J. Ignarro, *Prostaglandins*, **18**, 541–546 (1979).

364. J. Maclouf, P. Bernard, M. Rigaud, G. Rocquet, and J. C. Breton, *Biochem. Biophys. Res. Commun.*, **79**, 585–591 (1977).

365. J. Rogulski, T. Strzclecki, A. Pacanis, E. Kaminska, and S. Angielski, *Curr. Probl. Clin. Biochem.*, **4**, 106–110 1975.

366. Unpublished data, Health, Safety and Human Factors Laboratory, Eastman Kodak Co., Rochester, NY, 1973.

367. C. A. Winter and J. Trillius, *Am. J. Ophthalmol.*, **33**, 387–388 (1950).

368. E. Boyland, *Biochem. J.*, **34**, 1196–1201 (1940).

369. *Encyclopedia of Occupational Health and Safety* (International Labor Office, Geneva, Switzerland), **1**, 30 (1971).

370. M. Shimomura, *Physiol. Chem. Phys.*, **9**, 539–542 (1977).

371. M. Silverman and L. Haung, *Am. J. Physiol.*, **231**, 1024–1032 (1976).

372. V. J. Rosen, H. J. Kramer, and H. C. Gonick, *Lab. Invest.*, **28**, 446–455 (1973).

373. K. Schaerer, T. Yoshida, L. Voyer, S. Berlow, G. Pietra, and J. Metcoff, *Res. Exp. Med.*, **157** 136–152 (1972).

374. H. J. Kramer and H. C. Gonick, *J. Lab. Clin. Med.*, **76**, 799–808 (1970).

375. A. Gougoux, G. Lemieux, and N. Lavoie, *Am. J. Physiol.*, **231**, 1010–1017 (1976).
376. N. Chakravarty and H. J. Soerensen, *Acta Physiol. Scand.*, **88**, 401–411 (1973).
377. D. R. Mileski, H. R. Kaplan, M. H. Malone, and K. A. Nieforth, *J. Pharm. Sci.*, **54**, 295–298 (1965).
378. *Registry of Toxic Effects of Chemical Substances*, NIOSH ed., 1978.
379. J. M. Weiss, C. R. Downs, and H. P. Corson, *Ind. Eng. Chem.*, **15**, 628–630 (1923).
380. C. G. Smith, J. E. Grady, and J. I. Northrum, *Cancer Chemotherapy Rep.*, **30**, 9–12 (1963).
381. S. Levey, A. G. Lasichak, R. Brimi, J. M. Orten, C. J. Smyth, and A. H. Smith, *J. Am. Pharm. Assoc.*, **35**, 298–304 (1946).
382. K. Kuroda and M. Akao, *Arch. Int. Pharmacodyn. Ther.*, **226**, 324–330 (1977).
383. L. Mircevova and J. Bicanova, *J. Physiol. Bohemoslov.*, **14**, 289–293 (1965).
384. *Chem. Abstr.*, **71**, P26974m (1969).
385. *Chem. Abstr.*, **78**, P30808x (1973).
386. National Research Council Chemical—Biological Coordination Center. *Summary Tables of Biological Tests*, **6**, 147 (1954).
387. *Encyclopedia of Chemical Technology*, 2nd ed., Vol. 7, Wiley, New York, 1965, p. 415.
388. *Chem. Abstr.*, **71**, Abstract No. P92772v (1969).
389. *Chem. Abstr.*, **71**, Abstract No. 82292z (1969).
390. M. Finkelstein and H. Gold, *J. Am. Pharm. Assoc. Sci. Ed.*, **36**, 173–179 (1947).
391. A. N. Booth, J. Taylor, R. H. Wilson, and F. DeEds, *J. Biol. Chem.*, **195**, 697–702 (1952).
392. H. A. Lardy, *Methods Enzymol.*, **13**, 314–319 (1969).
393. C. T. Chopde, A. K. Dorle, and D. M. Brahmankar, *Indian J. Physiol. Pharmacol.*, **19**, 157–160 (1975).
394. U.S. Army Armament Research and Development Command, Chemical Systems Laboratory, NIOSH Exchange Chemicals, NX #00189.
395. G. Kallistratos and E. Fasske, *Folia Biochim. Biol. Graeca*, **13**, 94–107 (1976).
396. G. Kallistratos and U. Kallistratos, *Folia Biochim. Biol. Graeca*, **13**, 1–10 (1976).

Subject Index

Refer to the Chemical Index for specific compounds.

Chemical Index

Abate [*3383-96-8*], 4807-4809
Acephate [*30560-19-1*], 4809-4810
Acetaldehyde [*75-07-0*], 4024, 4151, 4419, 4549, 4553
Acetaldehyde cyanohydrin [*78-97-7*], 4876
Acetanilide hydroxylase [*9059-06-7*], 3954, 4732
Acethylene [*74-86-2*], 4335
Acetic acid [*64-19-7*], 4909-4911
Acetol phosphate [*926-43-2*], 3861
Acetone [*67-64-1*], 4566, 4718, 4720-4727
Acetone cyanohydrin [*75-86-5*], 4877-4878
Acetonitrile [*75-05-8*], 4458, 4846, 4868-4871
Acetonyl acetone, *see* 2,5-Hexanedione
Acetophenone [*98-86-2*], 4719, 4784-4785
Acetylacetone, *see* 2,4-Pentanedione
Acetylacetylene, *see* 3-Butyn-2-one
Acetylbenzene, *see* Acetophenone
Acetylcholine [*51-84-3*], 4091, 4802
Acetylcholinesterase [*9000-81-1*], 4073, 4111
Acetylene [*2143-69-3*], 4251
Acetylenecarboxylic acid [*471-25-0*], 4951-4954
Acetylenyl carbinol, *see* 2-Propyn-1-ol
Acetylethyne, *see* 3-Butyn-2-one
N-Acetyle-S-(3-hydroxypropyl)-L-cysteine [*23127-40-4*], 4669
Acetyl-β-methylcholine [*55-92-5*], 4802
2-Acetyl propane, *see* 2-Butanone, 3-methyl
Acetyl 2-propanone, *see* 2,4-Pentanedione
Acid phosphatase [*9001-77-8*], 4103, 4107, 4568
Aconitase [*9024-25-3*], 4975
Aconitic acid [*499-12-7*], 4974-4975
Acrilan, *see* Polyacrylonitrile
Acrolein [*107-02-8*], 4251, 4669
Acrylamide [*79-06-1*], 4352

Acrylic acid [*79-10-7*], 4954-4955
Acrylonitrile [*107-13-1*], 4258, 4324, 4458, 4846, 4864
Acrylonitrile-butadiene copolymer [*9003-18-3*], 4238, 4261, 4270-4271
Acrylonitrile-butadiene-styrene copolymer, [*9003-56-9*], 4238, 4331, 4343-4347
Acrylonitrile rubber, *see* Acrylonitrile-butadiene copolymer
Acrylonitrile-styrene copolymer [*9063-54-7*], 4236
Acrysol, *see* Polyacrylic acid
Adipic acid [*124-04-09*], 4945-4946
Adiponitrile [*111-69-3*], 4880-4881
Adipyl dinitrile [*111-69-3*], 4880-4881
Agar [*9002-18-0*], 4391
Alanine transaminase [*9000-86-6*], 4727
Alathon, *see* Polyethylene
Alcohol dehydrogenase [*9031-72-5*], 3827, 3829, 4549, 4553, 4560, 4669
Aldehyde oxidase [*9029-07-6*], 3827
Aldol [*107-89-1*], 3877
Algin [*9005-40-7*], 4391
Alkaline phosphatase [*9001-78-9*], 4568
Alkathene, *see* Polyethylene
Allylacetic acid [*591-80-0*], 4958
Allyl alcohol, *see* 2-Propen-1-ol
Amco, *see* Polypropylene
Amerfil, *see* Polypropylene
p-Aminobenzhydrazide-Terephthaloyl chloride copolymer [*59573-52-3*], 4414
p-Aminobenzoic acid [*1321-11-5*], 4062
3-Aminopropionitrile [*151-18-8*], 4884-4885
Aminopyrine demethylase [*9037-69-8*], 4732

Butylene glycol adipic acid polyester [*9080-04-0*], 4018

1,3-Butylene glycol diacrylate, *see* 1,3-Butanediol diacrylate

Butylene glycol methyl ether [*53778-73-7*] [*111-32-0*], 3975, 4003-4004

Butylene glycol mono-*n*-butyl ether [*4161-40-4*], 3975, 4006-4007

Butylene glycol monoethyl ether [*111-73-9*], 3975, 4004-4005

Butylene glycols [*25265-75-2*], 3872-3880

Butylene hydrate, *see* 2-Butanol

N-Butyl nitrite [*544-16-1*], 4198-4201

sec-Butyl nitrite [*942-43-6*], 4198-4201

tert-Butyl nitrite [*540-80-7*], 4198-4201

Butyl oxitol glycol, *see* Ethylene glycol mono-*n*-butyl ether

1-Butyn-3-ol, 3-methyl, [*115-9-5*], 4688

3-Butyn-2-ol [*2028-63-9*], 4688

3-Butyn-2-one [*1423-60-5*], 4718, 4733-4734

Butyraldehyde, 2-methyl [*96-17-3*], 4598

Butyraldehyde, 3-methyl [*590-86-3*], 4598

Butyric acid [*107-92-6*], 4913-4915

Butyric alcohol, *see* 1-Butanol

γ-Butyrolactone [*96-48-0*], 3879

Butyrone, *see* 4-Heptanone

N-Butyronitrile [*109-74-0*], 4872-4873

Butyrylcholine [*3922-86-9*], 4802

Calcium carbimide [*156-62-7*], 4856-4857

Calcium K-carrageenan [*62362-81-8*], 4391

Calcium λ-carrageenan [*62362-82-7*], 4391

Calcium cyanamide [*156-62-7*], 4856-4857

Calcium cyanide [*60448-22-8*], 4855-4856

Calcium oxalate [*563-72-4*], 3822, 3829, 3924, 3935, 4935

Calcium DL-tartrate [*15808-04-5*], 4944

Calcium L-tartrate [*3164-34-9*], 4944

Capric acid [*334-48-5*], 4927-4928

sec-Capric alcohol, *see* 2-Octanol

Caproic acid [*142-62-1*], 4919-4920

Capryl alcohol, *see* 2-Octanol

Caprylic acid [*124-07-2*], 4924

Carbapol, *see* Polyacrylic acid

Carbinol, diisobutyl [*108-82-7*], 4625-4629

Carbitol acetate, *see* Diethylene glycol monoethyl ether acetate

Carbitol solvent, *see* Diethylene glycol monoethyl ether

2-Carbomethoxy-1-methylvinyl dimethyl phosphate [*7786-34-7*], 4830-4831

Carbon dioxide [*124-38-9*], 4057, 4125-4126, 4335

Carbon monoxide [*630-08-0*], 4079, 4080, 4114-4124, 4251, 4335

Carbon tetrachloride [*56-23-5*], 4264, 4568, 4732

Carboxymethyl cellulose [*9000-11-7*], 4358-4359

Carrageenan [*9000-07-1*], 4391

ι-Carrageenan [*9062-07-1*], 4391

Castor oil acid [*141-22-0*], 4966-4967

Catalase [*9001-05-2*], 4060, 4560, 4727

Celcon, *see* Polyoxymethylene

Cellophane [*9005-8-16*], 4383-4384

Cellosolve acetate, *see* Ethylene glycol acetates; Ethylene glycol monoethyl ether acetate

Cellosolve solvent, *see* Ethylene glycol monoethyl ether

Cellulose [*9004-34-6*], 4356-4358

Cellulose, hydroxyethyl [*9004-62-0*], 4386-4389

Cellulose, hydroxypropyl [*9004-64-2*], 4386-4389

Cellulose, hydroxypropyl methyl [*9004-65-3*], 4386-4389

Cellulose, sodium carboxymethyl [*9004-32-4*], 4386-4389

Cellulose acetate [*9004-35-7*], 4358-4359, 4384-4385

Cellulose acetate butyrate [*9004-36-8*], 4385

Cellulose gum, *see* Cellulose, sodium carboxymethyl

Cellulose methyl [*9004-67-5*], 4386-4389

Cellulose nitrate [*9004-70-0*], 4358-4359, 4385-4386

Cellulose triacetate [*9012-09-3*], 4358-4359, 4384-4386

Cetylacetic acid [*57-11-4*], 4932-4934

Cetyl alcohol, *see* 1-Hexadecanol

Cetylic acid [*57-10-3*], 4931

Chlorine cyanide, *see* Cyanogen chloride

Chloroacetaldehyde [*107-20-0*], 4677

Chlorocyanogen, *see* Cyanogen chloride

2-Chloro-2-(diethylcarbamoyl)-1-methylvinyl dimethyl phosphite [*13171-21-6*], 4831-4832

β-Chloroethyl alcohol, *see* Ethanol, 2-chloro

1-Chloro-1-nitroethane [*598-92-5*], 4163

1-Chloro-1-nitropropane [*600-25-9*], 4147, 4163

2-Chloro-2-nitropropane [*594-71-8*], 4163

Chloropicrin, *see* Trichloronitromethane

3-Chloropropionitrile [*542-76-7*], 4883-4884

Chlorthion [*500-28-7*], 4811-4812

Cholinesterase [*9001-08-5*], 4107, 4108, 4574, 4727, 4804

Citraconic acid [*498-23-7*], 4973

Cumulative Index

The Cumulative Index combines all index entries appearing in Volume 2A (pages 1467-2878), Volume 2B (pages 2879-3816), and Volume 2C (pages 3817-5112). For easy reference, the inclusive page numbers of each volume are printed at the top of each page of the Cumulative Index.